Rudenko International Conference "Methodological Problems in Reliability Study of Large Energy Systems" (RSES 2023)

E3S Web of Conferences Volume 461 (2023)

Olkhon Island, Russia
9-15 July 2023

ISBN: 978-1-7138-8496-5

Printed from e-media with permission by:

Curran Associates, Inc.
57 Morehouse Lane
Red Hook, NY 12571

Some format issues inherent in the e-media version may also appear in this print version.

For additional information, please contact EDP Sciences – Web of Conferences at the address below.

EDP Sciences – Web of Conferences
17, Avenue du Hoggar
Parc d'Activité de Courtabœuf
BP 112
F-91944 Les Ulis Cedex A
France

Phone: +33 (0) 1 69 18 75 75
Fax: +33 (0) 1 69 28 84 91

contact-edps@webofconferences.org

Additional copies of this publication are available from:

Curran Associates, Inc.
57 Morehouse Lane
Red Hook, NY 12571 USA
Phone: 845-758-0400
Fax: 845-758-2633
Email: curran@proceedings.com
Web: www.proceedings.com

TABLE OF CONTENTS

A software platform for constructing a digital twin of the integrated energy system

Valery Stennikov[1,*], *Dmitry* Sokolov[1], *Evgeny* Barakhtenko[1], *Gleb* Mayorov[1]

[1]Melentiev Energy Systems Institute of Siberian Branch of the Russian Academy of Sciences (ESI SB RAS),

664033, Irkutsk, Russia.

Abstract. An integrated energy system is a complex technical object, digital twin technologies improve the quality of its design and management. The paper proposes a methodological approach to the development of a software platform for constructing a digital twin of the integrated energy system and solving a set of problems that arise during system design and management. The Model-Driven Engineering (MDE) concept is implemented to automate the stages of a digital twin construction based on the platform. A three-level architecture of a network multiuser software system is proposed. An algorithm used for automated construction of a digital twin of the integrated energy system is presented.

1 Introduction

Recently, an integrated approach has been developed to address the joint development and functioning of several separate systems (electricity, heat, gas, cooling, etc.) as part of an integrated energy system (IES) [1, 2]. At the same time, IES is understood as a system that includes a set of individual energy systems interconnected through common interdependent modes of operation when producing, consuming and transporting the corresponding types of energy, and with the integrated development of these systems.

Integration of individual energy systems into a single technological complex will ensure the implementation of new functionality, the use of more advanced technologies in management and the design of integrated centralized systems with coordinated control of their modes and active participation of consumers in the energy supply process. The transition to the construction of IES provides higher economic efficiency of the listed systems, high reliability and uninterrupted power supply to consumers.

The issue of optimal design of IES is very important, and its effective solution provides a significant reduction in economic costs and an acceptable quality of supplying the consumer with the necessary energy [1, 3, 4]. The use of modern approaches based on the concept of digital twins makes it possible to improve the adequacy of mathematical and computer models, provides the ability to solve design problems within a single software, which makes it possible to obtain a solution to a design problem of higher quality in comparison with the approaches of previous generations. At present, approaches to the design of complex technical systems based on the concept of digital twins [5] become widespread both in the energy sector [6] and in other industries [7-9].

The Melentiev Energy Systems Institute of the Siberian Branch of the Russian Academy of Sciences (ESI SB RAS) has formulated a new scientific problem of developing a unified and universal technology for building a digital twin of IES and developing a software tool for designing these systems. This paper proposes an original methodological approach to the design of a software platform for constructing digital twins of IESs and their design.

2 Formulation of the scientific problem

Based on the performed structuring of the problems and research of IES, a conclusion was made about the need to develop a unified software platform for modeling and designing IES. The platform should act as a basis for the automated construction of IES digital twins and solve the whole range of problems of computer simulation of IESs during their design.

The new platform must meet the following requirements:

- Ensure that all sub-problems are solved within a single software system.
- Link the solutions to individual sub-problems into a single solution.

* Corresponding author: sva@isem.irk.ru

• Flexibly adjust the computational procedure to the features of the equipment and mathematical models that describe its operation.

• Flexibly adjust the computational process by implementing various algorithms for solving the problem.

Platform elements should be developed in accordance with the component programming paradigm, which will ensure the possibility of presenting methods, algorithms and models in the form of reusable software fund elements. Software components should have standardized software interfaces within a single software tool with a description of subroutines, input and output data formats. In turn, the practical application of the component approach leads to a complex problem caused by the need to integrate a large number of software components used to model various system elements, which can be represented by a wide range of equipment and be in different states, into the software system. The high difficulty of maintaining the entire set of necessary software components up to date allows a conclusion that in order to implement the software platform, it is necessary to involve programming paradigms focused on automating the stages of software construction.

Modern approaches to software development are the subject of research by many specialists and are widely presented in the literature. The object-oriented approach [10, 11] is the basis for modeling complex software system. In [12-15], the Model-Driven Engineering (MDE) concept is proposed to automate the stages of software construction. The MDE concept is a set of methodological approaches to the automated construction of complex software systems based on pre-developed models. This concept of software development is currently actively developed; approaches based on it make it possible to successfully develop complex software systems [16-18]. Approaches to automating the stages of software construction based on metaprogramming are described in [19, 20].

3 A new methodological approach

The paper proposes an original methodological approach to the implementation of a software platform for the automated construction of digital twins of IES and the solution of practical problems arising in the design of these energy systems. The MDE concept is the basis for automated software construction based on this platform. A distinctive feature of the implementation of this concept within the framework of the developed approach is the wide use of modern metaprogramming technologies [19, 20], which provide high flexibility to dynamically design elements of a software system and make necessary changes to it.

The main components of the developed methodological approach are:

• Range of software components and their standardized software interfaces.

• Composition of the ontology systems and methods for their development.

• Three-level architecture of a multi-user software system.

• An original approach to the automated construction of a digital twin of the IES.

The main components of the developed software platform are the following: components that control the operation of other software components; components that implement mathematical methods and algorithms; components that provide access to databases and ontologies (an ontology of IES, which includes ontologies of individual energy systems; a software ontology, problem ontology). Java is the main programming language for implementing the platform. Individual software components can be implemented in C/C++, Fortran, Python, and other languages. To store ontologies, a domain-specific language developed on the basis of the XML data description language (eXtensible Markup Language) is used.

The three-level architecture of the network multiuser system proposed for the implementation of the software platform is shown in Fig. 1. The system consists of the following three subsystems:

1. Web application running on the server. This subsystem is responsible for interacting with users within a computer network and for solving computational and analytical problems on the server side.

2. The client part of the network system. It is designed to work with graphic computer models of IES and data on their elements. This subsystem is used to prepare initial data and view calculation results.

3. Data storage subsystem. It is intended for storage and multiple use of a computer model of the IES, initial data and calculation results, and graphical data on urban development.

The server part of the software system is developed in Java and runs under the Apache Tomcat servlet container. To store the IES model and transfer its configuration between subsystems of the software platform, the text data exchange format JSON (JavaScript Object Notation) is used. Computer models of IES created by users are stored by the platform in a database and can be repeatedly used in calculations. To interact with databases, JDBC (Java DataBase Connectivity) technology is used, which provides a single interface for interaction between the platform and database management systems (DBMS) using the SQL

(Structured Query Language). Depending on the scale of the problem to be solved, the number of one-time users and the specific features of the network load on the web application, JDBC provides the possibility of flexible integration into the platform of various multi-user relational DBMSs, such as Firebird, PostgreSQL, MS SQL Server, and others.

The interaction of the user of the software platform with the computer model of the IES is organized as work with a set of layers, each of which corresponds to one of the levels of the hierarchical model of the IES. At the same time, depending on the goals of modeling and the features of the problem being solved, one layer can represent both a separate energy system (as a whole) as part of the IES, and one of the levels of the hierarchical model of any subsystem of the IES. The IES hierarchical model can consist of models of individual energy systems of different scales (interregional networks, large urban backbone networks, regional distribution networks, networks of individual settlements, etc.). An IES of a small settlement, for example, can be connected to a large gas transmission system and an urban power system, while the heat supply system as part of this IES can be of a small scale and run on gas or coal.

We developed an original approach to designing a digital twin and organizing the computational process. In accordance with this approach, the digital twin and the computing subsystem are built in an automated mode.

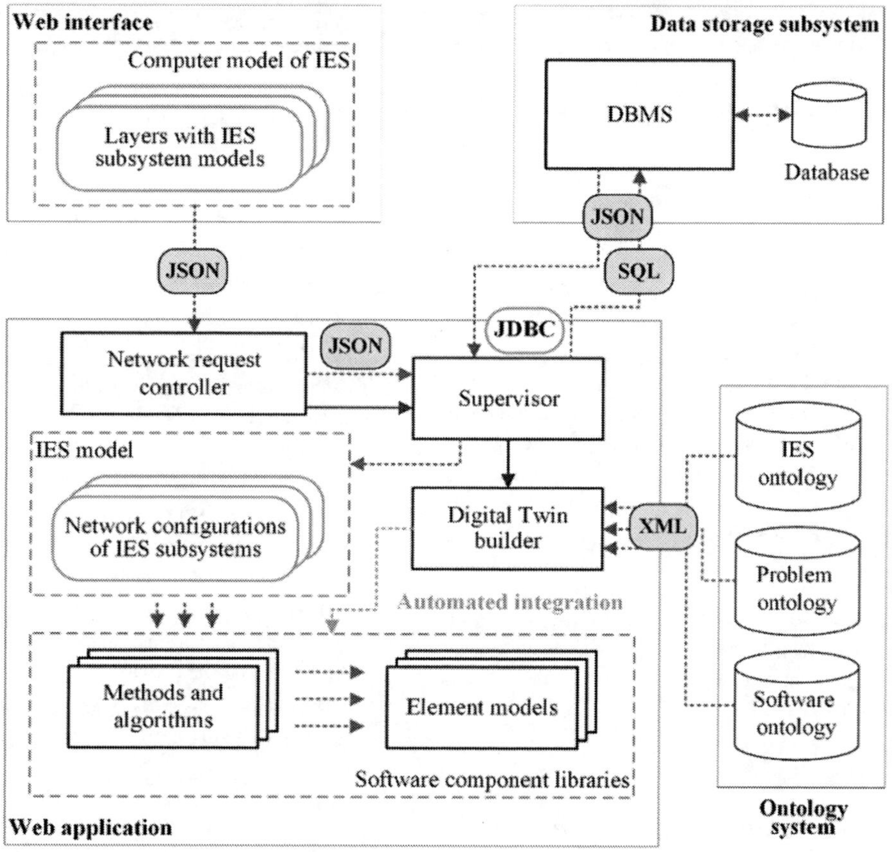

Fig. 1. Three-level architecture of a multi-user network software system.

The approach suggests that automated design is performed in accordance with the algorithm, the steps of which are presented in the following form.

1. Generate a list of equipment to be used based on the computer model of the IES created by the user.
2. Generate a list of mathematical models of the equipment used based on the computer model of the IES.
3. Generate a list of required software components based on the lists created in the previous steps.
4. Construct a description of a software component that implements an algorithm for solving an applied problem.
5. Construct a model of the designed subsystem of the software system.
6. Import the description of the required software components from the software ontology.

7. Link software components that implement algorithms and mathematical models of equipment to the software system.
8. Construct the data structures necessary for the functioning of the linked software components.

4 Conclusions

The transition to the modeling and design of modern energy systems such as IESs can significantly increase the efficiency of their work. IESs are complex technical objects that require the structuring of large amounts of data, the use of complex mathematical models and methods, and the organization of complex computational processes. Digital twin technologies make it possible to overcome these difficulties. The paper proposes a methodological approach to the development of a software platform for automated construction of digital twin of IES and solving a set of problems that arise when designing these systems. To automate the stages of the digital twin construction, the concept of MDE is implemented. A three-level architecture of a network multiuser software system is proposed. An approach to the automated construction of a digital twin is proposed and the algorithm used for the automated construction is presented.

The methodological support proposed in the paper can be used in organizations involved in the development of software for the analysis, design and optimization of IES for various purposes. The developed software, when implemented in design organizations, will improve the quality of the received design recommendations and reduce the complexity of obtaining them.

The research was performed at Melentiev Energy Systems Institute of Siberia Branch of the Russian Academy of Sciences under the support of Russian Science Foundation (Grant number 22-29-01611).

References

1. N. Voropai and V. Stennikov, Izvestiya RAN. Energetika **1**, 64–73 (2014).
2. N. Voropai, V. Stennikov, S. Senderov, E. Barakhtenko, O. Voitov and A. Ustinov, J. Energy Eng. **143**, 1-11 (2017).
3. V. Stennikov, E. Barakhtenko, D. Sokolov and B. Zhou, Energy Reports **8**, 10025–10036 (2022).
4. V. Stennikov, E. Barakhtenko, G. Mayorov, D. Sokolov and B. Zhou, Appl. Energy **309** (2022).
5. M. Grieves, *Digital twin: manufacturing excellence through virtual factory replication: White paper 1* (Institute of Technology, Florida, 2014), pp. 1-7.
6. A.K. Sleiti, J.S. Kapat and L. Vesely, Energy Reports **8**, 3704–3726 (2022).
7. D. Jones, Ch. Snider, A. Nassehi, J. Yon and B. Hicks, CIRP J. Manuf. Sci. Technol. **29**, 36–52 (2020).
8. M. Liu, Sh. Fang, H. Dong and C. Xu, J. Manuf. Syst. **58**, 346–361 (2021).
9. V. Zambrano, J. Mueller-Roemer, M. Sandberg, P. Talasila, D. Zanin, P. G. Larsen, E. Loeschner, W. Thronicke, D. Pietraroia, G. Landolfi, A. Fontana, M. Laspalas, J. Antony, V. Poser, T. Kiss, S. Bergweiler, S. P. Serna, S. Izquierdo, I. Viejo, A. Juan, F. Serrano and A. Stork, Array **14** (2022).
10. G. Booch, J. Rumbaugh, I. Jacobson, *The Unified Modeling Language User Guide* (Addison Wesley, Boston, 1999).
11. G. Booch, *Object-Oriented Analysis and Design with Applications*, 2nd. ed. (Addison Wesley, Boston, 1994).
12. A. R. da Silva, Computer Languages, Systems & Structures **43** (2015).
13. M. Brambilla, J. Cabot and M. Wimmer, *Model-driven software engineering in practice. Synthesis Lectures on Software Engineering* (Morgan & Claypool, Kentfield, CA, 2012).
14. J. Seixas, A. Ribeiro and A. Rodrigues da Silva, A Model-Driven Approach for Developing Responsive Web Apps, in: *Proceedings of the 14th International Conference ENASE 2019*, SciTePress, Setubal, pp. 257–264.
15. D. Akdur, V. Garousi and O. Demirörs, J. Syst. Archit. **91**, 62–82 (2018).
16. I. Boussaïd, P. Siarry and M. Ahmed-Nacer, Autom Softw Eng **24**, 233–294 (2017).
17. I. Al-Azzoni, J. Blank and N. Petrović, Algorithms **14** (2021).
18. E. de Araújo Silva, E. Valentin, J. R. H. Carvalho and R. da Silva Barreto, J. Comput. Lang. **62** (2021).
19. K. Hazzard and J. Bock, *Metaprogramming in .NET* (Manning Publications, NY, 2013).
20. R. Lämmel, *Software Languages: Syntax, Semantics, and Metaprogramming* (Springer, Cham, 2018).

 https://doi.org/10.1051/e3sconf/202346101002

Analysis of the prospects for the gas industry to meet Russia's needs for primary energy resources and assessment of the possibilities for exporting Russian gas

Sergey Senderov[1, 2*], *Viktor* Rabchuk[1]

[1] Melentiev Energy Systems Institute of Siberian Branch of the Russian Academy of Sciences, 130 Lermontov str., Irkutsk, Russia
[2] Federal State Budget Educational Institution of Higher Education «Irkutsk National Research Technical University», 83 Lermontov str., Irkutsk, Russia

Abstract. The capabilities of the Russian gas industry in meeting the country's internal needs with primary fuel and energy resources are predicted. These opportunities take into account the potential of the oil and coal industries, as well as all other sources (hydroelectric power plants, nuclear power plants, solar energy, wind energy, firewood, etc.) in the period until 2035, taking into account the effects of existing negative factors. After fulfilling the condition of fully meeting the country's internal gas needs, the prospects for Russian gas exports are assessed. These prospects are assessed both taking into account the implementation of foreign policy and foreign economic factors (sanctions restrictions and their consequences), and taking into account the possible volumes of Russian natural gas production.

1 Introduction

The article assesses the potential situation with the fulfillment of the most important requirement of energy security (ES) of Russia, regarding the reliable provision of domestic consumer demand for primary fuel and energy resources (FER) [1], taking into account the possibilities for export supplies of Russian gas until 2035. Moreover, Regarding gas exports under existing sanctions, we have to answer two main questions: what could be the possible volumes of Russian gas exports and will the promising technical potential of the Russian gas industry correspond to them?

Questions regarding the listed requirements (satisfying Russia's promising domestic demand for primary fuel and energy resources and the needs of Russian gas exports) will be considered for the following reference years: 2025, 2030 and 2035. In this case, we will take into account:

- actual production volumes of primary fuel and energy resources in Russia, as well as volumes of Russian gas exports;
- results of a qualitative analysis of factors influencing the implementation of threats to the Russian electronic security, incl. taking into account the effect of anti-Russian sanctions;
- requirements to increase the share of clean energy.

The most significant for the perspective under consideration, among the factors influencing the implementation of threats to Russia's energy security, are the following:

1. Lack of investment, incl. associated with a noticeable deterioration in the situation with businesses obtaining "long-term" cheap credit funds.
2. Increasing difficulties with the implementation of the best available technologies (BAT), incl. at gas industry enterprises.
3. Reduced opportunities to increase the share of renewable energy sources and hydrogen energy in the balance sheets of regions and the country as a whole, primarily due to difficulties with investment.
4. Reduced opportunities for the development of new gas and oil fields in Russia (especially in hard-to-reach, expensive areas) due to the worsening investment problem and reduced opportunities for implementing BAT.
5. A sharp decrease in the possibilities of importing: modern equipment for gas production and main gas transportation; means of intellectualization of control systems for the functioning of the main production facilities of the energy industries; modern machines and mechanization equipment for developing new gas fields and laying gas pipelines (especially in hard-to-reach areas), etc.
6. Slowdown in the development of the gas chemical industry in the country (due to difficulties with the implementation of BAT, lack of investment).

Today, the most important threat to energy security is the lack of investment in the energy sector. The situation here was significantly aggravated by large-scale changes in the conditions for the export of Russian hydrocarbons, which began in 2022. The threat of a significant reduction in their exports to European markets began to be realized with natural gas. It should be recognized that the emerging trends in energy-dependent industrialized countries are

* Corresponding author: ssm@isem.irk.ru

supported by active stimulation of scientific and technological progress in the field of energy, incl. in the development of nuclear energy, in the use of non-traditional types of fuel and energy resources and renewable energy sources. It is quite possible that the situation will develop with a significant reduction in the possibilities for exporting Russian hydrocarbons in the future. The implementation of these challenges will inevitably lead to a reduction in foreign exchange earnings from the export of Russian fuel and energy resources and thereby to a reduction in investments to ensure the sustainable functioning and development of industries in the Russian fuel and energy complex. First of all, this will affect the development of new expensive oil and gas areas.

In the current situation, including the establishment of a price ceiling for Russian oil from December 2022 [2] and for Russian petroleum products from February 2023 [3], the niche for the profitable export of Russian fuel and energy resources has narrowed significantly. There are serious concerns about a significant reduction in the profitability of alternative export destinations, for example, the potential redirection of natural gas flows to China. There is also an issue with the reorientation of the "freed up" volumes of oil and gas to the production of highly processed products (petroleum and gas chemistry). It is connected with the need for significant investments in the creation and development of these industries, as well as with the lack of a niche in world markets for the large-scale export of these highly processed products.

Against the backdrop of these trends, we can expect a significant decline in investment in the Russian energy sector with a corresponding decrease in its promising production capabilities. Already beyond 2025, Russia may experience a slight decrease in the potential to provide even "truncated" (due to sanctions) export supplies of fuel and energy resources, especially natural gas. It would seem that with restrictions on the export of hydrocarbons, the reserves of fuel and energy resources available in the subsoil are consumed more slowly. At the same time, a decrease in investment in energy industries pushes back the development of new areas of their production, thereby reducing their promising production capabilities. Under certain conditions, the question may arise not only of providing natural gas for export, but also of meeting domestic needs. The negative effects of the above factors will be especially noticeable until 2025. Further, until 2035, the situation should somehow stabilize with a decrease in sanctions pressure regarding the export of Russian fuel and energy resources, the introduction of BAT in the energy sectors and at least a partial return of opportunities for obtaining long-term cheap borrowed funds. Such changes must be accompanied by a corresponding increase in GDP and a decrease in its energy intensity associated with the diversification of the economic structure towards the development of the production of high-tech products and services.

Speaking about taking into account climate change factors when solving problems of ensuring Russia's energy supply today, it should be noted that for Russia their importance is still significantly less than the importance of the negative factors taking place today. To a certain extent, the reduction in the carbon footprint in Russia will be associated with the displacement of some volumes of coal from the consumption structure of fuel and energy resources due to reduced volumes of natural gas exports.

The article presents a quantitative forecast of the expected role of the gas industry in providing Russia with primary fuel and energy resources and the expected levels of Russian gas exports. Given the uncertainty of the situation, primarily with the export of Russian gas, it is advisable to make such a forecast over two time periods. The first - 2023-2025, the second - 2026-2035.

2 Forecast of the situation with primary FER in Russia and with the gas exports in 2023-2025.

Taking into account the current situation in Russia and in the world, it can be considered that today it is quite difficult to make a reasonable forecast of the values of annual production of oil, petroleum products and coal, at least until the situation with their exports stabilizes.

2.1 Natural gas

As for gas, already in 2022 the volume of its exports due to current economic sanctions decreased compared to 2021 from 244 to 184 bcm [4]. It is worth remembering here that the entire first quarter of 2022 gas exports were complete in accordance with the concluded contracts. In each subsequent quarter, exports were reduced by 20 bcm compared to the same quarter in 2021. Thus, most likely, the starting point for calculating the initial value of possible gas exports in 2023 will be approximately 165 bcm. It seems unlikely that there will be a further reduction in Russian gas exports, taking into account the expected increase in its supplies to China (only through the Power of Siberia gas pipeline, supplies should increase from 15 bcm in 2022 to 38 bcm in 2030. This will compensate the above reduction may lead to a slight increase in the total volume of gas exports in relation to 2022. From here, we can assume that in 2023 the volume of Russian gas exports could amount to about 170 bcm with some growth in 2024 and 2025 up to 180-190 bcm/year.

Domestic gas consumption may increase until 2025 due to some displacement of coal by gas released from exports (excess gas on the domestic market). In 2023, the volume of domestic gas consumption can be expected at the level of 480-490 bcm (484 bcm in 2022 [5]). It is quite acceptable to expect in 2024-2025 small (1.0-1.5% per year) growth of Russia's GDP. Based on this assumption, as well as due to the presence of excess gas production capacity (limited exports), as well as the implementation of gasification programs in the regions of the country, we should expect some increase in domestic gas consumption: in 2024 - 490-500 bcm, in 2025 year – 490-510 bcm.

Knowing the possible volumes of gas exports and the volumes of its domestic consumption, we can talk about the volumes of gas production required in this case in Russia: in 2023 – 650-660, in 2024 – 670-690 and in 2025 – 680-700 bcm.

In this analysis, the authors did not take into account the possibility of importing gas to Russia, because this import has not been very noticeable in recent years (at the level of 6-8 bcm per year). It is unlikely that this import, given the existing surplus of gas on the domestic market, can increase, but a reduction or cessation of this import is quite possible.

2.2 Oil

Total domestic consumption of oil (direct consumption) and petroleum products in 2021 and 2022. was (respectively) 225 and 224 million tce [6]. Until 2025 the annual consumption of these fuel and energy resources is unlikely to change noticeably. It may increase slightly (increasing demand for motor fuel from the military) or decrease slightly (increasing share of electric vehicles). We will assume that the consumption of oil and petroleum products in the forecast period under consideration until 2025 will remain within the same limits and amount to 220-230 million tce per year.

2.3 Coal

When assessing the volume of domestic coal consumption for the period up to 2025, it should be taken into account that the inevitable displacement of coal by gas from the fuel and energy resources balances (gas released when its exports are reduced) will be constrained by the difficulties of quickly replacing coal with gas (due to inertia). In connection with the above, most likely, in 2023, coal consumption will remain at the level of 2022 (173 million tce [7]) and will amount to 170-180 million tce with a further decrease in this value to 160 -170 million tce in 2024 and up to 150-160 million tce in 2025.

2.4 Hydroelectric power plants, nuclear power plants and other FER

The share of participation of hydroelectric power plants, nuclear power plants and other fuel and energy resources (wind energy, firewood, peat, etc.) in covering the country's internal needs for primary fuel and energy resources has not actually changed over the past 5-7 years and amounted to 156 and 157 million tce in 2021 and 2022, respectively [8]. It should be expected that until 2025 it will remain within the range of 150-160 million tce.

2.5 Primary fuel and energy resources in total

Amount for primary fuel and energy resources

Based on the above reasoning, the volumes of total domestic consumption in the country of primary fuel and energy resources, as well as the volumes of Russian gas exports for the period until 2025 and the actual values of these data for recent years are given in Table 1. At the same time, the domestic total consumption of primary fuel and energy resources in the country is calculated as the sum of the corresponding components. This value is presented for perspective as the average value of the corresponding possible range.

Table 1. Actual and projected volumes of consumption of primary FER within Russia until 2025, volumes of export and gas production

Indicators	Year				
	Fact		Forecast		
	2021	2022	2023	2024	2025
Domestic consumption of primary FER, total, million tce, incl.	1051	1071	1060	1070	1070
- natural and associated gas*, bcm	470	484	480-490	490-500	490-510
- oil, petroleum products, million tce	225	224	220-230	220-230	220-230
- coal, million t	155	173	170-180	160-170	150-160
- energy of the hydroelectric power stations, nuclear power plants, etc. FER, million tce	156	157	150-160	150-160	150-160
Gas exports, bcm	244	184	170-180	180-190	180-190
Production of natural and associated gas, total, bcm	764	674	650-670	670-690	680-700

* Does not include excess gas injection into underground gas storage facilities and losses.

3 Forecast of the situation with primary FER in Russia and with the gas exports in 2026-2035.

The forecast is made taking into account the data in table 1 considering that:

- by 2035, the situation with consequences for the Russian energy sector should become increasingly stabilized with a decrease in sanctions pressure;
- during this period, the country's economy should grow steadily with at least an average annual increase in GDP of 2-3% in 2026-2030 and 2.5-3.5% in 2031-2035. The increase in growth rates in the second five-year period is associated with expectations of a reduction in sanctions pressure;
- with sustainable economic growth, it becomes possible to further reduce the energy intensity of GDP at an annual rate of 1.5-2% in 2026-2030 and 2-3% in 2031-2035.

Prospective total annual domestic needs for primary FER are determined taking into account the current situation and the expected growth rates of GDP and a decrease in the specific energy intensity of the economy:

$$Q_t = Q_{t-1}(1 + K^t_{GDP} - K^t_{ei}), \quad (1)$$

where Q_{t-1} is the volume of consumption of these fuel and energy resources in the *t--st* year; K^t_{GDP} and K^t_{ei} are the coefficients of change in Russia's GDP and the specific energy intensity of GDP in the *t* - year relative to the *t-1st* year. When determining the internal requirements for primary fuel and energy resources, the following average annual values K^t_{GDP} and K^t_{ei}, table 2.

Table 2. Annual coefficients of change in Russia's GDP and specific energy intensity of GDP until 2035.

Period	K^t_{GDP}	K^t_{ei}
2026–2030	0,020–0,030	0,015–0,020
2031–2035	0,025–0,035	0,020–0,030

Using expression (1), the data in table 2 and the approximate volume of total consumption of primary FER in Russia in 2025 (1070 million tce) will be obtained for 2030 and 2035 these volumes are within 1110 million tce and 1140 million tce respectively.

When determining the future participation in the consumption structure of primary FER for 2030 and 2035, first of all, we will determine the possibilities of supplying all other FER, except gas. Then, we will accept gas as a closing resource and determine the requirements for the required volumes of gas supply. These volumes, together with the possible volumes of gas exports, will be the required volumes of gas production in the country for the corresponding years.

3.1 Oil

In 2023-2025, the expected participation of the oil industry in providing the country with primary FER is expected to be at the level of 220-230 million tce (Table 1). Starting from 2026 and until 2035, domestic needs for petroleum products, if they grow, will be no more than 0.5-1% per year. On the one hand, there is some displacement of vehicles using motor fuel by electric transport; on the other hand, there is a need to increase the volume of oil consumption as a raw material for petrochemicals, although due to difficulties with investments this increase will not be significant. In this case, we can expect the possible participation of the oil industry in providing the country with primary FER in the amount of 225-235 million tce by 2030, and 235-245 million tce by 2035.

3.2 Coal

The possible participation of the coal industry until 2030 will continue its downward trend (to 140-160 million tce, with 150-160 million tce expected in 2025). Then, until 2035, these levels should not change (including due to difficulties with investments in the gas industry).

3.3 Hydroelectric power plants, nuclear power plants and other FER

The total volumes of production of primary FER in Russia at hydroelectric power stations, nuclear power plants and through other FER in 2023-2025 are expected to amount to 150-160 million tce per year. Under the current sanctions, it can be assumed that this situation will continue until 2030. Further, in quantitative terms, the degree of participation of these energy resources in providing the country with primary FER should increase somewhat, primarily due to renewable energy sources (mitigation of the investment situation). In 2035, the production of FER at hydroelectric power stations, nuclear power plants and other sources may increase to 160-180 million tce.

3.4 Requirements for the gas industry.

The above-mentioned total needs of Russia for primary FER, as well as the volumes of possible participation in covering these total needs from the oil, coal industries, as well as hydroelectric power stations, nuclear power plants and other FER, make it possible to determine the requirements for the gas industry in terms of its participation in covering the indicated needs. In 2030 these volumes should be about 530-560 and in 2035 – 530-570 bcm. These volumes, together with the volumes of gas exports, limited by the conditions of the world gas market, determine the required volumes of gas production in the country for reference years.

Based on foreign policy and foreign economic factors, we can talk about the possibility of a slight increase in the niche for Russian gas exports from 180-190 bcm in 2025 to 180-200 bcm in 2030. It is difficult to expect further growth in its exports from -for part of lost markets, due to increased use of renewable energy sources, etc. The volume of Russian gas exports may decrease if the physical capabilities for gas production are insufficient to fulfill obligations for export supplies (while fully covering domestic needs). At the above export levels, the required volumes of gas production in the country should be at the level of 710-760 bcm in 2030 and 710-770 bcm in 2035.

All quantitative data obtained above for 2030 and 2035 are presented in table. 3. Actual data for 2022 and expected volumes for 2025 are also shown here (Table 1).

Table 3 presents the required gas production volumes in the country to meet its domestic needs, as well as the expected export volumes. Now it is necessary to determine the forecast possibilities for the production of this gas. To do this, it is necessary to consider the main conditions for the development and functioning of the Russian gas industry (objective factors and accepted prerequisites), and then analyze the situation in the main gas-producing regions of the country.

Table 3. Actual and projected volumes of primary FER consumption within Russia until 2035, export volumes and required gas production volumes

Indicators	Year			
	2022	2025	2030	2035
	Fact	Forecast		
Domestic consumption of primary FER, total, million tce, incl.	1071	1070	1110	1140
oil, petroleum products, million tce	*224*	*220-230*	*225-235*	*235-245*
coal, million t	*173*	*150-160*	*140-160*	*140-160*
energy of the hydroelectric power stations, nuclear power plants, etc. FER, million tce	*157*	*150-160*	*150-160*	*160-180*
Requirements for gas production to meet the internal needs for primary FER, bcm	484	500-520	530-560	530-570
Possible* export of Russian gas, bcm	184	180-190	180-200	180-200
Required volumes of gas production, bcm	668	680-710	710-760	710-770

* Export volume limited by foreign economic conditions.

The main conditions for the development and functioning of the Russian gas industry for the analyzed period are as follows:

1. Implementation of all the negative factors mentioned above (lack of investment, difficulties with the implementation of BAT, difficulties with obtaining borrowed funds, etc.).
2. The presence today (2023) of excess natural gas production capacity, which appeared after 2022 in the amount of about 100 bcm/year (Table 1). These surpluses must be used to ensure the necessary volumes of further gas production.
3. After using these surpluses, it is advisable to increase gas production in Yamal (operating fields and development of the Kharasaveyskoye field, bringing its annual production level to 30-40 bcm/year by 2035).
4. Decrease in the annual production level at Tyumen fields by 6-8 bcm/year [9]. Such a decline may occur mainly after 2030, when excess production capacity is exhausted.
5. Decrease in oil gas production levels due to the expected decrease in the export of Russian oil and oil products (and therefore due to a decrease in oil production levels in the country). According to the authors, the decrease in oil production during the analyzed period will be approximately 20-25% of the 2022 level (534 million t [10]). Accordingly, the level of oil gas production will decrease in the same proportions. Then we can expect that the level of annual total production of this gas in the period until 2035 will be 70-80 bcm.
6. Delay in the development of trillions gas fields on the shelf: Shtokman (Barents Sea), Leningrad and Rusanovskoye (Kara Sea) beyond 2035 due to investment problems.
7. Lack of growth in gas production levels in the Far East, at least until 2035 due to difficulties with investment.
8. Continued development of the Chayandinskoye and Kovyktinskoye fields, bringing the level of annual gas production to 40-50 bcm by 2035 (due to the need to fulfill the agreement on gas supplies to China). The results of the analysis of gas production capabilities are presented in Table. 4.

Table 4. Assessment of gas production capabilities in Russia for the period up to 2035, bcm

Indicators	Year			
	2022	2025	2030	2035
	Fact	Forecast		
Total capabilities (natural and petroleum gas), incl.	674	680-730	670-740	610-680
Production of natural gas, incl. by region	574	600-640	600-660	540-600
Yamal	*100*	*120-130*	*130-150*	*150-170*
Nadym–Pur–Tazovsky district	*389*	*390-400*	*370-390*	*300-320*
European part	*27*	*25-30*	*25-30*	*20-25*
Western Siberia incl. Khanty-Mansi Autonomous Okrug	*20*	*15-20*	*15-20*	*10-15*
Eastern Siberia	*6*	*25-30*	*25-30*	*25-30*
Far Eastern Federal District	*32*	*35-40*	*35-40*	*35-40*
Petroleum gas production	100	70-80	70-80	70-80

Now, understanding the possible benchmarks for gas production in Russia, as well as the possible volumes of its domestic consumption for the future until 2035 (data in Table 3), we will determine the production capabilities of the industry for the export of Russian gas without taking into account the situation on world gas markets and foreign economic conditions, Table 5. Here we present the volumes of possible gas exports from the perspective of expected foreign economic conditions.

Table 5. Analysis of production capabilities for Russian gas exports, bcm

Indicators	Year			
	2022	2025	2030	2035
	Fact	Forecast		
Total gas production capabilities	674	680-730	670-740	610-680
Domestic gas consumption volumes	484	500-520	530-560	530-570
Production capabilities for export of Russian gas	190	180-210	140-180	80-110
Possible export of Russian gas from the perspective of foreign economic conditions	184	180-190	180-200	180-200
Production capability margin for possible exports	6	10	- 30	- 100

The last row of Table 5 presents the discrepancy between the production capabilities of Russian gas exports and its possible volumes from the perspective of foreign economic conditions. For ease of perception, this discrepancy is represented by the difference in the average values of the given ranges. The table shows that for 2025 the expected level of gas production can, with a certain margin, ensure its potential export (with full satisfaction of domestic needs). But, in the conditions of the negative factors and threats discussed in the article, already approximately from 2030, the production capabilities of the gas industry may not be enough to ensure the export of Russian gas in the volumes possible in the context of the implementation of foreign economic sanctions and their consequences in the form of a reduction in the niche for Russian FER on the European market. By 2035, such a discrepancy may already amount to half of the theoretically possible export volumes.

4 Conclusion

An analysis of the promising capabilities of the Russian energy sector in the period up to 2035 in the context of negative factors and their potential consequences allowed us to formulate the following conclusions:

- despite the powerful sanction pressure, the level of production of Russian fuel and energy resources will fully satisfy domestic needs for primary FER throughout the entire period under review (until 2035);
- the share of gas in meeting the country's domestic needs for primary fuel and energy resources will increase slightly due to a decrease in the share of coal (at the beginning of the period under review, there is a surplus of gas on the domestic market and the need to fulfill the requirements to increase the share of clean energy);
- expected volumes of gas production and exports approximately until 2028-2029 will be determined by the sum of domestic gas needs and possible volumes of its exports, limited by sanctions and their consequences in the form of the loss of part of the European markets. Already from about 2030, the volumes of Russian gas exports may be lower than potentially possible due to limited production capabilities oil and gas complex of the country.

The research was carried out under State Assignment Project (no. FWEU-2021-0003) of the Fundamental Research Program of Russian Federation 2021-2030 using the resources of the High-Temperature Circuit Multi-Access Research Center (Ministry of Science and Higher Education of the Russian Federation, project no 13.CKP.21.0038).

References

1. Doctrine of Energy Security of the Russian Federation. http://publication.pravo.gov.ru/Document/View/0001201905140010?index=0&rangeSize=1
2. The EU has published a decision to introduce a cap on oil prices from Russia at $60. https://tass.ru/ekonomika/16495575
3. The EU Council approved the price ceiling for Russian petroleum products. https://tass.ru/mezhdunarodnaya-panorama/16962183
4. Gas exports from Russia in 2022 decreased by a quarter. https://newprospect.ru/news/aktualno-segodnya/eksport-gaza-iz-rossii-v-2022-godu-sokratilsya-na-chetvert
5. Russia in 2022 increased gas consumption by 3%, LNG exports by 8%. https://www.interfax.ru/business/880885
6. Neftegaz of Russia in 2023. https://www.cdu.ru/tek_russia/articles/1/1118/
7. Coal production in the Russian Federation in 2022 amounted to 443.6 million tons, exports - 210.9 million tons – Novak. https://www.interfax-russia.ru/main/dobycha-uglya-v-rf-v-2022-sostavila-443-6-mln-t-eksport-210-9-mln-t-novak
8. The volume of electrical generation in Russia in 2022 increased by 0.6%. https://www.akm.ru/news/obem_elektricheskoy_generatsii_v_rossii_v_2022_godu_vyros_na_0_6_/
9. Yamal will appear: the Ministry of Energy saw the risk of reducing gas production by a third in 17 years. Why this might happen and what the authorities are proposing. https://iz.ru/1489778/valerii-voronov/iamala-pokazhetsia-v-minenergo-uvideli-risk-snizheniia-dobychi-gaza-na-tret-cherez-17-le
10. Oil and gas condensate production in Russia in 2022 increased by 2.1%, gas production fell by 13.4%. https://neftegaz.ru/news/dobycha/768736-dobycha-nefti-i-gazovogo-kondensata-v-rossii-v-2022-g-vyrosla-na-2-1-gaza-upala-na-13-4/

Problems of searching for critically important combinations of gas industry objects from the position of ensuring energy security

Sergey Vorobev[1*] and *Alexey* Edelev[1]

[1]Melentiev Energy Systems Institute, 130 Lermontov str., Irkutsk, Russia

Abstract. The article is devoted to the problem of finding critical combinations of gas industry objects from the standpoint of ensuring energy security. A critically important object is the partial or complete failure of which can cause significant damage to the country from the fuel and energy complex as a whole or within a separate energy system. Gas transportation networks in Russia and Europe are very extensive, branched, and have many intersections and backup gas pipelines. Therefore, cases of simultaneous failure of several system objects are possible. The article presents the results of calculations in which the authors identify the most significant combinations of gas transportation network objects from the point of view of providing consumers with gas. The significance and influence of these combinations on the performance of the Russian gas industry is shown. Conclusions are drawn about the feasibility of searching for critical combinations of gas industry objects.

1 Introduction

Large, geographically distributed energy systems are exposed to a significant number of external threats. Among which the main ones can be identified are the aging and wear and tear of the linear part of fixed production assets, and the threat of deliberate impacts of various negative natures. Large-scale accidents in energy systems [1, 2], resulting from the failure of various critical objects of these systems, entail damage to consumers in the form of shortages of fuel and energy resources. Therefore, the search and determination of critical elements and their combinations in energy systems are important current problems.

2 Current state of the problem under study

In the world practice of studying the most important objects and their combinations in energy systems, the following points can be noted.

In [3, 4], the authors analyzed the gas transportation network to determine its most important components. The methodological approaches used are based on topological network analysis with an emphasis on studying issues of reliability and controllability. This analysis allows us to quantify the reliability of the gas transmission network and determine the role of each network component at different time intervals. As an example, the authors consider a real integrated gas transmission network of several EU countries. The article presents the results of an analysis of such critical infrastructure, showing the need to take into account physical characteristics, such as gas pipeline capacity limitations. To assess the impact of negative external influences on the normal supply of gas to consumers, a special flow model was developed. The vulnerability analysis was carried out from three perspectives: global vulnerability analysis, demand security analysis and critical analysis of gas pipelines. A global vulnerability analysis was carried out, taking into account possible disruptions in the operation of gas sources and gas pipelines.

In [5] presents a method for discovering and ranking critical components and sets of components in technical infrastructures. The criticality of a component or set of components is defined as the vulnerability of a system to failure when a particular component or set of components fails. This problem also addresses the problem of multiple simultaneous failures with synergistic consequences that complicate the problem. The proposed method solves this problem. A method for analyzing a gas distribution system in a Swedish municipality was presented as a case study.

In [6], the authors propose a comprehensive model for assessing the impact of interdependence between electric and gas systems on the reliability of power supply to consumers. The operation of the gas network was modeled using a number of constraints. Restrictions on gas supplies may affect changes in the operating

* Corresponding author: seregavorobev@isem.irk.ru

mode of the electric power industry. The case studies reviewed by the authors proved this.

An analysis of possible impacts in an integrated gas-energy network is presented in [7]. Failures of the gas supply system have been shown to be more critical to the integrated power supply system than failures in the power supply subsystem itself. Accordingly, the authors drew attention to possible control actions aimed at minimizing the negative impact of failures in the gas supply system.

Research [8, 9] is devoted to various issues of modeling energy systems as critical infrastructures. In [6], the authors propose an agent-based model of a typical regional power system that includes the characteristics of specific plant types and their cooling systems, which depend on an adequate water supply at appropriate temperatures to maintain full power operation. The study [9] presents a new approach to assessing the vulnerability of intracity gas distribution and road networks caused by gas pipeline accidents. The impacts of pipeline failures are quantified based on the gas supply areas they directly impact and the loss of traffic efficiency caused by traffic disruptions. This approach allows the identification of vulnerable links in the pipeline network, which can not only have a significant impact on the pipeline network, but can also have a significant impact on the road network.

The study [10] examined and analyzed the problem of the vulnerability of critical energy infrastructures to terrorist attacks on them. The study [11] proposes a risk analysis methodology for systems of interdependent critical infrastructures under various extreme weather events.

Study [12] presents a probabilistic approach to identify and rank important gas network components from a security of supply perspective. The authors conduct a probabilistic risk analysis of the regional European gas transportation network under selected attack scenarios. Results from 1 million Monte Carlo simulations of attack scenarios clearly indicate different impacts on gas supply. Thus, the authors obtained a list of the most important infrastructure components.

In these articles, the authors come closest to identifying critical objects of the energy system, in this case the gas transportation network. At the same time, they assign different indices to different objects of the system, which collectively determine the vulnerability of the system if the operation of this object is disrupted.

A number of studies were carried out at the ISEM SB RAS to identify critically important objects in the gas transportation network. A list of intersections of main gas pipelines in the Unified Gas Supply System of Russia has been determined, the disruption of which will lead to a relative shortage of daily gas supplies throughout the system as a whole of 5% or more [13, 14]. Research has been carried out to search for and determine combinations of individual sections of main gas pipelines, the simultaneous disruption of the functioning of which can lead to a significant shortage of daily gas supplies through the system (5% or more) [15, 16]. Works [17 - 19] are devoted to the analysis of critical objects in electric power systems.

Research has also been carried out, as a result of which critical combinations of gas industry facilities have been identified. Facilities have been found, the failure of which in combination with one of the air purifiers will lead to a much greater gas shortage among consumers than if these objects fail separately [20]. After identifying possible critical combinations of objects in each specific case, the possibilities of bypassing bottlenecks by short-term increasing the throughput capacity of individual sections of main gas pipelines were analyzed. This technological measure was used to minimize gas shortages among consumers. As a result of these measures, the significance of a number of potential critical asset combinations has been reduced. In this way, all possible critically important combinations of gas industry facilities were identified and a list of them, ranked by degree of importance, was formed.

At this stage of research, it is proposed to find and analyze combinations of gas industry objects of three elements. That is, within the framework of this study, three unrelated, independent objects are considered, the failure of which can lead to a significant gas shortage among consumers.

3 Determination of critical elements in the gas industry

In the ongoing study, both to determine the critical important objects themselves and to search for critically important combinations of objects, a flow model was used, which is the core of the software-computing complex "Oil and Gas of Russia" [21, 22]. The use of this internal control system makes it possible to determine the degree to which gas needs are met within the country and export supplies are ensured. In addition, the software-computing complex "Oil and Gas of Russia" makes it possible to identify bottlenecks - sections of the gas transportation network that, in some cases, limit the production capabilities of the system.

The flow distribution model in the Unified Gas Supply System of Russia, laid down in the software-computing complex "Oil and Gas of Russia", is intended to assess the production capabilities of the Unified Gas Supply System of Russia under conditions of various types of disturbances. The purpose of such studies is to minimize gas shortages at consumption points. The Unified Gas Supply System of Russia is presented in the model as a combination of three subsystems: gas sources, the main transport network and consumers.

When solving the problem of assessing the state of the system after a disturbance, the criterion for the optimal distribution of flows is the minimum gas deficit for the consumer at the minimum cost of delivering gas to consumers. This problem can be solved by finding the maximum flow through the network, followed by minimizing the cost of delivering gas to consumers [23]. The mathematical formulation of this problem is described in [24].

In the Unified Gas Supply System of Russia flow distribution model, as already mentioned, the Basaker-Gowen algorithm is used to calculate the maximum flow of minimum cost, which as a result allows us to determine the possible level of consumer satisfaction with gas. As a result of various emergency situations, gas shortages may occur among consumers, caused by a lack of throughput capacity of certain sections of gas pipelines. Bypassing such bottlenecks or areas of the system that limit production capabilities, in acceptable quantities, will reduce the gas shortage among consumers that arose in the situation under consideration.

If there is a gas shortage among consumers caused by a lack of capacity of the relevant gas pipelines, other branches of the main gas pipelines not affected by the violation in question take on increased volumes of gas. In such a situation, the congestion of the network changes and a lack of capacity may appear in other sections of the main gas pipelines.

Subsequent debottlenecking in the gas transportation network will minimize gas shortages among consumers and makes the assessment and identification of possible critical combinations of gas industry facilities as adequate as possible.

To solve the problem of bypassing the identified bottlenecks, the flow distribution model included the possibility of increasing the gas flow through the arcs within 10% of their throughput. Such a short-term increase in the throughput of a section of the main gas pipeline is possible with an increase in the operating power of compressors at large main compressor stations [25]. As a result of increasing the operating pressure in the gas pipeline, an increase in the throughput of the main gas pipeline section by up to 10% is achieved. As a result (by using the technical capabilities of the gas transportation network), the problem of minimizing gas shortages among consumers is solved.

4 Calculation results

The Unified Gas Supply System of Russia calculation scheme used to model the operation of the Unified Gas Supply System of Russia in this work takes into account all the main features of the functioning of the Unified Gas Supply System of Russia and contains:

- 378 nodes, including: 28 gas sources; 64 gas consumers (subjects of the Russian Federation); 24 underground gas storage facilities; 266 hub compressor stations;
- 486 arcs representing main gas pipelines and branches to gas distribution networks.

Initial data, such as daily volumes of gas consumption, production, export and import, and the throughput capacities of existing gas pipelines are taken in accordance with official statistics [26-28] for 2019. In a specially conducted study [29], a corresponding analysis was carried out, as a result of which 61 gas industry objects were assigned to the Unified Gas Supply System of Russia critical important objects. Among these objects: 25 arcs between nodal compressor stations and 36 nodes, including 30 nodal compressor stations, 5 head compressor stations at the outlets of large gas fields and a compressor station at one underground gas storage facility.

A calculation was carried out for the simultaneous shutdown of three unrelated elements of the design circuit. These calculations were carried out using a software package [30] that reflects in detail the functioning of the Russian gas transportation network and allows simulating various operating conditions of its facilities, including complete shutdown. Calculations were carried out using parallel and distributed computing technologies [31].

As a result, 17,808,318 combinations of three elements were obtained, the failure of which could lead to a gas shortage of 5% or more in the system as a whole. As previously assumed [32], such a volume of output information is difficult to analyze further, so combinations of elements with a total deficit of 15% for the entire system as a whole were accepted for further consideration. There were 3,655,963 such combinations. The obtained quantitative calculation results are summarized in Table 1.

Table 1. Results of determining the most important combinations of UGSS objects

Number of combinations	Gas deficit, %
102	> 50 %
2421	40-50 %
3655963	> 15 %
17808318	> 5 %

As shown in Table 1, in the current configuration of the Unified Gas Supply System of Russia there are 102 triplets of unrelated objects, the failure of which could lead to a gas shortage of more than 50% for the system as a whole. The most severe disruptions to gas supplies to consumers are 17 combinations, the failure of which could lead to a gas shortage of 52.8%. It should be noted that these combinations consist of the most significant critical facilities in the gas industry.

The calculation results show a synergistic effect from such disruptions in gas supplies to consumers. A detailed analysis of the combination leading to the occurrence of maximum gas deficiency showed the following. It consists of three critical facilities. Separately, the failure of these facilities using the same design scheme leads to the following gas shortages: 21.2%, 16% and 10%. These gas shortages add up to 47.2%, which is less than 52.8% if they fail together.

In general, from the results of the study it should be noted that disruption of the functioning of a combination of objects of three elements can lead to a significant gas shortage among consumers (> 50%).

4 Conclusion

This study examines the issues of searching for the most important combinations of gas industry facilities. The identification of such combinations of gas industry facilities was carried out, the failure of which could lead to a gas shortage in the system as a whole, comparable to the deficit that arose as a result of the failure of individual water treatment plants. The obtained main most important combinations of gas industry objects are analyzed in detail.

The studies have shown that in the modern configuration of the Unified Gas Supply System of Russia, situations are possible when, if three network facilities fail, the total gas shortage among consumers may exceed 50% of the total gas demand. Determining and taking into account such combinations of objects is necessary to increase the reliability of the Unified Gas Supply System of Russia during its development and reconstruction work. The main efforts should be aimed at reducing the conditional significance of the identified specific critical combinations of objects for the performance of the entire system.

The research was carried out within the framework of the state assignment project FWEU-2021-0003 (registration number AAAA-A21-121012090014-5) for fundamental research of the Siberian Branch of the Russian Academy of Sciences, using the resources of the Shared Use Center Siberian Supercomputer Center ICM&MG SB RAS.

The authors express their gratitude to the Irkutsk Supercomputer Center SB RAS for providing access to the high-performance cluster "Academician V.M. Sailors" (Irkutsk Supercomputer Center SB RAS, Irkutsk: IDSTU SB RAS; http://hpc.icc.ru).

References

1. Accidents on main gas pipelines in Russia in 2018-2019. URL: https://ria.ru/20190728/1556953028.html.
2. Top 5 largest and most destructive accidents on gas pipelines. URL: https://sila-sibiri-rabota.ru/avarii-na-gazoprovodax/.
3. Han F., Zio E., Kopustinskas V., Praks P. *Risk, Reliability and Safety: Innovating Theory and Practice*, 2016, pp. 2565-2571. DOI: 10.1201/9781315374987-389.
4. Su H., Zio E., Zhang J. Li X. (2018) Reliability Engineering & System Safety, **Vol. 175**, P. 79-91. DOI: 10.1016/j.ress.2018.03.006.
5. Jonsson H., Johansson J., Johansson H. (2008) J. of Risk and Reliability. **Vol. 222.** No. 2. P. 235-243. DOI: https://doi.org/10.1243% 2F1748006XJRR138.
6. Li T., Eremia M., Shahidehpour M., (2008) Interdependency of natural gas network and power system security, IEEE Transactions on Power Systems, **vol. 23**, 1817-1824. DOI: 10.1109/TPWRS. 2008.2004739.
7. Dokic S.B., Rajakovic N.Lj. (2018). Energy, DOI: 10.1016/j.energy. 2018.04.165
8. Thompson, J. R., Frezza, D., Necioglu, B., Cohen, M. L., Hoffman, K., Rosfjord, K. (2019). Int. Journal of Critical Infrastructure Protection **Vol. 24**, p. 144-165.
9. Kai, L., Ming, W., Weihua, Z., Jinshan, W., Xiaoyong, Y. (2018). Int. Journal of Critical Infrastructure Protection **Vol. 23**, p. 79-89.
10. Tichy, L. (2019). Int. Journal of Critical Infrastructure Protection, DOI: https://doi.org/10.1016/j.ijcip.2019.01.003.
11. Tsavdaroglou, M., Al-Jibouri, S. H.S., Bles, T., Halman, J. I.M., (2018). Int. Journal of Critical Infrastructure Protection **Vol. 21**, p. 57-71.
12. Praks P., and Kopustinskas V. (2019). CRITIS 2018, LNCS 11260, pp. 3–16. DOI: 10.1007/978-3-030-05849-4_1
13. Senderov S.M., Rabchuk V.I., Edelev A.V. (2016) News of the Russian Academy of Sciences. Energetika, **No. 1**, pp. 70-78. (in Russian)
14. Senderov S., Edelev A. (2017) Energy, DOI:10.1016/J.ENERGY.2017.11.063.
15. Vorobev S., Edelev A. Analysis of the importance of critical objects of the gas industry with the method of determining critical elements in networks of technical infrastructures / Management of Large-Scale System Development (MLSD), 2017 Tenth International Conference. IEEE, 2017. DOI: 10.1109/MLSD.2017.8109707.
16. Vorobev S., Edelev A., Smirnova E. (2017). E3S Web Conf. **Vol. 25**, DOI: 10.1051/e3sconf/20172501004.
17. Senderov S., Krupenev D. (2019) Energy Systems Research. **Vol.2.** № 2 (6). p. 41-50.
18. Krupenev D.S. Principles for identifying critical objects of electric power systems / Methodological issues in studying the reliability of large energy systems. International scientific seminar named after. Yu.N. Rudenko: In 2 books. Ed. Voropai N.I., 2018. pp. 329-337. (in Russian)
19. Krupenev D., Boyarkin D., Iakubovskii D. (2020) Reliability Engineering & System Safety. **Vol. 204.**
20. Senderov S., Vorobev S., Edelev A. (2018). E3S Web Conf. **Vol. 58**, DOI: 10.1051/e3sconf/20185803002.
21. Edelev A.V., Enikeeva S.M., Senderov S.M. (1999) Computational technologies, **Vol.4**, No. 5, pp. 30 – 35. (in Russian)
22. Vorobev S.V., Edelev A.V. (2014) Software Products and Systems, **No. 3**, pp. 174 – 177. (in Russian)
23. Ford L.R., Fulkerson D.R. *Flows in networks.* Princeton university press, Princeton, New Jersey, 1962. 276 p.
24. Vorobev S.V., Edelev A.V. (2016) Scientific Bulletin of NSTU **Vol. 62**, No. 1, p. 181-194. (in Russian)
25. Yu.P. Korotaeva, R.D. Margulova, *Extraction, preparation and transport of natural gas and condensate. Reference manual in 2 volumes. Volume II.* Nedra, 288 p., (1984). (in Russian)
26. *Exports of the most important goods to the Russian Federation in 2012 - 2020 (according to the Federal Customs Service of Russia)* http://customs.ru/index.php?option =com_ newsfts& view=category&id=52&Item id=1978&limitstart=60. (in Russian)
27. *InfoTEK Monthly oil and gas magazine.* **Vol. 1**, 2020, p. 150. (in Russian)

28. *Ministry of Energy of the Russian Federation. Statistics.* http://minenergo.gov.ru/activity/statistic. (in Russian)
29. Senderov S.M., Rabchuk V.I., Vorobev S.V. Formation of a list of critically important gas industry facilities from the standpoint of the country's energy security / Materials of the collection of reports Methodological issues in studying the reliability of large energy systems. 90th meeting "Reliability of developing energy systems". July 1-7, 2018, Irkutsk. (in Russian)
30. Feoktistov A., Gorsky S., Sidorov I., Kostromin R., Edelev A., Massel L. (2019) Communications in Computer and Information Science. **Vol. 965**. P. 289-300.
31. Edelev A.V., Beresneva N.M., Gorsky S.A. (2022) Modern science-intensive technologies. **No. 1**. – P. 47–52. DOI: 10.17513/snt.39008. (in Russian)
32. Vorobev S., Kolosnitsyn A., Minarchenko I. Energies. 2022; **15**(2):501. DOI: 10.3390/en15020501.

E3S Web of Conferences **461**, 01004 (2023) https://doi.org/10.1051/e3sconf/202346101004
RSES 2023

Application of the rating method in energy security assessments at the regional level

Elena Smirnova[1]* and *Sergey* Senderov[1]

[1]Melentiev Energy Systems Institute ESI SB RAS, Irkutsk, Russia

Abstract. The article presents a method for generating standardized values of energy security indicators and standardized qualitative assessments of the state of energy security, based on the use of a convolution apparatus for qualitative assessments of the state of the most important indicators of energy security. An analysis of the dynamics, main trends and scale of changes in the state of energy supply when using this method is presented using the example of an assessment of the constituent entities of the Russian Federation located on the territory of the Volga Federal District.

Introduction

The concept of energy security (ES) is interpreted as "the state of protection of citizens, society, the state, and the economy from threats of shortages in meeting their energy needs with economically available energy resources of acceptable quality, and from threats of disruption of the uninterrupted power supply" [1,2, etc.]. An indicative assessment of the level of energy security of a particular region of the country is carried out according to three, largely interconnected, blocks of indicators: production and resource availability of the fuel and energy supply system of the region; reliability of the fuel and energy supply system of the region; state of fixed production assets (BPA) of energy systems in the region (Table 1). Taking into account the characteristics of the energy supply of individual regions, previously [1,3,4, etc.], the threshold values of indicative indicators for different groups of constituent entities of the Russian Federation, as well as the relative shares of indicators in the overall system of their value, were expertly determined. Using the method of convolution of the obtained values of the analyzed indicators, integral assessments of the state of the economic security of the constituent entities of the Russian Federation were formed.

Research shows that individual indicators are measured in different units, and in order to obtain an integral assessment of a region's economic security, the principle of normalizing indicator values depending on the ratio of their values to threshold values can be applied. To do this, it is proposed to use a special normalization apparatus that allows you to work with both increasing and decreasing values of indicators, that is, with those indicators whose state improves as the value increases and with those whose state improves as their values decrease.

Table 1. The composition of the most important indicators of regional energy security.

1. The block of industrial and resource security of the fuel- and energy supply of the region
1.1. The ratio of the total available capacity of the region's power plants to the maximum electric load of consumers on its territory. 1.2. The ratio of the amount of available capacity of power plants and the capacity of inter-system connections of the region with neighboring consumers to the maximum electric load on its territory. 1.3. Opportunities to meet the needs of boiler heating oil (BHO) from the region's own sources.
2. The block of reliability of fuel and energy supply of the region
2.1. The share of the dominant resource in total consumption of BHO in the region. 2.2. Share of the largest power plant in the installed electric capacity of the region. 2.3. The level of potential supply of demand for fuel in the conditions of a sharp cooling (10% consumption of consumption) in the region.
3. Block of the state of the BPA of energy systems in the territory of the region
3.1. Degree of depreciation of the BPA in the energy sector of the region. 3.2. The ratio of the average annual input of installed capacity and reconstruction of power plants in the region over the previous 5-year period to the established capacity of the region.

The conversion of indicator values expressed in various units of measurement into normalized ones is carried out according to the following expression:

$$X_i^N = \frac{X_{PCi}-X_i^t}{X_{C,i}-X_{PC,i}}, \tag{1}$$

where X_i^N – is the normalized value of indicator i in the analyzed period, rel. units; X_i^t – is the actual value of the indicator in the system of initial units; $X_{PC,i}$, $X_{C,i}$ - respectively, the threshold values of the pre-

* Corresponding author: smirnova.e.m@isem.irk.ru

crisis and crisis states of the indicator i in the system of initial units.

In accordance with the calculation algorithm, the normalized threshold value $X^N_{PC,i}$ - is always equal to zero and is the starting point of the pre-crisis states of the indicator, and the normalized value $X^N_{C,i}$ - is equal to -1 and represents the boundary of transition to crisis status of indicators.

In [5], the method of normalizing indicative indicators was used using the example of the Central and Southern Federal Districts. Based on the results of the analysis of the obtained normalized indicators and their graphical display, it became necessary to further adjust the overlap of threshold values. And correlate the specific weights of each i-th normalized indicator to unity. The fact is that if the acceptable threshold value for some indicator, for example, overlaps by 100 percent or more, then the state of the indicator still remains acceptable, but this indicator pulls up the entire integral assessment and crisis and pre-crisis assessments may be lost other indicators. The same with crisis values. You should not take into account too great a depth of the crisis; it is enough to focus on crisis values that do not distort the overall integral assessment too much. In this regard, it was decided that any significant overlap of the threshold value of the indicator should be reduced to a sufficient margin: 25% for acceptable conditions and 50% for crisis ones.

Let us present the application of this approach using the example of constituent entities of the Russian Federation located on the territory of the Volga Federal District.

Let us first present the results obtained using conventional indicative analysis. Then, in order to more conveniently compare the results and compare the dynamics of the state of energy supply in different regions, we will apply the approach of normalizing the obtained quantitative estimates.

Results of indicative ES analysis by regions of the Volga Federal District

This section provides information on the qualitative state of energy security indicators for the constituent entities of the Russian Federation on the territory of the Volga Federal District, as well as a qualitative description of the state of energy security of these constituent entities for 5 years: from 2017 to 2021, Table. 2-5. The initial information for the study was taken in accordance with statistical information for 2017-2021. [6-8], as well as with the necessary information on specific regions. As a result of the analysis of relevant data for the subjects of the Volga Federal District, it is possible to judge the trends inherent in the energy sector of the studied territories in ensuring energy security. In table Table 2 presents information on assessing the status of indicators for the block of production and resource security of the fuel and energy supply system for 2017 and 2021. in the subjects of the Volga Federal District.

Table 2. The status of indicators on the territory of the subjects of the Volga Federal District of the district for the block of production and resource provision of the fuel and energy supply system for 2017, 2021.

Region, area	Indicator	Dimen-sion	The threshold values of the indicator		The meaning and status of the indicator, year			
			N	C	2017		2021	
Republic Bashkortostan	1.1	un.	0,5	0,3	1,2	N	1,3	N
	1.2	un.	1,5	1,2	2,4	N	2,5	N
	1.3	%	40	20	24,5	PC	21,6	PC
Republic Mari El	1.1	un.	0,5	0,3	0,5	PC	0,5	N
	1.2	un.	1,5	1,2	5,2	N	5,3	N
	1.3	%	40	20	57,7	N	63,7	N
Republic Mordovia	1.1	un.	0,5	0,3	0,7	N	0,7	N
	1.2	un.	1,5	1,2	2,7	N	2,7	N
	1.3	%	40	20	0	C	0	C
Republic Tatarstan	1.1	un.	0,5	0,3	1,6	N	1,7	N
	1.2	un.	1,5	1,2	3,5	N	3,4	N
	1.3	%	40	20	24,7	PC	22,8	PC
Republic Udmurtia	1.1	un.	0,5	0,3	0,4	PC	0,4	PC
	1.2	un.	1,5	1,2	2,9	N	3,1	N
	1.3	%	40	20	2,7	C	3,6	C
Chuvash Republic	1.1	un.	0,7	0,5	1,78	N	2,4	N
	1.2	un.	1,5	1,2	5,5	N	5,9	N
	1.3	%	40	20	0,4	C	0,1	C
Kirov region	1.1	un.	0,5	0,3	0,8	N	0,8	N
	1.2	un.	1,5	1,2	2,5	N	2,6	N
	1.3	%	40	20	3,1	C	3,3	C

Nizhny Novgorod region	1.1	un.	0,7	0,5	0,8	N	0,8	N
	1.2	un.	1,5	1,2	1,9	N	1,9	N
	1.3	%	40	20	59,8	N	71,8	N
Orenburgskaya region	1.1	un.	0,5	0,3	1,1	N	1,7	N
	1.2	un.	1,5	1,2	1,4	PC	2,1	N
	1.3	%	40	20	123,3	N	76,8	N
Penza region	1.1	un.	0,5	0,3	0,5	N	0,4	PC
	1.2	un.	1,5	1,2	2,6	N	2,5	N
	1.3	%	40	20	1,2	C	1	C
Permian edge	1.1	un.	0,5	0,3	2,1	N	2,3	N
	1.2	un.	1,5	1,2	2,8	N	3,2	N
	1.3	%	40	20	21,2	PC	22,1	PC
Samara region	1.1	un.	1,2	1,1	1,5	N	1,6	N
	1.2	un.	1,5	1,2	3,8	N	4,1	N
	1.3	%	40	20	48,1	N	44,1	N
Saratovskaya region	1.1	un.	0,5	0,3	3,2	N	3,2	N
	1.2	un.	1,5	1,2	7,1	N	7,1	N
	1.3	%	40	20	46,2	N	54,3	N
Ulyanovskaya region	1.1	un.	0,5	0,3	0,9	N	1,1	N
	1.2	un.	1,5	1,2	14,2	N	12,5	N
	1.3	%	40	20	1,3	C	1,8	C

According to indicator 1.1, the situation is acceptable in the Republic of Mari El, Nizhny Novgorod, Orenburg and Saratov regions. The maximum electrical load here is provided with a reserve of its own power generating capacities. As for electrical connections with neighboring regions (indicator 1.2), their capacity, in the event of large-scale emergency situations with energy supply, can be assessed as acceptable. Also, thanks to a sufficient amount of production of own BHO in certain regions (fuel oil in Nizhny Novgorod, Samara, Mari El, coal and gas in the Orenburg region, fuel oil and natural gas in the Saratov region), acceptable values of indicator 1.3 are provided. A crisis situation according to indicator 1.3 can be noted in Mordovia, Kirov, Chuvash and Ulyanovsk regions due to insufficient volumes of BHO production or its complete absence. The situation with respect to the block of indicators "Reliability of fuel and energy supply to the region" is presented in Table. 2.

Table 3. Characteristics of the state of indicators on the territory of the subjects of the Volga Federal District for the fuel and energy supply reliability block for 2017, 2021.

Region, area	Indica-tor	The threshold values of the indicator, %		The meaning and status of the indicator, year			
		N	C	2017		2021	
Republic Bashkortostan	2.1	40	70	94,1	C	95,3	C
	2.2	50	70	35,0	N	32,7	N
	2.3	100	<100	91	C	91	C
Republic Mari El	2.1	40	70	97,5	C	91,7	C
	2.2	50	70	31,4	N	31,7	N
	2.3	100	<100	91	C	91	C
Republic Mordovia	2.1	40	70	99,9	C	99,8	C
	2.2	50	70	89	C	87,6	C
	2.3	100	<100	94	C	94	C
Republic Tatarstan	2.1	40	70	99,5	C	99,4	C
	2.2	50	70	27,9	N	27,1	N
	2.3	100	<100	93	C	93	C
Republic Udmurtia	2.1	40	70	95,8	C	97,6	C
	2.2	50	70	52,5	PC	56,8	PC
	2.3	100	<100	91	C	91	C
Chuvash Republic	2.1	40	70	98,8	C	99,7	C
	2.2	50	70	62,8	PC	62,8	PC
	2.3	100	<100	91	C	91	C

 https://doi.org/10.1051/e3sconf/202346101004

Kirov region	2.1	40	70	90,7	C	93,6	C
	2.2	50	70	45,9	N	46,8	N
	2.3	100	<100	>100	N	>100	N
Nizhny Novgorod region	2.1	40	70	95,5	C	90,0	C
	2.2	50	70	45,9	N	46,8	N
	2.3	100	<100	>100	N	>100	N
Orenburgskaya region	2.1	90		87,3	N	91,7	PC
	2.2	50	70	63,2	PC	61,4	PC
	2.3	100	<100	>100	N	>100	N
Penza region	2.1	40	70	98,4	C	97,8	C
	2.2	50	70	77,8	C	82,8	C
	2.3	100	<100	>100	N	>100	N
Permian edge	2.1	40	70	97,0	C	95,0	C
	2.2	50	70	42,5	N	43,1	N
	2.3	100	<100	>100	N	>100	N
Samara region	2.1	40	70	97,7	C	97,5	C
	2.2	50	70	41,9	N	42,6	N
	2.3	100	<100	>100	N	>100	N
Saratovskaya region	2.1	40	70	98,2	C	97,3	C
	2.2	50	70	60,3	PC	60,8	PC
	2.3	100	<100	>100	N	>100	N
Ulyanovskaya region	2.1	40	70	98,3	C	99,2	C
	2.2	50	70	49,8	N	42,2	N
	2.3	100	<100	>100	N	>100	N

According to indicator 2.1, in all subjects of the Volga Federal District, the share of gas, as the dominant resource, in the total consumption of BHO is extremely high and amounts to 90-95%, which cannot be acceptable from an energy supply standpoint. As for the share of the largest power plant in the installed electrical capacity of the regions, in general the situation is acceptable. Exceptions are the Republic of Udmurtia (Izhevsk CHP-2), the Chuvash Republic (Cheboksary HPP), the Orenburg region (Iriklinskaya SDPP) and the Saratov region (Balakovo NPP). Here the share of the largest source is more than 60%, and the situation from an energy security point of view is assessed as pre-crisis. The crisis situation according to this indicator is in the Republic of Mordovia (the largest power plant Saranskaya CHP-2, which accounts for more than 70% of installed capacity and more than 60% of electricity generation), as well as in the Penza region (more than 80% is the share of Penza CHP-1, in addition, The Penza region is an energy-deficient region both in terms of electricity and power, and the deficiency is compensated by the flow of electricity from neighboring regions). The most important indicative indicators include indicator 2.3 (Table 3), reflecting the level of potential supply of demand for fuel and energy resources in conditions of a sharp cold snap (10% surge in consumption) in the territories of the region. It is estimated based on the results of model studies described in [2,3,9] as the amount of supply of boiler and furnace fuel to consumers during a possible peak cold snap, with an increase in CHP consumption by 10%. The subjects of the Volga Federal District belong to regions with a particularly cold climate with the average temperature of the coldest five days below minus 30°C [10]. The very high share of natural gas in certain regions, noted as a crisis from an energy supply point of view, determines a crisis situation with the possibility of providing consumers with fuel in conditions of increasing demand during sudden cold snaps in the Volga Federal District. Research shows that in such a situation the supply of the required amount of gas cannot always be ensured. At the same time, in the Kirov region, despite the fact that there is no own production of BHO, and the share of natural gas in the balance of BHO is 93%, the presence of powerful gas pipeline corridors in the region makes it possible to meet the peak increase in demand for fuel in the event of a sharp cold snap. The same can be said in the Perm region, Penza and Nizhny Novgorod regions. According to the indicators of the block "State of BPA of energy systems in the region", averaged data on wear and tear of BPA of energy industries were taken into account, Table. 4.

 https://doi.org/10.1051/e3sconf/202346101004

Table 4. Characteristics of the state of indicators on the territory of the subjects of the Volga Federal District in the block of the state of the BPA of energy systems for 2017, 2021.

Region, area	**Indica-tor**	**Dimen-sion**	**The threshold values of the indicator**		**The meaning and status of the indicator, year**			
			N	**C**	**2017**		**2021**	
Republic Bashkortostan	3.1	%	40	60	54,7	PC	51,9	PC
	3.2	%	2	1	1,9	PC	2,2	N
Republic Mari El	3.1	%	40	60	42,4	PC	47,5	PC
	3.2	%	2	1	0,9	C	0,9	C
Republic Mordovia	3.1	%	40	60	55,7	PC	58,6	PC
	3.2	%	2	1	0	C	1,3	PC
Republic Tatarstan	3.1	%	40	60	40,3	PC	41,8	PC
	3.2	%	2	1	2,6	N	1,7	PC
Republic Udmurtia	3.1	%	40	60	54,1	PC	58,5	PC
	3.2	%	2	1	5,9	N	1,0	C
Chuvash Republic	3.1	%	40	60	57,1	PC	60,9	C
	3.2	%	2	1	0,7	C	0,7	C
Kirov region	3.1	%	40	60	57,5	PC	61,7	C
	3.2	%	2	1	6,7	N	0,5	C
Nizhny Novgorod region	3.1	%	40	60	44,3	PC	49,6	PC
	3.2	%	2	1	2,4	N	0,2	C
Orenburgskaya region	3.1	%	40	60	61,6	C	63,9	C
	3.2	%	2	1	0,6	C	1,7	PC
Penza region	3.1	%	40	60	68	C	73,9	C
	3.2	%	2	1	0	C	0	C
Permian edge	3.1	%	40	60	47,3	PC	51,5	PC
	3.2	%	2	1	3,4	N	2,3	N
Samara region	3.1	%	40	60	50,4	PC	53,7	PC
	3.2	%	2	1	0,1	C	0,1	C
Saratovskaya region	3.1	%	40	60	63,6	C	0,1	C
	3.2	%	2	1	0,1	C	0,2	C
Ulyanovskaya region	3.1	%	40	60	55,7	PC	60,7	C
	3.2	%	2	1	0,8	C	1,9	PC

According to the third block of indicators, the condition in none of the subjects cannot be called completely acceptable. In terms of the degree of depreciation of the main production assets of the energy sector, as well as in terms of the average annual commissioning of installed capacity, the Chuvash Republic, Kirov, Penza and Saratov regions are in a state of crisis. In the Republic of Bashkortostan, according to indicator 3.2, the state moved from pre-crisis to acceptable due to the commissioning of a number of solar power plants with a total capacity of 50 MW in 2020-2021, as well as units at the Novo-Salavatskaya CHP and Ufa CHP-1. In the Orenburg region, the situation changed from crisis to pre-crisis due to the commissioning of solar power plants with a total capacity of 327 MW over a 5-year period.

Integral assessment of ES by regions of the Southern Federal District

Taking into account the previously presented and analyzed values of the main indicators, integral assessments of the level of energy security in the territories of the constituent entities of the Volga Federal District were obtained. To form these estimates, an approach was used based on convolution of indicator values, taking into account their specific weights. Qualitative characteristics of the state of all discussed indicators from table. 2-4, were collected from the corresponding territories and processed according to a specially developed methodology [3,4,9]. As a result, a qualitative integral assessment of the state of energy security in the regions of the Volga Federal District was obtained, presented in Table. 5.

Table 5. Integrated qualitative assessment of the state of energy security in the territory of the subjects of the Volga Federal District for 2016, 2020.

Years	The order numbers of the estimated ES indicators								The sum of the specific weights by state			Quality condition ES
	1.1	1.2	1.3	2.1	2.2	2.3	3.1	3.2	Boundaries of states			
	Specific weights of indicators								C[1]	PC	N[2]	
	0,104	0,138	0,133	0,120	0,079	0,170	0,127	0,129				
1	2	3	4	5	6	7	8	9	10	11	12	13
Republic Bashkortostan												
2017	N	N	PC	C	N	C	PC	PC	0,29	0,38	0,32	PC
2021	N	N	PC	C	N	C	PC	N	0,29	0,26	0,45	PC
Republic Mari El												
2017	PC	N	N	C	N	C	PC	C	0,41	0,23	0,35	C
2021	N	N	N	C	N	C	PC	C	0,41	0,12	0,45	C
Republic Mordovia												
2017	N	N	C	C	C	C	PC	C	0,63	0,12	0,24	C
2021	N	N	C	C	C	C	PC	PC	0,50	0,25	0,24	C
Republic Tatarstan												
2017	N	N	PC	C	N	C	PC	N	0,29	0,26	0,45	PC
2021	N	N	PC	C	N	C	PC	PC	0,29	0,38	0,32	PC
Republic Udmurtia												
2017	PC	N	C	C	PC	C	PC	N	0,42	0,31	0,26	C
2021	PC	N	C	C	PC	C	PC	C	0,55	0,31	0,13	C
Chuvash Republic												
2017	N	N	C	C	PC	PC	PC	C	0,55	0,20	0,24	C
2021	N	N	C	C	PC	PC	C	C	0,67	0,07	0,24	C
Kirov region												
2017	N	N	C	C	N	N	PC	N	0,25	0,12	0,62	N
2021	N	N	C	C	N	N	C	C	0,50	0	0,49	PC
Nizhny Novgorod region												
2017	N	N	N	C	N	N	PC	N	0,12	0,12	0,75	N
2021	N	N	N	C	N	N	PC	C	0,24	0,12	0,62	PC
Orenburgskaya region												
2017	N	PC	N	N	PC	N	C	C	0,25	0,21	0,52	PC
2021	N	N	N	PC	PC	N	C	PC	0,12	0,32	0,54	PC
Penza region												
2017	N	N	C	C	C	N	C	C	0,58	0	0,41	C
2021	PC	N	C	C	C	N	C	C	0,58	0,10	0,30	C
Permian edge												
2017	N	N	PC	C	N	N	PC	N	0,12	0,26	0,62	PC
2021	N	N	PC	C	N	N	PC	N	0,12	0,26	0,62	PC
Samara region												
2017	N	N	N	C	N	N	PC	C	0,24	0,12	0,62	PC
2021	N	N	N	C	N	N	PC	C	0,24	0,12	0,62	PC
Saratovskaya region												
2017	N	N	N	C	PC	N	C	C	0,37	0,07	0,54	PC
2021	N	N	N	C	PC	N	C	C	0,37	0,07	0,54	PC
Ulyanovskaya region												
2017	N	N	C	C	N	N	PC	C	0,38	0,12	0,49	PC
2021	N	N	C	C	N	N	C	PC	0,38	0,12	0,49	PC

[1] The state of ES in the region is recognized as a crisis if the sum of the shares of indicators in the state "C" exceeds 0,4

[2] The state of ES in the region is recognized as normal if the sum of the specific weights of the indicators in the "N" state exceeds 0,7

As shown by the analysis of data on the qualitative state of the most important indicators of energy security in the constituent entities of the Russian Federation on the territory of the Volga Federal District (Table 5), of the 14 subjects examined, an acceptable situation with the provision of energy security in the analyzed period was observed only in the Nizhny Novgorod region, and then only in 2017 ., and by 2021 it has moved to a pre-crisis state due to the deterioration of the situation with the renewal of the electric power industry's general fund. In seven regions of the district, the situation is also assessed as pre-crisis: the Republic of Bashkortostan, Tatarstan, Orenburg region, Perm region, Samara, Saratov and Ulyanovsk regions. In the remaining six regions, the situation is assessed as a crisis, including according to the indicators of the state of the BPA of the energy sector, due to the high share of gas in the balance of consumption of BHO and, accordingly, the insufficient ability to meet the needs for BHO in conditions of peak-increasing winter demand: Mari El, Mordovia, Udmurtia, Chuvash Republic, Penza and Kirov regions.

Normalization of ES indicator values

The next stage in the work was the normalization of the obtained indicator values for all subjects of the Volga Federal District, in accordance with expression (1). Based on the obtained values for a 5-year period, graphs of the state of normalized indicator values for the corresponding blocks were constructed, Fig. 1-3. The threshold values of the indicators were also taken into account here as normalized; where according to expression (1) "crisis" = - 1, "pre-crisis" = 0. Thus, all states located in the range of values below "-1" can be considered crisis, in the range from "-1" to "0" - pre-crisis and in the area above "0" - acceptable. The obtained results of normalized values for each indicator for the regions of the Volga Federal District are presented in graphs (Fig. 1-3). Comparing the results obtained in the graphs with those previously presented in the tables (Tables 2-4), one can note the clearly expressed dynamics of changes in the level of energy security in the regions, which allows us to examine the situation in more detail and take measures to improve it. For example, according to indicator 1.1 (Fig. 1), almost all subjects, except the Republic of Udmurtia, have a sufficient supply of available power. As for ind. 1.2, then, due to fairly high indicators, significantly exceeding the threshold values of the indicators (more than 100%), a restriction was applied to a sufficient margin of 25%. Therefore, on the graph, all these subjects of analysis "coincided" at around 1.25, with the exception of the Orenburg region, where there was a transition from the pre-crisis to an acceptable state according to this indicator, due to a slight increase in generating capacity by 2021. According to indicator 1.3 (Fig. 1), the analysis showed a positive trend in meeting the needs for BHO from own sources in the Nizhny Novgorod and Saratov regions. There is some deterioration in the situation, but still within the range of pre-crisis values for this indicator, in the Republics of Bashkortostan and Tatarstan. According to indicator 2.1 (Fig. 2) - "Share of the dominant resource in the total consumption of BHO", one can immediately note the absence of regions in the zone of an acceptable state from the point of view of energy security. According to indicator 2.2 (the share of the largest power plant in the installed capacity of the region), one can note the division of regions according to all three states of the indicative analysis - crisis, pre-crisis and acceptable.

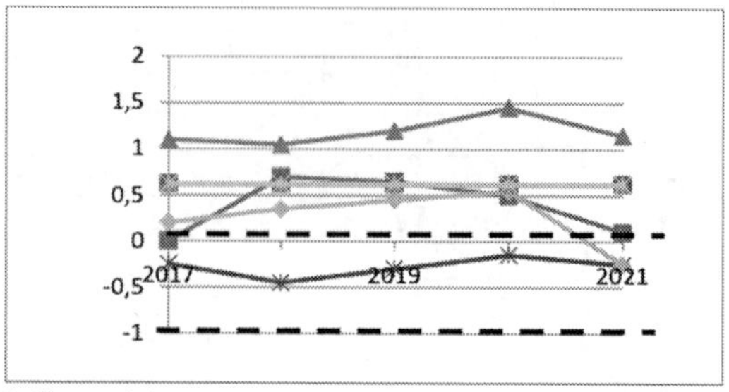

Indicator 1.1

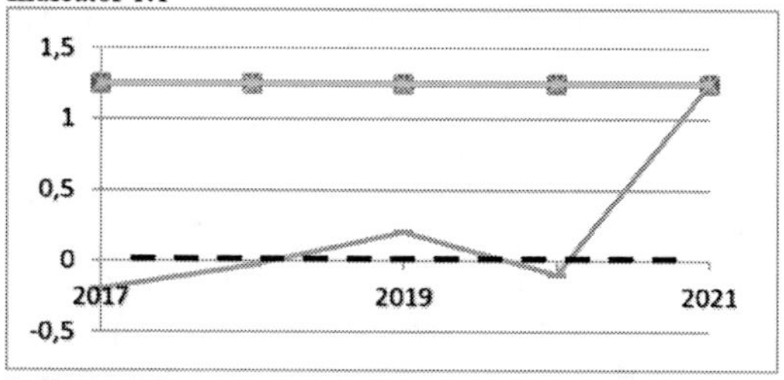

Indicator 1.2

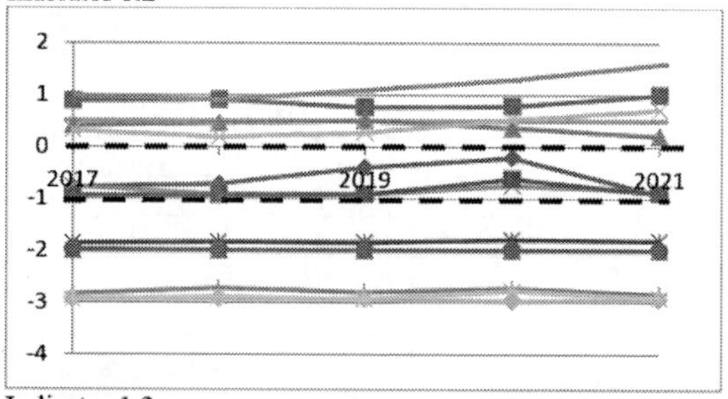

Indicator 1.3

Fig. 1. Analysis of the situation with the provision of ES requirements for the first block of indicators in the regions of the Volga Federal District.

A fairly high reserve of values for the index. 2.3 on the graph (Fig. 2) determines their reduction in this case and the accepted sufficient margin of 25% above the threshold value for regions in conditions of sudden cooling.

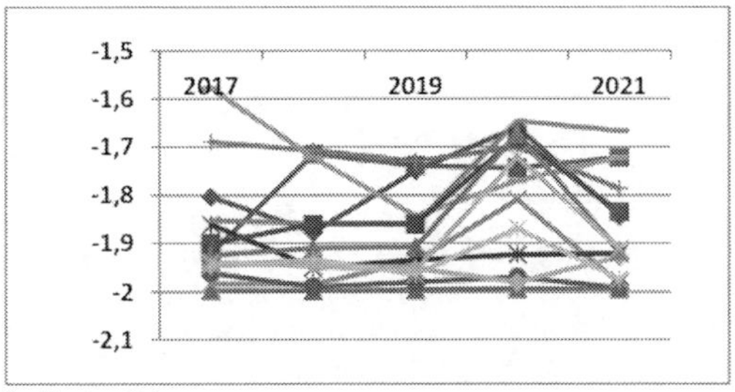

Indicator 2.1

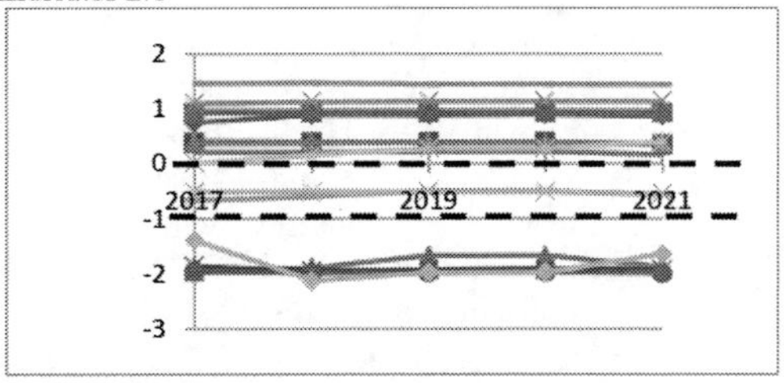

Indicator 2.2

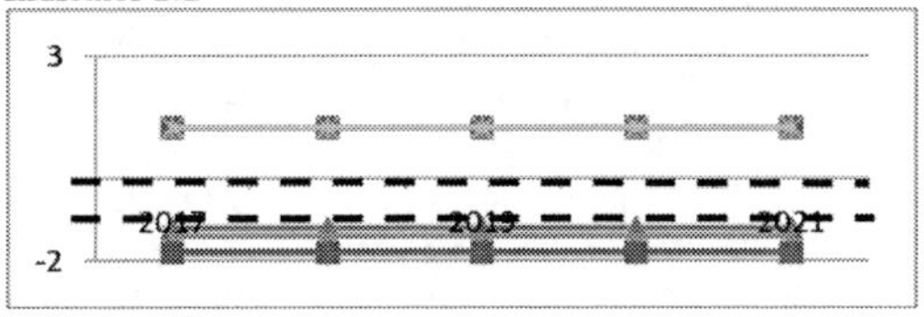

Indicator 2.3

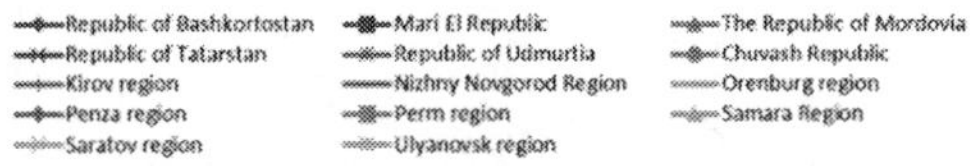

Fig. 2. Analysis of the situation with the provision of ES requirements for the second block of indicators in the regions of the Volga Federal District.

For the third block of indicators (Fig. 3), due to the insufficiently active policy in updating the BPA of the energy sector of the Volga Federal District, the values are located mainly in the zone of pre-crisis and crisis values. Sharp changes in index values. 3.2 (Fig. 3) are observed due to long breaks between the commissioning of new capacities and significant renewal of fixed production assets in the regions.

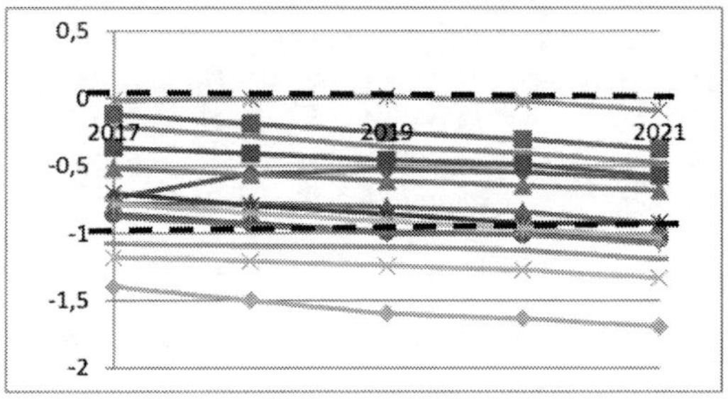

Indicator 3.1

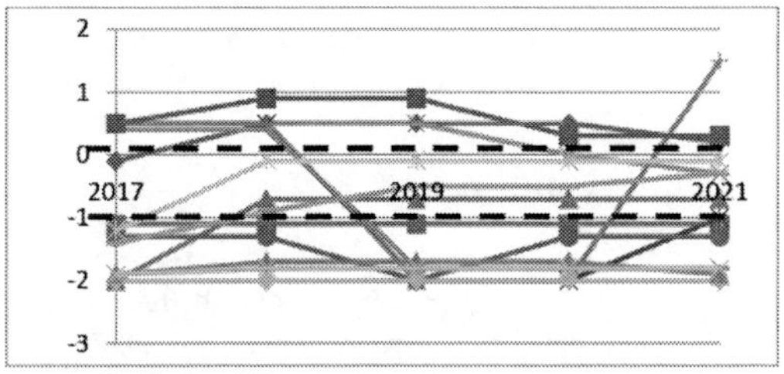

Indicator 3.2

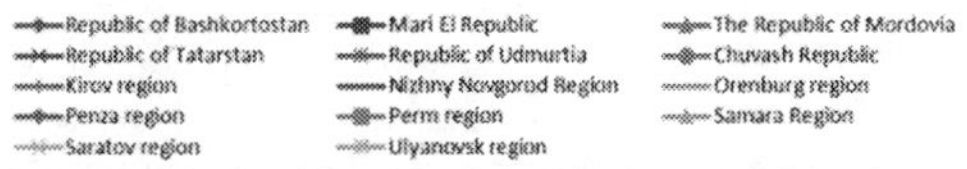

Fig. 3. Analysis of the situation with the provision of ES requirements for the third block of indicators in the regions of the Volga Federal District.

To obtain the final integral assessment of the level of ES, the normalized values of the indicators were convoluted taking into account their specific weights. The resulting complex integral assessment of the state of the energy supply of the subjects of the Volga Federal District is presented in Fig. 4. From the assessment presented in Fig. 4, as well as in Table 5, it is clear how the situation with the integral assessment of regional energy security has changed, but the use of normalized indicators allows us to visualize not only the main trends and dynamics in providing energy security, but also show proximity to acceptable or crisis state from the point of view of electronic security. Thus, as a result of the study, a slightly different picture was obtained on the integral assessment graph in comparison with the final table of indicators.

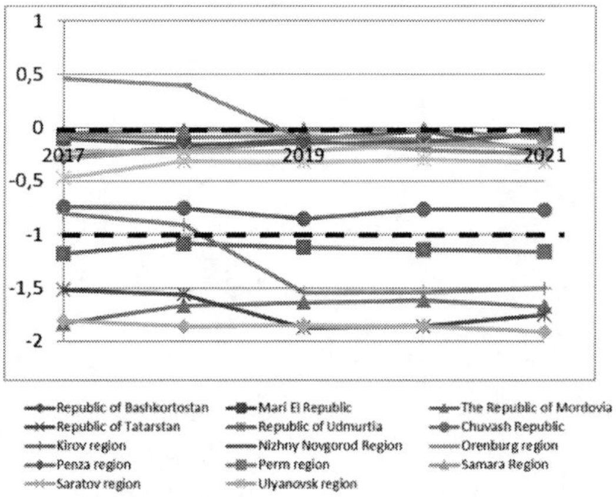

Fig. 4. Integrated qualitative assessment of the state of energy security in the regions of the Volga Federal District.

All subjects (Fig. 4) are located in the zone of pre-crisis values, and the Perm Territory and Nizhny Novgorod Region are in the acceptable zone, while according to the results of a qualitative assessment (Table 5), all subjects in 2021 were in crisis and pre-crisis states. The clarification occurred as a result of the application of the standardization method, which made it possible to more accurately determine the

https://doi.org/10.1051/e3sconf/202346101004

state of the regions on the scale of the crisis state from the point of view of economic security.

Conclusion

The approach discussed in the article differs from those used previously in that the integral assessment takes into account not only qualitative, but also quantitative assessments of the situation with ensuring energy security according to individual indicators. The use of this approach makes it possible to obtain normalized values of indicators, which in turn correctly reflect the obtained results of assessing the state of the regional energy security, and also more clearly and in detail show the dynamics of changes in the situation with the provision of energy security, both for each individual indicator and comprehensively for the subjects. The analysis showed that in the regions of the Volga Federal District, which largely do not have their own sources of fuel and energy resources, there is a crisis situation with the share of gas in the total consumption of BPA and with the share of the largest generating source. In general, it should be noted that in almost all regions the situation with the aging of the energy sector is rapidly deteriorating, and, consequently, with the danger of an increase in the number of emergencies with fuel and energy supply to consumers due to the failure of one or another equipment.

The study was carried out within the framework of the state assignment project FWEU-2021-0003 (registration number. AAAA-A21-121012090014-5) for fundamental research of the SB RAS.

References

1. S.M. Senderov, N.I. Pyatkova, V.I. Rabchuk, G.B. Slavin, S.V. Vorobyov, E.M. Smirnova *Methodology for monitoring the state of Russia's energy security at the regional level* (Irkutsk: ISEM SO RAN, 2014), 146 p.
2. V.V. Bushuev, N.I. Voropay, A.M. Mastepanov, Yu.K. Shafranik and other *Energy security of Russia.* (Novosibirsk: Science. Siberian Publishing Company RAS, 1998), 302 p.
3. Pyatkova N.I., Rabchuk V.I., Senderov S.M., Slavin G.B., Cheltsov M.B. *Energy security of Russia: problems and solutions.* – Novosibirsk: Publishing house SB RAS, 2011, 198 p.
4. Senderov S.M. *Assessment of the level of energy security of Russian regions and the basic principles of creating a system for monitoring energy security.* // Safety of fuel and energy complex facilities, 2012, No. 1(1), p. 125-130.
5. Senderov S.M., Smirnova E.M. *Using standardized values of indicators to determine the level of energy security using the example of the Central and Southern Federal Districts* // International Conference named after. Y.N. Rudenko "Methodological problems of studying the reliability of large energy systems" (RSES 2022), Vol. 384, No. 01026, 2023.
6. Statistical form of Rosstat *Information on balances, receipt and consumption of fuel and heat, collection and use of waste oil products,* 2017-2021.
7. Statistical form of Rosstat *Information on the use of fuel and energy resources*, 2017-2021
8. Statistical form of Rosstat *Technical and economic indicators of power plants, district boiler houses*, 2017-2021
9. Pyatkova N.I., Senderov S.M., Cheltsov M.B. and others. *Application of two-level research technology in solving problems of energy security* // Proceedings of the Russian Academy of Sciences. Energy, 2000, No. 6, p. 31-39.
10. Code of rules SP 131.13330.2020 *Construction climatology* SNIP 23-01-99 *, Moscow, 2020.

 https://doi.org/10.1051/e3sconf/202346101005

About the Planned Reliability of Electric Power Systems

Dmitry Krupenev[1,2*]

[1] Melentiev Energy Systems Institute of Siberian Branch of the Russian Academy of Sciences, Department of Energy Security, 664033 Irkutsk, Russia
[2] Irkutsk National Research Technical University, School of Power Engineering, 664074 Irkutsk, Russia

Abstract. The article discusses the issue of expanding the methodological foundations of system reliability of electric power systems as part of the development of a new type of system reliability - planned reliability. Planned reliability is a type of system reliability intended for the analysis of electric power systems when planning their operation for a period of up to 1-2 years. Planned reliability includes some signs of security and adequacy. The need for the development of methodological foundations for this type of system reliability is associated with the need to solve a number of important problems in electric power systems, such as planning repairs of power equipment, long-term planning of power system modes, etc.

1 Introduction

Reliability is a complex property of objects to perform specified functions to a given extent under certain operating conditions [1]. Projecting this definition onto the electric power system (EPS), we can conclude that the reliability of the electric power system is its property, which consists in the generation, transmission and distribution of electricity in the required volume and standardized quality, taking into account failures of power equipment, which are of a probabilistic nature, planned repairs of power equipment, probabilistic the nature of electricity consumption, disruption of the supply of primary energy resources to stations.

In the theory and practice of ensuring the reliability of EPS, such a concept as system reliability is used. System reliability is the reliability of a system as a complex technical or production facility (system reliability includes adequacy and security) [1]. Adequacy is the property of an object to satisfy consumer requirements within specified values and restrictions on the supply of energy resources, taking into account planned and unplanned interruptions in the operation of its elements and operational restrictions. If we make a composition of individual properties that form adequacy, then the subject of consideration can include reliability and maintainability (recoverability) [2]. Security is the property of an object to maintain specified operating states when conditions change, element failures and sudden disturbances. The composition of individual properties that make up operational reliability consists of reliability, stability, survivability, and controllability.

Both adequacy and security are used to solve specific problems at various stages of EPS control. Conventionally, the stages of management can be divided into the following: long-term design, work planning and operation. In terms of duration, the operation stage includes direct online control, planning of EPS operation is carried out for up to one year, long-term design is carried out for a period of up to 10-15 years.

The main goal of the security analysis is to assess the impact of possible failures of power equipment and other adverse phenomena in the EPS on its ability to maintain operating parameters within the required limits and uninterruptedly supply consumers with electricity and power in full. The stability of the EPS power system is the basis of security and, by analogy with static and dynamic stability, there are two subtypes of security: static security and dynamic security [3]. Various methods are used to analyze security [4-6], the main requirement for these methods is speed while maintaining the required accuracy. Indicators of operational reliability are indicators showing the probability of deviation of operational parameters from nominal values, the lack of resources in different places of the system in a specific operating states of the EPS. As part of the assessment of operational reliability, problems of online control of EPS are solved, and security is also assessed when solving some design problems. With online control of EPS, the main task of security is to determine the optimal composition of control actions to ensure reliability of power supply.

Adequacy is assessed when planning work and designing the development of EPS. The main goal of assessing adequacy is similar to the purpose of assessing operational reliability and is to analyse the ability of the system to supply consumers with electricity and power, taking into account random and non-random events: failures and repairs of power equipment, load fluctuations, etc. To analyse adequacy, a technique based on the method is mainly used Monte Carlo [2,7,8]. When assessing adequacy, only the balance of the active power of the EPS and the ability of the EPS elements to withstand active power loads

* Corresponding author: krupenev@isem.irk.ru

are taken into account; accordingly, operating parameters such as voltages in nodes and reactive power are not analysed and not taken into account. Not taking into account operating parameters is an assumption that affects the final result of determining reliability indicators. The main indicators obtained as a result of assessing adequacy are: probability of deficit-free operation, mathematical expectation undersupply of electricity, energy reliability characteristics. Based on these indicators, the problems of synthesizing adequacy are solved, mainly to justify the level of redundancy of generating capacity, the structure and capacity of electrical networks in the long-term design of development.

Among the many problems solved at all stages of the design and operation of EPS, there are a number of problems that do not correspond well to the use of the methodological apparatus of security and adequacy for their solution. First of all, this refers to the tasks that arise when planning the operation of an EPS. The main limitation, in methodological terms, on the part of security is the time limitation of reliability assessment, since in essence the reliability of a certain EPS state is assessed. As for adequacy, the main limitation here is the assumptions made in the assessment, which are fully justified in long-term development design, when there is objective high uncertainty in consumption forecasts and operating conditions of EPS and fairly rough estimates of EPS states are required.

Thus, in order to correctly solve problems that arise when planning the operation of EPS, it is necessary to develop a methodological basis for assessing system reliability. To do this, it is proposed to introduce a third type of system reliability – planned reliability. As a result, system reliability will consist of three components: security, planned reliability, adequacy (Fig. 1).

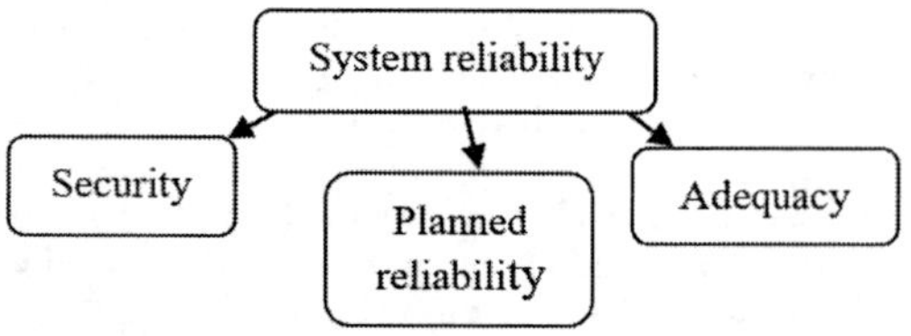

Fig. 1. Types of system reliability of EPS.

As part of the assessment of planned reliability, it is necessary to comply with the requirements for the dynamics of changes in random processes in the EPS over the required period of time and at the same time model the EPS states quite accurately. In essence, the problem of assessing the planned reliability of an EPS can be formulated as follows: for a given forecast of power consumption for a period of up to 1-2 years, a known structure and composition of EPS equipment (including taking into account plans for the input/output of equipment), it is necessary to determine reliability indicators characterizing the provision of electricity and the power of consumers, reflecting the "weaknesses" of the EPS and allowing rational decisions to be made when solving subsequent problems of synthesizing planned reliability, taking into account random processes in the behaviour of power equipment, the random nature of fuel supplies to power plants, random fluctuations in electricity consumption, as well as the most correct power flow distribution.

2 Problems solved based on the planned reliability of EPS

When planning the operation of EPS, a set of problems is solved that directly or indirectly affect their reliability. This may mean that to effectively solve such problems it is necessary to develop methods and approaches that will correctly take into account the factors affecting reliability. Based on the content of the problems, it is possible to formulate requirements for analyzing and determining the planned reliability of the EPS. By planned reliability we understand the property of an EPS, which consists in the ability to perform the required functions of supplying consumers with electricity and power in the required volume and standard quality and quickly respond to changes in electricity consumption and generation, taking into account random processes in the behavior of power equipment, the random nature of fuel supplies to power plants, random fluctuations in electricity consumption, as well as correct power flow distribution. The main idea of developing a methodological apparatus for analyzing planned reliability is a comprehensive analysis of existing probabilistic events and processes occurring in the EPS and determining their impact on the reliability of the supply of electricity and power to consumers. The composition of individual properties of the planned reliability of an EPS will combine reliability, maintainability (recoverability), stability, survivability and controllability.

The question of developing a methodological apparatus for planned reliability has arisen in recent decades in connection with the transformations taking place in the global electric power industry. Along with the existing problems of ensuring the reliability of the EPS, transformations in the electric power industry, contributing to changes in the structure of production and consumption of electricity, contribute to ensuring the reliability of the EPS. The ongoing changes make both a positive and negative contribution, but in any case it is necessary to be able to adequately model these changes in order to assess the reliability of the power supply. Thus, on the electricity production side, uncertainty in its production has increased due to the development of renewable energy sources (RES), which are characterized by a highly probabilistic nature of work. One can also note the development of distributed generation, which introduces additional specifics when ensuring the reliability of EPS [9]. Significant volumes of RES in the balance of power and electricity can lead to undersupplies of electricity and power to consumers and the development of system accidents [10-12]. On the electricity consumption side, a demand management mechanism can be identified, with the help of which it is possible to "smooth out" the power consumption schedule and thereby increase the reliability of the system. Undoubtedly, as part of the analysis of planned reliability, special attention must be paid to the modeling of energy storage systems (ESS). In modern energy systems, various types of energy storage systems are being developed, each with its own specific operation.

As noted above, modern trends in the electric power industry create requirements for expanding the methodological apparatus for assessing and ensuring system reliability. The following main problems can be identified that can be solved at a higher quality level when using planned reliability analysis:

1. Formation of schedules for repairs of power equipment during long-term planning of EPS operation.

Planning for repairs of EPS equipment depends on two main components. Firstly, the need for scheduled repairs, which can be determined based on operating hours and calendar time. Secondly, the ability of the EPS to bring power equipment into scheduled repairs with maximum efficiency and without increasing the risk of system accidents is above critical assessment.

2. Analysis of planned reliability for the period of planning the operation of the EPS.

A similar analysis of adequacy is carried out in various power systems [13,14]. A numerical assessment of planned reliability will make this analysis more in-depth. When identifying bottlenecks in the EPS, it is necessary to adjust control actions to ensure reliability of power supply and eliminate bottlenecks. In addition, in the process of assessing planned reliability, it becomes possible to take into account at a fairly detailed level the specifics and operating modes of power plants, including the risks of extreme water availability for hydroelectric power plants.

3. Assessment of EPS capabilities and efficiency in the integration of renewable energy sources (RES).

One of the main problems in the electric power industry in recent decades is assessing the effectiveness of RES integration. Such integration, due to the uneven nature of electricity and power generation from renewable energy sources, can lead to problems in ensuring the system reliability of EPS and the reliability of power supply. Assessing the planned reliability of EPS with adequate modeling of RES will make it possible to reflect the most critical random processes for EPS that characterize the operation of RES. The main issue in the problem under consideration is the ability of the EPS to compensate for the imbalance of power and electricity caused by the lack of a renewable source (sun, wind). Ultimately, the amount of generating capacity reserves, as well as the structure and capacity of the electrical network, will depend on the accuracy of RES modeling.

4. Analysis of the effectiveness of integration of energy saving systems (ESS) into electric power systems.

There are various types of ESS, which are characterized by such parameters as capacity, charge/discharge rate, operating principle. These parameters must be correctly reflected when assessing the reliability of the EPS. The specificity of the ESS operation lies in the continuity of the charge/discharge process. This process depends on the capabilities of the solar energy system and on the operating modes of the EPS, so it is important to reflect all influencing and limiting factors when modeling the solar energy system. In addition, the methodology for assessing the planned reliability of EPS is based on the analysis of random processes, which is important when modeling the operation of ESS.

5. Analysis of the need to attract electricity consumers to ensure the required level of planned reliability.

The main idea of this task, first of all, comes down to the selection of electricity and power consumers to manage demand. Demand management is one of the mechanisms for managing the reliability of the power system. In the process of selecting consumers, it is important to know which consumers can make the maximum contribution to ensuring the reliability of the EPS.

3 Methodology for assessing the planned reliability of EPS (content characteristics)

As noted above, the planned reliability of an EPS essentially integrates some aspects of assessing both security and adequacy, and also has a number of specific features. The most effective (in different positions) method for assessing system reliability is the Monte Carlo method [8]. It is on this basis that the methodology for assessing planned reliability is based.

The functioning of an EPS is a multidimensional random process in which many numerical parameters that characterize the operation of the power system change over time. By a random process of functioning of the EPS $X^n(t)$ we will understand a process whose value for any fixed $t = t_0$ is a random state of the system $X^n(t_0)$ [15,16]. During the operation of an EPS, various events (ω) occur: equipment failures, changes in consumer load, changes in the parameters of elements, which are analyzed when assessing reliability. Taking the latter into account, the random process of EPS functioning can be written as a function of two arguments:

$$X^n(t) = \varphi(t, \omega), \omega \in \Omega, t \in T, X^n \in \Xi,$$

where: ω – events occurring in the EPS; Ω – event space; T – time period of EPS reliability analysis; Ξ – set of possible values of the random process of functioning of the EPS $X^n(t)$.

The EPS consists of various elements with different operating features. The random process of EES functioning is a process with continuous time and with continuous states (in fact, the states are mixed, since there are both continuous and discrete ones; in the theory of random processes, such states are also called continuous). Continuous time is explained by the fact that for EPS, system transitions from states to other states can occur at any moment t of the observed period T. In the process of assessing the reliability of EPS based on the Monte Carlo method, the periods under consideration are discrete and equal to 1 hour. As for random states, as already noted, various states are considered, both discrete and continuous, for example, failures of power equipment form discrete states, and the random behavior of an EPS parameter, for example, voltage and a parameter that affects the reliability of the EPS, for example, speed winds or solar insolation form continuous conditions.

The flow of events for an electric power plant is extraordinary, since more than one event can occur at any given time, for example, the failure of a generating unit, the failure of a power transmission line, a change in wind speed, which affects the power output of a wind farm. If we consider the flow of events for a specific element of the EPS, then it can be accepted as ordinary, since events, in the case under consideration - failures and restorations, appear one by one. The flow of events for an EPS is a flow with consequences, since for any non-overlapping time periods the number of events (for example, failures) falling on one of them can depend on how many events fell on the previous one, for example, the failure of some elements of an EPS can lead to subsequent periods of time lead to overload of other elements, their failure and the occurrence of power shortages and undersupply of electricity. At the same time, the flow of events for a specific element will appear without consequences, since the events that form the flow appear at successive moments in time independently of each other, for example, failures and restorations. As for the stationarity of the flow of events in the EPS, in fact it is non-stationary, since the probability of the analyzed events can change over time depending on various factors.

4 Conclusion

Ensuring the reliability of electric power systems is the most important problem when managing them at different time stages. In the theory and practice of managing electrical power systems, the concept of system reliability is used. System reliability includes two components: security and adequacy. These components cover a number of problems solved when managing operations and designing development. However, these components do not fully cover a number of problems that arise when planning the operation of electric power systems. To solve problems that arise when planning the operation of electric power systems, it is proposed to use a new type of system reliability, which is called planned reliability. Based on planned reliability indicators, it becomes possible to solve a number of pressing problems in the electric power industry at a qualitatively new level.

Acknowledgements

The study was supported by the Russian Science Foundation grant No. 23-29-00435.

References

1. Reliability of energy systems. Collection of recommended terms, IAC "Energy",**192** (2007) (in Russian)
2. G. Kovalev, L. Lebedeva, Reliability of Power Systems, Springer Cham (2019)
3. K. Morison, Lei Wang, P. Kundur, Power system security assessment. IEEE Power and Energy Magazine, Volume **2**, Issue 5, (2004)
4. D.S. Krupenev, A.V. Domyshev, Assessment of security of electric power systems based on the Monte Carlo method, Electricity. **2**. (2015) (in Russian)
5. G.F. Kovalev, L.M. Lebedeva, Areas of use and limits of applicability of criterion N-1 in the formation of the structure and selection of parameters of EPS elements. Irkutsk, **69** (1999) (in Russian)
6. K. Teeparthi, D. M. V. Kumar Power System Security Assessment and Enhancement: A Bibliographical Survey. J. Inst. Eng. India Ser. B (2020)
7. R. Billinton, R.Allan, Reliability Evaluation of Power Systems, Plenum Press, (1996)
8. D. Krupenev, D. Boyarkin, D. Iakubovskii, Improvement in the computational efficiency of a technique for assessing the reliability of electric power systems based on the Monte Carlo method. Reliability Engineering and System Safety. Vol.**204** (2020)
9. P.V. Ilyushin, Prospects for application and problematic issues of integrating distributed energy sources into electrical networks: monograph. Electrical Engineering Library, No. 8 (260), **116** (2020) (in Russian)
10. 9 August 2019 power outage report, **50** (2020)
11. N.I. Voropai, D.S. Krupenev, S.V. Podkovalnikov, S.M. Senderov, Two energy collapses - in the state of Texas, USA and in the Primorsky Territory, Russia Electricity. Transmission and distribution. No. **4 (67)** (2021)
12. After oil and gas, Europe is now running out of wind (https://bigthink.com/strange-maps/europe-wind-drought/) (2022)
13. 2023 Summer Reliability Assessment (https://www.nerc.com/pa/RAPA/ra/Reliability%20Assessments%20DL/NERC_SRA_2023.pdf) (2022)
14. Mid-term adequacy report on the electricity supply-demand balance in France (https://assets.rte-france.com/prod/public/2020-09/2019_generation_adequacy_report.pdf) (2019)
15. E.S. Ventzel, L.A. Ovcharov, Theory of random processes and its engineering applications. M.: Science. Ch. Ed. Phys.-math. lit., **384** (1991) (in Russian)
16. A.D. Ventzel, Course on the theory of random processes. M.: Science. Ch. Ed. Phys.-math. lit., , **320** (1975) (in Russian)

Formation of power equipment repair schedules using indicators of planned reliability

Dmitry Krupenev [1*], *Denis* Boyarkin [1], and *Dmitrii* Iakubovskii [1]

[1] Melentiev Energy Systems Institute of Siberian Branch of the Russian Academy of Sciences, Lermontov str., 130, Irkutsk, 664033, Russia

Abstract. This article discusses the problem of planning maintenance of power equipment during long-term planning of the operation of electric power systems. Here we propose a mathematical model for the formation of maintenance schedules for power equipment, which takes into account two aspects: the current technical condition of power equipment and maximizing the planned reliability of electric power systems when carrying out maintenance of power equipment. Provided methodology for generating maintenance schedules for power equipment was tested on using the scheme of three reliability zones.

1 Introduction

Ensuring reliable power supply to consumers is an essential condition for their successful operation and development. Over the past decades, the operating conditions of electric power systems (EPS) have changed dramatically. New elements and technologies are integrated into EPS, such as renewable energy sources (RES), energy storage systems (ESS), distributed generation, active consumers [1]. It's also worth noting the increasing role of the uncertainty of natural and weather-related phenomena. For example, extreme water content, which influent to the production of electricity and power at hydroelectric power stations depends. The increase in the frequency of extreme phenomena: strong wind, snowfall, freezing rain and other phenomena, which affect determine the accident rate of power equipment at different periods of time, as well as the wear and tear of power equipment [2,3]. All of the above leads to a complication of the technological structure of power systems and the complexity of managing technological processes, which includes the process of planning maintenance of power equipment. In the current conditions, the role of timely and adequate comprehensive analysis of EPS and the development of informed decisions to manage their functioning is increasing. One of the aspects of such analysis is the assessment of the reliability of the EPS, taking into account the maximum number of influencing factors. In the case under consideration, we are talking about the planned reliability of the EPS, which is an integral part of system reliability. In case of assessing planned reliability, the regime and balance limitations of the EPS and their influence on each other are taken into account.

The problem under consideration, namely the problem of optimal formation of maintenance schedules for power equipment in EPS, is discrete, dynamic, and probabilistic. Direct mathematical methods for solving such problems do not yet exist. In modern power systems, where uncertainty due to the reasons indicated above increases, the use of specialized tools to solve the problem of planning maintenance becomes extremely necessary. Both the reliability of the functioning of such a system and economic efficiency directly depend on this. The creation of a methodological apparatus for solving this problem will improve the correctness of decision-making when planning the operation of EPS. As a result, with the practical application of the methodological apparatus, the risks of emergencies in energy systems and negative impacts on the economy, ecology and social sphere will be minimized.

The problem of optimal planning for maintenance of power equipment has existed since the advent of EPS. Throughout history, various criteria and methods have been developed and proposed for its solution. The problem under consideration became more and more relevant with the complication of the structure of the EPS, increasing uncertainty of operation and increasing dependence of economic efficiency and safety of electricity consumers on reliable power supply. The main user of the results of solving the problem under consideration are organizations that carry out technical management of the power system. In Russia, for example, the main users are the services of the System Operator and the United Dispatch Directorates (UDD). In modern conditions, at the UDD level, the maintenance planning process is practically not automated. The service for long-term planning of energy regimes and balances collects applications from generating companies for maintenance of generating equipment for the coming annual period, and then employees of this service, with minimal use of automated tools (special software), create schedules for maintenance of generating and network equipment, taking into account the main external influences conditions. The specialized software is the Information System "System for Automated Planning of Electric Power Regimes" (IS "SAPER") [4]. IS "SAPER" does

* Corresponding author: krupenev@isem.irk.ru

not provide a direct solution to the identified problem. This software acts as an assistant for performing variant calculations of power balances and checking the feasibility of monthly and annual maintenance schedules for main energy and electrical grid equipment according to the conditions of power balances, electricity and operating restrictions. Naturally, issues of system reliability are considered at a superficial expert level, the same can be said about issues of taking into account natural phenomena, for example, extreme water content at hydroelectric power station cascades, environmental and safety issues.

Reasonable scientific approaches to solving the problem under consideration began to be proposed in the 60-70s of the last century. The formulation of the problem under study may have several options. In fact, all options represent classical formulations of the reliability optimization problem, in which the objective function can act as a reliability criterion, for example, the minimum expected value undersupply of electricity, and an economic criterion, for example, minimum operating costs, while the reliability of the energy system is taken into account in the form of restrictions. It is worth noting that some works propose methods for planning maintenance of generating equipment only, and in some, generating and network equipment.

As for the methods for solving the problem, a sufficient number of them have been developed. We can highlight the following organizations in which work was carried out on setting the problem to be solved and methods for solving it: ESI SB RAS [5-7], MPEI [8,9], UrFU [10-12], etc.

In all the analyzed works, the system reliability of EPS is either not taken into account at a sufficient level or is completely absent, although the requirements for ensuring reliability are the main ones when solving the problem under consideration.

In general, the current state of affairs in the scientific field under consideration can be characterized as follows: both domestic and foreign researchers have made significant advances to competently solve the problem under consideration, but in all analysed studies there is no comprehensiveness in taking into account the impact of maintenance on the properties of power systems: economics, reliability , safety, ecology; Also, the methodological part concerning the system reliability of the EPS is not at the proper level, and in some cases is completely absent.

2 Statement of the problem of generating power equipment maintenance schedules

Maintenance schedules for power equipment in EPS are formed based on the need for various types of maintenance. Such a need can be determined as a result of a certain operating time of the power facility, if energy enterprises have adopted a planned preventive strategy for the prevention of power equipment [13,14], or as a result of an assessment of the technical condition of the power facility [14,15].

In terms of content, we formulate the task of generating optimal maintenance schedules for power equipment as follows: for the known structure of the EPS, technical (including reliability) and economic characteristics of power equipment, forecast data on the behavior of weather and environmental phenomena for the period of planning the operation of the EPS, the known need for maintenance of generating units and lines power transmission, it is necessary to create a schedule for maintenance of power equipment to ensure the required level of reliability of the power system and maximum economic efficiency.

It should be noted that it is only necessary to determine the sequence (order) of maintenance, and issues related to the availability of resources to perform these maintenance are not considered in this work. It is believed that there are a sufficient number of maintenance teams to complete maintenance within a certain time frame.

In essence, the problem being solved is the task of generating an optimal schedule, in the case under consideration for maintenance of power equipment. This means that its description will be valid when applying the methodological apparatus of scheduling theory [16-18].

We consider an EPS consisting of N power facilities (generating units, power transmission lines), and J of them are power facilities for which it is necessary to carry out maintenance in the upcoming period T. Each maintenance is characterized by a duration τ_j, $j = 1, \dots, J$. There are estimated values of planned reliability indicators for each sub-period t, $t = 1, \dots, T$.

Maintenance of any kind must be carried out completely without interrupting the deadlines for its completion; for this condition we introduce Boolean variables $y_j \in \{0,1\}, j = 1, \dots, J$.. If $y_j = 1$, then the maintenance is performed, otherwise $y_j = 0$. Let us denote G_i as the set of options for changing the composition of generating and network equipment at node i, $i = 1, \dots, I$ when it is taken out for maintenance.

When solving the problem of generating power equipment maintenance schedules, we write the objective function as follows:

$$\sum_{t=1}^{T}\sum_{i=1}^{I} E_{t,i} \underset{\Delta x_j}{\rightarrow} min, \quad (1)$$

given the limitations

$$\sum_{j=1}^{J} \Delta x_j \tau_j y_j \leq W, \quad (2)$$

$$\Delta x_j \in G, \quad (3)$$

$$y_j \in \{0,1\}, j = 1, \dots, J. \quad (4)$$

$$R_i(\Delta x, \xi) \geq R_i^{norm}, i = 1, \dots, I, \quad (5)$$

where: $E_{t,i}$ – value of expected value undersupply of electricity in EPS nodes in design intervals, MW; I – number of EPS nodes; Δx_j– operating characteristic (power, throughput) of energy equipment put out for scheduled maintenance, MW; W – volume of scheduled maintenance of all types, MW h; G – set of options for shutting down power equipment to perform scheduled maintenance at node i; R_i - actual value of the planned reliability indicator; R_i^{norm} – normalized value of the planned reliability indicator; ξ – random component, reflecting the probabilistic basis for assessing planned reliability.

3 Methodology for creating maintenance schedules

To solve the problem, a methodology was developed based on an analysis of the planned reliability indicators of the EPS. The developed methodology represents the sequence of the following actions:

1. Formation of the order of removal of generating equipment for maintenance (for example, according to the power weight criterion: the power of the unit multiplied by the number of days required for maintenance).
2. Assessment of the planned reliability of the initial version of the EPS for the period of work planning, without taking power equipment out of operation for scheduled maintenance.
3. Analysis of the values of the controlled reliability indicator in each day of the billing period and in each EPS node for compliance with the standard. Identification of the sequence of days with the smallest integral value for the period under consideration sum expected value non-supply of electricity for EPS and electricity consumption, the sequence of days should be equal to the number of days required to maintenance the generating unit.
4. Assessment of the planned reliability of the EPS, taking into account the maintenance option for the selected unit.
5. Repeating the second and third stages of the methodology until all power equipment requiring maintenance is taken into account, or the limitation that controls the standard level of reliability is violated.

If condition (5) is violated, and power equipment requiring maintenance is not taken into account in the maintenance schedule, then several options for solving this problem are possible, based on the prevailing conditions:

1. Changing of maintenance dates for equipment that requires maintenance, but is not taken into account in the current schedule, to later periods.
2. If the generated equipment maintenance schedule is relatively "tight" and there is no possibility of rescheduling maintenance, then for the EPS under consideration, as part of its operation planning, it is necessary to provide additional control actions to compensate for the removal of equipment for maintenance.

4 Experimental studies

To solve the problem, based on the proposed formulation of planning maintenance of power equipment of EPS (1)-(5) and testing the proposed solution algorithm, experimental studies of planning maintenance of generating equipment were carried out on a three-zone design scheme (Fig. 1). Tables 1-3 present the main initial data used in experimental studies.

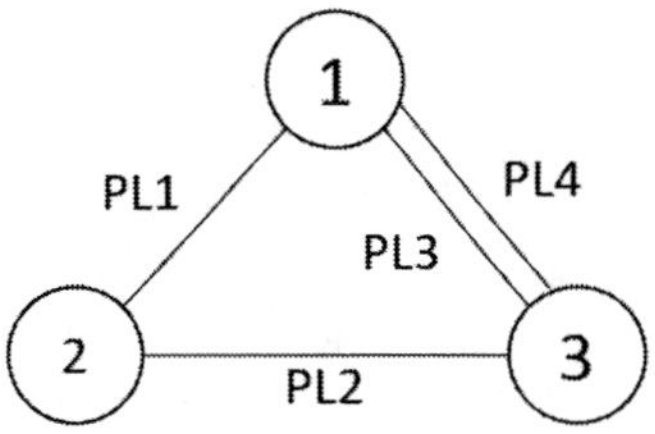

Fig. 1. Investigated system scheme. (PL – Power line)

Table 1. Maximum value of loading in reliability zones.

Name of node	Maximum value of loading MW
Node 1	17500
Node 2	17500
Node 3	17500

Table 2. Basic input data for generating units in reliability zones.

№	Power MW	Number of units	Accident rate, c.u.	Standards for maintenance , days	
				Overhaul	Refurbishment
1	50	50	0,02	37	1
2	100	50	0,045	40	2
3	200 (heating unit)	15	0,045	44	3
4	200 (hydro unit)	10	0,005	44	4
5	400	10	0,055	51	5
6	500	3	0,07	62	40

Table 3. List of repair requests for each of the reliability zones (number of units).

Power, MW	Reliability zone 1		Reliability zone 2		Reliability zone 3	
	Over.	Refu.	Over.	Refu.	Over.	Refu.
50	1		2		2	3
100	1	1	1	1	1	
200				1	2	1
400	1		1			
500	1					

As a result of solving the problem, a repair schedule for generating equipment was determined that satisfied all the specified criteria (Fig. 2).

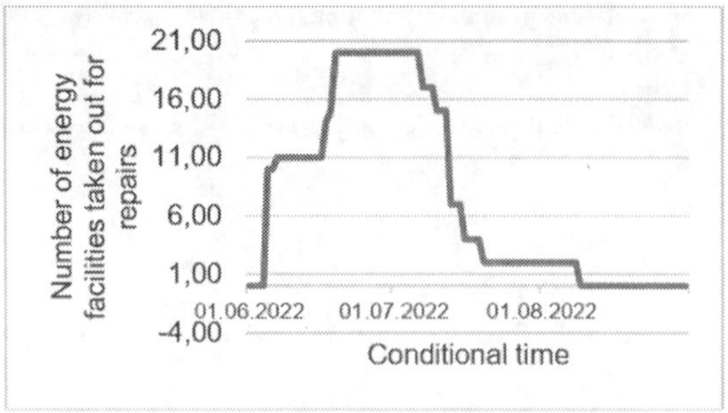

Fig. 2. Integrated schedule for repairs of generating equipment.

5 Conclusion

In while of planning the operation of electric power systems, a number of tasks are solved. One of the task is the formation of a schedule for repairs of power equipment. The solution to this problem depends on many factors, the main one of which is the reliability of the EPS. Thus, the solution to the identified problem must be carried out on the basis of an assessment of the planned reliability of EPS, which most fully reflects the specifics of the functioning of modern EPS. The article analyses the identified problem, proposes a substantive and mathematical formulation and an algorithm for solving it, and shows the results of experimental studies.

6 Acknowledgements

The study was supported by the Russian Science Foundation grant No. 23-29-00435.

References

1. N. I. Voropai, "Directions and problems of transformation of electric power systems," Electrichestvo, no.7, pp.12-21, 2020. (in Russian)
2. Voropai, N.I., Krupenev, D.S., Podkovalnikov, S.V., Senderov, S.M. (2021). Blackout in Texas, USA: Analysis and Some Conclusions. ECO. No. 9. Pp. 125-143. (in Russian) DOI: 10.30680/EC00131-7652-2021-9-125-143
3. Voropai N.I., Krupenev D.S., Podkovalnikov S.V., Senderov S.M. Two energy collapses - in the state of Texas, USA, and in the Primorsky Territory, Russia. Electricity. Transmission and distribution. 2021, No. 4 (67), p. 166-174 (in Russian).
4. Information system "System for automated planning of electric power modes" GENERAL DESCRIPTION OF THE SYSTEM. Moscow, 2021, 19 pages.
5. Lebedeva L.M., Fedotova G.A. Software-computing complex RESERVE - implementation of a methodology for the rational use of power reserves in EPS. Methodological issues in studying the reliability of large energy systems. Irkutsk: SEI SB AN USSR, 1981. (in Russian)
6. Dubitsky M.A., Rudenko Yu.N., Cheltsov M.B. Selection and use of generating capacity reserves in electrical power systems. M.A. Dubitsky, Yu.N. Rudenko, M.B. Cheltsov. M.: Energoatomizdat, 1988. (in Russian)
7. Voropai N.I., Fedotova G.A. Planning maintenance of power generating equipment in market environment with regard for reliability, Automation and Remote Control, 2010, Volume 71, Issue 7, Pages 1442–1446
8. Botvinnik M.M. Computers in Chess: Solving Inexact Search Problems. New-York: Springer-Verlag, 1984.
9. Botvinnik M.M. Chess method for solving exhaustive search problems. M: Soviet sport, 1989, 112 p. (in Russian)
10. Arzamastsev D.A., Oboskalov V.P. Determination of a plan for capital repairs of the main equipment of power systems using the coordinate-wise optimization method. Izvestiya VUZov. Energy, 1970, No. 8, p. 106–110. (in Russian)
11. Nesterenkov V.P., Oboskalov V.P. On the issue of optimal planning of major repairs of the main units of energy systems stations. Issues of optimization of development and operation of power systems, 1966, p. 83–90. (in Russian)
12. Gubin P.Y., Oboskalov V.P. Differential Evolution Method for Generation Maintenance Scheduling. News of the Russian Academy of Sciences. Energy, 2021, No. 2. p. 50-64. (in Russian)
13. Order of the Ministry of Energy of the Russian Federation dated October 25, 2017 N 1013 "On approval of requirements for ensuring the reliability of electric power systems, reliability and safety of electric power facilities and power receiving installations" Rules for the organization of technical maintenance and repair of electric power facilities. (in Russian)
14. Nazarychev A.N., Krupenev D.S. Reliability and Assessment of the Technical Condition of the Equipment of Power Supply Systems. Novosibirsk: Nauka, 2020, 224 p. (in Russian)
15. Nazarychev A.N., Pugachev A.A., Andreev D.A. Risk-based management of the operation of electrical equipment taking into account its technical condition. Electricity. Transmission and distribution, 2020. No. 6 (63), p. 134-135. (in Russian)
16. Tanaev V.S., Shkurba V.V. Introduction to scheduling theory. Main editorial office of physical and mathematical literature of the Nauka publishing house, 1975, 257 p. (in Russian)
17. Lazarev A.A., Gafarov E.R. Scheduling theory. Tasks and algorithms. Moscow: MSU, 2011, 222 p. (in Russian)
18. Brucker P. Scheduling algorithms. Springer Science & Business Media, 2004, 367 p.

Usage of electric energy storages to increase controllability and reliability of the Belarusian energy system

Mikhail Kashyn[1,*], *Nikolay* Novikov[2] and *Alexander* Novikov[3]

[1]RUE Belenergosetprojekt, Republic of Belarus
[2]JSC STC FGC UES, JIHT RAS, National Research University MPEI, Russia
[3]STC Energobezopasnost, Russia

Abstract. The paper provides an efficiency assessment of lithium-ion energy storage unit installation, including flattening the consumers daily load curve, reducing electricity losses and regulating voltage at the ESS installation point.

Introduction

The basic Energy System construction paradigm is a simultaneous and synchronous electricity production and consumption. Thus allowing to maintain one of the most important operating parameters – the frequency of the alternating current.

A daily load curve of both the entire Belarusian Energy System and its individual energy centers is characterized by a significant unevenness between the distinguished daytime peaks and the nighttime power dips. The unevenness coefficient of the daily load curve is 0.65-0.7.

When the second power unit of the Belarusian NPP is put into commercial operation in 2023, the capacity of the two units operating based on the daily load curves of the Belarusian Energy System will be about 40% of the maximum load of the Energy System. To comply with the self-balance operation condition, it is necessary to reduce the number of the powerful cycling power units of the condensing power plants (CPP) in hot reserve.

ESS connection to the electric energy system will make it possible to separate in time the processes of electricity generation and consumption (considering high efficiency of the ESS), which will lead to the load curve flattening of an individual power unit and the Energy System as a whole. This problem being solved, the ESS can become one of the key elements of the Smart electric power industry.

Potential areas of the ESS implementation in the Energy System also include: voltage and frequency regulation, hot reserve of the Energy System, emergency power supply to prevent system failures (when sectioning off the Energy System) and to recover the system after an accident, the electricity consumer emergency power supply. A special value of the ESS is its ability to simultaneously perform the above mentioned functions [1].

It should be noted that the renewable energy sources (RES) in the Republic of Belarus are not significantly distributed (the share of the power plants with RES in the installed capacity of the Belarusian Energy System is about 3%). However, the ESS usage is a key solution for the effective integration of a renewable energy source into the Energy System.

1 Analysis of ESS usage experience in the world

In the world, the Energy storage systems have been used for quite a long time. For example, galvanic cells were invented in the 1800s, and the first pumped storage projects were introduced in the early 1900s.

A relevant objective of using ESS in the Belarusian Energy System, minding a significant installed capacity of the Belarusian NPP, is to flatten the uneven daily load curves. ESS can be used to supply consumers with electricity during those periods of the day when the energy consumption exceeds its production at an economically efficient generating equipment (NPPs, large power units of TPPs), and for the accumulation of electricity during the periods when its production at those sources exceeds the consumption (resulting in the need to stop the operation of the generating equipment). Moreover, ESSs reduce the necessity to increase or decrease the load of the generating equipment in case of an emergency in the Energy System.

ESS energy intensity, their maximum output power during the discharge period, the duration of the discharge, and the efficiency of the storage unit shall be considered to determine the ESS efficiency.

The well-known energy storage technologies can be divided according to the type of stored energy into:

- mechanical (pumped storages, flywheels),
- electrochemical (accumulator batteries, flow batteries),
- chemical (fuel cells),

* Corresponding author: m.kashin@besp.by

https://doi.org/10.1051/e3sconf/202346101007

- electrical (capacitors, supercapacitors, superconducting magnetic accumulators),
- thermal (use of molten salts and hot water).

Depending on the technology, the duration of the energy storage can vary from less than 10 hours (e.g. some of the batteries) till the seasonal storage (weeks, months and years) (e.g. pumped storage systems).

Fig. 1 schematically presents the information about the power level and the ESS discharge time (duration) of various technologies [2].

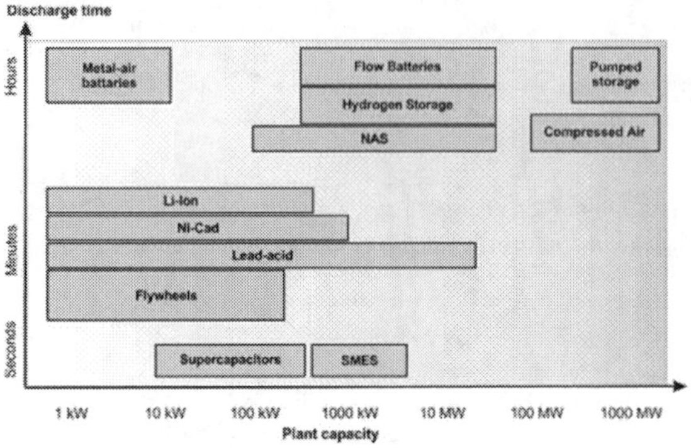

Fig. 1. Power levels and time (duration) of discharge of the various ESSs

The key technical characteristics of the main energy storage technologies are given in [2].

2 ESS technologies possible to use in the Belarusian Energy System

At present, the ESS technology level and the industrial designs maturity in the CIS countries enable us to draw the conclusions on the applicability of only several of the ESS described above in the Belarusian Energy System.

As it was mentioned above, the commissioning of the Belarusian NPP has prioritized the issue of covering the peak load and the dip in the daily load curve of the Belarusian Energy System, since NPP units usually operate in the base-load operation condition. A traditional means to solve this problem is to construct a **pumped-storage station (PSS)** together with a nuclear power plant. The pumped-storage station is both a highly controllable source of peak power and a regulating consumer. It is effectively used:

- to cover the peak part of the graph;
- for multiple short-term system load pick-ups;
- to optimally increase the night load of the TPP in terms of technical reliability;
- as a frequency maintaining quick-operating reserve and a short-term emergency reserve;
- as a network reactive power balance regulator.

In Belarus, due to the terrain peculiarities and the necessity to flood large land areas, the unit capacity of a pumped-storage station is limited to 400-570 MW. Therefore, to provide a reliable power backup of two power units of the Belarusian NPP, each having capacity of 1200 MW, the construction of several such stations is required. In addition, this infrastructure solution requires significant capital investments and a long implementation time, which is economically inadvisable at present.

At the same time, the world experience of using ESS in the energy sector demonstrates a significant interest in the use of **lithium-ion accumulator batteries (AB)**, since they have better characteristics compared to other AB technologies. The number of lithium-ion batteries used in power generation and in other spheres grows due to their long service period, high cell voltage, good low temperature performance, sufficient charge retention, and required depth of charge.

The project "Usage concepts of the energy storage systems based on lithium-ion batteries in the Belarusian Energy System", which provides for the integrated implementation and the use of ESS at the generating facilities of the State Production Association "Belenergo", in the electrical networks, and at the electric power facilities of the industrial enterprises, and at transport, was developed in 2022.

According to this document, technically available potential of the ESS installation in industrial consumers distribution electrical networks in the Belarusian Energy System ranges approximately 500 MWh and 100 MW.

3 Usage of ESS in the distribution networks up to 110 kV of the Belarusian Energy System

In 2022, RUE Belenergosetprojekt carried out an R&D work to assess feasibility and justify technical effectiveness of installing lithium-ion energy storage systems in order to flatten the daily load curve, reduce electricity losses and regulate voltage at the ESS installation point.

A number of substations, at which the transformers load was about 50% during the standard operation, were selected as "typical" 110/10 kV substations (SS). It should be stated that in the Belarusian Energy System the percentage of the substations with such a significant transformers load is less than 5%. They are basically located either in the capital city, or in the large regional cities, and in the large industrial centers of the Republic of Belarus. However, in accordance with the policy to increase electricity consumption for heating and hot water supply pursued in the Republic of Belarus, the contracted capacity increase of both domestic and industrial consumers is expected in the coming years. This will entail the need to solve the problem indicated below.

With the increase of the electrical load at those substations the efficiency of ESS was tested in two alternative options:

- the reconstruction with an increase in transformer capacity;
- the installation of ESS for flattening the daily load peaks and keeping the existing transformers in operation avoiding their overloading.

The initial data were:

- the reported measured values of the Electricity Commercial Metering Automated System (half-hour load measuring) on the transformers windings of the substation during four typical days (winter/summer, working day/day off);

- the electric load in accordance with the active technical specifications issued for the power supply of the consumers by a given substation;

- the schematic diagrams of the substation and of the supplying network.

Substation Korzyuki of RUE Minskenergo was chosen as a "typical" 110/10 kV substation with a dominating residential type of the load.

A theoretical substation was modeled as a "typical" one with a dominating industrial type of the load.

Two 110/10 kV transformers with the capacity of 16 MVA each are installed at the 110 kV Korzyuki substation. Fig. 2 displays a simulated daily load curve of a winter weekday of this substation, considering the load increase, which allows for the use of ESS, provided having the equal discharge/charge areas of the battery. The permissible maximum load is 28% higher than the power limit of the substation, in this case, the required operating capacity of the battery should be about 38.0 thousand kWh, and the rated power should be 6.5 MW.

The horizontal line on the graph shows the overload power limit of the substation (40% –1232 A), considering the withdrawal of one of the transformers in the case of the maintenance or the emergency shutdown.

Comparative cost calculation for 110/10 kV Korzyuki substation modernization options (transformer replacement or ESS installation) shows that the capital investment in 10 kV ESS installation 12 times exceeds the cost of the transformer replacement.

110/10 kV substations with dominating industrial type of the load and a continuous operation of the enterprise (oil refining, metallurgy, heavy engineering, etc.) were not considered in this work, since they have a daily load curve without distinguished peaks and dips in electricity consumption, during which the accumulator could be charged/discharged. The installation of ESS at such substations in order to reduce the peak load is inadvisable. This does not exclude the installation of ESS for the reasons of ensuring the reliability of power supply for the most critical units of the production process of this type of an enterprise.

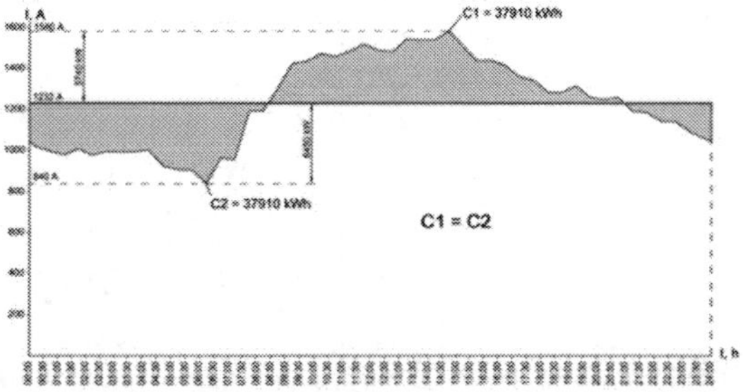

Fig. 2. Daily load curve (load current, A) of the 110/10 kV Korzyuki substation on a winter weekday, considering the equality of the discharge/charge areas of the ESS.

A 110/10 kV substation of the wood-processing industry with two transformers of 16 MVA capacity each was considered as a substation with a dominating industrial type of the load. Fig. 3 shows a load curve for such a substation, having equal battery discharge/charge areas, which allows for the usage of ESS.

The permissible maximum load exceeds the power limit of the substation by 62%, the required operating capacity of the battery is about 79.0 thousand kWh at a rated power of 11.5 MW.

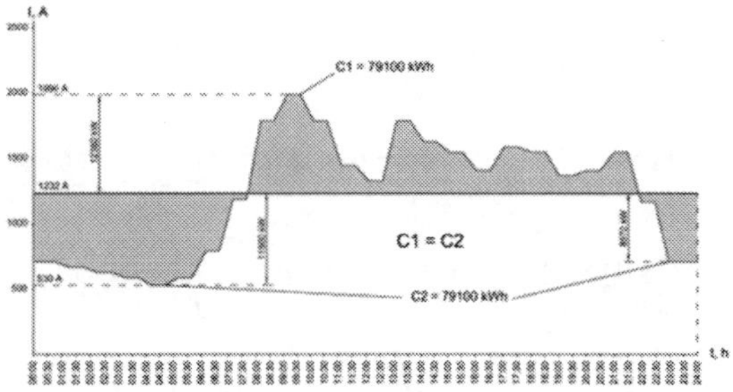

Fig. 3. Daily load curve (load current, A) of 110/10 kV substation of the wood-processing industry on a winter weekday, considering the equality of the discharge/charge areas of the ESS

Comparative cost calculation for the wood-processing industry 110/10 kV substation modernization options (transformer replacement or ESS installation) shows that the capital investment in 10 kV ESS installation 26 times exceeds the cost of the transformers replacement.

A series of calculations performed for the substations with two typical load curves showed that in order to achieve complete flattening of the load curves, it would be necessary to install an ESS of significant power and capacity – 12 MW and 80 thousand kWh (Fig. 4).

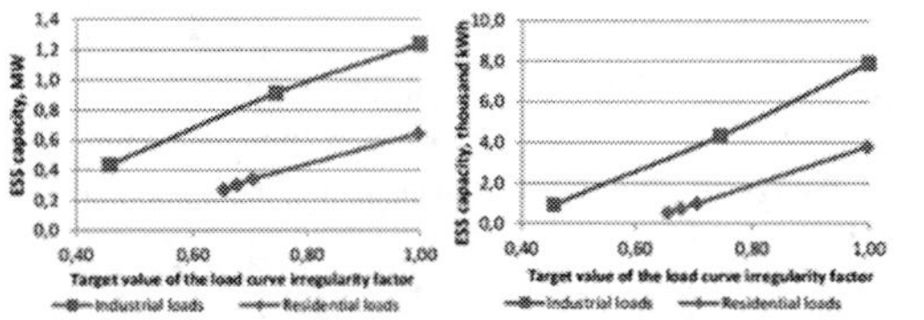

Fig. 4. Dependence of the required power and capacity of the ESS on the target coefficient of unevenness of the load curve.

4 Electricity losses in the network with an installed ESS

To assess the impact of ESS on the electricity losses, the changes in the electricity losses of the Minsk electrical network of 10 kV and above within a given period were calculated for the following options:

Option 1 - the ESS and a lower power transformer (16 MVA) were used;

Option 2 - a higher power transformer (25 MVA) was used.

The calculations were performed using the RASTRWIN program.

Additional annual electricity losses in the network of 10 kV and above were defined by the formula:

$$\Delta W = \Delta P \cdot \tau, \quad (1)$$

where ΔP – additional electricity losses in the network of 10 kV and above caused by increased load (without ESS);

τ – time of maximum losses, determined by the formula:

$$\tau = (0{,}124 + \frac{T_{MAX}}{10000})^2 \cdot 8760, \quad (2)$$

where T_{MAX} – time of the maximum load being used is taken equal to 5000 hours.

Electricity losses in the transformers are determined by the formula:

$$\Delta W_T = \Delta P_{NL} \cdot 8760 + (\frac{S_{MAX}}{S_R})^2 \cdot \Delta P_{SC} \cdot \tau, \quad (3)$$

where ΔP_{NL} – no-load operation losses of the transformer;

ΔP_{SC} – short circuit losses of the transformer;

S_{MAX} – the highest value of the total power flowing through the transformer;

S_R – rated power of the transformer.

Electricity loss in 10 kV ESS is determined based on the overall efficiency of the lithium-ion energy storage device that equals about 85%, according to the manufacturer's data.

The analysis of the calculation results proved that, when the ESS was used, the electricity loss in the ESS, the increase in the losses under the load in the lower power transformers, according to the option 1, were not compensated by the decrease in no-load operation losses of the lower power transformers according to the option 2. Therefore, in general, the electricity loss with the installed ESS is greater than without it. And, consequently, it should be considered in the feasibility studies.

5 Usage of ESS for voltage regulation

It was calculated in what way a 6.5 MW ESS affects the voltage level in the 10 kV network of the 110 kV Korzyuki substation. 10 kV ESS inverters are selected considering the possibility to generate the required reactive power in order to regulate the voltage in the 10 kV network. The power of the inverters should be about 7.2 MVA, if a load power factor equals 0.9. According to the ESS passport data, the power factor control range is from 0.1 to 1 p.u., provided that the inverter is not loaded with active power. Thus, a 7.2 MVA inverter can control reactive power in the range from -6.5 MVar to +6.5 MVar.

A change by 6.5 MVar in the reactive power on the 10 kV buses of 110 kV substation Korzyuki, causes a voltage level change by 0.6 kV in the 10 kV network. By means of the full range of regulation, a 7.2 MVA inverter is capable of regulating the voltage within the range of 1.2 kV (about 12%).

The 10 kV ESS are mainly used to prevent overloading of 110 kV substation transformers when one of the two 110/10 kV transformers is disconnected in the situations of the significant loads at the substation in winter. While both of the two transformers are in operation during the previously mentioned period or considering the operation of the summer loads, the active power of the ESS will not be used to its full potential. This circumstance makes it possible to use those systems as a means of voltage regulation (generation/consumption of reactive power) in a 10 kV network when consumption/generation of active power is below the nominal value of the installation.

The reactive power generation/consumption technical feasibility can be determined only after the required level of the inverter load by the active power for a specified substation load operation mode has been calculated. It should also be mentioned that for the voltage regulation on the 10 kV side, 110/10 kV transformers are equipped with on-load control devices, in this regard, the use of the ESS for 10 kV voltage regulation is not of the primary importance.

According to the preliminary estimates, the cost of an ESS being below 200 US dollars / kWh can be an economic feasibility criterion of the ESS wide usage in the 10-110 kV distribution networks of the Belarusian Energy System, in this regard the following factors should be considered:

- the service life and the degradation of ESS,
- the payback period of ESS, as an energy saving means, not exceeding 10 years.

It should be noted that when specifying the efficiency of the ESS usage in 10 kV distribution electrical network of an industrial enterprise, it is essential [3]:

- to determine specific requirements for the reliable work of the basic equipment of the enterprise;
- to clarify the cost-beneficial analysis based on the additional conditions of the contractual relations with the power-supplying organization (paying for the contracted capacity, taking part in the frequency regulation demand management, etc.);
- to clarify the cost indicators and the maintenance costs of the devices to be installed.

Conclusion

1. Usage of ESS at a 110/10 kV substation allows for compensating daily load unevenness, compensating daytime peak loads and increasing nighttime minimum loads (load dips), and avoiding the replacement of 110/10 kV transformers with higher power transformers.

2. ESS can be used to regulate the voltage in the 10 kV substation network, however, due to the use of 110/10 kV transformers with on-load tap-changers, this task is not of the primary importance.

3. The ESS installation cost significantly exceeds the facility reconstruction cost, which makes it indefinite and disputable to conclude the advisability of ESS widespread use in 10 kV distribution electrical networks.

References

1. Bushuev V.V., Novikov N.L. Infrastructure storage units in the electric power industry. Energy Policy, 2020, No. 10, p. 74-89.
2. Review of Energy Storage System for Microgrid. G.V. Brahmendra Kumar and K. Palanisamy, 2021.
3. Ilyushin P.V., Shavlovsky S.V. Payback mechanisms for investments in electric energy storage systems when used to reduce peak loads and power costs // Relay protection and automation, 2021, no. 3, p. 14-22.

Ensuring balance reliability in managing the development of electric power systems: condition, problems, information content

Iurii Chukreyev[1,*], *Mikhail* Chukreyev[1]

[1] Institute for Socio-Economic & Energy Problems of the North, Federal Research Center Komi Science Centre Ural Branch of the Russian Academy of Sciences, Syktyvkar, Russia

Abstract. The stages of improving the methodological and model support problem of balance reliability in recent years from the standpoint of developing new methodological instructions (MI) on the design of power systems are considered. The analysis of the developed and approved by the Russian Ministry of Energy and the impact of their application on the decisions made.

1 Introduction

The planning development of electric power systems (EPS) at all times has been given due attention. By the 90s of the last century, a hierarchical system was implemented, within the framework of which a strategy for the development of generating sources and power lines for the prospect of 5 to 20 years was developed. Methods implemented in software developments made it possible to solve problems related to development of the EPS development. Solving the problems of predicting the need for electric energy and power; optimization of the prospective development of generating capacities and the main electric network and today use the experience accumulated then. Unfortunately, this cannot be said about solving the problem of balance reliability (BR), from the standpoint of its influence on the creation of new or dismantling outdated generating sources in the management EPS of Russia development.

Balance reliability in the task of managing the EPS development always and today is understood to ensure the demand for electricity and the power of consumers within the specified restrictions on energy supply, taking into account the planned and unplanned breaks in the operation of its elements. It is important to note that only those whose solution is related to the need account for the system failures due to emergency damage equipment and random load deviations caused mainly by the temperature factor. The difference between BR and other (regime, structural) is that random conditions caused by unplanned breaks can last several tens of days (unscheduled equipment repair). At the same time, the depth of possible power deficiency can reach significant values (tens of GW), when several large generators go into emergency repairs.

Exactly 10 years ago, at the 2013 seminar, at the request of N.I.Voropaya, we presented a reviewing report "Methodological foundations of studying the balance reliability of electric power systems", published in section 1.2 of the monographs [1]. It noted that the BR task is complicated and multifaceted in our country, for known reasons from 1990 to 2010, it practically did not develop. The existing approaches to solving the problem of evaluating BR indicators and its synthesis, as well as the essence of the criteria for making managerial decisions on the substantiation of power reserves, were given. As a result, problematic issues were identified the tasks of ensuring BR in the new conditions of the transition from a centralized management system to the new paradigm of a multilateral process in the context of digitalization, multi -criteria and multiplicity of interests. They mainly came down to the following:

- the formation of models calculated scheme EPS of Russia to study various tasks of balance reliability;
- justification for new market conditions of criteria for making managerial decisions from the standpoint of ensuring balance reliability;
- the formation of probabilistic indicators of frequency and duration certain types of repairs (capital, medium, current and emergency) for serial equipment;
- the formation of probabilistic characteristics of a change in the load caused mainly by the temperature factor.

In the material below, the authors allowed themselves to do an analysis of what happened in solving the problem of providing BR over the past decade (approximately since 2011).

2 Stages of improving methodological support.

As already noted until 2010, the task of providing BR was not so relevant. This is due to a sharp decrease in electrical consumption (in different regions from 10 to 40 %) during the reform period of the industry (90s). The impetus for the development of BR tasks was the Decree Russian Federation Government No. 1172[2] regarding the procedure for long-term competitive power selection (CPS).

* Corresponding author: chukreev@iespn.komisc.ru

Table 1. Comparison the components of full power reserve for Russian UPS as a percentage of the maximum load.

Parameter	ISE&EPN FRC KSC UB RAS	Institute "EnergosetProject"	Offers «SO UPS»	Not approved MI 2012	MR 2003
1	2	3	4	5	6
UPS of Russia					
Normative reserve,	15,04 / 15,55	16,70	22,18	20,5	16,06
including excluding strategic	12,96 / 13,47	14,62	19,81	17,5	13,98
– Repair	4,04	5,56	10,51	6,8	4,04
– strategic	2,08	2,08	2,37	3,0	2,08
– operational	8,92 / 9,43	9,3	9,3	10,7	9,94
European part UPS of Russia					
Normative reserve	15,59 / 16,23	17,49	21,08	20,1	17,00
Siberia EPS					
Normative reserve	12,94 / 13,53	13,70	26,20	22,0	12,00

The procedure was focused on determining the paid generating power in the wholesale market, i.e. to identify ineffective generators. This process, a significant degree, depended on the parameter of demand for power in the power balance planned for the future period. The generating sources participating in the form and did not pass the selection were not payable when agreed with the Russian Federation Ministry of Energy, they could be dismantled.

After the publication of this resolution with the EPS, the first attempt was pre-adopted aimed at forming new approaches to the justification level of reservation in the EPS, which is one of the components demand in the power balance. In 2011, the Institute Energosetproekt ordered a research [3] on the improvement, but outdated methodological recommendations (MR) 2003 [4]. The percentage of the regulatory reserve of generating capacity (NRGC), which was obtained, with the assistance of our institute, did not arrange the customer, especially in terms of the repair reserve and were significantly increased to them. The stages of the change are given in table. 1 (column 1 - the results of our institute, column 2 - correction of the energy network project, column 3 - proposals for correction from the EPS, column 4 - proposed by the Ministry of Energy for approval). Methodological instructions (MI) 2012 were not approved for many reasons. Basically, they concerned use old information base (there was simply no other) and in the unreasonable overestimation of the repair component from 4.04 % (MR 2003) to 6.8 % in the amount of NRGC.

The analysis above results shows a significant increase in relation to the values NRGC in the 2003 MR, proposed with the EPS (22.18 %, column 4) and in unreasonable MI in 2012 (20.5 %, column 5). Basically, the overestimation is associated with the component of the power repair reserve, in offers from the EPS by 2.5 times, MI 2012 by 1.5 times. In the last two lines of the table. 1 shows the parameters of regulatory power reserve for CPS European and Siberian parts EPS of Russia in the procedure.

It can be seen that the differences in the values of power reserves are significant. During the CPS procedure, the value NRGC in the Siberian EPS increased due to possible restrictions on the energy supply of the hydroelectric power station, by another 8.55 % [5]. At the same time, NRGC was not accepted 12, as in the MR 2003, but 17 %. Given this, the increase NRGC for the CPS in the Siberian took a value of about 9 GW or 26.55 %.

From 2012 to 2014, our institute on assignment the SO UPS conducted research the problem of BR and, to a greater extent, of its component - the task of evaluating its indicators. By and large, they attempted to solve the problems posed in the mentioned monograph [1] and the introduction of this article. The tasks of information security in terms of probability information certain about unscheduled conclusions in the repair of generating equipment, chronological by hour of describing the power consumption modes and its random changes caused by temperature fluctuations. There were options for the formation models of calculated schemes EPS of Russia. Methods of evaluating random states were implemented, taking into account not only one, but also both law of Kirchhof [6].

In 2016, on instructions the SO UPS, the Institute completed research in a comparative analysis the level of operational power reserve [7] and new approaches to its substantiation [8]. In the first work, a lot of attention was paid to the analysis of methodological approaches and model-program developments used in domestic and foreign practice to evaluate BR indicators and justify the values of operational power reserve. The above table. 1 received in this work. It was given recommendations for the application of certain normative values to the BR indicators, depending on the accepted conditions for accounting for the electrical consumption mode [5]. Unfortunately, they did not find proper reflection in the approved regulatory values of BR indicators for territorial zones in the probability of deficiency work $_{pnorm}$ = 0.996 [10] or the probability of a power deficiency $J^{н}_{д}$ = 1 - p_{norm} = 0.004.

 https://doi.org/10.1051/e3sconf/202346101008

Table 2. Influence of various representations power consumption regime on the components of Normative power reserve

Parameter	Load schedule - December day		An hourly annual schedule	
	Taking into account the correlation	Excluding correlation	Taking into account the correlation	Excluding correlation
1	2	3	4	5
Normative reserve	18,86	15,73	15,61	13,53
– operational	9,99*	6,86*	6,74**	4,66**

*) – The results were obtained in the above research (link 6);
**) – The results obtained in this article.

The purpose "Justification of regulatory values", carried out in the same 2016, was to develop the recommendations corresponding to the name. We note two points. The first - the accounting of the power consumption mode was adopted by one December daily schedule, in the assumption of its actions on all working days of the year (as in the 2003 MR [4]). The second - the studies were conducted for the work EPS model presented by the customer SO UPS. The adoption of such a model of the calculated scheme, and not a model with fragmentation EPS and the achievement 50 territorial zones, as in the work of the Institute "Energosetproekt" can be explained by the results given in table. 1. In the column fraction column 2, the values of the full and operational power reserves are given for the number of territorial zones equal to the EPS number, in the denominator-equal to 50. As you can see the differences in the levels of power reserves are insignificant and do not exceed even 0.5 percent.

According to the generating equipment, an information was actualized about its accident rate and the influence significant excesses the generating power of planned repairs used in the conditions of the equipment presence as a state of equipment. Repair with such a policy increased from 4.04 % to 8.87 %. Under the power consumption regimes, random load changes caused by temperature variations, taking into account their correlation dependence within the territories covered by EPS, were subjected to actualization. The results of the developed recommendations are given in the 2nd and 3rd columns table. 2. In it, planned repairs on the recommendation with the EPS are adopted by 8.87 %. It can be seen that taking into account the correlation of the temperature factor leads to a significant increase in the value of the operational power reserve (9.99 and 6.86 %). The reduction in the percentage of the operational power reserve given in column 3 (6.86 %) in relation to the column 2, table. 1 (8.97 %). This is due to a decrease in the average accident rate of generating equipment with 3.68 % [11] to today 2.77 %, a change in random load deviations from a given value and a factor in the availability of excess capacity relations. Recall that the results of the table. 1 obtained with optimization of access capacities. The results of the table. 2 show that taking into account the correlation dependence of the temperature leads to a significant increase in the value of operational power reserve (by 3.13 %). It should be noted that the customer overestimation of the repair component on 4.83 % (from 4.04 % (table. 1) to 8.87 %) is leaning this increase. Given this, the value NRGC, obtained at the new information base, amounted to only 14.03 %.

In February 2018, a competition was opened on the basis of the «Nonprofit Partnership Council for Organizing Efficient System of Trading at Wholesale and Retail Electricity and Capacity Market» to develop a generating power reserve of the Russian EPS. In March of the same year, the winners of the competition were recognized by two Academic Institutes of the Russian Academy of Sciences: ISEM SB RAS and the Institute for Socio-Economic & Energy Problems of the North, Federal Research Center Komi Science Centre Ural Branch of the Russian Academy of Sciences (ISE&EPN FRC KSC UB RAS). In accordance with the terms of the contract, the work should have been completed within 6-8 months, but for unknown reasons were delayed. In December 2018, the ISE&EPN FRC KSC UB RAS, due to the reorganization and entry into the FRC of the Komi SC of RAS, left the contract (official version). Work with ISEM SB RAS continued and only by the end of 2020 the results were adopted by the customer of work. The results obtained by the ISEM of the SB RAS in terms of levels of NRGC for the EPS of Russia (approximately 13 %) [12] were agreed with the studies of the ISE&EPN FRC KSC UB RAS (table. 1, 2nd line, 2nd column). It should be assumed that the procedure for determining the value of the NRGC proposed by ISEM was supposed to become the basis of the new methodological instructions for the design development of energy systems [13].

Almost all worked out from 2012 to 2016 in the framework of the work with SO UPS on the methodological support of the BR task, including taking into account the territorial correlation of random changes caused by temperature fluctuations in one form or another, included in the national standard of the Russian Federation [6]. This can be attributed to the positive moments of the studies. It should be noted that the results of these studies were partially presented in the monograph [14] and work [15, 16]. Unfortunately, the studies of the ISE&EPN FRC KSC UB RAS and ISEM SB RAS did not find understanding SO UPS and the Russian Ministry of Energy on the substantiated values normative power reserve. It can be assumed that this is why they did not become the basis for the development of new MI.

3 Analysis the new MI 2022 from the standpoint of ensuring balance reliability.

In the new MI, in accordance with the Federal Law "On Electro-Energy" No. 35-Ф3 dated 03/26/2003, a two-level planning model was adopted aimed at developing two software documents-a general scheme for the placement of electric power facilities and the development program (SDP) electric power systems of Russia . The substantiation of decisions to ensure BR in the development of these software documents, one way or another, is associated with the features formation power balance and the methodological provisions of the estimate and provision of BR. Two chapters (V and VI) are devoted to these issues.

Power balances initially imply the presence of equality between the receipt (generation) and the consumable (consumption) of its parts. At the very beginning of the chapter, it is noted that when forming power balance sheets, a calculated reserve should be taken into account, the concept of which is not found in further points of this and subsequent chapters. Instead, the concept (paragraph 87) of the required power of power plants is introduced to ensure power balance ($P_{required}$). Its determination is carried out from the conditions for ensuring the regulatory indicator BR (ρ_{norm}) (paragraph 113). At the same time, it is quite obvious that the difference between $P_{required}$ and the need value for power ($P_{consumption}$) (paragraph 82) is precisely the same calculated or NRGC. It should be emphasized that the concept of the required power is introduced only in the framework development (updates) the general scheme. For the development of SDP, this concept, as well as the concepts of excess and power balance, are completely ignored.

A distinctive feature of the new MI is introduction chapter "Ensuring Balance Reliability". In previous editions of MI and MR BR, it was ensured by the fulfillment of the requirements for the values NRGC of the territorial zones EPS. Their values were determined by multivariate calculations EPS and were given by the percentage maximum load [11]. New MI is aimed at developing managerial decisions as a result of calculating BR indicators in the form of a probability deficit work (ρ). It is important to note that the need to increase the territorial zones power reserves and accessories of bonds is required when developing both software documents (paragraph 115). At the same time, the solution of the issues ineffective generating capacities (paragraph 116) and their dismantling (paragraph 117) can only be resolved only when developing or updating the general scheme that are relevant for modern electric power.

In the new MI algorithms for decision -making on the justification required capacity ($P_{required}$) and, therefore, the territorial zones NRGC EPS did not undergo changes in relation to the MR 2003. This cannot be said about the methodology for solving the problem of evaluating BR indicators. Changes are mainly associated with increasing information security. It should be noted that almost all probability defined information used to evaluate BR indicators is formed and used only by SO specialists and is not available to the scientific community. The alleged public discussions of the adopted management decisions, in our opinion, will be ineffective due to lack of the entire spectrum of source information among interested parties.

4 Assessment impact of changes introduced into new MI in the accounting of initial information on decisions made

The changes should include new approaches to accounting: electrical consumption modes (paragraphs 97, B, 106), Kirchhof laws (paragraph 109), models the EPS calculation scheme (paragraph 98), random power deviations caused by the temperature factor (paragraph 106). The influence of these changes are tested and partially published [6, 14]. At the same time, which included in the new MI related to the issues of accounting for planned equipment repairs (paragraph 102) and the representation the power of renewable power plants, including the hydroelectric power station (paragraph 83, D, G) were not tested and, moreover, they were not published. Below is briefly a analysis the influence of information on decisions the development and dismantling of equipment.

Presentation the power consumption mode. In many publications [9, 14, 16], the influence of various submission of the power consumption regime on decisions made was shown. Recall that in the MR 2003 and in the 2016 work report (link 5), the power consumption mode was presented by one daily load schedule of the month that lasts a calendar year. In the new MI, as in the national standard, the power consumption mode is represented by an annual hourly load schedule (8760 hours).

The normative indicator BR for the reliability territorial zones EPS in accordance with the order the Ministry of Energy No. 321 (link 6) remains unchanged and, as in the old MR 2003, equal to $\rho_{norm} = 0.996$. For this indicator in the table. 2 (columns 4 and 5) include calculations the value of operational power reserve. As you can see, the values of power reserve are significantly different from the principles of accounting for the regime of electrical consumption used in the 2003 MR (columns 2 and 3).

Model the estimated scheme EPS and taking into account the Kirchhof laws. The new MI proposes use of a model with a large number of reliability zones than in the development of 2003 MR and with the condition implementation of both laws of Kirchhof when calculating randomly formed states of generating power. At the same time, throughput abilities are determined by restrictions on the transmission of power by sections obtained by calculation. Studies given in a number of publications [14] show sufficient discrepancies in assessing the regime when using different distribution of power deficiency (DPD) models. At the same time, when optimizing $P_{required}$ these discrepancies become not so significant due to the possibility of redistributing generating power between reliability zones. Experimental calculations show that an increase in the reliability zones estimated EPS model can lead to an increase in operational power reserve by a value not exceeding more than 0,5 percent of the maximum load (the number and denominator of the column 2 table. 1), accounting of both Kirchhof laws maximum by 0,2 percent.

Accounting for scheduled equipment repairs. The approaches to the new MI to take into account the scheduled repairs of generating equipment are slightly different from the methodology used in 2003 MR. There are some explanations to certain types of generators. In any case, it can be stated that repairs should not exceed the average data. In accordance with [9], all types of planned repairs of generating equipment are 8.87 %. At the same time, approximately half falls on capital and average repairs, which is much higher than those in 2003 MR. This data was obtained during the work entire composition of generating equipment. They need to be adjusted to a really working composition, i.e. excessive excess. This accounting will reduce the amount of the repair reserve by about 2 %.

Presentation the power of renewable energy sources (RES), including hydroelectric power plants (HPP). In the new MI, as in the 2003 MR, the disposable capacities the HPP and RES are accepted equal to the established values, minus the restrictions caused by technical and regime parameters (paragraphs 83, D, G). There are no specific studies on such a representation of these stations types in the literature. During the CPS, NRGC in Siberia EPS increased by 8.55 % due to the restrictions of the hydroelectric power plants. The studies presented in [5] showed the need to reduce this value to 4.5 %. Studies on the RES South EPS [17] showed that the influence of these types of stations on the NRGC value must be taken into account with their share in the production of electricity exceeding the 5 percent barrier.

Based on the calculations presented in this section and the results (table. 2) following follows. The impact of changes in the presentation initial information in the new MI with existing regulatory requirements for BR indicators will lead to approximately the same parameters of the required power ($P_{required}$) or NRGC, which were also in the 2003 MR. Numerous calculations showed that the NRGC values for the UPS will fluctuate from 13.5 to 16.5 % (for our example - 6.74+0.5+0.2+8.87-2 = 14.31 %).

5 Conclusion

The introduction in the new MI section to ensure balance reliability will lead to negative consequences in the field of BR study. The reason lies in concentration of both the tools and the information component in one structure – SO UPS. Public discussions made from the SO UPS decisions on the industry development in this case are unlikely to be obtained. The reason is the impossibility of at least some verification of issues discussed in the discussion.

This is clearly seen from the projects accepted at the turn of 2022/23. New MI 2022 entered into force on January 1, 2023 two days before, i.e. on December 30, 2022, the general scheme was updated until 2035 [18]. It would seem that it was more logical to do this after the approval of the new MI. In our opinion, the adoption of such a decision was associated, in addition to other circumstances, by the requirements of new MI, in aspect the need to justify balance reliability.

In accordance with the above analysis, it is completely in vain that the BR indicators, with excessive power, presented in the materials actualized general scheme until 2035 and developed on its basis in 2023 of SDP materials at 2023-2028 [19] will significantly exceed the normative values. This contradicts the requirements of the new MI, in accordance with which, when developing a general scheme, it is necessary to bring the territorial zones BR indicators of the EPS to regulatory values (paragraph 116). This is achieved by multiple calculations of BR indicators with the withdrawal from the composition of the most emergency equipment. The derived generating equipment when fulfilling certain conditions paragraph 117, completely unrelated to the balance reliability conditions, should be subject to dismantling.

References

1. A.F. Dyakov, V.A. Stennikov, S.M. Sentorov et al. Reliability of energy systems: problems, models and methods for solving them, Novosibirsk: Science, 284 p. (2014)
2. Decree the Government of Russian Federation No. 1172, December 27, 2010 (as amended on 01/19/2018) "On approval the rules of wholesale electric energy and power market and on amendments to some acts the Government of Russian Federation on the organization functioning wholesale electric energy and power market"
3. Methodological instructions for the design development of energy systems / OJSC Institute Energosetproekt, 2011, the "STC UPS" Section is approved. "Technical regulation in the electric power industry", July 20, 2012, were not approved by the Russian Ministry of Energy.
4. Methodological recommendations for design development of energy systems. (Approved by order of the Ministry of Energy of Russia No. 281 of June 30, 2003).
5. Chukreev Yu., Chukreev M. Impact of the regulatory reserve and capacity demand on the process of justification of the generating sources in respect of the UES Russia management development / E3S Web of Conferences Volume 209, 06002 (2020) ENERGY-21 – Sustainable Development & Smart Management.
6. National Standard of the Russian Federation. A single energy system and isolated energy systems. Planning for the development of energy systems. Balance reliability calculations - norms and requirements. GOST R 58730-2019.
7. Report on research work Comparative analysis of differences in distribution the compensation reserve of power between the EPS in regulatory documents of various years approval. / Syktyvkar, 62 p. (2016)
8. Report on research work substantiation the normative values of components full power reserve in the EPS and UPS of Russia as a whole when planning their development. / Syktyvkar, 2016 - 66 p.
9. Chukreev Yu.Ya., Chukreev M.Yu. The indicators of balance reliability to justify the components of regulatory power reserve in terms of modern conditions for

the development EPS of Russia / news of the Russian Academy of Sciences. Energy. 2022. **Vol. 5**, pp. 22-35.

10. Order the Russian Ministry of Energy No. 321 dated 04.30.2021. On the establishment of a regulatory level of balance reliability for the unified energy system of Russia, used in assessing the possibility of withdrawing generating equipment from operation.
11. S.S. Rokotyan and I.M. Shapiro. M. Directory for the design of electric power systems . M.: Energy, 352 p. (1995)
12. http://isem.irk.ru/news/article-2021-03-19
13. Methodological instructions for the design development of energy systems. (Approved by order the Russian Ministry of Energy dated December 6, 2022, No. 1286).
14. Chukreev Yu.Ya., Chukreev M.Yu. Models for assessing the indicators of balance reliability in managing the development of electric power systems. Syktyvkar: Komi NC URS RAS, 2014, 207 p.
15. Chukreev Yu.Ya., Kucherov Yu.N. Problems of substantiating balance reliability in managing the development of electric power systems / Energy policy. **Vol. 2**, pp. 3-20 (2015)
16. Chukreev Yu.Ya. Indicators of balance reliability and their rationing in the management development of electric power systems: Information aspect / news of the Russian Academy of Sciences. Energy. **Vol. 5**, pp. 33-44 (2015)
17. Chukreev Yu.Ya. The influence of renewable energy sources on assessing and ensuring balance reliability in managing the development of electric power systems / Methodological issues of studying the reliability of large energy systems: **Vol. 73**. Irkutsk: ISEM SB RAS, pp. 240-249 (2022)
18. Order the Russian Federation Government at December 30, 2022 N 4384-r
19. Order the Russian Ministry of Energy dated 02.28.2023 "On the approval SDP EPS of Russia".

The influence of models assessing the state in task of ensuring balance reliability on managerial decisions when the ups planning

Iurii Chukreyev[1,*]

[1] Institute for Socio-Economic & Energy Problems of the North, Federal Research Center Komi Science Centre Ural Branch of the Russian Academy of Sciences, Syktyvkar, Russia

Abstract. Various models of assessment a random state are considered. The solution problem of substantiating power reserves is associated with the need to redistribute generating capacities between the EPS territorial zones. It is shown that this need leads to almost the same values of operational power reserves when using various models of assessing state. This is confirmed by experimental calculations on many models of estimated EPS schemes for various parameters.

1 Introduction

The task of ensuring the balance reliability of electro power systems (EPS) is inextricably linked with the solution to problem of evaluating balance reliability indicators (BRI). The mathematical models of BRI assessment existing in our country and abroad are mainly focused on use of statistical modeling methods to form random states of generating power and the EPS territorial zones (TZ) load and their assessment from the position of loading, taking into account the bandwidth of bonds. The so -called task of distributing power deficits (DPD). The significance the setting and solving problem of ensuring balance reliability is shown in the new methodological instructions (MI) for the design of development energy systems approved in December 2022 [1]. The values of RBI in complex EPS, and, consequently, the values of required generating power to ensure the proper regulatory level of reliability are affected by many factors and random events:

- model of estimated scheme UPS of Russia with allocated territorial zones and connecting them;
- reserves of access abilities of bonds between territorial zones and the limit maximum passing capacity (MPC) in power sections;
- the structure of generating capacities territorial zones;
- planned equipment repairs;
- graphs of changes in territorial zones loads in context the year and day;
- irregular fluctuations load and error of forecasting consumer demand;
- a decrease in the generating capacity territorial zones, accessories of ties and MPC sections caused by emergency exits of power equipment.

The number of calculated states the system when assessing the reliable BRI of multi-zone EPS is quite large. Therefore, the solution to the problem of evaluating their consequences for consumers EPS territorial zones (DPD block) should be carried out by effective methods. The resulting BRI for individual TZ, to a large extent depend on the setting and solving this problem. All above factors and approaches to solving the problem of DPD, of course, were previously taken into account when evaluating the BRI and were given in sufficient detail in many publications, for example, monographs [2, 3]. The issues formation the model calculated scheme of Russia UPS and the determination parameters of restrictions MPC lines and the MPC sections between the allocated TZ remained debatable. Below are solution to these issues from the standpoint of their influence in conjunction with the principles of solving problem assessing the state (DPD block) on accepted solutions to ensure balance reliability.

2 Formation model and parameters estimated scheme UPS of Russia for the tasks of substantiating balance reliability

In solving the problem of substantiating the balance reliability EPS and the UPS country in particular, at all times a number of various kinds assumptions were made to simplify the solution problem. This also related to the issues of aggregation actually existing diagram of electrical compounds EPS with the determination its parameters considered in this article. It is quite obvious that the model estimated scheme UPS of Russia for the tasks of balance reliability should be presented in a graph form, the vertices of which are zones same reliability, ribs - intersystem connections. The zones same reliability (TZ) mean a set of nodes included in it, network restrictions between which, as in the concentrated system, do not affect the BRI in it.

In the methodological recommendations (MR) of 2003 [4] and the Methodological Instructions (MI) not approved by the Russian Ministry of Energy 2012 [*], the country's model estimated scheme was formed on

* Corresponding author: chukreev@iespn.komisc.ru

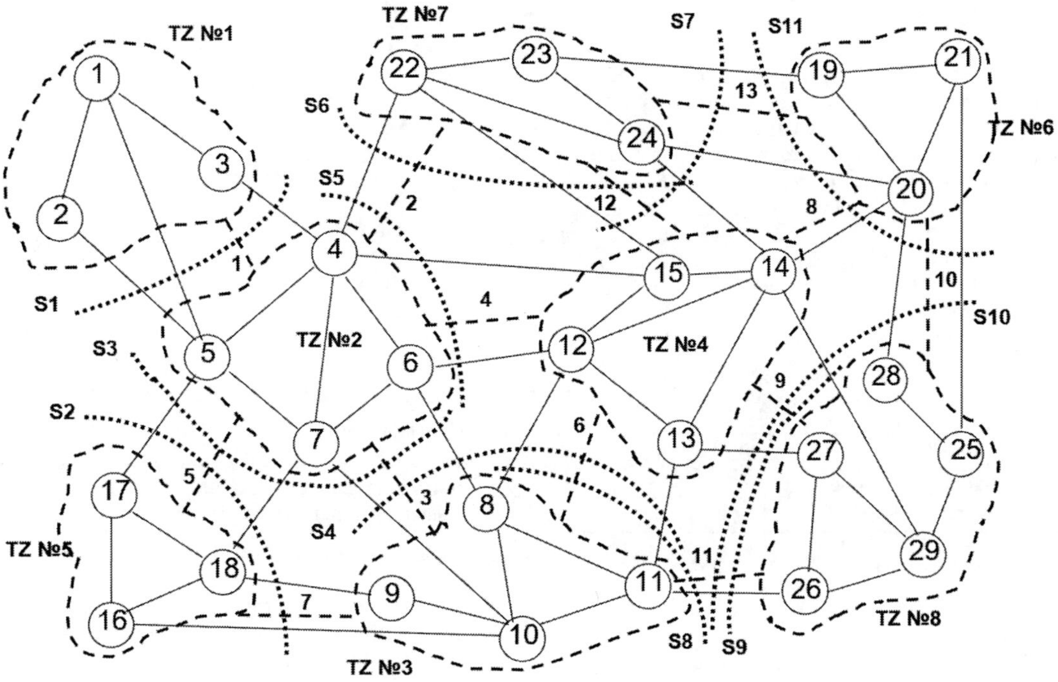

Fig. 1. Fragment hypothetical UPS with the release of territorial zones equal to reliability.

the basis an analysis capacity controlled sections actually existing circuit of electrical connections energy system under operating conditions - dispatch control . In the 2003 MR, a model was used with the release TZ in united EPS. In the uncontrolled MI 2012 model with the allocation 51st TZ, built by specialists of OJSC "Institute Energosetproekt" and the Energy research institute of the Russian Academy of Sciences [3, p. 137], based on the analysis MPC controlled sections in each country's EPS.

In section VI - ensuring the balance reliability in new MI 2022 [1] (paragraph 98) the boundaries of aggregated nodes are also determined by controlled sections of electric network with the bandwidth that limits the mutual reservation generating capacities. In the work [5], published even before the approval new MI 2022, it is noted that the independence of controlled sections allows us to simply select TZ equal to reliability. This is achieved by unification into TZ of those real nodes that do not have any specified restrictions in the form of controlled sections to transmit power. In Fig. 1 for a certain fragment of the hypothetical UPS (29 knots, 52 ties are given in thin lines), the process of allocation TZ and intersystem connections between them is shown. Based on the analysis 11th sections (S1-S11), 8 TZ and 13 intersystem connections were formed (in the figure are indicated by fat dashed lines). It can be seen that inside the allocated territorial zones there are no restrictions on controlled sections for the transfer of electricity between the nodes in it.

It is quite difficult to determine the bandwidth for formed and physically non -existing connections, for example, connections 5 and 7 between TZ with numbers 2, 3 and 5. On this subject, in the new MI 2022, it is noted that the throughput of intersystem relations between the reliability zones It is determined on the basis of MPC values in controlled sections in accordance with the methodological instructions for the stability of energy systems. From this it should be assumed that reserves of access abilities (5 and 7) in the controlled 2nd section (S2), as well as for connections of other controlled sections, must comply with the implementation 2nd Kirchhof Law during the distribution MPC. In accordance with this, the status assessment proposed in the model (DPD block), network coefficients (paragraph 109) should take into account these provisions. True, there are questions related to the receipt of network coefficients for territorial zones that somehow teach MPC sections independent from each other.

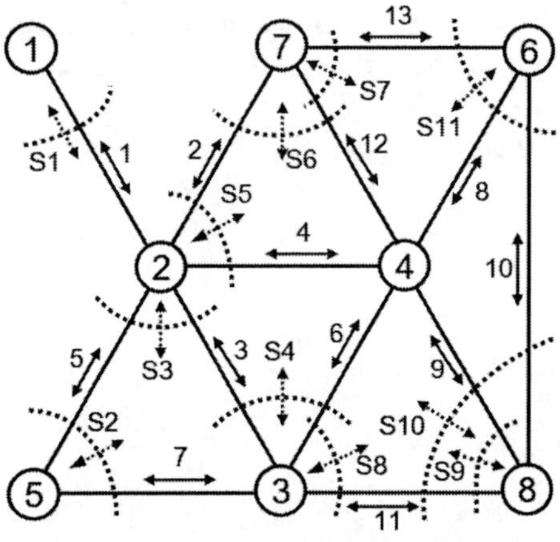

Fig. 2. Aggregated model the EPS calculation scheme

In Fig. 2 shows an aggregated model calculation scheme, EPS that meets the above provisions new MI 2022 on it for the selected controlled sections (S1 - S11, dashed lines) with the arrows in direct and reverse directions shows the restrictions on the MPC and the bandwidth calculated for connections or groups abilities (numbers 1-13 in arrows), taking into account the distribution of power in the sections under consideration, taking into account the implementation 2nd Kirchhof Law (Table 1).

Table 1. Parameters of access capacities and MPC the power of controlled sections in MW

Intersystemic lines			Controlled sections		
№	Binding nodes	Passing capacity	№	Numers of incoming lines	MPC
1	1 – 2	±700	1	1	±700
2	2 – 7	±260	2	5, 7	±810
3	2 – 3	±330	3	3, 5	±760
4	2 – 4	±330	4	3, 6	±680
5	2 – 5	±430	5	2, 4	±590
6	3 – 4	±350	6	2, 12	±440
7	3 – 5	±380	7	12, 13	±460
8	4 – 6	±330	8	6, 11	±680
9	4 – 8	±300	9	9, 11	±630
10	6 – 8	±320	10	9, 10, 11	±950
11	3 – 8	±330	11	8, 10, 13	±930
12	4 – 7	±180			
13	6 – 7	±280			

In fact, it is quite difficult to obtain the equality sums throughputs of selected intersystem ties in controlled sections, due to their independence from each other. Nevertheless, in the model assessment the state (DPD block) in new MI 2022, an approach was proposed, based only on restrictions the throughput abilities highlighted intersystem connections obtained on the basis analysis of restrictions on independent controlled sections.

3 Models for analyzing an assessment of random condition - block DPD

3.1 Balance model

During linear production, its solution is to minimize the power deficit [2. 3]:

$$\sum_{j=1}^{n} c_j^{\text{н.}}(\overline{P}_j^{\text{н.}} - P_j^{\text{н.}}) \rightarrow \min, \quad (1)$$

with restrictions:

$$P_j^{\text{н.}} - P_j^{\text{г.}} \pm \sum_{l(i,j)\in V} P_{l(i,j)}^L = 0, \quad j = 1, 2, ..., n; \quad (2)$$

$$0 \le P_j^{\text{н.}} \le \overline{P}_j^{\text{н.}}, \quad j = 1, 2, ..., n; \quad (3)$$

$$0 \le P_j^{\text{г.}} \le \overline{P}_j^{\text{г.}}, \quad j = 1, 2, ..., n; \quad (4)$$

$$\underline{P}_{l(i,j)}^L \le P_{l(i,j)}^L \le \overline{P}_{l(i,j)}^L \quad l = 1, 2, ..., m, \quad (5)$$

where $P_j^{\text{н.}}, \overline{P}_j^{\text{н.}}$, $P_j^{\text{г.}}, \overline{P}_j^{\text{г.}}$ - respectively covered with power and power of load demand, used and available generating power of the *j-th* TZ; $P_{l(i,j)}^L$, $\overline{P}_{l(i,j)}^L$, $\underline{P}_{l(i,j)}^L$ - a line power and its limit values in throughput in the direct (from the *i-th* TZ to *J-th*) and reverse directions; V - a lot of pairs TZ that have an intersystem connection with each other; *n(m)* - the number TZ (intersystem ties) association.

3.2 The model in idealization of direct current based on attachment the matrix of network coefficients

The target function remains the same as in the balance sheet model. The restrictions are restrictions (3) - (5) and restrictions on mode the species [1, 6]:

$$\sum_{j=1}^{n-1} S_{lj} P_j^{\text{нб.}} - P_l^L = \sum_{j=1}^{n-1} S_{lj}(P_j^{\text{н.}} - P_j^{\text{г.}}) = 0,$$

$$l = 1, 2, ..., m; \quad (6)$$

$$P_n^{\text{н.}} - P_n^{\text{г.}} \pm \sum_{l(i,j)\in V} P_{l(i,n)}^L = 0, \quad (7)$$

where S_{lj} - are the numerical values of the matrix of the network coefficients of each *i-th* communication for all *j-th* nodes; n - is the number of balancing node; $P_j^{\text{нб.}} = P_j^{\text{н.}} - P_j^{\text{г.}}$ - the injection power (non-balance) in the *j-th* node.

When using this approach, restrictions on bandwidth of relationships or group relationships are obtained on the basis formation a matrix of network coefficients. Unfortunately, the methodology for obtaining the latter for aggregated model EPS calculated scheme was not published and was not discussed anywhere. As noted, it is precisely this approach that is proposed to be used in new MI 2022 when assessing the BRI.

3.3 The balance model, taking into account controlled sections in the EPS model calculated scheme

The target function (1) and restrictions (2) - (4) remain unchanged. To take into account the MPC of controlled sections, restrictions on power flow through the intersystem connections (5) included in the section, bring to level the limiting on current loads and additionally introduce restrictions on power outlets along the controlled sections [3, 5]:

$$\underline{P}_k \le \sum_{l \in U} P_l^L \le \overline{P}k, \quad k = 1, 2, \ldots, K, \tag{8}$$

where P_l^L -calculated power on the *i-th* line; $\overline{P}_k$, $\underline{P}_k$ - DPD power for *k-th* controlled section in direct and reverse directions; U - a lot of *i-th* lines included in *k -th* section; K - is the number of controlled sections under consideration.

4 Experimental calculations

The calculations are made for the three models assessment state (DPD block) given in the previous section. As a calculated EPS scheme, a model was used the basis analysis of independent controlled sections shown in Fig. 2. The adequacy use of certain DPD blocks was estimated on basis of analysis the values operational power reserve and, in accordance with the new MI 2022, the values of required capacity to ensure the standard level of balance reliability ($P_{required}$). It should be noted that the latter are unambiguously related to the concept of regulatory power reserve in territorial zones that has established in the planning of EPS development. This connection for each *j-th* territorial zone is determined by the difference in the values of $P_{requiredj}$ and the maximum of the load.

As a tool for solving the problem of assessing BRI TZ EPS, the "Orion-M" software complex was used. The methodology for determining BRI implemented in it is based on the solution of two interconnected stages. At the first stage, the methods of statistical modeling on the analytically obtained functions reducing the generating power determine the generation power for all TZ EPS. The reduction in power is due to the withdrawal main repairs the main generating equipment. At the second stage, the methods of linear programming solve the problem of assessing the state the system (DPD block) from the standpoint of possibility, with given restrictions, to ensure the load in full. For these purposes, the software complex has been implemented in the previous section DPD model, including presented in the new MI 2022 [1].

The power of power plants ($P_{requiredj}$), taking into account the need to compensate for planned and emergency (unscheduled) repairs generating equipment, was determined based on the conditions for provision standard probability the deficit work in all *j-th* territorial zones [1]. This indicator in accordance with the order of Russian Ministry of Energy until 2027 was adopted by size of $\rho_{\text{норм.}j} = 0.996$.

To simplify calculations, the same daily load schedule for all TZ was used in assumption of its action all 365 days of the year. The ratio minimum load to the maximum is adopted by 0.75. For the same reason, random deviations load from their average value caused mainly by the temperature factor were not taken into account. It is quite obvious that use an annual daily graph will affect calculating results the operational power reserve in the direction of its reduction, as well as taking into account random deviations load, but in the direction of its overestimation. It is important to note that the adopted method of presenting the load will not affect the comparative characteristics use different DPD models.

In table 2 for TZ shown in Fig. 2, shows parameters the initial composition of generating equipment

Table 2. Calculation parameters and initial composition of generating equipment

Parameter		EPS in general	The number of the territorial zone							
			1	2	3	4	5	6	7	8
Generating power involved in the load covering, MW		57775	6075	5450	6500	9525	9475	6375	6000	8375
Single power, MW	Equity of equipment, relative units	The initial quantitative composition of generating equipment								
25,0	0,025	137	9	10	18	17	21	15	22	25
25,0	0,040	180	10	8	10	20	40	10	40	42
50,0	0,040	175	18	14	20	20	23	19	31	30
100,0	0,055	120	9	11	7	16	22	6	25	24
200,0	0,045	43	4	6	3	10	8	6	2	4
500,0	0,065	41	6	4	7	8	6	6	0	4

Table 3. Comparison of different models state assessment - DPD block when substantiating the regulatory BRI TZ EPS

№	Maximum load, MW	Planned repair, MW	Operational power reserve, MW / %			Required power of power plants ($R_{Req.}$), MW		
			Balance restrictions (p. 2.1)	Idealization on constant current (p. 2.2)	Controlled sections (p. 2.3)	Balance restrictions (p. 2.1)	Idealization on constant current (p. 2.2)	Controlled sections (p. 2.3)
1	2	3	4	5	6	7	8	9
1	5350	380	720 / 13,46	715 / 13,36	720 / 13,46	6450	6445	6450
2	5200	310	120 / 2,31	150 / 2,88	145 / 2,79	5630	5660	5655
3	5850	350	565/ 9,66	600 / 10,26	530 / 9,06	6765	6800	6730
4	8650	520	785 / 9,07	845 / 9,77	715 / 8,27	9955	10015	9885
5	8150	490	1205 / 14,79	1160 / 14,23	1215 / 14,91	9845	9800	9855
6	5750	345	640 / 11,13	615 / 10,70	640 / 11,13	6735	6710	6375
7	5700	340	245 / 4,30	285 / 5,00	245 / 4,30	6285	6325	6285
8	7600	455	820 / 10,79	815 / 10,72	875 / 11,51	8875	8870	8930
EPS	52250	3190	5100 / 9,76	5185 / 9,92	5085 / 9,73	60540	60625	60525

(without planned repairs) and the loads of generating capacities involved in the coating. The substantiation the value of operational power reserve in each *j-th* TZ was carried out proposed in 90s last century by the method of sequential approximations (iteration method) [2]. The probabilistic BRI ρ_j obtained on each iteration were compared with regulatory values $\rho_{\text{норм.}j} = 0.996$. Depending on this comparison, the power of power plants necessary to ensure the regulatory BRI TZ changed for each *j-th* TZ, and in model this is conditionally carried out by non-vital units. The process ends provided, when for all *j-th* TZ, the obtained probabilistic BRI (ρ_j) exceeds regulatory values ($\rho_{\text{норм.}j}$). Moreover, the excess should be extremely low, not more than 0.05 % the normalized value. It should be noted that in the new MI 2022, just such an approach was proposed.

In table 3 for each *j-th* TZ model EPS calculation scheme (Fig. 2), the parameters of maximum loads and planned repairs generating equipment are given. The latter for each *j-th* TZ are conditionally accepted 6 % the maximum their load. The introduction the values of planned repairs is necessary to determine the value a normative power reserve equal, as you know, the sum of operational, repair and strategic. The strategic reserve for modern conditions redundancy of energy systems is not taken into account. It is known [2, 3] that the values of planned repair do not affect the BRI TZ, since these indicators are determined for the generating power involved in the load coating, i.e. excluding, displayed for repair of generating equipment. In the same table, for each *j-th* TZ, the model of estimated circuit EPS (Fig. 2) shows the above components of regulatory power reserve (column 3-6). In addition, the column 7-9 tables give the generating power of the power plants necessary to ensure the normative BRI (as amended by the new MI 2022 $P_{requiredj}$).

Network coefficients were determined by calculation based on conditionally accepted ratios between the resistance of intersystem ties. At the same time, the throughput abilities of the intersystem ties included in one section were determined by the values of their MPC, taking into account the ratios of the accepted resistances. Different options for the resistance of relations and, accordingly, different matrices of network coefficients and throughput abilities of intersystem connections included in one or another section were considered. Below are the results only one of considered options for the formation network coefficients matrix, because for other options, the variation of these parameters did not affect the results of comparing the values operational power reserve when using different DPD models.

From the table 3 it can be seen that accounting as restrictions only by the MPC of controlled sections leads to somewhat smaller values the operational power reserve (column 6) for EPS as a whole in relation to other DPD models (4 and 5 columns). The same applies to the values required power on power plants necessary to ensure the normative BRI (column 9 in relation to 7 and 8). This fairly obvious conclusion is explained by the tightening of restrictions in transition from the accounting MPC of controlled sections (paragraph 2.3) to accounting for only the throughput abilities intersystem connections (paragraph 2.1). An even greater difference is observed with the introduction of more stringent restrictions on the accounting 2nd Kirchhof Law in form the network coefficients (paragraph 2.2). At the same time, it is necessary to emphasize once again that the receipt of network coefficients for an aggregated EPS model, based on the analysis of independent sections in operating conditions, is very problematic and is associated with certain assumptions associated with the consideration of independent controlled sections as dependent.

It should be noted that for the above example, a change in the operational power reserve for EPS as a whole a percentage of the maximum load for all three DPD models is extremely insignificant - less than 0.2 % (by power 100 MW). Moreover, the multiple experimental calculations with a change MPC of controlled sections showed sufficient stability of this conclusion. Differences grow on a percentage of interest with an increase MPC sections and are practically leveled with their small values.

It can be noted that the distribution operational power reserve, according to separate TZ EPS for the considered models DPD varies significantly. This can be explained by use restrictions of different stiffness (3-5), (6-7) and (8) in mentioned models. A fairly explanatory trend of increasing the values operational power reserve when using the DPD block with restrictions in form network coefficients (6-7) in TZ that have a larger number of interisystem connections (in example the TZ with numbers 3, 4 and 7) and a decrease in the rest is traced. But this is not always unambiguous, there may be exceptions that are explained by the ratios of network coefficients used DPD model in the idealization of direct current.

5 Conclusion

1. It is shown that in order to form a model estimated scheme EPS and the UPS of Russia, in particular, when solving the tasks of balance reliability, application the approach proposed in new MI 2022, based on assessment controlled sections used in dispatch management, is acceptable. At the same time, it is noted that it was precisely this approach that was used in the formation models of settlement schemes country UPS in development the 2003 MR and not approved MI 2012
2. Different models of assessment the random states of generating power and load (DPD blocks) are given when assessing BRI: Balance model, the model in idealization of direct current and model based on accounting for controlled sections. It is shown that obtaining restrictions on the throughput of intersystem ties is associated with certain assumptions associated with the independence controlled sections used in the

tasks of balance reliability of the design reliability model. These assumptions are especially characteristic of the first two models above.

3. The hypothetical scheme EPS shows the process of obtaining a model EPS calculation scheme for the tasks of balance reliability and comparative calculations were made to justify the value of operational power reserve and the required power of power plants to ensure the normative indicator of balance reliability. On the example of this scheme and on many other examples with various parameters, it is shown that the use three models of the random condition considered in the work - DPD blocks leads to almost the same results of values the operational power reserve. The differences in values of the operational power reserve for EPS as a whole do not exceed 0.2 % of the maximum load energy system.

References

1. Methodological instructions for the design development of energy systems. (Approved by order the Russian Ministry of Energy dated December 6, 2022, No. 1286).
2. Chukreev Yu.Ya. Models for ensuring the reliability of electric power systems. Syktyvkar: Komi NC URS RAS, 2014, 207 p.
3. Chukreev Yu.Ya., Chukreev M.Yu. Models for assessing the indicators of balance reliability in managing the development of electric power systems. Syktyvkar: Komi NC URS RAS, 2014, 207 p.
4. Methodological recommendations for design development of energy systems. (Approved by order of the Ministry of Energy of Russia No. 281 of June 30, 2003).
5. Chukreev M.Yu. Justification the model of calculated scheme UPS of Russia for the task of ensuring balance reliability / Methodological issues of studying the reliability of large energy systems: **Vol. 70**. Irkutsk: ISEM SB RAS, pp. 97-107 (2019)
6. Chukreev Yu.Ya. Evaluation of streaming in the idealization of constant current in the task of evaluating indicators and means ensuring reliability EPS / Methodological issues of studying the reliability of large energy systems: **Vol. 65**. Irkutsk: ISEM SB RAS, pp. 430-439 (2015)

Information content the task of balance reliability in managing the development of electric power systems

Mikhail Chukreyev[1,*]

[1] Institute for Socio-Economic & Energy Problems of the North, Federal Research Center Komi Science Centre Ural Branch of the Russian Academy of Sciences, Syktyvkar, Russia

Abstract. The issues of influence initial information on the obtained indicators of balance reliability are considered. The analysis new methodological instructions (MI) developed and approved by Russian Ministry of Energy from the position of possibilities information filling the task of assessing the indicators of balance reliability is given.

1 Introduction

Historically, the development of civilization directly depends on ability agricultural society to the constant presence of an excess food, thanks to which some people could master activities not related to agriculture, which, in turn, led to an increase in production, trade, population, social stability. The ability of population to provide themselves to necessary products was carried out due constant technological progress. Progress is possible in the presence of several components:

- information, intermediate goals located on some time segment and motivation;
- road map.

The information is constantly changing, it must be able to interpret and find the right one, but without it is not possible at all any development. You can't do something without experience, but experience is also information. Therefore, the presence of information is the most important resource that determines the possibility of progress, and its correct perception affects the effectiveness of development. A road map is an understanding of how to use time. The presence of both components gives the ability to move contrast. Any controlled action is power over it, therefore, there is an opportunity to establish their own rules.

In recent decades, the most significant catalyst for progress can be considered electric power. Its development directly affects all other sectors of our lives. In the vast majority of industries, electricity has a significant weight in the cost of product. In everyday life and social life, we also do not imagine ourselves without this resource.

To ensure all areas of our lives, strict planning for the development of electric power industry is necessary. In this direction, many organizations that form the planned level of load, generation by reliability zones, the bandwidth between them, etc. All this is consistent with the chosen reliability criteria. This generally reflects the essence of the task of balance reliability (the property object to satisfy the requirements of consumers within the specified values and restrictions on supply energy resources, taking into account the planned and unplanned breaks in the work of its elements and operational restrictions). And it is precisely its solution that determines the ability and speed of progress. It turns out that work on assessing balance reliability is the fundamental direction in the development of the state, society ... Therefore, organizations involved in these issues are responsible for the potential of this development. They have all components to obtain power, and it, in turn, can be used in different ways.

The development of electric power is along the path of creating unified energy systems (hereinafter - the UPS). An example is UPS of Russia, the unification energy systems of the European Union, and the unification energy systems of North America. The reliability power balance of such systems is always analyzed. In our country, in the framework UPS of Russia, zones are allocated in the form of united energy systems (hereinafter referred to as EPS) with their fragmentation into separate territories. The reliability power balance is ensured by both its own sources and the generation of related zones without any restrictions on the supply of power from them.

On the basis of this, fairly strict hierarchical systems are formed, within which effective methods implemented in software products are developed to solve the entire complex of problems related to development country's UPS for the prospect of 5 to 20 years, including the task of ensuring balance reliability.

At the present stage, the most acceptable indicators of balance reliability (BRI) from standpoint substantiating the means of ensuring reliability are considered [1.2]:

- in domestic practice, the probability of non-deficiency of territorial zones $\rho = 1 - J_д$ ($J_д$ - the integral probability of a power deficiency);
- in the foreign: Loss of Load Expectation (Lole) and Loss of Load Hours (Lolh).

* Corresponding author: chukreev@iespn.komisc.ru

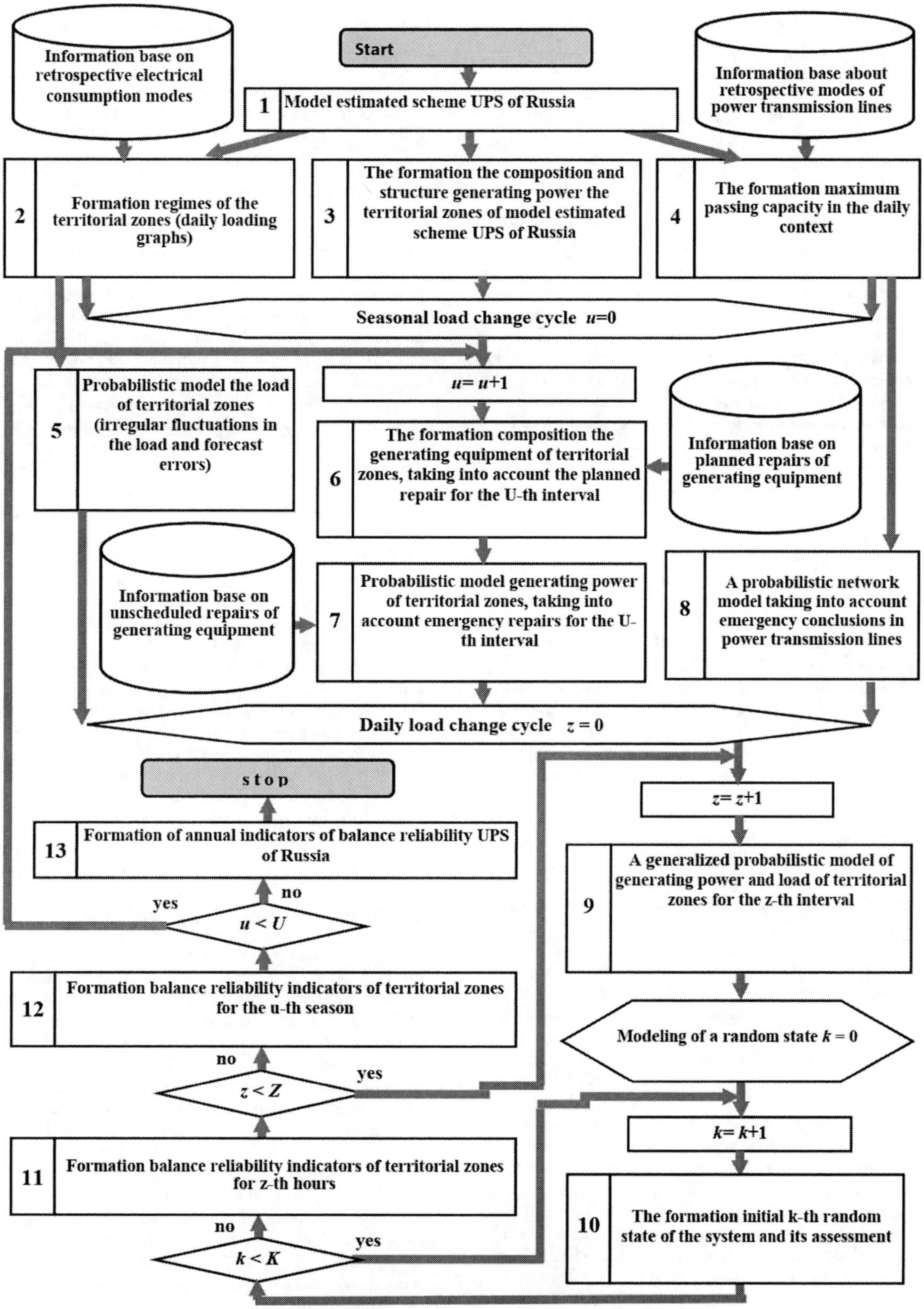

Fig. 1. Scheme the algorithm for assessment of BRI

2 Methodology for assessing the balance reliability EPS

The process determining the indicators balance reliability of complex EPS when using statistical modeling methods is to form a spectrum of random states the main generating and network equipment and their assessment for providing consumers with electricity. In the models developed in the pre-reform period [2, 3, 4, 5, 6, 7], the determination of balance reliability was carried out sequentially for all time intervals, during which the structure of generating capacities and system-forming lines, as well as the load, remained unchanged. After which they were summarized taking into account the probabilities existence of selected time intervals. In this aspect, taking into account market relations does not bring any cardinal changes to setting the task of evaluating balance reliability indicators. The main components of model-software should remain blocks (Fig. 1):

- the formation of random states of the system, caused by unreliable the generating and network equipment of the system;
- assessments the formed conditions for ensuring the load of territorial zones.

In accordance with the above components, the algorithm for assessing the indicators of balance reliability the unification EPS should include models formation the calculated states of the system and the model of their assessment (Fig. 1). The formation of estimated random conditions can be carried out in two ways: a complete overcoming of all possible states the generating power individual EPS of association and methods of statistical modeling. In the methodology implemented in the "Orion-M" software [3,8,9], this process is carried out by statistical modeling methods based on analytically obtained functions the distribution of deficits and excesses the generating power of territorial zones EPS association.

As can be seen from the enlarged block diagram determining the indicators of balance reliability EPS (Fig. 1) the first four blocks have a purely informational orientation and will be disclosed below. The next 4 blocks (from the 5th to the 8th) are aimed at solving the problem of forming the estimated states EPS, the 9th and 10th-to solve the problem of evaluating the consequences these states for consumers. The blocks 11th, 12th and 13th are aimed at determining the indicators of balance reliability EPS for various temporary slices. For the conditions for reforming the electric power industry, more detailed accounting of random states generating equipment of independent entities located in different territorial zones (blocks 3, 6) are required. In turn, this leads to a change in approaches the formation probabilistic models of generating power and the load of territorial zones (blocks 5 and 7), as well as the state of the maximum passing capacity of power in bandwidth (Block 8).

The purpose in solving the problem of balance reliability is to obtain data by means of its provision. The operational reserve can be considered the main one (the operation the "aggregator" and the presence of local energy systems also affect its value). The resulting amount of operational power reserve excites both consumers and suppliers. If you translate the above one question can be formulated as follows: why consider reserves and who needs it? Let's try to answer it based on the history of the last few decades.

2.1 Former USSR

In the period from 1972 to 1991, the formation scheme USSR EPS and all EPS until 1990 was developed, and after the scientific and technical program "Unified electric power system of the USSR" was developed. In connection with the termination of existence USSR, the program was not fully implemented, but it served as basis of the program "Development EPS of Russia". The leading organizations in this matter was the All-Union State Design and Research Institute of Energy Systems and Electric Networks "Energosetproekt" founded in 1962. The institute has become a head for the targeted comprehensive scientific and technical program "Further development USSR EPS in order to increase its efficiency, reliability and reduce electricity losses in electric networks".

During this period, all information was collected within the walls of organization "Energosetproekt", which had research status, NIIPT and ORGRES. This made it possible to receive information and roadmap to independent researchers, scientific and scientific and educational organizations for work on assessing the BRI.

In the electric power industry there was a monopoly structure of management. The state dictated the rules of the game, but at same time supported external research in the field of balance reliability. During this period, the answer to question posed can be formulated as: to whom to consumers, for the most part represented by the state, why - to control the dynamics of state development.

2.2 Reform of the electric power industry

A similar structure was preserved before the reform to grow in 2003. In February 2003, the State Duma of Russian Federation adopted the Law "On Electricity" [10]. If you see the adopted law shows that not one of the goals set in it is aimed at reducing electricity prices, ensuring energy security, sustainable functioning, and attracting investments. As a result of reform the electric power industry, market relations appeared. For the functioning of such a system in June 2002, the infrastructure organization of OJSC "System Operator - Central Dispatch Administration the UPS" was created, which was allocated from OJSC RAO UPS of Russia as an independent enterprise. Subsequently, the name was changed to the JSC "System Operator the UPS" (JSC "SO UPS"). In the framework of reorganization industry, the functions of developing software documents in the field of prospective development industry were entrusted: the general scheme for the placement electric power facilities for the long term, as well as schemes and programs development the EPS of Russia. Thus the System operator began to be responsible for the promising development EPS of Russia. It should be noted that the choice of reform model proposed by Anatoly Borisovich Chubais was primarily justified by the emphasis in it for the promising development Russian UPS.

At this time, despite the declared competition in the industry, SO a consistent movement to collect and store information at home begins. All works related to the tasks of determining reserves and, accordingly, the assessment BRI were carried out by SO or delegated as contractual work to profile organizations. The information provided for the implementation of these works was the intellectual property of OS and could not be published. But the work was carried out, partially with the aim of accumulating certain competencies inside SO, partly to solve problems in the interests of SO.

Based on the foregoing, we will try to answer the above question. Why consider reserves, That the SO as a subject responsible for the promising development EPS of Russia could stably ensure the proper level of reliability. Who needs it - consumers, who, in conditions of competition, both inside the country and beyond, should have the lowest possible tariff value. It turns out to consumers low prices are profitable, and this is a low reserve level, which is bad for SO, because it is more difficult to control the proper level of reliability. The income of generating companies, their ability to conduct investment activities, depends on the tariff, the SO itself lives on the tariff.

2.3 Present

In December 2022, with the direct participation new MI, they adopted. New MI suggest the development of two documents: the general scheme (GS) the placement of electric power facilities and the scheme and development program (SDP) Russian electric power system. Two chapters (V and VI) are devoted to the substantiation of balance reliability provision decisions.

The new MI introduces the chapter "Providing balance reliability". Earlier, balance reliability was ensured by the fulfillment of requirements for the normative power reserve (NPR) of EPS territorial zones [11] (percentage of maximum load, 17 % for the European part and 12 % for the Siberian). The introduction of this chapter allows you to develop managerial solutions for each territorial zone UPS. It is important to note that the need to increase power reserves in territorial zones and the access capacities of the bonds connecting them is required when developing both normative documents (paragraph 115).

In new MI, the methodology for solving problem of evaluating BRI has undergone changes in connection with an increase in information security.

The introduction of a separate section on ensuring balance reliability will lead to a decrease in research on the field of balance reliability. The reason is to concentrate both the tools and the information component in one structure - SO UPS. In addition, it should be noted that almost all probabilibly certain information used to evaluate BRI is formed and used only by specialists and not available to the scientific community. Thus, new MI implies the solution problem of evaluating balance reliability only within the framework of SO. Possible public discussions of managerial decisions, in our opinion, are ineffective due to the lack entire spectrum of source information among interested parties.

Let's try to ask again the question: why consider reserves and who needs it? And the answer will not change in comparison with the previous one, but now with itself determines what level of the reserve he needs. What contradicts the very idea of competition, consumers have lost all power over what is happening, they are simply confronted with a fact.

3 Conclusion

The intellectualization UPS of Russia allows you to use the entire range of necessary information in the tasks of evaluating BRI, which reduces the need to resort assumptions. Analysis the existing retrospective information allows you to identify the dependence of random load deviations between the individual EPS included in UPS of Russia and to form new indicators of the norms for non-wound and planned repairs generating equipment. Implementation this data is not serious problems. But the adoption new MI leads to an unreal scenario to realize this. According to this document, the SO is not obliged to provide information required for calculations receipt of BRI. This contributes to further giving the scientific community from participating in solving the problems of balance reliability. By summarizing the above, you can identify several questions:

- Is there any competition? If all information and roadmap in one player on the market.
- What will the motivation of this player and how to control whose interests he represents?
- How to check and evaluate the decisions made?

References

1. Chukreev Yu.Ya., Chukreev M.Yu. Models for assessing the indicators of balance reliability in managing the development of electric power systems. Syktyvkar: Komi NC URS RAS, 2014, 207 p.
2. Billinton R., Allan R.N. Reliability Evaluation of Power Systems. Second Edition. New York and London, Plenum Press, 1996. 509 p.
3. Yu. Ya. Chukreyev Models for ensuring the reliability of energy power systems. Komi SC UB RAS, Syktyvkar, 1995, 176 p.
4. Yu. N. Rudenko, I. A. Ushakov Reliability of energy systems. Science, Moscow, 1986, 252 p.
5. Kucherov Yu.N., Kucherova OM, Kapui L., Rudenko Yu.N. The reliability and effectiveness the functioning of large transnational EPS. Analysis methods: European measurement. Novosibirsk: Science. Siberian publishing company RAS, 1996. 380 p.
6. G. A. Volkov Optimization of reliability of electric power systems. Science, Moscow, 1986, 117 p
7. J. Enderne Modeling when calculating reliability in energy power systems, per. from English. Energoatomizdat, Moscow, 1983, 336 p.

8. Chukreev Yu.Ya. Methods and models for evaluating indicators and means of ensuring reliability in managing the development EPS / Izv.AS. Energy, **Vol. 4,** pp. 31-39 (1999).
9. Chukreev Yu.Ya., Chukreev M.Yu. Methodological features of assessing the reliability power balance of liberalized electric power systems. / Izvestia of universities. Problems of energy: **Vol. 11-12/1,** pp. 158-169 (2008).
10. The Federal Law "On Electricity" dated 03.26.2003 No. 35-03.
11. Methodological recommendations for design development of energy systems. (Approved by order of the Ministry of Energy of Russia No. 281 of June 30, 2003).

Accident analysis of 220, 500 kV overhead lines of the main power systems network

Irek Galiaskarov[1], *Misrikhan* Misrikhanov[2], *Vladimir* Ryabchenko[3*], and *Andrey* Shuntov[2]

[1]JSC «CEMC UES», 4, Belovezhskaya st., Moscow, Russia
[2] National research university «MPEI», 17, Krasnokazarmennaya st., Moscow, Russia
[3] JSC «R&D Center FGC UES», 22/3, Kashirskoe sh., Moscow, Russia

Abstract. The statistical analysis of overhead lines (OHL) of the unified national power grid (UNPG) of the Central European part of the country for the previous 15 years is subjected to statistical analysis. The structure of the failure flow parameter (failure frequency) of 220 and 500 kV overhead lines with a total length of 17.6 and 8.5 thousand kilometers, respectively, is compared. It is shown that the last decade is characterized by the most favorable reliability characteristics. At the same time, a tendency was detected, when the parameter of overhead line failure flow became insignificantly dependent on the nominal voltage of the network. This fact makes us doubt the existing belief of specialists, according to which the specific (per 100 km) failure rate of overhead transmission lines decreases with the increase of the nominal voltage of the network. The above trend once again demonstrates the objective possibility of purposeful management of electric networks reliability.

Introduction

The problem of ensuring the reliability of power systems has been and remains relevant. One of its aspects is the study of reliability indicators of power grid elements. The most important of them are the failure flow parameter (more precisely - the average failure flow parameter or failure rate), the average restoration time, as well as the frequency and duration of scheduled repairs. According to its functions, the power grid is subdivided into the main grid and the distribution grid [1]. The main grid includes the Unified National Power Grid (UNPG), which forms the backbone of the country's energy systems. It includes overhead power lines with voltage of 220 kV and higher. Reliability of overhead power lines attracts close attention because of their relatively high accident rate due to natural, climatic and socio-economic phenomena and events. As is known, these are the least reliable elements of power grids.

Failures of overhead transmission lines were analyzed in a large number of publications [2-7, etc.] - see Table 1, where references to literature sources are given in descending order of the chronological period of statistical sampling. In Table 1, the values (per 100 km of overhead line length) of the failure flow parameter are highlighted: unrecoverable ω_H; the sum of unrecoverable and recoverable by automatic reclosure ω_Σ. Characteristics are given per circuit or line. If the literature source differentiates the corresponding values depending on the type of supports (reinforced concrete or metal), the data on metal supports, as the most frequently used ones, were taken as a basis.

Accident analysis

Let us turn to the parameters characterizing the accident rate of overhead power lines. Among the data from [2-7] stands out the work [4], which, in fact, is the desktop book of the country's design engineers. It refers to the use of statistical data [6]. But in fact it reflects the values of ω_Σ. Thereby, the failure rate of overhead lines used to assess the reliability of electrical connection schemes is inappropriately overestimated. In [8] the data of more than half a century ago are presented. At the same time, for a number of voltage classes (220 and 330 kV) the values of ω_Σ have not changed much since then. As for 500 kV overhead power lines, it was directly stated in [8] that ω_Σ =2.5 1/(year 100 km) cannot be regarded as a natural reflection of their peculiarities: the data covered only the first experience of operation of overhead power lines from Volga HPPs to Moscow, when the «dancing» of wires was the main cause of emergency outages of these lines. The parameters of the flow of unrecoverable failures in [2, 3, 5, 6] are generally comparable, despite the fact that, say, the statistical sample [3] covered the time period from 2002-2007, and [5, 6] - the period up to the mid-1980s of the last century. The reason lies in the fact that failures in the core networks are subjected to thorough investigations with subsequent development of organizational and technical measures to prevent them.

Table 1 shows that the specific (per 100 km) failure rate of overhead lines steadily decreases with increasing nominal voltage of the network. This is due, as many believe, to the increasing overall dimensions of lines (phase- to-phase distances, distances relative to the ground, etc.) - see Table 2.

* Corresponding author: ryabchenko.vn@yandex.ru

Table 1. Failure flow parameter of overhead line

Voltage, kV	110	220	330	500	750	Source
ω_n, 1/(year·100 km)	1,28	0,56	0,45	0,37	–	[2]
ω_n, 1/(year·100 km)	0,81	0,42	0,33	0,31	–	[3]
ω_n, 1/(year·100 km)	3,9	1,7	1,3	0,6	0,6	[4]
ω_n, 1/(year·100 km)	1,07	0,36	0,55	0,24	–	[5]
ω_n, 1/(year·100 km)	1,28	0,5	0,55	0,21	0,20	[6]
ω_Σ, 1/(year·100 km)	1,32	1,34	1,29	2,5	–	[7]
ω_Σ, 1/(year·100 km)	4,49	2,98	2,07	0,77	–	[2]
ω_Σ, 1/(year·100 km)	4,68	1,51	1,37	0,69	–	[3]
ω_Σ, 1/(year·100 km)	5,3	2,0	2,2	0,58	0,56	[6]
ω_n/ω_Σ, relative units	0,29	0,19	0,21	0,48	–	[2]
ω_n/ω_Σ, relative units	0,17	0,28	0,24	0,45	–	[3]
ω_n/ω_Σ, relative units	0,24	0,25	0,25	0,36	0,36	[6]

Table 2. Parameters of 220 and 500 kV overhead lines

No	Name	Unit	220 kV	500 kV
1	Protection zone	m	20	30
2	The shortest distance from overhead line conductors to the ground surface in unpopulated and inaccessible areas	m	7	8
3	The shortest vertical distance from overhead line conductors to the groundsurface in populated areas	m	8	15,5
4	Smallest horizontal distance between overhead line conductors and tree crowns	m	5	6
5	Number of wires in phase	psc.	1	3–4
6	Geometric mean distance betweenphases	m	8	14
7	Wire cross sections	mm^2 (aluminum/ steel)	240/32 300/39 400/51 500/64	300/66 330/48 400/51 500/64

As is known, in the early 2000s, the organizational structure of management of the domestic electric power industry was changed. The 220 kV power grids referred to the UNPG were separated from regional power systems and transferred to the newly established Federal Grid Company. This ensured uniform rules for repair and maintenance of 220-750 kV grids. It turned out that these transformations hada noticeable impact on the reliability of these grids. The relevant arguments are given below.

The structure of unrecoverable failures of 220 and 500 kV overhead power lines with the total length of 17.6 and 8.5 thousand km in the central European part of the country for the period 2006-2022 was analyzed (Table 3). It follows from Table 3 that the predominant influence on overhead power line failures is caused by natural and climatic phenomena and events (items2, 3 and 5 - more than half of cases for 220 and 500 kV overhead power lines), as well as by conditionally social reasons (items 1, 4 and 6 - about 40% of failures for both voltage classes). As can be seen from Table 3, the structure of failures of 220 and 500 kV overhead lines is generally comparable. This peculiarity, as well as the unified since 2006 principles of repair and operational maintenance of UNPG, prompted the authors to compare the accident rate of 220 and 500 kV overhead lines.

Table 3. Organizational causes of overhead line failures for the period 2006-2022

No	Failures reason	Number of failures			
		500 kV		220 kV	
		piece	%	piece	%
1	Failure to meet deadlines, failure to perform the required amount of maintenance or repair of equipment anddevices, including:	36	9,00	80	9,54
1.1	Failure to identify and eliminate defects in a timely manner (breakage or unraveling ofwires and cables, contamination of insulation, destruction of garland, wear and tear of separate main elements)	18	4,50	71	8,46
1.2	Personnel errors	18	4,50	9	1,07
2	Impacts of birds	5	1,25	18	2,15
3	Exceedance of natural phenomena impactparameters relative to project conditions	17	4,25	41	4,89
4	Exposure to unauthorized persons andorganizations not involved in the technological process	131	32,75	205	24,43
5	Impacts of recurring natural events,including:	194	48,50	436	51,97
5.1	Impacts of recurring natural events,including:	18	4,50	10	1,19
5.2	Atmospheric overvoltages (thunderstorms)	68	17,00	111	13,23
5.3	Natural fires	55	13,75	49	5,84
5.4	Other impacts of unfavorable natural phenomena (falling trees, wild bush vegetation)	51	12,75	262	31,23
5.5	Animals	2	0,50	4	0,48
6	Defect(s) in design, installation,manufacturing	5	1,25	33	3,93
7	Undiscovered causes	12	3,00	26	3,10
8	Total	400	100	839	100

Fig. 1 shows the time series of the unrecoverable failure flow parameter of the above mentioned 220 and 500 kV overhead lines. As can be seen from Fig. 1, at the initial stage (2006) ω_n of 220 and 500 kV overhead lines was approximately 0.5 and 0.15 1/(year 100 km), respectively, which confirmed the data of Table 1. However, since 2012 the failure rate of overhead power lines has practically equalized, and in some cases it became more favorable for 220 kV overhead power lines compared to 500 kV: in 2020 the failure flow parameter of 220 kV overhead power lines was 0.08 1/(year 100 km), and that of 500 kV - 0.10 1/(year 100 km).

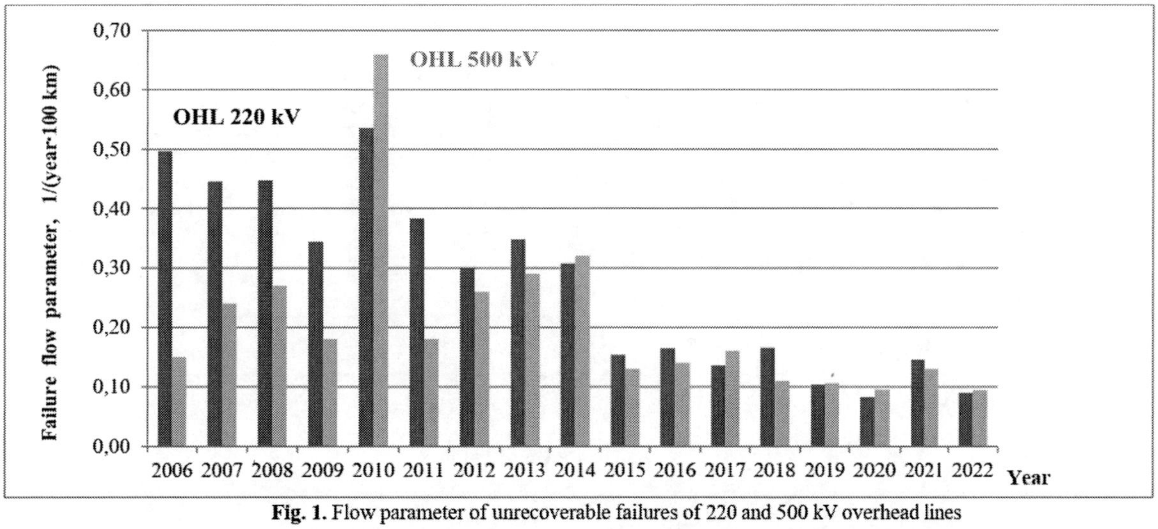

Fig. 1. Flow parameter of unrecoverable failures of 220 and 500 kV overhead lines

Thus, the uniform natural and climatic conditions and repair and maintenance resulted in the fact that the parameter of overhead line failure flow became little dependent on the nominal voltage of the network. Consequently, the identified trend once again demonstrates the objective possibility of managing the reliability of power grid elements.

As follows from Fig. 1, the failure rate of overhead transmission lines is subject to a certain variability from year to year. Therefore, it is not surprising that depending on the time period of the statistical sample, the overhead line reliability parameters, other things being equal, differ (Table 1). The maximumvalues ω_H 0.54 and 0.67 1/(year 100 km) at 220 and 500 kV respectively were observed in 2010 - the year of abnormal natural fires. At the same time,the last five-seven years have revealed a historical minimum of the accident rate of the considered overhead lines (0.1-0.15 1/(year 100 km)), despite the fact that about 40% of the lines have been in operation for more than 50 years (exceeding their standard service life).

At present, operating organizations are equipped with modern mechanisms and means for inspection and elimination of overhead power line failures. Analytical work is underway to collect and process statistical data on their accident rate in order to determine the causes and develop measures to prevent failures.

For example, starting from 2010, a program for the expansion of clearings has been developed and implemented. Trees in the overhead line protection zone that threaten to fall (leaning and dead trees), as well as trees and shrubs under the wires, are cut down annually. This significantly reduced the accident rate due to the fall of trees on wires and overlaps on tree and shrub vegetation (Fig. 2). For more visual dynamics of accident rate reduction in Fig. 2 and subsequent figures, both non-recoverable and recoverable by reclosure action failures are taken into account. As can be seen from Fig. 2, there was a multiple reduction in the accident rate of 220 and 500 kV overhead lines due to the considered cause.

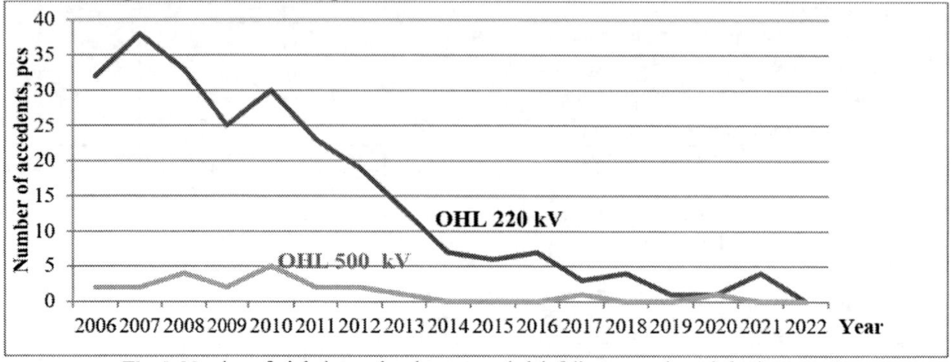

Fig. 2. Number of violations related to tree and slab falls on woody and shrub vegetation

Programs to improve lightning resistance of the most accident-proneoverhead power lines have also been implemented:

- inspection and, if necessary, replacement of the grounding circuit of supports;
- strengthening of lightning rods grounding by increasing the number of additional electrodes, changing the grounding scheme;
- increasing the length of the insulating suspension by adding insulators.

If there was no effect from increasing lightning resistance, surge arresters were additionally installed on the supports. As a result, the number of lightning-related accidents was significantly reduced (Fig. 3).

The situation is more complicated with accidents related to the impact of unauthorized persons and organizations not involved in the technological process. First of all, work was carried out to replace outdated and exhausted operational life of supports with new ones with improved characteristics (replacement of reinforced concrete supports with metal ones with increased dimensions). A program was implemented to reconstruct overhead line sections with wire-to-ground dimensions that do not meet modern requirements (Electrical installation rules, edition 7) due to the growth of settlements. A comprehensive system of notification and informing the population (by authorities, mass media, information and warning posters, signs on poles, day and night markings, etc.) about the inadmissibility and danger of violation of overhead line protection zones has been developed.

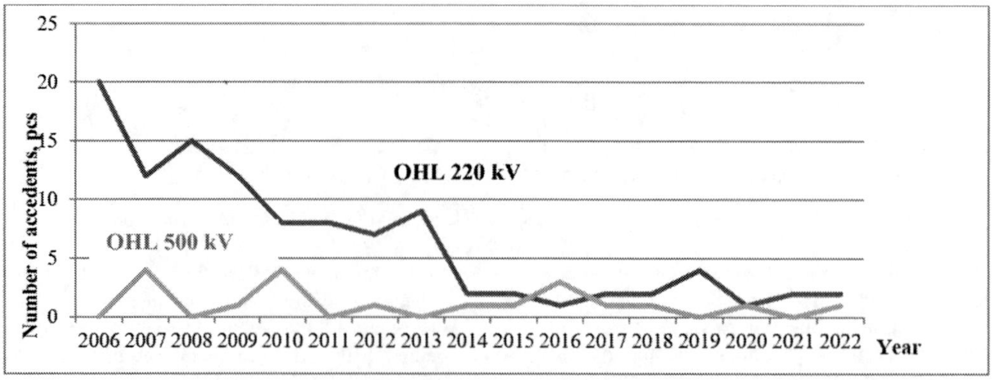

Fig. 3. Number of disturbances related to lightning impacts

 https://doi.org/10.1051/e3sconf/202346101011

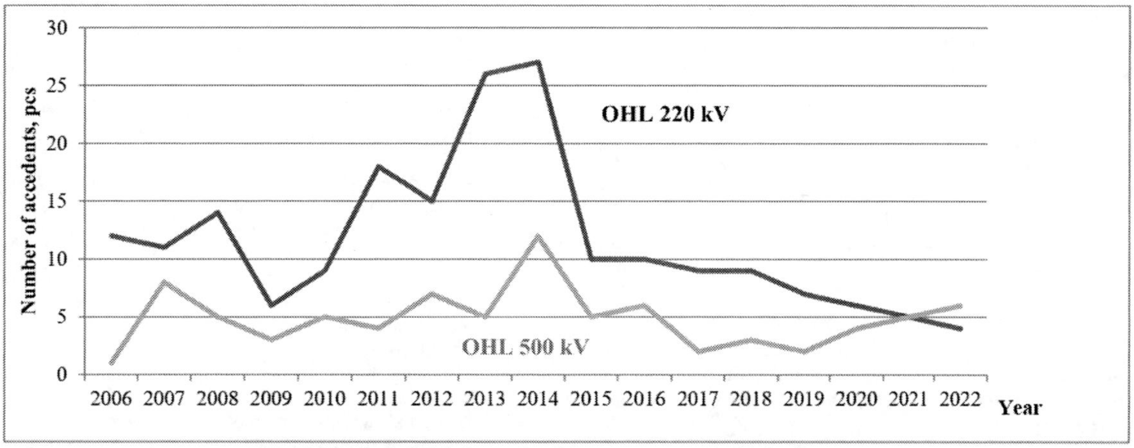

Fig. 4. Number of violations related to exposure to unauthorized persons

During 2009-2014 (Fig. 4), there was a «surge» of overhead line accidents. This is due to the rapid development of agricultural lands, increase in the number of industrial and social facilities, intensification of construction of large road interchanges, highways, as well as individual housing near overhead line protection zones in the region.

At that time, the existing regulations on interaction (during work) between electric grid and third-party organizations were not efficient enough; as a result, overhead line accidents increased. After the introduction of property and monetary agreements between these organizations, which provide for compensation of potentially caused damage to power grids, as well as the implementation of measures to fence territories, installation of information signs, inspection of overhead line protection zones by field crews, regulations for work in overhead line protection zones, etc., the number of accidents decreased (Fig. 4). Thus, if in 2006, 12 such failures were registered on 220 kV overhead lines (see Fig. 4), in 2022 - only four.

Table 1 shows the ratio $\omega / \omega_{H\sum}$, i.e. the number of unsuccessful failures divided by their total number. As can be seen from Table 1, as the nominal voltage of the network increases, the share of unsuccessful failures increases approximately 1.5-2 times and more: 0.19-0.28 and 0.36-0.48 at 220 and 500 kV, respectively. Fig. 5 shows the time series of the share of unrecoverable failures of 220 and 500 kV overhead power lines for the period under consideration 2006-2022. As can be seen, the ratio $\omega_n / \omega_{\sum}$ varies in wide ranges. Thus, in 2006 its value was about 0.45 for both voltage classes, and in 2022 - 0.18 and 0.34 for 220 and 500 kV, respectively.

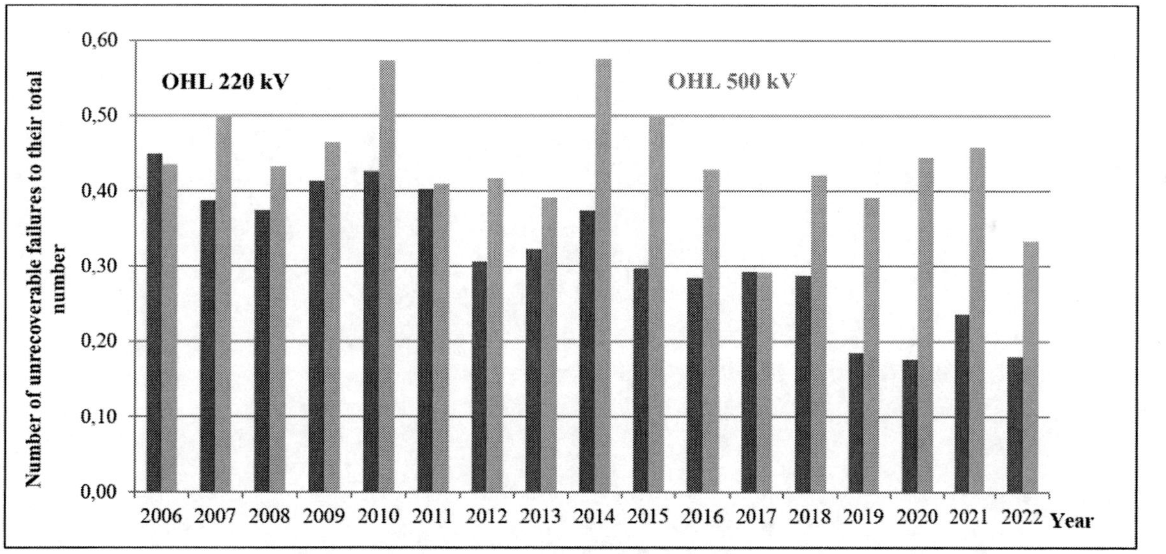

Fig. 5. Ratio of the number of unrecoverable failures to their total number to their total number

Thus, this indicator is subject to noticeable temporal variability. However, there is an increase in the efficiency of successful reclosure on 220 kV OHL. It should be assumed that this is due to the above-mentioned list of measures to improve their reliability.

Conclusion

For many decades there has been a stable opinion of specialists that with increasing nominal voltage of the network the specific (per 100 km) frequency of overhead line failures decreases. This fact is reflected in many literature sources.

The changes in the organizational structure of management of the domestic electric power industry that took place in the early 2000s provided uniform rules for repair and maintenance of 220-750 kV networks. This led to the fact that the parameter of overhead line failure flow became little dependent on the nominal voltage of the network.

Analysis of the organizational structure of failures of 220 and 500 kV overhead power lines in the vast region for the period 2006-2022 revealed the main factors influencing reliability, including their weight in the total accident rate: natural, climatic and social. Targeted impact of operation on thesefactors led to the historical in this time interval minimum accident rate of 220 and 500 kV overhead lines at the level of 0.1-0.15 1/(year 100 km). This is two to four times less than previously published statistical data.

The efficiency of reclosing on 220 and 500 kV overhead power lines is subject to significant temporal variability and is within the previously known limits. Never the less, the efficiency of reclosing on 220 kV overhead lines in the period 2006-2022 increased approximately twice, which, it should beassumed, is associated with the implementation of organizational and technical measures to improve reliability.

References

1. Methodical recommendations for designing the development of power systems. SO 153-34.20.118-2003. Approved by the order of the Ministry of Energy of Russia from 30.06.2003 **No. 281**. [Electronic resource] URL: http://docs.cntd.ru/ document/1200045669 (date of reference: 11.08.2023).
2. V.A. Nepomnyashchiy *Reliability of power systems equipment*. M.: Izd. of the journal "ELECTROENERGY. Transmission and Distribution", 2013,196 p.
3. A.M. Abdurakhmanov, M.S. Misrikhanov, K.V. Mozgalev, A.V. Shuntov, *About the switching resource of the switches at short circuits in the power systems* (in Russian) // Electrical power stations, 2008, **No. 10**, p. 59-62.
4. Reference book on designing of electric networks / Edited by D.L. Faibisovich. M.: Izd-vo NTs ENAS, 2005, 320 p.
5. I.G Barg., V.I. Edelman, *Air Power Transmission Lines*. Moscow: Energoatomizdat, 1985, 248 p.
6. Guidelines for application of reliability indicators of power system elements and operation of power units with steam turbine units. Moscow: SPO Soyuztehenergo, 1985, 18 p.
7. V.D. Katson, O.V. Nikitin, I.F. Polovoy, F.H. Khalilov, *Annual number of switching operations of lines and transformers of higher voltage classes* // Electrical Stations, 1970, **No. 10**, p. 42-44.
8. V.I. Edelman, I.G. Barg, S.V. Korobanov, T.P. Pripuskova *Statistical analysis of deliberate tripping of 35-500 kV overhead power lines* // Electrical Stations, 1979, **No. 11**, p. 49-54.

Synthesis of control and observation laws power system based on generalized formula Ackermann for MIMO systems

Irek Galiaskarov[1], *Misrikhan* Misrikhanov[2], *Vladimir* Ryabchenko[3*], *Eugene* Tregubov[3], and *Andrey* Shuntov[2]

[1] JSC «CEMC UES», 4, Belovezhskaya st., Moscow, Russia
[2] National research university «MPEI», 17, Krasnokazarmennaya st., Moscow, Russia
[3] JSC «R&D Center FGC UES», 22/3, Kashirskoe sh., Moscow, Russia

Abstract. For an electric power system (EPS), as a dynamic system with many inputs and many outputs (Multi Inputs Multi Outputs System – MIMO), compact analytical formulas are obtained for calculating the coefficients of the controller matrix and the observer matrix of the state of the solution of the synthesis problem, providing a given placement of eigenvalues along full state vector. These formulas are generalizations to MIMO systems of the well-known Ackermann formula used to design the control of systems with one input and many outputs (Single Input Multi Outputs System – SIMO). The approach is based on the transformations used in the original multi-step (multilevel) decomposition method, as well as a nondegenerate similarity transformation in the form of the Kalman controllability matrix. The obtained formulas are applicable to dynamic systems, for which the dimension of the state space is a multiple of the dimension of the inputs (controls). This limitation is removed by using the Yokoyama transform. These formulas differ in terms of parameterization of the set of equivalent laws. An example of the synthesis of a control law for a synchronous generator in a complex EPS is considered in order to preserve the existing modes of electromechanical oscillations and meet additional requirements (roughness with respect to disturbances and/or increased sensitivity to changes in controlled parameters in a given region or frequency band).

Introduction and problem statement

One of the most well-known explicit calculation formulas used for the synthesis of controllers and observers of linear stationary dynamic systems in the state space with one input and one output, including electric power systems (EPS), are the Ackermann and Bass-Gura formulas [0-4].

Let a fully controlled stationary linear dynamic SIMO-system (Single Input Multi Outputs System) be given

$$\sigma\mathbf{x} = \boldsymbol{A}\mathbf{x} + \mathbf{b}u, \tag{1}$$

where
$\mathbf{x} \in \mathbb{R}^n$ is the state vector, is the scalar input, $u \in \mathbb{R}$, $\mathbb{R}$ – the set of real numbers, $\sigma\mathbf{x}(t) \triangleq \dot{\mathbf{x}}(t)$ for the case of continuous time and discrete time $\sigma\mathbf{x}(t) \triangleq \mathbf{x}(t+1)$.

The condition for complete controllability of the system (1) or complete controllability of the pair corresponds to the nonsingularity of the Kalman controllability matrix $(\mathbf{A}, \mathbf{b})$

$$\Omega(\mathbf{A}, \mathbf{b}) = (\mathbf{b} \quad \boldsymbol{A}\mathbf{b} \quad \cdots \quad \boldsymbol{A}^{n-1}\mathbf{b}) \in \mathbb{R}^{n\times n}. \tag{2}$$

Let also the characteristic polynomial of the matrix $\boldsymbol{A}$ equals

$$det(\lambda\boldsymbol{I}_n - \boldsymbol{A}) = \lambda^n + \alpha_{n-1}\lambda^{n-1} + \cdots + \alpha_1\lambda + \alpha_0, \tag{3}$$

where

$\boldsymbol{I}_n$ is the identity matrix of order n, $\alpha_i \in \mathbb{R}$ are the coefficients of the characteristic polynomial, λ is the set of complex numbers $\mathbb{C}$.

Using the feedback law on state variables

$$u = -\mathbf{k}^T\mathbf{x} \tag{4}$$

required to provide a closed system
$\dot{\mathbf{x}} = (\boldsymbol{A} - \mathbf{b}\mathbf{k}^T)\mathbf{x}$,
the following characteristic polynomial:

$$det(\lambda\mathbf{I}_n - \boldsymbol{A} + \mathbf{b}\mathbf{k}^{\mathrm{T}}) = \lambda^n + \beta_{n-1}\lambda^{n-1} + \cdots + \beta_0. \tag{5}$$

If we introduce the notation for the matrix polynomial

$$\Xi(\mathbf{A}, \beta) = \boldsymbol{A}^n + \beta_{n-1}\boldsymbol{A}^{n-1} + \cdots + \beta_1\boldsymbol{A} + \beta_0\boldsymbol{I}_n, \tag{6}$$

then the calculation formula of Ackermann, which determines the vector of coefficients of the controller (dually, the observer of the state) in $\mathbf{k}^T$ (4) looks like [0- 4]

$$\mathbf{k}^T = (0 \quad \cdots \quad 0 \quad 1)\cdot(\mathbf{b} \quad \boldsymbol{A}\mathbf{b} \quad \cdots \quad \boldsymbol{A}^{n-1}\mathbf{b})^{-1} (\boldsymbol{A}^n + \beta_{n-1}\boldsymbol{A}^{n-1} + \cdots + \beta_1\boldsymbol{A} + \beta_0\boldsymbol{I}_n)$$

or taking into account(2), (6)

$$\mathbf{k}^T = (0 \quad \cdots \quad 0 \quad 1)\cdot\Omega^{-1}(\boldsymbol{A}, \mathbf{b})\cdot\Xi(\boldsymbol{A}, \beta). \tag{7}$$

The main advantage of the Ackermann calculation formula (7), is its explicit form, which allows using the known parameters of the system (1), its characteristic polynomial (3) and a given arrangement of poles (polynomial roots), expressed in coefficients of the charac-

* Corresponding author: ryabchenko.vn@yandex.ru

teristic polynomial (5), calculate the feedback that provides the closed system with the required location of the poles (setting the roots). The same applies to the Bass-Gura formula [4], which is not shown here.

In this paper, we consider a generalization of the Ackermann formula (7) for the case of a mathematical model of an EPS in the form of a dynamic system with many inputs (Multi Inputs Multi Outputs System – MIMO). This task is relevant, since there are no explicit calculation formulas for synthesizing the feedback law for MIMO systems similar to the Ackermann formula. Some exceptions are [5, 6]. In [5] to solve the problem, the Sylvester matrix equation and the associated inversion of the matrix polynomial (6), and in [6] is a special form of the control law, which significantly limits the area of its use.

In what follows, all transformations, taking into account duality, turn out to be valid for the problem of synthesis of the state observer.

Let a fully controlled MIMO system be given, which is understood as mathematical model of EPS,

$$\sigma \mathbf{x} = \boldsymbol{A}\mathbf{x} + \boldsymbol{B}\mathbf{u} \tag{8}$$

with state and control matrices of multiple dimensions
$\boldsymbol{A} \in \mathbb{R}^{k \cdot m}, \quad \boldsymbol{B} \in \mathbb{R}^{m}, \quad k, m \in \mathbb{N}$,
where the matrix

$$\Omega(\mathbf{A}, \mathbf{B}) = [\boldsymbol{B} \quad \boldsymbol{AB} \quad \dots \quad \boldsymbol{A}^{k-1}\boldsymbol{B}], \tag{9}$$

made up of the first k block columns of the Kalman controllability matrix is reversible. It is required to determine the matrices of regulators (observers) such that the state matrix "object - regulator" ("object - observer") $\boldsymbol{A} - \boldsymbol{BK}$ has given eigenvalues as $\boldsymbol{\Phi}^*$. At the same time, in the future we will compare the sets of solutions to this problem obtained by various analytical methods: using multilevel decomposition [7] and on the basis of a non-degenerate similarity transformation.

1. Solution using multilevel decomposition

Let us perform the following similarity transformation [8]:

$$\widetilde{\boldsymbol{A}} = \Omega^{-1}\boldsymbol{A}\Omega = \left[\begin{bmatrix} \mathbf{0}_{m\times(n-m)} \\ \boldsymbol{I}_{n-m} \end{bmatrix} \quad \Omega^{-1}\boldsymbol{A}^k\boldsymbol{B}\right],$$

$$\widetilde{\boldsymbol{B}} = \Omega^{-1}\boldsymbol{B} = \begin{bmatrix} \boldsymbol{I}_m \\ \mathbf{0}_{(n-m)\times m} \end{bmatrix}. \tag{10}$$

For the resulting pair of matrices, we synthesize the state controller with the matrix $\widetilde{\boldsymbol{K}}$.

In this case, when the condition(9), the number of decomposition levels of the pair of matrices $\widetilde{\boldsymbol{A}}, \widetilde{\boldsymbol{B}}$, (including the zero level) is k [7]. The levels do not degenerate (the rank of the control matrices at each of the levels is m). Decomposition is performed according to the formulas [7]

$\widetilde{\boldsymbol{A}}_0 = \widetilde{\boldsymbol{A}}, \ \widetilde{\boldsymbol{B}}_0 = \widetilde{\boldsymbol{B}},$
$\widetilde{\boldsymbol{A}}_{i+1} = \widetilde{\boldsymbol{B}}_i^{\perp L}\widetilde{\boldsymbol{A}}_i\widetilde{\boldsymbol{B}}_i^{\perp L}, \quad \widetilde{\boldsymbol{B}}_{i+1} = \widetilde{\boldsymbol{B}}_i^{\perp L}\widetilde{\boldsymbol{A}}_i\widetilde{\boldsymbol{B}}_i,$
$i = 0, 1, \dots, k-2,$
using semi-orthogonal left zero divisors

$\widetilde{\boldsymbol{B}}_i^{\perp L} = [\mathbf{0}_{(n-(i+1)m)\times m} \quad \boldsymbol{I}_{n-(i+1)m}], \quad i = 0, 1, \dots, k-2.$

The state matrices by decomposition levels take the form

$$\widetilde{\boldsymbol{A}}_i = \begin{bmatrix} \mathbf{0}_{m\times(n-(i+1)m)} & \widetilde{\boldsymbol{A}}_{(i+1,k)} \\ \boldsymbol{I}_{n-(i+1)m} & \widetilde{\boldsymbol{A}}_{(i+2:k,k)} \end{bmatrix} =$$
$$= \begin{bmatrix} \mathbf{0}_{m\times m} & \underbrace{[\mathbf{0}_{m\times m} \ \dots \ \mathbf{0}_{m\times m} \ \ \widetilde{\boldsymbol{A}}_{(i+1,k)}]}_{(k-1-i)\text{ blocks}} \\ \widetilde{\boldsymbol{B}}_{i+1} & \widetilde{\boldsymbol{A}}_{i+1} \end{bmatrix},$$
$i = 0, 1, \dots, k-2,$
$\widetilde{\boldsymbol{A}}_{k-1} = \widetilde{\boldsymbol{A}}_{(k,k)}.$

The control matrices by decomposition levels have the form

$$\widetilde{\boldsymbol{B}}_i = \begin{bmatrix} \boldsymbol{I}_m \\ \mathbf{0}_{(n-(i+1)m)\times m} \end{bmatrix} = \begin{bmatrix} \widetilde{\boldsymbol{B}}_{i+1} \\ \mathbf{0}_{m\times m} \end{bmatrix}, \quad i = 0, 1, \dots, k-2,$$
$\widetilde{\boldsymbol{B}}_{k-1} = \boldsymbol{I}_m.$

Here and below, the designation corresponds to the block located at the intersection $\boldsymbol{A}_{l(i,j)}$ i-th block line and j-th a block column of a matrix $\boldsymbol{A}_l$ divided into blocks of the same size $m \times m$.

Assign matrices with given eigenvalues by decomposition levels
$\boldsymbol{\Phi}_0, \boldsymbol{\Phi}_1, \dots, \boldsymbol{\Phi}_{k-1} \in \mathbb{C}^{m\times m},$
such that

$$\bigcup_{i=0}^{k-1} eig\,\boldsymbol{\Phi}_i = \boldsymbol{\Phi}^*.$$

The controller matrix at the uppermost (k–1)-th level has the form
$\widetilde{\boldsymbol{K}}_{k-1} = \widetilde{\boldsymbol{B}}_{k-1}^{-1}\widetilde{\boldsymbol{A}}_{k-1} - \boldsymbol{\Phi}_{k-1}\widetilde{\boldsymbol{B}}_{k-1}^{-1} = \widetilde{\boldsymbol{A}}_{(k,k)} - \boldsymbol{\Phi}_{k-1}$

Next, we sequentially perform the calculations of the controller matrices at the levels. Using pseudoinverse and auxiliary matrices
$i = k-2, k-3, \dots, 1, 0,$
$\widetilde{\boldsymbol{B}}_i^{+} = \widetilde{\boldsymbol{B}}_i^{T}, \quad \widetilde{\boldsymbol{B}}_i^{-} = \widetilde{\boldsymbol{B}}_i^{+} + \widetilde{\boldsymbol{K}}_{i+1}\widetilde{\boldsymbol{B}}_i^{\perp L} = [\boldsymbol{I}_m \quad \widetilde{\boldsymbol{K}}_{i+1}]$
on i-th decomposition level, define the controller matrix
$\widetilde{\boldsymbol{K}}_i = \widetilde{\boldsymbol{B}}_i^{-}\widetilde{\boldsymbol{A}}_i - \boldsymbol{\Phi}_i\widetilde{\boldsymbol{B}}_i^{-}$
$= [\widetilde{\boldsymbol{K}}_{i(1,1)} \quad \widetilde{\boldsymbol{K}}_{i(1,2)} \quad \dots \quad \widetilde{\boldsymbol{K}}_{i(1,k-i-1)} \quad \widetilde{\boldsymbol{K}}_{i(1,k-i)}],$
the blocks of which are written in terms of blocks of the controller matrix at the (i+1)-th level as follows:
$\widetilde{\boldsymbol{K}}_{i(1,1)} = \widetilde{\boldsymbol{K}}_{i+1(1,1)} - \boldsymbol{\Phi}_i$
$\widetilde{\boldsymbol{K}}_{i(1,2)} = \widetilde{\boldsymbol{K}}_{i+1(1,2)} - \boldsymbol{\Phi}_i\widetilde{\boldsymbol{K}}_{i+1(1,1)}$
...
$\widetilde{\boldsymbol{K}}_{i(1,k-i-1)} = \widetilde{\boldsymbol{K}}_{i+1(1,k-(i+1))} - \boldsymbol{\Phi}_i\widetilde{\boldsymbol{K}}_{i+1(1,k-(i+1)-1)}$

$$\widetilde{\boldsymbol{K}}_{i(1,k-i)} = \widetilde{\boldsymbol{A}}_{(i+1,k)} + \sum_{j=1}^{k-(i+1)} \left(\widetilde{\boldsymbol{K}}_{i+1(1,j)}\widetilde{\boldsymbol{A}}_{(i+1+j,k)}\right) - \boldsymbol{\Phi}_i\widetilde{\boldsymbol{K}}_{i+1(1,k-(i+1))}.$$

Let's rewrite the expressions for the blocks of the controller matrix
$\widetilde{\boldsymbol{K}}_{i(1,j)}$ $(j = 1, 2, \dots, k-i)(i = 0, 1, \dots, k-1)$ at the i-th level of decomposition $\widetilde{\boldsymbol{A}}$, $\widetilde{\boldsymbol{B}}$, in a recurrent form using the values of the known blocks of the same level, located to the left of the current block,

$$\widetilde{\boldsymbol{K}}_{i(1,j)} = \widetilde{\boldsymbol{S}}_{k-i,j} + \widetilde{\boldsymbol{A}}_{(k-(j-1),k)} + \sum_{r=1}^{j-1}\widetilde{\boldsymbol{K}}_{i(1,r)}\widetilde{\boldsymbol{A}}_{(k-(j-1)+r,k)}. \tag{11}$$

Here are the coefficients of the matrix polynomial $\boldsymbol{S}_{k-i,j}$

$$(\lambda \boldsymbol{I}_m - \boldsymbol{\Phi}_i)(\lambda \boldsymbol{I}_m - \boldsymbol{\Phi}_{i+1}) \dots (\lambda \boldsymbol{I}_m - \boldsymbol{\Phi}_{k-1}) = \\ = \lambda^{k-i}\boldsymbol{S}_{k-i,0} + \lambda^{k-i-1}\boldsymbol{S}_{k-i,1} + \lambda^{k-i-2}\boldsymbol{S}_{k-i,2} + \cdots \\ + \lambda \boldsymbol{S}_{k-i,k-i-1} + \boldsymbol{S}_{k-i,k-i}$$

Comparing the values of the blocks of the controller matrix at the lowest level with the values of the blocks in the lower k-th rows of matrices , we obtain the value of the controller matrix in explicit form

$$\widetilde{\boldsymbol{K}}_{0(1,j)}\widetilde{\boldsymbol{K}} = \widetilde{\boldsymbol{K}}_0 i = 0\widetilde{\boldsymbol{A}}^0, \widetilde{\boldsymbol{A}}^1, \dots, \widetilde{\boldsymbol{A}}^k$$

$$\widetilde{\boldsymbol{K}} = \left[\widetilde{\boldsymbol{K}}_{0(1,1)} \quad \widetilde{\boldsymbol{K}}_{0(1,2)} \quad \dots \quad \widetilde{\boldsymbol{K}}_{0(1,k)}\right] = \\ \sum_{j=0}^{k}\left(\boldsymbol{S}_{k,j}\underbrace{[\boldsymbol{0}_{m\times m} \quad \dots \quad \boldsymbol{0}_{m\times m} \quad \boldsymbol{I}_m]}_{k \text{ blocks}}\widetilde{\boldsymbol{A}}^{k-j}\right). \tag{12}$$

Let us carry out the reverse transition by the similarity transformation (10)

$$\boldsymbol{K} = \widetilde{\boldsymbol{K}}\Omega^{-1}$$

and write the result

$$\boldsymbol{K} = \\ \sum_{j=0}^{k}\left(\boldsymbol{S}_{k,j}\underbrace{[\boldsymbol{0}_{m\times m} \quad \dots \quad \boldsymbol{0}_{m\times m} \quad \boldsymbol{I}_m]}_{k \text{ blocks}}\Omega^{-1}\boldsymbol{A}^{k-j}\right). \tag{13}$$

Recall that here are the coefficients of the matrix polynomial$\boldsymbol{S}_{k,j}$ $(j = 1, 2, \dots, k)$

$$(\lambda \boldsymbol{I}_m - \boldsymbol{\Phi}_0)(\lambda \boldsymbol{I}_m - \boldsymbol{\Phi}_1) \dots (\lambda \boldsymbol{I}_m - \boldsymbol{\Phi}_{k-1}) = \\ = \lambda^k \boldsymbol{S}_{k,0} + \lambda^{k-1}\boldsymbol{S}_{k,1} + \lambda^{k-2}\boldsymbol{S}_{k,2} + \cdots + \lambda \boldsymbol{S}_{k,k-1} + \boldsymbol{S}_{k,k}, \tag{14}$$

from which (in accordance with the theorem on the equality of spectra in the method of multilevel decomposition [7]) the characteristic polynomial of the matrix is formed $\boldsymbol{A} - \boldsymbol{BK}$

$$\left|\lambda^k \boldsymbol{S}_{k,0} + \lambda^{k-1}\boldsymbol{S}_{k,1} + \lambda^{k-2}\boldsymbol{S}_{k,2} + \cdots + \lambda \boldsymbol{S}_{k,k-1} + \boldsymbol{S}_{k,k}\right| \\ = |\lambda \boldsymbol{I}_m - \boldsymbol{\Phi}_0||\lambda \boldsymbol{I}_m - \boldsymbol{\Phi}_1| \dots |\lambda \boldsymbol{I}_m \\ - \boldsymbol{\Phi}_{k-1}|.$$

Formula (13) describes the set of equivalent solutions (regulator matrices) of the modal synthesis problem under consideration, and the parameterization of this set (the construction of the set of solutions)is carried out by all possible matrix polynomials (14).

2. Solution based on a nondegenerate similarity transformation

For a pair of matrices $(\boldsymbol{A}, \boldsymbol{B})$ perform a similarity transformation [8]. Note that in the general case the Yokoyama transformation [7], which removes restrictions on the ratio (multiplicity) of the dimensions of the state vectors and inputs used in this work. However, to simplify calculations, we will use the transformation(10).

As a result of the transformation (10) we obtain an equivalent pair of matrices $(\widetilde{\boldsymbol{A}}, \widetilde{\boldsymbol{B}})$, for which we define a controller with matrix $\widetilde{\boldsymbol{K}}$. Next, we form the characteristic polynomial of the matrix $\widetilde{\boldsymbol{A}} - \widetilde{\boldsymbol{B}}\widetilde{\boldsymbol{K}}$:

$$|\lambda \boldsymbol{I}_n - (\widetilde{\boldsymbol{A}} - \widetilde{\boldsymbol{B}}\widetilde{\boldsymbol{K}})| \\ = \begin{vmatrix} \lambda \boldsymbol{I}_m + \widetilde{\boldsymbol{K}}_{(1,1)} & \widetilde{\boldsymbol{K}}_{(1,2)} & \widetilde{\boldsymbol{K}}_{(1,3)} & \widetilde{\boldsymbol{K}}_{(1,4)} & \cdots & \widetilde{\boldsymbol{K}}_{1,(k-1)} & \widetilde{\boldsymbol{K}}_{(1,k)} - \widetilde{\boldsymbol{A}}_{(1,k)} \\ -\boldsymbol{I}_m & \lambda \boldsymbol{I}_m & \boldsymbol{0}_{m\times m} & \boldsymbol{0}_{m\times m} & \cdots & \boldsymbol{0}_{m\times m} & -\widetilde{\boldsymbol{A}}_{(2,k)} \\ \boldsymbol{0}_{m\times m} & -\boldsymbol{I}_m & \lambda \boldsymbol{I}_m & \boldsymbol{0}_{m\times m} & \cdots & \boldsymbol{0}_{m\times m} & -\widetilde{\boldsymbol{A}}_{(3,k)} \\ \vdots & \ddots & \ddots & \ddots & \ddots & \vdots & \vdots \\ \boldsymbol{0}_{m\times m} & \cdots & \boldsymbol{0}_{m\times m} & -\boldsymbol{I}_m & \lambda \boldsymbol{I}_m & \boldsymbol{0}_{m\times m} & -\widetilde{\boldsymbol{A}}_{(k-2,k)} \\ \boldsymbol{0}_{m\times m} & \cdots & \boldsymbol{0}_{m\times m} & \boldsymbol{0}_{m\times m} & -\boldsymbol{I}_m & \lambda \boldsymbol{I}_m & -\widetilde{\boldsymbol{A}}_{(k-1,k)} \\ \boldsymbol{0}_{m\times m} & \cdots & \boldsymbol{0}_{m\times m} & \boldsymbol{0}_{m\times m} & \boldsymbol{0}_{m\times m} & -\boldsymbol{I}_m & \lambda \boldsymbol{I}_m - \widetilde{\boldsymbol{A}}_{(k,k)} \end{vmatrix}.$$

Here the determinant is divided by solid lines into blocks

$$|\lambda \boldsymbol{I}_n - (\widetilde{\boldsymbol{A}} - \widetilde{\boldsymbol{B}}\widetilde{\boldsymbol{K}})| = \begin{vmatrix} \boldsymbol{X}_{1,1} & \boldsymbol{X}_{1,2} \\ \boldsymbol{X}_{2,1} & \boldsymbol{X}_{2,2} \end{vmatrix}.$$

Since the block $\boldsymbol{X}_{2,1}$ is an invertible matrix, and , we write the equivalent characteristic equation$|\boldsymbol{X}_{2,1}| = (-1)^{n-m}$

$$\begin{vmatrix} \boldsymbol{X}_{1,1} & \boldsymbol{X}_{1,2} \\ \boldsymbol{X}_{2,1} & \boldsymbol{X}_{2,2} \end{vmatrix} = 0 \Leftrightarrow \left|\begin{bmatrix} \boldsymbol{X}_{1,1} & \boldsymbol{X}_{1,2} \\ \boldsymbol{X}_{2,1} & \boldsymbol{X}_{2,2} \end{bmatrix}\begin{bmatrix} -\boldsymbol{X}_{2,1}^{-1}\boldsymbol{X}_{2,2} & \boldsymbol{I}_{n-m} \\ \boldsymbol{I}_m & \boldsymbol{0}_{m\times(n-m)} \end{bmatrix}\right| = 0 \Leftrightarrow \\ \Leftrightarrow \begin{vmatrix} \boldsymbol{X}_{1,2} - \boldsymbol{X}_{1,1}\boldsymbol{X}_{2,1}^{-1}\boldsymbol{X}_{2,2} & \boldsymbol{X}_{1,1} \\ \boldsymbol{0}_{(n-m)\times m} & \boldsymbol{X}_{2,1} \end{vmatrix} = 0 \Leftrightarrow |\boldsymbol{X}_{2,1}||\boldsymbol{X}_{1,2} - \boldsymbol{X}_{1,1}\boldsymbol{X}_{2,1}^{-1}\boldsymbol{X}_{2,2}| = 0 \Leftrightarrow \\ \Leftrightarrow |\boldsymbol{X}_{1,2} - \boldsymbol{X}_{1,1}\boldsymbol{X}_{2,1}^{-1}\boldsymbol{X}_{2,2}| = 0. \tag{15}$$

Next, we define the inverse matrix

$$\boldsymbol{X}_{2,1}^{-1} \\ = -\begin{bmatrix} \boldsymbol{I}_m & \lambda \boldsymbol{I}_m & \lambda^2 \boldsymbol{I}_m & \lambda^3 \boldsymbol{I}_m & \cdots & \lambda^{k-2}\boldsymbol{I}_m \\ \boldsymbol{0}_{m\times m} & \boldsymbol{I}_m & \lambda \boldsymbol{I}_m & \lambda^2 \boldsymbol{I}_m & \cdots & \lambda^{k-3}\boldsymbol{I}_m \\ \vdots & \ddots & \ddots & \ddots & \ddots & \vdots \\ \boldsymbol{0}_{m\times m} & \cdots & \boldsymbol{0}_{m\times m} & \boldsymbol{I}_m & \lambda \boldsymbol{I}_m & \lambda^2 \boldsymbol{I}_m \\ \boldsymbol{0}_{m\times m} & \cdots & \boldsymbol{0}_{m\times m} & \boldsymbol{0}_{m\times m} & \boldsymbol{I}_m & \lambda \boldsymbol{I}_m \\ \boldsymbol{0}_{m\times m} & \cdots & \boldsymbol{0}_{m\times m} & \boldsymbol{0}_{m\times m} & \boldsymbol{0}_{m\times m} & \boldsymbol{I}_m \end{bmatrix}$$

and matrix polynomial

$$\boldsymbol{X}_{1,2} - \boldsymbol{X}_{1,1}\boldsymbol{X}_{2,1}^{-1}\boldsymbol{X}_{2,2} = \lambda^k \widetilde{\boldsymbol{S}}_{k,0} + \lambda^{k-1}\widetilde{\boldsymbol{S}}_{k,1} + \lambda^{k-2}\widetilde{\boldsymbol{S}}_{k,2} + \\ \cdots + \lambda \widetilde{\boldsymbol{S}}_{k,k-1}\phi + \widetilde{\boldsymbol{S}}_{k,k}, \tag{15}$$

where by grouping like terms in powers ϕ matrix coefficients are determined

$$\widetilde{\boldsymbol{S}}_{k,j} = \widetilde{\boldsymbol{K}}_{(1,j)} - \widetilde{\boldsymbol{A}}_{(k-(j-1),k)} - \sum_{r=1}^{j-1}\widetilde{\boldsymbol{K}}_{(1,r)}\widetilde{\boldsymbol{A}}_{(k-(j-1)+r,k)}.$$

From here one can express

$$\widetilde{\boldsymbol{K}}_{(1,j)} = \widetilde{\boldsymbol{S}}_{k,j} + \widetilde{\boldsymbol{A}}_{(k-(j-1),k)} + \sum_{r=1}^{j-1}\widetilde{\boldsymbol{K}}_{(1,r)}\widetilde{\boldsymbol{A}}_{k-(j-1)+r,k}.$$

The resulting formula is similar to the expression (11) for $i = 0$ up to coefficients , which are now replaced by . Therefore, by analogy with $\boldsymbol{S}_{k,j}\widetilde{\boldsymbol{S}}_{k,j}$(12) and (13) we get

$$\widetilde{\boldsymbol{K}} = \left[\widetilde{\boldsymbol{K}}_{(1,1)} \quad \widetilde{\boldsymbol{K}}_{(1,2)} \quad \dots \quad \widetilde{\boldsymbol{K}}_{(1,k)}\right] = \\ \sum_{i=0}^{k}\left(\widetilde{\boldsymbol{S}}_{k,i}\underbrace{[\boldsymbol{0}_{m\times m} \quad \dots \quad \boldsymbol{0}_{m\times m} \quad \boldsymbol{I}_m]}_{k \text{ blocks}}\widetilde{\boldsymbol{A}}^{k-i}\right) \tag{16}$$

and

$$\boldsymbol{K} = \sum_{j=0}^{k}\left(\widetilde{\boldsymbol{S}}_{k,j}\underbrace{[\boldsymbol{0}_{m\times m} \quad \dots \quad \boldsymbol{0}_{m\times m} \quad \boldsymbol{I}_m]}_{k \text{ blocks}}\Omega^{-1}\boldsymbol{A}^{k-j}\right). \tag{17}$$

Expression (18) describes the set of equivalent solutions (regulator matrices) of the problem under consideration, the parameterization of which is carried out by all possible matrix polynomials (16). The main difference between this solution and (13) is that the matrix polynomial (16) does not have to be decomposed into matrix factors, as is done in (14). The main thing is that the roots of the characteristic equation (15)

$$\left|\lambda^k \boldsymbol{I}_m + \lambda^{k-1}\tilde{\boldsymbol{S}}_{k,1} + \lambda^{k-2}\tilde{\boldsymbol{S}}_{k,2} + \cdots + \lambda\tilde{\boldsymbol{S}}_{k,k-1} + \tilde{\boldsymbol{S}}_{k,k}\right| = 0$$

coincided with the corresponding eigenvalues of the matrices $\boldsymbol{\Phi}^*$.

3. Example

Consider an example of practical synthesis, when the task is to modify the previously obtained control law for synchronous generators so that the modes of electromechanical oscillations provided by the original control laws are preserved, while the modified laws would satisfy some additional requirements. Such requirements may include "enhanced" roughness with respect to disturbances (robustness), or, conversely, increased sensitivity to changes in controlled parameters in a given region or frequency band, etc. These requirements can be met by varying the given matrices $\bigcup_{i=0}^{k-1} eig\,\boldsymbol{\Phi}_i = \boldsymbol{\Phi}^*$. that the desired solution of the modal control problem under consideration, which belongs to the set of possible solutions (18), is not contained in the set (13), obtained using multilevel decomposition. In other words, consider the case when the matrix polynomial (16) does not decompose into matrix factors by analogy with the controller matrix (14).

Let the model be given EES, including synchronous generators and represented by the following linearized equations $N/2$:

$$\frac{d\delta_i}{dt} = s_i, \frac{ds_i}{dt} = -\frac{1}{J_i}(\Delta P_i + D_i\Delta s_i - \Delta u_i),\ i = 1,2,\ldots,k = N/2. \quad (19)$$

Here D_i is the damping coefficient, J_i is the moment of inertia, ΔP_i is the change in active power, determined using the equations of the electrical network

$$\begin{bmatrix}\Delta \boldsymbol{P}\\ \Delta \boldsymbol{Q}\end{bmatrix} = \begin{bmatrix}\boldsymbol{H} & \boldsymbol{M}\\ \boldsymbol{N} & \mathbf{D}\end{bmatrix}\begin{bmatrix}\Delta \boldsymbol{\delta}\\ \Delta \boldsymbol{U}\end{bmatrix},$$

Where $\Delta\mathbf{P}$ is the active power increment vector in all system nodes, $\Delta\mathbf{Q}$ is the reactive power increment vector, $\Delta\mathbf{U}$ is the voltage amplitude increment vector, $\Delta\boldsymbol{\delta}$ is the voltage phase shift angle increment vector, $\Delta\mathbf{s}$ is the generator slip increment vector, $\Delta\mathbf{u}$ is the control vector (signals of PID-controllers). The elements of the Jacobi matrix are the corresponding partial derivatives

$$H_{ij} = \frac{\partial P_i}{\partial \delta_j}, M_{ij} = \frac{\partial P_i}{\partial U_j}, N_{ij} = \frac{\partial Q_i}{\partial \delta_j}, D_{ij} = \frac{\partial Q_i}{\partial U_j}.$$

Under the assumptions made, the matrix $\boldsymbol{A}$ and $\boldsymbol{B}$ in the vector equation of the model EES take the form

$$\boldsymbol{A} = \begin{bmatrix} 0 & 0 & \cdots & 0 & 1 & 0 & \cdots & 0 \\ 0 & 0 & \cdots & 0 & 0 & 1 & \cdots & 0 \\ \vdots & \vdots & \ddots & \vdots & \vdots & \vdots & \ddots & \vdots \\ 0 & 0 & \cdots & 0 & 0 & 0 & \cdots & 1 \\ -\frac{\bar{H}_{11}}{T_{J1}} & -\frac{\bar{H}_{12}}{T_{J1}} & \cdots & -\frac{\bar{H}_{1,k}}{T_{J1}} & -\frac{\bar{D}_{11}}{T_{J1}} & 0 & \cdots & 0 \\ -\frac{\bar{H}_{21}}{T_{J2}} & -\frac{\bar{H}_{22}}{T_{J2}} & \cdots & -\frac{\bar{H}_{2,k}}{T_{J2}} & 0 & -\frac{\bar{D}_{2}}{T_{J2}} & \cdots & 0 \\ \vdots & \vdots & \ddots & \vdots & \vdots & \vdots & \ddots & \vdots \\ -\frac{\bar{H}_{k-1,1}}{T_{Jk}} & -\frac{\bar{H}_{k-1,2}}{T_{Jk}} & \cdots & -\frac{\bar{H}_{k,k}}{T_{Jk}} & 0 & 0 & \cdots & -\frac{\bar{D}_{k}}{T_{Jk}} \end{bmatrix},$$

$$\boldsymbol{B} = \begin{bmatrix} 0 & 0 & \cdots & 0 \\ 0 & 0 & \cdots & 0 \\ \vdots & \vdots & \ddots & \vdots \\ 0 & 0 & \cdots & 0 \\ b_1 & 0 & \cdots & 0 \\ 0 & b_2 & \cdots & 0 \\ \vdots & \vdots & \ddots & \vdots \\ 0 & 0 & \cdots & b_k \end{bmatrix}.$$

Here, as you can see, the dimension of inputs is a multiple of the dimension of states, and phase angles and slips are used as outputs.

Let us consider the case of specifying the parameters of the control law in the PID-controllers, which provide the solution of the problem formulated earlier.

Let and be given an arbitrary non-zero real number $m = 2, k = 2, n = k \cdot m = 4\kappa$.

Let also the matrices from (14) have the form $\boldsymbol{\Phi}^* = \{\phi^*, \ldots, \phi^*\}$

$$\tilde{\boldsymbol{S}}_{2,1} = \begin{bmatrix}-2\phi^* & 0\\ 0 & -2\phi^*\end{bmatrix}, \quad \tilde{\boldsymbol{S}}_{2,2} = \begin{bmatrix}\phi^*\phi^* & \kappa\\ 0 & \phi^*\phi^*\end{bmatrix}.$$

Characteristic equation

$$\left|\lambda^2 \boldsymbol{I}_m + \lambda\tilde{\boldsymbol{S}}_{2,1} + \tilde{\boldsymbol{S}}_{2,2}\right| = \begin{vmatrix}(\lambda-\phi^*)^2 & \kappa\\ 0 & (\lambda-\phi^*)^2\end{vmatrix} = 0$$

has roots coinciding with given eigenvalues (given modes of electromechanical oscillations), but the system of equations

$$\begin{cases}\boldsymbol{\Phi}_0 + \boldsymbol{\Phi}_1 = -\tilde{\boldsymbol{S}}_{2,1}\\ \boldsymbol{\Phi}_0\boldsymbol{\Phi}_1 = \tilde{\boldsymbol{S}}_{2,2}\end{cases} \quad (18)$$

turns out to be unsolvable with respect to the matrices $\boldsymbol{\Phi}_0, \boldsymbol{\Phi}_1 \in \mathbb{C}$.

Let further and $\boldsymbol{\Phi}^* = \{\phi_x^*, \phi_x^*, \phi_y^*, \phi_y^*\}$

$$\boldsymbol{S}_{2,1} = \begin{bmatrix}-2\phi_x^* & \kappa\\ 0 & -2\phi_y^*\end{bmatrix}, \quad \boldsymbol{S}_{2,2} = \begin{bmatrix}\phi_x^*\phi_x^* & 0\\ 0 & \phi_y^*\phi_y^*\end{bmatrix}.$$

Characteristic equation

$$\left|\lambda^2 \boldsymbol{I}_m + \lambda\boldsymbol{S}_{2,1} + \boldsymbol{S}_{2,2}\right| = \begin{vmatrix}(\lambda-\phi_x^*)^2 & \kappa\phi\\ 0 & (\lambda-\phi_y^*)^2\end{vmatrix} = 0$$

has roots coinciding with the given eigenvalues, and the system of equations

$$\begin{cases}\boldsymbol{\Phi}_0 + \boldsymbol{\Phi}_1 = -\boldsymbol{S}_{2,1}\\ \boldsymbol{\Phi}_0\boldsymbol{\Phi}_1 = \boldsymbol{S}_{2,2}\end{cases} \quad (19)$$

is solvable with respect to matrices . One of the two possible solutions to system (21) has the following simple form:$\boldsymbol{\Phi}_0, \boldsymbol{\Phi}_1 \in \mathbb{C}$

$$\boldsymbol{\Phi}_0 = \begin{bmatrix}\phi_x^* & -\frac{\kappa\phi_y^*}{\phi_x^*-\phi_y^*}\\ 0 & \phi_y^*\end{bmatrix}, \quad \boldsymbol{\Phi}_1 = \begin{bmatrix}\phi_x^* & \frac{\kappa\phi_x^*}{\phi_x^*-\phi_y^*}\\ 0 & \phi_y^*\end{bmatrix}, \quad (20)$$

$$eig\,\boldsymbol{\Phi}_0 = \{\phi_x^*, \phi_y^*\}, \qquad eig\,\boldsymbol{\Phi}_1 = \{\phi_x^*, \phi_y^*\},$$

where, as can be seen, the eigenvalues are located on the main diagonal of the corresponding matrices. Triangular matrices (22) have the following features:

- firstly, to exclude division by zero in the expressions of off-diagonal elements in matrices (22), it is necessary to prevent the coincidence of the given eigenvalues (multiplicity of oscillation modes);

- secondly, the off-diagonal elements of matrices (22) contain the previously mentioned "adjusting" parameterκ, which makes it possible to provide the previously mentioned "enhanced" roughness with respect to disturbances, or increased sensitivity to changes in controlled parameters in a given region or frequency band (while maintaining the given oscillation modes).

4. Conclusion

For the EES, represented by a linear stationary dynamic system with many inputs and many outputs (MIMO-system), compact highly efficient analytical formulas for calculating the coefficients of the feedback law are obtained when solving the problem of modal control (given placement of eigenvalues) by the full state vector. The resulting analytical formulas are generalizations to MIMO-systems of the well-known Ackermann formula applied to systems with one input and many outputs (SIMO-systems). The transformations are based on a non-degenerate similarity transformation and the algorithm of the multi-step decomposition method. The obtained formulas are applicable to dynamic systems, in which the dimension of the state space is a multiple of the dimension of the system inputs. This limitation is removed by using the Yokoyama transform. The resulting formulas differ in terms of the parametrization of the set of equivalent laws. It is shown that there are cases when the set of equivalent control laws obtained using an analytical formula based on a non-degenerate similarity transformation is wider than the set of control laws formed on the basis of the multi-step decomposition method. The synthesis example demonstrated the possibility of changing the previously obtained control law for synchronous generators in order to preserve the previous oscillation modes and satisfy some additional requirements for robustness and sensitivity. These requirements are provided by varying the parameters of special matrices. The implementation of the obtained laws involves the use of synchronized vector measurements $\boldsymbol{\Phi}^{*}$.

References

1. T. Kailath *Linear Systems*. Prentice Hall. Englewood Cliffs. NJ. 1980.
2. R. Dorf, R. Bishop *Modern control systems*. Moscow: Basic Knowledge Laboratory, 2002.
3. M.G. Gadzhiev, M.Sh. Misrikhanov, V.N. Ryabchenko, Yu.V. Sharov *Matrix methods of analysis and control of transient processes in electric power systems*. Moscow: MPEI Publishing House, 2019.
4. B. Kuo *Theory and design of digital control systems*. M.: Mashinostroenie, 1986.
5. K. Nordström, H. Norlander *On the multi input pole placement control problem* // Proc. 36 IEEE Conf. decision and control. 1997. V. 5. P. 4288 - 4293.
6. N.E. Zubov, E.A. Vorobieva, E.A. Mikrin, M.Sh. Misrikhanov, V.N. Ryabchenko *Synthesis of stabilizing control of a spacecraft based on the generalized Ackermann formula* // Izvestiya RAN, Theory and Control Systems, 2011, **No. 1**, pp. 96 – 106.
7. N.E. Zubov, E.A. Mikrin, V.N. Ryabchenko *Matrix methods in the theory and practice of aircraft automatic control systems*. M.: Publishing House of Moscow State Technical University named after N.E. Bauman, 2016.
8. N.E. Zubov, E.A. Mikrin, V.N. Ryabchenko *Synthesis of astatic control of a linear system based on the generalized Ackermann formula* // Bulletin of the Moscow State Technical University. N.E. Bauman. Ser. Instrumentation, 2017, **No. 1 (112)**, pp. 67 – 74.

Methodological approach to assessing additional active power losses in overhead power lines at non-sinusoidal modes

Lidiia Kovernikova[1*]

[1]Melentiev Energy Systems Institute, Power Systems Department, 664033, Lermontov Str., Irkutsk, Russian Federation

Abstract. Non-sinusoidal voltage in electrical networks is one of the most pressing problems in the field of power quality. In non-sinusoidal modes in electrical networks, in addition to losses of active power at the fundamental frequency, i.e. at the 1-st harmonic, active power losses occur at frequencies that are multiples of the fundamental frequency, corresponding to harmonics from 2 to 40. Additional harmonic active power losses increase losses in electrical networks, creating economic damage for both energy companies and consumers. The paper provides a review of the literature on the topic under consideration. A methodological approach is proposed to assess additional active power losses in overhead power lines. Based on the methodological approach, an analysis of the power quality indicators characterizing the non-sinusoidal mode, measured at the point of connection of 110 kV transmission line to the electrical network, was carried out, additional active power losses were determined.

1 Introduction

In the last few years, the federal authorities have begun to actively discuss issues of power quality in Russian electrical networks and adopt legislative documents, which, of course, is caused by the low power quality and the associated economic damage. In 2008, the economic damage caused by the poor power quality, according to a minimum estimate, amounted to 25 billion dollars [1]. Government Decree No. 2425 dated December 23, 2021 [2] indicates that electrical energy is subject to mandatory certification for compliance with the requirements of GOST 32144-2013 [3]. At the beginning of 2022, the Section for Legislative Regulation of Energy Efficiency and Energy Saving of the Expert Council under the State Duma Committee on Energy proposed "Rosseti" PJSC to hold a seminar on amending the Federal Law "On Electric Power Industry" [4, 5]. One of the proposals concerned the establishment of requirements for the power quality and the responsibilities of electricity industry entities and consumers to comply with them. In 2022, the law on amendments to the Federal Law "On Electric Power Industry" was adopted [5]. According to experts from the State Duma, "resolving issues of the power quality will lead to a significant economic effect for power industry entities and consumers, a reduction in irrational spending of budget funds, and an increase in resources that can be directed to the technological modernization of electrical networks" [4]. High power quality is necessary for the digital economy, the task of creating which was set in [7]. The number of electronic equipment, which requires high power quality on the one hand, and, on the other hand, degrades it, is rapidly increasing [8–10]. One of the pressing problems in the field of power quality is voltage non-sinusoidality. The values of power quality indicators, characterizing the distortion of the voltage curve shape, exceed the standards established in [3]. One of the negative consequences of non-sinusoidal voltages is additional losses of active power and energy in power transmission lines. To assess their magnitude, the methodological approach is required. It includes: analysis of measured indicators of power quality and operating parameters characterizing the non-sinusoidal mode at the point of connection of the power transmission line to the electrical network, determination of the resistance of the power transmission line at harmonics from 2 to 40, taking into account changes in temperature environment; determination of the maximum, average and total values of additional active power losses in the power transmission line and other issues. The application of the methodological approach is illustrated by the assessment of additional active power losses in the three-phase 110 kV power transmission line "Mogocha-Davenda" in Trans-Baikal Territory.

2 Review of literature on the assessment of additional active power losses in overhead power lines under non-sinusoidal modes

Power lines are an element of electrical networks. In non-sinusoidal modes, harmonic currents flow in them not only of the 1-st, corresponding to the fundamental frequency, but also of harmonics that are multiples of it. As in any other electrical equipment, that is the element of electrical networks, harmonic currents in power lines cause additional active power losses. Many publications are devoted to determining the value of

* Corresponding author: kovernikova@isem.irk.ru

active power losses in power transmission lines under non-sinusoidal modes due to the relevance of the problem [10–14].

In [10] it is indicated that when a load with a nonlinear current-voltage characteristic (nonlinear load) is powered from a source with a sinusoidal voltage, the flow of active power of the 1-st harmonic is directed into it. Most of this power is consumed by the load to perform useful work, which in [11] was proposed to be called "working power". Part of the active power of the 1-st harmonic is converted by the nonlinear load into active powers of harmonics with numbers $n>1$ [10]. In [11], it is proposed to call such harmonic active powers "harmful powers".

In [9], the question is raised about additional financial losses of the supply network and consumers associated with active harmonic powers. In [12] the accounting of electrical energy in non-sinusoidal modes is considered and the misunderstandings that arise between energy supply organizations and consumers when paying for electrical energy are pointed out. Electric energy meters measure the active energy of not only the 1-st, but also other harmonics. For this reason, energy supply organizations and consumers incur additional costs when paying for electrical energy.

The authors of [13] note that additional harmonic active power losses in electrical system networks can amount to several percent of losses with sinusoidal voltage. In networks of industrial enterprises and in networks of electrified railway transport, losses can reach 10–15%. In [14] the values of active power losses at the fundamental frequency and additional losses at harmonic frequencies are given based on the values of current harmonics measured during the day in a 110 kV power transmission line at "Buryatenergo". At some points in time, additional losses amounted to 30% of losses at the fundamental frequency. The level of additional active power losses per day amounted to 13.1% of electrical energy losses at the 1-st harmonic.

The authors of [15–19] examined the influence of current harmonics and ambient temperature on the value of resistance of the power transmission line wire.

3 Methodological approach to assessing additional active power losses in overhead power lines

In non-sinusoidal modes, the voltage and current in one phase of the electrical network are represented by the expressions.

$$u(t) = \sum_{n=1}^{40} U_{mn} \sin(n\omega t + \varphi_{Un}), \tag{1}$$

$$i(t) = \sum_{n=1}^{40} I_{mn} \sin(n\omega t + \varphi_{In}), \tag{2}$$

where $u(t)$, $i(t)$ – instantaneous values of voltage and current; n – harmonic number; U_{mn}, I_{mn} – amplitude values of voltage and current of the n-th harmonic; $\omega=2\pi f$ – angular frequency, where f=50 Hz; φ_{Un}, φ_{In} – phase angles of voltage and current of the n-th harmonic.

Active power at non-sinusoidal voltage and current in one phase of the electrical network in accordance with [15] is represented as

$$P = U_1 I_1 \cos\varphi_1 + \sum_{n=2}^{40} U_n I_n \cos\varphi_n , \tag{3}$$

or

$$P = P_1 + \sum_{n=2}^{40} P_n = I_1^2 R_1 + \sum_{n=2}^{40} I_n^2 R_n , \tag{4}$$

where P_1 – the active power of the 1-st harmonic, P_n – the active power of the n-th harmonic; $\varphi_1 = \varphi_{U1} - \varphi_{I1}$, $\varphi_n = \varphi_{Un} - \varphi_{In}$ – the angles between the current and voltage of the 1-st and n-th harmonics; I_1, I_n – effective values of currents of the 1-st and n-th harmonics; R_1 and R_n – resistance of the wire for the 1-st and n-th harmonics.

Additional losses of active power transmitted through one phase of the power transmission line for one of the harmonics, including the 1-st, are determined by the expression

$$\Delta P_n = I_n^2 R , \tag{5}$$

where I_n – the effective value of the current flowing through the wire of one phase; R – the resistance of the wire. The total active losses at harmonics from 2 to 40 can be calculated by the expression

$$\Delta P_{n\Sigma} = \sum_{n=2}^{n=40} I_n^2 R_n . \tag{6}$$

The reference literature provides the values of the resistance of the wires for the 1-st harmonic currents. When determining active power losses in the power transmission line for harmonics with number $n>1$, it is necessary to take into account the change in the resistance of the wire caused by the electromagnetic field of different frequencies [20]. When transmitting electrical energy, alternating electric current is distributed unevenly across the cross-section of the wire. The current density has the highest values on the surface of the wire and decreases with distance from the surface into the depth of the wire. The phenomenon of surface effect occurs in the wire, as a result of which the resistance increases. The change in the value of resistance is characterized by the surface effect coefficient. Taking into account the surface effect, expression (6) takes the form

$$\Delta P_{n\Sigma} = \sum_{n=2}^{40} I_n^2 R k_{Rn} , \tag{7}$$

where k_{Rn} – the surface effect coefficient at the n-th harmonic.

There are various proposals in the literature for calculating the surface effect coefficient [13, 15–17]. In [13, 15, 16], the value of the surface effect coefficient is proposed to be determined as

$$k_{Rn} \approx 0.47\sqrt{n} . \tag{8}$$

In [17], under the assumption of a sharp manifestation of the surface effect, i.e. when the length of the electromagnetic wave in the wire is significantly less than the cross-sectional radius of the wire [20], it is proposed to accept that

$$k_{Rn} = \sqrt{n} . \tag{9}$$

In [10] it is noted that the resistance of the power transmission line is influenced by climatic conditions and line load. However, since power lines are only one of the elements of the electrical network, the parameters of which are constantly changing, "it is hardly necessary to calculate the ultra-precise value of the surface effect that has influence on the characteristics of a limited number of elements." The author proposes to take a simplified value of the coefficient for practical calculations. In [14], the surface effect is calculated using expression (9).

Publications [18, 19] note that when calculating the parameters of electrical network modes, it is necessary to take into account the influence of the ambient temperature, since the resistance of power transmission line wires changes with temperature changes. As the temperature increases, the resistance of the wire increases, and as the temperature decreases, it decreases. The reference literature provides information for determining the resistance of wires at a temperature of 20°C. For other temperature values, the value of the resistance of the wire is determined by the expression from [18, 19]

$$R_t = R_{20} + \alpha_T(t-20), \quad (10)$$

where R_{20} – the wire resistance at a temperature of 20°C; R_t – the wire resistance at ambient temperature; α_T – the temperature coefficient of wire resistance; t – the ambient temperature.

Thus, additional active power losses in power line wires under non-sinusoidal modes must be determined taking into account the influence of the surface effect and ambient temperature. The mathematical expression for determining the resistance of the wire, taking into account these two factors, takes the form

$$R = \sqrt{n}\,[R_{20} + \alpha_T(t-20)]\,. \quad (11)$$

To estimate the value of the total additional losses arising in non-sinusoidal modes, relative to the active power losses of the 1-st harmonic, it is proposed to calculate as

$$\Delta P_{n\%} = \Delta P_{n\Sigma} / \Delta P_{1\Sigma} * 100\%\,. \quad (12)$$

The sign Σ means the summation of active power losses, which are calculated from indicators and parameters measured in accordance with [3] at intervals of 10 minutes for seven days. Additional active power losses due to harmonics can be used to calculate possible electrical energy losses for one year as a forecast.

Thus, the proposed methodological approach for assessing and analyzing additional active power losses in the overhead power transmission line in non-sinusoidal mode should include the points below.

1. Analysis of measured indicators of power quality and mode parameters.

In accordance with [3], to assess the power quality, measurements are carried out within seven days. The analysis uses: K_U – total harmonic distortion in phases A, B, C, %; $K_{U(n)}$ – coefficient of the n-th harmonic component of voltage in phases A, B, C, %; $K_{I(n)}$ – coefficient of the n-th harmonic component of the current in phases A, B, C, %; I_1 – root mean square value of the positive sequence current of the 1-st harmonic, A.

The current values I_n for harmonics from 2 to 40 are determined as

$$I_n = \frac{K_{I(n)} I_1}{100}\,. \quad (13)$$

2. Determination of the resistance of the wire, taking into account the surface effect for harmonics from 2 to 40 and the range of changes in ambient temperature.

The values of resistances are calculated using expression (11). Resistance R_{20} is calculated as

$$R_{20} = r_0 L\,, \quad (14)$$

where r_0 – the resistance of the wire at the temperature of 20°C, Ohm/km; L – length of the power transmission line, km.

3. Calculation of additional active power losses for harmonics from 2 to 40 for the ambient temperature range.

Additional active power losses at each harmonic are calculated using expression (4). The resistance R of the wire used in (4) is determined by (10), the current I_n by (13).

4. Calculation of active power losses of the 1-st harmonic.

Additional losses of active power of the 1-st harmonic for each temperature of a given interval are calculated as

$$\Delta P_1 = I_1^2 R_1\,, \quad (15)$$

where I_1 – the measured root-mean-square value of the positive sequence current of the fundamental frequency, R_1 calculated by (10).

5. Calculation of the relative value of the additional active power losses.

The relative value of additional active power losses $\Delta P_{n\%}$ is determined by (12), where $\Delta P_{n\Sigma}$ is calculated by (6), $\Delta P_{1\Sigma}$ by (15).

6. Calculation of the amount of additional losses of electrical energy per year.

In accordance with [3], the measurement of one indicator of the power quality, characterizing the non-sinusoidal mode, is carried out over the time interval of 10 minutes, which represents 1/6 of an hour. The amount of electrical energy at one time interval is determined as

$$\Delta W_{(1)} = (1/6)\Delta P_{n\text{-}(1)}\,. \quad (16)$$

Within seven days, 1008 measurements of power quality indicators are made. The amount of electrical energy for seven days will be determined as

$$\Delta W_{1008} = (1/6) * (\sum_{i=1}^{1008} \Delta P_{n\text{-}(1)i})\,. \quad (17)$$

The predicted value of electrical energy losses for the year can be calculated using the expression

$$\Delta W = (52/6 * (\sum_{i=1}^{1008} \Delta P_{n\text{-}(1)i})\,. \quad (18)$$

Expression (17) can be used to calculate electrical energy in one phase and in three phases, using the corresponding values $\Delta P_{n\text{-}(1)}$, i.e. for one phase or for three phases.

https://doi.org/10.1051/e3sconf/202346101013

4 Assessment of additional active power losses in the 110 kV power transmission line

The analysis and assessment of additional active power losses in non-sinusoidal mode was carried out for the overhead power transmission line "Mogocha – Davenda", connected to the 110 kV buses of the Verhnyaya Davenda substation. A fragment of a network diagram with the power transmission line is shown in Fig. 1. Measurements of power quality indicators and mode parameters were carried out at node 73. The following designations are used in the figure: OSG – open switchgear, CSG – closed switchgear. The measurements were carried out with the Resurs-UF2M device [21]. The three-phase 110 kV power transmission line "Mogocha – Davenda" has a length of 41.5 km, made with the SA-95 wire [22]. The ambient temperature during the measurements varied in the range from +19°C to +23°C.

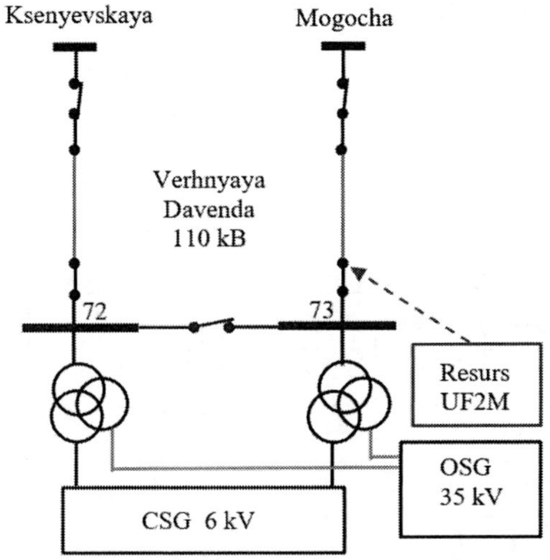

Fig. 1. Fragment of the network scheme of the Chita energy system.

Over the period of seven days, the following indicators were measured in three phases: K_U, $K_{U(n)}$, $K_{I(n)}$, I_1. In Fig. 2 shows the measured values of the indicator K_U and the standard values (K_{UH95}, K_{UH100}) established for it in [3] for 95% and 100% of the measurement time. From the given graphs, it is clear that in all phases the measured indicators exceed both standard values throughout the entire measurement time.

Figure 3 shows a diagram of the values in three phases for odd harmonics for 95% of the measurement time (95% A, 95% B, 95% C), maximum values (max A, max B, max C) and standard values ($K_{U(n)H95}$, $K_{U(n)H100}$) established for them in [3]. The diagram shows that the standard values are significantly exceeded at the 3-rd, 5-th, 7-th, 9-th, and 11-th harmonics.

Figure 4 shows the measured values of the indicator $K_{I(n)}$. Standard values for current indicators are not established in [3].

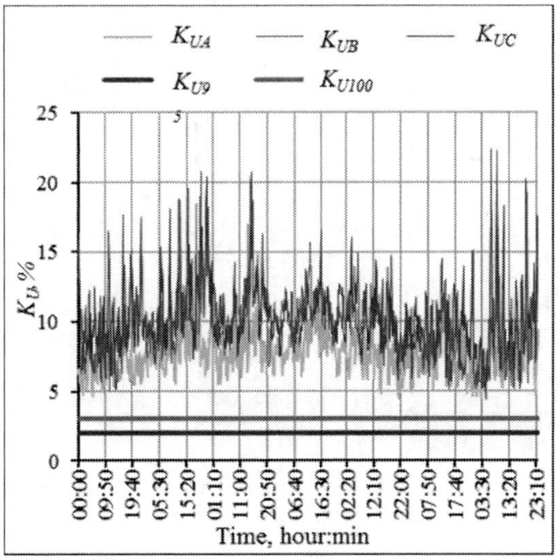

Fig. 2. Measured and standard values of K_U.

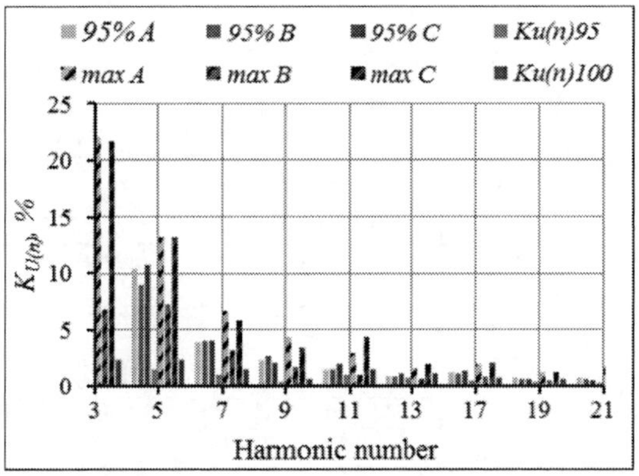

Fig. 3. Diagram of measured and standard values $K_{U(n)}$.

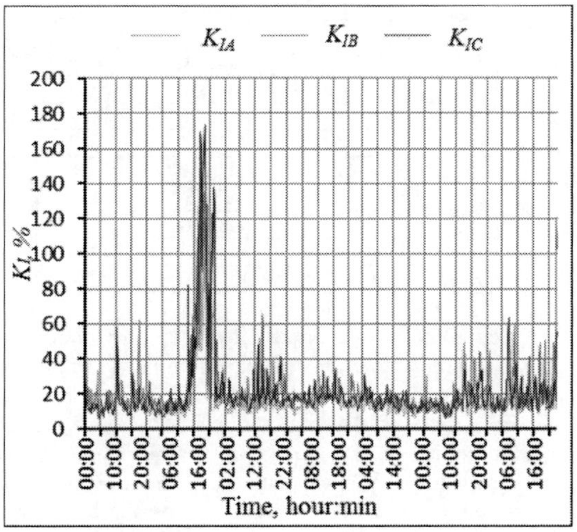

Fig. 4. Measured values $K_{I(n)}$.

Figure 5 shows the measured values of the positive sequence current of the 1-st harmonic.

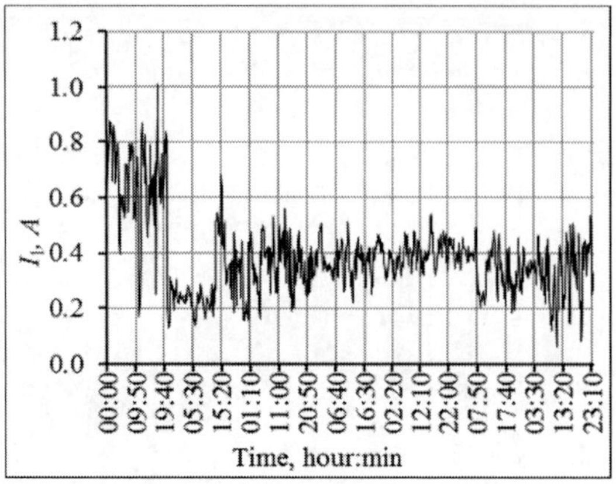

Fig. 5. Measured values I_1.

Based on the measured values of the indicators $K_{I(n)}$ and the current I_1, the values of the current harmonics in three phases were determined using expression (13).

The resistance of the wire of one phase of the power transmission line is determined for the ambient temperature range in the Trans-Baikal Territory from -40°C to +40°C with a step of 10° for harmonics from 2 to 40 according to expression (11).

The resistance in (11) is calculated using (14) for r_0=0.33 Ohm/km [22]. Table 1 shows the calculated values of the resistance of the power line for the analyzed temperature range with the step of 20° and for the odd harmonics from 1 to 17. The table shows that with increasing temperature and harmonic number, the resistance of the line increases.

Table 1. Resistance of the power line, Ohm.

n	**Ambient temperature, °C**				
	-40	**-20**	**0**	**+20**	**+40**
1	10.162	11.340	12.517	13.695	14.873
3	17.601	19.640	21.680	23.720	25.761
5	22.723	25.355	27.989	30.623	33.257
7	26.886	30.000	33.117	36.234	39.350
9	30.486	34.017	37.551	41.085	44.619
11	33.704	37.607	41.514	45.421	49.328
13	36.640	40.883	45.131	49.378	53.625
15	39.357	43.916	48.478	53.041	57.603
17	41.899	46.752	51.609	56.466	61.323

Table 2 presents the calculated additional power losses in three phases of the power transmission line $P_{n\Sigma}$, $P_{1\Sigma}$, $P_{n\Sigma}/P_{1\Sigma}$, as well as $P_{n\text{-}max}$, $P_{n\text{-}avg}$ – the maximum and average value of harmonic active power losses from the 2-nd to the 40-th of 1008 values for seven days.

Table 2. Additional active power losses, W, %.

Phase		**Ambient temperature, °C**								
		-40	-30	-20	-10	0	+10	+20	+30	+40
Phase A	$P_{n\text{-}max}$	4.6791	4.9504	5.2210	5.4920	5.7610	6.0350	6.3060	6.5770	6.8490
	$P_{n\text{-}ever}$	0.1637	0.1732	0.1820	0.1930	0.2016	0.2112	0.2210	0.2300	0.2400
	$P_{n\Sigma}$	165.070	174.640	184.200	194.100	203.260	212.900	222.470	232.030	241.590
	$P_{1\text{-}max}$	10.387	10.989	11.591	12.193	12.795	13.397	13.998	14.601	15.200
	$P_{1\text{-}ever}$	1.7300	1.8311	1.9310	2.0317	2.1320	2.2320	2.3326	2.4330	2.5300
	$P_{1\Sigma}$	1744.700	1845.80	1946.90	2048.00	2149.10	2250.20	2351.30	2452.40	2553.50
	$P_{n\Sigma}/P_{1\Sigma}$	9.4614	9.4615	9.4614	9.4776	9.4580	9.4610	9.4610	9.4610	9.4610
Phase B	$P_{n\text{-}max}$	7.4003	7.8293	8.2581	8.6881	9.1161	9.5450	9.9739	10.3970	10.8320
	$P_{n\text{-}ever}$	0.2815	0.2979	0.3141	0.3313	0.3471	0.3635	0.3798	0.3931	0.4125
	$P_{n\Sigma}$	283.750	300.200	316.640	333.930	349.910	366.380	382.840	396.290	415.750
	$P_{1\text{-}max}$	10.3870	10.9890	11.5910	12.1930	12.7950	13.3970	13.9990	14.6010	15.2030
	$P_{1\text{-}ever}$	1.3080	1.8311	1.93140	2.0317	2.1321	2.2323	2.3326	2.4329	2.5333
	$P_{1\Sigma}$	1744.700	1845.80	1946.90	2047.90	2149.10	2250.20	2351.30	2452.40	2553.50
	$P_{n\Sigma}/P_{1\Sigma}$	16.2640	16.2640	16.2640	16.3050	16.2820	16.2820	16.2820	16.1580	16.2820
Phase C	$P_{n\text{-}max}$	8.8280	8.7630	9.2429	9.7231	10.2020	10.6830	11.1630	11.6430	12.1230
	$P_{n\text{-}ever}$	0.5870	0.2737	0.2887	0.2947	0.3186	0.3233	0.3379	0.3524	0.3669
	$P_{n\Sigma}$	252.700	267.350	282.080	297.070	311.270	325.920	340.570	355.210	382.350
	$P_{1\text{-}max}$	10.387	10.989	11.591	12.193	12.795	13.397	13.999	14.601	15.203
	$P_{1\text{-}ever}$	1.7308	1.8311	1.9314	2.0317	2.1320	2.2323	2.3326	24329	2.5333
	$P_{1\Sigma}$	1744.70	1845.80	1946.90	2047.90	2149.10	2250.20	2351.30	2452.40	2571.30
	$P_{n\Sigma}/P_{1\Sigma}$	14.4840	14.4850	14.4890	14.5060	14.4840	14.4840	14.4840	14.4840	14.8700

Additional active power losses on harmonics in phase A amount to 9.5% of the active power losses on the 1-st harmonic, in phase B 16.3%, in phase C 17.5%. To predict the amount of electrical energy spent on additional active power losses per year, it is necessary to carry out calculations according to (17). At a temperature of +20°C, active power losses are equal to: $\Delta P_{n\Sigma A}$ = 222.47 W, $\Delta P_{n\Sigma B}$ = 382.84 W, $\Delta P_{n\Sigma C}$ = 340.57 W, total losses are 945.87 W. Electrical energy losses for the year will be 50.32 kW*hour. For the power transmission line 41.5 km long, losses are insignificant. If we determine the total losses of electrical energy due to harmonics in 110 kV overhead power lines in Russia, the length of which is almost 200 thousand km, and non-sinusoidal voltage occurs in all 110 kV electrical networks [9], then their value will be enormous.

Conclusion

Reducing additional active power losses in electrical networks is possible by improving the power quality, which requires the adoption of documents regulating the reduction of the generation of current harmonics by nonlinear loads in the electrical network.

At the design stage of power supply systems, it is already necessary to provide for the installation of equipment to eliminate the harmonics of nonlinear load currents.

The research was carried out under State Assignment Project (No. FWEU-2021-0001) of the Fundamental Research Program of Russian Federation 2021-2030 using the resources of the High-Temperature Circuit Multi-Access Research Centre (Ministry of Science and Higher Education of the Russian Federation, project no 13.CKP.21.0038).

References

1. L.N. Dobrusin L.N, *Investments in the Russian electric power industry and a program to increase their efficiency,* in Proceedings of the VI All-Russian Energy Forum "Russian Fuel and Energy Complex in the 21-st Century", 1-4 April, 2008, Moscow, Russian Federation (2008)
2. Decree of the Government of the Russian Federation dated December 23, 2021 No. 2425. http://publication.pravo.gov.ru/Document/View/0001202112300200?index=5&rangeSize=1
3. State Standard 32144-2013. Electric Energy. Electromagnetic compatibility of technical equipment. Power quality limits in public power supply systems (M.: Standartinform, 2014)
4. Decision of the State Duma Committee on Energy No. 3.25-5/22 dated April 6, 2022 Letter to the General Director of PJSC "Rosseti" A.V. Ryumin dated February 15, 2022 No. S70A-4/25
5. Federal Law No. 174-FZ "On Amendments to the Federal Law "On Electric Power Industry" and Certain Legislative Acts of the Russian Federation". https://www.garant.ru/products/ipo/prime/doc/404720347/
6. Federal Law "On Electric Power Industry" (version in force from January 1, 2023). https://docs.cntd.ru/document/901856089
7. Putin: The formation of a digital economy is a matter of national security of the Russian Federation. https://tass.ru/ekonomika/4389411
8. V.E. Vorotnitsky, Yu.I. Morzhin. Digital transformation of the Russian energy sector is a systemic task of the fourth industrial revolution, Energiya Yedinoy seti, No. 6(42), 12-21 (2018)
9. L.I. Kovernikova, V.V. Sudnova, R.G. Shamonov, et al., Power quality: current state, outstanding issues, and proposals to their solutions (Nauka, 2017)
10. J. Arrillaga, D. Bradley, P. Bodger, Power systems harmonics (Energoatomizdat, 1990)
11. L.S. Czarnecki, Working, reflected and detrimental active powers, (IET Generation, Transmission & Distribution, 2012)
12. I.V. Zhezhelenko, Higher harmonics in power supply systems of industrial enterprises (Energoatomizdat, 2010)
13. Kartashev I.I., Zuev E.N, Power quality in power supply systems. Ways to control and ensure it (Izdatelstwo MEI, 2001)
14. N.N. Kharlov, V.S. Borovikov, V.V. Litvak, A.V. Pogonin, V.A. Melnikov, Energy survey of non-sinusoidal modes of multi-wire power transmission lines, Elektrichestvo, No. **12**, 12-15 (2001)
15. A.E Emanuel, Power definitions and the physical mechanism of power flow. (John Wiley & Sons, Ltd, 2010)
16. A.K. Shidlovsky, V.G. Kuznetsov, Improving the power quality in electrical networks (Nauk. Dumka, 1985)
17. I.V. Zhezhelenko, Higher harmonics in power supply systems of industrial enterprises (Energoatomizdat, 1994)
18. O.N. Voitov, E.V. Popova, L.V. Semenova, Algorithm for calculating the flow distribution of electrical networks, Elektrichestvo, No. **12**, 19-26 (2013)
19. F.E. Evdokimov, Theoretical foundations of electrical engineering: Textbook for technical schools (M: Vyssh. shkola, 1981)
20. L.A. Neiman, K.S. Demirchan, Theoretical foundations of electrical engineering. Textbook for universities, **1** (Energoatomizdat, 1981)
21. Meters of electrical energy quality indicators "Resurs – UF2". https://www.entp.ru/
22. S.S. Rokotyan, I.M. Shapiro, Handbook of Electrical Power Systems Design (M., Energy, 1977)

Polyhedron circuit topology switchgear as a novel complex solution

Nikolai Grinev[1*]

[1]Everest LLC, 77 Shcherbakov str., Ekaterinburg, Russia

Abstract. The paper offers and analyzes a new complex solution – open switchgear with polyhedron circuit topology, with novel circuit and arrangement approaches including new electrical equipment. Decreasing numbers of disconnectors, better structural characteristics, elimination of vulnerabilities such as sections and busbar systems are the key characteristics of this solution. Due to a new circuit topology and equipment components, the open switchgear on the whole becomes more reliable, flexible, and less costly. Examples of arrangement solutions are given. We study the data on the 220 kB switchgear component outages in one of the largest regions of Russia in the period from 2018 to 2020. The use of SF6 circuit-breakers decreases the frequency of occurrence and time of the switchgear component outages in power grids. However, it is reasonable to use transfer-busbar circuits or double-breaker circuits.

1 Introduction

Intellectualized power grids, new types of equipment, reliability, compactness, environmental friendliness, and safety should be the main features of power systems in XXI century.

The density of electrical connections has greatly grown in the developed regions, thus n-2 criterion (as extension of n-1) is being increasingly used. This criterion considers all possible combinations of outages and repairs.

In terms of n-2 criterion, switchgears will inevitably become bottlenecks of power grids. They will remain so until they contain such centralization components as busbars sections/systems with bus-coupler breakers, since centralization is not conceptually compatible with survivability. Thus, power grid architecture should be developed.

There are high requirements to switchgears in terms of their design, functionality, user-friendliness. They are the framework of a power grid as their parameters are of strategic value.

Schemes, design and arrangements of switchgears were described in the 1960s-1980s [1,2], were studied in the 1990s-2000s [3-5]. The reported works are still relevant today. Switchgear circuits which are currently used in substations are found in [6]. The object of this study is 110-220 kV open switchgears.

According to [7], the way the power grids of the future will look like will depend on the conditions of their development and functioning. Scales of power grids grow, service areas increase, different power systems merge. Simultaneously, electricity consumption is distributed throughout the country. Decentralization trend is observed in terms of generation. Thus, current principles of development and construction of power grids will require new approaches.

The report offers ways to increase reliability of open-type switchgears due to a novel topology. The following issues were considered:

- a new solution and its key characteristics were described;
- possible realizations were studied;
- factual data on outages were analyzed, the effects of introduction of SF6 circuit-breakers (CBs) on repair characteristics of switchgears were assessed;
- a combination of outages of open switchgears and switching applications was modelled.

2 Methods and materials

Analogy, elements of systems theory, graph theory and reliability theory were used in the study. Power industry of the future will require novel approaches, thus there appeared an attractive hypothesis to use a topology of one of the most known biological structures – DNA molecule (Fig.1).

As can be seen, a DNA structure is topologically equivalent to that of a polyhedron, namely of polygonal prism. Polyhedron structure is highly cohesive, however decentralized. This structure can be realized as switchgear of a new type in the following way: each switching application is done through a junction of disconnectors, thus the switching application and its disconnectors form an edge. Circuit breakers are put in the vertices of a polyhedron. Each vertex connects no more than three edges. The switching application is broken with breakers of incident vertices.

*Corresponding author: grinevn@ya.ru

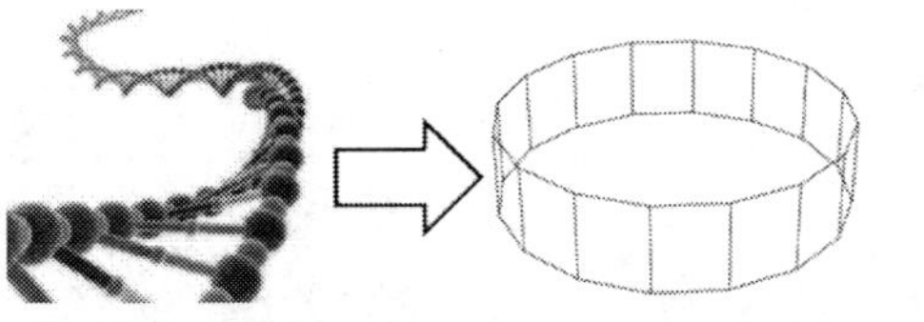

Fig. 1. DNA helix.

The switching application in the new-type switchgear is repaired by switching off two disconnectors. Breakers are repaired by switching off three disconnectors which are adjacent to the vertex, i.e. the polyhedron vertex as a whole is put under repair without an outage of the switching application.

According to the connection conditions in the three-edge vertex, at least two circuit breakers (CBs) or a special device – 3D-circuit-breaker (3D-CB) are put in each vertex.

3D-CB is a circuit breaker which has three primary terminals for different circuits. The main contacts of breaking units are connected following the star-scheme, whose arms have primary terminals thus making 3D-curcuit breaker. This circuit breaker can use elements of current commercial CBs. In addition, a 3D-CB has fewer elements than two conventional CBs do, for example, a support insulator in each pole, a support structure, a foundation and an operating mechanism are not necessary (Fig.2). This will help new equipment to avoid many problems. The cost of the live-tank type 3D-CB is predicted to be 150-190 % of the conventional commercial live-tank type CB cost.

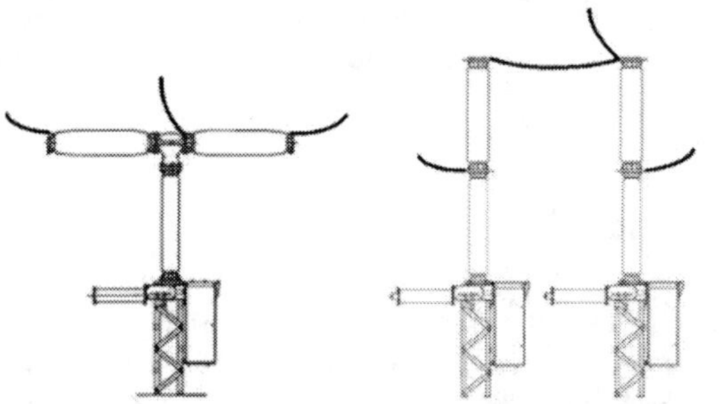

CB A couple of live-tank type CBs

Fig. 2. Comparison of a 3D-CB and a conventional CB.

The concept of the new electrical equipment has been developed together with a Russian plant which has all necessary facilities to implement it. By using polyhedron Euler theorem and handshaking lemma, one can calculate a required number of vertices N_v and that of disconnectors N_d depending on a number of switching applications N_c:

$$N_v = N_{sa} \cdot 2/3, N_d = N_{sa} \cdot 2 \quad (1)$$

Equation (1) shows that a number of disconnectors in the new solution is lower than that in the analogs. This will make the switchgear more cost-effective, taking into account the fact that the cost and a number of 3D-CBs will not change the costs on the breakers. Currently, the most unreliable and problematic element of the switchgear is disconnectors since further improving their reliability is limited [8]. Decreasing number of critical components is already a noticeable advantage.

Thus, new circuit topology and equipment components are the key characteristics of the novel solution, which is expected to improve reliability, flexibility and cost-effectiveness. This will be done not only by decreasing a number of disconnectors, but also by eliminating such vulnerabilities as sections and busbar systems, as well as by improving structural characteristics of the switchgear. This will be illustrated below.

Structural characteristics are a part of topological methods of reliability assessment and are not used in the design practice as such, since they do not allow further technical-and-economic assessment of options. These parameters are of value due to the fact that they have been obtained only on the base of the data on the element composition and their connections. Ref. [9] names the following structures applicable in power industry: series, ring, radial, complete graph, disconnected circuits. Characteristics of a cube as an example of polyhedron were calculated in comparison with the reported circuits (Table 1).

Table 1. Structural topological characteristics of different circuits.

Structure type	Parameter						
	R	K_C	ε^2	Q	d	δ	r_i
series	0	1	1,2	1	4	0,67	0,25
ring	0,25	2	0	0,5	2	0	0,2
radial	0	1	7,2	0,6	2	1	0,5
tree-like	0	1	3,2	0,75	3	0,76	0,38
complete graph	**1,5**	**2**	**0**	0	1	0	0,2
disconnected	-0,25	0	-	-	-	-	0,25
polyhedron	**0,71**	**2**	**0**	0,71	3	0	0,13

First three parameters are important for power industry objects. Parameter R (structure redundancy) is an indirect assessment of reliability and cost-effectiveness, it makes the system function and preserve its connections when some of its elements malfunction. The system with a higher redundancy is supposed to be more reliable. Parameter K_C (connectivity) is a characteristic of possible fault and repair of an element. Parameter ε^2 (quadratic deviation of vertex degree distribution) characterizes underutilization of the structure capacity and indirectly assesses cost-effectiveness.

The obtained parameters show that the polyhedron circuit is more cohesive, reliable, flexible, cost-effective, open and scalable. The polyhedron structure is very similar to the reference structure of complete graph in terms of parameters. Hence, the new-type switchgear parameters will not limit, but in contrast enhance intellectualization of electrical grids.

3 Implementation

Configurating a switchgear is arranging electrical equipment in a specific order in accordance with its functions and current requirements, and interconnecting its parts following the scheme. A switchgear typically consists of a number of analogous switching applications. Thus, it is sufficient to consider only one typical cubicle and the way these are connected with each other.

Convex polyhedron can always be shown as a flat graph with any number of switching applications and without intersecting edges.

A cube is a good example of the proposed structure. A simplified scheme of such switchgear is given in Figure 3. Eight vertices are required for twelve switching applications. In conventional circuits, a total number of CBs is 13-14.

Arranging the vertices in two rows along the passage for their maintenance is convenient and descriptive for an open switchgear with polyhedron topology (Fig.3).

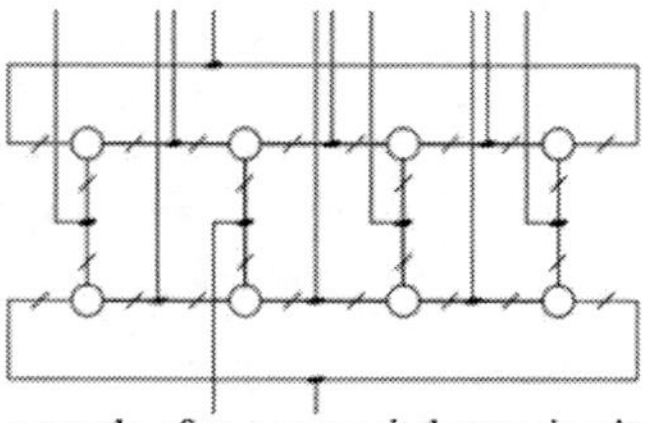

Fig. 3. An example of an open switchgear circuit with a cube topology

The switchgear of a new type has a modular structure. For 110kV, it is viable to make a polyhedron vertex as a module of block structure (Fig. 4). Fig. 5 shows an open-type 110 kV switchgear arrangement.

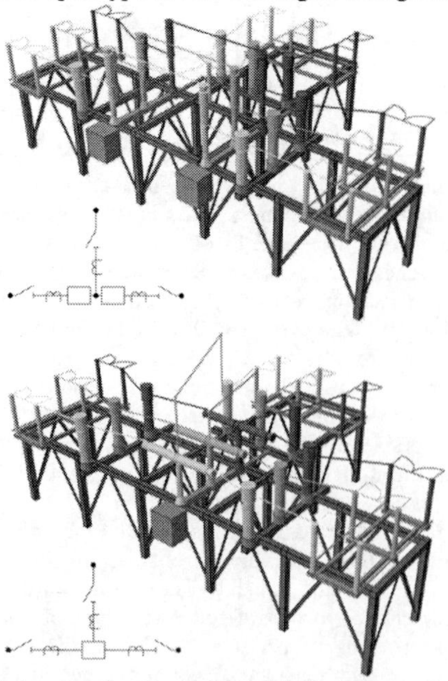

Fig. 4. An example of a module of block structure of a new-type open switchgear

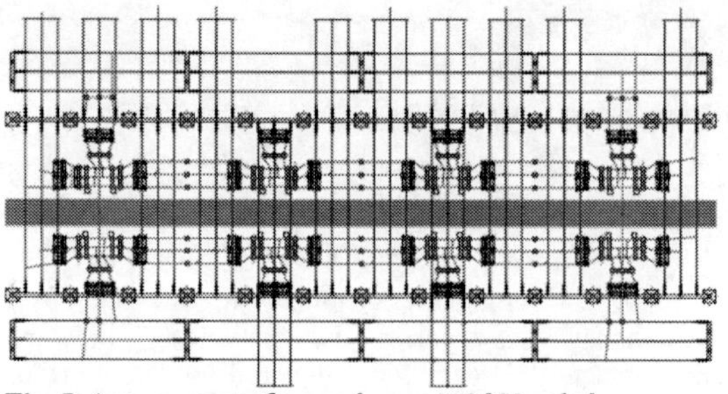

Fig. 5. Arrangement of a novel open 110 kV switchgear

An example of a 220 kV switchyard is shown in Fig.6. Primary terminals in the middle of a 3D-CB are connected via elements of a rigid busbar.

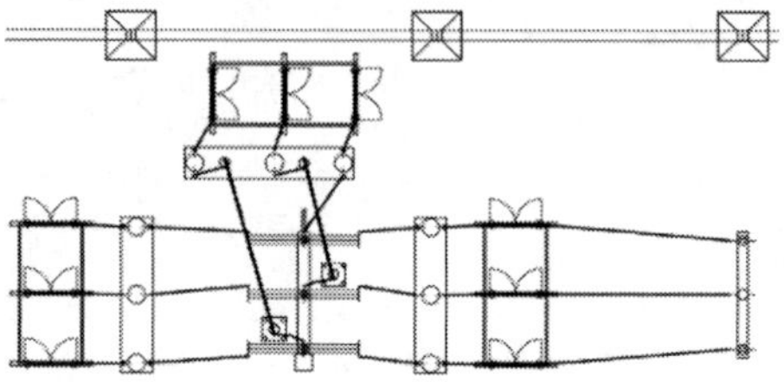

Fig. 6. 220 kV switchyard with a 3D-CB

Fig. 4-6 show that arrangement of the novel open switchgear is in agreement with the conventional rules, which meets all necessary conditions. In terms of compactness, it is better or at least not worse than its analogs. A polyhedron structure is not bulky.

Thus, an idea of a fewer number of connectors, absence of vulnerabilities, better structural characteristics of the novel open-type switchgear appears to be valid.

4 Discussion

For recent 30 years numerous SF6 CBs have been installed in the Russian power grid. At the early stage of their use, they were expected to decrease the frequency of occurrence and time of switchgear outages in such a way that they would be assimilated by switching application outages. Today we can see whether these expectations have come true.

Little attention is typically paid to outages, their data are often inconsistent and contradictory. For instance, the data from [10] and [11] have been used since the 1980s. Both sources refer to recommendations of ORGRES firm, the data however are inconsistent.

The author of this paper considered cases of repair of switchgear elements in one of the largest regions of Russia (over 150 substations, over 1,000 CBs in different climate zones) in 2018-20 – in total 3,777 outages in 220 kV open-type switchgears. Only minor maintenance and medium maintenance repairs in the switchgears where CBs were not changed in the reported period were analyzed. Major repairs were not considered since such repairs for ageing types of CBs

https://doi.org/10.1051/e3sconf/202346101014

are not made at present, and change of CBs is a part of such repairs. The obtained indicators are presented in Table 2.

The current indicators are in correspondence with those from [10], whereas they differ by 9-12 times from the ones in [11].

Advantages of SF6 CBs are remarkable. However, great changes by an order or higher has not been found. For instance, a mean time of an outage of a cubicle of an open 220 kV switchgear with a live-tank SF6 CB is 105 hours, that of the one with an oil CB is 137 hours. The obtained data are generally in agreement with those from [5, pp.43-45].

Table 2. Repair parameters of elements in an open 220 kV switchgear.

	[10]		[11]			Current indicators				
Breaker type	Air	Oil	Air		Dead-tank oil		Oil	Dead-tank oil	Dead-tank SF6	Live-tank SF6
μ, 1/year	2	2	0,2	0,14	0,14	1,77	1,73	1,61	1,18	1,13
T_{Π}, h	88	37	96			142	137	148	97	105

Outage data according to electrical equipment types are given as a fractional number in Table 3. The numerator of the fraction is a percentage of the total number of the operations done in all outages, the sum in it is 100 %. The denominator of the fraction is a percentage of the total number of outages, the sum in it is higher than 100 %. For example, CTs have 17 % of the total number of cases, whereas CTs are involved in 25 % of outages. In approximately 45 % of outages, the work was done simultaneously in several electrical devices.

Table 3. Outage data of open 220 kV switchgear elements.

CB	CT	BD	LD (TD)	BD	SC	Other	Total
37% / 53%	17% / 25%	19% / 28%	11% / 16%	2% / 2%	8% / 12%	6% / 9%	100% / 145%

Note: CT - current transformers, BD/LD/TD/BD – bus /line/transformer/by-pass disconnector, SC – secondary circuits.

Fig.7 shows distribution of repair time of open 220 kV switchgear elements. This graph is of interest because it is identical to a similar graph for overhead power transmission overhead lines from [13], both have exponential distribution which is found when a distribution of accident events occurs. This means that two independent, accidental in time occurrences should be brought into coincidence.

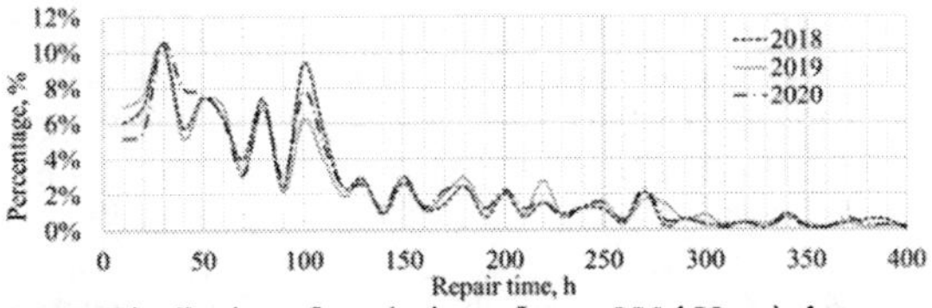

Fig. 7. Distribution of repair time of open 220 kV switchgear elements.

Independence of the occurrences of outages makes their synchronization very difficult. Based on this, synchronization was modelled (the occurrences were mixed and compared), the time of the total outage was taken as the maximum one from the two ones (the receiving end of a power transmission line was not considered by convention). For power transmission lines the repair time distribution was taken from [12], for open switchgear cubicles the data were collected in this work (for SF6 CBs).

The time of the synchronized outages was longer on average by 60 %. The results of modelling are in correspondence with the factual data in which switchgear outages and connection outages are brought into coincidence.

Thus, the use of SF6 CBs decreased the frequency of occurrence and time of outages of open switchgear elements in power grids on the whole. However, it is still not viable to synchronize outages, instead it is reasonable to use circuits with transfer busbar systems or schemes with two CBs on a switching application, an example of which is a polyhedron circuit.

5 Conclusion

A new complex solution – a switchgear with polyhedron circuit topology – includes novel scheme and arrangement approaches as well as new equipment – 3D-CBs. These breakers can be produced at one of the Russian factories. New-type switchgears are expected to be more reliable, flexible, cost-effective and compact.

For 3D-CBs to be used in new open switchgears, they should be designed, manufactured and factory-tested.

It is viable to study this new-type open switchgear and to do a pilot project in order to make a decision on further commercialization of this technical solution.

New-type switchgears can be used in constructing new and re-constructing current switchgears which have 5 or more switching applications. Due to cohesive structure of a polyhedron, it is reasonable to use it in circuits with parallel operation: in 110 kV grids and higher-voltage grids, as well as in generation switchgears. In separate operation grids (6-35 kV switchgears) such switchgears can become an alternative to busbar sections.

Arrangement of novel open switchgears is in agreement with the conventional rules, which meets all necessary conditions. In terms of compactness, they are better or at least not worse than their analogs.

We considered current indicators of outages of

open-type 220 kV switchgear elements in one of the largest regions of Russia in the period from 2018 to 2020. Even taking into account the use of SF6 CBs, which decreased the frequency of occurrence and time of outages of open-type switchgear elements, it is still viable to employ circuits without synchronizing switchgear outages and those of switching applications.

The offers presented in the report open new opportunities to study the important issue which can generally improve reliability of power grids.

References

1. L. Dvoskin, *Schemes and constructions of switchgear*, Energy, (1985)
2. G. Lisovskiy, M. Heifits, Eur. Main schemes and electrical equipment of 35-500 kV substations, Energy, (1977)
3. Y. Balakov, A. Vasilchikov, et al., Switch nodes in power grids, Energoatomizdat, (1997)
4. Y. Balakov, M. Misrihanov, et al., Power generation schemes of power plants : methodological aspects, Energoatomizdat, (2002)
5. Y. Balakov, M. Misrihanov, A. Shuntov, Designing schemes of power systems, Dom MEU, (2006)
6. GOST R 59279-2020, Electrical circuits of switchgears, (2021)
7. N. Voropay, A. Osak, Energy policy **5**, pp. 60-63 (2014)
8. A. Abdurahmanov, M. Lint, et al., Power plants **5**, pp. 50-55, (2008)
9. N. Voropay, Systems theory for power engineers, Nayka Sib. Izd. RAN, (2000)
10. V. Ershevich, A. Zeilinger, et al., Electrical system design manual, Energoatomizdat, (1985)
11. I. Karapetyan, D. Faibisovich, Power grid design manual, ENAS, (2012)
12. A. Abdurahmanov, S. Glushkin, et al. Electricity **8,** pp.12-17 (2018).

Multi-criteria comparison of 220 kV switchgears which have different circuits of electrical connections

Nikolai Grinev[1]*

[1]Everest LLC, 77 Shcherbakov str., Ekaterinburg, Russia

Abstract. The paper compares different open 220 kV switchgears in the context of a new complex solution offered by the author. The solution is a polyhedron circuit topology switchgear with novel scheme and arrangement approaches. Comparison is viewed here as a multi-criteria task with the following criteria: reliability, cost-effectiveness, maintenance activities, environment friendliness, compactness. To assess reliability, an unconventional calculation model is offered. The characteristics of open switchgears with a polyhedron circuit topology are better or at least not worse than the best analogs in each criterion. For example, such switchgears are highly reliable, cost-effective in terms of capital expenditures and highly cost-effective in terms of maintenance and balance-of-plant needs (carbon footprint decrease). In terms of all the characteristics together, open switchgears with a polyhedron circuit topology outperform all the existing analogs.

1 Introduction

The author offers a novel complex solution – an open switchgear with a polyhedron circuit topology. Regarding this, the report compares this new solution and the existing analogs. The new solution can consistently increase reliability of the Russian national power grid, which makes this study important.

The following criteria were considered in the work: reliability, cost-effectiveness, compactness, convenience of maintenance, maintenance activities, environment-friendliness. However, reliability was given a special emphasis following the topic of the seminar. An open switchgear with the lowest costs on construction and maintenance, without electric power supply interruptions, of the smallest area, and environment-friendly is the aim of the study.

Currently, there is no universal, best-in-all-criteria circuit of electrical connections (hereinafter, circuit). In Russia switchgear circuits which are in use today are assigned numbers for convenience. These circuits and their nomenclature as well as the circuit of a new switchgear are given in Fig.1.

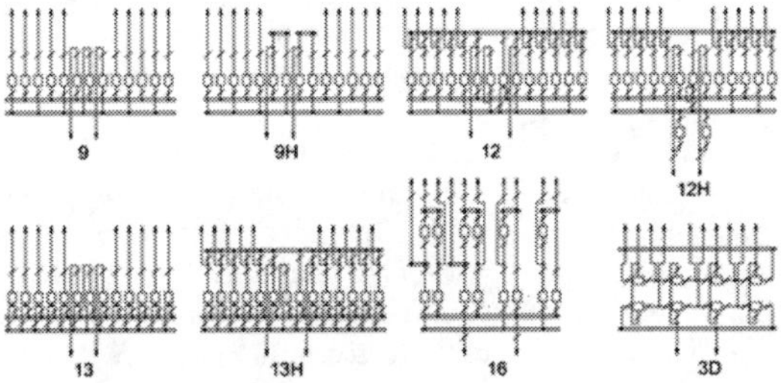

Fig. 1. Circuits and their nomenclature.

Principles of formalization of structure description are found in graph theory [1]. It is known that any convex polyhedron can be presented as a flat graph without intersecting edges. Regarding this, a polyhedron is not a solid figure and can lie in the earth surface plane.

2 Models, methods

Mathematic modelling, technical-economical comparison, elements of graph theory and reliability theory, Markov chains were used in the study.

We compared new-type switchgears with the existing analogs, thus it was viable to use the known proven methods. A task of comparing different switchgear circuits in terms of reliability had been repeatedly discussed at the meetings of *Methodoligical problems in reliability study of large energy systems* seminar, as well as in the monographs by F. I. Sinchugov (1971), M. N. Rozanov (1974), Yu. B. Guk (1976, al eds.), A. T. Shevchenko (1997, al eds.), M. Sh. Misrihanov and A. V. Shuntov (2002, 2004), V. A. Nepomniaschy (2013), etc. These methods were used with the currently relevant data on major failures [2-5] and the actual data on outages which were collected by the author in this work.

To solve the multi-criteria task, we used the ideal point method and graph method (Pareto optimization).

The ideal point method is an example of turning a multi-criteria task into a single-criterion one, where a distance to an ideal unachievable point is taken as a criterion. In the ideal point method, a target function of the combined effect can be expressed as:

$$C = \sqrt{C_1^2(x,t) + C_2^2(x,t) + \cdots + C_n^2(x,t)} \times (1+d)^{-t} \rightarrow min, x \in R \quad (1)$$

*Corresponding author: grinevn@ya.ru

https://doi.org/10.1051/e3sconf/202346101015

where $C_1, C_2, \ldots, C_n$ are quantitative indicators of the criteria; d is a discount rate; R is a constraint zone. Variables x are switchgear construction and maintenance activities per year t.

Equation (1) is an expression for calculating a vector magnitude from the central point of coordinates (the aim is to find the minimum, thus the point with coordinates $\{0; 0; 0; 0; 0\}$ is taken as the ideal one). The target function C is actually a vector in the n-dimensional space, where n is a number of the criteria under analysis (in this case – 5). Equation (1) reflects the principle of minimization of the given combined costs when the mandatory requirements to functioning are met.

The graph method (Pareto optimization) consists in the fact that the indicators of all the circuits are shown (as a radar chart) in the n-dimensional space. If there is a circuit which is superior to another one in all the indicators simultaneously, then the latter is considered to be an "underdog" and is dropped out.

Pareto optimization is a direct method, thus the best one since other methods are typically turning a multi-criteria task into a single-criterion one through different allowances, approaches and (or) priorities.

Both chosen methods require quantitative indicators in each criterion. The way in which they were obtained is shown below.

Criterion (C_1) is reliability. When choosing a circuit of a switchgear, failure effects and maintainability are of importance. Major failures (MaFs) and outages are usually described through occurrence indicators: frequency ω and μ, 1/year; time of recovery T_r и T_o, h, respectively.

A switchgear is typically decomposed into assembling units – cubicles. By using a double decomposition, we divided each cubicle into three zones in terms of MaF/repair effects:

Zone 1 is a switching application zone. A MaF here is handled by switching the circuit breaker (CB) off. When a failure occurs, the switching application is lost during recovery time. Repair is possible only together with the switching application.

Zone 2 is a zone of a CB, current transformers (CTs), protection relays and automation components. When a MaF occurs here, adjacent switching applications are lost during operational switchings. Loss of switching applications during repair is found only in some circuits.

Zone 3 is a busbar zone. A MaF here is handled by protecting busbars and switching the adjacent switching applications off. Busbar disconnectors are also found in this zone.

Thus, the system has five states: "0" is a basic state, all elements are in operation; "1…3" is a state at the transition into zones 1…3, respectively; "4" is a cubicle repair. As a result of occurrences (repairs and MaFs), the objects evolve from one state to another.

Such subdivision is in agreement with [5, 6].

Probability of occurrence of events (finding the system in one of these states) was calculated with Markov chains. The method is well-known, was employed in [7], and is easily presented in MS Excel. Instead of Markov chains another method can be used.

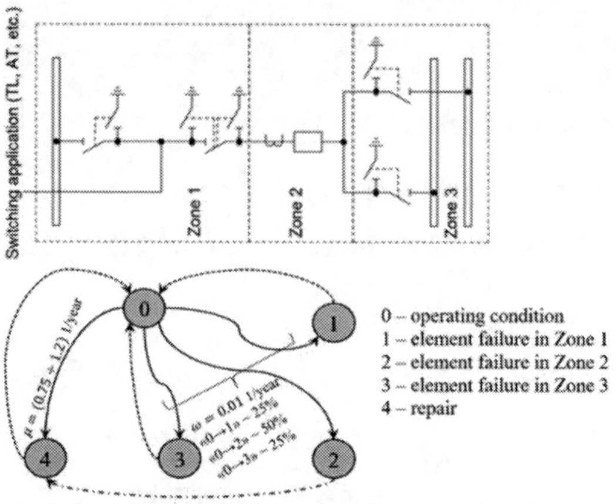

Fig. 2. A graph of a calculation model with Markov chains

A graph of the model is given in Fig.2. Occurrences of events in the model are described through trajectories of transitions between the states. A statistical model is described with Kolmogorov equations, the solution of which is a state probability vector $P(A) = \{P(A)_0, \ldots, P(A)_i\}$. To solve them, we took MaF indicators from [3,4,5,8] - $\omega = 0.01$ 1/year, repair indicators were collected by the author - $\mu = 0.75 \div 1.2$ 1/year.

However, the obtained probabilities are a data set, which does not allow them to be used as one quantitative indicator. Thus, we introduced an additional conditional indicator W'– an expected amount of lost electric energy W per year (to all K switchgear switching applications), – which was found by using the expression:

$$W' = \sum_{k=1}^{K} \sum_{i=1}^{N} Pmax_i \times P(A)_i \times 8760 \quad (2)$$

where K is a number of switching applications, $Pmax_i$ is the calculated power which is lost when a MaF or outage in zone i of switching application k occurs.

Not to forget about the conventional character of the indicator, a prime is used in it – W'. It is conventional because the loss of energy delivered to customers really occurs when two mutually redundant feed circuits are lost, since in the majority of cases electricity is supplied via two circuits.

Indicator W' is collective from n-1 criterion (it shows probabilities of total loss of energy delivery to customers). It is convenient since it does not depend on the circuit of the adjacent grid and allows differentiating the value of switching applications, for example of autotransformers (ATs), which are connected through a junction of CBs in 9H and 12H circuits.

The indicators for the other criteria $(C_2 - C_5)$ were obtained on the base of a number of cubicles of the open switchgear and characteristics of electrical equipment: SF6 CBs, disconnectors, CTs.

Criterion of cost-effectiveness (C_2). We used capital costs on the main equipment of an open switchgear as a quantitative indicator. We found these by using aggregated standards from [9, pp. 64-65]: three-phase unit of a 220 kV CT – 4.5 million roubles;

three-pole 220 kV disconnector – 6.03 million roubles; one-pole 220 kV disconnector – 2.09 million roubles; SF6 220 kV CB – 11.39 million roubles; SF6 3D-CB – 11.39 x 170 % = 19.36 million roubles. Hence:

$$C_2 = 11.39 \times N_S + 19.36 \times N_{S3} + 6.03 \times N_{D3} + 6.27 \times N_{D1} + 4.5 \times N_{CT} \quad (3)$$

where C_2 is capital costs on open switchgear equipment, million RUB.;
N_S and N_{S3} are numbers of SF6 CBs и SF6 3D-CBs, respectively;
N_{D3} and N_{D1} are numbers of three-pole and three-phase units of one-pole disconnectors, respectively;
N_{CT} is a number of CTs.

Criterion of convenience and maintenance activities (C_3). We used service maintenance activities of an open switchgear as a quantitative indicator. We obtained these by using specific values from [10]: a three-phase unit of SF6 220 kV CTs – 0.57 volumetric unit (VU); 220 kV disconnector – 1.09 VU; SF6 220 kV circuit-breaker – 1.18 VU; SF6 3D-breaker – 1.46 VU (SF6 500 kV CB as a prototype). Hence:

$$C_3 = 1.18 \times N_S + 1.46 \times N_{S3} + 1.18 \times (N_{D3} + N_{D1}) + 0.57 \times N_{CT} \quad (4)$$

where C_3 is amount of service maintenance activities of an open switchgear, volumetric unit (VU).

Criterion of environment-friendliness (C_4). Lowering environmental impact is an important issue. Open switchgears do not significantly affect atmospheric air, groundwaters, soil, etc. Thus, we consider carbon footprint as a quantitative indicator of environment-friendliness. It can be determined via power of heating systems. In this case we mean global warming effects. This is typically the main aspect for the equipment which consumes energy during lifetime. It is expressed with an equivalent emission of carbon dioxide CO_2. Environmental effects of consumption of energy which was produced by different power plants were estimated by OECD countries as 0.6265 kg of CO_2 per kWh [11]. Heating system powers at one of the Russian factories were the following: SF6 CB drive – (50+600*0.5) W, disconnector drive – (25+200*0.5) W, remote control unit of disconnector – (100*0.5) W. Hence:

$$C_4 = [350 \times (N_S + N_{S3}) + 175 \times N_{D3} + 425 \times N_{D1}] \times 8760 \times 0.6265 \times 10^{-6} \quad (5)$$

where C_4 is an equivalent CO_2 emission, kt of CO_2 per year.

Criterion of compactness (C_5). We used area of land plot as a quantitative indicator. It was found by using the equation below:

$$C_5 = [L \times a/10{,}000] \times N_{cubicle} \quad (6)$$

where C_5 is an area of land plot occupied by an open switchgear, ha; $L \times a$ is a cubicle size, m. The data were taken from standard projects or, for example, from [12, pp.413-423] – (Table 1).

Table 1. Open 220 kV switchgear cubicle sizes.

Switchgear circuit	9	9H	12	12H	13	13H	16	3D
L, m	76	95	95	105	76	95	154	76
a, m	15.4							

3 Results

To demonstrate the methods we took open 220 kV switchgear for 12 switching applications (2 ATs, 10 transmission lines (TL)) with circuits: 9; 9H; 12; 12H; 13; 13H; 16 (Fig. 1). The following indicators *Pmax* were used: TL – 40 MW; AT – 200 MW; busbar sections/systems – 200/500 MW.

The results of reliability (C_1) assessment are presented in Tables 2, 3. Similar calculations can be done for another switching application makeup.

In accordance with the final results, the obtained states can be distributed in the following way:

1) The lowest effect – "2". The effect is very low since it is rare. In case of MaF, the damaged element is removed out of service for repair.

2) The second effect – "1". It is a conditional constant for all circuits.

3) The third effect – "3". It is due to busbar sections/systems. It is more pronounced for the circuits with busbars.

4) The highest effect – "4". It is typical of the circuits in which a switching application is lost during outages (9, 9H, 13).

The obtained results allow the circuits to be ranked in terms of reliability (from the most reliable to the least reliable one): 3D; 16; 12H; 12; 13H; 9H; 9; 13. Changing W' by 2 orders is a solid ground to choose a more reliable solution. The obtained ranking is in correspondence with the known results, namely with the fact that circuit 16 is the most reliable among the existing ones, circuits 12H, 12, 13H are more often used than circuits 9, 9H.

The underlying cause behind the obtained results is high maintainability of the new open switchgear. Any electric device can be taken for repair without a power supply interruption for a switching application. All disconnectors in its circuit are linear (or transformer ones), which makes them similar to the existing analogs. The new open switchgear has no busbar systems and busbar disconnectors, which are responsible for 20 % of all outages and up to 25 % of MaFs in open switchgears. It is noteworthy that circuit breaker failure protection (CBFP) of a new-type open switchgear switches off only 2 "sound" switching applications, which makes them much superior to radial circuits and slightly inferior to ring ones in this indicator.

Table 2. State probability $P(A)$, unit per value (p.u.).

Switchgear circuit and its elements		State				
		«0»	«1»	«2»	«3»	«4»
9	TL, AT	9.87E-01	4.36E-05	1.13E-06	7.68E-05	1.32E-02
	BSB	9.86E-01	7.67E-05	1.13E-06	7.67E-05	1.35E-02
9H	TL	9.87E-01	4.36E-05	1.13E-06	7.68E-05	1.32E-02
	AT	9.75E-01	6.68E-05	1.22E-06	6.53E-05	2.48E-02
12	TL, AT	9.86E-01	5.63E-05	1.13E-06	7.67E-05	1.36E-02
	BSB	9.86E-01	7.67E-05	1.13E-06	7.67E-05	1.35E-02
12H	TL	9.86E-01	5.63E-05	1.13E-06	7.67E-05	1.36E-02
	AT	9.75E-01	6.68E-05	1.22E-06	6.53E-05	2.48E-02
13	TL, AT	9.87E-01	4.36E-05	1.13E-06	7.80E-05	1.32E-02
	BSB	9.86E-01	7.67E-05	1.13E-06	7.67E-05	1.35E-02
13H	TL	9.86E-01	5.63E-05	1.13E-06	6.73E-05	1.36E-02
	BCB	9.86E-01	7.67E-05	1.13E-06	7.67E-05	1.35E-02
16	TL - BC	9.86E-01	2.81E-05	1.13E-06	1.14E-04	1.36E-02
	TL - TL	9.86E-01	2.82E-05	1.13E-06	0	1.36E-02
3D	TL, AT	9.88E-01	2.82E-05	1.69E-06	0	1.19E-02

Note: BSB – bus-section breaker, BCB – bus-coupler breaker, BC – bus-coupler.

Table 3. Conditional loss of energy delivery W' (indicator C_1)

Switch-gear circuit	State				Total W'	
	Zone 1 (failure towards switching application)	Zone 2 (CBFP)	Zone 3 (busbar failure)	Zone 4 (repair)	MWh/year	p.u.
9	305.88	36.51	2,151.58	92,374.59	**94,868.56**	0.970
9H	386.96	40.84	1,802.11	46,187.29	**48,417.21**	0.495
12	394.52	36.49	2,150.82	0	**2,581.84**	0.026
12H	431.28	40.83	1,801.56	0	**2,273.68**	0.023
13	305.88	36.51	5,053.02	92,374.47	**97,769.87**	1.000
13H	394.52	36.49	4,688.23	0	**5,119.24**	0.052
16	138.09	23.67	1,601.42	0	**1,763.18**	0.018
3D	197.61	23.71	0	0	**221.32**	0.002

Estimations in terms of criteria of cost-effectiveness (C_2), maintenance activities (C_3), environment-friendliness (C_4), compactness (C_5) are shown in Table 4. A number of cubicles and their positions are conditionally shown in Fig.1.

Quantitative indicators of all criteria are normalized in summary table 5. Calculation results with the ideal point method [equation (1)] are also given there.

Table 4. Repair parameters of elements in an open 220 kV switchgear.

Switchgear circuit	9	9H	12	12H	13	13H	16	3D
Equipment elements								
N_S	13	14	14	15	13	14	14	-
N_{S3}	-	-	-	-	-	-	-	8
N_{D3}	26	28	41	43	38	39	40	24
N_{D1}	-	-	-	-	-	14	-	-
N_{CT}	13	14	14	15	13	14	14	24
$N_{cubicle}$	13	14	14	13	13	14	11	12
Quantitative indicators, actual values.								
C_2, mln. RUB.	363	391	470	498	436	545	464	408
C_3, VU	51	55	69	73	64	82	68	52
C_4, kt CO_2 e/y	50	53	66	70	61	96	65	38
C_5, ha	1.52	2.05	2.05	2.10	1.52	2.05	2.61	1.40

The ideal point method shows that 3D-circuit is the most universal one, whereas the most commonly used 13H circuit shows disadvantageous indicators. However, it is noteworthy that in this method criteria are equally significant, and mutual priorities are rather difficult issues which are not to be discussed here.

Table 5. Normalized criteria indicators and estimation by Equation 1.

Indi-cator, p.u.	Equa-tion	9	9H	12	12H	13	13H	16	3D
C_1	(2)	0.97	0.5	0.03	0.02	1	0.05	0.02	0
C_2	(3)	0.67	0.72	0.86	0.91	0.8	1	0.85	0.75
C_3	(4)	0.62	0.67	0.84	0.89	0.78	1	0.83	0.63
C_4	(5)	0.52	0.55	0.68	0.72	0.63	1	0.67	0.4
C_5	(6)	0.58	0.79	0.79	0.81	0.58	0.79	1	0.54
C	(1)	1.54	1.46	1.59	1.67	1.73	1.9	1.69	1.18

The solution obtained with the graph method is shown in Fig. 3. Numerical indicators of the constructed graphs are given in Table 4.

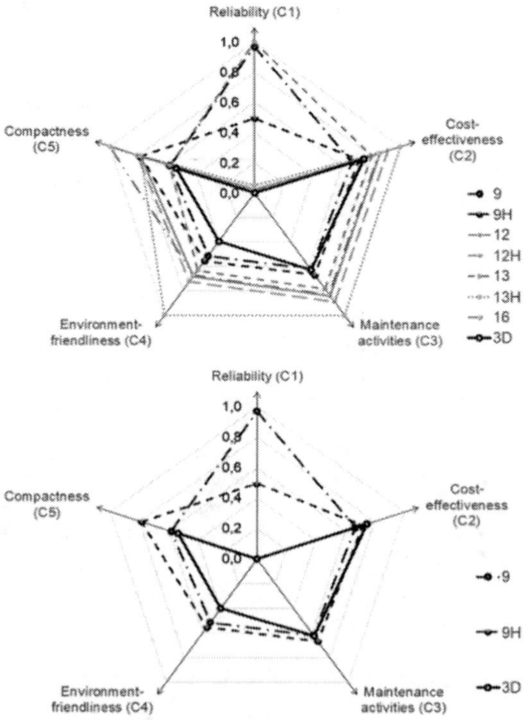

Fig. 5. Graph solution of the task

The graph should be interpreted in the following way: the lower the indicator is, the better it is. Thus, the smallest pentagon indicates the best circuit.

The left-hand part of Fig. 3 shows graphs for all the analyzed open switchgears (the "underdogs" are shown with grey color). The right-hand side of Fig. 3 features the ones which cannot be improved, the excluded circuits are not presented. New open switchgears do not put out of comparison the cheapest circuits (9, 9H) in terms of cost-effectiveness. This is due to a high cost of a 3D-CB in comparison with traditional SF6 CBs and many current transformers. A cost of new equipment for new open switchgears can be specified and lowered at the implementation stage.

The graph method shows that a new open switchgear is in each criterion superior to the best analogs or has the same indicators with them. All the indicators together creates a figure which is better than all the analogs.

If not to take into account a new-type open switchgear, the employed methods show that currently there is no best-in-all-criteria circuit. Thus, the results are in good agreement with practical usage. For example, circuits 9, 9H, which have such advantages as small sizes and low cost, are used less often than circuit 13H. This is due to such advantage of circuit 13H as reliability because of higher maintainability.

The use of new-type open switchgears will provide the advantages:

- higher reliability of customer power supply, since energy loss probability depends on the time of repair;
- increased staff safety, since all switching applications are done uniformly, a number of electrical devices is reduced;
- convenient repair schedule due to a circuit topology;
- decreased capital costs on the main equipment due to equipment elements;
- decreased operational costs on maintenance and balance-of-plant needs due to a number of electrical devices;
- increased carbon neutrality.

4 Conclusion

Multi-criteria comparison of open 220 kV switchgear which are currently used and new-type open switchgears shows that the latter are more balanced. The indicators of new open switchgears in each criterion are better or at least not worse than those of the best analogs. For example, they are highly reliable, very cost-effective in terms of capital and operational costs. In terms of all the characteristics together, open switchgears with a polyhedron circuit topology outperform all the existing analogs.

The offers presented in this work open new opportunities to study an important issue which can generally improve reliability of power grids.

Although the reported open switchgear requires further studies and improvements, the tasks to be solved can be set now:

1) to use combined switch-disconnector devices [11] and to stop using disconnectors;
2) to design 3D-current transformers.

References

1. F. Harary, Graph theory, (1973)
2. A. Abdurahmanov, M. Misrihanov, et al., Power plants, **4**, pp. 41-48 (2005)
3. A. Abdurahmanov, M. Misrihanov, et al., Power plants, **5**, pp. 54-57 (2005)
4. Yu. Dementiev, M. Misrihanov, et al., Power plants, **1**, pp. 51-54 (2011)
5. A. Abdurahmanov, M. Misrihanov, Sh. Shuntov, Electricity, **4**, pp. 2-11 (2007)
6. P. Grudinsky, S. Mandrykin, M. Ulitsky, Maintenance of the main electrical equipment of stations and substations, Energy, (1974)
7. IEC 61703:2016. Mathematical expressions for reliability, availability, maintainability
8. H. Ito, N. Uzelac, F. Richter, R. Le Roux, W. Pepper, L. Peng, A. Man-Im, I. Hategan on behalf of CIGRE AG A3.01, "CIGRE Reliability Survey on Equipment", CSE 023, pp. 1-11 (2021)
9. Order of the Ministry of Energy of Russia (January 17, 2019) No. 10 "Aggregated standards of costs of typical technological solutions of major construction of electric power facilities in electricity supply facilities"
10. Manual on labour costs calculations for operating personnel, (2008)
11. Disconnecting circuit breaker : manual. http://www05.abb.com/global/scot/scot245.nsf/veritydisplay/47a0a0579b 641143c12578a200222b19/$file/1HSM954323-03ru%20Disconnecting%20Circuit%20Breakers%20Buyers%20and%20Applica tion%20Guide.pdf
12. L. Rozhkov, L. Karneeva, T. Chirkova, Electrical equipment of power plants and substations, Izd. center Academia, (2013).

X-ray digital imaging detection system for high-voltage equipment inspection

Leonid Darian[1]*, *Roman* Obraztsov[1], Oleg *Nikitin*[2]

[1] JSC "Technical Inspection UES", 109012 Moscow, Russia
[2] FSUE «RFNC – VNIITF», 456770 Snezhinsk, Russia

Abstract. In 2019 a portable diagnostic X-ray system was developed and tested for the inspection of high-voltage equipment (HVE). The system operates based on discrete radiography and is referred by the authors as the first generation X-ray system. As a result of critical analysis of the first generation system, it was further proposed to create a second generation mobile scanning system for HVE digital radiography which is for the first time applied in the power engineering. A physical and mathematical model was developed to establish the relationship between the energy spectrum of the source X-ray radiation and the contribution of this radiation to the image generation during HVE radiography. Using this model, the authors assess the efficiency of the unique X-ray machine used in the first generation X-ray system.
The main technical requirements for the second generation X-ray system, implementing digital scanning radiography, for examining HVE in operation are given.

1 Introduction

To conduct the visual inspection of the internal elements of high-voltage equipment (HVE) it needs to be disassembled. And the disassembly of the equipment, minimum oil and SF6 circuit breakers (CB) in particular, requires down-conductor disconnection, oil drain or SF6 pumping, and its dismantling using lifting machines. After finishing the visual inspection, it is necessary to assemble and install back the HVE, refill it with an insulating medium, and reconnect down-conductors. In order to increase the efficiency of monitoring the HVE technical condition and reduce equipment downtime, new technologies that allow either to exclude some of these activities or to reduce the required time are needed [1, 2].

X-ray inspection, for example, of high-voltage CB allows you to quickly identify most of the defects associated with a change in the size of the internal elements and structural assemblies [3]:

– deformation, wear, damage to the moving (bending, burning, scuffing, etc.) and fixed contacts (damage to the housing, violation of the placement order of lamellas, etc.);

– damage to cables, bearings and springs of the moving contact control system;

– fastener defects and burnouts of inner shells;

– displaced or missing structural elements (fasteners, etc.);

– foreign bodies (chips, etc.).

The worldwide experience has been gained in HVE radiography using radioisotopes, constant and pulsed X-ray generators, and betatrons as sources of ionizing radiation. As a rule, X-ray films, photoluminescent storage screens, and digital flat-panel detectors are used as X-ray detectors [4–7]. At the same time, the experts in the field of X-ray nondestructive testing showed that thanks to X-ray digital imaging systems it is possible to obtain a better image quality compared to film systems [8, 9]. It was also confirmed that X-ray digital imaging technology allows to completely replace the film radiography in most practical application cases. It should be noted that all the above-mentioned X-ray systems for HVE inspection implement the so-called discrete radiography – when a full-size X-ray image is generated with simultaneous irradiation of the entire surface of the radiation receiver.

However, discrete radiography has two main disadvantages: distortion of images of extended elements on radiographs and an increased level of radiation hazard during HVE inspection. Moreover, the existing systems, though being portable, need special solutions for their positioning and reliable fixing, which usually requires more time to prepare for conducting HVE inspection.

Depending on how radiographs are generated, X-ray systems can be divided into two generations: the first generation X-ray systems are those that implement discrete radiography, and the second generation X-ray systems are based on scanning radiography.

The purpose of this work was to develop technical requirements for a mobile diagnostic X-ray system of the second generation, which can provide operational monitoring of the HVE technical condition and ensure the minimization of the radiation-hazardous zone.

* Corresponding author: daryan-la@ti-ees.ru

2 Portable diagnostic X-ray system of the first generation

In 2019 a portable diagnostic X-ray system was developed to examine the technical condition of the internal structural elements of 110 kV minimum oil CB. The system has been tested and put into operation at electric utilities [10]. The first generation X-ray system consists of the following main functional elements:

– X-ray machine with a collimator, power supply and control panels;
– digital flat panel detector (X-ray receiver);
– positioning system;
– software for processing digital X-ray images.

The main feature of the monoblock X-ray machine of constant action is a sectioned X-ray tube with a maximum voltage of 400 kV. This device is one of a kind in terms of penetrating power of radiation (60 mm in Fe-equivalent) considering its moderate weight (22 kg). A filter (copper plate) is installed at the exit window of the X-ray tube, which makes it possible to significantly reduce the level of radiation scattered by the test object and improve the contrast on the X-ray image of the internal elements of heterogeneous structures with massive porcelain insulators. The collimator mounted on the X-ray machine forms a 32°×24° rectangular pyramid beam [11].

It should be noted that the X-ray machine feasibility is determined by the energy of the radiation quanta that contribute the most to the resulting X-ray image.

3 Physical and mathematical model of HVE radiography

The physical and mathematical model of HVE radiography includes an X-ray machine with its own radiation spectrum, a test object and the medium in which the radiation propagates. The initial energy spectrum of the radiation emitted by the X-ray tube target (the distribution of the number of ionizing particles according to their energy) is described in the physical and mathematical model of HVE radiography in accordance with the Kramers equation [12]:

$$I(E) = k \cdot i \cdot Z \cdot (\frac{E_0}{E} - 1), \quad (1)$$

where k is the coefficient of proportionality ($k \approx 10^{-9}$); i is the current, carrying through the X-ray tube; Z is the atomic number of the anode material; E_0 is the maximum photon energy in the spectrum.

According to equation (1), if $E \to 0$ then $I_E \to \infty$, which means one can set lower threshold energy at 10 keV when calculating the energy spectra. Such a limitation does not affect the calculation results, since the emitters (monoblocks) of the X-ray machine almost completely absorb radiation with energy up to 15 keV.

The change in the energy spectrum of X-ray radiation during the object inspection is described in the physical and mathematical model of HVE radiography proposed by the authors using the Beer-Lambert law [12, 13]:

$$I_n = I_0 \cdot e^{-\eta\delta} \quad (2)$$

where I_0 is the intensity of radiation incident on the translucent medium; I_n is the intensity of radiation that has passed through a layer of matter with a thickness of δ; δ is the thickness of the absorber; $\eta \sim Z^3 \cdot \lambda^3$ – linear attenuation coefficient; Z is the atomic number of the element; λ is the wavelength of radiation.

Since the X-ray machine has an external filter, the source spectrum (Fig. 1a, graph 2, or Fig. 1b) contains a very small fraction of low-energy quanta (up to 120 keV). The fact that the voltage on the tube is very high for this type of X-ray machine makes it possible and efficient to apply the filtering of the initial radiation (Fig. 1a, graph 1). The ratio of the areas of graphs 2 and 1 corresponds to the fraction of X-ray machine radiation quanta transmitted by the external filter [14].

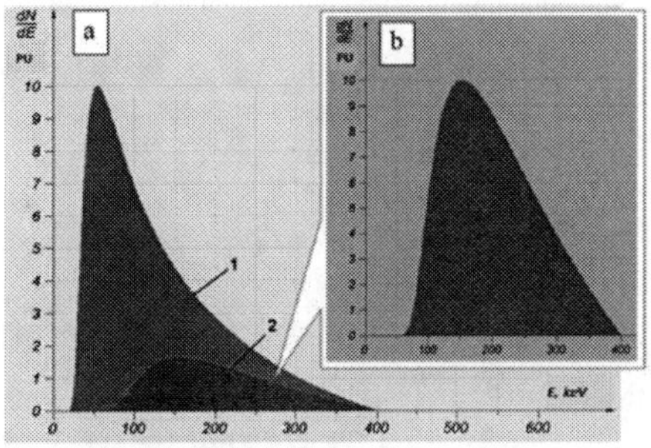

Fig. 1. Energy spectrum of X-ray machine radiation: 1 – energy spectrum of initial radiation; 2 – energy spectrum of radiation after passing through an external filter.

The efficiency of X-ray machine can be assessed by comparing its radiation spectra obtained without the test object and with it. To do so, firstly, the convolution of the spectra of the initial radiation and radiation emitted after passing through the test object should be calculated. Therefore, we need to obtain the product of the functions of these two energy emission spectra:

$$S(E) = I_0 \cdot I_n \quad (3)$$

where I_0 is the intensity of radiation that is incident on the translucent medium; I_n is the intensity of the radiation that has passed through the test object and is involved in the radiograph generation.

Secondly, we need to determine the fraction of the radiation energy that has passed through the test object and participates in the X-ray image generation. Thus, the percentage ratio of the convolution area of the radiation energy spectra to the area of the initial radiation energy spectrum is calculated:

$$P(E) = \int_{10}^{400} S(E)dE / \int_{10}^{400} I_0(E)dE. \quad (4)$$

Using the graph of the initial radiation energy spectrum combined with the convolution graph of the energy spectra of the initial radiation and radiation emitted after passing through the test object, it is possible to estimate the region of the X-ray machine radiation spectrum that most effectively produces an X-ray image. This region (Fig. 2) is located between the convolution peak of the energy spectrum of radiation (2) and point (3). It characterizes the energy of X-ray quanta that contributes the most to the resulting X-ray image.

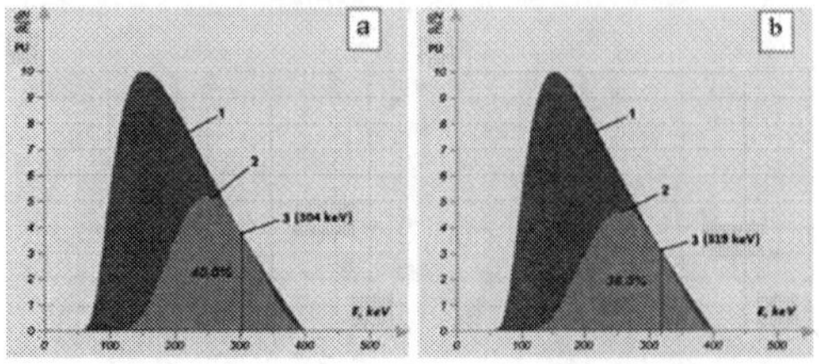

Fig. 2. Evaluation of the X-ray machine efficiency in inspecting VGT-110 (a) and VMT-110 (b) poles: region 1 – energy spectrum of the initial radiation; region 2 – convolution of the spectra of the initial radiation and radiation emitted after passing through the test object; point 3 – the upper limit of the radiation energy spectrum, the quanta of which contributes the most to the X-ray image generation.

Thus, analysis of the applied X-ray machine efficiency shows that the fractions of radiation energy involved in the generation of X-ray images of VGT-110 (Fig. 2a) and VMT-110 (Fig. 2b) are 40% and 36.5%, respectively. It is important to note that, according to [14], the efficiency of machines with 250 kV tube voltage in X-ray imaging of the same test objects is 8-9%.

4 Inspection of VMT-110 CB in the field

The results of X-ray inspection of VMT-110 CB in operation using the developed first generation X-ray system are presented. The X-ray system is mounted on the CB pole (Fig. 3). The positioning system (**S**) is attached to the CB frame. The X-ray machine (**M**) with a collimator (**C**) is mounted on the first pole of the positioning system, and a flat-panel digital detector (**D**) is mounted on the second pole.

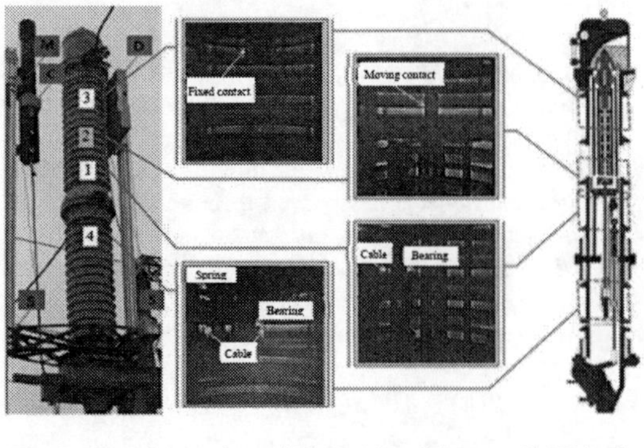

Fig. 3. X-ray inspection of VMT-110 minimum oil CB in the field.

The assessment of the technical condition of the contact rod of the CB moving contact was carried out using two X-ray images (Fig. 4a and 4b) obtained in two mutually perpendicular views (Fig. 3, position 2). According to the assessment results, there were no such defects as deformation, wear, burning of the contact surface, decrease in the height of the contact part of the moving contact [15].

At the end of radiography, the examined CB was disassembled (Fig. 4c). The further visual inspection of the CB internal elements confirmed the absence of the defects mentioned above.

Thus, it was confirmed that it is possible to assess the condition of the examined CB assemblies using X-ray images. Therefore, there is no need to disassemble the equipment, which can significantly save time and reduce labor forces.

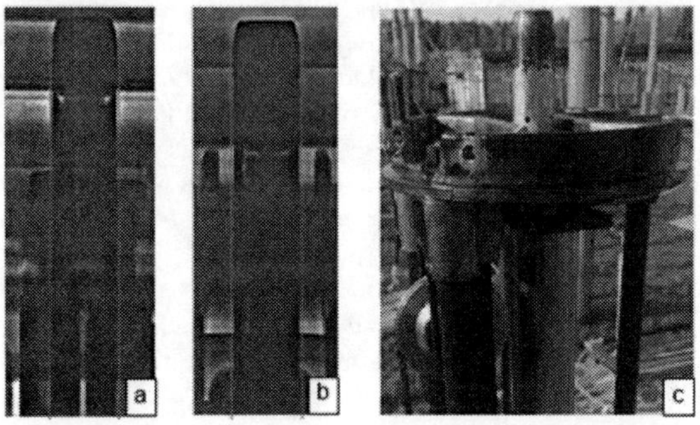

Fig. 4. Inspection of the condition of the contact rod of VMT-110 CB moving contact: (a) X-ray image obtained in the first view; (b) X-ray image obtained in the second view, perpendicular to the first one; (c) contact rod of the moving contact of the disassembled CB.

5 Technical requirements for mobile X-ray systems of the second generation

The further development of mobile X-ray systems is caused by the need to increase the efficiency of the HVE inspection and to improve the mobility and versatility of the applied technical solutions, as well as the need to significantly reduce the level of radiation hazard during radiography.

Taking into account the accumulated experience in the field of X-ray nondestructive testing and the results of assessing the X-ray machine efficiency, it seems appropriate to use the same X-ray machine that the first generation X-ray system was equipped with as an emitter for HVE radiography.

Unlike discrete radiography, the scanning system operates as follows: a narrowly collimated X-ray source and a linear radiographic recorder move synchronously along the object, obtaining its X-ray image line by line [16]. The data received from the linear recorders are digitally processed in the control unit of the scanning system and transferred to the software as pixels. When scanning is completed, a full-size X-ray image is generated from the received pixels. Compared to discrete radiography, the X-ray digital imaging detection systems have the following advantages [3]:

– reduction of the "noise" level on X-ray images by at least 4 times due to the radiation collimation;

– obtaining all lines of the image in one geometric projection, which makes it possible to avoid distortion of the images of object extended elements (for example, CB poles);

– expanding the possibilities of post-processing of the resulting image on a digital radiographic image;

– ensuring a reduced level of radiation hazard during HVE radiography in the field (reduction of the radiation-hazardous zone around the X-ray system during exposure by at least 2 times).

To increase the efficiency of the HVE inspection, the scanning speed needs to be at least 0.5 m/min. At the same time, the HVE can have a housing with a diameter of up to 1 m with a total radiation thickness of the test object up to 60 mm in Fe equivalent. The resolution of the X-ray image should be no more than 1 mm in the original scale of the equipment.

The selected vehicle for a mobile scanning X-ray system must meet the following technical requirements:

– being able of riding unpaved and public roads;

– having overall dimensions that provide high mobility on substations;

– having all-wheel drive and optimal carrying capacity to ensure the ability of riding muddy roads;

– having structural elements and auxiliaries that ensure the protection of the scanning system from the precipitation, dust and other external factors during transportation and storage;

– being equipped with a radiation-safe operator cabin;

– providing an automated workplace in the operator cabin.

The positioning of the scanning system is carried out by a crane-manipulator installed on the chassis of the X-ray system. The manipulator must provide positioning of the scanning system from one position of the parked X-ray system (from 1 m to 7 m vertically from the ground and 2 m horizontally to either side of the central position).

6 Conclusions

1. A critical analysis of the developed and tested portable diagnostic X-ray system of the first generation is presented. The choice of X-ray machine is vital for HVE radiography.

2. Based on the previously developed physical and mathematical model for HVE radiography, the feasibility of the applied X-ray machine was assessed. It is shown that with an increase in the tube voltage, the fraction of radiation energy involved in the formation of X-ray images increases as well.

3. The X-ray tube with the maximum photon energy of 400 keV, used in the first generation X-ray system, surpasses in its efficiency all known X-ray machines, taking into account its weight and size. Assessing the proposed X-ray machine showed it is reasonable to use the machine in a mobile scanning system of the second generation.

4. The main technical requirements for mobile diagnostic X-ray systems of the second generation, which implement digital scanning radiography and provide operational examination of the HVE while reducing the radiation-hazardous zone around the X-ray system, are proposed.

References

1. J. Michaelson. ABB Library (2016)
2. *Verify SF6 circuit breaker reliability while reducing maintenance costs,* Siemens (2017)
3. L.A. Darian, E.P. Grabchak, R.M. Obraztsov, et al. Russ. Electr. Engin., **91**, pp. 500-504 (2020)
4. L.A. Darian, E.P. Grabchak, R.M. Obraztsov, et al. ELECTRIC POWER. Transmission and distribution, **3(60)**, pp. 96-105 (2020) [in Russian]
5. L.A. Darian, R.M. Obraztsov. Energyexpert, **2(70)**, pp. 48-55 (2019) [in Russian]
6. X. Pang, Y. Qi, X. Li et al. *Research on Defect Pattern Recognition of GIS Equipment Based on X-ray Digital Imaging Technology*, in Proceedings of IEEE International Conference on High Voltage Engineering and Application, ICHVE, September 10–13, 2018, Athens, Greece (2018)
7. T. Li, X. Pang, B. Jia et al. Energies, **13**, art. № 661 (2020)
8. A.A. Mayorov. In the world of NDT, **3(45)**, (2009) [in Russian]
9. K.A. Bagaev, A.B. Spirkov. Industrial Safety, **5(30)**, (2013) [in Russian]
10. L.A. Darian, P.V. Golubev, R.M. Obraztsov, et al. ELECTRIC POWER. Transmission and distribution, **1(52)**, pp. 134-141 (2019) [in Russian]
11. L.A. Darian, P.V. Golubev, R.M. Obraztsov, et al. Patent № 189440 RU (2019)
12. A.M. Gurvich. Physical foundations of radiation monitoring and diagnostics, (1989) [in Russian]
13. K. Siegbahn Beta-and gamma-ray spectroscopy (1959) [Russian ed.]
14. L.A. Darian, R.M. Obraztsov, O.V. Ozerov, O.A. Nikitin. Energetik, **3**, pp. 3-12 (2021) [in Russian]
15. L.A. Darian, E.P. Grabchak, P.V. Golubev, R.M. Obraztsov. *Investigation of X-Ray Technology to Assess the Technical Condition of High-Voltage Minimum Oil Circuit Breakers*, in Proceedings of the 87-th International Conference of Doble Clients, March 8-13, 2020, Boston, MA, USA (2020)
16. L.A. Darian, P.V. Golubev, R.M. Obraztsov, E.P. Grabchak, V.I. Zverev et al. IEEE Trans. Dielectr. Electr. Insul., **1(27)**, pp. 172-180 (2020)

Seasonal changes in the concentrations of cellulose aging markers in transformer oil

Leonid Darian[1]*, *Roman* Obraztsov[1], *Alexey* Maksimchenko[1], *Vladimir* Polistchook[2], *Sergey* Asoskov[3]

[1] JSC "Technical Inspection UES", 109012 Moscow, Russia
[2] Joint Institute for High Temperatures of Russian Academy of Sciences, 125412 Moscow, Russia
[3] LLC «Gazprom Energo», 117574 Moscow, Russia

Abstract. The results of long-term monitoring of the content of aging markers of cellulose insulation – methanol, ethanol, and carbon oxides – dissolved in power transformer oil are presented. A correlation has been established between seasonal temperature extrema of transformer oil and the concentrations of aging markers. To improve the reliability of diagnostic conclusions based on measuring the concentration of aging markers, it is proposed to take into account their seasonal changes.

1 Introduction

It is known that during aging of the internal insulation of power transformers (PT) the deterioration of physical, chemical, mechanical, electrical, and other properties occurs which can lead to a heavy equipment fault. To prevent this, the assessment of the insulation aging degree and rate needs to be done. The rate of aging of the PT internal insulation can be estimated by the content of its decomposition products (aging markers) dissolved in transformer oil (TO). The most common aging markers for cellulose insulation are carbon oxides and water (first generation aging markers), furan derivatives (second generation aging markers), and alcohols – methanol and ethanol (third generation aging markers).

Recently, the interest in third generation aging markers has increased, since they allow to identify the processes of cellulose insulation degradation at an early stage of their development [1, 2]. Thus, methanol is predominantly produced during the cellulose hydrolysis reaction, i.e. at relatively low-temperature aging processes. Whereas the presence of ethanol in TO indicates local overheating of cellulose insulation, accompanied by a pyrolysis reaction [3, 4].

Aging markers, especially alcohols and water, are actively sorbed by cellulose insulation. From the point of view of physical and chemical diagnostics, this property, on the one hand, is very useful, since only a small amount of aging markers (dissolved in oil) is removed during degassing, regeneration, and oil change. On the other hand, the analysis of third generation aging markers requires quite a high sensitivity of instrumental and analytical software, since a relatively small amount of the markers is generated during insulation aging. For example, the lower limit of the detection for DGA is at the level of 10-4% vol. (1 ppm), while for third generation aging markers this limit should be at the level of 10-7% vol. (1 ppb). The distribution coefficient of aging markers in the oil-cellulose system is heavily dependent on temperature. Therefore, to interpret the measurement results it seems appropriate either to normalize the measured values of aging marker concentrations, or to calculate their total content in the internal insulation [2].

The sorption-desorption processes of insulation aging markers during temperature changes are currently insufficiently studied and are not often used for assessing the insulation technical condition of the equipment at electric power utilities. The only well-studied aging markers are water and furan derivatives, the distribution of which between TO and cellulose insulation is calculated on the basis of equilibrium curves in the operating temperature range [5]. Note that temperature changes in PT are associated with both load modes and seasonal changes in ambient temperature.

The purpose of the presented studies is to identify the relationship between seasonal changes in the temperature of TO and the concentrations of aging markers of PT cellulose insulation.

2 Research object and methods

The object of the studies was a three-phase PT (110 kV, 80 MVA, TO weight – 41 t) that was put into operation in 1986.

The PT is equipped with a monitoring system for diagnostics of nine gases (carbon oxides, hydrocarbons, hydrogen, oxygen, and nitrogen) dissolved in TO. The values of gas concentrations were measured at least once every 24 hours. According to the manufacturer of the monitoring system, the measurement error of gas concentrations is 10%. TO samples were taken about once a month to determine

* Corresponding author: daryan-la@ti-ees.ru

https://doi.org/10.1051/e3sconf/202346101017

the concentrations of methanol and ethanol. The analysis was carried out in the laboratory using a unique chromatography system and according to a certified method. The measurement range was from 10 ppb to 21 ppm, the accuracy index (the limits of the relative error at a confidence level of 0.95) did not exceed 15%, and the reproducibility index did not exceed 4%.

The monitoring of the concentrations of diagnostic gases and alcohols has been carried out for five years and is still going on.

The ambient temperature, the top and bottom oil temperature, and the load current of the PT are recorded by the monitoring system every hour. The temperature of the insulation hottest point was calculated based on the data of the ambient temperature and the PT load current according to the IEC standard method [6].

The change in the concentrations of monitored aging markers during preventive measures – current and major repairs – is of particular interest. The PT was taken out for overhaul and the oil was changed almost a year after the beginning of the monitoring aging markers. During the repair, the cellulose insulation was dried and the active part was revised. Inspection of the PT internals did not reveal any sign of deterioration in the condition of the cellulose insulation and the pressing of the windings.

The PT was also taken out for repair four years after the beginning of the monitoring to change the sealing gasket between the cover and the PT tank and to seal the welded joints. During this period, the measurements of alcohols' concentrations were carried out at least once a day.

3 Measurement results

The analysis of the measurement results showed that the difference in the top and bottom oil temperatures is usually about 10 °C. The maximum difference in these temperatures reached 30 °C. The ambient temperature differed from the bottom oil temperature by a maximum of 5 °C. The difference between the calculated values of the insulation hot spot temperature and the top oil temperature was approximately the same.

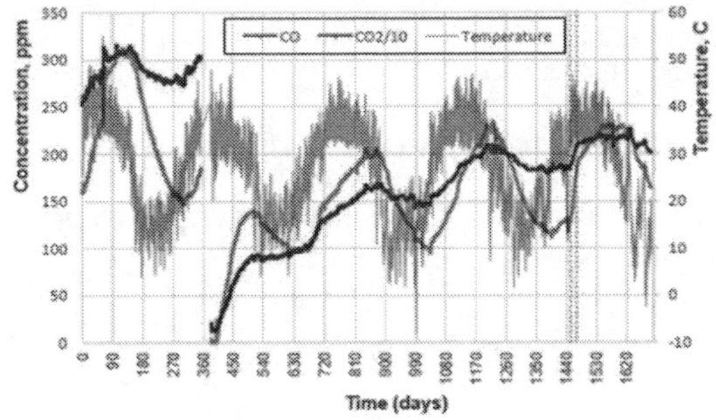

Fig. 1. Concentrations of carbon oxides.

The average value of the top and bottom oil temperature was taken as the typical temperature of the internal insulation. Fig. 1 shows the values of the typical temperature, concentrations of carbon oxide and carbon dioxide for the entire monitoring period.

Since these gases are shown simultaneously on the graph, the values of the concentration of carbon dioxide are reduced by 10 times to make it clear. A sharp decrease in the concentration of carbon oxides occurred after TO change during the overhaul.

Fig. 2 shows the values of methanol concentrations, as well as the PT load for the entire monitoring period. The PT load had been changing in a relatively small range – from 20% to 50%, or 30% in average. No influence of load on the seasonal changes in concentrations is observed. The period when the PT was taken out for TO change is marked in Fig. 2 by yellow dotted lines. The concentration of the points on this graph is due to the increase in the frequency of sampling and measurements.

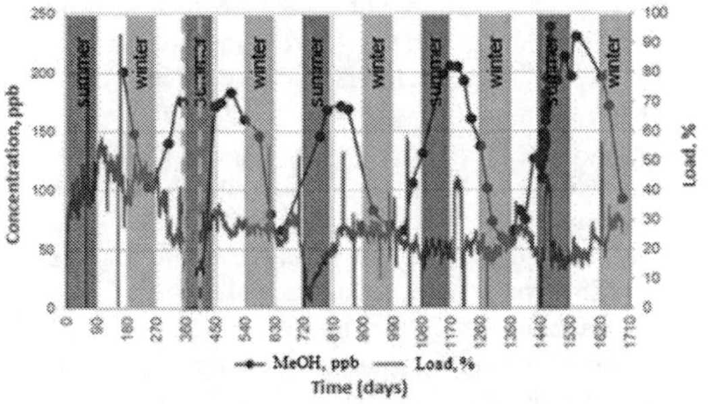

Fig. 2. Dependences of methanol concentration and PT load on time.

As in the case of carbon oxides, a sharp decrease in the methanol concentration occurred during the overhaul that included the TO change. However, while the concentrations of carbon oxides after the TO change decreased almost to zero (Fig. 1), the value of the methanol concentration remained quite noticeable – 31 ppb. This phenomenon, apparently, is due to the fact that oil sampling for analysis on methanol content was carried out not immediately after the oil change, but 20 days later. Fig. 1 also shows that the concentrations of carbon monoxide and carbon dioxide increased during this time. Thus, it can be assumed that after changing the oil, there was no methanol in it.

The monitoring of ethanol concentration began later than that of methanol, since the method for measuring the volume fraction of ethanol in TO was developed and implemented later. The qualitative dependence of the ethanol concentration on time was the same as that of methanol. It is interesting to note that after a dramatic growth of the PT load (the current rose from 75 to 160 A – green dotted lines in Fig. 1) the ethanol concentration increased more than 2 times from 15 to 35 ppb, and the maximum ethanol concentration was about 15% of the methanol concentration.

https://doi.org/10.1051/e3sconf/202346101017

4 Discussion of the results

As can be seen from Figures 1 and 2, the dependences of the concentrations of aging markers on time are periodic. However, after the PT overhaul the dependences were not periodic anymore since a significant decrease in the concentration of aging markers was noticed. Taking this into account, the dependence of the concentration of aging markers on time is qualitatively consistent with a similar dependence of the typical temperature of the PT internal insulation (Fig. 1). The maximum concentrations of aging markers for the studied PT were recorded in late summer and early autumn (August-October), and the minimum concentrations – in late winter and early spring (February-April). The load of the studied PT remained almost unchanged for a long time.

Considering that the generation rate of aging markers is low, the values of their concentration in TO are established as a result of the balance of sorption processes on the surface of cellulose insulation and diffusion of aging markers deep into the insulation. The desorption rate depends exponentially on the insulation surface temperature. Therefore, it may be concluded that there is a correlation between the dependences of the typical temperature of the insulation and the concentrations of aging markers on time. However, the dependencies shown in Figures 1 and 2 seem to be very non-trivial.

As can be seen from Fig. 1, a sharp decrease in the concentration of carbon oxides after the oil change was followed by a gradual increase while the seasonally oscillating concentration values could be seen. One can note a qualitative feature in the dependence of the carbon dioxide concentration: a monotonous increase in the concentration had been seen for 18 months after the oil change. During the same period of time, two extrema were recorded on the dependence of the concentration of carbon monoxide. Three years later, the extrema of carbon monoxide concentration (at a typical internal insulation temperature of about 45 °C) reached approximately 75% of the values recorded before the repair. Apparently, the content of carbon oxides in TO after the oil change began to recover due to their generation as a result of oxidative processes and desorption from cellulose insulation.

It was noted above that the qualitative dependences of the concentration of methanol (Fig. 2) and carbon monoxide (Fig. 1) on time are similar. At the same time, it is likely that the ratio of the amount of methanol contained in cellulose to its amount in TO was a bit higher than that of carbon monoxide and carbon dioxide. This is confirmed at least by the fact that the ratio of concentrations in the maxima and minima within 12 months for methanol varied from 3 to 5 (Fig. 2), and for carbon oxides this ratio was about 1.5-2 (Fig. 1) with an about 25 °C change in the typical temperature during season. The methanol concentration values returned to the original values approximately two years after the overhaul. This is presumably due to the low rate of methanol generation in PT.

It follows from Fig. 2 that the value of the first maximum of methanol concentration after repair turned out to be a bit less than before it. Then an increase in methanol concentration appeared – in two years it increased by about 35%. However, the actual increase in concentration could be less, since the determination error is about 15%. The almost constant value of the concentration for the last three minima indicates that the rate of methanol generation in the PT was low – its concentrations in the oil at the minima differed from each other by ≤8 ppb.

Comparing the data in Figures 1 and 2, there is a time lag between the extrema of the typical temperature and the concentrations of aging markers. The extrema of the aging marker concentrations and temperatures were determined according to the following algorithm: a time interval of 6-8 months was chosen near the extremum point, then the experimental points were approximated by a polynomial:

$$y = A \cdot x^3 + B \cdot x^2 + C \cdot x + D, \quad (1)$$

where y – aging marker concentration (or temperature); x – time (number of days from the beginning of the research).

The parameters of equation (1) for the given characteristics of TO are presented in Table 1. The degree of determination of the approximating curves is more than 0.8, which indicates a satisfactory approximation. The time (days) corresponding to the extrema is given in Table 1. These values are obtained by equating the first derivative of each approximating curves to zero.

Table 1. Parameters of equation (1).

$y(x)$	A	B	C	D	R^2	x if $y' = 0$
Typical TO temperature	$8 \cdot 10^{-7}$	-0,0031	3,4025	-1098,4	0,830	770
MeOH concentration	10^{-5}	-0,0386	39,472	-12733	0,996	823

In fact, the approximating curves were also used to smooth concentration and temperature values. From the obtained dependence, the value of the concentration of the aging marker at the extremum point was determined. Similarly, the value of the typical temperature of the internal insulation at the concentration extremum point was found. The time to reach the extremum of the typical temperature was determined in the same way.

Fig. 3 shows an example of approximation by equation (1) of the measured values of methanol concentration and typical temperature in the region of local maxima.

 https://doi.org/10.1051/e3sconf/202346101017

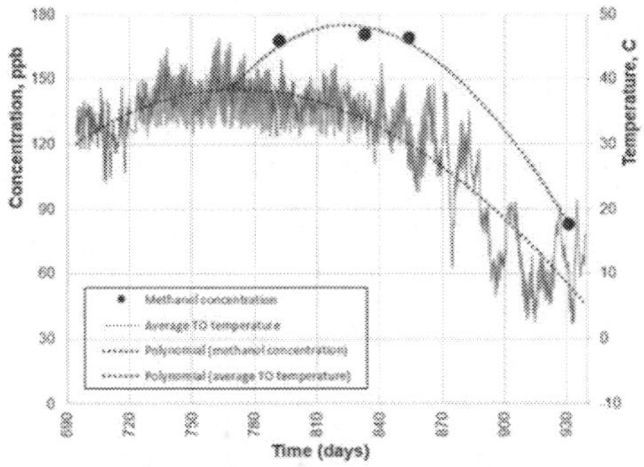

Fig. 3. Concentrations of carbon oxides

Table 2 shows the characteristics of methanol concentration extrema and previous typical temperature extrema depending on time over the entire monitoring period.

Table 2. Characteristics of methanol concentration extrema.

Extremum type	Day of MeOH concentration extremum	MeOH concentration, ppb	Days of the previous extremum of the typical temperature	Time lag, days
Min.	255	112	229	26
MAX.	481	190	438	43
Min.	659	64	584	75
MAX.	823	173	770	53
Min.	990	63	953	37
MAX.	1166	210	1129	37
Min.	1351	61	1323	28
MAX.	1544	234	1504	40

As can be seen, the extrema of the typical TO temperature (the average value of the top and bottom oil temperatures) "outran" the extrema of the methanol concentration by 1-2 months. When making estimations, it was assumed that the time lag between the extrema of the methanol concentration and the typical temperature is 45 days. The possibility of applying this assumption to other PT requires additional research.

5 Seasonal changes in methanol concentration

To normalize the concentrations of aging markers that depend on the internal insulation temperature, it is proposed to use the temperature coefficient (K(T)) [7], which is the ratio of the concentrations of the aging marker in TO at a temperature of 20 °C ($C_M(20)$) and at an arbitrary temperature $C_M(\mathrm{T})$.

$$\mathrm{K(T)} = C_M(20)/C_M(\mathrm{T}) \qquad (2)$$

In this case, the temperature coefficient K(T) can be described as:

$$\mathrm{K(T)} = a \cdot exp(b_M/(\mathrm{T}+273)) \qquad (3)$$

where a – pre-exponential factor; b_M – exponential factor; T – arbitrary temperature, °C.

The value of the exponential factor b_M, calculated for methanol based on the experimental data given in [7], is 3720 K.

Thus, using the temperature coefficient K(T), normalized concentrations of aging markers measured at arbitrary TO temperatures can be obtained.

As noted above, in order to assess the degree of insulation aging, the data on the total methanol content in the PT internal insulation is needed. Modifying equations (2) and (3) and considering the fact that methanol practically ceases to be sorbed by cellulose insulation at 90 °C [7], we obtain the following equation:

$$C_M(90) = C_M(\mathrm{T}_{av}) \cdot exp(b_M\left(\frac{1}{273+\mathrm{T}_{av}} - \frac{1}{273+90}\right)), \qquad (4)$$

where $C_M(90)$ – concentration of methanol in TO at a temperature of 90 °C; $C_M(\mathrm{T}_{av})$ – methanol concentration in TO at the typical temperature of the PT internal insulation; T_{av} – typical temperature of the PT internal insulation, equal to the average value of the top and bottom oil temperatures.

The obtained $C_M(90)$ value can be used to calculate the total amount of methanol Q_M in the PT internal insulation:

$$Q_M = C_M(90) \cdot V_{TO}, \qquad (5)$$

where V_{TO} – TO volume in PT.

The influence of the time lag between the typical temperature and methanol concentration on the results of calculating the $C_M(90)$ concentration is shown in Fig. 4. The calculation based on equation (4) was carried out without taking into account the time lag (blue columns). Thus, the typical temperature was considered as the average temperature of top and bottom oil on the day of sampling, taking into account the time lag of 45 days (green columns). So, when calculating according to equation (4), the typical temperature of the PT internal insulation, which was "45 days ago," was taken as T_{av} temperature. Considering that the rate of methanol generation in the PT is low, the calculated values of the $C_M(90)$ concentration should coincide for all measurements.

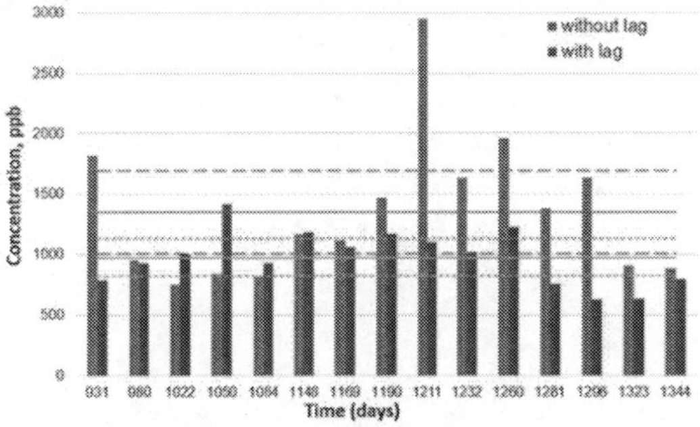

Fig. 4. Methanol concentration at the PT internal insulation typical temperature.

https://doi.org/10.1051/e3sconf/202346101017

Fig. 4 also shows the results of statistical processing (based on GOST [8]) of the $C_M(90)$ concentration values, calculated according to (4). The solid horizontal lines show the average values of the $C_M(90)$ concentration: blue line – without taking into account the time lag; green line – taking into account the time lag. Horizontal dotted lines of similar colors show confidence intervals at a confidence level of 0.95. According to GOST [9], this means that in a long series of samples, the relative frequency of cases where the confidence interval covers the true value of the population parameter is ≥0.95 for the given calculation.

It is clear that when the time lag is taken into account, the standard deviation of the $C_M(90)$ concentration values decreased by 2.6 times (from 592 to 228 ppb), and the confidence interval decreased by 2.1 times (from 339 to 159 ppb). Therefore, with the same confidence level, the interval that includes the true value of the $C_M(90)$ concentration narrows down, which increases the calculation accuracy. Thus, it can be stated that taking into account the time lag between the methanol concentration and the typical temperature makes it possible to increase the reliability of calculations of the total amount of the aging markers in the PT internal insulation.

6 Conclusions

1. According to the results of a long-term monitoring of aging markers' content in the oil of the PT in operation, there are seasonal extrema of their concentrations: the maximum concentrations of aging markers were recorded in late summer and early autumn, and the minimum concentrations – in late winter and early spring.

2. The extrema of methanol concentration lag behind the extrema of the typical temperature of the internal insulation by approximately 45 days.

3. The concentrations of third generation aging markers in TO can differ by 3-5 times with a change in the typical temperature of the internal insulation within 25 °C during a season, and the concentrations of carbon oxides under the same conditions – by 1.5-2 times. Thus, it seems appropriate to use either normalized concentration values or the total amount of aging markers in the PT internal insulation when assessing the condition of the PT internal insulation.

4. Taking into account the time lag when normalizing the concentration values or calculating the total amount of aging markers in the PT internal insulation increases the accuracy of calculations.

References

1. CIGRE TB 779. WG A2/D1-46. (2019)
2. L.A. Darian, A.V. Maksimchenko, R.M. Obraztsov, E.A. Goncharova. Patent RU 2647224. (2018) [in Russian]
3. E.M. Rodriguez-Celis, S. Duchesne, J. Jalbert and M. Ryadi. Cellulose, **22**, pp. 3225–3236, (2015)
4. E.M. Rodriguez-Celis, J. Jalbert, S. Duchesne, B. Noirhomme, and M-C. Lessard. CIGRE-083, CIGRE Canada Conference (2012)
5. T.V. Oommen. Proceedings of the Electrical/Electronics Insulation Conference, Chicago, IL, USA, pp. 162–166 (1983)
6. IEC 60076-7:2018 Power transformers – Part 7: Loading guide for mineral-oil-immersed power transformers (2018)
7. J. Jalbert, M.-C. Lessard, and M. Ryadi. IEEE Trans. Dielectr. Electr. Insul. **20(6)**, pp. 2287-2291 (2013)
8. GOST R 50779.22 2005. Statistical interpretation of test results. Estimation of the mean. Confidence interval (2005)
9. GOST R 50779.10-2000 (ISO 3534-1-93) Statistical methods. Probability and general statistical terms. Terms and definitions (2000)

Reliability of heating networks as a factor of sustainable development of thermal power industry

Irina Akhmetova[1]*, *Alina* Zalyalova[2], *Adelya* Nurislamova[1]

[1] Federal State Budgetary Educational Institution of Higher Education «Kazan State Power Engineering University», Kazan, Russia
[2] JSC "Kazenergo", Kazan, Russia

Abstract. The reliability of heating networks is the foremost principle for organizing relationships in the field of heat supply, as established by Federal Law. The overall reliability of heat supply across the Russian Federation is declining due to insufficient replacement of aging heating networks. This article explores the potential use of a program document, specifically a heat supply scheme, as a means to justify the allocation of budgetary funds for initiatives aimed at enhancing the reliability of heat supply or transitioning to a "price zone.".

1 Introduction

The climatic conditions in the Russian Federation, characterized by long periods of sub-zero outdoor temperatures across a vast area of the country, make the heat supply industry a key sector that ensures the normal functioning of the population.

Reliability of heat supply is one of the main principles governing the organization of relations in the field of heat supply, as established by Federal Law No. 190-FZ "On Heat Supply," dated July 27, 2010. The standard reliability of heat supply is essential for making sustainable development plans for the industry.

The heating network is a critical component of the heat supply system, determining its reliability. This is due to the presence of backup heat-generating and pumping equipment in thermal energy sources. In the event of a failure in one boiler or pump, it typically does not result in a disconnection of consumers. At worst, it may lead to a reduction in the parameters and quality of the supplied thermal energy.

In the event of accidents on heating networks, there is a possibility of a complete shutdown of heat supply to consumers. This can exacerbate the situation, particularly at low outside temperatures, due to the thawing of internal heating systems and sections of the heating networks.

2 Main part

Based on the report on the state of the heat power industry and centralized heat supply in the Russian Federation in 2020, prepared by the Ministry of Energy of the Russian Federation and the Federal State Budgetary Institution 'Russian Energy Agency' of the Ministry of Energy of Russia [1], it is evident that out of the total length of heating networks in the country, which is 167.4 thousand km, 51.51 thousand km (30.77%) require replacement, and 38.8 thousand km (23.18%) are in a dilapidated condition. Furthermore, over the past 5 years, the total length of heating networks in the Russian Federation has decreased by 4.15 thousand km, while the share of dilapidated heating networks has increased by 2.04 thousand km.

Insufficient replacement volumes serve as the main reason for the high wear and tear of heating network pipelines. In 2020, the relocation of heating networks in Russia as a whole amounted to only 3.371 thousand km, which represents merely 6.7% of the required replacement volume (Fig. 1).

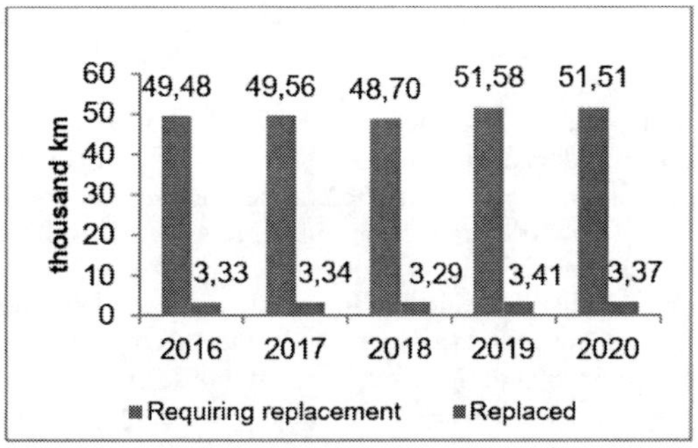

Fig.1 The ratio of the length of pipelines requiring replacement and those replaced in the Russian Federation in 2016-2020, thousand km.

An increase in the proportion of pipelines requiring replacement leads to a rise in the number of accidents in heating networks. From 2018 to 2020, the overall accident rate of heating networks in the Russian Federation increased by 22% [1].

The ongoing trend of escalating wear and tear in heating networks will inevitably result in a critical decline in the reliability of heat supply.

Simultaneously, the insufficient volume of replacement for heating networks can be attributed to both a shortage of funds allocated for reconstruction, modernization, and repairs accounted for in the thermal

* Corresponding author: irina_akhmetova@mail.ru

energy tariff, as well as unsatisfactory and poor-quality planning by heat supply organizations within their investment programs and repair plans.

It is worth noting that the Russian Government acknowledges the issue of diminishing reliability in utility infrastructure, including heating networks. Mechanisms have been developed to provide additional financing sources for replacing dilapidated networks. These programs are implemented with the support of the Territorial Development Fund:

- issuing subsidies to regions for the creation, reconstruction, and modernization of municipal infrastructure facilities (Resolution of the Government of the Russian Federation of December 26, 2015 No. 1451);
- preferential loans for the modernization of municipal infrastructure facilities (Resolution of the Government of the Russian Federation of February 2, 2022 No. 87);
- allocation of non-refundable budget funds for the reconstruction of linear facilities (Resolution of the Government of the Russian Federation dated December 8, 2022 No. 2253).

Programs implemented through DomRF Bank have also been developed:

- financing the construction (reconstruction) of infrastructure using bonds of SOPF Infrastructure Bonds LLC;
- provision of loans for the purpose of design, construction and (or) reconstruction of infrastructure facilities on the basis of a concession agreement or public-private partnership agreement, municipal-private partnership agreement.

Furthermore, it is important to highlight the solution proposed by the Government of the Russian Federation to address the problem as part of the transition to a new target model of the thermal energy market: the method of price zones or the 'Alternative Boiler House' (AK) method. This method of regulation was introduced by Decree of the Government of the Russian Federation No. 1562, issued on December 15, 2017, which determines the maximum price level for thermal energy (power) in heat supply price zones. It includes the indexation of the maximum price level for thermal energy (power) and the technical and economic operating parameters of boiler houses and heating networks used to calculate the maximum price level [2].

The concept of a new heat market model was initially announced in 2010 and has since undergone significant evolution, taking into account the minimization of all potential risks. This model involves transitioning from state regulation to establishing a maximum price level for thermal energy. Importantly, this maximum level is not determined based on the current costs of the heat energy supplier in the existing centralized heat supply system, but rather through a comparison (benchmarking) with the cost of the most effective alternative heat supply for the consumer, which is the price of an 'alternative boiler house.' The cost of centralized heat supply for the consumer should not exceed the cost of switching to their own sources of heat generation [3]. The methodology and procedure for calculating the maximum level for thermal energy have been approved [2] and are standardized nationwide.

Under the 'alternative boiler house' method, the price of thermal energy may differ from the actual established tariffs, both upwards and downwards. If the cost of the 'alternative boiler house' is lower, the tariffs will be frozen until they reach the price limit. If the cost is higher, a schedule will be developed to gradually bring it in line with the price of the 'alternative boiler house' over a period of 5-10 years.

Heat supply organizations are required to allocate additional funds from the sale of heat energy to measures that enhance the reliability, quality, and energy efficiency of heat supply if the price of the 'alternative boiler house' exceeds the tariffs for heat energy established prior to the transition to price zones.

When transitioning to the price zone method, a single heat supply organization (ETO) is designated for a populated area. This organization is responsible for all aspects of heat supply, including production, transportation, and sales of thermal energy. There may be multiple ETOs in one locality.

The ETO is granted expanded powers to manage the heat supply system, while control, including antimonopoly control, over its activities is strengthened. The ETO has increased responsibility to both consumers and authorities. It also has the opportunity to influence the optimization and development of the heat supply system. Prices for produced thermal energy and transmission services between the ETO and other companies in its area of operation are determined through mutual agreement, ensuring the direct economic interest of all heat supply participants in improving efficiency. The ETO enters into contracts with consumers, and fines, such as significant reductions in heat payments for consumers, are imposed on the ETO for heat supply shutdowns and violations of quality, such as low coolant temperatures [4].

Currently, 38 municipalities in Russia have transitioned to the new model for calculating tariffs, and the practical results demonstrate the effectiveness of the price zone method. Heat supply organizations that have adopted this new method are actively striving to expand its implementation to additional cities. The projected growth of investments under the new tariff calculation method exceeds the financing levels under the 'old' model. While the specifics of implementing the 'alternative boiler house' method may vary across regions, this does not hinder the overall goal of enhancing reliability and efficiency in heat supply.

However, the nationwide transition to the new model is not without its challenges. According to some experts, although the new method has proven to be a viable alternative to the current tariff regulation approach, it should not be considered the sole reference. They argue that adopting the AK method as the exclusive means of heat market regulation is misguided, as it may not be suitable for all municipalities. Moreover, scientists have identified negative consequences resulting from the transition to the new tariff regulation model. These include unjustified tariff increases, which could prompt consumers to abandon centralized heat supply systems, leading to a decrease in the heat load at combined heat and power (CHP) plants. The monopolistic structure of heat supply management does not encourage entrepreneurial initiative or foster competition for consumers. Additionally, higher payments for thermal energy may contribute to an increase in non-payment rates.

In response to these challenges, the Government of the Russian Federation has proposed an extensive infrastructure menu to attract investment aimed at improving the overall reliability of utility infrastructure, including heat supply.

However, it is important to acknowledge that the aforementioned solutions are either financed through budget funds or through an increase in thermal energy tariffs for consumers. Therefore, any decisions made by each heat supply enterprise must be accompanied by an analysis of the targeted use of existing tariff-based financing sources, ensuring that they are directed towards enhancing the reliability of heat supply. This analysis should consider the validity of investment programs and capital and current repair programs, as well as determine the minimum required financing to maintain the heat supply system in a standard condition.

A key task is to assess the reliability of heat supply, which involves evaluating not only current indicators but also future reliability under various scenarios for replacing heating networks. Reference [5] proposes a rapid assessment of the condition of heating networks without decommissioning them. Methods for calculating the reliability of heating networks, principles for generating failure statistics, and the influence of real initial data on the actual reliability indicator are discussed in [6]. Furthermore, [7] presents an assessment of the reliability of heating networks using cluster analysis.

In reference [8], the main features of decision-making regarding the selection of sections of heating networks for major repairs are examined, with a ranking based on various criteria.

Reference [9] provides proposals for monitoring the reliability of heating networks.

Despite the availability of numerous methods for assessing the reliability of heat supply systems, their practical implementation is not mandatory for heat supply organizations.

However, there is a document that must be developed and approved in accordance with current legislation - a heat supply scheme.

The heat supply scheme serves as the primary program document that outlines the long-term development of the heat supply system for settlements, urban districts, and federal cities. It is developed in compliance with the requirements established by [10, 11]. The heat supply scheme is subject to annual updates, and if the master plan is approved, a draft of a new heat supply scheme is formulated.

One of the crucial objectives in developing or updating a heat supply scheme is to ensure reliable heat supply in the most cost-effective manner while minimizing the negative impact on the environment. The heat supply scheme includes a dedicated chapter on assessing the reliability of heat supply to propose the reconstruction of heating networks that fail to meet standard reliability criteria. Reliability calculations are conducted using an electronic model of the heat supply system, ensuring adherence to the calculation requirements specified in [11].

The objectivity of calculating the reliability of the heat supply system relies on accurate and complete information regarding the historical damage records of heating networks and heat supply organizations for a retrospective period of at least 5 years, as well as information on the years of commissioning, reconstruction, and repairs of heating networks.

Accurate reliability calculations not only allow for an assessment of the current state of heating network reliability but also facilitate the development of long-term measures to bring the conditions of heating networks to standard values. Based on the reliability calculations, an objective evaluation of the necessary capital investments in heating networks can be made to ensure high-quality and reliable heat supply. In cases where there is insufficient funding accounted for in heat energy tariffs, the developer of the heat supply scheme should propose options for raising additional funds.

By the end of the long-term planning horizon specified in the heat supply scheme, the implementation of a set of measures should result in reliability indicators that meet standard values.

Unfortunately, in practice, the inclusion of a reliability chapter in the heat supply scheme is often a mere formality and fails to address the aforementioned issues.

The main reasons for the superficial development of the reliability chapter in heat supply schemes include:

- the problem of the lack of complete and reliable information for an objective calculation of the reliability of the heat supply system (for example, some organizations artificially underestimate the actual number of failures in heating networks, which, in turn, leads to erroneous attainment of standard performance indicators for the heat supply system). Additionally, there is a lack of information regarding the years of commissioning of heating networks;
- developers of the heat supply scheme take into account only measures proposed by heat supply organizations and secured by sources of financing, reliability calculations are artificially adjusted to standard values;
- to objectively perform a reliability calculation, significant labor costs are required for analyzing and entering a large array of data into an electronic model; when developing heat supply schemes for small settlements, taking into account the relatively small contract price for developing a heat supply scheme, the developer does not perform a reliability calculation.

The heat supply scheme is the most important tool for implementing state policy aimed at improving the reliability, quality, and energy efficiency of heat supply. Currently, in order to receive budget funding or

https://doi.org/10.1051/e3sconf/202346101018

transition to the 'price zone,' it is necessary to ensure the inclusion of planned activities in the heat supply scheme. However, only the latest update of the scheme is taken into account in this process. When comparing such schemes with previous updates, it becomes evident that the need for additional activities and sources of funding is only reflected in the latest update, while previous versions of the scheme lack this information. Updating heat supply schemes to meet the requirements of a specific project raises doubts about the objectivity of the results obtained.

To fully utilize heat supply schemes as a tool for long-term planning, high-quality reliability calculations, and the development of planned activities, it is proposed to legislate that, in order to justify the receipt of budget funding or the transition to the 'price zone,' the proposed activities must be aligned with at least three updated heat supply schemes.

This will ensure a more thorough development and updating of heat supply schemes and will make it possible to make maximum use of the results obtained in order to increase the reliability of heat supply.

3 Conclusions

1. The reliability of heat supply in the Russian Federation as a whole tends to decline as a result of insufficient replacement of heating networks. The breakdown rate of heating networks is increasing.

2. The Government of the Russian Federation has developed mechanisms for attracting additional investment in the industry.

3. Justification of the required volume of additional investments in the context of each heat supply organization is not objective enough.

4. The role of a tool for assessing the required volume of additional investments in the context of each heat supply organization can and should be performed by heat supply schemes when carrying out reliability calculations and proposing measures.

5. A significant part of heat supply schemes is made with formal, biased reliability calculations.

6. In order to improve the quality of development of heat supply schemes, the objectivity of the reliability calculations performed and the selection of measures, it is proposed to establish legislatively that in order to justify the receipt of budget funding or the transition to the "price zone", it is necessary to coordinate the proposed measures with at least three updates of heat supply schemes.

The research was carried out with the financial support of the Ministry of Science and Higher Education of the Russian Federation within the framework of state assignment No. 075-01262-22-00 dated December 28, 2021 (additional agreement No. 075-01262-22-01 dated January 28, 2022, No. 075-03 -2022-151/1 dated 01/31/2022, No. 075-03-2023-291/1 dated 02/28/2023).

References

1. Report on the state of the heat power industry and centralized heat supply in the Russian Federation in 2020, Ministry of Energy of the Russian Federation, Federal State Budgetary Institution "Russian Energy Agency" of the Ministry of Energy of Russia [Electronic resource]. Access mode: https://minenergo.gov.ru/node/22832.
2. Decree of the Government of the Russian Federation of December 15, 2017 N 1562 "On determining the maximum price level for thermal energy (power) in heat supply price zones, including indexation of the maximum price level for thermal energy (power), and technical and economic parameters of the operation of boiler houses and heating networks used to calculate the maximum price level for thermal energy (power)" [Electronic resource]. Access mode: https://docs.cntd.ru/document/556118651.
3. Bukharov, S.V. "Alternative boiler room" as a mechanism for solving existing heat supply problems, EKO, **№ 3 Vol. 537,** pp. 57-65 (2019)
4. Decree of the Government of the Russian Federation of 08.08.2012 N 808 "On the organization of heat supply in the Russian Federation and on amendments to some acts of the Government of the Russian Federation" [Electronic resource]. Access mode: https://docs.cntd.ru/document/902363976.
5. Chicherin S.V. Methodology for a general assessment of the condition of pipelines in heating networks. In the world of non-destructive testing, **4,** pp. 66-69 (2017)
6. Gorbunova T.G., Vankov Yu.V., Medyakov A.A. Reliability of heating networks: theory and practice. Reliability and safety of energy, **4,** pp. 275-279 (2017)
7. Ilyina M. N., Laryushkin D. V. Improvement of methods for assessing the reliability of heating networks of main pipeline objects. Science and technology of pipeline transport of oil and petroleum products, **2,** pp. 198-202 (2021)
8. Tsuverkalova O. F. Selection of heating network sections for major repairs in the implementation of heat supply schemes for territories. Science of Krasnoyarsk, **3-4,** pp. 173-182 (2020)
9. Sorokin A. M. Monitoring of heat supply systems. Bulletin of the Volgograd State University of Architecture and Civil Engineering. series: construction and architecture, **2, 87,** pp. 90-98 (2022)
10. Decree of the Government of the Russian Federation dated February 22, 2012 No. 154 "On the requirements for heat supply schemes, the procedure for their development and approval" [Electronic resource]. Access mode: https://docs.cntd.ru/document/902332189.
11. Order of the Ministry of Energy of Russia dated 03/05/2019 No. 212 "On approval of Guidelines for the development of heat supply schemes" [Electronic resource]. Access mode: https://docs.cntd.ru/document/553937025.

Methodological issues for the formation of reliable and efficient district-distributed heating systems

Ivan Postnikov[1,*]

[1]Melentiev Energy Systems Institute of Siberian Branch of the Russian Academy of Sciences, Laboratory of Heat Supply Systems, 664033 Irkutsk, Russia

Abstract. The general methodological issues for the formation and operation of district-distributed heating systems are formulated, including determining the areas of economic efficiency of district and distributed (prosumer sectors) heat supply to consumers, reliability analysis and synthesis, as well as economic analysis of failed conditions of the system (determination of damage from undersupply of thermal energy). Methodological provisions for solving these problems are proposed, structured in the form of a general algorithm with connection of problems, methods and results.

1 Introduction

The formation and operation of district heating systems (DHS) in the process of their development involves solving a wide range of scientific and methodological problems. The book [1] provides a fairly extensive methodological basis for solving main issues for existing DHSs. However, at the stage of the current energy transition in heat supply, quite transformations are taking place, considered first of all on the technological level of 4th and 5th generation of DHS [2–5].

One of the general directions of DHS development is associated with the formation of integrated energy systems [6], which involves the effective comprehensive use of various energy technologies (multi-resource systems), and on the other hand, the formation of a fundamentally different concept of development a system based on rational (in theory, optimal) application district heating with active demand for thermal energy. Active consumers or *prosumers* form subsystems (sectors) of distributed heating. Thus, at the present stage of their development, DHS are transformed into *district-distributed heating systems* (DDHS), effectively combining district and distributed generation of thermal energy [7, 8]. The implementation of prosumers with their own heating sources (HS) determines new properties of studied systems which requires the development of methodological support for solving the problems of their functioning and development.

This paper briefly provides number of methodological provisions for solving some problems of formatting and operating a DDHS with prosumers: scaling systems with determining zones of efficiency of district and distributed heating, the reliability analysis both stochastic and economic accounting damages from undersupply of thermal energy, some approaches to ensuring reliability when using prosumer reserves. Brief results of the computational experiment is considered.

2 Methodology

2.1 Scaling DDHS: determining the areas of district heating efficiency

The problem of effective scaling of DHSS (as well as traditional DHS) is to determine the boundaries of district heating efficiency based on the *effective heating radius* (EHR) criterion [9]. Constrains of many methods is the assumption of territorial uniformity of heat load distribution. The main principle of the proposed approach is that for each HS of studied system, the operating costs for the production and distribution of thermal energy are calculated for each section of transmission pipeline to consumers connected to it. The result of the analysis of the entire system is a map of district efficiency areas, which determines the scale of the necessary reconstruction and possible implementation of distributed sectors at these boundaries (prosumer subsystems). Methods and models to solve the problem of effective scaling of DDHS are discussed in detail in [9, 10].

2.2 Stochastic reliability analysis of DDHS based on the nodal indices

A comprehensive approach to the reliability analysis of the DDHS is proposed, which consists of determining the integral impact of all system objects (or subsystems) on the reliability of heating to consumers. The set of problems of reliability analysis and methods for solving them is divided into two main blocks: probabilistic and physical modeling of the functioning of the system. The set of system states is formed as a combination of states of its district and distributed subsystems accounting the technological possibility of their implementation. State probabilities are assessed accounting the specified reliability parameters of components (failure and restoration rates of facilities) based on a *markov random*

*Corresponding author: postnikov@isem.irk.ru

process model. Physical modeling consists of calculating the corresponding thermal-hydraulic conditions in the heat network using models of the theory of hydraulic circuits (THC) [11]. The final result is *nodal reliability indices* (RI), calculated using previously obtained results of probabilistic and physical modeling of states. The methodology for a comprehensive analysis of the reliability of heating systems and its practical applications are described in detail in [12–14].

2.3 Economical reliability analysis of DDHS

The purpose of the economical reliability analysis of DDHS is to assess the emergency conditions while minimizing operating costs accounting *expected damages* (economic losses from possible undersupplies of thermal energy) in the event of failures of system components (facilities). To obtain the parameters of the desired conditions, a methodological approach is proposed, which consists in the joint calculation of nodal prices for thermal energy and damage in the studied system based on the *Lagrange multiplier method*. This method with test calculations in relation to DHS is discussed in detail in [15]. The developed methods and models make it possible, based on the obtained indicators, to identify the most "bottleneck" points in the system, corresponding to the maximum economic losses from interruption of heating to consumers.

2.4 Ensuring the reliability of DDHS accounting the prosumer reserves

Achieving and maintaining a certain standard level of reliability for DDHS is ensured by means of redundancy both in the district and distributed (prosumer) sectors. Based on this, the research problem is formulated as the development of methods for ensuring the reliability of DDHS accounting the additional functional reserve of own prosumer HS (passive time redundancy, active thermal power). The formulation of the scientific and methodological problem consists in determining the optimal relation of reliability parameters of system components (failure and restoration rates) and the capacity of distributed HS of prosumers, which provide the required level of reliability of heating to consumers with minimal total costs for reliability and operation of prosumers within the acceptable parameter values. The study is a development of previous work [16].

3 General algorithm for formation reliable and effective DDHS

The methods discussed above are combined into a general algorithm presented in Fig. 1. The scheme is divided into three categories (problems, methods and results) and four stages corresponding to the previously discussed sections. The first stage (P1) corresponds to the problem of scaling of DHS (sect. 2.1), as a result of which the implementation areas of distributed heating are determined according to the EHR criterion.

In this case, the decision on the localization of the distributed sector is made under a combination of two conditions:

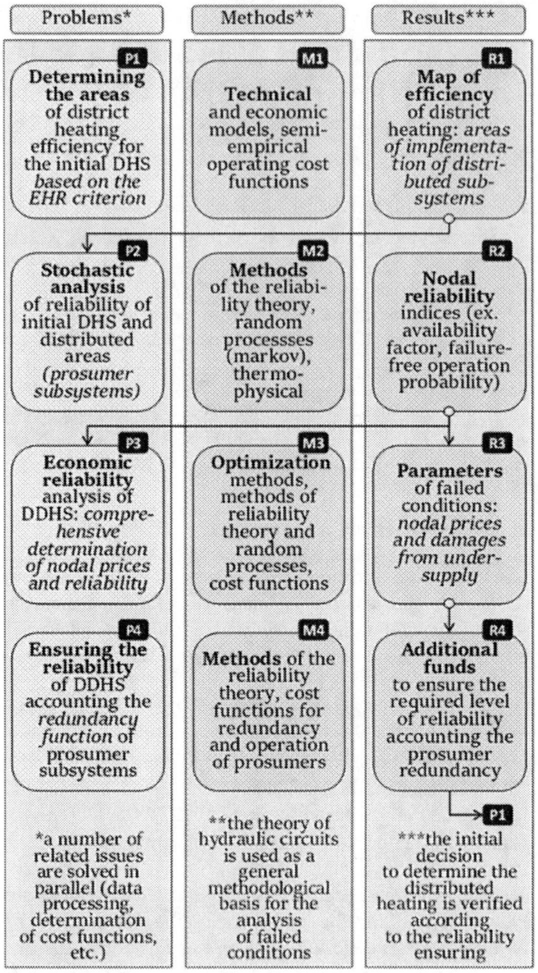

Fig. 1. Shorted diagram of the general algorithm for formation reliable and effective DDHS

1) part of the DHS scheme is located outside the boundaries of the EHR;
2) there is no possibility of structural solutions to expand the HER areas within the existing district system.

For a certain localized distributed sector, a prosumer subsystem is formed with a choice of HS type and heat supply parameters (temperature curve, heating period, etc.), subject to increased economic efficiency within this subsystem compared to the initial connection diagram. This problem is not considered here, but some results of such an analysis are given in [9].

At the second stage (P2), the transition to solving the problem of reliability analysis for the prosumer subsystem is carried out (sect. 2.2). The nodal RI obtained at this stage are used in next two stages. If there are unreliable part an economic assessment of the consequences of failures is carried out (stage P3), as a result of which the economic damages from undersupply of thermal energy are determined (sect. 2.3). To assess their possible reduction, the problem of ensuring (increasing) reliability is solved (sect. 2.4).

Increasing the reliability of the prosumer subsystem can be carried out both through internal redundancy (within the subsystem) and taking into account the possibility of using the reliability potential of the district system. The second option is possible in the case when the prosumer HS operates in parallel with district HS, covering part of the load of its consumer group. At stage P4 (see Fig. 1), the volume of additional funds is determined to ensure the required level of reliability in the considered distributed subsystem. The initial solution along the EHR boundaries is corrected according with the conditions for ensuring reliability (in the case of a partial increase, it is necessary to additionally take into account the expected damage from undersupply).

4 Brief results of computational case example study

Here are the main brief results of the application of the methodological support developed for solving the problems of effective and reliability of DHS accounting their transformation to DDHS with implementation of prosumers. The main issues were to determine areas of efficiency of district heating based on the EHR criterion, technical and economic assessment of the implementation of distributed HS of prosumers at the boundaries of EHR, analysis and ensuring the reliability.

For the studied DHS, the boundaries of the EHR were assessed in accordance with the given standard value of specific operating costs. It was determined that beyond the boundaries of the efficiency of district heating there are 14 branches (sectors) of the network with a total length of 14.8 km (38%) and 42 (46%) nodes of the initial system diagram with a total heat load of 357 GJ/h (39 %).

Next, to evaluate the effectiveness of implementing a prosumer with its own HS (distributed sector), one of the subsystem is selected, remaining actually connected to the district system, while having its own generation. Calculation of operating costs within the studied prosumer subsystem showed that specific cost at nodes within the subsystem is significantly reduced compared to the initial heating scheme from a district HS: the average cost reduction was 25%. The reliability level of the localized system when connecting a prosumer also increases significantly compared to the initial scheme: the increase is 44%. The distribution of the reliability effect on subsystem nodes is shown in Fig. 2. The total economic effect of introducing the prosumer subsystem is shown in Fig. 3 and is estimated at 93.85 million rub/year, which is 48.1% of the operating costs corresponding to the initial scheme. The main part of the effect is associated with a reduction in the cost of thermal energy and amounts to 87.53 million rub/year (44.8%) and only 6.32 million rub/year (3.2%) with a reduction in expected damage from undersupply thermal energy to consumers.

It is also important to note here that the effect is achieved with optimal control of the joint functioning of district and distributed sources. This problem represents the subject of a special study using bi-level programming methods (one level corresponds to the district system, the other to the distributed heating subsystem).

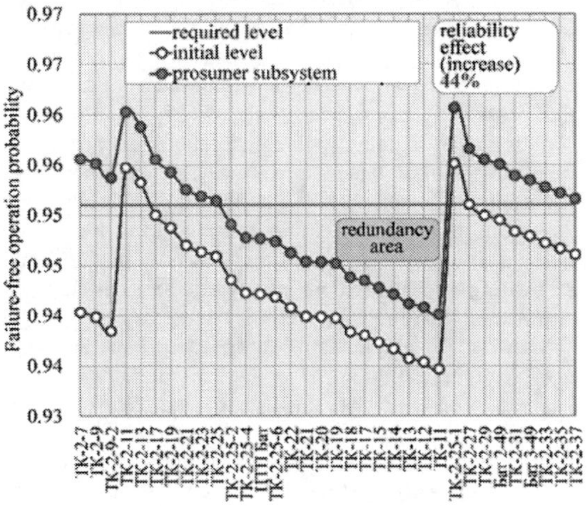

Fig. 2. Distribution of reliability effect on nodes of the prosumer subsystem (distributed sector in DHS)

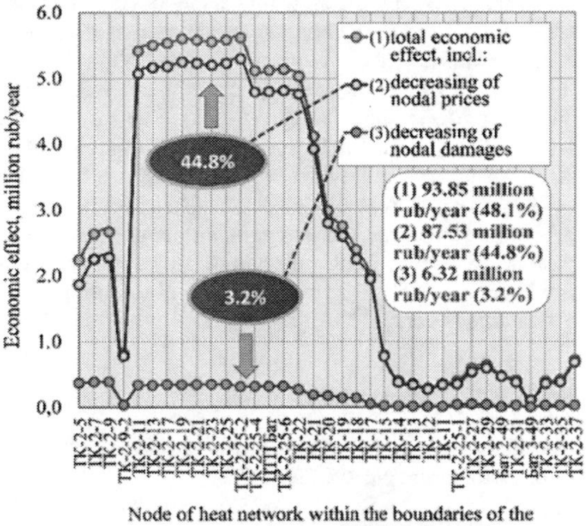

Fig. 3. Distribution of economic effect on nodes of the prosumer subsystem (distributed sector in DHS)

5 Conclusions

A possible version of the methodological basis is proposed to justify decisions on the transition of existing DHS to systems of a district-distributed type with the implementation of prosumers as sources of localized subsystems formed on the boundaries of the technical and economic efficiency of district heating. A number of methodological provisions are presented aimed at solving the problems of substantiating the effectiveness and reliability of decisions made. It is important to note that in addition to the considered issues, there are a number of related and quite complex problems associated with obtaining and analytical processing of data, allowing them to be used as objective functions (mainly, operating costs). As a result of the decision, DDHS should be formed with a higher level of efficiency and reliability of heating to consumers compared to the initial scheme. An important property of the proposed approaches is the ability to adjust the economic efficiency of the system according to reliability conditions accounting the technical and economic consequences of possible heat shortages and the creation of the necessary reserves to reduce them. The resulting solutions can be useful both in the design of heating systems and in justifying their development accounting the reliability factor.

Further development of the proposed methodology involves the solution of many methodological and applied problems. At this stage, we focus on the following priority direction: determination of the feedback for decisions obtained at different levels of the district and distributed heating system. A change in the specific cost of production and distribution of heat in a localized system of a prosumer (or several prosumers at once) will affect the value of the system-wide criterion for determining the EHR boundaries. In this case, the more load will be transferred to the distributed heating part, the more significant will be the deviation of the EHR from its initial value. According to the adjusted criterion, in turn, the previously obtained boundaries of the efficiency of district heating should be revised. When adjusting these boundaries, a reassessment of the technical and economic indices of the distributed sectors will also be required, which will again update the value of the EHR. As a result, to obtain the optimal district-distributed structure of the system, an iterative procedure with a "floating" EHR criterion is required, which is implemented up to some equilibrium solution with acceptable accuracy.

Acknowledgements

The study was supported by the Russian Science Foundation grant No. 22-29-01252, https://rscf.ru/project/22-29-01252/.

References

1. E. Sennova, V. Sidler, *Mathematical modeling and optimization of developing heat supply systems* (Nauka, Novosibirsk, 1987) [in Russian]
2. N. Novitskii, Z. Shalaginova, A. Alekseev, et al., Thermal Engineering **69** (2022)
3. H. Lund, P. Østergaard, M. Chang, et al., Energy **164** (2018)
4. I. Pakere, A. Gravelsins, D. Lauka, et al., Energy **234** (2021)
5. A. Revesz, P. Jones, C. Dunham, et al., Energy **201** (2020)
6. N. Voropai, A. Makarov, eds., *Systems studies in energy: energy transition* (ESI SB RAS, Irkutsk, 2022) [in Russian]
7. L. Brange, J. Englund, P. Lauenburg, Applied Energy **164** (2016)
8. M. Pipiciello, M. Caldera, M. Cozzini, et al., Energy **223** (2021)
9. V. Stennikov, E. Iakimetc, Energy **110** (2016)
10. V. Stennikov, E. Mednikova, I. Postnikov and A. Penkovskii, Environmental and Climate Technologies **23(2)** (2019)
11. A. Merenkov, V. Khasilev, *The theory of hydraulic circuits* (Nauka, Moscow, 1985) [in Russian]
12. V. Stennikov, I. Postnikov, Power Technology and Engineering **47(6)** (2014)
13. I. Postnikov, V. Stennikov, E. Mednikova and A. Penkovskii, Applied Energy **227** (2018)
14. I. Postnikov, Environmental and Climate Technologies **24(3)** (2020)
15. I. Postnikov, A. Penkovskii, Energy Reports **8(13)** (2022)
16. I. Postnikov, Energy Reports **9(1)** (2023)

E3S Web of Conferences **461**, 01020 (2023)
RSES 2023
https://doi.org/10.1051/e3sconf/202346101020

Forecasting changes in the parameters of the failure flow in heat networks at different temperature graphs

Yurii Vankov[1], *Pavel* Kuznetsov[1], *Evgeniya* Izmajlova[1]*, *Elena* Garnyshova[1], *Radik* Valiev[1]

[1] Kazan State Power Engineering University, 51 Krasnoselskaya str., Kazan, Russia

Abstract. The paper considers the transition from the existing temperature graph to a temperature graph with reduced coolant parameters in the city's district heating system and the results of calculating the predictive values of the failure flow parameter in heat networks.

1 Introduction

The stages of development of heat supply systems can be divided into four generations. Generations differ in the parameters of the coolant, sources of thermal energy, control systems, use of non-traditional and renewable energy sources [1]. To date, most countries have heat supply systems of the 2nd 3rd generation. A number of European countries have moved to the 4th generation, and development strategies are being adopted for the transition of their heat supply systems to the fifth. An example is the city of Copenhagen, the capital of the Kingdom of Denmark, in which the heat supply system operates in combination with cooling stations, waste incineration plants, thermal power plants and peak boilers [2].

Heat supply systems of older generations differ from the previous ones in low coolant temperatures, which leads to an increase in the energy efficiency of systems through the use of non-traditional heat sources and a reduction in losses during its transportation.

In the Russian Federation (RF), one of the factors affecting the magnitude of heat losses in networks is the physical deterioration of pipelines (corrosive wear of pipeline metal) [3-5], which results in an increase in the number of accidents associated with the loss of thermal energy with the coolant (leaks in emergency sections, drains coolant during emergency restoration work) [6].

The main reason for the high wear and tear of heating networks is the insufficient volume of pipelines [7]. Thus, the replacement of heating networks in 2020 amounted to 3.371 thousand km in Russia as a whole, which is 6.7% of the required volume of replacement of networks [8].

Over the past 5 years, the situation with the transfer has not fundamentally changed: on average, 3.3 thousand km of heating networks are transferred to the Russian Federation per year, and this volume is insufficient, because. there is a steady increase in the length of pipelines in need of replacement and, as a result, an increase in the number of accidents and the magnitude of heat losses (Fig. 1) [9].

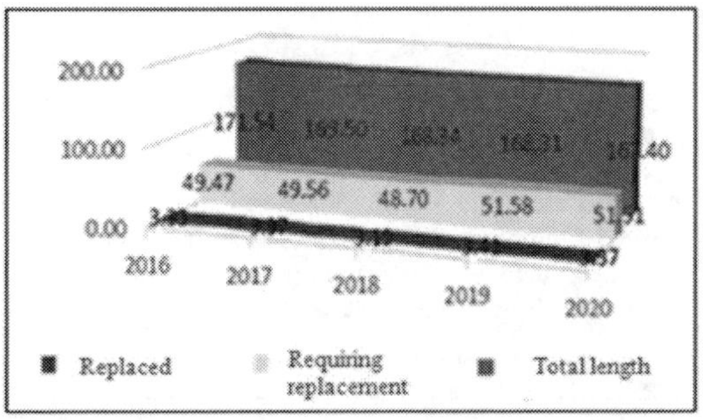

Fig. 1. The ratio of the total length of heating networks with the length of networks in need of replacement and replaced in the Russian Federation in 2016-2020, thousand km.

In the near future, the replacement of all dilapidated pipelines cannot be carried out for objective reasons [10]. A decrease in the temperature of the coolant will lead to a decrease in losses through insulation and a decrease in failure flows due to the fact that heating networks will switch to a more gentle operation mode.

The article deals with the reliability of the heat supply system of a settlement with a population of 65,000 people when it is transferred to a lower temperature schedule.

The design temperature of the outside air for the design of heating systems is assumed to be minus 33 °C. The duration of the heating period is 209 days.

2 State of the research question

2.1 Structure and characteristics of heat networks

The total length of heating networks is 242.3 km in one-pipe terms, including 55.3 km of main lines, 142.4 km of quarterly networks, and 44.6 km of hot water supply networks after heating points. The material characteristic of heat networks is 43.944 m^2. The

* Corresponding author: evgeniya-izmailova@yandex.ru

length is dominated by pipelines with diameters of 50, 100 and 150 mm.

The output from the heat source is carried out according to the scheme with a common collector by means of a branch to two heating mains. Return network water is returned to the boiler shop according to a similar scheme. Consumers are connected to the centralized heating system through their own heating points and heating units, and hot water supply is being prepared in central heating points. A four-pipe heating network has been laid from the central heating point to consumers [11-12].

The regulation of the supply from the source of thermal energy is carried out by a qualitative method. The heat carrier parameters set during the design of the heating network corresponded to the values of 150/70°C at the calculated outdoor air temperature. Heating networks operate according to a temperature graph of 150/70 °C with a cutoff at 130 °C and a lower threshold of 70 °C, to ensure the load on the needs of hot water supply [13].

2.2 Wear of the walls of pipelines of heating networks

The corrosion rate was determined using data from the analysis of corrosion indicators installed annually on the inner surface of pipelines [14]. The measurement results are shown in the Table 1.

Table 1. Rate of internal corrosion, mm/year.

Installation location	2019	2020	2021
TM-1, TK-119 (supply)	0.0743	0.0747	0.0607
TM-1, TK-119 (return)	0.0758	0.0783	0.0667
TM-2, TK-2405 (supply)	0.0748	0.0705	0.0611
TM-2, TK-2405 (return)	0.0791	0.0694	0.0625
TM-5, TK-530 (supply)	0.0700	0.0671	0.0591
TM-5, TK-530 (return)	0.0743	0.0690	0.0626
TM-8, TK-822 (supply)	0.0746	0.0684	0.0618
TM-8, TK-822 (return)	0.0723	0.0798	0.0637
TM-12, TK-12004 (supply)	0.0786	0.0753	0.0643
TM-12, TK-12004 (return)	0.0757	0.0767	0.0581

To assess the residual resource of sections of the heat network, the results of the presented measurements are extended to all types of heat pipelines of heat networks.

The measurement results were processed in accordance with GOST R 8.736 2011 [15]. The verification of the hypothesis that the measurement results belong to the normal distribution was carried out with a significance level of 5%.

There are no actual data on external corrosion, so the corrosion rate was taken depending on the type of gasket in the range from 0.01 to 0.03 mm/year.

2.3 Damage (failures) in sections of heating networks

Data on failures of each group of pipeline diameters are given in the Table 2.

Table 2. Failures in the heating networks of the city for 2015-2021

Diameter of the section of the heating network, mm	Average service life, years	Length (in one-pipe terms), m	Number of failures in 7 years	Failure rate, 1/year	Reduced failure rate, 1/km/year
700	39.8	1 778	0	0.0	0.00
600	27.7	3 897	6	0.9	0.22
500	26.2	10 760	19	2.7	0.25
400	23.6	12 397	11	1.6	0.13
350	43.9	7 215	6	0.9	0.12
300	37.3	8 872	9	1.3	0.14
250	33.3	8 963	27	3.9	0.43
200	30.7	12 303	25	3.6	0.29
150	33.0	33 770	55	7.9	0.23
100	34.5	42 233	88	12.6	0.30
80	29.0	12 548	26	3.7	0.30
70	32.8	13 267	22	3.1	0.24
50	36.7	17 057	32	4.6	0.27
Total		**185 060**	**326**	**46.6**	**0.25**

For 7 years of observations during the operation period, 326 failures were registered. Most of the failures were recorded on heat pipes with diameters of 250 mm or less [16].

The weighted average service life of heat pipelines, the reduced failure rate (the reduced damage rate) of the heating network of the central heating systems of the city was calculated using the methodology and formulas given in [17]. The failure rate for each group of sections of heat networks is shown in Fig. 2. The failure rate for all diameters exceeds the frequency of stable failures equal to 0.05 1/year/km, which characterizes the city's heat supply system as medium reliable, in which the effect of aging (degradation) is manifested.

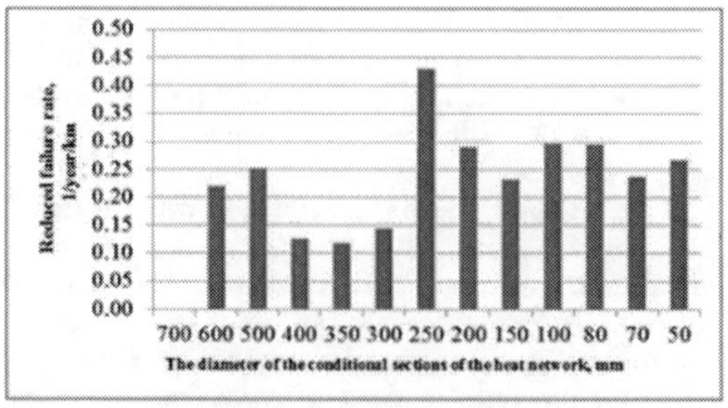

Fig. 2. Reduced failure rate of sections of the heat network of district heating systems.

3 Technical solution for the reconstruction of the heat supply system

The transition from a dependent to an independent connection scheme for all consumers in the city's centralized heat supply system remains unattainable in terms of the amount of required investments, sources and the timing of their return.

It is proposed to replace the existing dependent circuit for connecting consumer heating systems using elevator installations with a dependent circuit using corrective pumps [18]. The correction pump is installed parallel to the mixing line of the elevator and allows you to change the overall mixing ratio for the heat carrier supplied for heating from the heating network. The scheme for connecting the heating system at the subscriber input with a correction pump is shown in Fig. 3.

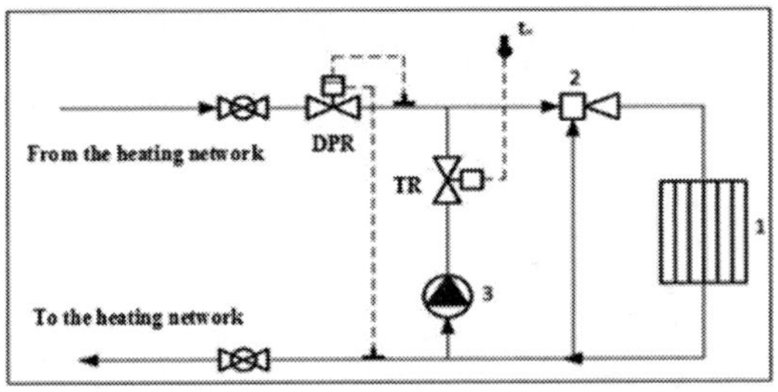

Fig. 3. Scheme of connection of the heating system with a corrective pump: 1 – heating system; 2 – elevator; 3 – corrective pump; DPR – differential pressure regulator; TR – temperature regulator.

The calculation of temperature graphs is made taking into account the implementation of a dependent scheme for connecting consumers using a corrective pump for a model heat supply system with a heat load of the heating system equal to 1 Gcal/h. It is assumed that the central regulation of heat supply to heating networks is carried out in accordance with the graph for heating heat load.

For completeness of the comparison conditions, in addition to the initial and reduced temperature graphs, a graph of 150/70 °C was also considered with the installation of corrective pumps:

- temperature graph 150/70 °C under initial conditions for connecting heating systems;
- temperature graph 150/70 °C with correction pump;
- temperature graph 130/70 °C with correction pump;
- temperature graph 110/70 °C with correction pump;
- temperature graph 95/70 °C with correction pump;

The conditions adopted in the calculation of indicators are given in the Table 3.

To compare regime indicators, calculations of temperature graphs, graphs of water flow from the heating network and through corrective pumps were carried out, and the dependence of heat supply on the outside air temperature was determined.

Table 3. Calculation conditions for various temperature graphs.

Temperature graph, °C	Estimated water consumption, t/h	Elevator mixing ratio
150/70 original	12.5	2.2
150/70	12.5	2.2
130/70	16.7	1.4
110/70	25.0	0.6
95/70	40.0	0.0

With a decrease in temperature graphs, the average annual average temperature in the supply and return pipelines decreases.

The average values of coolant flow rates increase with a decrease in the temperature graph, but this increase is significantly less than with a direct decrease in the temperature graph without the use of corrective pumps. Heat supply to consumers is the same for all temperature graphs. This is due to the operation of corrective pumps and the corresponding scheme of their automation.

4 Processing of data on failures in thermal networks

Data on failures in heat networks are presented only for the last seven years (from 2015 to 2021) and are not complete. The sample of data on failures in thermal networks refers to the samples "truncated on the left" and "censored on the right". The truncation on the left is formed due to the loss of data on failures of heat networks from the beginning of the service life until the start of observations. Censoring on the right is formed by taking into account sections of heating networks, the monitoring of which has been terminated without registering a failure.

To process data of this kind in models of lifetimes with the Weibull-Gnedenko distribution, the method of proportional intensities of Cox was used [19-20].

Evaluation of the impact of the transition to lower temperature graphs on the failure rate, the cost of emergency and major repairs is possible only in a comparable form. A comparable form is achieved by the estimated function of forecasting the number of failures per year for the entire complex of heating networks.

When switching to lower temperature graphs, heating networks need to be re-routed with an increase in their diameter in order to ensure standard pressure drops at subscriber inputs. Accounting for these shifts leads to an increase in the material characteristics of heat networks while maintaining its length unchanged and slightly affects the frequency of failures in them.

Data on the transfer of heat networks are given in the Table 4.

Table 4. The length of the transfers (km) of heating networks, to ensure the normative hydraulic regime of subscribers during the transition to a lower temperature graph.

Nominal diameter, mm	Temperature graph, °C				
	150/70	130/70	115/70	110/70	95/70
1000	0	0.00	0.00	0.00	0.50
800	0	0.00	0.00	0.00	0.01
700	0	0.00	0.00	0.00	2.49
600	0	0.00	1.34	1.88	2.28
500	0	0.00	0.26	0.36	1.85
450	0	0.00	0.00	0.00	0.94
400	0	0.00	0.26	0.26	0.39
350	0	0.00	0.00	0.09	0.56
300	0	0.00	0.01	0.10	0.81
250	0	0.02	0.12	0.59	2.48
200	0	0.40	1.75	1.94	5.26
150	0	0.27	0.86	1.31	4.68
125	0	0.58	2.03	2.25	3.76
100	0	0.32	0.92	0.87	3.17
80	0	0.14	0.33	0.62	1.52
70	0	0.27	0.56	0.58	1.53
40	0	0.00	0.00	0.00	0.00
50	0	0.00	0.23	0.23	0.07
Total, km	**0**	**2.01**	**8.68**	**11.07**	**32.30**

They characterize the length of the heating network transfers over existing sections with an increase in diameter.

For emergency repairs, it is assumed that all elements of the heat pipeline that have lost density during operation and hydraulic tests are replaced with new elements. It is assumed that the element is a section of the heat pipe with a length of 10 meters. For major repairs, it is assumed that the replacement of heating networks is carried out during the period of hydraulic tests.

Figure 4 shows the results of calculations of the predicted values of the failure rate parameter in the city's district heating networks for various temperature graphs.

An estimate of the costs of eliminating emerging failures was made based on the calculated values of the failure flow in heat networks (Fig. 4) at various calculated temperatures of the coolant in the supply heat pipeline of heat networks (temperature graphs) for all gradations of diameters of the heat pipelines of the existing heat network and data on the cost of eliminating these failures. The values of these costs for each temperature graph are taken into account in the financial and economic model.

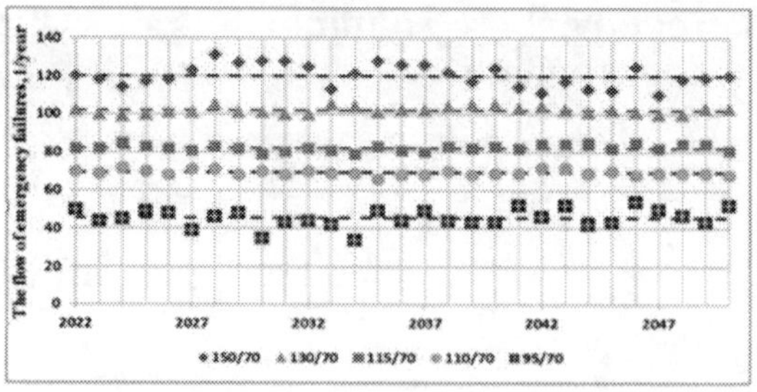

Fig. 4. Forecast values of failure flow parameters in city heat networks at different temperature graphs.

On Fig. 5. predicted values of heating network relocations are presented for various temperature graphs.

Based on the calculated values of the flow of overhauls in heat networks at various calculated temperatures of the coolant in the supply heat pipeline of heat networks (temperature graphs) for all gradations of diameters of heat pipelines of the existing heat network and the data of the approved values of the aggregated norms for the prices of construction of heat networks, an estimate was made of the costs of overhauls.

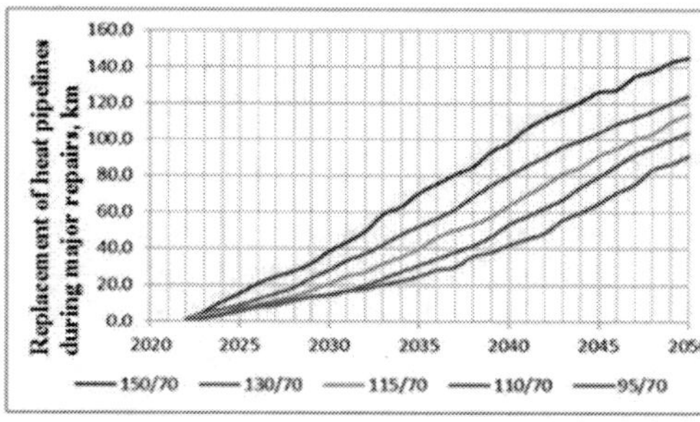

Fig. 5. Forecast values of heating network relocations as a cumulative total for various temperature graphs.

5 Conclusion

Based on the results of the work carried out, it was determined that the current temperature graph for the release of heat from the heat source is not optimal. EBITDA is taken as the target optimization function. To determine the value of EBITDA at different temperature graphs, a tariff-balance model was used, which makes it possible to predict the structure of heat supply costs, taking into account changes in the fuel and energy balance, operating and capital costs.

When determining the long-term price consequences and bringing capital investments in the implementation of projects to the prices of the corresponding years, macroeconomic parameters established by the Ministry of Economic Development of Russia were used:

- forecast of the social and economic development of the Russian Federation for 2022 and for the planned period of 2023 and 2024. (published November 9, 2021);
- forecast of socio-economic development of the Russian Federation for the period up to 2036

(published by the Ministry of Economic Development of the Russian Federation on November 28, 2018).

Assumptions adopted for the formation of the return on capital expenditures: capital expenditures are carried out at the expense of debt resources, return in annuity payments within 10 years, WACC = 12.5%.

It has been established that the maximum cost reduction is possible with the transition to a temperature graph of 130/70. When switching to lower temperature graphs, a significant increase in capital investments will be required, which increases the payback period and the burden on the end user in the form of an increase in payment tariffs.

Ultimately, when switching to the optimal temperature graphs for heat supply from a heat source, in the long term (a payback period of 10 years was considered), the purchase tariff will decrease below the level of 2022, which will make it possible to invest additional funds in the development of the city's heat supply system, replacing the heat exchanger equipment at the central heating point for the needs of hot water supply, as well as the use of more modern methods of control and automation.

The study was carried out within the framework of the state task of the Ministry of Science and Higher Education of the Russian Federation № 075 03 2023 91 dated January 16, 2023.

References

1. Enkhzharkhal H, Namkhaynam B., Batmand L., Bayarsaikhan G., Lkhamsuren Ts. Trends in the development of heat supply systems. Heat Supply News № 232 (01, 2023). pp. 16–25.
2. https://www.danfoss.com/en/markets/district-energy/
3. Vankov Yu.V, Ziganshin Sh.G., Izmailova E.V, Serov V.V. Determination of the oscillation frequencies of corrosion defects finite element methods in order to develop methods of acoustic monitoring of pipelines / IOP Conf. Series: Material Science and Engineering 86 (2015) 012035. doi: 10.1088/1757-899X/86/1/012035.
4. Elizarov S., Bardakov V., Shimanskiy A., Alyakritskiy A., Terentyev D., Barat V., Gogin A., Koltsov V. UNISCOPE: Instrument Integrating NDT Methods // Springer Proceedings in Physics. 2020. № 218. pp. 65–74.
5. Lim M.K.; Cao H. Combining multiple NDT methods to improve testing effectiveness. Constr. Build. Mater. 2013, 38, pp. 1310–1315.
6. D'orazio, T.; Leo, M.; Distante, A.; Guaragnella, C.; Pianese, V.; Cavaccini, G. Automatic ultrasonic inspection for internal defect detection in composite materials. NDT E Int. 2008, 41, pp. 145–154.
7. Broberg P., Sjödahl M., Runnemalm A. Comparison of NDT – methods for automatic inspection of weld defects. Int. J. Mater. Prod. Technol. 2015, 50, pp. 1–21.
8. Vanaei H.A., Eslami A., Egbewande A. A review on pipeline corrosion, in-line inspection (ILI), and corrosion growth rate models. Int. J. Press. Vessel. Pip. 2017, 149, pp. 43–54.
9. On the state of thermal power engineering and district heating in the Russian Federation in 2020. News of heat supply № 231(03, 2022). pp. 12–21.
10. Vankov Yu.V., Izmajlova E.V., Garnyshova E.V., Zagretdinov A.R. Monograph: Improving energy efficiency by controlling pipeline systems. – Kazan: Kazan University Press, 2022, 142 p.
11. Biserova K.A. Analysis of the main problems of modern district heating networks // Modern technologies in construction. Theory and practice. 2021. T. 1. pp. 294–296.
12. Rudig Wolfgang. Combined heat and power for districtheating. Phis. Technol. 1986. № 3.
13. Vankov Yu.V., Zapolskaya I.N., Gaponenko S.O., Muhametova L.R. Improving the reliability of heat energy transportation to consumers in the conditions of modernization of the hot water supply system // Bulletin of the Kazan State Power Engineering University. 2020. vol. 12. № 4 (48). pp. 29–37.
14. RD 153 34.1 17.465 00. Guidelines for assessing the intensity of internal corrosion processes in heat networks. JSC "VTI", 2001, 12 p.
15. GOST R 8.736 2011. Multiple direct measurements. Methods for processing measurement results. Basic provisions.
16. Coramik M., Ege Y. Discontinuity inspection in pipelines: A comparison review. Measurement 2017, 111, pp. 359–373.
17. RD 50 690 89. Methods for assessing reliability indicators based on experimental data. Methodical instructions. Reliability in technology. M.: Rosstandart, 1990, 133 p.
18. Baibakov S.A., Filatov K.V. On the possibilities of regulating the elevator units of heating systems. Heat Supply News № 7 (119), 2010.
19. Chimitova E.V. Mitrofanov I.M. Construction of probabilistic survival models based on data truncated on the left and censored on the right. Bulletin of ASTU. Ser.: Management, computer technology and informatics. 2016. № 1, pp.50-61. ISSN 2072 9502.
20. Cox D.R. Regression models and life tables (with Discussion), Journal of the Royal Statistical Society. Ser. B, 34 (1972), pp. 187–220.

Transformation of corporate governance in the face of modern challenges

Galina Sheveleva[1]*

[1]Melentiev Energy Systems Institute, Siberian Branch of the Russian Academy of Sciences, Irkutsk, Russia

Abstract. Russian generating companies in the power industry require a certain transformation of corporate governance to reduce the investment risks associated with today's challenges in order to increase investment attractiveness. For this purpose, the paper examines the rotation of ownership in these companies, the updated methodology of the Bank of Russia for assessing corporate governance, including in the context of the ESG agenda, HR strategy and growing digitalization. The results of a general assessment of the corporate governance of generating companies are presented; an additional analysis of the compliance of companies with the criteria for each of the sections of the Russian Corporate Governance Code was carried out; a comparative analysis of the structural characteristics of the boards of directors of Russian generating companies and foreign companies that are similar in terms of concentration of ownership (Germany) and the presence of the state in it (France) was carried out; the article presents the results of the assessment of internal documents. The generalization of the obtained data makes it possible to formulate proposals for further transformations of corporate governance in generating companies in the face of modern challenges, including new corporate governance standards with a certain set of additional compliance criteria, reorganization of boards of directors by updating the structural characteristics of the board, as well as the composition and content of internal documents.

Introduction

The problems of attracting investments in Russian electric power generating companies, which have become aggravated under the conditions of anti-Russian sanctions and the lack of alternatives to the domestic borrowing market, have not been resolved.

The purpose of the study is to search for further transformations of corporate governance in these companies in the interests of investors, taking into account modern challenges.

Mainly, wholesale and territorial generating companies of the electric power industry (hereinafter referred to as generating companies) were studied. The corporate governance assessment was performed for generating companies that published in their annual reports or reports on sustainable development on their official websites or the Interfax website reports on compliance with the principles and recommendations of the Corporate Governance Code (hereinafter referred to as the principles of the Code) [1]. This Code is based on OECD standards [2].

Publicly available information was used: reporting data and general corporate governance practices of these companies; Corporate Governance Code; Unified State Register of Legal Entities; information from the Bank of Russia; materials of information agencies Bigelectricpowernews and RosBusinessConsulting; business newspaper Kommersant; international company Spencer Stuart; National Council on Corporate Governance; French AFEP-MEDEF Code; JPMorgan Chase & Co. ESG reports; LECAP laboratories reports; authoritative ESG ratings (MSCI ESG rating, RAEX-Europe ESG ranking, ACRA); draft model methodology for ESG ratings of the Bank of Russia; Russian Institute of Directors, publications of domestic and foreign authors.

The rotation of ownership in generating companies at the end of 2022 and the accompanying transformation of corporate governance have been determined. The features of the updated methodology of the Bank of Russia for assessing corporate governance have been identified [3]. On its basis, the corporate practice of generating companies was assessed, which showed a relative deterioration, especially in section II of the Code (Board of Directors). A comparative analysis of the boards of directors of these companies and companies similar in terms of concentration of ownership (Germany) and the prevalence of the state in it (France) was carried out. By summarizing all the data obtained, the transformations of corporate governance are indicated. The transformations mainly concerning in updating corporate governance standards and criteria for assessing compliance with these standards, as well as structural reorganization of boards of directors, and updating internal documents.

* Corresponding author: sheveleva@isem.irk.ru

1 Property rotation

One of the current challenges is the departure of foreign owners and the accompanying rotation of the ownership of generating companies.

In 2022, the Italian Enel sold the shares of Enel Russia PJSC to Lukoil PJSC (26.9%) and the investment fund Gazprombank-Frezia (29.5%). The name of Enel Russia PJSC was changed to EL5-Energo PJSC. In 2023, 56.4% of the shares of EL5-Energo PJSC were transferred to Lukoil PJSC, 5.5% of this company was acquired by LLC PPIT-7 [4]. Finnish Fortum and German concern Uniper were not allowed to withdraw from Russian generating assets. However, Uniper, which owns Unipro, deconsolidated the assets of this company in 2022 and classified its activities as "discontinued" [5]. Finnish Fortum also excluded the Russian company Fortum PJSC from its new strategy [6]. The Russian assets of these two foreign companies were transferred under the external management of the Federal Agency for State Property Management. It is possible that the assets will be transferred to Rosneft. The appointment of their new general directors, who came from Rosneft, is evidence in favor of this possibility [7].

The main shareholder of Quadra PJSC changed in 2022. 82.47% of its shares were bought out by JSC Rusatom Infrastructure Solutions (a division of the state corporation Rosatom) [8]. Taking into account further transactions, this share increased to 95.02% [9]. 92.2% of the shares of TGC-14 PJSC was acquired by Far Eastern Management Company JSC [10]. The main shareholder of this company is the former top manager of Russian Railways K. Lyulchev. The shares of TGC-2 PJSC owned by its offshore companies, as well as Kores Invest LLC and Debt Agency LLC, were transferred to the trust management of SOVLINK LLC (57.8%). According to Rescore, its owner is a nominal holder [11]. En+Group O. Deripaska maximized his participation in EuroSibEnergo JSC (formerly Irkutskenergo JSC). All generating assets of SIBEKO JSC (formerly Novosibirskenergo JSC) were transferred to the Siberian Generating Company (A. Melnichenko). One of the current challenges is the departure of foreign owners and the accompanying rotation of the ownership of generating companies.

With an increase in of the state presence, large private owners and business groups among the owners of generating companies, a certain transformation of corporate governance and the associated investment risks are inevitable.

1.1 Risks of ownership rotation

When the state's presence expands, the number of former government employees on boards of directors and CEOs tends to increase. To vote on the most important issues, members of the board of directors receive certain directives "from above". Their responsibility (including independent directors) to shareholders and the company as a whole is reduced [12]. There are increasing risks of government promotion of social and strategic programs at the expense of shareholder value of companies. Most likely, this becomes possible due to the lack of a clear separation of the functions of the state as a regulator and as a shareholder. In the role of a regulator of corporate relations, the state has achieved some success in recent years, however, as an owner, it exercises only its own rights, ignoring the interests of other shareholders [13].

In the case the company is owned by large private owners, it was noted that they can usurp the powers delegated "de jure" to the boards of directors. In such companies, the majority shareholders, as a rule, are closely connected with the top management, and the boards of directors turn into formal participants in corporate relations, often into mere observers [13-14]. Related party transactions, expropriation of minority shareholders, overpayments for resources or asset purchases, underpriced products or asset sales, extension of loans to affiliates, or excessive compensation to management are not uncommon. Indirect evidence of this is a formal attitude to compliance with the principles of the Code [15] and a relatively low market value [16]. For large private owners or business groups, electricity generation, as a rule, is not a core activity, but performs the function of supplying the main industries under their control, where "points of profit" are concentrated. Investors in generating companies are at risk of missing dividends and low share prices.

The ownership structure of generating companies, in our opinion, should be included in the methodology of the Bank of Russia as a separate corporate governance standard, and not as an "inconspicuous" criterion of the Code in its Section IV [17-18]. Moreover, this is accepted by the international and national ESG agenda, authoritative rating agencies (MSCI ESG Rating, RAEX Europe ESG, and ACRA) along with the allocation of a separate standard "implementation of the principles of the Code" in the assessment of corporate governance [19-21].

2 Assessment of corporate governance

Certain steps to reduce investment risks in the face of modern challenges have been taken by the Bank of Russia. In part, they are defined in the updated (2021) methodology of the Bank of Russia (hereinafter referred to as the updated methodology) for assessing corporate governance.

2.1 Updated methodology of the Bank of Russia

The updated methodology retained 7 sections of the Code of the A part: I Shareholder rights and equality of conditions for shareholders exercising their rights; II. Board of directors of the company; III. Corporate secretary of the company; IV. System of remuneration due to members of the board of directors, the executive bodies, and other key managers of the company; V. Risk management and internal control system; VI. Disclosure of information about the company and its information policy; VII. Material corporate actions. The number (24) of subsections of the part B of the Code has not changed ("1.1." etc.). 79 principles have been preserved under three-digit numbers ("1.1.1." etc.). The status of the criteria for assessing compliance with the principles of

https://doi.org/10.1051/e3sconf/202346101021

the Code has remained ("complied with", "partially complied with", and "not complied with").

The distinctive features of the updated methodology are listed below.

• Increased the number of Code criteria (135 instead of 128).
• The criteria are more clearly articulated.
• Redistributed differently between principles.
• Part of the criteria has been excluded.
• Additional criteria have been introduced (from the Code, which were not included in the previous methodology of the Bank of Russia).
• New ones were formulated (they were not in the Code, nor in the previous methodology).
• Mandatory fixing of these criteria in the internal documents of companies has been introduced.

The criteria of the updated methodology Code were identified, reflecting modern challenges related to the ESG agenda (5.2.2., 6.2.2., and 6.1.1.), new technologies (1.1.5., 2.7.2.), and HR strategy (2.1.1., 2.3.1., 2.3.3., 2.9.1.).

Unfortunately, the updated methodology did not include the criteria identified during the additional analysis of the Code: an Internet forum was organized to collect opinions on candidates for the Board of Directors; a preliminary discussion by the shareholders of these candidates was held; information about the persons who nominated them was disclosed, as well as the nature of the relationship of candidates with affiliated persons of the company.

2.2 Implementing the principles of the Code

The assessment of compliance with the Code criteria by generating companies was carried out mainly on the basis of data from their annual reports "Report on compliance with the principles and recommendations of the Bank of Russia Corporate Governance Code" for 2021 using the updated methodology of the Bank of Russia.

Based on the results of the assessment, clear leaders were identified, which are companies with state participation - Inter RAO PJSC and RusHydro PJSC, as well as with foreign owners - the Italian Enel Russia PJSC and the German Unipro PJSC. Of the 135 criteria of the Code, they met 85.9%, 97.0%, 88.9% and 85.9%, respectively. Among the outsiders - TGC-2 PJSC, Quadra PJSC and T Plus PJSC, owned by domestic private owners, these companies meet 56.3%, 52.3%, 49.0% of the Code's criteria, respectively. Only in non-compliance with these criteria, outsider companies showed their leadership - 20.7%, 28.1%, and 31.0%.

An additional analysis of partially and non-complied criteria for each section of the Code showed a certain "vulnerability" of the boards of directors of generating companies.

Almost all generating companies included in the research did not meet more of the criteria of Section II (Board of Directors) of the Code. Failure to comply with the criteria specifically for this section (without diminishing the significance of other sections) can provoke significant risks for investors. Indeed, in domestic companies, boards of directors have been delegated the vast majority of all powers to ensure the rights and interests of shareholders, which is due to the lack of proper external control with the relative weakness of the institutional environment. In our opinion, in the situation that has arisen, a certain structural reorganization of the boards of directors of these companies is required.

3 Reorganization of boards of directors

In Russian generating companies, boards of directors "de facto" do not play the key role delegated to them "de jure" in ensuring the rights and interests of investors [22]. The overdue structural changes in the boards of directors of these companies are indicated in their comparative analysis with the boards of directors of France and Germany (at the beginning of 2021). The main material for this analysis was a typical set of structural characteristics of the Spencer Stuart Board Index, which provides data on the boards of directors of foreign companies [23].

3.1 Analysis of structural characteristics

The vector of development of the boards of directors of Russian generating companies is outlined below through a comparative analysis of their structural characteristics with the boards of directors of France and Germany.

3.1.1 Numbers

The average number of boards of directors of Russian generating companies (10.7) is below the level of France (13.7) and Germany (13.8).

3.1.2 Executive directors

The share of executive directors in the identified leading companies in corporate governance (Inter RAO PJSC, RusHydro PJSC, Enel Russia PJSC and Unipro PJSC) is comparable to that of France (8%). In outsider companies (Quadra PJSC and TGC-2 PJSC) it was overestimated (33%) even relative to the recommended values of the Federal Law "On Joint Stock Companies" (25%).

3.1.3 Independent directors

The share of independent directors (25.7%) is lower than recommended by the Code (30%). Its highest value was noted in Enel Russia PJSC (45.5%). Although this value is relatively small in comparison with its values in France (58%) and Germany (60%).

3.1.4 Serving on other boards of directors

Work on other boards of directors per director (1.8) is almost comparable to the French average (1.3).

Taking into account the additional structural characteristics proposed by the author (work in other organizations), which is typical for Russian companies, this average indicator has increased by 3 times (up to 5.4). The maximum presence of individual directors in other boards (without brackets) and other organizations (in brackets) is noted in RusHydro PJSC 17 (22), Enel Russia PJSC 16 (20) and PJSC Mosenergo 37 (37).

3.1.5 Proportion of women

The proportion of women is small (about 9%) compared to France (42.5%) and Germany (32%). The presence of women on boards of directors is highly valued in foreign companies: the microclimate of work is improving; the number of violations of laws and regulations is decreasing; monitoring of the actions of top management is intensified; and the number of cases of fraud, including transactions with related parties, is decreasing [24-26].

3.1.6 Foreign directors

Foreign directors are few in number. In purely Russian companies, there is a maximum of one (Inter RAO PJSC, TGC-1 PJSC). In companies with foreign owners, there are naturally more of them: Enel Russia PJSC (9), Unipro PJSC (6). Such directors are characterized by high competence and independence in decision-making. The disadvantages are mainly related to language and cultural barriers, lack of knowledge about the Russian specifics of doing business.

3.1.7 Age

The average age of board members (49.9) is lower than in France (58.9) and Germany (58). Unipro PJSC (60), Inter RAO PJSC (54), RusHydro PJSC board members (53) look relatively aged among Russian generating companies. Apparently, these companies value age diversity, which combines the relatively high technological level of young directors with the experience and discretion of older colleagues.

3.1.8 Term of office

The term of office of directors is not more than 7 years. It is slightly higher than the average values of this indicator in France (5.7) and Germany (5.7).

3.1.9 Number of meetings

The average number of meetings per year is 17.3. As a rule, all of them are in absentia, Unipro PJSC is an exception (4 in-person). However, a large number of meetings (especially in absentia) is not sufficient evidence of the effectiveness of corporate governance. The duration of the meetings and the quality of the discussions are valued. Meetings in absentia are most often “formal” [27]. In France, for comparison, boards of directors meet on average 9.1 times a year, in Germany - 6.8.

3.1.10 Self-evaluation and external evaluation

Unipro PJSC, Inter RAO PJSC, Enel Russia PJSC and RusHydro PJSC conducted an annual self-assessment and an external assessment (at least once every three years) of the work of the boards of directors. An independent company (PricewaterhouseCoopers Consulting LLC) was involved in the external evaluation. For comparison: in France, 75% of CAC 40 companies conduct an external evaluation of boards of directors. Current assessments of boards of directors, in our opinion, underestimate the importance of the personal and behavioral qualities of their members, although the regulator (the Bank of Russia) directly suggests paying attention to these criteria [28-29]. Unfortunately, in the recommendations of the Bank of Russia, these qualities are presented in a generalized form and require further development, for example, using foreign experience [30].

3.1.11 Remuneration

The remuneration of members of the boards of directors (for comparability, recalculated at the euro exchange rate as of December 31, 2020) in Russia was characterized by high values and disclosure of mainly the total amount, without division by person and their specific merits. The average amount of remuneration for non-executive directors in 2020, for example, at RusHydro PJSC was 44,173 euros, at Inter RAO PJSC - 37,755 euros. For comparison: in France, with a larger number of councils, each member accounted for 24,449 euros.

Conclusion

In the face of modern challenges, the problems of attracting investment in generating companies are becoming more acute. The reason is partly the investment risks of corporate governance in these companies. Ownership rotation, non-compliance of fundamental principles of corporate governance, relatively low structural characteristics of boards of directors and weak updating of internal documents reduce the level of investor confidence in generating companies. The paper proposes new corporate governance standards with a specific set of additional compliance criteria, which, along with updated structural characteristics of boards of directors and timely updating of internal documents, could improve the corporate governance of these companies in the interests of investors.

The research was carried out under State Assignment Project (No. FWEU-2021-0001) of the Fundamental Research Program of Russian Federation 2021-2030.

References

1. Corporate Governance Code. Regulations and up-to-date information of the Central Bank of the Russian Federation. Bank of Russia Bulletin, **40**, (2014)
2. G20/OECD Principles of corporate governance. OECD Publishing (2016) [Online]. URL:

http://dx.doi.org/10.1787/9789264252035-en (Accessed 15.01.2022)

3. Bank of Russia Information Letter dated 27.12.2021 No. **IN-06-28/102**. [Online] URL: https://cbr.ru/issuers_corporate/ acts/ (Accessed: 02.17.2022)
4. Lukoil sent an offer to buy back the shares of EL5-Energo at a price of 0.48 rubles per share. [Online] URL: https://bigpowernews.ru/print/107032.phtml (Accessed: 01.09.2023)
5. Unipro's sale under question, Russia's approval for the deal not received yet. [Online] URL: https://bigpowernews.ru/print/107665.phtml (Accessed: 02.17.2022)
6. Fortum has not included business in Russia in new strategy. [Online] URL: https://bigpowernews.ru/print/107803.phtml (Accessed: 03.02.2023)
7. Exchange me. [Online] URL: https://www.kommersant.ru/doc/5954363 (Accessed: 04.26.2023)
8. Rosatom bought Quadra from Prokhorov's Onexim. [Online] URL: https://bigpowernews.ru/print/101362.phtml (Accessed: 11.11.2022)
9. Rosatom increased its stake in Quadra to 95.02%. [Online] URL: https://bigpowernews.ru/print/103885.phtml (Accessed: 06.26.2022)
10. Russian Railways left the TGK-14; the Sinara group of Dmitry Pumpyansky could become the new owner of the generating company. [Online] URL: http://www.bigpowernews.ru/print/101071.phtml (Accessed: 01.11.2022)
11. SOVLINK LLC. [Online] URL: https://companies.rbc.ru/id/1027739102225-ooo-obschestvo-s-ogranichennoj-otvetstvennostyu-sovlink/ (Accessed: 05.11.2023)
12. Rostislav I. Kapelyushnikov. Russian Management Journal, **Vol. 4, 1**, pp. 3–28 (2006)
13. Belikov Igor V. *Board of Directors: a new approach.* Delibri, 580 p. (2019)
14. Pence Igor Sh., Furs Svetlana A. *Corporate management in the modern industry of Russia: the state and factors of improvement.* OOO Nedra Communications LTD, 120 p. (2008)
15. Galina I. Sheveleva. Global Energy Interconnection, **Vol. 5, 5**, pp. 512–523 (2022) DOI: 10.1016/j.gloei.2022.10.005.
16. Systems rresearch in energy: Energy transition. ISEM SO RAN, 594 p. (2021)
17. Galina I. Sheveleva. ECO, **7**, pp. 51-70 (2023)
18. Galina I. Sheveleva, *Corporate governance factors in the ESG agenda of Russian electric power companies*, in Proceedings of the International Conference on Management of Large-Scale System Development, MLSD, September 26-27, Moscow, Russia (2022)
19. Christophe Huet. Corporate governance in France: Main provisions of the AFEP-MEDEF Code as amended on Jan 30 2020. URL: https://nccg.ru/assets/files/issledovaniya/k.ue.-ku-vo-francii.pdf (Accessed 10.02.2021)
20. Environmental Social & Governance Report JPMorgan Chase & Co. [Online] URL: https://www.jpmorganchase.com/content/dam/jpmc/jpmorgan-chase-and-co/documents/jpmc-esg-report-2020.pdf (Accessed 24.02.2022)
21. ACRA - ratings. [Online] URL: https://www.acra-ratings.ru/upload/iblock/meth/20221118_ESG%20 Methodology_Appendix_Non-Financial.pdf (Accessed: 21.02.2023)
22. Valentina B. Kostyleva, Hector J. Lehuede. Board Formation: Nomination and Election in OECD Countries and Russia. [Online] URL: https://www.oecd.org/daf/ca/Board%20RUS.pdf (Accessed 29.03.2023)
23. Spencer Stuart Board Indexes. [Online] URL: http://www.spencerstuart.com/research-and-insight/board-indexes (Accessed 17.08.2021).
24. Melsa Ararat, Stijn Claessens, B. Burcin Yurtoglu. Emerging Markets Review, **Vol. 48**, 100767 (2021).
25. Jana Oehmichen. International Business Review, **Vol. 27, 2,** pp. 465–480 (2018)
26. National report on corporate governance. **Vol. 12** (2020) [Online] URL: https://nccg.ru/nacionalnyj-doklad-po-korporativnomu-upravleniyu/572.html (Accessed: 17.03.2022)
27. Russia Spencer Stuart Board Index. URL: https://www.spencerstuart.com/-/media/2021/june/russiabi2020/russiabi_2020.pdf (Accessed: 17.03.2022)
28. Bank of Russia Information Letter dated 26.04.2019 No. **IN-06-28/41.** [Online] URL: https://cbr.ru/StaticHtml/File/59420/20190426_in-06-28_41.pdf (Accessed: 17.01.2022)
29. Bank of Russia Information Letter dated 22.04.2020 No. **IN-06-28/80**. [Online] URL: http://www.cbr.ru/statichtml/file/59420/20200422_in_06_28-80.pdf (Accessed: 29.03.2023)
30. Leblanc Richard, Gillies James. *Inside the Boadroom: how boards really work and the coming revolution in corporate governance.* Wiley & Sons Canada, Ltd. 287 p. English edition. (trans: Alpina Business Books) (2006)

Multi-agent optimization in solving the problem of optimal power system control and reliability assessment

Alexander Domyshev[1,*]

[1]Melentiev Energy Systems Institute of Siberian Branch of the Russian Academy of Sciences (ESI SB RAS), 664033, Lermontov str., 130, Irkutsk, Russia

Abstract. The paper proposes a method for distributed optimal control of EPS with the presence of stochastic elements. The proposed technique is based on distributed optimization methods. The forecast of electric mode parameters is used as input data for optimization. The forecast horizon is divided into separate time steps at equal intervals. Distributed optimization methods are extended to the dynamic optimization problem, when not just independent states for each moment of time, but the process of their change in time is considered.

1 Introduction

This paper is a continuation of the work on automated optimal control of power system [1]. In the software complex for optimal control of voltage and reactive power in the automatic control cycle developed in ISEM SB RAS, forecasting and dynamic optimization of electrical modes are performed. The control algorithm is realized according to the hierarchical principle. Data from all substations are collected in a single control center, where state estimation and forecasting are automatically performed, based on which the optimal control actions and the time of their application are calculated.

The electric power system (EPS) is a complex dynamic object, the control of which requires the use of complex automatic systems. The developed hierarchical system of organization of automatic optimal control of EPS can be improved, if only because fully centralized control becomes ineffective due to the presence of large flows of information and significant time costs for its transfer to the center to calculate the control action. This issue is especially relevant in a power system with a significant number of sources of small generation or active load. The peculiarity of such systems is the stochastic nature of both load and generation. Optimal control in systems with a large share of uncertainty becomes even more complex, and hierarchical control, when it is required in a single center to quickly collect a model to optimize the entire network may no longer be appropriate. Multi-agent control can be a possible variant of optimal control system realization under the condition of stochasticity of EPS parameters.

The multi-agent approach is based on the concept of a mobile software agent, which is implemented and functions as an independent specialized computer program or an element of artificial intelligence. The paper proposes a method of distributed optimal control of EPS with the presence of stochastic elements. The proposed control system is aimed at application in systems with small and distributed generation, where the introduction of electricity storage systems is required, so it is important to take into account the fact that it is not always possible to organize a centralized optimal control system. Thus, when developing the model, the possibility of building a distributed multi-agent system is investigated at each stage.

2 Proposed solution

The proposed technique is based on distributed optimization methods [2, 3]. As input data for optimization, the forecast of changes in the regime parameters for a given time is used. The forecast horizon is divided into separate time steps at equal intervals.
The global target function of the distributed optimization problem can be formulated as the sum of target functions f_i for subsystems represented as individual computational agents.

$$\min \sum_{i=1}^{n} f_i(x). \quad (1)$$

Each of the functions f_i is known to only one agent. The admissible region for the values of the control vector x must take into account the constraints of each agent. Therefore, it is defined as the intersection of the corresponding admissible regions.

$$x \in X \triangleq \bigcap_{i=1}^{n} X_i. \quad (2)$$

The agents are connected by a directed graph $G = \langle V, E \rangle$ representing the available communication channels. A node $i \in V$ corresponds to an agent that

* Corresponding author: domyshev@isem.irk.ru

computes a function f_i. Each edge $(i, j) \in E$ corresponds to an element a_i^j of the incident matrix of the graph G.

The solution of the distributed optimization problem by the subgradient projection method for the case of convex functions $f_i(\cdot)$ is presented in [4]. At each k step at node i, the value of the vector of control actions is refined as follows:

$$x^i(k+1) = P_{X_i}\left[\sum_{j\in N_i(k)} a_j^i(k)x^j(k) - \alpha_k d_i(k)\right], \quad (3)$$

where P_{X_i} is the projection of the computed value of the control vector at node i onto its admissible area;
$a_j^i(k)$ is the coefficient of the incident matrix;
$N_i(k)$ - list of neighboring nodes for the i-th node;
$x^j(k)$ - value of the vector of control actions of the neighboring node j;
$d_i(k)$ - subgradient of the local target function $f_i(x)$;
α_k - step size at the calculation step k.

To ensure convergence of the distributed optimization problem, a necessary condition is that the switching graph is balanced, i.e., $a_j^i(k) \geq 0$ and $\sum_{j\in N_i(k)} a_j^i(k) = 1$.

When agents interact in the process of power system control, delays in information transmission are inevitable. Data received from neighboring control nodes may be outdated. In the presence of communication delays, the balance and symmetry of communication graphs will be broken, which makes algorithms that do not take into account such delays inapplicable. The loss or corruption of individual data can also be expressed in the form of data transmission delays. To solve the problem taking into account possible delays in information transmission, the following algorithm for computing the vector of control actions is proposed in [3]:

$$x^i(k+1) = P_{X_i}\left[\sigma x^i(k) - \sigma x^i(k-\rho) + \sum_{j\in N_i(k)} a_j^i(k)x^j(k-\rho) - \alpha_k d_i(k)\right] \quad (4)$$

The idea behind the algorithm is to use each agent's own state and information having a delay of size ρ to balance the impact of communication delays assuming an initially balanced graph. The coefficient σ is chosen such that the condition $\min_{j\in N_i(k)} a_j^i(k) \leq \sigma \leq a_i^i(k)$ is satisfied.

If we consider the problem of optimal control in an integral formulation, it is transformed into a dynamic optimization problem, where, unlike static optimization of a single mode, it is important to take into account the "cost" of control actions, depending not only on the system state vector, but also on the time of their implementation.

The cost of controlling one or another equipment depends on such factors as: the residual resource of the equipment; the priority of using the control action; the minimum permissible time between commutations by the same device. The optimization problem taking into account the cost of control action is written as:

$$min \sum_{t=1}^{T} f_t(X_t, t) = \sum_{t=1}^{T}\left(f_{d_t}(X_t) + \sum_{i=1}^{C} f_{c_i}(x_{t_i}, t)\right), \quad (5)$$

where X_t is the control actions available at time t;
f_{d_t} - static optimization function of each single state for time t;
f_{c_i} - monotonically decreasing function of the cost of the control action x_{t_i}, depending on the time of the actions that have been performed up to time t.

In the multi-agent formulation, the dynamic optimization problem is written as:

$$\min_X F(X,t) = \sum_{i=1}^{n} \int_{t=0}^{T} f_i(X_t, t).$$

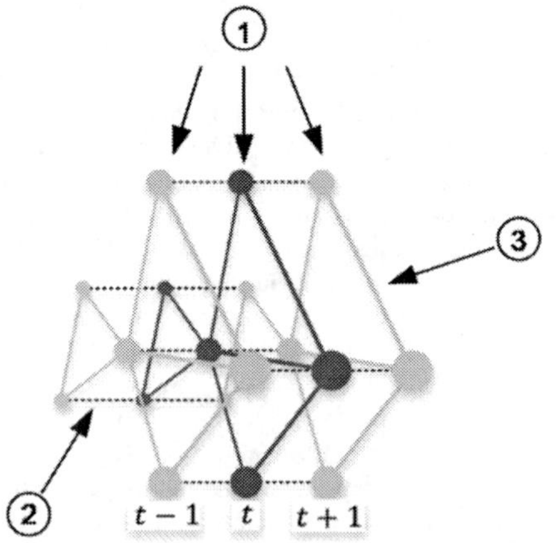

Fig. 1. Communication graph of dynamic multi-agent optimization. 1 – vertex corresponding to the agent at different moments of time; 2 – the edge corresponding to the relationship between different times of the same agent; 3 – the edge corresponding to the connection between agents.

Since time can be represented as discrete samples, each representing the time of data acquisition by one of the agents, $F(X,t)$ can be rewritten as:

$$\min_X F(X,t) = \sum_{i=1}^{n}\sum_{t=1}^{T} f_i(X_t, t) = \sum_{t=1}^{T}\sum_{i=1}^{n} f_i(X_t, t).$$

Having the sum of agents' target functions by time steps, the dynamic optimization problem is reduced to the problem (1), for the solution of which we use the subgradient method taking into account the information transmission delays (4). In this case, the communication graph G_Σ is a union of communication graphs G of the multi-agent interaction for each time step, supplemented by links between the corresponding agents at different time steps. In Fig. 1, these links are represented by dashed lines.

The component of the cost of the control action f_{c_i} that depends on the time of its application can be represented in the form of information delay ρ. However, it will already be time dependent.

$$x^i(k+1) = P_{X_i}\left[\sigma x^i(k) - \sigma x^i(k-\rho(t)) + \sum_{j\in N_i(k)} a_j^i(k)x^j(k-\rho) - \alpha_k d_i(k)\right].$$

The multi-agent dynamic optimization cycle uses multiple static optimization calculations. Therefore, they require fast electrical optimization software, which can be embedded in an external software package in the form of a library.

The paper presents such a library. The main optimization method implemented in it is the interior point method. The calculation of the Hessian in the optimization cycle is the most costly procedure, which must be performed at each iteration. To reduce the computation time, the Hessian is replaced by its approximation, refined iteratively according to the LBFGS algorithm. The use of the quasi-Newtonian algorithm speeds up the optimization calculation for each of the modes in the dynamic optimization process.

The interior point method [5 - 8] can be applied to solve the optimization problem both in the formulation (3 - 5) and in the formulation (7, 4, 5).

In general, the optimization problem for each time instant can be written as:

$$\begin{cases} \min f(x) \\ h(x) = 0 \\ \underline{g} \le g(x) \le \overline{g} \end{cases}$$

where $x = (p_g, u_g, k_t)$, h - equality constraints, g - inequality constraints.

Inequality constraints are reduced to equality constraints using auxiliary non-negative variables ω:

$$\begin{cases} g(x) - \underline{\omega} - \underline{g} = 0 \\ g(x) + \overline{\omega} - \overline{g} = 0 \\ \quad (\underline{\omega}, \overline{\omega}) \ge 0 \end{cases}$$

In this case, the dual problem is solved:

$$\min L(x, \lambda_h, \lambda_g, s),$$

where $L(.)$ is the Lagrangian function defined as:

$$L(x, \lambda_h, \lambda_g, s) = f(x) + \sum_i^n \lambda_{h_i} h_i(x) + \sum_j^m \lambda_{g_i}(g_i(x) + s_j) - \mu \sum_j^m \ln s_j.$$

This problem is solved by Newton's method, where a linearized system of equations is solved at each step:

$$\begin{bmatrix} H & -\nabla h(x)^T & -\nabla \underline{g}(x)^T & \nabla \overline{g}(x)^T & 0 & 0 \\ -\nabla h(x) & 0 & 0 & 0 & 0 & 0 \\ -\nabla \underline{g}(x) & 0 & 0 & 0 & I & 0 \\ \nabla \overline{g}(x) & 0 & 0 & 0 & 0 & I \\ 0 & 0 & \underline{W} & 0 & \Lambda_{\underline{g}} & 0 \\ 0 & 0 & 0 & \overline{W} & 0 & \Lambda_{\overline{g}} \end{bmatrix} \cdot \begin{bmatrix} \Delta x \\ \Delta \lambda \\ \Delta \lambda_{\underline{g}} \\ \Delta \lambda_{\overline{g}} \\ \Delta \underline{\omega} \\ \Delta \overline{\omega} \end{bmatrix} = \begin{bmatrix} \nabla_x L \\ \nabla_\lambda L \\ \nabla_{\underline{g}} L \\ \nabla_{\overline{g}} L \\ \nabla_{\underline{\omega}} L \\ \nabla_{\overline{\omega}} L \end{bmatrix}$$

where Hessian

$H = \nabla^2 f(x) - \sum_i^n \lambda_{h_i} \nabla^2 h_i(x) + \sum_i^m \lambda_{g_i} \nabla^2 g_i(x)$.

$\underline{W} = diag(\underline{\omega})$, $\overline{W} = diag(\overline{\omega})$, $\Lambda_{\underline{g}} = diag(\lambda_{\underline{g}})$,

$\Lambda_{\overline{g}} = diag(\lambda_{\overline{g}})$, I - unit matrix.

The calculation of the Hessian is the most costly procedure, which must be performed at each iteration. To reduce the computation time, the Hessian is replaced by its approximation, refined iteratively according to the LBFGS algorithm [9 - 14].

$$\widetilde{H}_{k+1} = V_k^T H_k V_k + \rho_k \delta_k \delta_k^T,$$

where $\rho_k = 1/\gamma_k^T \delta_k$ и $V_k = I - \rho_k \gamma_k \delta_k^T$,

δ_k is the change in the argument vector at step k:

$$\delta_k = \left(x_k, \lambda_{h_k}, \lambda_{g_k}, s_k\right) - \left(x_{k-1}, \lambda_{h_{k-1}}, \lambda_{g_{k-1}}, s_{k-1}\right)$$

γ_k is the change in the Lagrangian gradient.

$$\gamma_k = \nabla L\left(x_k, \lambda_{h_k}, \lambda_{g_k}, s_k\right) - \gamma_{k-1}.$$

The use of the quasi-Newton algorithm allows us to speed up the optimization calculation for each of the time steep in the dynamic optimization process. The optimization problems to be solved are actually not quite independent of each other. Indeed, the change of electrical parameters in time does not occur in a jump-like manner. Additional time speeding up of the calculation time is possible due to the use of the vector of control actions obtained as a result of optimization of the state for the previous moment of time as a starting point of optimization. At the same time, it is necessary to control the proximity of the previous state. For example, during switching operations, significant changes in the state may occur and the application of past optimal actions will not be effective. The proximity of the steady state can be estimated by the total deviation of active power flows on branches:

$$\sum_{j\in B,\, max\left(P_{j_k}, P_{j_{k-1}}\right)\neq 0} \left|\frac{P_{j_k} - P_{j_{k-1}}}{max(P_{j_k}, P_{j_{k-1}})}\right| \le \rho.$$

The value of the relative value of ρ is determined for each power system individually.

3 Implementation

The presented methodology is implemented as a C++ library and included in the ANARES software package. At the same time, an interface in Python language is provided, which allows using this library both in commercial solutions and in research projects. For use in research projects, the library is planned to be made publicly available in 2023

The proposed optimization algorithm was compared with the existing ones in the ANARES complex (gradient descent and LBFGSB). The calculation was performed on the Irkutsk power system model. The considered model of the power system includes power grids of the Irkutsk region with voltage of 500 kV - 110 kV. Generators in the model are mainly specified by nodes of the voltage at which the corresponding generators actually operate (6 kV - 15 kV). Some generation is represented as equivalent generators on high or medium voltage busbars. The load is mainly specified on the low side of transformers or on the tapping side of transmission lines. There are 486 generator nodes in the model and 485 load nodes. The total number of nodes in the node/branch model for this system is 1248, with 1481 branches. The calculations were performed on a workstation with the following parameters: Platform based on Intel Xeon W-2125 4 GHz processor with 32 GB DCU.

The transformation coefficients, as well as the change of reactive power generation of stations and synchronous compensators were considered as control parameters. The calculation was performed on the archive data obtained from the ANARES state estimation subsystem.

In a series of calculations from different initial conditions corresponding to different states from the archive, in comparison with the quasi-Newtonian algorithm LBFGSB showed the acceleration of calculations in the amount from 28% to 55%. These figures will also include data input from the model. On average, the optimization time per mode ranged from 212 ms to 437 ms. Whereas, in the LBFGSB implementation, the computation time ranged from 296 ms to 983 ms.

4 Conclusion

The presented optimization methodology was implemented in the research version of the ANARES software-computing complex. The calculations have shown high speed of the proposed methods with provision of the necessary accuracy of calculations.

The algorithm performance has been tested on the data of a real sufficiently large power system. The library for optimization of EPS modes with reformulated target function can be used in the cycle of mode reliability analysis.

References

1. A.V. Domyshev, A.B. Osak, D.A. Panasetsky Methods and approaches of building a complex of automatic hierarchical control of reactive power sources in normal and post-emergency modes of EES, Hierarchical modeling of energy systems, ed. by Voropai N.I., Stennikov V.A. - Novosibirsk: Academic Publishing House "Geo" 263-271, (2020), (in Russian).
2. A.V.Gasnikov Modern numerical methods of optimization. The method of universal gradient descent: Textbook. - 2nd edition, revised. M.: MCNMO, (2021), (in Russian).
3. P. Lin, W. Ren, Y. Song Distributed multi-agent optimization subject to noni-dentical constraints and communication delays, Automatica, **65**, 120-131, (2016).
4. A. Nedić, A. Ozdaglar Convergence rate for consensus with delays, Journal of Global Optimization, **47**(3), 437-456, (2010).
5. T. Morstyn, B. Hredzak, V.G. Agelidis Control strategies for microgrids with distributed energy storage systems: An overview, Transactions on Smart Grid, **9**(4), 3652-3666, (2016).
6. M.F. Zia, E. Elbouchikhi, M. Benbouzid Microgrids energy management systems: A critical review on methods, solutions, and prospects, Applied energy, **222**, 1033-1055, (2018).
7. F. Nejabatkhah, Y.W. Li Overview of power management strategies of hybrid AC/DC microgrid, IEEE Transactions on power electronics, **30**(12), 7072-7089, (2014).
8. H. Fontenot, B. Dong Modeling and control of building-integrated microgrids for optimal energy management – a review, Applied Energy, **254**, 113689, (2019).
9. D. Espín-Sarzosa, R. Palma-Behnke, O. Núñez-Mata Energy management systems for microgrids: Main existing trends in centralized control architectures, Energies, **13**(3), 547, (2020).
10. S.K. Rathor, D. Saxena Energy management system for smart grid: An overview and key issues, International Journal of Energy Research, **44**(6), 4067-4109, (2020).
11. O. Azeem et al. A comprehensive review on integration challenges, optimization techniques and control strategies of hybrid AC/DC Microgrid, Applied Sciences, **11**(14), 6242, (2021).
12. G. Chaudhary et al. Review of energy storage and energy management system control strategies in microgrids, Energies, **14**(16), 4929, (2021).
13. H. Zou et al. A survey of energy management in interconnected multi-microgrids, IEEE Access, **7**, 72158-72169, (2019).
14. S. Vuddanti, S.R. Salkuti Review of energy management system approaches in microgrids, Energies, **14**(17), 5459, (2021).

Investigation of the impact of integrating renewable sources with variable power generation, taking into account restrictions on the voltage stability limit

Nariman Rahmanov[1], *Viktor* Kurbatsky[2], *Huseyngulu* Guliyev[3*], *Nikita* Tomin[2], *Ayaz* Yagubov[1]

[1] Azerbaijan Scientific–Research and Designed–Prospecting Institute of Energetics, 94 H.Zardabi str., Baku, Azerbaijan
[2] Melentyev Energy Systems Institute of Siberian Branch of the Russian Academy of Sciences, 130 Lermontov str., Irkutsk, Russia
[3] Azerbaijan Technical University, 25 H.Javid str., Baku, Azerbaijan

Abstract. This article is devoted to the development of a methodology for estimating the amount of balancing power in a system with a large share of renewable energy sources (RES). For this purpose, an analysis was made of the influence of an increase in the installed capacity of wind power plants (WPPs) on the value of the balance power of the energy system, random characteristics of its variability for various combinations of the number of workers in the WPP system. The possibilities of realizing the distribution of balance power in characteristic critical loaded modes of the system are considered. Using the example of a real power system, the wind speed measurement data obtained for the geographic locations of wind stations on the power systems using the developed programs, simulations of the system modes were carried out for various combinations of these stations. The dependences for the balancing power are constructed, the conditions for reducing this power, as well as the conditions for its limitation on the stability of the voltage in the balancing part of the system, are determined.

Introduction

The growth in the share of generation of renewable sources has significantly changed the approaches to the technological implementation of power balancing in the modern energy system. As a result, in the modern energy system, along with traditional sources, with a specified nature of power generation, renewable sources also operate in parallel, a significant part of which are wind turbines and solar PV electrical systems with variable (randomly uncertain) generation. The choice of a balancing source (BS), under these conditions, should provide with a high probability the absolute range of power generation variability (MW), as well as the average value of the slew rate (MW/min) of power generation.

In addition, when choosing a balancing source, it is also important to ensure the stability of the system mode at each step of the variability of the BS generation power.

The generation of electrical energy by renewable sources depends on weather conditions and load [1-3].

Continuous spatiotemporal balancing of the magnitude of deviations between supply and demand in the energy system makes it possible for a stable and reliable power supply. Based only on measurements of random processes of power generation by wind and solar sources, it is not possible to accurately model the magnitude of their variability [4].

The increase in the number of renewable energy sources connected directly to the busbars of the load node of the electrical network of the power system leads to the uncertainty of the consumption forecast [5]. Sources of electrical energy storage (battery systems) are used when there is insufficient power generation to cover the demand of the load to the node to which they are connected [6].

The random and intermittent nature of power generation by wind (WPP) and solar (PV) stations requires the choice of a source for balancing (compensating) the random share of WPP and SPP generation. In most cases, one or two of the existing traditional stations are selected as a balancing source, in which part of the capacity from the total generation is allocated to compensate for the excess or deficit in power generation by the wind and solar stations of the system.

The choice of the value of the set power for the balancing station makes it possible to optimally distribute the load between the stations at different periods of the daily schedule, control the flow distribution in the electrical network of the power system and thereby effectively use the power generation of WPPs and SPPs.

Preliminary research results have shown that for the energy system of Azerbaijan, Shirvan and Sumgayit gas

* Corresponding author: huseyngulu@mail.ru

 https://doi.org/10.1051/e3sconf/202346101023

turbine stations can be considered as balancing stations, which are connected to 110 kV distribution networks, and are also connected in the supply network of the system at a voltage of 220 kV through lines. It has been established that the probability of overloading lines in distribution networks from fluctuations in the generation of power of a wind station and a solar station is small, which leads to a slight reduction in the generation of these stations.

Power systems with a high share of integrated RES [7-11] present a problem for dispatchers due to intermittent power generation and limitations in their predictability. At some point, the operator may be forced to allow a reduction in wind and solar generation relative to current generation. This amount of reduction in generation from conversion of wind and solar energy can be considered as an opportunity to reduce the total generation in the system. Thus, restrictions arise due to restrictions on power transmission through the supply network or balancing the power in the distribution network between incoming and consumed power [7,8,12]. Therefore, it is necessary to increase the flexibility of the power system in order to increase the production of VRE and at the same time compensate for the discontinuity in the generation of power from these sources [8,13,14]. Some traditional power plants cannot quickly adapt the management of their generation in the system due to technical or economic constraints [4].

Mode restrictions are associated with the speed of set, the duration of the start, as well as the value of the minimum load [8,13,14].

According to the above restrictions, power plants can be classified for use in operational control as not participating in the control process, used in certain situations, or fully accepting control in the tasks of operational maintenance of the system mode [15,16].

The power generation of WPPs and SPPs is intermittent and probabilistic in nature, which does not allow the system dispatchers to use in the operational management of the power system. Power plants not involved in the operational management of the power system, usually these include nuclear power plants, thermal power plants (TPP), which support the base part of the total demand of the power system. Power plants with traditional technology (gas turbines, hydropower plants), as well as all other renewable sources (except wind and solar PV) are involved in the regulation of the regime in the system.

The static stability of the power system in terms of voltage, which is mainly deteriorating due to the overload of backbone transmission lines or the insufficient supply of electric networks of existing traditional power systems with active and reactive power flow control, is facing new problems due to the large-scale integration of variable renewable generation. This article proposes a method for estimating the static stability of voltage in a system with a large share of power generation by wind and solar stations, examines the influence of the number of operating WPPs and SPPs on the mode of a balancing source in order to determine the amount of power generated by it to maintain voltage stability. An analytical criterion has been derived to ensure the stability of the voltage on the tires for connecting WPPs and SPPs to the system for various states of power generation by these stations and the shares of total demand at the current time points. Based on this model, a methodology for optimal system planning is proposed to minimize power generation by a balancing station due to the optimal distribution of active and reactive power generation between renewable sources.

1 Modeling the assessment of the value of the balancing power reserve in the energy system with RES.

Integration of RES into the energy system increases the uncertainty of the generated power and, together with the effect of the uncertainty of the load, makes it impossible to have a strict deterministic description of the value of the balance power.

In general, an estimate of the balancing power can be determined from the following equation:

$$0 = P^{t_i}_{trad.source} + P^{t_i}_{RES} - P^{t_i}_{load} - \Delta P^{t_i} + P^{t_i}_{b.s} \quad (1)$$

Where $P^{t_i}_{trad.source}$, $P^{t_i}_{RES}$ – capacity of traditional and renewable sources in the daily interval of the daily schedule; $P^{t_i}_{load}$ – load power in the balancing part of the power system in the interval i; ΔP^{t_i} – electrical network losses; $P^{t_i}_{b.s}$ – balancing power value.

In equation (1) $P_{trad.source}$, P_{RES}, P_{load} are given on the basis of the obtained measurements in a long period of system operation. The value of ΔP, $P_{b.s}$ - balancing power and losses in the network are the desired parameters.

To solve equations (1) with stochastic given input parameters, $P_{trad.source}$, P_{RES}, P_{load} the latter are given by the limits of their changes in the form of inequality equations, for example

$$P_{\min,RES} \leq P_{RES} \leq P_{\max,RES} \quad (2)$$

$P_{\min,RES}, P_{\max,RES}$ – are established at the stage of preliminary analysis of stochastic variables P_{RES}.

Thus, the estimation of the values of the balance power $P_{bs,i}$ in each interval $[0,i]$ is reduced to solving the stochastic system of equations

$$0 = P^{i}_{trad.source} + P^{i}_{RES} - P^{i}_{load} - \Delta P^{i} + P^{i}_{b.s}$$

Given the restrictions

$$P^{i}_{\min,RES} \leq P^{i}_{RES} \leq P^{i}_{\max,RES}$$

2 Research on the impact of renewable energy on the power balance in the Azerenerji system

The probabilistic approach in relation to the problem of determining the power reserve for the balance between coverage and demand is caused primarily by the stochastic nature of RES generation. The use of a stochastic approach when choosing a balance power is associated with a probabilistic analysis of observational data on wind speeds and solar radiation over a long observation period, in obtaining a stable assessment of the characteristics of their distribution, obtaining a predictive model for generating power by wind turbines in the geographical area under study and other parameters. At the same time, it is also important to have data on the nature of the stochastic variability of the system load.

Determining the balance power for steady post-accident modes (after failures of its main elements - generators, power lines) will require an analysis of the stability limits of the system state in these modes. The value of the balance during periods of critical conditions, in contrast to the existing deterministic approach, should have a probable estimate, taking into account the uncertainty of the expected emergency conditions. This paper proposes a method for statistical estimation of the limit of static stability of voltage for N-1, N-2 states of the system circuit.

The paper presents the results of calculating the power balance in the national energy system of Azerbaijan with a different combination of power plants with RES for normal and post-accident steady modes and various states of emergency shutdowns of circuit elements. Considering that the basic part of the structure of power plants is (80% of the installed capacity) thermal plants, where it is supposed to place a capacity reserve to cover the imbalance from renewable energy sources, therefore, a very important factor is the placement of a capacity reserve and the determination of flow distribution control strategies in the network.

In 2025, three wind farms are planned to be put into operation in the Azerenergy system: the Khyzi wind farm with a capacity of 240 MW (conditionally named Baku station); WPP Janub 220 MW (Lenkoran) and WPP Shimal 240 MW (Maraza).

To assess the balance capacity and analyze the regimes of the Azerenergy system in connection with the commissioning of the above VRE plants, modeling and computer tests were carried out for the following scheme options:

- WPP "Baku" is connected to the system;
- WPP "Maraza" is connected to the system;
- WPP "Lenkoran" is connected to the system;
- WPP "Baku" + WPP "Lenkoran" are connected to the system;
- WPP "Baku" + WPP "Maraza" are connected to the system;
- WPP "Baku" + WPP "Lenkoran" + WPP "Maraza" are connected to the system.

On Fig. 1 shows the structure of the parts of the systems - balancing and synchronous, in Fig. 2 shows the probability distribution function of the WPP power forecast error for different values of this power generation, and in Fig. 3, the reserve for power balancing in the energy system of Azerbaijan in the case of installing each individual wind farm at the locations of the WPP "Baku", WPP "Lenkoran" and WPP "Maraza" . Fig. 4 and Fig. 5 show the change in wind speed in the regions of Baku and Maraz.

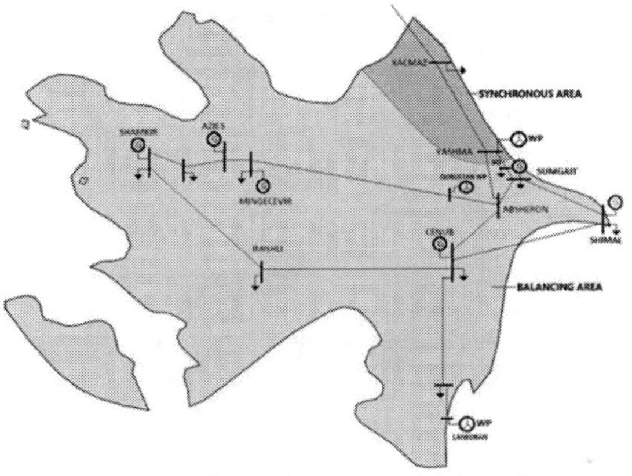

Fig. 1. The structure of parts of systems is balancing and synchronous for Azerenerji

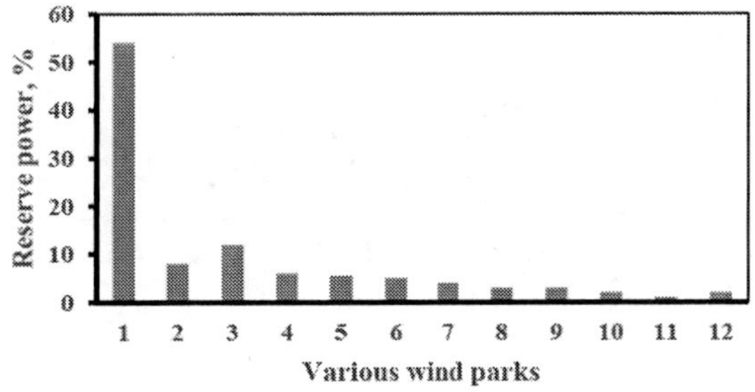

Fig.2. Power Prediction Error Probability Distribution WPP at different values of this power generation

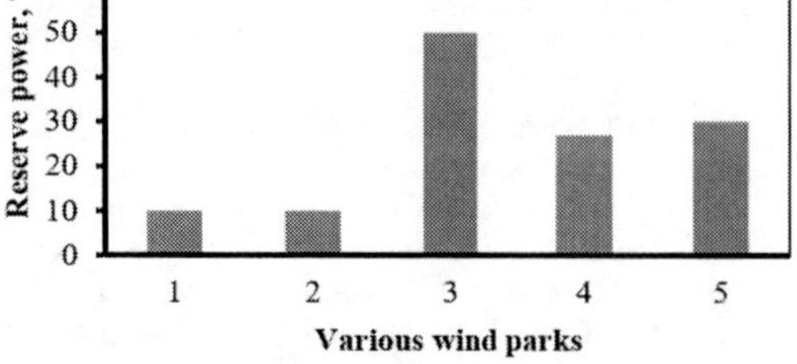

Fig.3. Reserve for power balancing in the energy system of Azerbaijan in cases of installation of each individual wind farm (Baku, Maraza, Lankaran) and their combinations: 1-Lankaran; 2-Maraza; 3-Baku; 4-Baku+Maraza; 5-Baku+Maraza+Lankaran

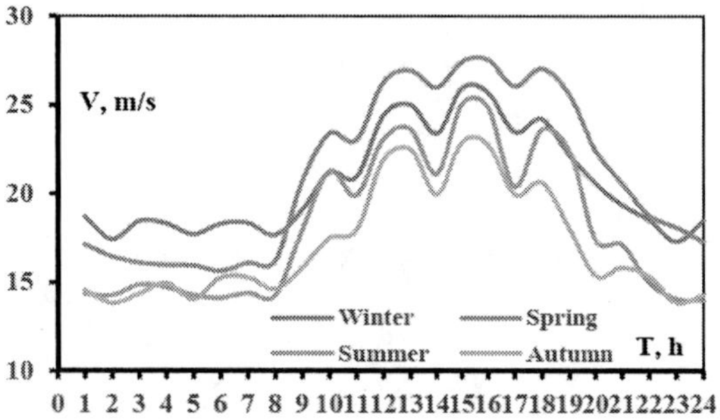

Fig.4. Hourly changes in wind speed in the area of Baku wind farm installation

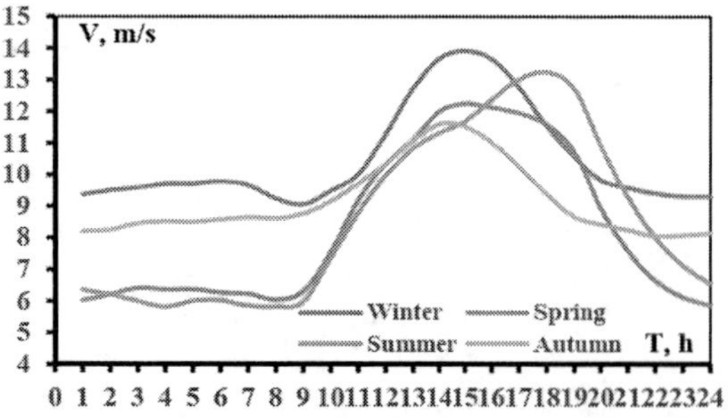

Fig.5. Hourly changes in wind speed in the area of installation of the Maraza wind farm

3 Impact on the statistical stability of RES during power balancing

The need to analyze the impact of RES on the static stability of the energy system of Azerbaijan under the conditions of operational management of the selected value of the balancing power reserve is caused by a large share of the power generated in the system from wind farms (up to 20% of that installed in the system of traditional plants). Long-term observations of climatic conditions show that in a number of regions of Azerbaijan, due to the characteristic features of the dynamics of the energy potential of wind speeds, the electrical energy produced by wind turbines in some periods of time can have a large amplitude. In this paper, to assess the limit of voltage stability in the balancing part of the power system (Fig. 1), at each stage of the implementation of the balance power reserve, the method of sequential mode weighting was applied [6].

Fig. 6 shows the dependence curves $U-P$ for power transmission through a 220 kV power transmission line during the hours of the evening maximum load. The total generation from all wind farms is about 340 MW, which is about 10% of the total system demand. As it was established earlier [3], with the joint operation of three wind farms (wind farms of Baku, Maraza, Lenkoran), the total reserve amounted to about 25% of the installed capacity of wind farms ($\approx 80-85\,MBm$).

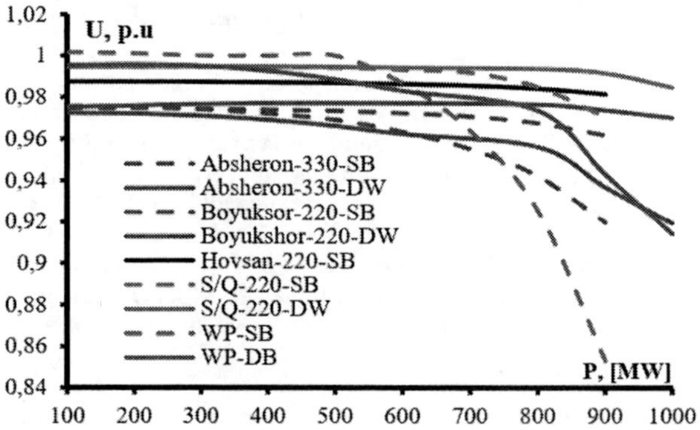

Fig. 6. Power Transfer Curves $U-P$ along power lines 220-330 kV during the hours of the evening maximum load

As can be seen from the curves in Fig. 5, if the total generation from wind farms is reduced by 30-40%, i.e. more than 85 MW, the power reserve will keep the voltage of the controlled power transmission within the allowable critical voltage.

Conclusion

1. The growth of RES integration increases the need for a power reserve to cover the deficit in the balance between supply and demand. The stochastic variability of wind energy sources leads to the need to take into account the influence of the following factors when choosing the reserve capacity to balance: forecast errors in power generation by wind turbines and solar PV installations, the influence of errors in the calculation model of balancing reserves, the influence on the marginal value of the reserve balancing deviation between total generation and consumption .

2. An approach is proposed for estimating the reserve for balancing active power in a system with a dominant

share of RES in the interval of 30-minute stochastic changes in generating capacity from wind turbines; stochastic load changes, random emergency failures of the main objects of the circuit - traditional generators and power lines of the supply network. The proposed approach makes it possible to take into account the dynamics of changes in the balancing power depending on the error in forecasting the discrepancy between the processes of RES generation and power consumption.

3. On the basis of computational and experimental studies in a real power system, it has been established that with an increase in the number of wind farms connected to the backbone circuit in its various nodes, the need for a power reserve can be significantly reduced.

The influence of wind farms of identical capacity in different places of the power system scheme has an unequal effect and the magnitude of the balance power required from their influence can vary significantly.

References

1. IEA (2014a) Advancing variable renewables-grid integration and the economics of flexible power system. International Energy Agency, Paris 2014a, 726 p., https://iea.blob.core.windows.net/assets/e6f58562-203e-474c-97a3-486f409aa7ff/WEO2014.pdf
2. Rakhmanov N.R., Kurbatsky V.G., Guliev H.B., Tomin N.V. *Short-term forecasting of power generation of wind power plants to ensure the reliability of electrical networks,* Proceedings of the International Scientific Seminar. Yu.N.Rudenko "Methodological issues of researching the reliability of large energy systems". Actual problems of reliability of energy systems. Issue 66, Minsk, 2015, pp. 397-404.
3. Rakhmanov N.R., Guliev H.B., Ibragimov F.Sh. *Modeling the generation process of a wind farm when predicting the load power coverage mode in the energy system in intra-daily cycles,* Problems of Energy, No. 3, Baku, 2015, pp. 3-11.
4. Grubb M. *Value of variable sources on power system. IEE,* Pros: Gener. Transm Distib.1991; 138 (2), p. 65-149.
5. 5.UNECE (Renewable Energy Status Report) 2017, pp.98, https://unece.org/DAM/energy/se/pp/renew/Renewable_energy_report_2017_web.pdf
6. 6. Hirth L., Ziegenhagen In. *Balancing power and variable Renewables: Three links,* Renewable and Sustainable Energy Reviews, 50 (2015), p.1035-1051, https://neon.energy/Hirth-Ziegenhagen-2015-Balancing-Power-Variable-Renewables-Links.pdf
7. Guerra K., Haro P., Gutierres R.E., Gomez - Barec . *Facing the high share of variable renewable energy in the Power System: flexibility and stability requirements,* Journal Applied Energy, Elsevier, Vol. 310(C), https://ideas.repec.org/a/eee/appene/v310y2022ics0306261922000460.html
8. European Commission , EG3 Report : *Regulatory recommendations for the development of flexibility,* Smart Grid Force, 2015, pp.93, https://projects2014-2020.interregeurope.eu/fileadmin/user_upload/tx_tevprojects/library/EG3%20Final%20-%20January%202015.pdf
9. Spiegel T. *"Impact of Renewable Energy Expancion to the Balancing Energy Demand of Different Balancing Groups",* Journal of Sustainable Development of Energy, Water and Environment Systems.(2018), Vol. 6., Issue 4 . pp. 784-799
10. Holttinen H., Milligan M., Ela E., Menemenlis N., Dobchinski J., Rawn B., Bessa R.J.,Flynn D., Gomes-Lazaro E., and Detlezsen N.K. *Methodologies to determine operating reserves due to increased wind power,* IEEE Transactions on Sustainable Energy, Vol.3, No. 4, pp. 713-723, 2012
11. Brener C., Engelhartdt C., and Moser A. "*Expectationbased reserve capacity demensioning in power system with an increasing intermittent feed-in",* in 2013 10th Internatinal Conference on the European Energy Market (EEM), IEEE ,2013, pp.1-7.
12. Matos M.A., and Bessa R.J., *Setting the operating reserve using probabilistic wing power forecasts,* IEEE Transactions on Power Systems, Vol. 26, Iss. 2, May 2011, p.594-603, https://ieeexplore.ieee.org/document/5565529
13. Cognet, Sylvain, Tracy Wilkison. *Balancing services in Europe.* HIS Cera decision brief ; 2013
14. Rakhmanov N.R., Guliyev H.B. *Application of a neural network model to assess the current value of the voltage stability limit of an electrical system,* Electricity, No. 4,2015, pp. 4-11.
15. Ajjarapu V., and A.Christy C. *The continuation power flow: A tool for steady state voltage stability analysis,* IEEE Transactions on Power Systems , Vol.7., No.1, Feb. 1992, pp. 416-423.
16. Holttinen H. et al. *Design and operation of power system with large amounts of wind power,* Final Resort, IEA wind Task 25, 2006-08, January 2009, Publisher: VTT Tiedotteita - Research Notes 2493, Editor: VTT/Julkaisiga (Ed.), ISBN: 9789513873097

On convergence of nonlinear topological algorithms for calculation of steady modes of electric networks

Dauren Akhmetbaev [1*], *Arman* Akhmetbaev [2]

[1]L.N. Gumilyov Eurasian National University, Astana, 010000, Kazakhstan
[2]JSC Kazakhtelecom, Almaty, 050057/A15H9D6, Kazakhstan

Abstract. The paper presents the results of a study of topological algorithms for the formation of a steady state of a complex electrical network, on the convergence of the iterative process. For a power transmission with distributed parameters, consisting of 5 nodes with a voltage of 500 kV, operating on a system of unlimited power, it is shown that there are two solutions and the need to introduce restrictions to obtain real solutions to nonlinear systems of equations. As the regime becomes heavier, the topological method ensures convergence and slow growth of the iterative process compared to the results obtained using the RASTR software package.
Key words: directed graph, tree, chord, current distribution coefficients, nodal voltages.

1 Introduction

The use of distribution coefficients of master currents as parameters of electrical networks makes it possible to simplify the performance of some complex calculations [1]. Therefore, the use of current distribution coefficients from the perspective of the topological theory of electrical networks is of particular interest.

Problems associated with the analytical method for determining the Z matrix of nodal resistances can be solved based on the following. With a known matrix C, it is always possible to find a unique correspondence to the response of the circuit of the electrical network under study to disturbances of the driving currents [2]. Current distribution coefficients, together with branch resistances, quite fully characterize the properties of electrical network circuits, which indicates a certain universality of their application. The complexity of the existing method for determining current distribution coefficients lies in determining the numerators of topological expressions by dividing the network into two parts in order to find two graph trees [3].

For the first time, the exact analytical dependence of the matrix of nodal resistances as a function of the matrix of current distribution coefficients was established by transforming the known matrix equations of the network state in [4]. This paper presents the results of an analysis of the convergence of topological algorithms for the formation of steady-state modes, developed on the basis of the properties of possible graph trees, without dividing the network into two parts [5,6].

2 Materials and methods

The analytical method for constructing the topological matrix of the electrical network of the power system was chosen as the research method. This method is described in detail in [7].

The topological method is applied to a well-known circuit, studied in detail in [2], IEEE test circuits http://energy.komisc.ru/dev/test_cases.

The conducted studies showed that the elements of the matrix of current distribution coefficients of a circuit of arbitrary complexity can be determined based on the natural parameters of the network according to the formula [9]:

$$C_{ij} = \frac{\sum F_{ij}}{\sum F} \quad (1)$$

F_{ij} – is specific tree containing the i-th branch relative to the j-th node; $\sum F$ – is the arithmetic sum of possible trees of the graph.

The steady state of complex networks is described by the matrix equation developed in [6]:

$$\dot{U} = U_0 + C^T Z_B C \hat{U}^{-1} \hat{S} \quad (2)$$

where C is the matrix of current distribution coefficients; Z_B – diagonal matrix of branch resistances; $\hat{U}$ – diagonal matrix of nodal conjugate stresses; $\hat{S}$ – matrix-column of conjugate powers of node loads and generators.

The matrix-column of conjugate powers of nodal loads and generators $\hat{S}$ depends on the nodal voltages:

$$S_k(U_k) = S_{0_k} + j \cdot \alpha_k \cdot |U_k|^2 \cdot 10^{-3} \quad (3)$$

where S_{0_k} – are given fixed values of generation power (load), coefficients α_k are expressed through the given values of capacitive conductivities B, in mSm and B_s, in mSm - the conductivities of shunt reactors.

* Corresponding author: abeked@mail.ru

3 Results and discussions

For a comparative assessment of the convergence of iterative processes for calculating a steady state in a topological model, let us consider the example discussed in Chapter 3 [2]:

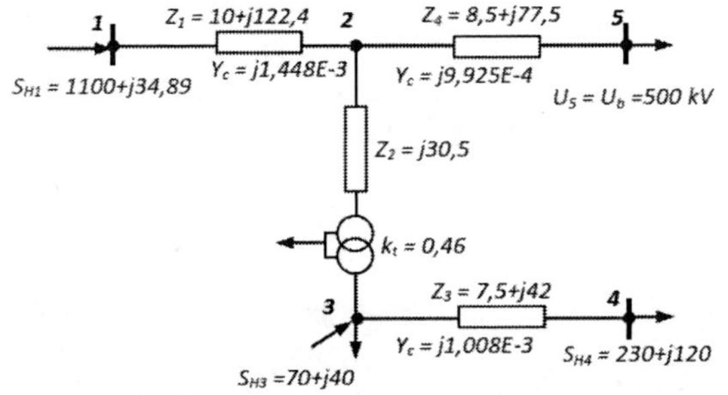

Fig 1. Equivalent circuit of the system under study (resistance in Ohm, conductivity in cm)

A comparison of the results of calculating the steady state performed by the method of simple iteration of the topological model and the Seidel method of the equation of nodal voltages in the form of a current balance is given in table1. It should be noted that the execution time of the proposed program was 10-2 seconds on a regular laptop. First of all, this is due to the fact that in this model there is no need to invert the conductivity matrix, which significantly reduces the number of computational operations.

Table 1. Comparative results of stress calculations with different accuracy (calculation accuracy 10^{-1} - 10^{-3}).

Calculation accuracy	Models and relative difference	Number of iterations	Program execution time (milliseconds)
10^{-1}	1	124	47
	2	95	35
	Δ%	23	25
10^{-2}	1	461	69
	2	328	47
	Δ%	28	31
10^{-3}	1	1313	115
	2	1001	85
	Δ%	23	26

Table 2. Comparative results of stress calculations with different accuracy (calculation accuracy 10^{-1} - 10^{-3}).

	10^{-1}			10^{-2}			10^{-3}		
Re(U1)	404	397	1,70	393	390	0,80	390	387	1,08
IM(U1)	350	351	0,22	351	351	0,09	351	352	0,11
|U1|	534	530	0,86	527	525	0,40	525	522	0,64
arg(U1),°	40,9	41,4	1,35	41,7	42	0,62	42	42,3	0,82
Re(U2)	509	506	0,59	504	503	0,27	503	501	0,35
IM(U2)	113	113	0,18	113	113	0,08	113	113	0,07
|U2|	521	518	0,55	517	515	0,25	515	513	0,32
arg(U2),°	12,5	12,6	0,73	12,7	12,7	0,34	12,7	12,8	0,42
Re(U3)	239	23	0,57	236	236	0,27	236	235	0,33
IM(U3)	44,5	44,5	0,10	44,5	44,6	0,03	44,6	44,6	0,03
|U3|	243	241	0,55	240	240	0,27	240	239	0,46
arg(U3),°	10,6	10,6	0,71	10,7	10,7	0,31	10,7	10,7	0,37
Re(U4)	212	210	0,81	209	208	0,38	208	207	0,48
IM(U4)	2,04	1,8	12,4	1,64	1,51	7,79	1,53	1,36	11,25
|U4|	212	210	0,81	209	208	0,38	208	207	0,48
arg(U4),°	0,55	0,49	11,3	0,45	0,41	7,56	0,42	0,37	10,71

Comparative table of calculations: 1-Seidel method for equations of nodal voltages in the form of current balance, 2-method of simple iteration of the topological model, Δ-relative difference in percentage.

Let us now consider the 5-node test circuit rucase_5_4.txt. The equivalent circuit has the form shown in Fig. 2.

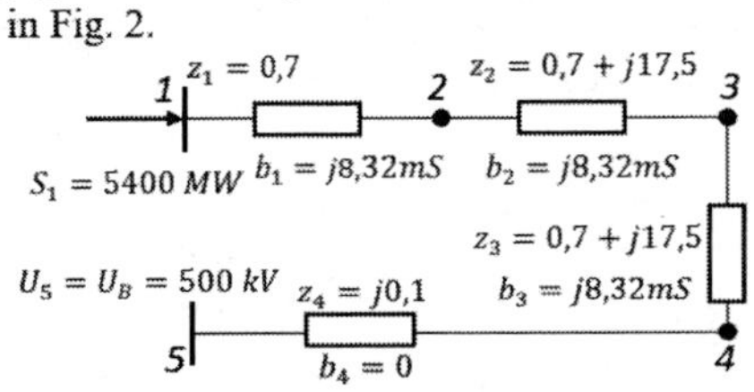

Fig. 2. IEEE test 5-node circuit rucase_5_4.txt.

The initial and calculated data are presented in the following tables.

Table 3. Calculated and initial data for nodes of the test circuit rucase_5_4.txt.

Nodes №	U nom	Voltage		Load power		Generation power	
	kV	phase, de	module, kV	P, MW	Q, MVar	P, MW	Q, MVar
1	500	68.73	484.5	0.0	0.00	5400	0.00
2	500	45.05	482.0	0.0	0.00	0.00	0.00
3	500	21.97	489.3	0.0	0.00	0.00	0.00
4	500	0.12	499.9	0.0	0.00	0.00	0.00
5	500	0.00	500.0	0.0	0.00	0.00	457

Table 4. Initial data for the branches of the test circuit rucase_5_4.txt.

Branch no.	start	end	R, Ohm	X, Ohm	B, mSm (cap+,ind-)
1	1	2	0.7000	17.5000	8.3200
2	2	3	0.7000	17.5000	8.3200
3	3	4	0.7000	17.5000	8.3200
4	4	5	0.0000	0.1000	0.0000

The calculations performed using both methods coincided exactly to the third decimal place. It should be noted that in this case there are several solutions to the vector equation (2), so it is necessary to take into account technical restrictions such as $|U_1| \leq 550$.

Moreover, the execution time and number of iterations, depending on the accuracy of the solution, are as follows:

Table 5. Calculated data for a 5-node circuit rucase_5_4.txt

Calculation accuracy	Number of iterations	Program execution time (milliseconds)
0,1	4	22
0,01	8	28
0,001	16	32
0,0001	18	48
1E-05	24	64

The required number of iterations for a 5-node scheme turned out to be greater than expected. This is because in this case we are considering a case close to the maximum power transmission power, in which additional "imaginary" solutions of the nonlinear system arise, to which the iteration converges very quickly, almost regardless of the required accuracy of the solution (no more than 5-6 iterations).

As noted above, to exclude these imaginary solutions, additional conditions are imposed on the solution to limit the voltage module to 525 kV. Therefore, the number of iterations increases. If we reduce the power transfer by 20%, that is, instead of 5400 MW we take, for example, 4400 MW, then the iteration converges to the desired solution quite quickly. Below are the calculation results for a generation power of 4400 MW.

Table 6. Calculated data for a 5-node circuit rucase_5_4.txt with a generation capacity of 4400 MW.

Calculation accuracy	Number of iterations	Program execution time (milliseconds)
0,1	4	8
0,01	8	8
0,001	5	12
0,0001	5	14
1E-05	6	18

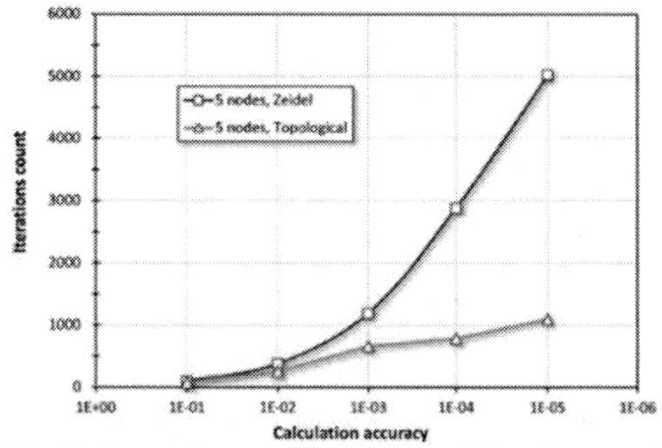

Fig. 3. Dependence of the number of iterations on the calculation accuracy in semi-logarithmic coordinates.

As can be seen from the graphs presented in Figure 3, calculations by the Seidel method become significantly more complicated with increasing accuracy. At the same time, the topological model method proposed by the authors shows a slight increase in the complexity of calculations with increasing accuracy, which is an extremely promising method for calculating complex circuits with high accuracy.

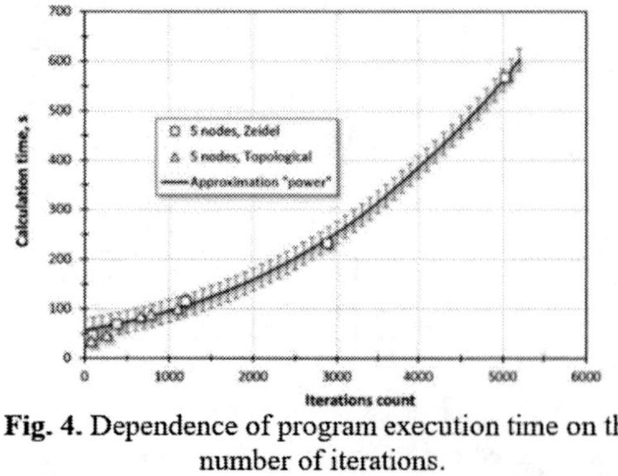

Fig. 4. Dependence of program execution time on the number of iterations.

Figure 4 shows the general dependence of the increase in calculation duration with an increase in the number of iterations for the 5-node circuit presented in Figure 2. As can be seen from the figure, with an increase in the number of iterations, the calculation time increases according to a power law, which is shown by the solid line.

The equation for the approximating curve was obtained according to the recommendations [10]:

$$T = 30 + 1.05 \cdot 10^{-9} \cdot (Ci + 2970)^3 (EQAdd.01)$$

This type of dependence allows us to describe the ratio of the time of the number of iterations with the coefficient of determination $R^2 = 0.999$. It should be noted that using the standard "Power" trend line in MS Excel gives the coefficient of determination value $R^2 = 0.88$, and for the "Exponential" trend line $R^2 = 0.93$, which is significantly lower than the value we obtained.

The vertical dashes on the solid line in Figure 3 show the standard error of approximation. As can be seen from the figure, it does not exceed 5%, which also confirms the correctness of the chosen form of the equation.

Thus, it is shown that the topological model makes it possible to increase the speed of calculations several times for complex circuits with high calculation accuracy compared to the Seidel method.

To confirm the conclusion obtained, the authors performed additional calculations of various electrical circuits, including more complex ones in composition and number of nodes.

Next, we present the results of applying the developed topological model to various test circuits.

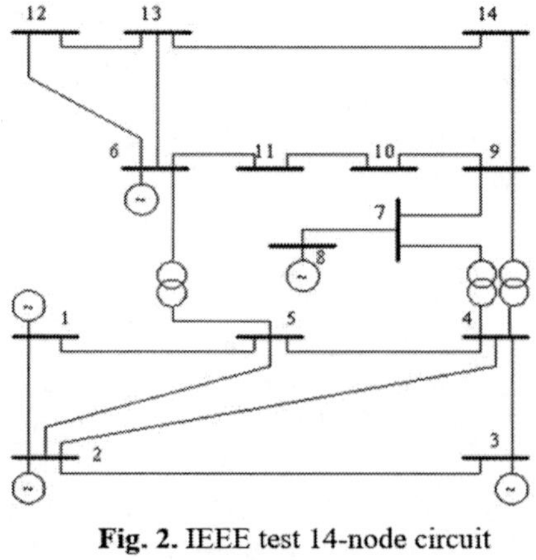

Fig. 2. IEEE test 14-node circuit

Table 7. Initial data for the branches of the test circuit

Branch no.	start	end	R, Ohm	X, Ohm	B, mSm (cap+, ind-)	k_t
1	1	2	10,25	31,30	0,0998	-
2	1	5	28,58	117,99	0,093	-
3	2	3	24,86	104,73	0,0828	-
4	2	4	30,74	93,27	0,0643	-
5	2	5	30,13	91,98	0,0654	-
6	3	4	35,45	90,47	0,0242	-
7	4	5	7,06	22,28	0	-
8	4	7	0	105,81	0	0,51125
9	4	9	0	276,26	0	0,516
10	5	6	0	115,80	0	0,53648
11	6	11	12,56	26,30	0	
12	6	12	16,25	33,83	0	
13	6	13	8,75	17,23	0	
14	7	8	0	23,29	0	-
15	7	9	0	14,55	0	-
16	9	10	4,21	11,17	0	-

Continuation of Table 7. Initial data for the branches of the test circuit

Branch no.	start	end	R, Ohm	X, Ohm	B, mSm (cap+, ind-)	k_t
17	9	14	16,81	35,76	0	-
18	10	11	10,85	25,40	0	-
19	12	13	29,22	26,43	0	-
20	13	14	22,6	46,02	0	-

Table 8. Comparative results of calculations.

Nod. no	Voltage calculation in the RASTR program		Voltage calculation in the topological model	
	phase, deg	module, kV	phase, deg	module, kV
1	0	243,8	0	243,8
2	4,983	240,35	4,973	240,141
3	12,725	232,3	12,73	231,813
4	10,313	234,064	10,291	233,361
5	8,774	234,488	8,772	234,375
6	14,221	123,05	14,056	124,538
7	13,36	122,075	13,424	120,36
8	13,36	125,35	13,424	126,882
9	14,939	121,432	15,152	120,033
10	15,097	120,863	15,284	120,878
11	14,791	121,544	14,83	120,802
12	15,076	121,347	15,025	119,867
13	15,156	120,794	15,152	119,395
14	16,034	119,086	16,2363	118,503

As we can see, the maximum deviation of the voltage modulus is 1.4%, and it is good results.

Table 9. Calculated data for the circuit

Calculation accuracy	Number of iterations	Program execution time (milliseconds)
0,1	6	22
0,01	8	24
0,001	10	28
0,0001	12	33
1E-05	13	38

When considering networks with a lot of nodes with possible random dynamic changes, associated, for example, with various types of emergency failures of nodes, it is necessary to consider random graphs. They can be described and studied using pseudo-finite models of standard graph theory. General laws for well-defined systems can be investigated using statistical and model-theoretic methods. From a model-theoretic point of view, one can approach the approximation [11]. The family of such pseudo-finite theories, as well as their topological properties, were studied in [12].

4 Conclusion

The considered iterative process has reliable and faster convergence than the Gauss-Seidel method using a matrix of nodal conductivities. The proposed method, unlike the method using Zy, does not require determining and storing a matrix of nodal resistances. A feature of the new topological method for calculating the mode of an electrical network is that the current distribution coefficients in a complex circuit can be determined at the stage of generating initial data on the network topology, which significantly reduces the computational costs of real calculations of a complex electrical network.

References

1. Kesavan, H. K.; Pai, M. A.; Bhat, M. V. (1972) Piecewise solution of the load-flow problem IEEE Transactions on Power Apparatus and Systems, PAS-91 (4). pp. 1382-1386. http://dx.doi.org/10.1109/TPAS.1972.293269
2. Gracheva E.I., Naumov O.V., Gorlov A.N. Problems of studying the equivalent resistance of low voltage electrical workshop networks // Bulletin of the Kazan State Energy University. – 2019. – T.11. – No.3(43). – P.37-45
3. Geraskin O.T. Methods for determining 2 graph trees in the topology of electrical networks // Izvestia of the USSR Academy of Sciences. Energy and transport. – 1967.– No. 5.– P. 106-112.
4. Egorov A., Semenov A., Bebikhov Yu.V., Sigaenko A. Assessment of the energy efficiency of the modernized main fan unit of the underground mine, Inter J Ener Clean Env. 20(2019)153–165. DOI: https://doi.org/10.1615/Inter J Ener Clean Env. 2019026505.
5. Akhmetbaev, D.S., Akhmetbaev, A.D., Dzhandigulov, A.R. New Approaches to the Topological Method of Analysis of Electrical Networks/ 2019 International Multi-Conference on Industrial Engineering and Modern Technologies, FarEastCon 2019, 8933870
6. Akhmetbaev D.S. Method for calculating steady-state modes of electrical networks based on current distribution coefficients // Electricity. – 2010.– No. 11.– P.23-27.
7. Akhmetbaev, D.S., Dzhandigulov, A.R. Development of algorithms for the formation of steady-state modes based on the topology of electric power systems/Journal of Physics: Conference Series, 2019, 1392(1), 012079
8. Akhmetbayev D.S., Dzhandigulov A.R., Bystrova S.V. Topological system method of formation of transformer transformation coefficients.(2020) E3SWebofConferences, 216, No. 01087, DOI:10.1051/e3sconf/202021601087
9. Akhmetbaev, D.S., Dzhandigulov, A.R., Akhmetbaev, A.D. Topological algorithm for forming nodal stresses of complex network energy systems/(2019) E3S Web of Conferences, 139, No. 01066. DOI: 10.1051/e3sconf/201913901066
10. Pinchuk V.A., Kuzmin A.V. Method of approximating experimental data on fuel combustion, InterJEnerCleanEnv.20(2019)261–272. DOI: https://doi.org/10.1615/InterJEnerCleanEnv.2019032466.
11. N. D. Markhabatov, "Approximations of acyclic graphs", Bulletin of Irkutsk State University, series "Mathematics", 40 (2022), 104–111
12. N. D. Markhabatov, S. V. Sudoplatov, "Topologies, ranks and closures of families of theories. I", Algebra and Logic, 59:6 (2021), 437–455, N. D. Markhabatov, S. V. Sudoplatov, "Topologies, ranks and closures of families of theories. II", Algebra and Logic, 60:1 (2021), 38–52

Simplified topological algorithm for forming a steady mode of electric power system

Abdygali Jandigulov [1,2*], *Korgaubek* Rauan[1], *Svetlana* Bystrova[3]

[1]L.N. Gumilyov Eurasian National University, Astana, 010000, Kazakhstan
[2]JSC «Astana IT University», Astana, 010000, Kazakhstan
[3]Ekibastuz Engineering and Technical Institute named after academician K. Satpaev, Ekibastuz, 141208, Kazakhstan.

Abstract. The paper presents the possibility of simplifying the algorithms for forming the parameters of the steady state mode of a complex electrical network, using the independence of the currents of its equivalent circuit. In this case, all independent currents of a complex circuit of an electrical network must be found in advance, which is, in the general case, a rather difficult task, since it requires additional calculations of the steady state. The matrix of current distribution coefficients, formalized on the basis of the topology of the electrical network, can be used to directly determine the currents in the branches of its equivalent circuit. This allows you to preliminarily estimate the values of pre-selected independent currents in the circuit branches of a complex network and determine its simplified design scheme. In this case, a calculation model is established, consisting of some trees of a complex electrical network graph.
Key words: directed graph, tree, chord, current distribution coefficients, nodal voltages.

1 Introduction

The conducted studies [1,2] proved that the difficulties of obtaining real solutions in the calculation of steady-state conditions can be significantly overcome if we proceed from the inverted form of the nodal equations. The difficulties of forming the inverted form of nodal equations, to a certain extent, were overcome using various methods of matrix factorization and iterative methods [3].

Calculations of electrical networks are greatly simplified if the distribution coefficients of driving currents are known [4]. The matrix of distribution coefficients can be formalized both by circuit conversion methods and by analytical methods, which are associated with known difficulties as the electrical network circuit becomes more complex.

Topological methods for forming a matrix of distribution coefficients were developed in [5], with known difficulties in finding and determining specific 2-trees of a graph of a complex electrical network. These difficulties were overcome in [6], where an analytical approach to determining the topological content of the driving current distribution coefficients was proposed. This led to a significant simplification of the technology of searching for specific trees without defining the 2-trees of a complex network graph. The proposed method significantly reduces the amount of work performed and provides fast and reliable convergence of the steady state [7, 8].

The paper proposes simplified topological models for estimating the parameters of the steady state, using the conditions of independent currents in the branches of the electric network circuit.

2 Formalization of the problem

The conditions for mutual independence of currents in the branches of the circuit can be determined on the basis of the matrix equation of the first Kirchhoff law [9]:

$$M \cdot \dot{I} = \dot{J} \quad (1)$$

where M is a rectangular first incidence matrix; $\dot{I}$ – column matrix of complex branch currents; $\dot{J}$ – column matrix of complex driving currents.

If we divide equation (1) into block matrices with respect to tree currents and chords, then the resulting block equation can be written as:

$$M_d \cdot \dot{I}_d + M_x \cdot \dot{I}_x = \dot{J}. \quad (2)$$

where $\dot{I}_d$ – is the column matrix of circuit tree currents; $\dot{I}_x$ – column matrix of circuit chord currents; $\dot{J}$ – initial matrix of setting currents.

If we take into account that the matrix M_d – is square and non-singular, then the matrix of currents $\dot{I}_d$ of the circuit tree can be determined by the formula:

$$\dot{I}_d = M_d^{-1}(\dot{J} - M_x \cdot \dot{I}_x). \quad (3)$$

This expression confirms that the predetermined currents in each chord of the circuit can be expressed as terms of the driving currents applied at its vertices.

Equations (3) determine the equivalent driving current in the form:

* Corresponding author: abeked@mail.ru

 https://doi.org/10.1051/e3sconf/202346101025

$$\dot{J}_э = \dot{J} - M_x \cdot \dot{I}_x. \qquad (4)$$

In this case, at the end vertex, the chord currents are taken into account with the same signs as the currents in the branches, and at the initial vertex, with opposite signs. This leads to a significant simplification of the computational model by removing all chords from the equivalent circuit, leaving only the circuit tree.

Thus, after selecting and determining the currents in the chords of a complex circuit of an electrical circuit based on the known matrix of current distribution coefficients, it is possible to significantly simplify the design circuit of a complex electrical circuit.

3 Formation of simplified steady state models

According to the known matrix of distribution coefficients, the currents in the branches of the circuit are determined [9]:

$$\dot{I} = C \cdot \dot{J}. \qquad (5)$$

It is known that all currents of circuit chords are independent and can be removed from the circuit after specification of the nodal driving currents according to formula (4).

After that, the nodal stresses are determined, according to a simplified scheme, in the form [10,11]:

$$\dot{U} = U_0 + C_d^T Z_B C_d \overline{U_D^{-1}} \bar{S} \qquad (6)$$

where $C_d = M^{-1}$ - current distribution coefficient matrix of the circuit tree, Z_B – resistance matrix of the circuit tree branch, $\bar{S}$– column vector of coupled powers of nodal loads and generators, T - matrix transposition sign, U_D - diagonal matrix of nodal coupled voltages

All transverse nodal conductivities, considering the no-load power of power transformers and the conductance of the power line, are represented by complex powers in the form:

$$\dot{S}_i = g_i + j \cdot (b_L - b_C)_i |U_i|^2 \qquad (6)$$

where g_i – is the active conductivity of the transformer and the line of the i-th node; b_L – inductive conductivity of the transformer of the i-th node; b_C – capacitive conductivity of the line of the i-th node.

4 Implementation of the algorithm

Preliminary calculations of the steady state of the system under study are performed on the basis of its complete model with the determination of the distribution coefficients of nodal currents. After choosing independent currents, the calculation model of the electrical network is determined, according to which the steady state mode of the complex electrical network of the power system is formed. As an example, a 110 kV network diagram is shown in Figure 1 with the parameters presented in tables 1 and 2.

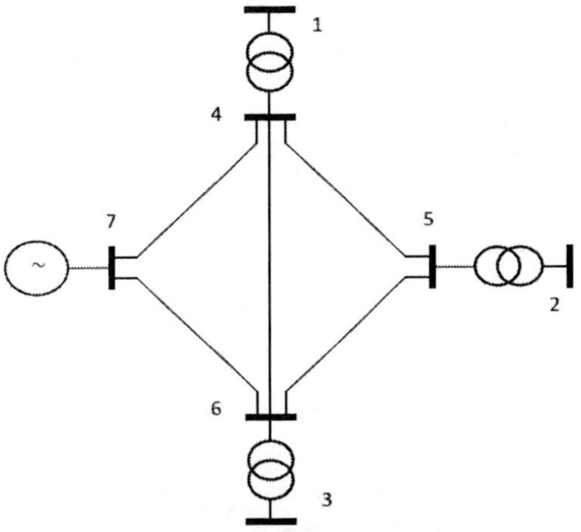

Fig.1. Network diagram 110 kV.

Table 1. Initial data by nodes.

Nodes №	U nom	Load power		Conductivity nodes	
	kV	P, MW	Q, MVar	g, mksm	b, mksm
1	10	60	40	-	-
2	10	40	35	-	-
3	10	25	15	-	-
4	110	-	-	27	162
5	110	-	-	21	-10
6	110	-	-	14	273
7	110	-	-	0.00	510

Table 2. Initial data by lines

Lines №	start	end	R, Om	X, Om	TC (transformation coefficients)
1	4	1	0,514	9,54	0,0956
2	5	2	0,81	13,4	0,0956
3	6	3	1,22	20,15	0,0956
4	7	4	6,6	19,9	-
5	4	5	16,2	25,2	-
6	7	6	6,8	16,24	-
7	6	5	24,3	37,8	-
8	6	4	13,5	21	-

The matrix of distribution coefficients is formed according to a specially developed program [13] based on the directed graph of the original network, shown in Fig.2.

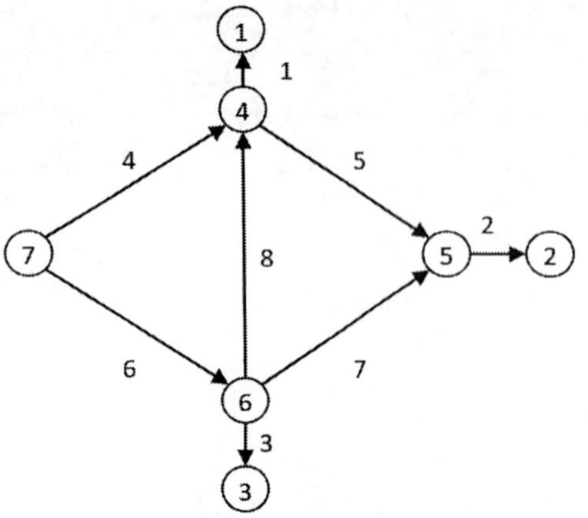

Fig.2. Directed graph of the scheme

The values of distribution coefficients are given in Table 3.

Table 3. Current distribution coefficient matrix

-1	0	0	0	0	0
0	-1	0	0	0	0
0	0	-1	0	0	0
-0.634+0.039i	-0.504+0.02i	-0.309-0.009i	-0.634+0.039i	-0.504+0.02i	-0.309-0.01i
0.091+0.01i	-0.576+0.005i	-0.077-0.002i	0.091+0.01i	-0.576+0.005i	-0.077-0.002i
-0.091-0.01i	-0.424-0.005i	0.077+0.002i	-0.091-0.01i	-0.424-0.005i	0.077+0.002i
-0.366-0.039i	-0.496-0.02i	-0.691+0.009i	-0.366-0.039i	-0.496-0.02i	-0.691+0.009i
-0.274-0.029i	-0.072-0.015i	0.232+0.007i	-0.274-0.029i	-0.072-0.015i	0.232+0.007i

The steady state mode of the electric power system is formed on the basis of a complete network model, the parameters of which are presented in Table 4, with the values of nodal loads [13,14]:

$$S = \begin{pmatrix} 60 + 40i \\ 40 + 35i \\ 25 + 15i \\ 0{,}389 - 2{,}333i \\ 0{,}302 + 0{,}144i \\ 0{,}202 + 0{,}391i \end{pmatrix}$$

Table 4. Steady state parameters for the complete network topological model.

Nodes number, j	U_j	$\lvert U_j \rvert$	$\arg(U_j)$
1	99,817–12,686i	100,62	-0,126
2	89,415–13,282i	90,396	-0,147
3	105,146–10,421i	105,661	-0,099
4	104,575–7,762i	104,862	-0,074
5	95,727–8,542i	96,107	-0,089
6	108,733–6,159i	108,907	-0,057

The values of currents in the branches of the circuit, determined by formula (5), are equal to:

$$I = \begin{pmatrix} 0{,}315 - 0{,}21i \\ 0{,}21 - 0{,}184i \\ 0{,}131 - 0{,}079i \\ 0{,}337 - 0{,}254i \\ 0{,}1 - 0{,}096i \\ 0{,}323 - 0{,}192i \\ 0{,}111 - 0{,}088i \\ 0.08 - 0.042i \end{pmatrix}$$

5 Simplified Models for Calculating Steady Mode Parameters

Suppose we know the independent currents in chords 7 and 8. Then, the equivalent driving currents in the nodes, determined according to expression (4), as well as the values of the nodal powers, respectively, will be equal to:

$$J = \begin{pmatrix} 0{,}315 - 0{,}21i \\ 0{,}21 - 0{,}184i \\ 0{,}131 - 0{,}079i \\ -0{,}078 + 0{,}053i \\ -0{,}11 + 0{,}087i \\ 0{,}192 - 0{,}113i \end{pmatrix} \quad S1 = \begin{pmatrix} 60 + 40i \\ 40 + 35i \\ 25 + 15i \\ -14{,}841 - 10{,}045i \\ -20{,}904 - 16{,}622i \\ 36{,}495 + 21{,}525i \end{pmatrix}$$

The steady state parameters are determined by (8), based on the simplified topological model shown in Fig.3.

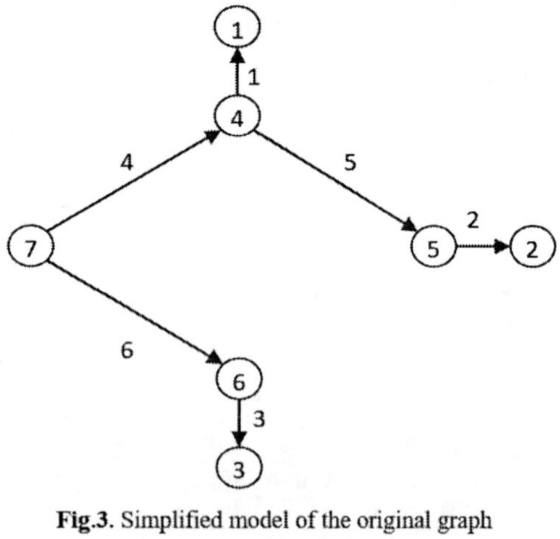

Fig.3. Simplified model of the original graph

The results of calculating the parameters of the steady state, obtained on the basis of a simplified model, are presented in Table 5. The errors of the calculation results are within the allowable values.

Table 5. Comparative results of steady state calculations

j	By full scheme		By simplified model		Relative error	
	$\lvert U_j \rvert$	$\arg(U_j)$	$\lvert U_j' \rvert$	$\arg(U_j')$	Δ_j, %	δ_j, %
1	100,62	-0,126	99,49	-0,128	1,12%	1,59%
2	90,4	-0,147	88,39	-0,147	2,22%	0,00%
3	105,66	-0,099	107,23	-0,097	1,49%	2,02%
4	104,86	-0,074	103,79	-0,074	1,02%	0,00%
5	96,11	-0,089	94,24	-0,086	1,94%	3,37%
6	108,91	-0,057	110,43	-0,056	1,40%	1,75%

If we assume that the currents in branches 5 and 8 are known, then the equivalent driving currents and, accordingly, the nodal powers will be equal to:

$$J = \begin{pmatrix} 0{,}315 - 0{,}21i \\ 0{,}21 - 0{,}184i \\ 0{,}131 - 0{,}079i \\ 0{,}022 - 0{,}044i \\ -0{,}099 + 0{,}096i \\ 0{,}081 - 0{,}025i \end{pmatrix} \quad S1 = \begin{pmatrix} 60 + 40i \\ 40 + 35i \\ 25 + 15i \\ 4{,}255 + 8{,}332i \\ -18{,}842 - 18{,}257i \\ 15{,}337 + 4{,}782i \end{pmatrix}$$

The steady state parameters determined from the simplified model shown in fig. 4 are shown in Table 6.

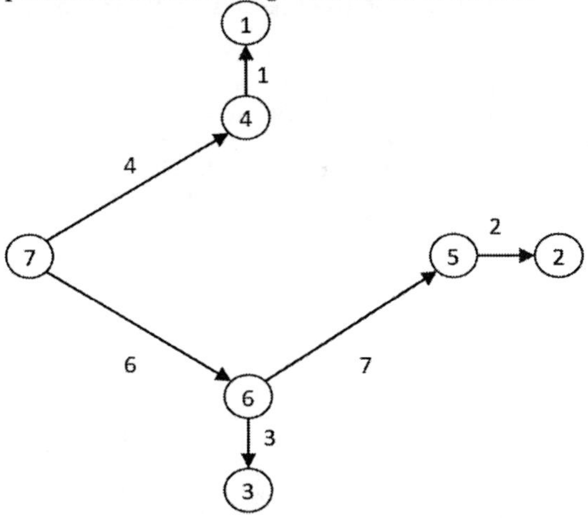

Fig.4. Simplified model of the original graph

Table 6. Comparative results of steady state calculations

j	By full scheme		By simplified model		Relative error	
	$\lvert U_j \rvert$	$\arg(U_j)$	$\lvert U_j' \rvert$	$\arg(U_j')$	Δ_j, %	δ_j, %
1	100,62	-0,126	101,42	-0,128	1,12%	1,59%
2	90,4	-0,147	88,84	-0,147	2,22%	0,00%
3	105,66	-0,099	105,56	-0,097	1,49%	2,02%
4	104,86	-0,074	105,63	-0,074	1,02%	0,00%
5	96,11	-0,089	94,66	-0,086	1,94%	3,37%
6	108,91	-0,057	108,81	-0,056	1,40%	1,75%

Table 6 shows that the results obtained based on the simplified model are in good agreement with the results of the calculation using the full model.

The conducted studies have shown that the same mode of the electrical network can be estimated on the basis of a different simplified model using possible independent currents of the electrical network.

6 Conclusion

1. The topological model of a complex electrical system allows you to set independent currents in the branches of its equivalent circuit.

2. To form the parameters of the steady state, simplified topological models of a complex network of the power system can be used.

References

1. Venikov V.A. Modeling of large power systems // Electricity. 1978.- No. 5. - S. 64-67.
2. Manusov B.Z., Lykin A.V., Sidorkin Yu.M. Algorithms of the Newton-Raphson method for solving nodal equations in inverted form. Izvestiya vuzov SSSR. Energy. 1974.- No. 9. - P.3-7.
3. Fazylov Kh.F., Nasyrov T.Kh. Steady Modes of Electric Power Systems and Their Optimization T.: Molniya. 1999.-370 p.
4. Melnikov N. A. The use of distribution coefficients in the calculation of complex equivalent circuits for electrical circuits / / Electricity No. 2.1961. pp. 9-13.
5. Geraskin O.T. Methods for determining 2 - graph trees in the topology of electrical networks // Izvestiya ANSSSR. Energy and transport. 1967.- No. 5. - P.106-112.
6. D. S. Akhmetbaev, A. D. Akhmetbaev, S. Zhumazhanov, B. Zhakishev. New Modeling of Steady-State Modes of Complex Electrical Grids of Power Systems MATEC Web of Conferences 155, 01043 (2018)
7. D. S. Akhmetbaev. Method for calculating steady-state modes of electric networks based on current distribution coefficients. Electricity, 2010 (11)
8. Akhmetbaev, D.S., Dzhandigulov, A.R., Akhmetbaev, A.D. Topological algorithm for forming nodal stresses of complex network energy systems/(2019) E3S Web of Conferences, 139, No. 01066. DOI: 10.1051/e3sconf/201913901066
9. Melnikov N.A. Matrix method for the analysis of electrical circuits. M.: Energy. 1972.-232 p.
10. Akhmetbaev, D.S., Akhmetbaev, A.D., Dzhandigulov, A.R. New Approaches to the Topological Method of Analysis of Electrical Networks/ 2019 International Multi-Conference on Industrial Engineering and Modern Technologies, FarEastCon 2019, 8933870
11. Akhmetbayev, D.S., Dzhandigulov, A.R., Bystrova, S.V. Topological system method of formation of transformer transformation coefficients. (2020) E3S Web of Conferences, 216, № 01087, DOI: 10.1051/e3sconf/202021601087
12. Akhmetbaev D.S., Dzhandigulov A.R Development of algorithms for the formation of steady-state modes based on the topology of electric power systems// Journal of Physics: Conference Series 1392(1),012079, 2019.
13. Jandigulov A.R., Akhmetbaev D.S. Implementation of a new topological algorithm for calculating current distribution coefficients in complex electrical networks. Computer program. Certificate of entering information into the state register of rights to objects protected by copyright. No. 1552 dated January 31, 2019 7.
14. Jandigulov A.R., Akhmetbaev D.S. Finding all spanning graphs of a given graph. Computer program. Certificate of entering information into the state register of rights to objects protected by copyright. No. 1551 dated January 31, 2019

A method for synthesizing and analyzing fault trees to improve the survivability of a demand response aggregator

Irina Kolosok[1]*, *Elena* Korkina[1]

[1] Melentiev Energy Systems Institute of Siberian Branch of the Russian Academy of Sciences, 130 Lermontov str., Irkutsk, Russia

Abstract. The paper is concerned with the application of fault tree and event tree methods to analyze and improve the survivability of the structure of the demand response aggregator (DR Aggregator). The approach used for this purpose was first called the Method of Fault Tree Synthesis and Analysis, combining the fault tree and event tree methods based on the mechanism and kinetics of undesired events affecting the object under study. With this approach, one can more accurately determine the most vulnerable components of the DR Aggregator structure, and ways to enhance their security.

1 Introduction

A highly researched topic for many years has been the survivability properties of large energy systems and their components [1]. Necessary conditions for maintaining the survivability of the system are the fulfillment of requirements for the trouble-free operation of its components, ensuring their redundancy or interchangeability, the use of adaptive algorithms for information interaction, and the organization of protective measures against undesired events (UE). This research focuses on the Demand Response (DR) Aggregator, which is a coordinating center in the demand-side management structure created to unite electric power entities not involved in the wholesale electricity market (WEM), which have their own excess energy resources for sale to the energy system [[2]-[5], etc.]. In today's context of the transition to intelligent energy systems (IES), each of these entities (a power plant, an industrial enterprise, a hotel, an apartment complex, and others) is a cyber-physical system (CPS), combining information, control, and technological components [6].

When studying the survivability of the DR Aggregator as an integration of several CPSs, it is necessary to take into consideration not only the survivability of the technical (physical) subsystem but also the information and control (cyber) subsystem of its participants. Therefore, when examining the undesired events in the analysis of the survivability of a technical (physical) system (major technical failures, emergency events, and others), cyber-attacks should be factored in and the survivability of the DR Aggregator should be analyzed from the standpoint of the cyber-physical stability of this complex structure. This study utilizes an original approach [7], combining FT and ET methods based on the mechanism and kinetics of undesired events, with the view to boosting the stability of the DR Aggregator.

2 Fault Tree and Event Tree Methods

Logical-graphic methods of FT and ET are used to identify cause-and-effect relationships between events [8]. The fault tree method underlies a logical-probabilistic model of cause-and-effect relationships between system failures and failures of its components and other undesired events [9]. Fault tree allows identifying all the paths that lead to the main event (an accident at the facility), and makes it possible to determine the minimum number of combinations leading to an accident. The Event Tree method is used to analyze the consequences of accidents that have occurred [10].

The disadvantages of the FT and ET methods include the need to involve the knowledge and experience of experts at some stage of the analysis, which reduces the universality of the methods. To systematize and formalize the knowledge of experts and increase the stability of the DR Aggregator operator, an original approach is used, which combines FT and ET based on the mechanism and kinetics of undesired events [7]. The essence of the method, called the Method of Fault Tree Synthesis and Analysis, is to superimpose undesired events on the branches of the fault tree, given the degree of degradation of each component (from a defect state to a complete failure), and in the case of a complex technical system (CTS) - to an emergency state (emergency event). The mechanism of change of the CTS is based on the properties (characteristics) of the components; the kinetics of events involves changes in the state of the components:

Defect (Def)->Damage (Dam)->Destruction (Des)->Failure (F).

Moreover, these states are analyzed for the components of both the technical and information subsystems of the cyber-physical system. With this approach, one can more accurately determine which component is most vulnerable in case of emergency

* Corresponding author: kolosok@isem.irk.ru

situations, and how to enhance the security of individual components and the entire CTS, in our case, the DR Aggregator.

3 Fault tree of DR Aggregator. Mechanisms of enhancing the DR Aggregator survivability

Figure 1 shows the structure of the DR Aggregator fault tree with three levels: I: CPS, consisting of control, technical, and information subsystems; II: components of these subsystems, III: measures to increase the fault tolerance of their components.

From the perspective of CPS, DR Aggregator consists of three subsystems - process, information-communication, and control one, which provides interaction of DR Aggregator with SO and ATN as part of the DSM pilot project. As shown in [11], a failure in the operation of one of the CPS subsystems can lead to the failure of the other subsystem, to a combined emergency state, and even to the failure of the entire structure.

A cluster structure consisting of the components, which are territorially and technologically integrated electricity consumers is proposed in [12] to increase the efficiency of the DSM mechanism. Cluster participants are electricity consumers represented by CPS facilities consisting of two subsystems. A failure in the operation of one of the cluster subsystems may lead to the failure of this cluster, and not the entire structure of the DR Aggregator.

Figure 2 shows a diagram of the FT-based cluster structure of the DR Aggregator and Active Consumers (AC) in the CPS format. The lower level of the FT is the components of the cyber and physical subsystems, and the Human Factor (HF) and smart sensors are included in both subsystems.

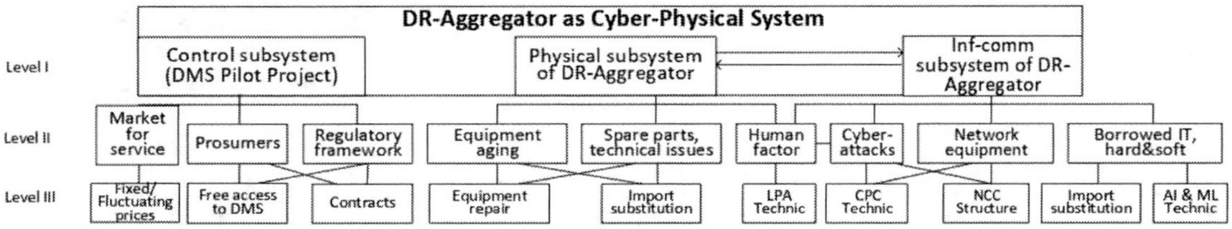

Fig. 1. Fault Tree of DR Aggregator survivability (DSM – demand-side management, IT - information technology, computer hardware/software, LPA - logical-probabilistic analysis, NCC - network-centric control, CPC - cyber-physical control, AI - artificial intelligence , ML - machine learning)

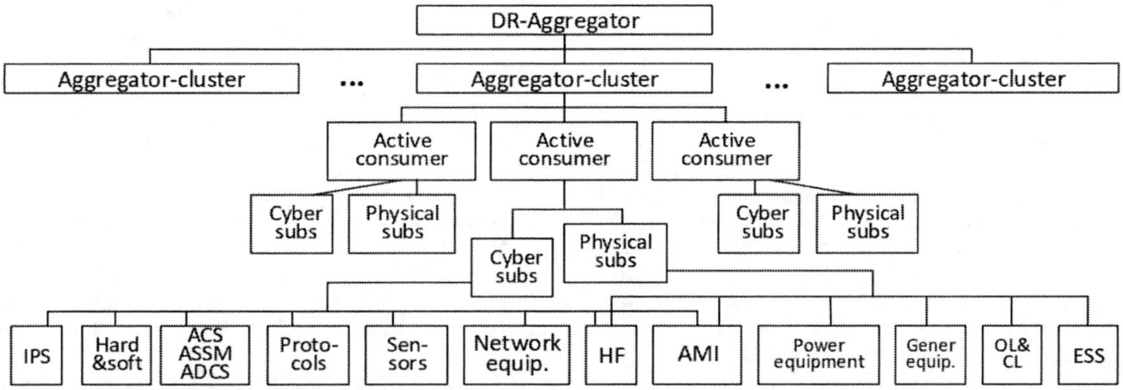

Fig. 2. Fault tree of the DR Aggregator cluster structure (APCS/ASSM/ADCS -automated process control systems / automated system of synchronized measurements / automated dispatch control system, HF – Human Factor, AMI – Automatic Measuring Infrastructure, OL&CL - Overhead and Cable lines, ESS – Energy Storage System)

A properly organized network-centric control (NCC) system for this structure can also contribute to boosting the survivability of the DR Aggregator [13]. In contrast to the traditional classical hierarchical structure inherent in any power system, with NCC, there are horizontal connections at each level of the hierarchy. Components located at the same level carry out information exchange, each with each other. As a result, the speed of decision-making and the reliability of the DR Aggregator structure increase. All levels of the network-centric control structure are managerial and manageable (except for the first) and only the lowest, process level, remains fully manageable.

Modern smart technologies allow lower-level components to also be involved in responding to changing conditions. This opportunity is provided to the lower level by cyber-physical control (CPC) [14],[15]. The implementation of the CPS becomes possible thanks to the combination of CPS and NCC system, when the participants of the DR Aggregator are considered as cyber-physical systems and can interact directly with each other.

4 Method of synthesis and analysis - development of a fault tree model for a complex technical system

Building a fault tree for a complex engineering system (stages 1-3 in Fig. 3) is not a difficult problem and can be implemented by computer systems, for example, [16]. However, formalizing the cause-and-effect chain of events that bring about a failure in structural components (level 4 in Fig. 3) is a problem, since it is performed exclusively by experts.

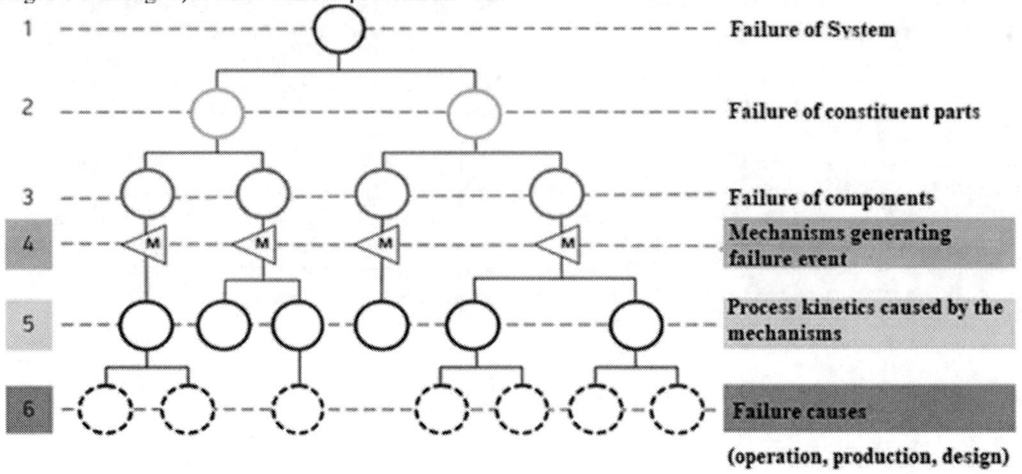

Fig. 3. Conditional diagram of constructing a fault tree [7].

To systematize and uniformly represent expert knowledge, enhance accuracy and achieve an unambiguous interpretation of fault trees, the research [7] proposes constructing level 4 based on structured knowledge about possible undesired/degradation processes and the events that constitute them.

The unique feature of the fault tree in terms of the mechanism and kinetics of undesired events (FT MKUE) [7] is the projection of physical and mechanical phenomena onto the operability state of the system and objects. **The mechanism of an undesired event** lies in the way each node (object) is represented, namely, in the description of its properties (qualities) and the listing of factors affecting it. **The kinetics of an undesired event** (the development of the degradation process) is characterized by a sequence of events gradually leading to failure. An event (triggering of an impact factor) at a vulnerable point in the characteristic of the object (its defect) is a condition for triggering the undesired event mechanism.

The state matrix for FT MKUE. Traditional fault tree shows the relationship between successive levels of hierarchy - object, components, and subcomponents, using the logical operators AND and OR. This structure is transparent, since its correctness can be verified using the technical documents on the configuration of the object (system).

FT MKUE contains not only the equipment, but also the events leading to failure. In order to clearly describe the FT MKUE of an object, it is first necessary to create a detailed matrix of object states from Defect to Failure. At this preliminary stage, events that are not related to the technical nature of failures, for example, the human error, are eliminated. The example of a mechanical system (MS) (Fig. 4, a) used in [7] demonstrates the successive steps leading to the failure of the entire mechanical system:

• Under the influence of a certain event, hidden defects in the part (D) begin to appear, leading to its damage; at this stage, another event contributes to the destruction of the damaged part; after this, the destroyed part becomes inoperable (state: failure of D).

• Failure of D leads to damage to the assembly unit (AU), then destruction and, finally, to failure of the AU.

• Failure of the AU through the stages of damage and destruction leads to failure of the mechanical system.

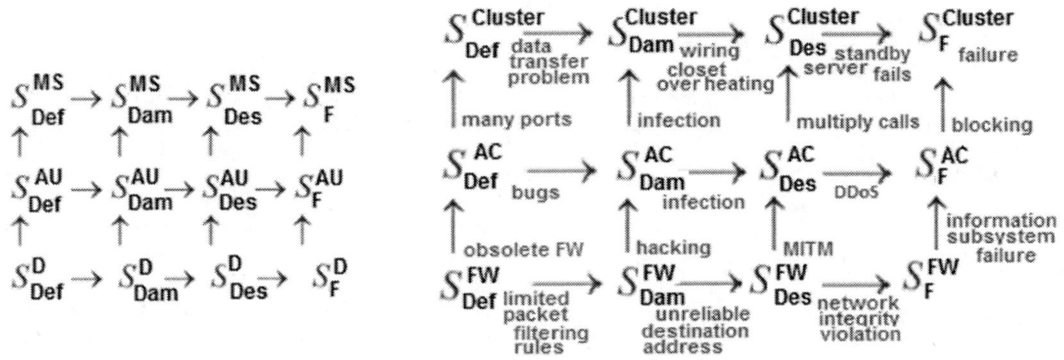

a) b)

Fig. 4. State transition matrix: a – mechanical system (MS) [7]; b - cluster structure of the DR Aggregator (Cluster - DR-A control unit, AC - active consumer, NF - network firewall).

Figure 4 b indicates such a state matrix compiled for the cluster structure of the DR Aggregator.

The cause-and-effect relationships of the formation of physical and mechanical events that cause failures used for the analysis and synthesis of a fault tree are presented as rules, which are one of the ways to represent knowledge in artificial intelligence. To enable representation of the information showing the mechanism of the process in the FT, an extension of the representation of generally accepted graphic symbols for constructing FT with a special symbol M was introduced in [7] to denote the mechanism of the process.

5 Application of FT MKUE to increase the survivability of the DR Aggregator

The next step aimed at boosting the survivability of the DR Aggregator is to build a FT MKUE by adding to the typical FT a layer of events that generate failure and a layer of factors influencing certain vulnerabilities present at each level of the FT. The levels considered following the designation in the traditional fault trees are (Fig. 5):

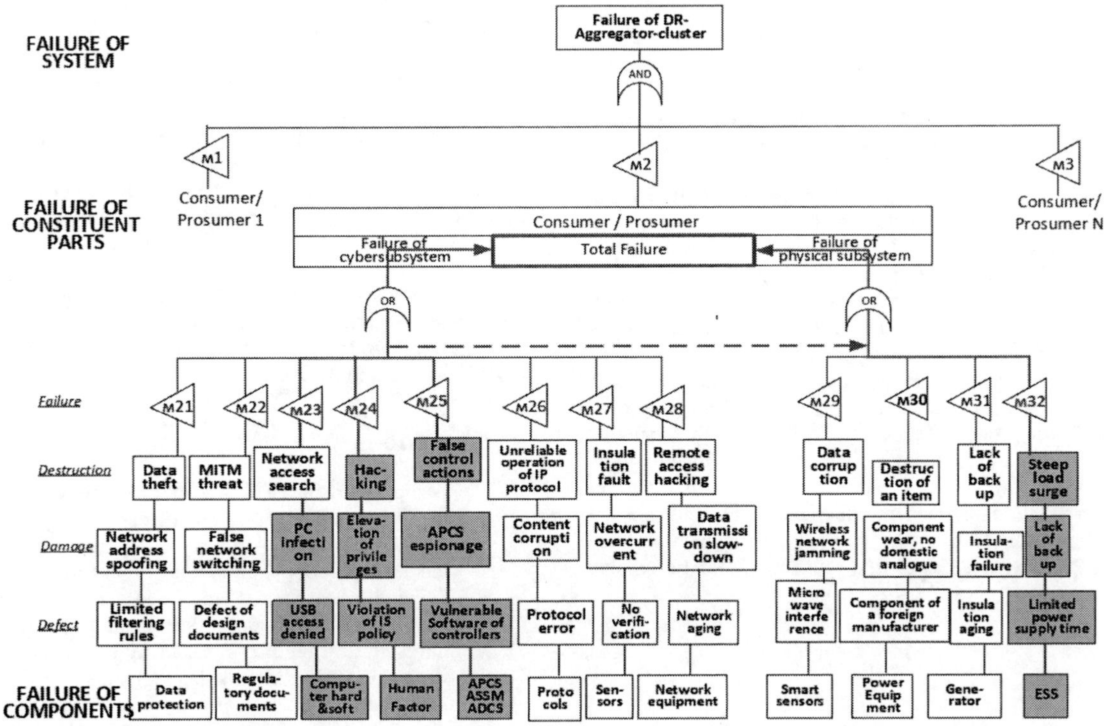

Fig. 5. A FT MKUE of the active consumer.

I. System level – DR Aggregator Cluster.
II. Basic component: Active consumer as CPS (cyber subsystem and physical subsystem).
III. Component: Information security (information security tools), Regulatory documents, Computer hardware/software, APCS/ADSS/ADCS, Protocol, Sensor, Network equipment, Smart sensor, Power equipment, Generator.

Three intermediate levels between the Constituent part and Component levels are specific state levels

Defect (Def)->Damage (Dam)->Destruction (Des).

Figure 5 shows a fault tree constructed with failures in both the information and physical subsystems:

In the information subsystem - 1) branch of computer hardware and software - a virus software was recorded on a USB memory card (defect), then the computer was infected and the attacker searched for an opportunity to get into the local network of an energy facility. 2) branch of HF - an employee of an energy facility violated strict rules for access to the network, increased the level of privileges, and the attacker's virus entered the network. 3) branch of APCS - through the APCS software, the virus reached the control controllers, which made it possible to spy on processes in the APCS and generate false signals at the controller input.

In physical subsystem - an active consumer uses a small-capacity ESS of poor quality assembly; there is no ESS reserve. The power output is limited in time, which caused a change in operating conditions with a steep increase in load. The ESS has entered a state of failure; the active consumer cannot fulfill the obligations assumed to the DR Aggregator cluster. Failures that occur simultaneously in both subsystems are interpreted as a combined failure.

Similarly, all mechanisms of undesired events are described based on an expert analysis of the chains of all events leading to failure. The construction of such a tree helps identify the root causes of failures, i.e., hidden defects that are present in each component and unexpectedly manifest themselves with some "trigger" undesired event or combination of several events.

In the theory of reliability of technical systems, there are standards and statistical data that allow determining the technical condition of an object or its part during its life cycle. A cyber-physical system is a new round of evolution of complex technical systems operating under the control of a rapidly changing information environment. It is too early to talk about norms and statistics regarding CPS. It is however essential to use expert experience and eradicate the slightest defects in the parts and components of the CPS now, thus laying the foundations for its reliable operation.

6 Conclusion

The survivability of the DR Aggregator is a new crucial issue which has become the focus of the research after gaining some experience in studying the structure of the DR Aggregator. In the context of "digitalization" of the energy sector, the DR Aggregator and the active consumer included in its structure acquire the properties of CPS, consisting of industrial-process and information-and-communication subsystems. Therefore, when analyzing the survivability of the DR Aggregator, it is necessary to take into account not only process failures, but also emergency conditions in information -communication subsystem and consider the survivability of the DR Aggregator from the standpoint of the cyber-physical stability of this sophisticated structure.

As was shown earlier in the studies by the authors, increasing the DR Aggregator survivability is facilitated by a rationally built cluster structure of the DR Aggregator, an optimally designed network-centric control system for this structure, and a system of measures to restore the DR Aggregator operation in the face of negative impacts.

This study examines the possibility of using the *Method of fault tree synthesis and analysis* to further increase the survivability of the DR Aggregator. This involves superimposing undesired events on the branches of the fault tree, considering the degree of degradation of each component from a defect state to a complete failure. These states are analyzed for both the components of the technical and information subsystems.

The considered methodology was used to construct a modified fault tree for the DR Aggregator cluster. This allowed the formalization of experts' knowledge regarding the cause-and-effect relationships of events leading to cluster failure, including the causes and mechanisms behind their occurrence. Furthermore, this method allows us to identify hidden defects in the components of the cluster structure, which significantly reduce its survivability and the survivability of the entire DR Aggregator.

This study was carried out within the framework of the State Assignment project (No. FWEU-2021-0001) of the Program for fundamental research of the Russian Federation for 2021-2030.

References

1. Voropai N.I. Survivability of EPS: methodological framework and methods of the research. Proceedings of USSR Academy of Sciences. Energy and Transport **6**, pp. 52-59 (1991) (in Russian).
2. Demand management in the Russian electric power industry: opportunities opening up. Expert and analytical report/ ed. Opadchy F., Kholkin D. Moscow, 2019. Available at: https://www.so-ups.ru/fileadmin/files/company/markets/dr/publication/EnergyNet_2019.pdf

3. Kuleshov M.A. (2019) Pilot project to create Demand Response aggregators. Available at: https://www.so-ups.ru/fileadmin/files/company/markets/dr/present/2019/present_041219_kuleshov.pdf.
4. Konceptsiya funktsionirovaniya agregatorov raspredelennykh energeticheskikh resursov v sostave Edinoj energeticheskoj sistemy Rossii. Agregatory upravleniya sprosom na elektroenergiyu [The concept of functioning of aggregators of distributed energy resources as part of the Unified Energy System of Russia. Aggregators of electricity demand management] Available at: https://www.so-ups.ru/fileadmin/files/company/markets/dr/docs/dr_agregator_concept.pdf
5. Byk F.L., Myshkina L.S. Aggregator – an element of digital transformation of the regional network. Proc. of 91st Int. Scientific Workshop on Methodological Problems in Reliability Study of Large Energy Systems **Vol. 70**, pp. 144-152 (2019) (in Russian).
6. Voropai N.I., Kolosok I.N., Korkina E.S., Osak A.B. Problemy uyazvimosti i zhivuchesti kiber-fizicheskih elektroenergeticheskikh sistem [Problems of vulnerability and survivability of cyber-physical electric power systems]. Energeticheskaya politika, **5**, pp. 53-61 (2018)
7. Berman A.F., Pavlov N.Y., Nikolaychuk O.A. the method for synthesis and analysis of "failure trees" based on the concepts of mechanism and events kinetics. Issues of Risk Analysis, **Vol. 15**, **3**, pp. 62-77 (2018) (in Russian) Available at: https://doi.org/10.32686/1812-5220-2018-15-3-62-77.
8. Guk Yu.B. Rashchet nadezhnosti electroenergeticheskih ustanovok, Leningrad, Energoatomizdat, 224 p. (1988)
9. Ryabinin IA On the relationship of mathematical logic with the theory of probability. Scientific notes. Saint-Petersburg **6**, pp.70-176 (2008)
10. Reliability of technical systems and man-made risk. Event tree. Available at: http://www.obzh.ru/nad/6-9.html
11. Diakov A.F., Stennikov V.A., Senderov S.M. et al.; Reliability of energy systems: Problems, models and methods for solving them/ ed. By N.I.Voropai // Novosibirsk: Nauka, 284 p. (2014)
12. Kolosok, I., Korkina, E. Selection of prosumers in the structure of the demand response aggregator by analytical methods. ELECTRIC POWER. Transmission and Distribution. **Vol. 3**, **66**, pp.22-28 (2021)
13. Kolosok, I., Korkina, E. Application of net-centric management to the structure of the electricity demand aggregator. Industrial power engineering **12**, pp.2-8 (2020).
14. Kudzh S.A., Tsvetkov V.Ya. Network-centric control and cyber-physical systems. Educational Resources and Technologies, **Vol. 2(19)**, pp. 86-92 (2017) (in Russian).
15. I. Kolosok, E. Korkina. Applying the principles of cyber-physical management to Demand Response Aggregator structure // IFAC Papers OnLine, **Vol. 50, 9**, p. 198-203 (2022)
16. Mozhaev A.S., Gladkova I.A. Software package for automated structural and logical modeling of complex systems (ПК АСМ 2001). Registration certificate № 2003611099

Estimation of acceptable failure rates for equipment of power distribution network

Artem Vanin [1*], *Dmitry* Gabdushev [1], *Vladimir* Tulsky [1]

[1]National Research University "Moscow Power Engineering Institute", Moscow, Russia

Abstract. Regulatory requirements for reliability indexes of power distribution networks are considered. The dynamics and structure of the reliability indexes of electricity distribution companies are analyzed. Criteria for determining the acceptable number of outages of power network equipment are proposed to calculate the acceptable frequency of outages. The proposed approach is illustrated by calculation of acceptable values of reliability indexes for the power distribution network of Moscow Region.

1 Introduction

Power distribution networks are very diverse in terms of areas covered, network structure, consumer composition, length and other characteristics. Therefore, it is impossible to develop uniform standards for the reliability level of power distribution networks. Both in Russia and in most countries of the world, the planned values of reliability indexes are set for each electricity distribution company separately. Reliability indexes *SAIDI* and *SAIFI* are used as standard indicators.

In the most countries, there is usually a regulatory body at the legislative level, which sets distribution network reliability targets, develops a methodology for their calculation, sets penalties and incentives for each electricity distribution company. In the UK, OFGEM sets targets and incentive rates for each network operator [1]. In Italy, the AEEG sets SAIDI and SAIFI targets for electricity distribution company, as well as penalties and bonuses [2]. In the Netherlands, the Netherlands Competition Authority is an independent government agency responsible for supervising compliance with general competition law and laws related to energy and transport [2]. In the USA, the electricity distribution companies are regulated by The Public Service Commission [3].

The reliability of different distribution networks can vary a lot. Many countries divide power distribution networks depending on the type of territory served, climatic conditions, load density and other factors [4]. Reliability standards are set for each type of network separately, based on historical data for the previous 3-5 years. In many countries, the maximum permissible duration of one power outage at the consumer level is set by law. This standard takes values from 4 to 24 hours.

In Russia, the methodology for calculation planned values of reliability indexes is determined by the order of the Ministry of Energy of the Russian Federation № 1256 dated 29.11.2016 [5]. For this purpose, the basic values of reliability indexes are calculated for groups of territorial power distribution networks with comparable characteristics and operating conditions. The base values of reliability indexes are given in the order of the Ministry of Energy of Russia № 976 dated 18.10.2017 [6]. Annual planned values of reliability indexes are determined based on the rate of improvement, which is determined in such a way that the power distribution network achieves the basic indexes in the long-term regulation period (5 years). For electric distribution companies whose reliability level is better than the baseline, the minimum rate of improvement (1.5% per year) is applied. The planned reliability indexes are achieved if the actual reliability indexes for the calculation period do not exceed them by more than 30%. Tariffs and company's profit directly depend on the fulfillment of the reliability standards [7].

2 Analysis of the reliability indexes of electric distribution company, case of Moscow Region

Figure 1 shows the planned reliability indexes of Moscow and Moscow Region for two long-term regulation periods with their allowable ranges [8-11]. Actual indicators for the first long-term regulation period are also shown.

* Corresponding author: vaninas@mpei.ru

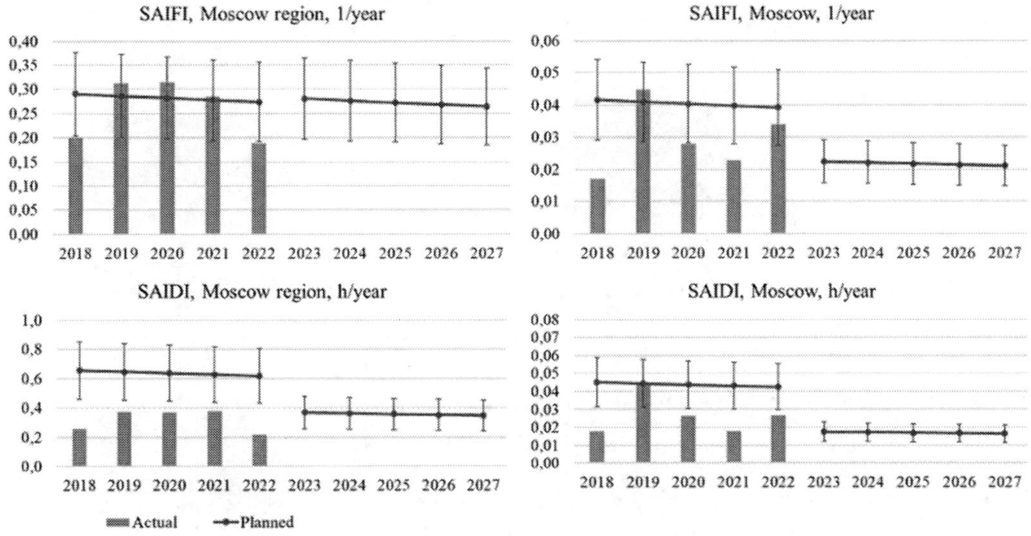

Fig. 1. Actual reliability indexes of Moscow and Moscow region for 2018-2022 and planned indicators for 2018-2022 and 2023-2027.

Figure 1 shows that current standards reduce reliability targets annually. At present, there are no regulatory limits for reduction the reliability planned indexes. Obviously, it is economically impractical and technically unfeasible to reduce reliability indexes to zero. Hence, there is an acceptable reliability level, when achieved, which should be further maintained without requiring improvement.

The reliability level of power distribution networks depends on the structure, composition of the network and reliability of equipment. The structure of the distribution network is determined by location and reliability category of the majority of consumers on served territory. The structure of the network in long-term operation does not change, therefore, reliability indexes are determined by the reliability of the equipment of this network.

Since the reliability indices *SAIFI*, *SAIDI* have additive character, they can be represented as a sum of components from failures of different equipment groups caused by various reasons:

$$SAIFI = \sum_{i=1}^{I} \sum_{j=1}^{J} SAIFI_{ij};$$
$$SAIDI = \sum_{i=1}^{I} \sum_{j=1}^{J} SAIDI_{ij} \quad (1)$$

where I – set of equipment types, J – set of failure causes.

Figure 2 shows the distribution of contributions to the reliability indexes of Moscow and Moscow region of different types of equipment: overhead line (OHL), cable line (CL), substation equipment, relay protection and automation devices (RPA). Figure 3 shows the contributions from different failure causes.

The most significant contribution to the reliability indexes is made by failures of overhead and cable transmission lines caused by inappropriate technical status, insufficient fault detection and effect of trees and vegetation.

3 Criteria for determining the acceptable frequency of equipment outages

Acceptable reliability indicators of the distribution network can be calculated based on the sample of acceptable failures. First of all, failures should be considered in accordance with the order of the Ministry of Energy of the Russian Federation № 1256 dated 29.11.2016 [5]. Intermittent failures eliminated by successful action of automatic reclosing and emergency repairs without power supply interruption are excluded from consideration.

Failures caused by personnel errors and non-compliance with the timing and scope of maintenance and repairs are excluded from the set of acceptable failures.

When calculating the frequency of acceptable failures, equipment with systematic failures are not taken into account. Overhead lines for which the number of failures per year caused by poor technical state or state of route is more than 3 is excluded [12]. Overhead lines are excluded for which the number of outages per year due to one meteorological phenomenon or another external factor is more than 3 [13]. Cable lines are excluded, for which failure rate caused by electrical breakdown is 30 or more failures per 100 km per year [14].

 https://doi.org/10.1051/e3sconf/202346101027

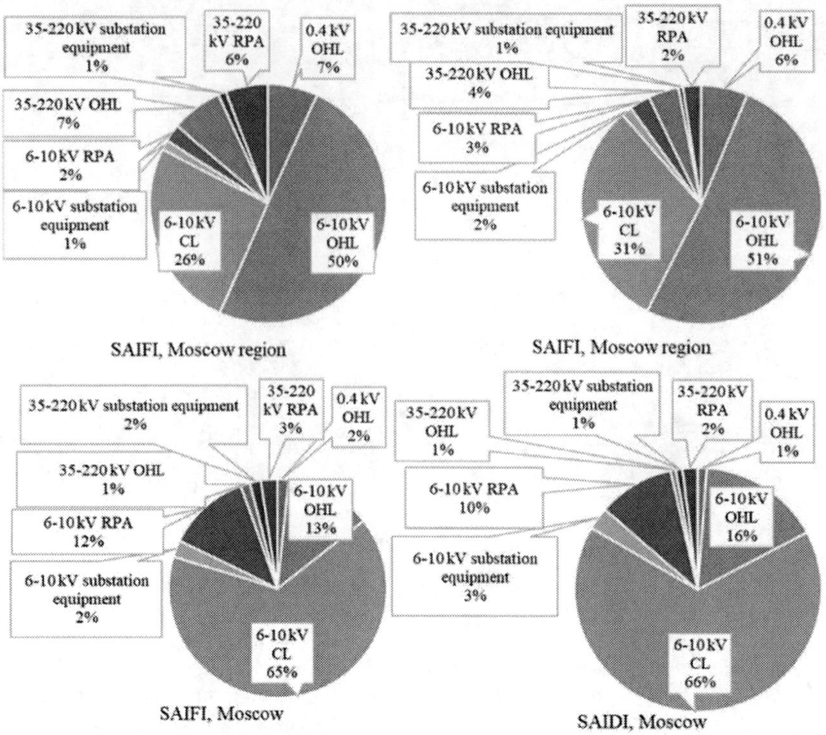

Fig. 2. Contribution of different equipment types to reliability indexes of Moscow and Moscow region

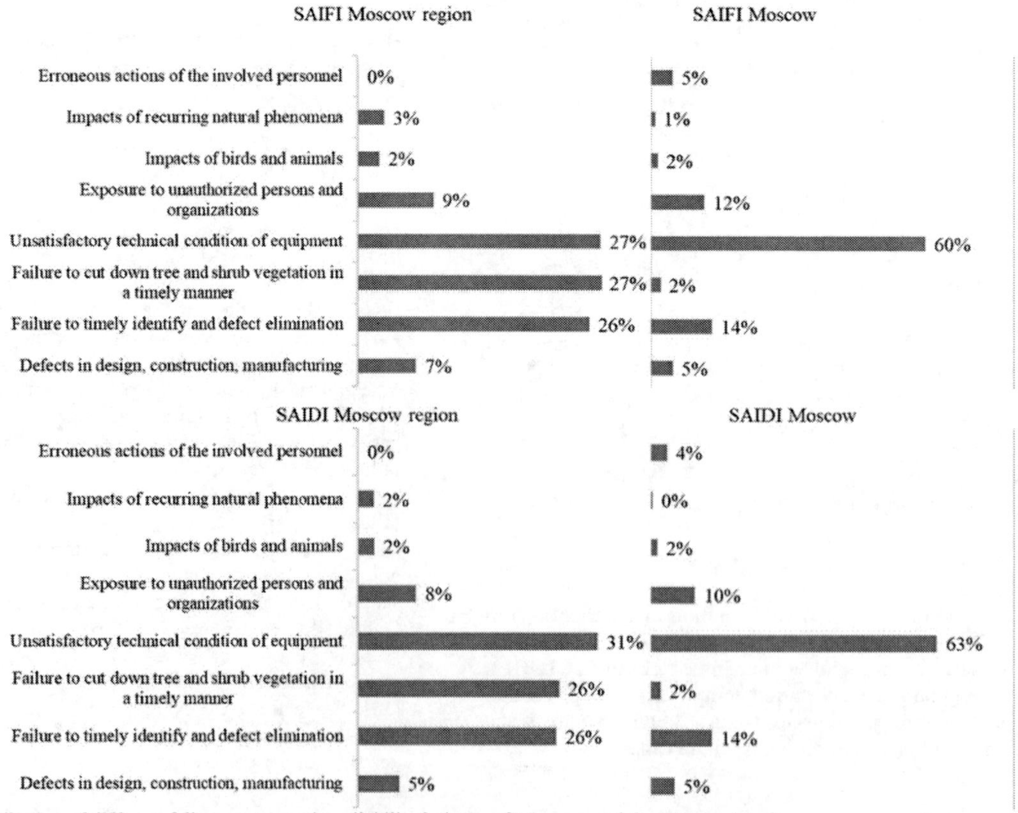

Fig. 3. Contribution of different failure causes to the reliability indexes of Moscow and the Moscow Region

Also excluded cable lines, which failures are caused by violation of cable laying requirements: violation of cable laying depth requirements, backfilling of the trench with soil with large elements of construction debris [15].

The remaining failures caused by external factors considered as an acceptable.

Failures caused the technical state or route condition (for OHLs), are further checked in compliance with the scope and regularity of maintenance.

The majority of OHL failures caused by undetected faults in the form of wire sagging, damage or contamination of insulators, heating and weakening of contact connections. Such defects may appear during the period between repairs, therefore, if the frequency of OHL inspections is observed, the failures are classified as acceptable. The standard frequency of inspections is at least once a year [12].

OHL failures due to falling branches or trees are usually associated with missing the detection of threatening trees outside the line route, therefore the frequency of inspections is also checked for these failures. Failures due to untimely cutting of trees and shrub vegetation within the OHL route should be classified as unacceptable, since the growth of vegetation under the OHL is observable and controllable.

For failures caused by the wear and tear of OHL elements, the regularity of major repairs is checked, which is 12 years for OHL on concrete, metal and composite poles and 6 years for OHL on wooden poles [16].

For 6-20 kV cable lines, the main type of maintenance is testing of insulation with increased voltage of ultra-low frequency 0.1 Hz [14]. The regularity of testing is once every 5 years for CL with paper insulation and once every 9 years for CL with cross-linked polyethylene insulation.

For 35-220 kV cable lines there are tests of cable sheathing with increased rectified voltage and infrared inspection once in 3 years [14].

Failures of circuit breakers are most often associated with defects of the actuator (misalignment, deformation, jamming) and defects of insulators (microcracks, chips, contamination). In accordance with [17], circuit breaker tests are carried out at least once every 5 years.

A significant factor of equipment failures at transformer substations 6-10 kV is a violation of the leak proofness of the roof with the subsequent ingress of moisture. For the equipment located in transformer substations 6-10 kV the periodicity of inspections is checked, which should be at least once every 6 months [14].

For power transformers, the requirements for failure-free operation are also specified as a standardized MTBF (Mean Time Between Failures). The acceptable failure rate caused by technical state is determined based on the standard MTBF.

In standards [18, 19], the probability of failure-free operation within the specified MTBF γ for transformers is not explicitly specified. In [20], the requirement is specified: the probability of failure-free operation over an operating time of 8800 h is at least 0.995. This probability corresponds to a probability of failure-free operation within the MTBF of 25,000 hours of operation of at least 0.985.

For the exponential distribution of operating hours, the expression for the acceptable failure rate is valid:

$$\omega_{ac} = \frac{-\ln(\gamma)}{MTBF / 8760} \quad (2)$$

Then for oil transformers with standard MTBF of at least 25000 hours, the permissible outage frequency is 0.005 1/year; for dry-type transformers with standard MTBF of at least 43800 hours, the permissible outage frequency is 0.003 1/year.

When calculating the acceptable number of failures, incorrect operation of relay protection systems associated with deficiencies and errors of operation are not taken into account. For the rest of cases, a check of compliance with the periodicity of maintenance in accordance with [21] is performed: preventive restoration or preventive inspection, technical inspection, testing - once every 4 years; technical inspection - once every 8 months; test inspection - once every 12 months.

For equipment failures due to workmanship and installation defects, the warranty periods for the equipment and electrical installation work are checked. The failures are classified as acceptable if the service life exceeds the warranty period.

4 Estimation of acceptable values of reliability indexes

The acceptable values of reliability indicators of electric distribution companies can be determined based on historical values through the ratio of acceptable and actual failure rates for each group of equipment and causes:

$$\begin{aligned} SAIFI_{perm} &= \sum_{i=1}^{I}\sum_{j=1}^{J}\frac{\omega_{perm\,ij}}{\omega_{ij}} SAIFI_{ij}; \\ SAIDI_{perm} &= \sum_{i=1}^{I}\sum_{j=1}^{J}\frac{\omega_{perm\,ij}}{\omega_{ij}} SAIDI_{ij} \end{aligned} \quad (3)$$

Based on the proposed criteria, acceptable failure rates for Moscow and Moscow Region equipment were determined.

The average values of SAIFI and SAIDI reliability indexes for the period 2021-2022 for each group of equipment were determined and the acceptable values were calculated. The final values of actual and acceptable reliability indexes SAIFI, SAIDI for Moscow and Moscow region are shown in Figure 4. Taking into account the probabilistic nature of equipment failure rates for the distribution network, confidence intervals of acceptable reliability indexes are also assessed. For most of the equipment the permissible frequency of outages is insignificantly less than the actual value. Figure 4 shows that current reliability indexes for Moscow are within the confidence intervals of acceptable values, for Moscow region current reliability indexes slightly exceed acceptable values. For Moscow and Moscow Region

 https://doi.org/10.1051/e3sconf/202346101027

there is practically no potential for reducing the frequency of power equipment failures. Therefore, it is reasonable to fix the planned reliability indicators for Moscow and Moscow Region at the current level.

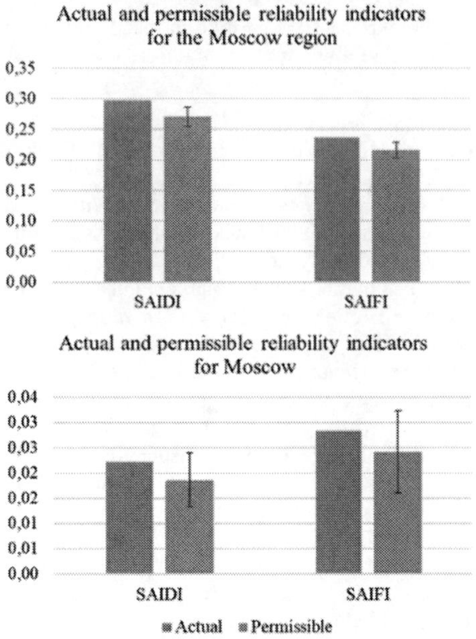

Fig. 4. Actual and acceptable values of reliability indexes for Moscow and the Moscow region

5 Conclusion

It is shown that currently there is a regulatory requirement for electric distribution companies in Russia to improve the level of reliability annually, while the limit level of improvements is not set. Based on the current standards for the power equipment were developed the criteria for determining the acceptable number and frequency of failures. On the example of power distribution network of Moscow region, the acceptable values of reliability indexes were calculated.

References

1. «Statutory security of supply report», Department for Business, Energy & Industrial Strategy, UK Government, 2018, HC 1644 doc. no., pp. 6– 6 URL: https://www.gov.uk/government/publications/statutory-security-of-supply-report-2021.
2. Approaches to setting electric distribution reliability standards and outcomes. The Brattle Group, Ltd. 2012.
3. Order Adopting Standards On Reliability and Quality of Electric Service was issued on July 2, 1991 (case 90-E-1119).
4. Brown R. E. Electric power distribution reliability. – CRC press, 2017.
5. Order of the Ministry of Energy of the Russian Federation dated November 29, 2016. N 1256 "On Approval of the Methodological Guidelines for Calculation of the Level of Reliability and Quality of Supplied Goods and Services for the Organization for Management of the Unified National (All-Russian) Electric Grid and Territorial Grid Organizations" (as amended and supplemented)
6. Order of the Ministry of Energy of Russia No. 976 dd. 18.10.2017 "On approval of base values of reliability indicators, values of coefficients of allowable deviations of actual values of reliability indicators from planned ones and maximum dynamics of improvement of planned reliability indicators for groups of territorial grid organizations having comparable economic and technical characteristics and (or) operating conditions with the use of the method of comparison of analogues"
7. Order of the Federal Tariff Service of October 26, 2010 N 254-e/1 "On Approval of the Methodological Guidelines for Calculation and Application of Reducing (Increasing) Coefficients to Ensure Compliance of the Level of Tariffs Established for Organizations Engaged in Regulated Activities with the Level of Reliability and Quality of Supplied Goods and Services"
8. Order of the Department of Economic Policy and Development of the City of Moscow No. 486-TR dd. 26.12.2017 "On Setting Uniform (Boiler) Tariffs for Electricity Transmission Services on the Moscow City Grids for 2018"
9. Order of the Committee for Prices and Tariffs of the Moscow Region No. 326-R dated 20.12.2017 "On Establishing Long-Term Regulatory Parameters for Territorial Grid Organizations for 2018-2022"
10. Order of the Department of Economic Policy and Development of Moscow No. 466-TR dd. 28.11.2022 "On Setting Long-Term Regulatory Parameters and Required Gross Revenue for 2023-2027 for Public Joint Stock Company Rosseti Moscow Region, in respect of which tariffs for electric power transmission services are set"
11. Mosoblkomtsen Order N 221-R dated 25.11.2022 "On setting individual tariffs for electric power transmission services for mutual settlements between PJSC Russian Grids Moscow Region and territorial grid organizations in the Moscow Region"
12. STO 34.01-24-002-2021 "Organization of maintenance and repair of electric grid facilities"
13. Order of the Ministry of Energy of Russia dated 04.10.2022 N 1070 "On Approval of the Rules for Technical Operation of Electric Power Plants and Grids of the Russian Federation and on Amendments to Orders of the Ministry of Energy of Russia dated September 13, 2018 N 757, dated July 12, 2018 N 548"
14. STO 34.01-23.1-001-2017 "Scope and norms of testing of electrical equipment".

https://doi.org/10.1051/e3sconf/202346101027

15. STO 34.01-21.1-001-2017 Distribution power grids with voltage of 0.4-110 kV. Requirements for technological design.
16. Order of the Ministry of Energy of Russia dated 04.10.2022 N 1070 "On Approval of the Rules for Technical Operation of Electric Power Plants and Grids of the Russian Federation and on Amendments to Orders of the Ministry of Energy of Russia dated September 13, 2018 N 757, dated July 12, 2018 N 548"
17. GOST R 52565-2006 "AC circuit breakers for voltages from 3 to 750 kV. General technical conditions"
18. GOST R 52719-2007 Power transformers. General technical conditions
19. STO 56947007-29.180.01.206-2015 Dry-type transformers for voltage 6-35 kV. Typical technical requirements
20. GOST 11677-85. Power transformers. General technical conditions.
21. STO 34.01-4.1-005-2017 "Rules for maintenance of relay protection, automation, remote control and signaling devices at electric grid facilities".

Development of methods for visual representation of low-frequency oscillation parameters in power systems

Alexander Popov[1*], *Kirill* Butin[1], and *Andrey* Rodionov[2]

[1]Engineering center «Energoservice», Russia
[2]Northern (Arctic) Federal University, Engineering center «Energoservice», Russia

Abstract. The development of wide area measurement system (WAMS) expands the analysis capabilities of the Unified power system based on computational processing of synchronized phasor measurements (SPM). The analysis of low-frequency oscillations (LFO) in the power system is one of the most relevant applications of WAMS data. The task of locating the source of forced LFO in real-time is highlighted. Due to the increasing complexity and heterogeneity of real power systems, a definitive solution to this problem does not currently exist. However, there are SPM data analysis methods that provide a high probability of identifying the source of oscillations. Informative representation of measurement signals, intermediate, and final LFO analysis results can serve as the basis for decision support tools for operational dispatch personnel. Direct visualization of raw data and computed parameters, such as a set of time function graphs, is inefficient due to the large volume. Therefore, compact high-level representations are in demand. This paper proposes some approaches to visualizing SPM data, computed LFO parameters, and the results of applying various oscillation source detection methods. The presented data visualization methods have been implemented by the authors in a LFO analysis software system.

1 Introduction

In Russia, the development of the Wide Area Measurement System (WAMS) [1] continues. The number of devices supporting Synchronized Phasor Measurement (SPM) technology installed in the components of the Unified Power System is increasing. As of now, approximately 1000 SPM data streams from phasor measurement units (PMUs) are available for online processing, with a data transmission frequency of 50 frames per second.

System Operator [1] outlined the current challenges associated with utilizing WAMS data. One of the top priorities is to enhance the awareness of the operational dispatch personnel regarding the power system's condition. Specifically, there is a demand for the development of tools that provide timely information about the presence of poorly damped low-frequency oscillations (LFO) [2] in the power system, their current and forecasted parameters, as well as potential sources of these oscillations. Effective data visualization tools play a crucial role here.

The analysis of LFO in a centralized setup involves a large number of input measurement signals. Even more signals are generated as a result of data processing. Therefore, compact and informative methods of representation are required for effective perception of LFO analysis data.

A simple set of time function graphs is not suitable for these purposes. Let's consider an example. Suppose there are 1000 sources of SPM data. The following dynamic parameters of oscillations are calculated: amplitude, phase angle, frequency, rate of change of mode magnitude, a measure of oscillation damping, and oscillation energy. Oscillatory components (modes) are extracted from frequency, current, voltage, active and reactive power signals. Monitoring is performed across 20 spectral bands. As a result, we have 600,000 signals. There is a trend towards increasing this number, particularly due to the deployment of more PMU devices.

This paper discusses some approaches to visually representing the results of SPM data processing that are applicable for LFO analysis at the power system level.

Researchers from Engineering center «Energoservice» are developing software for LFO analysis - ES Phasor [3, 4]. One of the main tasks of ES Phasor is to support decision-making in identifying the sources of forced LFO in the power system. The high-level data visualization methods discussed in this report are implemented in ES Phasor. During the development of visualization tools, the authors analyzed the experience of projects such as Phasor Point [5], Oscillation Source Locating [6], Monitoring Software for Synchronous Swing of Active Power [7], as well as joint research experience with the System Operator.

The structure of the paper is as follows. Firstly, methods of compactly representing arrays of raw signals in the time and frequency domains are discussed, along with additional visualizations to express common oscillation parameters and the system's involvement in the process. Then, an approach to visualize the dynamics of oscillation parameters using phase spaces is presented. Methods of interpreting the resulting images are discussed.

* Corresponding author: a.popov@ens.ru

Subsequently, the peculiarities of displaying time-varying amplitude-phase characteristics of the mode array are examined. Finally, a system of symbolic notations for visualizing the flow of oscillation energy on the power system's schematic is proposed. Screenshots of ES Phasor forms obtained during the analysis of real cases of LFO in the Unified Power System were used in the preparation of figures.

2 Visualization of arrays of initial data

Fig. 1 demonstrates one of the methods for visualizing an array of signals in the time domain, using the example of recording network frequency from approximately 100 PMUs during the start of LFO. The original signals are depicted as thin, semi-transparent lines of the same color. This creates a "statistical background" of the process. Additional plots include the mean signal value at each moment in time and the deviation from the mean by the standard deviation. This helps to understand the "commonality" of the oscillatory process.

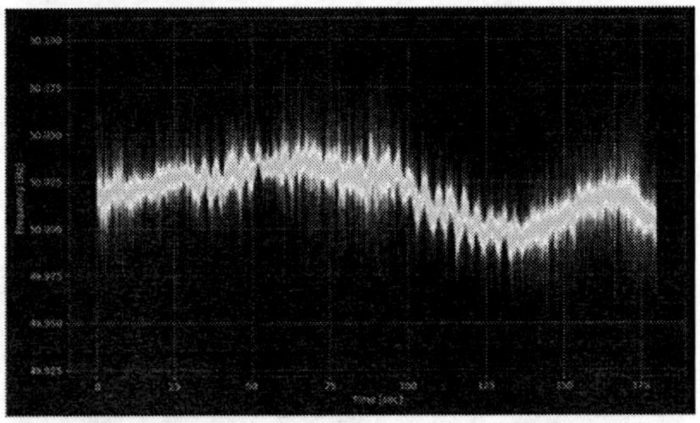

Fig. 1. Raw data in the time domain.

In the provided example, on the interval from 100 to 130 seconds, LFOs are observed in the average values. However, the range of values here is smaller than, for instance, on the interval from 60 to 100 seconds, where the LFO has a smaller amplitude.

It is also useful to examine the original signals in the frequency domain. In Fig. 2, amplitude spectra of active power measured on approximately 100 transmission lines during the LFO mode are shown as thin lines of the same color.

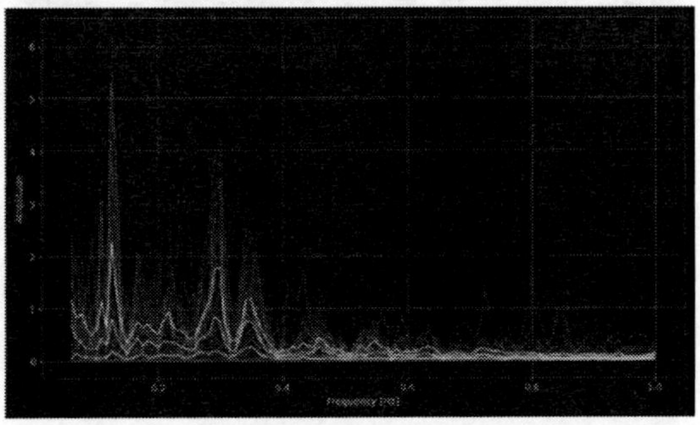

Fig. 2. Raw data in the frequency domain.

For each frequency value, the first, second (median), and third quartiles are calculated. They are represented as bold lines in the figure. Around 0.2 Hz, some signals exhibit an increase in their corresponding harmonic components. However, an analysis of the quartiles (especially the third quartile) indicates that the system's involvement in the oscillatory process as a whole in this case is less than, for example, around 0.1 Hz or 0.3 Hz.

It is also valuable to consider the aggregated spectrum (mean, median) over time. Fig. 3 displays the average amplitude spectrum of frequency at a certain moment in time and a spectrogram.

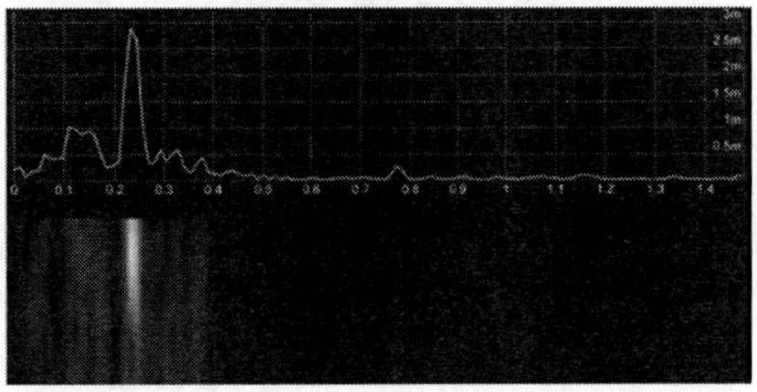

Fig. 3. The average spectrum over time.

In the spectrogram, the most recent values are located at the top. Stable oscillations are observed around 0.25 Hz and 0.8 Hz.

3 Visualization of oscillation dynamics

To represent the dynamic parameters of LFOs in the power system, it is proposed to use a set of moving points on the phase plane. Fig. 4 illustrates the schematic of the phase plane.

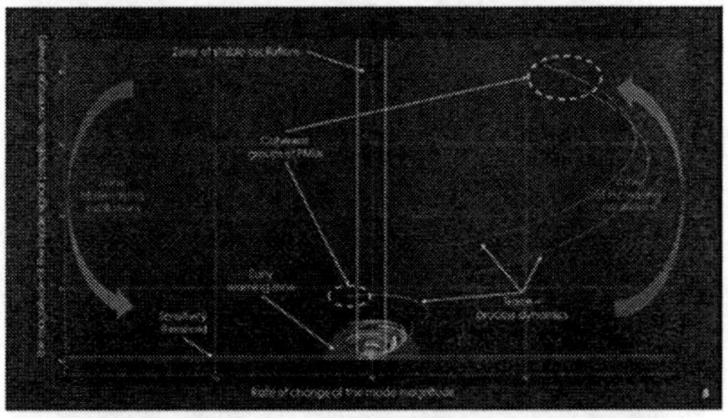

Fig. 4. Interpretation of the representation of LFOs on the phase plane.

The x-axis corresponds to the rate of change of the mode magnitude, while the y-axis represents the mode magnitude itself. Each point describes the evolution of the oscillation state over time. The movement of points to the right signifies the system's oscillatory growth, while movement to the left corresponds to damping of the oscillations. An important element of this representation is the "trace" left by the points. It reflects the recent history of changes in the mode magnitude and the velocity of point movement, allowing the observation of dynamic changes even on a static image.

Different regions on the phase plane correspond to different characteristics of LFO. The right half-plane represents the amplification of oscillations, while the

 https://doi.org/10.1051/e3sconf/202346101028

left half-plane corresponds to damping. The upper-middle part is the zone of low-damping, high-amplitude oscillations. A threshold value for the mode magnitude is also set, below which data is not displayed.The proposed representation of data on the phase plane provides feedback on the power system's state under control actions and is a valuable tool for monitoring LFOs.

The magnitude of the oscillations can be estimated in various ways. One approach involves using the amplitude (envelope of the signal), calculated with a Hilbert digital filter. Another approach is based on computing the average power of the signal over a sliding window. This integral value provides an averaged measure of the mode magnitude over a specific time interval and offers greater stability in noisy conditions. Additionally, average power can be calculated for the entire low-frequency range without isolating a specific mode. This approach provides a more general picture of the signal's energy distribution across frequencies and can be useful in situations where it is necessary to analyze the entire low-frequency range rather than individual modes.

Fig. 5 presents an example of phase space for a real case of LFOs.

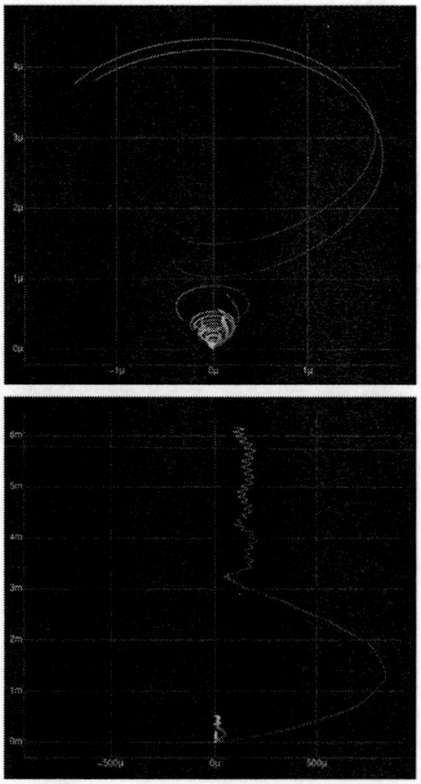

Fig. 5. An example of visualizing modes on the phase plane.

Two frames are depicted, describing the onset of oscillations (top) and their subsequent development (bottom). The mode is extracted from the frequency signal, and data from 80 sources are analyzed.

The process of increasing the magnitude of oscillations is evident. The values have changed by several orders of magnitude. The source of LFO is highlighted, which in this case is characterized by the maximum mode magnitude. On the bottom frame, there is also an attempt to dampen the oscillations when their growth accelerates significantly.

4 Visualization of amplitude-phase characteristics of modes

As another way to visualize the dataset on the scale of the power system, an amplitude-phase diagram is proposed. Fig. 6 presents an example of such representation for one of the cases of LFO. In this form, each mode is depicted as a point at a given moment in time. The distance from the point to the center of the diagram corresponds to the current amplitude, and the angle represents the full phase of the mode. A decaying trace is displayed behind each point.

This representation is associated with one of the common methods for identifying the source of LFO - Mode Shape Estimation (MSE) [8]. The idea behind MSE is that a mode with sufficient amplitude and leading phase belongs to the source of the oscillations. Modes can also group into clusters. In this case, you can talk about leading mode groups and leaders within those groups. Understanding the cluster's structure allows for determining the part of the system involved in the oscillations.

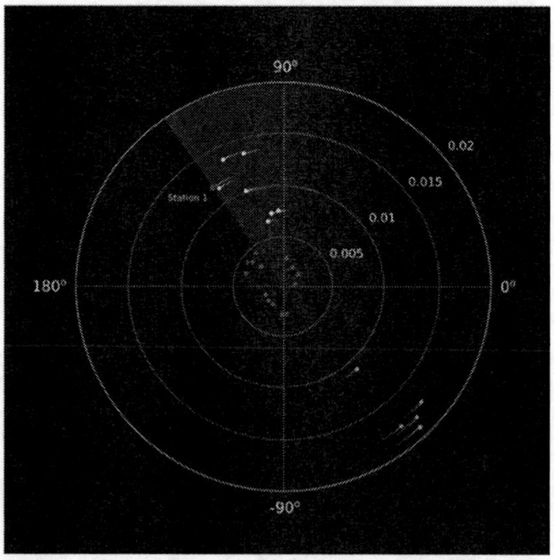

Fig. 6. The amplitudes and full phases of the mode set.

In the example shown in Fig. 6, three clusters of modes are identified: leading, lagging, and a group of modes with low amplitude. The presumed source of the LFOs is highlighted. It's worth noting that in this case, the amplitude of the oscillations from the source is not the highest.

5 Visualization of Oscillation Energy

For visualizing the high-level information about the state of measurements and the development of the

https://doi.org/10.1051/e3sconf/202346101028

oscillatory process, a visualization on the power system scheme is proposed. The cartographic representation includes three levels: the structure of the power system, the state of the WAMS, and the results of the LFO source identification. The visualization at the third level depends on the analysis methods used.

Let's take a closer look at the visualization of LFO energy flows. The assessment of the quantity, direction, and velocity of LFO energy movement is based on the Dissipating Energy Flow (DEF) method [9]. DEF provides an expression for calculating the amount of energy dissipated as a result of oscillations based on filtered synchrophasor data. The sign before the dissipated energy value provides information about the direction of LFO propagation.

The proposed notations are shown in Fig. 7. The power system is represented as a graph, where nodes correspond to power stations and substations, and edges represent power transmission lines. Nodes are tied to the geographical coordinates of power system objects, while the geography of the lines is not considered. Different line thicknesses are used to represent different voltage classes. Power transmission lines with voltage classes of 110 kV and higher are displayed.

Depending on the availability of reliable data in the WAMS, lines are considered as active or inactive. This, in turn, determines the presence of observable and unobservable objects in the power system, such as lines, power stations, and substations. For example, if at least one active PMU is missing on a line, the line is considered unobservable. Similarly, a substation is considered observable if it is connected to at least one observable line. Inactive PMUs and unobservable parts of the system are shown in gray.

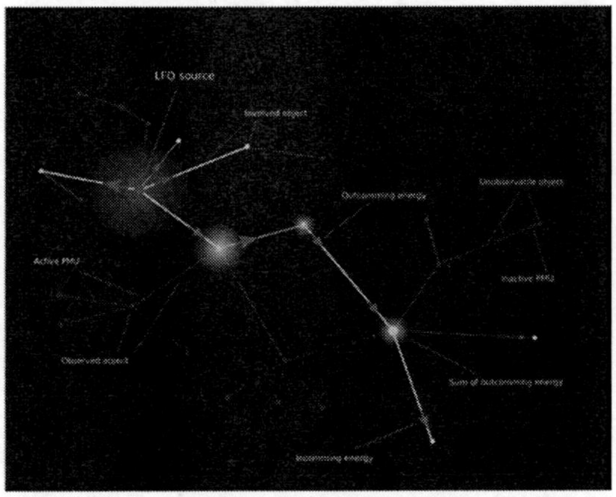

Fig. 7. Legend for DEF visualization on the map.

Dissipative energy is visualized by the brightness of the lights, the thickness of the lines, and the arrows indicating the flow direction of energy. The thickness of the arrows is proportional to the amount of energy from oscillations registered at the corresponding connections, and the total outgoing energy is represented by the glowing around the stations and substations.

Positive values of dissipative energy for PMUs are shown with red arrows pointing away from the stations. Negative values are represented by blue arrows pointing towards the stations. If PMUs are correctly installed on the same line, there should not be two arrows of the same color. The glowing indicates the total outgoing energy.

In Fig. 8, a frame for one of the real cases of large-scale LFOs is shown when the source was located in the power system of a neighboring country.

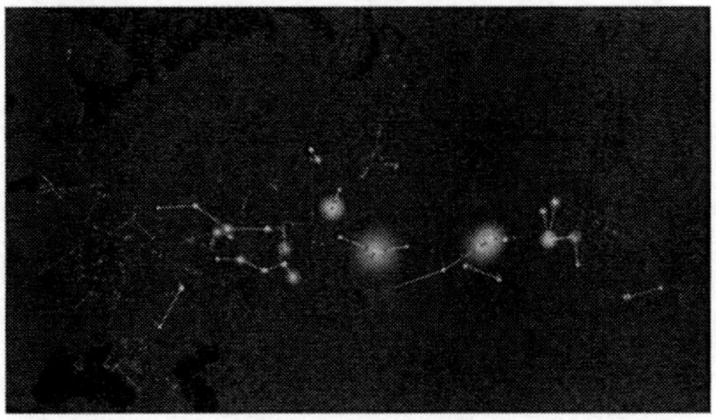

Fig. 8. Dissipative energy of large-scale LFOs.

The part of the power system involved in LFOs, the most disturbed part of it, and the source of LFOs are visible. Arrows at this scale have become dots, but thanks to the colors, the direction of the flow is visible. Similarly, the results of the modified MSE method [3], in which the frequency is calculated from the current and the leading mode is sought not in the entire power system, but at each station, are superimposed on the power system and WAMS, allowing the selection of an edge leading to the source of LFOs at each observed vertex of the graph.

6 Conclusion

The report discusses various approaches to compact and informative visualization of large volumes of data resulting from the analysis of LFOs in the power system. The proposed approaches have been implemented by the authors in the ES Phasor software. The images presented in the work are obtained from the analysis of real cases of LFOs in the Unified Power System.

The report highlights the challenges of high-level representation of datasets, dynamically calculated oscillation parameters, and data from various LFO source detection methods.

The proposed visualizations are designed to provide additional support for decision-making by operational and dispatch personnel and to assist in further research into LFO analysis methodologies.

The authors express their gratitude to the JSC «System Operator of the Unified Power System» for providing the SPM data.

References

1. A.V. Zhukov, E.I. Satsuk, D.M. Dubinin, O.V. Zhuravleva. "Development of Dispatch Control Technologies in Russia on the Synchronized Phasor Platform." Proceedings of the International

Scientific Seminar "Methodological Issues of Reliability Research in Large Energy Systems," **73** (2022)

2. L.G. Meegahapola et al. "Review on Oscillatory Stability in Power Grids with Renewable Energy Sources: Monitoring, Analysis, and Control using Synchrophasor Technology." IEEE Transactions on Industrial Electronics (2020)
3. A.I. Popov et al. "Examples of Processing Low-Frequency Oscillations in Russia and Ways to Improve the Analysis."International Conference on Smart Grid Synchronized Measurements and Analytics (SGSMA), (2022)
4. A.V. Rodionov, K.P. Butin, I.O. Komolykh, A.I. Popov. "Aspects of Implementing a System for Locating the Sources of Low-Frequency Oscillations. (Proceedings of the International Scientific and Technical Conference "Relay Protection and Automation of Power Systems – 2023)
5. PhasorPoint PC Software for Monitoring and Analysis of Power System Behavior Based on Synchronized Vector Measurements. URL: https://www.rtsoft.ru/catalog/programmnoe-obespechenie/prikladnoe-po-iprilozheniya/phasor-point/
6. Supplement to PhasorPoint for Low-Frequency Oscillation Source Detection. URL: https://www.iso-ne.com/
7. Software for Monitoring Synchronous Swings in Active Power. URL: https://www.ap-soft.ru/mskam
8. A.Monti., C. Muscas, F. Ponci. "Phasor Measurement Units and Wide Area Monitoring Systems." Academic Press (2016)
9. S. Maslennikov et al. "Dissipating Energy Flow Method for Locating the Source of Sustained Oscillations." International Journal of Electrical Power & Energy Systems 88 (2017)

Identification of electricity theft perpetrators in the low-voltage distribution network

Irina Golub[1]*, *Evgeny* Boloev[1, 2], *Alexander* Anfinogenov[3], *Anatoly* Buchinsky[3], *Andrey* Maslennikov[3]

[1] Melentiev Energy Systems Institute of Siberian Branch of the Russian Academy of Sciences, 130 Lermontov str., Irkutsk, Russia
[2] Federal State Budget Educational Institution of Higher Education «Irkutsk National Research Technical University», 83 Lermontov str., Irkutsk, Russia
[3] Regional State Unitary Enterprise «Oblkommunenergo», 54 Shiryamova St., Irkutsk, Russia

Abstract. Correlation analysis proves its effectiveness in pinpointing the culprits behind electricity theft, as demonstrated by the comprehensive data obtained from measurement snapshots taken in an actual distribution network over the course of a month. The analysis relied on various methods of theft simulation that correspond to attacks on meters, communication lines, and measurement data.

1 Introduction

The technical and non-technical losses of electricity in a low-voltage distribution network (DN) can be assessed using the phase measurements for active powers and voltage magnitudes. These values are measured by smart meters (SM) installed at the power supply node of the network and at the points where consumers are connected to it. A significant support in determining the actual losses is a balance smart meter installed at the power supply node of the distribution network, which measures the electricity transmitted to the phases. The comparison of the electricity transmitted with the total energy received by the consumers of each phase makes it possible to find out the unmeasured electricity consumption in the phases, which corresponds to the sum of technical and non-technical losses. Determining technical losses using the contribution method [1] by measuring active power and voltage of the consumer does not require information about the topology and parameters of the network. The difference between the actual and technical losses in the phases, corresponding to large non-technical losses, is an indicator of the electricity theft in the network.

According to the literature review [2, 3], the perpetrators of electricity theft are normally detected using clustering methods, state methods, and game theory methods. It is also possible to use correlation analysis [4 – 6], the performance of which was proven in [7, 8] when separating the SM measurements into phases corresponding to the phases of the balance meter.

The 30-minute measurement snapshots made in a real-world DN during a month exemplify the use of the correlation analysis method in detecting the perpetrators of electricity theft. The total number of simultaneously performed complex measurements, each of which includes the measurement of active power and voltage magnitude, is 3 measurements at the power supply node and 39 measurements at consumption points. Measurements were made by 18 single-phase and 7 three-phase smart meters placed at 10 DN supports and by a three-phase balance smart meter installed at the feeder power supply node. The information from the smart meters is transferred to the data collection and processing system by the automated metering system, which is a measuring infrastructure that connects the DN with a communication network and a modern measurement structure.

2 Determination of phases of complex measurements at the consumption points

The first step in the theft identification algorithm is to determine the phase of each complex measurement in which it was performed. This is done using the correlation analysis [7, 8], which involves the study of voltage magnitudes measured simultaneously over a long period of time in known phases of the power supply node $U_{\sup A}(t)$, $U_{\sup B}(t)$, $U_{\sup C}(t)$, $t=1,\dots,T$, and at the points where meters are connected to consumers, whose phases must be identified. In our case, the total number of consumers K in three phases is 39.

Phase X of the consumer connection is identified using the maximum value of the cross-correlation coefficients between the measurement vectors of the voltage magnitudes at the node of feeder supply and the k-th consumer $U_{\sup A}$ and U_{kX}, $U_{sup B}$ and U_{kX}, $U_{sup C}$ and U_{kX}. If the maximum correlation coefficient relates the first pair of vectors U_{supA} and U_{kX}, then X corresponds to phase A. Similar

* Corresponding author: golub@isem.irk.ru

conclusions can be obtained for measurements performed in phases *B* and *C*.

After the phase of smart meter connection is determined, the number and list of consumers are found out for each phase. For each consumer, the measurement vectors of the voltage magnitudes and active powers of the loads are generated with the number T of measurement snapshots. For the low-voltage DN considered in our case, 11 consumers were identified in phase A, and 14 consumers in phases B and C, in each.

3 Determination of technical energy losses by the contribution method

Consider the contribution method [1] for an open-loop network, which is the distribution network, for one of the measurement snapshots of phase f.

Measurements of active power of load P_n^f and voltage magnitude U_n^f of the n-th consumer, where the number of consumers in phase f, $n=1,\ldots,N_f$, allow determining the value of the current load

$$I_n^f = P_n^f / U_n^f . \qquad (1)$$

A current equal to the current load of the consumer flows to this consumer from the feeder supply node. The product of the current load of the n-th consumer and the voltage at the feeder supply node U_{sup}^f, measured by balance smart meter, determines the active power transmitted to the n-th consumer from the power supply node

$$P_{\text{sup}\ n}^{f\ \text{trans}} = P_n^f U_{\text{sup}}^f / U_n^f . \qquad (2)$$

The sum of the powers transferred from the feeder supply node to each of the N_f consumers determines the total power transmitted to all consumers of the phase

$$P_{\text{sup}}^{f\ \text{trans}} = \sum_{k=1}^{N_f} P_n^f U_{\text{sup}}^f / U_n^f . \qquad (3)$$

If the power of the n-th consumer load is subtracted from the power transmitted to this consumer from the feeder supply node, it is possible to find out the power losses that occur along the path of its transmission. These are the so-called technical losses, the sum of which for N_f consumers of the phase for each of the T measurement snapshots equals

$$\Delta P_t^f{}_{\text{tech}} = \sum_{n=1}^{N_f} \left(P_n^f U_{\text{sup}}^f / U_n^f - P_n^f \right)_t ,\ t = 1,\ldots,T . \qquad (4)$$

4 Determination of technical energy losses by the contribution method

For each of the measurement snapshots, a comparison of the power measured by the balance meter, which is transmitted to the feeder from the supply node $P_{\text{sup}}^{f\ \text{trans}}$, with the total load power of consumers $\sum_{n=1}^{N_f} P_n^f$ for this snapshot, allows estimating the total losses in the feeder, which are also called actual losses. These losses for each of T measurement snapshots can be represented as

$$\Delta P_t^f{}_{\text{act}} = \left(P_{sup}^f - \sum_{n=1}^{N_f} P_n^f \right)_t ,\ t = 1,\ldots,T . \qquad (5)$$

Non-technical energy losses, which are of the greatest interest in this study, are determined for each of the measurement snapshots as the difference between actual and technical losses

$$\Delta P_t^f{}_{\text{non-tech}} = \Delta P_t^f{}_{\text{act}} - \Delta P_t^f{}_{\text{tech}} ,\ t = 1,\ldots,T . \qquad (6)$$

Non-technical losses include unmetered electricity consumption, measurement errors, as well as errors in the method of determining technical losses.

5 Determination of the perpetrators of non-technical losses

In this work, using the recommendations given in [6] to detect the consumers stealing electricity, we calculate the absolute values of coefficients K_{xP}^f of correlation between the vector of non-technical losses and each of the load power vectors of the consumer. In doing so, it is assumed that the maximum coefficients correspond to consumers guilty of stealing electricity. The number of components of the non-technical loss vector $\Delta P_t^f{}_{\text{non-tech}}$ and each of the N_f load power vectors of consumer P_n^f, is equal to the number T of measurement snapshots.

Honest consumers can be identified using normalized correlation coefficients K_{hP}^f, equal to the relation $K_{hP}^f = K_{bP}^f / K_{xP}^f$, where K_{bP}^f is correlation coefficient between the vector of the balance meter measurements reduced by the value of technical losses $\left(P_b^f - \Delta P_{\text{non-tech}}^f \right)$ and vectors of measurements of each consumer P_n^f. The value of K_{hP}^f should be calculated in the case where correlation coefficients K_{xP}^f do not allow confidently identifying the culprits of theft.

Another, alternative way to determine the perpetrators of theft, which does not require the calculation of either technical or non-technical losses, can be the calculation of the maximum correlation coefficients K_{xI}^f between the current unbalance vector I_{unb}^f in phase f and each of N_f vectors of current loads I_n^f calculated using expressions (1).

Unbalance currents caused by the unmetered loads, including their theft, for each measurement snapshot are equal to the difference between the current supplied to the network from phase f of power supply source I_{sup}^f and the sum of consumer load currents

$$I_{unb}^{f} = I_{sup}^{f} - \sum_{n=1}^{N_f} I_n^f . \quad (7)$$

Honest consumers in this case can be identified using normalized correlation coefficients $K_{hI}^{f} = K_{bI}^{f} / K_{xI}^{f}$, where K_{bI}^{f} is correlation coefficients relating the vector of power supply source currents I_{sup}^{f} with each of N_f vectors of current loads I_n^f. Even in the absence of theft, the correlation coefficients K_{xP}^{f} and K_{xI}^{f}, which correspond to different consumers and do not exceed the values of weak correlation, can differ significantly from each other.

An important indicator of the absence of electricity theft in the network is the coincidence of the correlation coefficients K_1^f and K_2^f for each relation. The first correlation coefficient characterizes the relationship between the voltage loss vector and the vector of current in the lines, which is calculated considering the nodal currents that flow in the direction from the power supply node to the end nodes of the network. The second correlation coefficient establishes the relationship between the vector of voltage loss and the vector of current in the lines, which is calculated based on the nodal currents running in the opposite direction. To determine the nodal parameters, each support of the main feeder is assigned a load node. When several smart meters are installed at the support, the total load power and the average value of the voltage magnitude are calculated at this node.

6 Methods for simulating electricity theft

To confirm the effectiveness of the considered approaches to detecting the perpetrators of electricity theft, the research relied on the methods for simulating the perpetrators of theft, which are offered in various literature sources [6, 9, 10]. These methods involve:

1. Multiplying the power of consumer $P_n(t)$ by the randomly generated number α, within the range 0.1 – 0.9, for all snapshots of power measurements*.
2. Replacing $P_n(t)$ with β, if $P_n(t) > \beta$, for randomly chosen number β, minimum of the maximum meter readings $P_n(t)$.
3. Replacing $P_n(t)$ with $\left(P_n(t) - \beta\right)$, if $P_n(t) > \beta$, and replacing $P_n(t)$ with 0, if $P_n(t) \leq \beta$, for randomly chosen number β, minimum of the maximum meter readings $P_n(t)$.
4. Replacing meter readings of consumer $P_n(t)$ with zero values for a randomly determined period $t2 - t1 \geq 4$ hours a day.
5. Multiplying each of the power measurement snapshots of consumer $P_n(t)$ by a randomly generated number α, in the range 0.1 – 0.9.
6. Replacing each of the power measurement snapshots of consumer $P_n(t)$ with the product of the average load of the consumer and a randomly generated number α, in the range 0.1 – 0.9.
7. Replacing all power measurement snapshots of consumer $P_n(t)$ with the average load value over the analyzed period of time.

7 Illustration of theft detection methods

Figure 1 shows the daily graphs of active power of consumers in the phases of a real-world distribution network, which are built using 39 measurements taken from the automated metering system protocols. Table 1 indicates the values of the total energy consumption according to the measurements of the balance meter and consumer meters, as well as the values of the energy loss components for January 2023 for 1488 30-minute measurement snapshots.

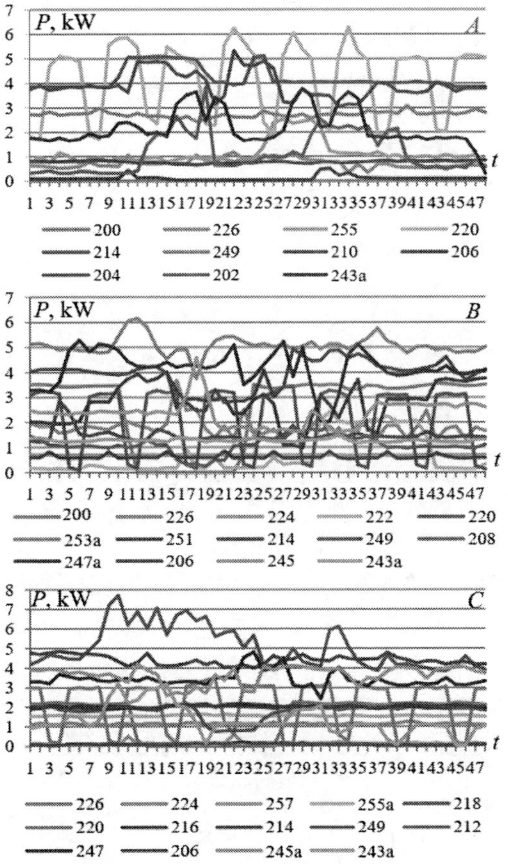

Fig. 1. Graphs of daily active power loads of consumers in phases *A*, *B*, *C*.

* The number of measurement snapshots in all cases $t = 1,.., T$

Table 1. Values of total energy consumption and losses in phases *A*, *B*, and *C* of the distribution network for the month of January 2023.

Phase	Total electricity consumption according to balance meter measurements, kWh	Total electricity consumption according to consumer meter measurements, kWh	Total actual electric energy losses, kWh /%	Total technical electric energy losses, kWh /%	Total non-technical electric energy losses, kWh /%
A	32 188.32	30 727.66	1 460.66/4.54	1 268.00/3.94	192.30/0.6
B	45 850.32	43 383.76	2 466.56/5.38	2 144.77/4.68	321.78/0.7
C	45 437.52	42 328.72	3 108.80/6.84	2 890.90/6.36	217.90/0.48

An assessment of the contribution of power of the consumer of each phase to non-technical losses, in the form of graphs of correlation coefficients between the vectors of non-technical losses and each of the consumer load vectors for the three phases of January 2023 (Figure 2) indicates a weak correlation. This gives us hope for a theft-free period at issue, which is additionally evidenced by the coincidence of the graphs of the correlation coefficients K_1^f and K_2^f, presented in Figure 3.

Let us simulate a decrease in the power of load for consumer 220 of phase *C* by 5%, 10%, 30%, 50%, 70%, 90%, corresponding to the first method of theft, for α equal to 0.95, 0.9, 0.7, 0.5, 0.3, and 0.1. Table 2 shows that in this case, for a constant value of the electricity supplied to the phase, equal to 45437.52 kWh, the total electricity consumption and technical losses go down, while the actual and non-technical losses go up.

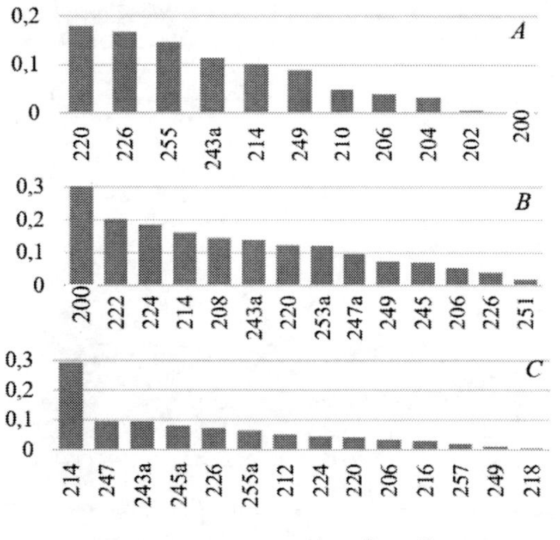

Fig. 2. Correlation coefficients K_{xP}^A, K_{xP}^B, K_{xP}^C in phases *A*, *B*, *C* for January 2023.

As for the accusation of the consumer of the electric energy theft, in all cases, the theft is proved by the maximum correlation coefficient K_{xP}^C, which rises with an increase in the share of stolen electricity in the range of 0.809 ÷ 0.998, Figure 4. An almost identical result, pointing to consumer 220 as the culprit of the electricity theft, was also obtained for the maximum values of the correlation coefficients K_{xI}^C of the vectors of total current unbalances and vectors of current loads of consumers.

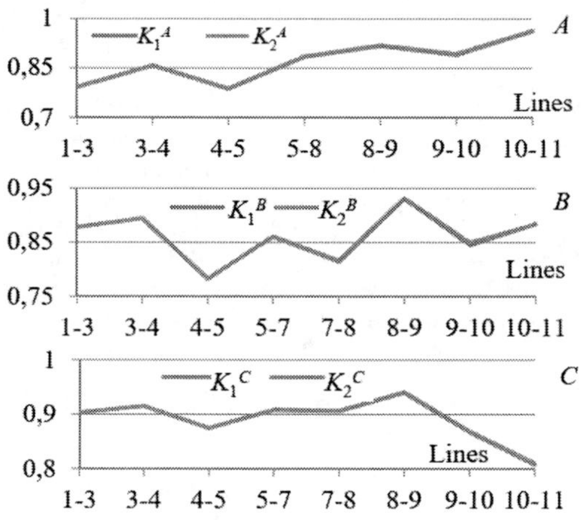

Fig. 3. Graphs of correlation coefficients K_1^A and K_2^A, K_1^B and K_2^B, K_1^C and K_2^C.

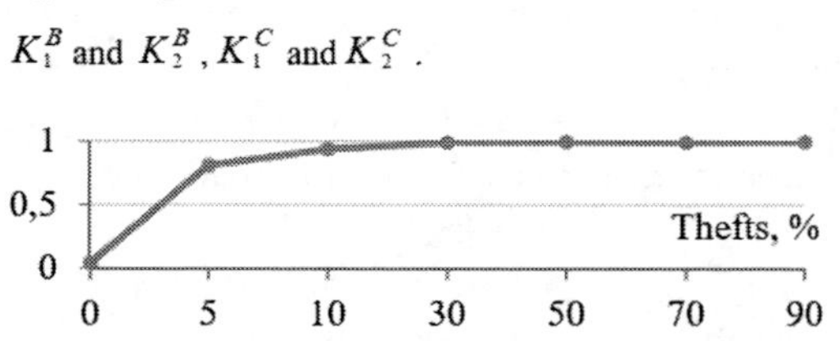

Fig. 4. Change in the correlation coefficient K_{xP}^C, with an increase in the percentage of electricity stolen by consumer 220.

Table 2. Influence of the percentage of electricity stolen by consumer 220 of phase *C* on the total consumption and losses.

Electricity theft, %	Total electricity consumption according to consumer meter measurements, kWh	Total actual electricity losses, kWh	Total technical electricity losses, kWh	Total non-technical electricity losses, kWh
0	42 328.72	3 108.80	2 890.90	217.90
5	42 195.56	3 241.961	2 884.671	357.2902
10	42 062.4	3 375.118	2 878.438	496.6792
30	41 529.78	3 907.745	2 853.51	1 054.235
50	40 997.15	4 440.372	2 828.58	1 611.791
70	40 464.52	4 972.299	2 803.652	2 169.347
90	39 931.15	5 505.626	2 778.723	2 726.903

 https://doi.org/10.1051/e3sconf/202346101029

Figure 5 demonstrates the graphs of correlation coefficients K_1^C and K_2^C, showing how, with an increase in the amount of electricity stolen by consumer the difference between the ordinates of the graphs rises. In addition, the intersection of the graphs corresponding to the fifth node or the fifth support, at which the smart meters of consumers 218 and 220 are located, indicates that at least one of them is associated with the electricity theft.

Two columns of graphs, Figure 6, illustrate seven ways of simulating theft with the load of consumer 220 as an example. The left column shows the daily load curve and the daily non-technical loss curve corresponding to the simulated theft. The right column shows graphs of correlation coefficients for phase *C* consumers. The maximum load of consumer 220 for 1,488 measurement snapshots in January is 4.112 kW, and the average load is 1.789 kW.

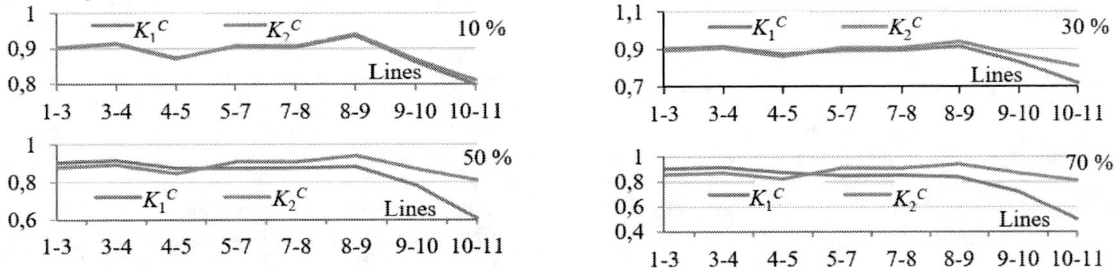

Fig. 5. Divergence of graphs of correlation coefficients K_1^C – 1 and K_2^C – 2 when simulating 10 %, 30 %, 50 % and 70 % load theft of consumer 220.

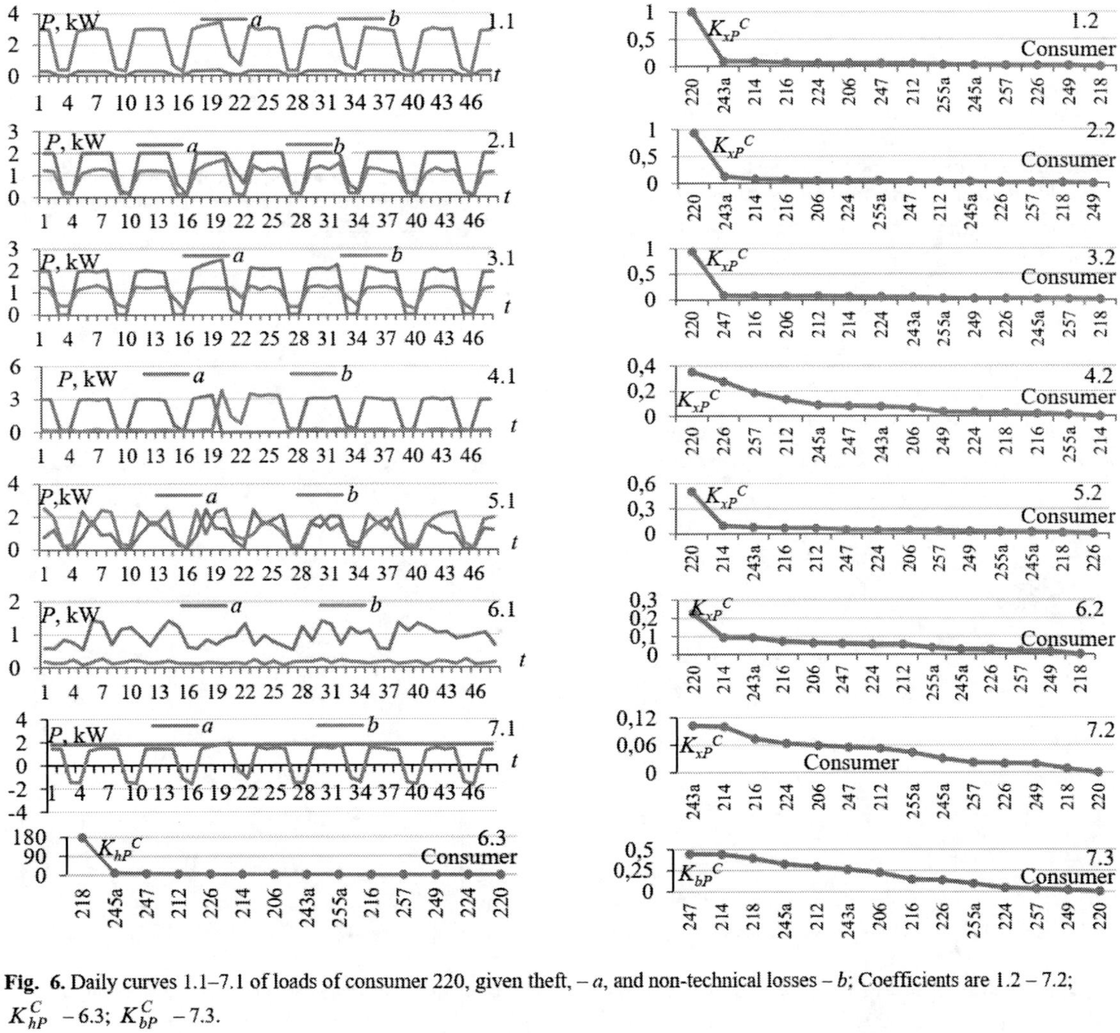

Fig. 6. Daily curves 1.1–7.1 of loads of consumer 220, given theft, – *a*, and non-technical losses – *b*; Coefficients are 1.2 – 7.2; K_{hP}^C –6.3; K_{bP}^C –7.3.

To comment on Figure 6, we will provide the values of the parameters and correlation coefficients used for each method of theft simulation:

1.1. $\alpha = 0.1$, 1.2. $K^{C}_{xPmax} = 0.998$;

2.1. $\beta = 2\,\mathrm{kW}$, 2.2. $K^{C}_{xPmax} = 0.934$;

3.1. $\beta = 2\,\mathrm{kW}$, 3.2. $K^{C}_{xPmax} = 0.935$;

4.1. $t_2 - t_1 = 4$ hours, 4.2. $K^{C}_{xPmax} = 0.352$;

5.1. $\alpha \approx (0.1 \div 0.9)$; 5.2. $K^{C}_{xPmax} = 0.502$;

6.1. $\alpha \approx (0.1 \div 0.9)$, $P^{aver}_{220} = 1.789$ kW;

6.2. $K^{C}_{xPmax} = 0.226$; 6.3. $K^{C}_{hPmin} = 0.123$.

The maximum correlation coefficient of 6.2, indicating the theft committed by consumer 220, is rather insignificant. Additional confirmation of the theft is the coefficient of 6.3 showing that consumer 220 is the least honest.

7.1. $P^{aver}_{220} = 1.789$ kW, 7.2. $K^{C}_{xPmin} = 1.75E-16$;

7.3. $K^{C}_{bPmin} = 5.7E-15$.

The proximity to zero of coefficients 7.2 and 7.3 denies the relation of the load of consumer 220 with both non-technical losses and measurements of the balance meter, which enables the possibility of accusing consumer 220 of electricity theft.

The study of simultaneous 30 % thefts of electricity, corresponding to the first method, by two consumers 220 and 247 and three consumers 220, 218, and 247 of phase *C*, as follows from the graphs of correlation coefficients K^{C}_{xP} and K^{C}_{xI} presented in Figure 7, allows in both cases of theft to successfully detect the thieves.

Figure 7 (1.3 and 2.3) shows scatter plots, with ordinates K^{C}_{hP}, and abscissas K^{C}_{xP}. The points with minimum ordinates and maximum abscissas, which are marked with arrows, correspond to consumers guilty of energy theft.

8 Conclusion

The problem of identifying the consumers stealing electricity in the phases of the low-voltage distribution network is solved considering technical losses that do not require knowledge of the topology and parameters of the network. The findings suggest that in order to detect theft, there is no need to analyze the correlation between the vector of non-technical losses and the vectors of load powers of consumers, instead, one should investigate the correlation of the vector of current unbalances with the vectors of current loads. To assess the absence of theft in the network, it is proposed to check the coincidence of the correlation coefficients between the voltage loss vector and the current vectors in the lines, when they are calculated considering the nodal currents that flow from the power node to the end of the feeder and in the opposite direction. The example of a real-world distribution network feeder illustrates how correlation analysis can effectively identify simulated perpetrators of different types of theft.

The research was carried out under State Assignment Project (no. FWEU-2021-0001) of the Fundamental Research Program of Russian Federation 2021-2030.

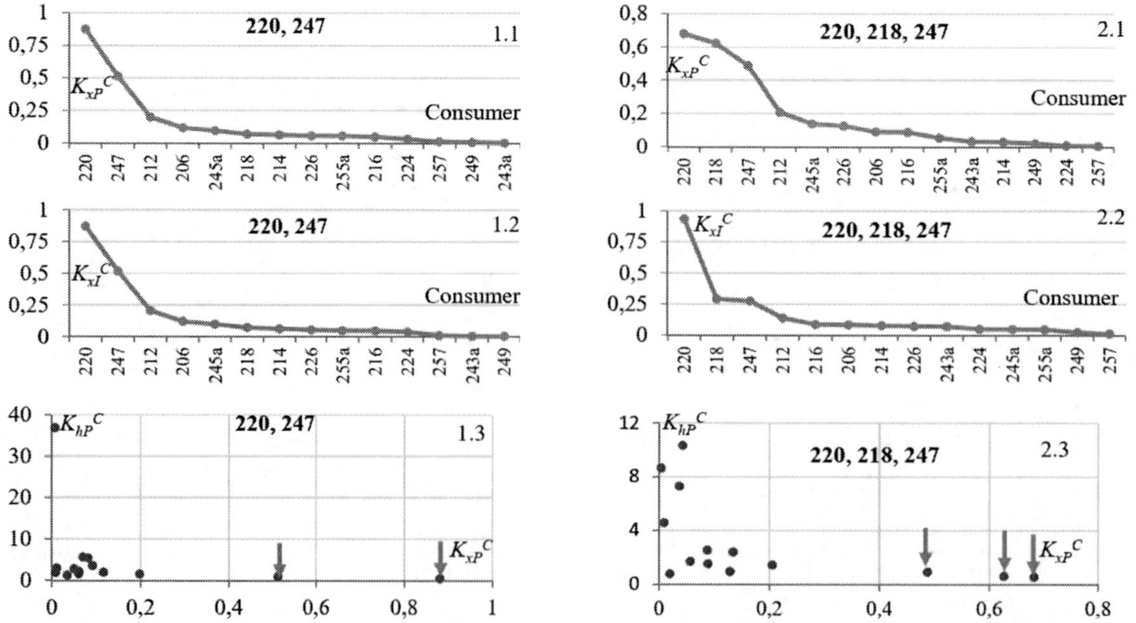

Fig. 7. Correlation coefficients K^{C}_{xP} and K^{C}_{xI} for the cases of two culprits of 30 % electricity thefts 1.1–1.3 and three culprits of 30 % electricity thefts 2.1–2.3.

References

1. I. Golub, E. Boloev *Determination of Losses in Distribution Networks by Smart Meter Measurements*, in Proceedings of the International Conference on Electrical, Communication, and Computer Engineering, ICEECCE, 12-13 June 2021, Kuala Lumpur, Malaysia (2021)
2. R. Jiang, R. Lu, Y. Wang, J. Luo, C. Shen, X. Shen, Tsinghua Sci Technol, **19** (2014)
3. B. K. Hammerschmitt, A. Abaide, F. Lucchese, C. Martins, A. Silveira, J. Rigodanzo, J. Castro, J. Rohr, *Non-Technical Losses Review and Possible Methodology Solutions*, in Proceedings of the 6th International Conference on Electric Power and Energy Conversion Systems, EPECS, 05-07 October 2020, Istanbul, Turkey (2020)
4. P. P. Biswas, H. Cai, B. Zhou, B. Chen, D. Mashima, V. W. Zheng, IEEE Transactions on Smart Grid, **11**, 4 (2020)
5. K. Zheng, Q. Chen, Y. Wang, C. Kang, Q. Xia, IEEE Transactions on Industrial Informatics, **15**, 3 (2019)
6. Y. Yang, R. Song, Y. Xue, P. Zhang, Y. Xu, J. Kang, H. Zhao, IEEE Access, **10** (2022)
7. I.I. Golub, E.V. Boloev, Ya.I. Kuzkina, *Using automated metering system measurements to check the topology and calculate the power flow of the secondary distribution network*, in Proceedings of the Rudenko International Conference Methodological Problems in Reliability Study of Large Energy Systems, RSES, 23-27 September 2019, Tashkent, Uzbekistan (2019)
8. I.I. Golub, Ya.I. Kuzkina, Proceedings of Irkutsk State Technical University, **24**, 1 (2020)
9. P. Jokar, N. Arianpoo, V. C. M. Leung, IEEE Transactions on Smart Grid, **7**, 1 (2016)
10. M. Zanetti, E. Jamhour, M. Pellenz, M. Penna, V. Zambenedetti, I. Chueiri, IEEE Transactions on Smart Grid, **10**, 1 (2019)

https://doi.org/10.1051/e3sconf/202346101030

Assessment of energy security levels objects using complexes data collection, processing and management

Mahsud Sultanov[1], *Ilya* Boldyrev[1], *Maksim* Ivanitckii[1]*, *Petr* Shamigulov[1]

[1]Branch of the Federal State Budgetary Educational Institution of Higher Education National Research University MPEI in Volzhsky, 69 Lenina str., Volzhsky, Russia

Abstract. The results of the development of a methodology for assessing the safety levels of energy facilities are presented. Energy security indicators are proposed, the values of which determine the degree of possible threat in the operation of the energy facility under consideration. The results obtained in the work can be used to determine the level of safety of the operation of TPP power units.

1 Introduction

In accordance with the requirements of the need to ensure the energy security of critical infrastructure, the functioning of generating systems is based on maintaining the operability of the resource-determining equipment of functional units and systems, taking into account their technical condition. Considerable attention should be paid to maintaining the observability and controllability of power equipment [1-3]. In the work [4], the formation of a complex indicator of energy security is considered, using the arithmetic mean value of point estimates depending on the nature of the situation, and recommendations are given for carrying out a set of measures to improve the economic security and reliability of the entire energy system. The works [5-8] describe methods using modern computational and experimental modeling tools to ensure the reliability and safety of power equipment operation. The main attention is paid to the physical wear of equipment elements. A high degree of depreciation of fixed assets, the availability of equipment that has exhausted its resource, reduce the energy security of enterprises. The analysis of the presented data shows that the allocation of certain calculated indicators in comparison with the threshold values will allow to identify an area with a risk of disruption of the normal functioning of the power system.

The purpose of the study is to develop a methodology for assessing the level of safety of power equipment, taking into account physical wear and tear, the technical condition of functional units and elements.

2 Main safety indicators of an energy facility

To date, there are many indicators characterizing energy security, reflecting both qualitative and quantitative characteristics of energy security [5, 6, 9, 10]. In this paper, the main safety indicators of an electric power complex facility are considered:

- the degree of depreciation of fixed assets of the enterprise;
- the coefficient of renewal of fixed assets;
- the share of electric power equipment that has exhausted its resource;
- availability of fuel and energy resources.

In accordance with the accepted safety indicators, the actual values of the indicators are determined. The degree of depreciation of fixed assets of the enterprise, %, is determined by the degree of equipment failure:

$$I = \frac{\sum_{1}^{n} I}{I_0} \cdot 100\%, \qquad (1)$$

where $\sum_{1}^{n} I$ – the amount of depreciation of fixed assets, rub.; I_0 – initial cost of fixed assets, rub.

The coefficient of renewal of fixed assets, %, is determined by the need for a complete renewal of fixed assets over their service life:

$$K_f = \frac{I_{ki}}{I_{ke}} \cdot 100\,\%, \qquad (2)$$

where I_{ki} – the cost of fixed assets introduced in a year, rub.; I_{ke} – the cost of fixed assets at the end of the year, rub.

The share of electric power equipment that has exhausted its resource depends on the availability of worn-out, exhausted equipment in power systems and reduces the reliability of power supply:

$$C_r = \frac{C_{cf}}{C_{tc}} \cdot 100\,\%, \qquad (3)$$

where C_{cf} – the cost of fixed assets of electrical equipment that has exhausted its resource, rub.; C_{tc} –

* Corresponding author: ivanickii@vfmei.ru

the total cost of fixed assets of electrical equipment, rub.

Availability of fuel and energy resources, years, which is determined by available resources:

$$R_f = \frac{B_{br}}{B_{vf}} \cdot 100\ \%, \qquad (4)$$

where B_{br} – balance reserves of fuel, tons; B_{vf} – the volume of fuel production per year, tons/year.

However, the achievement of such tasks is impossible without the development of a safety and risk management mechanism in the Russian energy complex, both individually for each energy enterprise and for the whole industry.

3 Comprehensive safety and energy efficiency indicator of an energy company

Modern design technologies are based on the use of digital models that allow numerical simulation of the operation of technological equipment operating taking into account the technological features of production.

The formation of a comprehensive indicator of the safety and energy efficiency of an energy company is carried out as an arithmetic mean of the point estimates of the nature of the situation for each energy security indicator presented in Table 1.

Table 1. Threshold values of energy security indicators.

№	Indicators	N (stable-normal situation)	PC (pre-crisis situation)	C (crisis situation)
1	2	3	4	5
1	The degree of depreciation of fixed assets of the enterprise	≤50	>50 <60	≥60
2	The coefficient of renewal of fixed assets	≥2	<2,0 >1,5	≤1,5
3	The degree of price balance in the market	≥100	<100 >90	≤90
4	The share of the energy component in the cost of production for energy security	≤3	>3 <5	≥5
5	The share of electric power equipment that has exhausted its resource	≤30	>30 <50	≥50
6	Availability of fuel and energy resources	≥30	<30 >20	≤20
7	Coefficient that takes into account climatic factors	≤0,05	>0,05 <0,10	≥0,10
8	A coefficient that takes into account the presence of technical supervision regulations in case of violation of industrial safety requirements	≤1	>1 <2	≥3
9	The coefficient characterizing the probability of trouble-free operation of the power unit	≥0,95	<0,95 >0,70	≤0,70
10	Coefficient that takes into account the under-production of electrical energy	≤0,05	>0,05 <0,30	≥0,30
11	Coefficient that takes into account the under-production of thermal energy	≤0,05	>0,05 <0,30	≥0,30
12	Information security coefficient	≤2	>2 <5	≥5
13	Complex (integral) indicator of the safety level of the power unit	≥0,89	<0,89 >0,12	≥0,12

In this regard, the methods currently being developed should be based on experimental methods for assessing the condition of equipment elements throughout the entire life cycle, as well as identifying possible damage at early stages of development.

Threshold values of enterprise safety indicators are proposed in order to determine their actual nature (normal (stable), pre-crisis, crisis) for each indicator presented in Table 2.

Table2. Color indication of the assessment of reliability and safety levels of power equipment.

Range of physical wear values	Visualization (color)	Reliability level (by actual wear)	Security level	Type of technical impact
>0,75	Red	Critical	Unacceptable (Crisis C)	Exploitation is unacceptable. Urgent impact on the equipment and (or) the electric power facility is required
>0,50 and ≤0,75	Orange	Unsatisfactory	Level of additional maintenance and technical re-equipment (pre-crisis PC level 3)	Additional maintenance and repair, enhanced monitoring of technical condition, technical re-equipment
>0,30 and ≤ 0,50	Yellow	Satisfactory	Enhanced control level (pre-crisis PC level 2)	Enhanced control of technical condition, major repairs, reconstruction
>0,15 and ≤0,30	Green	Good	The level of normal operation and planned diagnostics (pre-crisis PC level 1)	According to the results of the planned diagnosis
<0,15	Dark green	Very good	Required (stable-normal NS)	Routine diagnosis

For the convenience of processing the received information, visualization of the level of reliability and safety is used, taking into account the physical wear of the equipment. The proposed threshold values of enterprise safety indicators in order to determine their actual nature (normal (stable), pre-crisis, crisis) for each indicator are presented in Table 2.

In the modern conditions of the market functioning, the factors of efficiency and accuracy of obtaining a forecast of the reliability of equipment safety, characterizing the condition of its elements and components, are of great importance. To develop criteria for assessing the level of reliability and safety of power equipment K_i, it is necessary to identify the main indicators.

1. The wear coefficient of the equipment of the thermal power plants (TPP) power unit is determined by the formula:

$$K_w = 1 - ITC_i, \tag{5}$$

where ITC_i – the index of the technical condition of the equipment, which is evaluated on the basis of the methodology [11, 12].

Based on the methodology [11], the following parameters of functional nodes are evaluated:

– for a steam turbine: the maximum vibration value of bearing supports, bearing defects, the maximum temperature of the babbit of bearing liners (pads), corrosion damage to discs, the maximum temperature behind the oil cooler;

– for a steam boiler: the presence of restrictions on parameters / resource after industrial safety examination / technical diagnostics, the number of defects eliminated by welding during the entire period of operation, violations of the geometry of the boiler frame, suction into the furnace and gas path before exiting the superheater;

– for a turbo generator: the insulation resistance of the rotor winding, the temperature according to the results of the rotor winding heating tests, the breakdown of the rotor winding insulation during operation, the temperature of the stator winding rods according to the results of the generator heating tests, generator power limitations (due to increased heating of the stator winding), the difference in the values of the resistance of the windings to direct current, vibration of the stator core;

– for a power transformer with a voltage of 110 kV and higher: oil leakage, oil pressure, the degree of development of a defect in contact connections according to the results of thermal imaging control, breakdown voltage, moisture content, insulation resistance of the measuring output.

2. Coefficient of renewal of power unit equipment:

$$K_r = \frac{ITC_i}{ITC_n}, \tag{6}$$

where ITC_i – the index of the technical condition of the equipment at the current time; ITC_n – the normative index of technical condition, characterizing the type of technical condition at the lower limit of the «good» level, is not less than 70%.

3. Coefficient of the ratio of the electric power of the power unit to the electric power of the station:

$$K_n = \frac{N_{e.b}}{N_{e.st}}, \tag{7}$$

where $N_{e.b}$ – electric power of the power unit; $N_{e.st}$ – installed electric capacity of the station.

4. Fuel reserve ratio:

$$K_{fr} = \frac{B_i}{B_{max}}, \tag{8}$$

where B_i – fuel consumption at the current time; B_{max} – maximum fuel capacity at the power plant.

5. Coefficient that takes into account climatic factors:

$$K_c = \frac{24 \cdot \tau_i \cdot (t_i - \tau_{io})}{D_x \cdot (t_x - \tau_x)}, \tag{9}$$

where τ_i – the duration of the period under consideration at the current time; D_x – duration of the heating period, in days; $(t_i - \tau_{io})$ – the difference in air temperatures at the current time; $(t_x - \tau_x)$ – the difference in indoor and ambient temperatures for the coldest month of the year.

6. A coefficient that takes into account the presence of technical supervision regulations in case of violation of industrial safety requirements:

$$K_s = \frac{n}{\sum n_\kappa}, \quad (10)$$

where n – number of technical supervision regulations in the current year; $\sum n_r$ – average annual number of prescriptions for the last 5 years.

7. The coefficient characterizing the probability of trouble-free operation of the power unit:

$$K_t = (1 - \alpha \cdot t) / K_r, \quad (11)$$

where α=0,05 – a coefficient that takes into account the forecast of changes in the index of the technical condition of the functional unit of the main technological equipment; t - prediction time, years.

8. Coefficient that takes into account the underproduction of electrical energy:

$$K_{\Delta N} = (1 - p) \cdot \frac{N_i}{N_{max}}, \quad (12)$$

where N_i – release of electric energy at the current time, MW·h; N_{max} – maximum annual output of electric energy, MW·h.

9. Coefficient that takes into account the underproduction of thermal energy:

$$K_{\Delta Q} = (1 - p) \cdot \frac{Q_i}{Q_{max}}, \quad (13)$$

where Q_i – heat energy release at the current time, MJ; Q_{max} – maximum annual output of thermal energy, MJ.

10. Information security coefficient:

$$K_{is} = \frac{n}{\sum N}, \quad (14)$$

where n – the number of cyber attacks in the current year; $\sum N$ – the average annual number of cyber attacks over the past 5 years.

To obtain a comprehensive assessment of energy security for TPP power units, it is necessary to determine the weight index of each of the analyzed indicators, calculated by the formula:

$$K_a = \frac{K_{ai}}{\sum_{1}^{10} K_{ai}}, \quad (15)$$

where K_{ai} – the number of identified requests, determined by the number of requests in information sources; $\sum_{1}^{10} K_{ai}$ – the total number of identified requests, determined by the number of common requests in information sources.

The complex (integral) indicator of the safety level of the power unit is calculated by the expression:

$$K_{saf}^{e} = \sum_{i=1}^{n} (K_{wei} \cdot K_i). \quad (16)$$

It should be noted that in the case of assigning two or more safety indicators to the «crisis» level, despite the results of the calculated value K_{saf}^{e}, the complex (integral) indicator of the safety level of the power unit will also relate to the crisis threshold value of their actual nature.

4 Monitoring of safety indicators

Operational determination of security levels can be implemented on the basis of specialized digital data collection systems, including [13]. At the same time, it is necessary to take into account the importance of ensuring information security issues. The creation and operation of information systems for remote monitoring and diagnostics of power equipment is regulated by the requirements for ensuring information security described in the order of the order of the Ministry of Energy of the Russian Federation, 2018, №. 1015.

In terms of the detailed improvement of the management methodology of an energy enterprise of generating and energy systems, the authors consider the following methodological tasks.

1. Development of criteria for assessing the degree of use of digital technologies in ensuring the reliability and safety of energy enterprises.
2. Formalization of evaluation criteria.
3. Formation of a comprehensive criterion for the optimality of top-level control actions, from the point of view of ensuring maximum efficiency of an energy enterprise in terms of ensuring a given system-wide reliability and safety in the conditions of digital transformation.
4. Development of an algorithm for generating records in a distributed blockchain network about changes in the state of controlled power equipment and an algorithm for managing a database with information about controlled equipment and facilities [14].
5. Development of a control program that ensures the interaction of the database with neural network algorithms for analyzing and predicting the technical condition of equipment, security levels of energy enterprises and systems [13, 15].

Conclusion

The proposes a methodology for assessing the level of safety of TPP power units, which allows timely determining the degree of safety of equipment operation, forming a factor of physical impact on power equipment to normalize this level, if necessary, and taking managerial actions to prevent an emergency.

Safety indicators of generating systems are formulated, criteria for assessing the level of safety of TPP power units are defined.

The limits of the change of the complex (integral) indicator of the safety level of the power unit are determined depending on the results of the calculation of indicator indicators and their weighting coefficients.

The research is funded by Russian Federation public contract № FSWF-2020-0025 «Technique development and method analysis for ensuring power system object security and competitiveness based on the digital technologies» on the equipment of the collective use center «Testing range of technologies for the transmission of electricity and distributed intelligent power systems» NTI MPEI.

References

1. Federal Law № 116 of 21.07.97 «On Industrial Safety of hazardous production facilities». https://cornplus.ru/userfls/ufiles/fz-116
2. Energy strategy of Russia for the period up to 2035 (approved by the Decree of the Government of the Russian Federation, № 1715-r, 2017). https://minenergo.gov.ru/node/1026
3. Order of the Ministry of Energy of the Russian Federation, № 123, 2019, «On approval of methodological guidelines for calculating the probability of failure of a functional node and a unit of basic technological equipment and assessing the consequences of such a failure». https://www.garant.ru/products/ipo/prime/doc/72113770
4. Gaifullina M.M., Khaliullina D.R., Hafizova L.K. Assessment of energy security and energy efficiency of an oil company // Online journal «Science Studies», Vol. 9, 3, (2017)
5. Baranovsky V.V., Korotkova T.Yu., Konovalov M.Yu. The use of logical-probabilistic methods for assessing the safety and reliability of thermal power plants as structurally complex technical systems // Energy Security and Labor Protection, pp. 5-9 (2015)
6. Digital economy of Russia // TAdviser. URL: https://www.tadviser.ru/index.php
7. Gaev A.V., Danyushevsky I.A., Shevchuk R.E., Zhuravlev D.N. Development of a methodology for assessing the reliability and safety of thermal mechanical equipment // Reliability and safety of power engineering, pp. 65-69 (2015)
8. Sudakov A.V., Saikova M.S., Danyushevsky I.A. Accounting of defects in the examination of industrial safety and resource assessment of elements of boilers and pipelines // Berg-Collegia, (2010)
9. Grabchak E.P., Loginov E.L. Digital approaches to the management of electric and thermal power engineering facilities using intelligent cyberphysical systems // Reliability and safety of power engineering, Vol. 12, 3, pp. 172-176 (2019)
10. Veselov F.V., Dorofeev V.V. The intelligent energy system of Russia as a new stage of energy development in the digital economy // Energy Policy, pp. 43-52 (2018)
11. Order of the Ministry of Energy of the Russian Federation, № 676, 2017 «On approval of the methodology for assessing the technical condition of the main technological equipment and power transmission lines of power stations and electric networks». https://base.garant.ru/71779722
12. GOST R 55173-2012. Boiler installations. General technical requirements National Standard of the Russian Federation. Moscow: Standartinform, https://docs.cntd.ru/document/1200103141
13. Sultanov M. M., Boldyrev I. A., Gorban Yu.A., Smirnov A.A., Yurov V.A. A program for collecting, storing and processing data on operating modes of energy systems to ensure a high level of reliability. № 2022685054 (2022)
14. Sultanov, M.M. The principle and algorithm of distributing data on the reliability and operating modes of power equipment when using a blockchain network and a centralized data storage system // New in the Russian electric power industry. pp. 30-40 (2022)
15. Sultanov M.M., Boldyrev I.A., Shamigulov P.V. Application of models for calculating and predicting reliability and safety of power equipment using data collection, processing and management complexes // Methodological issues of research on the reliability of large energy systems. Materials of the 94th meeting of the International Scientific Seminar. Editor V.A. Stennikov. Irkutsk, pp. 573-582 (2022)

Cyber security risks of interconnected information systems in intelligent management of microgrid communities

Liudmila Gurina[1,*] and *Nikita* Tomin[1]

[1]Melentiev Energy Systems Institute of Siberian Branch of the Russian Academy of Sciences, 130 Lermontov str., Irkutsk, Russia

Abstract. The paper examines ways to form energy communities, analyzes various management structures for such communities, assesses and identifies possible threats and vulnerabilities of information systems (IS), possible failures and failures in IS during cyberattacks, which can lead to errors in the formation of control actions. An approach to reducing the cybersecurity risks of the information infrastructure of the microgrid community is proposed.

1 Introduction

Aggregation of microgrids in the form of an energy community promotes the efficient use of local energy resources between community members, curbing the growth of electricity tariffs, and uninterrupted and reliable energy supply to consumers [1, 2].

To implement the noted advantages, the microgrid community uses special digital automatic control systems and performs functions for optimal management of available energy sources. In this regard, such an energy community forms a cyber-physical energy system (CPES) with the introduction of advanced information and communication technologies (artificial intelligence, blockchain, Internet of things, etc.) [3].

As a rule, the energy system (ES) is strongly influenced by various dependencies that exist within the system itself, between the technological part and the information and communication infrastructure of the ES, as well as between the ES and other critical infrastructures or its environment [4]. In [5], interdependence is defined as "a bidirectional relationship between two infrastructures, whereby the state of each infrastructure influences or correlates with the state of the other".

When forming an energy community, as well as during the distributed control of energy facilities, interdependencies appear both between IS microgrids and technological interdependencies, due to which new potential vulnerabilities, common cause failures and other interdependent failures appear. Interdependencies also mean that microgrids are more susceptible to cyberattacks, even if such attacks do not directly target the microgrid itself. Cyber threats are constantly evolving, and there are many measures that can be taken to make IS more secure. However, many of the available measures are best suited for traditional ESs, but it may be more difficult to apply these measures to ESs

Microgrid management is actually a multi-objective task, spanning over different technical domains, time scales and physical layers. Multi-level management includes primary

in which distributed objects are closely interrelated. Thus, during cyberattacks, it is important to identify and analyze possible failures in the interconnected IS of microgrids in order to ensure the normal functioning of the energy community.

In this regard, a complex task arises for microgrid communities, which is to maintain the cybersecurity properties of IS management due to the growth of digital objects used and interconnection, which contribute to an increase in vulnerabilities to cyberattacks. Therefore, the purpose of the study is to develop an approach to ensuring cyber security of the information and communication infrastructure of energy community microgrids, the essence of which is to model energy community, simulate cyberattacks and develop methods and means of protecting interdependent information systems from cyberattacks.

2 Microgrid community management framework

Energy community management includes control of power converters, distribution of active/reactive power between distributed generation sources, control of charge level and charging/discharging power of energy storage systems, synchronization of multiple microgrids, maintaining voltage-frequency balance and generation-load balance in microgrids etc. The principles of construction and operating modes of microgrids are described in [6].

Microgrid management is actually a multi-objective task, spanning over different technical domains, time scales and physical layers. Multi-level management includes primary management, secondary management and tertiary management. Based on this hierarchy, the way to implement energy community management levels can be centralized, decentralized, distributed or hierarchical [7].

management, secondary management and tertiary management. Based on this hierarchy, the way to implement energy community management levels can be centralized,

* Corresponding author: gurina@isem.irk.ru

decentralized, distributed or hierarchical [7]. In a centralized structure, there is a central control unit that collects and transmits information to local generation sources. Decentralized and distributed structures do not require a central controller. Decentralized control, as defined in [8], performs regulation based on local measurements, and distributed control is based on both local measurement and neighboring communication [9]. The hierarchical control structure distributes control functions between local controllers and upper-level controllers.

Centralized control requires collecting data from all the main components of the microgrid [10]. Based on the collected information, monitoring and control procedures can be carried out in the controller to achieve correct and efficient operation. The benefits of centralized control include high observability and controllability across the entire energy community, as well as ease of implementation. However, this creates the problem of a single point of failure, so that failure of the central controller will lead to the loss of all functions [11].

Decentralized microgrid control belongs to control methods that do not require information from other parts of the system. The controller regulates the corresponding block using only local information. The advantage of decentralized schemes is that they do not require real-time communication, although the lack of coordination between local regulators limits the ability to achieve globally coordinated behavior.

The functions provided by the centralized control scheme can also be implemented by the distributed way. Information is transferred between controllers via communication lines, so that the necessary information is distributed between each local system to facilitate the coordinated behavior of all units. The main challenge of a fully distributed control scheme is the coordination among distributed units to accomplish control goals. The exchange of information between microgrids within an energy community allows controllers to find the optimal operating strategy for sustainable and efficient operation of the energy community.

Distributed secondary control, as a new control strategy, performs all the functions of a centralized controller with less communication and computational overhead, while being resilient to failures or unknown system parameters. The idea is to combine primary and secondary controls into one local controller. Unlike decentralized primary control, built-in secondary controllers must "talk" to their neighbors to operate properly. Each agent (i.e., converters, such as AC/DC inverters) exchanges information with other agents in the communication environment. Each local secondary controller makes a decision according to the information of its neighbors [12].

To ensure the cybersecurity of the energy community under centralized and distributed control, it is necessary to take into account the interdependencies of not only information, communication and physical subsystems, but also the interdependence of microgrid ISs within the energy community.

3 Threats and vulnerabilities that pose cybersecurity risks to the microgrid community

3.1 Microgrid vulnerabilities

Integrating the information and communication infrastructure with the physical (technological) subsystem provides benefits, but also creates a new set of vulnerabilities that can expose the system to various threats. Exploitation of such cyber vulnerabilities can lead to physical consequences. The microgrid, being a CPES, inherits their general cyber vulnerabilities, added to the vulnerabilities caused by the specifics of distributed energy. Reasons for vulnerabilities may include: use of wireless communications, use of heterogeneous communication technologies, increased exposure to external networks, exposure to the Internet, increased system automation, increased use of distributed control and automation devices, coexistence between legacy and new systems, use of multiple independent systems [13].

Thus, microgrids can be exposed to various cyber threats.

3.2 Potential cyberattacks on microgrids

The main target of cyberattacks is usually control and monitoring objects [14]. Attackers can exploit vulnerabilities across assets to gain access to multiple levels of control and impact the functioning of microgrid energy communities, thereby creating cybersecurity risks.

There are various types of cyberattacks on IS that can be implemented for microgrid energy communities: hardware and software manipulation, false data injection attacks (FDI attack), denial of service attacks (DoS attack), hijacking attack and etc. [15, 16].

An FDI attack can affect the integrity of information, DoS attacks will interrupt access to transmitted data, hijacking attacks will disrupt the control and "hijack" controllers of a distributed control system [16]. In this case, a hijack attack on the controller breaks the communication channel and replaces other data, therefore interrupting the process of updating the received signal. [17] shows that hijacking attacks can reduce the optimal performance of microgrids. Since such attacks replace the time-stamped measurement with a constant input, the linear consensus algorithm cannot update its reference state with respect to neighboring agents, which ultimately leads to an inevitable power imbalance.

In this regard, an approach is proposed to reduce the cybersecurity risks of interconnected IS microgrids, which is as follows:

- Modeling a microgrid energy community.
- Modeling cyberattacks, assessing the spread and impact of the consequences after an attack on interconnected microgrid ISs within the energy community.
- Development of possible measures to ensure the security of microgrid IS from cyber attacks.

In this regard, awareness of cyber-attacks and identification of their causes allows us to better assess the impact of failures on interconnected distributed objects.

4 Multi-agent distributed secondary voltage control system

In this paper, a multi-agent control system developed in [18] was used to implement distributed secondary voltage control in a microgrid community.

In this approach, the electrical network is considered as a multi-agent $G = (V, E)$, where each agent $i \in V$ interacts with its neighbors $N_i: \{j | \varepsilon_{ij} \in E\}$. The agent here is understood as a secondary control controller for a DC/AC inverter, used to connect DC distributed generation (DG) sources to the AC network.

In the presented structure, during the learning process, the agent "learns" a control strategy based on sub-global rewards, as well as local states and encoded communication messages from its neighbors (other agents). Each agent i in this model observes only part of the environment (its state and its neighbors), which leads to a partially Markov Decision Process. This problem is solved by the method of multi-agent reinforcement learning, for which the following key elements are defined:

- Action space: the control action for each agent is the secondary voltage control setpoint V_n. 10 discrete actions were used, evenly distributed between 1.02 and 1.12 p.u.
- State space: the state of each agent is chosen as $s_t = (\delta_i, P_i, Q_i, i_{odi}, i_{oqi}, i_{bdi}, i_{bqi}, v_{bdi}, v_{bqi})$ to characterize the CIGs modes, where $\delta_i$is the measured reference angle; P_i, Q_i are active and reactive power, respectively; $i_{odi}, i_{oqi}, i_{bdi}, i_{bqi}$are output currents d-q CIG i and directly connected buses, respectively; and v_{bdi}, v_{bqi} are the output voltages d-q of the connected bus, respectively.
- Observation space: it is assumed that each agent can only observe its local state, as well as messages from its neighbors, i.e., $o_{i,t} = S_{i,t} \cup m_{i,t}$, where $m_{i,t}$ is a communication message received from neighboring agents $j \in N_i$, which will be discussed in more detail below.
- Reward function: the goal of all agents is to maximize the total reward, $R_{i,t} = \sum_{k=o}^{T} \gamma^k \sum_{j \in v} \alpha(d_{i,j}) r_{i,t+k}$, where $\alpha(d_{i,j}) \in [0,1]$ is the spatial discounting function; $d_{i,j}$ is the distance between agent i and j; $r_{i,t}$is the reward of agent i at time step t. The function $r_{i,t}$is defined as follows, so that the voltages in the generator units quickly converge to the reference values (for example, 1 p.u.):

$$r_{i,t} = \begin{cases} 0.05 - |1 - v_i|, v_i \in |0{,}95; 1{,}05|, \\ -|1 - v_i|, v_i \in |0{,}8; 0{,}95| \cup |1{,}05; 1{,}25|, \\ -10. \mathit{Otherwise} \end{cases} \tag{1}$$

where $r_{i,t}$ is the reward of agent i at time step t.

In fact, we divide the voltage range into 3 working zones: normal mode zone (|0.95,1.05| p.u.), heavy duty zone |0.8,0.95|U|1.05,1.25| p.u.) and emergency zone (|0.0.8|U|1.25.∞| p.u.). With the formulated reward, an agent with "emergency" voltages will receive a large penalty, and an agent with a voltage close to 1 p.u. will receive a positive reward.

In the multi-agent structure under consideration, information from neighboring agents is used to improve learning efficiency. Thus, based on the structure proposed in [18], agent i updates hidden state, $h_{N,t-1}$ at each step t:

$$h_{i,t} = f_i\left(h_{i,t-1}, q_0\left(e_s(o_{i,t})\right), q_h(h_{N,t-1})\right), \tag{2}$$

where $h_{i,t-1}$ is the hidden state from the previous time step; $o_{i,t}$is the observation of agent i made at time t, i.e., its internal state of and the states of its neighbors; $h_{N,t-1}$is integrated state from neighbors; $e_s(o_{i,t})$ and q_h are differentiable functions for encoding and retrieving messages.

Instead of low-dimensional indicators, here the full states of the neighboring agent are included in the local observation $o_{i,t} = s_{i,t} \cup s_{N,t}$to improve the agent's observability. In this case, the received communication message $m_{i,t}$, *i-th* agent, is a combination of internal states and hidden states of its neighbors.

5 Case study

Consider several scenarios in which an attacker corrupts data, such as live data exchanged between agents, by injecting false data into communication channels or wireless communication channels. Consider an FDI attack on multiple agents. This attack modifies the measured information from neighboring agents by adding false data [16]. The actual current measurement (observation) $o_{i,t} = S_{i,t} \cup m_{i,t}$of neighboring agents during this attack is described as:

$$o_{i,t}^a = o_{i,t} - \alpha x_{i,t}^a, \tag{3}$$

where $x_{i,t}^a$ is the injected false data, specified as a random distribution in a certain range, $\alpha \in \{0,1\}$ is the data distortion coefficient, $\alpha = 1$ where corresponds to a full-fledged FDI attack.

Let's also consider a controller hijacking attack. In the event of such an attack, the attacker replaces the measurements with malicious data [20]:

$$o_{i,t}^c = (1 - \alpha) o_{i,t} - \alpha x_{i,t}^a, \tag{4}$$

where $o_{i,t}^c$ is modified observation of the agent; $x_{i,t}^a$is injected false data, specified as a random distribution in a certain range, and $\alpha = 1$ means a full-fledged attack on the inverter with complete replacement of correct observations.

To test a multi-agent system with simulated cyberattacks, by analogy with [20], we considered a model of a microgrid energy community with distributed power sources, obtained based on a modification of the IEEE34 scheme (Fig. 1).

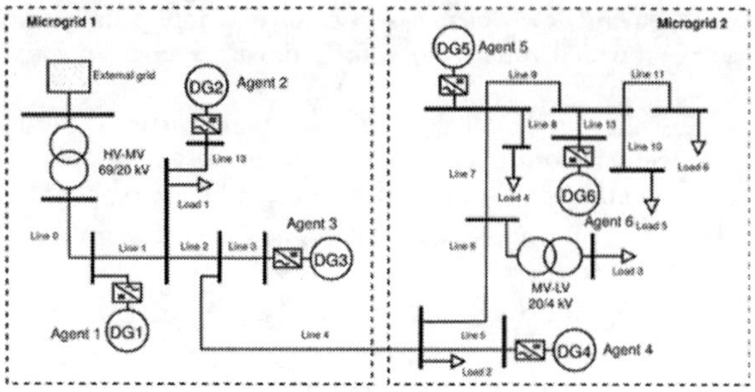

Fig. 1. Microgrid community test circuit.

To simulate heavy duty modes, random load changes were added throughout the network with deviations of ± 20% from the nominal values, as well as random disturbances in the range of ± 5% for each load. All agents in the considered schemes were monitored with a sampling time of 0.05 s, and each agent could communicate with its neighbors across local communication boundaries. Primary control of the lower level is implemented by analogy with [21].

For the circuit shown in Fig. 1, the scenario of an FDI attack and a Hijacking attack on agents 3, 5 and 6 was considered, according to (3) and (4). In Fig. 2 the results of such modeling are presented, which show the quality of voltage stabilization after a load disturbance for various scenarios.

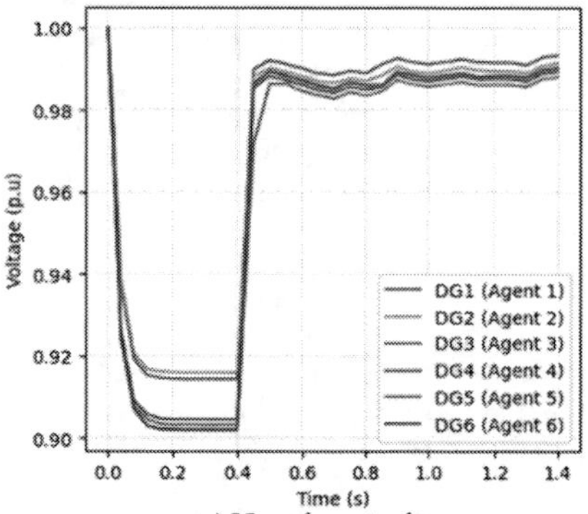

a) No cyber attacks

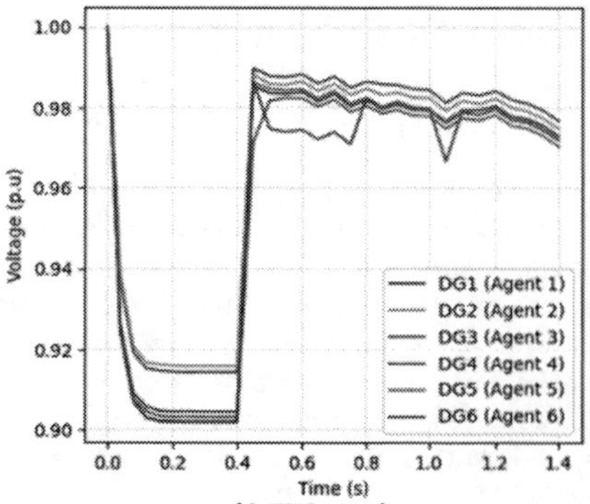

b) FDI attack

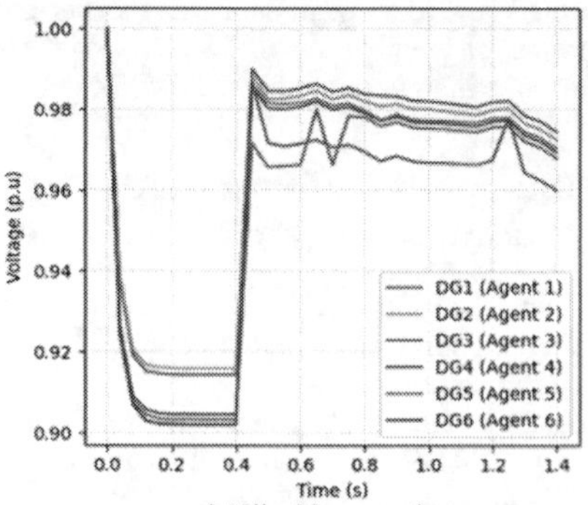

c) Hijacking attack

Fig. 2. Results of modeling the behavior of agents of a trained system voltage regulation during load disturbances in the absence and presence of cyberattacks

It is clearly seen that for the scenario in the absence of cyberattacks, inverter agents effectively cope with the task of coordinated voltage stabilization after a disturbance. However, when attackers interfere in information communications (Fig. 2b and 2c) through an FDI attack and Hijacking attack, the quality of voltage regulation deteriorates, especially in terms of voltage regulation by distributed generator inverter No. 6 (agent 6). At the same time, the hijack controller scenario seems to be the most "dramatic" in terms of stabilizing voltage profiles in generating nodes.

6 Conclusions

Various management structures for energy community microgrids are considered. It is shown that with centralized and distributed management of microgrid energy communities, the number of vulnerabilities to cyberattacks increases due to the interdependence of the IS.

An approach has been proposed to ensure the cybersecurity of the information and communication infrastructure of the energy community, which allows, when modeling cyberattacks on microgrid ISs, to analyze their distribution, assess the consequences for intelligent management and, in the future, develop measures to protect interdependent ISs from cyberattacks.

Acknowledgment

The research was carried out within the framework of the scientific project "Theoretical foundations, models and methods to control the expansion and operation of intelligent electric power systems", No. FWEU-2021-0001.

 https://doi.org/10.1051/e3sconf/202346101031

References

1. Gjorgievski V.Z., Cundeva S., Georghiou G.E., Renewable Energy. **169** (2021)
2. Warneryd M., Hakansson M., Karltorp K., Renewable and Sustainable Energy Reviews. **121** (2020)
3. The Microgrid Case Studies: Community Resilience for Natural Disasters, 2020. https://sepapower.org/resource/the-microgrid-case-studies-community-resilience-for-natural-disasters/
4. Voropay N.I., Electrichestvo, **7** (2020)
5. Rinaldi S.M., Peerenboom J.P., Kelly T.K., IEEE Control Systems Magazine. **21** , 6 (2001)
6. Ilyushin P.V., Volnyi V.S., Relay protection and automation. **1(50)** (2023)
7. Saha D., Bazmohammadi N., Vasquez J.C., Guerrero J.M., Energies. **16**, 2 (2023)
8. Davison E.J., Aghdam A.G., Decentralized Control of Large-Scale Systems. (2014)
9. Scattolini R., Process Control. **19**, 5 (2009)
10. Tsikalakis A.G., Hatziargyriou N.D., IEEE Trans. Energy Convers. **23**, 1 (2008)
11. Meng L. et al., IEEE Journal of Emerging and Selected Topics in Power Electronics. **5**, 3 (2017)
12. Nasirian V., Moayedi S., Davoudi A., Lewis F.L., IEEE Trans. Power Electron. **30**, 4 (2015)
13. Rekik M., Chtourou Z., Gransart C., Atieh A.A., 15th International Multi-Conference on Systems, Signals & Devices (SSD) (2018)
14. Kolosok I.N., Gurina L.A., Methodological issues in the study of the reliability of large energy systems (2019)
15. Kolosok I.N., Gurina L.A., Cybersecurity issues. **6(46)** (2021)
16. Abhinav S., Modares H., Lewis F.L., Ferrese F., Davoudi A., IEEE Trans. Smart Grid. **9**, 6 (2018)
17. Duan J., Zeng W., Chow M.Y., IECON 2016 – 42nd Annual Conference of the IEEE Industrial Electronics Society (2016)
18. Tomin N., Voropai N., Kurbatsky V., Rehtanz C., Energies. **14** (2021)
19. Zhang K., Yang Z., Liu H., Zhang T., Basar T., arXiv:1802.08757 (2018)
20. Sahoo S., Peng J.C.H., Mishra S., Dragicevic T., IEEE Trans. Power Electron. **35**, 7 (2020)
21. Bidram A., Davoudi A., Lewis F.L., Qu, Z., IET Gener. Transm. Distrib. 7 (2013)

E3S Web of Conferences **461**, 01032 (2023) https://doi.org/10.1051/e3sconf/202346101032
RSES 2023

Machine learning application for power systems reliability assessment

Dmitry Sorokin[1*]

[1] Research and Development Center at Federal Grid Company of Unified Energy System, 22 building 3, Kashirskoe sh., Moscow, Russia

Abstract. The paper presents the principles and features of the use of machine learning methods to assess the power system reliability. Based on the analysis of publications, the main approaches to the application of machine learning methods are given. A prototype of an automatic system has been developed to identify in real time potentially dangerous power system states, the occurrence of which can lead to the power system failures.

1 Introduction

Machine learning and artificial intelligence methods are among the key technologies of Industry 4.0 that underlie the digital transformation of the power systems [1]. The progress achieved over the past decades in the development of machine learning and artificial intelligence methods makes the use of these methods promising for solving many problems in various fields of science and technology. Such capabilities of machine learning methods as searching for patterns (characteristic sequences) in data, identifying various data features and characteristics, the ability to predict the appearance of certain values in data allow us to solve problems that either cannot be solved by classical methods, or have high computational complexity [2]. An additional driver for the development of machine learning methods (in addition to increasing the performance of computer technology and improving the machine learning methods) is the increasing availability and "quality" of big data (data reliability, structure, ability to process, etc.), which is the basis for training artificial intelligence models.

In power systems, machine learning methods show effectiveness in solving problems of short-term forecasting of electrical energy consumption [3, 4, 5], in problems of monitoring and predictive diagnostics of electrical equipment [6, 7, 8, 9], in problems of emergency control and relay protection [10], etc. It is also worth noting the experience of using machine learning methods in creating a software and hardware complex for recognizing the state and readings of electricity meters [11], the experience of developing electric power quality analyzers with a built-in artificial neural network [12], the development of the information system for identifying the power and electricity imbalances in electrical distribution networks [13]. This paper discusses the possibilities and features of using machine learning methods to solve the problem of ensuring the reliable functioning of electric power systems.

2 Review of the machine learning application to assess the power system reliability

There are quite a lot of publications devoted to assessing and ensuring the reliability of power systems based on the use of machine learning methods. A significant number of works have been published from the 2010s to the present. Publications present various aspects of the reliability of power systems (for example, from the point of view of power system adequacy [14], stability [15, 16], etc.). This paper provides a selection of references to publications devoted to assessing the reliability of power systems based on the parameters of steady-state (quasi-steady-state) electrical modes. A large number of publications have been analyzed and selected the papers that reflect the most common approaches to assessing the reliability of power systems. As a results, it is possible to identify the main approaches: using the Monte Carlo method to obtain a training sample and predicting the SAIDI, SAIFI indices [14, 17], assessing the power system state admissibility indices [18, 19, 20], assessing various aggregated indices taking into account the electrical network topology, including the L-index [21], SSI index (static security index) [22], the use of Markov chains for assessing the reliability of power systems, including to prevent cascading failures [23]. A number of publications reflect the features of the use of specific machine learning methods in assessing the reliability of power systems [24, 25]. Based on the results of the analysis of publications, the following general conclusions can be drawn:

1. The processes of creating machine learning models are fairly unified and, in general, consist of the following main stages: generating a training set from

* Corresponding author: sorokin_dv@ntc-power.ru

https://doi.org/10.1051/e3sconf/202346101032

the parameters of power system states (voltages, currents, etc.), training the machine learning model, assessing the reliability of the power system based on test data (at the stage of the model operation – according to available data, for example, from SCADA systems, PMU, etc.).

2. To obtain a basic training set, mathematical models of power systems are used. To form the required power system states, in most cases the following criteria are used: N-1 criterion (in the general case – N-k), Monte Carlo methods, special power system states that simulate the required situations (for example, the voltage collapse, etc.).

3. The reliability of the power system is assessed in most cases based on the calculation of various indices. The use of indices has certain limitations and specific application area. In addition, the calculation of different indices requires different amounts of information about the power system. For example, to calculate the L-index, information about the topology of the electrical network, the impedances of the branches and the voltages in the nodes are required [26].

4. In most cases, various supervised learning methods are considered: support vector machines, random forest methods, decision trees, ensemble algorithms, neural networks, etc. From the results obtained in the publications reviewed, it is impossible to conclude that any one is superior or several machine learning methods for assessing the reliability of power systems.

5. In all the publications reviewed, a high accuracy of classification of power system states (prediction of the indices used) was obtained. In most cases, the performance assessment of the algorithms used by the authors of publications is obtained for mathematical power system models of a fairly small dimension (for example, IEEE power system models with 57, 96 and 118 nodes are often used).

3 Prototype of an automatic system for assessing the power system reliability

To study the features and possibilities of using machine learning methods, a prototype of an automatic system has been developed to identify potentially dangerous power system states in real time, i.e. states, the occurrence of which is associated with the possibility of developing emergency processes and failures in the power systems. The prototype of the system is based on the use of machine learning methods to solve the problem of classifying power system states according to the reliability index values.

The generalized process of creating machine learning models for assessing the reliability of power systems is shown in Fig. 1. At the first stage, a training set from power system parameters is generated. To generate this sample, the electrical parameters of the power system are used (voltages, currents, power flows, etc.), obtained from the results of calculations of the steady states in mathematical power system model. This paper examines states formed according to the N-1 criterion and states corresponding to different levels of power consumption in the power system.

The second step involves training several different machine learning models. Based on the training results, the best model is selected – the model with the best performance indicator values. At the third stage, the reliability of power systems is assessed using test or real data. Depending on the formulation of the problem, the result of the machine learning model operation can be either the classification of the power system state to one of the categories (for example, normal, emergency, etc.), or the determination (prediction) of reliability index value.

At the stage of testing the final performance of the proposed machine learning models, the reliability of power systems is assessed for a test sample, i.e. according to data obtained from the results of power system simulation. At the stage of operation of the proposed algorithms in real time, the prediction of the reliability index value is carried out using available data (for example, received from SCADA systems, PMU, etc.).

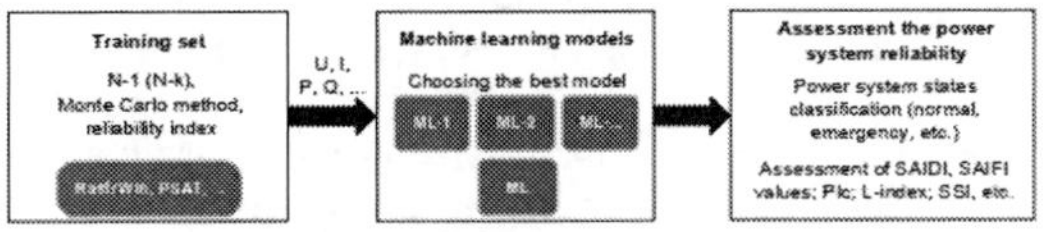

Fig. 1. A generalized process for generating machine learning models for assessing the power system reliability.

It should be noted that the presented prototype of the automatic system is a convenient tool (framework) for testing technologies for applying machine learning methods in electrical power systems. If necessary, the system prototype can implement various reliability indices, various scenarios for generating training samples, and apply various machine learning models to identify the model that is most appropriate for solving the specific problem.

4 Power system reliability indices

The aggregated index PI_c, defined as follows [19]:

$$PI_c = [\sum_i (q_{U_i}^{\max})^{2n} + \sum_i (q_{U_i}^{\min})^{2n} + \sum_j (q_{I_j})^{2n}]^{\frac{1}{2n}} \tag{1}$$

$$q_{U_i}^{\max} = \begin{cases} \dfrac{U_i - H_i^{\max}}{A_i^{\max} - H_i^{\max}}, \text{if } U_i > H_i^{\max}; \\ 0, \textit{otherwise}. \end{cases} \tag{2}$$

$$q_{U_i}^{\min} = \begin{cases} \dfrac{H_i^{\min} - U_i}{H_i^{\min} - A_i^{\min}}, \text{if } U_i < H_i^{\min}; \\ 0, \textit{otherwise}. \end{cases} \tag{3}$$

$$q_{I_j} = \begin{cases} \dfrac{I_j - I_{H_j}}{I_{A_j} - I_{H_j}}, \text{if } I_j > I_{H_j}; \\ 0, \text{otherwise}. \end{cases} \tag{4}$$

where i, j – the numbers of nodes and branches,
n – parameter that determines the exponent,
$q_{U_i}^{\max}$, $q_{U_i}^{\min}$ – voltage deviation indices,
q_{I_j} – current deviation indices,
I_j – current in the branch,
I_{A_j} – permissible current in the branch,
I_{H_j} – warning value of the current in the branch,
U_i – voltage in the node,

A_i^{min}, A_i^{max} – minimum and maximum permissible voltage levels in the node,
H_i^{min}, H_i^{max} – warning permissible voltage level in the node.

Index PI_c values are used to classify power system states. The paper presents two methods of classification: into two and three categories.

Classification of power system states into two categories:

1. $PI_c<1$ – for power system states, the parameters of which do not exceed the permissible ranges (normal state);
2. $PI_c\geq 1$ – for power system states, whose parameters have either reached or are outside the permissible ranges (emergency state).

Classification of power system states into three categories:

1. $PI_c=0$ – normal state;
2. $0< PI_c <1$ – for power system states, the parameters of which are outside the established warning ranges, but have not yet exceeded the permissible limits (normal state (warning));
3. $PI_c\geq 1$ – emergency state.

Thus, the value of the index PI_c value determines the admissibility of the power system state and is an indicator of the operational reliability of electric power system. Using the specified index, all electrical states can be classified into two classes: normal or emergency states, which allows the use of machine learning methods that carry out binary classification. Additionally, one more category of classification of states can be identified – potentially dangerous electrical states, the parameters of which go beyond the expertly established warning ranges, but have not yet exceeded permissible limits.

5 Power system model

To generate a training set and evaluate the performance of the machine learning models discussed in the paper, the IEEE-57 power system model was used. The model contains 57 nodes (including 7 generator nodes) and 80 branches. The following principles were used when selecting a test model of the power system. The power system model must be of a sufficiently large dimension, which makes it possible to obtain a large number of different electrical states and corresponds to models of real power systems. At the same time, based on the research nature of this paper, obtaining machine learning models should occur within an acceptable time and should not require the use of special hardware and software for storing and processing very large data sets. A single-line diagram of the power system model is shown in Fig. 2. The model was developed in the software RastrWin.

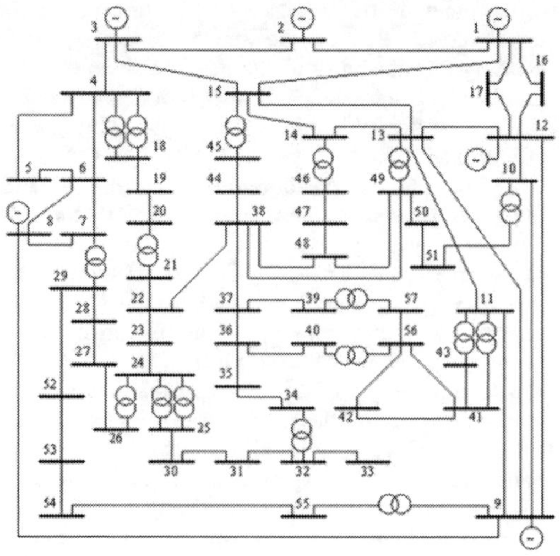

Fig. 2. IEEE-57 power system model.

6 Machine learning model performance indicators

To evaluate the performance of machine learning models in solving binary classification problems, the following indicators are used [2]:

$$Precision = \frac{TP}{TP+FP} \quad (5)$$

$$Recall = \frac{TP}{TP+FN} \quad (6)$$

$$F_1 = 2\cdot\frac{Precision\cdot Recall}{Precision+Recall} \quad (7)$$

where TP – the number of true positive predictions,
FP – the number of false positive predictions,
FN – the number of false negative predictions.

All three indicators can be obtained from the confusion matrix, which is one of the characteristics of machine learning models and is used in solving classification problems. In the context of the problem under consideration, the precision characterizes the classifier's ability not to classify emergency states as normal. In turn, the recall characterizes the classifier's ability not to classify normal states as emergency. The F_1 (f-score) indicator is interpreted as the harmonic mean value for the precision and recall values. The range of possible changes in the values of precision, recall and F_1 indicators is from 0 to 1. In this case, value 1 corresponds to the best values of the performance indicators of the machine learning model.

Performance indicators (5)-(7) are used when solving binary classification problems, i.e. classification into two categories. A common indicator for assessing the performance of machine learning models is the accuracy, which characterizes the number of true predictions relative to the total number of predictions. For classification problems, the accuracy indicator is defined as follows:

$$Accuracy = \frac{TP+TN}{TP+FP+TN+FN} \quad (8)$$

where TN is the number of true negative predictions.

7 Simulation results

Machine learning methods were applied to classify the electrical states of the power system model. Two groups of machine learning models were obtained: for classifying electrical modes into two categories (normal state, emergency state), and also for classifying electrical modes into three categories (normal state, normal state (warning), emergency state). The models are trained based on a sample consisting of parameters of 1182 electrical states. The following machine learning models were selected: decision trees, support vector machine, k-nearest neighbors, ensemble method (bagged trees) and neural network (multilayer perceptron). When training the models, 10-fold cross-validation was used. Characteristics of the performance of the resulting machine learning models for solving the binary classification problem are presented in Table 1. Comparative characteristics of the accuracy of machine learning models when predicting two and three categories of the power system (respectively, when solving the problem of binary classification and classification into three categories) are presented in Table 2.

As can be seen from the tables, all considered machine learning models have fairly high accuracy values. The ensemble method showed the highest accuracy on the training and test data used (accuracy 95.6% in the binary classification problem). As the number of classified categories increases, the prediction accuracy decreases. This may be due to the fact that with an increase in the number of classified categories, the training of the model for each category actually occurs over a smaller number of electrical states. A characteristic feature of the resulting machine learning models is their relatively high computational performance. Power system state classification is performed by the resulting machine learning models in about 50 ms (Intel Core i5, 8 GB RAM).

Table 1. The machine learning model performance to solve the problem of binary classification of power system states.

Machine learning model	Performance indicator, pu		
	Precision	Recall	F-score
Ensemble method	0,9601	0,9629	0,9614
Neural network	0,9514	0,9599	0,9556
Support vector machine	0,9419	0,9736	0,9575
k-nearest neighbors	0,9407	0,9658	0,9531
Decision trees	0,9478	0,9450	0,9464

Table 2. The accuracy of classification of power system states into two and three categories.

Machine learning model	Accuracy, %	
	Binary classification	Classification into three categories
Ensemble method	95,6	91,8
Neural network	94,9	91,2
Support vector machine	95,0	90,3
k-nearest neighbors	94,6	88,7
Decision trees	93,9	85,8

8 Conclusion

The paper discusses the main approaches to assessing the reliability of electric power systems based on the use of machine learning methods. Specific machine learning models have been developed to classify power system states according to the steady-state electrical parameters of power system. The high accuracy and performance of these models are shown. The presented models and the considered approaches can be used as a basis for creating an automatic system for assessing the reliability of electric power systems, namely, a system for identifying in real time potentially dangerous electrical states, the occurrence of which can lead to the development of emergency processes and failures in the power system.

References

1. Concept of Digital Transformation 2030, PJSC Rosseti. https://www.rosseti.ru/sustainable-development/digital-transformation-2030/
2. Steven W. Knox. Machine Learning: a Concise Introduction. – Wiley – 2018.
3. Kychnik A.V. and others. Forecasting power consumption in aggregated groups, ensemble models and selective choice models. – Energy of a single network, No. 1 (56). – 2021.
4. Tomin N.V., Kornilov V.N., Kurbatsky V.G. Increasing the efficiency of hourly forecasting of electricity consumption using machine learning models using the example of the Irkutsk energy system. - Electricity. Transmission and distribution. – No. 1(70), 2022.
5. Tanfilyev O.V., Sidorova A.V., Cheremnykh A.A. Application of ensembles of decision trees and linear regression for operational load forecasting. – Electricity. Transmission and distribution. – No. 6(69), 2021.
6. Goffman A.V. and others. Machine intelligence technologies for monitoring power transformers. – Energy of a single network. – No. 4 (53), 2020.
7. Zakharov O.A. Diagnostics of high-voltage rotating electrical equipment using the example of a generator. PRANA system for generators. – Energy of a single network. – No. 4 (53), 2020.
8. Khalyasmaa A.I. Application of machine learning methods to identify the technical condition of oil-filled instrument transformers. – Electricity. Transmission and distribution. – No. 6(63), 2020.
9. Kulikov A.L. Using machine learning and artificial neural networks to recognize turn faults in power transformers. - Electricity. – No. 10, 2022.
10. Kulikov A.L. and others. Formation of generalized information signs to increase the recognition of emergency modes by relay protection and automation. – Relay protection and automation. – No. 1, 2023.
11. Hardware and software complex for recognizing the state and readings of electricity meters based on artificial intelligence algorithms. – Electricity. Transmission and distribution. – No. 1(76), 2023.

12. Breakthrough forecasting technologies in the electric power industry using power quality analyzers with an artificial neural network. – Electricity. Transmission and distribution. – No. 4(23), 2021.
13. Turkina O.V., Voltov I.P., Ivanov D.S., Shcherbakov M.V. Information system for determining imbalances in the distribution network. – Energy of a single network. – No. 5-6 (60-61), 2021.
14. Boyarkin D.A., Krupenev D.S., Yakubovsky D.V. Using machine learning methods in assessing the reliability of electric power systems using the Monte Carlo method. – Bulletin of the South Ural State University. Series: Mathematical modeling and programming. – 2018.
15. Senyuk M. Power System Transient Stability Assessment Based on Machine Learning Algorithms and Grid Topology. – Modeling and Simulation for the Electrical Power System, No. 11(3), 2023.
16. Xin Li. Deep learning-based transient stability assessment frame-work for large-scale modern power system. – International Journal of Electrical Power & Energy Systems, No. 139(6), 2022.
17. Kamruzzaman Md. A convolutional neural network-based ap-proach to composite power system reliability evaluation. – International Journal of Electrical Power & Energy Systems, Vol. 135, 2022.
18. Tomin N., Sidorov D., Zhukov A. Machine Learning Techniques for Power System Security Assessment. – IFAC, 2016.
19. Yahui Li. Online Static Security Assessment of Power Systems Based on Lasso Algorithm. – Developing and Implementing Smart Grids: Novel Technologies, Techniques and Models, No. 8(9), 2018.
20. Mollaiee A. Novel continuous learning scheme for online static security assessment based on the weather-dependent security index. – IET Generation, Transmission & Distribution, 2022.
21. Zhukov A. Methods of classification and restoration of regression based on compositions of decision trees with application to the problem of assessing the operational reliability of EPS. – Dissertation for the degree of candidate of technical sciences. – Melentiev Energy Systems Institute of Siberian Branch of the Russian Academy of Sciences, 2021.
22. Ramirez-Gonzalez M. Convolutional neural nets with hyperparameter optimization and feature importance for power system static security assessment. – Electric Power Systems Research, 2022.
23. Wu W. A Markov chain-based model for forecasting power system cascading failures. – Automation of Electric Power Systems, No. 37(5), 2013.
24. Singh M. Tree-Based Ensemble Machine Learning Techniques for Power System Static Security Assessment. – Electric Power Components and Systems, No. 50, 2022.
25. Avouris N. Power Systems Contingency Analysis using Artificial Neural Networks. – Proceedings of the 4th international workshop on computer science and information technologies, 2002.
26. Elemary A. Power System Voltage Stability Index. – International Journal of Engineering Research & Technology (IJERT), Vol. 11, 2022.

 https://doi.org/10.1051/e3sconf/202346101033

Intelligent control system to increase the flexibility of power systems – the direction of development and a way to solve problems in the Irkutsk power system

Alexey Osak[1*], *Elena* Buzina[1]

[1] Melentiev Energy Systems Institute of Siberian Branch of the Russian Academy of Sciences, 130 Lermontov str., Irkutsk, Russia

Abstract. In the power system of the Irkutsk region since 2020, there has been a significant increase in the maximum electrical load, which amounted to more than 1.2 GW. As a result, the available reserves of power generating capacity and capacity of power grid were exhausted. In the coming years, situations of occurrence of an uncovered power shortage during periods of maximum loads are possible. It is proposed to solve the problems by increasing the flexibility of the power system, reducing the peak consumption and levelling the load schedule. Including through price-dependent demand response for electric energy and power through the development of regulatory legal acts in the field of electricity supply, technological connection and improving the payment system for electricity and power, including at the level of utility and household consumers.

1 Introduction

By 2023, the Irkutsk power system has exhausted all available reserves, and its development on traditional principles is no longer possible. Can be noted:

1) In the Irkutsk power system in the winter of 2021-2022, a historical maximum power consumption of 9,111 MW was reached (the previous historical maximum was 8,664 MW in 1989). The maximum consumption of power in the winter of 2021-2022 is 770 MW higher than in 2020-2021 [1]. In the winter of 2022-2023, a new maximum consumption of power is 9,421 MW was recorded [2]. The maximum consumption of power in winter 2022-2023 is 1.08 GW higher than in 2020-2021, i.e. 13% growth in two years.

2) In the power system of the Irkutsk region, all 110-500 kV power centers (central power substations) are "closed power centers" [3], i.e. there is no technical possibility of technological connection of consumers on general terms. Technological connection of any new consumers, with the exception of certain preferential categories, is possible only on individual projects.

3) There is a sharp decrease in the reliability of power supply to consumers due to an increase in accidents in distribution power grid [4-6]. By 2022, the capacity of the 500 kV Bratsk – Irkutsk power transmission, has been exhausted [1, 7].

2 Causes of problems and features of the power system

Solving these problems in the traditional way is impossible; a revision of approaches is required, which will solve many problems using available sources of financing for the industry. The main causes of problems:

1) A significant share of electric heating (severe sharply continental climate, lack of gasification and poor coverage of district heating in areas of low-rise housing). Electric heating is characterized by an extremely uneven schedule of electricity consumption throughout the year, which correlates with the ambient temperature (the lower the temperature, the higher the power consumption). The duration of peak consumption usually does not exceed 1-3 weeks per year, but a violation of heating during this period can pose a threat to the life and health of the population.

It is necessary to develop methods for forecasting peak consumption not only on the scale of the entire regional energy system, but also at the level of smaller districts, because weather conditions in different districts in the Irkutsk region are different (area: 767,900 km^2). To accurately account for the load on electric heating, a correct choice of the averaging period, depending on the thermal inertia of buildings, is necessary. For buildings with electric heating to be usually low-inertia (or even non-inertia).

2) Low power tariffs for the population, which makes electric heating the most cost-effective method of heating, leads to the unprofitability of many measures to improve the energy efficiency of buildings and premises. In itself, this phenomenon is not bad, because it increases the well-being of the population without special financial costs (an increase in energy efficiency requirements will bring down the level of well-being of the population, or will require significant budget subsidies for energy efficiency programs).

3) Significant internal migration processes in the Irkutsk region; housing and communal services are actively developing near Irkutsk, which leads to a

* Corresponding author: osakalexey@mail.ru

noticeable increase in power consumption in the southern part of the power system (which, both initially, and now is power deficient).

4) The appearance of "white" and "gray" cryptocurrency mining with a large value of powers consumption. Tight deadlines for commissioning of data centers, which are less than the deadlines for commissioning of any other types of production.

5) There is an extremely uneven level of use of the maximum connected power. At large aluminium plants, the average annual capacity is close enough to the maximum load and the maximum connected capacity. For small objects with electric heating, the average annual power is significantly lower than the maximum load and the maximum connected power. The load of the "miners" is close to the maximum connected capacity, and the periods of "downtime" are quite difficult to predict (equipment breakdowns or external events in the "cryptocurrency markets").

As a result, the biggest problems in the power system of the Irkutsk region are in the field of distribution grids and electricity supply to municipal consumers. At the same time, we traditionally have an extremely low level of automation in grids of 10 kV and below.

3 Analysis of load and peak consumption schedules

Let's analyse the load graphs, estimating the magnitude of the daily fluctuations in power consumption and peak consumption.

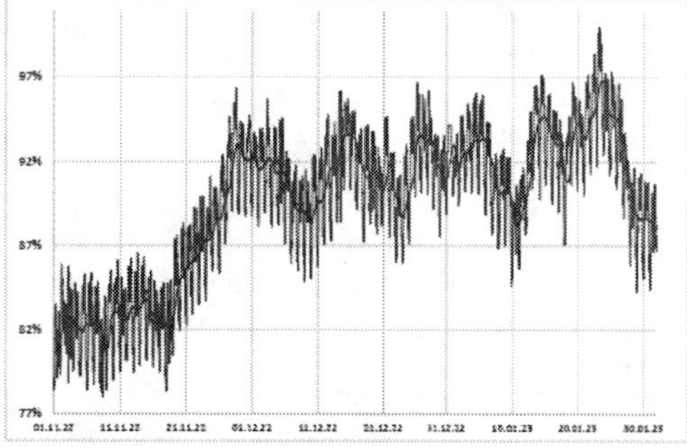

Fig. 1. Power load graph for the period from 01.11.22 to 31.01.23 (the coldest period of 2022/23).

Fig. 2. The period and value of limiting the consumption of the power system by up to 5% of the actual maximum (by up to 460 MW), which will require limiting the power consumption to 2.1%.

Analysis of the actual load schedule for 2020-2023 shows that problems with power shortage and lack of capacity of electric grids arise only for a short period of emergency repairs and maximum load, which is 2-5 weeks a year. Accordingly, automatic limitation of the load level (including price-dependent) during such periods will avoid excessive construction of power grids and power plant. To do this, it is proposed to amend the legislation and regulatory legal acts.

4 The legal side of electricity supply and proposals for their reform

The features of technological connection (TC) to power grids [8] and contractual relations for power supply:

1) The connected power determined in applications for TC, in technical specifications (TS) for TC and power supply contracts is the maximum permissible value of the power consumed simultaneously.

2) The connected capacity does not either characterize the average daily, average monthly, or average annual capacity, since these indicators are not the subject of contractual relations at all. For example, if a consumer on average consumes conditionally 2-3 kW of power, but once a year for 15 minutes he needs 50 kW at once for some purposes, then he must conclude an power supply contract for these 50 kW, while the order of connection of systematically and episodically demanded power is the same.

The need for the actual capacity of electric grids does not follow from the contracts for electricity supply, taking into account the factor of unevenness and non-simultaneous consumption. Statistical information on uneven and non-simultaneous consumption reflects the past, but is not an indication of the future, so these factors can only serve as estimated characteristics. And the methodological guidelines [9] lead to an underestimation of the capacity of power grids, as a result, the existing regulatory framework ensures the functioning and development of the power grids either with a minimum margin of capacity.

3) The current categories of reliability of power supply specified in clause 14.1 [8], characterize only the circuit of electrical connections, but does not characterize the actual reliability of power supply, nor the real needs of consumers in terms of quality and reliability of power supply. When determining reliability categories, neither real failure statistics, nor climatic factors, nor the length of power transmission lines, nor a host of other factors that really affect the reliability of power supply are taken into account. In conditions of strict regulation of the transmission and distribution of electricity, the incompleteness of reliability categorization plays a clearly negative role.

It should be noted that the current practice of "social norms" [10] of power consumption is essentially discriminatory, because on the one hand it creates a radical inequality of people depending on the place of residence, the presence/absence of district heating and the availability of gasification of the residential sector. If, due to the growth of tariffs, the introduction of "social norms", a significant reduction in electricity consumption by the utility sector is achieved, this will lead to a significant increase in unit operating costs (per 1 kWh). It is advisable to reform the power industry in terms of organizational, legal, contractual and technical aspects of TC and power supply to consumers.

A modern approach to ensuring the flexibility of power systems [11-13], including a modern approach to electricity demand response [14] are a possible way to solve the accumulated problems [15]. And if at the level of large consumers, emergency automation with the function of switching off the load has been operating effectively for many years [16-18], and in recent years the demand response system has been successfully functioning [19], then at the level of municipal consumers in the Russian Federation, these tasks, in fact, have not yet begun to be solved.

The modern level of digitalization ("government services", "personal cabinets", electronic document management) allows us to implement more complex and at the same time flexible contractual relations with household consumers than are available today. Instead of two fixed contractual positions (connected capacity and reliability category) and one variable (monthly volume of electricity consumed), it would be possible to introduce much more indicators.

5 Technical implementation

If there are appropriate regulatory and legal grounds, there are no problems in creating a relatively inexpensive device that combines the functions of a "smart meter" and a load disconnection/limitation device that allows you to implement all this complex and flexible system of prospective billing. And also, allowing one input (single- or three-phase) to be distributed to several outgoing lines with different modes of their operation, including with the possibility of limiting the load of part of the lines, both under contractual conditions and according to the commands of emergency control, including to realize the possibility of demand response at the household level.

It is proposed to implement an load control system for consumers power, including located both at the level of power centres and at the input power devices of end consumers. A system should solve the following tasks:

- at the level of the power center:
  - retransmission to end users of up-to-date pricing information (for flexible tariffs), both on the basis of system-wide data and on local parameters;
  - the functions of automation of voltage reduction limitation and automation of equipment overload limitation with the transmission of commands (via any available communication channels) load limitation to consumers, and if the required reduction volume is not realized, the shutdown of individual feeders as a whole;
- at the level of the input power device of the end power customers:
  - the functions of flexible changing of power consumed, and thereby realize the possibility of price-dependent reduction of power consumption;
  - the function of receiving a command to limit the amount of power load from the power center or dispatch center;
  - the functions of automatic switching on/off of individual lines (for providing power to individual electric receivers) depending on tariff criteria or in the presence/absence of a load limitation command.

Earlier, in the papers [20, 21], the authors considered a load disconnection system in distribution grids of 35 kV and below, for which it is proposed to use any existing communication channels [18], including cellular, Fig. 3. The device installed in the power center sends a shutdown command, simultaneously controlling the load reset in the feeder head. If the required amount of unloading has not occurred in the required time, then you can turn off the feeder entirely. Thus, the task of disconnecting the load will be solved in any case.

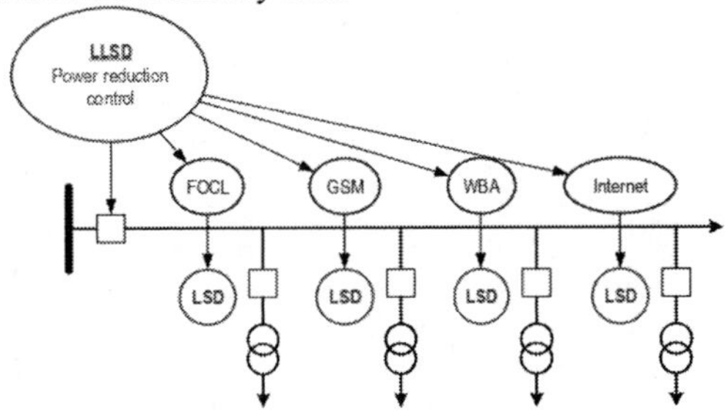

Fig. 3. Technology for disconnecting the load from emergency control devices in 0.4 – 35 kV grids without the use of expensive duplicated communication channels.

6 Conclusion

By 2023, the energy system of the Irkutsk region has exhausted all available reserves and reserves, and its development on traditional principles is no longer possible. The problem is reflected in the Scheme and program of development of electric power systems in Russia for 2023-2028 [7]. Solving the problem by traditional methods (construction of new power lines and substations) requires large capital investments, and therefore inefficient

It is proposed to solve the problems by increasing the flexibility of the power system, reducing the peak consumption and levelling the load schedule. Including through price-dependent demand response for electric energy and power due to the development of legal acts in the field of power supply, technological connection and improving the payment system for electricity and power, including at the level of utility consumers. On the technical side, in order to implement this approach, it is necessary to create an automatic load control system for utility consumers, located both at the level of power supply centers and at the input devices of end consumers. With the possibility of limiting the load of consumers, both under contractual conditions and according to the commands of regime and emergency automation upon overload of power lines, transformers, prevention of violations of stability, unacceptable voltage reduction.

The research was carried out under State Assignment Project (no.FWEU-2021-0001) of the Fundamental Research Program of Russian Federation 2021-2030 using the resources of the High-Temperature Circuit Multi-Access Research Center (Ministry of Science and Higher Education of the Russian Federation, project no 13.CKP.21.0038).

References

1. Mayakov D.V., Kornilov V.N. Irkutsk – the capital of mining: new challenges for the power system // Bigpower Daily. 16.12.2022. URL: https://www.bigpowernews.ru/interview/document106777.phtml
2. Power consumption in the power system of the Irkutsk region has reached a new record value. URL: https://www.so-ups.ru/odu-siberia/news/odu-siberia-news-view/news/20772/
3. Information about the availability of power available for technological connection. // The website of JSC "IESK". URL: https://www.irk-esk.ru/customers/attachment/availability
4. The volume of undelivered power energy as a result of emergency shutdowns // The website of JSC "IESK". URL: https://irk-esk.ru/customers/transmission/blackouts/accidents_values
5. Residents of the Irkutsk region complained of power outages // Information site of the Irkutsk region. URL: https://www.irk.ru/news/20221213/mining/
6. Decree of the Governor of the Irkutsk Region dated 04/28/2022 No. 71-ug "On approval of the scheme and program for the development of the electric power industry of the Irkutsk region for 2023-2027". URL: https://www.ogirk.ru/pravo/archives/law/330860
7. Order of the Ministry of Energy of the Russian Federation No. 108 dated 02/28/2023 "On approval of the scheme and program for the development of power systems of Russia for 2023-2028". URL: https://minenergo.gov.ru/node/24125
8. Rules for Technological Connection of Power Receiving devices of Electric energy Consumers, electric energy production facilities, as well as power grid facilities Belonging to grid organizations and Other Persons to power grids, approved by Decree of the Government of the Russian Federation No. 861 of 12/27/2004. URL: https://docs.cntd.ru/document/901919551
9. Methodological guidelines for determining the degree of loading of power grid facilities introduced after construction, as well as for determining and applying the coefficients of combining the maximum consumption of electric energy (capacity) when determining the degree of loading of such facilities, approved by Order of the Ministry of Energy of the Russian Federation No. 250 dated May 6, 2014. URL: https://docs.cntd.ru/document/499099225
10. Regulation on the establishment and application of the social norm of consumption of electric energy (capacity), approved by the Decree of the Government of the Russian Federation dated July 22, 2013 No. 614. URL: https://base.garant.ru/70422296/
11. Status of Power System Transformation 2018: Advanced Power Plant Flexibility. IEA. Paris. 2018. DOI:10.1787/9789264302006-en.
12. Voropai N., Rehtanz C. Flexibility and resiliency of electric power systems: Analysis of definitions and content // EPJ Web of Conferences. – EDP Sciences, 2019, Vol. 217, pp. 01018. DOI: 10.1051/epjconf/201921701018.
13. Voropay N.I. Elektrichestvo, 2020, № 7, p. 12-21. DOI: 10.24160/0013-5380-2020-7-12-21.
14. About the technology of price-dependent consumption. URL: https://www.so-ups.ru/?id=dr.
15. Osak A., Buzina E. Flexibility and security of power systems, methods of analysis, and criteria for their evaluation in the conditions of digital transformation of the power industry // AIP Conference Proceedings 2552, 040008 (2023). DOI: 10.1063/5.0114361.
16. GOST R 55105. Unified energy system and isolated power systems. Operational dispatch control. Automatic emergency control of power systems modes. Emergency automation of power systems. Norms and requirements. URL: https://meganorm.ru/Data2/1/4293724/4293724536.pdf
17. Voropai N.I., Efimov D.N., et. al. Proc. of 9th IFAC Symp. on Control of Power and Energy Systems, New Delhi, India, December 9-11, 2015, IFAC-PapersOnLine Volume 48, Issue 30, 2015, Pages 554–559. DOI: 10.1016/j.ifacol.2015.12.438
18. Voropai N.I., Efimov D.N., et. al. Adaptive algorithms for automatic distributed load shedding // Electrical stations. 2016. №11. p 27-35. URL: http://elst.energy-journals.ru/index.php/elst/article/view/610
19. Reviews on the functioning of the mechanism of price-dependent reduction of consumption for 2022. URL: https://www.so-ups.ru/functioning/markets/dr/reviews/2022/
20. Osak A.B., Panasetsky D.A., et. al. Centralizer emergency control system taking into account distributed generation and consumer characteristics // Reports of the international conference "Relay protection and automation of power systems – 2017", St. Petersburg, April 25-28, 2017. URL: https://elibrary.ru/zqjwur
21. Osak A.B., Panasetsky D.A., Buzina E.Ya. Consumer operation mode-wise intelligent emergency control // Proceedings of Irkutsk State Technical University, 2017, №9 (128). C.173-185. DOI: 10.21285/1814-3520-2017-9-173-184

 https://doi.org/10.1051/e3sconf/202346101034

Study of transient synchrophasors for the development of relay protection and control systems

Alexey Mokeev[1, 2*], *Sergey* Piskunov[1, 2]

[1]Engineering Centre "Energoservice", Kotlasskaya st. 26, Arkhangelsk, Russia
[2]Northern (Arctic) Federal University, Electroenergetics and Electrotechnics Department, Severnaya Dvina emb.17, Arkhangelsk, Russia

Abstract. The paper discusses issues of improving control and relay protection systems based on the study and analysis of voltage and current synchrophasors of electromechanical and electromagnetic transient processes. The authors consider methods for analyzing transient synchrophasors, the use of synchronized phasor measurement technology in control systems and for improving line distance protection. The paper provides examples of line impedance estimation using traditional protection algorithms and proposed algorithms based on phasor measurements.

1 Introduction

The development of control, protection and automation of power systems implies new requirements for indicators of the quality of their functioning. First, this concerns ensuring high performance of such systems and reliability of measurements.

The synchronized phasor measurement (SPM) technology is one of the promising areas for improving control and relay protection systems [1-2]. The development of protection and control algorithms based on SPM requires research into the equations of electromechanical transient processes in the values of voltage and current synchrophasors. Analysis of such equations makes it possible to justify the choice of certain components of transient synchrophasors for the implementation of signal processing algorithms in various control, protection and automation systems.

The paper presents the results of theoretical research, mathematical modeling of various electromechanical and electromagnetic transient processes, and an analysis of transient synchrophasors in order to improve control and relay protection systems.

2 Methods for analyzing transient synchrophasors

The most important advantage that SPM technology provides in protection and control devices is the accurate measurement of current and voltage synchrophasors during electromechanical transients [1-3]. The authors claim that the SPM, under certain conditions, also provides a reliable estimate of the power system equivalent circuit parameters during electromagnetic transients [4-7].

Effective use of SPM-enabled devices requires the study of equations in transient synchrophasors. However, existing methods and programs for calculating electromechanical and electromagnetic transient processes have solutions to differential equations in the form of instantaneous values of currents and voltages.

The authors have developed several methods for analyzing synchrophasors of transient processes in the power system (based on time-frequency representations, differential equation transformation and convolution integral) [8-9]. The development of special methods for analyzing transient synchrophasors, first, involves a more effective analysis of the functioning of IEDs with SPM support. The authors note that transient synchrophasors are often more complex functions than the synchrophasors assessed using PMU [9]. Therefore, it is important to establish which components of current and voltage synchrophasors need to be assessed using PMUs and other SPM-enabled devices [4-7].

Let us consider the application of a method for analyzing transient synchrophasors based on the transformation of differential equations. Below is an example of determining the line current and voltage synchrophasors for given power system and line parameters (Fig. 1). The line model is a simple RL equivalent circuit, and the power system EMF synchrophasors are functions of time.

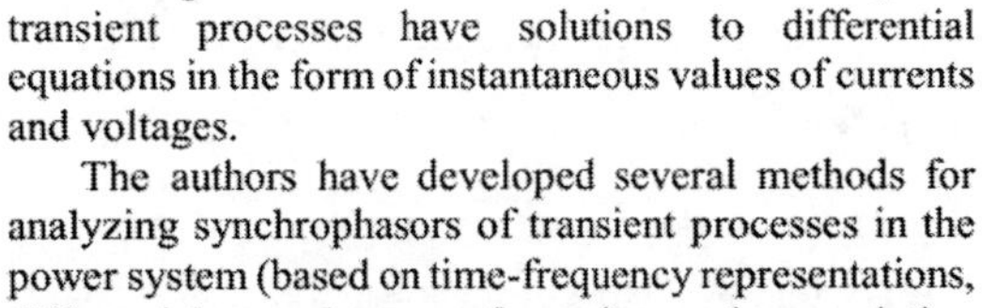

Fig. 1. Power system model № 1

Table 1 shows the solution of the line differential equation in transient synchrophasors. This solution involves a substitution $\Delta e_1(t) \to \Delta \dot{E}_1(t)e^{j\omega_0 t}$ and

*Corresponding author: a.mokeev@narfu.ru

$i_1(t) \rightarrow \dot{I}_1(t)e^{j\omega_0 t}$ that is widely used in the complex amplitude method. Thus, we replace the original differential equation (Table 1, par. 1) in the instantaneous values with the differential equation in the transient synchrophasors (Table 1, par. 2).

Table 1. Calculation of current and voltage synchrophasors.

№	Name	Expression
1	Differential equation	$\Delta e(t) = R_\Sigma i_1(t) + L_\Sigma \frac{di_1(t)}{dt}$, $\Delta e(t) = e_1(t) - e_2(t)$, $R_\Sigma = R + R_{e1} + R_{e2}$, $L_\Sigma = L + L_{e1} + L_{e2}$
2	Synchrophasor expressions	$\Delta\dot{E}(t) = \underline{z}_\Sigma \dot{I}_1(t) + L_\Sigma \frac{d\dot{I}_1(t)}{dt}$, $\underline{z}_\Sigma = R_\Sigma + j\omega_0 L_\Sigma$
3	Current synchrophasor $\dot{I}_1(t)$	$\dot{I}_1(t) = \frac{1}{L_\Sigma}\int_0^t \Delta\dot{E}(\tau)e^{p_1(t-\tau)}d\tau$, $p_1 = \beta_1 + j\omega_0$, $\beta_1 = R_\Sigma / L_\Sigma$, $\omega_0 = 2\pi 50$ rad/s
4	Voltage synchrophasor $\dot{U}_1(t)$	$\dot{U}_1(t) = \dot{E}_1(t) - \underline{z}_{e1}\dot{I}_1(t) - L_{e1}\frac{d\dot{I}_1(t)}{dt}$
5	Voltage synchrophasor $\dot{U}_2(t)$	$\dot{U}_2(t) = \dot{U}_1(t) - \underline{z}\dot{I}_1(t) - L\frac{d\dot{I}_1(t)}{dt}$, $\underline{z} = R + j\omega_0 L$

The transient current synchrophasor (Table 1, par. 3) includes two components that determine the forced and free components of the transient process and is a compact description of the instantaneous transient current $i_1(t) = \mathrm{Re}\left[\dot{I}_1(t)e^{j\omega_0 t}\right]$. There are several methods for obtaining forced and free components from a complete synchrophasor, for example, based on the antiderivative integral, preliminary determination of the free component of the transient current synchrophasor and subsequent determination of the forced component [9].

Similar dependencies are present in the study of transient synchrophasors for relay protection. For example, to determine the current synchrophasor from the side of the first EMF source during a metal three-phase short circuit, it is necessary to make a replacement $\Delta\dot{E}(t) \rightarrow \dot{E}_1(t)$ and similar replacements in the expressions for resistances, inductances and impedances (Table 1).

The considered approach is valid for more complex models of lines and power systems. In this case, we will receive a set of differential equations in which synchrophasors of EMF, voltage and current are present. If it is necessary to take into account nonlinear elements in equivalent circuits, for example, nonlinear arc resistance, nonlinear inductance of the magnetic circuit of a transformer during a magnetizing inrush current (MIC), we obtain a set of differential equations, some of which will be nonlinear. In this case, current and voltage synchrophasors will contain many forced and free components [5-7]. The decision to use certain synchrophasor components for the development of new protection and control devices depends on the results of research.

3 SPM applications in control systems

The presented method for analyzing voltage and current transient synchrophasors provides dependencies for solving the following problems: estimating line parameters, verifying PMU data, implementing virtual PMU, increasing the efficiency of power system dynamic state estimation, searching for the source of low-frequency oscillations.

Let us consider solving the problem of estimating the parameters of a power transmission line (Fig. 1). The initial data are measurements of voltage synchrophasors at substations $\dot{U}_1(t)$ and $\dot{U}_2(t)$, as well as current synchrophasors in the line $\dot{I}_1(t)$. The task is to estimate line parameters using an *RL* equivalent circuit.

Using an approach similar to Table 1 we obtain the following differential equation:

$$\Delta\dot{U}(t) = \underline{z}\dot{I}_1(t) + L\frac{d\dot{I}_1(t)}{dt}. \quad (1)$$

From equation (1) we obtain an expression for estimating the impedance of the power line:

$$\hat{\underline{z}}(t) = \frac{\Delta\dot{U}(t)}{\dot{I}_1(t) + \underline{k}_1\dot{I}_1'(t)}, \quad (2)$$

where $\Delta\dot{U}(t) = \dot{U}_1(t) - \dot{U}_2(t)$, $\dot{I}_1'(t) = \frac{d\dot{I}_1'(t)}{dt}$, $\underline{k}_1 = \frac{L_\mu}{\underline{z}_\mu}$, $\underline{z}_\mu = R_\mu + j\omega_0 L_\mu$, R_μ, L_μ - line resistance and inductance, per km.

The additional component in the denominator of expression (2) provides a correct estimation of the line impedance during electromechanical and electromagnetic transient processes.

The authors emphasize that the error of the measuring current and voltage transformers, the error of the PMU and some other factors affect the accuracy of the line impedance estimation (2). Therefore, this expression represents the basis for the development of algorithms for identifying power line parameters. In turn, the practical implementation of the algorithm requires a transition from an analog model to a discrete one. Thus, we can improve the accuracy of line impedance estimation by using optimization techniques and using additional information about the current and voltage synchrophasors.

Estimation of power line parameters based on RL equivalent circuit (2) requires the use of current and voltage synchrophasors and the derivative of the current synchrophasor. A more complex *U*-shaped line equivalent circuit requires solving a set of differential equations in which current and voltage synchrophasors and their derivatives are present. We can obtain similar

 https://doi.org/10.1051/e3sconf/202346101034

models for other elements of the power system. The development of such models opens up opportunities for improving control, protection and monitoring systems, in particular, for power system dynamic state estimation and implementing virtual PMUs. The latter is a way to estimate the current and voltage synchrophasors at an adjacent substation based on the data from the synchrophasors at the substation where there is a PMU.

The practical implementation of algorithms based on expression (2) makes it difficult to calculate the discrete derivative of the current synchrophasor. The most preferable is to calculate this derivative in the PMU by analogy with the calculation of the instantaneous frequency as a derivative of the voltage synchrophasor. The second option involves increasing the transmission frequency of synchrophasors and calculating the derivative in another device. In this regard, additional requirements arise for protection devices, in particular, high transmission frequency of synchrophasors, high speed of operation. On the other hand, protection devices have lower requirements for the accuracy of synchrophasor measurements compared to PMUs.

In many cases, substations have PMU only at some bays. The presence of PMU at adjacent substations makes it possible to evaluate synchrophasors at bays without PMU. This problem arises in various applications, for example, when searching for the source of low-frequency oscillations in a power system.

To solve this problem, we use a discrete analogue of the integral expression to determine the current synchrophasor (Table 1, par.3):

$$\dot{I}_1(k) = \frac{1}{L}\sum_{m=0}^{k} \Delta\dot{U}(k) e^{p_0(k-m)T}, \qquad (3)$$

where $p_0 = \beta_0 + j\omega_0$, $\beta_0 = R/L$, $k \cdot T$ - discrete time.

The problem of calculating the current synchrophasor has a problem, which lies in the presence of a rapidly oscillating function in expression (3). To solve this problem, we need to increase the transmission frequency of the synchrophasors. At the same time, algorithms based on a simplified method for analyzing transient synchrophasors [3-9] provide satisfactory results for estimating the current synchrophasor for a standard transmission frequency.

There are other possibilities for increasing the efficiency of searching for the source of low-frequency oscillations using current and voltage synchrophasors, in particular, the use of additional signs of recognition of low-frequency oscillations based on the instantaneous frequencies of current synchrophasors, analysis of models of such oscillations using transient synchrophasors.

Thus, the use of transient synchrophasors makes it possible to solve various problems for improving control systems. The most promising areas are additional estimation of synchrophasors, virtual PMU, verification of PMU data, power system dynamic state estimation.

4 SPM applications in relay protection

4.1 Distance protection

Synchrophasor technology has several promising directions for relay protection; in particular, the authors argue that the use of current and voltage synchrophasors is promising for improving distance protection algorithms.
Let us consider the use of methods for one-sided and two-sided measurement of current and voltage synchrophasors in distance protection of a 110-220 kV line (Fig. 2).

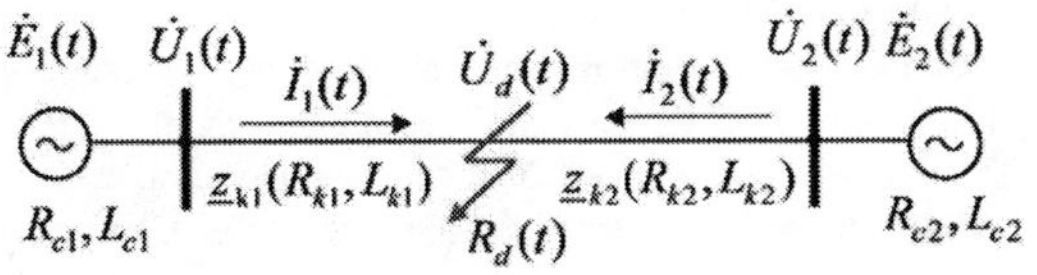

Fig. 2. Power system model № 2

Table 2 shows two line distance protection algorithms. The first algorithm involves one-sided synchrophasor measurements when we have data from only one protection device at the substation. The second algorithm uses two-sided measurements. This means that we have synchrophasor measurements from adjacent substations that the protected line connects.

Table 2. Estimation of short-circuit loop impedance.

№	Name	Expression
1	One-sided synchrophasor measurements	$\hat{\underline{z}}(t) = \frac{\dot{U}(t)}{\dot{I}(t) + \underline{k}\dot{I}'(t)}$, $\dot{I}'(t) = \frac{d\dot{I}(t)}{dt}$, $\underline{k} = L_\mu / \underline{z}_\mu$
2	Two-sided synchrophasor measurements	$\hat{\underline{z}}_1(t) = \frac{\Delta\dot{U}(t) + \underline{z}\dot{I}_2(t) + L\dot{I}_2'(t)}{\dot{I}(t) + \underline{k}\dot{I}'(t)}$, $\Delta\dot{U}(t) = \dot{U}_1(t) - \dot{U}_2(t)$, $\dot{I}(t) = \dot{I}_1(t) + \dot{I}_2(t)$, $\dot{I}'(t) = \dot{I}_1'(t) + \dot{I}_2'(t)$

We obtained the given algorithms (Table 2) based on the approach that this paper discusses in Section 3. These expressions have a more compact form compared to those previously published [4-6]. The one-sided line impedance estimation algorithm has disadvantages because it does not eliminate the influence of a nonlinear arc during a short circuit. The authors considered a more efficient one-sided measurement algorithm in [5, 7]. The algorithm for two-sided measurement of current and voltage synchrophasors eliminates the influence of a nonlinear arc [7] and ensures the correct operation of distance protection during electromechanical transient processes, including in asynchronous mode in the power system.

The proposed algorithms (Table 2) have a significant difference from traditional distance protection algorithms. This difference lies in the presence of

 https://doi.org/10.1051/e3sconf/202346101034

additional components with derivatives of current synchrophasors. These components can significantly improve the performance of the protection and ensure its correct operation during electromechanical and electromagnetic transients.

Distance protection depends on various factors that can have a negative impact on it [10-12]. One of these factors is the errors of current and voltage transformers. The saturation of the magnetic circuit of the current transformer has a particularly negative effect on distance protection. For this reason, the speed that the proposed algorithms (Table 2) can provide is an important tool for improving distance protection, since the time until saturation of current transformers is usually quite short.

Thus, the performance of distance protection algorithms may depend on the measurement system. The authors consider the use of digital optical current and voltage transformers [13] and digital current transformers based on a Rogowski coil [14] as a way to improve the efficiency of the proposed algorithms (Table 2). In this case, there is no saturation of the magnetic circuit of the current transformer, which ensures accurate current measurements (class 0.5) in the full range of its values. We propose the use of a combined digital current and voltage transformer ECIT for medium voltage switchgears (Engineering Center Energoservice LLC) [14]. ECIT has a Rogowski coil as a primary current converter and a capacitive voltage divider. Additionally, the ECIT has the function of measuring current and voltage synchrophasors. Thus, such a device provides more efficient operation of distance protection compared to electromagnetic current and voltage transformers.

4.2 Ground fault protection

Transient synchrophasors have promising applications for other types of protection, such as improving ground fault protection [15-16].

The zero sequence circuit of a three-phase network during a ground fault has distinctive features [17]. The authors analyzed differential equations for synchrophasors of zero sequence currents and voltages for various conditions of the occurrence of a ground fault. The results of the study allowed us to propose a new directional protection algorithm based on equivalent harmonic synchrophasors [7]. The advantage of such protection is the ability to work both locally and as part of centralized protection.

To improve ground fault protection, we can also use an approach based on line differential equation analysis similar to the short circuit mode (Section 4.1) [17-18]. However, for distance protection, it was sufficient to estimate the forced component of the current synchrophasor, and ground fault protection requires the estimation of synchrophasors taking into account the influence of free components. Therefore, this approach has a certain complexity of implementation.

The authors note that transient synchrophasors also have promise for other types of protection, for example, for differential protection of a power transformer [16]. The use of synchrophasors makes it possible to implement additional features for recognizing internal and external short circuit modes and the magnetizing current inrush mode.

5 Math modeling

We propose to consider several examples of mathematical modeling of the developed distance protection algorithms, which confirm their effectiveness.

Figure 3 shows a comparison of two distance protection options (Table 2). The power system model corresponds to Fig. 2. The upper graphs contain instantaneous values of currents and voltages, modules and phases of the corresponding synchrophasors. The lower graphs show the estimation of line impedance components based on the traditional (*t*) and proposed (*p*) algorithms. The graph with number 1 corresponds to the one-sided measurement algorithm (Table 2, par. 1). The graph with number 2 corresponds to two-sided measurements (Table 2, par. 2).

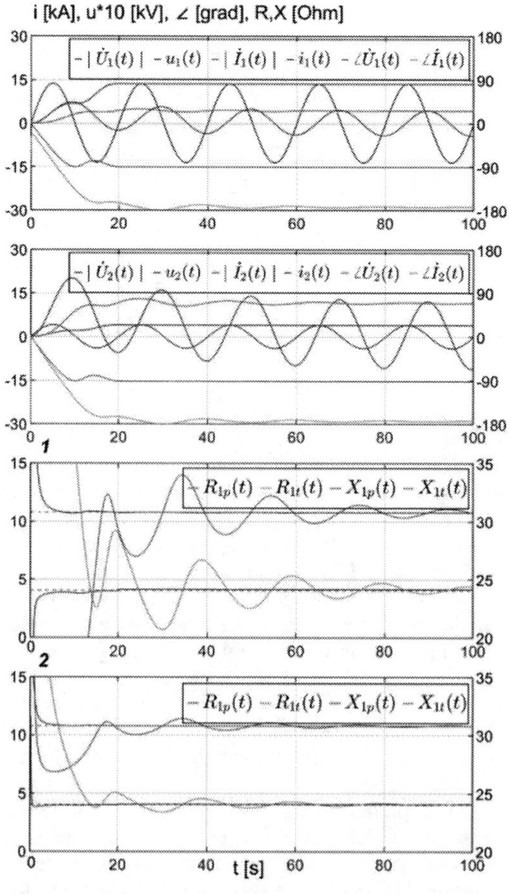

Fig. 3. One-sided (1) and two-sided (2) R, X estimation

The graphs (Fig. 3) confirm that the performance of distance protection based on the proposed algorithms (Table 2) significantly exceeds the performance of the traditional algorithm. The authors emphasize that the protection response time is a few milliseconds, despite

https://doi.org/10.1051/e3sconf/202346101034

the fact that the evaluation of current and voltage synchrophasors takes about 20 ms. This is possible because the proposed expressions (Table 2) take into account the parameters of the line model.

If we compare the one-sided and two-sided measurement algorithms, we can confirm that the two-sided measurement algorithm has advantages over one-sided measurement. Distance protection with two-sided measurements can operate stably under various electromagnetic and electromechanical transient processes, in particular, in asynchronous mode in the power system.

Fig. 4 shows plots similar to Fig. 3, but taking into account the current transformer saturation model.

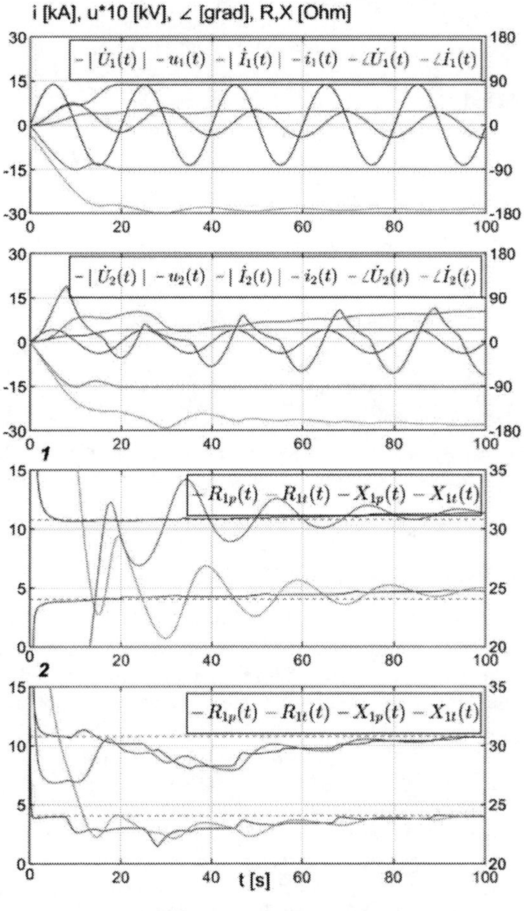

Fig. 4. R, X estimation in case of CT saturation

During a short circuit, rapid saturation of the current transformers occurs on the side of the second EMF source, which leads to a significant increase in the error in estimating the line impedance (Fig. 4). The authors note that the proposed distance protection algorithm (Table 2) provides high performance, so we obtain an accurate estimate of the line impedance before saturation of the current transformers (Fig. 4, plot 2).

Thus, the simulation results confirm that the use of transient synchrophasors is promising in relay protection systems.

6 Conclusion

The authors carried out a study concerning the use of current and voltage synchrophasors in protection and control systems. The paper examines several applications of transient synchrophasors. Analysis of voltage and current synchrophasors of electromechanical and electromagnetic transients allowed the development of distance protection algorithms (one-sided and two-sided measurement). The proposed algorithms provide accurate line impedance estimation under various conditions. In addition, the speed of the algorithms determines the correct operation of distance protection when the measuring current transformers are saturated.

Transient synchrophasors have promising applications in other protections, in particular, in transformer protection and ground fault protection.

References

1. A.G. Phadke, J.S. Thorp, Synchronized phasor measurements and their app., NY, Springer (2017).
2. M. Kezunovic, et. al., Application of time-synchronized measurements in power system transmission networks, NY, Springer (2014).
3. V.I. Antonov, et. al., Fundamentals of protection and automation of a smart electrical grid. Infra-Engineering (2023).
4. A.V. Mokeev, S.A. Piskunov, RSES, **93** (2022).
5. A.V. Mokeev, S.A. Piskunov, RPA, **3** (2022).
6. A.V. Mokeev, S.A. Piskunov, RPA, **4** (2022).
7. A.V. Mokeev, S.A. Piskunov, Releyshik, **1** (2023).
8. A.V. Mokeev, Márquez FPG (ed.) Digital Filters, Rijeka, InTech (2011).
9. A.V. Mokeev, Electrical equipment, **1** (2022).
10. E.M. Schneerson, Numerical distance protection, M., Energoizdat (2005).
11. G. Ziegler, Numerical distance protection: principles and applications, John Wiley & Sons (2011).
12. I.T. Titov, A.V. Bychkov, N.A. Doni. Proceedings of NSTU, **2** (2020).
13. N.A. Ivanov, R.I. Kanafeev, M.A. Yanin, CIGRE session, Paris (2022).
14. D.N. Ulyanov, et. al., Releyshik, **1** (2023).
15. S.A. Piskunov, A.V. Mokeev, RPA (2020).
16. S.A. Piskunov, A.V. Mokeev, REEPE (2021).
17. V.A. Shuin, et al., ISUE Bulletin, **4** (2018).
18. A. Wahlroos, et. al., CIRED, Stockholm (2013).

Application of WAMPAC principles in distribution networks

Sergey Piskunov[1, 2*], *Alexey* Mokeev[1, 2], *Dmitry* Ulyanov[1]

[1]Engineering Centre "Energoservice", Kotlasskaya st. 26, Arkhangelsk, Russia
[2]Northern (Arctic) Federal University, Electroenergetics and Electrotechnics Department, Severnaya Dvina emb.17, Arkhangelsk, Russia

Abstract. The paper examines the dissemination of principles for constructing WAMPAC systems in medium voltage networks. The authors provide the prerequisites for the use of WAMPAC in distribution networks and show the advantages of such systems over traditional ones. The first part of the paper considers directions for the development of monitoring, protection and automation systems in the distribution network. The second part is devoted to the localization of ground faults based on phasor measurement technology. The paper also presents the results of mathematical modeling and data on the experience of applying WAMPAC principles in 10 kV cable networks.

1 Introduction

The term WAMPAC (Wide Area Monitoring, Protection, Automation, and Control) appeared recently in the scientific community [1]. It gives a general designation for distributed monitoring, protection, automation and control systems implemented using synchronized phasor measurement (SPM) technology and PMUs. Currently, there are many areas of SPM application in the electric power industry to improve the reliability, stability and efficiency of power system management [2-3].

The initial concept of WAMPAC systems assumed strict compliance with the developed standard for PMU [4]. The authors emphasize that this approach limits both the development of the WAMPAC systems themselves and the areas of synchrophasor technology application.

Currently, there are several new approaches for WAMPAC systems, taking into account the development of their architecture and integration with other existing systems, in particular, SCADA [5]. The authors consider the issue of organizing several levels of WAMPAC management and their relationship with each other.

The authors suggest that the development of new technologies allows consider the WAMPAC application in both high and medium voltage networks. Synchrophasor technology has many promising areas for distribution networks, so the task of introducing WAMPAC is relevant here. In addition, there are other reasons for using WAMPAC in such networks (complex structure, distributed generation, non-linear load of consumers) [6].

Previously, the authors outlined general approaches to the implementation of wide area protection, monitoring, automation and control systems in medium voltage networks [7]. This paper is a continuation of research in the field of SPM application for distribution networks.

The paper has three main sections. Section 2 examines general issues of applying WAMPAC principles in medium voltage networks; the authors present relevant directions for automating such networks [8-10]. Section 3 describes a fault localization system based on equivalent harmonic synchrophasors (EHS). Section 4 considers the experience of applying WAMPAC principles in existing electrical networks [11].

2 WAMPAC applications

The reliability and quality of power supply to consumers of urban, rural and industrial distribution networks directly depends on the level of their automation. The challenges of developing medium voltage networks require rational and effective solutions based on the use of modern intelligent devices and technologies.

The following studies present the benefits of WAMPAC principles for power systems [2-3]. However, the authors limit the WAMPAC application to high voltage networks. At the same time, synchrophasor technology has prospects for use in medium voltage networks. We can solve this problem using new types of synchrophasor devices [7].

One of the main advantages of WAMPAC systems is their multi-level structure, which allows the implementation of central and several local levels of distribution network management.

In Fig. 1, the authors show the structure of the multi-level WAMPAC system in the distribution network.

*Corresponding author: s.piskunov@ens.ru

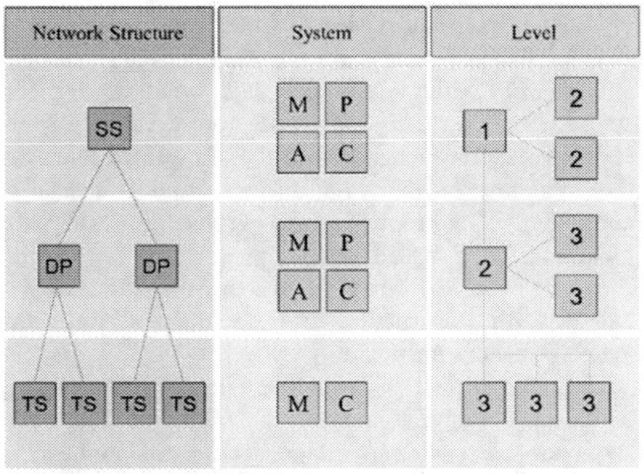

SS – Substation, DP – Distribution Point, TS – Transformer Substation 6(10)/0.4 kV, M – Monitoring, P – Protection, A – Automation, C – Control

Fig. 1. WAMPAC principles for distribution network

Fig. 1 has the following concept. Synchrophasor technology is an important part of WAMPAC, so based on it we can implement a centralized protection system for a substation or distribution point [8]. SPM technology opens up opportunities for the application of the differential principle of protection of substation buses and power transformer [9] and the distance principle for line protection [12-13]. The WAMPAC principles make it possible to ensure high performance protection in a centralized mode and maintain its reliability in a local mode when the communication network is partially or completely degraded.

We can apply a similar approach to WAMPAC in other distribution network systems. For example, a FLISR system (Fault Location, Isolation and Supply Restoration) that includes a fault localization subsystem may have several local control levels (transformer substation level) and a central level (substation/distribution point level) [11]. The presence of local levels allows for a phased modernization of the network, reducing the requirements for the technical characteristics of the system, in particular, for data transmission channels.

Let us consider an example that shows how the WAMPAC principles and synchrophasor technology make it possible to implement equipment state monitoring and network monitoring while reducing capital costs for the monitoring system (Fig. 2).

SPM provides measurements for transformer monitoring systems [9]. Such systems include monitoring of transformer electrical parameters (current and voltage synchrophasors, equivalent circuit parameters, power losses), which is an effective solution for distribution networks, since in many cases there are no other monitoring systems.

Fig. 2 shows two cases (1-2), which explain how, based on PMU, we organize monitoring of transformers at adjacent substations. Here, using the existing principles of PMU placement [14] and additional calculation of synchrophasors at an adjacent substation, we can implement monitoring of those transformers for which part of the synchrophasor measurements is missing.

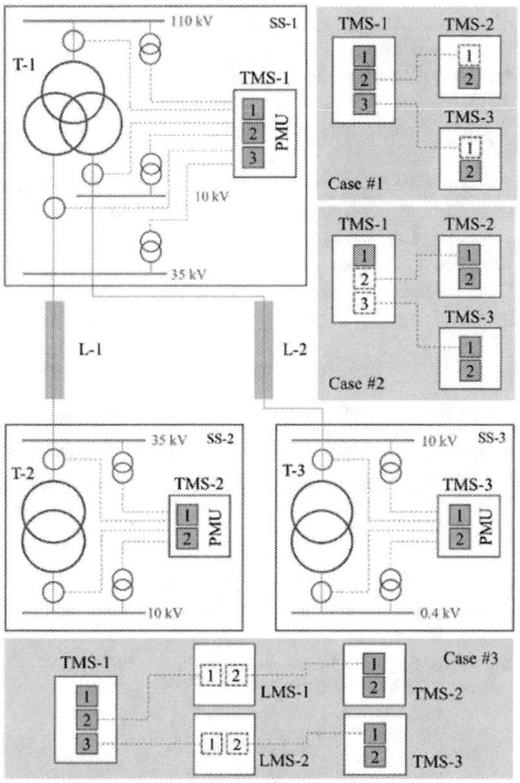

TMS – Transformer Monitoring System, LMS – Line Monitoring System

Fig. 2. Network and equipment monitoring

Case #3 (Fig. 2) also confirms that we can use PMU data to monitor other elements of the system, for example, to monitor power lines (L-1 and L-2). The authors suggest that a promising and effective solution is the integration of monitoring system functions in relay protection devices based on synchrophasor technology.

The WAMPAC principles are promising for the development of automation systems for networks with distributed generation and renewable energy sources [6]. Important tasks here are the advanced microgrid applications and operation, improved load shedding schemes, advanced distribution automation, real-time distribution system operation [15].

One of the most important and complex tasks for the automation of medium voltage distribution networks is fault localization, in particular the localization of single-phase ground faults (SGF). Let us consider in more detail the application of WAMPAC principles for the implementation of SGF localization system.

3 Fault localization

The authors presented methods for localizing short circuits and single-phase ground faults using synchrophasor technology in [11]. However, our work in this area is still ongoing, and we can present new research results in this paper.

Currently, there are several main methods for localizing SGF in medium voltage networks [16-17]. An urgent task is to develop new methods for networks with partial or full compensation of ground fault current. In this regard, the authors propose the use of equivalent zero sequence current and voltage synchrophasors [18].

A network with an isolated neutral makes it possible to implement maximum current and current directional SGF protection [17]. This is because in such a network the capacitive current of undamaged phases during a ground fault is directed to the fault location, so we can select the settings according to the theoretical values of the capacitive currents of the feeders. We can express this principle in zero sequence current synchrophasors:

$$\dot{I}_{0i}(t) = \dot{I}_{Ci}(t)\,, \tag{1}$$

$$\dot{I}_{0k}(t) = -\sum_{i=1}^{N} \dot{I}_{Ci}(t) + \dot{I}_{Ck}(t)\,, \tag{2}$$

where i – number of undamaged network feeder, k –feeder with ground fault, $\dot{I}_{0i}(t)$ – zero sequence current synchrophasor, $\dot{I}_{Ci}(t)$ – capacitive current.

Expressions (1) and (2) are valid for fundamental frequency synchrophasors and for the steady state SGF mode. The minus sign in expression (2) indicates that the currents and should have opposite directions in a network with an isolated neutral.

We can represent dependence (1) and (2) through an estimation of the feeder capacity:

$$C_i(t) = C_i\,, \tag{3}$$

$$C_k(t) = \sum_{i=1}^{N} C_i - C_k\,. \tag{4}$$

Expressions (1) - (4) are valid for a network with an isolated neutral and do not correspond to a network with compensation of the SGF current. In this case, we can use the active or harmonic components of the zero sequence current to identify the faulty feeder [17]. However, a network with full SGF current compensation allows only transient measurements to be used [16].

Measuring instantaneous values of zero sequence current and voltage in the transient SGF mode has certain difficulties due to the pulsed nature of such a process. In addition, methods that use zero sequence current harmonics to identify a faulty feeder require the development of specific harmonic filters. At the same time, there is no a priori data on the composition and level of harmonics in the zero sequence current, so the selection of harmonic filters is a complex task. In this regard, the authors propose the use of equivalent synchrophasors of zero sequence current and voltage harmonics (EHS) to improve algorithms for localizing faults in a network with a compensated neutral.

We previously reviewed the terms and theory of equivalent synchrophasors in [18]. Therefore, in this paper we use the following notations: $\dot{I}_{0e}(t)$, $\dot{U}_{0e}(t)$ - equivalent synchrophasors, including harmonic and fundamental frequency components, $\dot{I}_{0h}(t)$, $\dot{U}_{0h}(t)$ - equivalent harmonic synchrophasors (EHS).

Based on the provisions of the theory of electrical circuits, we can argue that in the transient SGF mode in a network with a compensated neutral, the arc suppression reactor does not compensate for the capacitive SGF current due to the significant inertia of the inductor. In this regard, we propose measuring equivalent synchrophasors to create a system for localizing SGF based on the principle of measuring transient electrical quantities. In the simplest case, we can use the estimation of fundamental frequency synchrophasors for this purpose.

Another direction for localizing faults in such networks involves measuring harmonics in zero sequence current and voltage. There are studies that confirm that distribution networks have significant harmonic components in the zero sequence circuit [17]. As we noted earlier, methods that involve measuring harmonics require the development of special filters. We propose the use of equivalent harmonic synchrophasors as the most rational option for solving this problem.

Both principles (measurement of equivalent synchrophasors of the transient SGF and measurement of equivalent harmonic synchrophasors) make it possible to implement a local and central control level of SGF localization system. This is in accordance with the WAMPAC principles in the distribution network. The local mode assumes that we can estimate the capacitances of individual feeders and determine the direction to the SGF location. The central control level allows determining a specific damaged section of the network.

The authors also propose another concept, which suggests that we can combine the considered methods of SGF localization to improve the efficiency of the system as a whole. In addition, we continue research to develop other methods for localizing SGF, in particular based on the transformation of the line differential equation.

Existing methods based on the transformation of the line differential equation use instantaneous values of zero sequence currents and voltages. The line capacity estimate in this case has the following expression:

$$3i_0(t) \approx C_0 \frac{du_0(t)}{dt}\,. \tag{5}$$

The authors emphasize that the use of instantaneous values of currents and voltages when solving the problem under consideration can lead to significant difficulties in implementing the SGF localization algorithm, since in the transient SGF mode the derivative of the zero sequence current and voltage can take on large absolute values. We consider that a much more effective approach is to estimate the capacitance from the integral values of the synchrophasors:

$$C_0(t) \approx \frac{3\dot{I}_0(t)}{j\omega_0 \dot{U}_0(t) + \dot{U}_0'(t)}\,, \tag{6}$$

where $\dot{U}_0'(t) = \dfrac{d\dot{U}_0(t)}{dt}$.

The authors obtained a more accurate estimate of the capacitance taking into account the conductivity of the network and in the presence of measurements of the current of the arc suppression reactor:

$$C_0(t) \approx \frac{1}{3}\frac{j\omega_0 \dot{I}_0(t) + \dot{I}_0'(t) + L_k^{-1}\dot{U}_0(t)}{\underline{a}_0\dot{U}_0(t) - \underline{a}_1\dot{U}_0'(t) - \dot{U}_0''(t)}, \qquad (7)$$

where $\dot{U}_0''(t) = \dfrac{d^2\dot{U}_0(t)}{dt^2}$, $\dot{I}_0'(t) = \dfrac{d\dot{I}_0(t)}{dt}$, $k = \dfrac{G_{0\mu}}{C_{0\mu}}$, $\underline{a}_1 = 2j\omega_0 + k$, $\underline{a}_0 = \omega_0^2 - j\omega_0 k$, $G_{0\mu}$ – feeder conductivity (per km), $C_{0\mu}$ – feeder capacity (per km), L_k – reactor inductance.

Thus, WAMPAC and synchrophasor technology make it possible to implement several new algorithms for localizing SGF in the distribution network.

4 SGF modeling

Let us consider an example of modeling the proposed algorithms for localizing SGF in Simulink. Fig. 3 shows a diagram of a network model with a compensated neutral. The network has two segments (damaged 1 and undamaged 2). The capacity of the undamaged segment 1 of the network is greater than the capacity of the damaged section 2 (in a ratio of 2:1). This makes it possible to compare the ratio of zero sequence currents across network segments in transient and steady state SGF modes.

Fig. 4 shows a comparison of equivalent synchrophasors and EHS of the zero-sequence current of a damaged and undamaged network segment. Fig. 5 shows the estimation of the segment's capacity using expression (6).

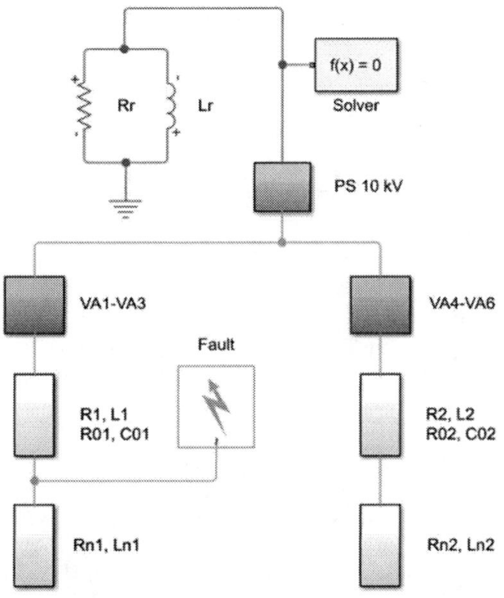

Fig. 3. Simulink test model

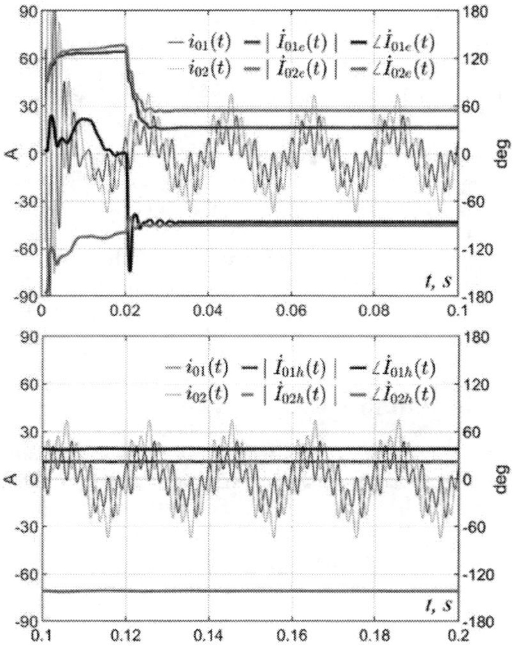

Fig. 4. Comparison of equivalent synchrophasors

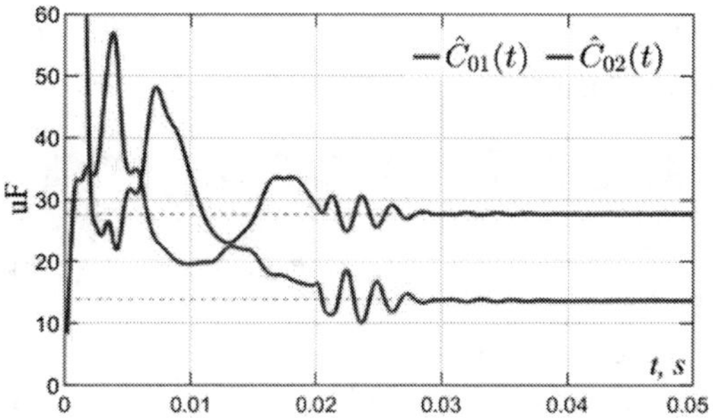

Fig. 5. Estimating the capacity of model segments

Fig. 4 confirms that the use of equivalent current synchrophasors in the transient mode (0-20 ms) and EHS in the steady state (> 20 ms) makes it possible to increase the efficiency of SGF localization algorithms by comparing the magnitude and phase of the corresponding synchrophasors of the damaged and undamaged area.

Fig. 5 confirms the effectiveness of expression (6) for estimating the capacity of network segments during SGF.

5 Practical experience

Currently, we have implemented several monitoring and control systems in the distribution network based on WAMPAC principles [11]. First, we would like to note the positive experience of using SPM for SGF localization in 10 kV cable networks. For this purpose, we have developed a special device with SPM support (ENLZ). The localization system operates at several operating facilities in Arkhangelsk and Cherepovets (Russia). The system measures and uses fundamental frequency zero sequence current and voltage synchrophasors. We obtained a cost-effective solution by reducing the requirements for measurement accuracy for the ENLZ device compared to PMU [4], as well as using split-core zero sequence current transformers.

Secondly, we can note the prospects for implementing monitoring systems for power transformers based on measurements of voltage and current synchrophasors. Currently, such a system is in pilot operation at two substations in Arkhangelsk (Russia) [11]. Important advantages of the system include the ability to evaluate the parameters of the equivalent circuit and power losses in the transformer in its operating mode.

6 Conclusion

The paper considers that the WAMPAC principles have promising applications in medium voltage networks. The authors show the advantages of a multi-level structure of automation, monitoring, protection and control systems. Using the example of an SGF localization system, we showed the effectiveness of using equivalent synchrophasors and EHS of zero sequence current and voltage. The paper presents the results of research using mathematical models and data from pilot industrial operation.

References

1. Terzija V. Wide area monitoring protection and control – WAMPAC (2007).
2. A.G. Phadke, J.S. Thorp, Synchronized phasor measurements and their app., NY, Springer (2021).
3. A. Monti, C. Muscas, F. Ponci, Phasor measurement units and wide area monitoring systems, Academic Press (2016).
4. IEEE C37.118.1-2011 Synchrophasor Measurements for Power Systems.
5. S. Gamboa, E. Orduña, IC ECC, IEEE (2015).
6. P.V. Ilyushin, A.L. Kulikov, RPA, **3** (2019).
7. S.A. Piskunov, A.V. Mokeev, D.N. Ulyanov, Electricity. Transmission and distribution, **2** (2023).
8. A.V. Mokeev, KSEU, **3** (2020).
9. S.A. Piskunov, A.V. Mokeev, REEPE, Moscow (2021).
10. S.A., Piskunov, A.V. Mokeev, D.N. Ulyanov, Relavexpo (2023).
11. S.A., Piskunov, A.V. Mokeev, D.N. Ulyanov, RPA, **4** (2021).
12. A.V. Mokeev, S.A. Piskunov, Releyshik, **1** (2023).
13. A.V. Mokeev, S.A. Piskunov, RPA, **4** (2022).
14. A. Pal, A.K.S. Vullikanti, S.S. Ravi, IEEE, T.32, **1** (2016).
15. NASPI Distribution Synchronized Measurements Roadmap Final Report (2021).
16. V.F. Lachugin, A.A. Ivanov, A.A. Belyanin, RPA, **3** (2012).
17. V.A. Shuin, et al., Electrical engineering, **1** (2018).
18. A.V. Mokeev, S.A. Piskunov, SIBCON, Kazan (2021).

Stability margins monitoring system performance evaluation for various methods of transfer capacity calculations with transient stability constraints

Vladimir Neuymin[1], *Eugene* Mashalov[1], *Aleksander* Aleksandrov[1], *Denis* Snegirev[2*], *Viktor* Klassen[2] and *Daniil* Teplukhin[2]

[1]JSC "STC UPS", Yekaterinburg, Sverdlovsk region, Russia
[2]JSC "STC UPS", UrFU, Yekaterinburg, Sverdlovsk region, Russia

Abstract. The purpose of this work is to determine ways to increase the performance of available power transfer capacity analysis with transient stability constraints, as well as to evaluate the accuracy of the proposed methods. A number of developed methods for performance improvement of available power transfer capacity (APTC) analysis with transient stability (TS) constraints in the stability margins monitoring system (SMMS) were discussed in this article, in particular: network reduction, transient stability analysis with implicit integration method, parallel computing of contingencies, and sequential TS simulations startup. In addition, evaluations of the method's accuracy of APTC calculations with TS constraints for various power transmissions were carried out.

1 Introduction

One of the most important properties that characterizes the SMMS is the calculation cycle duration required for the APTC. The longer this duration, the more likely the APTC calculated in the previous cycle will not correspond to the current power system state. This can cause incorrect dispatch control actions and secondary frequency control malfunctions. Reducing the duration of the APTC cycle calculation decreases the probability of this situation and brings the SMMS closer to the real-time system.

Considerable time of the calculations becomes noticeable for APTC evaluation with TS constraints. The main reason for this is the difference in analysis methods. Calculations of transient stability responses are used for large disturbances analysis, and steady state (SS) calculations are used for all other criteria.

The SMMS calculation cycle duration depends on many factors, including:

1. the server's computing performance;
2. the initial steady state before load flow continuation;
3. the power system model dimensions: nodes, branches, generators, and other elements count;
4. the number of power interfaces analyzed within the cycle;
5. the number and complexity of emergency processes (EP) – standard contingencies evaluated for each power interface;
6. the number and complexity of the simulated protection and control devices.

Some factors have a more significant impact than others. For example, the TS simulation time grows relatively faster than the SS calculation time with an increase in model dimension. Also, the complexity of emergency processes has a noticeable effect since it increases TS simulation time. The complex contingency can contain multiple automatic reclosing (AR) of the transmission lines in accordance with the requirements of the Stability of the Power Systems Guidelines [1]. The emergency process number may turn out to be about 10-15 also in accordance with these requirements.

The various approaches to accelerate APTC with TS constraints and limited computing performance are investigated in this paper.

2 The reduction of the network model

The most obvious way to shorten the time spent on TS simulation and noticeably speed up the APTC is to reduce dimension of the power system model.

It is important to keep the balance between the accuracy of the results and the simulation time while performing reduction. Accuracy deterioration is often associated with the wrong selection of elements to be reduced, or reduction method. Since the SMMS is an automatic system, it is also proposed to select the elements to be reduced automatically.

Network elements automatically selected to be reduced should not include buses, branches, and generators directly used in emergency process scenarios. In addition, reduction is not applicable to detailed generator models, such as Park's. The only generators modeled by the swing equation can be aggregated.

Various reduction methods and their performance were reported in [2]. The performance of methods in

* Corresponding author: snegirev@niipt-ems.ru

relation to the APTC problem with TS constraints was reported in [3]. It was determined that the combination of the Gauss and Zhukov methods has the best performance. The automatic selection of buses is based on their rank, i.e., the number of adjacent buses. Low rank buses are to be reduced. The automatic selection of aggregation groups is performed by combining several buses connected with low impedance branches (for example, circuit breakers) into one supernode.

3 The use of the implicit integration method for EMT calculations

TS analysis is based on the implicit integration method. It is implemented in the software package Raiden [4] and described in detail in [5]. Raiden is available as one of the backends for TS analysis for the APTC in the SMMS.

The former TS analysis software RUSTab [6] used the explicit integration method in the form:

$$y_n = F_E(y_{n-1}, h_n). \quad (1)$$

In this form, the integration step h_n is limited during the entire simulation process to maintain numerical stability and cannot be higher than the model minimum time constant, for example, the time constants of exciters or automatic voltage regulators (AVR).

For the implicit method used in Raiden the integration method has the following form:

$$y_n = F_I(y_n, y_{n-1}, h_n). \quad (2)$$

The calculation of y_n, in contrast to (1), requires the solution of a nonlinear system of equations. However, implicit methods have a much larger stability region, which allows choosing h_n that is significantly larger than the explicit methods step. As a result, the integration step can be significantly increased when the state variable oscillations decay during the simulation, reducing the number of steps required for the simulation without losing numerical stability.

The stability properties of the implicit integration method allow us to significantly reduce the time of the stable EMT simulation. The stability check for such an EP can take more than half of the total calculation time during the APTC estimation by the criterion of dynamic stability. Thus, TS simulation time can be reduced up to 3–5 times using the implicit method compared to the explicit method, depending on the case damping rate.

4 The sequential TS simulations startup

The former RUSTab package was used for the TS simulation using explicit integration methods and the APTC estimation with transient stability constraints. Its algorithm was developed for online user interaction and provides embedded tools to simplify the model data input. These tools include automatic equipment model parameter checks and corrections, and also model linking verification.

During the initialization of the EP analysis, the SS calculation, the TS differential-algebraic equations system (DAE) initial value calculation, and the updated simulation scenario setup are all completed. However, for each EP, the TS model structure setup, user data verification, and model link checks are only carried out once. The existing diagram of TS initialization in the SMMS is shown in Figure 1. The proposed procedure is displayed in Figure 2.

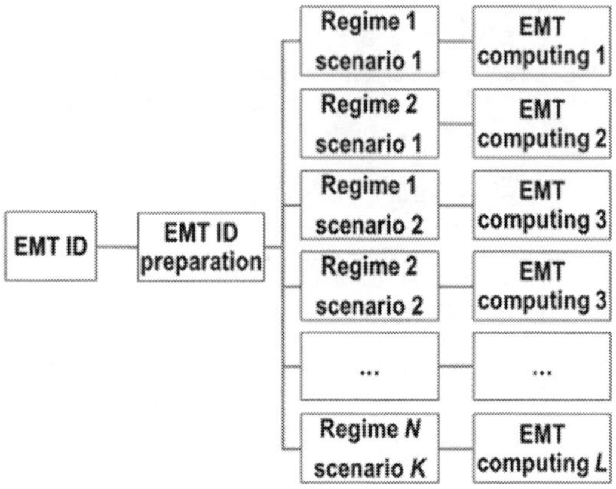

Fig. 1. Existing TS initialization

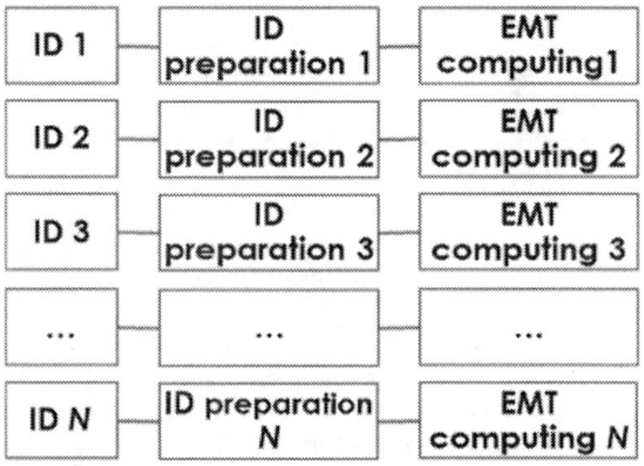

Fig. 2. Proposed TS initialization

5 Parallel EMT calculations

Parallel computing is a way of organizing computer calculations in which the problem to be solved is divided into subtasks distributed between interacting and simultaneously running computing processes [7].

The RUSTab algorithm for TS simulation already used in the SMMS supports parallel computing and vectorization, but only for the single TS case. Parallel computing efficiency can be increased by performing simultaneous TS simulations. Fig. 3 shows the existing TS simulation process. Fig. 4 shows the proposed process.

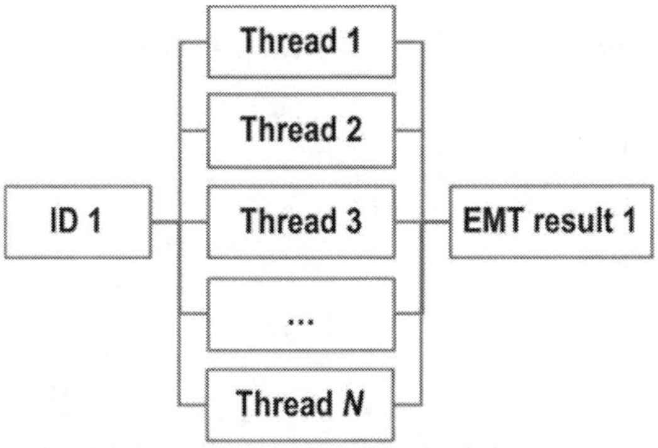

Fig. 3. Existing process of parallel TS simulations

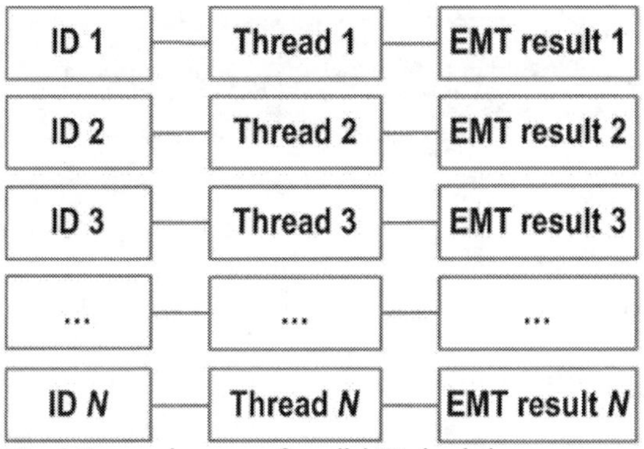

Fig. 4. Proposed process of parallel TS simulations

The existing process of parallel TS simulation can utilize up to 50-60% of computing power, spending the rest in synchronization.

In the proposed calculation procedure, the APTC state estimate creates a number of initial data variants (ID) for parallel simulations. The primary variations between these variants, in accordance with the provided continuation trajectory (CT), are the initial conditions and contingency situations. The computation of each of the prepared ID variants can be done independently in a separate thread. The calculation's outcome is whether there was a TS loss or not for each set of TS initial data. The following algorithm for parallel TS simulations is proposed to determine the APTC with transient stability constraints in the SMMS.

1. Formation of CT points.
2. Simulation tasks initialization.
3. Checking the CT points and EP combinations.
4. Aggregation of calculation results. Determining the available transfer capacity.

Fig. 5 illustrates this algorithm.

6 The acceleration methods effectiveness assessment

The following power system model is used as a test bed to assess the performance enhancements of the APTC with TS constraints: 1829 buses, 2966 branches, 438 generators with excitation systems (exciter, AVR, and discrete excitation controls, which match the particular Russian practice models), and also 66 swing equation models.

The performance improvement is the total APTC time reduction. The accuracy criterion selected is 5% proximity of APTC results in MW.

A number of computational experiments (CE) were conducted to determine the best approach to improving the performance of the APTC with TS. The following abbreviations are used to identify methods while performing computational experiments:

BC – baseline calculation used as a reference, with no proposed improvement methods;
EN – reduction with below N-rank buses exclusion;
IM – implicit integration method;
PC – parallel TS calculations;
S – sequential TS startup.

The TS simulation were conducted with the following settings: duration: 15s, minimum integration step for the explicit method: $5 \cdot 10^{-3}$s, and 10^{-8}s for the implicit method.

The APTC calculations results according with TS constraints using various acceleration methods for the considered power interfaces are given in tables 1, 2 and 3.

Table 1. The APTC calculations with TS constraints using acceleration methods for power transmission 1 results.

CE	APTC, MW	T, s	ΔAPTC, %	ΔT, %
BC	2 316.3	586.2	–	–
E1	2 314.0	582.3	–0.1	–0.7
E2	2 314.0	454.8	–0.1	–22.4
E3	2 111.3	415.1	–9.7	–29.2
IM	2 333.9	135.9	0.8	–76.8
PC	2 316.3	213.9	0.0	–63.5
S	2 316.3	563.7	0.0	–3.8
E2+S+PC	2 314.0	172.0	–0.1	–70.7
E2+IM+PC	2 330.4	56.7	0.6	–90.3

Table 2. The APTC with TS constraints using acceleration methods for power transmission 2 estimation results.

CE	APTC, MW	T, s	ΔAPTC, %	ΔT, %
BC	9	276.6	–	–
E1	1 935.6	276.4	–0.9	–0.1
E2	1 940.2	208.5	–0.6	–24.6
E3	1 765.2	188.3	–10.6	–31.9
IM	1 911.0	145.9	–2.1	–47.3
PC	1 952.3	146.6	0.0	–47.0
S	1 952.3	259.6	0.0	–6.1
E2+S+PC	1 940.2	155.0	–0.6	–44.0
E2+IM+PC	1 919.5	94.6	–1.7	–65.8

 https://doi.org/10.1051/e3sconf/202346101036

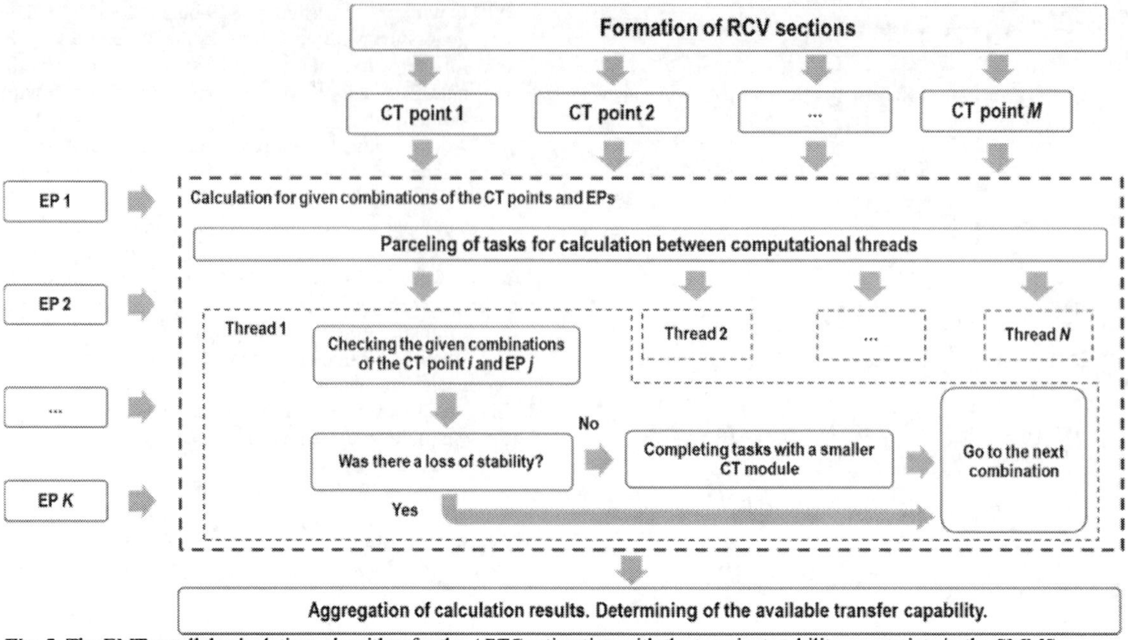

Fig. 5. The EMT parallel calculations algorithm for the APTC estimation with the transient stability constraints in the SMMS

Table 3. The APTC calculations with TS constraints using acceleration methods for power transmission 3 results.

CE	APTC, MW	T, s	ΔAPTC, %	ΔT, %
BC	2 017.2	427.9	–	–
E1	2 006.1	421.8	–0.6	–1.4
E2	2 012.1	333.6	–0.3	–22.1
E3	1 902.4	306.7	–6.0	–28.3
IM	2 017.2	134.3	0.0	–68.6
PC	2 017.2	187.8	0.0	–56.1
S	2 017.2	407.9	0.0	–4.7
E2+S+PC	2 012.1	141.5	–0.3	–66.9
E2+IM+PC	2 012.1	68.3	–0.3	–84.0

Analysis of the results in Tables 1, 2, and 3 leads to the following conclusions.

1. Eliminating buses with a rank of 2 or less gives the greatest computation time decrease effect. The APTC computation time fell by 22-24%, while the variance in the APTC result was less than 1%. There is no noticeable performance gain after the rank 1 bus reduction. The bus elimination for ranks 3 or lower saw the best computation time reduction (28–30%). Even though the APTC value discrepancy is 6-9%, that is unacceptable.

Fig. 6 shows the rotor angle transient response to the disturbance for the reduced and original models for the unstable case and the case which is close to the stability margin.

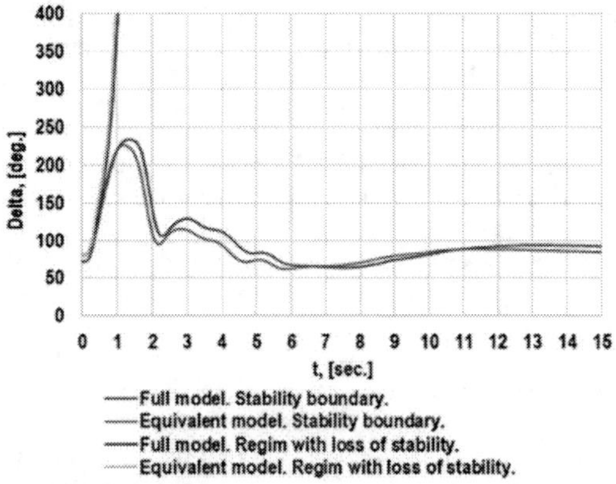

Fig. 6. Rotor angle transient response to the disturbance for the reduced and original models.

2. The use of the implicit integration method for solving the DAE in the TS simulation led to a significant time reduction that reached 45-75%, despite a significantly smaller minimum integration step. The best acceleration was achieved for the power interface with a large number of stable TS simulations. The difference in the APTC with TS constraints does not exceed 3% with the use of the implicit method.

The rotor angle transient response to the disturbance for explicit and implicit integration methods is shown in Fig. 7.

 https://doi.org/10.1051/e3sconf/202346101036

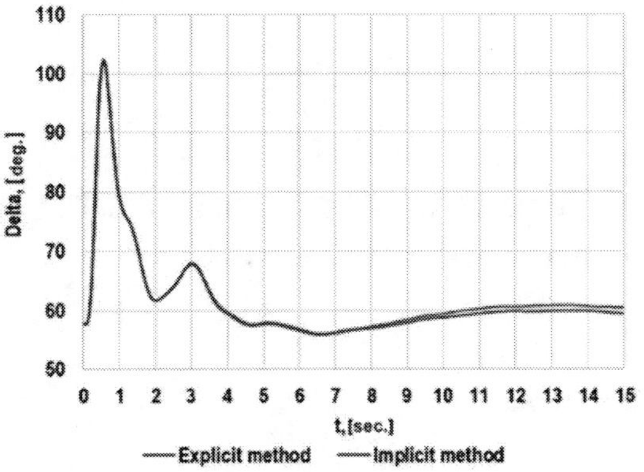

Fig. 7. Rotor angle transient response to the disturbance for explicit and implicit integration methods

3. Assuring the parallel TS simulations in accordance with the proposed approach resulted in a 45–60% reduction in calculation time. There were no differences in the APTC achieved. The time savings offered by parallel calculations became increasingly apparent as more emergency processes were examined during the APTC assessment.

4. Sequential TS initialization resulted in a 3-6% reduction in computation time with no changes in the APTC achieved. The performance improvement became more noticeable the more TS computations were carried out during the APTC with TS constraint estimation.

5. The use of a variety of acceleration techniques in combination leads to the highest efficiency for all the power interfaces taken into consideration.

7 Conclusion

This paper suggests several methods to enhance stability margin monitoring systems' analysis of available power transfer capacity with transient stability restrictions. Regarding the APTC estimation accuracy and the reduction in calculation time, these methods were compared when calculating the APTC for three different interfaces. It was shown that the best outcomes are obtained by combining the reduction of buses with rank 2 or below with the implicit integration method and parallel TS simulation. This combination enables a reduction in the APTC estimation time of up to three and, in some circumstances, even ten times.

For the stability margins monitoring system, the proposed approaches and their combinations could be used to accelerate calculation of the available power transfer capacity.

The focus of future research will be on refining the SMMS algorithm to achieve higher reductions in computation cycle duration. Additionally, the proposed approaches will be tested on an enhanced set of power system models.

References

1. Methodological guidelines on the stability of power systems (approved by the Order of the Ministry of Energy of Russia dated 03.08.2019 No. 630). Requirements for ensuring the reliability of electric power systems, reliability and safety of electric power facilities and power receiving installations. Moscow, 2018.
2. Mikhailenko A. F., Neuimin V. G., Satsuk E. I., Snegirev D. A. Comparison of the efficiency and accuracy of network equivalence algorithms for a sustainability inventory monitoring system. Izvestiya STC of Unified Power System. **No. 2(85).** pp. 5-16. 2021
3. Snegirev D. A., Neuimin V. G. Comparison of methods of equivalence of the computational model in determining permissible flows by the criterion of dynamic stability
4. Software package «Raiden». Available at: https://inorxl.com/raiden/
5. E. Mashalov, "Raiden. Power System Transient Stability Simulation Software Prototype," 2022 International Ural Conference on Electrical Power Engineering (UralCon), Magnitogorsk, Russian Federation, 2022, pp. 145-153, doi: 10.1109/UralCon54942.2022.9906601.
6. Software package «RUSTab». Available at: https://www.rastrwin.ru/rustab/
7. Almasi G. S., and Gottlieb A. Highly Parallel Computing, 2nd ed. Redwood City, CA: Benjamin/Cummings, 1994.

Model and experimental research of the transient process for the improvement of the traveling wave fault location

Rustem Khuziashev [1]*, *Iskander* Tukhfatullin[1], *Danil* Smirnov[1], *Iluza* Irkagalieva[1]

[1] Kazan State Power Engineering University, 51 Krasnoselskaya str., Kazan, Russia

Abstract. The results of modeling of transient signals in a single-wire line are considered. The parameters of voltage and current signals are compared. Signal oscillation's amplitude and period are given attention. The algorithm for determining the transient signal's onset time (TSOT) under the conditions of noise and oscillations of the signal's leading edge is presented. The results of experimentally recorded signals' analyses are given.

1 Introduction

The task of fault location in electric grids does not cease to be relevant due to the constant development of computing and communication technology, which allows to develop and implement more and more advanced algorithms for processing the recorded information.

Traveling wave fault location suite (TWFL) implements the measurement of TSOT, which is performed with an error of no more than 1 μs in the satellite-based navigational time scale. In homogeneous networks of a high voltage class, TWFL complexes determine the fault location in lines several hundred kilometers long [1]. For inhomogeneous networks of medium voltage class, the fault location is limited to line lengths of several tens of kilometers [2], which is due to the dispersion's destructive effect in a highly inhomogeneous line.

According to the TS parameters, in [3, 4] it is proposed to determine the correlation between the period of free oscillations and the distance to the fault location. When implementing the TWFL complex in distribution networks, it is important to increase the information value of the recorded signals and reduce the cost of the complex development, the dominant part of which is their software.

1.1 Model research of TS shaping by boundary inhomogeneities

TWFL hardware-software suite determines the place of TS occurrence. The structure of multimodal TS oscillations carries information about both the place of their occurrence and their cause. Without modelling of these processes, it is impossible to understand the mechanisms of TS shaping. The results of TS modeling were obtained with the PSCAD software package [5] in the simplest models of the electric network.

Fig. 1 shows the model of single-wire line with a return earth wire and concentrated capacitance; a source of alternating electromotive force with a 50 Hz frequency at the beginning. At the source of EMF with a phase voltage of 10 kV zero internal resistance is chosen in order to simplify the TS shaping's analysis. The line, 10 km long, is represented by 8 identical segments with distributed parameters of 1.25 km length. Line's wave impedance is 600 ohms.

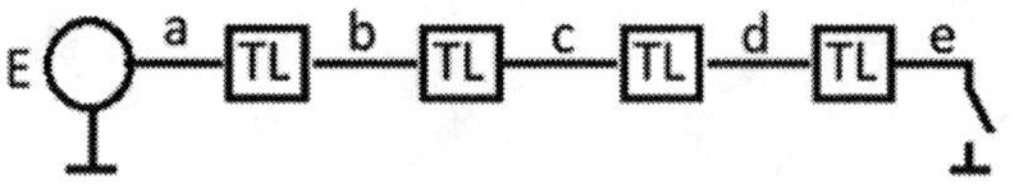

Fig. 1. Model of a homogeneous single-wire line.

Fig. 2 shows a voltage TS recorded in different parts of the line with a time sampling interval of 0.1 μs. At the end of the line a travelling wave (TW) is generated in the form of a voltage step equal to the instantaneous voltage amplitude at the moment of commutation. The TW propagates to the beginning of the line. At the beginning and at the end of the line, the voltage step's TW is reflected with reflection coefficient "-1", which corresponds to the boundary conditions of zero equality of falling and reflected waves' summed amplitudes. The superimposition of a multitude of reflected TW forms a standing voltage wave with nodes at the ends of the line.

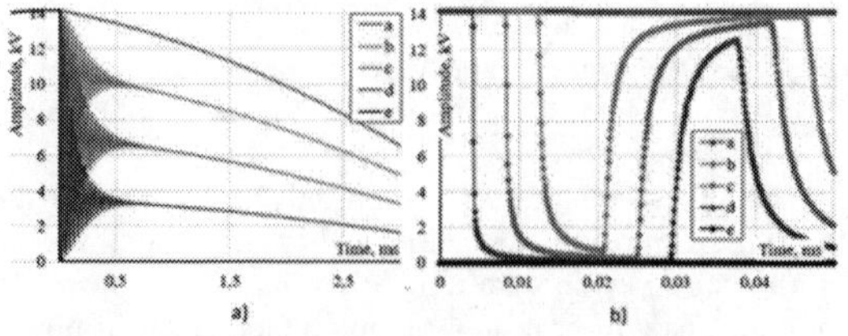

Fig. 2. Voltage TS with large- (a) and small-time scales (b) at different sections of the line. The sampling interval is 0.1 μs.

The oscillation's amplitude spread at TSOT (fig. 2) is equal to the amplitude of the phase voltage at the moment of breaker closure. As the TS amplitude

* Corresponding author: 142892@mail.ru

decreases, the voltage curve tends to become a steady-state emergency oscillation of industrial frequency of 50 Hz. The voltage oscillation amplitude's attenuation constant is 0.3 ms. The shape of voltage TW's leading edge illustrates an increase in its duration by about 1 μs for each kilometre travelled, in accordance with the dispersion effect on the TW voltage step's shape. The oscillation period of 36.2 μs corresponds to travelling twice the length of the line at a speed of 276 m/μs. This speed corresponds to the TW propagation in the ground or zero mode channel in a three-phase line.

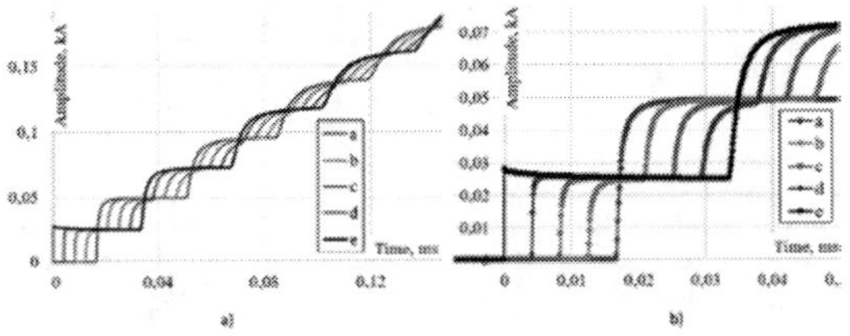

Fig. 3. Current TS with large- (a) and small-time scales (b) at different sections of the line. The sampling interval is 0.1 μs.

Fig. 3 shows a similar current TS. TW in the form of a current step is generated at the switching point and propagates to the beginning of the line. At the beginning and at the end of the line the current step's TW is reflected with reflection coefficient "+1", which corresponds to the boundary conditions of zero equality of falling and reflected waves' summed amplitudes. The superimposition of a multitude of reflected TW forms a rising standing current wave, which reaches a maximum value of 2 kA in 3 ms after the breaker closure. Current oscillation amplitude step spread for the line shown in fig. 1 on TSOT is equal to the ratio of the phase voltage to the value of line's wave resistance and is numerically equal to the value of 25 A.

It is the current TW's initial step that determines its onset time. Comparison with the voltage's initial step (fig. 2) indicates the advantage of voltage signals relative to current signals at the simplest threshold algorithms for determining the TSOT. This is due to the significantly leading edge's larger amplitude spread of the voltage signal compared to the current signal.

The TS shaping's research on the segment between the breaker's location and the end of the line is carried out (fig. 4).

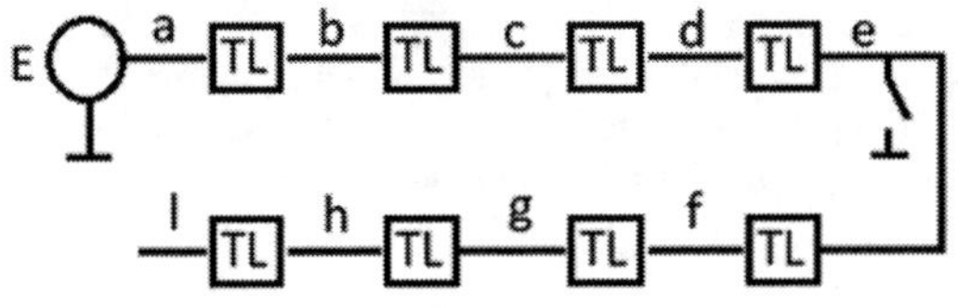

Fig. 4. Model of a single-wire line with a closing breaker in the middle of the line.

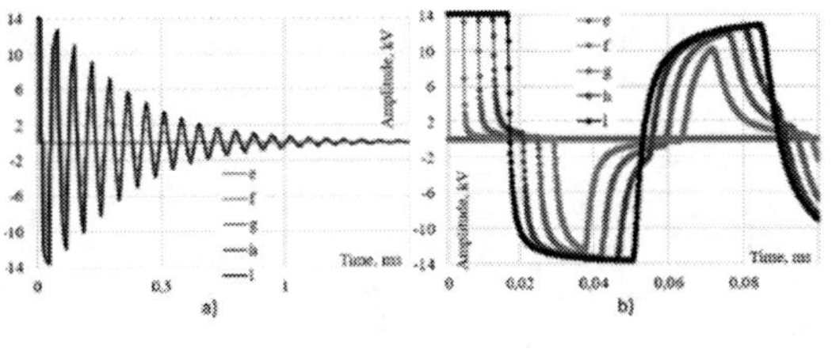

Fig. 5. Voltage TS with large (a) and small time scales (b) at different sections of the line.

The TS caused by breaker closure starts at the zero moment of time and causes zeroing of the voltage in node "e" (fig. 4, fig. 5). In this node a travelling wave (TW1) of negative polarity is formed with amplitude equal to the pre-fault value, propagating to the both line's ends (fig. 5b blue line - node "f", green line - node "g", red line - node "h", black line - node "l"). Upon reaching the end of the line with high impedance, TW1 is reflected with the coefficient "+1" and forms a reflected wave (RW1) of the same polarity, propagating to the short-circuit point. At the place of reflection, the amplitude spread of falling and reflected waves' sum is doubled. The further process of travelling and reflected waves' shaping is similar to the previously discussed one in fig. 3. The TS formed as a result of reflected waves' superimposition has free oscillations' period twice as long as the analogue period from fig. 1 for the signal to the left of the short-circuit location.

1.2 Model research of TS shaping by intermediate inhomogeneities

Equivalent circuit of all electric power facilities includes a parallel connection of complex impedances with a significant active component and a capacitor between their terminal leads having a non-zero area. According to the parameters used in the calculation models, the parasitic capacitance of the equipment is estimated to be 0.3-0.6 nF, the winding capacitance of a 220 kV autotransformer is approximately 2.7 nF, the capacitance of a plate insulator is 50 pF, and the capacitance of a reference pin insulator is 1-5 pF [6].

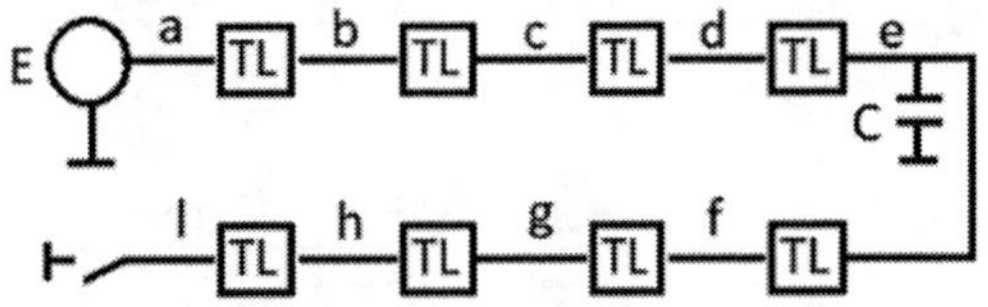

Fig. 6. Model of a single-wire line with concentrated inhomogeneity in the middle of the line.

At the initial moment of time the reflection of voltage step's TW from the capacitor occurs as from an inhomogeneity with zero resistance. This is due to the large charge current drawn by the capacitor at the initial moment of time. As the capacitor is charged, its resistance tends to infinity. Fig. 6 shows a diagram of a 10-kilometre single-wire line in the middle of which a capacitor with a variable capacitance is installed.

Fig. 7 shows the oscillogram of the voltage step's TW at node "f". TW has travelled the distance from node "l" to node "f" equal to 3.75 km in 12.4 μs, which is close to the speed of light. The wave reflected from the capacitance is registered at node "f" in 20.7-12.3 =8.4 μs, which corresponds to running a double distance from node "f" to node "e" and back from node "e" to node "f". The reflection coefficient is "-1", and the polarity of the reflected signal is similar to the situation illustrated in fig. 2.

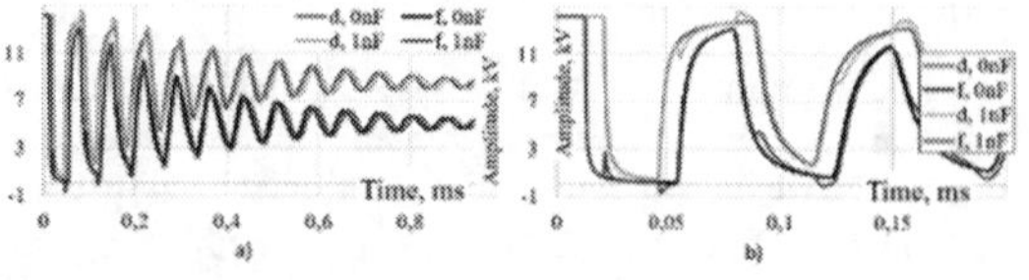

Fig. 7. Voltage TS with large- (a) and small-time scales (b) at different line sections with capacitor type inhomogeneity.

After 45.8-20.9=24.9 μs, a small wave was registered, which ran a distance of 1.25*6=7.5 km (from "f" to "l" and from "l" to "f") with a speed of 7.5/24.9=0.309 km/μs, and only after 52.3-12.3=40 μs, the main wave was registered, which ran 12.5 km (corresponding to the double running time from "f" to "a"). The decreasing distance between the TS registration point and the subsequent inhomogeneity is able to generate an oscillatory process at the TW's leading edge. This feature can be used in the machine analysis of the registered oscillograms, to determine the place of commutation inside or outside the area controlled by the TWFL complex. The decrease in TS amplitude even for a single inhomogeneity in the form of capacitance can be clearly seen. Increasing the number of inhomogeneity nodes will cause a proportional weakening of TS. The presence of an inhomogeneity in the form of a capacitor with a capacitance of 1 nF causes the appearance of high-frequency mode oscillations at the beginning of TS and a decrease in the frequency of low-frequency mode oscillations by about 1.5%.

Fig. 8 shows a diagram of a single 10-kilometre line with a branch of varying length connected to the middle of the line.

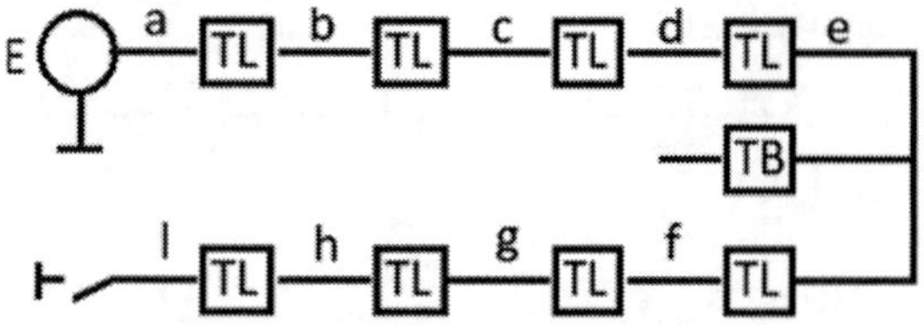

Fig. 8. Model of a simple single-wire line with a branch in the middle of the line.

Fig. 9 shows the TS oscillograms with large- and small-time scales, which clearly depict the effect of the branch on the parameters of the generated TS in comparison to its absence. A branch of any length causes the formation of high-frequency mode oscillations at the beginning of TS and an increase in the period of the dominant low-frequency mode oscillations throughout the TS's duration. The increase in period duration is 1.5 per cent for a 0.1 km branch, 10 per cent for a 1 km branch, and 100 per cent for a 10 km branch. The presence of a branch causes TS amplitude overvoltage relative to the normal phase voltage. Overvoltage's value depends on the ratio of branch length, line length and fault location.

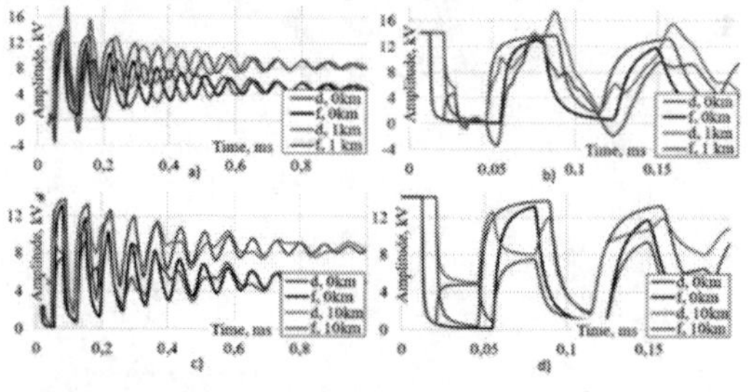

Fig. 9. Voltage TS with large- (a) and small-time scales (b) at different line sections with inhomogeneity of branching type with length of 1; 10 km.

2 Algorithm for determining the onset time of experimentally recorded TS

Fig. 10 shows typical shapes of experimentally recorded voltage TS's leading edge. The presence of oscillations with different polar amplitudes on the TS's leading edge, as shown in the previous sections, is explained by the inhomogeneous structure of the voltage step's TW propagation channel. These oscillations increase the error in determining the TSOT by threshold methods [7], two threshold methods with linear approximation to zero [8,9] and statistical methods based on high-order central moments [10, 11, 12]. Under conditions of lightning overvoltage, the amplitude of noise oscillations caused by streamer formation increases, which increases the error in determining the TSOT.

To illustrate the occurrence of the error in determining the TSOT by the statistical method based on the kurtosis function [11, 12], fig. 11a shows TS normalized to the maximum amplitude (fig. 10c, d) and calculated dispersion and kurtosis functions on the time window from the beginning of the oscillogram to the current value of the time series. We can see that statistical method for determining the TSOT requires the use of a non-zero threshold, the value of which decreases as the order of the central moment increases. The value of the threshold depends on the TS's noise value (fig. 10a, c), which is comparable to the amplitude of oscillations. Significant value of statistical method's threshold applied to the original TS causes an error in the determination of the TSOT (fig. 11a).

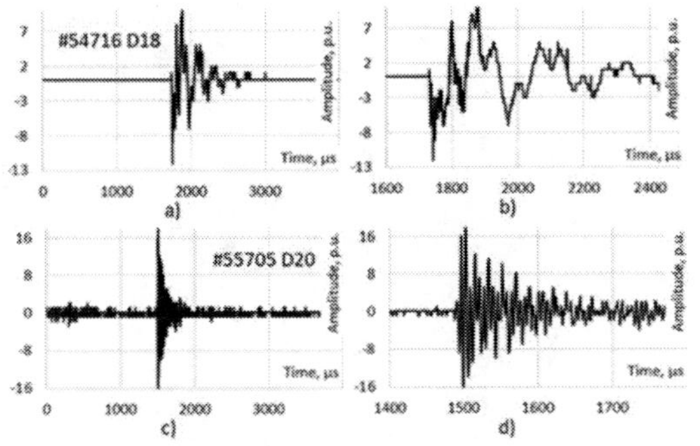

Fig. 10. Experimentally recorded TS with different shapes of leading edges. TS are depicted with large- (a, c) and small-time (b, d) scales.

In order to reduce the error of TSOT determination, an algorithm of TS's envelope construction was developed on the basis of replacement of instantaneous amplitudes by its modules, smoothing of emissions and dips of instantaneous amplitude without changing their polarity, smoothing of dips caused by oscillatory change of instantaneous amplitudes' polarity. This allowed us to approximate the TS's leading edge by a monotonic dependence and to reduce the number of noise emissions (fig. 11b).

The statistical functions of the second- and fourth-order central moments and the second-order initial moment were applied to the obtained envelope. The application of a zero threshold to the increments of these functions indicates the advantage of using the second-order initial moment from the TS envelope. The error of TSOT determination by competing functions is equal to 2 µs.

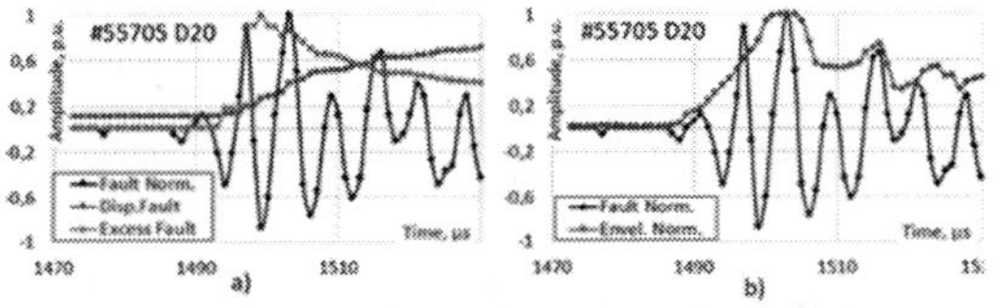

Fig. 11. Illustration of the algorithm for determining the TSOT based on the calculation of its envelope when the leading edge oscillates in the presence of noise.

When the number of noises is small (fig. 10 a-b), the second-order advantage of the initial moment vanishes, and all competing functions show the correct TSOT definition.

3 Conclusion

Medium voltage class distribution networks are significantly inhomogeneous. TS is shaped by superimposition of many reflected TW. All reflected TW in electrical network are formed by large and small inhomogeneities. Large inhomogeneities include the end and the beginning of the line. Small inhomogeneities include capacitances of substitution schemes of concentrated elements of electric power facilities (insulators, transformers) and nodes of branch connections. A three-phase electric line has three mode channels of TW propagation. Therefore, at TS registration point there is a superimposition of TW propagating in different mode channels, which greatly complicates the analysis. The paper demonstrated the mechanisms of TS shaping in a simple single-wire line, which is a substitution scheme of the ground mode channel of a three-phase line.

Model results indicate that the maximum voltage TS's spike is equal to or greater than the normal phase voltage of the line. The maximum current TS's spike is much smaller than the load current spike. This indicates the advantage of voltage TS's recording. Since standard voltage transformers have a smaller frequency bandwidth than standard current transformers, it is recommended to use input circuits based on capacitive dividers for TS recording.

This reduces the value of fault location methods based on TS period measurement.

The modelled TS shapes in the presence of a concentrated inhomogeneity indicate the occurrence of high-frequency mode oscillations at the beginning of the TS and an increase in the period of low-frequency mode oscillations proportional to the magnitude of the capacitance of the concentrated inhomogeneity or the length of the branch.

In order to reduce the error of TSOT calculation under the oscillatory nature of its leading edge and in the presence of noise, an algorithm based on TS envelope determination was developed.

References

1. Lvov A, Pinchukov P. Optimisation of search for fault locations of 220 kV overhead power lines located in the territory of the Republic of Sakha (Yakutia) using wave TWFL. // Electricity. Transmission and Distribution, **№S1(28)**, 2023, P.44-48.
2. Feoktistov A. V. Technical solutions for determination of the damage places in the distribution networks 6-35 kV. // Electricity. Transmission and Distribution, **№6(69)**, 2021, P. 94-95.
3. Kachesov, V.E. Method for determining the single-phase fault zone in distribution networks under operating voltage. // Electricity. 2005. **№ 6**. P. 9-18.
4. Shuin V.A., Filatova G.A. et al., Information parameters of electrical transient values for determining the ground fault location in distribution cable networks with voltage 6-10 kV. // Bulletin of ISPEU. **№2**, 2017, P.34-42.
5. PSCAD [Electronic resource] // URL: https://www.pscad.com (date of address: 10.09.23).
6. PAO "ROSSETI" STO 56947007-33.060.40.052-2010. Guidelines for the calculation of parameters and the selection of high-frequency carrier circuits for AC power lines of 35-750 kV. 2010. p. 143.
7. Dichev N.V. et al., Method for determining the time position of a slow-increasing echo impulse. // Bulletin of TPU. **№4**, 2010, C.146-149.
8. Soldatov A.I., Sorokin P.V., Makarov V.S., Determination of the temporal position of the acous-

tic pulse by approximation of the signal envelope. // Bulletin of SFedU. - Technical Sciences, **№ 10**, 2010, P. 178-184.
9. Schweitzer E.O., et al. Locating faults by the traveling waves they launch. // Proc. 67th annual conference for protective relay engineers. March 2014, P. 95-110.
10. Patent 2660403 Russian Federation, (06.07.18). Method of thresholdless automatic intellectual registration of acoustic emission signals by a nondestructive testing device. / Barat V.A., Shimansky A.G.
11. Lachugin, V.F.; Panfilov, D.I.; Smirnov, A.N. Implementation of the traveling wave fault location on power transmission lines using statistical methods of data analysis // Izvestia RAN. Energetika. **№ 6**, 2013, P.137-146.
12. Podshivalin, A.N. Research of kurtosis coefficient's properties for the purpose of the traveling wave fault location in transmission lines. // Relay protection and automation. №1, 2023, P.51-57.

E3S Web of Conferences 461, 01038 (2023) https://doi.org/10.1051/e3sconf/202346101038
RSES 2023

Problems of ensuring reliability and sustainable development of the United energy system of Central Asia under the conditions of energy consumption growth

Nasirov[1] T.Kh., Trofimov[2] G.G., Khamidov[1]* Sh.V., Tanirbergenov[1] R.M.

[1]Institute of Energy Problems of the Academy of Sciences of the Republic of Uzbekistan. 2-B Chingiz Aytmatov str., Tashkent
[2]Almaty University of Energy and Communications, 126 Baytursynova str., Almaty, Kazakhstan.

Abstract. The questions of development of the united power system of Central Asia at introduction in them of renewable energy sources (RES) defined by decisions of governments of the countries of the region are considered. The existing main scientific and technical problems related to the large-scale use of RES in the energy system and ways to overcome them are noted. The questions of reliability of power supply of the united energy system of Central Asia, joint solution of a complex of questions on introduction of RES and energy storages in the Association are analysed.

1 Introduction

In the Central Asian region, the problems of electricity shortages have recently become more acute [1]. Thus, Kazakhstan has been facing acute power shortages for the last two years, especially during peak load hours. In 2022, the electricity deficit was fixed for 8 months. According to the forecast of the Ministry of Energy of the Republic of Kazakhstan from 2023 to 2029, the volume of electricity used in the country will increase annually by an average of 3%. Levels of production and consumption in the power sector for the next seven years show that in 2023-2025 and in 2028 the balance will be negative. In Kyrgyzstan last winter, the country began cutting off electricity to reduce consumption. The shutdowns were caused by a shortage of electricity in the country, due to the lack of necessary capacity for production. In addition, this year the country may also face a significant shortage of electricity, and the deficit may amount to 1.9 billion kWh. The situation in Uzbekistan is similar. Due to energy shortages in the country, power outages of several hours are quite frequent. During peak loads in winter there is repeatedly observed a deficit of 18-20 mln kWh of electricity.

However, the forecast assessment shows that in the period up to 2030, the annual growth of electricity consumption in Uzbekistan will be about 6-7 percent [3]. The growth of electricity consumption requires accelerated construction and commissioning of new generating capacities and development of power grids in the region.

2 Main part

2.1 Energy return on Investment

Taking into account the fact that the construction of RES is planned everywhere in the region, under these conditions it is advisable to make a preliminary assessment of the efficiency of the energy resources used. As is known, the energy return on investment (EROEI, Energy Return on Investment) is one of the criteria for comparing the efficiency of energy resources and is used to assess the efficiency of energy production. EROI is defined as the ratio of the energy obtained from a certain resource to the energy used to produce that resource:

$$EROEI = Er / Es,$$

where Er - energy received, Es - energy spent on production.

In many respects, this indicator is methodologically unideal. Its value can vary significantly depending on the technologies used and the location of the power plant (insolation, wind strength, etc.). Nevertheless, it is useful for comparison in the dynamics of energy alternatives. If EROEI is greater than one, the resource is a net producer of energy, if it is less, it is a sink. The higher the EROEI, the

* Corresponding author: khamidov_sh@mail.ru

more useful the resource is, since it produces more energy per unit than others. It is believed that for the stable functioning and development of modern industrial society the energy payback index should be at least 5:1.

For a long time it was considered that RES (solar and wind power plants, SES and WPP) have a lower EROEI than conventional plants. For example, in [4] it is explicitly stated that nuclear, hydro, coal and gas power plants are an order of magnitude more efficient than photovoltaics and wind power.

It is also stated in [5] that for Russia alternative energy in the form of solar-wind power generation has completely discredited itself, and this is no longer in doubt. Russia is known to be a major producer of oil and gas, has significant hydro resources and therefore conventional power plants often operate on the basis of these natural and fossil resources. Not all CA countries have such resources.

However, the analysis has shown that for conventional power plants in the Central Asian region, such as thermal power plants based on fossil fuels (coal, oil, gas), the typical EROEI value is about 10:1 to 30:1. This means that each unit of energy expended in extracting and processing fossil fuels yields between 10 and 30 units of energy in the form of electricity. However, it is known that the EROEI of the main fossil fuels (oil and gas) is decreasing over time, which will cause the EROEI of fossil fuel based thermal power plants to decrease [6].

In the case of renewable energy sources, such as renewable energy sources (SES and WPP), EROEI values can vary depending on various factors, including technology, climatic conditions, choice of location and production methods. Moreover, the EROEI of wind and solar technologies are generally increasing [6].

Thus, for solar power plants, which are preferred in many regions of Central Asia, typical EROEI values range from about 10:1 to 30:1, although more precise values may vary depending on the type of solar technology (photovoltaic or tower-type plants) and their location. For example, the EROEI of silicon solar plants with an estimated lifetime of 25 years, located in Switzerland, where sunny days are much less than in Central Asia, is 9-10, respectively, the energy payback period is 2.5-2.8 years [7].

It should be noted that the reduction of material consumption per unit of output, in this case the consumption of raw materials, the production of which accounts for the main energy costs associated with solar energy, is the main factor in the growth of EROEI. In addition, the growth of energy efficiency in the production of photovoltaic cells and the increase in their service life also contributes to this.

For wind farms, EROEI values typically range from 15:1 to 40:1, although they can also vary depending on turbine size, wind speed and other factors.

Thus, the analysis shows that for the Central Asian region it is economically feasible to use SES and WES along with conventional energy sources.

Therefore, in all CA countries of the region included in the UES CA with development plans envisages the construction of not only new conventional power plants (NPP and hydroelectric power plant in Uzbekistan, planned NPP in Kazakhstan, large HPP in Kyrgyzstan), but also in the whole region is planned a wide integration of RES.

2.2 Direction of development of the Unified Energy System of CA with RES

The development of RES in the CA region has many advantages and can be very feasible. Here are some of the main reasons:

Resource potential. CA has a huge potential for RES development, making it ideal for intensive and efficient utilization of these sources.

Solar energy. The region has vast areas with high solar activity, long lasting sunshine. Therefore, SES can be an effective option for clean energy generation.

Wind energy. CA is also characterized by strong winds, especially in mountainous areas. Wind power plants can be installed in such locations to generate electricity.

In addition, there is significant potential for hydropower development in the region. The construction of large HPPs on the Amu Darya and Syr Darya rivers is expedient. Small and medium-sized hydropower plants may be in demand to meet local energy needs.

Energy independence. The development of natural resources will allow the region to diversify its economy, reduce the use of oil, gas and other traditional energy sources, and reduce the dependence of the region's countries on energy imports and strengthen their energy independence.

Environmental sustainability. The development of RES in CA will help reduce greenhouse gas emissions, limit negative environmental impact and counteract climate change.

Economic benefits. Energy production from RES can be economically advantageous compared to traditional energy sources, which is especially important if we take into account the resource potential of the region.

2.3 Global benchmarks for present value per unit cost of electricity

It should be noted here that Bloomberg NEF (BNEF), a world leader in strategic business and financial information, has published a new report [8] containing global benchmarks of the present value per unit of electricity (LCOE) for different power generation technologies in 2023 (Fig. 1).

It notes that in the last six months, onshore WPP and PV remain the cheapest new electricity generation technologies in countries that account for 82% of total global electricity generation. And the costs of wind power and battery storage projects have declined from 2022 levels and are the cheapest since BNEF began collecting project data in 2009, driven by market development.

 https://doi.org/10.1051/e3sconf/202346101038

Thus, the global average LCOE for onshore wind power is $42 per MW·h, for photovoltaic SES $44-48 (depending on the presence of trackers in the projects), for gas-fired power plants $92, and for coal-fired power plants $74 per MW·h. If the cost of carbon capture and storage (CCS) is added to gas and coal-based power generation technologies, their LCOE rises to $128 and $123, respectively.

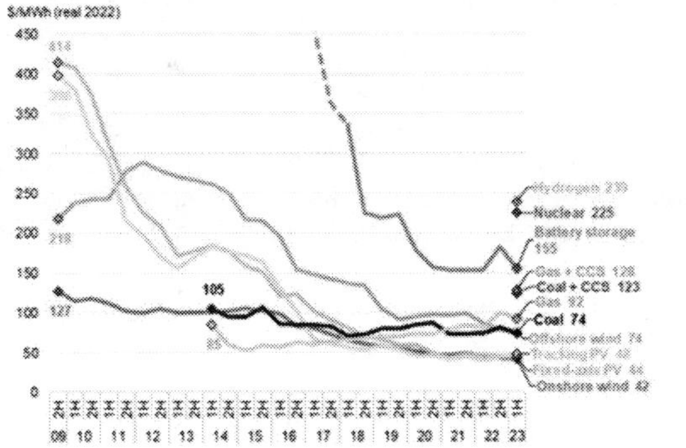

Fig.1. Global benchmarks for present value per unit cost of electricity

2.4 Capacity regulation in a power system with RES

The development of RES requires the development of effective policies, legislation and financing mechanisms, as significant investments are required to build RES power plants.

However, these sources are known to be intermittent and weather-dependent, which negatively affects the power system regime. In addition, in case of emergency situations in the power system, such as external short-circuits, the voltage on the busbars in the RES locations may decrease. If Solar power plant (SPP) and wind power plant (WPP) are not equipped with energy storages that can support it, the inverters included in RES are "overturned" and RES are disconnected from the grid by relay protection devices.

In this case, the available power of the system P_{AP} as a result of the accident will decrease by the amount of the disconnected power of SPP and WPP

$$P_{AP\_A} = P_{IC} - (\sum P_{SPP} + \sum P_{WPP}).$$

In this case, if there is insufficient operational ("hot, rotating") capacity reserve intended to compensate for the imbalance between power production P_{AP} and consumption P_C caused by failures or emergency shutdowns, the power system will have a capacity deficit equal to

$$\Delta P_{DEF} = P_C - P_{AP.}$$

As a result of the resulting deficit of available power P_{AC}, the frequency in the power system will decrease relative to the nominal value $f_{NOM} = 50$ Hz

$$f_{EF} = f_{NOM}\,(1 - \Delta P_{DEF} / P_C).$$

If a special Automatic frequency offloading or operational actions cannot prevent the frequency in the power system from decreasing and it is below some critical value

$$f_{CF} = (47\text{-}46{,}5 \text{ Hz}),$$

$$f_{EF} < f_{CF}$$

then low-inertia GT and HPPs will start to be disconnected from the grid, which will create an additional power deficit in the power system and can lead to a cascade development of a system accident.

In [9] the influence of the specific power weight of SPP and WPP in the presence of GT and GPP (Gas-piston power plant) in the UPS of CA was considered and it was shown that in the absence of energy storages in the power system the probability of violation of the normal operation mode of the power system and the damage caused by the deficit of generating capacity increases.

The above-mentioned issues related to power regulation will become even more actual in connection with construction in Uzbekistan of the first stage of nuclear power plant with two 1200 MW units designed for operation in the base mode.

The increased demand for natural gas and fuel oil in the country predetermines the need to diversify the supply of primary energy resources and to study the construction of the second stage of NPPs with similar units.

It should be noted that for solar-wind generation the installed capacity utilization factor (ICUF) can be relatively low due to the variability and inconstancy of renewable energy sources. Therefore, the ICUF for solar-wind generation typically ranges from 20% to 40% or even less, depending on specific operating conditions and technologies, and globally varies from 15% to 25% on average.

In contrast, conventional power plants such as coal, gas or nuclear can have higher ICUFs because they typically operate continuously at both constant and variable load. The ICUF for conventional power plants can be approximately 70-90% or more, depending on the mode of operation and maintenance. Therefore, to eliminate the above disadvantages of RES and improve their efficiency, it is advisable to equip RES with energy storage and accumulation systems.

It seems to us that in these difficult conditions energy storage systems can be used not only for wide integration of RES in the UPS of CA, which are becoming more and more important due to the presence of significant potential of solar and wind energy in the region, but also to improve the technical and economic performance of the energy system. Therefore, along with the construction of new power plants and integration of RES, it is important to install various energy storages in the energy system.

These accumulators will make it possible to store surplus energy and use it more efficiently during periods of absence or lack of sunlight and wind during the day, compensate for the variability of RES and ensure reliable energy supply. In addition, the use of energy storage and accumulation systems in the UPS of CA can be very expedient for the entire energy system, providing a number of technical and economic advantages for it.

Storages together with the system of power control are able to increase reliability of the energy system, smooth peak loads and balance fluctuations in power generation and consumption. Energy storage systems (ESS) with long-term energy storage capability will improve the energy security of the countries in the CA region, taking into account that some countries still have problems with the supply of energy resources.

Energy storage can reduce energy costs by using more efficient tariff schemes, e.g. energy can be stored during periods of low demand and used during peak demand periods when electricity prices are higher.

At the same time, the use of ESS can reduce the cost of building additional power plants to cover peak demand and overloaded transmission lines. Instead, surplus energy can be stored during low load periods and used during peak demand periods, thus optimizing the use of available resources and delaying the construction of new transmission lines.

ESSs can also affect the distribution of energy in the CA UPS between different countries, which may have different energy demand at different times, which will also contribute to the economic efficiency of electricity supply. Energy storage and accumulation allow efficient distribution of energy between countries according to their needs and reduce energy losses during long-distance transmission.

ESS will also contribute to improving environmental sustainability in the Central Asian region. The use of ESS along with the large-scale introduction of RES can help reduce the use of fossil fuels, reduce negative impacts on the climate and the environment by reducing greenhouse gas emissions.

In world practice, the most popular batteries are lithium-ion batteries, which, meanwhile, have a high degree of degradation - up to 5% per year and a short operating cycle.

In recent years, the technology of using liquefied air energy, as well as cryogenic energy storage technology has been rapidly developing [10]. Despite the possibility of returning only 50 percent of the energy consumed for preliminary accumulation, this SNE will be in high demand not only for reducing the peak part of the load schedule, but also as a means of long-term energy storage and improving the reliability of power supply.

However, decision-making on the introduction of ESSs in the UPS of CA should take into account many factors, including the potential for regional cooperation, the level of infrastructure development, social and environmental aspects, the cost of ESS, as well as the technical feasibility of various systems and the complexity of their implementation.

At the same time, it is necessary to determine by special studies the permissible share of RES in the composition of generating capacities, safe for maintaining sustainable operation of the energy system of each of the countries within the UPS CA, the necessary volumes of electric energy storage on the basis of promising types of storage systems and the places of their installation.

Usually it is said about installation of energy storage only in RES plants, but it is important to install energy storage in existing power plants to improve their operation and energy efficiency, as well as in distribution networks and at the consumer's end, which also improves technical and economic performance and contributes to the reliability of the power system.

In addition, it should be noted that strengthening the interconnectors between the CA countries in the region and with Russia will contribute most to ensuring large-scale integration of RES into the UPS.

2.5 Application of FACTS technology in UPS CA

Central Asian region purposefully strengthens the main power grids, increases generating capacities to meet the growing load of consumers. Calculations show that in UPS CA it is not always possible to ensure regime parameters in accordance with regulatory requirements. Under these conditions, the successfully developing theory and practice of using technologies of Flexible Alternative Current Transmission System (FACTS) allows to ensure stable and optimal operation of the power system. The use of FACTS technology allows obtaining various corrective actions depending on the conditions of a particular control task, reducing the gap between controlled and uncontrolled modes of uninterruptible power supply sources (UPS), providing dispatching personnel with additional degrees of freedom in controlling power flows and voltages to excessive and deficient sections of the power grid. The impact of FACTS technology on approaches to power plant equipment design and operation planning of transmission and distribution grids has been analyzed for UPS CA. As a result, it is shown that high technical capabilities of FACTS technology stimulate the development of mechanisms of innovation activity in the sphere of intellectualization of electric power industry of Uzbekistan and introduction of this technology. The efficiency of modern types of FACTS technology in control tasks of the united power system of Central Asia has been studied and analyzed, the effects of stability and mode controllability increase depending on the specific task have been obtained. The transfer of power grids from passive to active devices allows to regulate the mode parameters within wide limits. With the introduction of FACTS devices, power grids get an opportunity to change their electrical parameters in real time mode, to increase the limits of transmitted power and the stability of the power system to disturbances. Since the use of FACTS devices is equivalent to the construction of new transmission lines, an economic effect is achieved [11–13].

3 Conclusion

It should be noted that Europe's energy transition to zero-carbon electricity by 2050 in accordance with the Paris Agreement consists not only in the widespread construction of large numbers of solar panels and wind turbines, but is also accompanied by changes in infrastructure, with a special role for cross-border interconnectors, which significantly affect the cost-effectiveness of this process. Therefore, the CA region, which is 2.5 times smaller than Europe, should follow the same path to create a more flexible system that can balance geographical differences in wind and solar generation and minimize power fluctuations from RES. This pathway will be able to utilize the optimal locations of RES, their potentials, improve their generation efficiency and reduce the costs for countries in implementing decarbonization policies.

In addition, increased interconnection between countries in the region can incentivize investors by giving them the confidence to sell their generation both domestically and in foreign markets, which will also encourage the deployment of RES in the region.

At the same time, an important factor in maintaining the reliability of the energy system is the introduction of demand-side management practices, which involves actions by consumers to reduce their electricity consumption. The introduction of such consumer-regulators will mitigate overcoming emergency situations in the energy system, such as a sharp increase in electricity consumption during weather changes or emergency shutdowns of power units.

Bringing consumers within the allowed volume of consumed power during peak load hours and taking them out of operation during power shortages, introduction of automatic frequency and power regulation and stability reserve monitoring systems are also priority solutions for the energy system.

References

1. In a state of energy hunger: Central Asian countries face power supply shortages. https://ranking.kz/reviews /industries/v-sostoyanii-energeticheskogo-goloda-strany -tsentralnoy-azii-ispytyvayut-defitsit-elektrosnabzhe niya.html.
2. Zhanbolat Mamyshev. Kazakhstan predicted growth in electricity consumption. February 8, 2023. https://kz.kursiv.media/2023-02-08/zhnb-electrokazakh stan/#:~:text=В%202023-2025%20годы%20ежегод ный,(МЭА)%20в%20очередном%20отчете.
3. Concept of electricity supply to the Republic of Uzbekistan for 2020-2030. Ministry of Energy of RUz, 2020. Source: official website of the Ministry of Energy of the Republic of Uzbekistan - https://minenergy.uz/ru/lists/view/77.
4. D.Weißbach, G. Ruprecht, A. Huke, K. Czerski, S. Gottlieb, A. Hussein, Energy intensities, EROIs (energy returns on investment), and energy payback times of electricity generating power plants. Volume 52, April 1, 2013, pp.210-221. https://www.sciencedirect.com/scien ce/article/abs/pii/S0360544213000492. https://www.sciencedirect.com/science/article/abs/pii/S0 360544213000492
5. Masyutin S.A. Alternative Energy: Myths and Reality. 2021, Ruselprom. https://www.sciencedirect.com/science/article/pii/S0301 421516307066
6. Mohammad Vaferi, Kayvan Pazouki,Arjen Van Klink, Declines in EROI of Main Fuels and the Implications on Developing LNG as a Marine Fuel, 2020, file:///C:/Users/User/Downloads/jmse-08-00719.pdf.
7. Marco Raugei, Sgouris Sgouridis, David Murphy, Vasilis Fthenakis, Energy Return on Energy Invested (EROEI) for photovoltaic solar systems in regions of moderate insolation: A comprehensive response, Energy Policy, Volume 102, March 2017, pp.377-384 https://www.ruselprom.ru/support/obuchayushchie-ma terialy/obuchayushchie-prezentatsii/Альтернативная% 20энергетика% 20myths%20and%20reality.%20 Masy utin.pdf.
8. Cost of Clean Energy Technologies Drop as Expensive Debt Offset by Cooling Commodity Prices. June 7, 2023, https://about.bnef.com/blog/cost-of-clean-energy-tech nologies-drop-as-expensive-debt-offset-by-cooling-commodity-prices/
9. Nasirov T.H., Nepomnyashchiy V.A., Shamsiev H.A., Influence of wind and solar power plants on controllability and reliability of power systems functioning. // Scientific and Technical Journal "Problems of energy and resource saving". -Tashkent, 2021, No 3. -pp. 72-83.
10. Trofimov G.G., Petukhov Y. Prospects for the use of cryogenic technology in the energy system of Kazakhstan. // Energetika. Bulletin of the Union of Power Engineers of the Republic of Kazakhstan. 2022, №4 (83), pp. 65-68.
11. Misrikhanov, M.S.; Ryabchenko, V.N.; Sitnikov, V.F. FACTS devices (in Russian) // Elektro-info. 2008. № 1. C. 62 - 74.
12. Khamidov Sh.V., Shamsiev Kh.A., Shamsiev B.Kh. Technical and organizational aspects of increasing the reliability of the Unified Power System of Central Asian countries. // Rudenko International Conference "Methodological Problems in Reliability Study of Large Energy Systems" (RSES 2021). AIP Conference Proceedings 2552, 010001 (2023) https://doi.org/10.1063/12.0013114
13. Khamidov Shukhrat. Modeling of FACTS Devices and Their Application in Intersystem Tie Lines of the United Power System of Central Asia. // Energy Systems Research, Vol. 2, No. 3, 2019. -pp.51–54.

Cable technology as a complex, multicomponent intelligent system providing high accuracy, reliability and safety

Vera Ivanova[1]*, *Victoria* Tsypkina[1]

[1] Tashkent State Technical University, 2 University str., Tashkent, Uzbekistan

Abstract. Cable production consists of complex technological processes focused on the production of cable products with high operational reliability and fire safety. The article analyzes the main problematic issues that arise in the development of intelligent systems focused on information processing with advanced management functions, especially in the conditions of emerging technological uncertainties associated with the use of various production technologies. The classification of technological uncertainties of cable production is given and recommendations are developed for the selection of complex control systems integrated into existing technological processes to ensure modern requirements for the operational reliability of cable products. The main results on the development of a control system as a complex, multicomponent intelligent system that ensures high accuracy, reliability and safety of the production process of manufacturing cable and wire products are presented.

Introduction

Modern cable production is a combination of various technological processes for the manufacture of cable and wire products (CWP). Ideally, the production and technological process operating at a cable company is a closed technological cycle, which includes all technological operations starting from the production of a cable billet (copper or aluminum wire rod), followed by the manufacture of a finished cable product and ending with the recycling of cable production waste. Thus, it is a multidimensional production process, in which specialized cable making machines are engaged, ensuring the maintenance of a given technology and technological modes. At the same time, each individual technology is determined by various technological parameters (linear velocity, drawing, tension, pulling force, angular velocity, etc.) on the correctness of the choice, which determines the efficiency of process control and ultimately the quality of the finished cable product. Numerically, the value of these parameters is chosen in accordance with the regulatory, technical and design documentation for the production cycle under consideration.

The final result of the technological process is a finished cable product (cable, wire, cord), the structural elements of which are made of various materials and raw materials (Table-1).

Each material and raw materials used (Table-1) have their own technical and technological parameters, presented numerically with an established permissible range of variation, according to the technology requirements and features of cable making machines, as well as the design of the cable product itself: material density; material mass; melting point; softening temperature; fluidity; viscosity, etc.

Table 1. Distribution of materials and raw materials in structural elements of a cable product

Name of the structural element	Name of raw materials and materials
1	2
Conductive core	electrical copper and aluminum
Wire screen	
Insulation of the conductive core	insulating polymer materials, cable paper, impregnating compounds, bitumen
Cable/wire protective cover: hose, sheath	
Cable braid	polymer tape, cotton yarn
Cable armour	galvanized steel wire and tape, lead, aluminum

Each cable making machine also has its own technical parameters, depending on the features of the technology and the operation of the electromechanical system (EMS): billet drawing speed, drawing, tension, pulling force, input and output diameters of the billet and the finished cable product, intermediate diameters and parameters of the cable semi-finished product, design parameters of the cable product, etc.

Based on the above, it can be concluded that the process of the CWP manufacturing is a complex production system consisting of many interrelated technological operations, which ensures high quality and operational reliability of the cable product produced. The goal set by the production service of the cable company is achieved by a well-coordinated and proven technology, as well as the coordinated work of all cable manufacturing equipment involved in the process (main and auxiliary), including ensuring strict convergence of transient (interoperable) technological parameters within the technological chain: "the beginning of the operation – the end of the operation" (Fig.1) with an agreed range of regulation of

* Corresponding author: c-victory@yandex.com

https://doi.org/10.1051/e3sconf/202346101039

production indicators (speed and time of manufacture, % of output, volume of production, design parameters of the cable product). At the same time, it should be noted that during the production process there is a high probability of various (possible) minor deviations of technical and theological parameters from the nominal (established within the norm) values caused by the technical condition of both cable making machines, their service life, and violation of the manufacturing technology of cable products. These include: random external disturbances and impacts; interference of various nature; noise induced by EMS of technological equipment and instrumentation; signal disturbances (emergency modes of equipment operation); disturbances formed during abnormal situations of the technological process (conditions of extreme operation / modes exceeding the maximum and minimum values).

Ensuring the reliability of the technological process is a multi-component task, the solution of which is aimed not only at achieving high quality of the finished cable product, but also minimizing cable production waste, unplanned downtime of equipment, which is inextricably linked with the reliability, durability, maintainability and preservation of cable making machines during their operation, which does not always go according to the planned scenario. An adequate assessment of the operation of cable manufacturing equipment involved in the technological route (chain) of manufacturing a certain nomenclature of cable or wire will reduce or even eliminate the costly mechanism for carrying out restoration and repair work during the execution of the production order and prevent the failure of its production deadlines. This is ensured by maintaining the necessary level of reliability not only of performing a specific technological operation, but also ensuring the trouble-free operation of all coupled cable making machines involved in the manufacturing technology of the finished cable product.

The specifics of the work of cable production is based on the sequential manufacture of cable semi-finished products by cable making machines involved in the manufacturing technology, which work within the technological chain, according to the manufacturing route, in strict accordance with the sequence of technological operations involved. It should be noted that no cable company has the ability to fulfill only one (limited) order on cable making machines sequentially assigned to the technology. As a rule, there is a distribution of the production order (production of a certain volume and grade of the CWP) for the development on a variety of machines with sequential transfer of the volume of the produced, unfinished semi-finished product to another technology (Fig.1). For example, the technological operation - drawing, and consequently drawing machines, produce a drawn wire, which is the first stage (or the first structural element) in the production of any cable product, it is immediately supplied for insulation, stranding into a strand or for the manufacture of a conductive multi-wire core, etc. Thus, the failure or unplanned shutdown of at least one element of the system (cable manufacturing equipment) disables the entire production chain under consideration or, even worse, makes it inoperable (Fig.1), which leads to the disruption of the established production task.

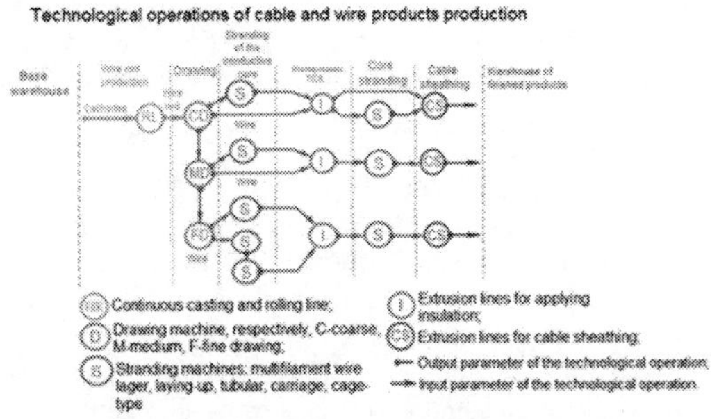

Fig. 1. The scheme of production routes of the technological process of manufacturing cable and wire products

To analyze the reliability state of the research object (technological process), a method has been applied based on the study of an object having a logical scheme of the dependence of states and their interactions described by a mathematical model based on the characteristics of its reliability, including input and output parameters of raw materials and materials, cable billet, finished cable product, as well as internal parameters of technological equipment, EMS and disturbing effects.

Each technological unit – a cable making machine, has a complex automatic control system (ACS), with the main task - to ensure the smooth functioning of all its systems and subsystems, as well as working units, their EMS within the established technology modes, subject to the action and occurrence of various kinds of uncertainties. However, this control system (CS) is an integral part of a single technological machine, which we can consider as a complex, multidimensional object of research. The synchronization of the work of all cable manufacturing equipment assigned to a certain technology shall ensure the speed, reliability and consistency of the work of all production units, which in fact is an unsolvable production task, since centralized control of the manufacturing technology of the CWP is coordinated by the production service, where there is a human factor existing in production management, distribution of the order volume across production workshops, with a high probability of delay or the choice of an erroneous action in making a decision [2, 3]. In this regard, the creation of a unified process control system (UPCS) for manufacture of the CWP will centrally ensure the execution of the production cycle, which includes all the main and auxiliary cable manufacturing equipment involved in the technological chain, including transitional (interoperation) processes, which confirms the relevance of the task we are solving through modern management theory [2, 4, 5]. At the same time, the UPCS shall be built as an intelligent control system (ICS) that ensures strict maintenance of the accuracy of the output design parameters of the finished CWP, all technological modes and deadlines, in various production conditions, taking into account the conditions of uncertainty, reliability and speed to various dynamic characteristics that arise when performing the technology [6, 7].

The solution of this issue begins with the design of the UPCS and is complicated by external and internal factors that operate under conditions of various definite, uncertain and inaccurate knowledge of the dynamic properties of the technological system we are considering:

— input and output parameters of the control system, both for a single cable making machine and for the entire technological operation;

— monitoring of input and output data and disturbances on the functioning of the control system of complex technological objects, their uncertainty, non-linearity, a small percentage of predictability.

Taking into account the wide variety of cable technologies connected into a single production process aimed at the manufacture of high-quality CWP, the variety of their operating modes and operating conditions, the emerging uncertainties in technology during the operation of cable units, the presence of uncontrolled external disturbances caused by changes in the properties of dynamic systems (DS), the subject of our research are models of complex multidimensional nonlinear DS optimization and control, which will allow to achieve the necessary level of reliability and stability by applying adaptive laws of the DS operation control.

Table 2. Top 15 digital technologies in industry in 2020 [1]

Technologies	Significance Index
1	2
Industrial robots	1,00
Artificial intelligence	0,86
Machine learning	0,68
Digital prototyping	0,56
Sensorics	0,42
Wireless connection	0,30
Blockchain	0,21
Big Data	0,20
Virtual and additional reality	0,12
Product as a service	0,09
Computer vision	0,03
Smart contact	0,03
Industrial Internet of Things	0,03
Digital twin	0,02
Smart factories	0,01

Many world scientific and practical laboratories are working on the problem of building complex, multicomponent intelligent systems that ensure high accuracy, reliability and safety of technology implementation, and the results obtained have been widely implemented (Table - 2). So currently, according to the analysis conducted by *"Marketsand Markets"*, the global Smart Manufacturing Market will grow from $ 214.7 billion in 2020 to $384.8 billion in 2025, with an average annual increase of 12.4%. The conducted research *"IT in Industry"* (CNews Analytics) showed that the overall value of digitalization for the industrial sector of Russia will increase by 14 times by 2030 [1].

Solution methods

The technological process of the CWP production, in fact, is a complex, rapidly changing DS, consisting of the equipment involved in the technology and its control system. The development of an intelligent process control system, as a technical system, cannot be represented as a solution of differential equations, and therefore have the form of a system described by a traditional mathematical modeling apparatus, since it is not able to cover the change in all technological and technical parameters of the system under consideration: raw materials, materials, cable manufacturing equipment (main and auxiliary), technological equipment and technological tools, as well as cable billets, semi-finished (unfinished) and finished cable products.

It should be noted that traditional neural network (NN) models describe such systems as a "black box", where the values of input and output parameters, including disturbing effects, are the resulting values of the experimental study of the technology. All this negatively affects the results of modeling, because it is a significant limitation in achieving the required level of complexity of the simulated multicomponent system. In the case we are considering, it is advisable to apply a hybrid neural network mathematical model, which includes differential equations that jointly describe the dynamics of the technology and the operation of all the cable manufacturing equipment involved.

Such an integrated approach will allow combining all the positive aspects of traditional models and compensating for the existing negative factors: ensuring adaptability, increasing the threshold of complexity of the task under consideration, as well as identifying the characteristics of a dynamically changing, multicomponent system.

The primary stage of the development of the automated control system for the CWP manufacture is based on working out the levels of interaction of all models of the system with the external environment, as well as disturbances of various nature and interference.

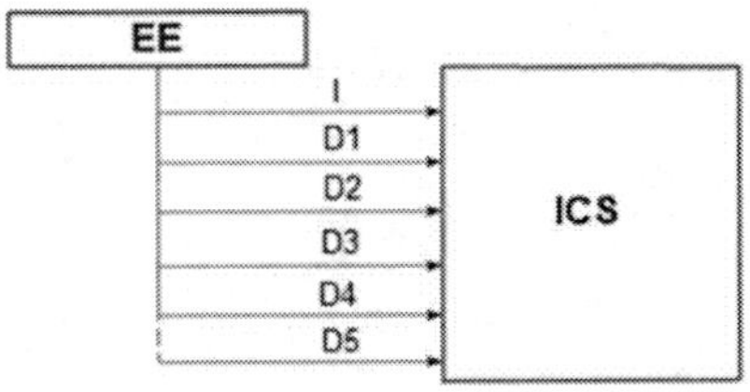

Fig.2. The primary structure of the ICS of the technological process of CWP manufacturing, built on information indicators: EE – the external environment; ICS – an intelligent control system; I – interference of the information and measuring systems and control and measuring system of cable making machines; D1 – disturbances, by technology and technological nature; D2 – disturbances caused by the operation electromechanical system and control system of cable making machines; D3 – disturbances caused by a deviation in the design of the cable product; D4 – disturbances caused by pre-emergency situations in the operation of cable making machines; D5 -other disturbances.

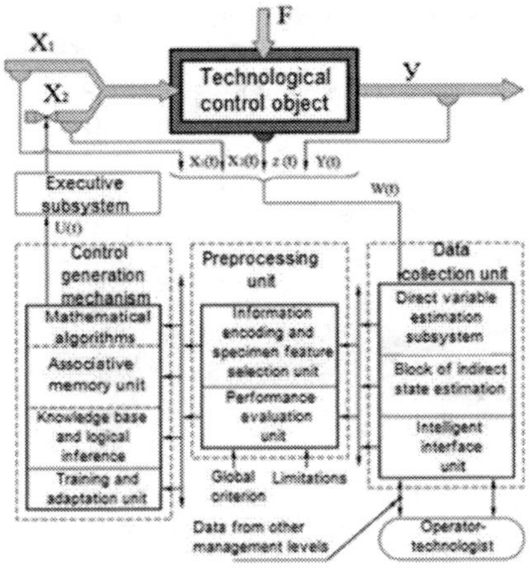

Fig.3. ICS structure: X1, X2 – input signals determined by the parameters and properties of raw materials and materials, according to the design of the cable product; F - disturbances and interference; Y - output signal.

However, in addition to possible disturbances and interference that act on the ICS, there are control actions (CA) formed by analyzing the results of system (regulatory) data and knowledge of the operation of cable making machines, technological modes and design parameters of cable products embedded in the regulatory, technical and design documentation accompanying a single technology (Fig.2). All CAs are "non-mathematical" data, but complex knowledge obtained in the process of manufacturing a cable product of a certain brand. When entering them into the ICS database (logical output) for subsequent associative memorization, the signals are initially encoded and decoded. At the same time, an algorithm of numerical, multi-valued (fuzzy) or symbolic form is being developed for the subsequent compilation of a mathematical model [2, 4, 5, 8].

The task of the mathematical model is to process the available information by implementing the main results of the corresponding numerical calculations. The working variants of the incoming information to the ICS, which are subject to system analysis, are divided into elements: accumulated knowledge; generalized and analyzed data (Fig.3). The developed control algorithm is focused on estimating dynamic parameters based on the formed production and technological basic values [8, 9, 10, 11,12,13, 14, 15].

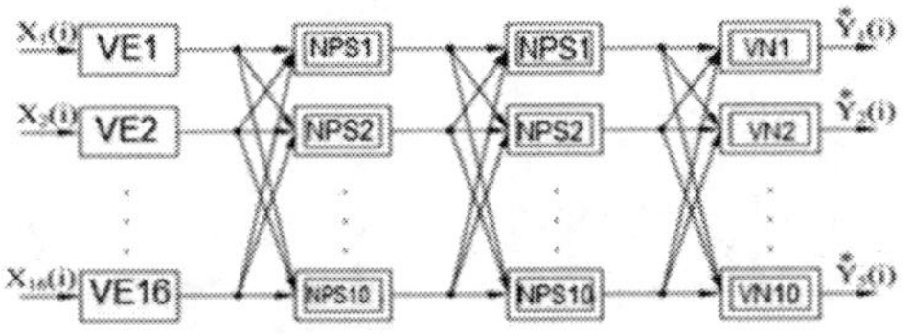

Fig. 4. Functional structure of the ICS: VE1, …. VE16 – input elements (normalizing and distribution converters); NPS1, ... NPS10 - neurons of the intermediate layer; VN1, ... VN5 – output neurons

The developed functional structure of the cable production control system is based on the multilayer structure of an artificial neural network (Fig. 4).

The initial stage of building a neural network model (NNM) of the technological process of manufacturing cable products is based on the development of a mathematical model that includes data from the analysis of the physical parameters of both technology and technological equipment, and is a system of differential equations that is solved by effective analytical and numerical methods.

It is a very difficult task to describe the generalized structure of the CWP production technology, because it consists of many working mechanisms and shall include specific parameters of the working systems of cable manufacturing equipment, including the initial values of variables, and difficult unobservable quantities that are derivatives of the joint operation of equipment, tooling and processed cable billet, including raw materials and materials involved in the technology.

One of the ways to solve the problem of creating a dynamic model of a neural network is its joint use with "classical" differential equations – objective functions (Fig. 5).

The construction of the NN reflecting the production process of CWP manufacturing includes neurons and their interconnection, the work of which, including the work of interneuronal connections, is set by means of a specially developed program. The functioning of the NN begins with a formal neuron, in which the initial moment of time (current time) along the dendrites receives a processed input signal coming from the structural units of the lower level and which are output signals for them. Next, the input multiplies the incoming signal to the neuron by the corresponding weighting factor and sums it with other signals, followed by multiplying it by the weighting factor of the available inputs, where the output signal is formed, which can be represented as follows:

$$S = \sum_{i=1}^{n} x_i w_i , \quad (1)$$

where, n - the number of neuron inputs, x_i - the value of the «I» neuron input; w_i - the weight of the «I» signal.

The mathematical model of a complex, multicomponent intelligent system of the technological process of cable product production has the form:

$$X(t) = \frac{X(t) - X^{min}}{X^{max} - X^{min}} \quad (2)$$

$$Y = \left[1 + exp\left(\sum_{m=0}^{n-1} W_{n-1}(t)\left[1 + exp\left(\sum_{m=0}^{n} W_n(t) \ldots \left[1 + exp\left[\sum_{m=0}^{1} W_1(t)\right]\right]\right)\right]\right)\right] \quad (3)$$

where, X_1, X_2 - input signals determined by the parameters and properties of raw materials and materials, according to the design of the cable product; W - a generalized signal of the system, including elements: both technological parameters and operating parameters of the cable manufacturing equipment involved in the technological process, including its technical condition and moral wear; Y - an output signal showing the quantitative indicators of the finished cable product.

Research results

The conducted complex of research works allowed to obtain the NNM DS of a complex, multicomponent, nonlinear, intelligent structure of the technological process of cable product production (Fig. 5) in real time, having the form of a connected (a) and optimized structure (b).

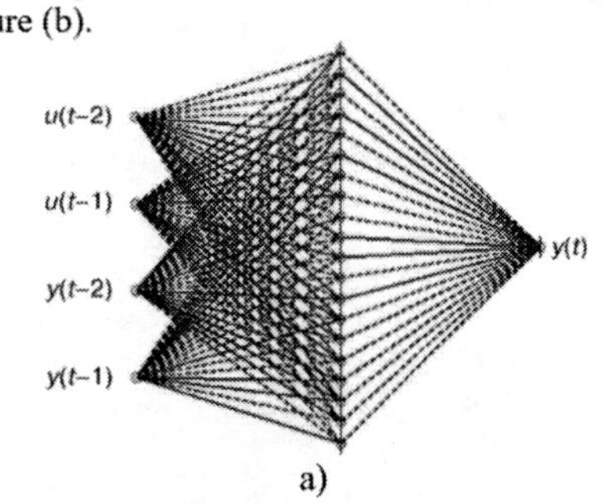

a)

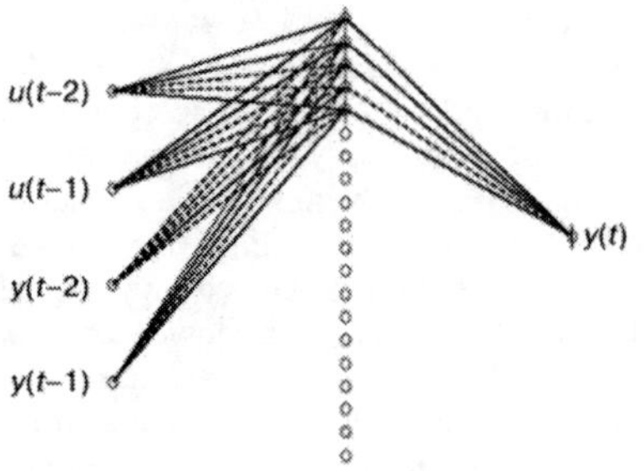

b)

Fig.5. Structural model of a neural network dynamic system: a) a neural network with a fully connected structure; b) a neural network with an optimized structure

The work of the NN is carried out by processing the corresponding input signals coming to some neurons and converting them into an output response signal determined by all the weighting coefficients of the corresponding technological process. The NN training process is based on the search with subsequent optimization of the values of weighting coefficients, the use of which minimizes the control error.

Fig. 6 shows the results of mathematical modeling for a neural network DS with a fully connected (a) and optimized structure (b) for a complex, multicomponent intelligent system of the technological process of cable product production. The given characteristics show that with random values of the flow of the technological process of cable product production, the ICS can work as unstable (Fig.6, a), and in the case of optimization of technologies operating within the production process, it behaves as a stable system (Fig.6, b). As an evaluation criterion, a technological parameter has been selected - tension, which is relevant for all types of cable manufacturing equipment and is defined as the force acting inside the object (cable product) that arose when the cable billet descended from the pay-off, passed through the working body of the cable making machine and received on the receiver as a finished (finished for this technology) product.

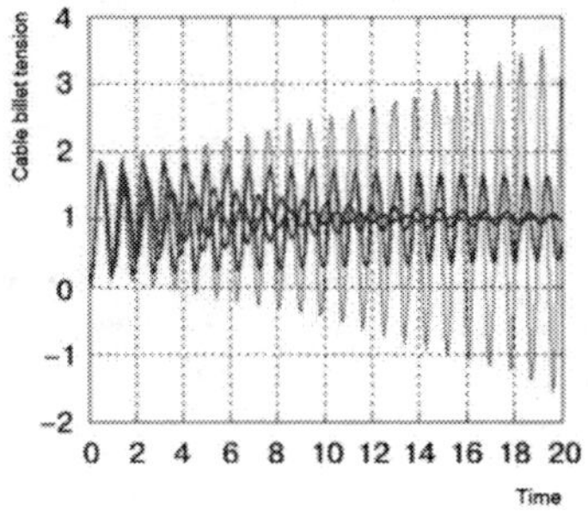

a)

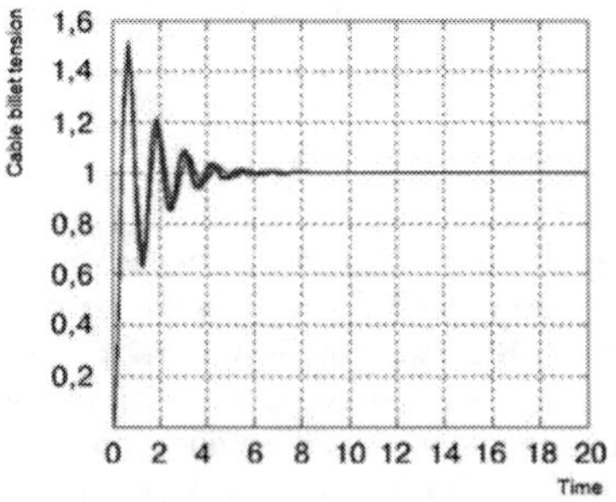

b)

Fig.6. Results of mathematical modeling for NN DS with a fully connected structure (a) – not steady–state mode and with an optimized structure (b) - steady-state mode.

Conclusion

Based on the above, it can be concluded that the proposed intellectualized system is adapted to the technological process of the CWP production. It is able to accumulate the information received, generalize and analyze the knowledge gained, solve complex problems of evaluating dynamic processes by using combined simulation models based on a combination of collected knowledge from the relevant subject area with practical experience and methods of teaching NN: technical processes, cable making machines with different operating principles, the design of cable products according to mark positions and mass of materials, from which structural elements of the cable product are manufactured.

The use of a hybrid NN model for cable production ensured the fulfillment of the research task by solving the problem of dimensionality of quantities and ensuring the specified accuracy of modeling by adapting the technology involved in real time. All this makes it possible to effectively evaluate cable production as a nonlinear controlled DS, capable of repeatedly increasing the accuracy of modeling, ensuring the reliability and safety of multicomponent CWP manufacturing technology.

Thus, there is a prospect of creating technical systems focused on cable technology, having expanded technical, technological and production capabilities.

References

1. Artificial intelligence technologies for industrial production automation: overview and prospects © Copyright 2015 - 2050 RPC INTEGRAL Ltd., 2022. https://integral-russia.ru/2022/08/26/tehnologii-iskusstvennogo-intellekta-dlya-avtomatizatsii-promyshlennogo-proizvodstva-obzor-i-perspektivy

2. Pupkov K.A. Intelligent systems / K.A. Pupkov, V.G. Konkov - M.: Publishing House of Bauman Moscow State Technical University, 2003. – p. 348.

3. Methods of classical and modern theory of automatic control: Textbook in 3 volumes. V.3: Methods of the modern theory of automatic control / Edited by N.D. Egupov - M.: Publishing House of the Bauman Moscow State Technical University, 2000. – p. 748.

4. Hangos K.M. Intelligent control systems: an introduction with examples / K.M. Hangos, R. Lakner, M. Gerzson - New York: Kluwer academic publishers, 2001. – p. 301.

5. Gaskarov D.V. Intelligent information systems / D.V. Gaskarov - M: Higher School, 2003. – p. 431.

6. Gavrilova T.A. Knowledge bases of intelligent systems / T.A. Gavrilova, V.F. Khoroshevsky - St. Petersburg: Pieter, 2001. – p. 384.

7. Bashmakov A.I. Intelligent information technologies: Textbook / A.I. Bashmakov, I.A. Bashmakov - M.: Publishing House of the Bauman Moscow State Technical University, 2005. – p. 304.

8. Modeling of a resource-saving method of drawing Victoria Tsypkina, Veronica Ivanova E3S Web Conf. 139 01073 (2019) DOI: 10.1051/e3sconf/201913901073.

9. Improving the reliability of cable lines operation in hot climates Daniyar Bakhtiyarovich Madrakhimov, Vera Pavlovna Ivanova, Victoria Vyacheslavovna Tsypkina E3S Web of Conf. 216 01151 (2020) DOI: 10.1051/e3sconf/202021601151.

10. Ivanova, Vera & Tsypkina, Victoria. (2020). Improving the reliability of power supply to active consumers by improving the technology for manufacturing cable product. E3S Web of Conferences. 216. 01152. 10.1051/e3sconf/202021601152.

11. Ivanova V.P., Tsypkina V.V., Turabekov A.U., Bokijonov U.A., Atajiyev Sh.M. Increasing the reliability of cable and wire products by improving the technology of applying insulation // Universum: Technical Sciences: Electronic Scientific Journal. Ivanova V.P. [et al.]. 2020. 11(80). URL: https://7universum.com/ru/tech/archive/item/10970 (reference date: 02.10.2023).

12. Development of generalized requirements for automated electric drive of cable equipment V.V. Tsypkina, V.P. Ivanova, D.N. Isamukhamedov, A.U. Turabekov, R.F. Atamukhamedova E3S Web of Conf. 401 02045 (2023) DOI: 10.1051/e3sconf/202340102045.

13. Improvement of the multifilament wire lager for cable production Olimjon Toirov, Vera Ivanova, Viktoriya Tsypkina, Dilnoza Jumaeva, Dilnoza Abdullaeva E3S Web Conf. 411 01041 (2023) DOI: 10.1051/e3sconf/202341101041.

14. Improving the reliability of power supply to active consumers by improving the technology for manufacturing cable product Vera Pavlovna Ivanova, Victoria Vyacheslavovna Tsypkina E3S Web of Conf. 216 01152 (2020) DOI: 10.1051/e3sconf/202021601152.

15. Cable conductor of cabling and wiring products based on composite materials for transport systems V. V. Tsypkina, V. P. Ivanova, K. K. Jurayeva E3S Web of Conf. 401 03036 (2023) DOI: 10.1051/e3sconf/202340103036.

https://doi.org/10.1051/e3sconf/202346101040

Algorithmic foundations of automated monitoring of commercial and technical power losses in distribution networks

T.T. Omorov[1*], *B.K.* Takyrbashev[1], *T.Dzh.* Koibagarov[1], *R.Ch.* Osmonova[1], *Zh.S.* Imanakunova[2], *T.O.* Zhanybaev[2], *A.T.* Asiev[2]

[1]National academy of sciences of the Kyrgyz Republic, Institute of Mechanical and Automation Engineering, Bishkek
[2] Kyrgyz State Technical University named after I. Razzakov

Abstract. It is known that significant power losses in power distribution networks (PDNs) caused by unauthorized power withdrawals (UPWs) lead to a decrease in their technical and economic performance. There are no digital technologies designed for separate assessment and monitoring of technical and commercial power losses in modern automatic system for commercial metering of power consumption (ASCMPC), which are currently widely used for automation and computerization of processes in distribution networks. The article proposes a new algorithm for solving the above problem on the basis of data obtained from electricity meters included in the ASCMPC structure. In order to ensure the solvability of the identification problem, the concept of a virtual network model characterizing the desired state of the PDN in the absence of UPW is introduced. On its basis, algebraic equations are obtained, the solution of which allows to identify technical and commercial power losses in three-phase networks. The obtained results are oriented on further improvement of modern ASCMPC and increase of their efficiency and reliability. **Key words:** distribution network, power losses, identification and monitoring algorithm.

Introduction

Automation of information processes in power distribution networks (PDNs) is currently carried out on the basis of implementation of automatic system for commercial metering of power consumption (ASCMPC) [1], which are elements of Smart Grid [2] technology. As it is known, ASCMPC hardware-software complexes belong to the class of information-measuring systems, the main function of which is commercial accounting of electric power in PDNs. The analysis of the functional structure of existing ASCMPCs shows that they do not solve the tasks of separate identification and operational monitoring of technical and commercial power losses in the network [3], as well as do not perform optimization tasks [4-7], which reduces their efficiency and technical and economic performance of distribution companies. In particular, this is due to the fact that to date, adequate and constructive methods for solving the above problems have not been sufficiently developed, taking into account such factors as the presence of unauthorized power withdrawals (UPW) in the network [8 -10], and the asymmetry of PDN operation modes [11-14]. The problem also lies in the fact that most PDNs belong to the class of large dynamic systems, which have a complex structure and function under conditions of incomplete information about their states and parameters. In [15, 16] some approaches to identification of power losses in 0.4 kV distribution networks are proposed. The article considers algorithmic problems related to the development of these approaches.

* Corresponding author: omorovtt@mail.ru

Task definition

We consider a four-conductor 0.4 kV PDN, the conditional design scheme of which is shown in Fig. 1, where k, ν - index variables, representing phase numbers, respectively A, B, C $(k = \overline{1,3})$ and the network's electrical circuits $(\nu = \overline{1,n})$; $\widetilde{E_k}$ – instantaneous electromotive force of power supply source for k – phase; $\tilde{I}'_{k\nu}$, $\tilde{U}'_{k\nu}$, $\dot{Z}_{k\nu}$ – sinusoidal instantaneous current, voltage and complex load resistance of the customer with coordinate (k, ν) correspondingly; $\tilde{\imath}'_{k\nu}, \dot{z}_{k\nu}$ – sinusoidal instantaneous current and complex resistance of ν - inter-customer site of k phase; $\tilde{J}'_\nu$, $\dot{z}_\nu$ – instantaneous current and complex resistance of ν – site of neutral conductor; $\tilde{I}'_k$, $\tilde{U}'_k$ – instantaneous currents and voltages at the inputs of the corresponding phases.

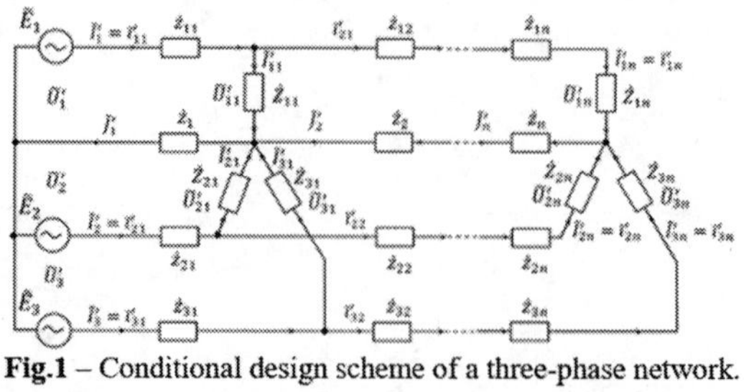

Fig.1 – Conditional design scheme of a three-phase network.

Let us assume that the following conditions are fulfilled:

1. The distribution network operates in asymmetrical mode;

2. There are unauthorized power withdrawals (UPW) in the network;

3. Cross-sections of phase and neutral conductors are assumed to be the same, i.e. $\dot{z}_{kv} = \dot{z}_v$;

4. At discrete time moment $\underline{t = t_\xi}$ synchronous measurements of the effective values of currents I'_k, I'_{kv} and voltages U'_k, U'_{kv} at the inputs of phases and loads of the network, respectively, as well as power factors $c_k = \cos\varphi'_k$ and $c_{kv} = \cos\varphi'_{kv}$ between them are performed using the ASCMPC electricity meters. The received information is recorded in the ASCMPC database and vectors of input phase currents $I' = [I'_1, I'_2, I'_3]$ are formed on their basis. According to the above data, the active P_κ, $P_{\nu\kappa}$ and reactive Q_κ, $Q_{\kappa v}$ powers consumed by the phases and customers of the network, respectively, are determined using known formulas.

At the moment $(t = t_\xi)$ of synchronous measurements of electricity meter readings included in the ASCMPC, the distribution network may be in operative (normal, desired) (C^o) or disturbed (C') state. There are no uncontrolled power losses in the network in the first case (state C^o). There are uncontrolled power losses in at least one of its phases in the disturbed state of the network (C'). In this case, the following balance ratios for capacities are valid:

$$\dot{S}_\kappa(\xi) = \dot{S}^a_\kappa(\xi) + \dot{S}^T_\kappa(\xi) + \Delta\dot{S}_\kappa(\xi), \quad \kappa = \overline{1,3}, \quad (1)$$

where $\dot{S}_\kappa(\xi)$ – total capacity, consumed by κ phase at the time moment $t = t_\xi$; $\dot{S}^a_\kappa(\xi)$ – total complex power consumed by customers of κ phase; $\dot{S}^T_\kappa(\xi)$ – technical power losses in κ phase; $\Delta\dot{S}_\kappa$- uncontrolled power losses in κ phase of the network, caused by the presence of UPWs in the PDN.

Thus powers $\dot{S}_\kappa(\xi)$ and $\dot{S}^a_\kappa(\xi)$ are known values, as in the existing ASCMPC it is possible to calculate their values according to the data of active and reactive powers received from the balance meter and customer meters using known formulas:

$$\dot{S}_\kappa(\xi) = P_\kappa(\xi) + jQ_\kappa(\xi), \quad (2)$$

$$\dot{S}^a_\kappa(\xi) = \sum_{\nu=1}^{n} \dot{S}_{\kappa v}, \quad k = \overline{1,3}, \quad (3)$$

where $\dot{S}_{kv} = P_{kv} + jQ_{kv}$; $j = \sqrt{-1}$ – imaginary value.

It shall be noted that technical power losses $\dot{S}^T_k(\xi)$ and commercial losses $\Delta\dot{S}_\kappa(\xi)$ in standard ASCMPC are not defined separately. Only the total $\dot{S}^\Sigma_k(\xi)$ power losses are estimated in existing ASCMPCs based on measurement data, i.e.

$$\dot{S}^\Sigma_k(\xi) = \dot{S}^T_k(\xi) + \Delta\dot{S}_\kappa(\xi) = \dot{S}_k(\xi) - \dot{S}^a_k(\xi), k = \overline{1,3}.$$

The task is to identify commercial $\Delta\dot{S}_\kappa(\xi)$ and technical $\dot{S}^T_k(\xi)$ power losses in real time based on measurement data obtained from ASCMPC electricity meters.

The solution of the generated identification problem includes the following main stages:

1. Building a virtual network (VN) model.
2. Identification of the current condition of the real network.
3. Formation of the general scheme of identification of power losses in PDN.
4. Estimation of input desired phase currents of the virtual network.
5. Estimation of currents of unauthorized customers.
6. Construction of the algorithm for identification of power losses in the network.

Building a virtual network (VN) model

It can be noted that the identification problem under consideration belongs to the class of those problems in which the initial data and conditions for its solution are not fully specified, i.e. the problem should be solved under conditions of insufficient information about the state of the real network. This uncertainty is due to the lack of data on unauthorized consumers, such as their coordinates and load parameters. Under these conditions, in order to solve the identification problem formulated above, we further introduce the concept of virtual network (VN) model, which describes the desired state of the real PDN under certain conditions. The analysis shows that for this purpose it is reasonable to introduce two types of virtual networks (Fig. 2):

1. Virtual network of the first type (BC_1), which is characterized by the fact that there are no uncontrolled power losses $\Delta\dot{S}_k$, and the inter-customer resistances $\dot{z}_{kv}$ have zero values (Fig. 2a):

$$\Delta\dot{S}_\kappa=0, \ k = \overline{1,3}, \ \dot{z}_v=0, \ \ v = \overline{1,n}. \quad (4)$$

2. Second virtual network (VN_2) - there are UPWs, i.e., the following conditions are fulfilled (Fig. 2b):

$$\Delta\dot{S}_\kappa \neq 0, k = \overline{1,3}, \quad (5)$$

As well as ratios (4). Here $\dot{E}_k$ – complex electromotive force of network supply source; $\dot{I}_{kv}$, $\dot{Z}_{kv}$ – complex current and resistance of the virtual customer load, having the coordinate (k, v); $\dot{I}_{k0}$, $\dot{I}_k$- complex current at the inputs of k –phase BC_1 and BC_2 respectively; $\Delta\dot{I}_k$, $\Delta\dot{Z}_k$ – complex current and resistance of load of unauthorized customer connected to the corresponding phase of the PDN. Next, without loss of generality of the task and for the sake of brevity, let us consider the k phase of the virtual network to which n customers are connected.

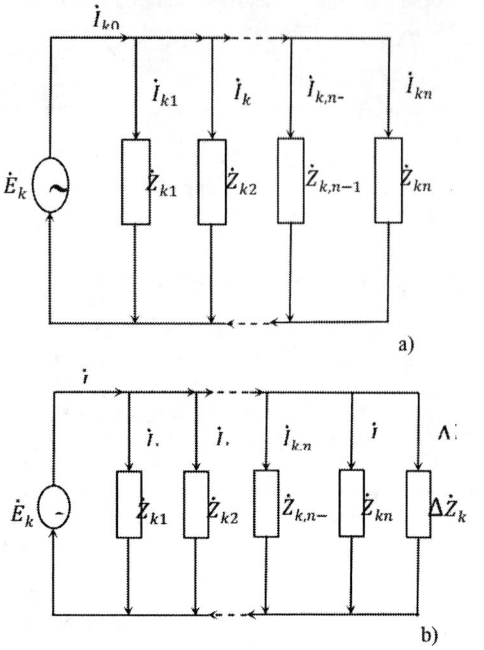

Fig. 2 – Virtual networks diagrams.

Input phase currents $\dot{I}_{k0}$ BC_1 characterize the normal operation mode of the considered real distribution network in the absence of uncontrolled currents $\Delta\dot{I}_k$. Therefore, hereafter we will call them the desired phase currents. A comparative analysis of the structures of the original PDN (Fig. 1) and virtual networks (Fig. 2) under conditions (4) and (5) shows that the following ratios for the complex currents $\dot{I}_k$, $\dot{I}_{k0}$ and $\Delta\dot{I}_k$ are valid:

$$\dot{I}_{k0} = \sum_{\nu=1}^{n} \dot{I}'_{\kappa\nu}, \quad k = \overline{1,3}, \tag{6}$$

$$\dot{I}_k = \dot{I}_{k0} + \Delta\dot{I}_k, \; k = \overline{1,3}, \tag{7}$$

where $\dot{I}_k = \dot{I}'_k$.

Current values $\dot{I}_{k\nu}$ and voltage values $\dot{U}_{k\nu}$, describing the state of the virtual networks differ from their corresponding values characterizing the state of the initial - real PDN (Fig.1.), i.e $\dot{I}_{k\nu} \neq \dot{I}'_{k\nu}$, $\dot{U}_{k\nu} \neq \dot{U}'_{k\nu}$. At the same time, the values of resistances $\dot{Z}_{k\nu}$ of the loads of customers of real and virtual networks have the same values. Further, we will represent the above network variables in the following complex form [17]:

$$\dot{I}_{k\nu} = I_{k\nu}\, e^{j(\beta_k + \alpha_{k\nu})}, \tag{8}$$

$$\dot{Z}_{k\nu} = Z_{k\nu}\, e^{j\varphi_{k\nu}}, \; k = \overline{1,3}, \nu = \overline{1,n},$$

where $I_{k\nu}, Z_{k\nu}$ - modules of the corresponding complex values; $\varphi_{k\nu}$ - resistance argument $\dot{Z}_{k\nu}$; $\propto_{k\nu}$ - deviations of phase shifts from their basic values β_k, defined by formulas $\beta_k = \frac{2(k-1)\pi}{3}$.

Complex resistances $\dot{Z}_{k\nu}$ of loads of network customers are known values which are calculated on the basis of readings (data) of customer electricity meters according to the formulas:

$$Z_{k\nu} = \frac{U'_{k\nu}}{I'_{k\nu}}, \; \varphi_{k\nu} = arccos(c_{k\nu}), \tag{9}$$

Relevant conductivities $\dot{Y}_{k\nu}$ of loads of network customers:

$$\dot{Y}_{k\nu} = \frac{1}{\dot{Z}_{k\nu}} = Y_{k\nu} e^{-j\varphi_{k\nu}}, k = \overline{1,3}, \nu = \overline{1,n}, \tag{10}$$

where $Y_{k\nu} = \frac{1}{Z_{k\nu}}$. At the same time, the powers $\dot{S}_{k\nu}$, consumed by the network customers and included in expressions (3) are determined by the formulas:

$$\dot{S}_{k\nu} = (I'_{k\nu})^2 \dot{Z}_{k\nu}, k = \overline{1,3}, \nu = \overline{1,n}. \tag{11}$$

Identification of the current state of the real network

Under the conditions of the task under consideration, the current values of phase currents $I'_k = I'_k(t_0)$ at the input of the initial real distribution network (Fig. 1) at time $t = t_0$ are measured by the balance three-phase meter, which are contained in the ASCMPC database in the form of vector $I' = [I'_1, I'_2, I'_3]$. The complex currents of unauthorized consumers $\Delta\dot{I}_k$ can be represented as:

$$\Delta\dot{I}_k = \Delta I_k\, e^{j(\beta_k + \Delta\alpha_k)}, \; k = \overline{1,3}, \tag{12}$$

where ΔI_k, $\Delta\alpha_k$ – modules and phases of complex currents $\Delta\dot{I}_k$ respectively. Estimates of the effective values of ΔI_k will be identified later.

The analysis shows that it is reasonable to use the following conditions to identify the current state of the PDN:

$$\Delta I_k \le \Delta I, \; k = \overline{1,3}, \tag{13}$$

where ΔI – maximum permissible error of current measurement in ASCMPC.

Obviously, if at least one of the ratios (13) is not fulfilled, there are UPW in the network, and their fulfillment means that the PDN operates in the normal mode. Thus, ratios (13) can be used as a criterion condition for identifying the current state of the PDN.

Formation of a general scheme for identification of capacity losses

We will consider the virtual network diagrams shown in Fig. 2. Let us denote by $\dot{S}'_k$ the powers consumed by the loads of the k-phase of BC2. The analysis shows that there are following balance ratios between the powers $\dot{S}'_k, \dot{S}^a_k$ and $\Delta\dot{S}_k$:

$$\dot{S}^a_k + \Delta\dot{S}_k = \dot{S}'_k, \; k = \overline{1,3}, \tag{14}$$

Where the total power $\dot{S}^a_k$ consumed by the loads of the k-phase is determined by formula (3). Each of these components, when conditions (4) and (5) are fulfilled, is determined by the following functional dependencies:

$$\dot{S}^a_k = I_{k0}{}^2 \dot{Z}_{k0}, \; \dot{S}'_k = I_k{}^2 \dot{Z}_k, \tag{15}$$

$$\Delta\dot{S}_k = \Delta I_k{}^2 \Delta\dot{Z}_k, \; k = \overline{1,3},$$

where $\dot{Z}_{k0}, \dot{Z}_k$ - total resistances of virtual phases shown on Fig 2a and Fig. 2b respectively; $\Delta\dot{Z}_k - load\ resistance$ of unauthorized consumer connected to the phase of number k. Estimates of unknown values included in ratios (15) will be defined later.

As can be seen from Figure 2, the corresponding conductivities are determined by the following formulas:

$$\dot{Y}_{k0} = \sum_{v=1}^{n} \dot{Y}_{kv}, \dot{Y}_k = \dot{Y}_{k0} + \Delta\dot{Y}_k, \quad (16)$$

$$\dot{Y}_{kv} = \frac{1}{\dot{Z}_{kv}}, \quad \Delta\dot{Y}_k = \frac{1}{\Delta\dot{Z}_k}.$$

Now, taking into account (15) and (16), ratios (14) can be represented in the form:

$$I_{k0}^2 \dot{Z}_{k0} + \frac{\Delta I_k^2}{\Delta \dot{Y}_k} = \frac{I_k^2}{\dot{Y}_{k0} + \Delta \dot{Y}_k}, \quad k = \overline{1,3}.$$

After simple transformations the last equations will be written in the form:

$$a_k(\Delta \dot{Y}_k)^2 + b_k\Delta\dot{Y}_k + c_k = 0, \ k = \overline{1,3}, \quad (17)$$

Where coefficients a_k, b_k, c_k are determined by the formulas:

$$a_k = \dot{S}_k^a,$$

$$b_k = \dot{S}_k^a \dot{Y}_k^0 + \Delta {I_k}^2 - {I_k}^2,$$

$$c_k = \Delta {I_k}^2 \dot{Y}_k^0.$$

To solve the quadratic equations (17), we can use the well-known Veta formula:

$$\Delta\dot{Y}_k = \frac{-b_k \pm \sqrt{{b_k}^2 - 4a_k c_k}}{2a_k}, \quad k = \overline{1,3}. \quad (18)$$

As a result, the required load resistances $\Delta\dot{Z}_k$ of unauthorized consumers will be determined:

$$\Delta\dot{Z}_k = \frac{1}{\Delta\dot{Y}_k}, \ k = \overline{1,3}. \quad (19)$$

Then the required values of power losses $\Delta\dot{S}_k$ caused by unauthorized withdrawals in the phases of the real network, taking into account (8), are calculated by the formulas:

$$\Delta\dot{S}_k = \frac{\Delta {I_k}^2}{\Delta\dot{Y}_k}, \ k = \overline{1,3}, \quad (20)$$

where the effective values of currents of unauthorized consumers ΔI_k will be defined further.

Now, based on the balance ratios (1), we determine the values of technical losses $\dot{S}_k^T$ of capacities in the phases of the real network:

$$\dot{S}_\kappa^T(\xi) = \dot{S}_\kappa(\xi) - \dot{S}_\kappa^a(\xi) - \Delta\dot{S}_k, \ k = \overline{1,3}, \quad (21)$$

Where $\dot{S}_\kappa(\xi)$ – complex capacity, consumed by k phase of the real network at the time moment $t = t_\xi$, which is calculated from the data of the balance three-phase electricity meter according to formulas (2).

Let us assume that data collection from the ASCMPC meters is performed at discrete moments of time $t = t_\xi, t = t_{\xi+1}, t = t_{\xi+2}, \ldots, t = t_{\xi+m}$. Then the use of the above computational procedure on the basis of these data makes it possible to perform automated operational control and monitoring of power losses in PDN in real time.

Estimation of input desired phase currents of the virtual network

To calculate the effective values of the input desired currents I_{k0} BC[1], we write the first ratios of the system (15) in the form:

$$\dot{S}_k^a = \frac{{I_{k0}}^2}{\dot{Y}_{k0}}, \quad k = \overline{1,3}. \quad (22)$$

Now let us represent the complex quantities $\dot{S}_k^a$ and $\dot{Y}_{k0}$ in exponential form:

$$\dot{S}_k^a = S_k^a e^{j\varphi_k^a}, \quad (23)$$

$$\dot{Y}_{k0} = Y_{k0}\, e^{-j\varphi_k^a}, k = \overline{1,3},$$

where $S_k^a, Y_{k0}, \varphi_k^a$ – modules and phases of complex values $\dot{S}_k^a$ and $\dot{Y}_{k0}$ respectively.

Ratios (22) taking into account (23) have the form:

$$S_k^a e^{j\varphi_k^a} = \frac{{I_{k0}}^2}{Y_{k0}} e^{j\varphi_k^a}, k = \overline{1,3}. \quad (24)$$

Based on equations (24), we obtain the numerical values of the desired input desired phase currents I_{k0} of the virtual network, since S_k^a and Y_{k0} are known values:

$$I_{k0} = \sqrt{S_k^a Y_k^a} = \sqrt{\frac{S_k^a}{Z_k^a}}, k = \overline{1,3}. \quad (25)$$

Estimation of currents of unauthorized consumers

For this purpose, based on ratios (7), let us write expressions for the complex currents $\Delta\dot{I}_k$ of unauthorized consumers:

$$\Delta\dot{I}_k = \dot{I}_k - \dot{I}_{k0}, \ k = \overline{1,3}. \quad (26)$$

Preliminarily these currents are represented in exponential form

$$\dot{I}_k = I_k e^{j(\beta_k + \alpha_k)},$$

$$\dot{I}_{k0} = I_{k0} e^{j(\beta_k + \alpha_k^0)},$$

where $I_k, I_{k0}, \alpha_k, \alpha_k^0$ - effective values and phase shifts of related complex currents.

It can be shown that the parameters of the indicated currents satisfy the following equations:

$$\Delta {I_k}^2 - {I_k}^2 - {I_{k0}}^2 + 2I_k I_{k0}\cos\theta_k = 0,$$

$${I_k}^2 - {I_{k0}}^2 - \Delta {I_k}^2 - 2\Delta I_k I_{k0}\cos\eta_k = 0, \quad (27)$$

$$\Delta {I_{k0}}^2 - \Delta {I_k}^2 - {I_k}^2 + 2\Delta I_k I_k \cos(\lambda_k) = 0,$$

where $\eta_k, \theta_k, \lambda_k$ – phase shift differences defined by the formulas:

$$\eta_k = \alpha_k^0 - \Delta\alpha_k, \lambda_k = \alpha_k - \Delta\alpha_k, \theta_k = \lambda_k - \eta_k, k = \overline{1,3}.$$

The aggregate of ratios (27) is a system of algebraic equations that includes three equations with three unknown values ΔI_k, η_k and λ_k. The analysis of ratios (27) has shown that the required currents ΔI_k are determined on the basis of the solution of the following equations:

$$\Delta I_k^4 + e_k \Delta I_k^2 + r_k^2$$

$$+ q_k \sqrt{\Delta I_k^8 + m_k \Delta I_k^6 + d_k \Delta I_k^4 + s_k \Delta I_k^2 + r_k^4} = 0,$$

$$k = \overline{1,3}, \quad (28)$$

Where their coefficients are known values, which are determined by the formulas:

$$e_k = -2f_k, f_k = I_k^2 + I_{k0}^2,$$

$$r_k = I_k^2 - I_{k0}^2, m_k = -4f_k,$$

$$d_k = 16 I_k^2 I_{k0}^2 + 6r_k^2, s_k = -4f_k r_k^2.$$

The choice of the sign function q_{kv} is carried out as follows:

$$q_{k1} = \begin{cases} +1, if\ sign(\Delta I_k^4 + e_k \Delta I_k^2 + r_k^2) < 0, \\ -1, if\ sign(\Delta I_k^4 + e_k \Delta I_k^2 + r_k^2) > 0. \end{cases}$$

The solution of the algebraic equation (28) for a given k does not present special difficulties on the basis of numerical methods [18-20], since its approximate solution ΔI_{k0} can be obtained on the basis of ratio (26), i.e., $\Delta I_{k0} = I_k - I_{k0}$.

General algorithm for identification of power losses in the network

Based on the obtained results, the following enlarged algorithm for identification and monitoring of power losses in the distribution network can be formulated.

1. Cyclic polling of the balance three-phase and ASCMPC customer electricity meters at a discrete moment of time$t = t_\xi$.
2. Recording of the received information into the ASCMPC database and formation of vectors $I' = [I_1', I_2', I_3']$.
3. Calculation of complex resistances $\dot{Z}_{kv}$ and conductances $\dot{Y}_{kv}$, as well as powers $\dot{S}_{kv}$, consumed by the network customers according to formulas (9) - (11), respectively.
4. Estimation of total powers $\dot{S}_k^a$ and resistances $\dot{Z}_k^a$ using formulas (3) and (24), respectively.
5. Determination of input desired phase currents I_{k0}, $k = \overline{1,3}$, by formulas (25).
6. Identification of uncontrolled current values $\Delta I_k, k = \overline{1,3}$ by solving algebraic equations (28).
7. Verification of the current state of the PDN by checking the conditions (13).
8. If ratios (13) are fulfilled, there are no UPWs in the network. Pass to item 1. Otherwise, proceed to item 9.
9. Determination of conductances $\Delta \dot{Y}_k$ by solving equations (17).
10. Estimation of commercial $\Delta \dot{S}_k$ and technical $\dot{S}_k^T$ $(k = \overline{1,3})$ power losses in the PDN by formulas (20) and (21), respectively, at time $t = t_\xi$.
11. Go to point 1 and repeat the computational procedure for consecutive time instants $t = t_{\xi+m}$, where $= 1, 2, 3, \dots$.

Conclusions

The practice of operation of modern ASCMPC used for control and metering of electric power in 0.4 kV distribution networks shows that they essentially lack digital technologies allowing to effectively solve the tasks of operational identification and monitoring of power losses with separate identification and estimation of such components as technical and commercial losses. A new computational algorithm for solving this problem based on the concept of a virtual model of a distribution network describing its desired state in the absence of external random factors, which, in particular, include unauthorized power withdrawals, is proposed. This utilizes measurement data received via communication channels from electricity meters installed in the transformer substation and at network customers. Systems of algebraic equations that describe functional relationships between variables and parameters of a virtual model of a three-phase network are obtained. Methods for solving these equations and a generalized algorithm for identifying and monitoring technical and commercial power losses in the distribution network are proposed. The obtained results are oriented for the development of algorithmic and special software for the subsystem of operational monitoring of power losses in PDN as part of ASCMPC.

References

1. Yakuishev K.V. Automated system of commercial electricity metering for the retail market // Informatization and management systems in industry. 2009. № 3(23). page 9-13.
2. El-Hawary, M.E. The Smart Grid—State-of-the-art and Future Trends. Electr. Power Compon. Syst. 2014, 42, pp 239–250.
3. Sapronov A.A., Kuzhekov S.L., Tynyanskiy V.G. Prompt detection of uncontrolled power consumption in power networks with voltage up to 1 kV //. Electromechanics. 2004. №1. page.55-58.
4. T.T.Omorov B.K.Takyrbashev1. K.E.Zakiriaev, T.Zh.Koibagarov. Digital control of electric power flows in unbalanced distribution networks as part of the automated metering and control system // Energy Sistems Reseach. 2021, vol.1, pp 38-46.
5. Omorov T.T., Takyrbashev B.K., Zakiryaev K.E., Imanakunova Zh.S., Koibagarov T.Zh., Asiev A.T. Ssues of development of systems for automation and digitalization of power distribution networks // E3S of Conferences, 2023, 384, 01026 .
6. Patent No. 2548656 (RF). Samokish V.V. Method of symmetry of phase currents of three-phase four-wire line and device for its realization // Bulletin No.11. 27.12.2013.
7. Omorov T. T., Takyrbashev B.K., Koibagarov T.D. Electricy losses management in distribution network as a composition of Automated Meter Reading and Control System (AMRCS)//Mekhatronika, Avtomatizatsiya, Upravlenie, 2021, 22(4), PP. 191-199.
8. Omorov T.T. Takyrbashev B.K., Koibagarov T.Dzh. Management of power losses in distribution networks as part of automated power control and metering systems // Mechatronics, automation, control. V. 22. №4. 2021. page 192 - 199.
9. Kirankumar T., Sri Madhu G.N. Power theft detection using probabilistic neural network classifier // International Research Journal of Engineering and Technology. 2018. Vol. 5, no. 8. Pp. 834–838.
10. Omorov T.T. Takyrbashev B.K., Osmonova R.Ch., Koibagarov T.Zh. Identification of current leakages in distribution networks based on ASCMPC data // Vestnik of South Ural State University. Energy Series. 2018. №2. page 48-54.
11. Omorov T., Koibagarov T., Zhanybaev T., Takyrbashev B., Boronin A. For the construction of

subsystem for power line diagnostic states of distribution networks as a part of AMRCS//E3S Web of Conferences, 2020, 216, 01042.

12. Kosouhov F.D., Vasiliev N.V., Filippov A.O. Reduction of losses from current asymmetry and improvement of power quality in 0.38 kV networks with household loads // Electrical engineering. 2014. №6. Page 8-12.

13. Omorov T.T., Osmonova R.Ch., Takyrbashev B.K., Imanakunova Zh.S. Methodology for identification of distribution network line parameters based on ASCMPC data // Vestnik of Kazan State Power University. 2021. Volume 13. №3 (51). page 168-177.

14. Voitov O.N., Mantrov V.A., Semenova L.V. Analysis of asymmetric modes of electric power systems and their control // Elektrichestvo.1999. №10. Page 2-18.

15 Omorov T., Zakiriaev K., Takyrbashev B., Imanakunova Zh. Automated control of unauthorized power take-offs in a distributed electrical network//Mechatronics, Automation, Control.2023.V.24.№1. p.24-32.

16 Omorov T., Takyrbashev B., Koibagarov T., ZhanybaevT. Identification and Monitoring of Power Losses in Distribution Networks as Composition of AMR // AIP Conference Proceedings, 2022, 2552, 020002.

17. Demirchyan K.S., Neyman L.R., Korovkin A.V. Theoretical fundamentals of electrical engineering. Volume 1. St. Petersburg: Peter, 2009. -512 page.

18. Bahvalov N.S., Zhidkov N.P., Kobelkov G.M. Numerical methods: Laboratory of basic skills, 2002. 632 page.

19. Aoki M. Аоки M. Introduction to optimization methods / M. Aoki. – M.: Science, 1977. – 334 page.

20 Omorov T.T., Kozhekova G.A. Synthesis of control system of multidimensional objects by criterion constraints // Izvestiya of Academy of Sciences of the Kyrgyz Republic 2009. №1. page 45-52.

Ensuring reliable operation of stranding machine for cable production through the use of intelligent technologies

Dilshod Isamukhamedov[1], *Artem* Ivanov[1*], *Aybek* Turabekov[1], *Diyora* Yokubjonova[1]

[1] Tashkent State Technical University, 2 University str., Tashkent, Uzbekistan

Abstract. The article discusses the reliability of cable manufacturing equipment by diagnosing its technical condition. The solution of this issue is provided by the use of intelligent systems with expert assessment and prediction of probable failures in operation. The built-in working environment for the management of cable manufacturing equipment makes it possible to solve the issues posed by the reliability of cable making machines with a long service life by integrating artificial intelligence elements into a system with limited resources. The implementation of the solution to the problem is carried out on the stranding machine for cable production, on the example of which the options are considered that provide the necessary level of not only the reliability of the equipment in the manufacturing technology, but also the reliability in operation of the finished cable product.

Introduction

The use of artificial intelligence in the operation of technological systems is an urgent task focused on centralized management of production processes provided with modern high-tech equipment.

Like any modern technological process, cable production is a flexible technology aimed at the production of finished cable products of various nomenclature, which is determined by the volume of the production order, depending on the contracts concluded by the sales service for the supply of cable or wire to consumers.

The main purpose of the production service is to ensure the preparation of production (PP) for the frequent changeability of the nomenclature of cable and wire products (CWP), taking into account the high productivity of the technological equipment involved. This is achieved by purchasing highly efficient in-line cable lines built on modern information technologies capable of providing the necessary level of automation of the entire production technology as a whole. However, the cable industry contains a large fleet of obsolete technological equipment that has been put into operation for more than 10 years (Table -1), which is not able to meet the requirements of technology: to maintain complex mechanization and automation at the proper level both for modern cable production and for a single technological process of manufacturing a single production order for a cable product. The new, modern technological equipment has input $x(t)$ and output $y(t)$ control signals focused on working in a single (centralized) control system for the production process of manufacturing finished products, which ensures the smooth introduction of digital technologies into the cable industry.

The cable company today is a production process of mastering new types of high-quality CWP, having a high level of operational reliability, with minimizing the set deadlines for both the PP and the manufacture of the finished product, ensuring the seriality of the products with a reduced level of labor intensity of the technological process and reducing the cost of the CWP. The task set by the production workers is solved by the acquisition of modern, high-tech production complexes, which include several cable lines that perform technological operations sequentially and are linked to a single automated system with numerical control. The peculiarity of modern cable making machines is that they freely adapt to a single automated process control system (APCS) within the entire cable plant, providing accurate and maximally predictable control of the manufacturing technology of the finished cable product in the context of the fulfillment of the general production order received from the sales service. Thus, the modern production of the CWP, oriented to new technological equipment, capable of providing complex mechanization and creating an efficiently functioning production structure with unconditional provision for rapid change of the manufactured range of cable products. All this is achievable only with the availability of a high technical level of the manufacturing technology used, which is equipped with new cable manufacturing equipment that provides the finished cable product at all stages of production: high quality, operational reliability, as well as energy and resource saving in general for the considered technological cycle of production.

However, the engineering services of cable companies do not seek to dismantle cable manufacturing equipment that has a long service life for several reasons: economic (lack of financial opportunity to upgrade the entire fleet of cable making

* Corresponding author: c-victory@yandex.com

machines at a time) and industrial (the technical condition of the equipment allows to fulfill a production order, even with zero automation). In this regard, the issue of increasing the efficiency of operation of previously installed technological equipment located at a cable company with a long operational life is particularly acute (Table-1).

Table 1. The service life structure of cable manufacturing equipment in the context of the JV OJSC "Uzkabel" (Uzbekistan, Tashkent)

Service life group	Type of cable manufacturing equipment							
	Continuous casting and rolling line	Wire-drawing machine	Extruding machine	Stranding machine	Winding machine	Inclined vulcanization line	Other equipment (auxiliary)	Total:
Up to 5 years	-	40	20	20	20	-	50	21,4
From 5 to 10 years	-	35	15	40	40	-	50	25,7
From 10 to 20 years	-	5	-	-	-	-	-	0,71
More than 20 years	100	20	65	40	40	100	-	52,1
Total:	100	100	100	100	100	100	100	100

It should be noted that the analysis of the state of the technological equipment fleet of cable companies of the Republic of Uzbekistan has a similar situation (Table-2) in terms of technical security of the production cycle.

Table 2. The share of cable manufacturing equipment installed at cable companies of the Republic of Uzbekistan having different service life.

Name of the cable company	Service life structure of the cable manufacturing equipment fleet, by year of commissioning, %			
	up to 5 years	from 5 to 10 years	from 10 to 20 years	more than 20 years
JV OJSC "Uzkabel"	21,4	25,7	0,71	52,1
"Andijan Kabel"	10	14,3	35,6	40,1
Yuqorichirchik metal invest	5	10	85	-

Table-1 demonstrates that the analysis of the technological equipment fleet of JV OJSC "Uzkabel" (Uzbekistan, Tashkent) showed that 58.8% of the cable making machines involved in the production process of the CWP manufacturing has a service life of more than 10 years and requires modernization not only of the EMC but also of the automated control system, and therefore the implementation of measures for the introduction of automated control systems is ineffective on the production areas under consideration. A similar situation is developing at other cable companies in Uzbekistan with extensive production experience in the CWP manufacture. The work on the introduction of full automation of cable production is effective only if measures are taken to improve not only the EMC of the above-mentioned technological equipment, but also its control system as a whole. Against the background of rapidly developing cable technologies, such machines belong to systems with limited resources, because they work within the framework of a separate technological operation and lose greatly in terms of productivity, energy efficiency, energy and resource conservation, which causes great production difficulties in fulfilling the established production task and meeting its deadlines. An individually designed built-in control environment for each technological unit will allow achieving the solution of the task by ensuring the reliability of cable making machines with a long service life, due to the integration of artificial intelligence elements into a system with limited resources. All of the above forms one common production problem – the creation of a single production process that coordinates the operation of the entire fleet of installed cable making machines with different commissioning dates and greatly differing in technological, technical and production parameters: the volume of production, production time of cable product manufacturing, estimated machine time, etc. The solution of the task is possible not only through the introduction of intelligent control systems (ICS) focused on the expert assessment of the technological parameters of each unit of cable making machines, but also ensuring the regulation of operating parameters in combination with the ability of independent analysis, as well as the learning process based on the prediction of probable failures in the operation of technological equipment, its EMC and possible failures of installed technological modes. All of the above confirms the relevance of the chosen research direction for the production process of the CWP manufacturing. At the same time, it is necessary to develop an individual system for a cable making machine that has a long service life and ensures its unconditional integration into a single production system on a par with modern cable manufacturing equipment, through its modernization.

Research methods

To date, there are a large number of directions for the implementation of intelligent systems for various types of technological objects. The most realistic, in terms of improvement and modernization of technological equipment with a long service life, is the development and implementation of a control system using artificial neural networks of technological parameters for a single cable making machine.

The cable making machine, which has a long service life, is focused on one technology (drawing, stranding, extrusion, etc.), which is carried out through the interaction of several units consisting of complex, interconnected mechanisms. During the entire period of operation for this group of machines, repair services carried out repeated, partial replacement of components and assemblies, which leads to a deterioration in the performance characteristics of the machine compared to the passport ones (Table - 3). During the technical diagnostics of the equipment (stranding machine of general stranding 630C 1-6-12-18-24), a discrepancy between the passport data and real indicators - output parameters was revealed. It should be noted that for the

 https://doi.org/10.1051/e3sconf/202346101041

other units of cable making machines, the same similar situation develops.

Table 3. The results of the technical diagnostics of the operating parameters of the 630C 1-6-12-18-24 stranding machine installed at JV OJSC "Uzkabel" (Tashkent, Republic of Uzbekistan)

Name of the parameter	Unit of measurement	Technical data	
		passport	actual
Linear speed of wire movement	m/min	30-55	20-35
Kinematic extraction	%	1,3÷1,25	1,15
Maximum rotation speed of the carriage	rpm	132	80-95

Solving the issues of modernization of cable making machines with a long service life is complicated by the fact that the existing element base corresponds to modern technologies, and therefore, obtaining performance characteristics close to the passport data for the machine in question is a complex engineering task. It is possible to carry out post-repair configuration of such equipment, as well as to maintain technological modes in conditions of strict compliance with technical regulations (regulatory and technical documentation) and production relationships only by developing an expert control system - an intelligent neural network (INN) operating for a specific cable making machine.

The considered type of INN refers to the RBF network, which has a target learning function focused on decision-making and technology management by approximating and predicting the technological parameters of both the EMC and the cable facility.

The creation of a digitalization system for cable manufacturing equipment is impossible without the development of a mathematical model that shall include a large array of input *x(t)* and output *y(t)* parameters linked to the production and technological indicators of not only the upgraded machine, but the input and output parameters of other cable making machines that are interfaced in the production chain as part of the overall production order – *"end-to-end vision"* of production technology. Thus, the mathematical model shall be based not only on the input and output parameters of the technological process, but also on the analytical data of equipment operation and technology execution in previous periods of operation (hour, shift, quarter, year), as well as disturbing signals *z(t)* - changes in external conditions (ambient temperature, humidity, dustiness, vibration), which is especially important for regions with elevated ambient temperatures, especially for Uzbekistan in the summer, when the temperature in the sun reaches 55-60°C).

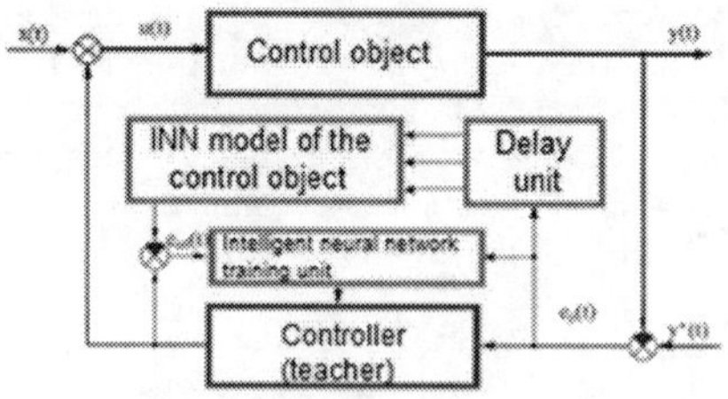

Fig.1. Block diagram of a generalized version of the INN: *x(t)* - input signal; $e_{об}(t)$- learning error, which is the difference between the output value and the control actions *u(t)*; *y(t)* - output action of the object.

A lot of research has been devoted to solving the problem under consideration on approaches to the formation of the INN, namely the inverse-direct (D. Psaltis) [1-4] and inverse-indirect (M. Kawato) [1-4] neural network model of the control object. The main task of integrating the INN into the control system (CS) of a technological object is to adjust for optimal control of the technological process through regulation of the main operating parameters of cable manufacturing equipment, followed by setting up the system for independent control of the entire technology of cable product manufacturing [5, 6], including the transition to a new production order. In a generalized version, the INN is an automatic control system with negative feedback, developed on the basis of a controller with a "teacher" function. Fig. 1 presents a block diagram of the learning process in the process of technology implementation by cable manufacturing equipment.

The implementation of the solution to the issue has been carried out using the example of stranding machine for cable production, within which it is necessary to consider options that ensure a given level of not only the reliability of the equipment in the manufacturing technology, but also the reliability in operation of the finished cable product.

As the object of the study, a rigid stranding machine (SM) of the disc type of cable production has been selected, which strands the conductive core (CC) of a cable product from a variety of previously cast (copper or aluminum) wires stranded according to the stranding system: *1+6+12+18+24+36*.

The developed mathematical model is presented as a system of differential equations in operator form, which describes the operation of the SM EMC with maintaining the tension of the cable billet, with the specified control parameters:

$$\left.\begin{aligned}
&M_{дин_i}(p)=M_{дв_i}(p)-P_i(p)\frac{R^{вх}_{б_i}}{j_{ред_i}}+Q_{(i-1)}(p)\frac{R^{вых}_{б_i}}{j_{ред_i}}-M_{хх_i}(p);\\
&\omega_{дв_i}(p)=M_{дин_i}(p)\frac{1}{J_{\Sigma_i}p};\\
&V^{вх}_{б_i}(p)=\omega_{дв_i}(p)\frac{R^{вх}_{б_i}}{j_{ред_i}};\\
&V^{вых}_{б_i}(p)=\omega_{дв_i}(p)\frac{R^{вых}_{б_i}}{j_{ред_i}};\\
&P_i(p)=\left(V^{вх}_{б_i}-V^{вых}_{в_i}\right)\frac{EF_i}{L_{P_i}p};\\
&V^{вых}_{в_i}=V^{вх}_{в_i}\mu;\\
&V^{вх}_{в_i}=Q_i(p)\frac{L_{Q_i}p}{EF_{(i-1)}}+V^{вых}_{б(i-1)};\\
&Q_i(p)=P_i(p)-N_i(p).
\end{aligned}\right\} \quad (1)$$

where, M - the moment acting in the system under consideration; ω - the angular velocity of carriage rotation, v - the linear velocity of workpiece movement; P - pulling force; Q - tension.

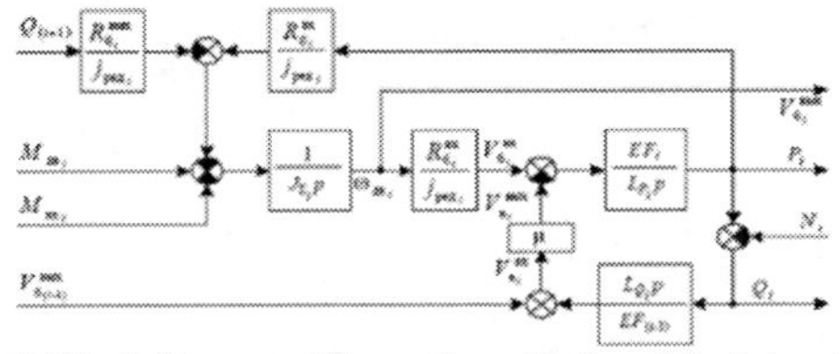

Fig.2. Block diagram of the mathematical model of stranding machine that provides tension maintenance.

The block diagram of the mathematical model, compiled in accordance with (2) is shown in Fig.2, reflects the essence of the process of forming forces during CC stranding, causing tension of the cable billet, as one of the main adjustable parameters in cable production.

The integrated INN in the SM EMC shall be maximally adapted in terms of the technological and production features of the technological operation "CC stranding" [7, 8]. Functionally, it is a multi-level control system that is built both on optimization and regulation of critical technological parameters (step and multiplicity of stranding, stranding ratio, compaction ratio, linear speed of the workpiece, rotation speed of the carriage, pulling force, tension, etc.), and on the forecast model - a recommendation service that works in real time (24/7). So in the real operating mode, the number of controlled system parameters is 125 pcs., each of which is considered important for the operation of the equipment, because it has a tightly controlled control range, which ultimately determines the quality of the technology performed. Therefore, the main intellectual load developed by the ICS is an expert assessment of the correctness of maintaining the technology, in order to fulfill the final production task – the manufacture of high-quality cable products.

All existing expert systems (ES) are based on computer programs, realize the main goal - to exclude the human factor (the opinion of the technologist) in making decisions in technology, as a separate (narrow) area of tasks (Fig. 3).

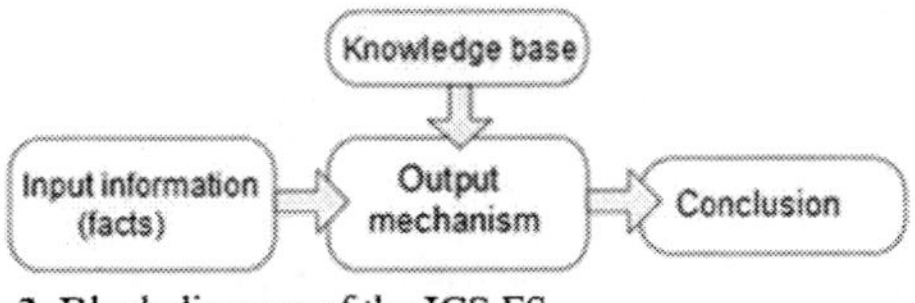

Fig. 3. Block diagram of the ICS ES.

The introduction of the ICS ES into the operation of the SM EMC will increase the speed of the SM operation by increasing the operating speeds, the correct selection of equipment operating modes, the amount of technological information processed, as well as the correct choice of solutions for extreme operating modes of equipment and the exclusion of the *"human factor"*.

The results obtained

The production process of CWP manufacturing is considered as a system mechanism that has a very tight time frame limited by the timing of the production order, and this complicates the task of performing the operation of the CC stranding due to time constraints associated with the execution of the production order.

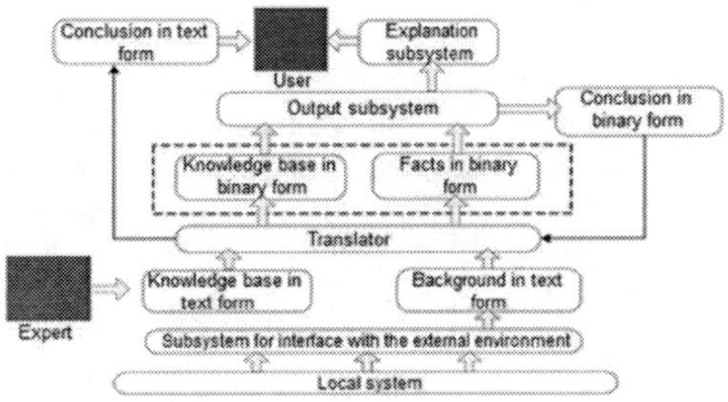

Fig. 4. Block diagram of the SM integrated control system.

Thus, the development of an algorithm for real-time ICS EC is a more complex task than the description of an automated control system having a static structure of equipment operation. Embedded and integrated into the general system, the ICS shall have many subsystems that are functionally connected and interfaced into a single system of interaction in terms of decision-making time and have strict limits on the amount of allocated memory, as well as execution time [8, 10-12]. At the same time, the integrated system shall also make the necessary edits independently and coordinate the calculation time of the control signal.

In addition, the SM ICS shall be adapted to work in a common network, i.e. it is compatible with other units of equipment included in the technological process of manufacturing a cable product, both within the production area, as well as the workshop and the company [7-12]. The block diagram of the integrated control system is shown in Fig. 4.

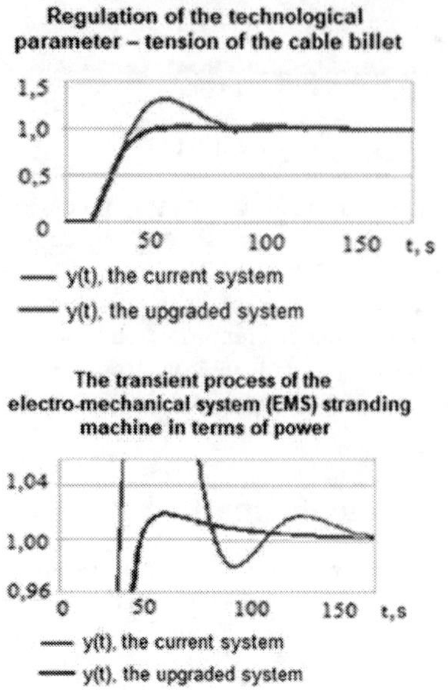

Fig. 5. The results of modeling the transition process in the current and upgraded system: *y(t)* - the output parameter of the system.

The results of mathematical modeling made it possible to test the operating and technological modes of cable manufacturing equipment – stranding machine, model "630C 1-6-12-18-24". The results of modeling the transition process in the current and upgraded system are shown in Fig. 5, both for systems in various situations and controlled by conditionally defined (selected) parameters.

Conclusion

Thus, the result of the research work done is to obtain, by mathematical modeling, the calculated characteristics of the operation of the technological equipment - the stranding machine, which allowed to conclude that the developed built–in INN is able to ensure the reliability of the stranding machine for cable production through the use of intelligent technologies. The implementation of the proposal was carried out at a specific stranding machine for cable production, and the necessary level of reliability of the technological equipment was obtained, but also the reliability of the finished cable product was increased.

References

1.Omatu S. Neuro-control and its applications / S. Omatu, M. Khalid, R. Yusof - London: Springer-Verlag, 1996. – p. 272.

2. Terekhov V.A. Neural network control systems / V.A. Terekhov, D.V. Efimov, I.Yu. Tyukin - M.: Higher School, 2002. – p. 183.

3. Modern applied management theory. New classes of technical system regulators. / Edited by A.A. Kolesnikov - Taganrog: TRTU, 2000. – p. 656.

4. Method of resource saving of cable production recyclables. Victoria Tsypkina and Daniyar Madrakhimov. E3S Web Conf.,139 01058 (2019) DOI: 10.1051/e3sconf/201913901058.

5. Improving the reliability of power supply to active consumers by improving the technology for manufacturing cable product, Vera Pavlovna Ivanova, Victoria Vyacheslavovna Tsypkina, E3S Web of Conf. 216 01152 (2020), DOI: 10.1051/e3sconf/202021601152

6. Improving the reliability of cable lines operation in hot climates, Daniyar Bakhtiyarovich Madrakhimov, Vera Pavlovna Ivanova, Victoria Vyacheslavovna Tsypkina, E3S Web of Conf. 216 01151 (2020), DOI: 10.1051/e3sconf/202021601151

7. Modeling of a resource-saving method of drawing, Victoria Tsypkina, Veronica Ivanova, E3S Web Conf. 139 01073 (2019), DOI: 10.1051/e3sconf/201913901073

8. Cable conductor of cabling and wiring products based on composite materials for transport systems, V.V. Tsypkina, V.P. Ivanova, K.K. Jurayeva, E3S Web of Conf. 401 03036 (2023), DOI: 10.1051/e3sconf/202340103036

9. Tsypkina, V., Ivanova, V., & Jurayeva, K. (2022). MATHEMATICAL MODELING OF THE CONVEYOR MOVEMENT FOR THE CABLE WASTE CRUSHING LINE. *Вестник КазАТК*, *122*(3), 255–261. https://doi.org/10.52167/1609-1817-2022-122-3-255-261.

10. Development of generalized requirements for automated electric drive of cable equipment, V.V. Tsypkina, V.P. Ivanova, D.N. Isamukhamedov, A.U. Turabekov, R.F. Atamukhamedova, E3S Web of Conf. 401 02045 (2023), DOI: 10.1051/e3sconf/202340102045

11. Improvement of the multifilament wire lager for cable production, Olimjon Toirov, Vera Ivanova, Viktoriya Tsypkina, Dilnoza Jumaeva, Dilnoza Abdullaeva, E3S Web Conf. 411 01041 (2023), DOI: 10.1051/e3sconf/202341101041.

Optimization of schemes and ways to expand the adjustment range for the power supply of combined heat and power plants

Elena Burdenkova[1] and *Michael* Garievsky[1*]

[1]Saratov Scientific Center of RAS, Department of Energy Problems, Saratov, Russia

Abstract. The article assesses the possibility of using low-flow operating modes at steam turbine with an additional boiler and steam and combined-cycle cogeneration plants for heat supply schemes, taking into account the storage properties of heat networks and buildings. It is shown that the use of low-flow operation modes of the turbine plant during the night period allows the combined heat and power plant to participate more effectively in covering the variable part of the electric load schedule. The dependence of the share of participation of an additional boiler in covering the heat load on the prices for fuel and electricity is determined. Comparison of the performance indicators of the steam turbine CHPP in low-flow modes with the release of the thermal energy necessary to the consumer from an additional boiler with the mode of its complete shutdown and the release of thermal energy from the peak hot water boiler made it possible, taking into account the accumulating properties of heat networks and buildings, to recommend a low-flow operation mode at night. For a combined-cycle cogeneration plant it is shown that it is advisable to reduce the electric power at night, while ensuring the nominal heat supply, by unloading the gas turbines with the transfer of the steam turbine to the motor mode.

1 Introduction

One of the important tasks in the development and operation of generating capacities is the reliable regulation of the variable part of the electrical load schedule of power systems, as well as the reliable and efficient supply of electricity demand. An increase in the unevenness of energy consumption schedules and limited possibilities for regulating the load in power systems has led to the fact that the operation modes of combined heat and power plants differ significantly from the basic ones, and low electricity prices in the wholesale electricity and power market during the night have led to the need for their deep unloading [1–5].

Power consumption modes change under the influence of various factors: technological features of production, climatic factors, etc. The dynamics of power consumption volumes in the considered period can be represented in the form of daily, weekly, seasonal, and annual load schedules. Of great importance for the operation of power plant equipment are the daily schedules of electrical loads.

In most of the energy systems of Russia, a significant number of combined heat and power plants with industrial and heating steam extractions are in operation. The possibilities of daily unloading on the electrical load of the CHPPs are limited by the load of regulated extractions [6, 7]. In winter, at low outdoor temperatures, the regulated extractions of a heating CHPPs are usually fully loaded and there are practically no control capabilities of the power plant [8–10]. At the same time, during these periods, the issues of ensuring the daily schedule of electrical loads are most acute [11–13].

Under these conditions, an urgent task is to assess the possibility of using low-flow modes of operation at steam turbine CHPPs (using the example of a T-100-130 cogeneration turbine with a capacity of 100 MW for heat supply schemes with an additional boiler plant) and combined-cycle cogeneration plant (using the example of CCGT-450T with a capacity of 450 MW) taking into account the storage properties of heat networks and buildings.

2 Materials and methods

For operation in a variable mode, several measures have been developed and are being used to expand the adjustment range limits of thermal power plants [14–18]. Such measures include the following: transfer of power units to the auxiliary load or to idling in case of generator shutdowns and load shedding; transfer of the steam turbine to the motor mode; CHPP unloading by transferring the thermal load of the turbine to a pressure-reducing cooling station or peak boilers.

In the practice of CHPPs operation, a forced reduction in electric power is used by throttling part of the live steam in the pressure-reducing cooling station. If there is a reserve of thermal power of peak hot water boilers, it is also possible to reduce the load of heating extractions by transferring it to these boilers. With forced unloading, a partial or complete transition from combined generation of electric and thermal energy to

[*] Corresponding author: garievsky.michael@gmail.com

separate is carried out, which is associated with a significant deterioration in the efficiency of the CHPPs, so it can only be used when the adjustment capabilities of the power plant have been exhausted.

In the motor mode, the main steam is not supplied to the steam turbine, and the electric generator is not disconnected from the network and operates as an engine, rotating the turbine rotor, consuming active power from the network to overcome ventilation, mechanical and other losses both in the turbine and in the generator. To cool the exhaust part of the turbine, a small amount of cooling steam is supplied to it through a pressure-reducing unit from constant sources (a general station auxiliary collector), and a normal vacuum is maintained in the condenser to condense the cooling steam. Such a result can also be achieved with a complete shutdown of the power unit, however, the motor mode has several advantages: switching the unit to motor mode takes less time; the power system is not deprived of a hot power reserve, since the unit operating in the motor mode is easily transferred to the generator mode; starting the unit from a warm and even hot state after stopping it for the night is much more difficult and lasts longer.

2.1 Comparative analysis of operating modes of the turbine T-100-130 heat plant

The need for electricity generated by the turbine at night decreases, during the day, on the contrary, the need for electricity increases. The papers [19–23] considered the efficiency of using thermal energy storage at night by heating networks and buildings in order to expand the regulation range of electricity generated by the turbine. To this end, only the steam needed for regeneration and minimal ventilation to the condenser was supplied to the turbine at night. Most of the steam, after its pressure reduction in the fast-response pressure-reducing station, entered the additional boiler.

In the proposed scheme, only steam is sent to the turbine at night only for a minimum ventilation pass to the condenser. Heaters of the regeneration system and network heaters are switched off.

At night during the winter period, it is proposed to generate heat at the heat plant that exceeds what is needed for the consumer. This will allow, thanks to the storage properties of heating networks and buildings, to generate less heat during the day, increase the steam flow to the condenser and generate peak electricity. Figure 1 shows a diagram of the heat plant of the T-100-130 turbine, where network water is heated in the first network heater (NH1) by steam of the 7th selection, in NH2 – by steam of the 6th selection, in an additional boiler – by fresh steam after a fast-response steam pressure temperature reducer, then in a peak water-heating boiler and is sent to the consumer.

Three variants for the operation of a heat plant at night are considered, taking into account the accumulation of heat by heating networks and buildings. In the first variant, network water is heated in a peak water-heating boiler and sent to the consumer. In this case, the turbine operates in motor mode, and the additional boiler installation is turned off. In the second option, network water is heated in an additional boiler. This happens as follows: fresh steam after a fast-response steam pressure temperature reducer is sent to an additional boiler, gives off its heat to the network water, condenses and then, using a drain pump, is sent to the mixing point of the condensate path before the deaerator. In the third variant, network water is heated in an additional boiler, and then in a peak water-heating boiler.

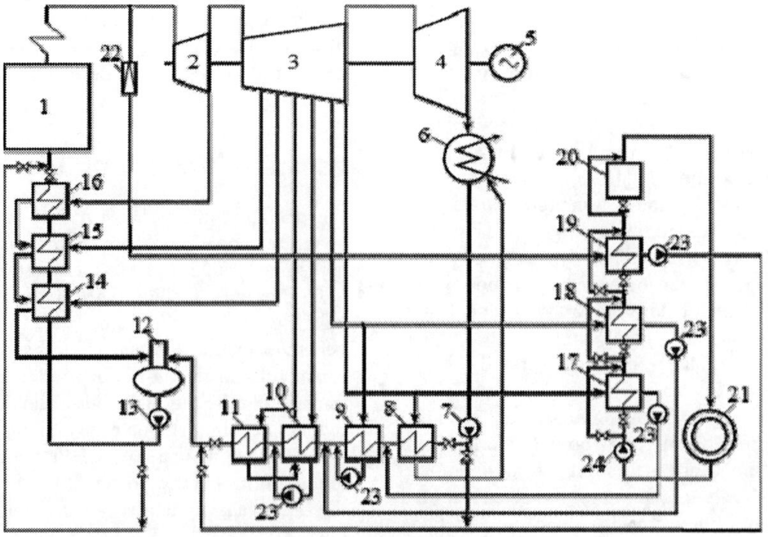

Fig. 1. Schematic diagram of the T-100-130 turbine heat plant with an additional boiler:
1 – boiler; 2, 3, 4 – high, medium and low pressure cylinders; 5 – electric generator; 6 – condenser; 7 – condensate pump; 8, 9, 10, 11 – low pressure heaters LPH-7, LPH-6, LPH-5 and LPH-4; 12 – deaerator; 13 – feed pump; 14, 15, 16 – high pressure heaters HPH-3, HPH-2 and HPH-1; 17, 18 – network water heater NH1 and NH2; 19 – additional boiler; 20 – peak water-heating boiler; 21 – consumer; 22 – fast-response steam pressure temperature reducer; 23 – drainage pumps; 24 – network pump.

For the accepted initial data (table 1), technical and economic calculations were carried out in order to determine the optimal mode of operation of the heat plant at night. Calculations were made for 3 price options for night and day electricity (table 1). Price 1 is the current one for the Volga region (data from www.atsenergo.ru), prices 2 and 3 are predictive. The ratio of the prices of night and day electricity for them is 0.86, 0.4 and 0.07, respectively.

In the first variant, the necessary steam flow rate for cooling the turbine and its seals is supplied from a neighboring boiler and is 0.56 kg/s, the electric power consumed by the turbine from the network is 2 MW. In all three variants, during the daytime, the steam flow from the boiler to the turbine is maximum and equals 127.78 kg/s. In the first variant, the temperature of the network water at the inlet to the peak hot water boiler at night is 48.5°C, in the third variant it is 90.3°C. In all variants during the day, the peak hot water boiler and the additional boiler are switched off, the network water is heated in the network heaters NH1 and NH2, the temperature at the inlet to NH1 is 50°C. In the second option, part of the main steam from the boiler is sent to a fast-response steam pressure temperature reducer, where its pressure is reduced to 0.12 MPa, after which it is sent to an additional boiler. In the third variant, the pressure of main steam in the fast-response steam pressure temperature reducer is reduced to 0.09 MPa, after which it is also sent to an additional boiler. Table 2 shows the results of calculations for three variants.

Table 1. Initial data.

Parameters		Variant 1	Variant 2	Variant 3
The amount of heat required by the consumer at night, GJ		5359.1		
Duration of night / day, h		8 / 16		
Duration of the heating period, days / year		198		
Fuel price, rubles/ths. m^3		5300		
Price for nighttime electricity / daytime electricity, rubles/kWh / ratio of nighttime and daytime electricity prices	price 1	1.16 / 1.35 / 0.86		
	price 2	0.64 / 1.61 / 0.4		
	price 3	0.26 / 3.73 / 0.07		
Temperature of network water at the outlet / at the inlet to the peak water-heating boiler at night, °C		100 / 48.5	– / –	100 / 90.3
Temperature of network water at the inlet to the lower network heater during the day, °C		50		
Steam consumption from the boiler to the turbine at night / day, kg/s		– / 127.78	2.79 / 127.78	2.79 / 127.78
Steam consumption for cooling the turbine and for seals in motor mode at night, kg/s		0.56	–	–
Electric capacity at night, consumed by the turbine in motor mode / generated by the turbine, MW		2 / –	– / 10	– / 10

Table 2. The results of calculations of the accumulation mode for three variants.

Parameters		Variant 1	Variant 2	Variant 3
Heat power of the peak water-heating boiler / additional boiler at night, MW		332.4 / –	62.5 / 269.9	– / 332.4
Share of power of an additional boiler in the heat plant at night		0	0.81	1.0
Steam consumption in the additional boiler at night, kg/s		–	87.41	123.7
Temperature of network water behind the peak hot water boiler / after the additional boiler at night, °C		100 / –	100 / 90.3	– / 100
Total fuel costs, million rubles/year		120.56	124.24	141.04
Heat power of the additional boiler at night, MW		332.39		
Excess heat generated per night, GJ		4213.6		
The required amount of night electricity per day / for the heating period, thousand kWh		16 / 3168	–	–
Costs of consumed night electricity for the heating period, million rubles/year	price 1	3.7	–	–
	price 2	2.0	–	–
	price 3	0.8	–	–

Thermal power of the lower network heater during the day, MW		112.9		
Temperature of network water during the day after the lower network heater, °C		67.5		
Steam consumption in the lower network heater / in the condenser during the day, kg/s		48.5 / 57.0		
Electric capacity at night/day, MW		– / 118.03	10 / 108.03	10 / 108.03
Electricity generated per night / per day / per 24-hour period, thousand kWh		– / 1888.4 / 1888.4	80 / 1888.5 / 1968.4	80 / 1888.5 /1 968.4
Generated amount of electricity during nights / days / total for the heating period, thousand kWh		– / 373910.3 / 373910.3	15840 / 373910.3 / 389750.3	15840 / 373910.3 / 389750.3
Adjustment range of electric power, MW		118.03	108.03	108.03
Income from the sale of electricity for the heating period: day / night / per 24-hour period / per 24-hour period, including fuel costs, million rubles/year	price 1	– / 504.8 /504.8/380.6	18.4 / 504.8 /523.2/398.9	18.4 / 504.8 /523.2/398.4
	price 2	– / 602.0 /602.0/479.4	10.1 / 602.0 /612.1/487.9	10.1 / 602.0 /612.1/487.4
	price 3	– / 1394.7 / 1394.7 / 1273.3	4.1 / 1394.7 / 1398.8 / 1274.6	4.1 / 1394.7 / 1398.8 / 1274.1

Table 2 shows that in the first variant, the thermal power of the peak water-heating boiler (heat plant) at night is 332.4 MW. This exceeds the required amount of heat by 78.63%. Excess heat is accumulated by heating networks and buildings. This allows to generate less heat during the day, turn off the upper network heater and reduce the power of the lower NH by 13%. Due to this, the steam flow to the condenser increases by 3.67 times. Electric power generated during the day increases by 18.03%. Income from the sale of daytime electricity in tariff options 1, 2 and 3 for night and day electricity for the heating period, taking into account fuel costs, amounted to 380.55, 479.41, 1273.31 million rubles/year, respectively.

In variant 2, the thermal power of the additional boiler at night is 269.9 MW, and the power of the peak hot water boiler is 62.49 MW. The total thermal power of the heat plant is 332.4 MW. Income from the sale of daytime electricity in tariff options 1, 2 and 3 for night and day electricity for the heating period, taking into account fuel costs, amounted to 398.4, 487.38, 1274.05 million rubles/year, respectively. In variant 3, the thermal power of the additional boiler, as well as the heat plant at night, is 332.39 MW. Income from the sale of daytime electricity in tariff options 1, 2 and 3 for night and day electricity for the heating period, taking into account fuel costs, amounted to 398.91, 487.89, 1274.56 million rubles/year, respectively. It should be noted that the adjustment range for the first, second and third variants is 118.03, 108.03 and 108.03 MW, respectively.

2.2 Comparative analysis of the operating modes of the CCGT-450T

The thermal scheme of CCGT-450T differs from traditional schemes of thermal power plants in a number of features. In connection with the task of expanding the adjustment range limits, the main ones are three electric generators (two with gas turbines and one with a steam turbine), two heat-recovery steam generators, the ability to operate the power unit with a different composition of operating equipment (work with two or one gas turbine), the dependence of CCGT efficiency on load (both for the power unit as a whole and for its main equipment) [24, 25].

The heating equipment of CCGT-450T at Severo-Zapadnaya CHPP consists of four levels of heating of network water in two horizontal (HNWH-1 and HNWH-2) and two vertical (VNWH-3 and VNWH-4) heaters of network water, which significantly expands the temperature range of the network water after the heat plant (the maximum temperature of direct network water after VNWH-4 can reach 153°C) [26]. VNWH-3 and VNWH-4 are connected in such a way that steam can be supplied to them both from steam turbine extractions and directly from heat-recovery steam generators, figure 2 [27].

If it is necessary to ensure the rated heat supply, unloading of the electric power of the cogeneration CCGT can be ensured only by switching to the GT based CHPP mode with the steam turbine shutting down or switching it to the motor mode [28]. The disadvantage of the GT based CHPP mode, including when the steam turbine is switched to the motor mode, is the need for capital investments in the modernization of the thermal scheme of the steam turbine and the heating system [29].

Based on the methodology [30–33], in this paper, using the example of a CCGT-450T, we compared two options for unloading at night: switching the CCGT to the GT based CHPP mode with the steam turbine shutting down and switching the steam turbine to the motor mode while providing a constant for 24-hour period of rated heat load 411.7 MW(t). CCGT in these modes is unloaded by electric power of 180.2 MW (from 450 MW to 269.8 MW).

3 Results

Figures 3*a* and 3*b* show the dependencies of fuel costs and additionally generated peak electricity on the share of heat produced by an additional boiler in the T-100-130 cogeneration plant.

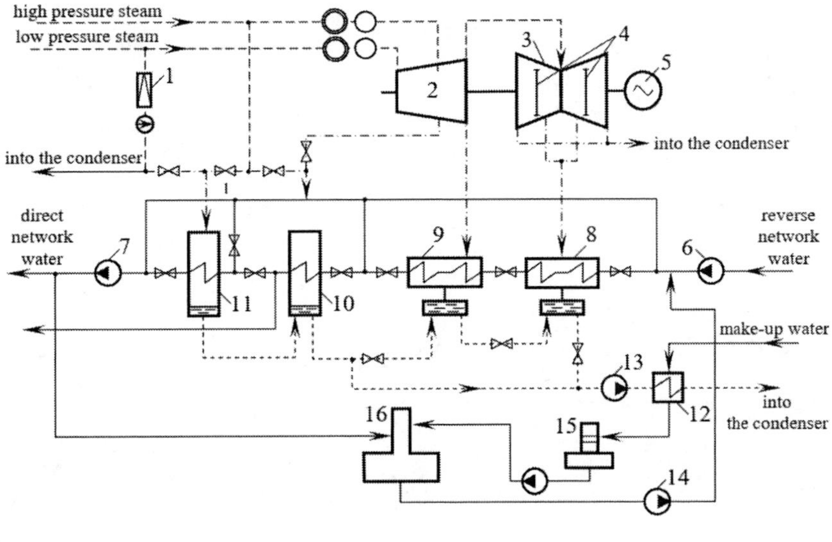

Fig. 2. Schematic diagram of the CCGT-450T network water heating system at Severo-Zapadnaya CHPP: 1 – fast-response steam pressure temperature reducer; 2 – high pressure cylinder; 3 – low pressure cylinder; 4 – regulating diaphragms; 5 – electric generator; 6, 7 – network pumps; 8, 9 – horizontal network water heaters HNWH-1 and HNWH-2; 10, 11 – vertical network water heaters VNWH-3 and VNWH-4; 12 – condensate cooler; 13 – drain pump; 14 – make-up pump; 15 – decarbonator; 16 – vacuum deaerator.

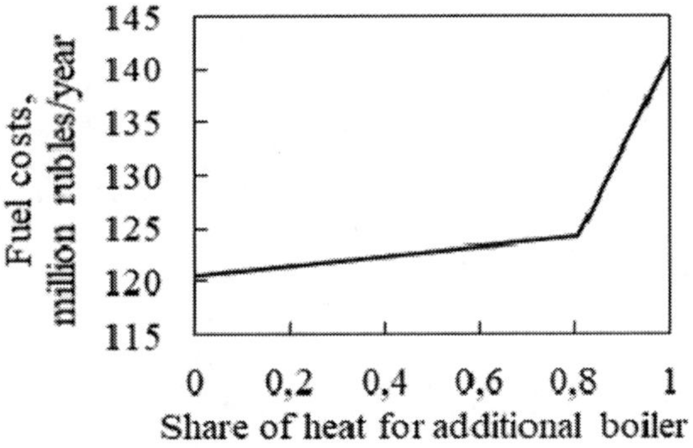

a)

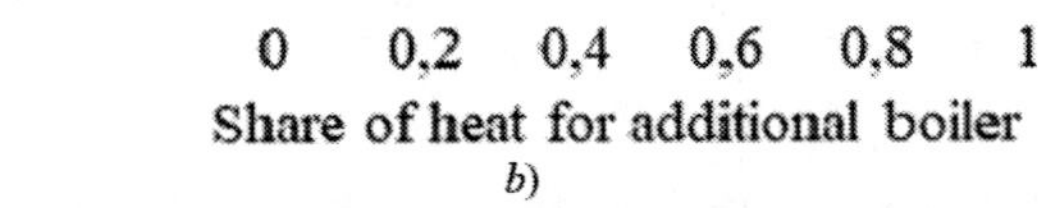

b)

Fig. 3. Dependences of fuel costs (*a*) and peak power increase (*b*) on the share of heat sent to the additional boiler.

Figure 3*a* shows that with an increase in the share of heat produced by an additional boiler in heat plant, fuel costs increase from 120.56 to 141.04 million rubles/year. From figure 3*b* it can be seen that the additional peak electric power with an increase in the proportion of heat produced by the additional boiler in the heat plant decreases from 18.03 to 8.03 MW.

Figures 4*a*, *b* and *c* show the dependence of the revenue from the sale of additional peak electricity, including fuel costs, on the share of heat produced by an additional boiler in a heat plant.

Figure 4 shows that with an increase in the share of heat produced by an additional boiler in a heat plant, the income from the sale of additional peak electricity, taking into account fuel costs, increases from 380.55 to 398.4 million rubles/year at actual tariffs, from 479.41 to 487.38 million. rubles/year (the ratio of the night-to-day price is 0.4) and from 1273.31 to 1274.05 (the ratio of the night-to-day price is 0.07) at forecasted prices.

Figure 5 shows the dependence of the cost of electricity in motor mode on the ratio of prices for night and day electricity.

Figure 5 shows that with an increase in the ratio of the night to day price for electricity, the prime cost of electricity in motor mode increases from 121.4 to 124.2 million rubles/year. The highest prime cost value corresponds to current electricity prices.

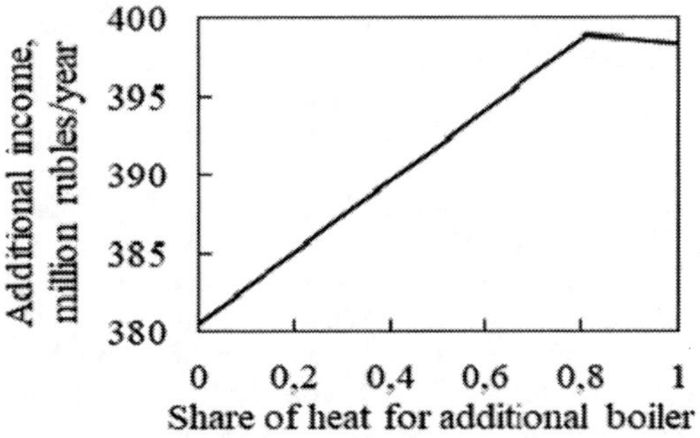

a)

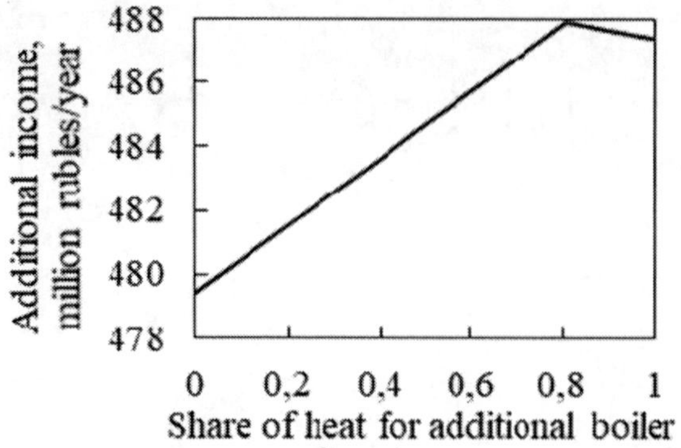

b)

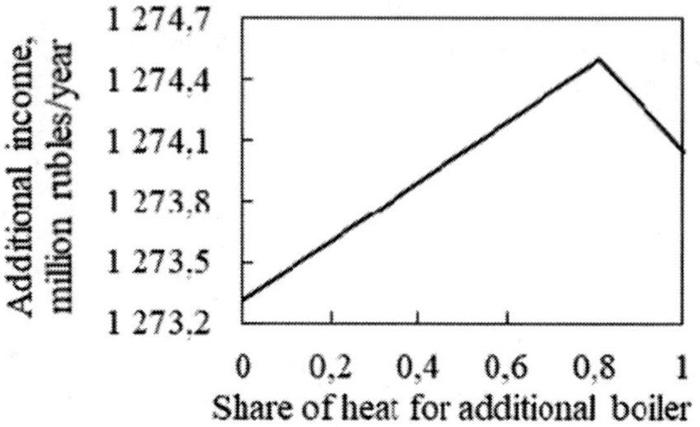

c)

Fig. 4. Dependences of additional revenues on the share of heat directed to an additional boiler with different price ratios for night and day electricity: *a* – 0.86; *b* – 0.4; *c* – 0.07.

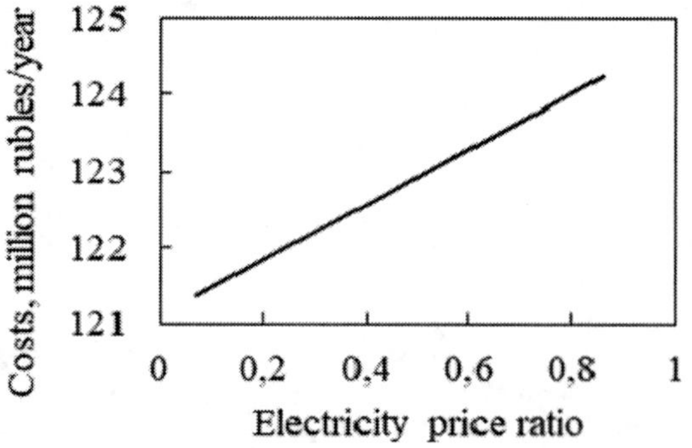

Fig. 5. Total costs for fuel and electricity, depending on the ratio of prices for night and day electricity, million rubles / year.

The results of calculating the costs of switching the steam turbine to the motor mode (instead of stopping), depending on the duration of unloading of the CCGT-450T, showed the following. Fuel costs are lower when unloading the CCGT with the transfer of the steam turbine to the motor mode with a duration of operation at reduced power of less than 5 hours, because in this case, less fuel consumption affects when loading the CCGT from the motor mode. With a longer duration of unloading of the CCGT, the fuel consumption for maintaining the motor mode turns out to be more than it is saved during loading. Repair costs and depreciation charges when switching a steam turbine to motor mode are lower compared to stopping a steam turbine. The revenue from the sale of electricity is higher in the motor mode compared to the shutdown of the steam turbine only when the CCGT is operating at a reduced power level during the hours of night load reduction less than 10 hours (figure 6).

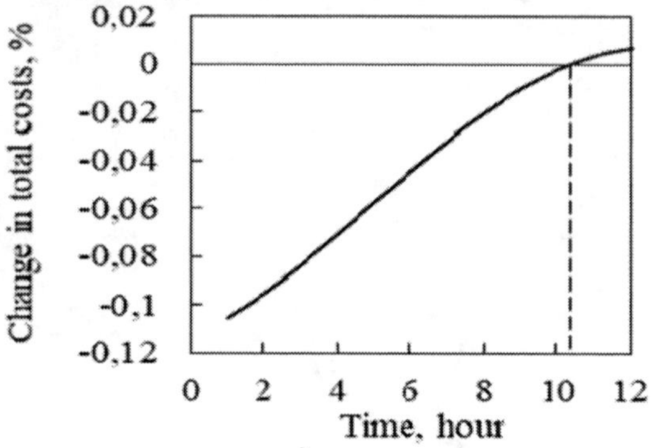

a)

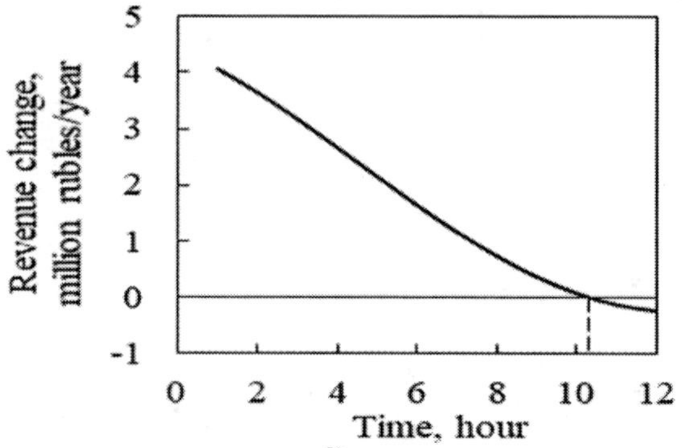

b)

Fig. 6. Changes in total costs (*a*) and revenues (*b*) when the steam turbine is switched to motor mode (instead of shutdown) depending on the duration of unloading CCGT-450T.

4 Conclusion

Three modes of operation of the heat plant scheme based on the T-100-130 turbine are considered, taking into account the accumulating properties of heat networks and buildings: 1) in the motor mode of operation of the turbine, when the heat load is fully provided by the peak water-heating boiler; 2) in the low-flow operation of the turbine, when the heat load is provided by an additional boiler and a peak water-heating boiler; 3) in the low-flow operation of the turbine, when the heat load is provided by an additional boiler. In all modes, regenerative and network heaters are disabled. As a result of calculations, it is shown that the largest expansion of the adjustment range for the supply of electricity (18.03%) corresponds to the motor mode, in low-flow modes it is 8.03%. The lowest fuel costs also correspond to the motor mode and amount to 120.56 million rubles/year, for the second mode they amount to 124.24 million rubles/year, for the third – 141 million rubles/year. Income from the sale of electricity for the heating period, taking into account the cost of fuel, is the highest for low-flow regimes and amounts to 398–1274 million rubles/year, depending on the price of electricity. In motor mode, it is less and amounts to 380-1273 million rubles/year, depending on the price of electricity, since in motor mode, electricity is not generated at night but is consumed from the network. Thus, in the motor mode of operation of the turbine, the mode of operation of the heat plant is optimal only in the case of a zero price of electricity. At modern prices for the supplied electricity, the modes of operation of the heat plant with low-flow turbine operation are preferable.

An analysis was made of ways to expand the adjustment range for the supply of electricity from combined heat and power plants. Using the example of a CCGT-450T, an assessment was made of the reduction in the service life of the main equipment and the economic efficiency of the combined-cycle cogeneration plant during unloading during the hours of the nighttime failure of the electrical load due to the transfer to the GT based CHP mode with the steam turbine turned off and with the transfer of the steam turbine to the motor mode at a constant (nominal) release of thermal energy. It is shown that if it is necessary to ensure the nominal heat supply, it is advisable to unload the CCGT-450T at night in terms of electric power for up to 10 hours by unloading gas turbines with the transfer of the steam turbine to the motor mode, with a longer unloading time – with the shutdown of the steam turbine.

References

1. J. Wang, S. You, Y. Zong, C. Træholt, Z. Y. Dong, and Y. Zhou, Applied Energy **252**, 113445 (2019)
2. J. Beiron, R. M. Montañés, F. Normann, and F. Johnsson, Applied Energy **278**, 115630 (2020)
3. V. V. Yanchuk, and V. N. Romaniuk, Energetika. Proceedings of CIS higher education institutions and power engineering associations **65**, 511 (2022)
4. B. D. Teplov and Yu. A. Radin, Therm. Eng. **66**, 323 (2019)
5. E.K. Arakelyan, A.V. Andryushin, Yu.Yu. Yagupova, and S.V. Mezin, MPEI Vestnik **4**, 122 (2022)
6. V. I. Nazarov, L. A. Tarasevich, and P. V. Nazarov, Energetika. Proceedings of CIS higher education institutions and power engineering associations **5**, 56 (2013)
7. E. K. Arakelyan, and E. D. Vasiliev, New in the Russian electric power industry **2**, 6 (2017)
8. E. K. Arakelyan, V. A. Makarch'yan, V. N. Samarenko, and E. I. Tazhiev, Therm. Eng. **48**, 296 (2001)
9. N. V. Goryachikh, A. G. Batukhtin, and S. A. Ivanov, Therm. Eng. **57**, 892 (2010)
10. D.Y. Tsypulev, E.K. Arakelyan, and V.A. Makarch'yan, Therm. Eng. **55**, 255 (2008)
11. V. N. Romaniuk, A. A. Bobich, N. A. Kolomytskaya, D. B. Muslina, and A. V. Romaniuk, Energetika. Proceedings of CIS higher education institutions and power engineering associations **2**, 67 (2012)
12. A.A. Mikhalevic, and U.A. Rak, Energetika. Proceedings of CIS higher education institutions and power engineering associations **64**, 5 (2021)
13. R. Z. Aminov, A. F. Shkret, E. Yu. Burdenkova, and M. V. Garievskii, Therm. Eng. **58**, 370 (2011)
14. A. A. Madoyan, Improving the maneuverability and efficiency of the use of thermal power plants (Energoatomizdat, Moscow, 1987)
15. A. A. Madoyan, B. L. Levchenko, E. K. Arakelyan [et al.] The use of motor mode at thermal power plants (Energia, Moscow, 1980)
16. G. G. Olkhovskiy, Y. A. Radin, O. N. Makarov [et al.] Electric Stations **3**, 2 (2015)
17. E. K. Arakelyan, A. V. Andryushin, S. Yu. Burtsev, and K. A. Andryushin, MPEI Vestnik **6**, 20 (2017)
18. E. K. Arakelyan, Power engineering: research, equipment, technology **25**, 58 (2023)
19. R. Z. Aminov, E. Yu. Burdenkova, and A. B. Moskalenko, Therm. Eng. **68**, 584 (2021)
20. R.Z. Aminov, and E. Yu. Burdenkova, Energy Safety and Energy Economy **1**, 22 (2023)
21. R. Z. Aminov, and E.Y. Burdenkova, Transactions of Academenergo **2**, 38 (2016)
22. R. Z. Aminov, and E.Y. Burdenkova, Transactions of Academenergo **2**, 45 (2018)
23. R. Z. Aminov and E. Y. Burdenkova, in *2019 International Multi-Conference on Industrial Engineering and Modern Technologies (FarEastCon)* (IEEE, Vladivostok, Russia, 2019), 19229068
24. Y. A. Radin, A. V. Davydov, A. V. Chugin, and I. N. Piskovatskov, Therm. Eng. **5**, 389 (2004)
25. S.V. Tsanev, V.D. Burov, D.V. Goncharenko, and N.V. Varaksina, Energy saving and water treatment **6**, 31 (2008)
26. Y. A. Radin, and A. V. Davydov, Electric stations **9**, 22 (2009)
27. A. D. Trukhniy, A.A. Makarov, and V.V. Klymenko, Fundamentals of modern energy. Part 1. Modern thermal power engineering (MEI Publishing House, Moscow, 2003)
28. E. K. Arakelyan, A. V. Andryushin, S. Yu. Burtsev, and K. A. Andryushin, Power Technol Eng **51**, 451 (2017)
29. E. K. Arakelyan, A. V. Andryushin, S. Yu. Burtsev, and K. A. Andryushin, Therm. Eng. **12**, 900 (2018)
30. R. Z. Aminov, A. F. Shkret, and M. V. Garievskii, Power Technol Eng **48**, 391 (2015)
31. R. Aminov and M. Garievskii, Power Technol Eng **53**, 479 (2019)
32. M. V. Garievskii, J. Phys.: Conf. Ser. **1683**, 042022 (2020)
33. M. V. Garievsky, Energy Safety and Energy Economy **2**, 19 (2022)

Resource modeling to provide the "flexibility" of the power system in its low-carbon redevelopment

Ruslan Alikin*

Energy research institute of Russian Academy of Science, 117186 Moscow, Nagornaya 31/2, Russia

Abstract. Initiatives to combat climate change imply the decarbonization of the economy. Transport and energy are the most promising sectors in terms of decarbonization's technological feasibility. However, the decarbonization of these sectors may cause several problems in terms of the sustainability of the energy system. This is because electric transport creates additional atypical demand in the energy system, while non-carbon energy sources such as nuclear power plants and renewable energy sources have generation limited by maneuverability characteristics or weather conditions. To avoid significant consequences from imbalances between demand and supply, before implementing decarbonization projects, it is necessary to conduct modeling of the energy system's flexibility. Solving such a task requires specialized modeling tools - optimization models of intra-annual commercial dispatching of generating and storage capacities. This article addresses critical requirements for a modeling tool, various applicable modeling approaches, and software options. The article also presents the results of modeling the conditions for balancing supply and demand in the UES of Russia at the level of 2050, considering a high level of transportation electrification and covering additional demand through various generation technologies, including carbon-neutral ones.

1 Introduction

In the context of global initiatives to combat climate change, such as the Paris Agreement [1], most countries around the world are focusing their attention on the decarbonization of their economies. Particular emphasis is placed on the energy sector, particularly the electricity industry. This is due to the unique technological capabilities of the industry in the use of various types of carbon-free resources as an alternative to traditional fossil fuels.

According to the International Energy Agency (IEA), the share of non-carbon sources in the total electricity production has increased by almost 3.5 times over the past 50 years [2], mainly due to an increase in the capacity of nuclear power plants (NPP) and renewable energy sources (RES).

For Russia, the task of decarbonizing the economy is also on the agenda. The President has outlined a long-term goal to achieve carbon neutrality by 2060 [3], and therefore, as part of the country's Development Strategy until 2050 with low greenhouse gas emissions, the Government of the Russian Federation [4] plans to significantly increase the capacity of carbon-free generation [5].

Within the framework of the Unified Energy System of Russia (UES), carbon-free power plants, primarily hydroelectric power plants (HPP) and nuclear power plants account for about 32% of installed capacity, providing 36% [6] of total electricity generation. At the same time, by 2035 it is planned to commission 12.2 GW of new NPP power units, 6.7 GW of new HPPs, and pumped storage power plants (PSPP) [5].

Unlike most European countries, the development of wind and solar energy in Russia is relatively small, despite the support measures taken by the government [7], the share of renewable energy is about 2.4% of the total installed capacity [8].

In addition to the energy sector, various countries, including Russia [9], are planning or already implementing substantial efforts towards decarbonization in other sectors, particularly the transportation sector, which also holds significant potential for replacing organic fuels with electric power [10] or hydrogen (a resource often produced using electric power).

As the use of carbon-free energy sources such as RES and NPPs increases, and more consumers switch to electricity, there is a need to adapt energy systems to cope with the growing imbalance between the changing daily profiles of electricity consumption and generation. This challenge is further amplified by the increasingly stochastic nature of electricity generation due to the impact of RES [11].

The adaptability of an energy system to such imbalances is often referred to as "flexibility" and can be ensured by having sufficient resources for reserving or rapidly adjusting both production and consumption loads.

Various solutions with varying degrees of complexity, effectiveness, and cost have been successfully implemented, contributing to the integration

* Corresponding author: ruslanalikin@bk.ru

of significant amounts of carbon-neutral energy generation in different energy systems. For instance, Denmark has implemented such solutions in parallel energy systems, Ireland in isolated energy systems, and King Island in Australia in small island systems [12, 13].

These solutions encompass geographic distribution of RES generation, restructuring markets for flexibility rewards, enhancing grid infrastructure, deploying advanced battery technologies, developing demand-side management programs, and much more [14].

Many of the aforementioned solutions do not require significant investments and can be considered as initial steps to unlock the potential of the energy system. However, for further progress toward achieving energy transition goals, it is crucial to take a comprehensive approach that weighs the costs of implementation against the achievable benefits of increased flexibility.

Early and long-term planning of the energy system development in the context of a low-carbon transition is necessary to avoid the implementation of costly urgent solutions when flexibility issues arise. For instance, a small system with limited maneuverable capacity and a high proportion of outdated equipment might experience flexibility shortages even with very low shares of NPP or RES, while a larger and more flexible system might encounter these challenges much later. In addition to the size of the power system, the structure and condition of existing capacity, the choice of future potential measures to increase the flexibility of the power system can also be influenced by the prospects for future growth in demand.

In several large energy systems, including the Russian Unified Energy System (UES) [15], a significant excess of capacity persists, many of which have limited maneuverability due to equipment conditions or co-generation of electricity and heat.

On the other hand, in larger energy systems, many thermal power plants have likely reached their payback periods, while newly constructed capacities in energy systems with rapidly growing demand might still have a considerable path to achieving profitability.

In newly established or fast-growing systems, it is usually easier to plan for flexibility as investments can be incorporated into market mechanisms and plans for introducing new generating and network capacities, instead of expensive upgrades to existing power plants and grid infrastructure [16].

Flexibility planning is a complex multi-stage process that requires consideration of a multitude of diverse factors and the solution of intricate mathematical problems.

2 Methods and tools for assessing energy system flexibility

Modeling is based on the creation of formalized models, which are simplified but still representative descriptions of reality. Modeling is widely used in scientific research to analyze complex phenomena and systems, as well as to facilitate decision-making under uncertainty. Energy systems are one of the traditional areas of application of various modeling methods.

Energy planning entails the use of mathematical methods, primarily optimization, and corresponding software to study the development and operation of energy systems. The results of energy modeling are indispensable for quantitatively justifying strategic decisions when formulating and implementing energy policies and programs, including in the context of low-carbon economic and energy transitions.

In modern energy planning practices, the study of the flexibility potential of an energy system and the assessment of the necessary measures are integral components of rationalizing the structure of generating capacities and prioritizing directions for technological upgrades in the power sector, particularly regarding decarbonization requirements.

Addressing such a task requires a specialized modeling tool that simulates the process of commercial dispatch of energy and meets several general requirements:

- The capability for co-modeling the functioning of various segments of the energy system (generation, consumption, inter-system energy flows, etc.),
- Sufficiently high levels of technological and temporal detailing for objective modeling of different energy technologies and their interactions,
- An adequate modeling horizon, spanning at least a year, to capture intra-annual (seasonal, weekly, daily, etc.) variations in consumer load profiles and the dispatch capabilities of various types of power plants,
- The ability to search for economically optimal solutions and analyze price-related implications

The last requirement is especially important to ensure that proposed solutions are justified from both technical and economic standpoints. Measures to enhance energy system flexibility during its low-carbon transformation are part of a comprehensive investment plan for altering the structure of generating capacities and developing the electrical grid. The task before experts is not to achieve maximum flexibility at any cost. On the contrary, given a predefined capacity development plan, it's necessary to determine a reasonable scale of flexibility enhancement and propose the least costly measures for achieving it. Moreover, the assessment of the cost of adapting the energy system to the new capacity structure could lead to adjustments in the initial plan of changing the generation capacity structure.

When exploring the flexibility potential of a power system, it's important to assess the price implications. This is especially true in a working classical competitive electricity market where pricing is based on the marginal short-term variable costs of suppliers. The intensive development of non-carbon generation with different intra-annual modes of power use significantly distorts the profile of the supply curve [11] and changes the levels of equilibrium prices (up to zero or negative values in some hours). The use of linear models of commercial dispatching with minimization of electricity production costs makes it possible to study the change in the spot price profile based on the solution of a dual linear programming problem. The shadow prices of the balance equations formed in the dual problem reflect the cost of producing an additional MWh of electricity over a given demand, that is, the spot price.

Existing tools and methods for flexibility assessment in global practice are intended for various purposes, ranging from visual comparisons to operational planning of energy systems of varying complexity. Simpler tools can be utilized for preliminary risk analysis of insufficient energy system flexibility in regions and countries lacking powerful computational tools required for detailed studies of renewable energy integration. They also serve to enhance the awareness of individuals involved in making strategic decisions and motivate a more detailed analysis of the problem. More complex and comprehensive tools can be employed for large-scale studies on the low-carbon transformation of energy systems, including their economic component (determining total costs, required revenue, and the impact on competitive price profiles).

Based on the complexity of the task and the level of detail (technological, regional, temporal, etc.), existing flexibility assessment approaches outlined in the literature [17] can be categorized into three levels:

• Level 1: Tools with minimal requirements for input data, for example, without time series. They can be based on data about generation structure, inter-system connections, and other potential flexibility sources and usually require expert assessment. A qualitative assessment can provide a quick comparison of different options for changes in the energy system structure and give recommendations for priority steps to enhance its flexibility. Such tools can be useful for understanding the depth of potential operational problems, but they won't provide much quantitative information for specific investment plans and projects.

• Level 2: Tools that assess a sufficient level of flexibility based on time series and more detailed individual data, usually with calculations performed without full optimization of the energy system's operation over the considered period. Time series (e.g., demand and generation, which must be synchronized with each other) are formed based on historical data and/or meteorological sources and transformed for future use. The modeling results can indicate when and to what extent additional flexibility may be required to avoid forced curtailment of RES, power shortages, or other undesirable phenomena.

• Level 3: Tools based on optimization models of power dispatch (possibly used in combination with models of investment planning for generation and network capacities). Third-level models are widely used in the operational management and planning of energy systems. Therefore, they provide the most detailed techno-economic basis for analyzing the sufficiency of flexibility. However, such models are often complex tools requiring vast amounts of input information. The quality of model solutions critically depends on the level of expertise and experience of researchers. Many of these models were developed for other purposes, and hence most of them require subsequent processing of results for flexibility analysis.

Based on this classification, Table 1 provides a brief overview of several tools and methodologies that can be used for flexibility assessment, along with their typical characteristics and limitations.

Among the considered tools, the most effective (but also difficult to use) are optimization models for intra-annual commercial dispatch of generation and storage capacities. One such model, called FlexTool, has been developed by the International Renewable Energy Agency (IRENA).

FlexTool is a linear optimization model that mathematically formalizes the real-world problem of optimizing hourly energy dispatch (including investment planning). In this model, the objective function and constraints have linear dependencies on variables. The model aims to find variable values that minimize the objective function (costs) while adhering to constraints defined by linear equations or inequalities.

This tool is currently being integrated into the ERI RAS SCANER model-information complex [18]. Input scenarios for changes in the generation structure (including potential emissions constraints from power plants and boilers, carbon pricing, and other carbon regulatory measures) are formed using another tool developed at ESI RAS, the Energy and Power Optimization Model (EPOS). Solutions obtained from EPOS (power structure determined through economic optimization over a significantly longer horizon than FlexTool) are tested for flexibility compliance for each hour using dispatch modeling.

FlexTool is capable of modeling systems of any size, provided that input data is sufficiently aggregated. Also, a significant advantage of FlexTool is that it has an open-source code, allowing users to understand its assumptions and operational principles, leading to more informed conclusions drawn from the obtained data.

3 Modeling the flexibility of the UES of Russia, taking into account the efforts to decarbonize the energy and transport sectors of the economy

As an example of flexibility modeling, five scenarios for the development of the UES of Russia for 2050 were modeled to assess the adaptability of the forecast structure of generating capacities to changing modes of power consumption due to the development of electric transport. In the considered scenarios, the basic demand (without electric transport) by 2050 will increase by 378.5 billion kWh or by 36% compared to the year 2020. At the same time, scenarios 2-5 provided for additional demand caused by the electrification of transport in the amount of 167.7 billion kWh or 10.5% of total demand. The scenarios considered various options to meet additional demand:

• In Scenario 1, the structure of installed capacity optimized in the EPOS model provides the predicted demand without the additional influence of electric transport.

Table 1. Methods and tools for assessing the flexibility of the energy system [17].

	Approach	Software	Brief description	Requirements and restrictions
Level 1	Expert review	NREL System Evaluation Tool	Provides a basis for evaluating characteristics that are important in terms of flexibility	Experience is required to evaluate power systems in terms of flexibility. Not based on evidence
	Visual comparison	GIVAR (IEA)	Provides a snapshot of the current situation with relevant information about system resilience.	Based on a limited set of data. Can only give a qualitative assessment
Level 2	Manoeuvrability valuation	FAST2 (IEA), IRRE	Simulates the dispatching of the system in an hourly manner. In case of an insufficient level of flexibility, it informs about the impossibility of changing the load according to the specified parameters	The main attention is paid to the increase (growth) of capacity and reserves
	Operational uncertainty	InFLEXion (EPRI)	Uses scheduling simulation results and historical data to assess potential weaknesses in flexibility in various situations	Post-processing tool. Preliminary calculations and results of the dispatching model are required
	Flexibility test	Flex Assessment (EDF), REFLEX (E3)	Assesses the need for flexibility within an hour during the planning phase. Ability to account for variables, operating constraints, and additional reserves	Pre-optimization tool. Requires a separate power system planning and generation balancing model
Level 3	Reserve valuation	FESTIV (NREL)	Tool for dispatching (operational) balancing of scenarios with a high share of RES in a relatively short time frame (from seconds to a day)	High level of detail. Doesn't optimize the capacity expansion plan, only operational management.
	System planning and operational modeling	REFLEX (NREL), RESOLVE (E3)	A tool for optimal scheduling, while taking into account operational constraints that are relevant in terms of flexibility. It also allows you to solve the problem of capacity expansion planning at the lowest cost.	REFLEX optimizes the scenario with a valley-filling algorithm, which can lead to inaccuracies.
	System planning and operational modeling	IRENA FlexTool	A tool for optimal capacity dispatching and/or for optimizing investment decisions can be used to determine whether a power system is sufficiently flexible and how to increase it	Plant characteristics, networks, and time series are required. Only linear optimization.

• In Scenario 2, the structure of installed capacity is optimized using the EPOS model, while demand is increased taking into account the electrification of transport.

• In Scenario 3, the additional demand for energy and capacity, caused by the electrification of transport, is provided exclusively by combined cycle gas turbines (CCGT).

• In Scenario 4, the additional demand is provided only by NPP.

• In Scenario 5, the additional demand is provided only by RES.

Detailed scenarios and assumptions are described in Article [19]. Table 2 reflects the aggregated types of power plants, their capacities, and average annual capacity utilization factors (CUFs), obtained through hourly aggregation. Changes in power plant CUFs and the extent of energy storage utilization, make it possible to obtain an integral characteristic of the flexibility level. Notably, changes in the CUFs of thermal power plants help determine shifts in their capacity utilization patterns as the share of non-carbon generation increases. Other characteristics of the scenarios are changes in annual CO_2 emissions, total costs of ensuring balance, as well as changes in the weighted average spot price of electricity.

Table 2. Analysis results [19].

Scenarios	*1*		2		3		4		5	
	Capacity utilization factor, %/ Capacity, GW									
Thermal power plants	*58.87*	*150,7*	59.00	*171,3*	58.13	*183,9*	51.00	*150,7*	65.46	*150,7*
Wind power plants	*27.97*	*15,7*	27.97	*15,6*	27.97	*15,7*	27.97	*15,7*	27.97	*37,3*
Solar power plants	*19.76*	*5,4*	19.81	*6,8*	19.76	*5,4*	19.76	*5,4*	19.39	*17,1*
Nuclear power plants	*90.85*	*42,1*	90.85	*48,2*	90.85	*42,1*	90.72	*75,3*	90.85	*42,1*
Hydropower plants	*44.04*	*50,2*	42.49	*52,1*	44.04	*50,2*	44.04	*50,2*	44.04	*50,2*
Energy storages	*13.67*	*2,5*	25.63	*5,7*	33.82	*2,5*	36.9	*2,5*	33.03	*2,5*
	Change relative to 1 scenario, %									
Annual total costs to ensure the balance of electricity (the value of the objective function of the model)	*0.00*		12.04		15.49		-8.60		12.88	
Annual CO_2 emissions	*0.00*		12.55		13.51		-13.20		12.37	
Weighted average spot price of electricity	*0.00*		0.18		-7.72		-14.17		15.50	

Scenario 1 is considered a reference scenario, where no additional growth in demand due to electric vehicles is foreseen. The primary source of "flexibility" in this scenario is HPP, whose generation throughout the day varies by a factor of 2.8. Thermal power plants (TPP) and energy storages also contribute to balancing the demand and generation but to a lesser extent. Energy storage units are charged during nighttime hours when electricity demand is minimal, thereby leveling the demand curve during the hours of the night dip. Discharge from energy storage units occurs during daytime and evening peak hours when electricity prices are at their maximum and the use of storage is economically justified. Similarly, with varying degrees of intensity, energy storage units' function in the other examined scenarios.

In scenarios with increased electrification of transportation, the total electricity demand by 2050 will grow by 10.5%, and the overall installed capacity of power plants will increase by 12.5%. The calculated parameters of scenarios 2 to 5 are compared with scenario 1.

In Scenario 2, the additional electricity demand is met through a combination of different types of power plants (thermal, nuclear, hydro, and renewable), based on the results of optimizing the capacity structure, which closely resembles the parameters of Scenario 1. Consequently, the overall costs of maintaining the electricity balance in Scenario 2 increase proportionally to the rise in demand relative to Scenario 1. Minor adjustments in the capacity structure also signify that the spot market supply curve and the market equilibrium point will closely align with Scenario 1. This assertion is supported by a marginal difference (0.2%) in the level of the average weighted spot electricity price. A change in the load profile due to electric transport will require more intensive use of storage devices - their storage capacity factor will increase by 12%, with a slight decrease in the HPP capacity factor by 1.5%.

In Scenario 3, where the additional demand is only by CCGT (with the capacity of all other types of power plants held constant at the level of Scenario 1), the higher fuel consumption leads to an increase in the cost of maintaining the electricity balance. This increase is higher than the demand growth itself, reaching 15.5% compared to Scenario 1. This scenario is characterized by the highest volume of CO_2 emissions. In this case, gas-fired power plants with lower specific fuel consumption partially replace less efficient stations in the balance, contributing to a certain reduction in the average weighted electricity price. In terms of flexibility, this scenario doesn't pose significant issues as CCGTs are among the most maneuverable energy sources with minimal operational limitations.

In Scenario 4, an intensive growth in the volume of electricity from NPPs displaces TPPs from the balance, which, in particular, can be seen from the decrease in their capacity factor in Table 2. To balance the low maneuverability of NPP, energy storages were more intensively utilized. Despite the increase in electricity consumption, both CO_2 emissions and the overall costs of maintaining the electricity balance decreased (by 8.7%). However, the most notable reduction occurs in the average weighted electricity price (by 14.2%). It's important to consider the conditional positive effect of reducing the spot price, as higher capacity payments might be necessary to ensure the financial viability of investments in new capacity (including NPP), leading to a substantial increase in the single tariff electricity price.

The last Scenario 5 assumes that the growth in demand is covered by RES. Model calculations revealed that this scenario places the highest demands on the energy system's flexibility. Due to the non-guaranteed generation of wind and solar power plants during certain hours, there are unacceptable reductions in reserves, and significant curtailment of RES generation occurs according to the balance conditions. In this variant, due to the low-capacity utilization factors of RES power plants, the forced solution is additional electricity generation from TPPs (with an increased capacity factor, as indicated in Table 2). Modeling this scenario showed an increase in the costs of maintaining the electricity balance, and a rise in the average weighted electricity price, and also revealed undesirable phenomena in terms of system flexibility, reflected in Figure 1.

4 Conclusions

Energy planning and flexibility modeling are essential components of finding an optimal combination of resources and technologies to meet energy demand sustainably, affordably, and reliably. Environmental considerations play an increasingly significant role, driving the transition towards a low-carbon technological structure in the energy sector, the replacement of fossil fuels with non-carbon sources, and electrification in end-use sectors, particularly transportation. Integrating large volumes of non-carbon sources and the emergence of new types of electricity consumers may require additional measures to enhance energy system flexibility. Special models of commercial dispatch are employed for investigating this need, typically with detailed intra-annual granularity and economic optimization criteria. The dual problem solution of linear optimization models allows the exploration of changes in the spot electricity price profile.

An analysis of the functioning of the UES of Russia on the horizon up to 2050, performed using the FlexTool linear optimization model, showed that electric vehicles are a fairly powerful factor influencing both the volume of demand and the requirements for flexibility of the UES of Russia. Scaling up the optimal structure of generating capacity to meet higher demand allows achieving it with comparable growth rates in total costs for maintaining the electricity balance and minimal changes in the average weighted price. Furthermore, the energy system possesses sufficient flexibility reserves to maneuver the capacities of power plants and storage systems.

From an economic and environmental standpoint, the most efficient scenario involves covering the additional electricity demand primarily through NPP. However, this scenario requires further development of storage systems, including pumped storage power plants. An alternative scenario based on a predominant expansion of renewable energy sources (RES) could lead to substantial increases in the overall costs for maintaining the electricity balance, a rise in the average weighted electricity price, and exhibit inadequate energy system flexibility, necessitating significant enhancement of maneuvering capabilities both in generation and among consumers.

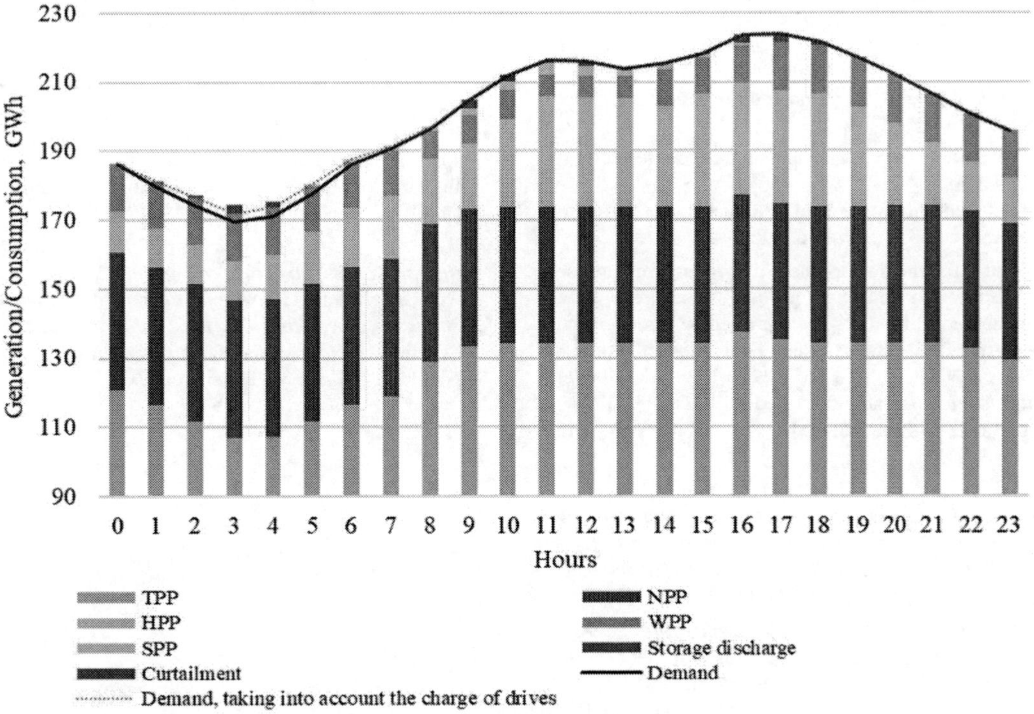

Fig. 1. Hourly electricity generation in 2050 for winter working day conditions (Scenario 5) [19].

References

1. *Paris Agreement*, in Proceedings of the UN Climate Change Conference (COP21), (2015) Available: https://unfccc.int/sites/default/files/english_paris_agreement.pdf, Accessed on: (July 6, 2023)
2. IEA, *Key World Energy Statistics*, (2021) Available: https://iea.blob.core.windows.net/assets/52f66a88-0b63-4ad2-94a5-29d36e864b82/KeyWorldEnergyStatistics2021.pdf, Accessed on: (July 6, 2023)
3. Plenary session of the international forum "Russian Energy Week", (2021) Available: http://www.kremlin.ru/events/president/news/66916, Accessed on: (July 6, 2023)
4. Decree of the Government of the Russian Federation of October 29, № 3052-p, (2021) Available: http://static.government.ru/media/files/ADKkCzp3fWO32e2yA0BhtIpyzWfHaiUa.pdf, Accessed on: (July 6, 2023)
5. *Russian fuel and energy complex 2022: challenges, results and prospects*, J. En. Pol. (2023) Available: https://energypolicy.ru/rossijskij-tek-2022-vyzovy-itogi-i-perspektivy/business/2023/12/13/, Accessed on: (July 6, 2023)
6. System operator of UES of Russia (2023) Available: https://www.so-ups.ru/functioning/ees/ups2022/, Accessed on: (July 6, 2023)
7. Ministry of Energy of the Russian Federation, *Decisions on the provision of subsidies from the federal budget for state support for the technological connection of generating facilities operating based on the use of renewable energy sources* (2020) Available: https://minenergo.gov.ru/node/489, Accessed on: (July 6, 2023)
8. Renewable Energy Development Association (2023) Available: https://rreda.ru/statistics_of_renewable_energy_in_russia, Accessed on: (July 6, 2023)
9. Concept for the development of production and use of electric road transport in the Russian Federation for the period up to 2030 (approved by Order of the Government of the Russian Federation of August 23, 2021, № 2290-p), Available: http://static.government.ru/media/files/bW9wGZ2rDs3BkeZHf7ZsaxnlbJzQbJJt.pdf, Accessed on: (July 6, 2023)
10. A.I. Solyanik, F.V. Veselov, R.O. Alikin, *Influence of electrification in the road transport sector on the level of electricity consumption and the daily load schedule in the UES of Russia*, J. Invest. of the Rus. Acad. of Sci., № 1, pp. 57-71 (2023)
11. S.P. Filippov, M.D. Dilman, *Renewable energy: System effects,* in Proceedings of the 12th international conference "Management of the development of large-scale systems", October 01–03, Moscow, Russia, pp. 558-567 (2019)
12. RTE, *Denmark on track to have 50% renewable energy by 2030*, (2018), Available: https://www.rte.ie/news/newslens/2018/0111/932573-denmark-wind-farm/, Accessed on: (July 6, 2023)
13. B. Kroposki, *Integrating high levels of variable renewable energy into electric power systems*, Journal of Modern Power Systems and Clean, (2017), Available: https://link.springer.com/article/10.1007/s40565-017-0339-3, Accessed on: (July 6, 2023)
14. IEA, *Status of Power System Transformation: Advanced Power Plant Flexibility*, (2018), Available: https://www.oecd.org/about/publishing/Corrigendum-StatusofPowerSystemTransformation2018.pdf, Accessed on: (July 6, 2023)
15. A.A. Titov, *Excessive power: It is not advisable to maintain output. Strategic choice of the location of the comma*, J. Energorynok, (2015), Available : https://www.so-ups.ru/fileadmin/news_import/e_market_0315_titov.pdf, Accessed on: (July 6, 2023)
16. IRENA, *Stranded Assets, and Renewables : How the energy transition affects the value of energy reserves, buildings, and capital stock*, (2017), Available: http://www.irena.org/publications/2017/Jul/Stranded-Assets-and-Renewables, Accessed on: (July 6, 2023)
17. IRENA, *Power system flexibility for the energy transition* (2018), Available: https://www.irena.org/publications/2018/Nov/Power-system-flexibility-for-the-energy-transition, Accessed on: (July 6, 2023)
18. F.V. Veselov, A.A. Khorshev, *Integrated modeling and information technology for strategic planning in the energy sector of Russia*. in Proceedings of the 11th International Conference on Application of Information and Communication Technologies (AICT), Moscow, Russia. pp 1-5 (2017)
19. R.O. Alikin, *Model assessment of the adaptability of the predictive structure of generating capacities with a growing share of carbon-free sources to changing power consumption modes*, in Proceedings of the School of Young Scientists, 16–17 November, Moscow, Russia, pp. 7-17. (2022)

Developing a three-phase converter to be used in a prototype of a virtual inertia system based on frequency and active power relationship

Maxim Burmeyster*, *Ilia* Berdyshev, *Ramis* Bulatov, and *Rinat* Nasyrov

Department of Electric Power Systems, National Research University "MPEI", Moscow, Russian Federation

Abstract. The integration of power plants based on renewable energy sources (RES) into existing energy systems causes a number of problems, one of which is the violation of the stability of the generators parallel operation. One of the possible methods for the above problem solving is the use of virtual inertia systems (VIS). In this paper, we review the process of a converter developing for further use in a VIS prototype. The developed device converts direct current from a laboratory power supply into a sinusoidal three-phase voltage of industrial frequency. The switching of the converter's power transistors is controlled through digital signal processor (DSP) TMS 320 F 28379 D via IR 2110 drivers. Using Altium Designer, a schematic diagram and layout of the printed circuit board of the converter were developed. Based on the manufactured printed circuit board, a converter was assembled, the correctness of its operation was checked, and a conclusion was made about the possibility of its use in the VIS prototype.

1 Introduction

Currently, renewable energy sources are widely used all over the world. The generation of electricity by wind power plants (WPP), solar power plants (SPP), etc. is actively introduced into electric power systems (EPS) and distribution networks. According to the report of the International Renewable Energy Agency, in the period from 2012 to 2022 the installed capacity of power plants based on RES more than doubled from 1443 to 3371 GW [1].

The RES integration into existing power systems causes a number of problems, among which one of the most important is the violation of the stability of the parallel operation of generators in the system. Most modern wind turbines operate with a variable speed of wind turbine rotation and are connected through semiconductor power converters, which does not provide inertial response. Photovoltaic cells and energy storage systems (ESS) are connected through inverter equipment and do not contribute to the overall inertia of the EPS either [2, 3]. The stability of the system directly depends on the reserve of kinetic energy in the generators. Hence, the sources connected through power converters do not increase the reserve of the system's kinetic energy, but at the same time, they act as sources of power. This leads to heavier transient processes in the EPS and a decrease in its stability during disturbances [4–7]. One innovative solution to the problems described above is the use of virtual inertia systems.

2 Virtual Inertia System Concept

A virtual inertia system can be characterized as a complex of devices and algorithms that converts direct current into alternating current of industrial frequency, simulating the inertial response of synchronous machines to disturbances in the EPS [8]. The main elements of the VIS are power converter equipment (converter), converter control system, and electric power storage system (Fig. 1).

The system is based on the virtual inertia algorithm. It allows sources of electricity connected to the network through converters acquire the properties of synchronous generators (SG). VIS determines the control signals for the converter transistors, so that energy generated by renewable energy sources has the characteristics of that produced by SGs – frequency maintenance, oscillation damping [9].

3 Development of the Functional Diagram of the Converter

The designed three-phase converter can be represented as four large functional blocks: a control system, a power section, current and voltage sensors, and a protection system. The developed functional diagram is shown in Fig. 2.

* Corresponding author: max.burmeyster@gmail.com

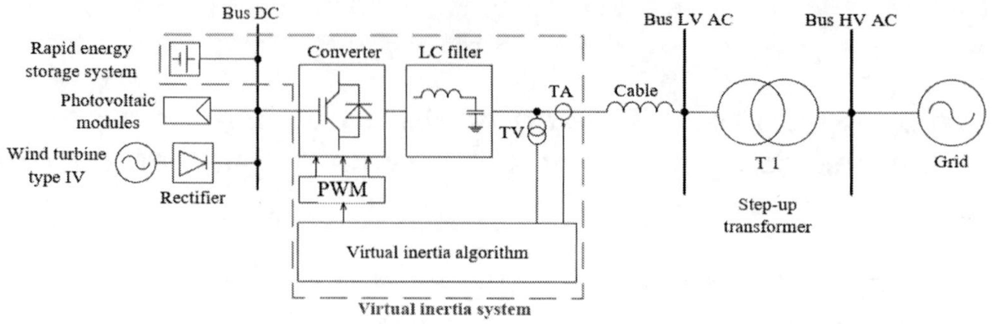

Fig. 1. An example of VIS use.

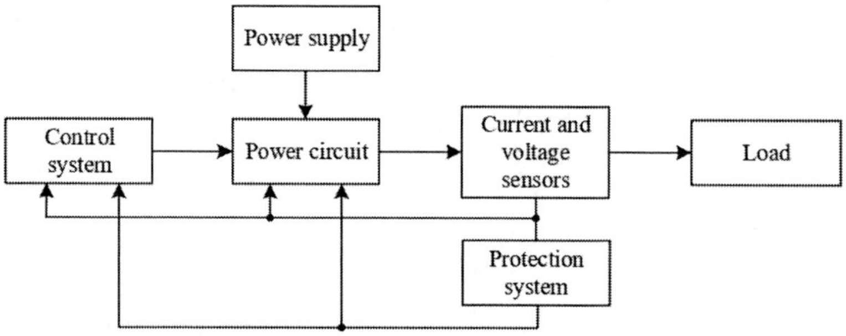

Fig. 2. Functional diagram of the developed converter.

The control system generates control signals for the power section of the converter (transistors) in accordance with the virtual inertia algorithm using a programmable DSP. The power section is a set of transistors connected in the form of a three-phase two-level inverter and their drivers. Current and voltage sensors measure the instantaneous values of current and voltage from the power supply and load side. After that, the measured values are transmitted to the protection and control systems to create feedback. The protection system is necessary to prevent emergencies on the converter, such as exceeding the permissible load or a short circuit. An alarm signal is sent to the control system and power section when an impermissible operating mode occurs. As a result, the control system blocks the PWM operation, and the transistor drivers turn off the power switches.

3.1 Developing control system

The control system was designed based on the LAUNCHXL-F28379D breadboard, with Texas Instruments DSP TMS320F28379D installed on it. TMS320F28379D is a powerful 32-bit floating-point microcontroller unit (MCU) designed to perform complex feedback control tasks in such areas as: industrial motor drives, solar inverters and digital energy, electric vehicles and transport, and signal processing.

To obtain sinusoidal voltage at the output of the converter, a method was chosen to control power transistors using pulse-width modulation (PWM). PWM signals are generated by the DSP using the developed software code.

3.2 Developing power section

Altium Designer software package was used to develop the functional diagram and the layout of the printed circuit board (Fig. 3). The elements were chosen, primarily, based on the technical limitations of the equipment available in the laboratory. The maximum current generated by laboratory power supplies and passing through the inverter is 10 A and the maximum voltage is 30 V.

MOSFETs were used in this design, which are the most widely used power devices due to their high switching speed, low gate drive power, and parallel connection capability. A MOSFET used in this work is an IRF 530 N with a drain current of 14 A at 25 °C, a drain-to-source voltage of 100 V (min.), a drain-to-source static resistance of 0.16 Ω (max.).

https://doi.org/10.1051/e3sconf/202346101044

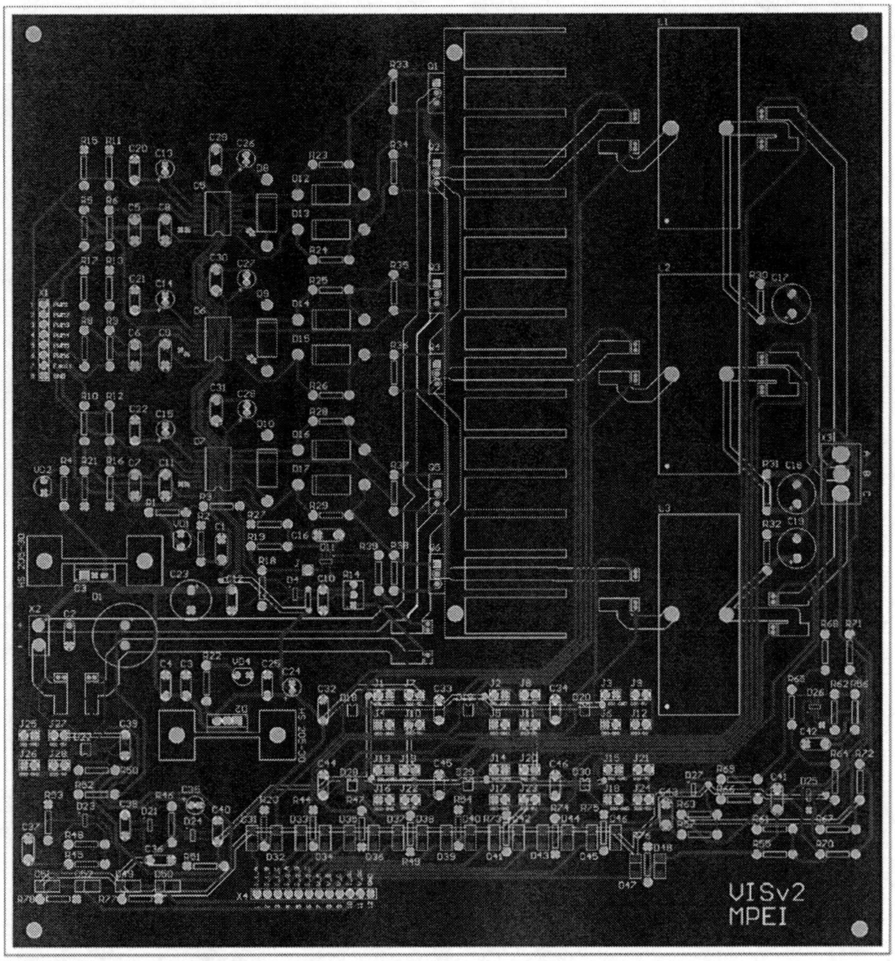

Fig. 3. PCB layout in Altium Designer.

The next step in the development of the converter was the coordination of key elements with the control system. To control the transistors, a voltage of 15 V and a pulsed current are required, which allows reducing switching losses. To turn on the upper transistor in the leg, it is necessary to apply voltage to its gate relative to its source, while the source potential changes in accordance with the voltage at the load. In other words, to unlock the upper transistor in the leg, the voltage at its gate must be Udc+15V, where Udc is the DC supply voltage. For this, half-bridge drivers IR 2110 are used, one for each phase (3 in total). This driver allows coordinating the 3.3 V control system and the power transistors of the converter, ensuring the correct functioning of the upper phase transistor, providing a delay between turning off the upper transistor and turning on the lower one.

3.3 Current and voltage sensors

To implement current and voltage feedback and protection operation, it is necessary to use appropriate sensors to measure the magnitude of currents and voltages at the converter output.

3.3.1 Voltage Sensing

The DC bus voltage and AC output voltages are measured differentially using resistor dividers and operational amplifiers, as shown in Figure 4. An offset voltage is added to the signal to allow measurement with an ADC that can only convert positive voltages.

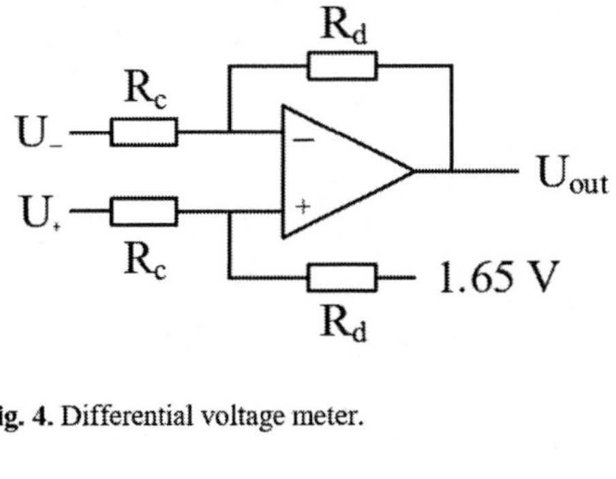

Fig. 4. Differential voltage meter.

https://doi.org/10.1051/e3sconf/202346101044

The voltage at the output of the differential voltage meter is calculated by formula (1):

$$U_{out} = \frac{R_d}{R_c}(U_+ - U_-) + 1.65 \qquad (1)$$

3.3.2 Current Sensing

Critical to building a feedback-control system is the accurate measurement of the inverter current. There are three main ways to measure current in power circuits, which differ significantly in their characteristics: current shunt, current transformer, and Hall sensor. This project uses shunt current measurement. The method was chosen because it is the simplest, cheapest and fastest way to measure current. At high currents, more power is dissipated on the shunt, which leads to a decrease in efficiency and the need to cool down the shunt. The absence of galvanic isolation when using a shunt in this project is not critical, since the operating voltage of the inverter does not exceed 15 V, and relatively small currents flowing through the shunt will not create major power losses. In this project, current is measured in three places: before, after the filter inductor, and in the DC circuit.

3.4 Protection systems

A system of protection against exceeding the maximum allowable current was developed for the converter. Overcurrent protection or short circuit protection is the most common type of protection because this type of malfunction can lead to the complete destruction of an electronic device. The installation of protection is especially important when developing a converter prototype due to possible flaws and errors in the printed circuit board layout.

A shunt with a 0.01 Ω resistance is included in the gap between the "ground" and the transistor emitters as a current sensor. An operational amplifier is connected to the shunt to amplify small voltage (from tens of millivolts to volts). The resulting amplified signal is compared with the reference signal using a comparator. If the measured voltage is above a certain threshold value, then the signal at the output of the comparator changes from "0" to "1", which disables the PWM signal to the transistor drivers. The trip current is set by a trimmer, which forms a voltage divider at the inverting input of the comparator. The functional diagram of the converter protection system is shown in Fig. 5.

4 Results

The work resulted in the development and manufacturing of a converter printed circuit board, a control system based on DSP TMS320F28379D, and a current protection system. To test the operation of the converter, a full-scale experiment was carried out with the connection of a resistive load and simulation in the MATLAB/Simulink software package to verify the results obtained.

For a full-scale experiment, a test bench was assembled (Fig. 6). A KORAD 3005D laboratory DC power supply was used as a power source and a RIGOL DS1054Z oscilloscope was used to record oscillograms. The main parameters of the test bench are shown in Table 1.

Table 1. Main parameters of the test bench.

Parameter	Value
DC supply voltage	20 V
Linear AC voltage at the output of the converter	12.25 V
Load resistance	6.8 Ω
Filter coil inductance	1.7 mH
Filter capacitor capacity	470 μF
Transistor switching frequency	10 kHz
AC frequency	50 Hz
Three-phase load power	24.852 W

After programming the DSP, the PWM signal generated at its outputs was checked. The PWM is tuned to 10 kHz frequency, with 50 Hz modulated signal frequency (Fig. 7).

Fig. 8 shows the output voltage oscillograms of the converter. Based on the data analysis, it was concluded that the inverter worked correctly, the voltage shape was close to a sinusoid, the total coefficient of harmonic components of the voltage at the output of the converter was 1.37%, as required by state standard GOST 32144-2013.

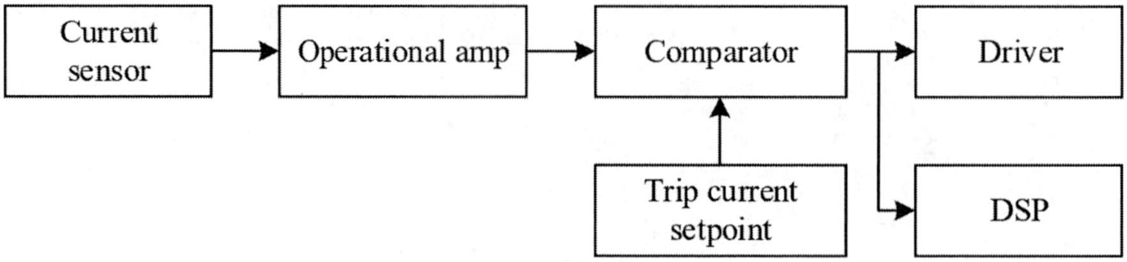

Fig. 5. Functional diagram of the converter protection system.

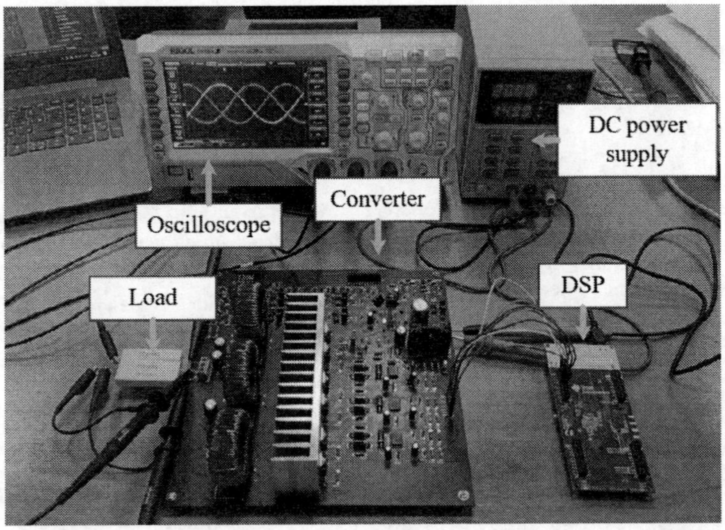

Fig. 6. Test bench for experiments.

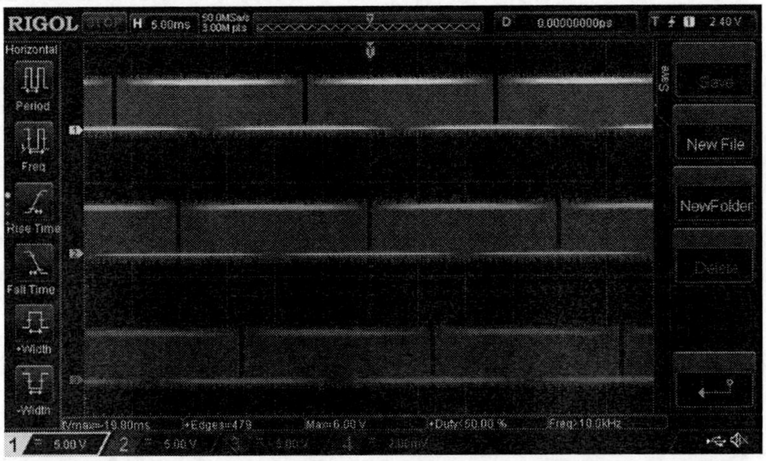

Fig. 7. PWM signal oscillograms of the upper keys of the developed phase converter A (yellow), B (blue), C (purple).

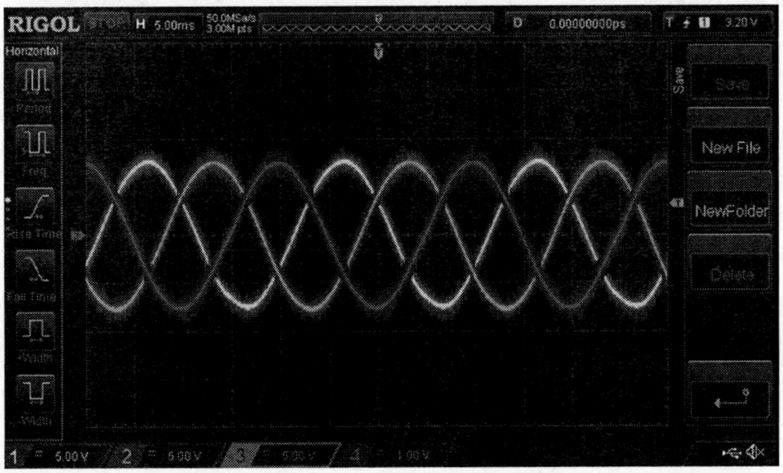

Fig. 8. Oscillograms of the output voltages of the developed phase converter A (yellow), B (blue), C (violet).

5 Conclusion

The work resulted in construction of a three-phase converter based on DSP TMS320F28379D, which was implemented using MOSFET power transistors. The converter generates 12 V three-phase sinusoidal linear voltage when powered by a 19.596 V power supply. For the DSP, the software code was written using the timers and interrupters built into the DSP, which allows for stable generation of control signals. In addition, as part of the work, a hardware current protection system was developed that disables drivers and the supply of control signals when the preset threshold is exceeded. To test the performance of the converter, an experiment was conducted with the connection of a 21.176 W resistive load. During the experiment, oscillograms of phase voltages at the output of the converter and control signals at the output of TMS 320F28379D were recorded.

The developed inverter is ready for use in the prototype of the virtual inertia system based on the frequency and active power connection. For this purpose, the converter is planned to be connected through a step-up transformer to an external network in the laboratory and finalize the software code of the virtual inertia algorithm for DSP.

The investigation has been carried out within the framework of the project "Development of a prototype virtual inertia system for use in solar and wind power plants as part of a microgrid" with the support of a subvention from the National Research University "MPEI" for implementation of the internal research program "Priority 2030: Future Technologies" in 2022-2024.

References

1. *Renewable capacity statistics 2023*, International Renewable Energy Agency, Abu Dhabi.
2. B. Kroposki, B. Johnson, Y. Zhang, V. Gevorgian, P. Denholm, B.M. Hodge, B. Hannegan. *Achieving a 100% Renewable Grid: Operating Electric Power Systems with Extremely High Levels of Variable Renewable Energy*. IEEE Power Energy Mag. 2017, **15**, pp. 61–73.
3. F. Gonzalez-Longatt, E. Chikuni, E. Rashayi. *Effects of the Synthetic Inertia from wind power on the total system inertia after a frequency disturbance.* In Proceedings of the IEEE International Conference on Industrial Technology (ICIT), Cape Town, South Africa, 25–28 February 2013; pp. 826–832.
4. F. Dorfler, S. Bolognani, J.W. Simpson-Porco, S. Grammatico, *Distributed control and optimization for autonomous power grids,* in: 2019 18th European Control Conference, ECC 2019, 2019, pp. 2436-2453.
5. T. Ackermann, T. Prevost, V. Vittal, A.J. Roscoe, J. Matevosyan, N. Miller, *Paving the way: a future without inertia is closer than you think*, IEEE Power Energy Mag. **15** (6), 2017, pp. 61–69.
6. L. Zhang, H.P. Nee, L. Harnefors, *Analysis of stability limitations of a VSC-HVDC link using power-synchronization control,* IEEE Trans. Power Syst. 26 (3), 2011, pp. 1326-1337.
7. C. Li, I. Cvetkovic, R. Burgos, D. Boroyevich, C. Engineering, V. Tech, *Assessment of virtual synchronous machine-based control in grid-tied power converters,* in: 2018 Int. Power Electron. Conf. (IPEC-Niigata 2018 -ECCE Asia), 2018, pp. 790–794.
8. M.V. Burmeyster, R.V. Bulatov, K.A. Blinova. "*The use of virtual inertia systems to improve the quality of transients in electric power systems*". Fedorov Readings - 2021: LI international scientific and practical conference with elements of a scientific school, Moscow, November 17–19, 2021. - Moscow: MPEI Publishing House, 2021. - pp. 318–324 (In Russian).
9. H. Bevrani, T. Ise, Y. Miura. *Virtual synchronous generators: A survey and new perspectives.* Int. J. Electr. Power Energy Syst. 2014, **54**, pp. 244–254.

Analysis of profitability of production of energy companies of the Irkutsk region

Evgeniy Filatov*

Irkutsk Scientific Center SB RAS, Department of Regional Economic and Social Problems, 664033 Irkutsk, Russia

Abstract. Energy is a life support industry that determines the state of the economy of the state as a whole. The energy sector is: 1. the infrastructure of all sectors of the economy and the life support system of the population; 2. business, which provides about 30% of GDP, 50-60% of the income of the country's budget system and up to 65% of all foreign exchange earnings; 3. the geopolitical factor determining the role of Russia in the international arena. Profitability of production is an important economic category of the market economy, knowledge of the nature of which allows the most successful regulation of the economy of individual enterprises, industries and the national economy as a whole. This indicator is often used in financial and investment analysis. The study of the effectiveness of the financial and economic activities of the organization, the identification of cause-and-effect relationships between the factors used and the results obtained create the basis for a scientifically based prediction of its state in the future when planning and forecasting. In the article, to carry out a factor analysis of the profitability of production, the author has developed a 3-factor model. The article reveals the influence of factors affecting the change in the profitability of production of energy companies of the Irkutsk region and gives the author's methodological approaches for its calculation (methods of factor analysis developed by Filatov E.A.). The article presents the author's analytical, systematized statistical material for the analysis of key indicators revealing the impact on the change in the profitability of production of energy companies of the Irkutsk region regions of the Russian Federation.

1 Introduction

The energy complex of the Irkutsk region is one of the largest in the Russian Federation. It is represented by objects of "big energy", which are mainly part of the structure of "EuroSibEnergo", as well as objects of "small energy", which was previously referred to as "communal". The structure of "small energy" includes distribution electric and thermal networks, municipal, one thermal power plant, boiler houses, diesel power plants, renewable energy sources with various forms of ownership. In addition to the listed facilities, the energy complex of the Irkutsk region also includes private and state-owned companies for the extraction and processing of fossil energy resources operating in various districts of the region [1].

In Russia, among the subjects of the federation, the Irkutsk region has the highest energy potential, where there are large power plants with a total capacity of 13.6 gigawatts (6-8% of the country's energy capacity). At the same time, electricity generation per capita in the Irkutsk region is one of the highest in the world. In terms of electricity generation, the Irkutsk power system ranks second in Siberia, behind Krasnoyarsk. About 20% of the generated electricity is transmitted outside the region, mainly to the Republic of Buryatia and the Trans-Baikal Territory. At the same time, due to the capacities of the Irkutsk power system, 20% of the total electricity demand of Transbaikalia is covered. About 7.5 billion kilowatts per hour are transmitted through the Krasnoyarsk power system for the needs of the unified power system of Siberia [2].

2 Research question

The issues of finding ways to achieve stability, economic reliability and sustainable development of the economy of electric power industries are devoted to the works of domestic scientists and specialists dealing with the problems of the development of the fuel and energy complex of Russia, such as: Shmatko S. I., Veselov F. V., Kurilov A.E., Makarova A.S. and others.

A lot of scientific works of leading domestic and foreign scientists-economists, such as: Bakanov M.I., Sheremet A.D., Safronov M.A., Lyubushin H.P., Volkov O.I., Kovalev V.V., Raitsky K.I., Shepelenko G.N., Krylov E.I. are devoted to the evaluation and analysis of the efficiency of companies and industries., Vlasova V.M., Zhuravleva I.V., Kaplan R., Peters T., King W., Heine P., Todaro M., Richard J., Maroshi M. and others.

The objective and most perfect measure of the degree of efficiency of production of energy companies is the analysis of profitability of production.

* Corresponding author: johnru@rambler.ru

The most important feature of energy production is that companies in this industry work directly for the consumer without creating inventory and accounting for work in progress, and the expenses incurred during the reporting period are completely written off to the cost of energy generated. Therefore, there is no need to distribute them between finished products and work in progress.

The coincidence in time of the phase of energy production with the phase of its consumption depends on the mode of energy consumption. In some periods, the energy demand may increase or decrease, which affects the operating mode of the stations. In order to maintain a constant correspondence between energy consumption and its production, to ensure uninterrupted supply of electricity to consumers at power plants, it is necessary to have reserve production capacities. The additional costs associated with their maintenance are included in the cost of energy. The cycle of energy production at the power plant ends with its transmission through the power grid lines to consumers. In the process of transmission, about 8-9% of electricity is spent on transportation and conversion and compensates for commercial losses.

The procedure for calculating the cost price is influenced by the nomenclature of the products produced. In the energy sector, some enterprises produce one type of product (electricity), while others produce two (electric and thermal energy). In the first case, the cost is determined by the total amount of expenses incurred, in the second – they must be distributed among individual types of energy [3-5].

The questions devoted to the comparative evaluation of various methods of evaluating the effectiveness of companies, and the urgent need to resolve them, indicate the relevance and practical significance of the chosen topic of the article.

3 Materials and methods

For factor analysis of the author's 3-factor model of profitability of production, four indicators were used: net revenue, net profit, value of total assets, cost of production.

Further, based on the methods of deterministic (functional) factor analysis developed by Filatov E.A. [6], we will assess the degree of influence of three factors on the change in the profitability of production of energy companies in the Irkutsk region of the Russian Federation.

The initial data for an alternative factor analysis of the profitability of the production of energy companies of the Irkutsk region of the Russian Federation are presented in Table 1 based on data from 4 statistical bulletins of Irkutskstat (1. Financial results of the organizations. Volume 2 – large and medium-sized organizations. 2019; 2. Financial results of the organizations' activities. Volume 3 – Small enterprises. 2019; 3. Financial results of the organizations' activities. Volume 2 – large and medium-sized organizations. 2020; 4. Financial results of the organizations' activities. Volume 3 – Small enterprises. 2020).

The initial formula derived by the author for factor analysis of profitability of production will have the following form (formula 1):

$$\boldsymbol{POP} = \frac{\boldsymbol{NP}}{\boldsymbol{TA}} * \frac{\boldsymbol{TA}}{\boldsymbol{NR}} * \frac{\boldsymbol{NR}}{\boldsymbol{COP}} = \boldsymbol{ROA} * \boldsymbol{CIA} * \boldsymbol{SOCR} = \boldsymbol{F_1} * \boldsymbol{F_2} * \boldsymbol{F_3} = \prod_{n=1}^{3} \boldsymbol{F_n} \quad (1)$$

The author's model of profitability of production consists of 3 factors:

F_1 – ***ROA*** – return on assets, calculated as the ratio of net profit (***NP***) to the cost of total capital (***TA***). This indicator is important for evaluating the effectiveness of all assets of the enterprise.

F_2 – ***CIA*** – capital intensity (resource intensity) assets is the ratio of the value of all assets (***TA***) per unit volume of sales of products, goods (***NR***). Capital intensity is an indicator that is the inverse of the asset turnover ratio (or resource efficiency ratio). This indicator is important for evaluating the effectiveness of operational and investment activities [7, 8].

Table 1. Initial data for factor analysis.

No.	Indicators	No. factor's	2019 year Plan (0) *	2020 year Fact (I) **	Deviation (Δ) ***
1	***NR*** – *Net Revenue*, thousand rubles		208011893	197975704	-10036189
2	***NP*** – *Net Profit*, thousand rubles		28224778	23713264	-4511514
3	***TA*** – *Total Assets*, thousand rubles		484065517	544974580	60909063
4	***COP*** – *Cost Of Production*, thousand rubles		166612982	157610414	-9002568
5	**POP** – *Profitability Of Production* (2/4) = 6 * 7 * 8		0.169403	0.150455	**-0.018948**
6	**ROA** – *Return On Assets* (2/3)	F_1	0.058308	0.043513	-0.014795
7	**CIA** – *Capital Intensity of Assets* (3/1)	F_2	2.327105	2.752735	0.425630
8	***SOCR*** – *Sales-On-Costs Ratio* (1/4)	F_3	1.248474	1.256108	0.007635

where: * 0 – past (base) period (year) taken as a basis for comparison; ** I – reporting (current) year; *** Δ – change for the period is calculated as the difference between the fact and the plan (I – 0).

F_3 – ***SOCR*** – the author introduced the indicator sales-on-costs ratio into scientific circulation, which is calculated as the ratio of sales income (revenue) (***NR***) per unit cost (***COP***). This indicator shows the ratio of the received revenue from sales (***NR***) to the amount of funds spent on its receipt (cost – ***COP***). This indicator is important for evaluating the efficiency of production.

The resulting indicator – ***POP*** – profitability of production is calculated as the ratio of net profit to cost. This indicator shows how much the company has net profit (***NP***) from one monetary unit of costs (***COP***).

Based on the data in Table 1, it can be seen that:

- the return on assets of energy companies of the Irkutsk region in 2019 was 5.83%, then in 2020 it was 4.35%, a decrease of -1.48 over the study period%;
- the capital intensity of the assets of energy companies of the Irkutsk region in 2019 was 232.71%, then in 2020 it was 275.27%, an increase of +42.56 over the study period%;
- the sales-on-costs ratio of energy companies of the Irkutsk region in 2019 was 124.85%, then in 2020 it was 125.61%, an increase of +0.76% over the study period.

The resulting indicator of the author's model of profitability of production of energy companies of the Irkutsk region in 2019 was 16.94%, then in 2020 it was 15.05%, a decrease of -1.89% over the study period.

The cumulative deviation of the resulting indicator (***ΔPOP***) is determined by the formula 2:

$$\Delta POP = \sum_{n=1}^{3} POP(F_n) = \\ = \Delta POP(F_1) + \Delta POP(F_2) + \Delta POP(F_3) \quad (2)$$

The algebraic sum of the influence of factors must necessarily be equal to the total increase in the effective indicator. The absence of such equality indicates errors in calculations.

Next, we will consider 10 methods of deterministic factor analysis developed by Filatov E.A.

Auxiliary data on the author's comparative coefficients for factor analysis are presented in Tables 2, 3.

Table 2. Multiple comparative coefficients for one factor.

Comparison of factors	The designation of comparative coefficient	Value	The product of coefficients (value)
$F_{1(I)} / F_{1(0)}$	A_1	0.746257	1.00
$F_{1(0)} / F_{1(I)}$	A_2	1.340020	
$F_{2(I)} / F_{2(0)}$	A_3	1.182901	1.00
$F_{2(0)} / F_{2(I)}$	A_4	0.845379	
$F_{3(I)} / F_{3(0)}$	A_5	1.006115	1.00
$F_{3(0)} / F_{3(I)}$	A_6	0.993922	

Table 3. Multiplicative comparative coefficients for two factors.

Comparison of factors	The designation of comparative coefficient	Factorial factors	Value
$(F_{1(I)} * F_{2(I)}) / (F_{1(0)} * F_{2(0)})$	B_1	$A_1 * A_3$	0.882749
$(F_{2(0)} * F_{3(0)}) / (F_{2(I)} * F_{3(I)})$	B_2	$A_4 * A_6$	0.840241

10 author's methods of factor analysis are presented in Table 4, in which the result is equal to the product of the main part of the formula and the corresponding correction coefficients.

Table 4. Methods of alternative factor analysis using comparative coefficients.

No. methods	No. formulae	formulas / calculations: the main part of the formula	formulas / calculations: adjustment factors
1.1	1.1	$\Delta POP(F_1) = POP_0*(A_1) - POP_0$	–
	1.2	$\Delta POP(F_2) = (POP_0*(A_3) - POP_0)*$	A_1
	1.3	$\Delta POP(F_3) = (POP_0*(A_5) - POP_0)*$	$(A_1 * A_3)$ or B_1
1.2	2.1	$\Delta POP(F_1) = (POP_I - POP_I * (A_2))*$	$(A_6 * A_4)$ or B_2
	2.2	$\Delta POP(F_2) = (POP_I - POP_I * (A_4))*$	A_6
	2.3	$\Delta POP(F_3) = POP_I - POP_I * (A_6)$	–
2.1	3.1	$\Delta POP(F_1) = (\Delta F_1 / F_{1(0)}) * POP_0$	–
	3.2	$\Delta POP(F_2) = (\Delta F_2 / F_{2(0)}) * POP_0)*$	A_1
	3.3	$\Delta POP(F_3) = ((\Delta F_3 / F_{3(0)}) * POP_0)*$	$(A_1 * A_3)$ or B_1
2.2	4.1	$\Delta POP(F_1) = ((\Delta F_1/ F_{1(I)}) * POP_I)*$	$(A_6 * A_4)$ or B_2
	4.2	$\Delta POP(F_2) = ((\Delta F_2/ F_{2(I)}) * POP_I)*$	A_6
	4.3	$\Delta POP(F_3) = ((\Delta F_3/ F_{3(I)}) * POP_I$	–
3.1	5.1	$\Delta POP(F_1) = (POP_I*A_4*A_6) - POP_0$	–
	5.2	$\Delta POP(F_2) = ((POP_I*A_2*A_6) - POP_0)*$	A_1
	5.3	$\Delta POP(F_3) = ((POP_I*A_2*A_4) - POP_0)*$	$(A_1 * A_3)$ or B_1
3.2	6.1	$\Delta POP(F_1) = (POP_I - (POP_0*A_3*A_5))*$	$(A_6 * A_4)$ or B_2
	6.2	$\Delta POP(F_2) = (POP_I - (POP_0*A_1*A_5))*$	A_6

https://doi.org/10.1051/e3sconf/202346101045

	6.3	Δ POP (F_3) = POP_I – (POP_0*A_1*A_3)	–
	7.1	Δ POP (F_1) = Δ POP – (POP_I – (POP_0*A_1)	–
4.1	7.2	Δ POP (F_2) = Δ POP – (POP_I – (POP_0*A_3))*	A_1
	7.3	Δ POP (F_3) = Δ POP – (POP_I – (POP_0*A_5))*	(A_1 * A_3) or B_1
	8.1	Δ POP (F_1) = Δ POP – ((POP_I *A_2) – POP_0)*	(A_6 * A_4) or B_2
4.2	8.2	Δ POP (F_2) = Δ POP – ((POP_I *A_4) – POP_0)*	A_6
	8.3	Δ POP (F_3) = Δ POP – ((POP_I *A_6) – POP_0)	–
	9.1	Δ POP (F_1) = Δ POP – (POP_I – (POP_I*A_4*A_6))	–
5.1	9.2	Δ POP (F_2) = Δ POP – (POP_I – (POP_I*A_2*A_6))	A_1
	9.3	Δ POP (F_3) = Δ POP – (POP_I – (POP_I*A_2*A_4))*	(A_1 * A_3) or B_1
	10.1	Δ POP (F_1) = Δ POP – (POP_0*A_5*A_3) – POP_0)*	(A_6 * A_4) or B_2
5.2	10.2	Δ POP (F_2) = Δ POP – ((POP_0*A_5*A_1) – POP_0)*	A_6
	10.3	Δ POP (F_3) = Δ POP – ((POP_0*A_3*A_1) – POP_0)	–

As can be seen from Table 4, the author's methods 1.1 and 1.2, 2.1 and 2.2, 3.1 and 3.2, 4.1 and 4.2, 5.1 and 5.2, based on the influence of correction coefficients, are mirror to each other.

4 Results

The result for methods 1.1, 2.1, 3.1, 4.1, 5.1 is presented in Table 5, the result for methods 1.2, 2.2, 3.2, 4.2, 5.2 is presented in Table 6.

Table 5. Results for methods 1.1, 2.1, 3.1, 4.1, 5.1.

No.	the main part of the formula	adjustment factors		result
1	Δ POP (F_1) = -0.042985	-		-0.042985
2	Δ POP (F_2) = 0,030984	0.746257	A_1	0.023122
3	Δ POP (F_3) = 0,001036	0.882749	A_1 * A_3	0.000914
	-0.010965			**-0.018948**

Table 6. Results for methods 1.2, 2.2, 3.2, 4.2, 5.2.

No.	the main part of the formula	adjustment factors		result
1	Δ POP (F_1) = -0.051158	0.840241	A_6 * A_4	-0.042985
2	Δ POP (F_2) = 0.023263	0.993922	A_6	0.023122
3	Δ POP (F_3) = 0.000914	-		0.000914
	-0.026980			**-0.018948**

As can be seen from the final result of Tables 1, 5, 6, the purpose of the analysis is achieved – the determination of the influence of factors is disclosed without deviations.

The final change in the profitability of the production of energy companies of the Irkutsk region had a positive impact:

- an increase in the capital intensity of assets by +42.56%, caused an increase in the studied indicator by **+2.31%**;

- an increase in the sales-on-costs ratio by +0.76%, caused an increase in the studied indicator by **+0.09%**.

The final change in the profitability of the production of energy companies of the Irkutsk region had a negative impact:

- a decrease in the return on assets by -1.48% caused a decrease in the studied indicator by **-4.29%**.

The combined influence of three factors led to a decrease in the profitability of the production of energy companies in the Irkutsk region by **-1.89%**.

5 Discussion

The initial data for analyzing the structure of the key indicators of the energy industry of the Irkutsk region of the Russian Federation in the context of types of organizations are presented in Table 7 based on data from 4 statistical bulletins of Irkutskstat (1. Financial results of organizations. Volume 2 – large and medium-sized organizations. 2019; 2. Financial results of the organizations' activities. Volume 3 – Small enterprises. 2019; 3. Financial results of the organizations' activities. Volume 2 – large and medium-sized organizations. 2020; 4. Financial results of the organizations' activities. Volume 3 – Small enterprises. 2020).

Table 7. The structure of key indicators of the energy industry of the Irkutsk region in the context of types of organizations, %.

No.	Indicators	2019 year	2020 year	Δ
1	NP – Net Profit	100.00	100.00	0.00
	including:			
	- large and medium-sized organizations	100.11	100.47	0.36
	- small enterprises	-0.11	-0.47	-0.36
2	NR – Net Revenue	100.00	100.00	0.00

	including:			
	- large and medium-sized organizations	96.89	96.61	-0.28
	- small enterprises	3.11	3.39	0.28
3	TA – Total Assets	100.00	100.00	0.00
	including:			
	- large and medium-sized organizations	99.09	99.06	-0.03
	- small enterprises	0.91	0.94	0.03
4	COP – Cost Of Production	100.00	100.00	0.00
	including:			
	- large and medium-sized organizations	95.56	94.89	-0.67
	- small enterprises	4.44	5.11	0.67

During the studied period of time, the final financial result of the aggregate of small enterprises of the energy industry of the Irkutsk region was negative – a slight loss was obtained – expenses slightly exceeded revenues.

In general, the share of small enterprises in the energy sector of the Irkutsk region is only 3%.

The size of the total assets of the Irkutsk region's energy industry increased by +12.6% over the period under study, with official inflation of +4.9%.

As a result of the pandemic that began in 2020 and the subsequent lockdowns, the main financial indicators of large and medium-sized organizations of the energy industry of the Irkutsk region have decreased:

- revenues by -5.1%;
- final financial result by -15.7%;
- cost of production by -6.1%.

In general, the structure of key indicators in the context of the types of organizations of the energy industry of the Irkutsk region remained unchanged over the period under study (changes were less than %).

Based on the statistical bulletins of Irkutskstat, we will calculate the final profitability of the production of energy companies of the Irkutsk region in the context of types of organizations (Table 8).

Table 8. Profitability of production of companies in the energy sector of the Irkutsk region in the context of types of organizations, %.

No.	Indicators	2019 year	2020 year	Δ
1	all organizations	16.94	15.05	**-1.89**
2	large and medium-sized organizations	17.75	15.93	-1.82
3	small enterprises	-0.41	-1.38	-0.97

Table 8 clearly shows that large and medium-sized companies in the energy sector of the Irkutsk region are significantly more efficient than small enterprises.

6 Conclusion

The pandemic of the Covid-19 coronavirus infection caused not only panic among the population, but also a quite tangible crisis in the economy. The energy industry is particularly worth considering. Energy consumption in the considered period of time in the world has significantly decreased [9-11].

The energy crisis due to the pandemic has not affected the Russian Federation as much as the countries of Europe and Asia. The Ministry of Energy of the Russian Federation reported that consumption decreased in the sector of mechanical engineering, metallurgy, railway industry, in small and medium-sized businesses, while electricity generation at hydroelectric power plants increased. But it is worth noting that the overall demand in the period under consideration still decreased, which affected wholesale prices and reduced the revenue of generating companies.

Competition between various sources of electricity generation has intensified abroad, in which renewable energy sources have been able to take an advantageous position. In the Russian electric power industry, the Covid-19 coronavirus pandemic has created a number of problems for the energy industry, including falling electricity prices, the risk of increased non-payments and difficulties with the implementation of new projects [12].

If we take the Irkutsk region, then as can be seen from the presented analytical tables 5-8, the situation in the energy industry was negative, but not critical.

Outdated equipment, high costs, shortage of working capital are the main negative internal factors that reduce the efficiency of energy sales activities. Falling demand, inflation, high taxation, lack of payment discipline are the main external factors that put companies in a difficult economic situation. In addition, the results of the activities of energy marketing companies are influenced by the climatic conditions prevailing in the region. To ensure the effective operation of companies, it is necessary to take into account the above external and internal factors in order to develop measures to reduce their negative impact.

It should be noted that the observed significant difference in the levels of electricity costs for individual power systems is largely explained by a number of objective conditions:

- differences in the technical level of energy enterprises;
- the structure of generating capacities;
- features of natural and geographical conditions;
- the density of electrical and thermal loads;
- technical and economic characteristics of fuel and energy resources;
- the level of fuel transportation costs;
- share of purchased energy.

The article was carried out within the framework of the scientific project of the Siberian Branch of the Russian Academy of Sciences "Socio-economic development of a resource region in the conditions of changing external factors (on the example of the Irkutsk region)" (Research Work of the ISC SB RAS No. 121021800157-8).

References

1. V.A. Stennikov, V.O. Golovshchikov, *Power engineering of the Irkutsk region: trends, challenges and threats in modern conditions*, Energy policy, **12 (178)**, 56-71 (2022)
2. O.L. Zuenok, *Power engineering of the Irkutsk region, prospects of development*, Youth Bulletin of IrSTU, **4**, 21 (2014)
3. P.P. Bezrukikh, *The role of renewable energy in energy conservation in the world and Russia*, Electrics, **4**, 3-5 (2004)
4. K.S. Beisenov, *Problems and prospects of electric power industry development in modern conditions*, Young scientist, **20 (154)**, 235-236 (2017)
5. V.Ya. Ushakov, Renewable and alternative energy: resource conservation and environmental protection (Tomsk, SibGrafix Publishing House, 2011)
6. E.A. Filatov, *Analysis of profitability of production of enterprises in the field of transportation and storage of the Irkutsk region*, Transportation Research Procedia, **63**, 518-524 (2022) DOI: 10.1016/j.trpro.2022.06.043
7. E.V. Zubova, L.V. Koyava, *The cost of capital as a factor in assessing the value of a company*, Innovative development, **1 (6)**, 47-50 (2017)
8. Z.R. Israilova, M.A. Barzaeva, *Problems of the efficiency of the use of the company's capital*, Achievements of science and education, **10 (11)**, 31-32 (2018)
9. V.E. Abaimov, A.A. Shavkatovna, *Influence of COVID-19 on the electric power industry of the world*, Innovative Science, **11**, 79-83 (2020)
10. M. Keay, D. Robinson, *COVID-19: glimpses of the energy future?*, Forum, COVID-19 and the energy transition, **123**, 29-32 (2020)
11. F. Laverón, *The role of the electricity sector in the energy transition after COVID-19*, Forum, COVID-19 and the energy transition, **123**, 36-39 (2020)
12. Yu.O. Daneeva, *Impact of the novel coronavirus pandemic on energy transition*, The Eurasian Scientific Journal, [online], **4(12)**, (2020)

Problems of the current state of power quality and proposals for their solution

Lidiya Kovernikova*

Melentiev Energy Systems Institute of the Siberian Branch of the Russian Academy of Sciences, Irkutsk, Russian Federation

Abstract. The power quality in Russian electrical networks does not meet the requirements of regulatory documents. Big problems occur in all areas of people's lives for this reason. Economic damage is one of the serious consequences. In the context of digitalization of the economy, the requirements for the power quality are increasing, because digital electrical equipment requires high power quality. The paper presents the history of solving the problem of thepower quality in Russia, including three periods - before the adoption of the Federal Law "On Electric Power Industry", after its adoption and after the adoption of the resolution of the Russian Government on the creation of a digital economy. Examples of the consequences of low power quality. To achieve a high level of power quality, it is proposed to create a power quality management system that meets the requirements of the emerging intelligent electrical power systems and digital economy.

1 Introduction

Electrical energy is used in all spheres of people's lives. In the Decree of the Government of the Russian Federation No. 1013 of August 13, 1997 "On approval of the list of goods subject to mandatory certification and the list of works and services subject to mandatory certification" [1] it was recognized as a product. Electrical energy is characterized by the coincidence in time of the processes of production, transmission, distribution, variable mode of consumption and can be converted into other types of energy. It cannot be returned to the seller if the quality is poor. The concept of product quality can be applied to electrical energy. State standard GOST 15467-79 [2] defines the concept of product quality as a set of product properties that determine its suitability to meet needs in accordance with its intended purpose.

The paper "On the issue of the power quality" was published in the journal "Elektrichestvo" in 1968 [3]. It was devoted to the state standard for the power quality, which was developed in our country. This standard was the first standard for the power quality in the world – GOST 13109-67 [4]. The author of the article wrote: "Electric energy has found application in modern society in all spheres of human activity. Therefore, its quality directly affects the living conditions and activities of people. The power quality affects also the technical and economic performance of individual devices and power supply systems as a whole. By improving the power quality, these indicators can be increased, and by worsening, they can be reduced".

Currently, the current state standard for the power quality is GOST 32144-2013 [5]. Poor power quality causes additional losses of power and energy in electrical networks, damage and reduction in service life of electrical equipment, decreased productivity of technological equipment of industrial enterprises, and economic damage [6-10].

The power quality in Russian electrical networks does not meet the requirements of GOST 32144-2013 [11]. The indicators δU(-) и δU(+) (positive and negative deviations of the voltage value), KU и KU(n) (indicators characterizing the distortion of the sinusoidal shape of the voltage curve), K2U (indicator assessing the degree of asymmetry of the three-phase voltage) exceed the norms GOST 32144-2013 [5].

Currently, the Russian economy is undergoing a digital transformation, as a result of which the amount of electronic equipment is increasing [12, 13]. Electronic equipment requires high power quality, but by consuming non-sinusoidal current, it introduces distortions into the electrical network. Federal authorities have been actively discussing the low power quality in Russian electrical networks and have been adopting legislative documents [14-18] over the past few years to reduce the economic damage caused by the low power quality. Economic damage was minimally estimated at $25 billion in 2008 [19]. The paper presents proposals for solving the problem of power quality in the Russian electric power industry, which should become mandatory elements of the power quality management system.

2 History of the issue of power quality in Russia

* Corresponding author: kovernikova@isem.irk.ru

The power quality has been studied for many years. The nature of the attitude towards the power quality can be divided into three periods: the first – until 2003, which ended with the adoption of the Federal Law "On Electric Power Industry" in 2003 [20]; the second – from 2003 to 2017, ending with the approval of the Russian Digital Economy program in 2017 [14]. The third period began in 2017.

The first period is characterized by active work in the field of power quality. Before the adoption of the Federal Law "On Electric Power Industry" [20], it was believed that the parameters of the modes of electric power systems should ensure the economical operation of both energy supply organizations and consumers, which is determined by the power quality, ensures the reliability of electrical equipment, and its functioning in accordance with its intended purpose. Numerous studies conducted to assess the impact on electrical equipment of deviations in voltage and frequency, voltage curve shape, and voltage symmetry from nominal values have confirmed the negative impact of low power quality on electrical equipment. For example, in [21] the results of studies of the influence of non-sinusoidal voltage on a power transformer are presented, in [22] of asymmetrical and non-sinusoidal voltage on asynchronous motors, which showed the negative impact of non-sinusoidal and asymmetrical voltages on electrical equipment. As a result, the world's first state standard for the power quality was developed and put into effect on January 1, 1968 - GOST 13109-67 [4]. The author [3] noted, that "the release of a state standard for the power quality will not only lead to an increase in the technical and economic indicators of energy systems, but will also increase our interest and attention to providing optimal solutions – both during design and in operating conditions". Work in the area of power quality continued. In 1989, on January 1, 1989, a new GOST 13109-87 [23] was put into effect, replacing GOST 13109-67, and in 1999, on January 1, GOST 13109-97 [24] was put into effect. In addition to standards, regulatory and technical documents were put into effect [25-29], which made it possible to create an economic mechanism for managing the quality of electrical energy. The "Rules" [27, 28] made it possible to determine the culprit of the distortion of the power quality, its contribution to the distortion, the degree of influence of the consumer on the power quality as a result of comparison with the permissible calculated contribution, an assessment of the excess of the actual contribution of the consumer to the distortion of the power quality relative to acceptable control contribution. If the culprit for the deterioration in the power quality was the energy supply organization, then a discount on the electricity tariff was provided for the consumer; if the culprit was the consumer, then a surcharge on the tariff.

The amounts of discounts and allowances were calculated taking into account [29]. The conditions formulated in the "Rules" were included in contracts for the use of electrical energy. Work on the creation of instruments for measuring the power quality continued. If the power quality indicators during measurements in electrical networks exceeded the regulatory values of the state standard, then issues of power quality between energy supply organizations and consumers were resolved in the courts. For example, employees of the Institute Energy Systems (ISEM SB RAS) took part as experts in the arbitration trial, which considered the claims of "Amurenergo" against the Trans-Baikal Railway for violation of the contract regarding the power quality [30].

The second period is associated with the adoption in 2003 of the Federal Law "On Electric Power Industry" [20], according to which the Unified Energy System of Russia was divided into a large number of entities that independently decided whether to deal with the power quality or not. This provision stemmed from the fact that the Law "On Electric Power Industry", the Civil Code [31], the Federal Law "On the Protection of Consumer Rights" [32] contained provisions on the need to ensure the required power quality and responsibility for it, but mandatory quality requirements electricity have not been established at the legislative level. State standards were considered documents of voluntary application. The "Rules," which were essentially an economic mechanism for managing the power quality, were abolished. During the second period, the power quality deteriorated. In the monograph [11], written based on the materials of the section "Power quality" of the conference "Russian Energy in the 21st Century. Innovative development and management", held in 2015, it was noted that violations of the requirements of GOST 32144-2013 in Russian energy systems are widespread and systematic. Despite the current situation, experts in the field of power quality have developed many new standards, including in order to harmonize the requirements for the power quality in Russian regulatory documents with international ones. The method for determining the source of distortion in the power quality and its contribution along the directions of flow of negative and zero sequences electrical energy and the n-th harmonic component electrical energy and their values was developed and approved in 2014 [33]. The development of technical means that normalize the power quality [34-36], instruments for monitoring and analyzing the power quality [37] continued. The concept of a smart electric power system with an active-adaptive network was developed in 2011 [38]. It proposed to consider the reliability of electricity supply and the power quality "as a service (product), taking into account the reasonable costs of the electric grid company to maintain electric networks at a level that ensures adequate reliability and power quality, and the likely damage to the consumer at a certain level of reliability." Consumers were given the opportunity to choose the level of reliability and power quality. The concept noted "problems in legal, regulatory and technical support for the quality of power supply at the energy system-consumer border", the lack of "clear responsibility of power grid companies for ensuring the quality of power supply and responsibility of consumers for the negative impact on the quality of power supply", virtual absence of centralized control of the power quality. A new period in the development of industrial production and changes in all spheres of society began in 2017, which required a new attitude to the power quality, since the digital economy requires high digital power quality.

Government Decree No. 2425 [15], confirming the 1997 Government Decree [1] on mandatory certification of electrical energy for compliance with the requirements of GOST 32144-2013, was adopted in 2021. The section on legislative regulation of energy efficiency and energy saving of the Expert Council under the State Duma Committee on Energy in 2022 proposed Rosseti PJSC to hold a seminar on amending the Federal Law "On Electric Power Industry" [16]. One of the changes confirms the requirements for the power quality and the obligation to comply with them by electricity industry entities and consumers. The Law on Amendments to the Federal Law "On Electric Power Industry" was adopted in 2022 [17]. The changes took effect in 2023 on January 1 [18].

3 Examples of the consequences of poor power quality

Non-sinusoidal voltage in electrical networks is a pressing problem nowadays [11], as it was many years ago. It was noted in [7] that the occurrence of harmonic components of voltages in electrical networks is the most important problem of electromagnetic compatibility, caused by the development of new technologies using electronic convert-ers that consume non-sinusoidal current from the network. At the same time, electronic automatic control systems and electronic devices are sensitive to non-sinusoidal voltage, and their operation is disrupted [39].

Harmonic components of currents create additional power and energy losses in electrical networks and electrical equipment of networks and consumers. An assessment of active energy losses of the fundamental frequency and additional losses caused by current harmonics in a 110 kV line during the day is given in [11]. Additional losses of electrical energy amounted to 13.1% of the main losses of electrical energy. In addition to harmonics, electrical net-works contain interharmonics, i.e. components with frequencies that are not multiples of the fundamental frequency. In [40] it is noted that their influence on the electrical network may be greater than the influence of harmonics. Elec-trical networks also contain supraharmonics, the frequency of which is a multiple of the fundamental frequency, but above 2 kHz. The authors of [41] determined additional active power losses in a 20 kV cable line at supraharmonics in the frequency range from 2 to 9 kHz. Additional power losses due to supraharmonics exceeded 10% of power losses caused by current harmonics. Asynchronous motors have been widely used in the past and are still in use to-day. If the power quality is low, their service life is reduced. They are damaged and fail, which has been confirmed by the practice of their operation for many years. At the processing plant of one of the mines in the Transbaikalia region, the asynchronous motor of the mill failed due to the poor power quality [10]. Nowadays [11], as many years ago [42], the source of distortion in the power quality in Transbaikalia is the electrified railway. The asynchronous motor bearings were damage three times in 2014. The cause of the damage was three-phase voltage asymmetry, which caused vibration and, therefore, accelerated wear of the bearings. Information on permissible vibration levels for an asynchronous motor is given in [43]. In accordance with [44], depending on the vibration speed, an asynchro-nous motor can be in three modes, in one of them - for a limited time, since voltage asymmetry increases the vibra-tion speed of an asynchronous motor, then with voltage asymmetry, the asynchronous motor must operate for a lim-ited time. The authors of [45] noted that when voltage unbalance occurs, bearings can suffer mechanical damage and that voltage unbalance consistently exceeding 2% can lead to damage to an asynchronous motor. The result of poor quality power supply to the enrichment plant during operation from 2014 to 2017 was the loss of working time due to interruptions in power supply and poor power quality, amounting to 333 hours, which led to economic damage of 239631 thousand rubles. In 2018, due to the poor power quality, electronic equipment failed at the "BAIKALSEA Company" water bottling plant in Irkutsk: Ethernet switch, uninterruptible power supply, optical module.

4 Proposals for solving problems with the power quality in Russian electrical networks

One of the tasks of managing the functioning of electrical power systems is the task of ensuring the standard power quality. It can be solved using a set of organizational, technical and economic measures that will represent a system for managing the power quality. Taking into account the experience of previous years, it should include:

- government agency for managing the power quality;
- complex of legislative documents regulating the rights and responsibilities of producers, suppliers and consumers of electrical energy;
- complex of regulatory and technical documents;
- complex of regulatory requirements for standard design of electrical power systems, taking into account ensuring the quality of electrical energy;
- economic mechanism for managing the power quality;
- technical means to ensure the power quality;
- system for continuous monitoring of power quality indicators in electrical networks;
- software tools that allow, based on the results of periodic monitoring or continuous monitoring of mode parameters and indicators of the power quality, to solve problems of analysis, assessment, and forecasting of the power quality in various modes;
- software tools for selecting technical devices to ensure the standard power quality in electrical networks.

4.1 About the government agency dealing with the power quality

The State Energy Supervision (Gosenergonadzor) under the Ministry of Fuel and Energy of the Russian Federation, which dealt with the power quality, was created by Decree of the Government of the Russian Federation No. 938 of 08/12/98 [46, 47]. The main task of Gosenergonadzor was to supervise the rational and efficient use of electrical energy. Gosenergonadzor included structural units for managing energy supervision of the central office of the Ministry of Fuel and Energy, regional departments and departments in the subjects of the country. Gosenergonadzor authorities were engaged in inspection of enterprises and electrical networks in terms of the power quality. Gosenergonadzor was abolished in 2003 after the adoption of the Law "On Electric Power Industry". Supervision over the power quality was transferred to Rostechnadzor, whose functions were significantly reduced. At present, when electronic equipment is being introduced not only in the electric power industry, but also in the country's economy, it is necessary to create a government agency that will ensure the digital power quality and the efficient use of electrical energy.

4.2 A complex of legislative and legal documents on the power quality

The need to maintain the power quality in accordance with GOST 32144-2013 is indicated in three legislative documents. Article 38 of the Law "On Electric Power Industry" [18] notes that power industry entities supplying electrical energy to consumers "are responsible to consumers of electrical energy for the reliability of supplying them with electrical energy and its quality in accordance with the requirements of this Federal Law and other manda-tory requirements". Article 542 of the Civil Code [31] states that "the quality of the supplied energy must comply with the requirements established in accordance with the legislation of the Russian Federation, including mandatory rules, or stipulated by the energy supply contract". Government Decree No. 442 "On the functioning of retail mar-kets for electrical energy, full and (or) partial limitation of the mode of consumption of electrical energy" [48] states that "all subjects of the electric power industry supplying electric energy to consumers, when fulfilling their obliga-tions under contracts on the wholesale and retail markets must ensure the power quality". In practice, reliability ob-ligations are sometimes violated, and the power quality is not fulfilled [10, 49]. In [11] it is noted that the reason for this is the lack of methods for determining the amount of civil liability for violation of requirements for the power quality. It is necessary to develop legal documents on the responsibility of consumers whose electrical equipment introduces distortions into the electrical network, and as a result, the power quality deteriorates. A document is re-quired on the liability of consumers who distort the power quality to the energy supply organization for the damage caused.

4.3 A complex of regulatory and technical documents on the power quality

Regulatory and technical documents include regulations, standards, rules, and methods. In accordance with the Federal Law "On Technical Regulation" No. 184-FL as amended in 2002 and 2021 [50, 51], a technical regulation is a document that establishes mandatory requirements for the application and execution of technical regulation ob-jects (products), i.e. has the status of law. In the latest edition of the Law "On Technical Regulation", technical regu-lations can be adopted by decree of the President of the Russian Federation, or by decree of the Government of the Russian Federation. Currently in force is the Technical Regulation of the Customs Union "Electromagnetic Compati-bility of Technical Equipment" [52], in which changes in voltage characteristics taken into account by GOST 32144-2013 when assessing the power quality are considered only as electromagnetic interference, including voltage devia-tion, frequency deviation, non-sinusoidal voltage, etc. When assessing the power quality, the levels of electromag-netic interference are assessed - the values of frequency and voltage deviations, the degree of distortion of the volt-age curve shape, etc. Issues of power quality are not presented in [52]. Since "the power quality determines the de-gree of danger of this product for the life and health of citizens, property and the natural environment" [11], it is necessary to develop technical regulations that establish mandatory requirements for the power quality and forms of mandatory confirmation of compliance with these requirements. This was discussed in Federal Law No. 184-FL "On Technical Regulation" as amended in 2002 [50].

There are a large number of standards in the field of power quality. Many of them are developed on the basis of international ones. To develop regulatory and technical support in the field of electrical energy quality, it is neces-sary to improve existing standards, taking into account the accumulated experience of their application, as well as the development of a regulatory document establishing indicators and standards for networks with voltages above 220 kV. It is necessary to develop standards for the emission into the supply network of harmonic currents, negative sequence currents by distorting consumers, interharmonics of voltage and current, and for permissible modes of ab-ruptly variable loads that cause voltage fluctuations at the point of connection of the consumer to the supply net-work. At present, the issue of the amount of payment for electrical energy taking into account its quality has not been resolved. An attempt to introduce discounts (surcharges) to tariffs when the power quality deteriorated was made in the 90-s of the last century in the developed "Rules" [27, 28] and "Price List" [29]. In [11], it is proposed to include in the energy supply contract conditions on the obligation of the energy supplying organization to pay a penalty for the supply of electrical energy that does not meet the established requirements. To do this, it is necessary to develop guidelines for determining the size of the penalty, which takes into account the reduction in the service life of devices, equipment, and the costs of consumers for installing special equipment to improve the power quality. It is also necessary to develop a methodology for assessing economic damage when supplying consumers with poor power quality.

4.4 A complex of regulatory requirements for standard design of electrical power systems

Currently, when designing electrical networks in Russia, voltage deviations at network nodes are considered. Non-sinusoidality, asymmetry, fluctuations and voltage dips are analyzed only when special work is carried out for areas with low power quality. This is due to the lack of information on the power quality indicators in electrical networks of different voltages and guidelines for their calculation, the unregulated distribution of responsibility between the subjects of the electric power industry for the implementation of measures to improve the power quality based on the design results. Failure to take into account the power quality when designing electrical power facilities leads to its deterioration during the development of the power system when connecting distorting consumers or equipment of electrical networks, for example, of static capacitor banks. When a static capacitor bank is turned on in an electrical network with a non-sinusoidal voltage, the amplitudes of voltage and current harmonics increase due to its resonance with the network at harmonic frequencies [52].

Regulatory design requirements must provide answers to many questions:

– what replacement circuits for elements of electrical power systems should be used when calculating indicators of the power quality,

– what operating modes of the electrical network and loads should be considered when analyzing indicators of the power quality and developing measures to improve the power quality,

– how to distribute measures to improve the power quality between several consumers and the electric grid company.

It is necessary to develop standard methods for calculating indicators of the power quality in electrical networks, characterizing the distortion of sinusoidality and symmetry, voltage fluctuations and dips, selecting parameters and installation locations for devices to ensure the power quality, assessing the economic effect of installing technical means to normalize the power quality.

4.5 Economic mechanism for managing the power quality

At the end of the last century, an economic mechanism for managing the power quality already was created in our country. It included "Rules for connecting consumers to a general purpose network according to the conditions affecting the power quality" [27], "Rules for applying discounts and surcharges to tariffs for the quality of electricity" [28], Price list "Tariffs for electric and thermal energy [29]. In [54] it was noted that "The Rules incorporated everything that was allowed by the existing means of measuring the power quality at that time". In 2014, a technique was developed for determining the source of distortion in the power quality [33] in the direction of power flows of the fundamental frequency and distorting powers caused by non-sinusoidality and asymmetry of currents and voltages. When directing distorting power from the network to the load (*P*+), the source of distortion is the electrical network; when directing distorting power from the load to the network (*P*-), the source of distortion is the load. Currently, the magnitudes and directions of distorting powers can be determined using instruments for measuring power quality indicators [55]. The technique can be used to develop an economic mechanism for managing the power quality.

As an example, in Fig. 1 and 2 show graphs of active powers directed from the network to the load (*P*+) and from the load to the network (*P*-) at the point of transmission of electrical energy under the power supply contract.

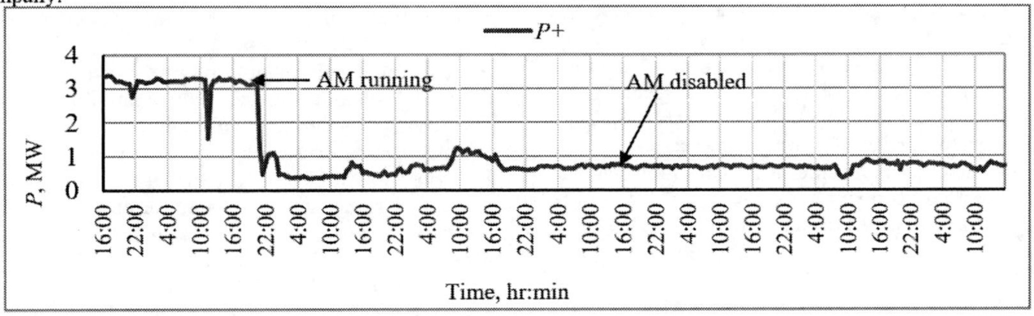

Fig. 1. Active power from the network to the load.

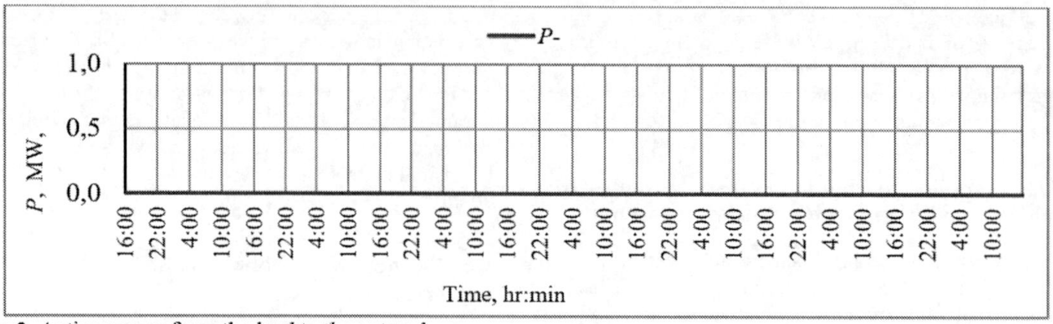

Fig. 2. Active power from the load to the network.

The figures show the moment of failure of the asynchronous motor (AM) of the mill of the processing plant, which was discussed above. A device for measuring power quality indicators "Resurs" [57] measured active powers. Graph in Fig. 2 shows that the active power supplied from the load to the network is zero.

It is necessary to develop an economic mechanism for managing the power quality, taking into account the capabilities of modern instruments for measuring indicators of the power quality and a methodology for determining the contribution of the distorting consumer to the distortion of the power quality.

4.6 Technical means to ensure the power quality

The concept of a smart power grid proposes 22 technical devices for normalizing the power quality in electrical networks [38]. The devices are divided into four groups depending on the method of connection to the electrical network: transverse, longitudinal, longitudinal-transverse devices, devices for combining electrical power systems and filters. Of the devices presented in the concept, a small number are used in electrical networks. The most commonly used of them are capacitor banks, controlled shunt reactors with magnetization, and passive harmonic filters. Currently, new effective technical means are being used and developed to comprehensively solve problems of electrical energy quality. Among them are active filter-balancing devices of the shunt type (transverse) and the serial type (longitudinal) [56]. They generate reactive power to reduce voltage sags, make the voltage symmetrical, and filter harmonics in a dynamic mode when the harmonics of the distortion source change rapidly. Modified magnetization-controlled shunt reactors are presented in [57]. They do not have a negative impact on the quality of electrical energy, unlike previously produced and used magnetization-controlled shunt reactors. A pilot project of a converter device is presented in [58]. It is an active filter for installation in electrical networks of 10-220 kV in order to reduce asymmetry and non-sinusoidality of currents and voltages, stabilize voltage, reduce dips and voltage fluctuations. To solve problems with the power quality using special technical means, it is necessary to create and use standard devices of various capacities that compensate for the asymmetry and non-sinusoidality of currents and voltages in electrical networks of all voltages.

4.7 Periodic control and monitoring of power quality indicators

Period control and monitoring of the power quality is carried out in accordance with the current GOST 33073–2014 [59]. Periodic control includes certification and arbitration tests of electrical energy, inspection supervision of certified electrical energy, control during the implementation of state control and supervision of consumer complaints, implementation of the energy supply contract in terms of the power quality, etc. Periodic measurements of power quality indicators in electrical networks have the disadvantage that events that cause a decrease in the power quality occur not only during the measurements. Therefore, the creation of a system for continuous monitoring of power quality indicators in the Russian electric power system is an actual task. Information received from the system for monitoring power quality indicators can be used to solve the following problems:

– assessment of the current state of the power quality when checking the compliance of measured indicators with established requirements in regulatory documents, energy supply contracts, etc.;

– obtaining information about the power quality in case of consumer complaints;

– providing information on the power quality to consumers when they connect and taking it into account when concluding energy supply contracts;

– assessment of the impact of low power quality on the service life of electrical equipment of consumers and power grid companies, generating electrical equipment, power and energy losses in asymmetrical and non-sinusoidal modes;

– selection and development of measures and technical means to normalize the power quality;

– assessment of the impact of distributed generation and new types of consumer electrical equipment on the power quality;

– planning the development of electrical networks, possible changes in the network structure, connection of consumers;

– forecasting, using special software, the power quality in various operating modes of electrical distribution networks and power supply systems;

– obtaining information about trends in the electrical network and individual areas in the field of power quality.

With the help of information received from the power quality monitoring system, in addition to the listed problems, many others can be solved. It is difficult to confirm the connection between consumer damage and poor power quality if, at the time of disruption of the enterprise's technological process, during equipment failure, or for a long time before that, measurements of power quality indicators were not carried out. Currently, there are electricity meters that, in addition to commercial metering of electrical energy, record indicators of its quality. Such meters must have an event log with the date and time of voltage deviation from the norm, no voltage or voltage below a specified threshold in each phase, recording the time of voltage loss and restoration, etc. It is necessary to formalize the need to install a system for commercial metering of electrical energy with control of all indicators of the power quality at the initiative of the consumer or the electric grid company.

5 Conclusion

In Russia, it is necessary to create a system for managing the power quality in smart power systems in order to provide the digital economy with the proper power quality. To create a full-fledged power quality management sys-tem, it is necessary, first of all, to develop a concept for power quality management, taking into account the positive experience of previous years.

The research was carried out under State Assignment Project (No. FWEU-2021-0001) of the Fundamental Research Program of Russian Federation 2021-2030.

References

1. "On approval of the list of goods subject to mandatory certification and the list of works and services subject to mandatory certification," Decree of the Government of the Russian Federation No. 1013 of August 13, 1997.(in Russian)
2. *Product quality management. Basic concepts. Terms and Definitions.* GOST 15467-79. Moscow: Standardinform, 2009. (in Russian)
3. N.A. Melnikov, "On the issue of the power quality," *Elektrichestvo*, no. 5, pp. 1-6. 1968.
4. *Electric Power. Standards for power quality indicators at its receivers connected to general-purpose electrical networks.* GOST 13109-67. Moscow: Publishing house of standards, 1985. (in Russian)
5. *Electric power. Electromagnetic compatibility of technical means. Quality regulation of electric power in general-purpose electric power supply systems.* GOST 32144–2013, Moscow: Standardinform, 2014. (in Russian)
6. A.K. Shidlovsky, A.K. Kuznetsov, *Improving the power quality in electrical networks*. Kyiv: Nauk. Dumka, 1985, 268 p. (in Russian)
7. J. Arrilaga, D. Bradley, P. Bodger, *Power System Harmonics:* Translated from the English, Moscow: Energoatomizdat, 1990, 320 p. (in Russian)
8. I.I. Kartashev, E.N. Zuev, *Power quality in power supply systems. Ways to control and ensure it.* M: Izdatelstwo MEI, 2001, 120 p. (in Russian)
9. I.V. Zhezhelenko, *Higher harmonics in power supply systems of industrial enterprises.* M: Energoatomizdat, 2010, 375 p. (in Russian)
10. O.V. Zapanov, L.I. Kovernikova, "On the power quality supplied to joint stock company "Aleksandrovsky mine"". https://www.e3s-conferences.org/artcles /e3sconf/abs/2020/69/ e3sconf_energy12020_07012/e3sconf_energy-212020_07012.html.
11. L.I. Kovernikova, V.V. Sudnova, R.G. Shamonov et al., *Power quality: current state, outstanding issues, and proposals to their solutions.* Novosibirsk: Nauka, 2017, 219 p. (in Russian)
12. V.E. Vorotnitsky, Yu.I. Morzhin, "Digital transformation of the Russian energy sector is systemic task of the fourth industrial revolution", *Energiya Yedinoy seti*, no. 6(42), pp. 12-21. 2018. (in Russian)
13. Putin: "The formation of a digital economy is a matter of national security of the Russian Federation." https://tass.ru/ekonomika/438941114. (in Russian)
14. "On approval of the Digital Economy of the Russian Federation program," Government Order of the Russian Federation No. 1632-q of July 28, 2017. (in Russian)
15. Government Decree of the Russian Federation No. 2425 of December 23, 2021. (in Russian)
16. Decision of the State Duma Committee on Energy No. 3.25-5/22 of April 6, 2022. Letter to the General Director of PJSC "Rosseti" A.V. Ryumin No. C70A-4/25 of February 15, 2022. (in Russian)
17. "On Amendments to the Federal Law "On Electric Power Industry" and Certain Legislative Acts of the Russian Federation", Federal Law No. 174-FL. (in Russian)
18. "On Electric Power Industry", Federal Law of the Russian Federation No. 35-FL of January 1, 2023. (in Russian)
19. L.N. Dobrusin, "Investments in the Russian electric power industry and a program to increase their efficiency", in Proceedings of the VI All-Russian Energy Forum "Russian Fuel and Energy Complex in the 21-st Century", 1-4 April 2008, Moscow. Final materials. (in Russian)

20. "On Electric Power Industry," Federal Law of the Russian Federation No. 35-FL of March 26, 2003. (in Russian)
21. E.A. Mankin, "Eddy current losses in transformer windings with non-sinusoidal current." *Elektrichestvo*, no. 12, pp. 48-52. 1955. (in Russian)
22. A.I. Tserazov, N.I. Yakimenko, "Information materials No. 70. Study of the influence of voltage asymmetry and non-sinusoidality on the operation of asynchronous motors", Moscow: Gosenergoizdat, 1963, 120 p. (in Russian)
23. *Electric Power. Requirements for the power quality in general purpose electrical networks.* GOST 13109-87. Moscow: Publishing house of standards, 1988. (in Russian)
24. *Electric Power. Electromagnetic compatibility of technical equipment. Standards for the power quality in general-purpose electrical networks.* GOST 13109-97. Moscow: Standartinform, 1998. (in Russian)
25. "On the rules for certification of electrical equipment", Resolution of the State Standard of the Russian Federation No. 36 of July 16, 1999. (in Russian)
26. "On introducing changes and additions to the Nomenclature of products and services (works), in respect of which the legislative acts of the Russian Federation provide for their mandatory certification," Resolution of the State Standard of the Russian Federation No. 74 of August 14, 2001. (in Russian)
27. "Rules for connecting consumers to a general purpose network according to the conditions for influencing the quality of electricity," *Promyshlennaya energetika*, no. 8, pp. 45-48. 1991. (in Russian)
28. "Rules for applying discounts and surcharges to tariffs for the quality of electricity," *Promyshlennaya energetika*, no. 8, pp. 49-51. 1991. (in Russian)
29. "Tariffs for electric and thermal energy", Price list No. 09-01. 1991. (in Russian)
30. Forensic electrical examination in case No. A78-1996/03-C1-26/48 of the Arbitration Court of the Chita Region. (in Russian)
31. Civil Code of the Russian Federation of October 21, No. 14-FL. 1994. (Of July 3, 2016). (in Russian)
32. "On the Protection of Consumer Rights," Federal Law of the Russian Federation of 02/07/1992 No. 2300-1. (Of 07/03/2016). (in Russian)
33. "Methodology for determining the source (direction to the source) of distortions in power quality parameters/approved," Deputy Chairman of the Board - Chief Engineer of JSC FGC UES Dikim V.P., of November 20, 2014. Included in the Federal Information Fund for Ensuring the Uniformity of Measurements under No. FR.1.34.2015.20724. (in Russian)
34. G. M. Mustafa, S.I. Gusev, "Experience in using active filter-compensating devices of shunt and serial type in electrical networks," *in Proc. of the International Scientific and Practical Conference "Electric Energy Quality Management"*, Moscow, 2016, pp. 67-77. (in Russian)
35. A.I. Nenakhov, S.I. Gamazin, "Application of a low-power StatCom device in a system with an asymmetric load," *Promyshlennaya energetika*, no. 1, pp. 30-36. 2017. (in Russian)
36. G.M. Mustafa, S.I. Gusev, "Features of the use of modular multi-level converters for normalizing voltage quality indicators of the electrical network," *Elektroenergiya. Peredacha i raspredeleniye*, no. 4(49), pp. 58-65. 2018. (in Russian)
37. E.V. Ilyashenko, Yu.V. Kolyuzhko, "Measuring instruments for monitoring and analyzing the quality of electrical energy," *in Proc. of the International Scientific and Practical Conference "Electric Energy Quality Management"*, Moscow, 2016, pp. 134-143. (in Russian)
38. "Report on the agreement "Development of the Concept of an intelligent electric power system with an active-adaptive network" Agreement No. I-11-11/10," Open Joint Stock Company "STC Electric Power Industry", Moscow, 2011. (in Russian)
39. R. Hartungi, L. Jiang, "Investigation of power quality in health care facility", *in Proc. of the International Conference on Renewable Energies and Power Quality*, Granada, Spain, 23-25 March, 2010.
40. I.V. Zhezhelenko, Yu. L. Saenko, T. K. Baranenko, *"Selected issues of non-sinusoidal modes in electrical networks of enterprises,"* Moscow: Energoatomizdat, 2007. 297 p. (in Russian)
41. A. Novitskiy, S. Schlegel, D. Westermann, "Estimation of Power Losses caused by Supraharmonics," *Energy Systems Research*, vol. 6, no. 4, pp. 28-36. 2020.
42. S.P. Gladkikh, L.I. Kovernikova, A.V. Kostin et al., "Higher harmonics in connection nodes of traction substations (on the example of the East Siberian Railway)", Irkutsk: ISEM SB RAS, 2002. 59 p. (in Russian)
43. G. R. Bossio, C. H. De Angelo, Donolo P. D. et al., "Effects of voltage unbalance on IM power, torque and vibrations", *In Proc. of the 2009 IEEE International Symposium on Diagnostics for Electric Machines, Power Electronics and Drives*, Cargese, France, 2009.
44. *Mechanical vibration – Evaluation of machine vibration by measurements on non-rotating parts. Part I. General guidelines*, ISO 10816-1:1995, International Standards Organization. 1995.
45. "Economic Framework for Power Quality," *Technical Brochure CIGRE*, no. 467. 2011.

46. "In Glavgosenergonadzor of the Russian Federation," *Promyshlennaya energetika*, no. 6, p. 49. 1999. (in Russian)
47. "Regulations on state energy supervision in the Russian Federation," *Promyshlennaya energetika*, no. 6, pp. 50-52. 1999. (in Russian)
48. "On the functioning of retail electricity markets, complete and (or) partial restrictions on the consumption of electrical energy," Decree of the Government of the Russian Federation No. 442 of May 4, 2012. (in Russian)
49. "A state of emergency was introduced in the Mogochinsky district due to the failure of 17 engines and a pump in boiler houses", URL: https://www.rosteplo.ru/ news/2017/12/12/ 1513024429-rejim-chs-vveli-v-mogochinskom-rajone-iz-za-sboya-17 01.03.2022). (in Russian)
50. "On Technical Regulation," Federal Law No. 184-FL of December 27, 2002. (in Russian)
51. "On Technical Regulation," Federal Law No. 184-FL of December 23, 2021. (in Russian)
52. "Electromagnetic Compatibility," Technical Regulations of the Customs Union TRTS 020/2011, Approved by decision of the Customs Union Commission No. 879 of December 9, 2011. (in Russian)
53. R.G. Shamonov, "Assessment of the influence of banks of static capacitors on the higher harmonic components of voltage in main electrical networks," *Energiya yedinoy* set, no. 2 (19), pp. 22-29. 2015.
54. "In Glavgosenergonadzor of the USSR Ministry of Energy," *Promyshlennaya energetika*, no. 11, p. 52. 1990.
55. "Measuring instruments," *Research and production association "Energotekhnika,"* https://www.entp.ru/.(in Russian)
56. G.M. Mustafa, S.I. Gusev, A.M. Ershov et al., "Calculation of the power of an active filter-balancing device for normalizing the voltage on the buses of the 220 kV Skovorodino substation", *Elektricheskiye stantsii*, no. 3, pp. 46-53. 2015. (in Russian)
57. A.M. Bryantsev, M.A. Bryantsev, M.A. Makarova, "Modified series of controlled shunt reactors and reactive power sources," *Elektroenergiya. Peredacha i raspredeleniye*, no. 4 (49), pp. 94-100. 2018. (in Russian)
58. R.G. Shamonov, N.A. Alekseev, A.M. Matinyan, A.V. Antonov, "Experience and prospects for using a high-voltage active filter of the MPU series to improve the power quality," *Elektroenergiya. Peredacha i raspredeleniye*, no. 2 (65), pp. 76-82. 2021. (in Russian)
59. *Electric Energy. Electromagnetic compatibility of technical equipment. Control and monitoring of the power quality in general purpose power supply systems.* GOST 33073-2014, Moscow: Standardinform, 2014(in Russian)

Application of the state estimation procedure to improve the cyber-physical stability of the smart grid

Irina Kolosok*, *Elena* Korkina, and *Ivan* Demidov

Melentiev Energy Systems Institute Siberian Branch of Russian Academy of Science, 664033 Irkutsk, Russia

Abstract. Current Smart Grids and their facilities acquire stable features of cyber-physical systems, the components of which are physical (technological) and information - communication subsystems. In a cyber-physical system, a failure in one subsystem can cause a failure in an other one, and combined emergency states can occur as a result, so there is a problem of ensuring the Smart Grid resilience to physical and information disturbances - the problem of cyber-physical stability (resilience). The report considers the issues of applying the State Estimation procedure to analyze and increase the resilience of Smart Grid and its facilities. It is shown that the State Estimation procedure serves as an effective mean of identifying deliberate impacts or cyber-attacks on the information-communication subsystem and eliminating their consequences on technological subsystem operation. Ways to improve the efficiency of the State Estimation procedure for solving this task are considered.

1 Introduction

The development of the Russian electric power industry, based on the concept of Smart Grids, leads to a radical change in the structure and properties of power systems, giving rise to new problems in their functioning. Modern electric power systems (EPS) are acquiring stable features of cyber-physical systems (CPS), the components of which are the physical and information and communication subsystems. As the EPS "digitalizes", both subsystems become equally complex, equally significant and significantly influencing each other. The introduction of complex technical, information and communication equipment leads to an increased vulnerability of the Smart Grids as a whole and its individual subsystems to various failures and disturbances, including deliberate ones. Therefore, at present, when developing conceptual models and projects for the development of EPS, much attention is paid to the problem of ensuring cyber-physical stability, i.e. stability of the Smart Grid and its objects to physical and information disturbances [1-5].

The most vulnerable and dangerous in terms of the consequences of realized failures, disturbances, deliberate actions or cyber attacks for Smart Grids are the objects of the information and communication control subsystem, such as SCADA systems that collect telemetry data, Wide Area Measurement System (WAMS), registering Synchronized Phasor Measurements (SPM), Wide Area Control System (WACS), etc. [6,7].

As a result of cyber-attacks on SCADA and WAMS aimed at distorting or stealing data, the tasks of managing the technological subsystem (TS) may receive distorted information about the operating parameters. If special measures are not taken to identify these distortions, serious errors may occur in the decisions made during the dispatching control of the TS. Therefore, to obtain a qualitative result of the state estimation (SE), the measurements used must be checked for the presence of distorted data. The SCADA data and SPM are processed in the SE problem, in which the power flow calculation is performed according to the measurement data [8,9].

The report considers the issues of applying the SE procedure for analyzing and improving the cyber-physical stability of the EPS and its facilities. It is shown that procedures detecting and compensating erroneous measurements serve as an effective means of identifying cyber attacks on SCADA and WAMS systems and eliminating their consequences, can significantly reduce the probability of distortion of the current state and improve the results of solving the tasks of dispatching control of the technological subsystem of the EPS.

2 Reliability and cyber-physical stability of Smart Grids

Security is one of the most important properties of an EPS, which ensures that the specified modes of the power system are maintained when operating conditions change, element failures and sudden disturbances. The problem of ensuring the security of electric power systems (EPS) is traditionally relevant for the electric power industry. Currently, a number of factors associated with the reform of the Russian electric power industry and the transition to market relations have led to

* Corresponding author: kolosok@isem.irk.ru

a decrease in the reliability and efficiency of the operation of the EPS.

To ensure a high level of system reliability and reliable power supply to consumers in normal modes of operation of the power system and to prevent emergency situations in a number of advanced countries of the world, including Russia, the process of building Smart Grids is underway. The use of the latest information technologies in the transition to Smart Grids creates new vulnerabilities in terms of cybersecurity, in this regard, deliberate intrusions or cyber attacks that threaten a large-scale power outage should also be considered as disturbances.

Existing approaches to the analysis of security relate mainly to the TS. Transitions from the normal state of a TS to an emergency state have been well studied by power engineers for many years [10,11, etc.].

At the same time, emergency conditions in the information and communication subsystem (ICS) due to failures and deliberate disturbances on its components can lead to a decrease in operational reliability and serious emergency conditions in the physical subsystem of the Smart Grids. Therefore, to analyze the stability of the Smart Grids and its objects to physical and information disturbances, i.e. Cyber-physical sustainability requires the development of new approaches to consider not only the technological, but also the information and communication subsystem, as well as their close relationship.

In a number of works [12-14], the reliability of the CPS of an electric power facility, consisting of two subsystems, is considered. The technological (physical) subsystem includes technical devices of primary equipment such as transformers, generators, reactors, bus coupler circuit-breakers, switches, disconnectors, relays, common auxiliaries, capacitor banks, etc., also includes electronic devices - PLC, IED, without which the operation of modern technical equipment is impossible. The ICS includes computing and network equipment, intelligent devices, software systems - everything that provides the collection, processing and transmission of information in digital form from physical components to the Control Center, as well as means of protection information.

As shown in [15] (Fig. 1), in the CPS, combined emergency conditions can occur, when a failure in the operation of one subsystem causes a failure in the second. The lack of electricity causes failure of the components of the information and communication subsystem (ICS) or the entire control unit, and vice versa, the erroneous results of the computing modules form incorrect control actions, causing, in turn, emergency events in the technological subsystem (TS).

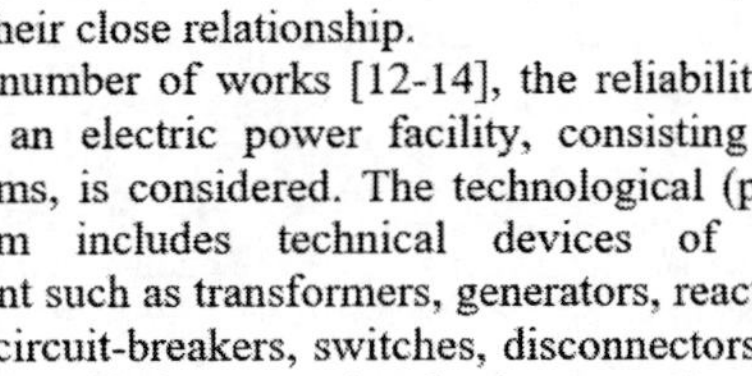

Fig. 1. States and transitions of the cyber-physical system [15].

One of the reasons leading to combined emergency events in the Smart Grids is malicious interference in its trouble-free operation. An analysis of events in the development of a number of systemic accidents in various countries, performed in domestic and foreign works, showed the presence of the mutual influence of failures and disturbances in the technological and information-communication subsystems of the EPS during the development of accidents (arrows of transitions to states C and D in Fig. 1). One of the main recommendations formulated in these works is the need to have a reliable and secure information and communication subsystem in the EPS that provides dispatch control tasks with information about the operating parameters obtained by estimating the state according to SCADA and WAMS data.

3 The role of the SE problem in providing the cyber-physical stability of the EPS

The organization of monitoring the state of the power facility in order to detect a sharp change in its state and subsequent response to this change will protect the power facility from making wrong decisions during management. For these purposes, programs are used that analyze operation parameters and can distinguish reliable information from false information and identify situations of malicious interference in the operation of power facilities. The category of such programs-analyzers includes software tools that implement the procedure Estimating the state of the EPS.

State estimation (SE) is a statistical data processing method used to check the reliability of data, filter measurement errors and recalculate unmeasured parameters. This task works on-line, detects the appearance of gross and systematic errors in the measurements of the operating parameters, identifies them and, if possible, determines the reasons for their occurrence. Locating the location of the error can help to find its source and cause.

The ESI SB RAS has developed a method of control equations for detecting erroneous data and assessing the state of the EPS [16]. The method was developed to detect incorrect data in SCADA measurements, then adapted to verify WAMS measurements and analyze the cybersecurity of data collection and processing systems - SCADA and WAMS - belonging to the information and communication infrastructure of the Smart Grid [17-18].

If deliberately false information comes from single recorders, then the SE task can independently cope with this problem, a priori detecting an erroneous measurement and replacing it with a pseudo-measurement. In the case when the attack is carried out on a large number of registrars, the number of erroneous measurements can lead to a divergence of the SE computational process. In such cases, the blocking of the SE task is provided, about which a message is issued to the dispatching personnel.

A large number of rejected measurements can lead to the loss of observability of the calculation scheme for the

SE. In this case, the SE is executed for the observed areas. In the unobserved part of the circuit, you can use the estimates from the previous cycle. One of the measures to combat the loss of SE observability is the installation of backup devices at power facilities for registering SPM, then when specific measurements are rejected, it will be possible to replace them with the values obtained from these backup PMUs.

To localize the impact site of a cyberattack, an SE algorithm was proposed in [19] based on the selection in the calculation scheme of areas observed by PMU. Performing a local SE for such areas separately according to SCADA and WAMS data, it is possible, based on the results of the SE, to determine which of these systems was attacked by a cyberattack.

In addition, the software itself, which implements the functions of the SE, can be subjected to cyberattacks. The main functions of the SE are the formation of the current settlement scheme according to TV signals; observability check; detection of gross errors in measurements and calculation of estimates of measured and non-measured parameters. In reliable SE software, all these functions should be duplicated [20].

Obviously, methods for detecting erroneous data and SE can give useful results in the event of emergency conditions in the information and communication subsystem and transitions from these states to the normal mode of the technological subsystem (transition B, shown in Fig. 1 by a red arrow). If the SE procedure cannot perform its functions due to disturbances in the ICS, then transitions to a combined emergency state occur (transition C, shown in Fig. 1 by a black arrow).

Based on the studies performed, Table 1 was compiled (with reference to Fig. 1.), which shows the possible transitions during the impact of a cyberattack on the ICS of a certain power facility and the functions that the SE procedure can provide to mitigate the consequences of such transitions.

Table 1. Interaction of cybernetic and physical subsystems of a power facility.

Transitions (Fig.1)	Information and communication subsystem (ICS)		Technological subsystem (TS)		The result of mutual influence
	Occasion	**Functions of the SE procedure**	**Occasion**	**State**	
B	Failure of the telemetry sensor or PMU	SE using additional pseudo-measurements	Regular mode	Regular mode	Both substations are functioning normally
	Failure of several telemetry sensors or PMU	Partial loss of observability. Distributed SE by monitored areas. Using estimates of the unobserved part of the scheme from the previous cycle.	Distributed SE by monitored areas. Using estimates of the unobserved part of the scheme from the previous cycle.	Regular mode	Both substations are functioning normally
	Cyber attack on SE software	Launch of duplicate algorithms of authentication blocks and SE.	Informing staff	Regular mode	Restoration of the OS program. TS in normal mode
	Cyber attack that led to a falsified break in the communication line	The division of the design scheme into subsystems, SE is performed using the decomposition algorithm	Regular mode	Regular mode	Both substations are functioning normally
	Cyber attack due to penetration into the local network, which did not lead to failures in the TS	Temporary shutdown of the ICS in the absence of a data protection tool	Regular mode	Regular mode	Recovery in the ICS. "Manual" control of the technologist / dispatcher in the work of TS is required
C	Complete non-receipt of SCADA and WAMS data	SE task in standby mode	The failure of the ICS led to the loss of information for the operator about the need for control action in the TS	Denial of TS	Recovery in the ICS, then in the TS Or "Manual" control action mode in TS
	Measurement out of sync	Incorrect results of calculations or divergence of SE computational process	SCADA and WAMS data are marked as FAILURE	Emergency mode	Requires "manual" control of the technologist / dispatcher at work TS Recovery in the ICS
	Cyber attack on SE task software (in the absence of	Failures in the computing process, emergency stop of the SE task	SCADA and WAMS data are marked as FAILURE	Emergency mode	Requires "manual" control of the technologist /

	duplicate SE algorithms)				dispatcher at work TS Recovery in the ICS
	Human factor	Making an incorrect decision on control action (contrary to SE results)	Dependent on staff	Deterioration in the TS	Recovery of a temporary failure of the TS. Troubleshooting by staff
	Penetration of a hacker into the ICP (theft or distortion of data, traffic overflow leading to failures in the TS)	Unable to run SE task	Dependent on external influence emergency event	Emergency state in TS	Independent recovery in each of the substations

As the obtained results showed, the application of the SE procedure to increase the cyber-physical stability of the Smart Grid and its objects requires a significant development and improvement of the methods and algorithms of the SE, which ensures the cybersecurity of the information and communication structure of the Smart Grid. As a development strategy, it is proposed to increase the redundancy of SCADA measurements, supplement them with WAMS measurements, to combine various methods for detecting erroneous data – a priori (Test equations method), a posteriori (analysis of evaluation residuals), robust criteria. Also it is proposed to use of criteria for the maximum probability of bad data detection when PMU is placed and to save the observability of the scheme when individual measuring devices failed.

4 Conclusion

In the process of "digitization", both electric power systems and their objects acquire the properties of cyber-physical systems, in connection with this, the problem of ensuring the stability of the Smart Grid and its objects to physical and information disturbances or cyber-physical stability arises. The procedures for detecting and compensating for erroneous measurements in the EPS SE serve as effective means of identifying technical failures and intentional effects on information and communication subsystems and eliminating the consequences of failures on the results of solving dispatching control problems. This makes it possible to prevent the occurrence of complex combined emergency conditions and increase the reliability of the functioning of the cyber-physical system of the EPS and its objects.

However, as shown in [21], the transition from a "safe network" to a "stable network" requires the implementation of strategies to improve situational awareness of personnel for monitoring, forecasting, detecting and mitigating the consequences of extreme events, to increase network fault tolerance and develop tools and methods to detect emerging situations and correct them in a timely manner when minimal impact on the operation of the system. An important role here is assigned to the strategy of significant development and improvement of information collection systems and methods and algorithms for their processing.

This study was carried out within the framework of the state assignment project (No. FWEU-2021-0001) of the program of fundamental research of the Russian Federation for 2021-2030.

References

1. N.I. Voropai, Electrichestvo **7**, 12-21 (2020)
2. N.I. Voropai, I.N. Kolosok, E.S. Korkina, A.B.Osak, Energy Policy, **5**, 53-61 (2018)
3. B.V. Papkov, A.L. Kulikov, V.L. Osokin, Library of Electrical Engineering, **9**, 1-96 (2017)
4. D. Faquir, N. Chouliaras, V. Sofia, K. Olga, L. Maglaras, AIMS Electron. Electr, **5**, 24-37 (2021)
5. D. Westlund, *The Essential Role of Cyber Security in the Smart Grid*. Available: https://electricenergyonline.com/energy/magazine/312/article/ (accessed on 20 January 2023).
6. N.I. Voropai, I.N. Kolosok, E.S. Korkina, A.B.Osak, ESR, **3**, 19-28 (2020)
7. I.N. Kolosok, L.A. Gurina, *Information and mathematical technologies in science and management*, **2**, 40-51 (2019)
8. A.Z. Gamm, Yu.N. Kucherov, S.I. Palamarchuk et al, Nauka, *Methods of solving real-time problems in the electric power industry* (1991)
9. A. Abur A., A. G. Exposito, Marcel Dekker, Inc: New York-Bazel, *Power System State Estimation.Theory and Implementation* (2004)
10. N.I. Voropai, Nauka, *Reliability of power supply systems* (1991)
11. N.I. Voropai, Avtomatiz. IT Energet, **3**, 11 (2011)
12. D. Obychaiko, V. Shikhin, G. Chrysostomou, *Reliability Analysis of Cyber-Physical Systems*, in Proceedings of Intern. Conf. on Industrial Engineering, Applications and Manufacturing (ICIEAM), Moscow, Russia, 1-6 (2018)
13. B.V. Papkov B.V., P.V. Ilyushin, A.L. Kulikov, Scientific Publishing Center "XXI century", *Reliability and efficiency of modern power supply* (2021)
14. N.I. Voropai, I.N. Kolosok, E.S. Korkina, Relay protection and automation, **34**, 78-83 (2019)

15. A. F.D'yakov, V. A. Stennikov, S. M. Senderov et al, Nauka, *Reliability of energy systems: problems, models and methods of their solution* (2014)
16. A.Z. Gamm, I.N. Kolosok, Nauka, *Bad data detection in measurements in electric power systems* (2000)
17. E.S. Korkina. Abstract of the PhD Thesis, Development of methods for assessing the state of electric power systems based on the integration of SCADA and PMU data (2009)
18. A.M. Glazunova, I.N. Kolosok, *Solving problems of dispatching control of intelligent electric power systems on the basis of state assessment methods*, in Proceedings of Conf. "Power Engineering of Russia in the XXI century. Innovative Development and Management", Saint Petersburg, Russia, (2018)
19. I. N. Kolosok, E. S. Korkina, *Decomposition of power system state estimation problem as a method to tackle cyberattacks*, in Proceedings of the 1st IEEE Industrial Cyber-Physical Systems, ICPS, **58** (2018)
20. N.I. Voropai, I.N. Kolosok, E.S. Korkina, *Resilience assessment of the state estimation software under cyberattacks*, in Proceedings of E3S Web of Conferences (2018) DOI: https://doi.org/10.1051/e3sconf/20185802013
21. S. Basumallik, P. Chatterjee, *A. Grid resiliency against cyber threats and wildfires*, Available: https://energycentral.com/topics/tags/special-issue-2022-10-grid-reliability-resilience (accessed on 20 January 2023)

Thermodynamic analysis of the processes of interaction and formation of nitrogen and sulfur compounds during the combustion of Kansk-Achinsk coals

Vladimir Elsukov* and *Svetlana* Latushkina

Bratsk State University, 40 Makarenko st., Bratsk, Russia

Abstract. The paper identifies and examines the mechanisms of the interaction of nitrogen and sulfur compounds during the combustion of Kansk-Achinsk coals (KAC). The researches were carried out on the basis of the model of extreme intermediate states (MEIS) developed at The L.A. Melentiev Institute of Energy Systems and presented in a number of publications. The KAC of various deposits characterized by different content of sulfur and mineral components are considered. On the basis of computational experiments, the influence of alkaline elements of the mineral part (Ca, Na, K) on the formation of NO and SO_2 was determined. The temperature interval of the transition of condensed calcium sulfate to sulfur dioxide during the combustion of KAC of various (including new) deposits has been clarified. Based on the modeling of thermodynamic states, the results of industrial tests are considered, in which inversely proportional dependences of NO and SO_2 on the resulting calcium sulfate have been recorded. These dependences can be explained by multidirectional energy flows during the formation of $CaSO_4/_c$ and NO_x.

1 Introduction

It is known that nitrogen oxides (NO) and sulfur oxides (SO_2) are the main gaseous pollutants during solid fuel combustion. Currently, there is insufficient data on the mechanisms of possible influence of these pollutants. The possibility of the influence of sulfur binding processes on the generation of NO was noted in particular by American specialists [1].

2 The study

The present study was carried out using the thermodynamic model of extreme intermediate states (MEIS), developed at the Institute of Energy Systems SB RAS (Irkutsk), B.M. Kaganovich and collaborators [2,3,4]. In [5], a modification of the model is presented, which includes restrictions on macroscopic kinetics and is written by the authors, in relation to these studies, in the form (for fixed pressure, temperature and initial composition of reagents):

find:

$$\max x_j = F(x^{ext}), \quad (1)$$

under the conditions:

$$Ax = b, \quad (2)$$

$$D_t(y) = \{x: x \leq y\}, \quad (3)$$

$$G(x) = \sum_j G_j(x)x_j\,, \quad (4)$$

$$x_r \leq \varphi, \ r \in R, \quad (5)$$

$$x_j \geq 0, \quad (6)$$

where $x=(x_1,.....x_n)^T$ is the vector of the quantities of moles of the components of the reaction mixture; $y=(y_1,.....y_b)^T$ is the vector of the number of moles of the initial reagents, $y \subset x$; ext is the index of the equilibrium state corresponding to the extremum of the objective function; $A - m \times n$ is the matrix of the contents of elements in the components of the system; b is the vector of the number of moles of elements; $D_t(y)$ is the region (set) of thermodynamic reachability from y; G – Gibbs energy; $j \in J^{ext}$ – is the set of indices of components, the extreme concentration of a set of which must be determined; R – sets of indices of constraints on macroscopic kinetics; φ is a constant.

The objective function (1) represents the desired maximum concentration of the harmful substance that we are interested in. Matrix equation (2) represents the condition for preserving the quantities of moles of elements. Expression (3) defines the domain of thermodynamic reachability. The sign «≤» inside the curly brackets has a thermodynamic interpretation: $x \leq y$, if it is possible to move from y to x along a continuous trajectory along which the characteristic function (in this case G) does not monotonically increase.

The authors included conditions (5) in the block of MEIS restrictions. Restrictions (5) on the maximum concentrations of individual substances were set taking into account experimental data and (or) the results of thermodynamic calculations at individual stages.

* Corresponding author: elswk@mail.ru

The paper considers the KAC of three deposits, the parameters of which are presented in Table 1.

Table 1. The main parameters of the KAC deposits to be taken under consideration.

Parameter	Kansk-Achinsk coals (KAC) of the following deposits:		
	Irsha-Borodinskoye	Bogotolskoye	Berezovskoye
Q_i^r, kJ/kg	15659	10140	15650
S^r, %	0,18	0,49	0,20
A^r,%	6,00	6,72	4,70
CaO,%	25,80	37,30	60,00
Na_2O,%	0,50	0,70	0,80
K_2O,%	0,50	0,30	1,20

At the first stage of the research, the influence of various components of the mineral part of fuels on the generation of the pollutants in question (NO_x и SO_2) was considered. It was noted in [6] that the alkali metals contained in the mineral part of coal (Ca, Na, K) can interact during the combustion and (or) with sulfur and nitrogen, changing the concentrations of the above pollutants.

Figure 1-2 shows the results of calculations for (1)-(6) burning of Berezovsky coal in two variants. In these variants, the composition of the vector x varied, including (in the first variant in Fig.1) 137 components formed by 15 chemical elements. In the second variant (Fig. 2), condensed calcium sulfate ($Na_2SO_4/_c$) was excluded from vector x. Berezovsky coal was selected for this computational experiment as the richest in the content of alkali metals (Na_2O=0,8% отA^d=4%). In both variants, the equilibrium values of nitrogen compounds (x_j^{eq}) were calculated. The characteristics shown in Fig. 1,2, are close to similar dependences based on the results of thermodynamic calculations of the combustion of the KAC Irsha-Borodinsky deposit in the works [2,4]. It can be seen from Fig. 1,2 that at flare combustion temperatures of KAC (1500K), mainly nitrogen oxide (NO) is formed in the furnaces. In the temperature range 1400-1700K, along with NO, the flue gases contain noticeable concentrations of NO_2, $NaNO_2$, KNO_2, N_2O, HNO_2. Moreover, the concentrations of $NaNO_2$ and KNO_2, unlike other nitrogen compounds, are maximum at a temperature of ≈ 1500K. It should be mentioned that in accordance with hygienic standards [7] sodium and potassium nitrites are poisonous and 20 times more dangerous than nitrogen dioxide (NO_2). The maximum permissible concentrations (MPC) in the air of the working area, respectively, are 0.1 and 2 mg/ m^3. It follows from the above that when burning KAC, along with various nitrogen oxides (NO_x), it is necessary to control the emissions of $NaNO_2$ and KNO_2. The exclusion of $Na_2SO_4/_c$ from the vector x led to an increase in the concentration of $NaNO_2$ (Fig.2) and a slight decrease in NO_2. Such exclusion of $Na_2SO_4/_c$ is possible, for example, if all the sulfur of the fuel is bound in $CaSO_4/_c$ during combustion in the fluidized bed. It should be mentioned that the concentration of calcium in the mixture is three or more times higher than the value required for stoichiometric binding of fuel sulfur. Fluctuations in the concentrations of sodium and potassium nitrite emissions are also possible with changes in the content of Na and K in fuel, which occurs in practice.

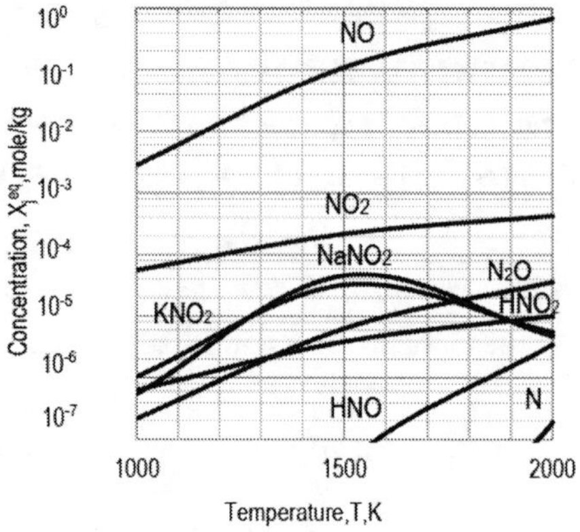

Fig.1. Equilibrium concentrations of nitrogen compounds in the combustion products of the KAC of Berezovsky deposit at α=1,2 (with the inclusion of the $Na_2SO_4/_c$ component in the vector x).

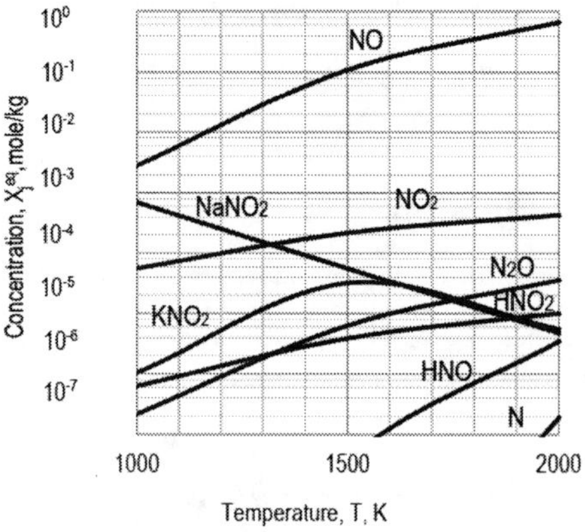

Fig.2. Equilibrium concentrations of nitrogen compounds in the combustion products of the KAC of Berezovsky deposit at α=1,2 (without the inclusion of the Na2SO4/c component in the vector x).

Thus, the amount and type of emissions of nitrogen compounds depend on the conditions of interaction of alkali metals (Ca, Na, K) with both nitrogen oxides and sulfur compounds.

At the second stage of the research, the conditions for the formation of sulfur compounds during the combustion of KAC of the four above-mentioned deposits were considered. The values of temperatures (T) and excess air (α) varied. The results of calculations based on (1)-(6) are presented in Fig.3-4. Fig.3 shows the equilibrium (x^{eq}) values at α = 1,2 and temperatures in the range 1000÷1600K. From these figures it can be seen that for all KAC, the temperature interval of transition $x^{eq}_{CaSO_4/c}$ in $x^{eq}_{SO_2}$ remains approximately constant and is ≈ 1400÷1500K. In the same temperature range, sodium and potassium sulfates are converted into sulfur dioxide and nitrogen compounds. The values of sulfur in the fuel (vector y) determine the equilibrium concentrations of the components formed $x^{eq}_{CaSO_4/c}$ and $x^{eq}_{SO_2}$, which are greatest during the combustion of the KAC Bogotolskoye deposit. Extreme values $x^{eq}_{CaSO_4/c}$ and $x^{eq}_{SO_2}$ do not depend on temperature and are determined only by the material balance (2).

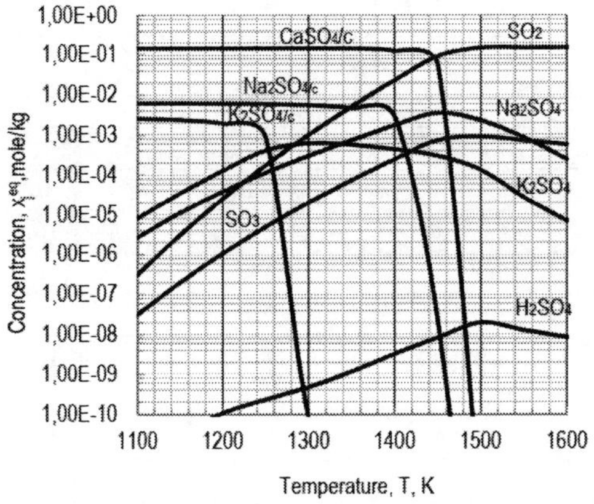

Fig.3. Equilibrium concentrations of sulfur compounds in the combustion products of the KAC of Bogotolskoye deposit (α=1,2; p=0,1 MPa).

Figure 4 shows the effect of excess air (α) on NO и SO_2 emissions and temperature transitions $x^{eq}_{CaSO_4/c}$ in $x^{eq}_{SO_2}$ for the Irsha-Borodinsky deposit. The calculations were performed at two values of α, respectively 1,1 and 1,2. It can be seen from the figures that at these values of α, the sulfur compounds in question change somewhat only in the temperature range (1400÷1500 K), and outside of it their concentrations are approximately constant (these compounds are only diluted with excess air). Large values of α correspond to large concentrations of $CaSO_4/_c$ and NO (and at all design temperatures).

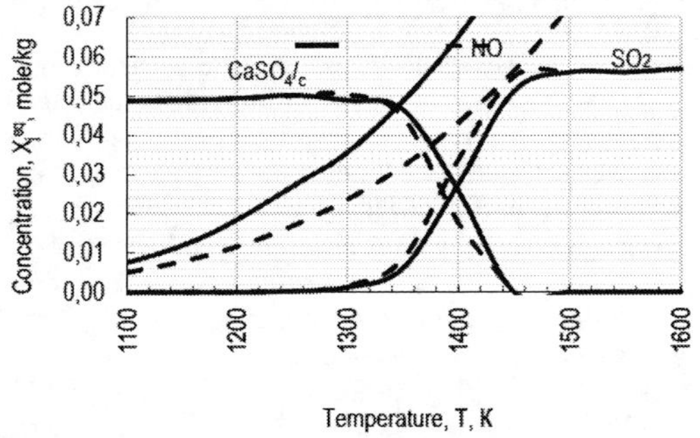

Fig.4. The effect of excess air on emissions NO and SO_2 during the combustion of the Irsha-Borodinsky KAC deposit (- α=1,2; - α=1,1).

In the work [8], the authors of the presented study performed an analysis of industrial tests of boilers with flare combustion of the KAC Irsha-Borodinsky deposit. Conclusions were drawn about the correspondence of experimental data with the calculated temperature range of calcium sulfate formation (1400-1500K) and the influence of α values on its value. The authors explained this by the fact that the time of presence of the combustion products in the region with T=1400÷1500K decreased with the increase of α and the proportion of SO_2 transition to $CaSO_4/_c$ decreased. By reaction:

$$CaO+SO_2+0,5O_2 \rightarrow CaSO_4/_c+500KJ, \quad (7)$$

Thus, for the considered KAC, the temperature interval of the transition of $CaSO_4/_c$ to SO_2 is determined by the final equilibrium of the system (combustion products) in accordance with (1)-(6) and is 1400–1500K, and the value of the transition fraction is non-thermodynamic factors (kinetics, diffusion, etc.).

At the last stage of the research, the possibility and conditions of a simultaneous increase in NO concentrations and a decrease in SO_2 were tested at values close to those measured by Japanese specialists in industrial experiments with a fluidized bed [9].

In these experiments, the researchers consistently increased the molar ratio of Ca/S (0; 2 and 4) in the layer at a sulfur content of S^d = 0.62% and recorded a simultaneous increase in NO_x from about 30 to 75 ppm and a decrease in SO_2 from 140 to 17 ppm. The obtained dependencies are not explained by the researchers.

The authors of the presented work suggested that such dependences can be explained by the multidirectional energy flows of the reactions of the formation of $CaSO_4/_c$ and NO. From (7) it can be seen that when a mole of $CaSO_4/_c$ is formed, 500KJ is released, and for NO, on the contrary, 90KJ is absorbed. It is clear that with an increase in the [9] molar ratio of Ca/S, a decrease in the concentration of SO_2 will lead to the formation of $CaSO_4/_c$ and, accordingly, the release of energy necessary for the generation of NO. To verify this assumption, the authors modeled the conditions of the above experiments in relation to the KAC of the Bogotolskoye deposit (closest to [9]). The objectives of the simulation are to determine the enthalpies (their changes) of the system during the above experiments. $Na2SO_4/_c$was chosen as the target function (1), since the conditions of its formation are close to $CaSO4_4/_c$ (as can be seen from Fig.3,4), but its effect on the concentrations of other sulfur-containing components is insignificant (due to the small content of Na).

The concentration of NO was set from experiments [9] taking into account the conversion of NO_x to NO ($NO=0.13 \cdot NO_x$) by the formula (5) in the form:

$$x_{NO}=\psi, \quad (8)$$

where ψ is the number of moles of NO, which increased sequentially in three calculated experiments (0; 2; 4) (states). The numbering coincides with the mole relations of Ca/S in [9] (for convenience of comparison).

The concentration of SO_2 was also set from experiments [9] by the formula (5) in the form:

$$x_{SO_2}=\varphi, \quad (9)$$

where φ is the number of moles of SO_2, which was consistently reduced in computational experiments in accordance with [9].

Table 2. The results of modeling the combustion process of the Bogotolskoye deposit under thermodynamic and kinetic constraints (T=1073K, α =1,2).

Parameters of the mixture	State of the system						
	y	$x^{ext}_{Na_2SO_4/c}$					x^{eq}
	State numbers						
	1	0	2	4	3	5	6
		Type of restrictions					
		(1)-(6), (8), (9) at			(1)-(4), (6)	(1)-(6), (8), (9) at	(2)-(4), (6)
		ψ=0,00082 φ=29,3	ψ=0,0012; φ=0,0115	ψ=0,0024 φ=0,0036	ψ=0; φ=0	ψ=0,0016; φ=0,0115	
	Gibbs energy, enthalpy, kJ/kg; concentrations, mole/kg						
G(x)	$-8,506 \cdot 10^3$	$-9,107 \cdot 10^3$	$-1,126 \cdot 10^4$	$-9,102 \cdot 10^3$	$-1,156 \cdot 10^4$	$-9,103 \cdot 10^3$	$-1,662 \cdot 10^4$
H(x)	$2,279 \cdot 10^2$	$-2,053 \cdot 10^2$	$-2,166 \cdot 10^3$	$-2,078 \cdot 10^2$	$-2,425 \cdot 10^3$	$-2,078 \cdot 10^2$	$-2,525 \cdot 10^3$
CO		$1,982 \cdot 10^1$	$5,636 \cdot 10^0$	$1,978 \cdot 10^1$	$2,067 \cdot 10^0$	$1,978 \cdot 10^1$	$1,060 \cdot 10^{-7}$
CO_2	$5,148 \cdot 10^{-3}$	$8,418 \cdot 10^0$	$2,260 \cdot 10^1$	$8,426 \cdot 10^0$	$2,617 \cdot 10^1$	$8,426 \cdot 10^0$	$2,824 \cdot 10^1$
O_2	$3,595 \cdot 10^1$	$2,092 \cdot 10^{-2}$	$3,949 \cdot 10^0$	$2,174 \cdot 10^{-2}$	$6,065 \cdot 10^0$	$2,174 \cdot 10^{-2}$	$5,495 \cdot 10^0$
SO_2		$2,930 \cdot 10^{-2}$	$1,150 \cdot 10^{-2}$	$3,600 \cdot 10^{-3}$	$8,227 \cdot 10^{-3}$	$3,600 \cdot 10^{-3}$	$8,249 \cdot 10^{-8}$
$CaSO_4/_c$		$1,166 \cdot 10^{-1}$	$1,334 \cdot 10^{-1}$	$1,344 \cdot 10^{-1}$	$1,367 \cdot 10^{-1}$	$1,344 \cdot 10^{-1}$	$1,448 \cdot 10^{-1}$
$Na_2SO_4/_c$		$6,233 \cdot 10^{-3}$	$6,233 \cdot 10^{-3}$	$6,233 \cdot 10^{-3}$	$6,233 \cdot 10^{-3}$	$6,233 \cdot 10^{-3}$	$6,224 \cdot 10^{-3}$
$K_2SO_4/_c$		$1,474 \cdot 10^{-3}$	$2,578 \cdot 10^{-3}$	$7,586 \cdot 10^{-4}$	$2,541 \cdot 10^{-3}$	$7,586 \cdot 10^{-4}$	$2,608 \cdot 10^{-3}$
NO		$8,200 \cdot 10^{-4}$	$1,200 \cdot 10^{-3}$	$2,400 \cdot 10^{-3}$	$5,064 \cdot 10^{-3}$	$1,600 \cdot 10^{-3}$	$8,200 \cdot 10^{-4}$
NO_2		$2,309 \cdot 10^{-5}$	$1,343 \cdot 10^{-3}$	$2,397 \cdot 10^{-5}$	$7,976 \cdot 10^{-4}$	$2,397 \cdot 10^{-5}$	$5,339 \cdot 10^{-5}$
$NaNO_2$		$1,709 \cdot 10^{-11}$	$1,711 \cdot 10^{-11}$	$1,704 \cdot 10^{-11}$	$1,695 \cdot 10^{-11}$	$1,704 \cdot 10^{-11}$	$1,493 \cdot 10^{-6}$
KNO_2		$3,756 \cdot 10^{-5}$	$2,914 \cdot 10^{-5}$	$1,700 \cdot 10^{-7}$	$1,679 \cdot 10^{-4}$	$1,700 \cdot 10^{-7}$	$2,882 \cdot 10^{-6}$

According to (7), the change in the number of moles of SO_2 is equal to the change in the number of moles of $CaSO4_4/_c$, which was taken into account when setting constraints (9). According to the calculation results, Table 2 is compiled, which presents 7 states of the system (from 0 to 6): 1 – characterizes the initial composition (y); 0, 2, 4 – modeling a system with constraints for the formation of NO and SO_2; 3 – without restrictions; 5 – close to state 2, but with a change in ψ; 6 – the state of final equilibrium.

Comparison of states 0, 2, 4 shows that with a decrease in SO_2, $CaSO_4/_c$ increases and the O_2 content decreases. From the comparison of the results, it can be seen that the enthalpy (H) in states 0, 4, 5 remains approximately constant, while in state 2 it decreases noticeably. Close values of H indicate that an increase in NO is compensated by a decrease in SO_2 (an increase in $CaSO_4/_c$). The exception from this series state 2 can be explained by various reasons: inaccuracy of measurements, fluctuations in the material and energy balances of the combustion zone, etc. Even small changes in concentrations of either NO or SO_2 ($CaSO_4/_c$) can greatly change the enthalpy of the system (compare states 2 and 5). State 5 is conditional, because it has a value of φ 25% greater than in state 2.

Thus, the results obtained confirm the authors' assumption that the inverse dependences of NO and SO_2 on the molar ratio Ca/S [9] are explained by the multidirectional energy flows of the reactions of $CaSO_4/_c$ and NO formation.

3 Conclusions

It is found that the alkali metals of the mineral part of the fuel (Ca, Na, K) can interact with both nitrogen and sulfur, affecting the formation of both NO and SO_2.

It is shown that further study and control of emissions of not only NO_x, but also sodium and potassium nitrites is necessary during the combustion of KAC.

It is clarified that for all the considered KAC (regardless of the content of sulfur and alkali metals), the temperature interval of the transition of condensed calcium sulfate to sulfur dioxide remains approximately constant (1400-1500K), and the transition process thermodynamically does not significantly depend on α.

A conclusion was made about the thermodynamic possibility of a simultaneous increase in NO_x formation and a decrease in SO_2, which is due to multidirectional energy flows during the formation of $CaSO_4/_c$ and NO.

The conclusions obtained, of course, require additional verification.

References

1. R. Pagne, *Natural gas reburning applied to three coal – fired utility boilers*/ R.Pagne, B. Folsom, T.Simmer, in Proceedings of the International gas – reburn technology workshop, 7-9 February 1995, Malmö, Sweden (1995)
2. B. M. Kaganovich and S. P. Filippov, Equilibrium thermodynamics and mathematical programming (Novosibirsk: Nauka: Siberian Publishing Company RAS), 236 (1995)
3. A. N. Gorban, B. M. Kaganovich, S. P. Filippov, Thermodynamic equilibria and extremes: Analysis of achievability areas and partial equilibria in physico-chemical and technical systems (Novosibirsk: Nauka), 296 (2001)
4. A. N. Gorban, B. M. Kaganovich, S. P. Filippov, A.V. Keiko, V.A. Shamansky, I.A. Shirkalin, Termodinamic Equilibria and Extrema Analisis of Attainability Regions and Partial Equilibria (Springer) 282 (2006)
5. B. M. Kaganovich, A.V. Keiko, V. A. Shamansky, Equilibrium thermodynamic modeling of dissipative macroscopic systems (Irkutsk: ISEM SB RAS), 76 (2007)
6. V. K. Elsukov, Complex researches of power sources efficiency in systems of centralized heat supply: autoabstract of dis. for the degree of doctor of technical Sciences, Irkutsk, (2012)
7. SanPiN 1.2.3685-21 "Hygienic standards and requirements for ensuring the safety and (or) harmlessness of environmental factors for humans"
8. V. Elsukov and S. Latushkina, *Features of formation and reduction of sulfur dioxide emissions when burning brown coal in boilers with liquid slag removal* (E3S Web of Conferences 209, 03012 (2020)
9. Ed. P. Basu, M. Hozio, M. Hasatani, Circulating Fluidized Bed Technology **III**, (Pergamon Press), (1990).

E3S Web of Conferences **461**, 01049 (2023) https://doi.org/10.1051/e3sconf/202346101049
RSES 2023

Prospects of low-carbon development in Russia: the role of renewable energy and challenges of sanctions

Liudmila Nefedova[1]*, *Dmitry*.Solovyev[2,3], *Michael* Berezkin[1], and *Kirill* Degtyarev[1]

[1]Lomonosov Moscow State University, 1/19 Leninskiye Gory, 119991, Moscow, Russia
[2]Joint Institute for High Temperatures of the Russian Academy of Sciences,13/2, Izhorskaya str., 125412, Moscow, Russia
[3]Shirshov Institute of Oceanology of the Russian Academy of Sciences, 36, Nakhimovskiy prospect, 117997, Moscow, Russia

Abstract. The article analyses the volume of electricity production in Russia at solar and wind power plants. The conducted studies allowed the authors to assess the potential for reducing carbon dioxide (CO2) emissions by using these renewable energy sources. The role of renewable energy in the energy supply of Russian regions and in the reduction of CO2 emissions is assessed. The regions of Russia were ranked by the level of decarbonisation due to use of renewable energy sources, and there was assessed the role of renewable energy in the energy balances of the country's regions. In 2023 due to the operation of power facilities without the use of fossil fuels it is possible to avoid emissions of 4.75 million tons. The study revealed a positive trend in reducing CO2 emissions in all leading in RE regions of Russia from 2020 to 2030. According to the forecast using neural networks the total reduction in CO2 emissions due to the use of renewable energy sources in all the regions under consideration may amount to approximately 0.46% of the total CO2 emissions in Russia in 2022 by 2030. This shows not so significant prospects for renewable energy in terms of decarbonisation while maintaining the current place of increasing the capacity of their use in Russia. However, for some regions located in the Southern Federal District, the development of renewable energy sources is of great importance for improving the environmental situation and shows the need for further development and support of domestic technologies.

1 Introduction

Global energy development trends indicate a transition to the fourth stage of energy development with an emphasis on low-carbon development. Many countries are adopting programs to reduce carbon dioxide emissions in the energy sector, and Russia is not an exception. After Russia joined the Paris Agreement, a number of other documents has been adopted. The key of which were the presidential decree on reducing GHG emissions, the Strategy for Low-Carbon Development of the Russian Economy and the Federal Law on Limiting GHG Emissions [1]. Despite the aggravation of relations with Western countries since February 2022 the course towards low-carbon development has been maintained. Additional measures are being taken to encourage low-carbon development. A number of resolutions has been adopted already in March-May 2022. They are mainly related to the implementation of climate projects and carbon regulation. As the main market and financial mechanisms adopted in recent years and stimulating the transition to a low-carbon development path on a voluntary basis one should mention climate projects, green certificates and green bonds - the same or similar instruments have been used in Western countries over the past 1-2 decades. Also a new program to improve the energy efficiency of Russian economy joined into force with the purpose of achieving carbon neutrality by 2060 [2]. In this context the development of renewable energy (RE) becomes especially important as an alternative to traditional fuel energy sources.

As of 07/01/2023 within the framework of the Capacity Supply Agreement (CSA) RES-1 Program. One hundred renewable energy generation facilities with a total capacity of 4.219 MW were operated on Russia's wholesale electricity and capacity market: solar power plants - 1.788 MW (70 facilities), Wind farms - 2 360 MW (25 facilities), SHPP - 71 MW (5 facilities). As of the beginning of the second half of 2023, the total capacity of all renewable energy facilities in Russia is about 2.4% of the total capacity of generating facilities in the energy system of the Russian Federation. At the same time, the capacity of power facilities under the CSA RES program is 1.7% in the country. Moreover, in the United Energy System of the South - SPP and WPP are as of 1.07.23 already 12.1% [3].

Some problems have arisen in the development of wind energy in Russia. Sanctions measures by western countries and the withdrawal of companies such as Vestas and Simens Gamesa from the Russian market have become an incentive to search for alternative suppliers and strategic partnerships in the technologies

* Corresponding author: nefludmila@mail.ru

and components for renewable energy and opened up prospects for the development of domestic production of wind turbine components. Nova Wind JSC (a division of Rosatom State Corporation) put into operation Berestovskaya WPP (60 MW - December 2022) and Kuzminskaya WPP (160 MW - June 2023). It is especially important that since May 2023, despite the sanctions and thanks to the efforts of Russian engineers and builders, Kola Wind farm (202.35 MW) in Murmansk region began to operate in full - the first wind farm in the UES of Russia connected to an automatic frequency control system.

State support measures for the RE industry, such as CSA mechanisms and free bilateral agreements promote investments in low-carbon projects and the development of long-term partnerships. The development of public-private partnership mechanisms and free contracts in RE can create favourable conditions for concentration of production resources and development of domestic production of components for renewable energy. This path contributes to the sustainable reduction of CO_2 emissions and low-carbon development in Russia.

2 Materials and methods

This paper presents the results of studies aimed at regional assessment of the role of renewable energy industry in Russia in reduction of carbon dioxide emission at grid-connected SPPs and WPPs built under the CSA RES program in 2023. The volume of electricity generation at these stations and levels of utilization of the installed capacities are analysed, since a considerable number of solar and wind plants has been constructed in Russia in recent years. The analysis performed with respect to commissioned photovoltaic and wind power stations in Russia allowed us to estimate reduction in emission volumes of CO_2 based on the calculated volumes of power generation at these eco-friendly power facilities. The estimates of reduction in greenhouse gas emissions due to expansion of renewable generation capacities were made assuming that in federal districts of Russia electricity generation is accompanied by different average emissions of CO_2 [4]. Analysing data on commissioned solar and wind power plants in Russia, we were able to estimate the annual reduction in CO_2 emissions in various regions. Estimation of annual CO_2 emissions reduction in regions took into account data on volumes of carbon dioxide emission associated with the complete "life cycle" of SPPs and WPPs. That is photovoltaic modules manufactured from highly purified silicon as well as metal frames and supports for the solar panels; steelmaking for WPP tower construction; recovery and recycling of panels and blades from WPPs etc. Based on the analysis of publications on this matter, the mean value that may be adopted for emissions is 35 g CO_2-eq/kWh of generated electricity for SPPs and 25 g CO_2-eq/kWh for WPPs [5-7]. The WPP and SPP facilities were built mainly as an addition to the grid connected Power Stations in regions [8]. Therefore, the estimates of "CO2 non-emissions" were made taking into account the calculated CO_2 emissions by category of power facilities in the production of electricity in Russia (Table 1).

Table 1. Carbon dioxide emissions reducing due to the operation of RES power plants as an alternative to fuel power plants in the Federal Districts of Russia.

	WPP g CO_2/kWh per year	SPP g CO_2/kWh per year
Southern FD	577	567
North Caucasian FD	474	464
Volga FD	418	408
North-Western FD	510	500

Compiled by the authors using data from [4-7].

For a more accurate assessment of the impact of WPPs and SPPs on the reduction of CO_2 emissions in the regions of Russia we applied the neural forecasting technique [9]. This method is based on training artificial neural networks on the available data on electricity generation at wind farms and solar power plants. The neural networks were trained on a large amount of data, including information about weather conditions, time of day, and other factors that affect electricity production and CO_2 emissions.

In this research the authors employed statistical and predictive analytics methods, economic and geographical analysis, and expert assesments. The estimations were based on materials provided by power producers and RPSO JSC, as well as statistical data from the Federal State Statistics Service of Russia, NP Sovet Rynka (Non-commercial partnership the Market Council), and the Ministry of Natural Resources and Environment of Russia.

3 Results and Discussion

Based on the calculations of the total installed capacity of solar photovoltaic and wind power plants operating in Russia at the 1/07/2023 for regions with a total SPP capacity of more than 5.0 MW and all grid connected WPP built under the CSA RES Program they were ranked according to the level of development of renewable energy. The largest capacities of 96.6% of wind power plants (except for wind farms in the Ulyanovsk region) are installed in the Southern and North Caucasian Federal Districts, which are part of the United Power Plant of the South of Russia. For calculations of electricity generation in the regions, the average annual value of the capacity factor (CF) was in our calculation according to the data of the System Operator of the Unified Power System of the Russian Federation for 2021: SPP –14.3%. WPP –29.8%; for 2022: SPP – 13.9%.WPP -32.1% [10, 11]. The calculation results are in Tables 2 and 3. These tables also compare the estimated annual electricity energy (EE) production by SPPs and WPPs of Russia with the volume of EE production in the regions for 2022 [12].

Table 2. Assessment of electricity generation and CO_2 emission reductions at operating Solar Power Plants in Russian regions as of 07/01/2023.

№	Federal Districts (FD), subjects of the Russian Federation	Installed capacity of SPPs as of 1/07/2023, MW	Electricity generation at SPPs over a year, mln KWh	Total electricity production in the region in 2022, mln kWh	Share of SPPs electricity in total EE prod. in the region	CO2 emission reducing per year, thousand ton
	Volga FD including:	*643.80*	*783.92*	*105694.80*	*0.74%*	*319.84*
1	Orenburg Reg.	360.00	438.35	12884.00	3.40%	178.85
2	Rep. of Bashkorstan	103.80	126.39	27283.50	0.46%	51.57
3	Samara Reg.	75.00	91.32	20328.90	0.45%	37.26
4	Saratov Reg	105.00	127.85	45198.40	0.28%	52.16
	Southern FD.includ.	*684.40*	*833.35*	*33967.50*	*2.45%*	*472.51*
1	AstrakhanReg.	275.00	334.85	4407.40	7.60%	189.86
2	Rep.of Kalmykiya	234.10	285.05	976.5	29.19%	161.62
3	Volgograd Reg.	120.00	146.12	17043.10	0.86%	82.85
4	Krasnodar Terr.	46.40	56.50	11421.60	0.49%	32.03
5	Rep. of Adygea	8.90	10.84	547.70	1.98%	6.14
	North Caucasian FD. including:	*120.00*	*146.12*	*23437.00*	*0.62%*	*67.80*
1	Stavropol Terr.	100.00	121.76	17165.60	0.71%	56.50
2	Chechen Rep.	5.00	6.09	1586.60	0.38%	2.82
	Rep. of Daghestan	15.00	18.26	4684.80	0.39%	8.47
	Siberian FD. including:	*185.20*	*225.51*	*25911.80*	*0.87%*	*204.08*
1	Republic of Altai	120.00	146.12	150.30	97.22%	132.24
2	Omsk Reg.	60.00	73.06	6509.30	1.12%	66.12
3	Rep. of Khakassiya	5.20	6.33	19252.20	0.03%	5.73
	Far Eastern FD. including:	*215.00*	*261.79*	*13807.40*	*1.90%*	*232.21*
1	Rep. of Buryatiya	115.00	140.03	6199.90	2.26%	124.21
2	Zabaikal'skii Territiry	100.00	121.76	7607.50	1.60%	108.00
	Total in RF as of 1/07/2023	*1848.40*	*2250.69*	*202818.50*	*1.11%*	*1282.89*

Compiled by the authors using data from [3, 11-12].

Table 3. Assessment of electricity generation and CO_2 emission reducing at operating Wind Power Plants in Russian regions as of 07/01/2023.

№	Federal Districts (FD). subjects of the Russian Federation	Installed capacity of WPPs as of 1/07/2023. MW	Electricity generation at SPPs over a year. mln KWh	Total electricity production in the region in 2022, mln kWh	Share of SPPs electricity in total EE prod. in the region	CO_2 emission reducing per year, thousand ton
	North Caucasian FD. including:					
1	Stavropol Terr.	670	1884.01	17165.6	10.98%	953.02
	Southern FD.includ.					
1.	Rostov Reg.	607.29	1707.68	48699.8	3.51%	985.33
2.	Astrakhan Reg.	340.2	956.63	4407.40	21.69%	552.04
	Rep.of Kalmykiya	216.6	609.000	976.5	62.37%	351.40
	Rep. of Adygea	150	421.29	547.70	77.01%	243.32
	Northwestern FD					

	Murmansk Reg.*	202.3	749.68	18 227.0	4.11%	382.31
	Volga FD including:					
	Volgograd Reg.	88.2	248.000	17043.10	1.46%	143.10
	Ulyanovsk Region	85.4	240.14	2 685.0	8.94%	100.38
	Total in RF as of 1/07/2023	1997.69	5617.42	202818.50	0.48%	3650.92

*Data for the Murmansk region. (Kola wind farm) - planned.
Compiled by the authors using data from [3, 11–12].

The total reduction of CO_2 emission per year in 2023 in RE in Russia is in Wind Energy – 3.65 mln t and in Solar Energy – 1.28 mln t. A regional analysis of the of carbon dioxide emissions reduction in all subjects of the Russian Federation, where grid connected power plants operate on renewable energy sources, has been carried out. The regions are identified, mainly in the Southern and North Caucasian Federal Districts, with the highest installed capacity of WPPs and SPPs, in these regions that SPPs and WPPs operate with the highest efficiency. The maximum effect in terms of decarbonisation due to development of solar energy was achieved in the Astrakhan and Orenburg regions - 189.86 thousand tons and 178.85 thousand tons per year, respectively. Wind power provides the maximum reduction of carbon dioxide emissions in the Rostov Region and Stavropol Territory (985.33 thousand tons and 893.02 thousand tons per year). The ranking of Russian regions by the level of decarbonisation by the use of energy facilities for RES made it possible to single out the seven leading regions and assessed the share of calculated values of the total EE production at SPPs and WPPs in the total electricity energy generated in 2022 in these regions of Russia (Table 4).

Fig. 1 shows a neural forecast for reducing CO_2 emissions due to renewable energy facilities (SPPs and WPPs) in the leading regions of Russia in terms of the level of decarbonisation based on the performed ranking. This forecast shows how ~~the~~ development of renewable energy sources in different regions of Russia helps to reduce carbon dioxide (CO_2) emissions into the atmosphere. The total reduction of CO_2 emissions per year, in thousands of tons, for each region and the total reduction for all seven regions under consideration are shown. Forecast data are given for the period from 2023 to 2030. All regions show a positive trend in CO_2 emissions the reduction due to renewable energy sources for the entire period. The largest increase is achieved in the Republic of Adygea (+263%), and the smallest - in the Ulyanovsk region (+228%). The highest level of reduction in CO_2 emissions in 2030 is achieved in the Stavropol Territory (1838.59 thousand tons), and the lowest - in the Ulyanovsk Region (328.25 thousand tons). This is explained by the different scale and speed of RES development in these regions. Among all regions, the largest jump in CO_2 reduction occurs in 2029 (+20.6% compared to 2028), and the smallest in 2024 (+17.3% compared to 2023). This may be due to various factors such as the economic situation, technological progress and climatic conditions. According to the forecast made using neural networks, the total reduction in CO_2 emissions due to the use of renewable energy sources in all the regions under consideration may amount to approximately 0.46% of the total CO_2 emissions in Russia in 2022 by 2030 [13].

We also calculated the investment costs for achieving carbon neutrality by 2050 without taking into account absorption, based on the objective function - reducing to zero the actual anthropogenic emission. The calculation methodology is based on the assumption that the absence of anthropogenic greenhouse gas emissions is possible if all the required amount of energy is supplied in the form of electricity generated at power plants operating on non-carbon sources - nuclear energy and renewable energy sources. We carried out calculations of the needs for generating capacities to provide the required amount of energy and, based on the available data on specific investment costs (per unit of installed capacity), estimates of the total investment costs were made.

Table 4. Ranking of Russian regions by the level of reduction of CO_2 emissions due to RE facilities (SPP and WPP) as of 01.07.2023.

	Subjects of the Russian Federation	Share of SPPs & WPP electricity in total EE prod. in the region in 2022	CO_2 emission reducing per year, thousand ton		Total CO_2 emission reducing per year, thousand ton
			SPP	WPP	
1	Stavropol Terr.	11,69%	56.50	893.02	1009.52
2	Rostov Reg.	3,51%	-	985.33	985.33
3	Astrahan Reg.	29,56%	189.86	552.08	741.94
4	Resp. of Kalmykiya	91.56%	161.62	351.41	513.03
5	Rep. of Adygea	78.99%	-	243.32	243.32
6	Volgograd Reg	2.32%	82.85	143.10	225.95
7	Ulyanovsk Region	8.94%	-	100.38	100.38

Compiled by the authors based on the performed calculations.

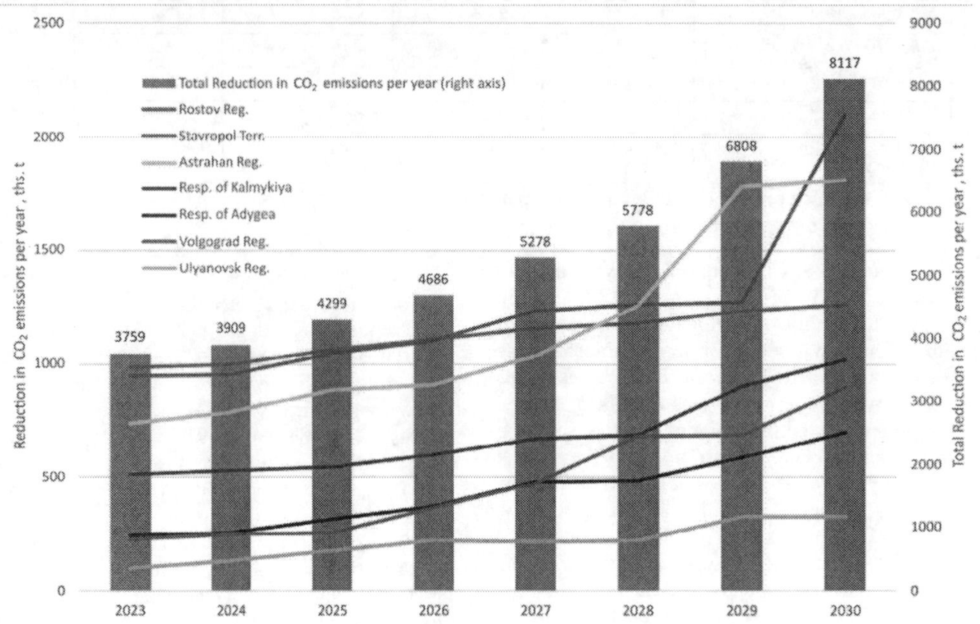

Fig.1. The Forecast of CO_2 emission reduction due to renewable energy facilities (SPP and WPP) in Russia's leading regions in RE decarbonisation. Compiled by the authors based on the calculations performed using a neural model [9].

We have considered nine scenarios with different values of energy consumption and ratios of energy production volumes from different sources. Depending on the scenario, in Russia the costs of the transition to full carbon neutrality, based on the reduction of anthropogenic greenhouse gas emissions to zero, range from 200 to 350 trillion roubles [14]. This is 3-5 times higher than the estimates given in the Energy Strategy of Russia because at the same time, the Strategy assumes that only 30% of the total reduction in net greenhouse gas emissions will come from reducing anthropogenic emissions.

4 Conclusion

According to the calculations, wind power facilities put into operation under the CDA RES-1 Program in total in Russia on the beginning of the second half of 2023 already provide a reduction in CO_2 emissions by 3.65 million tons per year, and in solar energy - 1.28 million tons per year. The maximum effect in terms of reducing CO2 emissions due the use of solar resources was achieved in the Astrakhan and Orenburg regions - 189.89 thousand tons and 178.75 thousand tons per year, respectively. Wind power provides the maximum reduction of carbon dioxide emissions in the Stavropol Territory and the Rostov Region – 893.02 thousand tons and 985.33 thousand tons per year. The ranking of the regions of Russian by the level of decarbonisation due the use of renewable energy carried out by the authors made it possible to single out seven regions - leaders with the Stavropol Territory in the first line. Regional assessments were made of the share of electricity generated from renewable energy facilities in the energy consumption of the constituent entities of the Russian Federation. The maximum regional values were obtained: for solar energy facilities - Republic of Altai (97.2%) and Republic of Kalmykia (29.2%), Astrakhan Region (7.6%) and the for wind energy facilities, respectively: Republic of Kalmykia (62.4%), Republic of Adygea (77.0%), Stavropol Territory (11.0%).

The resulting forecast for CO_2 emission reducing due to renewable energy facilities (SPPs and WPPs) in the regions - leaders in Russia in terms of the level of decarbonisation- allows us to conclude that all these regions show a positive trend in reducing of CO_2 emissions due to renewable energy for the entire period from 2023 to 2030. This indicates that the regions are actively introducing renewable energy sources into their energy system and striving to improve the environment in the country and the world.

The work was carried out within the framework of State Assignment No. 121051400082-4 and with the support of the Ministry of Education and Science of the Russian Federation State Assignment No. FMWE-2023-0002 (construction of a neural model).

References

1. K. Degtyarev, O. Sinyugin, M. Berezkin, *Journal of environmental Earth and Energy Study*, **2** (18). (2023).
2. O.S. Popel', A.B. Tarasenko A.B, *Applied Solar Energy*. **58** (2022).
3. NP Market Council. List of qualified energy facilities operating on the basis of RES, https://www.npsr.ru/ru/market/vie/index.htm

4. B. Saneev, I. Ivanova, A. Izhbuldin. E. Maysyuk, *Energy Policy*, **177** (2022).
5. M. Tawalbeha, A. Al-Othman, F. Kafiah et al. *Sci. Total Environ.*, **759** (2021).
6. S. Gerbinet, S. Belboom, A.Léonard. *Renewable Sustainable Energy Rev*, **38** (2014).
7. Carbon Neutrality in the UNECE Region: Integrated Life-cycle.EC for Europe UN. 2022.
8. L. Nefedova, D. Solovyev. *Applied Solar Energy*, **59**. 1 (2023).
9. L. Nefedova, D. Solovyev, V. Bushuev, V.M. Zaicheneko, *Journal of Physics: Conference Series*. 1(2096) (2021).
10. Russian Power System Operator, Report on the Functioning of the UES of Russia in 2021. Minenergo, 2022, https://www.so-ups.ru
11. Renewable energy market in Russia: current status and development prospects. RREDA. 2023.
12. Official website of the Federal State Statistics Service (Rosstat). Electric balance and electricity consumption in the Russian Federation from 2005–2022, https://rosstat.gov.ru/enterprise_industrial
13. BP. Statistical Review of World Energy 2022, https://www.bp.com/en/global/corporate/energy-economics/statistical-review-of-world-energy.html
14. K. Degtyarev, D. Solovyev, M. Berezkin, *Energy Policy*, **6** (2023).

Application of bang-bang control for frequency control in grids with high number of power electronics devices

Oleg Khamisov*, *Ildar* Idrisov, and *Stepan* Vasilev

Skolkovo Institute of Science and Technology, Moscow, Russia

Abstract. Frequency control is one of the key aspects connected with usage of power electronic devices. On the one hand, replacement of synchronous generation with power electronic interfaces removes kinetic energy of rotor's rotating mass, making system dynamics more volatile. On the other hand, fast response to control signals calls for a shift control paradigm. In this work a bang-bang primary frequency control is proposed. Its effectiveness is supported by both theoretical results and detailed numerical experiments.

1 Introduction

Effectiveness of frequency regulation in power grids greatly depends on system inertia as well as response time of the corresponding controller. Traditionally, proportional-integral controller is used in order to keep frequency at the nominal value. Its speed is limited by characteristic time of the generators' turbines, which normally lies within 3-7 seconds interval [1]. Due to this limitation, large control gains lead to positive feedback and as a consequence, stability loss [2]. Moreover, it is argued that large control oscillations increase wear of turbines and synchronous machines [3]. At the same time, controller speed remains sufficient to limit frequency oscillations due to system inertia, introduced mostly by synchronous generators. It prevents rate of change of frequency (RoCoF) form reaching large values, thus allowing the control to counter power disturbances within the acceptable timeframe, despite control gains limitations. Physical properties of power electronic devices differ significantly from synchronous generation: they do not have physical rotating mass and do not possess corresponding dynamical properties. As a result, shift towards power electronic interfaces leads to two significant changes:

1. RoCoF is no longer limited by system inertia. Therefore, even a small disturbance can lead to unacceptable frequency deviations.
2. Controller is no longer limited by turbines characteristic times. It allows an introduction of fast control systems.

Within this paradigm, there exists several approaches aimed to increase effectiveness of frequency control in low-inertia systems. Majority of them are focused on virtual inertia approach. The downside of it is the necessity to evaluate RoCoF, which contains significant amount of noise.

In this paper a novel primary frequency control algorithm is proposed. Here, the algorithm does not rely on RoCoF and instead utilizes discontinues feedback function in order to maximize efficiency of fast power electronics response. The control operates as a standard proportional control, if frequency deviations are small and shifts to bang-bang mode, if frequency deviations are significant. The control algorithm firstly derived for simplified power system model and then is substituted for more detailed one. The theoretical results are supported by case study. The numerical tests are done in RTDS NovaCor 1.0. hardware-in-the-loop complex.

The novelty of the paper is the following:

1. The controller is proven to deliver optimal solution for simplified model: is it shown, that no other controller could minimize integral of frequency deviation squared better than the one, presented in this paper.
2. For the detailed model, it is shown that the control remains effective via detailed case studies.

2 Problem statement

In this section two problem statements with different models are considered:

1. Simplified model. Here dynamics of converters are considered to be instant and power grid is aggregated into a single bus.
2. Detailed model. Here it is assumed that dynamics of synchronous generator and converter satisfy some general assumptions. Beyond those, no explicit model definition is required for the control derivation.

This separation is utilized to do a proof of concept on simplified model and then generalize it to the case, when system dynamics are nonlinear and system is non-observable.

* Corresponding author: o.khamisov@skoltech.ru

2.1 Simplified model

Here grid dynamics are aggregated to a single generator swing equation. It is assumed that mechanical power injections are not affected by the developed controller and converter dynamics are instant: it can output exact amount of power required by the control. Within these assumptions the following optimal control problem is formulated:

$$J(u) = \frac{1}{2}\int_0^T \omega^2(t)dt \to \min, \quad (1)$$

$$u \in [\underline{u}, \overline{u}], \quad (2)$$

$$M\dot{\omega} = -D\omega + p^m - p^e + u, \omega(0) = \omega^0 \quad (3)$$

Here u is converter control signal, ω is frequency deviation from the nominal value, M is inertia coefficient, D is damping coefficient, p^m is mechanical power injection, p^e is electrical load.

2.2 Detailed model

In the detailed model inverter response to control signal is no longer instant. Universal two-stage model of three-phase invertor is considered. Control loops consist of phase-lock-loop (PLL) and current control loop (CCL). Additionally, synchronous generator now receives mechanical power injections from three stage single reheat steam turbine. The objective function and control limits remain unchanged. As a result, detailed optimal control problem has the following form:

$$J(u) = \frac{1}{2}\int_0^T \omega^2(t)dt \to \min, \quad (4)$$

$$u \in [\underline{u}, \overline{u}], \quad (5)$$

$$\begin{pmatrix} \dot{\omega} \\ \dot{p}^{inv} \\ \dot{y} \end{pmatrix} = f\left(\begin{pmatrix} \omega \\ p^{inv} \\ y \end{pmatrix}, u, t\right), \begin{pmatrix} \omega \\ p^{inv} \\ y \end{pmatrix} = \begin{pmatrix} \omega^0 \\ 0 \\ 0 \end{pmatrix}, \quad (6)$$

where F has the following properties:

1. $f: \mathbb{R}^n \to \mathbb{R}^m$ is absolutely continues.
2. If $\lim_{t\to\infty} u(t) = u^*$ then $\lim_{t\to\infty} p^{inv}(t) = u^*$.
3. Frequency dynamics are described by a swing equation:
4. $M\dot{\omega} = -D\omega + g(y) + p^{inv}$.(7)
5. Control u is in only one f_n and f_n is proportional in u.

3 Control derivation

Here optimal control is derived for the simplified model and then obtained results are extended to the detailed one.

3.1 Simplified model

Application of Pontryagin maximum principle to the problem (1)-(4) gives the following set of equations:

$$0 = \dot{\omega} + \frac{1}{M}(D\omega - p^m + p^e - u), \quad (8)$$

$$0 = \dot{\psi}^s - \frac{M}{D}\psi^s + \omega, \psi^s(T) = 0, \quad (9)$$

$$0 \in \frac{1}{M}\psi^s + N_u(u), \quad (10)$$

where ψ is dual trajectory for the simplified model and

$$N_u(u) = \begin{cases} 0, & u \in [\underline{u}, \overline{u}], \\ \{l: l(v-u) \le 0, \forall v \in [\underline{u}, \overline{u}]\}, & u \in [\underline{u}, \overline{u}]. \end{cases} \quad (11)$$

Solving system (4)-(6) gives optimal control for the simplified model:

$$u^s(t) = \begin{cases} \underline{u}, & \psi(t) \ge 0, \\ \overline{u}, & \psi(t) \le 0, \\ p^e - p^m, & \psi(t) = 0. \end{cases} \quad (12)$$

Since $\psi^s(T) = 0$, for any $t \ge 0$ the inequality $\psi(t) \ge 0$ if and only if $\omega(\tau_0) < 0$ for some $\tau_0 \in [t, T]$. Let us now show that for any $\tau \in [t, T]$ trajectories $\omega(\tau)$ and $\psi(\tau)$ have different signs. By contradiction, if there exists τ_1 such that $\omega(\tau_1)\psi(\tau_1) > 0$, then there exists $\tau_2 \in [\tau_0, \tau_1]$, such that $\omega(\tau_2) = 0$. However, from Pontruyagin maximum principle, optimal control in τ_2 is equal to $u(\tau_2) = p^e - p^m$. Substitution of this control into (3) gives $\dot{\omega}(t) = 0$ for all $t \ge \tau_2$, which contradicts to the initial assumptions. As a result, optimal control can be represented in the following form:

$$u^s(t) = \begin{cases} \underline{u}, & \omega(t) \le 0, \\ \overline{u}, & \omega(t) \ge 0, \\ p^e - p^m, & \omega(t) = 0. \end{cases} \quad (13)$$

3.2 Detailed model

It can be seen that objective functional for the detailed model remains that same as for the simplified one. Therefore, optimal control has bang-bang structure with sliding mode. However, discontinuity points will differ from the control of simplified model u^s. Hamiltonian for the detailed model is given by

$$H(x, \psi, u, t) = \psi^\top f(x, u, t) + \frac{1}{2}\psi^0\omega^2, \quad (14)$$

Pontryagin maximum principle for the optimal control with detailed model has the following form:

$$\dot{\psi} = -\frac{\partial H(x, \psi, u, t)}{\partial x} = -\left(\frac{\partial f(x, u, t)}{\partial x}\right)^\top \psi - \begin{pmatrix} \psi_1^0 \\ 0 \\ 0 \end{pmatrix}, \quad (15)$$

$$\dot{x} = f(x, u, t), \quad (16)$$

$$H(x, \psi, u, t) = \max_{v \in [\underline{u}, \overline{u}]} H(x, \psi, v, t), \quad (17)$$

Note that according to the property 4 of the function f

$$\begin{aligned} \arg\max_{v \in [\underline{u}, \overline{u}]} H(x, \psi, v, t) &= \arg\max_{v \in [\underline{u}, \overline{u}]} \psi^\top f(x, u, t) = \\ &= \arg\max_{v \in [\underline{u}, \overline{u}]} \psi_n f_n(x, \psi, v, t) = \operatorname{sign} \psi_n \end{aligned}, \quad (18)$$

From the equations of Pontryagin maximum principle optimal control is given by

$$u(t) = \begin{cases} \underline{u}, & \psi_n(t) \ge 0, \\ \overline{u}, & \psi_n(t) \le 0, \\ p^e - p^m, & \psi_n(t) = 0. \end{cases} \quad (19)$$

Due to non-observability it is not possible to find moments t, such that $\psi_n(t) = 0$. Nevertheless, due to the properties 2 and 3 of the function f, it is reasonable to assume that ψ_n acts similarly to ψ^s from the simplified system, if $|\omega(t)| > \epsilon$ for some $\epsilon > 0$. Therefore $u(t) = u^s(t)$ for $|\omega(t)| \ge \epsilon$. Since equation

 https://doi.org/10.1051/e3sconf/202346101050

$\psi_n(t) = 0$ cannot realistically be solved with respect to t, optimal $u(t)$ for $|\omega(t)| < \epsilon$ is replaced with standard proportional controller. Therefore, final control form is defined by the equation

$$u(t) = \begin{cases} \underline{u}, & \omega(t) \leq -\epsilon, \\ \overline{u}, & \omega(t) \geq \epsilon, \\ -K\omega(t), & |\omega(t)| < \epsilon. \end{cases} \quad (20)$$

4 Case study

The IEEE 14-bus system is a well-established standard in the field of power system research, widely employed as a benchmark for various studies. This system encompasses five synchronous generators, 14 bus points, 20 transmission lines, and 11 load points [4]. In the context of this paper, we embark on a modification of this conventional system. Our adjustments primarily revolve around the introduction of two energy storage systems (ESS) on bus 3, each with power capacities of 1 MW and 30 MW, respectively.

The ESSs are connected to the power grid through an inverter. The overarching structural scheme governing the ESS integration into the power system is given in [5]. This scheme encompasses several pivotal components, including a lithium-ion battery, a three-phase transistor bridge conversion circuit, decoupling components that assume dual roles in harmonics filtering and decoupling functions, an ESS switch positioned at the network connection juncture, and the AC power system itself. The ESS model input receives active and reactive power setpoints. These values are subsequently translated into current setpoints, serving as inputs to a Proportional-Integral (PI) controller, functioning within the framework of a rotating direct-quadrature (dq) coordinate system. Precise control of the transistor conversion circuit is facilitated through the implementation of pulse width modulation. The output of the PI controller yields voltage modulation levels, expressed in dq coordinates. Then these coordinates are transformed into reference sinusoidal waveforms, which are supplied to the transistor circuit after conversion to a three-phase coordinate system.

We use two distinct ESS models: one allows to scrutinize the high-frequency switching dynamics of semiconductor switches, other interfaces with the network via an averaged inverter model, which operates based on a controlled alternating current source. The incorporation of these two distinct ESS models serves the purpose of minimizing the potential impact of errors on our measurement outcomes. The resulting modified 14-bus system configuration is depicted in Fig. 1.

In the endeavour to simulate the dynamic behaviour of the adapted 14-bus system, the Real Time Digital Simulator (RTDS) is employed. Recognized as a state-of-the-art simulation platform renowned for its real-time accuracy, RTDS excels in the precise modelling of power system components and their intricate interactions [6]. This powerful tool provides meticulousle assessment and validation of the converter-interfaced generation systems performances. Moreover, it provides detailed simulations for their accompanying control strategies, within a realistic operational environment. For the construction and execution of simulation models, RSCAD FX software [7] is used.

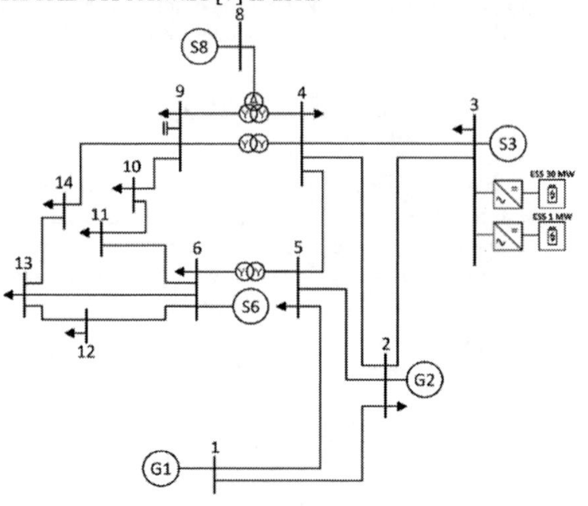

Fig. 1. Modified IEEE 14-bus system.

Simulation Scenario:

1. At the time t = 2.1 seconds, a load experiences a notable increase, rising from 29.5 MW to 89.5 MW.
2. This change in load conditions initiates the activation of the Underfrequency Load Shedding (ULS) mechanism, resulting in the disconnection of load 4, which has a rating of 47.8 MW.
3. Two distinct ESS regulation modes are considered: the conventional proportional mode and developed control.

In Fig. 2, frequency dynamics for different controls are presented. In the absence of ULS (depicted by the blue line) in conventional regulation mode, the frequency nadir stabilized at 59.4 Hz. Frequency restoration is very slow. However, upon the activation of ULS at t = 3 seconds, load 4 was disconnected, leading to a gradual frequency recovery, but with the overshooting 60.1 Hz (illustrated by the green line). The breaker tripping signal is also visible (indicated by the purple line).

Notably, when the developed control is engaged (as shown by the red line), the frequency nadir is even lower, specifically at 59.47 Hz. This prevents the initiation of underfrequency load shedding, as evidenced by the absence of a breaker tripping signal (highlighted in yellow).

Additionally, examining the generator torque graph (Fig. 3) reveals that, under the influence of the developed control, the load remains connected, and the generator increases its output, thereby continuously contributing to meeting the demands of all consumers (as indicated by the red line).

 https://doi.org/10.1051/e3sconf/202346101050

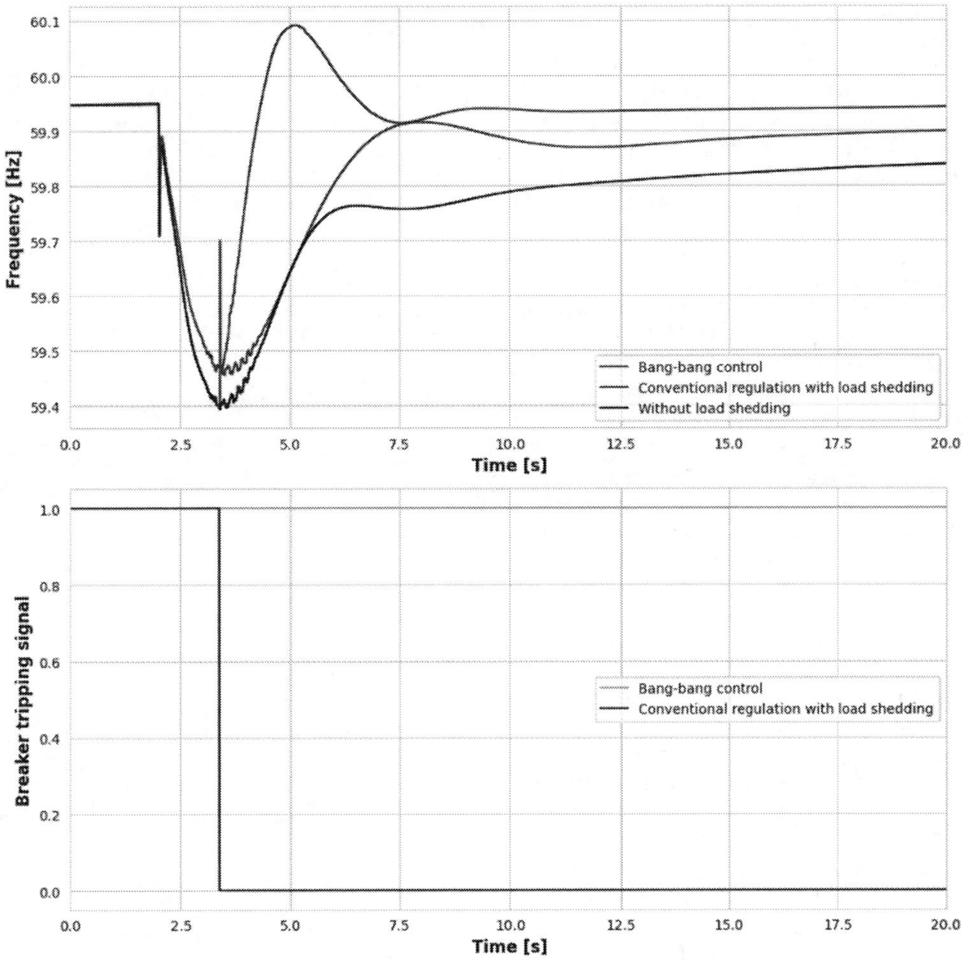

Fig. 2. Frequency and breaker tripping.

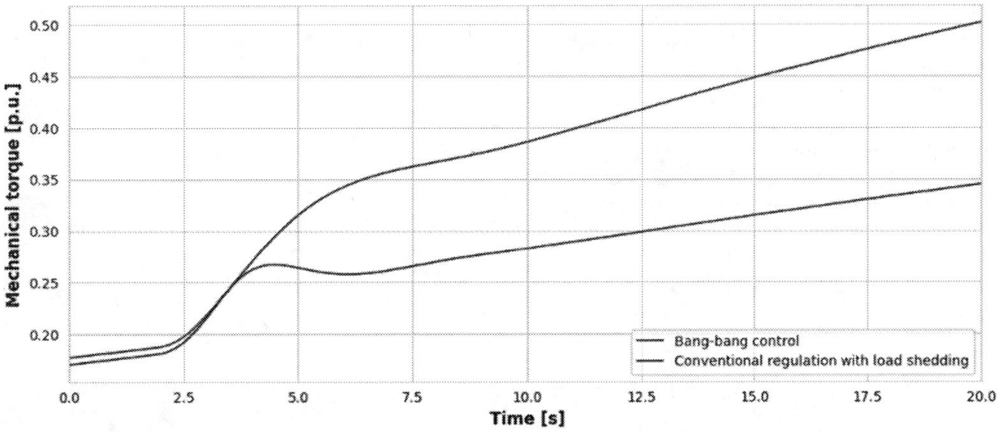

Fig. 3. Generator mechanical torque.

Based on the findings of this case study, it is evident that the developed control excels in its ability to sustain load connectivity within the power system and significantly curtails the extent of frequency overshoot during the restoration process.

5 Conclusions

In the presented work a novel frequency control algorithm is developed. Its main advantage is capability to fully utilize quick dynamics of power electronic interfaces, while using only frequency measurements and not measurements of RoCoF. The optimality of the control is proven for the simplified system. For full power system model exact optimal control cannot be applied due to non-observability. Therefore, an approximation of an optimal control is derived. Its effectiveness is demonstrated via detailed case study for IEEE 14-bus system using RTDS NovaCor 1.0 hardware-in-the-loop facility.

References

1. O. O. Khamsiov and V. Terzija, Distributed Frequency Control and Congestion Management with Coupling Inter-Area Constraints, 2023 IEEE Belgrade PowerTech, Belgrade, Serbia, 2023, pp. 1-6.
2. O.O. Khamisov, T. Chernova, and J. W. Bialek, Comparison of two schemes for closed-loop decentralized frequency control and overload alleviation. In 2019 IEEE Milan PowerTech Conference proceedings, pp. 1-6.
3. Y. Jiang, E. Cohn, P. Vorobev, E. Mallada, Storage-based frequency shaping control. IEEE transactions on Power Systems, volume V, issue **36**, pp. 5006-5019, 2012.
4. P.M. Anderson and A.A. Fouad, Power System Control and Stability, 2nd ed., IEEE Press, 2003.
5. A. Yazdani and R. Iravani. Voltage-sourced converters in power systems: modelling, control, and applications. Wiley-IEEE Press. 2010.
6. P. Forsyth and R. Kuffel, "Utility applications of a RTDS® Simulator," 2007 International Power Engineering Conference (IPEC 2007), Singapore, 2007, pp. 112-117.
7. RTDS Technologies Inc., "RSCAD FX: Real-Time Simulation Software Package," https://knowledge.rtds.com/hc/en-us/articles/360046352893-RSCAD-FX-Real-Time-Simulation-Software-Package, accessed on Oct. 1, 2023.

Development of simulation model of smart phase selector device

E. Khojiakbar[1*], *I.Kh.* Kholiddinov[1], *A.* Eraliev[1], *S.* Tukhtasinov[1], *S.* Komolddinov[1]

[1]Fergana Polytechnic Institute, 86 Fergana, Fergana, Uzbekistan

Abstract. In the quest for efficient and reliable power distribution systems, voltage asymmetry has emerged as a significant challenge. Voltage imbalances across different phases can lead to increased energy losses, compromised power quality, and reduced equipment lifespan. To address these issues, smart phase selector devices have gained attention as advanced electronic solutions for monitoring and managing voltage imbalances in low-voltage networks. This paper presents the development of a simulation model of a smart phase selector device using the Proteus software platform. Proteus offers a versatile environment for simulating electronic circuits and systems, making it an ideal choice for developing and evaluating the performance of the smart phase selector device. The primary objective of this research is to create a comprehensive simulation model that accurately represents the behavior and functionality of the smart phase selector device. Through a thorough literature review, the key principles and concepts underlying smart phase selector devices, voltage asymmetry, and related simulation techniques are explored. The developed simulation model incorporates the necessary circuit components, algorithms, and control mechanisms to faithfully simulate the operation of the smart phase selector device.

1 Introduction

In recent years, the demand for efficient and reliable power distribution systems has increased significantly. One of the challenges faced by power distribution networks is voltage asymmetry, which occurs when there are imbalances in voltage levels across different phases. Voltage asymmetry can lead to several issues, including increased energy losses, reduced equipment lifespan, and compromised power quality [1–5].

To achieve this objective, we will first conduct a thorough review of the existing literature on smart phase selector devices, voltage asymmetry, and related simulation techniques. This literature review will provide a solid foundation for understanding the key principles and concepts underlying the development of the simulation model.

Gana, [8] presents a "Single phase automatic changeover switch" which has power supply from just one phase then the generator phase as reference phase. This design was realized using atmeg8 microcontroller where the program for the logic control part of the device was coded by suitable assembly language. The circuit checks if the mains supply is available, if not, it activates the kick starter of the generator and energizes the generator contactor supplying power to the load. As soon as the mains supply is available while running on generator, the circuit energizes the mains contactor, and thereby turns off the generator in a delay time of 4seconds. The issue in this case is that, once that phase is faulty, there would be no alternative which is a limitation on his work. Oduobuk, et al [9] reported that, the automatic phase changer was made from several electronic components which include; operational amplifiers, diodes, resistors, capacitors, Zener diodes, transformers, relays and fuses. Results obtained during the test showed that whenever the system senses a higher voltage across at least one of the three inputs, it then engages the load. The system was designed to handle low power loads and not high-power loads. Also, system reliability, compatibility, and durability in this work were not considered. Nweke et al [10] noted that the automatic three phase power system selector was designed and constructed to automatically switch over to the alternative phase that has current when there is power outage or extremely low current in the phase which the load is connected without the power being off. The selector links the load and the other phases and relay switches allowing the usage of the remaining phases where there is outage on the mains source without disturbing or interrupting the load. It maintains constant power supply to the load by automatically activating the phases when the need arises. This safeguards the electronics system from being damaged and burnout as a result of voltage instability, collapse, insistent outages which are paramount in under developed and developing countries. Ofualagba, et al [11] proposed the design and construction of automatic phase selector and changeover

* Corresponding author: eraliyevhojiakbar@gmail.com

https://doi.org/10.1051/e3sconf/202346101051

switch for 3-phase power supply to provide a means of switching from one phase of AC mains to another in the case of failure in the existing phase. It also changes over to generator if there is failure in all the three phases of the AC mains. The circuit also senses the restoration of any or all the three phases of the mains and changeover without any notice of power outage. The project was aimed at improving the existing types of electromechanical device that has being in use over the years. The proposed system was realized by the use of 1-of - 4 analogue multiplexers (CD4052), analogue to digital converter (ADC0804), AT89C51 microcontroller and relay switches.

To address these challenges, the development of smart phase selector devices has gained significant attention. Smart phase selectors are advanced electronic devices designed to monitor and manage voltage imbalances in low-voltage networks. These devices intelligently analyze the voltage levels of each phase and automatically switch to the phase with the most optimal voltage, thereby minimizing voltage asymmetry and its associated problems. This paper presents the development of a simulation model for a smart phase selector device using Proteus 8 Professional. The simulation results demonstrate the effectiveness of the device in reducing voltage asymmetry and improving the performance of low-voltage networks. The device accurately detects and measures the voltage levels of each phase, intelligently selects the phase with the most optimal voltage, and dynamically balances the voltages across all phases. The simulation findings highlight the device's potential to enhance power quality, minimize energy losses, and increase equipment reliability. The paper's contribution lies in providing valuable insights for power system engineers, researchers, and practitioners involved in power distribution network planning and optimization. The simulation model serves as a powerful tool for evaluating the performance of the device and can guide the implementation of smart phase selector devices in real-world power distribution systems. The research contributes to the body of knowledge on smart phase selector devices and paves the way for further advancements in power distribution systems, promising more efficient and reliable energy infrastructure in the future.

2 Experimental research

A Smart Phase Selector Device is an advanced electrical device used in power distribution systems to automatically select the optimal phase for power supply. It is designed to ensure efficient power distribution and mitigate the issues caused by unbalanced loads. In electrical systems, power is typically distributed through multiple phases, such as three-phase or single-phase configurations. These phases carry electrical currents that need to be balanced for stable operation and optimal utilization of resources. However, due to various factors, such as unevenly distributed loads or faulty connections, the current distribution among the phases can become imbalanced. A Smart Phase Selector Device monitors the current flowing through each phase and analyzes the load distribution. It continuously measures and compares the currents in each phase, identifying any imbalances. Based on this analysis, the device intelligently selects the phase with the most balanced load to supply power to the connected devices or equipment.

The benefits of a Smart Phase Selector Device include:

• Load balancing: By automatically selecting the phase with the most balanced load, the device helps distribute the electrical load evenly across all phases, preventing overloading and reducing the risk of equipment failures.

• Energy efficiency: Balanced load distribution leads to improved energy efficiency as it ensures that each phase operates at its optimum capacity, minimizing energy wastage.

• Voltage stability: When the load is balanced, voltage stability is enhanced, promoting a consistent and reliable power supply to connected devices.

• Equipment protection: By preventing overloading and optimizing load distribution, a Smart Phase Selector Device helps protect electrical equipment from damage, extending their lifespan and reducing maintenance costs.

• Real-time monitoring: These devices often come with monitoring capabilities, allowing users to track power consumption, load distribution, and other relevant electrical parameters in real-time. This data can help identify patterns, troubleshoot issues, and optimize system performance.

Overall, a Smart Phase Selector Device provides an intelligent solution for managing power distribution in electrical systems, improving efficiency, and ensuring reliable operation of connected equipment.

It can be seen that the research of [1] and [2] are mainly focused on choosing the active phase in the network. This mainly ensures continuity of power supply but does not sufficiently eliminate voltage asymmetry. The research of [3] and [4] are aimed at choosing the optimal phase and connecting an additional generator when there is no power supply in all three phases. In the course of our research, it can be seen that there are many frequent changes in the difference between voltages in the network due to load asymmetry in low-voltage power networks as shown in Fig. 1.

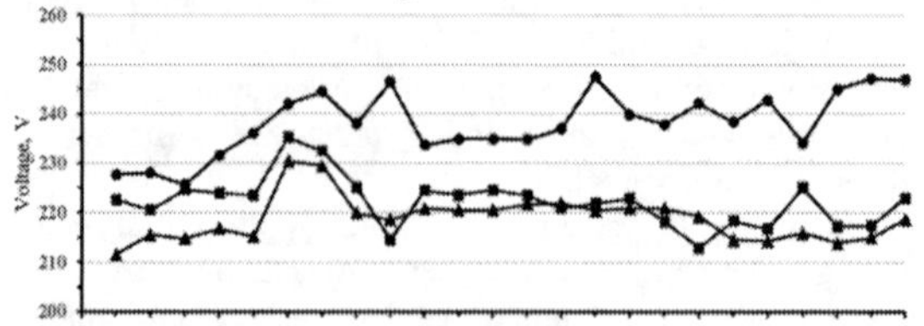

Fig. 1. Daily voltage asymmetry obtained in experimental research

In this case, the above solutions cannot sufficiently eliminate the voltage asymmetry in the network. In the design of the Smart phase selector device that we offer, the voltage difference and waiting times during the transition from the current phase to the optimal phase of the consumer are included. This dramatically reduces the number of redundant multiple switching switches at the output of the device and extends the life of the device.

Taking into account the considerations mentioned above, this work focuses on implementing a switching process using a microcontroller and leveraging the specific characteristics of electronic elements to create an automatic phase selection and replacement project. In the network, the phase with the highest load typically has the lowest voltage, while the phase with the lowest load usually has the highest voltage. This device serves as an effective system that employs programmed logic to accurately select the optimal voltage phase for each household from the power supply. It also switches to another optimal phase only when the power supply is completely disrupted. The device offers one phase through three available channels. The microcontroller, acting as the logical core of the system, continuously measures and compares the voltage of the three phases with voltage reduction. It is important to note that this programmed switching and phase selection process occurs within milliseconds without causing any power outages.

3 Research results

The Smart Phase Selector device is specifically designed for single-phase consumers. It incorporates automatic disconnections between the three phases and the load, utilizing various electrical and electronic components in its design. The ATMEGA16 microcontroller was employed in developing this device, and the software was written in the C programming language. Its purpose is to measure the voltage across the network phases, transfer the load from one phase to another, and display the voltage at the device's output.

Upon initial startup, the monitor displays the message "SMART PHASE SELECTOR" (Fig. 2 a). After a two-second delay, it measures the voltages of the three network phases and connects the load to the phase with the highest voltage, provided it falls within the voltage limit values. If the voltage value in all phases exceeds or falls below the threshold value, an "ERROR!" message is displayed on the monitor (Fig. 2 a, b). In such cases, the load is not connected to any phase, and the device awaits at least one phase voltage to enter the threshold value.

If the voltage value in all phases remains within the threshold values, the device connects the load to the phase with the highest voltage in the network. The values of all phase voltages and the phase to which the load is connected are displayed on the monitor as "ACTIVE-A," "ACTIVE-B," and "ACTIVE-C" (Fig. 2 d).

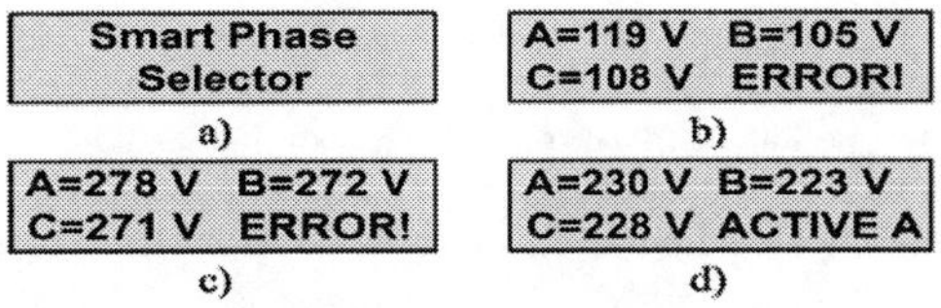

Fig. 2. Monitor statuses of the Smart Phase Selector device during start-up.

The device performs voltage measurements on the network phases and immediately compares them. It follows a specific algorithm to determine the optimal phase for load connection. Here is a breakdown of the algorithm:

1. The device starts by comparing the voltage of phase A with the voltage of phase B.
2. If the voltage of phase A is higher than or equal to the voltage of phase B, it proceeds to compare the voltage of phase A with the voltage of phase C.
3. However, if the voltage of phase A is lower than the voltage of phase B, the device compares the voltage of phase B with the voltage of phase C.
4. If the voltage of phase A is higher than or equal to the voltage of phase C, the device connects the load to phase A.
5. Conversely, if the voltage of phase C is higher than the voltage of phase A, the device connects the load to phase C.
6. In the case where the voltage of phase B is higher than or equal to the voltage of phase C, the device connects the load to phase B.
7. Conversely, if the voltage of phase C is higher than the voltage of phase B, the device connects the load to phase C.

Fig. 3 illustrates the flowchart of the startup algorithm of the Smart phase selector device, considering the aforementioned conditions.

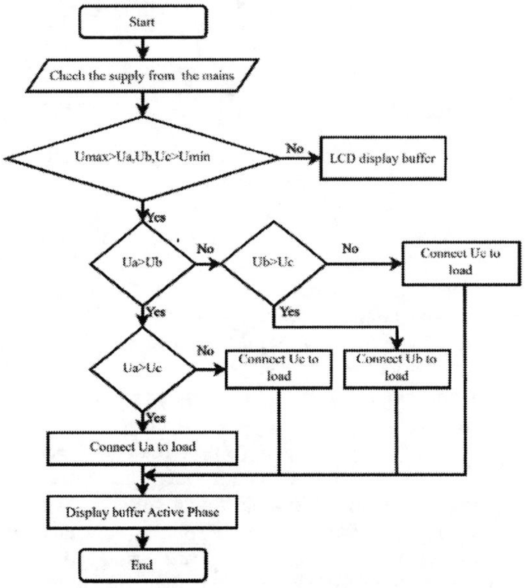

Fig. 3. Startup algorithm of the smart phase selector device

To validate the effectiveness of the developed algorithm, a simulation model was created, and experimental simulations were conducted. For the simulation, the actual components used in the circuit were implemented using components from the Proteus library.

The simulation model was built based on the device's schematic and operational algorithm, using components available in the Proteus 8 Professional software library. Proteus Professional is a comprehensive software package designed for computer-aided design of electronic circuits. It provides a graphical editor for circuit assembly, allows for the simulation of circuit operation, and facilitates the design of printed circuit boards, including three-dimensional visualization. In addition, Proteus supports SPICE models often provided by electronic component manufacturers and is compatible with various digital and analog device models. Specifically, Proteus Professional allows for the simulation of microcontroller operations, including popular models such as 8051, ARM7, ARM Cortex-M3, AVR, Texas Instruments, Motorola, PIC, and Basic Stamp. By utilizing the simulation capabilities of Proteus Professional and implementing the developed algorithm within the software, it becomes possible to evaluate and analyze the performance of the Smart Phase Selector device in a virtual environment before its practical implementation.

The overall scheme of the simulation model of the proposed smart phase selector device is illustrated in Fig. 4.

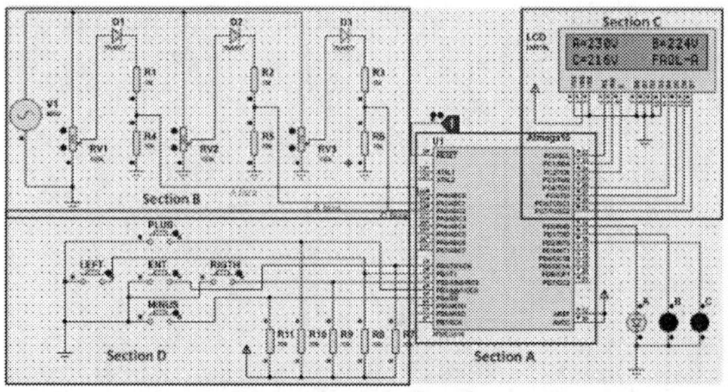

Fig. 4. Simulation model of the proposed smart phase selector device

In the Proteus 8 Professional software, a simulation model of the Smart Phase Selector device was developed using an Atmega16 microcontroller and other discrete components. Section A in Fig. 4 illustrates the inputs and outputs of the Atmega16 microcontroller. To measure the voltage values of each phase in the network, the alternating current of 0.4 kV is converted to a constant current using a 1N4007 diode. The voltage value is then reduced to 5 volts using a voltage divider circuit and transmitted to the microcontroller. This process is depicted in Section B. By adjusting a variable resistor to change the input voltage, the best phase is selected, and the voltage values of all phases are displayed on the monitor. A 16x4 LCD display, available in the Proteus library as LM016L, is used as the device monitor. Section C illustrates the connection of the LCD display to the Atmega16 microcontroller. To control the operating mode of the device and adjust threshold parameters, five buttons are utilized. These buttons are connected to the microcontroller through 10 kOhm resistors, as shown in Section D. LED lights serve as outputs for the switching part of the device, indicating the selected phase. Overall, the simulation model provides a virtual environment to test and analyze the functionality and performance of the Smart Phase Selector device, using the Atmega16 microcontroller and various components simulated within Proteus 8 Professional.

The development of a simulation model of a smart phase selector device in Proteus has provided valuable insights into the effectiveness and benefits of implementing such devices in power distribution systems. The simulation results have demonstrated that the smart phase selector device successfully mitigates voltage asymmetry, improves power quality, reduces energy losses, and enhances equipment reliability in low-voltage networks. Through the simulation model, it was observed that the device accurately detects and measures the voltage levels of each phase, intelligently selects the phase with the most optimal voltage, and dynamically balances the voltages across the phases. This intelligent decision-making process results in improved voltage balance, reduced energy wastage, enhanced power factor, and minimized harmonic distortions.

4 Conclusion

1. The development of a simulation model of a smart phase selector device in Proteus has demonstrated the device's effectiveness in reducing voltage asymmetry and improving the performance of low-voltage networks. The simulation results have highlighted the device's potential to enhance power quality, reduce energy losses, and increase equipment reliability. These findings contribute to the body of knowledge on smart phase selector devices and provide valuable insights for power system engineers, researchers, and practitioners involved in power distribution network planning and optimization. Further research and advancements in this area hold promise for even more efficient and reliable power distribution systems in the future.

2. The development of a simulation model for the smart phase selector device in Proteus has provided compelling evidence of its effectiveness in addressing voltage asymmetry and optimizing the performance of low-voltage networks. The simulation results have clearly demonstrated the device's ability to enhance power quality, minimize energy losses, and improve the reliability of equipment. Through the simulation model, it was observed that the smart phase selector device accurately detects and measures the voltage levels of each phase. It intelligently selects the phase with the most optimal voltage and dynamically balances the voltages across all phases. This intelligent decision-making process leads to improved voltage balance, reduced energy wastage, enhanced power factor, and minimized harmonic distortions. These findings

contribute to the existing body of knowledge on smart phase selector devices and offer valuable insights for power system engineers, researchers, and practitioners involved in power distribution network planning and optimization. The simulation model serves as a powerful tool for evaluating the performance of the device and can guide the implementation of smart phase selector devices in real-world power distribution systems.

3. The positive outcomes observed in the simulation highlight the potential for further research and advancements in this area. Continued exploration and innovation in smart phase selector technology holds promise for even more efficient and reliable power distribution systems in the future. By leveraging the capabilities of smart phase selector devices, power grids can achieve improved energy efficiency, reduced losses, and enhanced overall system stability.

References

1. S.Hao, Z. Fu, and others. 2016 International Symposium on Computer, Consumer and Control (IS3C), 16248424, (2016), https://doi.org/10.1109/IS3C.2016.35
2. K.I. Khosiljonovich, A. Saleem, and others. 2023 4th International Conference on Computing, Mathematics and Engineering Technologies: Sustainable Technologies for Socio-Economic Development, (iCoMET) 22963449, (2023), https://doi.org/10.1109/ICOMET57998.2023.10099113
3. T. Wang, M. Meskin, and others. 2017 Ninth Annual IEEE Green Technologies Conference (GreenTech), 16868340, (2017), https://doi.org/10.1109/GREENTECH.2017.50
4. Y. Liu and Y. Teng, 2019 12th International Conference on Intelligent Computation Technology and Automation (ICICTA), 19430909, (2019), https://doi.org/10.1109/ICICTA49267.2019.00138
5. F. Ni and J. Yu, 2010 International Conference on Electrical and Control Engineering, 11643070, (2010), https://doi.org/10.1109/ICECE.2010.1290
6. Allaev K., Toshov J., Toirov O. E3S Web of Conferences 401. 05090. (2023). https://doi.org/10.1051/e3sconf/202340105090
7. Abdi Yonis S., Yusupov Z., Habbal A., Toirov O. Advances in Electrical and Electronic Engineering.Vol 21. No 3 (2023). https://doi.org/10.15598/aeee.v21i3.5149
8. Toirov O., Khalikov S., Sodikjon Khalikov, F. Sharopov. E3S Web of Conferences 410. 05015. (2023). https://doi.org/10.1051/e3sconf/202341005015
9. Jumaeva D., Raximov U., Toirov O., Ergashev O., Abdyrakhimov A. E3S Web of Conferences 425. 04003. (2023). https://doi.org/10.1051/e3sconf/202342504003
10. Toirov O., Ivanova V., Tsypkina V., Jumaeva D., Abdullaeva D. E3S Web of Conferences 411. 01041. (2023). https://doi.org/10.1051/e3sconf/202341101041
11. S. Su, G. Liu, and X. Luo, 2008 International Conference on Intelligent Computation Technology and Automation (ICICTA). 10385736, (2008), https://doi.org/10.1109/ICICTA.2008.414
12. Z. Zhang and others. "2020 13th International Conference on Intelligent Computation Technology and Automation (ICICTA), 21139101, (2020), https://doi.org/10.1109/ICICTA51737.2020.00153
13. A. O. Ezenagu, "Design and implementation of automatic three phase selector using lm358 intergrated circuit," Project, Jan. (2018)
14. Augustine Okechukwu Ezenagu, Electrical Power Engineering, Project, (2018) https://www.academia.edu/43747488/
15. Enobong Oduobuk, Journal of International Journal of Engineering and Technology Research. (2014) https://www.researchgate.net/publication/361523645
16. Awunde, P. I., Ashigwuike, E. C. and Emmanuel, B. S. J Design, simulation and construction of automatic three-phase power selector system. (2021) https://www.researchgate.net/publication/355702310
17. E. Udoha. Journal of International Journal of Novel Research in Electrical and Mechanical Engineering. ISSN 2394-9678, (2017) https://www.researchgate.net/publication/323202875
18. Ofoegbu E.O, Ogbe Joy.O International Journal of Research and Innovation in Applied Science (IJRIAS) | Volume IV, Issue X, October 2019|ISSN 2454-6194, (2019). https://www.rsisinternational.org/journals/ijrias/DigitalLibrary/Vol.4&Issue10/15-21.pdf
19. Nnochiri Ifeoma U., Iroegbu Chibuisi, International Journal of Advances in Engineering and Management (IJAEM), Volume 3, Issue 8 Aug 2021, pp: 524-533 ISSN: 2395-5252, (2021). https://doi.org/10.35629/5252-0308524533

Device for reducing asymmetry

Eldor Usmanov[1*], *Bakhriddin* Kholikhmatov[1], *Bakhodir* Rikhsitillaev[2], *Kamoliddin* Nimatov[3]

[1]Tashkent State Technical University named after Islam Karimov, 100095, Uzbekistan, Tashkent, University St. 2A
[2]Nuclear energy development agency under the Ministry of Energy of the Republic of Uzbekistan, Tashkent, 100025, Uzbekistan
[3]Karshi Institute of Engineering Economics, 225 Independence Avenue, Karshi, Uzbekistan

Abstract. This article focuses on the development and analysis of a device designed to reduce voltage asymmetry in electrical systems. Voltage asymmetry can lead to negative consequences, such as equipment overloads, increased energy losses, and unstable power supply. The article discusses the fundamental principles of the device, its components, and control methods to achieve an optimal level of voltage symmetry. The presented research results emphasize the device's effectiveness in improving the quality of power supply and ensuring the stable operation of electrical grids.

1 Introduction

In 0.4 kV networks, their operation is often asymmetrical, i.e. Voltages in different phases may differ from each other. Asymmetrical modes in electrical networks arise for the following reasons: 1) unequal loads in different phases; 2) partial-phase operation of lines or other elements in the network; 3) different parameters of lines in different phases.

Most often, voltage asymmetry occurs due to inequality of phase loads. In urban and rural networks of 0.38 kV, voltage asymmetry is caused mainly by the connection of single-phase lighting and low-power household electrical receivers. The number of such single-phase power receivers is large, and they need to be evenly distributed among the phases to reduce asymmetry. However, nowadays, due to the rapid development of household electrical appliances, such a solution is practically difficult [1-3].

2 Experimental research

In high voltage networks, asymmetry is caused, as a rule, by the presence of powerful single-phase power receivers, and in some cases, three-phase power receivers with unequal consumption in the phases. These consumers include arc steel-smelting furnaces. The main sources of asymmetry in industrial networks of 0.38-10 kV are single-phase thermal installations, ore-smelting furnaces, induction smelting furnaces, resistance furnaces and various heating installations. In addition, asymmetrical electrical receivers are welding machines of varying power. Traction substations of AC-electrified railway transport are a powerful source of asymmetry, since electric locomotives have single-phase power receivers [4-9].

Asymmetry negatively affects the operating and technical and economic characteristics of rotating electrical machines. The positive sequence current in the stator creates a magnetic field that rotates at a synchronous frequency in the direction of rotation of the rotor. Negative sequence currents in the stator create a magnetic field that rotates relative to the rotor at twice the synchronous frequency in the direction opposite to rotation. Due to these double-frequency currents, a braking electromagnetic torque and additional heating occur in the electric machine, mainly in the rotor, leading to a reduction in the service life of the insulation.

In asynchronous motors, additional losses occur in the stator. In a number of cases, it is necessary to increase the rated power of electric motors during design, unless special measures are taken to balance the voltage. In synchronous machines, in addition to additional losses and heating of the stator and rotor, dangerous vibrations can begin. Due to asymmetry, the service life of transformer insulation is reduced, synchronous motors and capacitor banks reduce the production of reactive power.

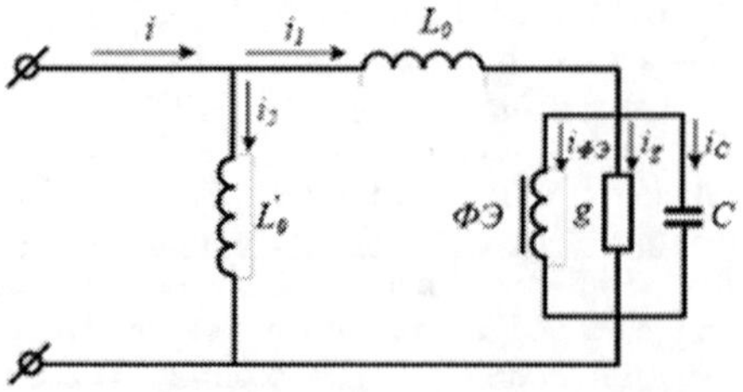

Fig.1. An equivalent circuit is a parallel oscillating circuit connected in series with a linear inductance and in parallel connected with a second linear inductance.

* Corresponding author: ilider1987@yandex.ru

Research conducted by the authors when creating a balun showed that a parallel oscillatory circuit with a falling section of the amplitude characteristic can be used as a sensitive organ of the stabilizer.

Let us consider the phase relationships of a parallel ferroresonant circuit connected in series with a linear inductance and connected in parallel with the second linear inductance (Fig. 1) of a linear choke, due to their smallness [10].

To carry out a theoretical analysis, we will accept the same assumptions as for the previously considered circuit. In addition, we neglect losses in the core.

Let's determine the amplitude-phase relationship between voltage U_{BX} and current i_1

The study is carried out using the method of slowly varying amplitudes. The circuit under study is described by the following differential equation:

$$u = w\frac{d\phi}{dt} + L_0\frac{di}{dt} \quad (1)$$

where

$$i = i_C + i_g + i_{\Phi Э} \quad (2)$$

here

$$i_C = wC\frac{d^2\phi}{dt^2} \quad (3)$$

$$i_g = wg\frac{d\phi}{dt}; \quad (4)$$

$$i_{\Phi Э} = \frac{K}{w}\phi^7 \quad (5)$$

Substituting the corresponding current values according to (3), (4), (5) into equation (2) from equality (1) we obtain [11,12]:

$$u = w\frac{d\phi}{dt} + wCL_0\frac{d^2\phi}{dt^2} + wgL_0\frac{d^2\phi}{dt} + \frac{KL_0}{w}\frac{d\phi^7}{dt} \quad (6)$$

Let us introduce the basic and dimensionless quantities corresponding to equation (6):

$$y = \frac{u}{U_б}; x = \frac{\phi}{\Phi_б}; \Phi_б = \sqrt[6]{\frac{64\omega^2 w^2 C}{35K}}; U_б = \omega w\Phi_б; \tau = \omega t. \quad (7)$$

Taking into account expressions (7), after simple transformations, equation (7) will have the following form:

$$y = \frac{dx}{d\tau} + \omega^2 CL_0\frac{d^3x}{d\tau^3} + \omega gL_0\frac{d^2x}{d\tau^2} + \omega^2 CL_0\frac{64}{35}\frac{dx^7}{d\tau} \quad (8)$$

Taking dimensionless coefficients:

$$\beta = \omega^2 CL_0; \gamma = \omega L_0 g \quad (9)$$

we have

$$y = \frac{dx}{d\tau} + \beta\frac{d^3x}{d\tau^3} + \gamma\frac{d^2x}{d\tau^2} + \beta\frac{64}{35}\frac{d\phi^7}{d\tau} \quad (10)$$

Let's integrate this equation:

$$\int y\mathrm{d}x + c = x + \beta\frac{d^2x}{d\tau^2} + \gamma\frac{dx}{d\tau} + \frac{64}{35}\frac{d\phi^7}{d\tau}\beta x^7 \quad (11)$$

where ***c*** is the integration constant.

We will look for solution (11) in the form:

$$x = X_m Sin(\tau + \psi) \quad \text{при} \quad y = Y_m Sin\tau \quad (12)$$

Derivative of ***x***; has the form:

$$\frac{dx}{d\tau} = \frac{dX_m}{d\tau}Sin(\tau + \psi) + X_m Cos(\tau + \psi) + + \frac{d\psi}{d\tau}X_m Cos(\tau + \psi) \quad (13)$$

Taking into account the fact that $X_m >> \frac{dX_m}{d\tau}$ and $X_m >> \frac{d\psi}{d\tau}X_m$ can be taken as a first approximation:

$$\frac{dx}{d\tau} = X_m Cos(\tau + \psi) \quad (13)$$

$$\frac{d^2x}{d\tau^2} = \frac{d^2X_m}{d\tau^2}Sin(\tau+\psi) + \frac{dX_m}{d\tau}Cos(\tau+\psi) + \frac{dX_m}{d\tau}\frac{d\psi}{d\tau}X_m Cos(\tau+\psi) + \\ + \frac{dX_m}{d\tau}Cos(\tau+\psi) \quad X_m Sin(\tau+\psi) \quad X_m\frac{d\psi}{d\tau}Sin(\tau+\psi) + \\ + \frac{dX_m}{d\tau}\frac{d\psi}{d\tau}Cos(\tau+\psi) + X_m\frac{d^2\psi}{d\tau^2}Cos(\tau+\psi) \quad X_m\frac{d\psi}{d\tau}Sin(\tau+\psi) \\ X_m \quad \frac{d\psi}{d\tau}^{2} \quad Sin(\tau+\psi) \quad (13)$$

Substituting (13) and (4) into equation (11) we get:

$$Y_m Cos\tau = -X_m Sin(\tau + \psi) - 2\beta\frac{dX_m}{d\tau}Cos(\tau + \psi) + 2\beta\frac{d\psi}{d\tau}Sin(\tau + \psi) + \beta X_m Sin(\tau + \psi) - \gamma X_m Cos(\tau + \psi) - \beta X_m^7 Sin(\tau + \psi) \quad (14)$$

After simple trigonometric transformations, we group the coefficients for the same trigonometric functions:

$$Y_m Sin\psi = -X_m + 2\beta X_m\frac{d\psi}{d\tau} + \beta X_m - \beta X_m^7 \quad (15)$$

$$Y_m Cos\psi = -\gamma X_m - \beta\frac{dX_m}{d\tau} \quad (16)$$

For steady state:

$$Y_m Sin\psi = \beta X_m + \beta X_m - X_m^7 \quad (17)$$

$$Y_m Cos\psi = -\gamma X_m \quad (18)$$

The joint solution of equations (17), (18) gives the amplitude value of the input voltage, in relative units, and the phase angle between the power source voltage and the ***FE*** magnetic flux.

$$Y_m = \sqrt{[\beta(1 - X_m^6]^2 - \gamma^2} \quad (19)$$

$$tg\psi = -\frac{\beta(1-X_m^6)-1}{\gamma} \quad (20)$$

To build circuit ***Current–Voltage characteristic*** and determine its amplitude-phase characteristics, it is necessary to determine dependencies $I = f(\Phi)$ and $\Phi = f(U_{вх})$.

Let us rewrite equation (2) taking into account (3), (4) and (5):

$$i = wC\frac{d^2\Phi}{dt^2} + wg\frac{d\Phi}{dt} + \frac{K}{w}\Phi^7 \quad (21)$$

introducing new dimensionless and basic quantities:

$z = \frac{i}{I_б}$; $I_б = \varpi^2 wC\Phi$

and taking into account expressions (7), (9) we rewrite equation (21) in the form:

$$z = \frac{d^2x}{d\tau^2} + \frac{g}{\omega C}\frac{dx}{d\tau} + \frac{64}{35}x^7 \quad (22)$$

or

$$z = \frac{d^2x}{d\tau^2} + \frac{g}{\delta}\frac{dx}{d\tau} + \frac{64}{35}x^7 \quad (23)$$

Let's assume that

$$z = Z_m(Sin(\tau - \varphi) \quad (24)$$

Let's substitute expressions (11), (12) and (25) into equation (24) and taking into account the fundamental harmonic of the magnetic flux we have:

$$Z_m Sin(\tau - \varphi) == -X_m Sin(\tau + \psi) + \frac{\xi}{\delta}X_m Cos(\tau + \psi) + X_m^7 Sin(\tau + \psi) \quad (25)$$

$$Z_m Sin(\tau - \varphi) == -X_m Sin(\tau + \psi) + \frac{\xi}{\delta}X_m Cos(\tau + \psi) + X_m^7 Sin(\tau + \psi) \quad (25)$$

After simple transformations and grouping of coefficients for the same trigonometric functions, we have [13-19]:

$$Z_m Sin\varphi = -(X_m^6 - 1)X_m Sin\psi - \frac{\gamma}{\beta}X_m Cos\psi \quad (26)$$

$$Z_m Cos\varphi = (X_m^6 - 1)X_m Cos\psi - \frac{\gamma}{\beta}X_m Sin\psi \quad (27)$$

A joint solution of equations (26) and (27) gives the amplitude value of the load current in relative units and the phase angle between the load current and the supply voltage:

$$Z_m = X_m\sqrt{(1 - X_m^6)^2 - \left(\frac{\gamma}{\beta}\right)^2} \quad (28)$$

$$tg\phi = -\frac{(X_m^6-1)tg\psi - \frac{\gamma}{\beta}}{(X_m^6-1) - \frac{\gamma}{\beta}tg\psi} \quad (29)$$

The voltage across inductance L_0' is determined from the formula [20-26]:

$$u = L_0'\frac{di}{dt} \quad (29)$$

From here

$$i_2 = \frac{1}{L_0'}\int U_m Sin\omega t dt = -\frac{U_m}{\omega L_0'}Cos\omega t \quad (29)$$

The current in an unbranched section of the circuit is determined as

$$i = i_1 + i_2 \quad (30)$$

or in relative units

$$z = z_1 + z_2 = Z_{1m}Sin(\tau - \psi_1) - \frac{y_m}{\beta'}Cos\tau,$$

where $\beta' = \omega^2 CL_0'$

$$Z_m Sin(\tau - \psi) = Z_{1m}Sin(\tau - \psi) - \frac{y_m}{\beta'}Cos\tau,$$

$$Z_m Sin\tau Cos\varphi - Z_m Cos\tau Sin\varphi == Z_{1m}Sin\tau Cos\varphi_1 - Z_m Cos\tau Sin\varphi_1 - \frac{y_m}{\beta'}Cos\tau$$

Let's highlight the sine and cosine components:

$$Z_m Sin\varphi = Z_{1m}Sin\varphi_1 + \frac{y_m}{\beta'}(31)$$

A joint solution of equations (31) and (32) gives the amplitude value of the current, in relative units, in an unbranched section of the circuit and the phase angle between this current and the supply voltage [27-32]:

$$Z_m = \sqrt{\left(\frac{Z_{1m}tg\varphi_1}{1+tg^2\varphi_1}\right)^2 + \left(\frac{Z_{1m}}{\sqrt{1+tg^2\varphi_1}}\right)^2} \quad (33)$$

$$tg\varphi = tg\varphi_1 + \frac{y_m}{\beta' Z_{1m}} \quad (34)$$

3 Research results

The results obtained from formula (34) have both positive and negative meanings:

– a negative value of the angle φ corresponds to the phase advance of the load current from the supply voltage;

– a positive value of the angle φ corresponds to a phase lag of the load current from the power source voltage.

Thus, if a transreactor is connected to the unbranched section of the considered circuit, then from its secondary windings you can receive a signal to control the thyristors connected to the arms of the diode bridge. This connection will allow you to obtain a rectified voltage, the value of which varies in inverse proportion to the voltage of the supply circuit [33, 34].

In Fig. 2 shows a diagram of connecting the proposed device to a three-phase circuit.

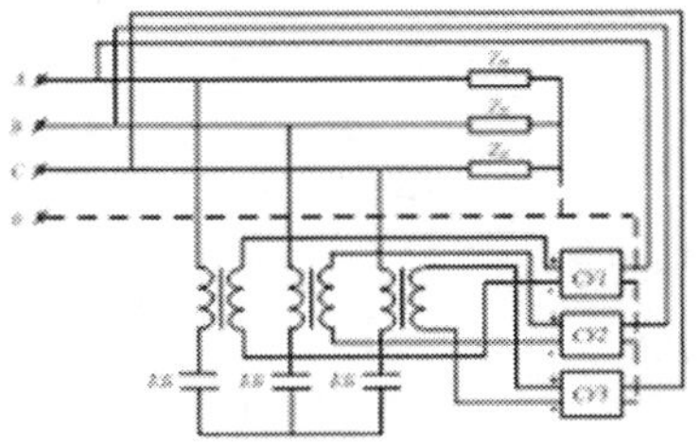

Fig.2. Single-line diagram of a balancing device.

4 Conclusion

As can be seen from this diagram, each control system is respectively connected to the same phase of the four-wire circuit as the capacitors that regulate the power. Therefore, when the voltage changes in phase "A", "СУ1" will operate, receiving power from phase "A" and turn on the "КБ" connected to this phase [35,36].

Thus, using a parallel oscillatory circuit in the capacitor bank control system, it is possible to create a system for smoothly controlling the power generated by the capacitor banks into the supply network [37].

References

1. Usmanov, E.G. E3S Web of Conferences, 384, 01034, (2023), https://doi.org/10.1051/e3sconf/202338401034
2. Usmanov, E. AIP Conference Proceedings, 2552, 050020, (2022), https://doi.org/10.1063/5.0111537
3. Usmanov, E.G., Khusanov, B.M. E3S Web of Conferences, 216, 01161, (2020), https://doi.org/10.1051/e3sconf/202021601161
4. Usmanov, E.G. E3S Web of Conferences, 216, 01160, (2020), https://doi.org/10.1051/e3sconf/202021601160
5. Karimov, R.C., Bobojanov, M.K., Rasulov, A.N., Usmanov, E.G. E3S Web of Conferences, 139, 01039, (2019), https://doi.org/10.1051/e3sconf/201913901039
6. Usmanov, E.G., Rasulov, A.N., Bobojanov, M.K., Karimov, R.C. E3S Web of Conferences, 139, 01079, (2019), https://doi.org/10.1051/e3sconf/201913901079
7. Hoshimov F.A., Rakhmonov I.U., Niyozov N.N., Omonov F.B. AIP Conference Proceedings 2647. 030025. (2022). https://doi.org/10.1063/5.0112388
8. Rakhmonov I.U., Najimova A.M., and Reymov K.M. AIP Conference Proceedings 2647. 030010. (2022). https://doi.org/10.1063/5.0104788
9. Rakhmonov I.U., Najimova A.M. AIP Conference Proceedings 2647. 030011. (2022). https://doi.org/10.1063/5.0104791
10. Rakhmonov I.U., Najimova, A.M., Esemuratova Sh.M., Koptileuov T.T. AIP Conference Proceedings 2647. 070024. (2022). https://doi.org/10.1063/5.0104793
11. Rakhmonov I.U., Hoshimov F.A., Kurbonov N.N., Jalilova D.A. AIP Conference Proceedings 2647. 050022. (2022). https://doi.org/10.1063/5.0112391
12. Rakhmonov I.U., Ganiev Sh.R., Alibekova T.Sh., Nematov L.A. E3S Web of Conferences 384. 01029. (2023). https://doi.org/10.1051/e3sconf/202338401029
13. Rakhmonov I.U., Niyozov N.N., Kurbonov N.N., Umarov B.S. E3S Web of Conferences 384. 01030. (2023). https://doi.org/10.1051/e3sconf/202338401030
14. Niyozov N.N., Rikhsitillaev B.Kh., Jalilova D.A., Kuatbaev P.M. E3S Web of Conferences 384. 01031. (2023). https://doi.org/10.1051/e3sconf/202338401031
15. Usmanov E.G. E3S Web of Conferences 384. 01034. (2023). https://doi.org/10.1051/e3sconf/202338401034
16. Rasulov A.N., Paxratdinov A.Dj. E3S Web of Conferences 384. 01035. (2023). https://doi.org/10.1051/e3sconf/202338401035
17. Taslimov A.D. E3S Web of Conferences 384. 01036. (2023). https://doi.org/10.1051/e3sconf/202338401036
18. Taslimov A.D., Rakhimov F., Rakhimov Feruz. E3S Web of Conferences 384. 01037. (2023). https://doi.org/10.1051/e3sconf/202338401037
19. Fayziyev M., Tuychiev F., Mustayev R., Ochilov Y. E3S Web of Conferences 384. 01038. (2023). https://doi.org/10.1051/e3sconf/202338401038
20. Saidkhodjaev A.G., Bobojonov Y.M. E3S Web of Conferences 384. 01040. (2023). https://doi.org/10.1051/e3sconf/202338401040
21. Bobojanov M., Mahmutkhonov S. E3S Web of Conferences 384. 01041. (2023). https://doi.org/10.1051/e3sconf/202338401041
22. Rasulov A.N., Ruzinazarov M.R. E3S Web of Conferences 384. 01042. (2023). https://doi.org/10.1051/e3sconf/202338401042
23. Rashidov N., Rozmetov Kh., Rismukhamedov S., Peysenov M. E3S Web of Conferences 384. 01043. (2023). https://doi.org/10.1051/e3sconf/202338401043
24. Tuychiev F, Haqberdiev A. E3S Web of Conferences 384. 01044. (2023). https://doi.org/10.1051/e3sconf/202338401044
25. Bobojanov M., Torayev S. E3S Web of Conferences 384. 01045. (2023). https://doi.org/10.1051/e3sconf/202338401045
26. Reymov K.M., Esemuratova Sh.M., Khusanov B.M., Mytnikov A.V. E3S Web of Conferences 384. 01047. (2023). https://doi.org/10.1051/e3sconf/202338401047
27. Kholikhmatov B.B., Samiev Sh.S., Erejepov M.T., Nematov L.A. E3S Web of Conferences 384. 01032. (2023). https://doi.org/10.1051/e3sconf/202338401032
28. Melikuziev M.V. E3S Web of Conferences 384. 01033. (2023). https://doi.org/10.1051/e3sconf/202338401033
29. Bobojonov Y.M., Saidkhodjaev A.G. E3S Web of Conferences 384. 01048. (2023). https://doi.org/10.1051/e3sconf/202338401048
30. Hoshimov, F.A., Najimova A.M., Rafikova G.R., Bakhadirov I.I. E3S Web of Conferences 384. 01049. (2023). https://doi.org/10.1051/e3sconf/202338401049
31. Ruzinazarov M.R. E3S Web of Conferences 384. 01050. (2023). https://doi.org/10.1051/e3sconf/202338401050
32. Bobojanov M.K., Rismukhamedov D.A., Tuychiev F.N., Shamsutdinov Kh.F. E3S Web of Conferences 365. 04024. (2023). https://doi.org/10.1051/e3sconf/202336504024
33. Ushakov, V.Y., Mytnikov, A.V., Rakhmonov, I.U. Past, Present and Future of Electric Power Systems (Brief Excursion). In: High-Voltage Equipment of Power

Systems. Power Systems. Springer, Cham. (2023). https://doi.org/10.1007/978-3-031-38252-9_1
34. Sattarov Kh., Sapayev M., Suyarov A., Turaev, A. Improving efficiency in a distribution network with asymmetric load due to connected solar panels with a phase relationship. E3S Web of Conferences, 2023, 401, 04001, https://doi.org/10.1051/e3sconf/202340104001
35. Jurayeva K., Sattarov Kh. Issues of investigation of the dependence of static characteristics of magnetoelastic converters of mechanical quantities on the influence of external. E3S Web of Conferences, 2023, 389, 01060, https://doi.org/10.1051/e3sconf/202338901060
36. Sattarov Kh., Safarov A., Kamolov A., Egamberdiev M., Raxmanova G. A device for detecting current asymmetry in a three-phase alternating current system. E3S Web of Conferences, 2023, 383, 01028, https://doi.org/10.1051/e3sconf/202338301028

Distribution network planning with DG units considering the network reconfiguration and reliability

Mansur Khasanov[1,2*], *Salah* Kamel[3], *Jasur* Abdubannaev[4], *Abror* Kurbanov[2], *Elnur* Abdullayev[2], *Urinboy* Jalilov[2]

[1]Tashkent State Technical University named after Islam Karimov, 100095, Uzbekistan, Tashkent, University St. 2A
[2]Department of Energy, Jizzakh Polytechnic Institute, 130100 Jizzakh, Uzbekistan
[3]Department of Electrical Engineering, Faculty of Engineering, Aswan University, 81542 Aswan, Egypt
[4]Zhuzhou Even Control Science and Technology Development Co. Ltd, 410000 Changsha, China

Abstract. The failure statistics of most utilities show that distribution networks (DN) make the most considerable individual contribution to the lack of supply to consumers. There are several advantages to optimizing distributed generation (DG) allocation in DNs, including loss reduction, voltage profile improvements, and reliability. A traditional objective function of optimal DG allocation is to minimize the total power loss and improve the voltage profile. Due to their ability to supply loads locally, DG units enhance the load-carrying capability of lines and, therefore, serve the same purpose as redundant lines. A DN's reliability indices can also be improved by optimal DG allocation. As such, it is interesting to investigate how network reconfiguration (NR) may be further enhanced to improve the achievement of these critical objectives and ensure the system's efficient operation in this context. The purpose of this paper is to present an intensive investigation of the reliability and performance improvement that NR can achieve when using DG units in modern DNs. An efficient metaheuristic rider optimizer algorithm (ROA) is applied to the complex combinatorial NR and DG allocation problem with stringent constraints. Analyses are conducted on benchmark IEEE 33-bus test DNs with explicit presentation of results.

1 Introduction

Planning distribution networks (DNs) with distributed generation (DG) units is a critical task in power systems engineering. As the electricity demand continues to grow and the integration of renewable energy sources (RES) becomes more prevalent, efficient and reliable DNs are essential for meeting the power needs of consumers. In this context, distribution network reconfiguration (NR) and reliability are crucial in ensuring optimal performance and minimizing power disruptions [1,2].

NR involves strategically altering the topology and configuration of the DN to optimize its operation, minimize power losses and improve reliability. By considering the placement and sizing of DG units, NR aims to improve the system's overall efficiency, voltage stability, and power quality. The planning process considers various factors, such as load profiles, generation capacity, and DN constraints, to determine the optimal configuration that maximizes the benefits of DG integration.

Reliability is another key aspect in the planning of DN with DG units. It refers to the ability of the system to deliver power consistently and without interruptions to consumers. The presence of DG units can enhance the reliability of the DN by reducing dependence on a centralized power source and providing localized generation close to the load centers. However, DG units' optimal placement and operation must be carefully considered to ensure a reliable and resilient power supply under normal and contingency conditions.

Due to simple protection and control mechanisms and lower short circuit current, DNs typically operate in radial topology [3]. There is only one source of supply per consumer in a radial topology. However, DNs are frequently the main reason for power system outages [4]. According to technical literature, issues in DNs account for 80% of consumer interruptions [5]. According to North American Electric Reliability Corporation records, 22% of power outages are caused by system equipment failures [6]. Overloaded systems can also cause outages. The frequency and severity of these power outages have increased in recent years. According to [7], from 2000 to 2013, the average number of grid failure events in the United States grew from 2.5 to 14.5 per month. According to [8], power outages could become more frequent and more severe in the future. Both electricity customers and electric utilities may also be affected by power outages. Consumers are primarily affected financially by an unpredictable power supply, which might also disrupt other operations that depend on electricity [9]. However, electrical utility firms experience a loss in energy sales and must cover additional operating and maintenance costs for service

* Corresponding author: hasanov6654525@mail.ru

restoration [10]. Distribution utilities invest in switching and protecting equipment to avoid prolonged outages and minimize the number of customers isolated by faults. With each outage, the quality of the power supply is degraded, so it's critical to measure and avoid them accurately. Reliability indices can quantify a network's ability to sustain continuous steady-state operation over a specific period. A system average interruption frequency index (SAIFI), a system average interruption duration index (SAIDI), and an energy not supplied (ENS) index are the most commonly used indices by utilities.

It was suggested in [11] to minimize load cost loss and average interruption indices by using a reliability-based NR technique. Switch exchange is based on the minimum path method, whereas the best configuration is found using meta-heuristic methods, including Tabu Search (TS), Evolution Strategies (ES), and Differential Evolution (DE). Simulating annealing (SA) redesigned the DN to increase reliability [12]. One needs to determine the optimal network topology to achieve the lowest disruption cost for customers. The cost of customer interruption is calculated using the reliability indices of the load points and the damage functions of customers, which correlate customer class with outage duration. A non-dominated sorting genetic algorithm (NSGA II) has been used to solve the reconfiguration problem of a large system containing DG units [13]. An ant colony optimization (ACO) approach combined with fuzzy multi-objective optimization has been used to minimize power loss and improve voltage profile by optimal simultaneous allocation of the distribution static synchronous shunt compensator (DSTATCOM), DG units and NR in a DN [14].

It is important to note that when DG units are connected to a distribution feeder, they alter both the active power flow and the reactive power flow within the feeder, whereas when NR units are connected to a distribution feeder, they alter both the active and reactive power flows. DNs' reliability and performance are directly affected by current flow since failure chances are primarily determined by current flow. Thus, it is essential to study how the simultaneous execution of NR and DG units affects DN performance, reliability, and efficiency. It has not yet been thoroughly explored in literature. The present research assesses whether NR and DG allocation can increase DN reliability and performance. To analyse the impact of DGs and NR on the reliability indexes, the failure statistics of feeders are proposed to be a function of feeder currents. This study uses a metaheuristic rider optimizer algorithm (ROA) to solve the complex combinatorial NR and DG allocation problem. The algorithm considers multiple reliability and performance enhancement objectives separately. An exhaustive analysis of DG allocation and NR in the IEEE 33-bus test DN is carried out by simulating some scenarios. The study's results and analyses are presented and analyzed extensively.

The remainder of this paper will be organized as follows: The problem formulation is presented in Section II. A detailed overview of the solution method is presented in Section III. The simulation results are presented in Section IV. Conclusion of the paper given in Section V.

2 Problem formulation

Due to increasing load demands, DNs suffer from power losses, low voltage levels, high currents, and low reliability. The optimal DG unit allocation considering optimal NR in the DN has been proven to improve reliability, reduce power loss, and improve voltage profile. As a result, to make the DN function effectively, it is essential to determine the optimal radial topology and the locations and sizes of the DGs to maintain a proper radial topology.

2.1 Objective function

The proposed technique has been used to perform four single-objective function optimizations. ENS, SAIFI, SAIDI, and power loss are four conflicting objectives. Through the use of suitable DG allocation and NR, the conflicting objectives have been minimized.

The first objective is power loss minimization, and it is formulated as:

$$Minimize \quad F = P_{loss} = \sum_{i}^{N_{br}} R_i \frac{P_i^2 + Q_i^2}{V_i^2} \tag{1}$$

where, P_i and Q_i are active and reactive power of bus i. V_i is voltage magnitude of bus i. R_i is resistance of branch i. N_{br} is number of branch in test system.

The second objective is system average interruption frequency index (SAIFI) minimization, and it is formulated as:

$$Minimize \quad F = SAIFI \tag{2}$$

$$SAIFI = \frac{\text{The total number of user outages}}{\text{total number of users}} = \frac{\sum \lambda_i N_i}{\sum N_i} \tag{3}$$

where, λ_i is the average failure rate of load point i, N_i is the number of users of load point i.

Minimization of SAIDI has been considered as the third objective, and it is formulated as:

$$Minimize \quad F = SAIDI \tag{4}$$

$$SAIDI = \frac{\text{The sum of user outage duration}}{\text{total number of users}} = \frac{\sum U_i N_i}{\sum N_i} \tag{5}$$

where, Ui is the annual average outage time of load point i.

The fourth objective is the minimization of expected energy not supplied EENS and its formulated as:

$$Minimize \quad F = ENS \tag{6}$$

$$ENS = \text{Total energy not supplied by the system} = \sum U_i L_i \tag{7}$$

where, Li is the average load of load point i.

2.2 Constraints

a) Power balance constraints

There must be a balance between the total power supplied to the network and the total power demand with loss [15].

$$P_{system} = \sum_{i=1}^{N_{br}} P_{brloss}(i) + \sum_{j=1}^{M} P_d(j) \quad (8)$$

$$Q_{system} = \sum_{i=1}^{N_{br}} Q_{brloss}(i) + \sum_{j=1}^{M} Q_d(j) \quad (9)$$

where, P_{system} and Q_{system} are active and reactive system power, P_{brloss} and Q_{brloss} are active and reactive power losses, P_d and Q_d are active and reactive power demand, M is total system bus number.

b)Bus voltage constraint

$$V_{\min} \le |V_i| \le V_{\max} \quad (10)$$

where $V_{\min}$ and $V_{\max}$ represent the minimum and maximum voltage values, respectively.

c)Line current constraints

The line current at any n line must meet the following constraints:

$$I_{Line,n} \le I_{Line,n(\max)} \quad (11)$$

where, $I_{Line,n(\max)}$ is maximum allowed current of line n.

d)Installed DG power size and power factor (PF) constraints

The DG power size in the DN is restricted as [1]:

$$P_{DG}^{\min} \le P_{DG}(n) \le P_{DG}^{\max} \quad (12)$$

$$Q_{DG}^{\min} \le Q_{DG}(n) \le Q_{DG}^{\max} \quad (13)$$

$$PF_{DG,\min} \le PF_{DG,n} \le PF_{DG,\max} \quad (14)$$

where, $P_{DG}^{\max}$, $Q_{DG}^{\max}$ and $PF_{DG,\max}$ represent maximum active, maximum reactive, and maximum PF of DG units. $P_{DG}^{\min}$, $Q_{DG}^{\min}$ and $PF_{DG,\min}$ represent minimum active, minimum reactive and minimum PF of DG units.

e)Constraints of radial topology

No insulated buses should exist in the DN topology, and the topology should be radial.

$$\det[BusInM] = 1 \, or -1 \quad (for \quad radial \quad network)$$
$$\det[BusInM] = 0 \quad (for \quad nonradial \quad network) \quad (15)$$

where, $BusInM$ is network incidence matrix.

3 Overview of solution technique and proposed algorithm

To solve the optimal planning of DG and NR problem, this study applies a recently developed metaheuristic rider optimizer algorithm (ROA).

3.1 Network reconfiguration method

NR is one of the feasible methods in which power flow is varied by ON or OFF the switches on the feeders. It is incorporated by opening a sectionalizing switch and closing a tie switch to protect the feeder radial structure. The proposed algorithm searches for better-switching combinations that further improve reliability indexes, voltage stability and minimize losses by considering a tie switch and its neighboring sectionalizing switches one at a time. It continues the search process until there is no further improvement in voltage stability, active power losses and reliability. NR affects only tie/sectionalized switches but not other equipment's location, properties, or status. Only feeder section failures are simulated for reliability evaluation in this study.

3.2 Feeder failure rate model

For reliability calculations, failure statistics must be available for all network components. Component failure statistics are mainly based on their historical behavior, but they remain valid as long as the components are subjected to similar working conditions. Unlike the conventional DN in which the utility grid supply is distributed to end customers, there are now several DG sources in addition to the utility grid supply available in the modern deregulated and competitive energy market [16]; therefore, the current derived from the grid has decreased significantly. In addition to localized generation, NR is also contributing to a reduction in the magnitude of the feeder current, thus changing the operating conditions of the switches. In underground cables, the operating lifetime of insulation reduces at an exponential rate with a rise in temperature with current flowing through it [16]. This is quite important since a conductor's current has a significant relationship with its temperature [16-17]. As a result, conductors that generate less heat are expected to last longer and be less prone to failure due to their reduced heat generation. Therefore, feeder failure statistics must account for this change in the current through feeders. According to the model presented in this paper, the feeder failure rate of an ith feeder is calculated as a linear function of how much current flowed through it (I_i). As an example, if a the failure rate is λ_{ibase} of an ith feeder carrying I_{ibase} current before any compensation and it changes to $\lambda_{icompensated}$ after full current compensation, then its failure rate (λ_i)for any arbitrary current I_i flowing through and its formulated as follows [16]:

$$\lambda_i = m * I_i + \lambda_{icompensated} \quad (16)$$

where, m is the compensation degree and defined by the following formula:

$$m = \frac{\lambda_{ibase} - \lambda_{icompensated}}{I_{ibase}} \quad (17)$$

As a result of this relationship, network reliability is affected because feeder conductors that carry lesser current limit the temperature of the conductor, increasing its lifespan and as the probability of failure is higher for an older conductor, reducing the current reduces the likelihood of failure.

3.3 Reliability evaluation in the integration of DG units

In recent years, the DNs have become more self-sufficient due to the integration of DG units that add to the system's reliability. These DG units supply the local demand during emergency periods. During a fault on a particular feeder, a small downstream segment of the network may be isolated from the grid based on the location and type of protection devices [16]. A faulty feeder in an isolated network bus may cause an outage for a duration equal to the repair or maneuver time of the feeder, depending on whether the supply can be restored after fault clearance by NR [16,18]. It is possible to form an island if a DG unit capacity S_{DG} is located in an isolated network segment consisting of d buses, each of which exhibits a demand of S_i and thus the condition must be satisfied for that island to form.

$$S_{DG} = \sum_{i=1}^{d} S_i + S_{loss} \tag{18}$$

Due to the localized nature of the generation, losses in the island (S_{loss}) are negligible. A minimal number of loads are shed if localized generation is insufficient to sustain the island under condition (18) and prevent unstable operations.

3.4 Rider optimizer algorithm (ROA)

In recent years, metaheuristic optimization techniques such as the rider optimizer algorithm (ROA) have gained popularity. The ROA is inspired by a group of riders [19]. There are four groups based on the total group number. There are equal numbers of riders in each group. To arrive at a final solution, the groups take different approaches.

This algorithm can be summarized in the following way in terms of its basic steps:

Step 1. Randomly initialize group position and rider parameters (steering angle, brake, acceleration, gear).

$$X_i^0 \quad U(X_{\min}, X_{\max}) \tag{3.27}$$

where, $X_{\min}$ and $X_{\max}$ are the lower and upper limits of the optimization dimension.

Step 2. The success rate of riders is calculated as follows:

$$R_i = \frac{1}{\left\| X_i - l_j \right\|} \tag{3.36}$$

where, X_i is position of ith rider. l_j is target position.

Step 3. Finding the leader rider according to the maximum rider's success rate.

Step 4. Update positions of riders.

Bypass rider updates its position as follows:

$$X_{t+1}^{bypass}(i,n) = \delta\left[X_t(\eta,n) * \beta(n) + \right. \\ \left. + X_t(\xi,n) * \left[1-\beta(n)\right]\right] \tag{3.37}$$

where, δ and β are random numbers within [0,1], η and ξ are random numbers within [1, G]. G is riders' total number.

The follower rider updates its position as follows:

$$X_{t+1}^{follower}(i,j) = X^{leader}(l,j) + \\ + \left[\cos(T_{i,j}^t) * X^{leader}(l,j) * d_i^t\right] \tag{3.38}$$

where, i is selector of coordinate, X^{leader} is position of leader rider, l is its index, $T_{i,j}^t$ is steering angle in the jth coordinate, d_i^t is travelled distance of the ith rider.

Over taker rider updated its position as follows:

$$X_{t+1}^{overtaker}(i,j) = X_t(i,j) + \\ + \left[Q_t^k(i) * X^{leader}(l,j)\right] \tag{3.39}$$

where $X_t(i,j)$ is rider position in the jth coordinate, Q_t^k is rider direction indicator at time t.

The attacker rider updates its position as follows:

$$X_{t+1}^{attacker}(i,k) = X^{leader}(l,k) + \\ + \left[\cos(T_{i,k}^t) * X^{leader}(l,k)\right] + d_i^t \tag{3.40}$$

where, $X^{leader}(l,k)$ is the leader rider position in k^{th} coordinate.

Step 5. According to the calculated objective function, the leader rider position is updated.

Step 6. This step updates riders' parameters (steering angle, brake, acceleration, gear).

Riders' parameters are updated following the success rate using the activity counter.

Step 7. Time to ride off. The rider determines a race winner in the lead at the end of the race.

A summary of the proposed method's parameters and the study's limitations is presented in Table 1 [1-2,15].

Table 1.

Parameter	Values
	33-bus DN
Total rider number	20
Iteration number	50
Buses voltage limits	$0.93 p.u. \le V_m \le 1.05 p.u.$
Power generation limits of DG units	$0.2MW \le P_{DG,m} \le 3MW$
DG units' PF limits	$0.7 \le PF_{DG,m} \le 1$

4 Research results

The proposed method has been carried out using MATLAB R2021b software on a PC having a core i5-6200U CPU @ 2.30GHz processor and 8 GB RAM, and its effectiveness is evaluated for IEEE 33-bus DN. The single-line diagram of the test network is shown in

Figure 1. The power flow calculations were performed based on the forward-backward sweep (BFS) method. A DG unit is assumed to be 100% reliable regardless of the network configuration and operated at its maximum generation capacity. A 100% availability assumption is made for disconnects, substations, and fuses. The test network's feeder failure rates and repair/maneuver time can be found in [20]. When DG units are integrated or NR performed, the failure rates will be modified following equations (16) and (17), and performance and reliability indices will be evaluated, as described in equations (1–7). An 85% reduction in feeder failure rate is assumed with total current compensation. The extent of support provided by DGs in the event of feeder faults is determined by equation (18) and the location of the protection devices within the network (Fig. 1). The test DN is expected to perform under balanced load conditions. A constant power is expected in nature for all loads. Bus and line data of DN are taken from [21].

The following scenarios are used to demonstrate the effectiveness of the proposed approach:
Scenario 1: DG placement, Network Reconfiguration, and Simultaneously Reconfiguration with DG placement carried out for power loss reduction when compared to a Base configuration
Scenario 2: Base configuration, DG placement, and Simultaneously Reconfiguration with DG placement carried out for reliability improvement when compared to Base configuration

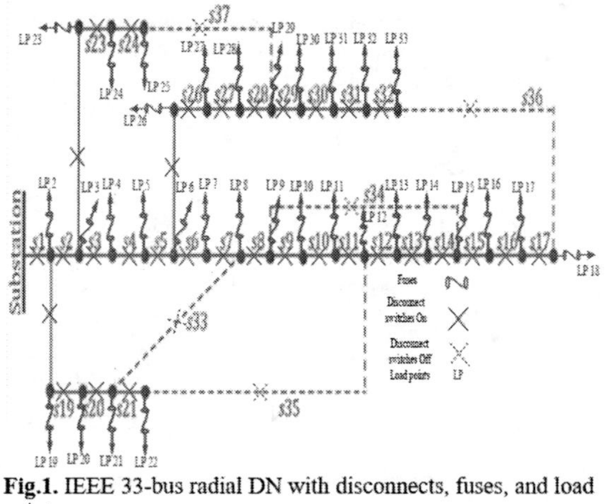

Fig.1. IEEE 33-bus radial DN with disconnects, fuses, and load points.

4.1.Scenario 1: Power loss reduction

Optimal DG allocation: In Table 2, the obtained results of the proposed technique for identifying the optimal size, location, PF, and simultaneously NR with DG allocation for single and multiple DG units in a 33-bus system are presented. According to this table, active power loss reductions for one, two and three PV-based DG units and simultaneously NR with one, two and three PV-based DG allocation in DN are 47.3761 %, 58.6854 %, 65.5013%, 61.28%, 68.15% and 73.80 %. As a result of the integration of one, two, and three WT-based DG units as well as simultaneous NR with one, two, and three WT-based DG allocation in the test DN, active power loss reductions are 67.83, 86.42, 94.43, 82.41%, 90.61%, and 95.07%, respectively. Due to the ability to generate reactive power, WT-based DG is also more efficient than PV-based DG in terms of power loss reduction. The impact of the DG integration and their simultaneous NR with DG allocation on voltage profile is shown in Figs. 2-4.

Table 2.

Cases	Bus №,(size (kW)/PF) and tie switch	Power loss (kW)	Loss reduction (%)
Base case	s33, s34, s35, s36, s37	210.9824	-
Network reconfiguration	s9, s7, s37, s14, s32	139.5513	33.85
One PV	s33, s34, s35, s36, s37	111.027	47.37
One PV+DNR	30, [1628/1] s13, s15, s7, s9, s28	81.6878	61.28
One WT	s33, s34, s35, s36, s37	67.83	67.83
One WT+DNR	30, [1594/0.82] s13, s9, s34, s33, s27	37.1050	82.41
Two PV	s33, s34, s35, s36, s37	87.165	58.68
Two PV+DNR	30,12, [1572/1,855/1] s8, s14, s7, s25, s16	67.1821	68.15
Two WT	s33, s34, s35, s36, s37	28.50	86.42
Two WT+DNR	10,30, [715/0.7,1725/0.7] s15, s35, s33, s27, s34	19.7920	90.61
Three PV	s33, s34, s35, s36, s37	72.786	65.50
Three PV+DNR	14, 25, 27, [1008/1, 1301/1, 707/1] s8, s9, s33, s28, s31	55.2626	73.80
Three WT	s33, s34, s35, s36, s37	11.740	94.43
Three WT+DNR	8,31,29, [1265/0.7,1004/0.7,1220/0.7] s2, s15, s14, s11, s22	10.40	95.07

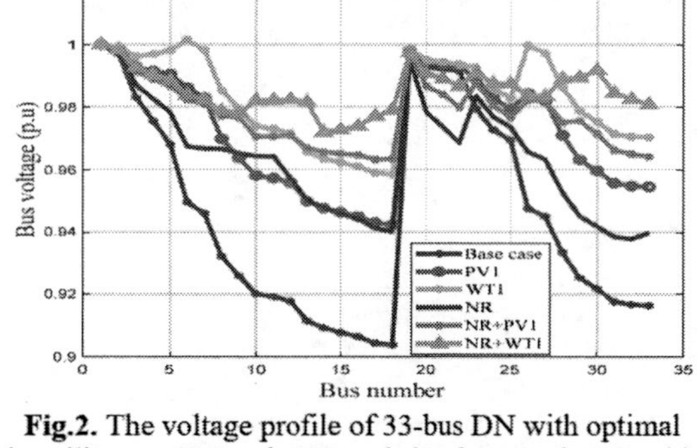

Fig.2. The voltage profile of 33-bus DN with optimal installing one DG unit, NR and simultaneously NR with one DG unit.

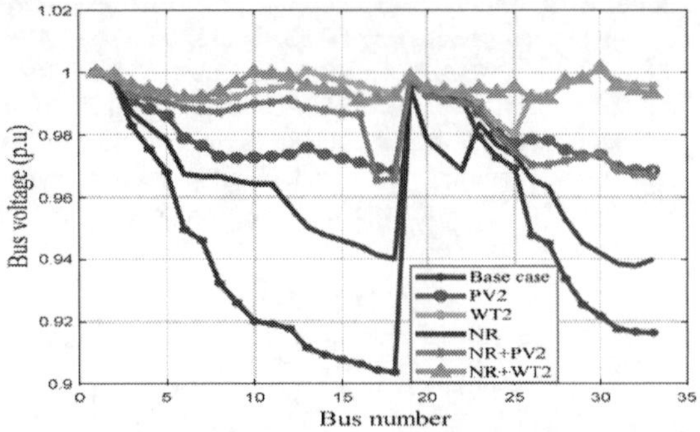

Fig.3. The voltage profile of 33-bus DN with optimal installing two DG units, NR and simultaneously NR with two DG units.

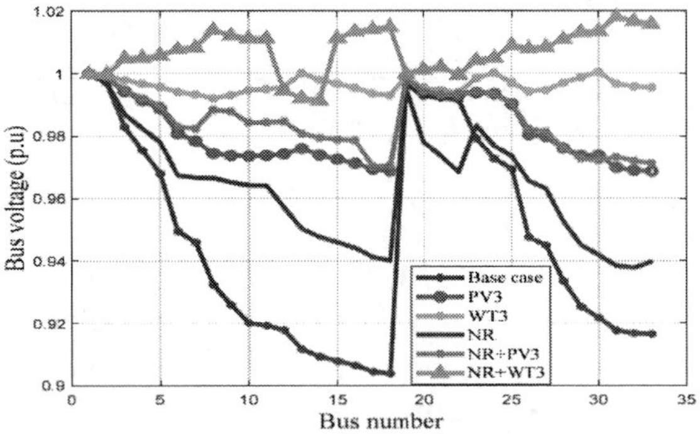

Fig.1. The voltage profile of 33-bus DN with optimal installing three DG units, NR and simultaneously NR with three DG units.

4.2. Scenario 2: Reliability improvement

Figure 1 shows the test network for this case. A set of links, namely s33-s37, represents the initial molecular structure, i.e., network configuration. In Table 3, the results obtained by NR are explicitly presented.

As shown in Table 1, by opening the switches s10, s15, and s27 and closing the tie switches s35, s36, and s37, the ENS decreased from 7097.95 kWh/year to 5057.0238 kWh/year, which can be reduced to 28.75% of the base case. As shown in Table 3, changing the radial structure appropriately will minimize SAIDI and SAIFI by 27.12 % and 15.74%, respectively.

Table 3.

Objective minimized (optimal configuration)	Reliability indices		
	SAIDI hours/customer year	SAIFI (interruptions/ customer year)	EENS (kWh/year)
Base case (s33, s34, s35, s36, s37)	2.0436	2.4126	7097.95
SAIDI (s10, s33, s34, s16, s27)	**1.4893**	1.588	1.4977
SAIFI (s27, s17, s33, s14, s35)	2.2164	**2.0328**	2.2363
EENS (s10, s33, s34, s15, s27)	5063.315	5366.7843	**5057.0238**

The test network is updated for this analysis by optimal allocation of three PV-based DG units' size and location as mentioned in Table 8 (DG location and size table) and for minimizing the objectives (2,4 and 6) separately. Under normal operating conditions, all DGs are assumed to have a unity PF. Table 4 summarizes the obtained results for this case.

All three reliability indices improved at base configuration when three PV-based DG units were integrated optimally. There is a significant reduction in SAIDI, SAIFI, and ENS, respectively, by 6.34%, 5.66%, and 6.1% as a result of optimally allocating three PV-based DG units.

Table 4.

Objective minimized (optimal configuration)	Reliability indices		
	SAIDI hours/customer year	SAIFI (interruptions/ customer year)	EENS (kWh/year)
Base case (s33, s34, s35, s36, s37)	2.0436	2.4126	7097.95
SAIDI (s33, s34, s35, s36, s37)	**1.914**	1.924	1.9205
SAIFI (s33, s34, s35, s36, s37)	2.2801	**2.2759**	2.2794
EENS (s33, s34, s35, s36, s37)	6687.3039	6689.5829	**6664.6792**

This analysis depicts the modern case analysis of DN by optimal integration of three PV-based DG units with

NR simultaneously, as given in Table 8. Table 5 summarizes the obtained results for this case.

From the table, it can be seen that three PV-based DG units with NR significantly improve network reliability. When obtained results in Table 8 were placed in the test network, SAIDI, SAIFI and ENS were reduced by 33.08%, 20.26%, and 34.27%, respectively. As a result of the reduction in feeder failure rates and the real power support provided by PV-based DG units during emergencies, the index was reduced by 34.27% when using NR with three PV-based DG units simultaneously. NR with three PV-based DG allocations also significantly improves the other reliability indices.

Table 5.

Objective minimized (optimal configuration)	Reliability indices		
	SAIDI hours/customer year	SAIFI (interruptions/ customer year)	EENS (kWh/year)
Base case (s33, s34, s35, s36, s37)	2.0436	2.4126	7097.95
SAIDI (s34, s33, s14, s9, s27)	**1.3675**	1.4782	1.3885
SAIFI (s35, s33, s27, s12, s17)	2.1024	**1.9236**	2.0497
EENS (s34, s27, s15, s9, s33)	4735.3695	4995.2572	**4665.1392**

The test network is updated for this analysis by optimal allocation of three WT-based DG units as mentioned in Table 8 (DG location and size table) and for minimizing the objectives (2,4 and 6) separately. Under normal operating conditions, all DGs are assumed to have an optimal PF. Table 6 summarizes the obtained results for this case.

All three reliability indices improved at base configuration when three WT-based DG units were integrated optimally. There is a significant reduction in SAIDI, SAIFI, and ENS by 10.48%, 9.62% and 9.61% respectively. Unlike three PV-based DG units, three WT-based DG units generated active and reactive power. Therefore, active and reactive currents are significantly reduced through feeder sections compared to three PV-based DG units.

Table 6.

Objective minimized (optimal configuration)	Reliability indices		
	SAIDI hours/customer year	SAIFI (interruptions/ customer year)	EENS (kWh/year)
Base case (s33, s34, s35, s36, s37)	2.0436	2.4126	7097.95
SAIDI (s33, s34, s35, s36, s37)	**1.8293**	1.8417	1.8439
SAIFI (s33, s34, s35, s36, s37)	2.1916	**2.1804**	2.2109
EENS (s33, s34, s35, s36, s37)	6429.5596	6373.4449	**6415.7992**

Following this case, the optimal integration of three WT-based DG units with NR was carried out simultaneously, as given in Table 8. Table 7 summarizes the obtained results for this case.

From the table, it can be seen that three WT-based DG units with NR significantly improve network reliability. When obtained results were placed in the test network at the base configuration, SAIDI, SAIFI and ENS were reduced by 40.18%, 27.11% and 40.20%, respectively. As a result of the reduction in feeder failure rates and the active and reactive power support provided by WT-based DG units during emergencies, the index was reduced by 40.20% when using NR with three WT-based DG units simultaneously. Additionally, NR with three WT-based DG allocations significantly improves the other reliability indices.

Table 7.

Objective minimized (optimal configuration)	Reliability indices		
	SAIDI hours/customer year	SAIFI (interruptions/ customer year)	EENS (kWh/year)
Base case (s33, s34, s35, s36, s37)	2.0436	2.4126	7097.95
SAIDI (s27, s9, s7, s34, s16)	**1.2223**	1.3843	1.2367
SAIFI (s17, s27, s33, s12, s35)	1.9199	**1.7585**	1.7629
EENS (s10, s34, s7, s36, s27)	4264.8434	4667.8849	**4243.8856**

Table 8.

Scenario	Bus location	Maximum generation capacity (MW/PF)
Three PV	24,30,14	1073/1,1058/1, 7961
Three PV+DNR	18,26,24	981/1, 747/1, 1384/1
Three WT	30,24,14	1176/0.7,982/0.7,677/0.70113
Three WT+DNR	29,27,11	1743/0.7,356/0.7,719/0.70146

5 Conclusion

In this paper, an effective optimizer called ROA with the BFS method developed here for radial distribution NR with optimal DG allocation has been developed. To check the effectiveness of the proposed method, it is applied to standard 33 bus DN. The first scenario was carried out to minimize power loss of the test network with optimal integration DG units with NR and analyzed impact of size and locations of DG units and tie switches on the reduction of power loss and voltage profile has been obtained. NR, base configuration with DG installation and simultaneous NR with DG installation cases are analyzed with MATLAB programming. The results show that NR showed better results than the base configuration with the installations of DG. Then, the improvement of reliability from base case to NR case, base case to base with DG, and base case to simultaneously NR with DG has been observed.

As a result of this research, researchers can develop an algorithm and model that will reduce power loss and improve the voltage profile and reliability of DNs.

References

1. Khasanov M, Kamel S, Rahmann C, Hasanien HM, Al-Durra A. Optimal distributed generation and battery energy stor- age units integration in distribution systems considering power generation uncertainty. IET Generation, Transmission & Distribution. 2021;15(24):3400–3422.
2. Khasanov M, Kamel S, Houssein EH, Rahmann C, Hashim FA. Optimal allocation strategy of photovoltaic- and wind turbine-based distributed generation units in radial distribution networks considering uncertainty. Neural Computing and Applications. 2022;p. 1–26.
3. Lavorato M. Imposing radiality constraints in distribution network optimization problems. IEEE Transactions on Power Systems. 2011;27:172–180.
4. Yin, Shih-An CN, Lu. Distribution feeder scheduling considering variable load profile and outage costs. IEEE Transactions. 2009;2:652–660.
5. Billinton R, Allan RN. Introduction. In: Reliability Evaluation of Power Systems. Boston, MA: Springer; 1996. Available from: https://doi.org/10.1007/978-1-4899-1860-4_1.
6. Hines P, Balasubramaniam K, Sanchez EC. Cascading failures in power grids. Ieee Potentials. 2009;28:24–30.
7. Wirfs-Brock J; 2016. Available from: https://insideenergy.org/2014/08/18/power-outages-on-the-rise-across-the-u-s/.
8. Byrd, H., & Matthewman, S. Exergy and the City: The Technology and Sociology of Power (Failure). Journal of Urban Technology. 2014;21(3):85–102.
9. Campbell RJ, Lowry S. Weather-related power outages and electric system resiliency. Washington, DC: Congressional Research Service, Library of Congress; 2012.
10. Cebeci ME, Tor OB, Oprea SV, Bara A. Consecutive market and network simulations to optimize investment and opera- tional decisions under different RES penetration scenarios. IEEE Transactions on Sustainable Energy. 2018;10(4):2152– 2162.
11. Silva D, Cassula AML, Sacramento AM, Nascimento CE, Ávila LC, F A. Network reconfiguration of distribution systems using metaheuristics and reliability measures. 15th International Conference on Intelligent System Applications to Power Systems. 2009;p. 1–8.
12. Skoonpong A, Sirisumrannukul S. Network reconfiguration for reliability worth enhancement in distribution systems by simulated annealing. 5th International Conference on Electrical Engineering/Electronics, Computer, Telecommunications and Information Technology. 2008;2:937–940.
13. Hivziefendić, J., Hadžimehmedović, A., & Tešanović, M. Optimal network reconfiguration with distributed generation using NSGA II algorithm. (IJARAI) International Journal of Advanced Research in Artificial Intelligence, 2016; 5(2):6-11.
14. Tolabi H, Bagheri, Ali MMH, Rizwan. Simultaneous reconfiguration, optimal placement of DSTATCOM, and photovoltaic array in a distribution network based on fuzzy-ACO approach. IEEE Transactions on sustainable Energy. 2014;6:210–218.
15. Khasanov M, Xie K, Kamel S, Abdubannaev J, Kurbanov A, Jalilov U. Optimal radial distribution network reconfiguration to minimize power loss by using mayfly algorithm. AIP Conference Proceedings. 2023;2612:50016–50016.
16. Sharma P, Gupta N, Niazi KR, Swarnkar A. Investigation of network reconfiguration on the reliability and performance of distribution systems using CRO. 6th International Conference on Computer Applications In Electrical Engineering-Recent Advances (CERA). 2017;p. 80–85.
17. Lewin PL, Theed JE, Davies AE, Larsen ST. Method for rating power cables buried in surface troughs. IEE Proceedings- Generation, Transmission and Distribution. 1999;146(4):360–364.
18. Carpaneto E, Chicco G, Prunotto A. Reliability of reconfigurable distribution systems including distributed generation. 2006 International Conference on Probabilistic Methods Applied to Power Systems. 2006;p. 1–6.
19. Binu D, Kariyappa BS. RideNN: A new rider optimization algorithm-based neural network for fault diagnosis in analog circuits. IEEE Transactions on Instrumentation and Measurement. 2018;68(1):2–26.

20. Gupta N, Swarnkar A, Niazi KR. Distribution network reconfiguration for power quality and reliability improvement using Genetic Algorithms. International Journal of Electrical Power & Energy Systems. 2014;54:664–671.
21. Baran ME, Wu FF. Network reconfiguration in distribution systems for loss reduction and load balancing. IEEE Transac- tions on Power delivery. 1989;4(2):1401–1407.

Research of energy-saving composite materials for electric motors

Utkirbek Sulaymonov[1], *Aleksandr* Jelutkevich[1], *Malika* Nabiyevna[1], *Oybek* Sayfullayev[1], *Ulug'bek* Berdiyorov[2*]

[1]Tashkent State Transport University, Tashkent, Uzbekistan, Tashkent, Temiryolchilar St. 1
[2]Tashkent State Technical University named after Islam Karimov, 100095, Uzbekistan, Tashkent, University St. 2A

Abstract. This article discusses samples of magnetic cores based on isolated powders of magnetic materials, and also investigates the main magnetic characteristics of magnetic cores using ASC100.29 iron powders.

1 Introduction

Recently, in various devices operating in alternating magnetic fields, cores made of composite powder materials with the necessary magnetic characteristics are widely used. At the same time, they have relatively small eddy current losses and high electrical resistivity [1, 2, 4]. The required characteristics are achieved by doping soft magnetic materials based on iron with phosphorus or silicon with the introduction of special dielectric layers of organic and inorganic additives. This makes them competitive with conventional laminated electrical steel cores. The use of powder metallurgy methods in the formation of products of complex shape, in addition, makes it possible to obtain isotropic materials that change the nature of the magnetization, as well as the direction and distribution of the magnetic fluxAn analysis of literature data shows that the developed materials are subjected to heat treatment at temperatures below 800 °C [2, 5]. This is apparently due to the destruction of the material of the interlayers of thermosetting inorganic materials. To increase the magnetic permeability of the metal matrix, recrystallization annealing is best carried out at higher temperatures using high temperature resistant dielectrics. Most authors use glass-containing compositions, enamels, etc. as such materials. Such materials are in a liquid state at high temperatures, which contributes to the appearance of properties characteristic of these phases. The arising capillary forces can ensure the movement of liquid micro volumes, similarly to impregnation processes, into antiparticle voids [4, 6].

Composite materials with the necessary set of performance characteristics are widely used in the nodes of various mechanisms. In any technical applications, certain properties of solids are used: electrical, magnetic, optical, thermal, mechanical, corrosion-resistant, etc. The production of magnetic materials with low energy losses during magnetization reversal is one of the urgent problems of industry today [1, 5, 8]. Despite the fact that research and development of such materials have been carried out since the beginning of the last century, the study of the mechanism of magnetization reversal and improvement of the quality of these materials is still relevant today. This is due to the fact that magnetic materials are widely used in various electrical devices (generators, electric motors, measuring devices, inductors, etc.).

2 Experimental research

The technology for manufacturing experimental samples of magnetic cores based on isolated powders of magnetic materials is a multi-stage process [5, 8].

For research, samples were made using ASC100.29 iron powder with particle sizes $d > 0.1$ mm and $d < 0.1$ mm and with an average insulation thickness based on phosphorus oxide ~ 1 nm, as well as samples based on ASC100.29 ($d < 0.05$mm) with insulating layer thicknesses of 10, 20, 30 nm.

Measurements of the magnetic properties of samples of experimental components for the stator and rotor were carried out on an express magnetometer (Figure 1), where losses and other magnetic parameters were determined from the magnetization reversal curves of the samples [6, 7].

Measurement of the magnetic characteristics of magnetic circuits in a wide range of magnetic fields, demagnetizations frequencies and temperatures was

* Corresponding author: ilider1987@yandex.ru

carried out using an express magnetometer developed and manufactured at the Center [2, 5]. To normalize the magnetometer, the magnetic characteristics were first measured on a certified flux meter. An F5050 microwebermeter was also used to normalize the magnetometer. Figure 1 shows the main blocks of the express magnetometer, created in the laboratory of the physics of magnetic materials of the State Scientific and Practical Center of the National Academy of Sciences of Belarus for Materials Science, and Figure 2 shows the results of data processing in the express fluxmeter and magnetometer mode.

Studies of the main magnetic characteristics of magnetic cores using ASC100.29 iron powders (China, Hoganas) were compared with similar parameters for E320 electrical steel. For this purpose, cores 24 × 13 × 10 mm were made from strip steel and composite magnetic materials with a density of ρ = 7.6 g/cm3. Their static and dynamic magnetic characteristics have been studied, and the value of hysteresis losses has been determined [2, 4, 7]. Samples of magnetic cores made of composite magnetic material were annealed in vacuum at a temperature of 400°C for 3 hours. The static characteristics were studied using an F5050 microwebermeter, and the dynamic characteristics and losses were studied with a flux meter and an express magnetometer in the frequency range up to 10 kHz and magnetic fields up to 30 kA/m.

Table 1

Composition of air-sprayed Laiwu Taidong Powder (China) and water-sprayed ASC100.29 powder (Sweden) (in %)

	Laiwu Taidong Powder (China)	ASC100.29 (Sweden)
Fe	98,69	99,5
Mn	0,35	0,08
Si	0,1	0,04
C	0,022	0,08
P	0,028	0,01

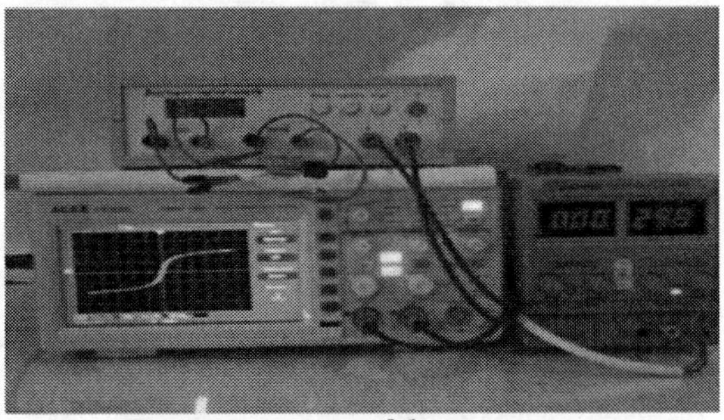

Fig 1. Appearance of the magnetometer

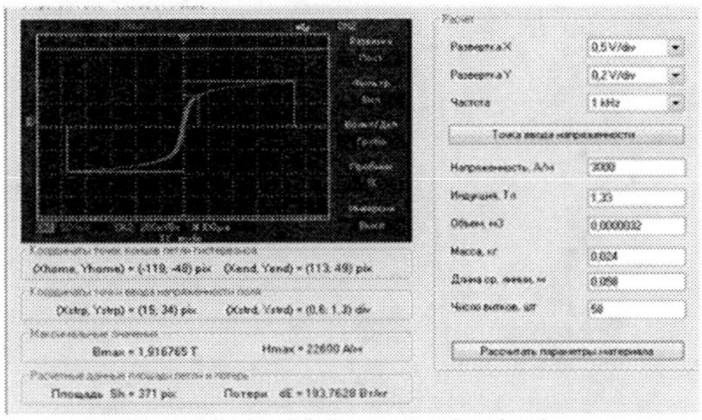

Fig. 2. Results of data processing in the express flux meter mode (a) and in the express magnetometer mode

The field dependences of magnetization and losses at a frequency of 1 kHz for a low-frequency sample of a magnetic circuit based on ASC100.29 powder upon annealing in a vacuum of 350°C for three hours are shown in Figure 3 [3, 6]. It follows from the figure that a linear dependence is preserved in the change in magnetic induction and the magnitude of losses.

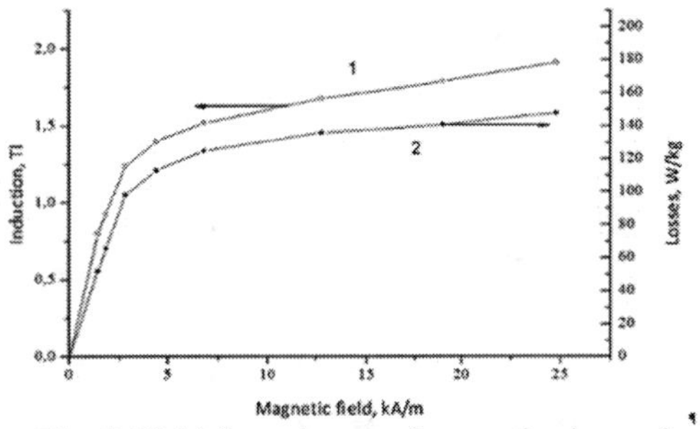

Fig. 3. Field dependences of magnetization and losses at a frequency of 1 kHz for a composite material based on ASC100.29 (1) and Laiwu Taidong Powder (China) (2)

3 Research results

The study of losses during magnetization reversal of samples of magnetic circuits showed that hysteresis losses are 25-30% higher for samples based on powders sprayed with air compared to losses for magnetic circuits based on powders sprayed with water (Figure 4) [7, 9]. On the whole, comparative studies have shown a certain advantage in the value of magnetization reversal losses of purer iron powders obtained by the water spray method in comparison with powders sprayed with air [3, 5, 8].

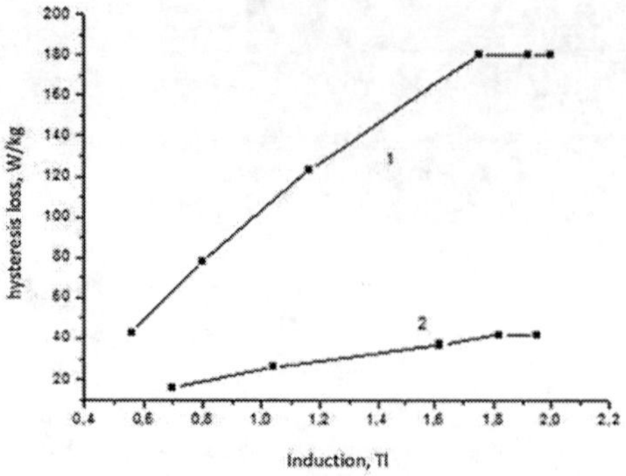

Fig. 4. Dependence of losses on the value of induction for a low-frequency composite material based on ASC100.29 during magnetization reversal in a full cycle - 1 and in a quotient from the zero point - 2 at a frequency of 1 kHz

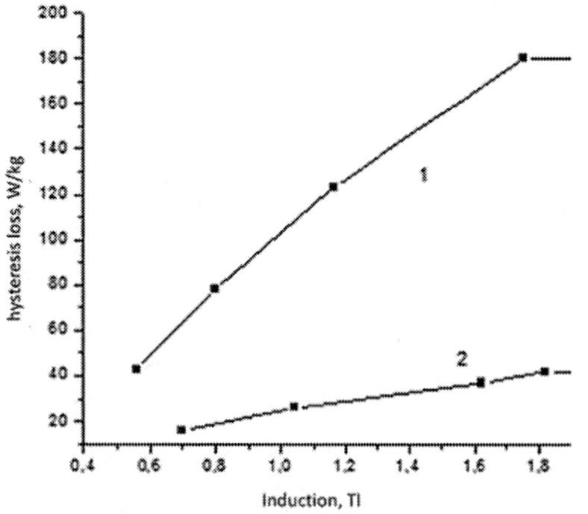

Fig. 5. Dependence of losses on the induction value for a low-frequency composite material based on ASC100.29 during magnetization reversal in a full cycle - 1 and in a quotient from the zero point - 2 at a frequency of 1 kHz

The main parameter that determines the possibility of using products in electrical devices is the magnitude of electromagnetic losses at different frequencies [5, 8-20]. The results of preliminary studies of losses from the magnitude of induction are shown in Figure 5. In general, such studies have shown an advantage in the magnitude of losses for magnetization reversal of components made on the basis of composite materials compared to sheet electrical steel.

4 Conclusion

Thus, the conducted studies have shown an advantage in the value of losses for the remagnetization of components made on the basis of composite materials, compared with sheet electrical steel. In general, comparative studies have shown some advantage in the magnitude of remagnetization losses of purer iron powders obtained by the water spray method in comparison with powders sprayed with air.

References

1. Vityaz, P. A. Powder materials based on iron and copper. Atlas of structures / P.A. Vityaz [and others]. – Minsk: Belarus. science, 2008. – 155 p.
2. Govor, G.A. Features of magnetic characteristics of new composite materials based on iron powders / G.A. Govor, A.K. Evening, K.I. Yanushkevich // Advanced materials and technologies: ed. V.V. Klubovich - Vitebsk, Educational Institution "VSTU". – Vitebsk, 2017. – T. 2, Ch. 15 – pp. 278–299.
3. A.K. Evening, G.A. Govor, K.I. Yanushkevich, U.T. Berdiev, F.F. Khasanov / Study of the influence of synthesis conditions on the magnetic characteristics of composite materials based on iron powders // Article submitted to the journal "Vestsi of the National Academy of Sciences of Belarus, gray physics and technical sciences".
4. Govor, G.A. Composite soft magnetic materials based on iron powders and prospects for their application in technology / G.A. Govor, A.K. Evening, V.I. Mityuk // Metal physics and newest technologies. – 2012. – T. 34, No. 4. – P. 439–444.
5. Soft magnetic materials based on iron used in electrical engineering / A.K. Evening, G.A. Govor, W.T. Berdiev, F.F. Khasanov // Bulletin of TashIIT. – 2019. – No. 3. – P. 212–217.
6. Govor, G.A. Composite soft magnetic material based on iron powders with oxide nanocoating / G.A. Govor, A.K. Evening // Current problems of solid state physics FTT-2016: collection. report VII Int. scientific Conf., Minsk, November 22-25, 2016: in 3 volumes / Publisher: Kovcheg; ed. coll.: N.M. Olekhnovich (prev.) [and others]. – Minsk, 2016. – T. 1 – P. 50–52.
7. Optimization of the method of oxide coating of metallic iron powder particles. Usan Berdiyev, Olga Demedenko, Mirjalol Ashurov, FF Hasanov and UB Sulaymonov. E3S Web Conf., 383 (2023) 04039. DOI: https://doi.org/10.1051/e3sconf/202338304039
8. Berdiev, UT , Kolesnikov, IK , Tuychieva, MN , Khasanov, FF , Sulaymonov, UB Methods of new technological developments of electric motors based on soft magnetic materials . E3S Web of Conf. Volume 401, 2023. 03038. https://doi.org/10.1051/e3sconf/202340103038
9. Berdiev U.T., Tuychieva M.N., Iksar E.V. Estimation of thermal parameters of frequency-controlled asynchronous traction motors when powered from asymmetrical sources. // Computer

Science and Energy problems. Uzbekistan Journal, №1 2022 July. 30-35 b.
10. Berdiev U.T., Burkhankhodzhaev A.M., Tuychieva M.N., Iksar E.V. Investigation of energy indicators with asymmetry of the voltage of the power source of mainline electric locomotives of alternating current. // International Scientific Conference Construction Mechanics, Hydraulics & Water Resources Engineering. CONMECHYDRO 2021 AUTUMN SEASON. Tashkent/Uzbekistan on September 7-9, 2021.
11. Pirmatov, N., Panoev, A., Samatova, G., Berdiyorov, U. Determination of methods of achieving the energy savings through mathematical modeling of static and dynamic modes of electromagnetic energy conversion in asynchronous motors used in feed crushers. E3S Web of Conferences, 2023, 383, 04046.
12. Pirmatov, N., Muminov, M., Akberdiev, M., Abdullayev, J. Use of Hybrid Renewable Energy Systems to Excite Autonomous Synchronous Machines. AIP Conference Proceedings, 2022, 2432, 020020.
13. Pirmatov, N.B., Madrakhimov, D.B., Ivanova, V.P., Ortikova, M.J., Atamukhamedova, R.F. Determination of the Degree of Digitalization at the Cable Enterprise Level. AIP Conference Proceedings, 2022, 2552, 070009.
14. Pirmatov, N., Panoev, A. Frequency control of asynchronous motors of looms of textile enterprises. E3S Web of Conferences, 2020, 216, 01120.
15. Kh Sapaev, Sh Umarov. Two approaches for automating analysis of electromagnetic processes in non-linear circuits with valves. E3S Web of Conferences 139, 01085 (2019)
16. M. M. Khamud-khanov, K. H. B. Sapaev, Umarov Sh. B. Multi-motor drive with common inverter for pumping units. MIP: Engineering-2020.IOP Conf. Series: Materials Science and Engineering 862 (2020) 062035 IOP Publishing. doi:10.1088/1757-899X/862/6/062035.
17. Islombek Abdullabekov, Sapaev Khushnud. An Energy Efficient Control System for Water Lifting Units of the Ramadan Pumping Station Based on Frequency Controlled Electric Drives. AIP Conference Proceedings 2552, 040023 (2023).
18. Sapaev Khushnud, Umarov Shukhrat, Abdullabekov Islombek, Khamudkhanova Nargiza and Nazarov Maxamanazar. Scheme of effective regulation of pumping station productivity. AIP Conference Proceedings 2402, 060016 (2021).
19. Pirmatov, N., Toshev, S. Overvoltage in the free phase of the stator winding in the case of asymmetric short circuit implicit pole synchronous generator biaxial excitation. E3S Web of Conferences, 2019, 139, 01030.
20. Pirmatov, N., Tosheva, S., Toshev, S. Best overall dimensions of synchronous generator with permanent magnets for small power wind plants and micro hydropower plants. E3S Web of Conferences, 2019, 139, 01027.
21. Abduraxman Akhmedov, Galina Sauchuk, NatalliaYurkevich, Sardorbek Khudoyberganov, MahammatyakubBazarov, KarimberdiKarshiev; The influence of production conditions on the electrophysical parameters of piezoceramics for different applications. E3S Web Conf. 264 04020 (2021), DOI: 10.1051/e3sconf/202126404020
22. R. Mirsaatov, S. Khudoyberganov, A. Akhmedov; Uncertainty estimation in determination of Cocoons silkiness by thickness of their shell. AIP Conference Proceedings 15 March 2023; 2612(1):050010. https://doi.org/10.1063/5.0114683

Development of a unified digital transport and logistics intelligent platform based on the National Operator

G.B. Yusupkhodjaeva[1]*

[1]Tashkent State Technical University named after Islam Karimov, 100095, Uzbekistan, Tashkent, University St. 2A

Abstract. This article is devoted to the research and development of a single digital transport and logistics intelligent platform based on the infrastructure of the National Operator. In the context of a rapidly developing transport and logistics industry, the creation of such a platform becomes a key strategic initiative. The abstract analyzes the main components and benefits of the platform, such as data integration, route and load optimization, demand forecasting, warehouse management, cargo tracking, improved planning and coordination, and analytics and reporting.

1 Introduction

The development of a unified digital transport and logistics intelligent platform based on the National Operator is an important and strategically significant initiative. Such a platform can have many benefits, including better management and coordination of transport and logistics processes, improved forecasting and planning, and increased transparency and cost reduction in transport and logistics.

In order to take measures to implement the concept of ensuring road safety in the Republic of Uzbekistan for 2018 - 2022, as well as within the framework of the Decree of the President of the Republic of Uzbekistan dated July 11, 2017 No. PP-3127 "On measures to further improve the system for ensuring road safety" and Resolution of the Cabinet of Ministers dated May 19, 2018 No. 377 “On measures to further improve the road safety system in the Republic of Uzbekistan”, foreign experience in the creation and implementation of intelligent systems for road safety and navigation (transport monitoring) was studied, consultations were held with potential international investors, local developments in this direction are studied.

2 Experimental research

Foreign experience in the creation and implementation of intelligent systems for road safety and navigation includes many innovative solutions and technologies aimed at improving the safety, efficiency and manageability of the transport infrastructure:

1. Automatic driving assistance systems (Advanced Driver Assistance Systems, ADAS): These systems include a variety of functions such as lane control, automatic braking in the event of obstacles, collision warning systems, etc. Examples include an automatic parking system, blind spot monitoring systems, adaptive cruise controls, and others.

2. Infrastructural traffic support systems: The introduction of intelligent systems on the roads, such as sensors, cameras and traffic lights, linked into "smart" roads, allows you to monitor and manage traffic flows, optimize traffic lights, detect violations and emergency situations.

3. Navigation and GPS systems: Very popular and widespread, these systems provide drivers with accurate maps, traffic information, speed limits and other important data. In recent years, navigation systems have also been integrated with weather, road conditions and other parameters.

4. Communication systems between vehicles (V2V) and with traffic lights (V2I): V2V and V2I systems allow vehicles to communicate with each other and with traffic infrastructure. This can help alert you to near emergencies, provide smoother driving, and so on.

5. Violation and Accident Detection Systems: Many countries are implementing systems for automatic photo- video recording of violations, such as speeding, running a red light, etc. This makes it possible to effectively control traffic and reduce violations.

5. Data integration for road infrastructure management: Large amounts of data collected from

* Corresponding author: raxmatillo82@mail.ru

various sources can be used to analyze and manage transport infrastructure. For example, traffic flow analysis can help optimize the distribution of traffic signals.

Examples of countries successfully implementing such systems are Sweden (road equipment and infrastructure systems), Germany (ADAS development), USA (V2V and V2I technologies), Japan (GPS and navigation systems development), Singapore (traffic management systems) and a lot others.

3 Research results

Such intelligent systems greatly improve traffic safety, efficiency and convenience, but also require attention to data, privacy and cybersecurity aspects. The main components and benefits of developing such a platform may include:

Data integration: Collecting, aggregating, and analyzing data from various sources (transport companies, warehouses, ports, weather, and other relevant data) will help you create a complete picture of traffic flows, available resources, and the current situation in the logistics network. Integration of data from various sources, such as transport companies, warehouses, ports, weather data and other relevant sources, plays an important role in optimizing and improving logistics operations. Here's how it works and why it matters:

Data Collection: First you need to collect data from different sources. This data can include information about cargo, routes, delivery times, warehouse conditions, weather forecasts, vehicle information, and many other parameters.

Data Aggregation: The collected data is then aggregated in a centralized system or database. This allows you to combine information from different sources and create a single data access point.

Data Analysis: After data aggregation, you can perform analysis to create a complete picture of the logistics network. Data analysis allows you to identify trends, problems, opportunities for optimization and make more informed decisions.

Route and loading optimization: Using optimization algorithms, the platform can offer optimal routes and loading methods for vehicles, taking into account parameters such as distance, travel time, road capacity, and other constraints. Route Optimization: Analyzing data about traffic flows and available resources helps you optimize delivery routes, which can save time and resources.

Improved forecasting: Integration of weather and other external factors data allows you to more accurately predict delays and other unexpected events, which contributes to more reliable delivery.

Inventory management: Analysis of data on the state of warehouses and expected deliveries helps you manage inventory effectively, avoiding a shortage or excess stock.

Improve customer service: Data integration allows you to quickly respond to changes in the logistics network and provide a higher level of customer service.

Save resources: By integrating data, companies can make better use of their resources, which can save money and reduce the negative impact on the environment.

Demand forecasting: Analysis of historical data and current trends will allow you to more accurately predict demand for transport and logistics services, which in turn will help you plan resources more efficiently. Demand forecasting is an important aspect in inventory management, logistics, and business in general. Demand forecasting allows companies to adapt their operations, optimize production, manage inventory, and provide a better level of customer service. Here are some key aspects of demand forecasting:

Data collection: The first step in forecasting demand is to collect data. This includes historical data on sales, trends, seasonal changes, as well as external factors that can influence demand, such as weather, economic events, etc.

Choice of forecasting methods: There are many forecasting methods available, including statistical methods (such as exponential smoothing, time series), machine learning, and artificial intelligence. The choice of method depends on the available data, the nature of the business, and the accuracy of the forecast that is required.

Data processing: Data may require pre-processing, such as eliminating outliers, filling in missing values, and normalizing data.

Modeling: Forecasts are made using the selected methods. Models can be trained on historical data and used to predict future values.

Accuracy assessment: It is important to evaluate the accuracy of forecasts by comparing the predicted values with the actual results. This allows you to improve models and forecasting methods.

Update forecasts: Forecasts are not static and may change depending on new data and changes in the business environment. Therefore, it is important to update your forecasts periodically.

Decision Making: Based on forecasts, you can make decisions about purchasing, production, inventory management, distribution, and marketing strategies.

Demand forecasting helps companies avoid shortage or excess inventory, optimize production and logistics, and provide more accurate promises on delivery times and customer service. This is a key element of successful business management and reducing transaction costs.

Warehouse Management: The platform can help optimize inventory management by offering data on current inventory levels, predicting needs, and optimal times to move to orders.

Inventory Monitoring: The platform allows companies to track current inventory levels in a warehouse. This includes information about the number of products, their location, expiration dates, and other characteristics.

Demand forecasting: The platform uses data on past sales and other factors (such as seasonality, marketing campaigns, changes in market conditions) to predict future demand. This helps you determine how many items you need to keep in stock.

Order management: The platform optimizes the order process by calculating the best moments for ordering goods from suppliers. This may include automating orders, optimizing order batches, and taking into account factors such as minimum orders from suppliers.

Optimization of distribution: The platform helps optimize the location of goods in the warehouse to reduce pick-up time пикерованd simplify the process of shipment of goods.

Expiration Date Management: For products with expiration dates, the platform can provide information about upcoming expiration dates, which allows you to manage inventory more efficiently and avoid losses.

Data Analysis: The platform can provide analytical tools that enable companies to analyze warehouse performance, identify bottlenecks, and optimize operations.

Cargo tracking: Monitoring and tracking systems will allow real-time monitoring of cargo movement, which will improve supply management and provide customers with up-to-date information, improve logistics operations management, improve customer service, and help companies be more efficient and competitive in the market.

Improved planning and coordination: The system can automate planning and coordination tasks, reducing the risk of errors and conflicts in the transport and logistics chain. Improved logistics planning and coordination are key components of successful logistics operations. Modern technologies and methods can significantly improve planning and coordination processes. Here are some of the ways this can be achieved: As mentioned earlier, integrating data from various sources allows you to create a complete picture of the logistics network. This makes it possible to plan and coordinate deliveries more precisely, taking into account many external factors. Transport management systems: Modern transport management systems (TMS) provide the ability to efficiently plan and route vehicles, as well as coordinate deliveries. They take into account route restrictions, traffic jams, and delivery time intervals. Machine learning and analytics allow you to predict demand, optimize routes, and take into account many variables, such as seasonality and weather conditions. Electronic marketplaces and platforms connect suppliers and buyers, making it easier to coordinate orders and deliveries. Tracking and Monitoring Systems: As mentioned earlier, real-time cargo tracking systems allow companies to more accurately coordinate deliveries and respond to real-time changes. The use of cloud technologies allows different participants in the logistics chain to work with shared data and provides easy access to information. Warehouse management systems (WMS) help to optimize processes in warehouses, which contributes to more efficient coordination and preparation of goods for delivery. Improved communication between participants in the logistics chain, including suppliers, carriers, and customers, plays an important role in effective coordination. Planning and coordination in logistics helps reduce delays, optimize costs, increase customer service, and make logistics operations more flexible and reactive. It also helps to reduce the negative impact on the environment through more efficient use of resources.

Improved transparency and security: The platform can provide greater transparency of the entire process of moving goods, which helps to improve security and prevent losses. Improved logistics planning and coordination are key components of successful logistics operations. Modern technologies and methods can significantly improve planning and coordination processes. Here are some of the ways this can be achieved:

Analytics and Reporting: Generating analytical reports and statistics helps identify trends, bottlenecks, and areas for improvement in transport and logistics operations.

However, the development and implementation of such a platform also involves a number of difficulties, including technical, organizational, security and legal aspects. It is necessary to take into account data confidentiality, provide protection against cyber threats, and ensure consistency and cooperation between different participants in the transport and logistics network.

It is also important to establish a regulatory framework and standards governing the use and development of the platform, as well as ensure the participation of all stakeholders, including transport companies, logistics operators, government agencies and other market participants, to ensure the effectiveness and success of this initiative.

Since the beginning of 2019, a similar system has been launched in the Republic of Kazakhstan (EVAC). Currently, work is underway to link emergency response systems in Russia and Kazakhstan. In addition, emergency response systems have been put into operation since April 2018 on the territory of 28 states of the European Union as part of the implementation of the ECall program. All these systems have an identical principle of operation, which consists in installing special equipment on vehicles, which automatically sends detailed information about the location of the vehicle and the nature of damage to the emergency response center at the time of a road accident. This significantly reduces the response time of emergency services and assistance to victims, and the presence of an intercom device in the car also allows you to provide psychological assistance before rescuers arrive.

The transport connectivity of the Republic of Uzbekistan with the Republic of Kazakhstan and the Russian Federation, as well as the increasing traffic flow between these states, make it necessary to create an emergency response system in case of accidents on highways of the Republic of Uzbekistan with subsequent integration with the ERA-GLONASS and EVAK systems, which will significantly expand the unified ecosystem of transport security and create conditions for security of citizens of the Republic of Uzbekistan, both on the territory of their own state and in neighboring territories.

Tasks of the national operator: One of the main tasks of the national operator will be the creation and operation of the state information system for emergency response in case of accidents in transport (traffic accidents) (hereinafter referred to as the Emergency Response System in case of an accident), an analogue of the European E-Call system and the Russian GAIS "ERA-GLONASS". The principle of operation of the European and Russian systems is identical and consists in installing special equipment on each vehicle, which automatically sends detailed information about the location and nature of damage to the emergency response center at the time of a traffic accident. Due to this, the response time of emergency ambulance services, the Ministry of Emergency Situations and the Ministry of Internal Affairs is significantly reduced, and the presence of an intercom in the car allows you to provide psychological support to persons injured in an accident before the arrival of rescue services.

Also, one of the key activities of the national operator should be the solution of state tasks related to the control of the location of vehicles of certain categories that carry out passenger and freight transportation.

4 Conclusion

Development of a unified digital transport and logistics intelligent platform based on the National Operator The emergency response system in case of an accident can also be used to create and ensure the functioning of a system for controlling road and rail transportation, which provides for the use of identification tools (seals) operating on the basis of GLONASS technology:

- ensuring the control of the integrity of the identification means (seals) installed on the vehicle, functioning on the basis of GLONASS technology (hereinafter referred to as the seals), with a control system with the transfer of information about the application, removal and violation of the seal to the authorized state authorities;
- checking the performance and identification of seals in the control system;
- providing access to the control system in 24x7 mode for the purpose of obtaining information about the location of vehicles and the safety of cargo.

Any types of cargo spaces (compartments) of vehicles and railway rolling stock, premises, containers and other places where goods are or may be located during their transportation through the territory of the Republic of Uzbekistan can act as mobile objects of protection. It can be various types of containers, wagons, tanks, car trailers, refrigerators, thermoses.

Market volumes (initial data for revenue estimation) If a decision is made on the mandatory equipping of vehicles put into circulation with emergency call devices, taking into account the current forecast for the volume of production of vehicles on the territory of the Republic of Uzbekistan, as well as the volume of exports and imports of vehicles, the predicted number of equipped vehicles by 2024 will be about 1.3 million units

According to international recommendations, the indicators of the commercial (financial) efficiency of the project "as a whole" take into account the financial consequences of its implementation for the participants implementing the investment project, assuming that they incur all the costs necessary for the implementation of the project at their own expense and use all its results. When constructing the model, an assumption was made that the cost of funds of the GIK services operator is assumed to be equal to the pledged rate of return and is 20%. Thus, the weighted average cost of the operator's funds is 20%.

References

1. Allaeva G.J. Problems of formation and innovative potential use of fec enterprises in Uzbekistan EDITOR COORDINATOR, 2021
2. Samatov G.A. Sarvirova N.S. Kamalov A.S. Evolution of the paradigm of logistics and supply chains through the system of concepts of philosophy and mission
3. https:// statistika Uzbekistan -2022
4. Yuldasheva, S. A., & Burikhodzhaeva, M. S. (2022, September). Improving the efficiency of customs activities in the development of international economic cooperation. In international scientific and practical conference" the time of scientific progress" (Vol. 1, No. 1, pp. 87-91).
5. Muzafarova, K. Z. (2019). The role of foreign investment in the development of the national economy of Uzbekistan. Economics and Entrepreneurship, (1), 411-414.
6. G B Yusupkhodjaeva JSC "GLONASS" innovative navigation and information technologies in Uzbekistan E3S Web of Conferences 289, 07031 (2021)

Research algorithm and dynamic characteristics of three-phase electromagnetic current transducers in filter-compensation devices

Mokhirbek Makhsudov[1*], *Dilyorbek* Karimjonov[1]

[1]Andijan Machine - Building Institute, 170019, Uzbekistan, Andijan, Bobur shakh St. 56

Abstract. The article presents a three-phase current transducer for filter-compensation devices, representing the value of the stator current of an asynchronous motor. An algorithm for determining the values of the output electric quantities in them was developed based on the methods of interconnecting the loops of the sensitive element of the three-phase electromagnetic current transducer arranged in accordance with the stator slots of the asynchronous motor. In addition, based on this algorithm, the time-dependent dynamic characteristics of the output signal of the sensitive element loops are determined according to the connection method. Based on the dynamic description, it is possible to see the time for the output voltage signal of the three-phase current transducer to reach stability and the effect of high harmonic currents affecting it. Theoretical studies were carried out in Matlab program, and practical studies were carried out using CassyLab device. The theoretical and practical results determined because of research are compared with each other.

1 Introduction

It is important to research three-phase electromagnetic current transducers for the control and management of filter-compensation devices of asynchronous motor reactive power based on the time-dependent characteristics of the values of the output signals. The parameters of the asynchronous motor and the sensitive element, as well as the time-dependent change of the output voltage signals, are determined by dynamic descriptions [1-3].

Dynamic descriptions combine the expressions formed by the graph model of the electromagnetic current transducer, FTEs and descriptions of the electromagnetic current transducer in different operating modes of the asynchronous motor. By researching the dynamic characteristics of an asynchronous motor electromagnetic current transducer, it is important to study the properties and characteristics of nonlinear signals caused by transient processes, because nonlinear and unbalanced signals have a negative effect on the operation of the control and management device.

Taking into account the interaction of the quantities with different characteristics affecting the electromagnetic current transducer, the dynamic characteristics of the voltage signal coming out of the sensitive element are studied, the differential equations characterizing the output signals and representing the transition processes are created. Determining the dynamic characteristics of an asynchronous motor electromagnetic current transducer is carried out on the basis of theoretical calculations, i.e., the formed graph model, its analytical expressions, simulation models developed using the Matlab program, and practical results determined by modern technologies [4].

A complete study of dynamic processes is achieved based on the analysis of the electrical and electromagnetic processes of the asynchronous motor based on the number and connection scheme of the sensitive elements of the electromagnetic current transducer. It is important to determine the dynamic characteristics of series, parallel and differentially connected electromagnetic current transducers with one sensitive element loop or two sensitive element loops and to compare them [5-7].

2 Experimental research

For the control and management of asynchronous motor filter-compensation devices, the number of sensitive element coils suitable for each phase of three-phase electromagnetic current transducers is taken at an equal value, as a result, it is possible to determine the amplitude and phase asymmetries generated between phases [8,10].

In the study of dynamic characteristics, a 4A63 type asynchronous motor with a rated power of P=250 W, with stator windings connected in a star form, was taken. The active resistance of the stator winding R_1=38.51 Ohm, the inductive resistance X_1=21.05 Ohm, the active resistance of the rotor $R`_2$=35.94 Ohm, the inductive resistance $X`_2$=21.05 Ohm, the

* Corresponding author: mohirbek2702@mail.ru

magnetizing resistance X_m=359.43 Ohm, the number of windings of the stator windings w_1= 169, the number of nominal revolutions n=1380 rpm.

It was found that at the nominal values of the asynchronous motor, the stator currents reach stability in the time interval t=0.07÷0.075 sec. The time of stator currents reaching stability can be seen through the simulation model created in the Matlab program and the oscillograms determined practically using the CassyLab device (Fig. 1 and Fig. 2).

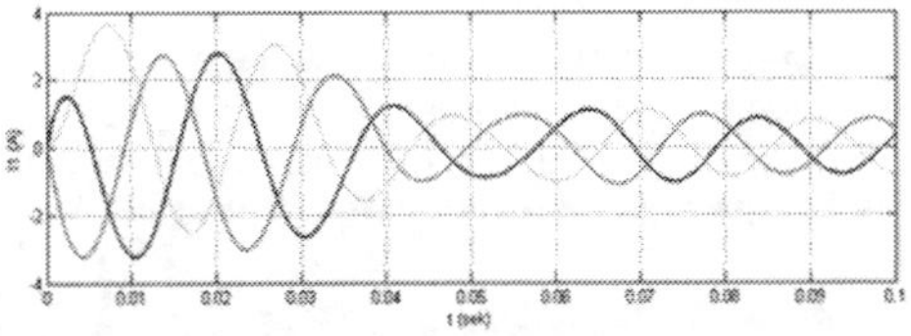

Fig. 1. Dynamic description of the stator currents determined by Matlab software through the switching scheme of an induction motor three-phase electromagnetic current transducer.

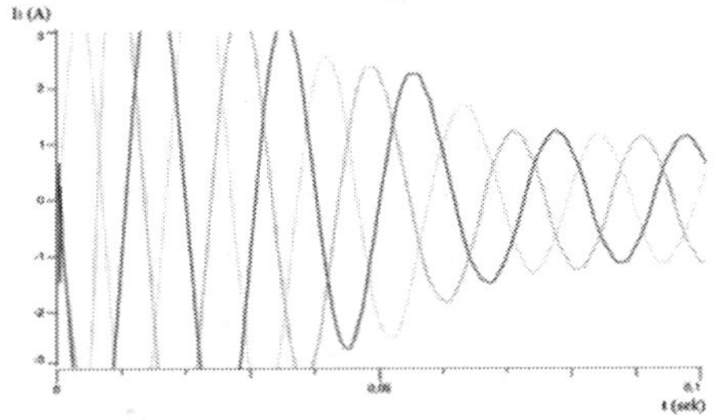

Fig. 2. Dynamic characterization of stator currents practically determined by the CassyLab device.

Mathematical expressions for determining the dynamic characteristics of the output voltages through the graph model of an asynchronous motor three-phase electromagnetic current transducer with one sensitive element loop suitable for each phase are given below:

$$\begin{cases} U_{a.out} = K_{U_{a.out}F_\mu} W(F_{\mu c}, F_{\mu x}) K_{I_A F_\mu}\left(I_{A.d} \sin \omega t + I_{A.n} e^{-\frac{t}{T}} \right) \\ U_{b.out} = K_{U_{b.out}F_\mu} W(F_{\mu c}, F_{\mu x}) K_{I_B F_\mu}\left(I_{B.d} \sin(\omega t + 120^0) + I_{B.n} e^{-\frac{t}{T}} \right) \\ U_{c.out} = K_{U_{c.out}F_\mu} W(F_{\mu c}, F_{\mu x}) K_{I_C F_\mu}\left(I_{C.d} \sin(\omega t - 120^0) + I_{C.n} e^{-\frac{t}{T}} \right) \end{cases}$$

where $I_{A.d}$, $I_{B.d}$, $I_{C.d}$, $I_{A.n}$, $I_{B.n}$, $I_{C.n}$ – are the periodic and non-periodic values of each phase current of the stator.

The dynamic description of the output voltages is determined using mathematical expressions defined by the graph model of an induction motor three-phase electromagnetic current transducer with two sensitive element loops suitable for each phase is given below:

$$\begin{cases} U'_{a.out} = K_{U'_{a.out}F_\mu} W(F'_{\mu c}, F'_{\mu x}) K_{I_A F_\mu}\left(I_{A.d} \sin \omega t + I_{A.n} e^{-\frac{t}{T}} \right) \\ U''_{a.out} = K_{U''_{a.out}F_\mu} W(F''_{\mu c}, F''_{\mu x}) K_{I_A F_\mu}\left(I_{A.d} \sin(\omega t - 180^0) + I_{A.n} e^{-\frac{t}{T}} \right) \\ U'_{b.out} = K_{U'_{b.out}F_\mu} W(F'_{\mu c}, F'_{\mu x}) K_{I_B F_\mu}\left(I_{B.d} \sin(\omega t + 120^0) + I_{B.n} e^{-\frac{t}{T}} \right) \\ U''_{b.out} = K_{U''_{b.out}F_\mu} W(F''_{\mu c}, F''_{\mu x}) K_{I_B F_\mu}\left(I_{B.d} \sin(\omega t - 60^0) + I_{B.n} e^{-\frac{t}{T}} \right) \\ U'_{c.out} = K_{U'_{c.out}F_\mu} W(F'_{\mu c}, F'_{\mu x}) K_{I_C F_\mu}\left(I_{C.d} \sin(\omega t - 120^0) + I_{C.n} e^{-\frac{t}{T}} \right) \\ U''_{c.out} = K_{U''_{c.out}F_\mu} W(F''_{\mu c}, F''_{\mu x}) K_{I_C F_\mu}\left(I_{C.d} \sin(\omega t - 300^0) + I_{C.n} e^{-\frac{t}{T}} \right) \end{cases}$$

where $I_{A.d}$, $I_{B.d}$, $I_{C.d}$, $I_{A.n}$, $I_{B.n}$, $I_{C.n}$ – are the periodic and non-periodic values of each phase current of the stator.

In the transient process of an asynchronous motor, its reactive power is also unstable. The output voltage of the three-phase electromagnetic current transducer is affected by the electrical and electromagnetic mechanical parameters of the asynchronous motor. The number of loops of sensitive elements placed in the stator slots, their interconnection methods and their location directly evaluate the sensitivity of the electromagnetic current transducer and the accuracy of the output signals [9].

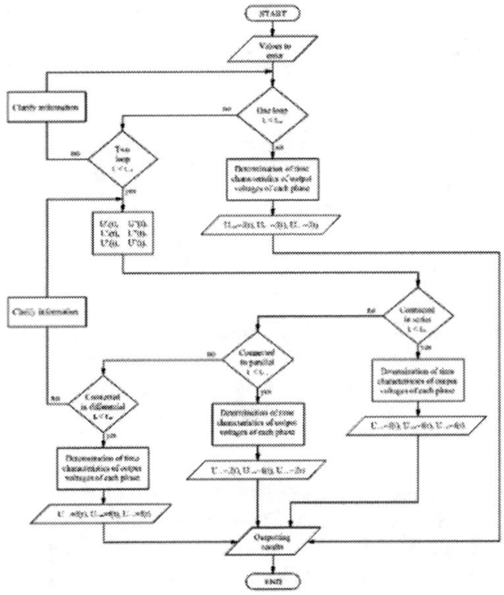

Fig. 3. Algorithm for studying the dynamic characteristics of the three-phase electromagnetic current transducer of an asynchronous motor.

Dynamic characteristics of three-phase electromagnetic current transducers with one sensitive element loop or two sensitive element loops suitable for each phase of the stator coils are studied based on the algorithm showing dynamic characteristics. The dynamic characteristics of series, parallel and differentially connected three-phase electromagnetic current transducers with one sensitive element loop suitable for each phase and two sensitive element loops are determined by the time of output signals reaching

https://doi.org/10.1051/e3sconf/202346101056

stability, location of loops in the stator wedges and connection methods (Fig. 3) [11].

3 Research results

Based on the above-mentioned expressions and algorithm, the simulation model of the three-phase electromagnetic current transducer with one sensitive element suitable for each phase, based on the commutation scheme, developed in the Matlab program and the dynamic descriptions determined by them are presented below (Fig. 4 and Fig. 5);

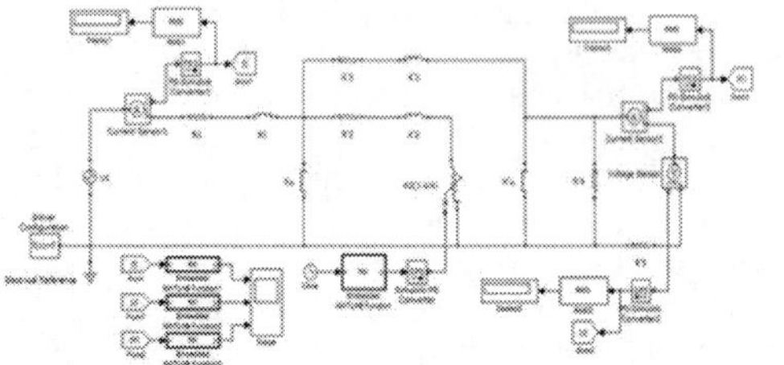

Fig. 4. A model for researching the dynamic characteristics of a three-phase electromagnetic current transducer with one sensitive element for each phase using the Matlab program.

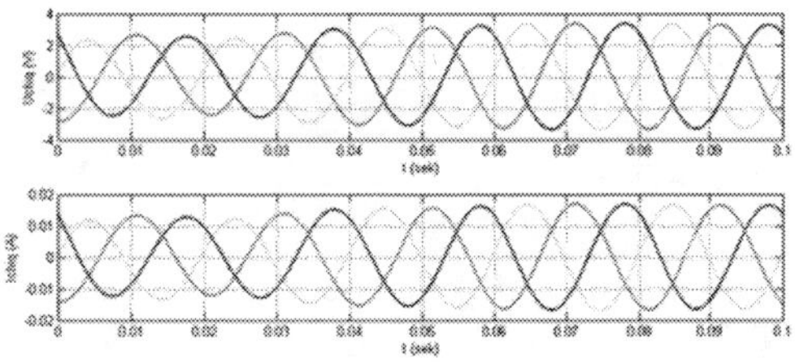

Fig. 5. Dynamic characteristics of a three-phase electromagnetic current transducer with one sensitive element for each phase determined using the Matlab program.

The simulation model for the study of three-phase electromagnetic current transducers developed using the Matlab program, with two sensitive element loops placed between the poles and in line with the stator coils, connected in series and parallel, and placed opposite to the slots where the stator coils are located and differentially connected, and their dynamic descriptions are presented below (Fig. 6, Fig. 7, Fig. 8 and Fig. 9);

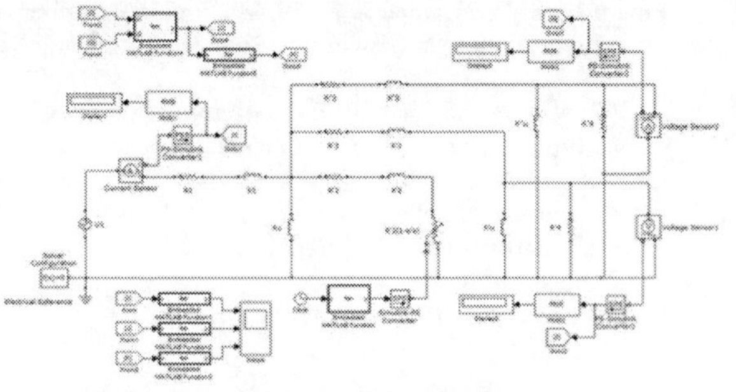

Fig. 6. A model for studying the dynamic characteristics of a three-phase electromagnetic current transducer with two sensitive element loops suitable for each phase using Matlab software.

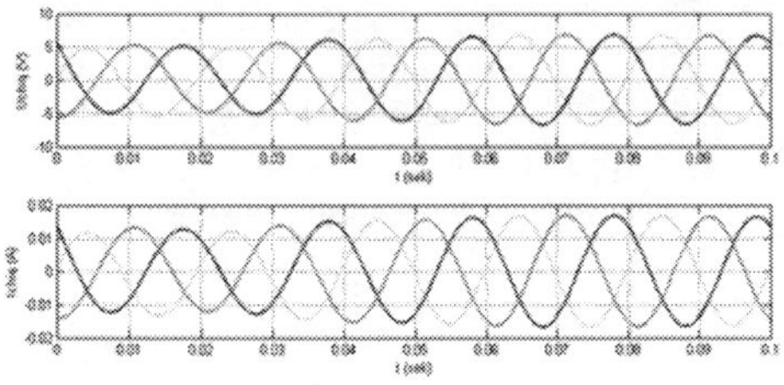

Fig. 7. Dynamic characteristics of a three-phase electromagnetic current transducer with series-connected sensitive element loops determined using the Matlab program.

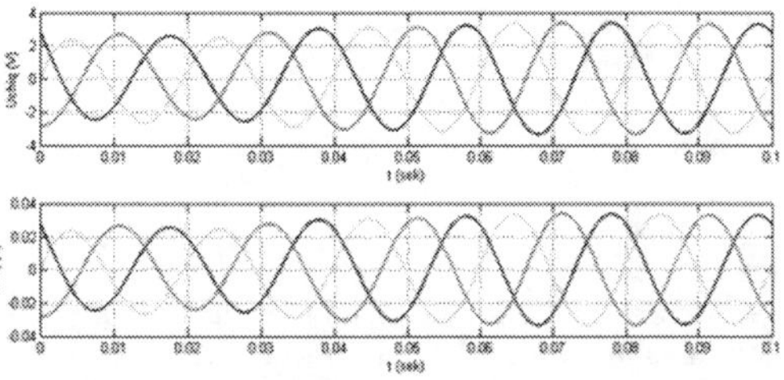

Fig. 8. Dynamic characteristics of a three-phase electromagnetic current transducer with parallel-connected sensitive element loops determined using the Matlab program.

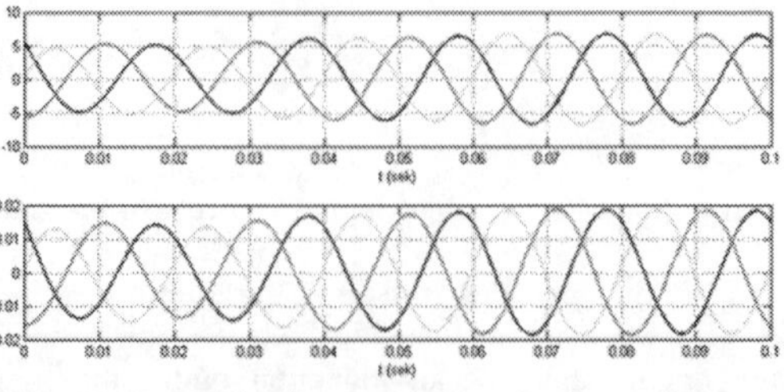

Fig. 9. Dynamic characteristics of a three-phase electromagnetic current transducer with differentially connected sensitive element loops determined using the Matlab program.

A graph of the time dependence of the output voltages determined by the CassyLab device based on the nominal values of a three-phase electromagnetic current transducer with one loop, suitable for each phase placed on the stator wedges, is presented (Fig. 10).

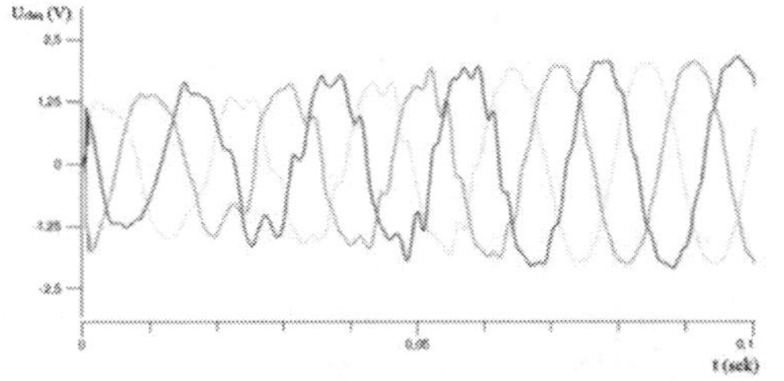

Fig. 10. A plot of the output voltages versus time of a three-phase electromagnetic current converter with sensitive element rings, determined practically by the CassyLab device.

Asynchronous motor three-phase electromagnetic current transducer loops with matching and two sensitive element loops for each phase placed in the stator slots are connected in series, parallel and differential and based on the nominal values, the graphs of the time dependence of the output voltages determined by the CassyLab device are presented below (Fig. 11, Fig. 12, and Fig. 13);

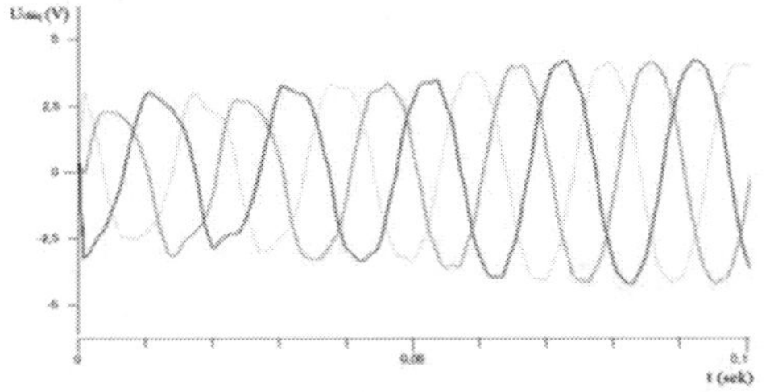

Fig. 11. Time-dependent graph of the output voltages of a three-phase electromagnetic current transducer with series-connected sensitive element loops, determined practically by the CassyLab device.

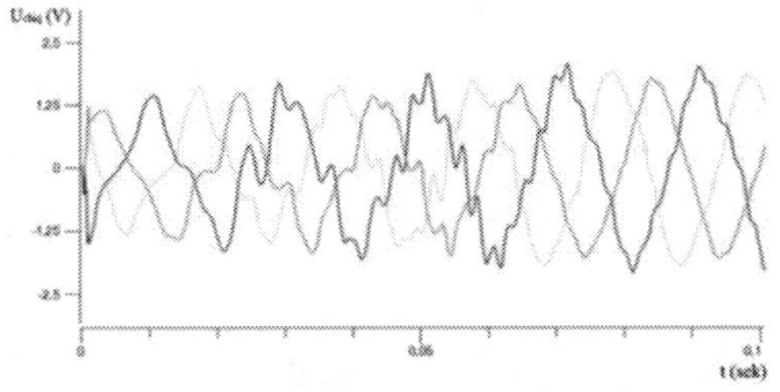

Fig. 12. Time-dependent graph of the output voltages of the three-phase electromagnetic current transducer with parallel-connected sensitive element loops, determined practically by the CassyLab device.

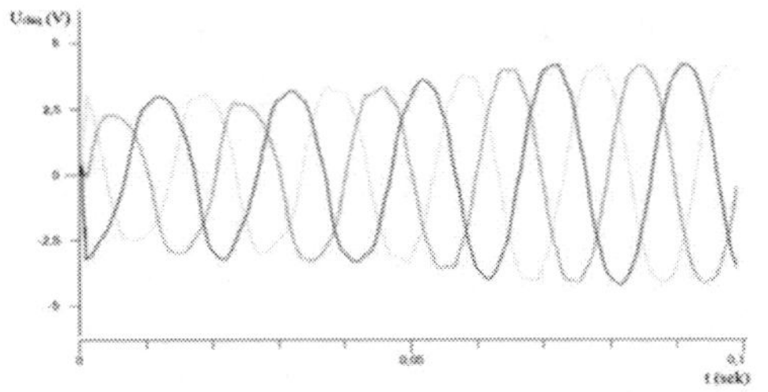

Fig. 13. Time-dependent graph of the output voltages of a three-phase electromagnetic current transducer with differentially connected sensitive element loops, determined practically by the CassyLab device.

During the start of the asynchronous motor, the number of rotations of the rotor and the inductance of the stator coils increase the stator current several times, as a result of the number of rotor revolutions approaching the nominal value and the interaction of magnetic currents, the stator currents become sinusoidal [12].

4 Conclusion

1. On the basis of theoretical calculations, simulation model of Matlab program and CassyLab device, based on practical results and graphs, asynchronous motor three-phase electromagnetic current transducer with one sensitive element loop and two sensitive element loops with serial, parallel and differential connected sensitive elements. voltages U_{out}=f(t) and stator currents I_1=f(t) based on the time dependence characteristics, it was determined that the start-up time of an asynchronous motor at nominal load reaches a steady state after t=0.07-0.075 seconds, depending on the parameters of the asynchronous motor.

2. Dynamic description of a three-phase electromagnetic current transducer with one sensitive element suitable for each phase, according to Figure 10, we can see that the output voltages are quantitatively small and the influence of high harmonics on the sinusoidal shape of the output signal, and the time for the output voltage to reach stability is t=0.068 seconds.

3. The dynamic description of the three-phase electromagnetic current transducer formed by the series connection of two sensitive element loops suitable for each phase, determined practically by the CassyLab device, according to Figure 11, the output voltages are close to the standard 5 V, the time for the output voltage to reach stability t =0.04 seconds, and we can see that the influence of higher harmonics on the output signal of the three-phase electromagnetic current transducer duloop the starting of the asynchronous motor is less than that of the single-phase current transducer.

4. Practical dynamic description of the output signal of an electromagnetic current transducer with two sensitive element loops placed on the interpolar heads

of the stator and connected in parallel. Based on Figure 12, the output voltage is equal to the output voltage of an electromagnetic current transducer with one loop, the effect of higher harmonics on the output signal and achieving stability It was determined that the time is t=0.072 seconds.

5. Based on Figure 13, the determined dynamic description of the output signal of the three-phase electromagnetic current transducer, which has loops of two sensitive elements placed in the slots where the stator coils are located and differentially connected, the value of the output voltages, the equality of the loops with the signals of the current transducer connected in series, according to the output signals of high harmonics we can see that there is little mystery and the time to reach stability is t=0.033 seconds.

6. Based on the results of research, we come to the conclusion that the three-phase electromagnetic current transducer with two sensitive element loops, placed opposite the wedges where the stator windings are located and differentially connected, has a short time to reach the stability of the output signals, and it was found that the influence of high harmonic currents on the output voltage signals is small. Since the output voltage signals of the investigated three-phase electromagnetic current transducers represent reactive power, it is effective to use three-phase currents as electromagnetic transducers for the control and management of filter-compensation devices of the reactive power of an asynchronous motor.

References

1. K.Siddikov, MA Anarbaev, and others. 2019 International Conference on Information Science and Communications Technologies (ICISCT), 2019, doi: 10.1109/ICISCT47635.2019.9011936.
2. Siddikov I.Kh., Karimjonov D.D., Abdigapirov A.A. Chemical technology. Control and management. International scientific and technical journal, ISSN 1815-4840, Toshkent, 2023. № (1)109, pp. 35-45.
3. Karimjonov D.D., Siddikov I.X., and others. IOP Conference Series: Earth and Environmental Science, ICECAE-2022, doi:10.1088/1755-1315/1142/1/012023.
4. Karimjonov D.D. Novateur publications. International journal of innovations in engineering research and technology [IJIERT] ISSN: 2394-3696 volume 10, issue 1, jan. -2023.
5. Siddikov , I.Kh. , Malikov , A., and others. AIP Conference Proceedings 2432, 020003 (2022); doi.org/10.1063/5.0089681 .
6. IK Siddikov , AA Abdumalikov and MT Makhsudov. 2020 International Conference on Information Science and Communications Technologies (ICISCT), doi: 10.1109/ICISCT50599.2020.9351482.
7. J. Abdubannaev , Y. Sun, and others. 2019 IEEE 3rd Conference on Energy Internet and Energy System Integration (EI2), doi: 10.1109/EI247390.2019.9062002.
8. Petrova, I., Zaripova, V. and others. E3S Web of Conferences, 97, 01023, 2019. doi: 10.1051/e3sconf/20199701023.
9. Safarov, A.M., Jumaboyev, S.K., Sattarov, K.A. IOP Conference Series: Materials Science and Engineering, 734 (1), 012196. doi: 10.1088/1757-899X/734/1/012196.
10. Siddikov, I.Kh., Sattarov, Kh.A., Khujamatov, Kh.E. 2017 International Conference on Information Science and Communications Technologies, doi: 10.1109/ICISCT.2017.8188581.
11. I.K.Siddikov, K.A. Sattarov and K. E. Khujamatov. 2017 International Conference on Information Science and Communications Technologies (ICISCT), doi: 10.1109/ICISCT.2017.8188581.
12. I.K.Siddikov and A.A.Abdumalikov. 2020 International Conference on Information Science and Communications Technologies (ICISCT), doi: 10.1109/ICISCT50599.2020.9351458.

Theoretical study of characteristics and mathematical model of convective drying of foods

Dj.E. Saparov[1]*, *S.A.* Sultonova[1], *E.C.* Guven[2], *D.I.* Samandarov[1], *A.M.* Rakhimov[1]

[1]Tashkent State Technical University named after Islam Karimov, 100095, Uzbekistan, Tashkent, University st. 2a
[2]Istanbul Technical University, Turkey.

Abstract. The characteristics of convective drying of potatoes at different temperatures, air velocity and slice thickness are investigated and a mathematical model is developed. The results show that the drying time decreases with increasing drying temperature. Compared with air velocity, drying temperature and slice thickness have a greater influence on the drying rate, and the period of velocity drop is the main stage of the convective drying process. As a model for convective drying of potato, the equations for 6 drying models are fitted and the optimal model is selected.

1 Introduction

Potato (*Solanum tuberosum L.*) is a root vegetable belonging to the nightshade family. After maize, wheat and rice, it is the most produced and consumed food in the world. It is reported that the major potato producing countries in the world are China, India, USA, Central Asia and Ukraine. Potato is defined as a food rich in vitamins, dietary fibre, minerals and essential amino acids. Potato, like many other vegetables, has a very short shelf life after harvest due to its high water content [1, 2].

Drying is one of the most commonly used methods of potato storage and processing in the world. Among drying methods, one of the most favoured methods for drying agricultural products is the hot air drying method. This method is based on the removal of water from food products using heated air, which reduces water activity and microbial growth, increases shelf life and prevents chemical reactions to a large extent. It also provides less packaging and lower transport and storage costs with reduced weight and volume [3, 4].

2 Materials and methods

2.1. Materials and equipment. Potatoes grown in Uzbekistan were chosen as the object of research. And for drying experiments, a laboratory convective drying unit developed at TSTU was used [5].

2.2. Test method. Single-factor experimental design. Hot air drying temperature, wind speed and slice thickness were selected as the study parameters, and an appropriate mathematical model was developed to study the effect of drying characteristics.

Each group of experiments was repeated three times.

1) drying temperature: to investigate the convective drying performance of potatoes, hot air drying temperatures of 50, 60 and 70 °C were set.

2) hot air drying speed: to investigate the potato drying performance, the wind speed was selected in three variations: 1.0, 1.5 and 2.0 m/s.

3) Slice thickness: the thickness of potato slices of 0.5, 1.0 and 1.5 mm to study the characteristics of convective drying of potatoes.

Measurement of indicators.

1) dry basis moisture content and drying rate: according to literature data [6].

2) Moisture content coefficient: according to literature data [7].

3) effective moisture diffusion coefficient: according to literature data [8].

4) activation energy: according to literature data [9].

Thin layer drying model. Thin-layer drying model, usually for materials less than 20 mm thick, fully exposed to the drying medium, is used to respond to the drying process of moisture with a change in equation time, through painstaking research on the former, a number of theoretical, semi-theoretical and empirical models have been summarised. In this experiment, 6 thin layer drying models were selected for analysis, as shown in Table 1 [10].

Table 1

Models of thin layer drying

№	Model name	Modelling equation
1	Verma	$MR = a\,exp(-kt) + (1-a)exp(-gt)$
2	Page	$MR = exp(-kt^n)$
3	Two-term	$MR = a\,exp(-kt) + b\,exp(-k_1 t)$
4	Wang and Singh	$MR = 1 + at + bt^2$
5	Midilli	$MR = a\,exp(-kt^n) + bt$
6	Henderson and Pabis	$MR = a\,exp(-kt)$

* Corresponding author: jasursafarov@yahoo.com

Data processing. The analysis and processing of the data obtained during the experiment were carried out using the state-of-the-art Comsol Multhipysics 5.5 software.

3 Results and analyses

3.1. Effect of temperature on convective drying characteristics of potatoes. As shown in Fig. 1(a), the drying time of potatoes to final moisture content was 210, 180 and 150 minutes at air velocity of 1.5 m/s, slice thickness of 1.5 mm and temperature of 50, 60 and 70 °C, respectively, and the drying time at 70 °C was reduced by 29.08% compared to 50 °C. The higher the temperature, the greater the change in the moisture content of the dry substrate and the shorter the time to the end of drying. From Fig. 1(b), it can be seen that the drying process of compound noodles has only a rate reduction stage associated with internal diffusion. At 70 °C, the drying rates are maximum and they have a positive correlation. In the pre-drying stage, the difference in drying rate is large, in the middle and late drying stages, the difference in drying rate obviously decreases, and in the late stage, there is almost no change.

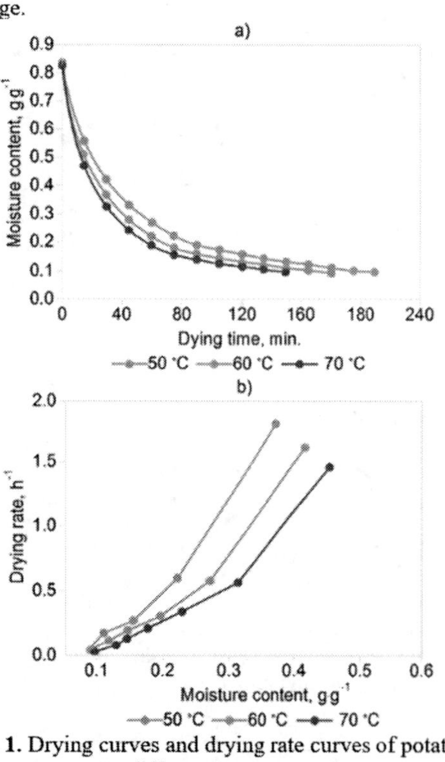

Fig. 1. Drying curves and drying rate curves of potatoes at different temperatures

Drying of materials refers to complex mass and heat transfer processes. In the early stage of drying, the water content of the liquid phase of the material is high and is in a continuous state, and the water transfer is mainly through liquid diffusion, heat flow and capillary flow, and the water in the surface layer of the compound noodle will evaporate first, and the water in the surface layer and inside will form a certain pressure gradient, and as the temperature of the hot air rises, the greater the momentum for mass transfer, the faster the drying speed. Inside and outside at the same time temperature difference will cause the chemical potential difference of water, that is, heat flow, this chemical potential difference will push the water from inside to outside along the flow, the greater the temperature difference, the greater the driving force [11].

At the same time, the temperature of hot air increases, so the internal temperature of the object increases, the surface tension of water decreases, resulting in increased capillary flow, increased mass transfer rate and increased drying rate. In the middle and late stages of drying, the moisture content of the object obviously decreases, at this time, the mass transfer is dominated by gas diffusion, the effect of temperature on it is insignificant, and the rate in the middle and late stages is practically unchanged. Thus, the temperature of hot air drying has a significant influence on the drying characteristics of potato, and the increase of temperature can obviously reduce the drying time.

3.2. Influence of air velocity on drying characteristics. As shown in Fig. 2(a), the drying times to final moisture point were 165, 150 and 135 min under conditions of 70 °C, slice thickness of 1.5 mm and speeds of 0.5, 1.0 and 1.5. The drying time was 12.5% shorter than at 0.5 m/s speed. The drying rate of potato increased with increasing air velocity. It can be seen from Fig. 2(b) that the drying rate of the object increased with increasing air velocity. In the hot air drying process, hot air can be used both as a heat transfer medium to transfer heat and as a moisture transfer medium to remove moisture from the material.

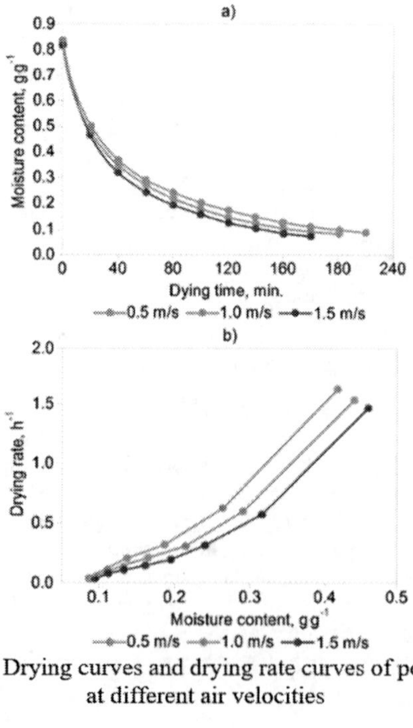

Fig. 2. Drying curves and drying rate curves of potatoes at different air velocities

 https://doi.org/10.1051/e3sconf/202346101057

Potato is a solid-phase material, its porosity is small, wind velocity can not penetrate into the interior, so when it is used as a heat transfer medium, heat is transferred to the material mainly by convection and acts only on the surface of the material, and the change in wind velocity of hot air has little effect on the heat exchange between the interior of the material and the solid frame. When it is used as a moisture carrier, the water evaporated from the material is mainly carried away by convection, and the wind velocity of the hot air cannot carry the moisture away from the material. When used as a moisture carrier, the moisture evaporated from the material is also carried away mainly by convection, and the wind velocity of the hot air cannot carry the moisture inside the material, and the mass transfer effect is not ideal. In addition, as the wind speed increases, the airflow turbulence on the surface layer of the material becomes larger, the boundary layer on the surface layer of the material becomes thinner, the diffusion resistance becomes smaller, the mass heat transfer is accelerated, and the drying rate increases. However, changing the air velocity can only change the drying rate of the surface moisture of the material, but has little effect on the internal moisture. Since potato is an internally diffusion driven material, the internal diffusion resistance is much greater than the surface evaporation resistance, so the effect of wind speed on drying rate is not as great as the effect of temperature [12].

3.3. Effect of potato slice thickness on drying characteristics. As can be seen from Fig. 3 (a), the drying times to final moisture content were 90, 105 and 135, min at a temperature of 70 °C, air velocity of 1.5 m/s and thicknesses of 0.5, 1.0 and 1.5 mm. The drying time of 0.5 mm thick material was 33.3% less than that of 1.5 mm thick potatoes. The smaller the thickness of the dried material, the greater the change in dry basis moisture content and the shorter the time to reach the final drying point.

From Fig. 3(b) it can be seen that the drying rate is maximum at a slice thickness of 0.5 mm, and as the thickness of the material increases, the drying rate decreases. In the drying process, the mass and heat transfer rate is also largely controlled by the drying process, and the thickness of the material changes the moisture transfer path. If the thickness is small, the heat brought by the hot air takes a shorter path to penetrate into the material, and thus the distance for moisture migration from the material is shortened and the drying time becomes shorter. If the material is thicker, the heat does not penetrate the material as easily, the path for water evaporation becomes longer and the internal diffusion resistance is greater, and the drying time increases.

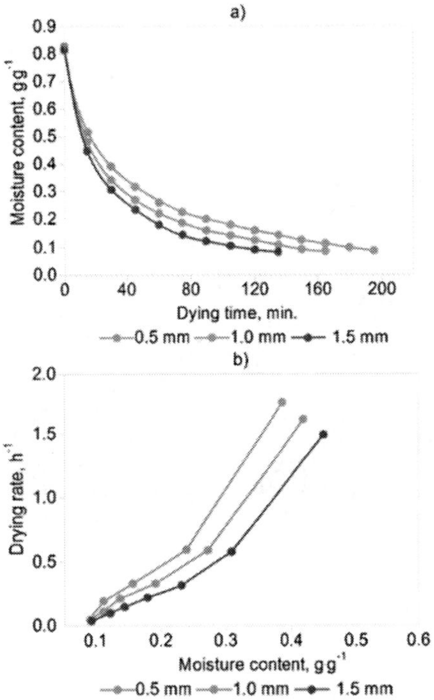

Fig. 3. Drying curves and drying rate curves of potatoes at different thicknesses

3.4. Development of mathematical models of convective drying. Selection of drying models. In this experiment, 6 thin layer drying models at 70 °C, air velocity of 1.5 m/s and thickness of 1.5 mm were selected to fit the data, and the corresponding values of the parameters R^2, χ^2 and RMSE are shown in Table 2. As shown in Table 2, the Midilli model had the highest R2 value of 0.9999 and the lowest output and RMSE of 8.4984×10-6 and 0.0012, respectively, which provided the best fit, which has the highest degree of fit, so this model was selected and validated.

Table 2

Drying parameters and modelling coefficients for different drying models

Models	R^2	χ^2	RMSE
Verma	0.9997	3.0118×10^{-5}	0.0049
Page	0.9995	4.1945×10^{-5}	0.0068
Two-term	0.9997	2.5910×10^{-5}	0.0045
Wang and Singh	0.8452	0.01513	0.0935
Midilli	0.9999	8.5071×10^{-6}	0.0012
Henderson and Pabis	0.9932	6.1623×10^{-4}	0.0284

Validation of the drying model. To ensure the accuracy of the selected model, experimental data different from the model were selected and fitted, and the results are presented in Fig. 4-6. From Fig. 4-6, it can be seen that the experimental data different from the model and the model prediction are mostly in agreement, indicating that the model fit is high, and hot air drying of potato can be predicted by the Midilli model. The relationship between drying time and moisture ratio can be traced by the model, which provides basic data for potato drying process optimisation and process control, and avoid unnecessary energy loss.

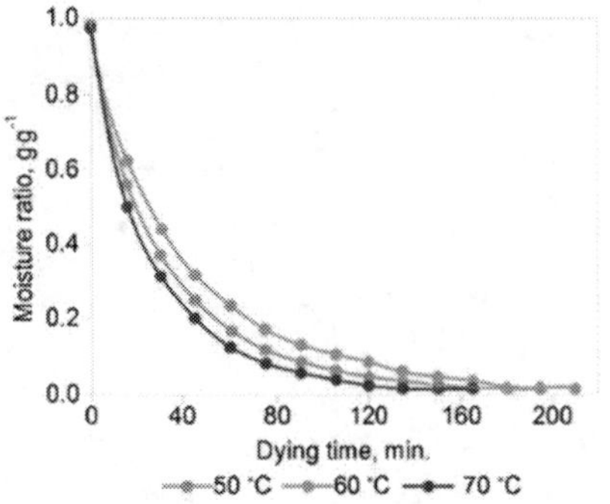

Fig. 4. Verification of the Midilli model of hot air drying of potatoes at different temperatures

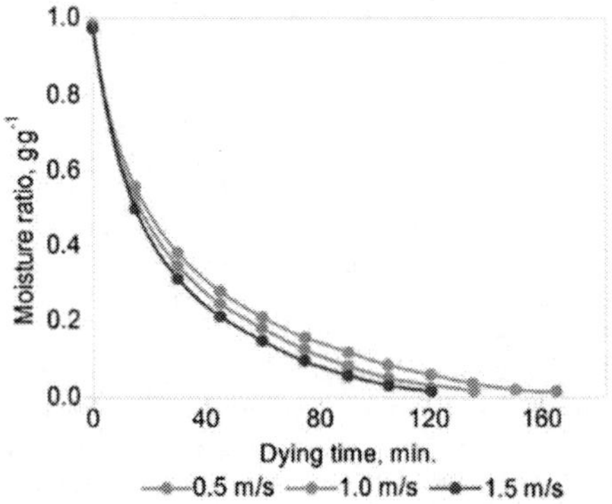

Fig. 5. Verification of the Midilli model of hot air drying of potatoes at different air velocities

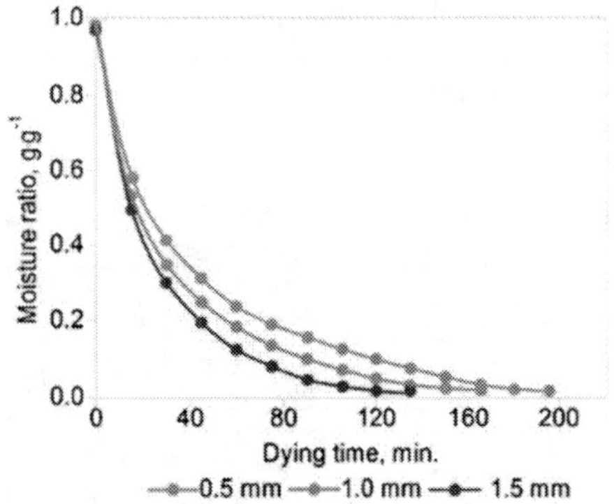

Fig. 6. Verification of the Midilli model of hot air drying of potato under different thickness conditions

Table 3

Effective potato moisture diffusion coefficients at different temperatures

Hot air drying temperature (°C)	Effective moisture diffusion coefficient ($m^2 \cdot s^{-1}$)
50	2.6718×10^{-10}
60	3.2271×10^{-10}
70	4.1933×10^{-10}

Table 4

Effective diffusion coefficients of potato moisture at different air velocities

Air velocity for hot air drying ($m \cdot s^{-1}$)	Effective moisture diffusion coefficient ($m^2 \cdot s^{-1}$)
0.5	3.2215×10^{-10}
1.0	3.8625×10^{-10}
1.5	4.1866×10^{-10}

Table 5

Potato effective moisture diffusion coefficients under different thickness conditions

Thickness of slices (mm)	Effective moisture diffusion coefficient ($m^2 \cdot s^{-1}$)
0.5	7.3209×10^{-10}
1.0	5.7275×10^{-10}
1.5	4.5115×10^{-10}

The temperature was 70 °C and the wind speed was 1.5 m/s. When the velocity was increased from 0.5 m/s to 2.0 m/s, the $D_{eff.}$ coefficient increased. This shows that increasing the drying temperature, air velocity and decreasing the thickness of the slices can enhance the mass-heat transfer properties of potatoes during hot air drying and thus increase the effective moisture diffusion coefficient of the material, but among the three influences of hot air drying, the influence of temperature and thickness is more significant, and the influence of air velocity is relatively small. The activation energy of potato drying was calculated from the slope of the straight line by linear fitting of lnD and 1/T curves and was 52.32 kJ/mol.

4 Conclusion

The results of this experiment show that hot air temperature and potato slice thickness have a significant effect on the drying rate, while air velocity has less influence. When the hot air temperature is 70 °C and the thickness of the slices is 0.5 mm, the drying speed obviously increases, while too high temperature may lead to the destruction of nutrients, so we can choose 70 °C and 1.5 mm thickness as the drying conditions. Thus, a drying temperature of 70 °C and a thickness of 1.5 mm can be selected as the drying conditions. The air velocity has little influence on drying and therefore 1.0 m/s is selected as the optimum wind speed condition considering the energy consumption. After fitting equations for 6 drying models, Midilli model was selected as the hot air drying model for potato. The effective moisture diffusion coefficient of potato $D_{eff.}$ increased with increasing drying temperature and wind speed, and with decreasing thickness. This experiment investigated the characteristics of potato hot air drying and provided a theoretical basis for its production drying control.

References

1. Wu B., Guo Y. et al. Journal of Food Engineering, **237**, 249-255, (2018), doi.org/10.1016/j.jfoodeng.2018.05.030.
2. Wang H., Liu Z.L. et al. Drying Technology, **39(3)**, 418-431, (2020), doi.org/10.1080/07373937.2020.1818254.
3. Tunckal C., Doymaz İ. Renewable Energy, **150**, 918-923, (2020), doi.org/10.1016/j.renene.2020.01.040.
4. Huang D., Men K. et al. Ultrasonics Sonochemistry, 63, **104950**, (2020), doi.org/10.1016/j.ultsonch.2019.104950.
5. Tarawade A., Samandarov D., Safarov J., Sultanova Sh. ICTACS, 825-830, (2022), doi: 10.1109/ICTACS56270.2022.9988323.
6. Liu Y.H. et al. Journal of Food Science and Technology, **52(8)**, 4955-4964, (2015), doi: 10.12691/ajfst-7-6-10.
7. Hi Q.L. et al. Journal of Food Engineering, **84(1)**,12-20, (2008), doi.org/10.1016/j.jfoodeng.2007.04.012.
8. Doymaz I. Journal of Food Processing and Preservation, **36(4)**, 310-319, (2011) doi.org/10.1111/j.1745-4549.2011.00594.x.
9. Guedes J.S. et al. Food Structure, 29, **100202**, (221) doi.org/10.1016/j.foostr.2021.100202.
10. 10. Hiranvarachat B. et al. Food and Bioproducts Processing, **89(2)**, 116-127, (2011), doi.org/10.1016/j.fbp.2010.03.010.
11. Tarawade A., Samandarov D.I., Azimov T.D., Sultanova Sh.A., Safarov J.E. IOP Conference Series: Earth and Environmental Science, 1112(1), **012098**, (2022), doi. 10.1088/1755-1315/1112/1/012098.
12. Sultanova Sh.A., Safarov J.E., Usenov A.B., Samandarov D.I., Azimov T.D. AIP Conference Proceedings 2507, **050005**, (2021), doi.10.1063/5.0110524.

Development of magneto elastic transducers of mechanical quantities invariant to destabilizing factors

Sulton Amirov[1]*, *Kamila* Jurayeva[1], *Zamira* Nazirova[1]

[1]Tashkent State Transport University, 100167, Uzbekistan, Tashkent, Temiryo'lchilar St. 1

Abstract. This article analyzes the transducers of mechanical quantities developed by various authors, invariant to destabilizing factors and ways to reduce the sources of errors affecting the main characteristics. At the same time, special attention is paid to the causes of various errors associated with the properties of the magnetic material, such as magneto elastic hysteresis and the nonlinearity of the transformation function. Additional errors caused by temperature changes and their studies have shown the possibility of reducing the temperature error by selecting the material of the magnetic circuit and choosing the optimal power supply modes. Fluctuations in the frequency and amplitude of the supply voltage are factors that significantly affect the sensitivity, while causing additional error, and the use of the classic Royer multivibrator scheme increases sensitivity.

1 Introduction

The development of automation makes it possible to create effective technical systems, which include control and control devices. An example of such systems are continuous production process control systems in the metallurgical, chemical, transport and other industries, where there is a process of weighing goods, sorting and storing products in the production process. In this regard, the main direction in the design of weighing and dosing means is the creation of weighing equipment capable of providing not only mass measurement – weighing with the necessary accuracy and speed, but also automatic control of technological processes, as well as their control and management.

Existing developments in instrument-making technology allow for the conversion of the measured value with a very high degree of accuracy. However, if during the weighing process the measured value is perceived by an elastic sensing element (SE) of poor quality, then, no matter how high the accuracy of further transformations, the characteristics of the elastic element will limit the accuracy of the device as a whole. Although elastic elements are simple mechanical parts, and many types of elastic elements have been known and widely used for many decades, their performance characteristics often do not meet the requirements imposed on them, which hinders the creation of devices of high accuracy classes. The growing requirements for a magnetoelastic transducers (MT) makes it relevant to develop converters invariant to destabilizing factors. At the same time, the solution to the problem lies in the study of external factors that explicitly or implicitly affect the accuracy characteristics. Therefore, special attention should be paid to the development of computational and experimental methods that will change the methodology of designing MT.

In magnetoelastic transducers (MT) using a sensing element (SE) that converts mechanical quantities (force, moment, pressure) into an electrical signal, external destabilizing factors (DF) act mainly on the SE, changing its physical characteristics. However, accurate measurement of the MT output signal based on the SE does not eliminate the influence of DF on the measurement result. Consequently, in order to improve the accuracy of the MT, it was necessary to develop transformations related to the processing of auxiliary measurements of the DF.

In transducers using various research methods (in particular, the energy-information model of circuits (EIMC)), a significant contribution to the overall error is made by the methodological component resulting from the difference between the converter conversion function and the mathematical model used, which is the basis of the principle of operation of the converter [1].

The enumerated disadvantages, without their elimination, limit the possibilities of the EIMC method. Therefore, the development of the theoretical foundations of the EIMC method in the field of control and control of mechanical quantities and the creation of MT invariant to DF on their basis is an important scientific problem.

* Corresponding author: ilider1987@yandex.ru

2 Problem statement

The development of transducers with high accuracy and stability in hard operating conditions, over a wide range of measured values and over long periods of time, is necessary both for converters based on the principle of operation on magneto elastic effects, and for information-measuring and control systems in general based on them.

3 Solving the problem

To solve this problem, it is necessary to identify the sources of errors and explore ways to reduce them. [2].

Analysis of existing MT has shown that the main sources of the basic error are: magneto elastic hysteresis and the nonlinearity of the transformation function.

Magneto elastic hysteresis appears with increasing and decreasing load at the output of the MT. The causes of this phenomenon are magnetic hysteresis and mechanical hysteresis. To reduce the first component, magnetically soft materials are used, and to reduce the second, materials with a large elastic limit are used. The hysteresis of the MT with stacked magnetic cores is 4-5% in the case of a tightened magnetic core and decreases to 1.5-2% when gluing the plates. A sharp decrease in hysteresis to 0.5% is observed when the MT is powered with a rectangular voltage of increased frequency, this phenomenon is accompanied by ultrasonic shaking [3].

According to the degree of nonlinearity of the characteristics, magneto elastic transducers have a significant value and are due to the fundamental nonlinearity of the dependence of induction on the field strength and on the magnitude of mechanical stresses. [4].

According to [5], the linearity error of MT made from various batches of metal or with insufficiently accurate compliance with heat treatment modes can vary within ± (1.5 – 2.5)%, in some cases, reaching values of the order of 5-7% or more.

Recently, much attention has been paid to the issues of reducing the zero signal and eliminating ambiguity [6].

Various circuit variants of compensation for the nonlinearity of the initial section of the transformation function are proposed [7].

A detailed analysis of the effect of the initial anisotropy of the magnetic core material was carried out in [8]. It is proposed to perform MT magnetic circuits with two pairs of protrusions (from all side faces) and an experimental determination of the direction of load application, which ensures unambiguity of the characteristic. However, this method is not always convenient due to the increase in the dimensions of the device and is completely unacceptable in the case of multi-element structures.

There are also works on the study of the influence of manufacturing inaccuracies (displacement of the centers of holes) on the course of the conversion function and improvements of the MT by manufacturing additional holes and introducing additional windings are proposed. It is revealed that a decrease in the influence of the initial anisotropy of the MT according to the proposed method also contributes to a decrease in the temperature error. The elimination of the nonlinearity of the initial section can also be achieved by preloading the converter [9].

The development of methods to eliminate the ambiguity of the initial section of the transformation function is not a complete task of the theory and practice of MT, because its solution contributes to a significant expansion of the working range of controlled efforts. At the same time, the most radical way to solve the problem is the constructive improvements of the MT, and not the complication of the output circuits.

At the same time, the dependence between the input and output values is ensured by the immutability of the sensitivity of the MT. If the sensitivity is constant ($S = const$), then each value of the input value corresponds to a single value of the output value y_i. The elastic element under this condition has no error. However, the condition ($S = const$) is never provided in practice [10], since the sensitivity of the MT is affected by various factors: climatic, mechanical loads, various fields, including gravity, time factor, etc. The measured value also affects the sensitivity of the MT. The action of each influencing factor causes the appearance of a corresponding error: 1) errors from non-linearity Δ_n (non-linearity of the MT characteristic). The reasons for the nonlinearity may be an ambiguous perception of the input value when it changes and elastic imperfections of the MT material. Elastic imperfections of the material also lead to an error from hysteresis (variation of readings); 2) multiplicative temperature error ΔS – sensitivity change due to temperature change. When the temperature changes, the geometric dimensions of the MT and the modulus of elasticity of the material change, which leads to a change in the sensitivity of the MT; 3) the error from the effect of linear accelerations Δ_l is manifested in the deformation of the MT under the action of inertia forces in the presence of linear accelerations. This error is additive in nature, since it shifts the transformation function (TF) parallel to itself by the amount of Δ_l; 4) the error from the effects of vibrations Δ_f. In this case, the corresponding vibration signals are superimposed on the output signal. This error is also additive in nature; 5) the time error Δ_B, which manifests itself in a change in the properties of the material from which the MT is made. As time passes, various kinds of phase transitions occur in the MT material, as a result of which both the modulus of elasticity, and hence the slope of the TF, and the geometric dimensions may change, which leads to a shift of the TF (a change in the initial level); 6) other errors from the effects of radiation, humidity, ambient pressure, etc. As a result of the appearance of these errors, the TF can have any value lying in the range of possible values. The width of this area and its position relative to the x and y coordinates determines the accuracy properties of the MT.

Due to errors, the unambiguity between the input and output values is violated. A single value of the input value x_i can correspond to many values of the output y_i. The actual TF in the presence of errors has the following form [11]:

$$y = \Delta_0 + S \cdot (1 + \delta_s) \cdot x$$

where, Δ_0 - is the additive error; δ_s - is the relative multiplicative error, x - is the factors (input parameters) of the technological process, S - is the sensitivity of the elastic sensing element.

Thus, the nominal input value, the nominal output value, the basic error and the components of the basic error completely determine the actual TF and are the metrological characteristics of the MT.

A significant proportion of the MT error is caused by additional errors due to temperature changes. According to [10], the change in the magnetization of the main types of electrical alloys is of the order of 2% for every 10 degrees and depends on the magnetic field strength and the magnitude of mechanical loads, most often to compensate for such a significant coefficient of influence, subtraction of the signal of the loaded converter from the signal of an unloaded compensation converter is used.

The output signal of the MT, which is a kind of differential transformer circuits, partially compensated by temperature, but nevertheless its changes with temperature have a noticeable value, so for example, in the case of using electrical steel, they are 1.3-1.7% by 10 degrees. It is impossible to use such a converter in a wide temperature range without taking additional measures. Some researchers consider the unevenness of the thermal expansion of plates with different rolling orientation to be the main source of the temperature error of the set MT [11].

Experimental studies of the temperature stability of monolithic magnetoanisotropic converters [12, 13, 14] have shown the possibility of reducing the temperature error by selecting the material of the magnetic core and choosing the optimal power supply modes up to (0.5 - 0.3)% by 10 degrees. However, the results achieved are still isolated, and even they cannot be considered satisfactory for operation in a wide temperature range. Recent studies show that when using the compensation method, the temperature error of the converter for every 10 degrees of temperature change varies within the following limits: (0-1.04)% for the 50KFVI alloy, (1.2-3.0)% for the 35KNM alloy, (0-2.6)% for the Yu12 alloy and (1.4 – 3.0)% for alloy 0X27Y5.

The magnitude of the error largely depends on the field strength and the magnitude of mechanical stresses, while the increase in the controlled load corresponds to an increase in the error. Due to the fact that the analysis of MTs of various types can be performed from a single position, the results obtained are of a general nature and are applicable to the construction of any varieties of MTs.

Other factors that significantly affect the additional error of the MT are fluctuations in the frequency and amplitude of the supply voltage. The relative error of the MT is 2-3 times greater than the relative fluctuations in the amplitude and frequency of the supply voltage, so the MT is always powered from special sources, in particular, from auto generators. At the same time, the additional error from the frequency change does not exceed (1-1.5)% by 10% of the frequency change.

Increasing the sensitivity of a low-effort MT can be achieved by using structures with a high load factor, using materials with low losses (Yu12, Yu14) and increasing the power frequency, etc. Two variants of MT have been studied: analog and frequency. In the analog version, the power supply is provided by a stabilized alternating voltage of rectangular shape. In the frequency variant, the classical scheme of the Royer multivibrator was selected from all possible circuit diagrams of autogenerators, which provides a linear dependence of frequency on the magnetic resistance of the core, the study was performed for CBM compressive and tensile forces in the range from 0 to 13.1 kgf.

The use of MTs in automatic control systems places increased demands on their dynamic characteristics [11]. The process of converting the input value - the controlled force - into the output electrical value is carried out in the MT sequentially in three stages: converting the measured force of the X_{input} into the mechanical voltage σ in the magnetic circuit, converting the mechanical stresses that have arisen in the magnetic circuit σ into a change in its magnetic characteristics $\Delta\mu$, and finally, converting changes in magnetic properties into an electrical signal of the X_{output}.

Therefore, the transfer function of the MT can be defined as the product of the transfer functions of the 1st, 2nd and 3rd links corresponding to three stages of transformation: mechanical, magneto elastic and electromagnetic. It was revealed [15] that the third stage of the transformation, electromagnetic, has the most significant effect on the speed of the MT, whereas the influence of the first two stages can practically be neglected.

4 Conclusion

1. The use of the energy-information model of the circuit for error analysis is difficult due to the complexity of the mathematical form of recording the transfer function and does not allow to express explicitly the influence of individual factors on the error of the MT.

2. The number of factors affecting the systematic error of the MT is large and diverse, it is not possible to obtain an equation of the dependence of the MT error on their total effect.

3. Errors of FT such as: errors from nonlinearity; multiplicative temperature; error from the effects of linear accelerations can have any value lying in the range of possible values. The width of this area and its position relative to the x and y coordinates determines the accuracy properties of the MT.

4. The mathematical model of the MT obtained with the help of EIMC is a convenient means of analyzing the properties, in the particular case of errors, of measuring instruments (transformations) with a complex transformation function. At the same time, it turns out to be quite simple to take into account various factors that affect the final error of the measuring instrument. The detection of a factor that introduces a large error in the measurement result is a signal indicating the need to make a correction in the structure of the measuring instrument or to tighten the requirements for the component of the measuring instrument that introduces a large measurement error.

References

1. Michnik Yu.O. Magneto elastic effect in the energy information model. Caspian Journal: Management and High Technologies No. 2 (2) 2008.

2. Jurayeva K.K., Nazirova Z.G., Mirasadov M.J. Identification and analysis of sources of errors of the magnetoelastic sensor. AIP Conference Proceedings, 2021, 2402, 060019

3. Antonets I.V., Tereshenok A.P. Methods of calculation and modeling of elastic elements. Textbook. Ulyanovsk UlSTU 2013 – 121p.

4. Jurayeva K.K., Sattarov Kh. Issues of investigation of the dependence of static characteristics of magnetoelastic converters of mechanical quantities on the influence of external. E3S Web of Conferences, 2023, 389, 01060

5. Zavarykin B.S., Nekhorosheva L.V., Murashkin S.I., Pavlov V.V. Elements of automation systems: textbook - Krasnoyarsk, 2005. - 172p.

6. Jurayeva K.K., Nazirova Z.G., Mamadaliyev U.S. Increasing the sensitivity of magnetoelastic sensors using the energy-information method. AIP Conference Proceedings, 2023, 2476, 020013

7. Piavchenko O.N., Udod E.V. Errors of intelligent pressure sensors. – Taganrog: Izvestiya TRTU, No. 3, 2007. – 131p.

8. Pleshev D.A. Investigation of nonlinear magnetoelastic dynamics in single- and multilayer magnetic films. Dissertation for the competition of the PhD. Syktyvkar – 2017.-128p.

9. Lopatin S.S., Panich A.E. Intelligent piezoelectric sensors in process control systems. – Taganrog: Izvestiya TRTU, No. 2, 2007. – 169p.

10. Timofeev I.A. Development of technologies for manufacturing powdered magnetic materials for electrical products. Abstract of the dissertation for the degree of Doctor of Technical Sciences. Moscow, 2009 – 62p.

11. Sokol-Kutylovsky O.L. Investigation of magnetoelastic interaction in amorphous ferromagnets for their application in magnetic and mechanical sensors. Abstract of the dissertation for the degree of Doctor of Technical Sciences. Yekaterinburg - 1997, 68p.

12. Dormidontov A.G., Lukin A.A., Perevoshchikov P.S., Shabalina E.V. Features of the formation of a highly coercive state in sintered magnets. – Abstracts of the XV International Conference on Permanent Magnets. M. 2005, p.118.

13. Lukin A.A., Dormidontov A.G., Shabalina E.V. The effect of heat treatment on the magnetic properties of permanent magnets of the NdRFeMB type. – Mining information and analytical bulletin. Thematic appendix: Functional materials. – Moscow: Publishing House of Moscow State University; 2005, pp. 421-431.

14. Gavrilyuk A.A., Semenov A.Yu., Mokhovikov A.L. Magnetic and magnetoelastic properties of amorphous ferromagnetic alloys treated with electric current/ ZhTF. – 2006. - Vol.76. - V.6. - pp.64-71.

15. Jurayeva K.K. Magneto elastic sensors for systems of monitoring and control of railway transport facilities. Dissertation for the competition of the PhD. Tashkent – 2019.

Control of modes of energy-saving vibro-drying devices

Sultonova *S.A.*[1*], Saparov *Dj.E.*[1], Mambetsheripova *A.A.*[2], Usenov *A.B.*[1], Alimova *D.Q.*[1]

[1]Tashkent State Technical University named after Islam Karimov, 100095, Uzbekistan, Tashkent, University St. 2A.
[2] Karakalpak State University, 742012, Uzbekistan, Republic of Karakalpakstan, Nukus, Ch. Abdirova St. 1.

Abstract. This paper proposes a method for optimizing the drying process of pectin hydrolyzate in the vibrofluidized bed of inert material, which allows to determine the values of the parameters of the regime at which the drying process continues with minimal costs for thermal and heat. It is based on the experimentally obtained relationship between the specific productivity of the drying equipment and the following parameters of the drying process: air movement speed in the drying chamber, initial air temperature, specific load depending on the drying area, gas distribution network , the initial concentration of dry matter. A complex criterion for the optimization of the vibrofluidized drying process has been developed, its minimization allows to increase the efficiency of the drying process. Restrictions on the ranges of changing the parameters of the drying process have been developed.

1 Introduction

The technological process of pectin production is energy-intensive and environmentally unsafe. Such a process traditionally includes the stages of washing apple pomace, extracting pectin by acid hydrolysis, separating the solid fraction, followed by treatment with ethyl alcohol, and drying [1]. In industrial conditions, large volumes of ethyl alcohol are required, as well as significant energy costs for its regeneration. One of the promising directions of the technological process of pectin production is the drying of the hydrolyzate directly before its precipitation with ethyl alcohol.

A promising direction for the modernization of the technological process of pectin production is the use of an intensive method of drying pectin hydrolyzate in a vibrating fluidized bed of inert granules [2]. Carrying out the drying process in a vibrofluidized layer of an inert material is associated with increased costs for thermal and electrical energy. In this regard, the urgent task is to find the optimal values of the regime parameters of the pectin hydrolyzate drying process in a vibrofluidized bed, which minimize the reduced costs, including the cost of steam and the cost of electricity. At the same time, in the optimization process, it is necessary to take into account a number of restrictions on the regime parameters of drying, due to the specifics of the technological process [3–6].

The purpose of the studies carried out in this work was to develop a methodology for optimizing the process of drying pectin hydrolyzate in a vibrofluidized bed, based on the criterion of minimizing the reduced costs for energy carriers. The proposed method is based on the dependence obtained experimentally, which determines the restrictions on changing the regime parameters of the drying process. The developed technique made it possible to determine the values of the regime parameters of the drying process, at which the minimum energy costs are achieved.

2 Methodology and research results

The limiting productivity of the vibrofluidized bed drying plant depends significantly on the parameters of the pectin hydrolyzate, including its initial concentration in terms of dry matter, and changes with varying temperature and air velocity in the drying chamber [3, 4]. Also, an important factor affecting the performance of the installation is its specific load. In order to identify the relationship connecting the listed mode parameters of the drying process, an experimental drying installation was developed, which includes the following elements (Fig. 1): drying chamber 1, fan 2, eccentric vibration drive 3, heater 4. The vibration drive provides vertical oscillatory movements of the drying chamber. Cubic fluoroplastic granules are placed on the gas distribution grate of the drying chamber. The fluidization of these granules is ensured both by fluctuations in the gas distribution grid and by supplying hot air to the drying chamber. Temperature control and regulation is carried out by an electronic potentiometer complete with a fast-response thermocouple. Before the product is fed into the drying chamber, it is preheated using a heat exchanger 6. From the container 8, the pectin hydrolyzate is pumped by a pump 7 into a pneumatic nozzle spraying the hydrolyzate in a layer of granules. The air consumption for spraying in all experiments was 11 m3/h. The product is deposited on the surface

* Corresponding author: sh.sultanova@yahoo.com

of the inert granules, forming a thin film. To reduce the viscosity of the product, before it was fed into the layer of inert material, it was preliminarily heated in heat exchanger 6 with hot water coming from a thermostat. After drying, the product loses its connection with the inert granules and moves with the air into the cyclone 5, which ensures the separation of the dried product from the air stream.

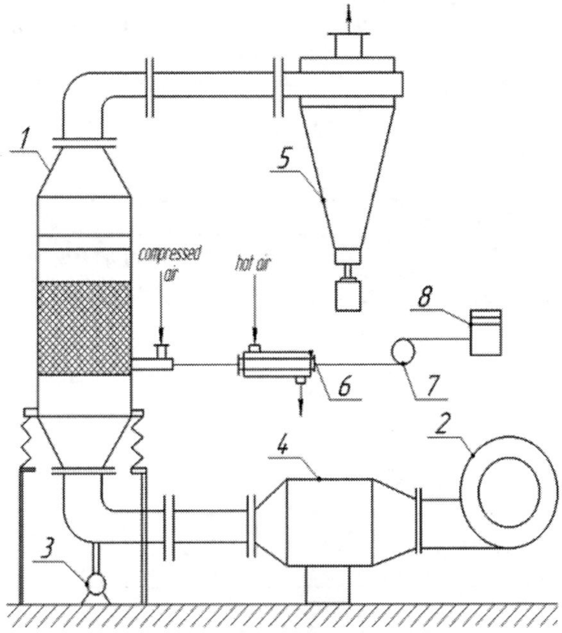

Fig . 1. Scheme of the experimental drying plant

In the course of a preliminary experimental study, it was shown that the process of drying pectin hydrolyzate in a vibrofluidized bed proceeds continuously. With an increase in the product flow rate, the nonuniformity of the boiling of the fluidized bed increases and sticky granules are observed, while the surface of the granules is not completely covered with the product. Therefore, to increase productivity, a vibrating gas distribution grate was used, which makes it possible to break agglomerates of granules. In the experiments performed , the vertical vibration frequency was 7.5 Hz, the amplitude was 8 mm. Drying was started with a low productivity with a further stepwise increase in the supply of pectin hydrolyzate with an interval of 30 minutes.

In the course of the experiments, the following regime parameters of the drying process acted as factors: air velocity in the drying chamber v , m/s, initial air temperature t 1, oC , specific load G , kg, related to the area of the gas distribution grid Fp , m2, initial dry matter concentration C, %. Accordingly, when conducting experiments, an orthogonal plan of the four listed factors, varied at four levels, was used. The plan is built on the basis of two ordered and two Latin squares of size 4 x 4 [4, 5]. Based on the results of these experiments, an empirical relationship was obtained between the performance of the installation and the controlled process variables

$$U_p = 1{,}47\cdot 10^{-7}(1{,}15G_p - 0{,}23)(0{,}38v - 0{,}026)(1{,}9t_1 - 41{,}63)(-3{,}12C_1 + 210{,}8) \quad (1)$$

where U_p=U/F_p , U is the ultimate performance of the plant, kg/h; F_p is the area of the gas distribution grid, m2; Gp = G / F_p,

G is the specific load, kg; v is the speed of air movement, m/s; t_1 - initial air temperature, ° C ; C_1 - initial concentration for dry matter, %.

The obtained dependence (1), as well as the recommendations outlined in [6, 7], allow us to formalize the following two indicators of the efficiency of the pectin hydrolyzate drying process in a vibrofluidized bed.

The first indicator is the reduced cost of thermal energy consumed by the vibrofluidized bed installation:

$$J_1 = 3600P_sC_a\rho_a \times \frac{(100-W_2-C_1)(t_1-t_0)}{U_p\eta_k r_s} \to min; \quad (2)$$

where $P_s$is the cost of steam, rub./kg; C_a- specific heat capacity of air, kJ/(kg/K); ρ_a- air density at the entrance to the drying chamber, kg/m ^{3}; W_2- humidity of dry product, %; t_0- ambient temperature, ^{0} C; η_k- efficiency of the heater; r_s- specific heat of phase transformation of steam, kJ/kg.

The second indicator is the reduced cost of electricity consumed by the vibrofluidized bed installation:

$$J_2 = P_e v(100 - W_2 - C_1)(G_p g - \Delta p)/U_p\eta_p r_d \quad (3)$$

where $P_e$is the cost of electricity, sum ./kW*h; g - free fall acceleration, m/s ^{2}; Δp- resistance of the air path of the dryer, Pa; η_p- fan efficiency; r_d- the same electric motor.

3 Research results

In accordance with the scalar convolution method [8, 9], we combine (2) and (3) into the following complex criterion for optimizing the drying process in a vibrofluidized bed:

$$J = J_1 + J_2 \to min \quad (4)$$

Thus, taking into account (4), the problem of optimizing the drying process in a vibrofluidized bed is reduced to finding the values of the regime parameters included in formula (1), which provide a minimum cost for thermal and electrical energy, with a maximum specific productivity per unit area of the gas distribution grid [10-19].

When minimizing the optimization criterion (4) , the following restrictions are imposed on the ranges of change of the regime parameters of the drying process in the vibrofluidized bed t_1, v, G_p and C_1:

air temperature at the inlet to the drying chamber

$$t_{1min} \le t_{1max} \le t_{1max} \quad (5)$$

where t_{1min}- 100 ° C; t_{1max}- 120 ° C - minimum and maximum air temperatures at the inlet to the drying chamber, the values of which are selected according to the conditions given in [2, 19-23];

air velocity in vibrofluidized bed apparatus

$$v_{min} \le v \le v_{max} \quad (6)$$

where v_{min}- 4.0 m/s; v_{max}- 5.5 m/s – minimum and maximum air velocities in the apparatus of the vibrofluidized bed, the values are selected according to the data of [2, 20];

– specific load of inert material granules

$$G_{p\ min} \le G_p \le G_{p\ max} \quad (7)$$

Where$G_{p\ min}$ - 140 kg / m ^{2} ;$G_{p\ max}$ - 170 kg/m ^{2} - minimum and maximum specific loads of cubic fluoroplastic granules with a fin size of 4 mm;

– dry matter concentration

$$C_{1\ min} \le C_1 \le C_{1\ max} \quad (8)$$

Where$C_{1\ min}$ - 5%; $C_{1\ max}$- 15% - minimum and maximum concentrations for dry matter according to [7].

Additionally, a limitation is introduced on the final moisture content of the dry product W_2

$$W_{2\ min} \le W_2 \le W_{2\ max} \quad (9)$$

where $W_{2\ max}$- 9%;$W_{2\ min}$ - 5% - the minimum and maximum allowable dry product moisture content according to technological conditions.

Also, according to technological conditions, the temperature of the product during drying should not exceed 75. Considering that the temperature of the product during drying in a fluidized bed practically coincides with the temperature of the exhaust air, the condition t_2 - $t_{2\ d\ occurs}$, where t_{2d} = 75 °C. In this case, the exhaust air temperature t_2 is related to the air temperature at the inlet to the drying chamber t_1 by the heat balance equation. In this equation, it is permissible to neglect the heat costs for heating the hydrolyzate because of their smallness in comparison with the costs for moisture evaporation. The equation for such a heat balance has the form

$$U_p = 3600 K_p \rho_a C_a v (t_1 - t_2)/r \quad (10)$$

Where K_p - heat loss coefficient; r is the specific heat of vaporization.

The value of the specific heat of evaporation is determined by the formula

$$r = r_0 + C_s t_2 - C_w \theta + \Delta r \quad (11)$$

where $C_w$is the specific heat capacity of water, kJ/(kgK); θ– product temperature, K; C_s– specific heat capacity of steam, kJ/(k*K); Δr is the specific heat of overcoming the bond between moisture and material, kJ/kg; r_0= 2500 kJ/kg.

Table 1. Optimal values of regime parameters of the drying process

Mode parameter	Optimal value mode parameter
Air velocity in the drying chamber v , m/s	5.5
Specific load of inert material granules G / F_p, kg/m ^{2}	170
The initial concentration of pectin on dry matter C_1,%	5
Exhaust air temperature t_2, ^{0}C	74.76

Thus, the problem of optimizing the mode of the drying process in a vibrofluidized bed is reduced to minimizing the quality criterion (4) under constraints (5)–(10). The problem posed was solved numerically by the method of sequential quadratic programming [10, 11, 19] using the MATLAB software package. Fig. 2 shows the analysis of the drying process to obtain pectin based on the experiments.

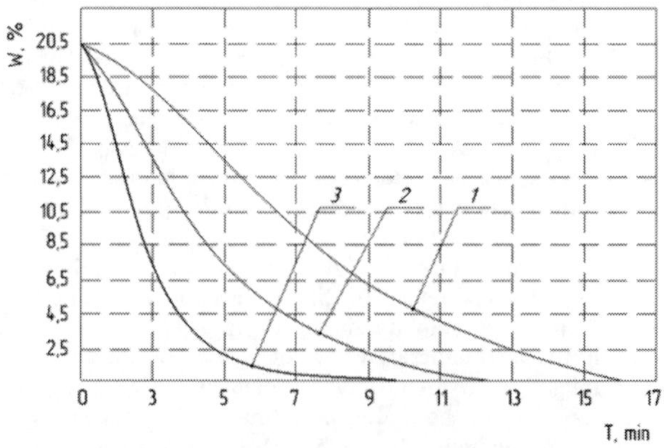

Fig. 2. Kinetic characteristics of the drying process in a vibrofluidized bed apparatus at a drying agent temperature of 60 °C(1), 70 °C(2), 80 °C(3).

Based on the results of solving the optimization problem, the values of regime parameters of the drying process were obtained (Table 1), at which the complex criterion (4) reaches a minimum.

4 Conclusion

A technique for optimizing the process of drying pectin hydrolyzate in a vibrofluidized layer of an inert material is proposed, which makes it possible to reduce the energy intensity of this process. An experimental dependence has been obtained that determines the restrictions on changing the regime parameters of the drying process. The optimization of the drying process based on the minimization of the complex criterion of reduced energy costs has been carried out. On the basis of the proposed method, the values of the regime parameters of the drying process are determined, at which the costs of thermal and electrical energy can be minimized.

References

1. K.G.Abidov, O.O.Zaripov, and ofters. AIP Conference Proceedings, 2552, **030023**, (2022), https://doi.org/10.1063/5.0112385
2. Liu Y.H. et al. Journal of Food Science and Technology, **52(8)**, 4955-4964, (2015), doi: 10.12691/ajfst-7-6-10.
3. Tunckal C., Doymaz İ. Renewable Energy, **150**, 918-923, (2020), doi.org/10.1016/j.renene.2020.01.040.
4. Hi Q.L. et al. Journal of Food Engineering, **84(1)**,12-20, (2008), doi.org/10.1016/j.jfoodeng.2007.04.012.
5. F.Westermann, J. S.Wei, M.Ringner, L.Saal, F.Berthold, M.Schwab, C.Peterson, P.Meltzer, J.Khan. GBM Annual Fall Meeting Halle (2002) https://doi.org/10.1240/sav_gbm_2002_h_000061
6. Arfan, M., Khan, R., Rybarczyk, A., & Amarowicz, R, 2012, International Journal of Molecular Sciences, **13(2)**, 2472–2480.
7. M.E.Adabi, A.M.Nikbakht, A.Motevali et al. Journal of Agricultural Science and Technology, (2012), **15(1)**: 23-34.
8. A.M.Castro, E.Y.Mayorga, F.L.Moreno. Journal of Food Engineering **223** (2018). pp.152-167.
9. S.Ebrahim, A.Ali Haghighi, M.Kamyar. Food Sci Nutr. (2019). **(7)**. pp.3589–3606.
10. M.Fazaeli, S.Yousefi, Z.Emam-Djomeh. Food Research Inte rnational, (2013), **50(2)**: 568-573.
11. Sultanova Sh.A., Safarov J.E., Usenov A.B., Samandarov D.I., Azimov T.D. AIP Conference Proceedings 2507, **050005**, (2021), doi.10.1063/5.0110524.
12. Guedes J.S. et al. Food Structure, 29, **100202**, (221) doi.org/10.1016/j.foostr.2021.100202.
13. S. Zhuang, G. Ravichandran, D. E. Grady. Journal of the Mechanics and Physics of Solids 51 (2003), pp. 245-265.
14. V. Bharti, Xu HS, G. Shanti, QM. Zhang, K.Liang. J Appl Phys (2000) **87** 452.
15. Y. Saito, H. Takao, T. Tani, T. Nonoyama, K.Takatori, T. Homma. Nature (2004) **432** 84–7.
16. S. Kalpat, X. Du, IR Abothu, A. Akiba, H. Goto, K. Uchino. Jpn J Appl Phys (2001) **40** 713.
17. Leriche, A. Aït-Kaddour, Foods **9(6)**, 724. (2020).
18. Rakhmonov, A. Berdishev, B. Khusanov, U. Khaliknazarov, U. Utegenov. IOP Conference Series: Materials Science and Engineering. **883(1)**, (2020).
19. Verma Sunil, A.B. Usenov, D.U. Sobirova, S.A. Sultonova, J.E.Safarov. IOP Conference Series: Earth and Environmental Science 17551307 (2022), doi: 10.1088/1755-1315/1112/1/012012
20. N.N. Merenkova, E.V. Sennova, V.A. Stennikov. Electronic Modeling, **6**, pp. 76-82, (1982).
21. S. Mitra, L Sun., I.E. Grossmann. Energy, **54**, 194-211 (2013).
22. S. M. Turabdjanov, H. Hasanov, A. Boboev, L. Yotova, International Journal BIOautomation (2011) **15(1)**, pp. 5-12.
23. N. R. Yusupbekov, A. R. Marahimov, Sh. M. Gulyamov, , H. Z. Igamberdiev. 4th International Conference on Application of Information and Communication Technologies, AICT 2010, 5612015, (2010) DOI: 10.1109/ICAICT.2010.5612015

E3S Web of Conferences **461**, 01060 (2023) https://doi.org/10.1051/e3sconf/202346101060
RSES 2023

Intelligent approach to power system management during well drilling in terms of environmental impact

Makhmudov *N.N.*[1*], Akhmedova *N.A.*[1], Sulitanofu *K.M.*[1]

[1]Tashkent State Technical University named after Islam Karimov, 100095, Uzbekistan, Tashkent, University St. 2A

Abstract. An autonomous hybrid power system is an intelligent power installation that provides electricity to isolated consumers without being connected to the power grid. The introduction of autonomous hybrid power systems in the oil and gas industry is considered as an option to improve energy and environmental security. As a rule, oil and gas fields are located remotely and off the grid, and therefore electricity is generated using diesel generators. This article discusses the possibility of reducing emissions of pollutants into the atmosphere by using an autonomous hybrid installation as a source of electrical energy.

1 Introduction

The energy for drilling is often provided by diesel generators. On average, the need for diesel fuel for drilling and construction of conventional oil wells usually ranges from 70 to 100 tons or more. Renewable generation can reduce or eliminate the need for a generator, which can lead to significant fuel savings.

2 Experimental research

On the example of the construction of two different groups of wells (Kashkadarya region, Mubarek Oil and Gas Production Department (OGPD)) using diesel units and power lines (power transmission lines) as a source of electricity, when analyzing the dispersion of pollutants in the atmosphere, it was found that hydrocarbon emissions were reduced and greenhouse gas emissions - carbon monoxide, nitrogen oxides, sulfur dioxide, as well as soot and formaldehyde (Figures 1,2).

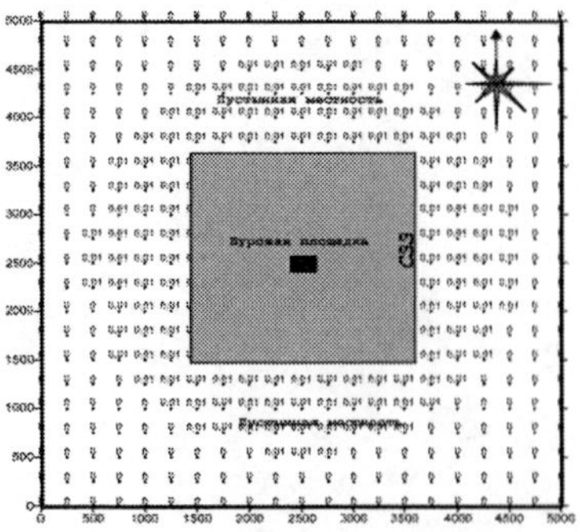

Fig. 1. Dispersion of NO_2 emissions during diesel-driven well construction

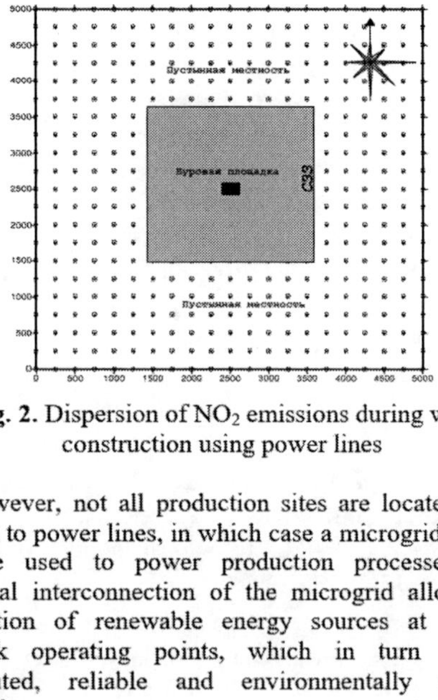

Fig. 2. Dispersion of NO_2 emissions during well construction using power lines

However, not all production sites are located close enough to power lines, in which case a microgrid system can be used to power production processes. The electrical interconnection of the microgrid allows the integration of renewable energy sources at various network operating points, which in turn ensures distributed, reliable and environmentally friendly generation.

The source of the microgrid is both power lines and its own generation sources, such as solar panels and wind generators, which can be installed on the roofs of residential cottages and dormitories for workers or administrative buildings. In addition, high-capacity batteries, as well as turbo expanders and other similar equipment of oil fields well pads and gas distribution lines can be used to convert mechanical energy into electricity with its further transmission to a microgrid with a controller that optimizes several sources, including environmentally friendly energy sources.

* Corresponding author: karolinasu@list.ru

Table 1.

Stratigraphic unit	Depth, m		Standard description of the rock: full name, characteristic features (structure, texture, mineral composition, etc.)
	from (roof)	to (bottom)	
Neogene + Quaternary	0	~150	Variegated continental clays, loams, limestone, sands, sandstones
Paleogene:	150	~300	
Suzak layers	150	~210	White clays, layered sandstones
Bukhara layers	210	~300	Cavernous light gray limestone
Cretaceous:	300	~1400	
Upper Cretaceous Senonian	300 300	~1400 ~800	Sandy, dense gray clays, layered with interlayers of sandstones, siltstones
Turonian VIII horizon	800 900	~1200 ~1100	Fine-grained and highly argillaceous sandstones. The clays are sandy, grayish-green with interlayers of shell rock and siltstone.
Cenomanian IX horizon X horizon	1200 1200 1360	~1400 ~1300 ~1400	Sandstones, siltstones with interlayers of clays, gray, dark gray, in the lower part. Sandstones, layers of clay and siltstones. Alternation of sandstones and siltstones with interlayers of clays.
Lower Cretaceous:	1400	~2300	
Albian XI horizon	1400 1450	~1800 ~1600	In the lower part, the clays are light gray, gray, silty; in the upper part, there is an interbedding of dense, inequigranular gray sandstones, clays, and siltstones.
Neocomian – Aptian	1700	~2200	In the lower part there are sandstones, clays, siltstones, brown red ones. In the middle part there are interlayers of sandstones and clays. In the upper part there are dark gray sandstones, fine-grained with interlayers of clays.
XII horizon	1750	~1950	
XIII horizon	2050	~2100	
XIV horizon	2100	~2250	
Jurassic:	2250	~2950	
Tithonian	2250	~2760	Salt-anhydrite stratum is represented by:
Upper anhydrites	2250	~2260	Dense whitish-gray anhydrites.
Upper salts	2260	~2640	Gray halite with an admixture of red-colored clays.
Medium anhydrites	2640	~2680	
Lower salts	2680	~2745	Dense anhydrites of gray color.
Lower anhydrites	2745	~2760	Halite, light gray.
Oxfordian-Kimmeridgian	2760	~3000	Gray, dark gray limestone, polythomorphic, fine-grained.

Let's consider an example of construction of 8 wells in the Alan field. Administratively, the Alan field belongs to the Mirishkor region and is located in the eastern part of the Kashkadarya region.

The field involves Jurassic, Cretaceous, Paleogene and Neogene-Quaternary sediments in its geological structure (Table 1).

The main gas-bearing horizon of the Alan field is the Jurassic sediments (Oxfordian-Kimmeridgian), consisting of fine-grained porous-fractured limestone with high productivity. The gas-water contact is located at a depth of about 3000 m. The porosity of reservoir rocks varies over a wide range - from 8% to 25%. The average porosity according to the study of core material and field geophysics is 17%. The coefficient of permeability of reservoir rocks varies within 0.1 - 0.3, which indicates rather high filtration properties of reservoir rocks.

Exact information on the depths of the formations is not provided due to an agreement on non-disclosure of data for official use.

Currently, intensive development of hydrocarbon deposits is underway in this region, which has a certain technogenic impact on the environment.

The area is located in a desert, waterless, uninhabited area. Due to the lack of an extensive irrigation network in the area under consideration and the lack of fresh water, crops are not currently sown. The lands are managed by karakul farms and are used as pastures for livestock.

A number of gas-oil and gas-condensate fields are located near the studied field: Kokdumalak, Zevardy, Kultak, Southern Pamuk, etc. The nearby territory is crossed by a network of gas pipelines going to the Mubarek Gas Processing Plant (GPP) from the Urtabulak-Dengizkul, Kokdumalak, Pamuk, Zevardy and Kultak fields.

In the north, the Northern Urtabulak-Oil loading platform "Sulfur Plant" oil pipeline passes, as well as a water pipeline and a communication cable.

The Alan field itself houses gas production wells, well production collection and transportation systems, and process equipment, which are the main sources of pollutant emissions into the atmosphere. In accordance with the regulatory data of the industrial sites of Mubarekneftegaz LLC (Kashkadarya region), the total atmospheric emission from the Alan field is 61.783746 t/y. An analysis of the surface concentration fields at the Alan field showed that there is no excess of established standards for ingredients emitted by existing sources, however, with the expansion of the developed area, measures to reduce emissions will be required.

Next, two options for managing energy resources when drilling wells are considered.

3 Research results

Option 1 – drilling 8 wells using a ZJ 30 DBS drilling rig.

The well construction cycle consists of the following successive technological operations: construction and installation, preparatory work for drilling; drilling and well casing; well development (testing); dismantling of drilling equipment.

Type of drive of drilling rig ZJ 30 DBS – diesel. Power supply sources will be a 292 kW diesel generator and a 100 kW diesel generator for power supply of amenity premises.

Combustive-lubricating materials (fuels and lubricants) will be used at the stages: rig installation works, preparatory work for drilling, drilling and well casing, well development.

Calculation of the daily need for fuel and lubricants is calculated by the formula:

$$\frac{N \times K \times \rho \times t}{1000}; \quad (1)$$

where:

N – consumption rate of fuel and lubricants;

K – the coefficient of fuel and lubricants use over time;

ρ – the density of diesel fuel, g/cm3;

t – time, h.

The need for fuel and lubricants for various operations is shown in Table 2.

Table 2.

Demand for fuels and lubricants for drilling rig engines, tons			
Total	including		
	fuel	oil	lubricant
1. Derrick installation works			
2.7	2.66	0.04	-
2. Preparatory work for drilling			
1.503	1.44	0.063	-
3. Drilling and well casing			
87.52	83.62	3.9	-
4. Well development			
4.70	4.54	0.16	-

In total, the construction of wells will require about 96.4 tons of fuel and lubricants.

The calculations were carried out in accordance with NGH 39.0-140:2012 "Methodology for calculating emissions of pollutants into the atmosphere for oil and gas producing and oil refineries" JSC "O'ZLITINEFTGAZ" [9].

The result of the calculation of pollutant emissions is shown in Table 3.

Table 3.

Ingredient name	Emissions			
	From 1 well		From 8 wells	
	g/s	t/y	g/s	t/y
Carbon monoxide	1.122	5.031	8.972	40.249
Nitrogen dioxide	0.784	3.517	6.272	28.135
Hydrocarbons	0.489	2.185	3.908	17.481
Sulphur dioxide	0.120	0.5370	0.958	4.298
Soot	0.076	0.342	0.610	2.735
Formaldehyde	0.016	0.073	0.131	0.586
Inorganic dust	0.003	0.002	0.023	0.017
Iron dioxide	0.001	0.001	0.006	0.010
Fluorides	0.0001	0.0002	0.0009	0.001
Silicon dioxide	0.0001	0.0002	0.0009	0.001
Hydrogen fluoride	0.0001	0.0001	0.001	0.001
Manganese dioxide	0.00003	0.0001	0.0003	0.0005
Benzopyrene	0.0000	0.000005	0.00001	0.00004
Total:	2.610	11.690	20.884	93.517

Thus, the input of pollutants during the operation of the designed facility will be about 93.517 tons of pollutants of 13 types. The main contribution is made by carbon monoxide, nitrogen dioxide and hydrocarbons.

Option 2 – drilling 8 wells using an off-grid microgrid hybrid power plant, allowing a combination of renewable energy sources, in this case solar panels and diesel-powered units.

The rig mainly provides power to the drilling rig during drilling, casing and completion of the well. The amount of energy consumed will be 548,700 kWh.

In the second option, fuels and lubricants will be used mainly in technical units to reduce the effect of friction during rigging, drilling and well casing. The total consumption of fuels and lubricants will be 14.52 tons for 8 wells (table 4).

Table 4.

Demand for fuels and lubricants for drilling rig engines, tons			
Total	including		
	fuel	oil	lubricant
14.52	14.29	0.23	-

The result of the calculation of pollutant emissions is shown in Table 5.

Table 5.

Ingredient name	Emissions			
	From 1 well		From 1 well	
	g/s	t/y	g/s	t/y
Inorganic dust	0.002	0.003	0.023	0.026
Hydrocarbons	0.052	0.424	0.413	3.391
Iron dioxide	0.001	0.001	0.007	0.010
Manganese dioxide	0.00004	0.0001	0.0003	0.0005
Silicon dioxide	0.0001	0.0002	0.001	0.001
Fluorides	0.0001	0.0002	0.001	0.001
Hydrogen fluoride	0.0001	0.0001	0.001	0.001
Total:	0.056	0.429	0.446	3.431

Thus, during the construction of wells according to the second option, emissions will amount to about 3.431 tons of pollutants of 7 types and the reduce will amount to 96.3% compared to the first option.

The next fundamental step is modeling and optimizing the operation of an autonomous hybrid power plant. As an example, Figure 3 shows a microgrid diagram created in the HOMER program. The system uses two diesel generators (operating and standby) rated at 292 kW, a photovoltaic generator and a wind turbine. It's also necessary to consider AC/DC converters and inverters when connecting generators to the microgrid. The costs for the required power electronics were easier to model with this arrangement. Due to the relatively weak wind potential of the site, the wind power plant can be replaced by solar panels.

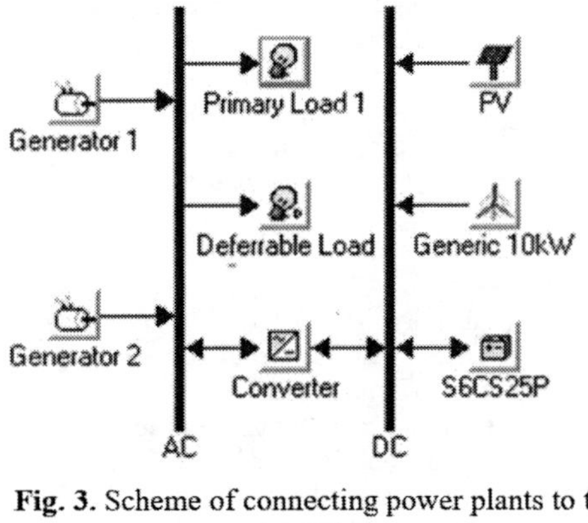

Fig. 3. Scheme of connecting power plants to the microgrid

4 Conclusion

To date, the approach under consideration is successfully implemented at remote bases, in settlements and in isolated areas, as it makes it possible to adapt activities to the industrial process, taking into account known energy needs and available energy resources. An intelligent approach to power system management improves the environmental situation at field sites, ensures reliable power supply and optimizes operation at a lower cost.

References

1. Energy and sustainable development issues. Review. UNDP, United Nations Department of Economic and Social Affairs, World Energy Council. USA, 2000.
2. Scoping studies on the development of a national strategy for the development of renewable energy in Uzbekistan. UNDP. - Tashkent, 2005.
3. UNFCCC (2008). Kyoto Protocol Reference Manual. Bonn, 2008.
4. Kulikova I. Yu., Dzerzhinskaya I. S. Protection of the environment in the oil and gas complex. - 2008. - Issue. 5. — P. 24–29.
5. N. Matchanov - Development of renewable energy in Uzbekistan: current state, problems and ways to solve them - 2019 - P. 9-10.
6. Report on the research work of Uztransgaz JSC. Review and analysis of the environmental performance of the facilities of Uztransgaz JSC for 2017-2019 Inventory and assessment of greenhouse gas emissions. Measures to reduce greenhouse gas emissions. - 2019. - P.14-16.
7. Halabi, M, A Al-Qattan, and A Al-Otaibi. 2015. "Applications of Solar Energy in the Oil Industry--Current Status and Future Prospects." Renewable and Sustainable Energy Reviews 43: 296-314.
8. Karras, Greg. 2010. "Combustion Emissions from Refining Lower Quality Oil: What is the Global Warming Potential." Environmental Science 44: 9584-9589.
9. Decree of the Cabinet of Ministers of the Republic of Uzbekistan No. 541 dated September 7, 2020 "On further improvement of the environmental impact assessment mechanism".
10. NGH 39.0-140:2012 "Methodology for calculating emissions of pollutants into the atmosphere for oil and gas producing and oil refineries." JSC "O'ZLITINEFTGAZ".
11. The Sanitary Regulations and Standards No. 0293-11 "Hygienic standards list of maximum permissible concentrations (MPC) of pollutants in the atmospheric air of populated areas on the territory of the Republic of Uzbekistan". Tashkent, 2011.
12. Regulations on the procedure for the development and approval of draft environmental standards. Approved by order of the Cabinet of Ministers of the Republic of Uzbekistan dated January 21, 2014 No. 14. Tashkent.
13. SanPiN No. 0179-04. List of maximum allowable concentrations (MPC) of pollutants in the atmospheric air of populated areas on the territory of the Republic of Uzbekistan.
14. KMK 2.01.01-94 "Climatic and physical-geological data for design. State Committee of the Republic of Uzbekistan for Architecture and Construction. Tashkent, 1994.
15. IPCC Guidelines for Compiling National Greenhouse Gas Inventories. Reporting Instructions for the Inventory of Greenhouse Gases. - 1996, Volume 1.

E3S Web of Conferences **461**, 01061 (2023) https://doi.org/10.1051/e3sconf/202346101061
RSES 2023

The role of energy security in forming the foundations for sustainable development of fuel and energy complex enterprises

Gulchekhra Allaeva[1*]

[1]Tashkent State Technical University named after Islam Karimov, 100095, Uzbekistan, Tashkent, University St. 2A

Abstract. The article considers energy security as a factor in the sustainable development of the fuel and energy complex of the Republic of Uzbekistan. The main types of threats at energy enterprises and the types of sources of risks and threats to the sustainable development of the fuel and energy complex are considered. A comprehensive mechanism for monitoring and preventing threats to sustainable development and energy security of the fuel and energy complex is proposed. Evidence-based proposals have been developed for the efficient and safe sustainable development of the sectors of the energy sector of the national economy.

1 Introduction

The implementation of the security strategy involves the identification, systematization and classification of events that may pose a threat to the fuel and energy complex, the regional economy and the population. The topic of efficient use of energy resources in all types of economic activity is currently extremely relevant for the Republic of Uzbekistan. One of the main priorities of the Concept for providing electricity to the Republic of Uzbekistan for 2020-2030, approved by the Government of the Republic of Uzbekistan on May 4, 2020, is to increase the energy efficiency of the economy of Uzbekistan and reduce the energy intensity of GDP in the context of increased energy security.

2 Experimental research

The main task of the fuel and energy complex is to ensure the sustainable functioning of the economy and support the livelihoods of the population of the region. The threat in the theory of energy security plays an important role. A threat is a potential risk that arises from a set of known factors and conditions. Awareness of the risk can lead to economic damage and environmental pollution. A threat is the possibility of developing an unfavorable situation or a series of events that can disrupt the normal operation of enterprises in the fuel and energy complex or cause them any damage (Figure 1).

The state of equilibrium of the system is influenced by various factors and conditions. An important task is to identify the causes (factors) of threats. To date, the following threats and risks have been identified that affect the sustainable development of the fuel and energy sector:

1) people (various criminal organizations, terrorist groups, etc.);

2) enterprises of the fuel and energy complex, facilities of other industries;

3) various force majeure situations and natural phenomena.

Realization of energy threats leads to the following.

- supply of fuel and energy resources to consumers with interruptions or stoppage of the supply of fuel resources;
- failure of production facilities;
- Significant deterioration of living conditions;
- various cascade accidents;
- economic losses;
- various socio-economic upheavals up to the occurrence of force majeure situations in the form of revolutions, protest marches, etc.

Our studies have shown that all the emerging risks and threats to the sustainable development of the fuel and energy complex can roughly be divided into two areas. One group of threats arises under the influence of destabilizing factors that transfer the fuel and energy complex to an emergency (force majeure) mode of operation. During the normal operation of the fuel and energy complex, a group of threats (in real or potential form) with varying degrees of risk may be present. Note that all groups of threats and risks can lead to various kinds of damage to

* Corresponding author: ilider1987@yandex.ru

https://doi.org/10.1051/e3sconf/202346101061

consumers due to insufficient supply of fuel and energy resources.

The main conclusion of our research is that any system (social, technical, economic) must work stably despite the impact of random circumstances on it. That is, in general, the stability of the system is the ability of the system to return to its basic state after the elimination of all negative effects. We believe that in order to maintain stability in a constantly changing market environment, the balance of the fuel and energy complex should be dynamic. However, the balance can be disturbed at any moment and an imbalance of the main components of the system can occur. In a normal situation, the fuel and energy complex should be in a stable or neutral state. Therefore, we can conclude that the balance and sustainability of the development of the fuel and energy complex is determined by the positive parity of internal and external circumstances.

Of course, today various processes and events affect the sustainability of the development of the fuel and energy complex. If you already have the tools, tactics, and strategies for emergencies, you can adapt to them. As a result, the threats are fully or partially equated. However, threat factors can be controlled (managed) or uncontrolled (unmanaged).

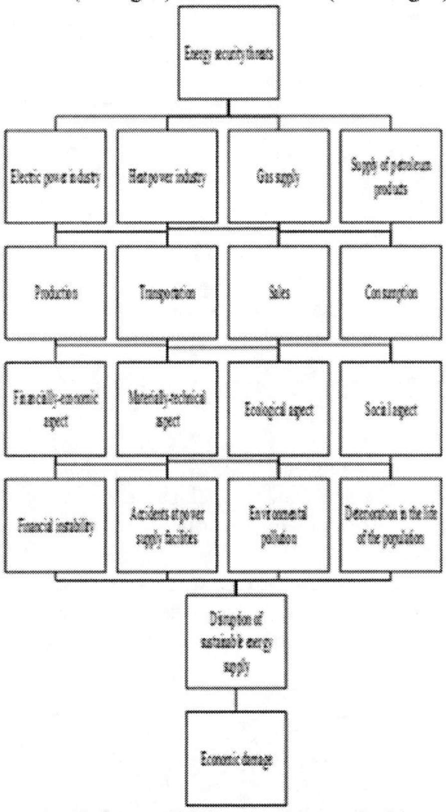

Fig. 1. Types of threats to the sustainable development of fuel and energy enterprises[1]

[1] Developed by author

3 Research results

Uncontrollable thesis includes threats and risks affecting from the outside - socio-political, natural, partial management shortcomings. The system can be adapted to the impact of uncontrolled risks. Controlled factors include threats arising in the fuel and energy complex itself. The main types of threat sources in the fuel and energy complex are shown in Figure 2.

Let's classify the types of threats and risks affecting enterprises in the fuel and energy complex.

Economic threats include:

- lack of competition in the energy market, monopoly of suppliers and producers of fuel and energy resources;
- irrational use of fuel and energy resources;
- financial instability of fuel and energy enterprises;
- insufficient investment in the modernization, renewal of fixed production assets, as well as the unsatisfactory situation in the field of innovation processes in the fuel and energy complex;
- high energy prices.

Although the energy intensity of Uzbekistan's GDP has been declining in recent years, this indicator remains well above the level of developed countries. Today, the average value of the energy intensity of GDP around the world is 240 kg.o.e./1000 US dollars. The energy intensity of Uzbekistan's GDP is almost 4 times higher than in the EU and 2 times higher than the global average. As a result, Uzbekistan is included in the group of countries with a fairly high level of CO2 per unit of gross domestic product.[2]

Social-political threats. Achieving an acceptable standard of living and sustainable economic development can be achieved by managing socio-political threats. This type of threat is directly related to the standard of living of a society and can be exacerbated by other threats.

Natural hazards. Measures to level out threatening natural disasters are provided at the design and construction stage of an energy facility. Nevertheless, natural disasters can lead to emergencies, taking into account all natural factors. However, the problem is that due to chance, the real course of natural phenomena does not always correspond to the forecast. Natural disasters are often of diverse origin, severity, negative impact, scale and frequency of them have a wide range: from several days of heavy rains to nine-magnitude earthquakes.

Artificial threats. Malfunctions and accidents can occur in any technical system for various reasons. In this regard, the concept of man-made threat is

[2] Sh. Zokirov. Renewable energy for sustainable development. Research Center "Scientific Foundations and Problems of Development of the Economy of Uzbekistan"

introduced. Large-scale fires and ports in the fuel and energy complex pose a serious threat to the population and the environment.

The main causes of man-made threats are:

– significant depreciation of fixed assets of the fuel and energy complex;

- low quality of installed equipment, as well as a low level of repair or construction work;

– violation of the rules for the operation of power facilities;

- Irrational placement of productive forces, leading to the accumulation of high-risk facilities located close to each other and near settlements. In this regard, an accident at one facility can cause a fire or explosion at another facility.[3]

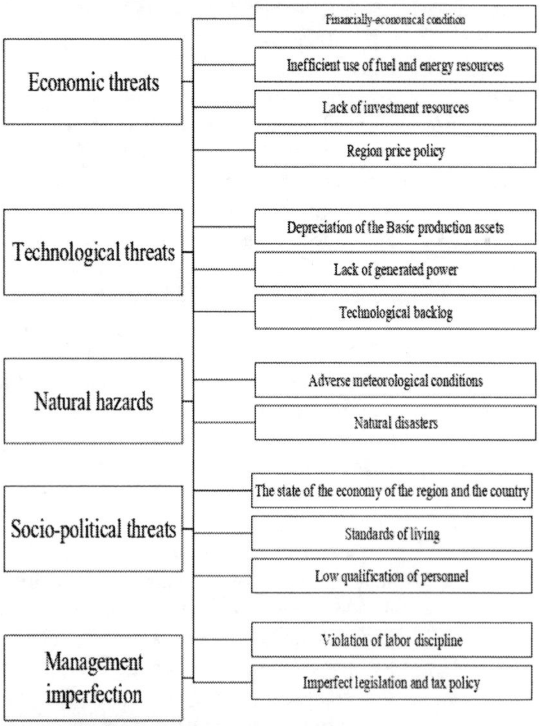

Fig. 2. Types of sources of risks and threats to the sustainable development of the fuel and energy complex[4]

Unfortunately, man-made accidents primarily cause great damage to the population. During the accident at the Takhiatash thermal power plant (TPP) on July 18, 2019, Khorezm region and part of Karakalpakstan were left without electricity for several days. The 8th power unit was switched off by the technological protection system. After 12 seconds, the system also turned off the 7th power unit and the 1st turbine unit. On January 10, 2021, a failure at the Talimzhan thermal power plant (TPP) in the Kashkadarya region led to a power outage in a number of districts and cities. Since, in the first place, such social facilities as enterprises with a continuous production cycle, kindergartens, hospitals, schools, the above-mentioned important facilities are transferred to backup power sources. Batteries and electric generators are usually always provided at such facilities.

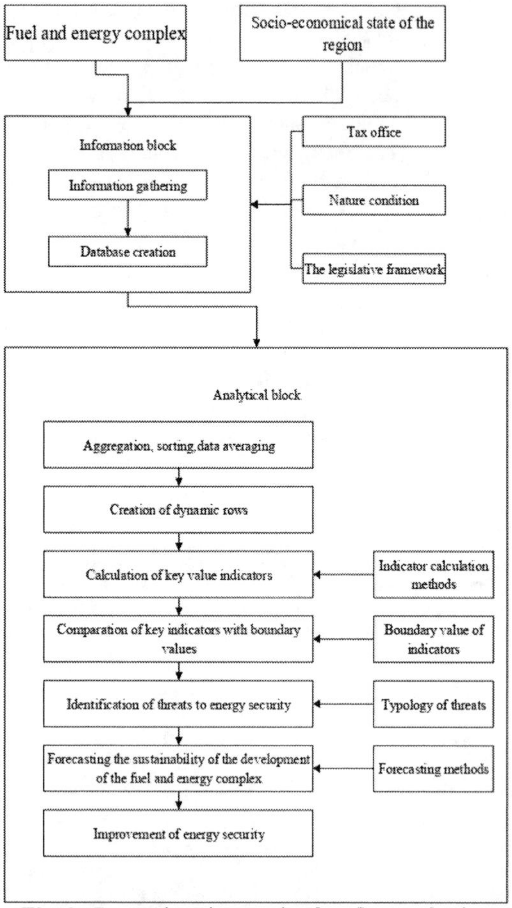

Fig. 3. Comprehensive mechanism for monitoring and preventing threats to sustainable development and energy security of the fuel and energy complex[5]

The proposed comprehensive mechanism for monitoring and preventing threats to sustainable development and energy security of the fuel and energy complex (Fig. 3) at first glance, has all the necessary elements, however, a more detailed study of each of its elements, and most importantly, the effectiveness of the relationship of these elements, reveals the presence and preservation of a number of problems, the solution of which is important for the sustainable development of energy sectors.

[3] Orsoeva, M.V. Energy security as a factor in the sustainable development of the region: dissertation of PHd Sciences: 08.00.05: / Orsoeva Maria Vasilievna; [Place of doctoral defense: Baikal. state University of Economics and Law]. - Irkutsk, 2005. - 194 p.

[4] Developed by author

[5] Developed by author

Science-based suggestions. Analysis of the factors limiting the efficient and safe sustainable development of the sectors of the energy sector of the national economy allowed us to formulate the most promising directions for the development of the fuel and energy complex, which determine the main strategic goals for the near and long term:

- monitoring the situation on the world energy market;
- expanding the use of energy efficient technologies not only in the energy sector, but also in the economy as a whole;
- the relationship between the growth of the gross domestic product and the energy intensity of the national economy with the development of demand for energy resources in Uzbekistan;
- creation of an optimal investment environment in the republic, in order to create a favorable climate in the tax, customs and price spheres for enterprises in the fuel and energy industry;
- accelerated innovative development and improvement of the fuel and energy complex, modernization of existing capacities, and also, in the long term, the development and implementation of such promising areas in the energy sector as renewable energy sources, nuclear and hydrogen energy and other types of energy of the future;
- carrying out measures to level the negative impact of the activities of fuel and energy complex enterprises on the environment.

4 Conclusion

Thus, the considered threats and risks of sustainable development of the sectors of the energy complex showed that a financially stable, economically sustainable, developing in accordance with the requirements of international standards, environmentally efficient, provided with innovative scientific and technical developments and qualified non-energy personnel is a complex non-energy necessity, the requirement of the time, but will also raise the economy of the republic to a qualitatively new level of development.

Based on the foregoing, we believe that it is necessary to develop and implement, at the state level, such measures that will allow:

- in the short term - to effectively use the existing production and economic potential at the enterprises of the fuel and energy complex, to form the basis for the further sustainable development of the national fuel and energy complex in the context of the formation of the digital economy;
- in the long term - to solve the problem of uninterrupted supply of all segments of the population of the republic and the economy itself with all available energy resources, to develop a long-term strategy for the innovative development of the fuel and energy complex, allowing the national fuel and energy complex to be a highly competitive, efficient and sustainable structure that meets all modern challenges for energy sectors of the economy.

References

1. Decree of the President of the Republic of Uzbekistan No. PD-4947 dated February 7, 2017 "On the strategy of actions for the further development of the Republic of Uzbekistan".
2. Decree of the President of the Republic of Uzbekistan on the Concept for the development of nuclear energy in the Republic of Uzbekistan for the period 2019-2029. No. PD-4165 dated February 7, 2019
3. No. PD-5544 dated September 21, 2018 "On the strategy of innovative development of the Republic of Uzbekistan for 2019-2021"
4. Gurieva L. Strategy for sustainable development of the region /L. Guriev// Problems of theory and practice of management. - 2012. - No. 2.
5. Economic security of Russia. General course: textbook / ed. V. K. Senchagova. - 6th ed. - Moscow: Knowledge Laboratory, 2020. - 818 p.
6. Dudarev A.S. Factors of sustainable economic development /A.S. Dudarev // Entrepreneur. - 2013. - No. 8. - P. 56-59.
7. Xodiev B.Yu., Mahmudov N.M. Factors and trends of sustainable development of leading sectors of the national economy. - T: Economics, 2018. - 344 p.
8. Ryasin, V.I. Energy security of the region in the context of reforming the electric power industry: diss. doc. economy Sciences: 08.00.05: / Ryasin, Vladimir Igorevich; - Ivanovo, 2006. - 363 p.
9. Kucherova E.N. Modern approach to sustainable development of the enterprise / Bulletin of the Orenburg State University. 2007. - No. 9 (73). – P. 76–81.
10. Burrows, M. A Strategic View of the Energy Future./ M. Burrows, G. Treverton –Washington: GPO, 2008. – 142 p. International Energy Outlook 2019. U.S. Energy Information Administration Office of Energy Analysis U.S. Department of Energy. - Washington, DC 20585, 2019. - 85 p.
11. Developing Sustainable Infrastructure for the Low-Carbon Transition in Central Asia and the Caucasus: Mapping the Situation of Potentially High-Impact Infrastructure Projects and Needs Assessment. Strategic infrastructure planning for sustainable development in Uzbekistan. OECD Ad Hoc Working Group on Green Action. 30.09 – 01.10. 2019, Paris.
12. Ministry of Energy: goals, objectives, plans and achievements.
13. Kondrakov O.V. Increasing energy security based on economically sustainable development of

the fuel and energy complex. Dissertation for the degree of Doctor of Economics. Tambov 2019

14. Akhmedov T.M., Muinov D.A., Shoahmedov Sh.Sh. Industrial strategy in Uzbekistan: aspects of problems and formation / Ed. Muinova D. A. - T .: CEEP, 2005. - P. 120.

15. Allaeva G.Zh. Prospects for sustainable development of enterprises of the fuel and energy complex of the Republic of Uzbekistan. //Monograph. - Tashkent, "Fan va texnologiyalar", 2021. 160 pages.

16. Allaeva G.Zh. Sustainable development of enterprises of the fuel and energy complex of the Republic of Uzbekistan in the digital economy. Sanoat iqtisodiyoti va management. Republic of ilmiy-amaliy anzhumanni materiallari - T: 2021, April 29. B. 368-372

17. Allaeva G.Zh., Makhkamova M.A. Promising technologies in the electric power industry are the basis of energy saving and energy security in Uzbekistan. // Spec. issue of the journal "Problems of Energy and Resource Saving". Tashkent, 2019. - B. 208-212.

Application of design methods for precision energy supply systems in the educational process using information technology

Tsybov *N.N.*[1*], Galbaev *Zh.T.*[1]

[1] Kyrgyz State Technical University n. a. I. Razzakov, Bishkek, Kyrgyz Republic

Abstract. Teaching electrical engineering disciplines in technical universities involves the use of information technology in modeling processes in electrical circuits and in organizing the educational process. The main problem of the existing automated training systems for technical universities is their low didactic efficiency. This paper presents the application of a training system that takes into account the psycho-factors of the participants in the educational process and the personal characteristics of students. The initial data for the operation of the automated system is information about the fundamental pedagogical concepts and information reflecting the personal characteristics of the participants in the educational process. The color design of the means of visual display of information blocks on monitors is tuned in accordance with the user's psychotype. The proposed automated training system contains a set of training electronic devices with interchangeable functional units, which, based on the results of testing academic training and testing personal characteristics, are offered to students for design and stage-by-stage modeling. The paper shows the design of a precision DC voltage stabilization system. This study proposes technical solutions that implement the principles of the combined use of negative and positive feedback circuits. In the proposed technical implementation of the device, a method is proposed for fully compensating for the instability of the stabilizer output voltage when the load current changes from zero to the nominal value.

1 Introduction

The study of electrical disciplines in technical universities is not complete without the use of new information technologies. Information technologies are used not only in modeling processes in electrical circuits, but also in automating the learning process itself using information training systems. When building informational training systems, a negative trend is to automate only the subject area of the studied technical discipline. With this approach, the system does not take into account modern pedagogical concepts and teaching methods, and therefore such software and hardware teaching tools have low didactic effectiveness. This paper presents a variant of the application of new information technologies in the educational process using automated information training systems that take into account not only academic training, but also the psycho-factors of the participants in the educational process [1, 2]. The identified psychofactors are used to implement a student-centered approach in teaching, and at the same time, psychofactors are new additional components of the system analysis of pedagogical situations.

Software and hardware learning tools are inherently information systems for monitoring, analyzing and managing the educational process. The database of information systems contain information resources for both students and teachers.

A feature of the proposed educational information system is the ability to quickly adjust the forms of presentation of educational material in accordance with the identified personal characteristics of the student.

In this paper, we consider a variant of designing a powerful precision system of stabilized power supply using virtual models of training devices that are part of the training system.

The functioning of the training system is shown on the example of the use of a precision DC voltage stabilizer included in its training virtual training device, made according to the author's patent KG 2029 [3].

Stage-by-stage calculation of the ratings of the electronic components of the stabilizer is given by us in [4], therefore, we will consider only the functioning of the system and the main features of technical solutions.

* Corresponding author: nikolay_research@mail.ru

The relevance of designing powerful precision power supply systems is as follows.

The increase in the functional complexity of new electronic devices for monitoring power supply systems during their creation implies an increase in the requirements for the quality and stability of stabilized power sources. Conducting experimental scientific work in the design of devices for monitoring and controlling power supply processes increases the requirements for stability and accuracy of the supply voltage parameters by dozens of times.

Industrially produced DC voltage stabilization systems, in terms of their characteristics, are quite consistent with the requirements of standard devices during their operation, but when designing such devices, developers need precision power supply systems.

Modern industrially produced stabilized power systems have an instability coefficient of at least 1%, but during the experimental development of new systems, sometimes the required stabilization accuracy can exceed 0.01%.

Due to the high requirements for the weight and size indicators of the power supply system, when creating electronic devices, pulsed power supply systems are mainly used. But if the designed power supply system has high requirements for the stability of the output voltage, for ripples, for radiated interference, linear stabilized power systems are widely used [5].

In addition to being used in precision power supply systems, linear DC voltage stabilization systems are now widely used in the construction of two-stage DC power supply systems [6]. In such systems, linear regulators with a low voltage drop (LDO - low dropout) at the output stage are used as the second stage [7]. But industrially produced integrated LDO stabilizers are produced for small load currents up to 5 A and for scientific research have a low stabilization coefficient of the order of 1%. In this regard, there is a need to design both precision reference voltage sources and powerful stabilized power supply systems..

2 Experimental research and results

In the process of designing an automated learning system, a synergistic approach was used to formalize educational components.

When formalized, the automated educational environment is a synergistic ergatic model "student - electronic educational tools". Memory is chosen as the "order parameter" of a self-organizing ergatic system. The description of the ergatic model with memory can be described by the following equation [8].

$$\frac{dY_T(t)}{dt} + X_T Y_T(t - \tau_T) = P_T$$

where: Y_T (t) – parameter of the state of the training system;

X_T – parameter coefficient of knowledge perception;

t – time;

τ_T – delay parameter;

P_T – portion of knowledge.

When automating the components of the educational process, the functioning algorithms of the proposed learning system reflect modern theories and concepts of learning. The educational system in its composition contains a module for psycho-diagnostics of the qualities of the participants in the educational process.

The automated training system considered in this paper includes virtual models of real electronic devices used as training devices.

The training system and all training electronic devices are protected by invention patents. Each training device has a main base part and a set of duplicate functional modules of varying complexity.

The educational information system, in accordance with the preliminary testing of the student's academic preparation and his personal characteristics, corrects the complexity of the designed device presented for study and provides him with modules of the appropriate complexity for designing.

The visual display system of the educational system, in accordance with the student's psychotype, adjusts the color representation of information blocks on the monitor screen, since choleric, sanguine and melancholic people react differently to the color display on the monitor screen.

In this case, the student is tasked with calculating the parameters of the components of a precision linear stabilization system with an output voltage of 12 V, 50 A with an instability coefficient for load current no worse than 0.001%.

To create a power supply system with a stabilization accuracy of 0.001% when the load current changes from zero to the nominal value, a precision reference voltage source is required [9, 10, 11]. Integrated reference voltage modules cannot provide such accuracy. In the proposed technical solution, the reference voltage source is made according to the scheme of two back-to-back current stabilizers (see Fig. 1).

The main precision zener diode D2 is powered by a back-to-back current regulator on the transistor VT2. In turn, the reference zener diode of the current stabilizer on the transistor VT2 is powered by the current stabilizer on the transistor VT1. Therefore, when supplying the reference zener diode D2 with a stable current, the internal resistance of the main reference zener diode has practically no effect on the value of the output reference voltage.

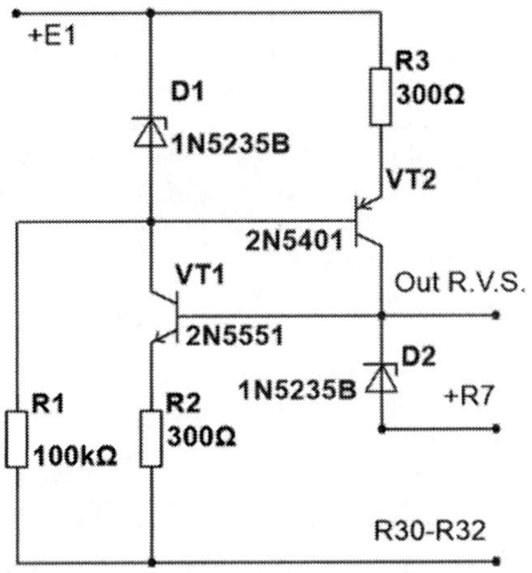

Fig. 1. Reference voltage source

To achieve the stability of the output DC voltage of the stabilizer, as a rule, the gain of the mismatch amplifier of the error signal is increased. But a significant increase in the gain of the error signal mismatch amplifier leads the power supply system to parasitic generation.

In this circuit solution, in order to significantly increase the stabilization factor with changes in the load current, in addition to circuits with negative feedback, circuits with positive feedback are used. To do this, a current sensor is inserted between the negative terminal of the main reference zener diode and the common wire of the voltage stabilizer - a low-resistance resistance R7, through which the load current passes (see Fig. 4). In this technical solution, at a load current of 50 A, the resistance value of the current sensor R7 is 0.01 mOhm. When the load current flows through the current sensor R7, a voltage drop is created on it, which is added to the reference voltage and is fed to the input of the mismatch amplifier (see Fig. 2).

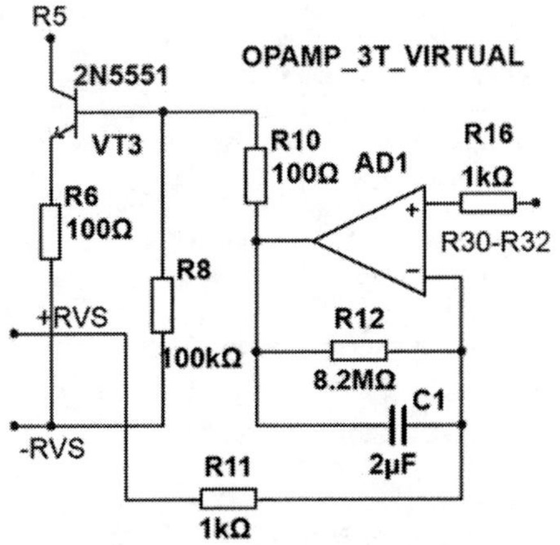

Fig. 2. Error amplifier

When debugging the stabilizer, by adjusting the resistance of the current sensor R7, you can completely compensate for the change in the output voltage of the stabilizer when the load current changes from zero to the nominal value.

The mismatch amplifier of the error signal in this technical solution is made on the operational amplifier AD1 and the matching transistor VT3.

The gain of the mismatch amplifier is set by resistors R10 and R12. The error signal from the output voltage divider is applied to the non-inverting input of the operational amplifier AD1, and the reference voltage is applied to the inverting input AD1. The mismatch signal isolated by the operational amplifier is compared with the reference voltage and the corrective signal is fed from the collector of the transistor VT3 through the resistor R5 to the base of the pre-terminal stage VT4 (see Fig. 2 and Fig. 3).

The final control stage is made according to the scheme of a composite stage with current-equalizing emitter resistors (see Fig. 3). At a load current of 50 A, the composite cascade includes 5 2SC5200 transistors.

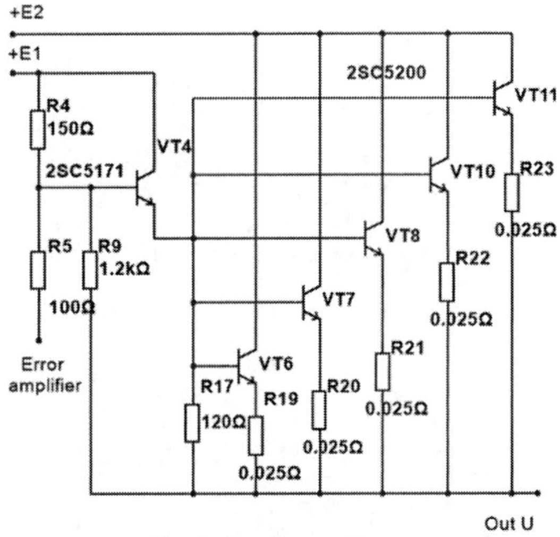

Fig. 3. Output control stage

In order to reduce losses on equalizing emitter resistances, their nominal value is reduced to 0.025 Ohm. At a load current of 50 A, a current of 10 A will flow through each transistor. At a current of 10 A, the voltage drop across the equalizing resistors will be 0.25 V, which is a quarter of the opening operating voltage at the bases of the transistors of the composite cascade. Such a voltage drop is quite sufficient for accurate equalization of the currents of the composite output control stage [12].

In order to reduce losses during power dissipation in the composite control stage, an additional low-voltage voltage boost source E1 is introduced into the circuit, which is connected in series with the main power source E2 (see Fig. 4).

The final output control stage is powered by the main power supply E2, while the reference voltage source and the final stage are powered by the sum of the two voltages of the sources E1 and E2. Such a circuit solution allows you to open the transistors of the output composite stage to saturation and increase the stabilization range with a decrease in the input supply voltage.

On fig. 4 shows a general block diagram of a precision DC voltage regulator.

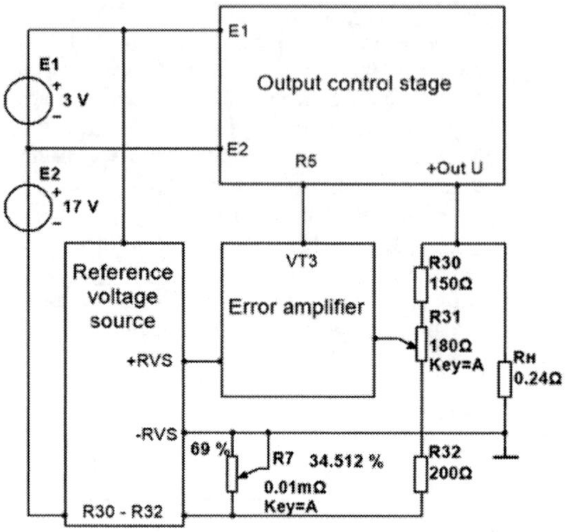

Fig. 4. Precision voltage regulator

If a student is offered to design a voltage regulator for 12 V and 8 A with a stability of 1%, then the student does not need to use a precision reference voltage source on two counter-connected current stabilizers. In this case, it will be sufficient to use a parametric stabilizer circuit on a single precision zener diode as a reference voltage source (see Fig. 5 b).

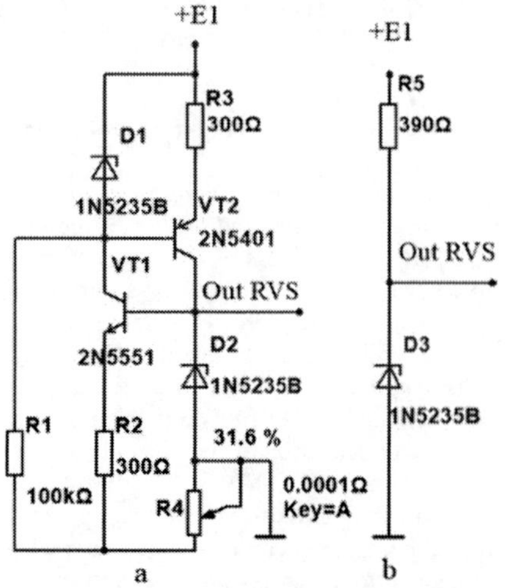

Fig. 5. Precision reference voltage source (a) and simple reference voltage source (b)

When designing a stabilizer for 12 V and 8 A as an output control stage at a load current of 8 A, a composite stage is not required. In this case, a cascade on one 2SC5200 transistor will be enough (see Fig. 6a), and in the case of a load current of 0.2 A, there will be no need for a pre-terminal cascade and one 2SC5171 transistor will be enough as a final control cascade (see Fig. 6 b).

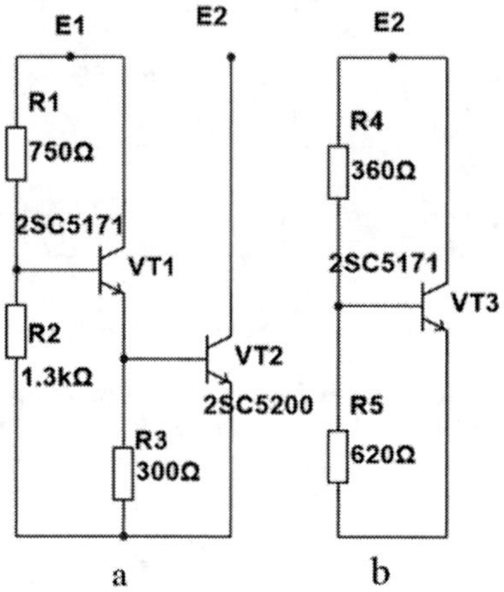

Fig. 6. Output control stage 8 A (a) and output control stage 0.2 A (a)

3 Conclusion

1. The use of negative and positive feedback circuits in the design of stabilized power supply systems makes it possible to fully compensate for the dependence of the stabilizer output voltage on changes in load current.
2. The designed information training systems in their functioning algorithms must contain the fundamental concepts and teaching methods, as well as modules for psychodiagnostics of participants in the educational process.
3. The identified psychofactors of participants in the educational process can be used as new elements of system analysis in solving pedagogical situations.
4. When formalizing the educational components of automated systems, synergistic ergatic models are the most effective.
5. In order to increase the effectiveness of the presentation of educational material, the means of visual display of information blocks on monitors, teaching systems must have the function of color alignment with the student's personal characteristics.

References

1. Tsybov, N. N., Galbaev Zh. T., Burmeyster M. V. Features of Approaches to Information Educational Components Design for Energy Universities // 2023 5th Int. Youth Conf. on Radio Electronics, Electrical

and Power Engineering (REEPE). – M., 2023. – P. 1–8. – doi: 10.1109/REEPE57272.2023.10086744.

2. Tsybov, N. N. Didactic features of the application of teaching systems in technical universities. Education and self-development. T. 17, No. 4.p.p. 267–287. (2022).

3. Tsybov N.N. Precision thermostable DC voltage stabilizer with compensation of internal resistance. Patent for invention KG 2029 Kyrgyz Republic, IPC G05F 1/56. No. 20170075.1; Application. 20.06.17; Republ. 28.02.18, Intellectualdyk menchik rasmii bul. No. 2. - 2 p.: ill. (2018).

4. Tsybov, N. N. Modeling of processes in electrical circuits in tasks of intelligent learning systems. Prakt. power electronics. no. 4 (72). p.p. 47–56. (2018).

5. Bityukov V.K., Ivanov A.A., Mikhnevich N.G., Petrov V.A. Measurement of the ripple suppression factor by linear voltage stabilizers. Measuring technology. no. 11.p.p. 52-56. (2016).

6. Bityukov V.K., Ivanov A.A., Mikhnevich N.G., Petrov V.A. Measurement of the ripple suppression factor by linear voltage stabilizers. Measuring technology. no. 11.p.p. 52-56. (2016).

7. Mikushin A.V. Features of the use of LDO stabilizers. In the collection: Modern problems of telecommunications. Materials of the Russian scientific and technical conference. Novosibirsk, p.p. 503-509. (2022).

8. Khramov, B. V., Vitchenko, O. V., Tkachuk, E. O., Golubenko, E. V. Intelligent methods, models and algorithms of educational process organization in modern universities. Rostov n/D: FGBOU VO RSUPS, 152. (2016).

9. Nedashkovskaya A.E. Comparative analysis of reference voltage sources for the CMOS process with a design standard of 45 nanometers. Scientific horizons. no. 12 (40). p.p. 100. (2020).

10. Verkhulevsky K. Precision reference voltage sources from apex microtechnology. Components and technologies. no. 2 (139). p.p. 30-34. (2013).

11. Smorodinov A. S. Reference voltage sources: basic parameters and design principles. Components and technologies. no. 9(170). p.p. 90-92. (2015).

12. Chzhu M., Go F. Parallel connection of Analog Devices LT3033 linear regulators for increased output current and uniform heat distribution. Components and technologies. no. 7 (228). p.p. 54-58. (2020).

Optimization of energy consumption in cotton ginning enterprises using neural network method

J. Tolipov[1*], *M.* Rakhmatova[1], *A.* Mannonov[1], *O.* Ishnazarov[1]

[1]Institute of Energy Problems of the Academy of Sciences, 100084 Tashkent, Uzbekistan

Abstract. In the modern world, energy consumption optimization has become a critical concern across various industries due to environmental considerations and economic efficiency. Cotton ginning enterprises, which play a pivotal role in the textile supply chain, are no exception. This article explores applying neural network methods to optimize energy consumption in cotton ginning enterprises. We delve into the challenges faced by the industry, introduce the concept of neural networks, and discuss their potential to enhance energy efficiency. A case study demonstrates the practical implementation of the neural network approach in a cotton ginning setting, showcasing the potential benefits and providing insights into future directions for sustainable energy practices.

1 Introduction

Cotton ginning, the process of separating cotton fibers from their seeds, is a resource-intensive operation within the textile industry. Energy consumption is a significant cost factor for these enterprises, making energy optimization a crucial objective for both economic and environmental reasons. In recent years, advancements in machine learning techniques, particularly neural networks, have shown promise in tackling complex optimization challenges. This article explores the integration of neural network methods into the energy consumption optimization of cotton ginning enterprises. Early in the decade, research primarily centered on conventional optimization techniques for energy consumption in cotton ginning enterprises. Studies such as Ismail et al. [1] and Dube et al. [2] and explored process reengineering and control strategies to minimize energy usage. These studies highlighted the importance of improving mechanical efficiency and adopting energy-efficient technologies.

As machine learning gained prominence, researchers began investigating its applicability to energy optimization. Sahu et al. [3] introduced a predictive model using regression analysis to optimize energy consumption in cotton ginning processes. This marked an initial step toward incorporating data-driven techniques.

The mid-2010s saw a shift towards neural network-based methods for energy optimization. Researchers recognized the ability of neural networks to capture intricate patterns in energy consumption data. Shanbeh et al. [4] introduced a study where neural networks were used to predict energy demand in ginning operations. This study laid the foundation for future research on the application of neural networks in the cotton ginning context.

As the decade progressed, researchers began exploring integrated systems and advanced algorithms to optimize energy consumption comprehensively. Fue et al. [5] proposed a model integrating artificial intelligence techniques, including neural networks, to optimize the entire ginning process. This holistic approach highlighted the potential of combining multiple strategies for enhanced results.

Towards the end of the decade, attention shifted towards practical implementation and case studies. Hardin et al. [6] presented a case study on a real-world cotton ginning plant, demonstrating the effectiveness of a neural network-based predictive model for energy optimization. This study provided valuable insights into the feasibility and benefits of implementing such methods in actual industry settings.

While significant progress was made during the decade, challenges remained. Data availability, model interpretability, and scalability were identified as key issues. As highlighted by Patel et al. (2020), efforts were needed to develop more robust neural network architectures specifically tailored to the cotton ginning industry.

The years between 2010 and 2023 marked a significant evolution in the approach to energy consumption optimization in cotton ginning enterprises. From traditional methods to the emergence of neural network-based techniques, researchers demonstrated the potential of data-driven approaches in enhancing energy efficiency. Case studies and integrated systems showcased the applicability of these techniques in real-world scenarios [7,8]. As the decade concluded, the stage was set for future research to address challenges and continue refining these methods for sustainable energy practices in cotton ginning enterprises.

* Corresponding author: tolipovjamshidn@gmail.com

This project is aimed at optimizing energy consumption in cotton ginning enterprises through the utilization of neural network methodologies.

2 Materials and Methods

Historical energy consumption data from diverse cotton ginning enterprises were gathered to form the basis for this study. The dataset encompassed a comprehensive array of ginning processes, encompassing distinct operational conditions and energy usage patterns.

In this research, a Bayesian neural network (BNN) architecture was adopted to optimize energy consumption in cotton ginning enterprises. The BNN framework integrates uncertainty estimation, offering a more comprehensive understanding of predictive model outcomes. The network comprised an input layer, multiple hidden layers, and an output layer. Notably, Bayesian principles were embedded into the network's structure to provide probabilistic predictions.

The Levenberg-Marquardt algorithm, renowned for its convergence speed in non-linear optimization problems, was employed for training the Bayesian neural network. This algorithm adjusts the neural network's weights by minimizing the difference between predicted and actual energy consumption values while considering the network's uncertainty estimates.

A feedforward neural network architecture was chosen for its ability to capture complex relationships within the data. The network consisted of an input layer, hidden layers, and an output layer. The number of neurons in the hidden layers was determined through experimentation to achieve optimal performance.

Relevant operational parameters such as ginning speed, humidity levels, and machine configurations were selected as input features. These features were normalized to ensure consistent scaling within the neural network.

2.1. *Training and Validation*

The dataset was divided into training and validation subsets. The neural network was trained using backpropagation, a gradient-based optimization technique. Mean Squared Error (MSE) was employed as the loss function to minimize the difference between predicted and actual energy consumption values.

The training of the Bayesian neural network (BNN) commenced with the initialization of weights and biases in the network's architecture. The Levenberg-Marquardt optimization algorithm was then employed to iteratively adjust these weights to minimize the difference between the predicted and actual energy consumption values while accounting for the uncertainty estimates. The Levenberg-Marquardt algorithm incorporates the curvature of the error surface, facilitating convergence even in the presence of non-linearities.

2.1.1. *Bayesian Uncertainty Estimation*

Integral to the training process of the BNN was the incorporation of Bayesian uncertainty estimation. This entailed the assignment of probability distributions to the weights and biases of the neural network, effectively quantifying the uncertainty associated with each parameter. By considering these uncertainties, the BNN not only produced point predictions but also provided a probabilistic range of predictions, allowing for a more comprehensive understanding of the model's confidence in its output.

2.1.2. *Loss Function for Training*

The objective of training was to minimize the negative log-likelihood of the observed energy consumption given the predicted energy consumption and its associated uncertainty. The loss function, denoted as L, was defined as follows:

$$L = -\frac{1}{2}\sum_{i=1}^{N}\left[log(\sigma_i^2) + \frac{(y_{actual,i} - y_{predicted,i})}{\sigma_i^2} + \right.$$
$$\left. +2log(2\pi)\right] \quad (1)$$

Where:

N is the number of data points; $y_{actual,i}$ is the actual energy consumption value for the ith data point; $y_{predicted,i}$ is the predicted energy consumption value for the ith data point; σ_i is the uncertainty estimate for the ith data point.

The optimization process aimed to minimize this loss function by adjusting the weights and biases of the BNN.

2.1.3. *Validation and Performance Metrics*

To assess the trained Bayesian neural network's performance, a separate validation dataset was employed. The model's predictions were compared to the actual energy consumption values in this dataset. The following performance metrics were calculated:

Mean Absolute Error (MAE)

$$MAE = \frac{1}{N}\sum_{i=1}^{N}\left|y_{actual,i} - y_{predicted,i}\right|. \quad (2)$$

Root Mean Squared Error (RMSE)

$$MAE = \sqrt{\frac{1}{N}\sum_{i=1}^{N}\left(y_{actual,i} - y_{predicted,i}\right)^2}. \quad (3)$$

The solution to this problem is a very complicated and time-consuming process. The technological processes of the considered cotton ginning enterprise are considered to be very complex and difficult. A prediction algorithm based on a common method of correlation and regression analysis consists of a mathematical description.

Planned consumption of electricity:

$$W_{Plan} = k_{month} \cdot Q_{Plan}, \quad (4)$$

 https://doi.org/10.1051/e3sconf/202346101063

where k_{month} – rate of electricity consumption per product unit for per month, kW·h/t; Q_{Plan} – planned cost of production, t.

The resolution to these challenges could potentially involve the creation and application of a predictive framework for estimating electric power usage, utilizing artificial neural networks (ANNs). This predictive technique relies on employing multilayered artificial neural networks organized in a sequential configuration. ANNs represent a computational paradigm inspired by basic biological mechanisms reminiscent of human brain functions. A fundamental unit within these architectures is an artificial neuron, denoted for its likeness to its biological counterpart.

The training and prediction of the artificial neural network (ANN) involve utilizing input data from the commercial electricity accounting automated system's database. This dataset is complemented by additional parameters specific to the electricity consumption of each enterprise, such as year days, air temperature, humidity, atmospheric pressure, days of the week, and time of day. Tailoring individual predictions for each entity involves the crucial task of discerning pertinent variables while disregarding extraneous ones. To address this, the ANN employs the Bayesian regularization method, as depicted in Figure 1. This technique guides the neuron's training to align with network operations, enabling the determination of weight coefficients.

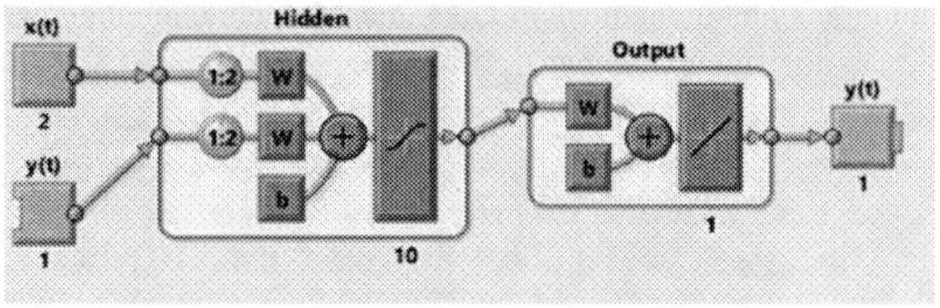

Figure 1. Neural Network Architecture

Figures 2 and 3 illustrate the outcome of the assessment of indicator correlations using the ANN approach. The findings demonstrate the existence of notable and moderate connections between input and output indicators.

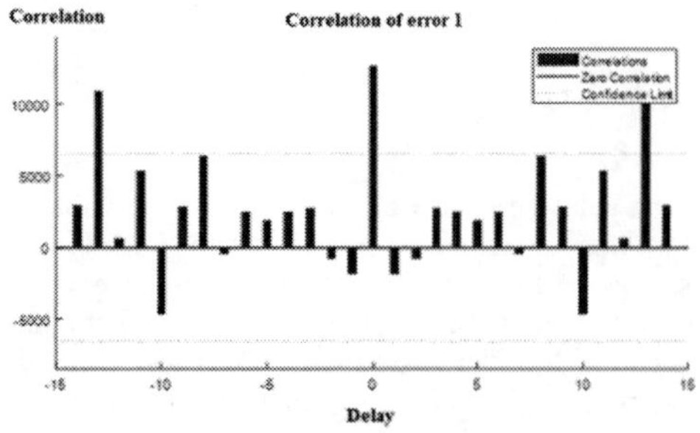

Figure 2. Autocorrelation of error 1

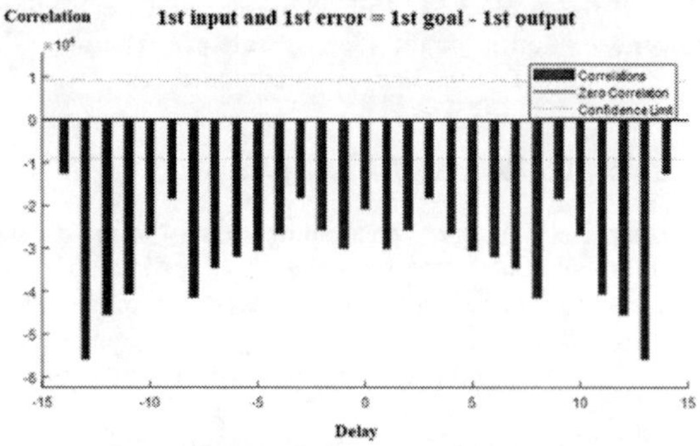

Figure 3. Correlation of model indicators

3 Results and Discussion

The utilization of Bayesian regularization mandates the incorporation of the Hessian matrix of the target function. In the context of ranking based on the ANN value and the summation of squared weights, the Hessian matrix adopts a quadratic structure that can be approximated through the application of the Levenberg-Marquardt algorithm (Gill and Murray, 1978) [9,10]. The formulation of the target function is outlined below

$$F = \alpha E_W + \beta E_D, \tag{5}$$

where α and β are the objective functional parameters.

Within the context of the Bayesian paradigm, the network's weights are viewed as stochastic variables. Consequently, the probability density function pertaining to the network's ensemble, encompassing its weight distribution, can be established through the subsequent function

$$f(w|D,\alpha,\beta,M) = \frac{f(w|D,\alpha,\beta,M) f(w|\alpha,M)}{f(D|\alpha,\beta,M)}, \tag{6}$$

where M - exact neural network model; $f(w|\alpha,M)$- this is the previous density, previous weights information; $f(D|\alpha,\beta,M)$ - is a probability function that is the probability that the data will occur given the weights; $f(D|\alpha,\beta,M)$ - normalization factor, which guarantees that the total probability is equal to 1 [11]

Depending on the Gaussian noise, there is a probability w of the given parameters

$$f(w|D,\alpha,\beta,M) = \frac{exp(-(\beta E_D + \alpha E_W))}{Z_W(\alpha) Z_D(\beta)} = \frac{exp(-F(w))}{Z_f(\alpha,\beta)}, \tag{7}$$

here $Z_f(\alpha,\beta) = \int exp(-F)dw$. Within this Bayesian context, the optimal weights are sought to optimize the conditional likelihood.

Figure 4 displays the outcome of the training process for the neural network constructed using the Bayesian regularization technique.

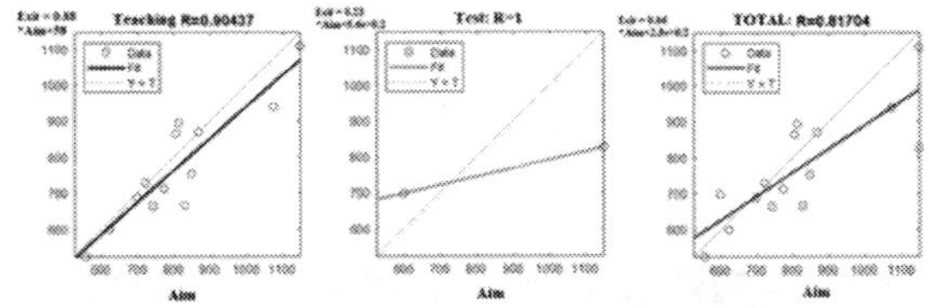

Figure 4. Evolution of neural network regression training over 1000 epochs, attaining maximum iteration

Figure 5 depict the training distribution error and the time series output element outcome.

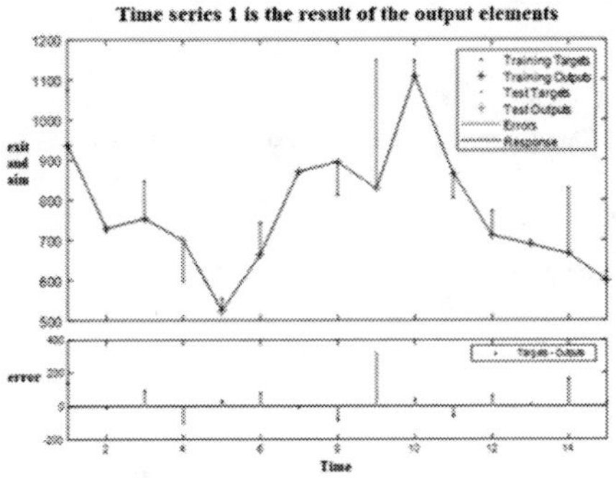

Figure 5. Time Series Output Element Outcome

4 Conclusions

1. The results obtained reveal that the ANN approach exhibits substantial promise in crafting predictive models aimed at enhancing precision. The optimal predictive model is characterized by straightforward training, rapid development, flexibility in accommodating environmental shifts, convenient adaptability, and consistent performance even within unstable conditions.

2. Consequently, the integration of this method alongside prediction systems based on ANNs presents a substantial opportunity to curtail expenses associated with fines arising from discrepancies between ordered and actual electricity consumption. To achieve this objective, we confined ourselves to mathematical representations of the neural network. As a consequence, significant enhancements in prediction accuracy can be achieved through the adoption of this technique.

5 References

1. Ismail, S. A., Chen, G., Baillie, C., and Symes, T. (2011). Energy uses for cotton ginning in Australia. Biosystems Engineering, 109(2), 140–147. https://doi.org/10.1016/j.biosystemseng.2011.02.010.
2. Dube K. J., Ingale L. T., and Ingle S. T. (2013). Respiratory Impairment in Cotton-Ginning Workers Exposed to Cotton Dust. International Journal of Occupational Safety and Ergonomics, 19(4), 551–560. https://doi.org/10.1080/10803548.2013.11077009
3. Sahu S., and Pramanik K. (2015). Delignification of cotton gin waste and its optimization by using white rot fungus Pycnoporus cinnabarinus. Journal of Environmental Biology, 36(3), 661.
4. Shanbeh M., Hasani H., and Tabatabaei S. A. (2011). Modelling and predicting the breaking strength and mass irregularity of cotton rotor-spun yarns containing cotton fiber recovered from ginning process by using artificial neural network algorithm. Modelling and Simulation in Engineering, 2011, 1-8.
5. Fue K. G., Porter W. M., Barnes E. M., and Rains, G. C. (2020). An extensive review of mobile agricultural robotics for field operations: focus on cotton harvesting. AgriEngineering, 2(1).
6. Hardin I.V, Barnes R. G., Delhom E. M., Wanjura C. D., and Ward J. K. (2022). Internet of things: Cotton harvesting and processing. Computers and Electronics in Agriculture, 202, 107294
7. Muratov Kh., Ishnazarov O., Isakov A., and Nuritov I. (2023) Assessment of effect of lining material on energy efficiency of starting up ball mills. E3S Web of Conferences, 365, 04005. DOI: https://doi.org/10.1051/e3sconf/202336504005
8. Khushiev S., Ishnazarov O., Juraev S., Izzatillaev J., and Karakulov A. Construction of an electric drive system for borehole pumps with frequency control. AIP Conference Proceedings 5 December 2022; 2686 (1): 020018. https://doi.org/10.1063/5.0114016
9. Sariev E., and Germano G. (2020) Bayesian regularized artificial neural networks for the estimation of the probability of default. Quantitative Finance, 20(2), 311-328, DOI: 10.1080/14697688.2019.1633014
10. Neal R. Bayesian learning for neural networks (1996). PhD Thesis, University of Toronto, Toronto, Canada
11. Sariev E. and Germano G. (2019) An innovative feature selection method for support vector machines and its test on the estimation of thecredit risk of default. Rev. Financ. Econ., 37, 404–427.

Digital protection of electrical equipment in railway transport

Komolbek Turdibekov[1]*, *Dilshod* Rustamov[1], *Sarvar* Xalikov[1]

[1]Tashkent State Transport University, 100167, Tashkent, Uzbekistan, Temiryulchilar street, 1 st

Abstract. In this scientific study, an analysis of the modernization of the digital protection of AC traction substation electrical equipment in operation, in particular power transformers, the implementation of the basic provisions for the implementation of digital protection of power transformers, finding algorithms for the implementation of both measuring and logic elements of digital protection, mathematical models of digital protection of power transformers are obtained, a method for blocking current protection is implemented, a method is selected finding the setpoints of the stage, obtaining differential protection, the schemes of secondary switching of digital protection are considered, the results of the scientific research carried out in the microprocessor terminal of protection of power transformers.

1 Introduction

The development of modern technologies in the railway power supply system for digital protection of electrical equipment of an alternating current traction substation, in particular power transformers, provides for the following tasks: differential current protection, maximum voltage current protection, external protections that control the state of the power transformer, automatic re-activation of the power transformer, automatic switch control with the function of a switch failure backup device. One of the responsible and difficult to implement stages of protection of power transformers is differential current protection, as well as maximum current protection.

Analysis of microprocessor protection of a power transformer provides the following set of functions: differential current protection, which includes a differential current stage with braking and a differential current stage without braking; eight stages of maximum current protection with voltage starting capability; directional action and connection to any side of the transformer, six stages from increasing the zero or negative sequence current with the possibility of starting by voltage, directional action and connection to any side of the transformer, four stages of maximum and four stages of minimum voltage with the possibility of return at the rate; four stages of maximum and four stages of minimum frequency with the possibility of return at the rate; sixteen external protections that monitor the state of the transformer using external signals; automatic transfer of reserve, automatic re-switching of the transformer.

Modernization of these stages can significantly increase the reliability of digital protection of the power transformer, increase the sensitivity of the magnetization current to voltage, algorithms for adjusting differential current protection from equilibrium currents of external short circuits based on correct transform intervals, in particular, develop a criterion that reacts to external signs. no matter a short circuit of the current size, which does not require the calculation of setpoints.

2 Experimental research

The formation of the path of orthogonal components of the digital microprocessor protection consists of an input sensor, a 2nd order analog filter on an operational amplifier, an analog-to-digital converter, as well as a digital filter, the filtration frequency is calculated by the following formula.

$$f(k) = \frac{k}{N} f_c \qquad (1)$$

where k is the harmonic number
N – the number of samples
f_c – sampling rate
The algorithm formula for the kth harmonic has the form

* Corresponding author: ilider1987@yandex.ru

$$X(t) = \left[\cos\left(\frac{2\pi}{N}k\right)\vartheta_{N-1} - \vartheta_{N-2} \right] - \sin\left(\frac{2\pi}{N}k\right)\vartheta_{N-1} \quad (2)$$

$$\vartheta_1 = x_1 + 2\cos\left(\frac{2\pi}{N}k\right)\vartheta_{N-1} - \vartheta_{N-2} \quad (3)$$

Where x_1 is the *i*th sample of the signal $0 \leq i \leq N-1$

Coefficients $\vartheta_2 - \vartheta_1 = 0$

A study using a computational experiment method showed that the path for forming controlled quantities makes it possible to implement filtering of the 1st, 2nd and 5th harmonics during one period of industrial frequency, which is sufficient for the implementation of measuring elements of transformer protection stages. The 1st harmonic filter is precisely tuned to a frequency of 50 Hz, but in the operating frequency range (47-52 Hz) it has an error of up to 3%. To ensure sufficient filtering accuracy, it is necessary to change the number of samples per period when the network frequency changes. The 2nd harmonic filter completely suppresses all harmonics, producing a 100 Hz output signal without attenuation or amplification. When filtering the 5th harmonic, the signal is attenuated to 97% due to the action of the analog filter. When implementing differential current protection of a transformer, the attenuation of the 5th harmonic must be taken into account using a correction factor of 1.03.

From false alarms of differential current protection against external short circuits, a new blocking method has been developed, which implements the ratio of increments of differential and braking currents according to the evaluation. This method is relatively easy to implement while maintaining the necessary accuracy and does not require special attention to the computing power of the digital protection terminal.

The functional scheme of blocking differential protection in case of external short circuits is given in Figure 1.

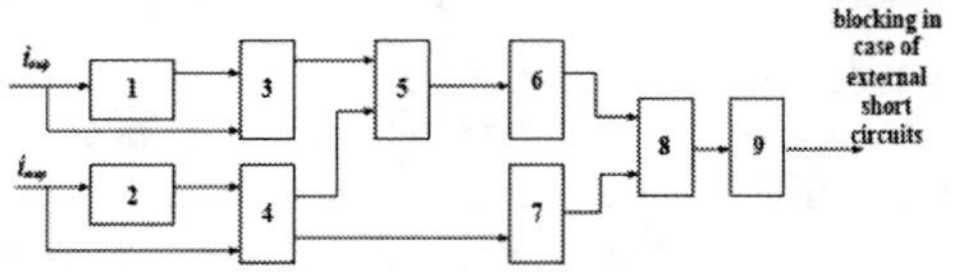

Figure 1. Functional diagram of blocking differential current protection in case of external short circuits

Blocks 1 and 2 implement the memorization and storage of the two previous samples of the differential i_{dif}. and i_{brk}. braking currents.

Blocks 3 and 4 determine the increment on two sampling periods of differential and braking currents. Block 5 calculates the ratio of increments of differential and braking currents on two sampling periods. Block 6 compares the above ratio with the coefficient K on two sampling periods. The coefficient K is selected from the range 0.2-0.3.

The functional scheme of blocking current protections against magnetization current surges (Figure 2) contains: frequency filter 1 of the first harmonic of the current, frequency filter 2 of the second harmonic of the current, filters 3 and 5 of the direct sequence of the current, filter 4 of the reverse sequence of the current, block 6 of blocking based on current control I_{1np}, block 7 of monitoring the content of the reverse current sequences of the first harmonic, block 8 of calculation and control of the parameter when the magnetization current is thrown I_{2np}. / I_{1np}., element 9 logical "And", element 10 "Prohibition".

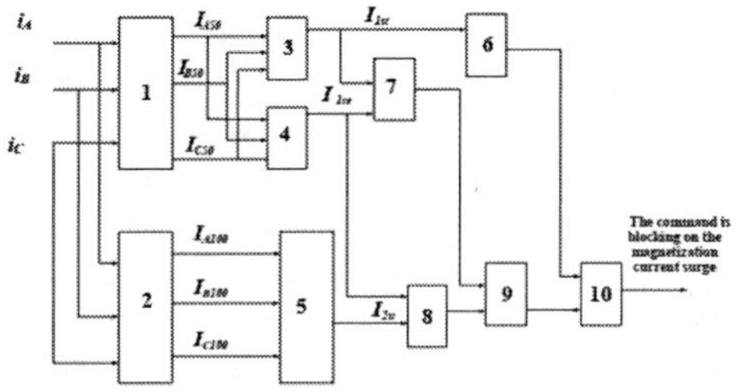

Figure 2. Functional diagram of blocking current protections during a surge of magnetization current.

As a result of each experiment of switching on an unloaded power transformer, the time dependences of the blocking parameters were obtained.

Blocking during magnetizing current surges is implemented in phase-by-phase or cross-phase modes; in the second case, the activation of the blocking in any phase leads to blocking of the other two.

The range of changes in the minimum value of the blocking parameter was 67.2-112.2%, the range of changes in the minimum value of the cross-blocking parameter was 40.3-117.1%. Thus, the blocking parameter is more stable and more effective at detecting inrush currents than the cross-blocking parameter.

In the initial period, during a short circuit, the transformer operates in saturation mode, and 10 ms after the occurrence of a short circuit, the measuring element of the differential stage is activated with braking. This is due to the phase reduction of the low side currents. Blocking signals for external short circuits, occurring 2 ms after the start of the short circuit, prohibit the operation of the differential current protection.

Blocking signals exist throughout the entire duration of the external short circuit, which ensures reliable detuning from unbalance currents.

Studies of differential current protection in external short circuit modes have revealed: differential current protection does not operate in the specified modes, the response time of the blocking algorithm in the event of an external short circuit is no more than 2 ms.

The use of a blocking algorithm for magnetizing current surges ensures unblocking of differential current protection in the event of significant fault currents.

Studies of differential current protection in internal short circuit modes have shown: differential current protection of a transformer operates reliably in internal short circuit modes, while its own response time does not exceed 30 ms; the use of a blocking algorithm during magnetizing current surges does not slow down the operation of differential current protection during internal short circuits. damage.

The mathematical model of differential current protection includes models of input current sensors and a second order analog filter, phase and amplitude reduction of currents, calculation of braking and differential currents, blocking from overexcitation of a power transformer, calculation of differential cutoff, blocking algorithm from external short circuits, blocking algorithm from magnetization current surges.

The mathematical model of the maximum current protection includes models of input current sensors and a second order analog filter, a model of a digital filter of orthogonal components, a blocking algorithm when a magnetization current is thrown.

3 Research result

The study of the maximum current protection in short circuit modes shows:

-the protection of the power transformer reliably operates in short-circuit mode, while its own response time does not exceed 30 ms;

-the application of the blocking algorithm during magnetization current surges does not slow down the operation of the maximum current protection in case of internal damage.

4 Conclusion

The functional scheme of the microprocessor protection terminal of the power transformer has been developed, which includes 12 current, 4 voltage and 26 discrete inputs and provides control of up to three arms of the protected power transformer. The input current and voltage signals are converted on the input sensors into voltage signals, which are converted into a digital signal on a 16-bit analog-to-digital converter.

Based on the data obtained, the operation of differential current protection and maximum current protection is realized.

Application. The developed digital protections were put into trial operation at the Tashkent power supply distance ECH -1 of Uzbekistan Temir Yullari AO.

References

1. Loman M.S. Development of a microprocessor terminal of a power transformer. //Power supply is the most important condition for the innovative development of the agro-industrial complex: materials of the International Scientific and Technical Conference, Minsk, November 24-25, 2001. https://studylib.net/doc/6938394

2. F.A. Romanyuk, S.P. Korolev, M.S. Loman. Microprocessor protection of power transformers. //Energy-News of higher educational institutions. 2001. No.5https://energy.bntu.by/jour/article/view/257?locale=en_US

3. S. Amirov, D. Rustamov, N.Yuldashev, U.Mamadaliev, M.Kurbanova, Study on the Electromagnetic current sensor for traction electro supply devices control systems, IOP Conf. Series: Earth and Environmental Science 939, 012009 (2021). https://www.researchgate.net/publication/356818951

4. S. Amirov, M. Yakubov, K. Turdibekov, A. Sulliev, Resource-saving maintenance and repair of the Traction transformer based on its diagnostics, Int J Advanced Science and Technology 29(5), 1500-1504 (2020). http://sersc.org/journals/index.php/IJAST/article/view/10060

5. F.A. Romanyuk, M.S. Loman., A.S. Gvozditsky. Investigation of the algorithm for blocking current protections of a power transformer in the mode of magnetization current surge. //Energy-News of higher educational institutions. 2014. https://www.researchgate.net/publication/314832192

6. Loman M.S., //Investigation of microprocessor transformer overcurrent protection in fault codes. Science and Technics-2014.-№3. https://paperity.org/p/213711766

7. Romanyuk F. Validation of mathematical model of differential protection / F. Romanyuk, I. Novash, M. Loman, P. Wegierek, M. Szrot // Przeglad electrotechniczny. – 2014. – № 3. – P. 187–190. https://www.semanticscholar.org/paper/Validation-of-mathematical-model-of-differential-Romanyuk-Novash/4fc440fac8b2017dba577031dd4d9cf3dd5417cc

8. K.H.Turdibekov, A.Kh.Sulliev, I.B. Qurbanov, S.B.Samatov, Voltage symmetration in high speed transport power supply systems, AIP Conference Proceedings 2432, 030084 (2022), https://pubs.aip.org/aip/acp/article-abstract/2432/1/030084/2823165

9. M. Yakubov, K. Turdibekov, A. Sulliev, I. Karimov, S. Saydivaliyev, S. Xalikov, Improvement of the informationmeasuring complex for diagnostics of traction power supply objects at high-speed traffic, E3S Web of Conferences 304, 02014 (2021) https://www.researchgate.net/publication/356720056

10. K.H.Turdibekov, A.Kh.Sulliev, I.B. Qurbanov, S.B.Samatov, Voltage symmetration in high speed transport power supply systems, AIP Conference Proceedings 2432, 030084 (2022). https://pubs.aip.org/aip/acp/article-abstract/2432/1/030084/2823165

11. S.F. Amirov, A.Kh. Sulliev, Two-parameter resonant sensors with distributed parameters, J Sensors and Systems 10, 41-43 (2008). https://www.researchgate.net/publication/369755211_Investigation_of_biparametric_resonance_sensors_with_distributed_parameters

12. Sulton Amirov, Islom Karimov and Iroda Abduazimova. Application of the method of direct mathematical modeling to diamond-shaped contact lines. E3S Web of Conf. Volume 401, 2023 V International Scientific Conference "Construction Mechanics, Hydraulics and Water Resources Engineering" (CONMECHYDRO - 2023). https://doi.org/10.1051/e3sconf/202340102060

13. Yakubov Mirdjalil, Karimov Islom, Abdullaev D. Structure and distribution of function between the stages of the information and measurement system for diagnostics of traction substations and contact network installations. 2023 International Scientific Conference Transport Technologies in the 21st Century, TT21C 2023. https://doi.org/10.1051/e3sconf/202338301027

14. Salihjan Khalikov; Sarvar Khalikov; Dilmurod Rahmatov; Shokhrukh Atazhiyev; Abdulla Bakizhanov. Issues of reliability and safety of an electrical substation //Rudenko international conference "methodological problems in reliability study of large energy systems" (rses2021) https://pubs.aip.org/aip/acp/article-abstract/2552/1/060016/2875676.

15. Sarvar Khalikov; Ubaydulla Boqijonov; Humoyun Muminov; Nematulloche Ikromov; Elmurod Yuldashev. Operational management of substation safety power supply // Rudenko international conference "methodological problems in reliability study of large energy systems" (rses2021) https://pubs.aip.org/aip/acp/article-abstract/2552/1/060017/2875673.

16. S. Amirov, K. Turdibekov D. Rustamov, S/ Saydivaliev. Mathematical models of magnetic circuits of high currents induction sensors for electric power supply systems devices of electric transport july 2023 e3s web ofconferences 401(3) doi:10.1051/e3sconf/202340102036

Ion sputtering as methods for generation of cluster particles

Sh.T. Khozhiev[1,2*], *S.E.* Maksimov[2], *Sh.K.* Kuchkanov[2,3], *B.B.* Gaibnazarov[3], *I.A.* Yuldoshev[3], *I.O.* Kosimov[1], *Sh.Yu.* Menglieva[4]

[1]Institute of bioorganic chemistry named after O.S. Sodikov, Academy of Sciences of the Republic of Uzbekistan, 100125, M.Ulugbek Str. 83, Tashkent, Uzbekistan;
[2]Arifov Institute of ion plasma and laser technologies, Academy of Sciences of the Republic of Uzbekistan, 100125, Durmon Yuli Str. 33, Tashkent, Uzbekistan;
[3]Tashkent State Technical University named after Islam Karimov, 100095, Tashkent, University St. 2A Uzbekistan
[4]Tashkent Pharmaceutical Institute, 100015, Oybek street, 45, Tashkent, Uzbekistan.

Abstract: The investigations of emission and fragmentation of silicon oxide clusters sputtered from Si surface have been performed. It has been shown that the processes of formation of these clusters can be qualitatively described within the framework of modern concepts, and the main channels of their formation are determined in accordance with the mechanism of combinatorial synthesis.

1 Introduction

The sputtering of targets by beams of accelerated primary ions is one of the most productive methods for generating cluster [1]. It is known that some of the sputtered particles correspond to cluster ions. Cluster ions are formed in vibrationally excited states and decay on their way from the target to the detector [2]. The discovery of fragmentation of sputtered clusters [2] has significantly complicated the understanding of the cluster emission mechanism, since their decays change the measured mass spectra, kinetic energy spectra and the distribution of their internal energies. Investigation of the processes of cluster decay makes it possible to obtain information both on the emission of clusters and on the chemical and physical properties of these particles. Physical and chemical properties of a cluster, varying from the properties of atoms to the properties of a solid and depending on the size and structure of the cluster, can serve as a basis for the synthesis of crystals with unique properties that promising for solving actual problems of modern nanotechnology [3]. Recently studies, considerable attention is paid to clusters of metal oxides, which play an essential role in modern technology of micro- and nanoelectronics, surface chemistry, catalysis processes, as well as the search for methods of their synthesis and study of fundamental properties. The study of silicon oxide, which plays an important role in these fields is especially important.

The purpose of this work is an experimental study by secondary ion mass spectrometry (SIMS) of the emission and fragmentation of heteronuclear silicon oxide clusters under ion sputtering and analysis of the obtained results from the point of view of existing theoretical concepts. These results are important to understand the nature of the formation of sputtered clusters. From a practical point of view, the obtained results can be essential for solving the problems of the power industry to obtain cheap and environmentally friendly hydrogen fuel.

2 Experimental research

We used the unique capabilities of an ion microanalyzer with double focusing and reverse geometry of Niro - Johnson type (Fig.1) [4] to study the formation and fundamental properties of stable hydrogenated silicon oxide clusters $Si_nO_mH_x^-$.

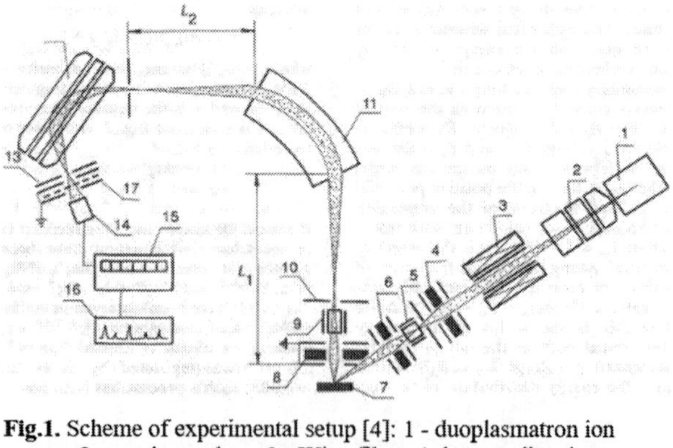

Fig.1. Scheme of experimental setup [4]: 1 - duoplasmatron ion source; 2 - condensor lens; 3 - Wien filter; 4 -intermediate lens; 5 - deflector; 6 - objective lens; 7 sample; 8 - immersion lens; 9 - dynamic transfer system; 10 - entrance slit; 11 - electromagnet; 12 - electrostatic analyser; 13 - energy slit; 14 - electron multiplier; 15 - counter; 16 - plotter register; 17 - grid system; L_1 and L_2 – field-free zones.

* Corresponding author: Khojiev69@bk.ru.

The clusters were synthesized by ion sputtering of the silicon surface by O_2^+ ions while simultaneously exposing the surface to the atmosphere of water wapour. The energy of primary beams was 18,5 keV. To study the emission of $Si_nO_mH_k$ clusters, we used the technique to one described earlier in [5,6], which consists in introducing oxygen or H_2O vapor onto the bombarded target.The dissociation of H_2O upon interaction with the silicon surface with the formation of atomic hydrogen will make it possible to synthesize the silicon oxide clusters and hydrogen atoms in the ion impact zone and generate stable cluster configurations $Si_nO_mH_k^-$. In our experiments, H_2O vapors were injected through a specially designed injection system directly into the bombardment area. The pressure range varied from $P=2*10^{-6}$ Pa (residual vacuum) to $P=4\text{-}5*10^{-3}$ Pa (the maximum allowable pressure of our vacuum system) The method of investigation of fragmentation processes of sputtered cluster ions in field-free zones L_1 and L_2 is described in [4].

3 Research results

The results of study of yields of $Si_nO_{2n+1}^-$ clusters sputtered from the Si surface by O_2^+ ions (Fig.2,3) at different pressures of oxygen and water vapor in the bombardment chamber, showed that $Si_nO_{2n}^-$ and $Si_nO_{2n+1}^-$ have an increased intensity in the mass spectra for all methods of their generation in agreement with [5,6].

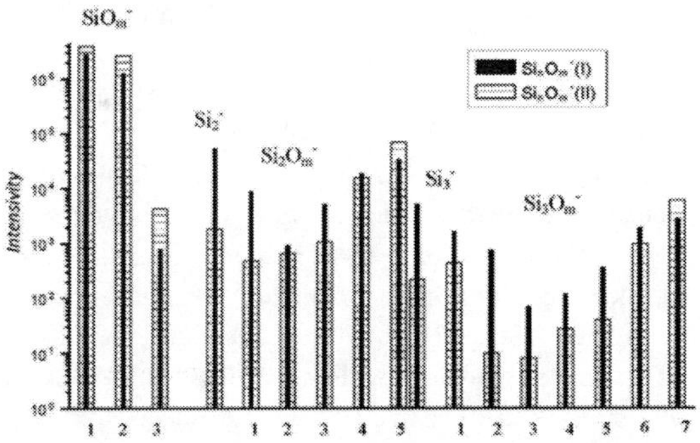

Fig.2. Diagram of yields of $Si_nO_m^-$ clusters (n = 1-3, m = 1… 2n + 1) obtained at a current of O_2^+ I=100 nA on the surface of Si target; I - residual vacuum $P=6.5 * 10^{-6}$ Pa, the surface was cleaned with an ion beam after H_2O was injected into the chamber; II - inlet H_2O into the chamber up to $P=2.5 * 10^{-3}$ Pa.

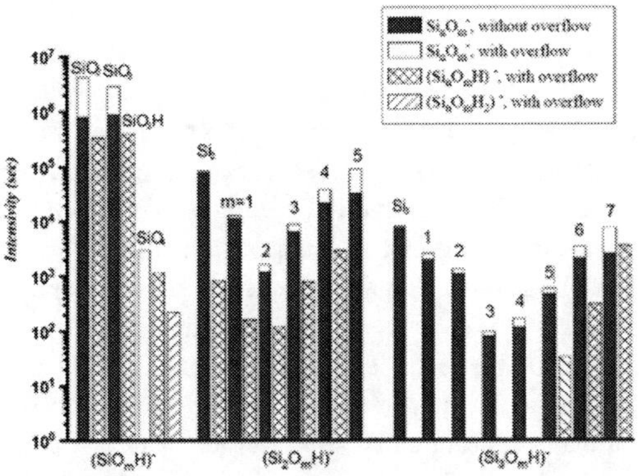

Fig.3. Diagram of yields of $Si_nO_m^-$ and $Si_nO_mH^-$ clusters under bombardment of Si target by O_2^+ ions with H_2O vapor in chamber.

Observation of the change in yields of "magic" [5,6] $Si_nO_{2n+1}^-$ clusters showed that with increasing pressure their intensities increase, and at pressure $P=4\text{-}5*10^{-3}$ Pa the yields of these clusters are maximum. In this case, the absolute values of their intensities when oxygen and water vapor are admitted into the chamber at the same pressure differ little. The most significant difference is that when H_2O is injected, additional intense peaks of $Si_nO_{2n}H_k^-$ and $Si_nO_{2n+1}H_k^-$ (k=1-3) clusters appear in the mass spectrum of sputtered clusters. Fig.2 shows the overall change in the yield of Si_n^- and $Si_nO_{2m}^-$ clusters (n=1-3) when water vapor is admitted into the bombardment chamber at a current of primary O_2^+ ions I_0=100 nA.

As can be seen from Fig.2, the presence of additional oxygen atoms on the sputtered surface significantly changes the mass spectrum of emitted clusters. The peak intensities of both homonuclear Sin- and heteronuclear $Si_nO_m^-$ clusters with m<2n decrease significantly (by one or two orders of magnitude) when H_2O is injected to the surface. The yields of $Si_nO_{2n}^-$ and $Si_nO_{2n+1}^-$ clusters also decrease several times. The increase in the yields of $Si_nO_{2n}^-$ and $Si_nO_{2n+1}^-$ with H_2O vapors can be associated, firstly, with an increase in the concentration of oxygen atoms in the ion bombardment zone, and secondly, with an increased stability of $Si_nO_{2n}^-$ and $Si_nO_{2n+1}^-$ clusters. When operating in the dynamic matching mode [4], the time of re-arrival of the primary beam at a given point of the sputtered surface is approximately 0.02 sec. Obviously, at a pressure in the bombardment chamber $P=4*10^{-3}$ Pa, this time is quite sufficient for the formation of at least one monolayer of gas molecules on the surface. In the case of water entering the surface, the decomposition of H_2O molecules on the radiation-damaged target surface into hydrogen and a hydroxyl group is also possible. Later, with the development of the sputtering cascade caused by the incident ion, oxygen atoms located on the surface are captured into the resulting cluster structures, and a sufficient number of them ensures an increase in the yields of $Si_nO_{2n}^-$ and $Si_nO_{2n+1}^-$ clusters. It is also interesting to note that the total decrease in the yields of homonuclear Si_n^- and heteronuclear $Si_nO_m^-$ clusters with m<2n in each cluster series n with an accuracy of several tens of percent (i.e., practically with an accuracy determined by the measurement technique) is equal to the corresponding increase in the yields of $Si_nO_{2n}^-$ and $Si_nO_{2n+1}^-$ clusters with the same number of silicon atoms n. Thus, an increase in the yield of $Si_nO_{2n}^-$ and $Si_nO_{2n+1}^-$ clusters at an optimal concentration of oxygen atoms on the bombarded surface indicates an increased stability of these clusters. In the case of water being injected onto the bombarded target, hydrogen atoms apparently saturate the existing free bonds in the cluster and stabilize the resulting cluster structures, leading to the effective formation of hydrogenated clusters $Si_nO_{2n+1}H_k^-$ (k=1-3) (Fig.3.).

The values of the dissociation energies for the $Si_nO_{2n+1}^-$ cluster with n = (2-7), obtained by us according to the above-described method from the experiment, are in the range 2.8-4.8 eV [5]. Based on these data, the calculated values of the excitation energies were obtained, which lie in the range 3.68-17.58 eV for clusters with n=2-7 and 0.26-0.35 eV, respectively, per one oscillator. Analisys of ion yields illustrated that 1) the yields of the $Si_nO_{2n+1}^-$ magic cluster with H_2O and O_2 inject are increased compared to the spectrum without injecing, 2) hydrated oxides of the type $Si_nO_{2n+1}H^-$, $Si_nO_{2n+1}H_2^-$, $Si_nO_{2n+1}H_2^-$, and the latter begins to appear only in magic clusters with n≥2. In magic clusters, the peaks of one hydrated (hydride) oxide are somewhat higher than the previous peak of $Si_nO_{2n+1}^-$ oxide.

Hydrogenation is a simple method for stabilizing the silicon surface against oxidation and thus is important in microelectronics. Reaching an inactive silicon surface, ideally ending in hydrogen (hydrogen), has been accomplished in the last 40 years. However, oxidation still takes place even on hydrogen terminated atomistic flat silicon surfaces. Electrochemical hydrogen passivation for easily emitting silicon pores is an essential but unstable process. Recently silicon nanowires have been produced for large scale synthesis. An essential requirement for their widespread use has been fulfilled. This is both technological and scientific importance of finding ways to stabilize them, so as to skillfully avoid the problems of degradation and low photoluminescence. To achieve this goal, the study of hydrogenated silicon clusters in relation to their certain local structural stability should be urgent, but so far, such information is not attainable. In this experiment, we studied the reaction of an aqueous molecule on different hydrogenated silicon clusters in order to relate stability to the local hydrogen configuration, and to shed light on the way to achieve stability, non-reactivity of hydrogenated silicon structures. Since size effects often appear for properties such as energy gap when the size of cluster structures reaches nanometers, we focus our study on the effect of size on oxidation and thus on stability. In [7], models of hydrogenated surfaces Si (001) and Si (111) - Si_9H_{14} - SiH_2 and $Si_{10}H_{15}$ - SiH_3 were obtained, as well as smaller Si_2H_6 - SiH_2 and Si_5H_{10} - SiH_2, SiH_3 - SiH_3 and Si_4H_9 - SiH_3. These reactions were studied to develop a dimensional dependence of reactivity with water and the stability of local structures. The reaction starts from both sides and then a transition state (TS) is formed. The reaction is eventually completed by an H_2 molecule that is released and forms OH attached to the silicon cluster. It turns out that the reaction can go through:

$$Si_9H_{14} - SiH_2 + H_2O \rightarrow IC \rightarrow TS \rightarrow Si_9H_{14} - SiHOH + H_2$$

The appearance of a weak intermediate complex of a similar reaction - IC, which is 1.4 kcal/mol lower than the reaction. In this case, the O-Si1 distance = 4.054 Å. Thus, it is similar to the fact that the forces for this weak chelator are the dipole-dipole interaction. The system then has to go through a transient state TS in order to reach the product. In the case of TS with the O-Si1 bond, the distance between the attached O of the H_2O atom and the attached silicone atom on the surface of 1.855Å is slightly longer than in the product; accordingly, the formed H_1 - H_2 bond is much longer than the regular value in the product. Meanwhile, one of the two hydrogen atoms in water, H1, moves away from the oxygen atom, resulting in an H_1 - O distance of 1.110Å, which is drawn longer than IC. The hydrogen atom H_2 is 1.842110Å, away from the silicon atom. The energy barrier to this reaction is 44.5 kcal/mol in relation to the reactants. From the point of view of thermodynamics, this reaction is beneficial, because it is exothermic at 14.9 kcal/mol. For other complexes - similar reactions. It is noticed that each of these reactions includes an intermediate complex with an energy of almost 2.3 kcal/mol. With an expansion in the bunch size of the response of each sort of silicon-hydrogen arrangement, the energy hindrance diminishes, everything shows an increment in the reactivity of the framework. Responses with SiH3 show a more modest energy obstruction than with SiH_2. To give more positive help for the reactivity pattern, rate constants were determined at 1 atm. for a watery response inside different hydrogenated setups. For reactions on a dihydride, the calculated reaction rate for medium clusters shows a much larger increase in relation to small clusters, but can only increase slightly with increasing cluster size. Confirmation of the above may also appear from the analysis of their boundary orbitals. It has previously been found that the overlap between the highest occupied molecular orbitals (HOMO) of a molecule and the lowest unoccupied molecular orbitals (LUMO) can be determined by the nature of the chemical reaction. A smaller energy difference between HOMO of one molecule (electron donor) and LUMO of another (electron acceptor) might show a more preferable reaction. Thus, the tendency is the close ratio of HOMO and LUMO of individual hydrogen-silicon clusters, for which, with an increase in the cluster size, their HOMO generally moves forward and LUMO falls, as a result of which the energy gap (gap) decreases. Since the energy barrier size effect is well known for silicon clusters when dimensional changes are in the nanometer range, the correlation of reactivity with the well documented energy barrier size effect could provide an important implication for the size dependent reactivity found here. Based on small silicon clusters, it is expected that the reactivity and rate constants for large clusters will also stabilize for a given temperature and pressure.

4 Conclusion

The performed experiments indicate that ion sputtering is an effective method for generating heteronuclear oxide clusters $Si_nO_m^-$ and $Si_nO_mH_k^-$ of various sizes. Qualitatively, the processes of formation of these heteronuclear clusters can be described in terms of modern concepts [8], and the main channels for their formation are determined in accordance with the mechanism of combinatorial synthesis [6,9-11].

References

1. A Wucher. //Matematisk-fysiske Meddelelser. Det Kongelige Danske Videnskabernes Selskab. Copenhagen, 2006. V.52. P.405-432.

2. N.Kh.Dzhemilev, U.Kh.Rasulev, S.V.Verkhoturov. //Nucl. Instr. Meth. B. 1987. V.29. P.531-536. https://doi.org/10.1016/0168-583X(87)90065-6

3. B. Hagenhoff, in J. S. Vickerman, D. Briggs (Eds.), ToF-SIMS: Surf. Analys. By Mass Spectr., IM Publ. And Surf. Spectra Lim., Manchester and Chichester. 2001. P. 285-308.

4. A.D.Bekkerman, N.Kh.Dzemilev, V.M. Rotshteyn. //Surf. and Interf. Anal. 1990. V.15. P.587-590.

5. N.Kh.Dzhemilev, L.F.Lifanova, S.E.Maksimov, Sh.T.Khojiev. // Poverkhnost. 2006. № 7. P. 68-71 (in Russian).

6. N.Kh.Dzhemilev. //Journal of Surface Investigation. X-ray, Synchrotron and Neutron Techniques. 2012. V.6. №4. P.254–258. DOI: 10.1134/S1027451012080083

7. I.A. Wojciechowski, U. Kutliev, Sh. Sun, Ch. Szakal, N. Winograd, B. J. Garrison. //Appl. Surf. Scienc. 2004. V.231-232. P.72-77.

8. Sh.T.Khozhiev, Kh.B.Ashurov, S.E.Maksimov, B.L.Oksengendler, N.Yu.Turaev. //Reports of the Academy of Sciences of the Republic of Uzbekistan. 2021. №4. P.21-25 (in Russian).

9. Dzhemilev N.Kh., Maksimov S.E., Khozhiev Sh.T. Statistical Mechanism of the Formation of the Energy Spectra of Sputtered Molecular Clusters. // Journal of Surface Investigation. X-ray, Synchrotron and Neutron Techniques. – 2014. – vol. 8. – No 5. – pp. 1063-1067.

10. Dzhemilev N.Kh., Kovalenko S.F., Maksimov S.E., Tukfatullin O.F., Khojiev Sh.T. Study of the formation and unimolecular fragmentation of $Si_nO_m^+$ clusters under ion bombardment. // Journal of Surface Investigation: X-ray, Synchrotron and Neutron Techniques. - 2017. - vol. 11. - No 3. - pp. 490–495.

11. Khojiev Sh.T., Tashmukhamedova. D.A., Rakhimov I.T., Gulbekov B.H. Processes of formation and mechanisms of formation of metal clusters of $Nb_nO_m^+$. // Technical science and innovation. – 2020. – No 1. – pp.142-147.

Soft start of induction electric motors using rezistor and denistor devices

Ikram Davletov[1], *Nemat* Khakimov[1], *Alibek* Qodirov[1], *Munisbek* Akhmedov[1], *Jurabek* Sadullayev[1], *Ogabek* Khamraev[1]*

[1]Urganch State University, Urganch city, Uzbekistan

Abstract. In this article, it is experimentally shown that by equipping asynchronous electric motorswith a new rezistor and denistor soft starting device, it is possible to effectively limit the current. In addition, as a result of using rezistor and denistor soft starting device in practice, it was determined that electrical energy wastage is saved and mechanical breakdowns are prevented.

1 Introduction

It is known that the real essence of the problem is that in the process of starting asynchronous electric motors, that is, when connecting it to the source, more power than the nominal power is required. This law of physics is valid for all objects in the world - after all, it is always more difficult to start a movement than to continue it. Today, asynchronous electric motors with a short-circuited rotor are the most widely used devices in production enterprises. Therefore, prevention of energy wastage and mechanical damage caused by overcurrent during the start-up and shutdown of these devices is one of the urgent problems of the traditional energy system. Scientists have published many scientific articles aimed at solving the problem of smooth start-up of asynchronous electric motors.

2 Experimental research

During the soft start-up of electric motors, the current slowly reaches the nominal value based on the non-linear dependence on time and decreases from the nominal value to zero value during shutdown. Figure 1 shows the time dependence of the current in the process of starting and stopping the asynchronous electric motor.

In fact, the current should ideally be as shown in Figure 1 a). But we can see in Fig. 1 b) that this is not the case in reality. As can be seen from the graph, the starting current of the asynchronous electric motor is more than the nominal current. So, it can be seen that the current is much higher than the nominal value when encountering inductive resistance. In this case, the question arises as to how the starting current of electric motors is determined.

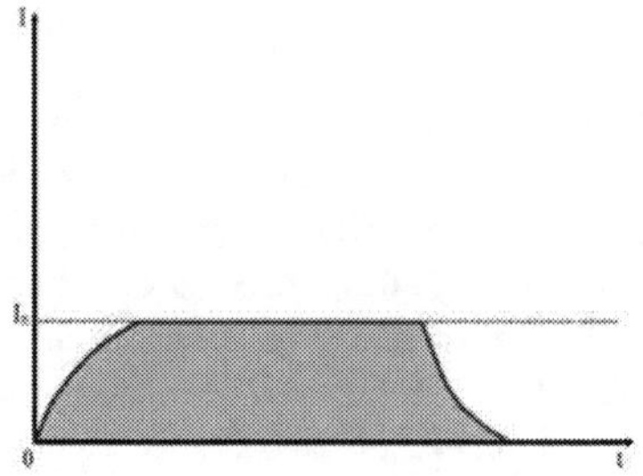

a)

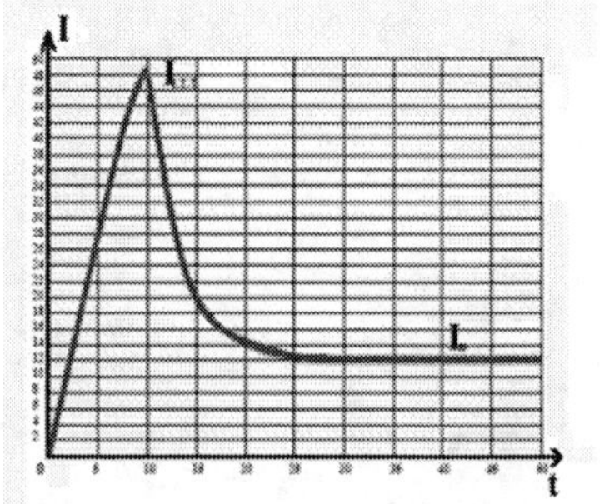

b)

Fig. 1. Time dependence of the current in the processof starting and stopping the asynchronous electric motor.
(a-ideal state, b-real state)

We can determine and calculate the starting current of the electric motor with the following formula.

$$I_{ish.t} = I_n * K \quad (1)$$

* Corresponding author: ogabekxamrayev678@gmail.com

 https://doi.org/10.1051/e3sconf/202346101066

Here $I_{ish.t}$ - the value of the current during the start-up process of the electric motor, I_n- pre-calculated rated current, K – the duration of the electric motor starting current (obtained from the passport data of the electric motor). If we know the power values of the electricmotor in the passport, it will not be difficult to calculate its current values. For example, let us know the nominal current of a 45 kW electric motor. In this case, we determine the current of the electric motor by power as follows. If the electric motor is connected to a 380 V three-phase system, its current can be fully determined:

$$I_n = P_n/\sqrt{3}(U_n * \eta * cos\varphi) \quad (2)$$

I_n= 45000/√3(380*0,92*0,85) =87,43A

I_n – current power of an asynchronous electric motor P_n – The nominal power of the electric motor is 45 kW
√3 – square root of one third = 1,73205
U_n – voltage in the circuit 380 V
η – useful work coefficient 92% (when calculating 0,92)

If the FIK and power factor of the electric motor are unknown, we can find its rated current with a small error of two amperes per kilowatt. We determine the current of the electric motor by the following formula:

$$I_n = P * 2 \quad (3)$$

Now let's talk about the consequences of overcurrent in the process of starting asynchronous electric motors. This is quite a problem and can lead to:

- leads to overloading of the electrical network, heating (up to the melting of contacts) and voltage drop;
- Overloading and overheating of the engine leads to its premature wear. Some manufacturers indicate the maximum number of starts per hour or day due to overheating among the parameters of the electric motor;
- premature wear and overloading of mechanical mechanisms, especially when moving with a large moment of inertia (bearings, reducers, belts);
- the noise caused by the addition of contactors, which is transmitted not only through wires, but also through the electromagnetic field;
- problems with technology - it is impossible to start many processes at once.

It is at this critical moment that the "weak link" may appear. In addition, many consumers of electricity in this network face problems such as dimming of light bulbs due to low voltage, and contactors may stop due tostrong noise. From this we can conclude that the current in the process of starting and stopping is an unavoidable problem. If this condition is not treated using effective methods, it can lead to bad consequences.

This article talks about the fact that the rezistor and denistor soft starting device for asynchronous electric motors is a device that effectively limits the starting current compared to other methods, as well as the types of soft starting devices for asynchronous electric motors used in industrial enterprises and their characteristics. In addition, it has been experimentally shown that by using rezistor and denistor soft starting device, electricity waste can be saved as a result of current reduction and mechanical breakdowns can be prevented.

Before proceeding to the discussion of the results obtained in the work, we will dwell on the features and shortcomings of devices for reducing excess currents in the startup process that have been used so far.

In the first method, a reduced voltage is supplied to the electric motor, and then, when the electric motor accelerates, it is necessary to increase the voltage and rotation speed to the nominal value. This method is used in electronic devices for starting electric motors - soft starters (SCP) and frequency inverters (frequency).Figure 2 shows the time dependence of the starting current of electric motors when using the softstarter method. It can be seen from the graph that the starting current is much higher than the nominal value.

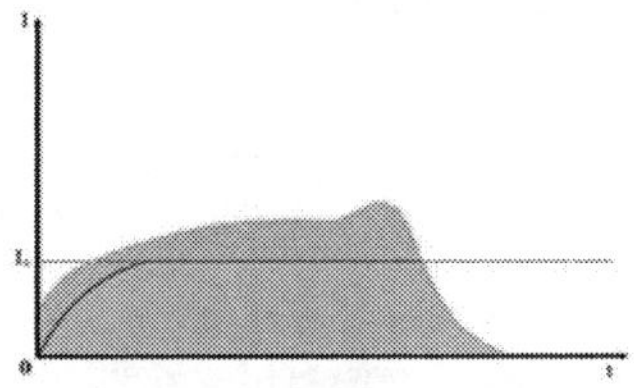

Fig. 2. Time dependence graph of electric motor startingcurrent when using the first method

This is shown in red. The disadvantages of this method are as follows:

- first of all, the electronic part for high-power asynchronous electric motors has not been developed today;
-secondly, as a result of the influence of the external environment, the parameters of the electric motor change and lead to limitations in the management of large capacities;
- thirdly, the price of the device is very high.

The second method. It is necessary to use current limiters during the start-up process. At first, the electric motor is powered by powerful rezistors, and then it is transferred to the nominal value by a timer. The resistance of the rezistors corresponds to the resistanceof the starter coils. This device is very easy to make (contactor and time relay). The advantage is that theelectrical circuit of the device is very simple.

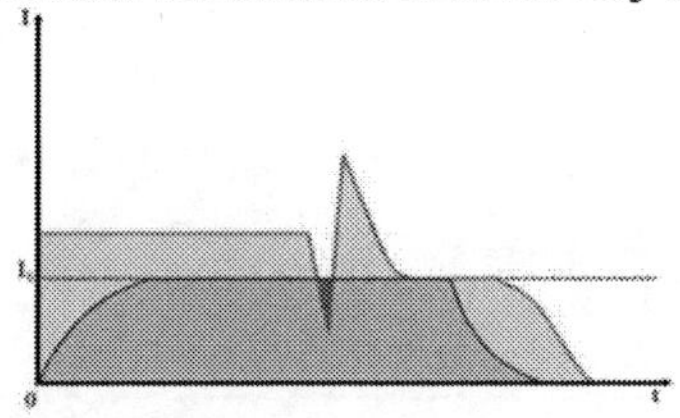

Fig. 3. Time dependence graph of electric motor starting current when using the second method

The third method. In this method, the electric motor is given full voltage at once, but first it is necessary to connect the coils in such a way that the electric motor does not rotate at full power. This scheme is called the Triangle-Star method. The disadvantages of this method are as follows:

- firstly, not all asynchronous electric motors can be connected in this way;

-secondly, it is necessary to carry two cables to connect the asynchronous electric motors to the network;

-three contactors are used to switch an asynchronous electric motor from a star connection to a delta connection.

3 Research results

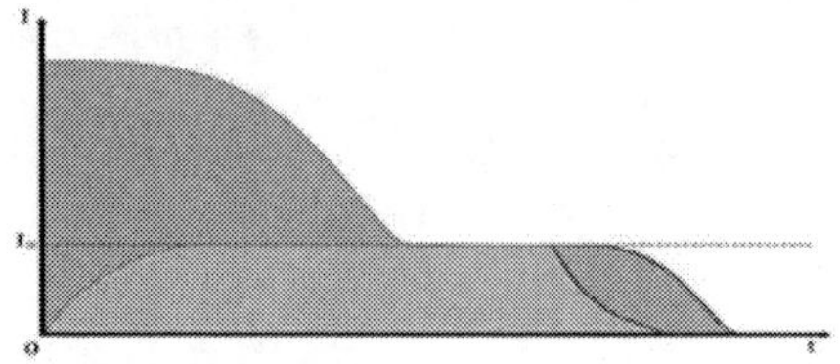

Fig. 4. Time dependence graph of electric motor start-up current when using the third method

The cheapest and most energy-efficient device among the above three connection methods is the contactor and time relay connection.

In this work, using the connection method we chose, instead of the time relay in the electrical circuit, the start-up current was brought to the minimum value by using an electronic switch (denistor) that connects depending on the voltage change in the internal electrical processes.

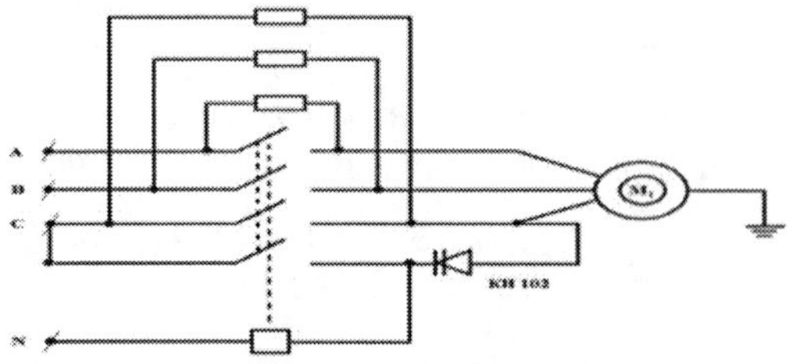

Fig. 5. Electrical diagram of the proposed rezistorand denistor soft starter

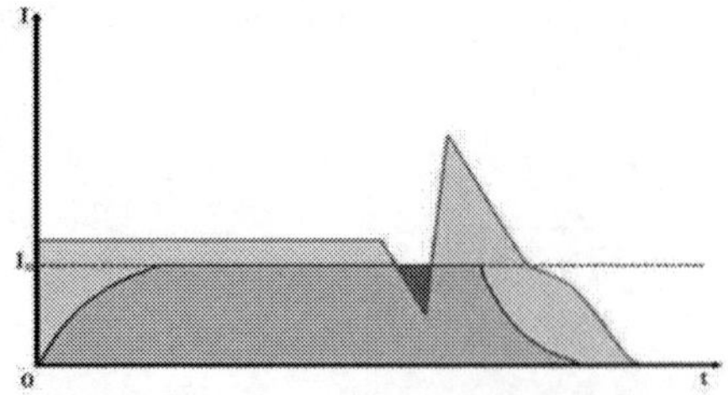

Fig. 6. Plot of current versus time during starting using the proposed rezistor and denistor soft starter

From the above graph, it can be seen that the rezistor and denistor soft starting device that we have chosen is more effective than other devices and methods. That is, the surface of the work performed by wasted electrical energy is the smallest. This is the surface shown in red inthe graph.

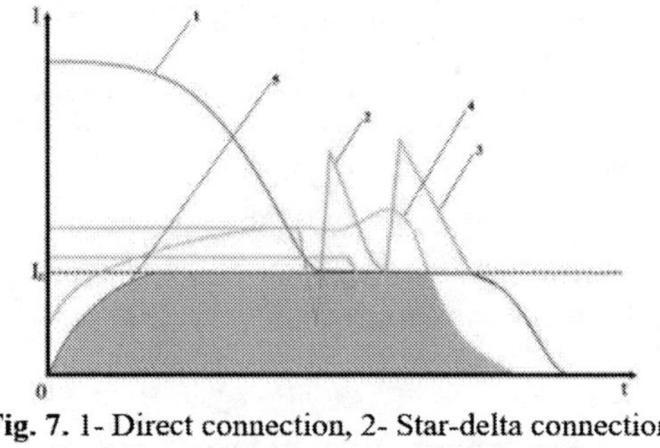

Fig. 7. 1- Direct connection, 2- Star-delta connection, 3- Rezistor and denistor connection, 4- Soft start (softstarter) connection

Conclusion

Several times the starting current of asynchronous electric motors is considered a big problem and has negative consequences. It can cause overloading, heatingand voltage drop of the electric network, wear and overheating of electric motors, wear and overload of mechanical mechanisms (bearings, reducers, belts), noise caused by connecting contactors, and other similar consequences. it is. A soft starting device for asynchronous electric motors with a rezistor and a denistor is effective in solving this problem, and the sharp increase in the current during the start-up is eliminated. As a result, excess electrical energy losses and mechanical breakdowns occurring in the system are prevented.

Here's a basic overview of how to achieve soft starting using resistor and reactor devices:

Primary Components:

Motor: The induction motor you want to start softly. Resistor: A power resistor that's connected in series with the motor during the starting process. Reactor (Denistor): Also known as an autotransformer, a reactor is connected in series with the motor to limit the starting current.

Start-up Sequence:

When the motor is initially energized, the resistor is connected in series with the motor. This effectively reduces the voltage applied to the motor. As the motor accelerates, the resistor is gradually bypassed or removed from the circuit. This allows the motor to receive the full voltage, and it reaches its rated speed.

Benefits:

Reduced current and torque peaks during motor

startup, which can be beneficial for the power supply and the connected equipment. Minimized mechanical and thermal stresses on the motor and the driven machinery.

Control Mechanism:

Typically, a controller or starter, often based on microprocessor technology, is used to automate the soft starting process. The controller is programmed to control the gradual reduction of the resistance in the circuit.

Adjustability:

Soft starters can be adjusted to vary the rate at which the resistance is reduced, allowing for fine-tuning based on the specific motor and load requirements.

Considerations:

The resistor can dissipate a significant amount of energy as heat, so it's important to size it correctly and choose a suitable type (e.g., wire-wound resistors). Denistor or reactors must be appropriately sized and rated to handle the motor's current and voltage requirements.

Soft starting methods can enhance the operational lifespan of the motor and reduce wear and tear on mechanical components. They are often used in applications where starting the motor abruptly could lead to damage or cause voltage disturbances in the power supply network.

It's essential to consult with a qualified electrical engineer or technician when designing and implementing a soft starting system for induction motors, as the specific requirements can vary depending on the motor's characteristics and the application.

References

1. Savvin N. Y., Riylov I. V., Ratushnyak V. R., Kaydalov M. V. Asinkhronniy elektrodvigatel // fundamentalniye i prikladniyye nauchniye issledovaniya: Aktualniye voprosiy, dostijeniya i innovatsii : statya v sbornike trudov konferensii. - OOO "Nauka i prosvesheniye", Penza, 2021.

2. Leontyev G.A., Zenina Ye.G. Issledovaniye asinkhronniykh dvigateley s korotkozamknutiyam i fazniym rotorom. - Volgograd.: Volgogradskiy gos. tekh.un-t., 2000.

3. Tuchkevich V. M., Grekhov I. V. // Noviyye printsipiy kommutatsii bolshikh moshnostey poluprovodnikoviymi priborami// L.:Nauka: Leningr. otdeleniye AN SSSR, ISBN 5-02-024559-3 (1988)

4. "Plavniy pusk". machinedesign.com 2014-07-16

5. Petrov L.P., Ladenzon V.A., Obukhovskiy M.P., Podzol R.G.. Asinkhronniy elektroprivod s tiristorniymi kommutatorami. — Moskva: Energiya, 1970.

6. Sovremenniy uroven razrabotok asinkhronniykh tiristorniykh elektroprivodov s fazoviym upravleniyem / obzornaya informatsiya. — Moskva: Informelektro,1975.

7. L.P. Petrov, O.A. Andryushenko, V.I. Kapinos. Tiristorniyye preobrazovateli napryajeniya dlya asinkhronnogo elektroprivoda. - Moskva: Energoatomizdat, 1986.

8. Djans, Tomas M.; Ouen, Edvard L. (yanvar 2001). "Privodiy peremennogo toka s reguliruyemoy chastotoy vrasheniya v epokhu tysyacheletiya: kak miy syuda popali?". IEEE Transactions po silovoy elektronike. 16 (1): 17

9. I.Ya. Braslavskiy, Z.SH. Ishmatov, V.N. Polyakov. Energo - sberegayushiy asinkhronniy elektroprivod: uchebnoye posobiye/. — Moskva: Akademiya, 2004.

10. K.N. Lebedev, S.A. Buzun. Adaptivniyye softstarteriy dlya pogruzhniykh elektronasosniykh agregatov: Monografiya. — Zernograd: Azovo-Chernomorskaya gosudarstvennaya agroinjenernaya akademiya, 2012.

11. Bartos, Frenk Dj. "Privodiy peremennogo toka ostayutsya jiznenno vajniymi dlya 21-go veka". Injeneriya upravleniya. (2004-09-01).

12. Eisenbrown, Robert E. "Privodiy peremennogo toka, istoricheskiye i budushiye perspektiviy innovatsiy i rosta". Osnovnaya prezentatsiya k 25-letiyu Viskonsinskogo konsortsiuma elektricheskikh mashin i silovoy elektroniki (WEMPEC). Universitet Viskonsina, Medison, Viskonsin, USA: WEMPEC. (2008-05-18).

E3S Web of Conferences **461**, 01067 (2023) https://doi.org/10.1051/e3sconf/202346101067
RSES 2023

Problems of ensuring security in open – pipe coal mines in the republic of Uzbekistan

Gulnora Gulomova[1*], *Urxiya* Turabekova[1]., *Dilrabo* Nizamova[1]., *Rayhon* Mudarisova[2]

[1]Tashkent State Technical University named after Islam Karimov, 100095, Uzbekistan, Tashkent, University St. 2A.

[2]Uzbek State University of World Languages, Tashkent, block G-9a, Kichik Khalka Yuli st., 21a.

Abstract. This article provides information on ensuring safety at coal mines by analyzing existing coal mining facilities. Information on the Angren coal mine is also given. It is shown that there are no accurate methods for determining the location of fires in order to increase the efficiency of spontaneous combustion prevention by supplying liquid compositions such as water and clay pulp flowing down the formation soil. Measures have been proposed to suppress spontaneous combustion centers, nitrogen is being increasingly used, the supply of which ensures volumetric processing of the collapsed mass and allows reducing the oxygen concentration in the mined-out space.

1 Introduction

The coal industry of the country, integrating a complex of various production processes, is the most dangerous injury not only due to the aging of the mine fund and the deterioration of the mining and geological conditions of coal mining, but also due to violation of the rules for the operation of coal deposits and coal mining technologies [4].

Coal mining has a huge negative impact on the environment, has one of the highest injury rates. One of the dangerous phenomena in the coal mining industry is endogenous fires. Endogenous fires associated with spontaneous combustion of seams pose a great danger to the life and health of miners. In addition, they lead to long shutdowns of mining operations, conservation of coal reserves and, subsequently, to a deterioration in the technical and economic performance of the enterprise.

Operation of obsolete equipment, low level of engineering and technical solutions to ensure production safety, unsatisfactory operational control by the departments of labor protection and safety, violations of the compliance of actual technological parameters with regulatory documents, low technological and production discipline, that is, the lack of a system for ensuring effective preventive measures development of industrial hazards, contributes to the manifestation and implementation of the latter, leading to accidents, explosions, fires, occupational diseases, injury, death of people and the failure of the cut for a long time. Therefore, the study of the causes and the development of preventive measures to reduce the fire hazard of coal seams is a very urgent task [1].

2 Experimental research

Insufficient knowledge of the mechanism for the implementation of industrial hazards and causes, the lack of a systematic analysis of the determining factors and causes, comprehensive criteria for the quantitative assessment of industrial hazards, workable methods for predicting the level of safety at the design stage and organization of production makes it difficult to develop and implement measures that ensure the effectiveness of decisions taken to minimize the level accident and injury rates in coal mines and cuts.

Based on the foregoing, it can be argued that scientific research in the direction of developing a system for ensuring the safety of the main production processes when performing mining operations in a coal mine is very relevant. And also relevant is the establishment of patterns of influence of a complex of mining-geological and technological factors on the characteristics of industrial hazard for the development of a system for ensuring the safety of the main production processes during mining operations, which allows developing measures to effectively reduce injuries in coal mines

It must be said that virtually all mining enterprises produce dust, not only coal mines or coal transportation and transshipment facilities. Just black dust is the most common and most noticeable. The task of dust suppression is a common one for many industrial companies [2].

The idea of the work is to comprehensively take into account the mining and geological conditions and technological parameters of production when predicting the level of industrial hazard in order to make decisions

* Corresponding author: raxmatillo82@mail.ru

that reduce the likelihood of injury (risk) for miners in the main professions.

Fig. 1. Coal excavation for vehicles.

Therefore, it is necessary to take appropriate measures to reduce the negative impact of the coal mining process on the environment and nearby settlements by adopting environmental technologies to prevent the formation of coal dust and flue gases from coal fires.

When coal is transported by large dump trucks of the Belaz type, a huge amount of dust also rises into the air, which can be extinguished by adding special pastes to the road soil.

Fig 2. General view of the Angren section.

As can be seen from the figure, the section has a huge territory. During open-pit coal mining, collapses, landslides, formation of cracks and destruction in the walls of the deposit are possible due to a violation of the mining technology.

Fig 3. Extraction of a thick coal seam.

Such undesirable phenomena in a coal mine can be constant causes of various accidents, accompanying the appearance of injury among the working personnel of the mine.

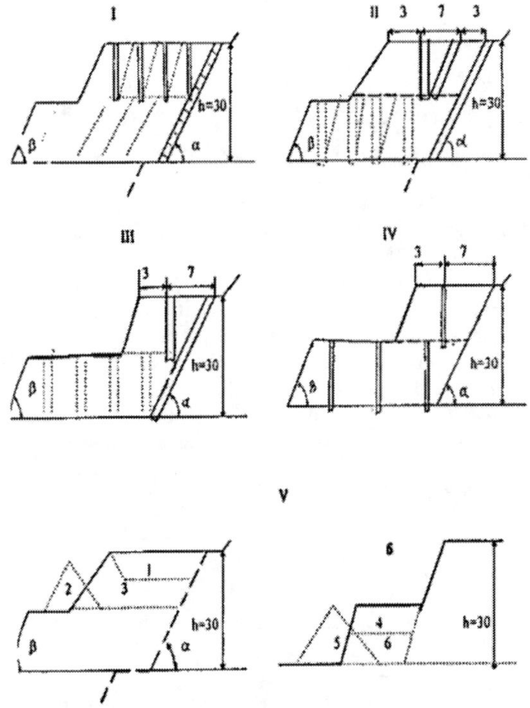

Fig 4. Standard rules for technical operation of mineral resources deposits during open-open mining.

Therefore, we have set the goal of analyzing and monitoring cases of accidents and gross violations of the relevant established coal mining technologies, which in turn attract casualties and injuries among the working personnel, as well as developing special measures and measures to ensure the safety of the coal mining process.

3 Research results

The Angren coal deposit is a brown coal deposit in the Akhangaron valley of the Tashkent region, located at the foot of the Kurama and Chatkal mountains. Exploration work began in 1934. The first mine was built in 1940, which was launched the same year. The coal basin covers about 70 km2. Proved reserves are estimated at 1.9 billion tons. Located between the Jurassic deposits, the formation is very loose, with a thickness of 20 m at the surface and 130 m at depth. Coal belongs to brand B2. Calorific value 13.9 MJ/kg. Coal is mined mainly by open pit mining. In addition, at the "Erostigas" station, coal is processed into combustible gas by coal gasification. Along with coal, kaolin is also mined. Portland cement, flux, shoulder and many rocks are laid between the layers, which can be used for the production of sewer pipes. More than 85% of the total coal produced in Uzbekistan comes from the Angren coal mine.

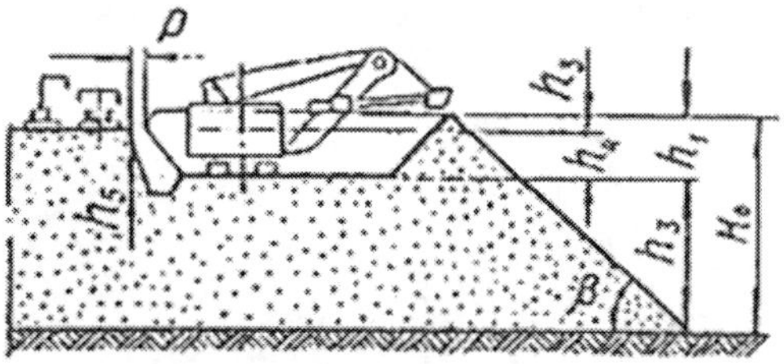

Fig 5. Open pit mining.

However, there are a number of problems with coal mining. The most important of these is environmental pollution. When coal is mined, especially in an open pit, a large amount of waste rock is formed in the form of coal dust and artificial embankments. Tailings of coal in dumps sometimes ignite spontaneously, while the atmospheric air is charged with flue gases. In addition, when coal is transported by heavy dump trucks of the Belaz type, a large amount of coal and soil dust rises into the atmospheric air. All this negatively affects the environment and the health of workers, as well as the nearby communities.

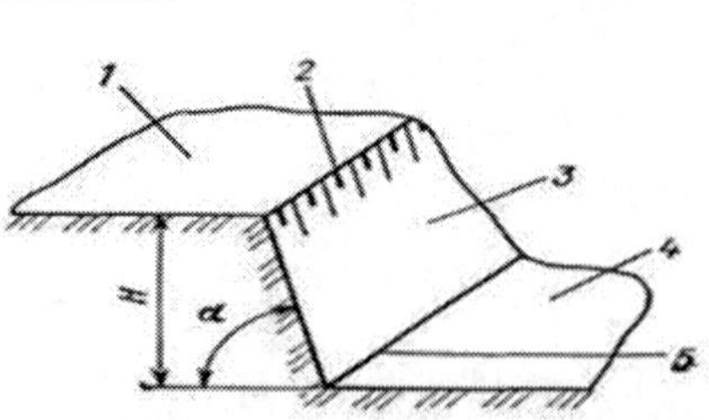

Fig 6. Main elements of the ledge: 1-upper platform, 2-upper edge, 3-slope, 4-lower platform, α-angle of slope of the ledge, *H*-height of the ledge.

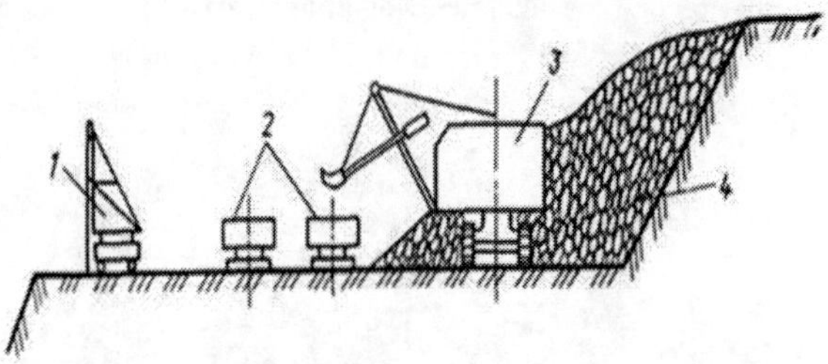

Fig 7. Bench working area: 1-drill rig, 2-vehicles, 3-excavator, 4-blasted rock.

In order to identify the highest risk HIFs, territories and industries, a classifier of the degree of accident risk has been developed using the developed risk-oriented indicators and criteria for categorizing HIFs given in table 1.

Table 1. Criteria for the degree of danger of accidents at on-site (on-site) hazardous production facilities

Comparative degree of danger of an accident at a component of a hazardous facility	The maximum possible number of victims whose life or health may be harmed as a result of an accident at the facility or component of the educational program
Small	Less - 10
Average	10-74
High	75-300
Extremely high	More - 300

To establish the hazard category of a production facility based on the level of accident risk, the hazard indicators calculated for hazardous production facilities are compared with the general criteria for categorizing emergency hazards. The upper values of the ranges of accident hazard indicators at hazardous production facilities with a high accident risk (Table 1) characterize the possibility of major accidents, and at hazardous production facilities with an extremely high accident risk, especially large accidents [3]

Dust suppression methods in quarries. The production process of open-cast coal mining is associated with a high level of occupational risk, which is formed mainly due to the intense inhalation exposure to coal dust against the backdrop of an unfavorable microclimate [8]. Dust control in mining operations includes dust prevention, dust suppression and dust collection. If it is impossible to completely exclude dust formation, along with dust collection at mining enterprises, in various technological processes, dust suppression of dust in the atmosphere (coagulation) and deposition directly at the places of dust formation, as well as dedusting (artificial) ventilation - liquefaction and removal of a dust cloud are widely used [7].

Technical measures do not always provide a full-fledged healthy dust environment in the workplace (reduction of dust content in the air to maximum permissible concentrations), in such cases it is mandatory to use personal respiratory protection equipment.

However, many coal mines and cuts operate in regions with a high temperature regime. In this regard, at these enterprises it is necessary to use special personal protective equipment that takes into account the negative ambient temperature.

4 Conclusion

Thus, it has been established that unfavorable labor factors at mining facilities include: sharply changing temperatures of air and rocks; high humidity; often - high speeds of air movement along the main workings; dust, explosive and exhaust gases; noise, vibration, souffle emissions of toxic gases; increased risk of injury and the development of occupational diseases. Dust remains the main unfavorable professional factor at the majority of mining facilities in the North, the concentration of which depends on the development technology, type of mineral, equipment used, production operation, and implementation of anti-dust measures.

In each case, these issues are resolved individually. The set of measures to reduce the factors of the working environment affecting miners should include the use of personal protective equipment. However, it must be understood that they are not able to replace the use of radical and complex technical measures to combat dust, such as dust collection and dust suppression.

References

1. G.M. Gulomova. Theoretical and practical aspects of studying Angren coal for the production of sorbents // International Conference of Academic Sciences, Novosibirsk, October 2021, pp. 92-97.
2. G.M. Gulyamova., M.N. Musaev. Problems of sludge waste disposal in the oil and gas industry. Collection of materials of the VII International Correspondence Scientific and Practical Conference, June 5, 2020, pp. 98-100. Minsk.
3. G.M. Gulomova, U.M. Turabekova. Problems of ensuring safety at open coal mines in the republic of Uzbekistan. TECHNICAL SCIENCE AND INNOVATION №3 ISSN: 2181-0400, (2022)
4. G.M.Gulomova, O.T.Tuxtamurodov, and ofters. AIP Conference Proceedings, 224-230, (2023) https://zenodo.org/record/7985791
5. G.M. Gulomova, E.B. Abdurakhmanov. Energy of adsorption of benzene vapors by activated carbon. Joutral. №11-1 (89), 100-102, (2021) https://scholar.google.com/citations?view
6. G.M Gulomova, E.B Abduraxmonov, Adsorption energy of toluene vapors into an adsorbent obtained on the basis of local angrene coal. Harvard Educational and Scientific Review, (2021) https://www.journals.company/index.php/article/view/14
7. G.M.Gulomova, M.N.Musaev, Problems of Ensuring Safety at Open Pits in the Republic of Uzbekistan. JournalNX, 314-319, (2020) https://scholar.google.com/citations
8. D.Rakhmatova, M.Musaev, G.Gulomova, M.Aripkhadzhaeva, D.Nizamova, Development of safe products from local waste of oil and fat production E3S Web of Conferences, 371, (2023) https://scholar.google.com/citations .
9. Gulomova G.M., Bozorov Z. P., Talipova Yu. A. The influence of the environment on the strength of building structures. Novateur publication, Indian magazine NX. May 25 and 26, 2021, pp.733-737.
10. A.I. Grazhdankin. Features of analysis and assessment of the consequences of accidents at hazardous production facilities // Labor Safety in Industry. – 2003. – No. 08. – P.45-46.
11. A.I. Grazhdankin. Accident risk and assessment of unwanted losses / A.I. Grazhdankin, D.V. Degtyarev, M.V. Lisanov, A.S. Pecherkin//Life safety. –2002. – No.
12. Aripkhodzhaeva M.B., Suleimanov A.A. Drones for monitoring the operational situation in emergency situations. Far Eastern Spring 2019, international scientific and practical conference on environmental and safety issues, Komsomolsk-on-Amur. 2019, - S. 173-175. ISBN 978-5-7765-1406-7.
13. Suleimanov A.A. The activities of the consolidated headquarters created in case of extreme and emergency events. UzR Milliy Guard Harbiy Technician Institute Harbiy-Technician Institute Akhborotnomashi. - Tashkent, 2018, - p. 158-159.
14. Bodrova A.S., Bezdenezhnykh S.I. Prospects for the development and use of complexes with unmanned aerial vehicles: Conf. Kolomna, 2016. 274 p.
15. Organization of protection of the population and territories from emergency situations: a textbook. - 3rd ed., revised. and additional / V. A. Sednev, S. I. Voronov, I. A. Lisenko, E. I. Koshevaya, N. A. Savchenko, N. I. Sedikh.
16. Emergency situations of peacetime and wartime. Characteristics of emergency zones: methodological development / [compiled by V. A. Gorishny, V. B. Chernetsov, L. N. Borisenko]. - N. Novgorod, NSTU, 2006.
17. Forecasting emergency situations of a natural nature: B.A. Khramsov, T.G. Bolotskikh, A.M. Yurev. Methodical instructions. –Belgrod: Publishing house of BSTU im. V.G. Shukhova, 2006. - 25p.
18. Madzhidov I.U., Ibragimov B.T., Suleymanov A.A. Composições modernas de isolação térmica para o aumento da resistência térmica das estruturas de concreto reforçadas Ciência, engenharia e tecnologia 4 de julho, SÃO PAULO, BRASIL 2019 Publicado por Science WA Ltd pp 78-86.
19. Levitt R.E., Kunz J.C. "Using Artificial Intelligence Techniques to Support Project Management." Artificial Intelligence in Engineering Design, Analysis and Manufacturing. Vol. 1, No. 1 (2017). pp.3-24.
20. Vambol S.O., Kondratenko O.M., Mishchenko I.V., Koloskov V.Yu. (2018). Investigati-on of hydraulic jets during creation of ecological safety management system of high-risk objects: Monograph, Kharkiv, Publ. NUCD of Ukraine, 204 p. ISBN 978-617-7555-58-1.

Efficient and economical energy use technology in the processing of domestic coarse wool fiber

Odil Toshbekov[1*], *Mustafokul* Urazov[1], *Shavkat* Ermatov[1], *G'ayrat* Arapov[1], *Maftuna* Khamraeva[1]

[1]Termez State University, 190111, Uzbekistan, Termez, Barkamol avlod St. 43

Abstract. There are many types of fabrics produced in the textile industry, and the scale of production and industrial use of non-woven fabrics is increasing with the help of textile technology. The production of non-woven products is considered a relatively new branch of the textile industry, and this type of products is characterized by low cost, originality of quality, and variety of production methods. Many types of fabrics have the potential to be used as short-term fabric substitutes in the textile, engineering, construction, automotive, shipbuilding, aviation, and medical industries.

1 Introduction

The essence of the method of determining the limit of deformation stages of non-woven fabric from coarse wool fiber is that the disproportion of non-linear diagrams of textile materials on stretching is the result of changes in their mechanical properties under stretching conditions. In the process of stretching, the initial structure of the material changes, as well as density, elasticity, plastic modulus, etc.

Based on these characteristics, the determination of the change of elastic, visco-elastic, visco-plastic modules of non-woven fabrics is somewhat the same, and the viscoelastic properties of the fabric appear at all stages of deformation. According to experiments, elastic deformation is also involved in the elastic and elastic-plastic deformation stage, so in [1-2] it was proposed to combine all the names of these modules under the general name of the deformation module of non-woven fabrics.

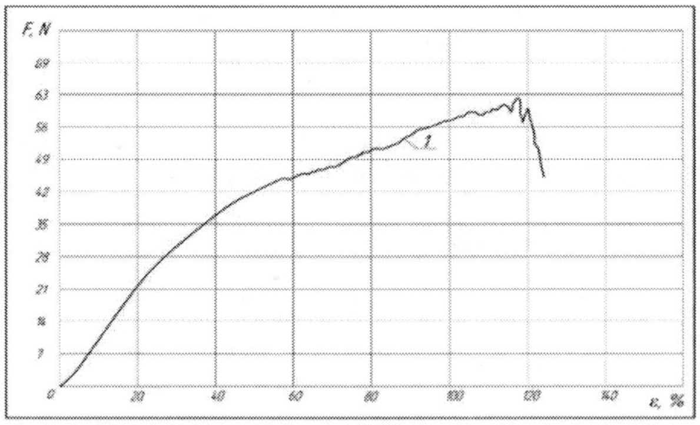

Figure 1. A diagram of the elongation of the non-woven fabric according to the 1st option

2 Experimental research

In the process of stretching deformation of non-woven fabric in width a different situation is shown (Fig. 2,4,6). In this case, changes take place along a different trajectory than in $E(\varepsilon)$ (Figure 1,3,5). Here, in the process of stretching, as well as when stretched along the length, the indicators of the deformation modulus decrease to the value E_*. In addition, there is an increase along the E curved trajectory during stretching. After reaching $E = E_k$, the fabric $\varepsilon = \varepsilon_k$ is torn [3-4].

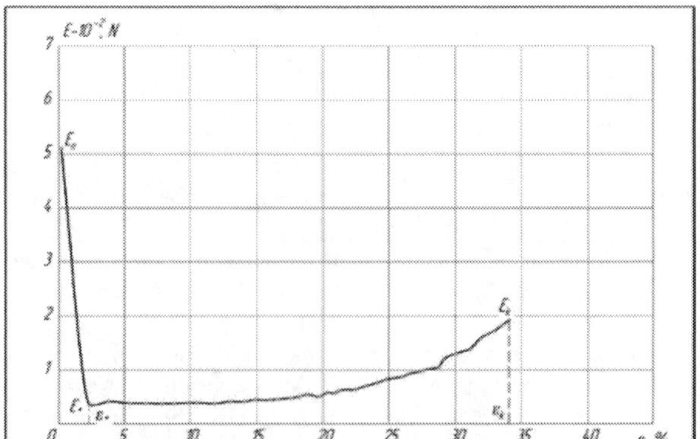

Figure 2. №1 Dependence of the modulus of deformation in stretching on the width of non-woven fabric on the relative deformation.

---

* Corresponding author: otoshbekov17@mail.ru

 https://doi.org/10.1051/e3sconf/202346101068

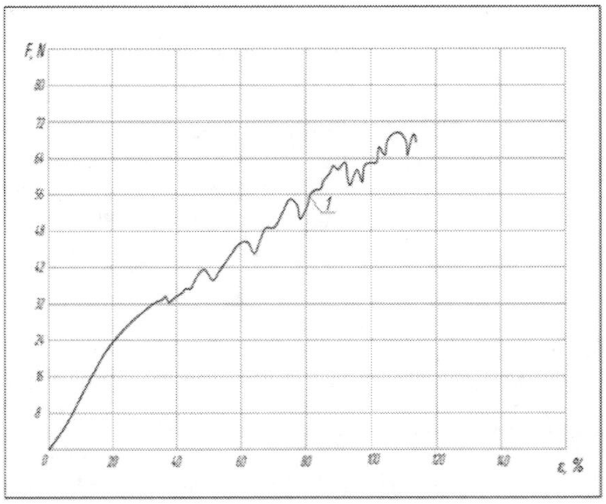

Figure 3. Option 2 is a diagram of the lengthwise stretching of non-woven fabric.

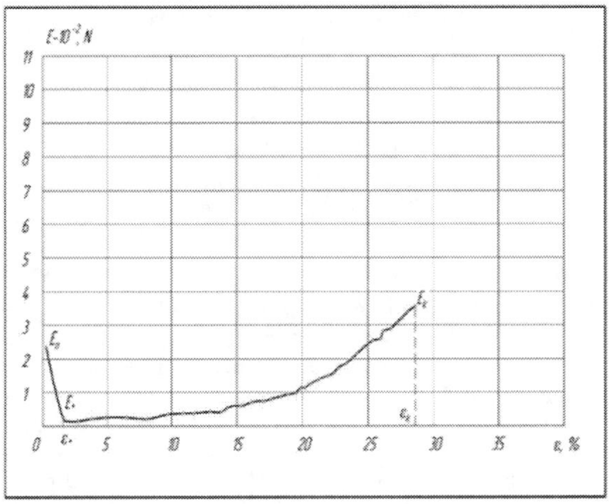

Figure 4. Dependence of the modulus of deformation in the widthwise elongation of non-woven fabric No. 2 on the relative deformation

During the stretching of the nonwoven fabric in width, as well as at the beginning of the deformation process, the soft structure of the nonwoven is densified without much resistance, therefore, the intensity of the ε_* deformation modulus decreases to a value of $E = E_*$ to 0. During stretching, the structure of the non-woven fabric becomes denser, the deformation modulus indicator increases $\varepsilon = \varepsilon_k$ and is torn at $E = E_k$ This $E(\varepsilon)$ change is explained by the dependence on the specific structural structure of the non-woven fabric in terms of width. Here, the deformation process along the trajectory is visco-plastic.

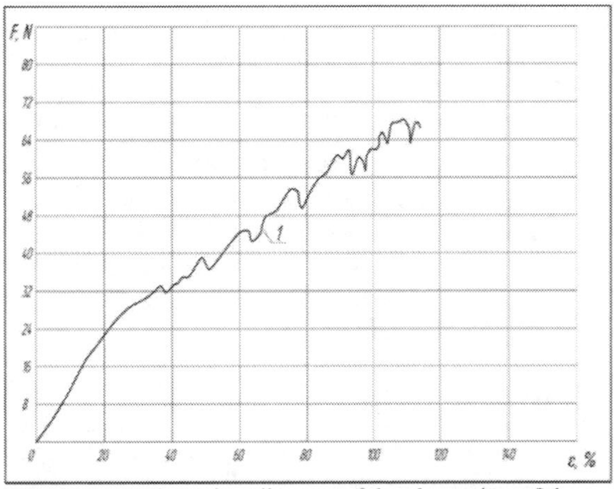

Figure 5. Option 3 is a diagram of the elongation of the non-woven fabric lengthwise

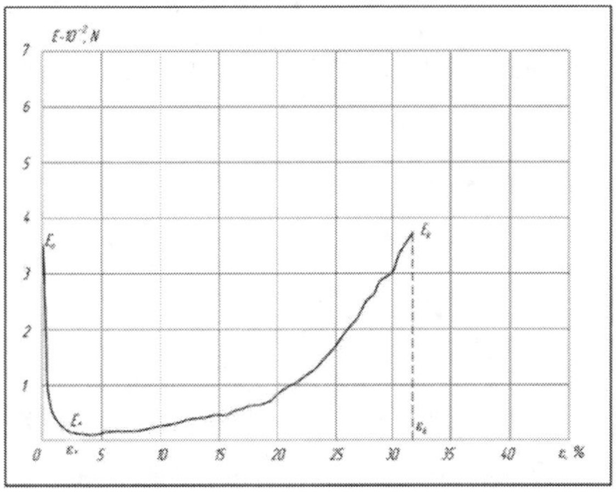

Figure 6. Dependence of the deformation modulus of the non-woven fabric widthwise on the relative strain

The variation of parameter value, the variation of the function of elongation $E(\varepsilon)$ along the width of the non-woven fabric is presented in Table 1. Here, too, E=EN is defined by the indicator $\varepsilon_N \approx 0$.

Table 1. Values of modulus of deformation at different stages of deformation in widthwise stretching of non-woven fabric

Deformation parameters and their indicators	Samples of non-woven fabric		
	№1	№2	№3
E_N, N	490,0	220,0	340,0
E_*, N	30,0	16,0	13,0
ε_*, %	4,0	1,8	3,8
E_k, N	180,0	320,0	370,0
ε_k, %	28,0	27,5	31,0

As can be seen from Table 1, an increase in the amount of wool in the structure of the non-woven fabric leads to an increase in the deformation modulus during tearing in $\varepsilon = \varepsilon_k$ and $E = E_k$. In all samples of non-tissue, the values of parameters ε_* and ε_k remain approximately the same between experiments.

The results presented in Figures 2, 4, 6 and Table 1 show that non-woven fabrics exhibit high stiffness in width stretching and nonlinear belt deformation during all deformation processes [5-6].

3 Research results

As can be seen from the above results, $F(\varepsilon)$ the dependences $E(\varepsilon)$ obtained in the experimental diagrams more accurately reflect the deformation stage of the non-woven fabric. It should be noted that the method of determining and evaluating the elastic, elastic plastic, and viscous stages of deformation proposed in is an effective and visual method for studying the deformation state and mechanical properties of nonwoven fabrics.

The results of the experimental and theoretical studies presented above allow us to determine the limits of the deformation stages of non-woven fabrics and its maturity indicators. To do this, we use the deformation law of (1) nonwoven fabric and rewrite it as follows:

$$\frac{d\sigma}{dt}+\sigma\frac{E_D(\varepsilon)}{E_S(\varepsilon)}\mu(\varepsilon)=E_D(\varepsilon)\frac{d\varepsilon}{dt}+\mu(\varepsilon)E_D(\varepsilon)\varepsilon \quad (1)$$

According to the experiments, we first consider the static deformation process, that is, we accept $d\varepsilon/dt \approx 0$. Then equation (2) takes the following form.

$$\frac{d\sigma}{dt}+\sigma\frac{E_D(\varepsilon)}{E_S(\varepsilon)}\mu(\varepsilon)=\mu(\varepsilon)E_D(\varepsilon)\varepsilon \quad (2)$$

In addition, we believe that the indicator $E_S(\varepsilon)$ is known, $\varepsilon = \varepsilon_k$ or $\varepsilon = \varepsilon_m$ is consistent with the results presented in **Fig. 2,4,6**. Accordingly, we assume that $E_D(\varepsilon) = \gamma E_S(\varepsilon)$ and the value $\gamma = 2$ are equal to 2, and the value of the parameter $\mu(\varepsilon)$ is known for the specified deformation values $\varepsilon = \varepsilon_m$ or $\varepsilon = \varepsilon_k$, $E_S = E_m$ or (10) [7-8].

Then the equation (2) takes the following form.

$$\frac{d\sigma}{dt}+\mu\gamma\sigma=\mu\gamma E_S\varepsilon_m \quad (3)$$

Integrating (3) we obtain the following

$$\sigma = E_m\varepsilon_m\left(1-e^{-\mu\gamma t_m}\right) \quad (4)$$

$\varepsilon = \varepsilon_k$ ва $E_S = E_k$ for points

$$\sigma = E_k\varepsilon_k\left(1-e^{-\mu\gamma t_k}\right) \quad (5)$$

Formulas (4) and (5) make it possible to determine the value of the load applied to the non-woven fabric, in which the non-woven fabric undergoes elastic deformation, that is, after the load is removed, the material returns to its original state [9].

When the indicators are $\varepsilon_m = 12\%$ or $\varepsilon = 0{,}12$ according to formula F_m (4) is equal to $F_m = 15{,}8\ H$.

In this case $\gamma = 2$ and $\mu = 10\ c^{-1}$ are accepted. If the parameters γ and μ have these parameters, according to (5) $\varepsilon_k = 1{,}3$ and $E_k = 46\ H$, we get $F_k = 47{,}1\ H$.

According to the diagram in Figure 4.1, indicators F_m and F_k are respectively equal to $F_m = 15\ H$ and $F_k = 47{,}28\ H$.

This (4) indicates that equations [3-10] are suitable for predicting the deformation of nonwoven fabric.

Based on this, if there is a deformation index ε_m^{cp} = 17,5% in the length of the non-woven fabric, the non-woven fabric will be deformed. And when stretching along the width, the strap deformation is kept at ε_k^{cp} = 1,08% deformation. Depending on the parameters of the deformation module E_S and E_D, as well as the parameters γ, μ and t_k, the allowable load can be determined by formulas (4) and (5). In this case, the non-woven fabric retains its original shape and structure.

In order to obtain non-woven fabrics, a new type of non-woven fabric based on softening, bleaching, drying process and needle pinning, gluing, and pressing of coarse wool fibers through cleaning, washing and chemical treatment was obtained, and effective production technologies were proposed in practice at the same time as the effectiveness of their application.

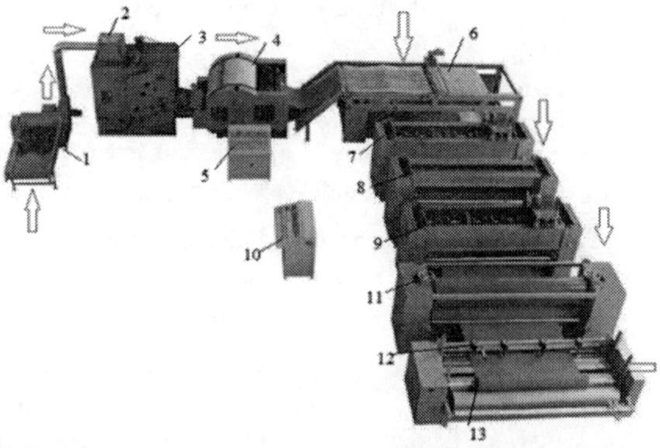

Figure 6. Chine liseng L-4 180906 principle technological scheme of non-woven fabric production

1-spinning of wool fiber, 2-pneumatic conveyor, 3-cleaning and mixing, 4-carding process, 5-monitor, 6-boarding process, 7,9-pinning process, 8-pressing process, 10-speed, temperature monitoring, 11- hot pressing drum, 12- cooling process, 13- finished product.

Based on the results of our research, non-woven fabric was obtained at the enterprise "JIZZAX JUN" in Jizzax city, and the technology of production of non-woven fabric of three different thicknesses used for the automobile industry is recommended. Chine liseng L-4 180906, a new non-woven fabric manufacturer based on coarse wool fiber for the automotive industry, is characterized by the fact that the non-woven fabric is produced by softening the coarse wool fibers by combing, combing, needling and heating at high temperature. The non-woven fabric production technology was carried out in a more simplified manner. In addition, non-woven fabrics are obtained mainly on the basis of local natural wool fibers. The technology of production of new non-woven fabrics is presented in Figure 5 below.

4 Conclusion

1. It was determined from the experimental diagrams of non-woven fabric stretching, depending on the change of the deformation parameters in the stretching of the fabric along the width and length of the deformation modulus of the non-woven fabric. It was found that the change of the deformation modulus depending on the value of the deformation is mainly nonlinear.

2. It has been found that non-woven fabrics for the automotive industry are the best flexible material because they are relatively elastic and have the ability to compensate for small-scale errors and retain their shape when deformed.

3. When the deformation index of the non-woven fabric made of recommended coarse wool fiber is E_m=17,5, it is found that the non-woven fabric is elastically deformed, and when it is stretched along the width, the elastic deformation is 1.08%.

References

1. Shcherbakov V.P. Applied and structural mechanics of fibrous materials. – M.: Tisoprint, 2013. - 304 p.

2. Sultanov K.S., Ismoilova S.I. Structural strength of textile threads. Monograph. – Tashkent: Fan, 2017. - 256 p.

3. Toshbekov O.A., Urozov M.K. Methodology for predicting, evaluating and determining deformation characteristics of non-woven fabrics // Евразийский журнал академических исследований. 2023. Vol 3, № 4 R. 7–9.

4. Toshbekov O.A., Ermatov Sh.Q. Research of the production technology of non-woven fabrics based on coarse wool fiber // Fergana Polytechnic Institute. Scientific-technical journal. 2023. T-27, № 3. 95-100 b. [05.00.00; №20].

5. Toshbekov O., Ermatov Sh., Qarshiyev B. Research of properties of wool fibers obtained from sheep breeds grown in our republic.// Agro science-a scientific and practical journal of agriculture and water management of Uzbekistan. 2023. № 2. 55-58 b. [06.00.00, №1].

6. Urozov M.K., Toshbekov O.A., Raximova K., Bobomurodov E. Determination of wool fiber diameter and roughness.Eurasian Journal Of Academic Research. 2022. Vol 2, № 13. R. 789–791; [01.00.00, №51].

7. Urozov M.K., Toshbekov O.A., Raximova K. Methods of testing wool thickness.Eurasian Journal Of Academic Research. 2022. Vol 2, № 13. R. 784–788; [01.00.00, №51].

8. Toshbekov O.A., Urozov M.K., Qulmuminov O. Determination of noise reduction and high flexibility properties of non-woven fabrics // Eurasian Journal Of Academic Research. 2022. Vol 2, №13 R 772-777; [01.00.00, №51].

9. Toshbekov O.A., Urozov M.K., Boltayeva I.B., Hamrayeva M.F. Use of wool fabrics, classification and coding of wool fabrics // World Bulletin of Public Health. 2022. T 11, C. 68-71.

10. Toshbekov O.A., Urozov M.K., Baymurova N.R., Hamrayeva M.F. Processes of bleaching and discolouring of wool fibers // International journal of social science & Interdisciplinary research issn. 2022. T 11, № 6. C. 231-235.

Algorithms for synthesis of control systems for dynamic objects based on the modal method

Jasur Sevinov[1*], *Oqila* Boeva[2]

[1]Tashkent State Technical University named after Islam Karimov, 100095, Uzbekistan, Tashkent, University St. 2A
[2]Navoi State Mining and Technology University, 210100, Uzbekistan, Navoi, St. Galaba, 76V

Abstract. Algorithms for the synthesis of control systems for dynamic objects based on the modal method are given. Algorithms have been developed for determining the placement of poles in multidimensional systems with proportional-differential feedback on the output, providing the desired distributions on the complex plane of the eigenvalues of the matrix of a closed system. Algorithms for selecting feedback in control systems of dynamic objects are proposed, which allow for stable identification of the coefficients of the equations of the object and the observed values of the object used in the control device. Adaptive algorithms have been developed to place poles of non-minimal-phase stochastic systems in the presence of white noise at the output, which makes it possible to exclude the operation of matrix inversion, which requires significant computational costs. Adaptive control algorithms based on the modal method have been developed for dynamic objects with a delay under conditions of a priori uncertainty and unsteadiness of technological process parameters, providing exponential dissipativity of the control process. Based on the developed regular algorithms, an adaptive control system for the synthesis of ammonia is proposed.

1 Introduction

The emergence of new efficient technological processes and the intensification of existing ones place ever higher demands on their management systems. The efficiency of the production management system is largely determined by the quality of regulation and control of technological parameters [1-3]. This is because modern technological complexes and installations are characterized by stressful conditions of the processes occurring in them. At the same time, for the synthesis of control systems, as a rule, there is not enough complete a priori information about processes and it is not always possible to obtain it. Most control objects are complex dynamic systems with many degrees of freedom, subject to the action of many external and parametric disturbances; processes in these objects are relatively little studied, and work on their identification and control is fraught with great difficulties; working conditions are predictable with low probability. This is determined by the complexity of the processes, their dynamics, the impossibility of obtaining any sufficiently adequate mathematical description of them, the physical impossibility of direct measurement of all parameters necessary for effective control, as well as the influence of controlled and uncontrolled disturbances, the nature of which may also change during the technological process and the functioning of the control system. In such cases, it is necessary to resort to automatic control with adaptation, which allows for stable and efficient operation of the entire system as a whole [3-9]. To solve these issues, it is required to develop adaptive control systems for linear or linearizable multidimensional objects with modal control using discrete identifiers and with disturbance compensation, which can improve the quality of control by choosing the shortest duration and amplitude of the identification signal when the system is out of order. In this regard, it is very tempting to consider various possible approaches to solving problems of increasing the accuracy of calculating the states of dynamic systems based on modal methods and to identify the most promising for practical use methods and algorithms for solving ill-posed problems [9–15].

2 Objects and Methods

- **Development of algorithms for the synthesis of control systems for dynamic objects based on modal methods.**

Consider the problem of choosing from several observable quantities, such that the system has minimal complexity in order. The solution to this problem is obtained by the operator-polynomial method [13, 16, 17], which provides the possibility of choosing not only the poles but also the zeros of the system according to the setting and disturbing influences.

We will assume that the synthesized system consists of a control device and a given part [13,16], the equations of which have the form:

$$A(p)y = B_0(p)u + B_1(p)f_1, \quad (1)$$

$$A(p)\tilde{v}_i = \tilde{V}_i(p)u + \tilde{Z}_i(p)f_1, i = (\overline{1, M}). \quad (2)$$

* Corresponding author: sevinovjasur@gmail.com

$$R(p)u = Q_0(p)g - \sum_{i=1}^{q} L_i(p)v_i + Q_1(p)f_1, \quad (3)$$

where y is a controlled quantity (not observed), u is control; f_1 is an observed disturbance; $\tilde{v}_i$ is output quantities that can be observed, M is their number; $A(p)$ is a proper operator [2,18]; $B_j(p)$, $\tilde{V}_i(p)$, $\tilde{Z}_i(p)$ – input operators of a given part, $j = 0,1$; g – the setting effect; $R(p)$, $Q_0(p)$, $L_i(p)$, $Q_i(p)$ – some polynomials from p; v_i – the observed magnitudes of the object, q – their number, $q \le M$.

Operators are polynomials from $p = d/dt$ with constant coefficients, we will agree to denote the degree of the polynomial: $n[A] = n$, $n[B_j] = m_j$, $n[\tilde{V}_i] = \tilde{n}_i$, $n[\tilde{Z}_i] = \tilde{s}_i$. Then in the equations (1), (2) $m_j \le n$, $\tilde{n}_i \le n$, $\tilde{s}_i \le n$. The magnitudes v_i are selected from the number of magnitudes $\tilde{v}_1, \tilde{v}_2, ..., \tilde{v}_M$, and after selection of them, as well as the corresponding ones according to (2), the polynomials $\tilde{V}_i(p)$ and $\tilde{Z}_i(p)$ are renumbered so that $n_{i+1} \le n_i$, $i = (\overline{1, q-1})$. The value is $\mu_y = \min\{r - r_j, r - l_i; j = 0,1; i = (\overline{1,q})\}$, where $r_j = n[Q_j]$, $l_i = n[L_i]$, $r = n[R]$, are the indices of the control device.

Let us pose the problem of choosing the variables v_i included in equation (3), as well as the degrees and coefficients of the polynomials of this equation so that, with the minimum possible order of the control device, i.e. at $r \to \min$, the following conditions were fulfilled:

$$\mu_y \ge \mu_y^*, H(p) = H^*(p),$$
$$H_j(p) = H_j^*(p)\tilde{H}_j(p), \; j = 0,1, \quad (4)$$

where $H(p)$ is the own operator of the synthesized system; $H_0(p)$, $H_1(p)$ are input operators of the system.

The polynomials of equation (3) under conditions (4) are determined by the solutions of the systems of the following algebraic equations [12, 13]:

$$A(p)R(p) + \sum_{i=1}^{q} L_i(p)V_i(p) = H(p), \quad (5)$$
$$B_0(p)Q_0(p) = H_0(p),$$
$$B_0(p)Q_1(p) + B_1(p)R(p) + \sum_{i=1}^{q} L_i(p)C_i(p) = H_1(p)$$

So polynomials $R(p)$ and $L_i(p)$, $i = (\overline{1,q})$ are determined by the solution of the system:

$$Gd = h \quad (6)$$

where $d = [\lambda_{q0} \ldots \lambda_{ql_q} \mid \cdots \mid \lambda_{10} \ldots \lambda_{1l_1} \mid \rho_0 \ldots \rho_r]^T$, $h = [\eta_0 \eta_1 \ldots \eta_{r+n}]^T$. Here $\lambda_{ik}, \rho_k, \eta_k$ are the coefficients at p^k polynomials $L_i(p)$, $R(p)$, $H^*(p)$ respectively; the matrix G is composed of coefficients α_k and v_{ik} with p^k polynomials $A(p)$ and $V_i(p)$, $i = (\overline{1,q})$.

When finding a solution to equation (6), computational difficulties arise because when solving the equation, it is necessary to use regularization methods [10, 11]. When solving equation (6), we take the approximation conditions in the form $\|h - \bar{h}\| \le \delta$, where $\bar{h}$ is the exact value of the right side of the equation. To regularize the solution of this equation, we will use the method of M.M. Lavrentiev [10, 11]:

$$\alpha d + Gd = h, \quad (7)$$

where $\alpha > 0$ is the regularization parameter; $R_\alpha = \theta_\alpha(G) = (\alpha + G)^{-1}$ if conditions $\lim_{\alpha \to 0} \|R_\alpha Gd - d_{(1)}\| = 0$, $d \perp \ker G$, (6) are satisfied leads to a limited approximation of the problem [11].

The solution of equation (7) can be expressed in the form: $d_{(\alpha)} = (\alpha I + G)^{-1}h = \theta_\alpha(G)h$, where $\theta_\alpha(\lambda)$ is the generating system of functions $\theta_\alpha(\lambda) = (\alpha + \lambda)^{-1}$; λ - spectral parameter $0 \le \lambda < \infty$; I is the identical operator; $\langle(\alpha I + G)d, d\rangle \ge \alpha \|d\|^2$, $\forall d \in H$. Thus, problem (7) is correct.

When solving equation (6), one can also use the concepts of pseudo-inversion [11, 19-21]. In case $Gd \ne h$, i.e. measure of inconsistency $\mu_G = \inf_{d \in D} \|Gd - h\|_F > 0$, $d \in D : \|Gd - h\| = \mu_G$, the solutions of this equation do not converge at $\alpha \to 0$ to the pseudo solution $d = G^+h$. In this case, it is advisable to use the following relation: $d_\alpha = d_\alpha + \alpha \frac{dz_\alpha}{d\alpha}$, where $\frac{dz_\alpha}{d\alpha}$ is the derivative of the element z_α as a function of the parameter α. Then one can write [11]:

$$d_\alpha = (\alpha I + G)^{-2} Gh = (\alpha I + G)^{-1} Gz_\alpha,$$

i.e. d_α is the solution of the equation regularized especially: $(\alpha I + G)^2 d_\alpha = Gh$. Taking into account that in this case, the pseudo-solution d_h satisfies the equation $G^2 d_h = Gh$ in the usual sense, we will have the:

$$d_\alpha = (\alpha I + G)^{-2} G^2 d_h \text{ and}$$
$$d_h - d_\alpha = \alpha(\alpha I + 2G)(\alpha I + G)^{-2} d_h.$$

Then, following [10,11], we can show the validity of the following limit relation: $\lim_{\alpha \to 0} \|d_h - d_\alpha\|_H = 0$, i.e. the considered method is regularizing.

The above expressions make it possible to implement a stable procedure for selecting feedback in control systems for dynamic objects and, thereby, improve the quality of control processes.

For a completely controllable system, the number of nontrivial invariant polynomials of the system matrix is less than the dimension of the control vector; in this case, the problem of constructing a control that leads to the required spectrum is not solved uniquely. When the system is completely controllable and observable with a one-dimensional input and a multidimensional output, various approaches to determining matrices in controllable, observable linear, multidimensional systems are considered, and these approaches allow arbitrary assignment of the poles of a closed system, i.e. determination of eigenvalues of matrices.

- **Development of adaptive-modal control algorithms for dynamic objects**

Assessment of the fulfillment of the conditions of controllability, observability, and non-degeneracy, as well as a quantitative assessment of their degree, are, of course, important stages in the analysis of the properties of objects in the construction of modal control systems with state or input-output controllers. However, it is known [3,4,11,15] that even under the condition of a high degree of controllability and observability of some objects, situations may arise when, as a result of solving the problem of synthesis of a modal control system, solutions with low parametric roughness are obtained.

When managing such objects under conditions of their stability and minimum-phase nature, certain contradictions may arise between the design requirements and the capabilities of the system, which are resolved only due to the appearance of non-minimum-phase links in its structure or the formation of positive feedback on the output coordinate and its derivatives, essentially worsening the robust properties of the ACS [4,8,22].

Consider a linear system with the following difference equation:

$$A(z^{-1})y_t = B(z^{-1})u_t + C(z^{-1})e_t, \qquad (8)$$

where A, B, C are polynomials for which $A(0)$ and $C(0)$ are equal to one, $B(0)$ is equal to zero, and z^{-1} is the unit inverse shift operator. Signals correspond to an output, a controlled input, and a zero-mean white noise disturbance.

If we introduce a controller corresponding to the equation [22-24]:

$$Fu(t) + Gy(t) = 0, \qquad (9)$$

for a closed loop, you can get the equation:

$$T\,y(t) = F\,e(t), \qquad (10)$$

that is, the pole system for a closed loop is determined by the zeros of the selected polynomial T, provided that F and G satisfy the polynomial identity

$$AF + BG = CT, \qquad (11)$$

at $F(0)$ and $T(0)$ equal to one.

Identity (11) corresponds to a unique solution for F and C, provided that A and B are coprime, and n_f, n_g and n_t satisfy the relations: $n_f = n_b - 1$, $n_g = n_a - 1$; $n_t \le n_a + n_b - n_c - 1$.

Under these conditions, the solution to equation (11) is obtained by inverting a matrix of order $n_a + n_b - 1$. This is one of the disadvantages of this technique, when the calculations must be repeated at each discrete time moment: $F = T - PB$, $G = PA$, which satisfies identity (11), for any polynomial P.

The expression (10) is replaced by:

$$ATy(t) = FCe(t).$$

In the case when the parameters are unknown, the controller (9), (11) should be synthesized based on the parameter estimates for the selected model structure [2, 12, 23]. The above algorithms turn out to be effective in solving applied problems of identification and synthesis of adaptive control systems for technological objects.

Consider the problem of designing a modal control system for a multidimensional lumped dynamical system. Let a multidimensional system with lumped parameters be described in a vector-matrix form [13, 15, 17, 24]:

$$\dot{x}(t) = Ax(t) + Bu(t), \quad y(t) = Cx(t),$$

where $x(t) \in R^n$ is the state vector of the object; $u(t) \in R^m$ - control vector; $y(t) \in R^p$ is the observed output vector; $A \in R^{n\times n}$, $B \in R^{n\times m}$, $C \in Rp^{p\times n}$ are constant matrices such that pair (A, B) is controllable, pair (A, C) is observable, and $\operatorname{rank}B = m$, $\operatorname{rank}C = p$ $(2 \le m,\ p < n)$.

It is required to obtain the general form of matrices $K \in R^{m\times p}$ in the control law

$$u(t) = Ky(t), \qquad (12)$$

which satisfy the condition:

$$\sigma(A + BKC) = \Lambda.$$

Here $A + BKC$ is a matrix with a simple structure [13, 24], $\sigma(\square)$ is the spectrum of the matrix, and $\Lambda = \{\lambda_1, \lambda_2, ..., \lambda_n\}$ is a given set of numbers.

Let matrix $K \in R^{m\times p}$ satisfy condition (12). Then for the matrix $A + BKC$ it is possible to specify n right $v_1, v_2, ..., v_n$ and n left $w_1^T, w_2^T, ..., w_n^T$ eigenvectors (T is the transposition symbol), which correspond to the eigenvalues $\lambda_1, \lambda_2, ..., \lambda_n$ and satisfy the conditions [19, 20]:

$$v_i \in R^n,\ w_i \in R^n \text{ at } \lambda_i \in R^1;$$
$$v_i = \bar{v}_j,\ w_i = \bar{w}_j \text{ at } \lambda_i = \bar{\lambda}_j; \qquad (13)$$
$$w_i^T v_j = \delta_{ij}\ (i, j = \overline{1, n}),$$

here δ_{ij} is the Kronecker symbol.

Writing equalities:

$$(A + BKC - \lambda_i I_n)v_i = (A - \lambda_i I_n, B)\begin{bmatrix} v_i \\ KCv_i \end{bmatrix} = 0_n, \qquad (14)$$

$$(A^T + C^T K^T B^T - \lambda_i I_n)w_i = (A^T - \lambda_i I_n, C^T)\begin{bmatrix} w_i \\ K^T B^T w_i \end{bmatrix} = 0_n,$$

where I_n is the identity $n\times n$ matrix, 0_n is the zero n vector.

We get vectors $(v_i^T, v_i^T C^T K^T)^T$, $(w_i^T, w_i^T BK)^T$, which are solutions of the equations, respectively:

$$(A - \lambda_i I_n, B)x' = 0_n,\ (A^T - \lambda_i I_n, C^T)x'' = 0_n. \qquad (15)$$

Let us define matrices P_i, Q_i, whose columns form the bases of subspaces of solutions to the first and second equations from (15), respectively. Since the matrices A, B, C are real and the numbers $\lambda_i \in \Lambda$ are symmetrical about the real axis, it follows that P_i, Q_i, can be chosen so that

$$P_i \in R^{(n+m)\times m},\ Q_i \in R^{(n+p)\times p} \text{ at } \lambda_i \in R^1;$$

$$P_i = \overline{P}_j,\ Q_i = \overline{Q}_j \text{ at } \lambda_i = \overline{\lambda}_j. \qquad (16)$$

Taking into account the method of obtaining matrices P_i, Q_i, for vectors $(v_i^T, v_i^T C^T K^T)^T$, $(w_i^T, w_i^T BK)^T$ from (14) we can write:

$$\begin{bmatrix} v_i \\ KCv_i \end{bmatrix} = P_i f_i,\ \begin{bmatrix} w_i \\ K^T B^T w_i \end{bmatrix} = Q_i g_i \qquad (17)$$

Here m-vectors f_i and p-vectors $g_i\ (i = \overline{1, n})$, according to (13), (16), are such that

$$f_i \in R^m,\ g_i \in R^p \text{ at } \lambda_i \in R^1;$$

$$f_i = \overline{f}_j,\ g_i = \overline{g}_j \text{ at } \lambda_i = \overline{\lambda}_j$$

Let's represent the matrices P_i, Q_i in the form:

$$P_i = \begin{bmatrix} S_i \\ M_i \end{bmatrix} \begin{matrix} \}n \\ \}m \end{matrix},\ Q_i = \begin{bmatrix} T_i \\ N_i \end{bmatrix} \begin{matrix} \}n \\ \}p \end{matrix} \qquad (18)$$

Then, substituting (18) into (17), we have:

$$v_i = S_i f_i,\ w_i^T = q_i^T T_i^T, \qquad (19)$$

$$KCv_i = M_i f_i,\ w_i^T BK = g_i^T N_i^T. \qquad (20)$$

Since the comoving matrices for $A + BKC$ are related to the eigenvectors v_i, w_i^T by formulas $H_i = v_i w_i^T\ (i = \overline{1, n})$ [19], then, taking into account (19), we can write for them that $H_i = S_i f_i g_i^T T_i^T$ or

$$H_i = S_i Z_i T_i^T,\ (i = 1, n), \qquad (21)$$

where $Z_i = f_i g_i^T$.

In the case when the matrices C and B are a matrix of incomplete rank, then the problem under consideration is ill-posed. To give numerical stability to the procedure of pseudo-inversion of matrices C and B, it is expedient here to use the concepts of regular methods [10, 11, 21].

Let us consider algorithms for stable computation C^+ that use certain matrix decompositions C [15,19,25]. Whereas if $\text{rank}C = p$ ($C \in R^{p\times n}$ with $p \le n$), then the pseudoinverse of C is a C^+ matrix defined by the second Gaussian transformation:

$$C^+ = C^T (CC^T)^{-1}. \qquad (22)$$

The validity of expression (22) is since any matrix $C \in R^{p\times n}$ can be represented as a "skeletal" decomposition [10,19,]:

$$C = U \cdot V,$$

with matrices $U \in R^{p\times r}$ and $V \in R^{r\times n}$, where $r = \text{rank}C \le \min(p,n)$.

Let's put it now

$$C^+ = V^+ \cdot U^+,$$

where, according to (22) and by the property of the pseudoinverse matrix [19], we can come to the following relations: $V^+ = V^T (VV^T)^{-1}$, $U^+ = (U^T U)^{-1} U^T$.

Then $CC^+C = UVV^T(VV^T)^{-1}(U^TU)^{-1}U^TUV = UV = C$. If we accept $CC^+C = UVV^T(VV^T)^{-1}(U^TU)^{-1}U^TUV = UV = C$, then we can show that $UC^T = V^+$.

Equation $C^+ = C^T\Omega$ with $\Omega = U(U^TU)^{-1}(VV^T)^{-1}(U^TU)^{-1}U$ is also true.

To find the pseudoinverse matrix B, one can also use the partition of the matrix into blocks [11,26,27]. Let's represent the rectangular matrix B in the form:

$$B = \left[\begin{array}{c|c} J & P \\ \hline \Phi & \Theta \end{array}\right],$$

where J is a nonsingular square matrix $(|J| \ne 0)$. Then the equality $\Theta = \Phi J^{-1} P$ is true and therefore $B = \begin{pmatrix} J \\ \Phi \end{pmatrix} J^{-1}(JP)$.

Since this formula is the result of two successive skeletal expansions:

$$B = \begin{pmatrix} J \\ P \end{pmatrix}\begin{pmatrix} E & J^{-1}P \end{pmatrix}, \begin{pmatrix} E & J^{-1}P \end{pmatrix} = J^{-1}(JP), \text{ to}$$

$$B^+ = (JP)^+ (J^{-1})^+ \begin{pmatrix} J \\ \Phi \end{pmatrix}^+ = (JP)^+ J \begin{pmatrix} J \\ \Phi \end{pmatrix}^+.$$

Then you can come to the expression:

$$B^+ = \begin{pmatrix} J^T \\ P^T \end{pmatrix} (JJ^T + PP^T)^{-1} J (J^T J + \Phi^T \Phi)^{-1} (J^T \Phi^T), \qquad (23)$$

Formula (23) gives an explicit expression for the pseudoinverse matrix B^+ in terms of blocks J, P, Φ.

The above algorithms make it possible to regularize the considered problem of modal control of multidimensional dynamical objects based on regular methods.

3 Objects and Methods

- **Application of the developed synthesis algorithms in the tasks of automation and control of the technological process of ammonia production**

Based on the analysis of the technological process, it was found that the synthesis of ammonia as a control object is characterized by the interaction of controlled parameters [3, 28]. The process occurring in the column is associated with the release of a significant amount of heat and the slightest changes in the control effects greatly affect the temperature profile of the column. The specifications for reactor temperature control are very stringent: high static accuracy is required, fast temperature equalization, i.e. the speed is high and the dynamic accuracy is significant. In addition, the temperature in the catalyst layers is affected by various disturbing influences, in which the concentration of inert compounds, the ammonia concentration at the reactor inlet, and the ratio of nitrogen and hydrogen are of great importance [28–30].

With this in mind, the following can be considered as the main process variables: control parameters

$u = \left[u^{[1]}, u^{[2]}, u^{[3]}, u^{[4]}\right]$, where $u^{[1]} \div u^{[4]}$ are the degrees of opening of dampers on bypass flows; controlled parameters $y = \left[y^{[1]}, y^{[2]}, y^{[3]}, y^{[4]}\right]$, where $y^{[1]} \div y^{[4]}$ are the temperatures in the catalyst layers; disturbing influences $f = \left[f^{[1]}, f^{[2]}, f^{[3]}\right]$, where $f^{[1]}$ is the concentration of ammonia, $f^{[2]}$ is the concentration of inert gases, $f^{[3]}$ is the hydrogen/nitrogen ratio.

Therefore, from the point of view of management, this process is an object with interdependent parameters. The topology of these relationships is shown in fig. 1.

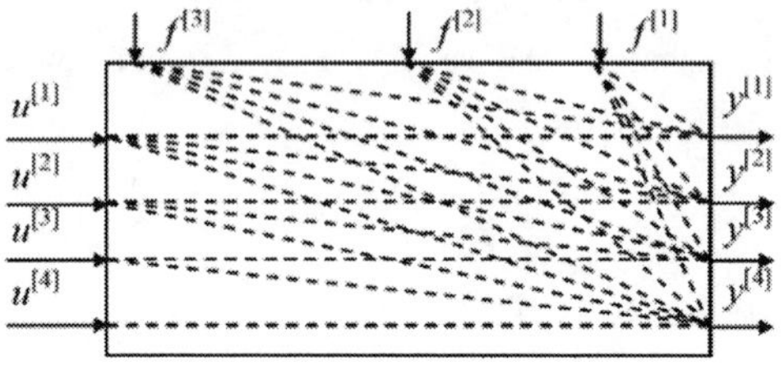

Fig.1. Block diagram of the object

We will describe the dynamics of the main and cross channels of the control object, channels of perturbation using finite-difference equations. After that, in the universal record of the differential equation based on the indexing variable, taking into account the total (measured) outputs, the channel coupling equations are described using the discrete transfer functions *W(z)* of the plant:

$$y^{u[k][j]}(z) = W_o^{u[k][j]}(z) \cdot u^{[k]}(z),\ k = \overline{1,4},\ j = \overline{k,4},$$

$$y^{f[h][j]}(z) = W_o^{f[h][j]}(z) \cdot f^{[h]}(z),\ h = \overline{1,3},\ j = \overline{1,4}, \quad (24)$$

$$y^{[j]}(z) = \sum_{k=1}^{j} y^{u[k][j]}(z) + \sum_{h=1}^{3} y^{f[h][j]}(z),\ \ j = \overline{1,4},$$

where $u^{[k]}(z)$ – control action; $y^{[j]}(z)$ – object output; $y^{u[k][j]}(z)$ – output via main (*k=j*) or cross (*k≠j*) channel; $f^{[h]}(z)$ – external disturbance; $y^{f[h][j]}(z)$ – output through the perturbation channel; $W_0^{u[k][j]}(z)$ – discrete transfer function for the main or cross channel, $k = \overline{1,4}$, $j = \overline{k,4}$; $W_0^{f[h][j]}(z)$ – discrete transfer function for the perturbation channel $h = \overline{1,3}$, $j = \overline{1,4}$; k – input number (bypass flow number); h is the serial number of the disturbance (the concentrations of ammonia, inert gases, the ratio of hydrogen and nitrogen); j – exit number.

Let us write the system of equations (24) in the vector-matrix form:

$$y = W_0^u \cdot u + W_0^f \cdot f, \quad (25)$$

where $y = \left[y^{[1]}(z), y^{[2]}(z), y^{[3]}(z), y^{[4]}(z)\right]^T$ is the vector of measured outputs of the control object; $u = \left[u^{[1]}(z), u^{[2]}(z), u^{[3]}(z), u^{[4]}(z)\right]^T$ – vector of control actions; $f = \left[f^{[1]}(z), f^{[2]}(z), f^{[3]}(z)\right]^T$ – disturbance vector; W_0^u – triangular matrix of discrete transfer functions of the object in the main and cross channels; W_0^f – matrix of discrete transfer functions of the control object through disturbance channels.

Thus, the developed structure of the discrete mathematical model (25) takes into account the one-sided influence of the degree of opening of the dampers in the bypass flows on the temperature in the corresponding and underlying catalyst layers, as well as the influence on the temperature in each layer of disturbing influences: the concentration of ammonia, inert impurities and the ratio of hydrogen and nitrogen . Active experimental methods [31, 32] were used to determine the dynamic properties of the object under study for the main signal transmission channels. Based on the obtained experimental dynamic characteristics of a four-stage ammonia synthesis reactor by the least squares method (quantization cycle T_0=10 sec.) and differential equations of the first order, a parametric determination of the transfer functions of the main and cross channels of the object was carried out.

The ammonia synthesis reactor is characterized by non-stationarity of dynamic properties, which is associated with a change in catalyst activity and thermophysical properties of the reactor over time. Synthesis of the system requires providing the necessary quality of control based on an adaptive approach. The parameters of the main digital controllers were calculated based on the algorithms developed in the dissertation work. The parameters of discrete transfer functions of cross-connect compensators are calculated.

Based on the analysis of scalar coupling equations, the steps of the current algorithm for identifying the parameters of the models of the main and cross channels of a multi-link control object in a closed control scheme have been developed. Taking into account the developed identification algorithms [6,33-36], taking into account for the non-stationarity of the dynamic characteristics of the reactor, a functional-structural diagram of an adaptive control system for the ammonia production process was proposed (Fig. 2).

The adaptive invariant system [28,36,37] works as follows: the current values of the temperature sensors 2-5 are sent to the block for identifying the parameters of a closed system 8. In this block, the parameters of closed systems are identified. Then the values of the parameters of closed systems are transferred to the block for determining the parameters of the model of the main and cross channels of the control object 9, where the parameters of the control object are determined based on the developed software. The evaluation of the parameters of the object and uncontrolled disturbances is transmitted to the reconfiguration unit of control part 10. In this case, the signals from the sensors for the concentration of ammonia 14, the concentration of inert gases 15, the ratio of the concentration of nitrogen and hydrogen 16, and the flow rate of the circulating gas 13 are fed to the disturbance compensation unit 7, and then to the unit reconfiguration of the control part 10. Based on the analysis of the current values of the control actions, the

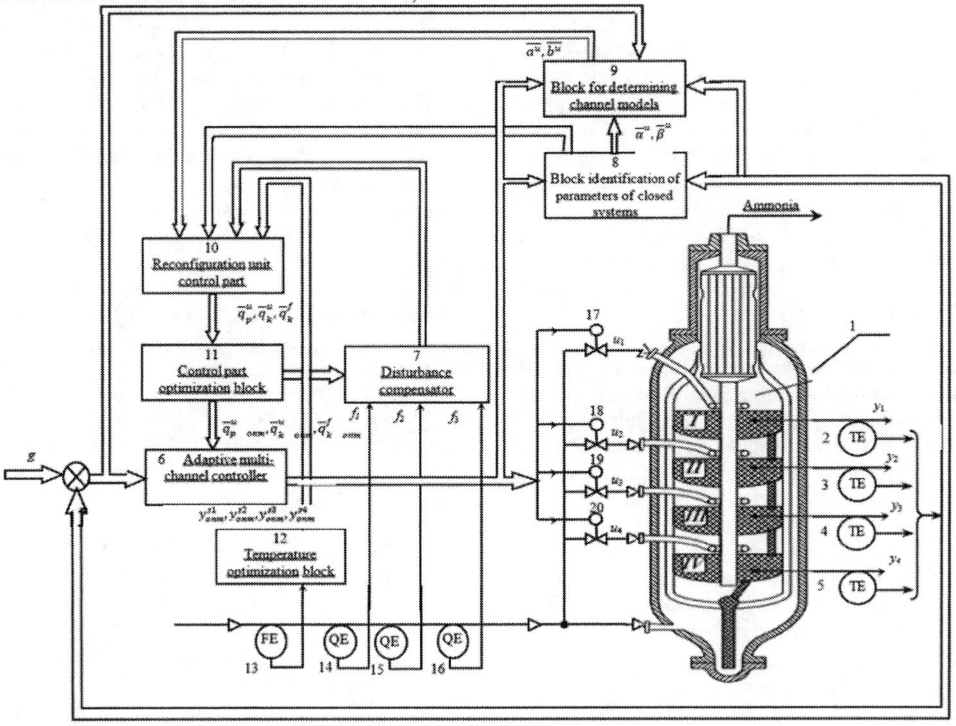

Fig.2. Functional diagram of an adaptive-invariant control system for the ammonia production process

reconfiguration unit of control part 10 determines the linearization interval of the control system for a given mode and transmits this information to the optimization unit of control part 11. Also, the flow rate of the initial nitrogen-hydrogen mixture (NHM) 13 enters the unit temperature optimization 12, where the optimal temperature of the synthesis reaction is determined, which ensures the maximum production of ammonia at a certain NHM flow rate.

Through optimization unit 11 of the control part, the optimal temperature values are fed as control actions to the multichannel adaptive controller 6. In the reconfiguration unit 10, taking into account the nonlinear dependence of the temperature in the first catalyst layer on the degree of damper opening, on based on piecewise linear approximation, one or another piecewise-range model of direct channels for the transmission of influences. Thus, the reconfiguration unit of control part 10, taking into account the information received, calculates the settings of the multi-channel adaptive controller 6, which in turn regulates the bypass flow dampers 17-20.

A simulation block diagram of the ammonia production process control system has been developed. The results of the computer simulation of invariant and adaptive-invariant control systems are presented in fig. 3.

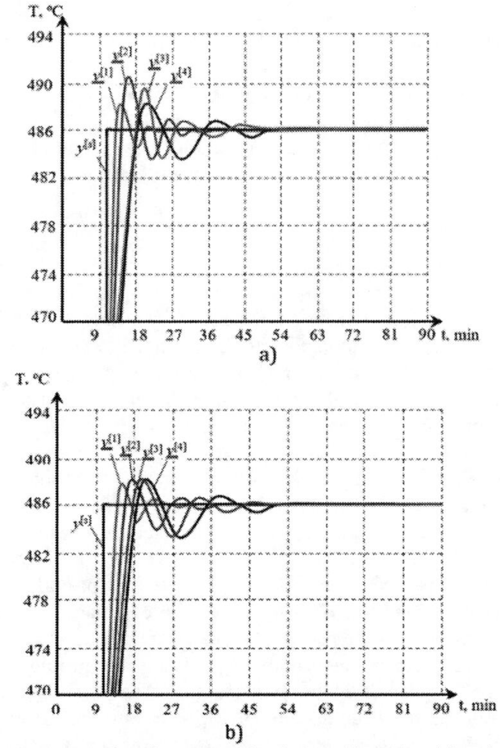

Fig.3. Results of computer simulation of invariant (a) and adaptive-invariant (b) systems of control systems

A comparative analysis of transient processes of invariant and adaptive-invariant systems, as well as control quality indicators, allows us to draw the following conclusions: for the first channel, the improvement in quality indicators for the integral square error is 0.5%, for the second - 11.7, for the third and fourth channels - 2.97 and 0.17%. Thus, the use of an adaptive control system makes it possible to stabilize the technological regime of the process under consideration, compensate for the change in the parameters of the object and the influence of uncertain disturbances on the dynamics of the system, and improve the quality indicators of control processes.

Thus, algorithms have been developed for determining the placement of poles in multidimensional systems with proportional-differential feedback on the output, providing the desired distributions on the complex plane of the eigenvalues of the matrix of a closed system. Adaptive algorithms for pole placement of non-minimum-phase stochastic systems in the presence of white noise at the output are developed, which makes it possible to exclude the operation of matrix inversion, which requires significant computational costs. Algorithms for selecting feedback in control systems of dynamic objects have been developed, which allow for stable identification of the coefficients of the object equations and the observed values of the object used in the control device. Algorithms for adaptive control based on the modal method for dynamic objects with delay under conditions of a priori uncertainty and non-stationarity of the process parameters are developed, which provide exponential dissipativity of the control process. Based on the developed regular algorithms, an adaptive control system for the ammonia synthesis process is proposed. The proposed adaptive control system based on modal methods makes it possible to stabilize the technological regimes of the process and increase the efficiency of its operation.

4 Conclusions

Based on the methods of system analysis, identification, the theory of automatic control, evaluation, adaptive-modal control, processing of experimental results and methods for solving ill-posed problems, constructive algorithms for the synthesis of control systems for dynamic objects based on the modal method were developed in the article and, as a result, the following scientific results were obtained:

Algorithms for determining the placement of poles in multidimensional systems with proportional-differential output feedback are developed, which provide the desired distributions on the complex plane of the eigenvalues of the matrix of a closed system.

An efficient algorithm for solving the problem of placing poles of multidimensional invariant systems with feedback has been developed. This algorithm is efficient for desired eigenvalues that are real, complex conjugate, and repeating poles.

We propose adaptive algorithms for pole placement of non-minimum-phase stochastic systems in the presence of white noise at the output, which make it possible to exclude the operation of matrix inversion, which requires significant computational costs.

Algorithms for adaptive control based on the modal method for dynamic objects with delay under conditions of a priori uncertainty and non-stationarity of the process parameters are developed, which provide exponential dissipativity of the control process. Based on the developed regular algorithms, an adaptive control system for the ammonia synthesis process is proposed. The proposed adaptive control system based on modal methods makes it possible to stabilize the technological regimes of the process and increase the efficiency of its operation.

References

1. Besekersky, V.A., Popov, E.P. (2003). Theory of automatic control systems (S-Pb: Publishing house "Profession", Ed. 4th) p752.
2. Bobtsov, A.A., Miroshnik, I.V. (2001). Linear automatic control systems (S-Pb: SPbSU ITMO) p245.
3. Egupov N.D., Pupkov, K.A. (2004). Methods of the classical and modern theory of automatic control. Textbook in 5 vol., Publishing House of Bauman Moscow State Technical University.
4. Tsykunov, A.M. (2009). Adaptive and robust control of dynamic objects by output. Fizmatlit, p268.
5. Dorf, R. Modern control systems / R. Dorf, R. Bishop., translated from English by B.I. Kopylova, Laboratory of Basic Knowledge, 2002. pp. 831–832.
6. Yusupbekov, N.R., Igamberdiev, H.Z., Sevinov, J.U. (2020). Journal of Advanced Research in Dynamical and Control Systems, Volume 12, Issue 6s, Pp: 77-88. DOI:10.5373/JARDCS/V12SP6/SP20201009.
7. Abdurakhmanova Y.M., Sevinov J.U. (2021). Algorithms to Synthesis for Adaptive Sub-Optimal Control of Dynamic Objects Based on Regular Methods. International Conference on Information Science and Communications Technologies (ICISCT). Pp. 1-3, DOI: 10.1109/ICISCT52966.2021.9670234.
8. Igamberdiev, H.Z., Boeva, O.H., Sevinov, J.U. (2020). Sustainable algorithms for selecting feedback in dynamic object management systems. Journal of Advanced Research in Dynamical and Control Systems. 12(7 Special Issue). -pp. 2162–2166.
9. Derevitsky D.P., Fradkov A.A. (1981). Applied theory of discrete adaptive control systems (Moscow: Nauka). p216.
10. Tikhonov A.N., Arsenin V.Ya. (1986). Methods of solving ill–posed problems (Moscow: Nauka). p288.
11. Morozov V.A. (1987). Regular methods for solving incorrectly set tasks (Moscow: Nauka).
12. Kuzovkov N.T. (1976). Model control and monitoring devices (Moscow: Mechanical Engineering).
13. Grigoriev V.V., Zhuravleva N.V., Lukyanova G.V. and Sergeev K.A. (2007). Synthesis of automatic control systems by the modal control method (S-Pb: St. Petersburg State University ITMO). p 108.

14. Sevinov J.U., Boeva O.H. (2022). Synthesis Algorithms for Adaptive-Modal Control Systems for Technological Objects With Delays. AIP Conference Proceedings 2647, 030007. https://doi.org/10.1063/5.0104891.

15. Yusupbekov, N., Igamberdiev, H., Mamirov, U. (2022). Algorithms for Robust Stabilization of a Linear Uncertain Dynamic Object Based on an Iterative Algorithm. Lecture Notes in Netstatement and Systems, 307, 225-232. DOI: 10.1007/978-3-030-85626-7_28.

16. Gaiduk A.R. (1987). On the limitations caused by a given part of the system and the control device (Moscow: Izv. vuzov. Instrumentation, №5). Pp. 11-16.

17. Sevinov J.U., Boeva O.H. (2021). Algorithms for Determining the Placement of Poles in Multivariate Systems With Proportional-Differential Output Feedback, International Journal of Advanced Research in Science, Engineering and Technology, Vol. 8, Issue 3, pp. 16896-16902.

18. Miroshnik I.V. (2005). Theory of automatic control. Linear systems (S-Pb: Peter) p336.

19. Gantmacher F.R. (1988). Matrix theory (Moscow: 4th ed. Nauka. Fiz. mat. lit.) 2. p552.

20. Paraev Yu.I., Perepelkin E.A. (2000). Linear matrix equations in the problems of analysis and synthesis of multiconnected dynamical systems (Barnaul: Publishing house of AltSTU) p117.

21. Zhdanov A.I. (2006). Introduction to methods of solving ill–posed problems (Ed. Samara State Aerospace University) p87.

22. Fomin V.N., Fradkov A.L., Yakubovich V.A. (1981). Adaptive control of dynamic objects (M.: Nauka) p448.

23. Tyutikov V.V., Tararykin S.V. (2006). Robust modal control of technological objects (State Educational Institution "Ivanovo State Power Engineering University named after V.I. Lenin". Ivanovo) p256.

24. Zubov N.E., Mikrin E.A. and Ryabchenko V. N. (2016). Matrix methods in the theory and practice of automatic control systems of aircraft (Moscow: Publishing House of Bauman Moscow State Technical University) p666.

25. Yusupbekov, N.R., Igamberdiev, H.Z., Mamirov, U.F. (2021). Stable algorithms for adaptive control and adaptation of uncertain dynamic objects based on reference models. CEUR Workshop Proceedings, 2965, Pp: 296-302.

26. Sevinov J.U., Mallaev A.R. and Xusanov S.N. (2021). Advances in Intelligent Systems and Computing, 1323. Springer, Cham. https://doi.org/10.1007/978-3-030-68004-69.

27. Alisher Mallayev, Jasur Sevinov, Suban Xusanov, Okhunjon Boborayimov (2022). Algorithms for the synthesis of gradient controllers in a nonlinear control system. AIP Conference Proceedingsthis link is disabled (CAMSTech-II), 2467, 030003. https://doi.org/10.1063/5.0093749.

28. Kudryashov V.S. (2006). Calculation of a digital system for multi-connected control of the ammonia synthesis process. Materials of the 3rd scientific and technical. conf. "Mechatronics, automation, control". – St. Petersburg. Pp: 325-328.

29. Sokolov R.S. (2003). Chemical Technology (M.: Humanit. ed. center VLADOS) p368.

30. Ivanov M.E., Olevsky V.M., Polyakov N.N. (1990). Production of ammonia capacity (M.: Chemistry) p288.

31. Faddeev M.A. (2008). Elementary processing of the results of the experiment (S-Pb: Lan) p128.

32. Alifanov O.M., Artyukhin E.A., Rumyantsev S.V. (1988). Experimental methods for solving ill-posed problems (M.: Nauka) p288.

33. Boeva O.H. (2021). Software for determining the parameters of the multi-channel modal adjuster of the stabilization system of column operation modes in ammonia synthesis. Certificate of official registration of the program created for computer program. DGU 12506, 27.09.2021.

34. Sevinov J.U., Boeva O.H. (2021). Software for solving problems of modal control using the optimal algorithm of the digital state controller. Certificate of official registration of the computer program, DGU 11396 dated April 22, 2021.

35. Yusupbekov, A.N., Sevinov, J.U., Mamirov, U.F., Botirov, T.V. (2021). Synthesis Algorithms for Neural Network Regulator of Dynamic System Control. Advances in Intelligent Systems and Computing, 1306, pp. 723–730. DOI: 10.1007/978-3-030-64058-3_90.

36. Igamberdiyev H.Z., Zaripov O.O., Yusupbekov A.N. (2014). Steady Estimation Algorithms of the Dynamic Systems Condition on the Basis of Concepts of the Adaptive Filtration and Control, ICTACT Journal on Soft Computing, Vol. 4, Issue 4, pp. 796-803. DOI 10.21917/ijsc.2014.0114.

37. Misrikhanov M.S. (2007). Invariant control of multidimensional systems. An algebraic approach (Moscow: Nauka) p398.

E3S Web of Conferences **461**, 01070 (2023) https://doi.org/10.1051/e3sconf/202346101070
RSES 2023

Photoelectric characteristics of Si-photocetitors

Abdumalik Gaibov[1]*, *Abdugani* Eshkulov[1], *Utkur* Juraev[1], *Kutbidin* Vakhobov[1]

[1]Tashkent State Technical University named after Islam Karimov, 100095, Uzbekistan, Tashkent, University St. 2A

Abstract. The work examines the influence of ultrasonic fields on the photoelectric and spectral characteristics of diffusion Si-n-p receivers of electromagnetic radiation. It has been established that ultrasonic irradiation increases the lifetime and diffusion length of carriers and, as a consequence, increases the efficiency of carrier collection at the electrical contacts of Si-n-p receivers. As a result of these processes, an increase in the short-circuit current is observed, which causes an increase in the open-circuit voltage of the Si-n-p structure operating in the photoconversion mode.

1 Introduction

The creation of solar energy photoconverters based on semiconductor materials is a priority in modern science and technology. The leading position in this area is occupied by the technology of creating solar cells based on silicon. The efficiency of solar cells based on monocrystalline silicon has currently reached a record value of more than 25%.

This work examines in detail the behavior of the photoelectric and spectral characteristics of Si-n-p optical radiation receivers before and after ultrasonic exposure. The information obtained is useful and, from a practical point of view, can be used in production technologies not only for semiconductor photodetectors operating on the basis of charge carrier transfer, but also in the production of photoconverters. The relevance of such research is indicated by the fact that there have been works showing improved characteristics of photoconverters based on Si single crystals after ultrasonic treatment, but detailed studies of the characteristics have not been carried out [1].

2 Experimental research

The experiments examined n-p structures made of low-resistance silicon, which had an area $S=5\times5$ mm^2 and a thickness $d=250\div50$ μm. The base p-Si layer had a resistivity of 1.8 Ωcm<ρ<3 Ωcm. To create a diffusion n^+ layer in the base with a thickness of $0.15\div5$ μm, phosphorus was used as a dopant. The concentration of N dopant for various Si samples ranged from $7\cdot10^{18}$cm^{-3} to $5.5\cdot10^{19}$cm^{-3}. The N concentration was determined by the four-probe method. A sawn-off layer of gold with a density $K = 30\div50$ μg/cm^2 was used as a frontal contact. The rear contact was a layer of sawn aluminum (Al). Figure 1 *a*) shows the structure and research methodology for photosensitive Si-n-p structures for these experiments. The Si photodetector (1) is located on a thick copper plate (2), which is also the lower electrical contact, and also serves to supply cold using a copper rod (3), which is a cold conductor and immersed in a Dewar vessel (4) with liquid nitrogen (5).

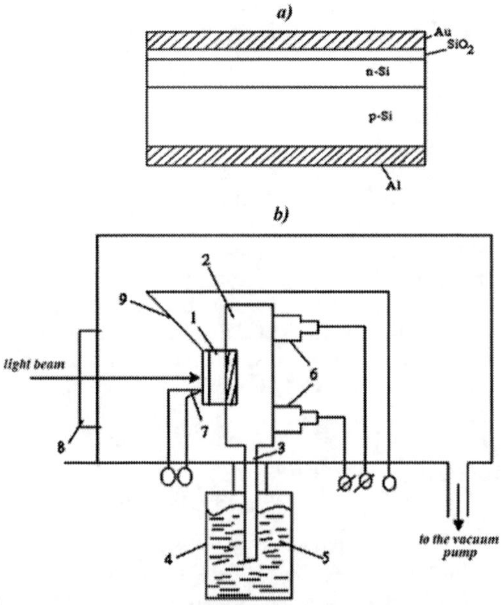

Fig, 1. Structure (*a*)and methodology (*b*) for studying photosensitive Si-n-p structures.

The samples were heated using D818 zener diodes (6), through which a direct current was passed. The temperature was controlled using a thermocouple (7)

* Corresponding author: raxmatillo82@mail.ru

with an accuracy of $\Delta T = \pm 0.1$ K. The sample was placed in a vacuum chamber that had a glass window (8). The electrical signal from the p^+-Au contact was recorded using a clamping contact (9). Electrophysical, photoelectric and spectral characteristics were measured in a standard way using spectrophotometers SF-4 and SF-20, as well as lighting sources in the form of xenon lamps with a power of 100 W to 150 W and infrared filters. Ultrasonic processing of the Si-n-p photodetector was carried out in a standard manner using a liquid acoustic conductor located between the Si photodetector and the ultrasonic emitter.

Figure 2 demonstrates a family of current-voltage characteristics of a Si photodetector, obtained and studied by us when illuminating the latter with light at different illumination levels before and after ultrasonic exposure. The Si photodetectors under study operated as photoconverters and at T = 300 K had an open-circuit voltage V_{xx} within the range of 0.5 V ÷ 0.53 V.

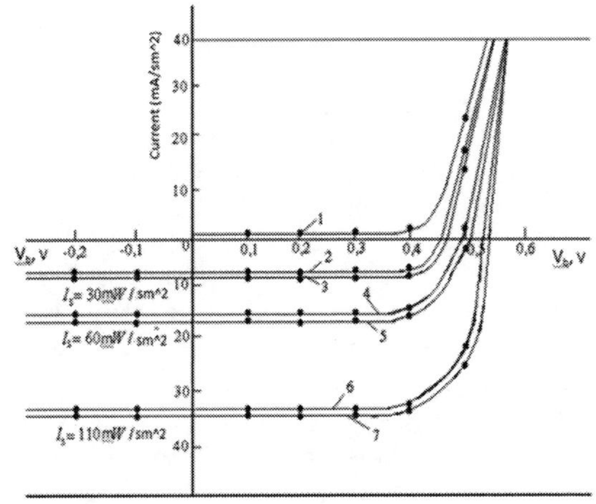

Fig. 2. Current-voltage characteristics of a Si-n-p photodetector at different illumination levels IS before (curves 2, 4, 6) and after (curves 3, 5, 7) ultrasonic irradiation of the photodetector with ultrasonic waves with intensity I* = 0.25 W/cm2, f =25 MHz for t=40 min, at T=300K. Curve 1 – darkened.

The usual dependences of I(V) for a Si photodetector are shown in Fig. 3. The filling factor ff, determined by the ratio of the maximum specific power P_M to the value of the product $V_{xx} \cdot I_{k3}$ (I_{k3} -short circuit current) for this photodetector, has the following value: ff ≅ 0,72. It should be noted that the dependence of Ikz on the power of the lighting level is linear throughout the entire measurement interval, and the dependence of V_{xx} on the lighting power is almost logarithmic.

Thus, it can be shown that the dependence $\log I_{k3}(V_{xx})$ approaches linear; this is observed from Fig. 3. Theoretical analysis of the behavior of the main characteristics V_{xx}, I_{k3}, ff depending on the influence of external factors (temperature, illumination) is quite well described in various works, for example [4], but the influence of ultrasonic fields on these characteristics has not been studied previously.

In these studies, we use Si photodetectors as models of solar energy photoconverters irradiated with ultrasonic waves.

Analysis of the experimental dependence $I_{k3}(V_{xx})$, presented in Fig. 3. shows that the coefficient "n"≈1,51÷1,52 and A ≈25,8 V^{-1}÷26 V^{-1} at a short-circuit current value $I_{k3} \approx (1{,}2 \div 1{,}4) \cdot 10^{-3}$ A/cm^2 /cm^2 and $I_0 \approx 1{,}2 \cdot 10^{-7}$ A/cm^2.

It is also clearly visible that an increase in illumination power even at 10 times solar illumination leads to a deviation of the dependence of the current-voltage characteristic from linearity in the region of high bias voltage values Vb > 0.45 V (Fig. 3). This effect is explained by the saturation of recombination centers with carriers, an increase in the lifetime and, as a consequence, leads to an increase in the efficiency of collecting nonequilibrium charge carriers at the current-collecting contacts of the Si photodetector [2, 3]. In the same Figure 2, curves 3.4 of the I(Vb) dependence are shown, measured after the ultrasonic wave passed through the Si photodetector.

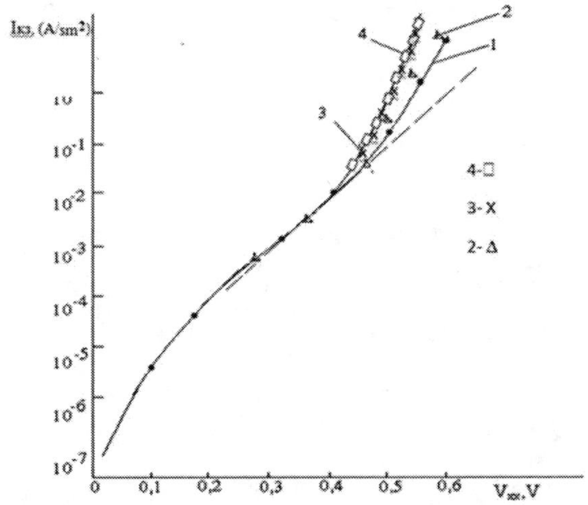

Fig.3. Dependence of the short circuit current on the open circuit voltage under single and tenfold solar illumination, measured before and after ultrasonic treatment of the Si-n-p photodetector at T=300 K. Before ultrasonic treatment: • - curve 1 (single solar illumination); Δ - curve 2 (10 times solar illumination). After ultrasonic treatment (I*=0.25 W/cm^2, f=25MHz, t=40min): * - curve 3 (single solar illumination); □ - curve 4 (10 times solar illumination). The dotted line is the theory.

The temperature dependences of the main characteristics of the Si photodetector under study, as well as the spectral dependence of the photoresponse before and after ultrasonic treatment, are presented in Figures 4 and 5, respectively. It is of interest to us to explain the observed acoustically stimulated changes in these characteristics, which has not been previously studied. Noteworthy is the fact that all the studied characteristics, to one degree or another, experienced a slight increase in their values. Since all the characteristics are interconnected by known relationships, there is no need to recall that they are all determined by the efficiency of collection of nonequilibrium charge carriers, which decisively depends on the degree of perfection of the crystal lattice of the Si photodetector.

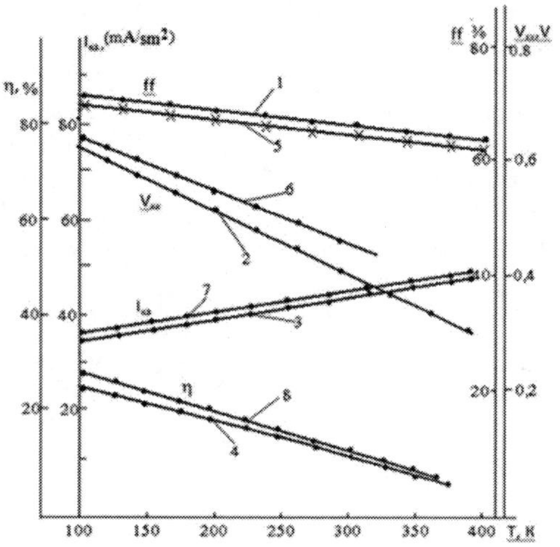

Fig.4. Temperature dependences of the parameters of the Si-n-p photodetector (ff, V_{XX}, $I_{к.з.}$, η), measured at illumination intensity I_S = 130 mW/cm^2 before (curves 1, 2, 3, 4) and after (curves 5, 6, 7, 8). Ultrasound irradiation with intensity I* = 0.25 W/cm^2 and frequency f = 25 MHz for t = 40 min at T = 300K.

3 Research results

As is known, the degree of perfection of a crystal lattice is determined by the concentration of defects. Defects can be electrically inactive or active. The latter affect the lifetime of minority carriers in sensitive areas of photodetectors. When studying the leading edges of charge pulses, a decrease in discharge was discovered, i.e. there was an increase in the lifetime τ of charge carriers by an average of 10÷15 %. As is known, when considering the processes of transfer of charge carriers (electrons), the diffusion equation is used:

$$I_n = qD_n\left[\frac{\partial(n_p - n_{po})}{\partial x}\right]; \quad (1)$$

where D_n is the electron diffusion coefficient related to the mobility of carriers μ_n by the Einstein relation $D_n = kT\mu_n/q$.

The value of D determines the diffusion length of charge carriers L through the equation:

$$L = \left(\frac{\mu kT\tau}{q}\right)^{1/2}; \quad (2)$$

Thus, the dependence $I_{кз}(V_c)$ is mainly due to an increase in the lifetime τ, which is influenced by the relative position of the energy levels of the capture and recombination centers. The theory gives the following well-known expression for the value τ, see for example [7-9]:

$$\tau = \frac{1}{N\sigma_n V_t}; \quad (3)$$

These acoustically stimulated processes lead to a smoothing of the potential relief of the sensitive region of the photodetector and, as a consequence, to an increase in the diffusion length L, an increase in the efficiency of carrier collection and radiation detection. It should be borne in mind that under certain conditions, such as, for example, in the case of introducing gold atoms into the sensitive region, as noted above, the value of τ decreases. A similar decrease in τ can also occur at threshold powers I* of ultrasound ≥ 2÷5 W/cm2, when there is intense generation of defects reducing τ. The nature of such defects has not yet been sufficiently studied [5, 6].

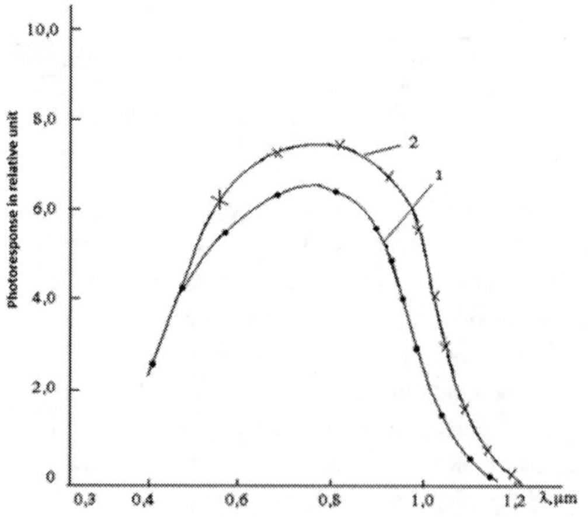

Fig.5. Spectral dependence of the photoresponse of the Si-n-p photodetector before (1) and after (2) ultrasonic treatment.

The presented spectral dependence of the photoresponse after ultrasonic treatment (fig. 5. curve 2) in the region λ_0 = 0,6 μm ÷ 0,8 μm shows its increase compared to the original curve. this indicates that surface recombination has a low speed and is suppressed by ultrasound in the near-surface layers of the n-p junction of the si photodetector. in the wavelength range λ_0 = 0,7 μm ÷ 0,8 μm, an increase in photosensitivity of approximately 15÷20 % is observed. the increase in the photoresponse is also clearly manifested in the longer wavelength region of the spectrum $\lambda_0 \geq$ 0.8 μm ÷ 1.2 μm, that is, the collection of nonequilibrium charge carriers formed during the absorption of photons of lower energies is facilitated due to the acoustically stimulated increase in the diffusion length l of carriers in silicon.

4 Conclusion

It has been established that ultrasonic irradiation of such receivers from the side of the "entrance" window leads to a change in the concentration of electrically active centers in the sensitive region of the receiver and smoothing of the impurity relief through the mechanism of acoustically stimulated diffusion. A positive change in the defective structure of the sensitive region leads to a change in its transport properties, which determine the processes of collection of nonequilibrium charge carriers. In this case, there is an increase in the lifetime and diffusion length of carriers and, as a consequence, an increase in the efficiency of collecting carriers at the electrical contacts of Si-n-p receivers.

As a result of these processes, an increase in the short-circuit current is observed, which causes an increase in the open-circuit voltage and the efficiency of such a diffusion Si-n-p structure operating in the photoconversion mode. It is noted that the surface recombination of carriers decreases in the near-surface layers of the Si-n-p photodetector under the influence of ultrasonic waves.

References

1. Gaibov A.G., Zaveryukhin B.N., Krevchik V.D. and others. Acoustic-stimulated processes in Si(Li)-p-i-n detectors. Letters to ZhTF. 1984, issue. 10, vol. 10, pp. 616-620.
2. A.A. Fahrenbruch, R.B. Bueb. Solar cells. Theory and experiment. Solar cells. Theory and experiment. 1987. Publishing house "Energoatomizdat", p. 278.
3. Zaveryukhin B.N., Zaveryukhina N.N. and others. Change in the reflectivity of radiation from the surface of semiconductors in the spectral range λ_0=0,4μm÷1μm under the influence of ultraviolet waves. ultrasonic waves. Letters to ZhTF. 2002, issue. 28, vol. 18, pp. 1–12.
4. Mirsagatov Sh.A., Sapaev I.B., Nazarov Zh.T. "Ultrasonic annealing of surface states in the heterojunction of a p-Si/n-CdS/n+-CdS injection photodiode." // Inorganic materials 2015, volume 51, No. 1 p.3-6
5. Zaveryukhina N.N., Vakhobov K.I., Gaibov A.G. Acoustic-stimulated diffusion of gold in Au-Si–n-p structures. Reports of the Academy of Sciences of the Republic of Uzbekistan. 2005, Vol. 6, pp. 20-23.
6. Gaibov A.G. Vakhobov K.I., Juraev U. E., Ibragimova B.V., Rasulova D. T. Influence of ultrasonic waves on current-voltage characteristics and polarization effects of si-n-p radiation receivers. Materials Science Forum, ISSN: 1662-9752, Vol. 1049,2022, pp. 317-324. [DOI:10.4028/www.scientific.net/MSF.1049.317].
7. Gaibov A.G. The influence of ultrasonic waves on the characteristics of the radiation detector. Modern Innovations, Systems and Technologies, 2022, 2(1) pp. 62-68.
8. Gaibov A.G., Vakhobov K.I. Acoustostimulated diffusion of gold atoms into highly doped N + -Layers of Silicon. International Journal of Advanced Research in Science, Engineering and Technology, 2020. Vol.7, Issue 7,p.14455-14461. https://ijarset.com/upload/2020/july/28-gaibov-35.pdf.
9. Gaibov A.G., Vakhobov K.I. (2020). Photoelectric and spectral characteristics of SI photodetectors. *International Journal of Advanced Science and Technology*, *29*(11s), 1431 - 1437. Retrieved from http://sersc.org/journals/index.php/IJAST/article/view/21010

To the adsorbent obtained on the basis of local Angren coal analysis of adsorption energy properties of benzene vapors

Gulnora Gulomova[1*], *Musurmon* Barotov[1]

[1]Tashkent State Technical University named after Islam Karimov, 100095, Uzbekistan, Tashkent, University St. 2A.

Abstract. This article provides information on ensuring safety at coal mines by analyzing existing coal mining facilities. Information on the Angren coal mine is also given. It is shown that there are no accurate methods for determining the location of fires in order to increase the efficiency of spontaneous combustion prevention by supplying liquid compositions such as water and clay pulp flowing down the formation soil. Measures have been proposed to suppress spontaneous combustion centers, nitrogen is being increasingly used, the supply of which ensures volumetric processing of the collapsed mass and allows reducing the oxygen concentration in the mined-out space.

1 Introduction

At present, activated carbon adsorbents are widely used in the world for purification of various organic compounds, heavy metal ions and life-threatening radioactive metals. Today, hydrophobic carbon adsorbents are mainly modified on the basis of lignite, coal, coke and stems and barks of various plants. Activated carbon adsorbents obtained on the basis of coal and oil coke are important in cleaning of organic, aromatic compounds and heavy metal ions in oil refining waste water [1].

At present, scientific and research work is being carried out in the world on obtaining activated carbon adsorbents using local coal and oil waste products from mines. In this regard, studying the composition of industrial waste; to determine optimal conditions for obtaining carbon adsorbents based on them; absorbing organic substances into the obtained carbon adsorbents, determining the number and location of their micro-, meso-, and macropores, the complete molecular mechanisms of sorption processes occurring in adsorbents; Special attention is being paid to the creation of industrial wastewater treatment technology using carbon adsorbents [2].

Today, preventing the accumulation of coal dust and residues in our Republic and directing it to targeted production is one of the urgent issues of today. By enriching coal residues with additional substances, obtaining adsorbents and fully determining the adsorption properties of porous coal adsorbents, we will have the opportunity to determine the full thermodynamic properties of the obtained porous coal adsorbents using capillary and volumetric methods in a high-vacuum adsorption device, high-sensitivity DAK 1-1 calorimetric apparatus.

By determining the adsorption isotherm, differential heat, entropy and equilibrium time of benzene on the obtained coal, the adsorption capacity of the new adsorbent, location of active centers, and pore sizes can be described in detail [3].

2 Experimental research

Carbon microporous adsorbents (activated carbons, active fibers, and fabrics) are a class of high molecular weight solid porous carbon materials that absorb substances of various chemical natures (in gaseous, vapor, and liquid media) onto certain surfaces [1]. According to the classification of M.M. Dubinin, based on the sorption mechanisms occurring in the adsorption pores and capillary phenomena, the pores of various adsorbents are divided into the following types: micropores, macropores (equivalent radius r greater than 100-200 nm), mesopores (1.5 -1, 6 <r <r <1.5-1.6 nm).

Carbon microporous adsorbents are significantly different from materials such as coke, pumice, graphite because they contain micropores and supermicropores. In the same way, they differ from coal, which have a relatively high surface area due to small particle size (up to 100 m2/g and above) and belong to non-porous carbon adsorbents. Micropores and supermicropores are actually adsorbed into their pores, as they are crucial for the adsorption of gases and vapors and, in many cases, for liquid phase adsorption. It depends on the structure and geometric surface of adsorbates.

Its value is ~400-1400 m^2/g according to, i.e. 15-55% of the theoretically possible value of 2630 m^2/g obtained for a hypothetical model of pure aromatic carbon of hexagonal form with a condensed flat band structure. The volume of micropores and supermicropores is usually in the range of 0.2 - 0.6 cm^3/g For the best

* Corresponding author: raxmatillo82@mail.ru

 https://doi.org/10.1051/e3sconf/202346101071

samples of carbon adsorbents, it can reach 1.3 - 1.5 cm^3/g [1].

A formula for calculating the geometric volume of micropores from the value of adsorption energy of standard benzene vapor was proposed. The possibility of redescription of adsorption isotherms for several substances (benzene, ethyl chloride, carbon tetrachloride, etc.) using the equations of the theory of volume saturation of micropores (MHTN) was shown.

Thus, according to the MHTN equations and the parameters of the porous structure of activated carbon calculated from the adsorption isotherms of standard benzene vapor, it is possible to evaluate their adsorption properties with respect to vapors of a wide class of substances. However, it should be noted that in some cases there is a discrepancy between the parameters determined by the adsorption of benzene and other substances.

At the Institute of General and Inorganic Chemistry of the Russian Academy of Sciences, they are conducting quantitative research on the adsorption of polar and non-polar molecules on surfaces, microporous adsorbents, in a high-vacuum adsorption device using a microcalorimeter [4].

3 Research results

The adsorption isotherm of benzene on the resulting adsorbent was determined in the pressure determination section of the high-vacuum adsorption device. The differential heat of adsorption is measured by theoretical calculation of the pressure values, and the energy output from the DAK 1-1 calorimeter in the Tian-Calve model. Volumetric and capillary (liquid) methods were used to determine the adsorption isotherm. The adsorption isotherm has an accuracy of 0.1% and a heat of 1% [5].

The benzene obtained as an adsorbate was purified in a mixture of zeolites and under vacuum before use in sorption. Additional gases are expelled until the vapor pressure of the absorbing material is the same as the vapor pressure data given in the tables for absolute benzene.

Adsorption of benzene was carried out at 303 K on a sample of activated carbon adsorbent by activating local anthracite coal.

Results obtained and discussion. Local coal was mixed with asphaltene, a solvent for residual (powder) and small particles. After the mixtures are thoroughly mixed, they are kept until they become thick and elastic. Then they are prepared for granulation. After granulation, it is heated at 6000C. As a result of heating, additional gases in the composition are released. After that, activated carbon is obtained by the pyrolysis method.

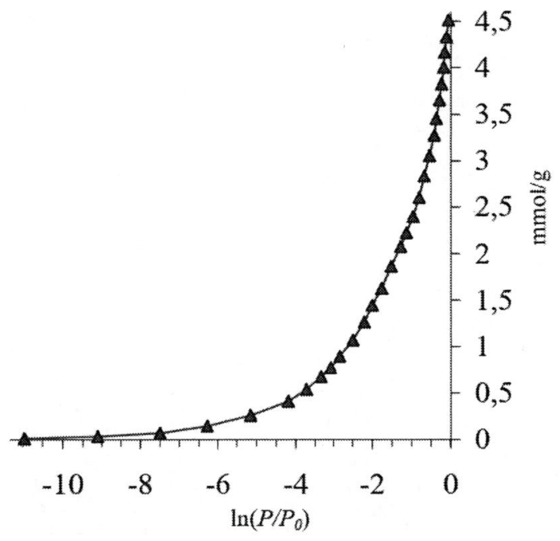

Fig. 1. Isotherm of the adsorption of benzene vapors on the adsorbent obtained from local coal residues at a temperature of 303 K

The theoretical study of the sorption properties of the obtained activated carbon is based on the results obtained from experiments conducted in a high vacuum adsorption device. Figure 1 shows the adsorption isotherm of benzene on activated carbon from local coal mines at 303 K. The adsorption isotherm initially shows $\ln(P/P_0)$=-11.

This means that coal is being adsorbed in the micropores. Considering that the size of the benzene molecule is 0.6 nm, it indicates the presence of micropores in our obtained adsorbent [6]. .

Adsorption of benzene at values, up to about 1 mmol/g is $\ln(P/P_0)$=-7.

This means that benzene molecules are adsorbed in the internal micropores of coal. When adsorption reaches 4.5 mmol/g, the pressure in the device is 119.38 mm.sim. above will be equal. This value is equal to the saturation vapor pressure of benzene at a temperature of 30^0C, and coal is saturated with benzene.

$$a = 0{,}64\exp[(A/13{,}64)^2] + 2{,}48\exp[(A/4{,}93)^2] + 0{,}99\exp[(A/0{,}97)^1],$$

where a is the adsorption value in a-micropores, mmol/g; a - is the work of free energy, kJ/mol, A = RTln(P/P_0).

 https://doi.org/10.1051/e3sconf/202346101071

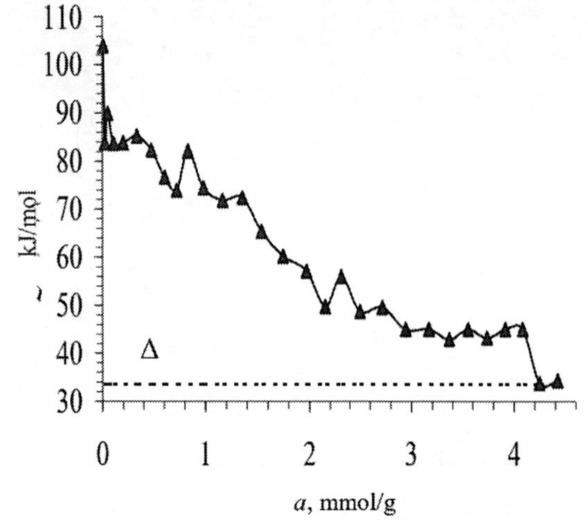

Fig 2. Differential heat of adsorption of benzene vapors on an adsorbent derived from native coal residues at a temperature of 303 K. The horizontal dashed line is the heat of condensation of benzene vapor at 303 K

The differential heat of adsorption of benzene vapors on the coal adsorbent obtained for research work at a temperature of 303 K is presented in (Fig. 2).

At first, the heat of adsorption of benzene is equal to 105.05 kJ/mol when α=0.055 mmol/g at the initial saturation, and it is observed that the heat value of adsorption of benzene molecules decreases slightly. Q_d=85.16 kJ/mol at α=0.52 mmol/g [7].

One of the main reasons for the high heat of adsorption of benzene molecules at low saturations is explained by their adsorption into micropores. Metal cations located in the micropores of the coal sample form π-complexes of $(Me^{n+}S_6N_6)n$ type. It was further observed that the adsorption exhibits a single maximum heat at a differential heat value of α=1.0 mmol/g and increases to Q=87.66 kJ/mol. In the subsequent stages of adsorption, the differential heat decreases. After a=4.2 mmol/g, the heat of adsorption becomes equal to the heat of condensation [8].

This decrease in the adsorption differential heat values is absorbed into the micro-, and mesopores of the coal, and the heats of adsorption decrease with increasing benzene saturation. The value of condensation heat of adsorption of benzene vapors on the coal obtained for research work is close to 33.8 kJ/mol, and a total of 4.6 mmol/g of benzene is adsorbed [9].

Adsorption entropy describes the state of motion of absorbed adsorbate molecules in the adsorbent. The entropy curve of the adsorbent obtained from local coal residues -benzene system had a wavy appearance according to the amount of adsorbent saturation (Fig. 3).

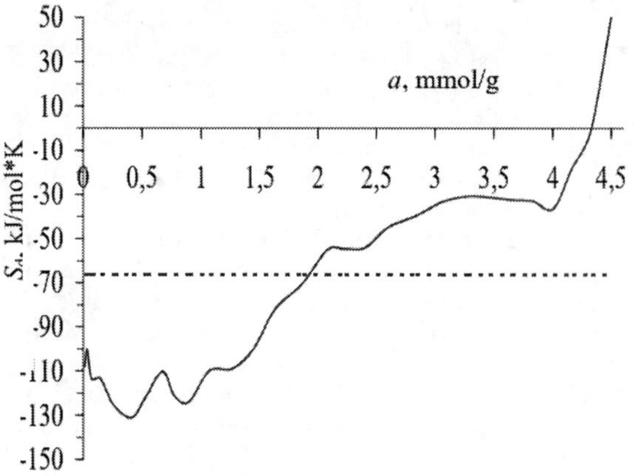

Fig.3. Molar integral entropy of adsorption of benzene vapors on an adsorbent obtained from native coal residues at a temperature of 303 K

It is known from the entropy graph that at values - up to 2 mmol/g, benzene is strongly adsorbed, that is, it is located below the average integral entropy line.

After 4.3 mmol/g, it is higher than the standard state benzene entropy value. Adsorption entropy in the process up to 4.3 mmol/g indicates strong localization in the pores of the adsorbent. In the initial stages of saturation, benzene molecules are strongly adsorbed in the micropores of coal, that is, benzene molecules are in an unexcited state, during the adsorption process up to -68.7 J/mol*K, benzene goes to the solid state entropy value. Then, benzene slowly changes to the entropy of the liquid state. The average integral entropy of benzene adsorption on coal obtained in the research work is -68.7 J/mol*K [10].

4 Conclusion

In our research work, an adsorbent activated from local coal residues was obtained and its adsorption properties were studied. The adsorption differential heat, isotherm, molar entropy and thermokinetics of benzene molecules were calculated and it was determined that 4.5 mmol/g of benzene and 3.5 mmol/g of toluene are adsorbed on the activated carbon adsorbent. It was observed that benzene adsorption on the obtained activated carbon adsorbent starts with 104 kJ/mol and toluene with 90 kJ/mol, and the differential heat gradually decreases.

Based on the adsorption isotherm values of benzene and toluene on the activated carbon adsorbent, recalculated using the volume saturation theory equation of micropores, 84.5% of benzene and 94.2% of toluene were adsorbed on the micropores of the adsorbent.

The main features of the obtained adsorbent are that the raw materials necessary for the production industry are obtained by effectively using the residual products of coal.

References

1. Gulomova G.M., Abdurakhmanov E.B. Adsorption energy of toluene vapors into an adsorbent obtained on the basis of local igneous coal //Harvard Educational and Scientific Review, International Agency for Development of Culture, Education and Science USA. Vol.1. Iss. 1. P. 118–128. DOI: 10.5281/zenodo.5670204

2. G.M. Gulomova. Theoretical and practical aspects of studying Angren coal for the production of sorbents // International Conference of Academic Sciences, Novosibirsk, October 2021, pp. 92-97.

3. Gulomova G.M., Shomansurov S.S., Rakhmatova D.M. Development of methods for obtaining carbon adsorbents based on Angren coal and oil residues. Occupational Safety in Industry, № 3 2023, p.68-74 https://www.scopus.com/2.0-85163386699

4. G.M. Gulyamova., M.N. Musaev. Problems of sludge waste disposal in the oil and gas industry. Collection of materials of the VII International Correspondence Scientific and Practical Conference, June 5, 2020, pp. 98-100. Minsk.

5. G.M. Gulomova, U.M. Turabekova. Problems of ensuring safety at open coal mines in the republic of Uzbekistan. TECHNICAL SCIENCE AND INNOVATION №3 ISSN: 2181-0400, (2022)

6. G.M.Gulomova, O.T.Tuxtamurodov, and ofters. AIP Conference Proceedings, 224-230, (2023) https://zenodo.org/record/7985791

7. G.M. Gulomova, E.B. Abdurakhmanov. Energy of adsorption of benzene vapors by activated carbon. Joutral. №11-1 (89), 100-102, (2021) https://scholar.google.com/

8. G.M Gulomova, E.B Abduraxmonov, Adsorption energy of toluene vapors into an adsorbent obtained on the basis of local angrene coal. Harvard Educational and Scientific Review, (2021) https://www.journals.company/index.php/article/view/14

9. G.M. Gulomova, M.N.Musaev, Problems of Ensuring Safety at Open Pits in the Republic of Uzbekistan. JournalNX, 314-319, (2020) https://scholar.google.com/citations

10. D.Rakhmatova, M.Musaev, G.Gulomova, M.Aripkhadzhaeva, D.Nizamova, Development of safe products from local waste of oil and fat production E3S Web of Conferences, 371, (2023) https://scholar.google.com/citations .

11. Gulomova G.M., Bozorov Z. P., Talipova Yu. A. The influence of the environment on the strength of building structures. Novateur publication, Indian magazine NX. May 25 and 26, 2021, pp.733-737.

10. Khudayberganova N., Rizaev A., Abduraxmonov E.B. Adsorption properties of benzene vapors on activated carbon from coke and asphalt//E3S Web of Conferences 264, 01022 (2021) CONMECHYDRO – 2021 https://doi.org/10.1051/e3sconf /202126401022

12. M.Khudoyberganov, F.Rakhmatkarieva, E.Abdurakhmonov, I.Tojiboeva, Kh.Tadjieva, Thermodynamics of water adsorption on local kaolin modified microporous sorbents//The 1st international conference on problems and perspectives of Modern science, ICPPMS, 10-11 June, 2021, Tashkent, Uzbekistan (2021), AIP Conference Proceedings 2432, 050001 (2022); https://doi.org/10.1063/5.0090736

12. F.Rakhmatkarieva, O.Ergashev, O.Davlatova, M.Kokhkharov, M.Khudoyberganov, E.Abdurakhmonov, T.Abdulkhaev, NaITA energy in molecular sieve guest-guest interactions energetics of host-guest interactions in NaITA molecular electrics// Journal of Physics: Conference Series, IV International Conference on Applied Physics, Information Technologies and Engineering 2022 (APITECH-IV 2022) 05-08 October, Bukhara, Uzbekistan (2022), J.Phys.: Conf. Ser. 2388 012175, https://doi:10.1088/1742-6596/2388/1/012175

13. E. Abdurakhmonov, I.B. Sapaev, B. Abdullaeva, F. Rakhmatkarieva, N.N. Dekhkanova. Thermodynamics of hydrogen sulfide adsorption in Zeolite LiX// E3S Web of Conferences 383, 04018 (2023) TT21C-2023 https://doi.org/10.1051/e3sconf/202338304018

14. Firuza Rakhmatkarieva, Mansur Xudoyberganov, Sergei Lyapin, Eldor Abdurakhmonov, Mirzakhid Kokharov, and Tolib Abdulkhaev. Energy, structure and localization of H_2O-Ba^{2+} and H_2O- Na^+ complexes in BaNaY molecular sieve// AIP conference proceedings « Problems in the Textile and Light Industry in the Context of Integration of Science and Industry and Ways to Solve Them (PTLICISIWS-2022)» AIP Conference Proceedings 2789, 020022 (2023) https://doi.org/10.1063/5.0145638

15. Nigora Dekhkanova, Eldor Abdurakhmonov, Feruza Rakhmatkarieva and Nodirabegim Jamoliddinova Thermodynamics of hydrogen sulfide adsorption in NaX zeolite // E3S Web of Conferences 402, 14037 (2023) TransSiberia 2023 https://doi.org/10.1051/e3sconf/202340214037

16 Eldor Abdurakhmonov, Nigora Dekhkanova/ Thermodynamics of hydrogen sulfide adsorption in Zeolite LiX// E3S Web of Conferences 413, 04004 (2023) INTERAGROMASH 2023 https://doi.org/10.1051/e3sconf/202341304004

17. Ziyada Djumanova, Eldor Abdurakhmonov. Adsorption energetics of benzene vapors on activated adsorbent based on glycyrrhiza glabra root extract residues// E3S Web of Conferences 413, 04005 (2023) INTERAGROMASH 2023 https://doi.org/10.1051/e3sconf/202341304005

18. Muxlis Oydinov, Tolibjon Abdulxayev, Mirzohid Kokhkharov, Eldor Abdurakhmonov, Firuza Rakhmatkarieva Thermodynamics of Carbon (IV)-Oxide adsorption on NaA Zeolite obtained by modification of angren kaolin// E3S Web of Conferences 413, 04007 (2023) INTERAGROMASH 2023 https://doi.org/10.1051/e3sconf/202341304007

19. Muminjon Ismoilov, Bosit Khamidov, G. Yakubova, Eldor Abdurakhmonov andShavkat Khamidov Oil production from local oil// BIO Web of Conferences 65, 03003 (2023) EBWFF-2023 https://doi.org/10.1051/bioconf/20236503003

Transition to alternative sources of electricity supply: diversifying the energy strategy in the Republic of Uzbekistan

Indiraxon Ayubova[1]*

[1]Tashkent State Technical University named after Islam Karimov, 100095, Uzbekistan, Tashkent, University St. 2A

Abstract. In today's world, the problem of energy security is becoming increasingly important. The issue of increasing energy independence and diversification of energy supply sources is becoming an important component of this problem. Transition from classical to alternative energy sources allows not only to reduce dependence on external suppliers, but also to improve the environmental situation. The article is devoted to analyzing the process of transition to alternative sources of electric energy in the Republic of Uzbekistan. In the conditions of constantly growing demand for electricity and irreversible depletion of natural resources, the importance of this topic can hardly be overestimated. The purpose of this study is to examine the theoretical and practical aspects of this transition, including not only changes in the energy sector, but also socio-economic, technological and environmental consequences of these changes for the country.

1 Introduction

The current electricity supply system in Uzbekistan relies heavily on conventional energy sources, predominantly natural gas. According to the World Bank, about 85% of the total electricity supply is generated by gas-fired power plants. At the same time, the country has faced a number of challenges, including systemic technical and financial constraints that increasingly emphasize the importance of diversifying energy sources [4].

At the same time, Uzbekistan has a significant potential for the use of renewable energy sources. Specialists of the International Renewable Energy Agency (IRENA) estimate the country's solar and wind energy potential at 51.3 GW and 520 GW, respectively [5].

In addition, public authorities are currently making efforts to increase the share of alternative energy sources in the total electricity supply balance, setting a target to increase their share to 25% by 2030, which implies a build-up of renewable capacity to 5 GW in 10 years [3].

2 Research

The strategic vision of energy management in the Republic of Uzbekistan is changing significantly in light of the growing need to diversify the sources of electricity supply and move towards greater use of renewable energy. This correlates with global trends and a response to the need to ensure sustainability, reliability and affordability of energy for the population and economy of the country.

The Republic of Uzbekistan has initiated several plans over the years to transition towards alternative sources of electricity supply as part of its long-term strategy to diversify its energy sector.

Uzbekistan is blessed with abundant resources of renewable energy, especially solar and hydro power, and the government is making efforts to tap into these resources.

As part of its strategy to modernize its energy infrastructure, the Republic of Uzbekistan has been working with international organisations such as the Asian Development Bank, the European Bank for Reconstruction and Development, the United Nations Development Programme, and the World Bank.

Key initiatives include:

1. Solar Power: Uzbekistan is one of the sunniest countries in the world and has the potential to generate more than 51 billion kWh per year from solar energy. A series of large-scale photovoltaic power plants are planned with the help of international partners.
2. Hydro Power: Hydropower contributes a significant percentage of Uzbekistan's electricity supply. There are plans to upgrade existing hydropower plants and construct new ones.
3. Wind Energy: Although not as extensive as solar and hydropower, there are plans underway to explore the potential of wind energy, with several feasibility studies in place.
4. Energy Efficiency: There are also considerable efforts made towards improving energy efficiency in the country. This includes modernizing current energy infrastructure, introducing energy-saving technologies

* Corresponding author: ilider1987@yandex.ru

and driving a cultural shift towards more prudent consumption of power.

5. Legislation: The Uzbek government has also made changes to the legal and regulatory framework to encourage the use of renewable energy. This includes incentives to local and foreign investors to invest in the renewable energy sector.

According to the "Action Strategy in the Five Priority Areas of Uzbekistan's Development 2017-2021", the government aims to implement reforms in the oil and gas sector, including the development of renewable energy [3]. It is planned that by 2030, the share of renewable energy in the country's energy balance will increase to 25%. (Fig.1.)

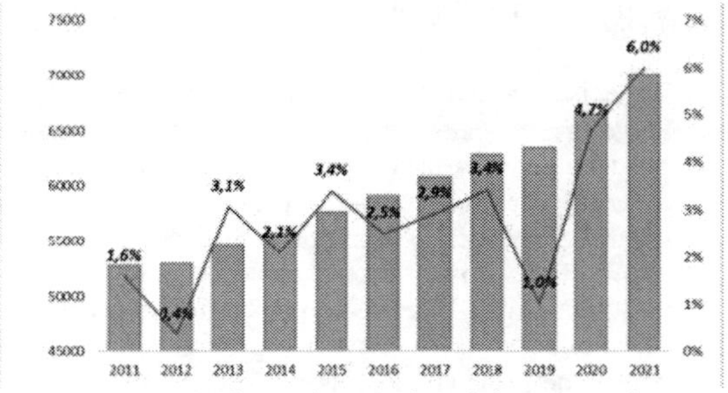

Fig. 1. Information on electricity generation

Steps in this direction are already being taken. In June 2020, the government of Uzbekistan commissioned the first major solar power plant in Zafarabad district of Jizzak region with a total installed capacity of 100 MW [4]. This is the first major solar power project implemented in the country, and it becomes a significant step towards achieving the goal of increasing the use of renewable energy.

However, the task of transition to renewable energy sources is long-term and requires serious work and significant investments. Taking into account the country's huge potential in the use of solar and wind energy, it is planned to create the necessary infrastructure and regulatory environment to attract domestic and foreign investments in this sector [5].

In addition, emphasis should also be placed on increasing energy efficiency and saving on energy use. In this context, energy demand management and industrial energy policy, respect for green building and urbanization principles become an integral part of the country's energy development strategy.

In addition to the above, a phased transition to alternative energy sources is a key focus of Uzbekistan's energy development strategy. The main elements of this strategy include decisions on infrastructure, financing, regulation, industrial policy and interaction with foreign partners.

Diversification of electricity supply sources through the active use of alternative energy sources is a sustainable solution that provides both economic and environmental benefits for Uzbekistan.

From an economic point of view, the use of renewable energy sources helps to reduce dependence on the international hydrocarbon market, thus avoiding unpredictable fluctuations in the prices of these resources. In the long term, it increases the energy security of the country and the stability of its energy system.

The development of alternative energy also contributes to the country's economic development by creating new jobs, attracting foreign and domestic investment and expanding opportunities for scientific and technological development. Dentons Law Firm's 2020 estimates show a significant growth of the renewable energy sector in the country, attracting more and more investments. In the long term, this will attract new investments and create new jobs, thus stimulating the economic development of the country [1].

Transformation to renewable energy can also help Uzbekistan address energy challenges, such as lack of access to electricity in hard-to-reach or remote locations. Small-scale renewable projects such as solar panels and wind turbines can provide for these needs at the local level, increasing access to electricity for a wider range of people.

From an environmental point of view, the use of renewable energy sources leads to a reduction in greenhouse gas emissions and pollution. Given the scale of the global climate crisis and the potentially threatening consequences for Uzbekistan, as 80% of its territory is arid, this advantage cannot be underestimated.

The report of the Institute for Strategy and Global Policy of the U.S. House of Representatives Standing Committee clearly points to the importance of transition to clean energy as the main factor in the fight against climate change, which is certainly relevant for Uzbekistan [2].

Uzbekistan's energy strategy:

1. International Investments: Uzbekistan has been seeking foreign investments and has offered attractive concessions to international companies for setting up renewable energy projects. For instance, the UAE-based company Masdar Clean Energy has signed an agreement for a 400 MW wind farm, marking it as one of the biggest investments of its kind in the region.

2. Nuclear Power: The country has also initiated plans to develop nuclear power to diversify its energy supply. A deal struck with Russia aims to build a nuclear power plant in the country by 2030.

3. Bioenergy: Uzbekistan is a significant producer of Cotton, a crop that leaves abundant agricultural residue that can be utilized to produce bioenergy. Developing bioenergy can simultaneously solve waste management issues and generate power.

4. Renewable Energy Project: As of 2020, Uzbekistan is executing a major $1 billion project, "Scaling Solar", to develop renewable energy resources, particularly solar. The project includes making infrastructural and policy changes to enable this and attract private investment.

5. Energy Transition: In 2020, the government approved the 'Concept for the Development of the Electric Power Industry of the Republic of Uzbekistan until 2030.' This key document identifies the power sector's transition to a green economy, increased energy

efficiency, and the reduction of environmental impact as crucial areas.

Though Uzbekistan currently relies significantly on fossil fuels, it's making substantial efforts to diversify its energy sources. The nation is particularly focusing on harnessing its abundant renewable energy resources as well as exploring the use of nuclear power. However, creating the necessary infrastructure and putting suitable regulations and incentives in place remains a ongoing task.

The energy transition from traditional fossil fuels to renewable and alternative sources is a major undertaking, requiring significant investment in infrastructure, technology, and education. It also requires policy and regulatory support to encourage investment and to make the renewable energy sector competitive. Here are more details about Uzbekistan's transition to alternative electricity supply:

Technological Support: Uzbekistan is using modern technology to improve its energy sector. For instance, they have implemented "smart grids" to improve the management of electricity supply and demand. They're also researching cutting-edge storage technologies and power plant designs.

Geothermal Power: Uzbekistan has started exploring its potential for geothermal power, looking to tap into the high-temperature water resources in its underground reservoirs for power generation.

Decentralized Power Generation: The government is encouraging the adoption of decentralized power generation systems, betting particularly on Solar Home Systems (SHS) and green mini-grids. This way, individual homeowners can produce their own electricity instead of relying solely on the national grid.

Natural Gas: In addition to renewable sources, natural gas also plays a significant role in Uzbekistan's energy strategy. The natural gas industry has been relatively developed. In the future, natural gas could potentially serve as a "bridge" between the current energy supply dominated by coal and a more sustainable, renewable-focused mix.

Institutional support: To oversee and accelerate the energy transition, the government created new institutions and bodies, such as the Ministry of Energy established in 2019. They are tasked with the design and implementation of energy policies.

International Co-operations: Uzbekistan has joined the International Renewable Energy Agency (IRENA) and the Central Asia Regional Economic Cooperation (CAREC) Energy Strategy 2030 to collaborate with global partners on renewable energy development and to share experiences and expertise.

These initiatives demonstrate a revolutionary shift in Uzbekistan's approach to energy production and consumption. However, significant challenges still exist, including a need for increased technical and institutional capacity, ensuring energy affordability, and overcoming infrastructure gaps. The successful transition towards a diversified electricity supply will require sustained efforts across all these areas.

3 Research results

Mathematically simulating the process of self-starting and setting the duration of a power interruption at different time intervals, we determine the maximum duration at

There is a sharp increase in pressure *H*, at point c M_g prevails, M_f and M_{din} becomes negative, the acceleration of the pumping unit at this point slows down sharply.Then M_{el} prevails and the speed of the pumping unit reaches the nominal level, while $M_{din}=0$.In the interval *c-b* on the pressure curve, a process of increasing pressure is observed, which has a wave character, which is explained by the action of a direct wave of hydraulic shock in the pressure pipeline.

Uzbekistan's goal to transition to renewable energy:

Solar Power Development: Uzbekistan has 300 sunny days in a year, offering high potential for the development of solar power. There are plans for the construction of multiple solar power projects, including a 100 MW Solar Power Project in Samarkand and the Quart Wind Power Plant that aims to produce 1GW by 2031. The latter is in partnership with United Arab Emirates renewable energy group Masdar.

Biomass Energy: As one of the world's leading cotton producers, Uzbekistan generates a significant amount of agricultural waste that can be converted into biomass energy. Recognizing this potential, several initiatives are being undertaken to develop the country's biomass energy capacity.

Innovation in Energy Solutions: Uzbekistan also acknowledges that innovation in energy solutions is key to exploiting renewable energy sources effectively. They are adapting new technologies to capture, store and distribute energy, whilst also aspiring to lead in the development of these technologies.

Energy Conservation: Efforts are being made to improve energy conservation and efficiency, with the government implementing several energy-saving programs and modernizing existing infrastructures. The government is also encouraging residents to use energy-efficient appliances.

Training and Education: Recognizing the need for a skilled workforce to support its energy transition, Uzbekistan has initiated various training and education programs in renewable energy technologies. This includes partnerships with international entities to provide technical training and capacity building.

Carbon Capture and Renewable Natural Gas: Uzbekistan is considering the development of carbon capture and storage technologies and the possibility of producing renewable natural gas to further diversify its energy sources and reduce greenhouse gas emissions.

Environmental and Social Safeguards: Uzbekistan is also considering the environmental and social impacts of renewable energy projects and aims to develop those projects in a manner that promotes sustainable development.

Development of a Regulatory Framework: Conducive regulatory frameworks are being developed to encourage private investment and promote competition in the energy sector. This includes formulation of policies that foster the use of renewable energy and environmental protection.

By taking these steps, Uzbekistan is demonstrating commitment to transition to a more sustainable and diversified energy future. The ultimate goal is to reduce dependence on fossil fuels, reduce carbon emissions, and have a positive impact on its economy and environment.

Uzbekistan's energy efforts:

Foreign Cooperation: Uzbekistan is cooperating with international organizations and foreign companies to develop renewable energy projects. This includes collaborations with Asian Development Bank, the European Bank for Reconstruction and Development, the International Finance Corporation, and organizations from countries like France, Russia, etc.

Power Purchase Agreements (PPAs): The country has enacted favorable legislation for Power Purchase Agreements (PPAs) to attract private sector investment in the renewable energy sector. Under these agreements, the government or a utility company agrees to buy power from a private power producer for a specified period, which can provide income stability for renewable energy projects.

Emission Reduction Goals: Uzbekistan aims to reduce greenhouse gas emissions by 10% by 2030. Development of renewable energy infrastructure plays a critical role in achieving this goal since the energy sector contributes a significant proportion of the country's emissions.

Hydropower: Sustainable utilization of its robust hydropower resources is another element of Uzbekistan's energy strategy. As Uzbekistan has impressive hydropower potential, upgrading and developing new hydropower capacities is a part of its overall renewable energy plan.

Public Engagement: The government has undertaken various initiatives to promote public engagement and awareness about energy conservation and renewables. This plays a crucial role in influencing household energy consumption habits and encouraging the adoption of renewable technologies.

Energy Efficiency Standards: Energy efficiency in buildings, appliances, and vehicles is also a key part of the strategy, with efforts to set, implement, and enforce energy efficiency standards throughout the country.

Research & Development: To foster innovation and technological advancement, the country aims to expand its research and development capabilities within renewable energy and energy efficiency.

Implementing these measures and reaching the outlined goals constitute a significant amount of work. There are considerable challenges to overcome—the need for significant capital investment, the need for building trained workforces, developing technological capabilities and infrastructures—however, the rewards in terms of improved energy security, economic stability, environmental protection, and public health justify these efforts.

4 Conclusion

Diversification of the energy sector will not only reduce dependence on fossil fuel sources, but also reduce environmental pollution, ensuring sustainable and reliable energy supply.

Utilizing sources such as solar and wind energy will help reduce Uzbekistan's carbon footprint and make the energy system more sustainable and reliable. It will also open new business opportunities and create jobs, which will contribute to the country's economic development.

Despite high initial costs, the cost-effectiveness and long-term benefits of alternative energy sources make them an important investment advantage. Strengthening energy infrastructure and education programs will also be key factors in ensuring an effective transition.

However, progress cannot be achieved without overcoming obstacles. State and international structures should be actively involved in overcoming these challenges and continue to support the development of alternative energy sources in Uzbekistan.

In contrast to traditional energy supply, the switch to alternative sources represents an important and inevitable solution for the future. This is truly an innovative approach that can make a significant contribution to the overall sustainability and well-being of the country. Ultimately, an effective diversification of Uzbekistan's energy strategy will have a positive impact on both the national and global arenas.

It's important to understand that the transition to renewable energy is a long-term process that requires considerable planning, investment, and international cooperation. Yet, as Uzbekistan progresses in these initiatives, it stands to greatly benefit from increased energy independence and decreased environmental impacts.

References

1. "Investing in Renewable Energy Projects in Europe". Dentons, 2020. https://www.dentons.com/en/insights/guides-reports-and-whitepapers/2020/june/30/investing-in-renewable-energy-projects-in-europe
2. "Climate Change and National Security". USA National Intelligence Council, 2017. https://www.dni.gov/files/documents/Newsroom/Reports%20and%20Pubs/Implications_for_US_National_Security_of_Anticipated_Climate_Change.pdf
3. "Decree of the President of the Republic of Uzbekistan 'Strategy of the Republic of Uzbekistan for 2017-2021'". Norma.uz, February 2017. https://www.norma.uz
4. "World Bank Support Diversification of Energy Sources in Uzbekistan". World Bank, 24th July 2020. https://www.worldbank.org/en/news/press-release/2020/07/24/world-bank-supports-diversification-of-energy-sources-in-Uzbekistan
5. "IRENA Renewable Readiness Assessment: The Republic of Uzbekistan". IRENA, April 2020. https://www.irena.org/publications/2020/Apr/Renewable-Readiness-Assessment-Uzbekistan

Cumulative generation of electricity from the wind of passing vehicles and natural wind in the Bukhara region of the Republic of Uzbekistan

Abduraxmon Akhmedov[1]*, *Sardorbek* Khudoyberganov[1]

[1]Tashkent State Transport University, 100167, Uzbekistan, Tashkent, Temiryolchilar Street, 1

Abstract. Wind energy occupies are a rather important place in the balance of global "green" generation. Wind energy directly depends on the wind speed in the region under consideration. The highest average wind speed during the year is observed in the Bukhara region of the Republic of Uzbekistan. Vertical wind generators are widely used to collect the energy of natural wind and air movement generated by traffic into electricity. In order to effectively use the power of natural wind and air turbulence that occurs during the movement of vehicles, it is proposed to install vertical wind generators on the side of highways of international importance in the Bukhara region at certain intervals of the road.

1 Introduction

When vehicles move, pressure waves and air rarefaction occur. Wind energy occupies a fairly important place in the balance of global "green" generation, but until now turbines are installed mainly in coastal areas, where the wind is constant and stronger. For this purpose, it is advisable to use vertical wind generators. Advantages of vertical wind generators over traditional ones:

• Application of innovative silent and vibration-free technologies;

• Application of highly efficient methods for obtaining and converting wind energy into electrical energy;

• The optimal profile of the wind wheel blade allows achieving wing efficiency close to ideal, regardless of the wind direction (independent "guidance" to the wind direction);

• The vertical wind generator does not require routine maintenance or repair. The design does not contain parts with rubbing surfaces, with the exception of the wind wing thrust bearing, which has a three-hundred-fold safety margin;

• Highly resistant to strong winds, stable enough to withstand hurricane force winds;

• The controller-converting system allows you to charge the battery at very low generator speeds. This makes it possible to consume previously generated energy during periods of calm;

• Requires a minimum of space for placement, is absolutely harmless due to the absence of radiation, vibration and noise pollution;

• Possibility of installation without damaging landscape views, bird-safe design;

• Fast installation and maintenance;

• The main advantage of wind turbines is its independence from main energy networks, autonomy of production and consumption of electricity. The relative simplicity of the device, the versatility of the equipment, and the accessibility of transportation and installation make it possible to build wind power stations in the most inaccessible areas remote from power supplies.

Among vertical wind generators, the following groups of rotors can be distinguished: orthogonal, Savonius, Darrieus, Helicoid, multi-bladed with a guide vane. The main advantage of vertical wind generators is that there is no need to orient them towards the wind.

Orthogonal wind generators. Orthogonal vertical wind generators have a vertical axis of rotation and several blades parallel to it, located at a certain distance from it. The advantages of orthogonal wind generators are: there is no need to use guiding mechanisms in their design, since the operation of these installations does not depend on the direction of the wind; Due to the vertically located main shaft, the drive equipment can be located at ground level, which greatly simplifies its operation.

Wind generators with Savonius rotor. Two or more half-cylinders are used as blades in the Savonius rotor. The Savonius rotor is characterized by high starting torques, operation at relatively low speeds and relatively high manufacturability of its production. Currently, wind generators with a Savonius rotor are produced in a power range of up to 5 kW. The Savonius rotor is also often combined with a Darrieus rotor to provide higher starting torques for the Darrieus rotor.

Wind generators with Daria rotor. Wind generators with a Darrieus rotor have a vertical axis of rotation and two or three blades, which are a flat strip that does not have a characteristic aerodynamic profile. The advantages of the Daria rotor are: the absence of a wind

* Corresponding author: ahmedov _1950@ inbox . ru

orientation system; technological simplicity of manufacturing blades; the ability to place drive equipment at ground level, which greatly simplifies its maintenance.

Wind generators with a helicoidal rotor. The helicoidal rotor or Gorlov rotor (its second name) is a modification of the orthogonal rotor. Due to the twisting of the blades, the rotation of the rotor is more uniform, which significantly reduces the dynamic loads on the support units and, thereby, increases their service life, compared to the support units of orthogonal rotors; however, the production technology of twisted blades becomes much more complicated, which affects the increase their cost.

Wind generators with a multi-blade rotor with a guide vane. Multi-blade wind generators with a guide vane are a modification of an orthogonal rotor. They have two rows of blades, the first row is stationary, it is a guide vane, the purpose of which is to capture the wind flow, compress it with increasing speed, and supply the wind flow at an optimal angle of attack to the second row of blades, which is a rotating rotor. The advantage of this type of rotor is its higher operating efficiency compared to other vertical wind generators; work at low wind speeds. The disadvantage of this rotor is its higher cost due to the use of a large number of profiled blades.

The Turkish company Deveci Tech has fundamentally changed the approach to the issue, showing that wind energy can be collected directly on city streets or intercity highways. Engineers from Istanbul have created a vertical wind turbine called ENLIL (named after the Mesopotamian god of wind and storms), which is powered by air currents created by fast-moving cars. The turbulence from passing heavy vehicles and buses causes the windmill to rotate even more, and the vertical arrangement of the long blades ensures maximum flow capture [1-5].

Enlil turbines take up very little space on the ground and are easy to assemble, disassemble and operate. The turbine is connected to a generator, and the energy produced can be fed into the grid or stored in batteries until it is needed. In addition to roadsides and railway tracks, such turbines can be installed on embankments, rooftops and parks. But according to the developers, the location next to busy traffic is optimal, since the almost constant flow of cars ensures uninterrupted operation of the installation [6-9].

The development of ENLIL turbines was carried out by Istanbul Technical University and the technology company Devecitech.

Fig.1. Enlil turbines installed in Istanbul

Fig.2. Wind turbines installed in Istanbul.

Wind generators have been installed in Istanbul, which produce energy using the wind flow generated by the movement of cars.

2 Experimental research

Average wind speed values for regions of the Republic of Uzbekistan during the year are given in Table 1

Table 1. Average wind speed values in the regions of the Republic of Uzbekistan during the year

No.	Area name	Average wind speeds throughout the year	
		km/hour	m/sec
1	Republic of Karakalpakstan	15.73	4.37
2	Andijan region	7.68	2.13
3	Bukhara region	15.85	4.40
4	Jizzakh region	9.16	2.54
5	Kashkadarya region	13.44	3.73
6	Navoi region	15.63	4.34
7	Namangan region	9.73	2.70
8	Samarkand region	11.90	3.31
9	Surkhandarya region	11.21	3.11
10	Syrdarya region	10.35	2.88
11	Tashkent region	9.75	2.71
12	Fergana region	8.47	2.35
13	Khorezm region	15.48	4.30

As can be seen from this table, the average wind speed throughout the year in the Bukhara region of the Republic of Uzbekistan meets the requirements for the installation and operation of wind power plants [10-13].

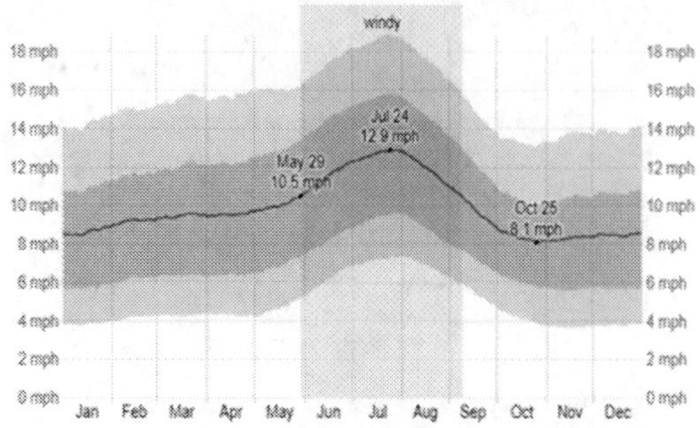

Fig.3. Average wind speed throughout the year in Bukhara region.

It is advisable to install wind turbines on the sides of highways in the region. The total length of roads of international importance in the Bukhara region is 540 km, and of national importance 1155 km.

Since the intensity of vehicle traffic on highways of international importance is higher than on other roads, we will consider issues related only to highways of international importance. It is proposed to use the Chinese wind generator Axies Wind SR - 600/800 as a wind turbine. The efficiency of such wind generators is very high [14-20].

3 Research results

The power of the air flow created by natural wind is:

$$P = \frac{dE}{dt} = \frac{C_p \cdot m \cdot v^3}{2} = \frac{C_p \cdot \rho \cdot S \cdot v^3}{2}, Bm$$

where Cp is the wind energy utilization coefficient. For the selected wind turbine we have Cp = 0.19; v - air flow speed, m/s; S - swept area, m2. For the selected wind generator S = 2 m2.

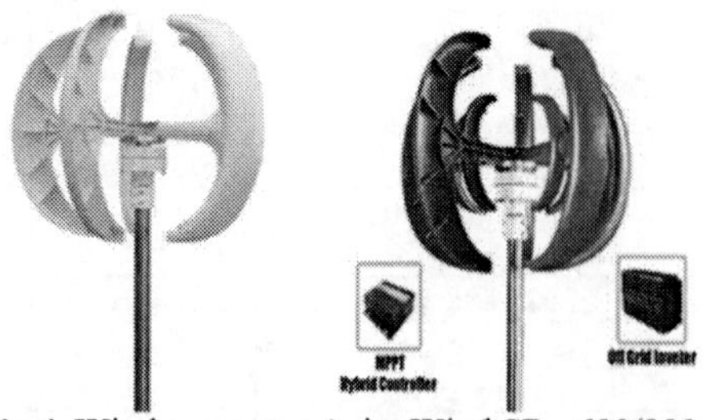

Fig.4. Wind generator Axies Wind SR - 600/800

Table 3 shows the technical characteristics and price of Axies Wind wind generators of two modifications.

Table 3. Technical characteristics and price of Axies Wind wind generators.

Manufacturer	Voltage, V	Power, W	Cost, in soums
Axies Wind (China)	12	800	2 840 187
Axies Wind (China)	24	1000	3 036 138

They are very sensitive and begin to generate energy at a speed of 2 m/s. It is also designed to be connected to the network. Assume that the wind turbines operate normally due to natural winds, and assume that the natural generation is due to the wind turbines. According to the results of the study, the speed of the air flow created by a passing car is about 60 km/h or 16.67 m/s

$$P = \frac{0.19 \cdot 1.226 \cdot 2 \cdot 16.67^2}{2} = 1371.12\ W$$

During the travel time of the car, that is, 0.267 seconds, we have the actual electrical energy received:

$$E = P\ t\ ,\ W \cdot s$$
$$E = 1371.12 \cdot 0.267 = 366\ W \cdot s$$

Then we get 152.5 Wh of energy from 1500 cars per day from the intersection, and this energy per year is equal to:

$$E = 152.5 \cdot 365 = 55.66\ kW \cdot h.$$

For example, when installing 2 devices on both sides of an intersection, we get the following energy output:

$$E_{flow} = 2 \cdot 55.66 = 111.3\ kW \cdot h.$$

The power of the air flow created by natural wind is:

$$P = \frac{0.19 \cdot 1.226 \cdot 2 \cdot 4.4^3}{2} = 19.84\ W$$

Electricity produced by one wind turbine per year due to natural wind:

$$E_{nat} = 19.84 \cdot 24 \cdot 365 = 173.8\ kW \cdot h$$

If wind turbines are placed one on each side of the road, then the total energy output on both sides of the road will be:

$$E_{nat} = 2 \cdot 173.8 = 347.6\ kW \cdot h$$

The total energy generated by one pair of wind turbines per year is:

$$E_{total} = E_{nat} + E_{flow} = 458.9\ kW \cdot h$$

Then the amount of electricity generated by one pair of wind turbines per year is:

$$S_{sum} = S_{nat} + S_{flow} = 600 \cdot 458.9 \text{ sum} = 275340 \text{ sum}$$

If we install one pair of wind turbines for every 100 meters of international highways in the Bukhara region, we will get a wind farm of 1080 wind turbines.

Then the total energy generated by the wind farm of wind turbines per year is:

$$E_{wind\ farm} = 5400\ (E_{nat} + E_{flow}) = 2.478\ \text{MW·h}$$

And the amount of electricity generated by a wind farm of wind turbines per year is:

$$S_{wind\ farm} = 5400(S_{nat} + S_{flow}) = 600 \cdot 2.478 \cdot 103 \text{ sum} = 1\ 486\ 836\ 000 \text{ sum}$$

Wind generator

Fig.5. Wind turbines installed along the road.

4 Conclusion

1. In order to effectively use the air turbulence that occurs when vehicles move, it is proposed to install wind generators along the side of the road at a certain interval of the road.
2. It is proposed to use the Chinese-made Axies Wind wind generator as a wind generator.
3. Electric energy generated by the method of obtaining electricity from natural wind and from the wind of passing vehicles can be used to illuminate highways and bus stops.

References

1. https://gisprofi.com/gd/documents/innovatsionnyj-vetrogenerator-enlil-rabotaet-ot-proezzhayushchih-mimo.html
2. https://lenta.ru/news/2021/08/24/woooooo/
3. https://ua.news/ru/technologies/v-turtsyy-yspolzuyut-veter-transporta-dlya-vyrabotky-energyy
4. AU Gapparov and GA Govor and UT Berdiyev and FF Hasanov and AM Kurbanov. Magnetic-soft materials based on iron for electromechanical engineering. IOP Conference Series: Earth and Environmental Science, 2020. volume 614, no. 1, p. 012048b doi 10.1088/1755-315/614/1/012048
5. Abduraxman Akhmedov, Galina Sauchuk, Natallia Yurkevich, Sardorbek Khudoyberganov, Mahammatyakub Bazarov, Karimberdi Karshiev; The influence of production conditions on the electrophysical parameters of piezoceramics for different applications. E3S Web Conf. 264 04020 (2021), DOI: 10.1051/e3sconf/202126404020
6. R. Mirsaatov, S. Khudoyberganov, A. Akhmedov; Uncertainty estimation in determination of Cocoons silkiness by thickness of their shell. AIP Conference Proceedings 15 March 2023; 2612(1):050010. https://doi.org/10.1063/5.0114683
7. Ravshanbek Mirsaatov, Sardorbek Khudoyberganov; Development of a non-destructive method of determination of the maturity of Mulberry cocoons. AIP Conf. Proc. 16 June 2022; 2432(1):040018. https://doi.org/10.1063/5.0089646
8. Burkhanov, SD, Mirsaatov, RM, Khudoyberganov, SB, Kadyrov, BH Relationship of parameters that characterize the quality of living cocoons. IOP Conference Series: Earth and Environmental Science, 2021, 677(4), 042032. DOI 10.1088/1755-1315/677/4/042032
9. Halina Sauchuk, Natallia Yurkevich, Abduraxman Akhmedov, Sardorbek Khudoyberganov, Sayfulla Kayumov and Ulmasbek Berdiyarov. Dielectric and microwave properties of ceramics of the Bi-Ti-O system. E3S Web of Conf., 401 (2023) 05076. DOI: https://doi.org/10.1051/e3sconf/202340105076
10. Optimization of the method of oxide coating of metallic iron powder particles. Usan Berdiyev, Olga Demedenko, Mirjalol Ashurov, FF Hasanov and UB Sulaymonov. E3S Web Conf., 383 (2023) 04039. DOI: https://doi.org/10.1051/e3sconf/202338304039
11. UT Berdiyev, AK Vetcher, FF Hasanov, BB Avazov; Soft magnetic materials for electric machine construction. AIP Conference Proceedings 15 March 2023; 2612(1):050014. https://doi.org/10.1063/5.0117785
12. Fayzullayev, JS , Jurayeva, KK The transfer function of a traction asynchronous motor controlled by a four-square converter IOP Conference Series: Materials Science and Engineering , 2020, 734(1), 012195
13. Juraeva, KK , Fayzullayev, JS Mathematical models of magnetic circuits of sensors of functional diagnostic systems of electric carriers IOP Conference Series: Materials Science and Engineering , 2019, 537(6), 062026
14. Kolesnikov, I. , Yaronova, N. , Ametova, A. Building a "Smart" Electric Network Using 5G and Unmanned Aircraft AIP Conference Proceedings , 2022, 2432, 030037
15. Kolesnikov, I. , Yaronova, N. , Kurbanov, J. , Khusnidinova, N. Intelligent railway transport radio

communication based on neural networks E3S Web of Conferences, 2021, 264, 05044
16. Yusupov D.T.Kutbidinov O.M., Samadov Sh.A.,Abdullaev E.S. Investigation of winding faults of traction transformers IOP Conference Series: Earth and Environmental Science , 2021, 173024
17. S. F. Amirov, A Kh. Sulliev, A T. Sanbetova, and I. Kurbonov, Study on the biparametrical transudations circuits with distributed parameters. IOP Conf. Series: Earth and Environmental Science 939 (2021) doi:10.1088/1755-1315/939/1/012008
18. Eldor Abdullaev, Odiljon Kutbidinov, Maxamadaziz Rasulmukhamedov, Dilmurod Yusupov; Transformer oils testing laboratories development of electronic process recording system. AIP Conf. Proc. 23 June 2023; 2789(1):040100. https://doi.org/10.1063/5.0145608
19. Burkhankhodjaev, A. , Tuychieva, M. , Iksar, E. , Kholbutayeva, K. , Nurmatov, B. Investigation of the Energy Performance of Electric Locomotives in Asymmetric Modes . AIP Conf. Proc. 2552, 010001 (2023) https://doi.org/10.1063/12.0013114
20. Berdiev, UT , Kolesnikov, IK , Tuychieva, MN , Khasanov, FF , Sulaymonov, UB Methods of new technological developments of electric motors based on soft magnetic materials . E3S Web of Conf. Volume 401, 2023. 03038. https://doi.org/10.1051/e3sconf/202340103038

Formation of investment activities of energy enterprises

Saodat Ibragimova[1]*, *Khilola* Bakhodirova[1]

[1]Tashkent State Technical University named after Islam Karimov, 100095, Uzbekistan, Tashkent, University St. 2A

Abstract. This article discusses the formation of the investment process in energy enterprises. Its functions and criteria. As well as approaches to the conduct of investment policy of energy enterprises and the role of regional authorities in monitoring the development of the electric power industry in the region to ensure energy security. And the importance of the business reputation of energy enterprises.

1 Introduction

Electric power industry in all periods and spheres of humanity has always been one of the most important levers for the development of the country. It is impossible to imagine the functioning of the national economy and maintaining the modern quality of life of the population without ensuring the production of electricity.

In addition, the electric power industry is one of the main consumers of mechanical engineering products and the construction complex, and at the same time creates a wide field for scientific and technological progress and the application of new technologies in the processes of thermal, nuclear and hydro generation, and transmission of electrical energy. Therefore, the task of developing the energy complex for the medium and long term is the most important topic for the economy of any state. In a nutshell, investment activity in energy enterprises is necessary to ensure uninterrupted supply to consumers, obtain long-term profits and competitive advantages.

2 Experimental research

In the context of developing competition in the electric power industry, the task of organizing the investment activities of energy companies becomes urgent. Because the electric power industry is a basic sector of the economy of Uzbekistan, strategically important for the state. The corresponding growth rates of other sectors of the economy, the stability of their work and the power supply depend on its condition and development. [1]

Reforms of the structure of the electric power industry lack an understanding of the standard state of the industry, which organizations operating in it could strive to achieve. As a result of the reform carried out by separating monopoly and potentially competitive sectors of the electric power industry for domestic energy companies in order to ensure uninterrupted production and distribution of electricity, issues of increasing the operational and investment efficiency of companies were added. Investment programs of energy enterprises are aimed at stabilizing power supply, eliminating energy shortages, meeting growing future consumer demand for electricity, increasing competitiveness, and reducing operating costs.

General data on the parameters of investment potential and the level of investment risk determine how attractive an economic system is for investment. A holistic approach is required when considering it. Which will allow us to describe the role of investment activity and its place in the system of social relations, as well as reveal a complex of influencing factors that determine the degree of development of the system. A holistic approach is required when considering it. Which will allow us to describe the role of investment activity and its place in the system of social relations, as well as reveal a complex of influencing factors that determine the degree of development of the system. An analysis of the efficiency of the functioning of real economic systems in industry allows us to conclude that the constant elimination of this need using only our own means leads to a slowdown in growth rates.

The fact is that the subject, concentrating attention on an activity unusual for him, involves the main resources in it. Therefore, for an industrial enterprise that has a limited amount of its own funds, we can recommend choosing a specific object that has the required properties that will make up for the specified deficit with minimal investment. Moreover, if in this case it is impossible to fully satisfy the needs of the enterprise, then there is a need for external participation, and, consequently, to attract investment. When studying investment activity, one should first of all take into account its heterogeneity, due to the presence of various combinations of parameters characteristic of its objects and subjects. Therefore, the desire to maximally cover all objects and subjects for unification, acquiring homogeneity and optimizing their properties determines the multiplicity of processes occurring during its implementation. At the same time, optimization is focused on maximizing the positive manifestations of these properties, which, in our opinion, can be identified

* Corresponding author: ilider1987@yandex.ru

with the overall development of the system. It is important that the overall dynamics of the development of society is determined precisely by the development of individual systems. In the event of a sharp increase in growth dynamics in a system based on breakthrough technologies in any area, an inevitable chain reaction of rapid development of adjacent systems is initiated.

In other words, as a result of effective investment activities, the prerequisites are created for accelerating social development based on the generation of chain reactions of accelerated development in related systems. Taking into account the specifics of investment activity in industry, the process of managing it should be based both on the implementation of regulatory procedures and on the use of economic and mathematical models that make it possible to justify the choice of directions and specific activities that contribute to achieving the goals of the functioning of the system (economic entity). Traditionally used scientific approaches that reveal the essence of investment activity as an equilibrium, linear and closed system have the disadvantage of narrowness and lack of perspective. They are not adequate to the real state of the economic environment, which is characterized by a significant level of instability, disequilibrium and a high probability of crisis phenomena. Traditionally used scientific approaches that reveal the essence of investment activity as an equilibrium, linear and closed system have the disadvantage of narrowness and lack of perspective. They are not adequate to the real state of the economic environment, which is characterized by a significant level of instability, disequilibrium and a high probability of crisis phenomena.

This special role was determined by the strong investment process in the industry. Investment is the most important activity in the electric power industry, ensuring reliable and efficient long-term energy supply to consumers through the development and technical renovation of production capacities. Today, the unity of the methodological and information base is necessary for the development of investment projects, and at the same time allows us to streamline the preparatory process and facilitates the monitoring of these projects and programs. In addition, reforming the electric power industry involves demonopolization and the development of competition in all spheres of the economy. Consequently, investment projects can be developed by a variety of organizations and the existence of a unified methodological and information basis makes it possible to compare these projects [2].

Investment policy in the electric power industry should be aimed at ensuring its sustainable development, developing energy saving, and also provide for attracting investments in all areas of the electric power industry. The basis of the investment strategy of the electric power industry is to facilitate the attraction of investments into the enterprise by creating a favorable investment climate, creating stable conditions for business activities, using innovative tools to attract investments, and providing economic incentives for the introduction of new highly efficient technologies into production.

Initiator of the investment project most often acts as an energy enterprise that has scientific and technical groundwork, the implementation of which will improve product quality (for example, through the introduction of efficient equipment), reduce tariffs (for example, by reducing losses), expand production (for example, due to construction of new capacities). In this situation the investor becomes dependent on the initiator of the project in choosing investment project options, implemented in cooperation.

There are situations when the initiator of the project is the state military intermediary or representative of regional authorities, which determined as a result of the completed marketing research promising measures measures to maintain energy securityness. Production and technological factors. These are the factors must answer the question about the fundamental possibility of implementing an investment project with using the accumulated resources of the enterprise scientific, technical and production progress. In this situation, as the business non-plan of the investment project, the remaining facts tori will also be taken into account.

In cases where the project initiator is a third party investor, for example, a credit institution or foreign creditor. In this case, the main thing is monitoring of energy indicators arises safety in combination with environmental indicators ical safety.

No matter which party acts as the initiator of transactions, investment activities and the introduction of innovations into the activities of the enterprise are profitable cooperation for all participants.

Making a profit from a strategic perspective is a determining factor influencing decision-making on transactions in the investment market. If we consider investment activity from the point of view of the system, we can highlight a number of the following characteristic features: unstable and intermittent influence of external factors; lack of influence of the type and level of the economic system on the nature and structure of mutual relations, which are also distinguished by a high degree of complexity, between components and factors; multiplicity of descriptive parameters, functions performed, as well as implementation results; adequacy to the goals and interests of business entities.

The circulation of funds of a business entity is largely determined by effective investment activity. The basis of absolutely any type of entrepreneurial activity is the processes of core and investment activity, since at the investment and production stage the prerequisites for its next cycle are formed. Thus, the importance of investments in the economy of an industrial enterprise is difficult to overestimate, since they have a direct impact on the growth of employment levels and the volume of output or provision of services, structural changes in the development of industrial complexes, reducing the level of costs of various types of resources, as well as ensuring the competitiveness of manufactured products.

The most important factor determining the innovative development of the country's economy is the economic, in particular investment policy of the state. [3].

State regulation of investment activities is a necessary effective tool. Based on the experience of countries with developed market economies, it can be seen that the state, in the process of regulating investment activities, assumes several functions.

— regulation of the total volume of capital investments of private businesses. In fact, this is the main macroeconomic function, carried out through interest, monetary, tax and depreciation policies.

— selective stimulation of capital investment through credit and tax incentives.

— administrative intervention for the purpose of entering or withdrawing certain production capacities.

In addition, regional authorities should monitor the development of the electric power industry in the region to ensure energy security. The main goal of this analysis should be to identify development trends and evaluate them from the point of view of compliance of the adopted investment programs with both the needs of electricity consumers and the requirements for stable and efficient development of the electric power industry.

The investment attractiveness of the regional electric power industry depends on the level of information openness of energy enterprises, which is determined by the reliability, objectivity, efficiency and completeness of information. It is necessary to observe the principle of de-bureaucratization of the investment process to reduce unnecessary interference of regional administrations in the economic activities of electric power companies. And also to form a new regulatory and control mechanism in this sector that is adequate to the conditions and results of liberalization in the electric power industry. De-bureaucratization processes should also simplify the mechanism for obtaining loans for the implementation of investment projects. In addition, assessing the competence of management is one of the main criteria for assessing the investment attractiveness of companies when determining the possible risk of investors investing in it.

One of the most basic functions is the creation of complete information support for the investment process, necessary for both the investor and the region. For example, forecasts of prices for energy services, demand, resources, the formation of clear regional priorities in the development of energy facilities, the provision of transparent risk insurance schemes for investors, a transparent taxation system and tax incentives for commissioned energy facilities.

The prospects of any enterprise are determined by its business activity. It is this that guarantees stable development and economic growth. An indicator of the business activity of an enterprise is its investment and innovation activities.

Improving the flow of investments comes from the innovative and production potential of the enterprise, which directly depend on factors such as the development of the technical and technological base, improvement of financial and labor resources, proper organization of activities and the creation of a base of all necessary material resources. A distinctive feature of innovation potential from production potential is the role of labor resources, or rather their qualifications. Therefore, the low price of labor is not only a positive factor in the process of innovation and investment development. Based on the experience of developed foreign countries, we can say that educating personnel in the spirit of creative thinking, accepting new innovative ideas and increasing the level of innovative culture is a priority task. All this, of course, dictates completely different conditions for assessing the workforce.

Studying the path of energy development in the world over the past two decades, it is not difficult to notice an increase in the interest of energy giants in issues of energy saving and energy efficiency. In general, the energy industry system itself has undergone significant changes during this period. All new trends and innovative ways of development were directly or indirectly aimed at solving the problems of decentralization of energy networks and increasing the share of alternative energy sources in the total electricity generation of the world. In addition, the most important tasks of recent years include the introduction and use of smart, intelligent devices in the activities of enterprises in the energy industry and, of course, the desire of all countries for energy security.

Such changes in the orientation of the energy industry pose completely new challenges for energy enterprises in organizing creative business activities, the achievement of which is impossible without the introduction of innovations that correspond to market trends. In this process, the role of the state as a regulator of the industry is important for the formation of the necessary institutional frameworks that serve the implementation, and subsequently the development, of the country's heat and power potential.

The electricity sector is now attracting more investment than the oil and gas sector, as investment is needed as the generation mix changes and aging infrastructure is modernized[4].

It should be noted that in the world today there are more than 50 scenarios and models for the development of global energy, developed by various energy organizations. The World Energy Model (a large-scale simulation model designed to replicate how energy markets function) [5] notes that the International Energy Agency (IEA) also provides medium- and long-term energy forecasts using the World Energy Model (WEM).

The New Policy Scenario (NPS) assumes that the policy goals and objectives in the energy field that are officially announced and recorded today will be realized. The Sustainable Development Scenario (SDS) shows what the global energy sector could look like if climate targets (as set out in the Paris Agreement) are met.

Based on all of the above, we can say that one of the most important factors in the prospects of any enterprise is its business activity.

If we talk about factors with a negative impact, then according to experts, the focus on increasing dividends and maximizing current profits is often achieved to the detriment of long-term development trends and can lead to a decrease in the investment activity of the enterprise [6]. The move to strategic management emphasizes ensuring continued profitability in the long term rather than simply maximizing profits in the short term. The focus of management on maximizing profits in the short term can lead to poor adaptation to changing operating

https://doi.org/10.1051/e3sconf/202346101074

conditions and, as a result, to a deterioration in the market position.

In this regard, many experts rightfully criticize policies aimed at maximizing current profits. In this approach, the ideology of maximizing current profits as a target function for managing electric power production comes to the fore and the main purpose of the electric power industry is relegated to the background - ensuring reliable and uninterrupted power supply to all types of consumers of the national economy.

3 Research results

When solving this problem, energy enterprises must proceed from the fact that only by ensuring reliable energy supply to consumers can commercial goals be realized. Consequently, the requirement to ensure an appropriate level of reliability becomes the main condition limiting the investment resources allocated for the commercial purposes of an energy enterprise.

If the analysis shows an insufficiently high level of reliability of energy supply, then among the promising areas, investment in measures to ensure reliability should have the highest priority. When analyzing the reliability of energy supply, assessing the state of the fixed assets of an energy enterprise is of utmost importance.

Consequently, the main goal of the investment activity of an energy enterprise should be to ensure reliable and uninterrupted energy supply to consumers. This is especially true in conditions of high rates of equipment wear and declining reliability of power supply. Another important goal of investment activity, determined by the process of reforming the domestic energy sector, is to ensure constant profitability potential in the long term.

The goals of the investment activity of an energy enterprise are as follows [7]:

1. Ensuring reliable and uninterrupted power supply to consumers. (reducing the level of wear and tear of power equipment; ensuring the necessary reserve capacity; ensuring the required amount of generating capacity to cover peak loads; reducing technological risks).

2. Ensuring constant profitability potential in the long term. (ensuring competitive advantages of the energy enterprise; increasing profits while ensuring the financial stability of the enterprise; increasing the market capitalization of the enterprise).

Business reputation is one of the general factors that determines the investment attractiveness of an enterprise. This factor characterizes the background of the economic activity of the enterprise. The company's stable reputation as a reliable partner is the key to its correct behavior in the future. The main source of information about the image of an enterprise is the media. However, the analysis of this information must be approached with caution, since there is also unreliable information.

4 Conclusion

The overall unity of the parameters of investment potential and the level of investment risk determine how attractive an economic system is for investment. Investment activity is never self-sufficient, therefore, a holistic approach is required when considering it, which will make it possible to describe its role and place in the system of social relations, as well as to reveal a complex of influencing factors that determine the degree of development of the system.

Moreover, the ability to develop is present in absolutely any subject, which necessitates the continuous reproduction of specific resources and the modification of specific properties. An analysis of the efficiency of the functioning of real economic systems in industry allows us to conclude that the constant elimination of this need using only our own means leads to a slowdown in growth rates. The fact is that the subject, concentrating attention on an activity unusual for him, involves the main resources in it.[8]

Therefore, for an industrial enterprise that has a limited amount of its own funds, we can recommend choosing a specific object that has the required properties that will make up for the specified deficit with minimal investment. Moreover, if in this case it is impossible to fully satisfy the needs of the enterprise, then there is a need for external participation, and, consequently, to attract investment.

It should be noted that the above goal is closely related to ensuring a constant potential for profitability in the long term, which can be achieved through the creation and development of competitive advantages, ensuring the profitability of the energy enterprise while maintaining its financial stability, as well as increasing the market capitalization of the enterprise. Undoubtedly, these factors in the analysis of investment activity of energy enterprises are not exhaustive, since within different regions the situation with energy security will develop according to its own scenarios. However, the problems highlighted above, in our opinion, are typical for many countries.

References

1. Allaev K.R. “Modern energy and prospects for its development.” Tashkent 2021.

2. Petr Aleksandrovich Rybkin, mechanism for managing investment activities in industrial integrated structures: a synergetic approach. Federal State Educational Institution of Higher Professional Education "State University - Educational, Research and Production Complex". Eagle 2014.

3. Cherkasov S.I. Improving the management of the process of innovative activity of an enterprise. Abstract of the dissertation for the degree of Candidate of Economic Sciences. Moscow-2007.

4. The Gold standart of energy analysis. World Energy Outlook. 2018 (WEO). (http://www.iea.org/weo2018/scenarios)

World energy Model. Documentation. 2019 Version. 88 p (http://www.iea.org/policiesandmeasures/).

5. Ansoff, I. Strategic management. Classic edition / I. Ansoff. – St. Petersburg: Peter, 2009. – 344 p.

6. Valinurova, L.S. Investing: textbook for universities / L.S. Valinurova, O.B. Kazakova. – M.: Wolters Kluwer, 2010. –438 p.

8. A. N. Melnik, R. R. Khabibrakhmanov. Indicators for assessing the effectiveness of investment activities of energy enterprises. udk 338.24.021.8.

https://doi.org/10.1051/e3sconf/202346101075

Development of a mathematical model for the analysis of different load modes of operation of induction motors

Uchkun Mirzaev[1]*, *Elnur* Abdullaev[1], *Botir* Kholdarov[1], *Bakhodir* Mamatkulov[1], *Abduvokhid* Mustafoev[1]

[1]Jizzakh polytechnic institute,130100, I. Karimov St. 4, Jizzakh. Uzbekistan

Abstract. This paper explores the conditions for creating and using mathematical models for the analysis of various modes of operation of the load of induction motors. Widely used in industrial plants, these motors play an important role in the world's energy consumption. Improving their energy efficiency is essential to reduce energy consumption, waste and improve sustainability. The paper also provides information on the introduction of modern mathematical models for accurate analysis and optimization of energy saving modes in asynchronous motors.

1 Introduction

Generally, induction motors (IM) are widely used in various industrial and commercial applications due to their simplicity of construction, robustness and economic efficiency. These motors are used in pumps, fans, conveyor belts, compressors, lifting mechanisms and many other mechanisms.

Global concerns about energy sustainability, the need to reduce greenhouse gases and optimize energy use, have led to increased efforts to improve the efficiency of induction motors. These efforts are not only in line with environmental protection goals, but also help to save costs and improve the operational efficiency of the industry. The widespread use of IMs leads to the generation of significant energy waste, and this is the reason for conducting research on improving their energy efficiency.

In particular, many foreign and domestic scientists are conducting research on increasing the energy efficiency of IM electric systems [3,4]. These studies show that today experiments are being conducted in the following areas of energy saving in IM electrical systems:

- Increase energy efficiency of IMs;
- promotion of frequency adjustment technology of speed;
- permanent magnet electric motors.

Increase the energy efficiency of induction motors.

As mentioned above, induction motors are the most energy-consuming electrical equipment. In order to achieve energy efficiency in induction motors, it is necessary to reduce the amount of internal waste. For this, the following measures should be taken:

1. Use efficient materials to reduce waste in steel and steel.
2. Implementation of magnetically conductive materials and technological processes to reduce waste in steel.
3. Change the structure of the wings to reduce ventilation waste.
4. Optimizing motor design and technological processes to reduce waste.
5. Optimizing motor design using computer technology to reduce waste.
6. Reform casting technology to reduce rotor wastage, for example using a copper rotor.

New design and effective materials apply through Can reduce losses in IM. For example, using copper coils in the rotor can reduce resistance, and silicon-laminated magnet conductors can reduce wastage in steel. But all the measures mentioned above are made of motors release increases the price [1,4].

Promotion of frequency adjustment technology. Since the 1980s, the frequency converter has been widely used in various fields. Frequency adjustment technology is mainly used in three-phase IM and synchronous motors, and it is the ideal and most promising method of speed control. The frequency converter has both excellent speed adjustment performance and excellent energy saving performance, making it an ideal speed adjustment equipment for manufacturing plants. Variable frequency drive technology can make motors operate at the most optimal energy-saving speed by precisely controlling the rotating speed [3,5].

Permanent magnet electric motors. Permanent magnet induction motors whose rotors are made of permanent magnet materials can increase efficiency by 3-5% by reducing waste by 20-30%. It has several advantages over ordinary IM. First, it is smaller in size, only two-thirds the size of a regular IM with the same power and poles. Second, it has a high starting torque, three times the normal rated torque. Third, the power

* Corresponding author: Uchqun8822@mail.ru

factor of the electric motor can reach 0.95. Fourth, when the pole number is less than 12, it can rotate slowly with high energy efficiency and power factor. Finally, it has a good speed adjustment performance because energy-efficient permanent magnet motors have appeared with the development of electronic control technology [7,8].

Energy saving means not only reducing the energy used to produce electricity, but also reducing the release of harmful gases such as CO_2 and SO_2 into nature. This, in turn, serves to slow down global warming. Therefore, energy saving in induction motors has become an important issue on a global scale [4,6].

2 Experimental research

In this paper, the issue of optimization of energy consumption by mathematical modeling of the nominal operating mode of asynchronous motors in the area of energy efficiency improvement of IM is considered.

An induction machine is a nonlinear, strongly coupled, high-order multiparameter system. The stator and rotor are connected by the magnetic flux generated by the windings, which is also called mutual induction [7,11]. It is important to find a simple and convenient mathematical model to investigate the operation of an induction machine under different conditions, because it is accurate enough to account for anomalous conditions. First, let's consider the mathematical models of the induction machine along the axes abc and dq.

According to the principle of operation of IM, the rotating magnetic field of the stator crosses the rotor windings and creates an electromotive force in them, which leads to the formation of a current flowing through the rotor windings. The interaction of this current with the stator magnetic flux creates an electromagnetic torque that causes the rotor to rotate. It can be seen that in order to generate a rotor rotating moment, the rotation frequency (speed) of the rotor must be less than the rotation frequency of the stator magnetic flux ω_r, and ωr=Zp ω, where ω is the mechanical speed of rotation of the rotor, Z_p is the torque of the motor number of poles $\Delta\omega=\omega_s-\omega_r$ is called the absolute slip of the motor and $s=(\omega_s-\omega_r)/\omega_s$ is called the relative slip, where ωs is the rated electric frequency of the stator source.

The equivalent circuit of the IM is shown in Fig. 1.

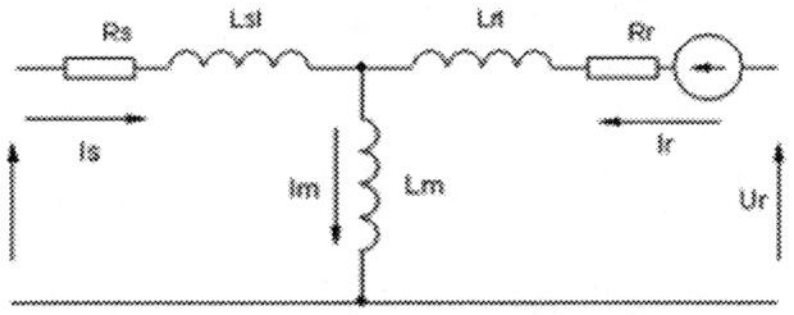

Fig 1. Equivalent circuit of IM.

IM is a nonlinear multidimensional object with a very complex structure. In its mathematical description, the correctness of the assumptions made in each specific case is of great importance. The diversity of IM mathematical models is related to the content of assumptions, the coordinate system in which the mathematical description is performed, the content of input and output signals of the model, and the system of relative values adopted.

Depending on the adopted coordinate system, the following IM models were developed:

- IM models in the three-phase coordinate system;
- IM models in two-phase orthogonal coordinate systems;
- single-phase models based on the static characteristics of IM with a simplified calculation of electromagnetic processes (on the basis of the equivalent circuit of IM in steady state).

The choice of the coordinate system and model configuration, that is, the composition of input and output signals, depends on the structure of the control system, in particular, the composition of the feedback signals and the characteristics of the IM power source.

The successful selection of a mathematical model is important not only in the analysis, but also in the synthesis of an automatic control system. This is important in the development of the model in the coordinate system directed along the rotor current connection, which is the basis for creating high-quality vector control systems.

A generalized vector concept is used to pass from the mathematical description of harmonic signals in polyphase coordinates to the mathematical description in orthogonal two-phase coordinates.

Such a generalized vector is a vector whose projections on the phase axes are equal to the instantaneous values of the corresponding phase quantities at a certain time.

To represent this connection, the generalized vector in the rotation angle φ_k is rotated counterclockwise with angular velocity.

$$\omega_v = \frac{d\phi_v}{dt}, \qquad (1)$$

and corresponds to the angular velocity of the corresponding periodic signal and has a modulus equal to the maximum value of the phase value.

$$|\tilde{v}| = v_{max}, \qquad (2)$$

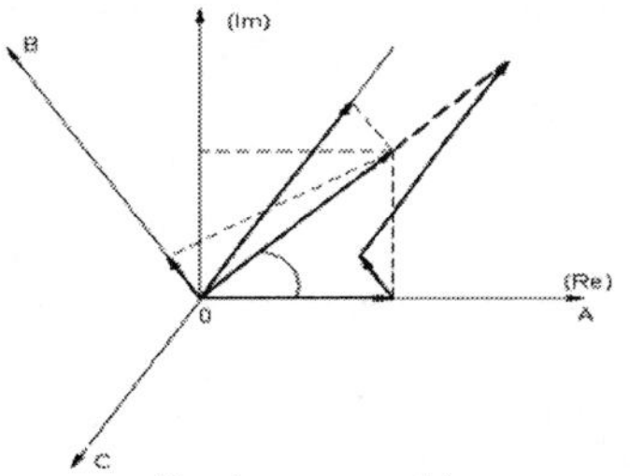

Fig 2. Generalized vector and its components.

Figure 2 shows the generalized vector $\tilde{V}_r$ of any physical quantity of the rotor, its projections $\tilde{V}_A, \tilde{V}_B, \tilde{V}_C$ on the physical axes are the vector sum of these projections $\tilde{V}_A + \tilde{V}_B + \tilde{V}_C$ and the projections of the generalized vector on the real d (Re) and abstract q (Im) axes , the first (vd) coincides with the axis A, the second (vq) is perpendicular to the first and rotated 90˚ counter-clockwise.

In the literature [2] and [3], it is shown that the sum of vectors $\widetilde{V}_A, \widetilde{V}_B, \widetilde{V}_C$, is a vector that corresponds to the generalized vector in direction and exceeds its amplitude by 1.5 times. With this in mind, we can write:

$$\underset{abc}{\tilde{v}_r} = 1{,}5(\tilde{v}_A + \tilde{v}_B + \tilde{v}_C) = 1{,}5(\tilde{v}_a + \tilde{v}_b e^{j2\pi/3} + \tilde{v}_c e^{j4\pi/3}) = \frac{2}{3} a v_r \, , \quad (3)$$

$$\underset{abc}{\tilde{v}_r} = \underset{dq}{\tilde{v}_r} = v_{r\max} r^{j\varphi_{vr}} = v_{r\max}(Cos\varphi_{vr} + jSin\varphi_{vr}) = v_d + jv_q \, , \quad (4)$$

where, $a = [1 \bar{a} \bar{a}^2]$; $\bar{a} = e^{j2\pi/3} = -\frac{1}{2} + j\frac{\sqrt{3}}{2}$; $\bar{a}^2 = e^{j4\pi/3} = -\frac{1}{2} - j\frac{\sqrt{3}}{2}$;

$$v_r = \begin{bmatrix} v_a \\ v_b \\ v_c \end{bmatrix} = \begin{bmatrix} |\tilde{v}_a| \\ |\tilde{v}_b| \\ |\tilde{v}_c| \end{bmatrix} = \begin{bmatrix} v_{r\max} Cos(\varphi_{vr}(t)) \\ v_{r\max} Cos(\varphi_{vr}(t) - 2\pi/3) \\ v_{r\max} Cos(\varphi_{vr}(t) - 4\pi/3) \end{bmatrix}$$

φ_{vr} is the angle between the direction of the generalized vector of the rotor and the d-axis.

$$\frac{d\varphi_{vr}}{dt} = \omega_{vr} - Z_p \omega$$

Thus, formulas (3) and (4) describe the generalized rotor vector in three-phase (abc) and orthogonal (dq) coordinate systems rotating at speed ω_r.

$$\omega_r = Z_p \omega \, , \qquad \frac{d\varphi_{vr}}{dt} = \omega_r$$

In order for the direction of the real axis of the orthogonal coordinate system to coincide with the direction of the generalized vector ψ_r of the coupling of the rotor current, the coordinate system taken as a basis must rotate synchronously with this vector. In this case, the rotor current vector in it will have only a real component, since the projection of this vector on the abstract q-axis will be zero. Then the conditions for mathematical description of IM in these coordinates will have the following form

$$\omega_k = \omega_{\psi_r} \, , \quad (5)$$

$$\Psi_{rq} = 0; \Psi_{rd} = |\widetilde{\Psi}_r| = \Psi_r = \Psi_{rm}, \quad (6)$$

The model in the dq coordinate system became the basis for the development of a vector control system for an induction motor with a short-circuited rotor. Therefore, in addition to conditions (5), (6), we add the condition that the rotor voltage is equal to zero

$$U_{rd} = U_{rq} = 0 \, , \quad (7)$$

When the motor stator windings are powered by a voltage source, the vector control system has feedback on the components of the stator magnetic flux and the relation of the rotor magnetic flux. Therefore, another condition for the development of the IM model is the presence of the listed i_{sd}, i_{sq}, ψ_r signals in it.

We write the equations describing the processes in IM in the form of generalized vectors, taking into account the condition in formula (7):

$$\begin{cases} \tilde{U}_s = R_s \tilde{i}_s + \dfrac{d\tilde{\Psi}_s}{dt} + j\omega_k \tilde{\Psi}_s, \\ 0 = R'_r \tilde{i}_r + \dfrac{d\tilde{\Psi}_r}{dt} + j(\omega_k - \omega_r)\tilde{\Psi}_r, \end{cases} \quad (8)$$

where ω_k is the rotation frequency of the coordinate system;

ω_r is the electrical frequency of rotor rotation;

U_s, i_s - voltage and current of the stator;

R_s is the resistance of the stator coil;

i_r - current strength in the rotor coil;

R'_r is the applied resistance of the rotor coil;

ψ_r is the rotor magnetic flux.

For an induction motor with a short-circuited rotor, the vectors differ from each other due to the presence of magnetic scattering of the stator and rotor windings. Therefore, we can write them in the following form:

$$\tilde{\Psi}_s = (L_m + L_{sl})\tilde{i}_s + L_m \tilde{i}_r; \qquad (9)$$
$$\tilde{\Psi}_r = (L_m + L_{rl})\tilde{i}_r + L_m \tilde{i}_s,$$

where L_m is the main leakage inductance;

L_{sl}, L_{rl} - leakage inductance of coils.

For L_s and L_r, we have:

$$L_s = L_m + L_{sl};$$
$$L_r = L_m + L_{rl}. \qquad (10)$$

In order to ensure the composition of the sample signals mentioned above, we find the generalized vectors of the dependence of the rotor current and the stator magnetic flux:

$$\tilde{i}_r = \frac{1}{L_r}\tilde{\Psi}_r - \frac{L_m}{L_r}\tilde{i}_s = \frac{1}{L_r}\tilde{\Psi}_r - k_r \tilde{i}_s, \qquad (11)$$

$$\tilde{\Psi}_s = L_s \tilde{i}_s - \frac{L_m^2}{L_r}\tilde{i}_s + \frac{L_m}{L_r}\tilde{\Psi}_r = \sigma L_s \tilde{i}_s + k_r \tilde{\Psi}_r, \qquad (12)$$

where $k_r = \frac{L_m}{L_r}$; $\sigma = 1 - \frac{L_m^2}{L_s \cdot L_r}$.

If we apply equations (11), (12) to the system of equations (8), then the following system of equations is formed:

$$\begin{cases} \tilde{U}_s = R_s \tilde{i}_s + \sigma L_s \dfrac{d\tilde{i}_s}{dt} + k_r \dfrac{d\tilde{\Psi}_r}{dt} + j\omega_k(\sigma L_s \tilde{i}_s + k_r \tilde{\Psi}_r); \\ 0 = \dfrac{1}{T_r}\tilde{\Psi}_r - \dfrac{L_m}{T_r}\tilde{i}_s + \dfrac{d\tilde{\Psi}_r}{dt} + j(\omega_k - Z_p \omega_r)\tilde{\Psi}_r, \end{cases} \quad (13)$$

where $T_r = \frac{L_r}{R'_r}$ - the electromagnetic time constant.

In order for the first equation from the system of equations (13) to have only one signal derivative, we derive the rotor magnetic flux coupling in the second equation of the same system and replace the resulting expression with the first one:

$$\frac{d\tilde{\Psi}_r}{dt} = \frac{L_m}{T_r}\tilde{i}_s - \frac{1}{T_r}\tilde{\Psi}_r - j(\omega_k - Z_p\omega_r)\tilde{\Psi}_r = k_r R'_r \tilde{i}_s - \frac{1}{T_r}\tilde{\Psi}_r - j(\omega_k - Z_p\omega_r)\tilde{\Psi}_r;$$

$$\tilde{U}_s = (R_s + k_r^2 R'_r)\tilde{i}_s + \sigma L_s \frac{d\tilde{i}_s}{dt} - \frac{k_r}{T_r}\tilde{\Psi}_r + j\omega_k \sigma L_s \tilde{i}_s - jZ_p\omega k_r \tilde{\Psi}_r. \quad (14),(15).$$

We rewrite (14) and (15) taking into account the projections on the d and q axes, and the following system of equations is formed:

$$\begin{cases} U_{sd} = (R_s + k_r^2 R'_r) i_{sd} + \sigma L_s \frac{di_{sd}}{dt} - \frac{k_r}{T_r}\Psi_r - \omega_k \sigma L_s i_{sq}; \\ U_{sq} = (R_s + k_r^2 R'_r) i_{sq} + \sigma L_s \frac{di_{sq}}{dt} + \omega_k \sigma L_s i_{sd} + Z_p \omega k_r \Psi_r; \\ \frac{d\Psi_r}{dt} = \frac{L_m}{T_r} i_{sd} - \frac{1}{T_r}\Psi_r; \\ 0 = k_r R'_r i_{sq} - (\omega_k - Z_p\omega)\Psi_r. \end{cases} ,(16)$$

and we rewrite the differential equation (16) in operator form:

$$\begin{cases} U_{sd} + \frac{k_r}{T_r}\Psi_r + \omega_k \sigma L_s i_{sq} = i_{sd} R_{sr}(T_{sr} p + 1); \\ U_{sq} - \omega_k \sigma L_s i_{sd} - Z_p \omega k_r \Psi_r = i_{sq} R_{sr}(T_{sr} p + 1); \\ L_m i_{sd} = \Psi_r (T_r p + 1); \\ \omega_k = k_r R'_r \frac{i_{sq}}{\Psi_r} + Z_p \omega. \end{cases} , (17)$$

In the last equation of the system of equations (17), we define the following:

$$\omega_r = Z_p\omega; \; \Delta\omega_r = k_r R'_r i_{sq} / \Psi_r = \omega_k - Z_p\omega, \quad (18)$$

We express the equation of the electromagnetic torque of the motor as follows:

$$M = \frac{3}{2} Z_p k_r \Psi_r i_{sq}, \quad (19)$$

In this case, the mechanical equilibrium equation is written in the usual form:

$$M - M_c = J\frac{d\omega}{dt}, \quad (20)$$

where J is the given moment of inertia of electric drive.

Using the equations obtained above in the Matlab package, we construct a short-circuit AD model [3, 9, 10].

For the research, we use the 4A 16 0S4 type 15 kW asynchronous motor installed on the aspirator in the elevator department of "Jizzaxdonmahsulotlari" JSC - the parameters listed in table 1.

The results of calculating the parameters of the IM model in the MatLab program according to the IM catalog data

Table 1

IM type	4 A 180 M 2	4 A 180 S 2	4 A 1 6 0 S 2
IM parametrs according to information			
P_n ,W	30000	22000	15000
U_n ,V	380	380	380
f_n ,Hz	50	50	50
S_n	0.025	0.027	0.030
COP ή	0.905	0.895	0.89
cosφ	0.88	0.88	0.89
k_{ms}	1.7	1.7	1.8
k_{mmax}	2.7	2.7	2.7
k_{is}	7.5	7.0	7.0
Results after calculation			
I_n, A	57.23	42.44	28.77
w, rad/s	306.31	305.68	304.73
$T_{e.n}$, Nm	97.94	71.97	49.22
$T_{m.n}$, Nm	93.17	68.09	46.44
R_s , Ohm	0.19 27	0.2980	0.46 02
L_s, Gn	0.4e-3	0.578e-3	0.853e-3
R_r, Ohm	0.0 795	0.1147	0.1896
L_r, Gn	0.4e-3	0.578e-3	0.853e-3
L_m, Gn	25.44e-3	34.26e-3	52.7e-3
J,(kg*m^2)	0.4 5	0.4 1	0.2 1
F_f, N m s	0.0 2	0.0 2	0.0 2
p	2	2	2

3 Research results

We create a block diagram of a 3-phase short-circuited rotor IM in the Matlab program using the above-mentioned IM mathematical models (Fig. 3). Using these schemes, we analyze the characteristics of IM under 4 different loads. We use parameters from 2 table for analysis and create graphs and oscillograms in different operating modes.

Table 2

Load on shaft(%)	0	100	50	25
Process duration (s)	3	3	3	3

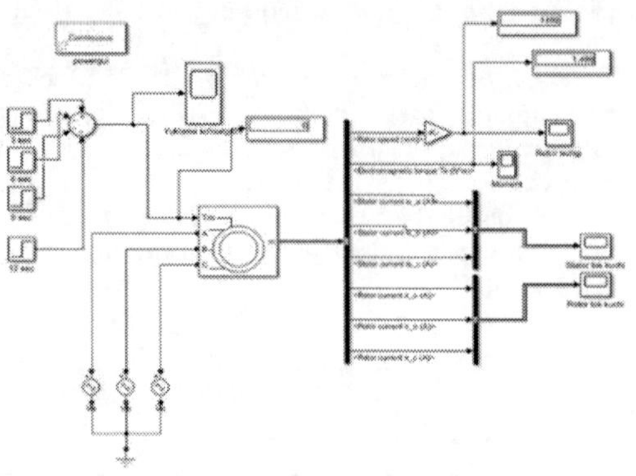

Fig. 3. Structural structure of IM in Matlab

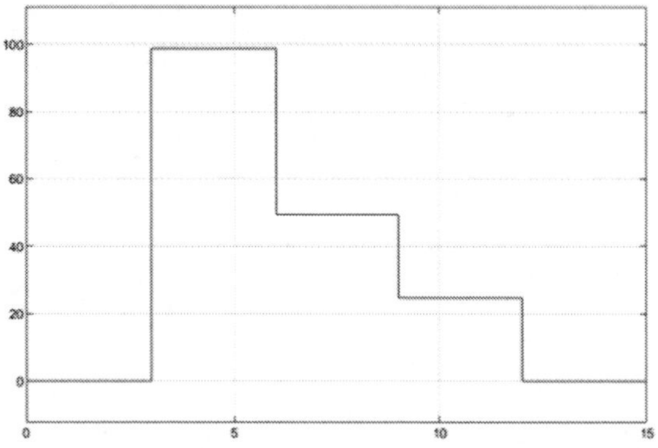

Fig. 4. Torque graph of the selected IM shaft under 4 different loads

Fig. 4 shows the values of the mechanical torque on the studied IM shaft in the considered operating modes graphically. It can be seen from the graph that the torque value is linearly related to the load value.

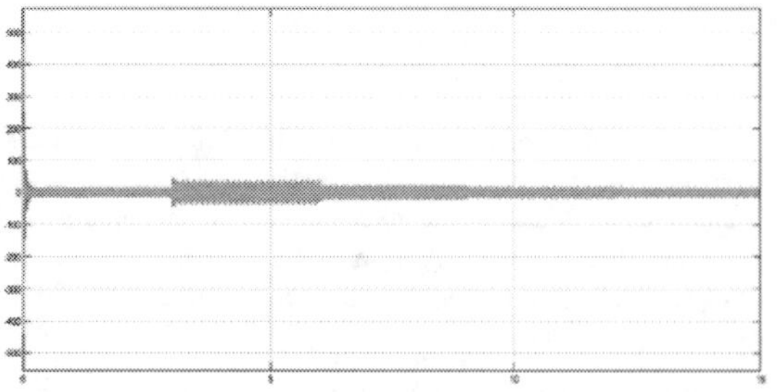

Fig. 5. Oscillogram of stator currents at 4 different loads on selected IM.

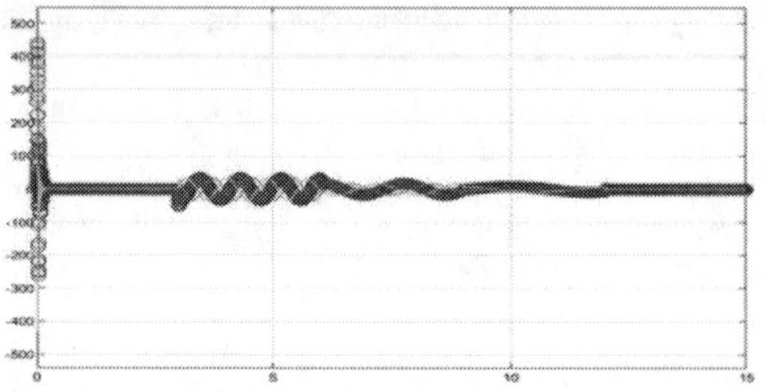

Fig. 6. Oscillogram of rotor currents at 4 different loads at selected IM.

Figures 5 and 6 show the oscillogram of the values of current forces in the studied IM stator and rotor in the considered operating modes. It can be seen from the graph that the change of the currents changed depending on the value of the applied load, and transient processes occurred in the initial phase of the change of operating modes.

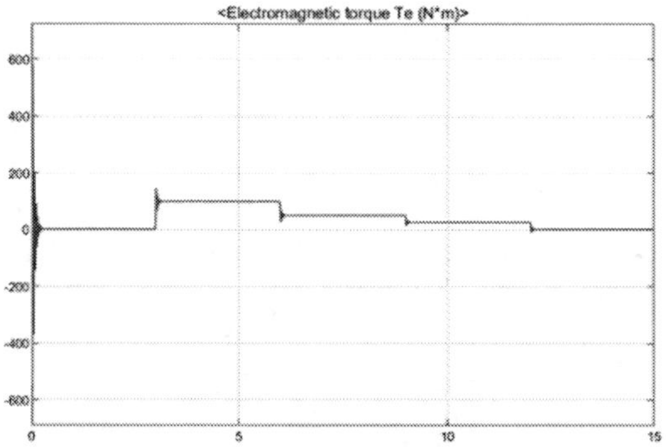

Fig. 7. Mechanical torque oscillogram on rotor shaft at 4 different loads at selected IM

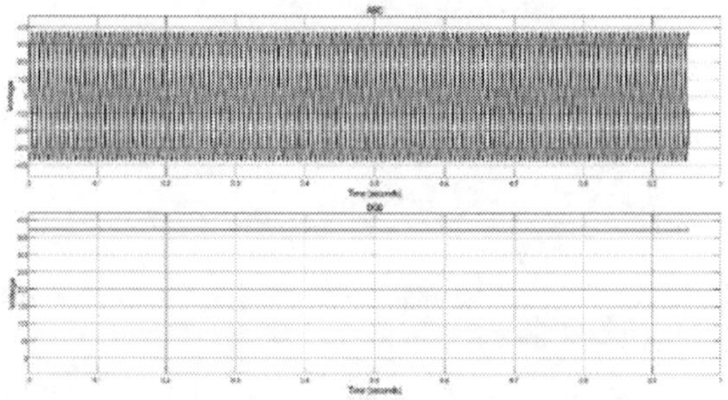

Fig. 8. oscillograms of voltages in the coordinate system abc and qd0 during start-up at nominal load IM.

4 Conclusion

Using the Matlab program, the graph of the mechanical torque on the rotor shaft, the oscillograms of the stator and rotor current in different operating modes, and the oscillograms of the voltages in the stator coils according to the abc and dq coordinate system were obtained. It can be concluded from these oscillograms that when starting and using asynchronous motors directly connected to the network, when the load value on the rotor shaft is used at a value equal to or close to the nominal torque of the rotor, its useful efficiency is high. and it is justified that it is economically efficient.

References

1. Nurali Pirmatov and Uchkun Mirzaev. "Ways to Improve the Energy Efficiency of an Electric Drive with Asynchronous Motors." International Journal of Engineering and Information Systems (IJEAIS) ISSN (2021): 230-233.
2. Bellure, Anand, and M. S. Aspalli. "Dynamic dq model of induction motor using simulink." International Journal of Engineering Trends and Technology (IJETT) 24.5 (2015): 252-257.
3. Liu, Yupeng. Induction Machine Modeling and Emulation for Asymmetrical Conditions. Diss. Concordia University, 2022.
4. Ma, Wenzhong, and Lianping Bai. Energy-Saving Principles and Technologies for Induction Motors. Hoboken, NJ, USA: Wiley, 2018.
5. Duvlonov, Jaloliddin. "Analysis Of Frequency Control Systems For Asynchronous Electric Drive." International Journal of Engineering and Information Systems (IJEAIS) ISSN (2021): 102-104.
6. Mirzaev Uchkun, and Hiloliddinov Diyor. "Methodology Of Energy Audit In Uzbekistan." International Journal of Engineering and Information Systems (IJEAIS) ISSN (2021): 197-199.
7. Orzikul Nurullaev. "Analysis of Energy Saving In Enterprises." International Journal of Engineering and Information Systems (IJEAIS) ISSN (2020): 130-134.
8. Xoshimov F.A., Taslimov A.D. Energiya tejamkorligi asoslari. O'quv qo'llanma. - T.: Voris, 2014.
9. ABID MANSURI (2023). Dynamic modeling of Three-phase Induction Motor (https://www.mathworks.com/matlabcentral/fileexchange/71885-dynamic-modeling-of-three-phase-induction-motor), MATLAB Central File Exchange. Retrieved August 24, 2023.
10. Baratov, R., Pirmatov, N., Panoev, A., Chulliyev, Y., Ruziyev, S., & Mustafoqulov, A. (2021). Achievement of electric energy savings through controlling frequency convertor in the operation process of asynchronous motors in textile enterprises. In IOP Conference Series: Materials Science and Engineering (Vol. 1030, No. 1, p. 012161). IOP Publishing.
11. Pirmatov, N., & Panoev, A. (2020). Frequency control of asynchronous motors of looms of textile enterprises. In E3S Web of Conferences (Vol. 216, p. 01120). EDP Sciences.

History of tectonic development of the southeastern part of the Turan plate

Ravshan Zokirov[1*], *Irmukhamat* Khalismatov[1], *Feruza* Zakirova[1], *Oybek* Bakhtiyorov[1], *Guzal* Kasimova[1]

[1]Tashkent State Technical University named after Islam Karimov, 100095, Uzbekistan, Tashkent, University St. 2A

Abstract. The history of the tectonic development of the territory of the Turan Plate in the pre-Jurassic and Lower-Middle Jurassic times is considered. The development of stages of rifting on the territory of the Amudarya syneclise is analyzed. The structural elements of the rift are examined. The tectonic development of the eastern part of the Turanian plate in Neogene-Anthropocene time is analyzed in more detail. Its large structural elements are substantiated with detailed descriptions of its characteristic features.

1 Introduction

The history of the tectonic development of the described territory was discussed by many researchers (A.A. Akramkhodzhaev, G.I. Amursky, A.A. Abidov, A.G. Babaev, T. Babadzhanov, A.A. Bakirov, R.G. Gavrilov, R.G. Garetsky, Sh. Peive, O.A. Ryzhkov, B.B. Sitdikov, B.B. Tal-Virsky, V.I. Troitsky, V.S. Shein, A.L. Yanshin, M.E. Egamberdyev, etc.).

The need to consider this issue arises from a number of reasons. Firstly, in recent years, new factual drilling and geophysical material has accumulated, significantly supplementing our knowledge of the paleotectonics of the region. Secondly, there are contradictory statements, for example, about the relationship between structural plans in the sedimentary cover and pre-Jurassic layers, etc.

The history of the tectonic development of the study area is relatively complete starting from pre-Jurassic deposits. Pre-Jurassic (preplate, intermediate or transitional complex) deposits on the territory of the Amudarya syneclise are confined mainly to the Khiva and Zaunguz (Izmail-Bagadzhinsky) grabens, delimited by the Central Amudarya horst. Here they are divided into Lower-Middle Carboniferous terrigenous-carbonate-volcanogenic, Upper Carboniferous-Lower Permian volcanogenic-sedimentary, Upper Permian-Lo-wer Triassic volcanogenic-terrigenous strata (according to A.M.Akramkhodzhaev, A.A.Bakirov, V.A.Bush, R.G. Garetsky, V.S.Knyazev, A.E.Starobinets, Kh.Uzakov and others.

The lower and middle formations are essentially marine, while the upper is continental. The latter forms an orogenic structural-tectonic stage, and the first two form a quasi-platform [6]

Within the Khiva graben, deposits of three formations are developed, up to 4-5 km thick. In the area of the Zaunguz graben, the thickness of pre-Jurassic rocks most often does not exceed 1-0.5 km. In the area of the Central Amudarya horst, pre-Jurassic layers are not recorded.

2 Methodology for analyzing tectonic development

The development of extension structures (grabens) along pre-Jurassic deposits, which were essentially filled with volcanics, indicates that a rift appeared on the territory of the Amudarya syneclise during Carboniferous - Early Triassic time. The rift was located southwest of the Chardzhou step that emerged later, within the eastern part of the Karakum microplate; extended submeridionally, had a length of about 350 km and a width in the middle part of up to 200 km, and had a wedge-shaped shape, expanding to the south (Fig. 1).

During the Northern Revolution, it most likely formed within the shelf located in the east of the Karakum microplate, where calcareous, clayey or volcanogenic formations accumulated, and in some places reefs and others formed.

The width of this coastal zone reached 200-250 km. In the east, the shelf ended with the Chardzhou geomorphologically and tectonically expressed fault (ledge) of a strike-slip type, behind which there was either a continental slope or an ocean (ocean crust). The shelf was structurally a system of megablocks, divided into macroblocks, articulated with each other and manifested with varying intensities. Intense volcanic activity has occurred here several times. The possibility of manifestation within the shelf of continental spreading (?), which was formed during the Riphean formation of the earth's crust and underwent spreading, cannot be ruled out. This conclusion arises from the fact that the volcano-plutonic complex reaches a thickness, at least in the northern part of the rift, where the Chardzhou,

* Corresponding author: raxmatillo82@mail.ru

Bukhara, Elan and Central Ustyurt faults meet, of about or more than 1000 m.

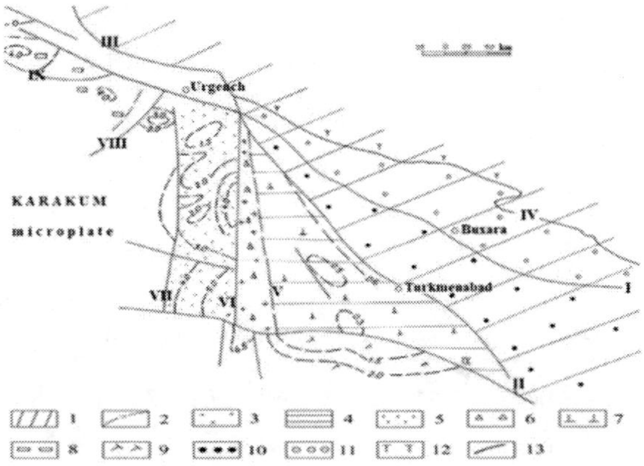

Fig 1. Amudarya rifts. (Ryzhkov O.A., Zakirov R.T., based on materials by A.Alan, G.E. Dikenshtein, K.N. Kravchenko, A.V.Peive, A.E. Starobintsev and others.)

Legend: 1.Paleotis: South Tien Shan geosynclinal region, pre-Jurassic (first generation) paleorift; 2.Isopachs of pre-Jurassic (Carboniferous - Lower Triassic sediments, in km); 3. Absence of pre-Jurassic deposits according to KMPV data; 4. The thickness of the Upper Paleozoic - Lower Triassic deposits is less than 0.5 km (up to their complete pinchout) - according to A.E. and M.E. Starbintsev; 5. Khiva graben; 6. Central - Amudarya horst; 7.Zaunguz (Izmail-Bagadzhinsky) graben; 8. Mangyshlak - Central Ustyurt rift zone; 9.Murgab rift; 10. Chardzhou shoulder; 11. Bukhara shoulder; 12. Kyzylkum uplift; 13. Faults (Flexural-fault zones): I – Chardzhou, II-Bukhara, III- Takhtakair-Sultanuizdag, IV-marginal Tien Shan orogen, V-Mergen, VI-Elan (Khorezm-Murghab), VII-Beurdeshik, VIII-Central Karakum, IX South Daryalyk.

Subsequently (R-T), the third stage, the stage of rift orogenesis, appeared. It is characterized by the formation of block mountains (mainly due to vertical movements of blocks), the manifestation of intense volcanism, moderate dissection of the relief, and the accumulation of relatively thick volcanic molasse.

The considered Amudarya rift of S-T time, corresponding to the rifting of the first generation in this area, went through three stages in its development: initiation, continental spreading (?) and rift orogenesis. The Amudarya rift of the first generation is located along the continental margin; its development was accompanied by active volcanic activity at all stages of its manifestation. It ceased its development at the end of the Late Permian-Early Triassic time. Its regional structures - the Beurdeshik shoulder, the Khiva and Zaunguz grabens and the Central Amudarya horst were revived and inherited continued to develop at the beginning of the Early-Middle Jurassic time; postumnian movements of the Central Amudarya horst are recorded even in the Jurassic and early Cretaceous times; then an inversion occurred, and the central part of the Amudarya syneclise began to form in this zone [4].

In the Early-Middle Jurassic (during the Cimmerian tectonic era), the second generation of the rift appeared on the territory of the Amudarya depression. It is associated with the general development of rifting at this time within the Eurasian continent [2]. If the first generation of the rift in the territory under consideration occurred on the eastern margin of the Karakum Plate, and this rift was replaced in the east by the Paleotethys, then its second generation appeared within the Eurasian continent. The Early-Middle Jurassic rift was formed not only on the territory of the first generation rift, but also significantly expanded to the east due to the involvement of adjacent parts of the continental crust of the Late Hercynian formation in its area; here the Chardzhou, Bukhara and Central Kyzylkum shoulders arose [5,7]. In the area of these rift shoulders, compression of the earth's crust sharply manifested itself in pre-Jurassic times. From the beginning of the Jurassic (possibly from the Late Triassic), it was replaced by extension, which led to the origin and development of these shoulders, as well as the formation of not only the slit-like Kimirek mesorift (O. Ryzhkov et al., 1985), but also the Zekrinsky one within the same Bukhara FRZ. In the Early and Middle Jurassic times, the Amudarya rift was filled with relatively thick, mainly terrigenous sediments - alternating gray, inequigranular, calcareous sandstones, gray siltstones, less often mudstones; tuffs were found among these rocks; interlayers of conglomerates and limestones are noted [6]. Volcanic rocks of alkaline composition are quite common in the Koshabulak region. Thus, the blocks of the Amudarya rift in the Cimmerian tectonic era were subjected to relatively intense spreading, penetration of the roots of deep faults into the "basalt" layer and, apparently, the upper mantle, from where volcanics came to the surface of the earth. This cycle of rifting began to degenerate by the end of the Middle Jurassic, as evidenced, firstly, by the cessation of activity of some of the previously manifested faults delimiting individual blocks (Beurdeshiksky, Mergensky), which indicates a weakening of the transverse movement, and secondly, the absence of the upper Jurassic terrigenous volcanic formations. The paleorift territory under consideration was located on the southern flank of the South Tien Shan uplift of the Eurasian continent.

3 Analysis of the tectonic development of the territory in Neogene-Anthropogenic times

At the end of the Oligocene, tectonic activity sharply intensified within the Turanian plate, which included the territory of our research. It flowed more intensely both from the northeast and to the southeast. As is known, the post-platform Tien Shan orogen arose here and further east on the plate. Within most of the Chardzhou and Bukhara steps, part of the Turan plate continued to develop. In the area under consideration, the orogen was manifested by an uplift - the southwestern subsidence of the Gissar meganticline and subsidence - by the foothill (marginal) Beshkent-Kashkadarya depression, which is an element of the foothills of the Newest Tien Shan [5,8]; the depression is also considered as a megablock. The south-eastern limit of the depression is the Lyangara-Karail FRZ, the north-western limit is the isopach of Neogene-anthropogenic layers of 500 m due to the fact that to the north-west of it the thickness of the sediments in question gradually decreases, and then fluctuates between 100-200 m, while as to the southeast of the 500m isopach it increases relatively quickly, reaching up to 2000m or more in individual blocks (Fig. 2).

The foothill trough is superimposed on the part of the Turanian plate involved in the subsidence (Fig. 2). It is relatively strongly divided into separate blocks, characterized by their own unique structure, the thickness of the Cenozoic molasse, and even the sections of the Jurassic terrigenous formation. Fault tectonics manifested itself more often in the form of reverse faults, less often faults (mainly in its northwestern part); The blocks developing here are articulated.

In general, the foothill basin can be considered as a megablock, which smoothly merges with the Turanian plate in the southwest and northwest, and gradually articulates with the uplifting part of the orogen in the northeast; the depression developed in the southeastern parts of the Bukhara and Chardzhou steps that had previously appeared here.

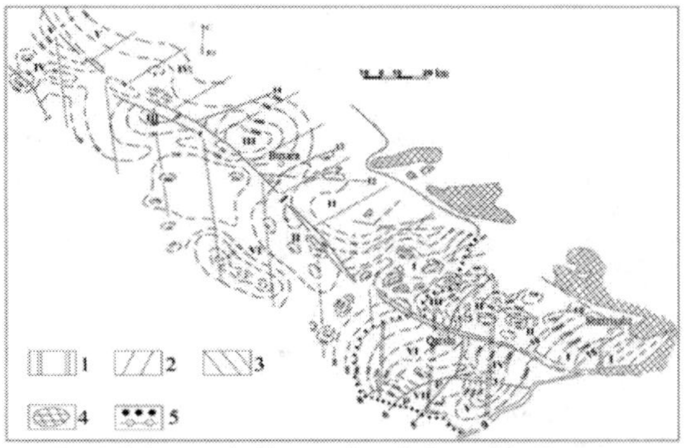

Fig. 2. Paleotectonic diagram along the base of the Neogene-Anthropogen at the time of completion of their accumulation. (Ryzhkov O.A., Zakirov R.T., Khalismatov I.)

Legend: 1. Beshkent-Kashkadarya foothill depression: I. Shakhrizyab mesoblock, II. Tashly-Uvidi mesoblock, III. Bayburak-Rudaksai mesoblock, IV. Zafar mesoblock, V. Shurtan mesoblock, VI. Yangikent mesoblock, VII. Alyaudinsky-Ilimsky mesoblock 2. Bukhara stage: I. Mubarek monoclinal slope, II. Kagan macrouplift, III. Rometan macrotrough, IV. Gazli structural nose, V. Tuzkoy monoclinal slope. 3. Chardzhou degree: I. Dengizkul macro-uplift, II. Ispanly-Chandyr macro-uplift, III. South Kimire fault trough, IV. Khodzhikazgan uplift V. North - Eastern Gugurtlinsky fault trough, VI. Kushabsky trough. 4. Outcrops of Paleozoic sediments on the surface. 5. Boundary of the latest Tianyshan orogen with Turan. Main regional faults: 1. Bukhara FRZ, 2. Chardzhou FRZ. Intrastage faults: 3. North-Shurtansky, 4. South-Kimireksky, 5. Kuruksai-Mangitsky, 6. Alyaudinsky, 7. Beshkent-Kamashinsky, 8. Nishansky, 9. Lyangara-Karailsky, 10. Uvadinsky-Sarychinsky.11 .Andabazar-Karaktai. 12. Proletar-Abad, 13. Kuyumazar, 14. Atbakar, 15. Yakkasaray, 16. Chirakchin.

The megablock, mainly by a system of faults that caused a sharp change in the thickness of the newest accumulations within them, is divided into three macroblocks: northern (Shakhrizyab-Tashlyk), central (Zafar) and southern (Shurtan-Yangikent) (Fig. 2).

The Shakhrizyab-Tashlyk macroblock consists of three mesoblocks (from southeast to northwest) Kashkadarya, Tashly-Uvadinsky and Baiburak-Rudaksay. The second structure is a horst complicated by folds. Within the Kashka-Darya structure, the maximum thicknesses of the rocks under consideration throughout the entire trough have accumulated - up to 2000 m or more. The Baiburak-Rudaksai mesoblock is characterized by a decrease in the thickness of accumulations to the northwest. It is also complicated by local folds and faults (Fig. 2).

The middle part of the megablock is formed by the Zafar macroblock. It is enclosed between the Bukhara FRF in the north and the North Shurtan fault in the south, and the northern part of the Kuruksai-Mangit fault in the west.

The macroblock is a graben inclined in an easterly direction (Fig. 2). The southwestern part of the macroblock is broken by several faults of local importance, forming the North Shurtan tectonic wedge - graben.

The southern macroblock at least consists of the Shurtan, Yangikent mesoblocks and the Alyaudinsky-Ilimsky wedge-shaped horst.

The Shurtan mesoblock relative to the Zafar macroblock is elevated in places by up to 1000 m. In regional terms, it is a near-fault hemicycline with a maximum thickness of the deposits in question more than 700 m. In the southwest it is distinguished by a near-fault uplift, which to the west is replaced by the Alyaudinsky-Ilimsky horst, where these deposits are absent.

The Yangikent mesoblock is located south of the Bukhara FRZ, west of the northern part of the Kuruksai-Mangitsky and west of the Alyaudinsky faults. Regionally, it is expressed by two near-fault hemisynclines with a sediment thickness of up to and more than 900 m. To the west of them, the thickness of the deposits in question gradually decreases.

Thus, the considered foothill depression was paleostructurally connected by Lower Jurassic-Lower Callovian deposits with the Shurtan near-fault macrotrough, the Kultak-Zafar macroblock, the Shakhrizyab and Azlyartepa troughs; for Kimmeridgian-Tithonian deposits - with the Shurtan, Nishan, Beshkent-Kamashin and eastern part of the Urtabulak-Aizavat mesoblocks and the southeastern part of the Bukhara stage; along the Cretaceous deposits - with the Shurtan mesoblock, the Kultak-Zafar macroblock and the Alyaudin-Ilim graben and the southeastern part of the Bukhara step.

All this indicates a serious reworking of the structure of the earth's crust here during the accumulation of sedimentary cover, especially in recent times, when independent blocks of various sizes and structures arose, articulated with each other. The foothill trough originated at the base of the Tien Shan orogenic uplift and thus developed on platform pre-Neogene structures.

North-west of the Shakhrizyab-Tashlyk macroblock, on the area of the Turan plate, the Mubarek monoclinal slope stands out. Within the uplift, numerous folds are outlined, broken by faults, mainly of north-eastern strike. The thickness of the rocks in question in the raised wings is 150-200m. North-west of the Mubarek uplift is the Kagan macrouplift. The uplift from the northwest is limited by the Proletarabad fault. To the north-west, between the Kuyumazar and Atbakor faults, the Rometan trough is located. In the central part, the thickness of these deposits is more than 500 m. Even further to the north-west there is a weakly expressed Gazli structural nose and the Tuzkoy monocline.

On the territory of the Chardzhou stage, based on Neogene-Anthropocene deposits, the Dengizkul uplift is distinguished with the absence of these deposits on the arches of the folds complicating it; also the Ispanly-Chandyr fault uplift. North-west of the Dengizkul and Ispanly-Chandyr uplifts, the Karakul trough is distinguished; the Kushab trough is also recorded between the Dengizkul and Ispanly-Chandyr uplifts. The South Kimirek near-fault trough, the folds of the Gugurtlinsky macro-uplift and the northeast located near-fault trough are well identified. In the area of the Khadzhikazgan uplift, the southwestern wing of the Chardzhou FRZ is lowered by approximately 100 - 150 m, forming a near-fault trough, as in the layers of the terrigenous Jurassic, and in the Cretaceous, this wing was raised by 400 m compared to the northeastern one.

The Bukhara FRZ is fixed by its previously described elements: the Prigugurtlinsky-Yangikazgan - near-fault trough, the Pritaikyrsky - elevated southern wing, the Zekrinsky-Kassantausky - graben or stepped faults, the Kassantausky - Kanallinsky - horst or large-amplitude step.

4 Conclusion

Based on the results of tectonic analysis, the following were identified:

1. The study of the modern structure and paleotectonics of the region indicates the development within its boundaries of both folded and discontinuous (block) structures, which manifested themselves in different places with unequal severity. The modern structure of the study area includes the Beshkent-Kashkadarya, Bukhara and Chardzhou megablocks and the Bukhara flexural fault zone.

2. As a result of paleotectonic analysis, Lower-Middle Jurassic rifting was revealed and it was expressed in the territory by the Bukhara slot-like mesorift, Chardzhou and Bukhara shoulders.

3. It has been established that tectonics affects both reserves and the distribution of oil and gas deposits in section and area. In this case, the intensity of manifestation, regional distribution, and the depth of active faults are of particular importance.

References

1. Dikenshtein G.Kh. and others. Oil and gas fields of Central Asia. Nedra, M., 1965 p.257

2. Zonenshain L.P., Gorodnitsky A.M. Paleozoic and Mesozoic reconstructions of continents and oceans: Art. 1-2 // Geotectonics. 1977: Art. 1. Early and Middle Paleozoic reconstructions. No. 2. P. 3-23; Art. 2. Late Paleozoic and Mesozoic reconstructions. No. 3. P. 3-24.

3. Kuvshinova M.F. Geological and geophysical model of the Uchbash-Karsha flexure-fracture zone and oil and gas potential prospects. Oil and gas geology. Theory and practice. -2022. -T.17. No. 4. - http://www.ngtp.ru/rub/2022/31_2022.html. DOI: https://doi.org/ 10.17353/2070-5379/31_2022. Page 1-10

4. Ryzhkov O.A., Shein V.S. and others. Inversion development of the Amudarya depression. Express information, M, VNIGazprom. 1972. P.37-43

5. Ryzhkov O.A., Zakirov R.T. Rifting on the territory of the Amudarya syneclise. Tr. IGIRNIGM, issue 60 Tashkent. 1987. pp. 33-41

6. Yanshin A.L., Garetsky R.L. Young platforms, their tectonics and oil and gas prospects: (Materials of the meeting on problems of tectonics in Moscow). M.: Nauka, 1965. 226 p.

7. Kholismatov, I.Kh., Tursunova, T.M., Zokirov, R.T., Hayitov, O.G. Prospects of Oil and Gas Bearing Capacity of Middle Lower Jurassic Terrigenous Sediments of the Baysun Trough. (2022) AIP Conference Proceedings, https://www.scopus.com/inward/record.uri?eid=2-s2.0-85133003719&doi=10.1063%2f5.0090927&partnerID=40&md5=da0d4f7c DOI: 10.1063/5.0090927

8. Xayitov, O.G., Zokirov, R.T., Agzamov, O.A., Gafurov, S.O., Umirzoqov, A.A. Classification of hydrocarbon deposits in the south-eastern part of the Bukhara-Khiva region, justification of its methodology and analysis of the results (2022) News of the National Academy of Sciences of the Republic of Kazakhstan, Series of Geology and Technical Sciences, 2022 (1), pp.46-52. https://www.scopus.com/inward/record.uri?eid=2-s2.085129923215&doi=10.32014%2f2022.2518-170X.139&partnerID=40&md
DOI: 10.32014/2022.2518-170X.139

Energy-saving technology for storage of cherry

Safarov *J.E.*[1*], Sultanova *Sh.A.*[1], Polatov *M.M.*[1], Ponasenko *A.S.*[1]

[1]Tashkent State Technical University named after Islam Karimov, 100095, Uzbekistan, Tashkent, Universitetskaya Street 2A

Abstract. Results of studies on extension of storage (shelf) life of cherries by treatment with ultraviolet light are presented. The most important indicators characterizing quality of cherry are considered. Analysis of variance of weight of cherries treated with ultraviolet light had been investigated. The Duncan Test for weight loss of cherries treated with ultraviolet rays was performed. Further studies and tests to determine quality of cherry had been carried out as well.

1 Introduction

Various food preservation methods are aimed at increasing useful life of foods during storage by applying methods that allow preventing microbiological changes while maintaining quality. Effectiveness of these methods depends mainly on following hygiene during its production, the purpose of which is to reduce the microbial burden and prevent its development. For this purpose, many products are subjected to heat treatment, which often changes both organoleptic characteristics (texture, taste and color) and nutritional properties (mainly, loss of vitamins) of the food. Because of these adverse effects of high-temperature processing, (some) non-thermal preservation processes, also called soft technologies, are being developed. These methods are not very aggressive and have an advantage of offering products that are similar to fresh ones and, therefore, meet current market requirements and never lose their security guarantees [1-2].

2 Experimental research

When vegetables and fruits are exposed to ultraviolet (UV-C) radiation, the initial microbial burden on their surface is reduced.

During treatment of cherries with UV-C and storage at 4 degrees Centigrade (°C) for 15 days, they showed a weight loss of 9.72% for samples treated at 4 kJ/m^2 (No.3) and of 10.22% for cherries treated at 2 kJ/m^2 (No.2). The control sample's weight loss has made 10.79% at 0 kJ/m^2 (No.1).

Water is the most common component of fruits. Its content ranges from 89 to 94%. Fruits are very sensitive to dehydration, and those of them that show a high rate of respiration cause water loss, which implies wrinkling, weight loss and deterioration of organoleptic qualities, which (in turn) affects texture and juiciness of the fruit.

The rate of transpiration depends on biological factors of each product and environmental conditions in which they are stored. It is a process in which water vapor from tissues is transferred to the ambient air surrounding the product. That represents one of degradation factors commonly seen during post-harvest processing due to a high water content, resulting in unattractive appearance of the fruit and significant weight loss which reduces its quality and commercial value.

Table 1 shows the analysis of variance of weight loss of cherries treated with UV-C which showed a significant effect at confidence level of 95% of radiation power and storage time.

* Corresponding author: raxmatillo82@mail.ru

Table 1.

Analysis of Variance of Weight Loss of cherries treated with UV-C

Variable	Variation Source	Sum of Squares	Degrees of Freedom	Mean Squares	F	p
Weight loss	Power	1.9	2	0,9	6.5	0.004
	Time	615.6	4	153.9	1,014.05	0.0
	Power Time	0,9	8	0.1	0,7	0.6
	Error	4.5	30	0.1		
	Common	623.1	44			

Table 2 shows results of Duncan Multiple Comparison Test, where it is seen that there is a significant difference between treatment methods marked by formation of subsets. In Subgroup 5 treatments under 4 kJ/m^2 in 15 days of storage and under 2 kJ/m^2 in 15 days of storage were observed, which showed a lower weight loss of 9.7% and 10.2%, accordingly. These processing methods are statistically equal.

Table 2.

Duncan's Weight Loss Test of cherry treated with UV-C

Parameters	Result, %					
	1	2	3	4	5	6
0 kJ/m^2 - 0 days	0.0					
2 kJ/m^2 - 0 days	0.0					
4 kJ/m^2 - 0 days	0.0					
4 kJ/m^2 - 3 days		2.8				
2 kJ/m^2 - 3 days		3.1				
0 kJ/m^2 - 3 days		3.2				
4 kJ/m^2 - 7 days			5,7			
2 kJ/m^2 - 7 days			5.8			
0 kJ/m^2 - 7 days			6.2			
4 kJ/m^2 - 10 days				8.2		
2 kJ/m^2 - 10 days				8.6		
0 kJ/m^2 - 10 days				8.8		
4 kJ/m^2 - 15 days					9.7	
2 kJ/m^2 - 15 days					10.2	10.2
0 kJ/m^2 - 15 days						10.7

Duncan's Criterion. In statistics, the Duncan Multiple Range Test is a multiple comparison procedure developed by B. Duncan in 1955. The Duncan Test belongs to a common class of multiple comparison procedures that use the q_r range statistics to compare sets of means. The Duncan Test is particularly protective against False-Negative Errors (Type II) at the cost of a greater risk of False-Positive Errors (Type I). The Duncan Test is commonly used in agronomy and other agricultural studies. The result of the test is a set of means subsets, where in each subset the mean values do not differ from each other [3].

Let us denote $Q_{(p,v,\gamma(p,\alpha))}$ as γ_α range distribution uantile with p observations and v degrees of freedom for the second sample. And now, let us denote $\eta_{(p,v\alpha)}$ as the standardized critical value determined by the following rule:

If p=2

$$\eta_{(p,va)}=Q_{(p,v,\gamma(p,\alpha))} \quad (1)$$

More

$$\eta_{(p,va)}=max(Q_{(p,v,\gamma(p,\alpha))}, \eta_{(p,va)}) \quad (2)$$

The shortest critical range (the actual critical value of the test) is calculated as follows: $R(p,v,\alpha)=\sigma_m\eta_{(p,va)}$. The denotations for Q and R are not same in literature, where Q is sometimes denoted as the Shortest Significant Interval, and R as the Significant Quantile for the Student's Distribution Range [3].

Effect of UV-C radiation and storage time on soluble solids content. The cherries exposed to a 4 kJ/m^2 (No. 3) radiation had the lowest soluble solids content when stored for 15 days at 14.23 °Brix, followed by samples under 2 kJ/m^2 radiation at 14 .30°Brix. In the Control Sample (0 kJ/m^2), the greatest increase was observed with an indicator of 14.43 °Brix, which shows an advance in ripening process of the cherries, as well as a loss of fruit weight as storage days increase [4].

Table 3 shows the analysis of variance of soluble solids content of cherries which demonstrates a significant effect at confidence level of 95% of radiation power and storage time.

Table 3.

Analysis of Variance of Soluble Solids Content of cherries treated with UV-C

Variable	Variation Source	Sum of Squares	Degrees of Freedom	Mean Squares	F	p
Soluble solids	Power	0,2	2	0.1	4.4	0.02
	Time	4.7	4	1.1	45.3	0.0
	Power Time	0.01	8	0.002	0,07	1.0
	Error	0,7	30	0.02		
	Common	5,7	44			

Table 4 shows the results of the Duncan Multiple Comparisons Test, in which a significant difference between treatments was observed, marked by formation of subsets. Similarly, the Subgroup 6 has treatment under 4 kJ/m^2 at 15 days of storage which showed the lowest soluble solids content at 14.2 °Brix.

Table 4.

The Duncan Test of Soluble Solids Content of cherries treated with UV-C

Parameters	Result, °Brix							
	1	2	3	4	5	6	7	8
4 kJ/m^2 - 0 days	13.3							
2 kJ/m^2 - 0 days	13.4	13.4						
0 kJ/m^2 - 0 days	13.4	13.4						
4 kJ/m^2 - 3 days	13.4	13.4						
2 kJ/m^2 - 3 days	13.5	13.5	13.5					
4 kJ/m^2 - 7 days	13.6	13.6	13.6					
0 kJ/m^2 - 3 days		13.7	13.7	13.7				
2 kJ/m^2 - 7 days		13.7	13.7	13.7				
0 kJ/m^2 - 7 days			13.8	13.8	13.8			
4 kJ/m^2 - 10 days				13.9	13.9	13.9		
2 kJ/m^2 - 10 days					14.02	14.02	14.02	
0 kJ/m^2 - 10 days					14.1	14.1	14.1	
4 kJ/m^2 – 15 days						14.2	14.2	14.2
2 kJ/m^2 - 15 days							14.3	14.3
0 kJ/m^2 - 15 days								14.4

Effect of UV-C radiation and storage time on durability. Cherries treated under 4 kJ/m^2 (#3) showed greater hardness on day 15 when stored at 1.56 N, followed by the sample treated under 2 kJ/m^2 (#2) at 1.52 N. Likewise, we found that the control sample (#1) demonstrated the smallest value of 1.48 N.

As fruits ripen, one of the most noticeable changes is softening associated with biochemical changes in the cell wall, middle lamina, and the membrane level. Firmness of minimally processed fruits is an important quality indicator which is influenced by these enzymes causing softening of fruit tissue and poor skin turgor due to loss of water.

Poor turgor is one of the most significant physico-chemical changes and is associated with loss of water generated during transpiration and respiration of the fruit, resulting in wilting and loss of the fruit's consistency.

Hardness is an important property of cherries, since these fruits are very perishable. Firm cherries are a sign of freshness and provide a longer storage life and faster delivery to the destination point. And, conversely, it is known that softening increases susceptibility to mechanical damage and exposure to pathogens.

Table 5 shows the analysis of variance of hardness of the cherries treated with UV-C where a significant effect is noticed at confidence level of 95% of radiation power and storage time.

Table 5.

Analysis of Variance of hardness of cherries treated with UV-C

Variable	Variation Source	Sum of Squares	Degrees of Freedom	Mean Squares	F	p
Hardness	Power	0.3	2	0.1	36.7	0.0
	Time	4.3	4	1,09	201.02	0.0
	Power Time	0.02	8	0.003	0.5	0.8
	Error	0.1	30	0.005		
	Common	4.9	44			

Table 6 shows results of Duncan Multiple Comparison Test, where it can be indicated that there is a significant difference between the treatments marked by formation of subsets. Similarly, In Subgroup 2 treatments under 2 kJ/m^2 of 15 days of storage and 4 kJ/m^2 in 15 days of storage were observed, which showed highest durability with 1.46 N and 1.54 N, accordingly. These treatment methods are statistically equal.

Table 6.

Duncan's Test for hardness of cherry treated with UV-C

Parameters	Result, N							
	1	2	3	4	5	6	7	8
0 kJ/m^2 - 15 days	1.35							
0 kJ/m^2 - 10 days	1.45	1.45						
2 kJ/m^2 - 15 days	1.46	1.46						
4 kJ/m^2 - 15 days		1.54	1.54					
2 kJ/m^2 - 10 days		1.56	1.56					
0 kJ/m^2 - 7 days			1.61	1.61				
4 kJ/m^2 - 10 days			1.64	1.64				
2 kJ/m^2 - 7 days				1.72	1.72			
4 kJ/m^2 - 7 days					1.81	1.81		
0 kJ/m^2 - 3 days					1.82	1.82		
2 kJ/m^2 - 3 days						1.94		
4 kJ/m^2 - 3 days							2.11	
0 kJ/m^2 - 0 days							2.16	
2 kJ/m^2 - 0 days								2.35
4 kJ/m^2 - 0 days								2.45

3 Conclusion

Food quality is determined by a combination of such parameters as appearance, texture, taste, aroma and nutritional value. Impact of each attribute depends on the food, although, in purchasing, visual impact will be a determining factor.

Treatment with irradiation power of 4 kJ/m^2 showed the highest brightness, durability, titratable acidity and total anthocyanin content; the lowest content of soluble substances, weight loss and the quantity of molds and yeasts in cherries in 15 days of storage at 4 °C (degrees Centigrade).

References

1.Bezlepkin A.I., Perevodchikov V.I., Shlifer E.D. Design of a plant for liquid and solid-phase objects disinfection by combined effects of microwave frequency and UV radiation, and ozone. "Vysokovoltnaya i preobrazovatel'naya tekhnika" ("High-Voltage and Converting Technology.") All-Russia Electrotechnical Institute Publishing House 2001. Pp. 137–143.

2. Wasserman A.L. Design and operation of ultraviolet bactericidal plants / Ed. Y.B. Popovsky. M., 2009, 56 p.

3.https://en.wikipedia.org/wiki/Duncan%27s_new_multiple_range_test

4. Safarov J.E., Sultanova Sh.A., Polatov M.M. A method for storing cherry fruits. III International scientific and technical conference "Problems and prospects of innovative equipment and technologies in agricultural and food sector". Tashkent, 2023. Part 1. Pp. 367-368.

5. Bartosh A. Peculiarities of choosing ultraviolet lamps for growing plants and their use. [E-Resource]: Lamp Expert. -URL: https://lampaexpert.ru/vidy-i-tipy-lamp/kvartsevye-i-ultrafioletovye/uf-lampa-dla-rastenij (accessed on 03/06/2022).

6. Kuryleva A.G., Kondatyeva N.P. Effectiveness of ultraviolet irradiation of grain crop seeds. [E-Resource]: Cyberleninka.-URL: https://cyberleninka.ru/article/n/effektivnost-ultrafioletovogo-oblucheniya-semyan-zernovyh-kultur/viewer (accessed on 03/06/2022)

7. Tudupova D. B. Using UV radiation as a pre-sowing treatment of flax seeds for growing at home / D. B. Tudupova, A. O. Shardina, V. S. Soldatkin // Role of Agricultural Sciences in sustainable development of rural areas: Proceedings of V All-Russia (national) scientific conference, Novosibirsk, December 18, 2020 - Novosibirsk: "Zolotoy Kolos" (The Gold Head), the Publishing Center of Novosibirsk State Agricultural University, 2020. – Pp. 335- 339.

8. Safarov J.E., Sultanova Sh.A., Polatov M.M. STUDY OF UV RADIATION ON ABSORPTION OF VEGETABLES AND FRUITS // Universum: Technical Sciences: a scientific e-magazine. 2023, 4(109). URL: https://7universum.com/ru/tech/archive/item/15256 (accessed on: 08/23/2023). DOI - 10.32743/UniTech.2023.109.4.15256

E3S Web of Conferences **461**, 01078 (2023) https://doi.org/10.1051/e3sconf/202346101078
RSES 2023

Heat-pump system on natural refrigerant CO_2

Kudrat Karimov [1*], *Asatilla* Karabayev[1], *Taxir* Nurmatov[1], Abdullakhuja Norkhujayev[1] and *Mirolim* Abdullayev[1]

[1]Tashkent State Technical University, University st., 2, 100095, Tashkent, Uzbekistan

Abstract. Modern science is closely dealing with conservation of non-renewable resources and development of systems capable of converting the energy of inexhaustible sources – the sun, air, earth, water – into resources for human life activity, such as heating and water supply. Active introduction of new methods takes place in all civilized countries of the world. One of the modern devices that allow to consume heat and hot water with minimal energy consumption is a heat pump. The study and analysis of a heat pump system using pure refrigerant CO_2 (R744) is of both scientific and wide practical interest.

1 Introduction

Energy conservation is one of the main problems being solved by the world community at the present time, it has two goals: preservation of non-renewable energy resources and the reduction of emissions of combustion products into the atmosphere, which, in particular, is the main factor in global warming. At present, the issue of using non-renewable energy sources is quite critical. Rising energy prices, environmental issues and the flexibility of heating and water supply systems entail the need for new technologies.

The problem of meeting the growing needs of the republic in fuel and energy resources includes a set of tasks to find and develop alternative and use renewable energy sources and introduce rational ways to reduce fuel costs. Heating, ventilation and air conditioning systems consume up to 40% of the extracted fuel and up to 10% of the generated electrical energy. In this regard, energy saving in this area gives the most tangible effect. First of all, it is necessary to apply methods and means of energy saving with the help of which the highest energy and environmental effect is achieved at minimal additional costs.

One of the effective measures to save fuel, as well as to protect the environment, is the widespread use of heat pump installations that convert the natural low-grade heat of the surrounding air, water, solar energy and heat waste into heat of a higher temperature, suitable, in particular, for heat supply to buildings and premises.

2 Analysis of operation of heat pump system on CO_2

In foreign scientific, technical, patent and advertising literature of recent years, the number of publications on the issue of reducing the consumption of electricity and fuel in the processes of heat and cold supply of residential and administrative buildings, as well as in industrial and agricultural technological processes through the use of heat pump systems (HPS) has drastically increased.

HPS is most effectively used in heating systems of residential and administrative premises that are not connected to a combined heat and power plant, are located away from gas mains, and in areas with low building density.

The use of HPS for heating was constrained by competition from less expensive gas and oil boilers until the price of liquid and gaseous fuels increased.

A sharp increase in oil and gas prices and a continuous outpacing of their growth rates compared to the growth rates of electricity prices requires the need for further improvement of HPS for heating and air conditioning systems, a feasibility study for the feasibility of their use.

Such systems operate without the use of fuel and do not produce harmful emissions into the atmosphere, in addition, they can significantly reduce operating costs, saving up to 75-80% of the energy used for heating and hot water preparation. Currently, about 40% of the total amount of fuel burned is spent on heat supply. Heat pumps, carrying out a reverse thermodynamic cycle on a low-boiling working substance, utilize the low-grade heat of natural, industrial or domestic sources and generate high-potential heat, while spending 1.2-2.3 times less primary energy than with direct fuel combustion. In a number of countries, heat pump installations have become a common attribute of life. For example, in Israel, hot water supply to 80% of all residential buildings is provided by solar water heaters, which saves more than 5% of electricity produced in the country.

* Corresponding author: kudrat_kf@list.ru

The use of HPP for heating and hot water supply instead of local boilers operating on gas or oil fuel, moreover, does not pollute the environment of residential areas with fuel combustion products and reduces the global level of thermal pollution, which is not an unimportant factor in favor of their use. The successes achieved in the world in the field of refrigeration engineering are the basis for the creation of more advanced heat pumps. As a result, the industry of the leading countries producing heat pumps has recently experienced rapid growth.

Heat consumption for domestic hot water supply of residential buildings is determined by consumption rates (SNiP 2.04.01-85).

Average daily heat consumption for hot water supply during the heating period:

$$Q_w^{av.d} = 1.2 \cdot m \cdot a(t_h - t_c)c_p^{av},$$

where m – design number of consumers; a – water consumption rate per 1 person, kg/day

(for apartment buildings with improved public services and amenities a = 105 to 130 kg/day); c_p^{av} is specific heat of water, kJ/kg-K.

Average load of hot water supply during the heating period:

$$Q_w^{av.d} = 1.2 \cdot m \cdot 120(55-5)4.187 = 0.349m = 0.35\ kW/person$$

Design maximum heat load of hot water supply:

$$Q_w^d = \chi \cdot Q_w^{av.d} = 2.4 \cdot 0.349 = 0.84\ kW/person$$

where χ - design coefficient of hourly variation for residential buildings.

Average heat consumption for heating during the heating season:

$$Q_c^{av} = Q_o \frac{t_{wd} - t_{av.c}}{t_{wd} - t_{\kappa c}} = Q_c \frac{18 + 2{,}4}{18 + 15} = 0.62Q_c$$

Average heat consumption for hot water supply in summer:

$$Q_{hs}^{av} = Q_c^{av} \frac{t_h - t_{cw}}{t_h - t_{cz}} = Q_c^{av} \frac{55-15}{55-5} = 0{,}22\text{kW/person}$$

Annual heat consumption for hot water supply:

$$Q_c^y = Q_c^{av} \cdot n_o + Q_{hs}^{av}(n - n_o),$$

where n is the duration of the hot water supply - 8400 h.

Annual heat consumption per inhabitant for heating and ventilation of residential buildings:

$$Q^{year} = n \cdot 3{,}85\ \text{YJ},$$

hot water supply:

$$Q^{year} = n \cdot 8{,}15\ \text{YJ}.$$

Estimated heat load per inhabitant for heating and ventilation of residential buildings, kW:

$$n = 0{,}82,$$

hot water supply:

$$n = 0{,}32.$$

Heating performance of residential buildings:

$$q_0 = 0{,}49\ \text{W/м}^3\text{K}.$$

Heat loss in residential buildings:

$$Q = q_o \cdot V(t_{in} - t_{out}).$$

The volume of residential buildings per person is 55 m^3. The calculated heat consumption for ventilation is 10% of the calculated heat consumption for heating.

As is known, the operation of a HPS requires not only energy costs for the drive, but also the energy of additional heat sources, the temperature level of which does not allow them to be used in the usual way, i.e. temperatures below 40 °C. Such sources of heat include waste heat and the energy of the surrounding space.

The source of energy from the surrounding space is the soil, ground and surface water, the energy of the sun and the surrounding air. The lower limit of the use of water sources is the freezing point of water, and for air, the temperature of frost formation on the surface of the air cooler (heat receiver).

In the conditions of our Republic, the use of outdoor air as a low-potential source for heat pumps is the most appropriate in view of its versatility (unlimitedness, availability, no additional costs for its use, etc.).

When using outdoor air as a source of heat, one should take into account not only temperature fluctuations during the day, but also the fact that the lowest temperature and the maximum required heating capacity coincide in time, electricity consumption increases in cold weather conditions, due to a decrease in the conversion factor.

When the air temperature is below +5 °C, the evaporator freezes, and in this case it is necessary to provide a defrosting device.

Heat pump refrigerants are in principle subject to the same requirements for physical and chemical properties as for refrigeration machines. Special requirements arise due to higher evaporating and condensing temperatures. In general, these temperatures are in the same range as chillers operating in harsh environments (air-cooled air conditioners). Therefore, they can be used in heat pumps.

In recent years, the choice of refrigerant for heat pumps has become one of the key issues. This is due to the negative environmental impact of CFC and HCFC refrigerants.

In the 80s of the last century, it was proved that CFCs and HCFCs (CFCs to a greater extent, and HCFCs to a lesser extent) are capable of destroying the ozone layer that protects the flora and fauna of the Earth from the harmful effects of solar ultraviolet radiation.

In accordance with the 1987 Montreal Protocol for the Protection of the Earth's Ozone Layer and subsequent amendments to it (Uzbekistan ratified the Montreal Protocol and its subsequent amendments), the production of CFCs in developed countries has been stopped since 1996, and in developing countries it should be stopped by January 1, 2010, and HCFC production in developed countries by 2020 and in developing countries by 2030.

The second negative factor in the impact of refrigerants on the Earth's atmosphere is the greenhouse effect. Absolutely all refrigerants, including non-ozone-depleting substances, have it. This effect occurs due to the fact that certain gases absorb infrared radiation and trap it in the Earth's atmosphere. As a result, the Earth's surface maintains a temperature suitable for the origin and development of life. Water vapor, carbon dioxide and other gases have this ability to absorb.

An increase in their concentration as a result of human activity leads to an increase in the temperature of the Earth's atmosphere (to global warming). The greenhouse properties of CFCs and HCFCs are thousands of times greater than those of carbon dioxide.

Thus, when choosing a refrigerant, it is necessary to take into account, along with thermodynamic characteristics, a complex environmental indicator, including not only its effect on the ozone layer, but also on the level of global warming.

CO_2 belongs to so-called "natural" refrigerants such as ammonia, propane, butane or water. Each of them has its drawbacks: ammonia is toxic, propane is flammable, water has a limited scope. Unlike them, CO_2 is non-toxic and non-flammable, although its impact on the environment is not unambiguous. On the one hand, CO_2 is contained in the air and is necessary for the flow of life processes. On the other hand, it is believed that a large concentration of carbon dioxide in the air is one of the causes of global warming.

The initiative to return to the use of CO_2 belongs to the Scandinavian countries, in particular Denmark. Legislation in Denmark severely restricts the use of HFC and HCFC refrigerants and plans to phase them out in the future. Ammonia has traditionally been used as a refrigerant for industrial installations, but its quantity in the system is very limited. This is not a problem for refrigeration units operating at high and medium temperatures (up to -15/-25 °C), where the amount of ammonia can be reduced by using a secondary coolant. But for lower temperatures, the use of a secondary coolant is inefficient due to large losses due to the temperature difference; in this case, it makes sense to use CO_2 as a coolant.

3 Research results

The main heat pumps in Uzbekistan in recent years have been used mainly in the housing sector. The winter-summer air conditioner is a typical air-to-air heat pump, in which the source of low potential is outdoor air, and the heated medium is indoor air.

The Department of Refrigeration and Cryogenic Engineering of the Tashkent State Technical University is working on the development and implementation of heat pumps. Thus, HPP was introduced at a number of dairy farms in the Tashkent region, which ensured the process of its pasteurization and the need for hot water for sanitary needs by utilizing the heat of cooled milk. In 2005, a heat pump for a mineral water bottling line was introduced at the Istiklol Fayz Private Firm in Tashkent, in which the heat of the cooled water is used to heat water for a bottle washing machine. A project was also developed for a heat pump station to supply heat to the boarding house of the Tashselmash plant on Lake Issik-Kul, using the heat of the lake as a source of low potential.

One of such recent works is the development and research of a modern HPI operating on CO_2 to produce hot water.

The main limitation in CO_2 operation is the high operating pressure. Thus, a "condensation" temperature of 40°C corresponds to a pressure of 85 bar. At the same time, one can speak about condensation rather conditionally. since this point lies above the critical one.

The study and analysis of a heat pump installation using pure CO_2 refrigerant (R744) is of both *scientific* and wide *practical* interest.

To compare the operating parameters of an experimental heat pump installation when using an environmentally friendly refrigerant in it, as well as to compare the values determined by the exergy method, the bench was tested, which was developed on the basis of a SANDEN heat pump, which belongs to the environmental organization ERP France (Fig. 1).

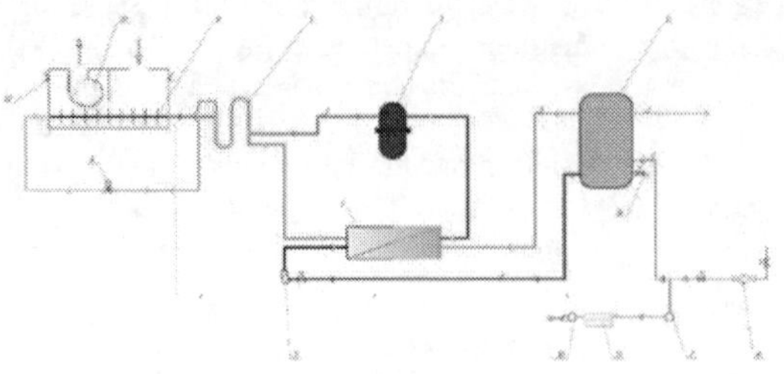

Fig. 1. Experimental stand for research

The bench works as follows:

The unit is a single-stage machine with systems for supplying water to the gas cooler and air to the evaporator. The heat pump consists of: a compressor 1, a plate-and-tube evaporator 2, a water two-pipe gas cooler 3, an expansion valve 4 and a heat exchanger 5.

The water supply system to the condenser consists of a hot water storage tank 6, a water pump 7. Fresh water for heating is supplied through a rotameter 8. This part of the installation has a protection system with water bypass into the drainage tank 9, 10.

The air supply system to the evaporator consists of centrifugal fan 11 and guide channels 12.

Compressor 1 sucks in CO_2 vapors, compresses them and pumps them into the annular space of the gas cooler 3, where the vapors are cooled, giving off heat to the heated water. The cooled vapors enter the heat exchanger 5, where they give off heat to the cold vapors coming from the evaporator 2. Then, after passing through the expansion valve 4, they are throttled and liquefied. Liquid CO_2 enters evaporator 2. In the evaporator, carbon dioxide boils in the tube space, cooling the outside air entering through the ducts. The resulting R744 refrigerant vapor enters the heat exchanger 5, overheats in it, due to heat exchange with the refrigerant leaving the gas cooler 3, and then is sucked in by the compressor and the cycle is repeated anew.

Accumulator tank with a capacity of 150 liters for storing sanitary hot water. It has 3 hydraulic connections on the bottom:

- city water supply
- to CO_2-Water heat exchanger
- drain.

And 2 hydraulic connections on the top:

- exit from the CO_2-Water heat exchanger;
- hot water outlet for the user.

The city water pressure at the inlet to the heat pump must be greater than 1 bar. The pressure regulator is set to 3.7 bar, and is effective until it reaches a maximum pressure of 10 bar at the top.

The maximum city water pressure recommended by SANDEN at the inlet reducer is 7 bar. The safety valve is set at 4.40 bar.

In the steady state mode of the heat pump, the following parameters of the installation were measured:

Temperature, °C

t_1 – of refrigerant at compressor suction;
t_2 – of refrigerant at compressor discharge;
t_3 – of refrigerant at condenser outlet;
t_4 – of refrigerant before thermostatic expansion valve;
t_5 – of refrigerant at evaporator inlet;
t_6 – of refrigerant at evaporator outlet;
t_7 – of water at gas cooler inlet;
t_8 – of water at gas cooler outlet;
t'_w – of urban water at accumulator tank inlet;
t''_w – of urban water at accumulator tank outlet;

Pressure, MPa

P_0 – of refrigerant at compressor suction;
P_k – of refrigerant at discharge to gas cooler;

Electric power, W

N – compressor power consumption;

Flow rate, kg/s

G_w – water through gas heat exchanger.

Measurement of refrigerant temperatures at the installation points (t_1, t_2, …, t_8) was carried out using a 12-channel thermocouple sensor of Datalogger – model TM500. The thermocouples were terminated on the surface of the pipes.

The water flow was measured using a magnetic-inductive flow sensor SM6000.

The electrical parameters were measured with a K-505 measuring complex, a D-522 wattmeter of class 0.5 according to GOST 8711-60 for the range of 150–600 W and class 0.1 for 100 W.

The operating parameters of the experimental setup in the experiments were as follows:

- Outside air temperature $t_{воз(нар)} = (-14\ to + 21)$ °C;
- Incoming water temperature $t_W = (10\ to\ 45)$ °C;
- Water temperature at the unit outlet was constant: $t'_W = 65$ °C.

Heating capacity Q_T and electric power of compressor $N_{эл}$, obtained after processing the experimental values of the installation, were depicted graphically. On Fig. 2 – 4, the dependencies are given $Q_к = f(t_{воз})$, $N_{эл} = f(t_{воз})$ and coefficient of performance or (COP) on outside air temperature $COP = f(t_{воз})$ at water temperatures $10°C$, $30°C$ and $45°C$

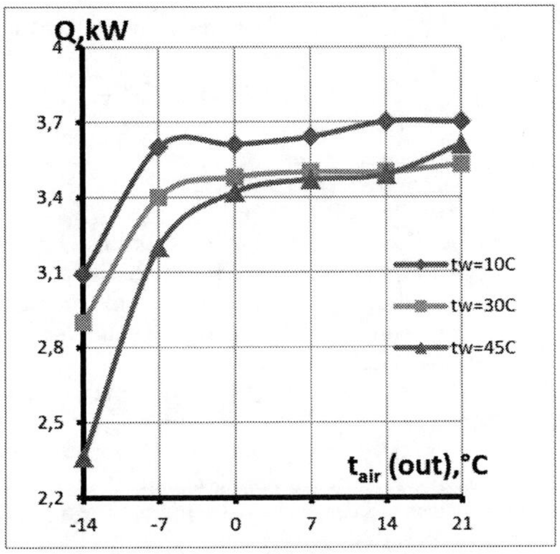

Fig. 2. Dependence of the heat output of the installation on the outside air temperature at fixed inlet water temperatures t_W.

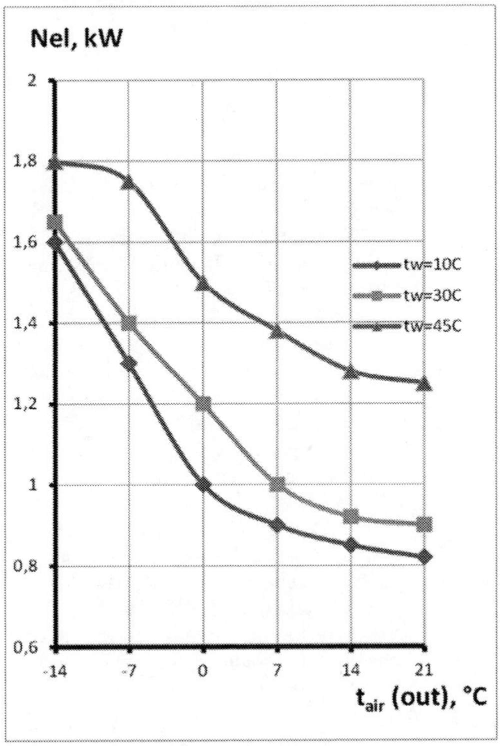

Fig. 3. Dependence of the electrical power of the installation on the outdoor temperature at fixed inlet water temperatures t_W.

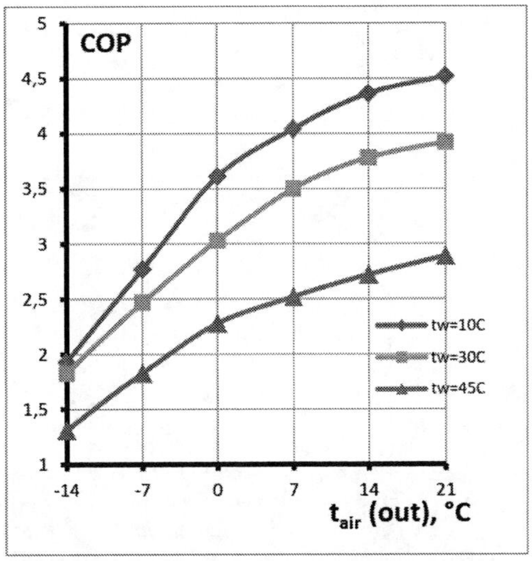

Fig. 4. Dependence of COP of the installation on the outdoor air temperature at fixed inlet water temperatures t_W.

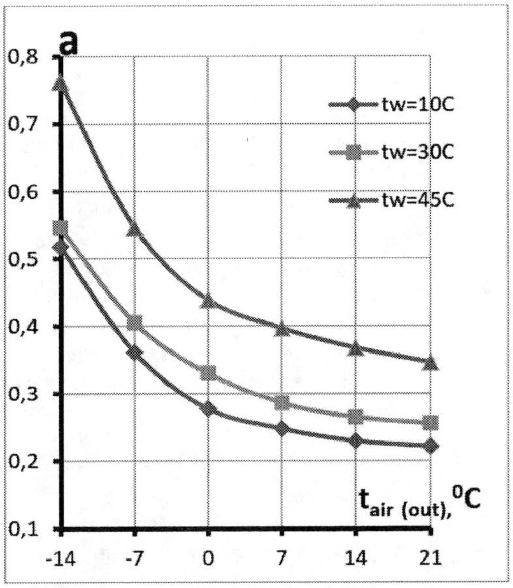

Fig. 5. Dependence of the unit's specific power consumption on the outdoor air temperature at fixed inlet water temperatures t_W.

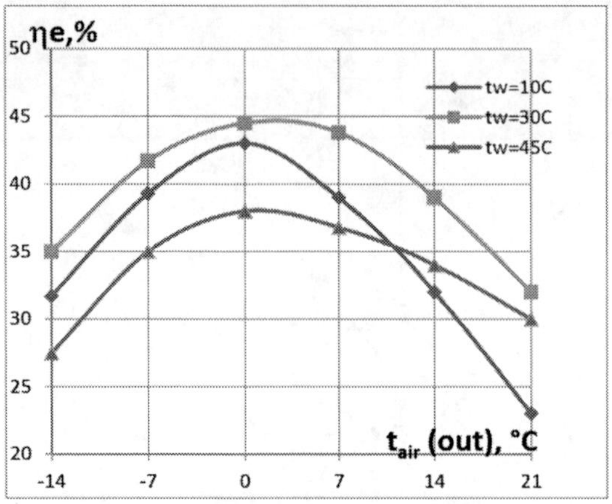

Fig. 6. The dependence of the exergy efficiency of the installation on the outdoor

The specific power consumption of the installation *a* decreased with increasing ambient temperature. In the region of negative temperatures, *a* has higher values than at positive temperatures.

4 Conclusion

When the ambient temperature changes from -14 °C to +21 °C, the heating capacity of the installation increases:

- by 19% at water temperature of 10 °C;

- the best result of the COP value was when the unit was operating on colder water (10 °C) – 4.5, which is 55% more than when operating on water with the temperature of 45 °C.

Costs (specific) at a water temperature of 10 °C were in average 48 - 59% less than when using water with a temperature of 45 °C. And when water enters the installation with the temperature of 30 °C, the costs are 4.8 that is 14% more than at 10 °C.

The dependence of the exergy efficiency of the unit on the outdoor air temperature at fixed inlet water temperatures gave the following results:

- the temperature region of the maximum values of the exergy efficiency falls on the interval -5 to + 7 °C at all temperatures of the water entering the gas cooler of the plant;

- absolutely high efficiency was observed when water with the temperature of 30 °C was supplied. In addition, it was in this case that a wider temperature range was observed (-10 to +12 °C), where the efficiency value exceeded 40%.

- in the mode below the air temperature +10 °C, the unit while operating with water t_W = 45 °C was inferior to the operating mode at t_W = 10 °C. After that temperature, the efficiency has become higher.

Based on the above, we can conclude that from the point of view of the 1st and 2nd laws of thermodynamics, the optimal mode in this case is:

- water temperature at the unit inlet t_W = 30 °C;

- temperature range at t_W = 30 °C should be -10 to +12 °C.

References

1. Azizov D.H., Norkhujaev A.S., Research of split air conditioner in environmentally friendly cooling agent // Scientific and technical journal of Chemistry and Chemical Technology. №2, 2017 - pages 75-78
2. Azizov D.H., Norkhojayev A.S., Research of a split air conditioner on environmentally friendly refrigerants // Collection of scientific papers international scientific and technical on-line conference. - Tashkent. TashSTU, 2020. pp. 285-288
3. Koraboev A.S., Norkhujaev A.S., Maxmudov O.O., Study of compressors with split conditioners with low pressure // Texnika i teknologii mashinostroeniya. Materials of the VII International scientific-technical conference. Omsk, May 21-23, 2018. pp. 174-177
4. Nurmatov T.B., Norkhujaev A.S., Khasanov B.B., Testing of split-conditioners with low pressure with the use of refrigerants R22 and R410A // Texnika i teknologii mashinostroeniya. Materials of the VII International scientific-technical conference. Omsk, May 21-23, 2018 C. 178-180
5. Svetkov O.B., Laptev Yu.A. Tablisы svoystv xolodilnых agentov: Ucheb.-metod. posobie. – SPb.: NIU ITMO; IXiBT, 2013. – 52 p.
6. Luneva S. K., Chistovich A. S., Emirov I. X. K voprosu primeneniya teplovix nasosov. // Jurnal «Texniko-texnologicheskie problemi servisa», 2013.
7. Kazakov V. G., Lukanin P. V., Smirnova O. S. Eksergeticheskie metodы otsenki effektivnosti teplotexnologicheskix ustanovok. — SPb.: Sankt-Peterburg. gos. texnol. un-t rastitelnых polimerov, 2013. — 63 s.
8. Azizov D., Saydiev F. Osnovы xolodilnoy texniki i texnicheskogo obslujivaniya xolodilnix sistem - Tashkent : Baktria press, 2017. - 176 s.
9. Azizov D. X., Saloxiddinov B. I. Povishenie effektivnosti i energosberejeniya v xolodilnix mashinax // Molodoy ucheniy. — 2016. — №9. — pp. 93-96

Textile process and electrical equipment used in the development and production of the textile industry in Uzbekistan

Ikromjon Rakhmonov[1]*, *Mukhammadkhon* Toirov[1], *Kamoliddin* Obidov[2], *Navruz* Uzakov[3]

[1]Tashkent State Technical University named after Islam Karimov, 100095, Uzbekistan, Tashkent, University St. 2A

[2]"TIIAME" NRU Bukhara Institute of Natural Resources Management, Bukhara, 105009, Uzbekistan

[3]"TIIAME" Karshi Institute of Irrigation and Agricultural Technologies of the National Research University, Karshi, Uzbekistan

Abstract. The article identifies ways to increase energy efficiency in textile industry. The power consumption modes of the electrical process devices involved in the technological processes were analyzed and the most efficient operating modes were identified. In addition, the technological process in the textile industrial enterprise was analyzed, adapted to the consumption of the most electricity, and the sequence of energy consumption was compiled and analyzed.

1 Introduction

We know that textiles are one of the biggest drivers of our economy and development. Previously, Uzbekistan ranked 6th and 3rd in terms of cotton exports, respectively. However, over the last 10 years, exports of raw materials have been declining, with an emphasis on exports of processed products. This is because the price of 1 kg of fiber in the world market is around $ 1.5, while the price of the finished product is around $ 25. In other words, we can see that the existence of a large economic force in the textile sector is reflected in prices.

2 Experimental research

There is a lot of legal support in our country, which is reflected in the following. In other words, in accordance with the Resolution of the Cabinet of Ministers of the Republic of Uzbekistan No. 211 of July 25, 2011, the Ministry of Economy of the Republic of Uzbekistan together with the World Bank launched the project "Improving energy efficiency in industrial enterprises." Another example is the Decree of the President of the Republic of Uzbekistan No. PF-5282 dated 14.12.2017 on the completion of JSC "Uzbekyengilsanoat" and the establishment of the Association "Uztextile Industry". Also, on 22.08.2019, the President of the Republic of Uzbekistan signed a resolution "On urgent measures to increase energy efficiency in the economy and social spheres, the introduction of energy-saving technologies and the development of renewable energy sources." The above reforms, well-chosen strategies and clear management have shown a positive growth trend for the country's textile companies. An example of this is the results of January-October 2019 and 2020 in the textile industry in our country. In January-October 2020, a total of 4,886 textile enterprises in the Republic of Uzbekistan produced goods worth 26.9 trillion sums. Compared to January-October 2019, the physical volume index was 112.3%. In January-October 2020, exports of textile products amounted to $ 1524.7 million, accounting for 11.4% of total exports of the Republic, an increase of 15.9% over the same period last year. Also, at the beginning of the year, 522 different types of textiles were exported to 65 countries [1-7].

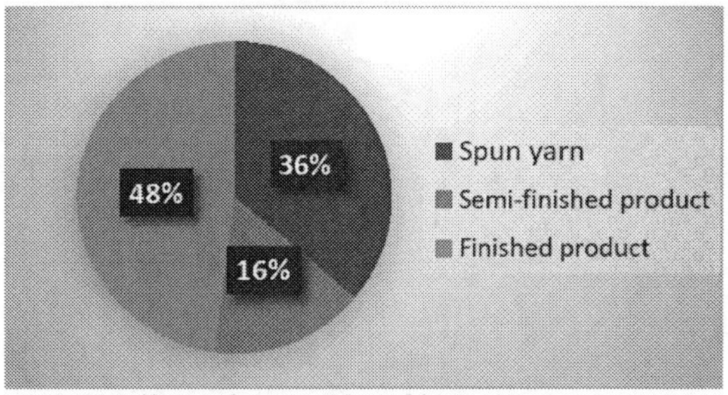

Fig1. Textile products produced in Uzbekistan in 2020

Fig.1 shows the share of spun yarn, semi-finished products and finished textile products in Uzbekistan in 2020.

* Corresponding author: ilider1987@yandex.com

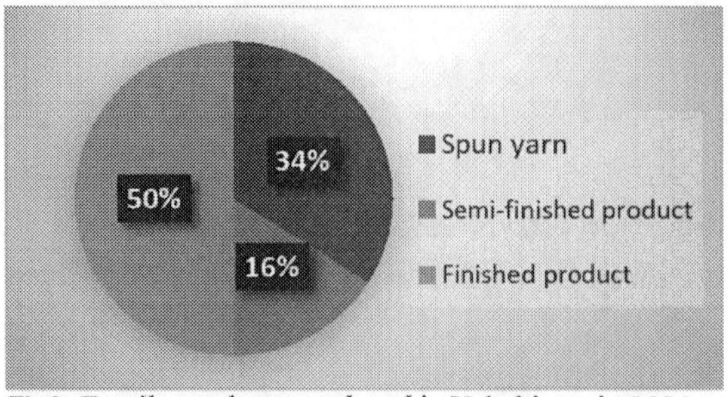

Fig2. Textile products produced in Uzbekistan in 2021

Fig.2. shows the share of spun yarn, semi-finished products and finished textile products in Uzbekistan in 2021.

The role of ready-made products in the field of Textiles is growing, as far as we are concerned.

From the above data, it can be seen that the importance of the technological process of textiles is reflected in the figures. This technological process consists of the following: The process of production in textiles is divided into 5 main sequences. These include: cleaning, spinning, weaving, wetting, and garment manufacturing [8-16].

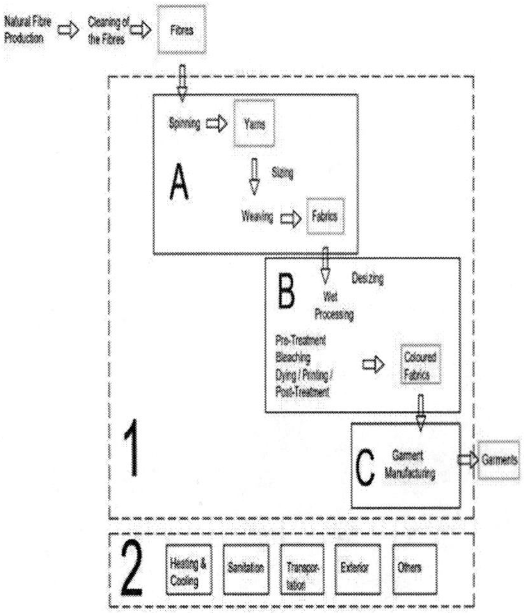

Fig.3. The process of converting cotton into a finished raw material in general

Number 1 and 2 show primary and secondary energy consumption. That is, A, B, and C are the main consumers in the spinning, measuring, weaving, wetting, and garment manufacturing processes. Heating and cooling, sanitation, transportation, outdoor work, etc. are secondary energy consumers. Each stage has its own individual energy consumption, for some need a lot of electricity, others need heat. The spinning process consumes more electricity with different steps. The amount of electricity consumed in this process varies depending on the type of spinning system, the different and two-part operation of the machine, the preparation of the yarn and the characteristics of the raw material. On average, in the production of 1 kg of individual yarn, the spinning process and the various processes of the machine account for 80% of energy consumption. Dimensioning is the main process between spinning and weaving. The measurement process often requires steam generated by gas, oil, or electricity before bending. The weaving process requires only electricity. The room temperature is controlled at 25°C and humidity 65% [17-25, 45].

The humidification process requires the consumption of hot water, steam, and hot air. The mechanical part of the humidification process consumes electricity. This process consumes less electricity than other processes.

The garment manufacturing process includes stocking, cutting, sewing, vacuuming, ironing and transportation. Only heating and ironing may require steam and hot air.

Textiles are fabrics made of woven yarn. Fabric production is not an easy process, so scientists and devices are being developed by scientists around the world to facilitate this process. The following is information on equipment used in the textile industry [26-35, 44].

The spinning process is the basis of fabric production, because it consumes the most electricity. This is the process by which a long yarn is formed - a yarn woven from short fibers. This production process is done on a spinning machine. This modern machine has a capacity of 18 kW, weighs 12 tons and has 240 rotating shafts and up to 1,800 revolutions per minute, and costs $ 60,000.

Fig.4. Spinning machine

The weaving machine is designed for the production of textile fabrics, it can be loosing and without loosing, round and flat, wide and narrow. Knitting machines are selected depending on what fabric should be produced: linen, silk, cotton or wool. The following is a modern weaving machine called ST-ERL-808 with a maximum speed of 800 RPM. The width of the fabric is from 600 mm to 1200 mm. Moreover, there are several fabrics getting produced in textile manufacturing industries including woven fabrics, non-woven fabrics, and knitted fabrics. To produce all those fabrics, heavy machinery is used by the textile manufacturing industries in many countries. Presently, there are several weaving machines

 https://doi.org/10.1051/e3sconf/202346101079

that are dominating the textile market. However, majorly they are divided into three basic types- Circular, Shuttle, and Narrow Fabric. a) Circular loom: This machine contains several circulating shuttles that move around the periphery disseminating the weft. The major concern of this machine is producing tubular fabric and thereby getting used to produce seamless fabric products such as clothing, fire hose, hosiery, sacks, etc. The modern circular machine uses electromagnets for high-speed fabric production. The speed of this loom is manageable and hence preferred by many of the textile manufacturing industries in many countries. a) Shuttle loom: This machine is mainly controlled via electronic and used to weave warp and weft patterns. This loom is a kind of micro wood or plastic that contains bobbins that are filled with yarn. It is fluttered across the loom back and forth to form interlacement that helps in producing fabric easily and more efficiently. This is also easy to operate and worker-friendly. c) Ribbon loom: A ribbon loom is a kind of narrow fabric loom that is highly efficient to weave more than forty pieces together. This loom has a jacquard fitted at the top of them, which is controlled by a stack of perforated cards. The weaving machine & jacquard machine are perfectly integrated into a seamless single control panel to provide ease of operation while ensuring error-free weaving. This fabric loom also helps in the effective processing of complex patterns like brocade and matelassé, etc. [36-43].

Fig.5. Weaving machine

The rolling machine is used to automatically roll the resulting fabric into a roll or roll using a rotating roller. A properly machined knurling machine works more efficiently by weavers, especially on a production scale than by hand wrapping. Furthermore, rolling machine machine is an industrial mechanical pressing tool used to bend sheets and most sheets (including sheet metal) by clamping the workpiece. Industrial bending machines not only improve work efficiency but also help small businesses reduce manufacturing costs [42-44].

Fig.6. Rolling machine

The dye line and printing machine allow fabrics to be dyed with natural or synthetic dyes. The printer applies color prints with ink or melts the screen pattern onto a ready-dyed fabric. The following modern machine is a high-speed 6-color combination and the rollers work synchronously with low noise. The rollers rotate at 120 meters per minute, while also printing shapes on material up to 8-100 meters per minute. In the industrial set up of the economy, the printing machines are growing in large numbers. Different printing techniques and processes are used by every industry for effective communication. Any machine used in printing ink on the substrate (printing medium) which can be cloth, paper or plastic, is referred to as the printing machinery. By applying pressure to the substrate, the printing machinery transfers the ink on the printing medium resting on an inked platform composed of movable type. The development of industrial printing machines changed a lot of things in the world. It would make transfer and preservation of historical records, scientific findings, and knowledge easy and possible. There's a great advancement in the printing machinery introduced in the market today given the improvement in science and technology

Fig.7. Dye line and printing machine

Washing and measuring machines are identified.The washing machine washes and dries woven fabrics after printing or dyeing, and inspection equipment is used to check the quality of the finished woven product, its length, width and density. Moreover, Washing in the general sense means washing something. In the case of textiles, washing is the process of removing excess chemicals from fabric or products after a wet processing process. There are different chemicals for different wet processing. So the chemicals used in a process can cause serious damage to the next process. So after the completion of any wet process, the product must be washed. [24-29].

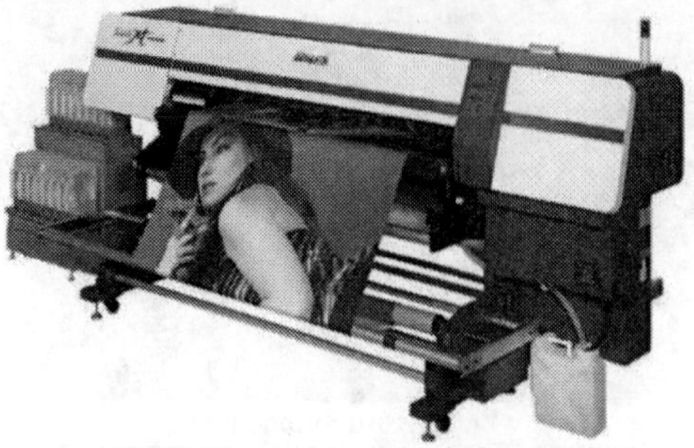

Fig.8. Measuring and printing machine

They also can cut prepared materials which are needed in textile industry. This machine can work faster and more efficient than previous one. An important field of application is the calibrated length measurement in textile fabric inspection, when the finished product is sold by length or by square meter. Other applications include length measurement for incoming and outgoing goods, as well as in production machines and slitter rewinders, where it is necessary to obtain precise information about material lengths. In these cases more efficient machines should be used in thhe process [10-12, 45].

3 Research results

Taking into account the fact that the textile industry has its place almost all over the world, it was considered how important it is in Uzbekistan. in this process, the sequence of actions for the development of the textile industry in Uzbekistan was presented. In this case, it was determined that if the textile process consists of cleaning, spinning, wetting, dyeing, and printing, the process of spinning the product is considered to be the process that requires more electricity consumption. If almost any scientific work is done to increase the efficiency of electricity consumption in a textile industrial enterprise, this process should be directed to the spinning part. In our article, the spinning machine was replaced with a more efficient one, and the results were compared and acknowledged.

Table 1

Spinning machine	Power	Weight	Number of rotating shafts	Number of revolution 1 min
Present	18kVt	12 tons	240	1800
Previous	20kVt	15 tons	260	1600

This table 1 shows the former product spinning machine in a textile industry and the current more efficient modern spinning machine. in this case, a spinning machine with a power of 2 kw, a weight of 3 tons less and 20 more rotating parts was selected.

4 Conclusion

1. Based on the above, it can be concluded that the following processes are considered as the points of the textile industry with high energy consumption: spinning, wetting, garment production.

2. The operating conditions of electrical equipment that cause these cases have been studied, and on the basis of these studies, ways to achieve their energy efficiency through the optimization of power consumption regimes, the use of energy-saving devices in the technological process have been identified.

3. It was found that the most electricity consumption is required in the process of spinning the product, among the processes that have a significant role in the consumption of electricity in the textile industrial enterprise.

4. Secondary electricity consumers can also take small measures to improve energy efficiency. These actions can be applied to product heating, cooling, transportation and other activities.

References

1. Rakhimov S. Modernization of Higher Education in Uzbekistan: Current Reforms and Future Prospects. Central Asian Journal of Education.2020.
2. Turaeva M., and Mukhamedova Z. The Role of Technology in Uzbekistan's Education System Transformation. International Journal of Emerging Technologies in Learning (iJET), 13(2).2018
3. Allaev K., Toshov J., Toirov O. E3S Web of Conferences 401. 05090. (2023). https://doi.org/10.1051/e3sconf/202340105090
4. Abdi Yonis S., Yusupov Z., Habbal A., Toirov O. Advances in Electrical and Electronic Engineering.Vol 21. No 3 (2023). https://doi.org/10.15598/aeee.v21i3.5149
5. Toirov O., Khalikov S., Sodikjon Khalikov, F. Sharopov. E3S Web of Conferences 410. 05015. (2023). https://doi.org/10.1051/e3sconf/202341005015
6. Hoshimov F.A., Rakhmonov I.U., Niyozov N.N., Omonov F.B. AIP Conference Proceedings 2647. 030025. (2022). https://doi.org/10.1063/5.0112388
7. Rakhmonov I.U., Najimova A.M., and Reymov K.M. AIP Conference Proceedings 2647. 030010. (2022). https://doi.org/10.1063/5.0104788
8. Rakhmonov I.U., Najimova A.M. AIP Conference Proceedings 2647. 030011. (2022). https://doi.org/10.1063/5.0104791
9. Rakhmonov I.U., Najimova, A.M., Esemuratova Sh.M., Koptileuov T.T. AIP Conference Proceedings 2647. 070024. (2022). https://doi.org/10.1063/5.0104793
10. Rakhmonov I.U., Hoshimov F.A., Kurbonov N.N., Jalilova D.A. AIP Conference Proceedings 2647. 050022. (2022). https://doi.org/10.1063/5.0112391
11. Rakhmonov I.U., Ganiev Sh.R., Alibekova T.Sh., Nematov L.A. E3S Web of Conferences 384. 01029. (2023). https://doi.org/10.1051/e3sconf/202338401029
12. Rakhmonov I.U., Niyozov N.N., Kurbonov N.N., Umarov B.S. E3S Web of Conferences 384. 01030. (2023). https://doi.org/10.1051/e3sconf/202338401030

13. Heikkinen, E. P., & Fabritius, T. (2012). Modelling of the Refining Processes in the Production of Ferrochrome and Stainless Steel. Recent Researches in Metallurgical Engineering-From Extraction to Forming. InTech, 186, 65-88.
14. Niyozov N.N., Rikhsitillaev B.Kh., Jalilova D.A., Kuatbaev P.M. E3S Web of Conferences 384. 01031. (2023). https://doi.org/10.1051/e3sconf/202338401031.
15. Usmanov E.G. E3S Web of Conferences 384. 01034. (2023). https://doi.org/10.1051/e3sconf/202338401034
16. Rasulov A.N., Paxratdinov A.Dj. E3S Web of Conferences 384. 01035. (2023). https://doi.org/10.1051/e3sconf/202338401035
17. Taslimov A.D. E3S Web of Conferences 384. 01036. (2023). https://doi.org/10.1051/e3sconf/202338401036
18. Taslimov A.D., Rakhimov F., Rakhimov Feruz. E3S Web of Conferences 384. 01037. (2023). https://doi.org/10.1051/e3sconf/202338401037
19. Fayziyev M., Tuychiev F., Mustayev R., Ochilov Y. E3S Web of Conferences 384. 01038. (2023). https://doi.org/10.1051/e3sconf/202338401038
20. Saidkhodjaev A.G., Bobojonov Y.M. E3S Web of Conferences 384. 01040. (2023). https://doi.org/10.1051/e3sconf/202338401040
21. Bobojanov M., Mahmutkhonov S. E3S Web of Conferences 384. 01041. (2023). https://doi.org/10.1051/e3sconf/202338401041
22. Rasulov A.N., Ruzinazarov M.R. E3S Web of Conferences 384. 01042. (2023). https://doi.org/10.1051/e3sconf/202338401042
23. Rashidov N., Rozmetov Kh., Rismukhamedov S., Peysenov M. E3S Web of Conferences 384. 01043. (2023). https://doi.org/10.1051/e3sconf/202338401043
24. Tuychiev F, Haqberdiev A. E3S Web of Conferences 384. 01044. (2023). https://doi.org/10.1051/e3sconf/202338401044
25. Bobojanov M., Torayev S. E3S Web of Conferences 384. 01045. (2023). https://doi.org/10.1051/e3sconf/202338401045
26. Palanichamy K.C., Sundar Babu N. Second stage energy conservation experience with a textile industry. Energy Policy 2005.
27. Reymov K.M., Esemuratova Sh.M., Khusanov B.M., Mytnikov A.V. E3S Web of Conferences 384. 01047. (2023). https://doi.org/10.1051/e3sconf/202338401047
28. Kholikhmatov B.B., Samiev Sh.S., Erejepov M.T., Nematov L.A. E3S Web of Conferences 384. 01032. (2023). https://doi.org/10.1051/e3sconf/202338401032
29. Melikuziev M.V. E3S Web of Conferences 384. 01033. (2023). https://doi.org/10.1051/e3sconf/202338401033
30. Bobojonov Y.M., Saidkhodjaev A.G. E3S Web of Conferences 384. 01048. (2023). https://doi.org/10.1051/e3sconf/202338401048
31. Hoshimov, F.A., Najimova A.M., Rafikova G.R., Bakhadirov I.I. E3S Web of Conferences 384. 01049. (2023). https://doi.org/10.1051/e3sconf/202338401049
32. Ruzinazarov M.R. E3S Web of Conferences 384. 01050. (2023). https://doi.org/10.1051/e3sconf/202338401050
33. Bobojanov M.K., Rismukhamedov D.A., Tuychiev F.N., Shamsutdinov Kh.F. E3S Web of Conferences 365. 04024. (2023). https://doi.org/10.1051/e3sconf/202336504024
34. Ushakov, V.Y., Mytnikov, A.V., Rakhmonov, I.U. Past, Present and Future of Electric Power Systems (Brief Excursion). In: High-Voltage Equipment of Power Systems. Power Systems. Springer, Cham. (2023). https://doi.org/10.1007/978-3-031-38252-9_1
35. Ushakov, V.Y., Mytnikov, A.V., Rakhmonov, I.U. Characteristics of the Main Elements of Electric Power Systems. In: High-Voltage Equipment of Power Systems. Power Systems. Springer, Cham. (2023). https://doi.org/10.1007/978-3-031-38252-9_2
36. Ushakov, V.Y., Mytnikov, A.V., Rakhmonov, I.U. Insulating Materials and Media Used in High-Voltage Elements of Electric Power Systems. In: High-Voltage Equipment of Power Systems. Power Systems. Springer, Cham. (2023). https://doi.org/10.1007/978-3-031-38252-9_3
37. Ushakov, V.Y., Mytnikov, A.V., Rakhmonov, I.U. Effects on Equipment Causing Insulation Aging and Failure. In: High-Voltage Equipment of Power Systems. Power Systems. Springer, Cham. (2023). https://doi.org/10.1007/978-3-031-38252-9_4
38. Maslov D.V., Development of algorithms and control systems for arc steel-melting furnaces that reduce electrode breakdowns. PhD Thesis, 2014.
39. Hoshimov F.A., Rakhmonov I.U., Niyozov N.N. E3S Web of Conferences 209. 07017. (2020). https://doi.org/10.1051/e3sconf/202020907017
40. Sattarov Kh., Sapayev M., Suyarov A., Turaev, A. E3S Web of Conferences, 401, 04001, (2023). https://doi.org/10.1051/e3sconf/202340104001
41. Jurayeva K., Sattarov Kh. E3S Web of Conferences, 389, 01060. (2023). https://doi.org/10.1051/e3sconf/202338901060
42. Sattarov Kh., Safarov A., Kamolov A., Egamberdiev M., Raxmanova G. E3S Web of Conferences, 383, 01028. (2023) https://doi.org/10.1051/e3sconf/202338301028
43. Tsypkina, V., Ivanova, V. E3S Web of Conferences, 139, 01073. (2019). https://doi.org/10.1051/e3sconf/201913901073
44. Ivanova, V., Madrakhimov, D. E3S Web of Conferences, 139, 01057. (2019). https://doi.org/10.1051/e3sconf/201913901057
45. Pirmatov N. B., Madrakhimov D. B., Ivanova V. P., Tsypkina V. V., Ortikova M. J., Atamukhamedova R. F. AIP Conference Proceedingsthis link is disabled, 2552, 070009. (2022). https://doi.org/10.1063/5.0117025

Automatic control system of greenhouse soil moisture using solar photoelectric systems research

Bakhramzhan Rasakhodzhaev[1,3*], *Zhao* Mingzhi[2], *Usmonjon* Akhmadjonov[1], *Yung* Jie Liu[2], *Yu* Rong[2], *Jasur* Shoimov[1]

[1]National Research Institute of Renewable Energy Sources 100084, Uzbekistan, Tashkent, Chingiz Aytmatov St. 2B, Building 2.

[2]Inner Mongolia University of Technology, People's Republic of China, Hohhot, Ai Min Street, No. 49, P.R.C

[3]Tashkent State Technical University, 100095, Uzbekistan, Tashkent, Almazar district, Universitetskaya st. 2A.

Abstract. This research work is devoted to the study of the system of automatic control of soil moisture in the greenhouse using photoelectric systems. In this regard, work continues to modernise existing and develop new designs of heliothermal greenhouses, in particular, energy-efficient greenhouses with transformable enclosure, in which the volume can be changed by transforming the enclosure. The main objective of the research work is to create an efficient way to save energy and water consumption in greenhouses. The use of solar water pump sprinklers in the soil irrigation system inside the greenhouse can reduce water consumption and electricity consumption. A special sprinkler device is used to spray water on the irrigated area to reduce water consumption. The combined sprinkler is an environmentally friendly and economical device used to minimise electricity and water consumption. Improved systems for saving and rational use of electricity and water resources were used in our research work.

1 Introduction

Ensuring the rational use of water, its quality and the safety of water management facilities in our country, as well as reducing water consumption, involve collective (complex) measures to introduce modern innovative accounting systems, along with consistent reforms in the field of groundwater extraction.New structures of water resource management are being created to effectively regulate water use [1].

Water pumps have a significant role in water production and storage. Nowadays, water pumps are widely utilized for water extraction and storage. They find extensive use in pumping water from wells, filtering water bodies and transporting water over long distances. The working principle of the water pump primarily relies on the principle of positive water displacement and kinetic energy [2].

In this regard, work is continuing on modernising existing and developing new heliothermal greenhouse designs, particularly in energy-efficient greenhouses with a transformable enclosure, in which the volume can be changed by transforming the enclosure.

The use of solar water pump sprinklers in the soil irrigation system helps minimize water consumption and reduce electricity usage. Sprinklers are utilized to spray water on the irrigation field, thus reducing water consumption. Combined sprinklers are environmentally friendly and economical devices aimed at minimizing electricity and water consumption [3]. The effect of the soil heat field on changes in the air inside the greenhouse was explored in scientific works [4,5]. Alterations in the air inside the greenhouse directly affect the soil temperature. Soil temperature changes were studied in the upper layers at 5, 10, 15, and 20 cm depths [6]. In scientific research, the impact of soil temperature on the water heating device placed on the soil was also considered.

Studies of radiation and thermal engineering calculations of solar greenhouses are considered in [5].

In [6], a thermal model of ground-mounted solar water-heating collectors is developed. To determine them, a non-stationary unit thermal model of solar water heater is proposed, which allows to study the dynamics of heating temperatures of the main elements of SVC (transparent enclosure, receiver, heat-insulating ground bottom).

In the work [7] presented research, the choice of heating system on the basis of calculations of economic efficiency and payback period of alternative systems, solar greenhouse with a transformable enclosure. Depending on the shape of the greenhouse, the total costs, economic efficiency and payback period are determined. The conducted research work shows that the developed greenhouse with transformable (adjustable) body is quite acceptable for its successful use among farmers and private farms in the Republic of Uzbekistan in terms of construction cost and material consumption. Calculation of economic efficiency and payback period of greenhouses with transformable enclosure

* Corresponding author: rashodjaev@mail.ru

https://doi.org/10.1051/e3sconf/202346101080

allows to choose the most acceptable heating system and technical characteristics of alternative systems acceptable for climatic conditions.

Scientists from the National Research Institute of Renewable Energy are conducting numerous scientific studies on the development and implementation of energy-efficient water management systems. Our institute's researchers have designed a 6 kW photoelectric module system for replenishing water supply reservoirs for consumers. By implementing this system, the electricity consumption of water supply facilities can be reduced by up to 70%, resulting in high efficiency [8].

The purpose of this research work is to develop an automated system for irrigating crops in a reenhouse, utilizing power from an autonomous photoelectric station, and effectively controlling the water quantity for watering plants while saving electricity.

2 Methods and materials

Electricity is supplied to the automated system through a photovoltaic station (FES). The electricity generated by the photovoltaic plant is used to charge batteries, ensuring the system's operation during the day and providing continuous power supply at night. Moisture sensors are employed to measure soil moisture levels and transmit signals to the control unit. To achieve effective crop irrigation, the soil composition is monitored using a moisture sensor until it reaches the desired moisture level.

According to KMK 2.09.08.07 "Greenhouses and Polytunnels" Table 5, recommended conditions for growing vegetable crops in greenhouses are as follows: cucumbers (autumn turnover) - overcast day, daytime temperature 22-23 °C, nighttime temperature 17-18 °C; tomatoes (winter-spring turnover) - overcast day, daytime temperature 19-20 °C, nighttime temperature 16-17 °C [8,9,10,11].

Such thermal regimes can be widely used on farms engaged in the cultivation of plants or vegetables with the assistance of greenhouses equipped with heating systems.

This system was developed for use in the cultivation of medicinal and valuable plants. One such plant is BLUEBERRY, which has a rich vitamin C content, along with numerous other beneficial vitamins and trace elements [12]. The average price of blueberries, a fruit-bearing bush that grows in northern swamps, is 49,000 soums for a 150-gram package, which translates to approximately 327,000 soums per kilogram. In other words, at the current exchange rate, 1 kilogram is nearly equivalent to 37.6 US dollars. However, its retail price in Ukraine is much cheaper, ranging from 9 to 12 per kilogram [5]. Therefore, there is potential to expand the cultivation of this fruit on a local scale. To achieve this, seeds should be germinated with watering at a soil temperature of 23-25°C and soil moisture levels up to 40%. Maintaining a moderate air temperature and soil moisture is of great importance when caring for seedlings. We propose the following innovative solutions for the reliable utilization of this system.

Variation of soil temperature fluctuation amplitude depending on depth, namely the reduction of soil temperature fluctuation amplitude [13,14,15,16,17], has been studied in scientific works. As a result, it was determined that the influence of the variation of the fluctuation amplitude of the temperature of the soil on the devices located on the upper part of the soil is very small.

Controlling soil moisture in a greenhouse and having an effective irrigation system are of great importance in influencing the crops grown in the greenhouse.

Below is a block diagram of an automated irrigation system based on a photoelectric system in Figure 1.

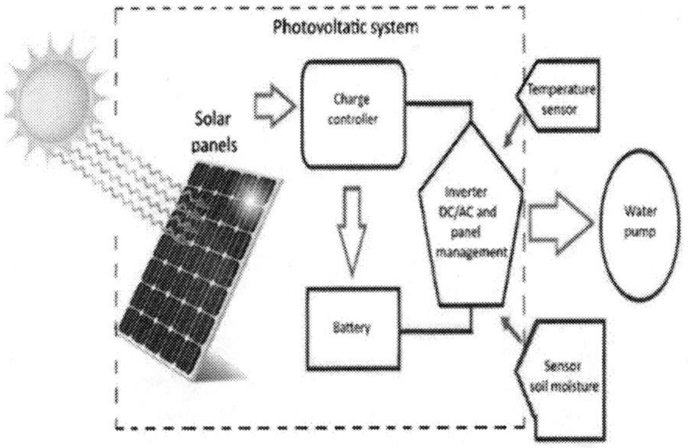

Figure 1. Automatic irrigation based on photoelectric modules block diagram of the system.

As can be seen from Figure 1, the water pump is controlled by electronic devices located in the control panel of the photoelectric system. The water pump irrigates the allocated area, and the uniform distribution of water is based on the data received from the moisture sensors. Thus, the automated system, based on the photoelectric station, is supplied with electricity, and with the help of an automated block diagram, continuous water is supplied to the irrigated area [25,26,27] .

Names and specifications of the required devices are listed in Table 1.

Table 1

№	Name of devices	number, piece	Characteristic
1.	Photoelectric modules	6	410 W
2.	1-phase machine	3	220V/32A
3.	Inverter (hybrid)	1	2 kW
4.	Humidity measuring sensor (XN-M214)	2	4...20mA, DC 12V, RH 20-90% RH ±1.3%
5.	Microcontroller	1	Atmega 328
6.	Limiter (relay)	1	220V
7.	Water pump	1	220V, 50 Nz, 1.5 kW
8.	Water storage capacity	1	500 liters

Figure 2 below shows the principle diagram of the electronic devices placed in the control panel of the Photoelectric system.

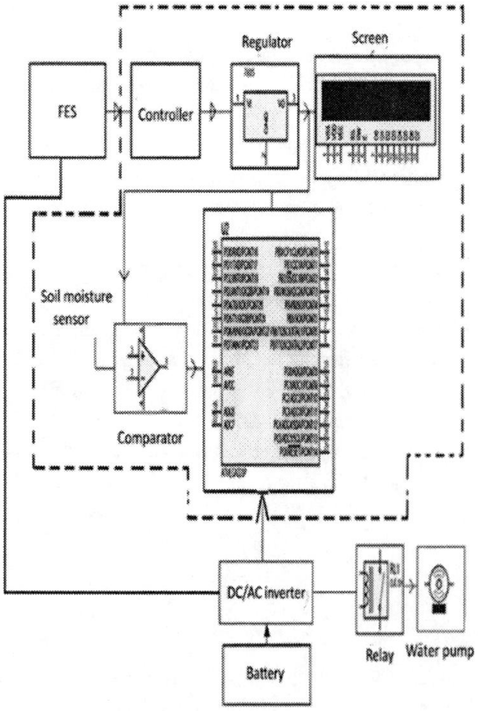

Figure 2. The principle diagram of the electronic devices placed in the control panel of the photoelectric system.

Figure 2 shows that the FES, the control circuit, and the water pump are connected in series. The elements listed in the control scheme: voltage and current control, data output screen, comparator (microcircuit), inverter, relay block - serve to ensure the continuity of water supply to the irrigated area.

The difference between the control scheme of this automated irrigation system and other similar irrigation systems is that it uses an autonomous photoelectric system, which provides continuous energy and water supply, as well as drip irrigation to achieve energy and water savings of up to 70%. As a result, irrigating crops in the greenhouse with the help of an autonomous photoelectric system that automatically controls soil moisture reduces water wastage and saves electricity, ensuring uninterrupted electricity and water supply.

3 Conclusion

The following conclusions can be drawn based on the conducted scientific research:

1. The use of solar water pump sprinklers in the greenhouse soil irrigation system is employed to minimize water consumption and reduce electricity usage.
2. Integrated (combined) systems, such as sprinklers, are environmentally friendly and cost-effective, serving to minimize both electricity and water consumption.
3. The automated control system, based on a photoelectric plant, effectively ensures the continuity of water supply to the irrigated area.

In summary, this research work has led to the development of an automated system powered by a photoelectric station, taking soil moisture into account for greenhouse crop irrigation. This system allows for the prevention of water wastage, efficient water management during plant irrigation, and electricity savings.

References

1. https://lex.uz/docs/-4611198.
2. S.Senthil Kumar Chidambaranathan Bibin K. Akash K. Aravindan M. Kishore G. Magesh Materials "Solar powered water pumping systems for irrigation: A comprehensive review on developments and prospects towards a green energy approach" Proceedings Volume 33, Part 1, 2020, Pages 303-307.
3. Mansur Aliyua Ghassan Hassanab Syed A. Saida Muhammad U. Siddiquic Ali T. Alawamid Ibrahim M. Elamind "A review of solar-powered water pumping systems"Renewable and Sustainable Energy Reviews Volume 87, May 2018, Pages 61-76.
4. Ismanzhanov, AI,Klychev, Sh.I.,Rasakhodzhaev, BS,Tokonova, TS,Borodin, DVThe heat efficiency of solar ground water-heating collectors.Applied Solar Energy (English translation of Geliotekhnika), 2011, 47(4), str. 318–320.

5. Ismanzhanov, AI,Rasakhodzhaev, BSInvestigation of the soil thermal field in a soil solar water heating collector,Applied Solar Energy (English translation of Geliotekhnika), 2011, 47(2), pp. 134–136.
6. Klychev, SI,Rasakhodzhaev, BS,Akhadov, ZZ,Akhmadjonov, UZ,Adylov, CAStudy of the Thermal Regime of Solar Greenhouses for Individual Purposes for Their Design Features.Applied Solar Energy (English translation of Geliotekhnika), 2022, 58(1), pp. 121–126.
7. Khalimov A.G., Khairiddinov B.E., Kim V.D. Radiation characteristics of the solar greenhouse //Heliotekhnika.2009. No. 1. S.22-25.
8. Ismanzhanov, A.I., Klychev, Sh.I., Rasakhodzhaev, B.S., Tokonova, T.S., Borodin, D.V. The heat efficiency of solar ground water-heating collectors. Applied Solar Energy (English translation of Geliotekhnika), 2011, 47(4), pp. 318–320.
9. Rasakhodzhaev, B., Makhmudov, S., Muminov, F. Selection of a heating system based on climatic conditions of Uzbekistan and on calculations of the technical and economic indicators of alternative systems: A case study of the solar greenhouse with a transformable building. IOP Conference Series: Earth and Environmental Sciencethis link is disabled, 2021, 939(1), 012003.
10. J.Z. Akhadov, B.S. Rasakhodjaev, U.Z. Akhmadjonov "Research of water extraction facilities using solar photoelectric systems" The role of advanced innovative technologies and education in the solution of automation and energy problems aimed at increasing the energy efficiency of economic sectors and the social sector. Namangan, 2021. S 760.
11. KMK 2.01.01-94 RUz.Klimaticheskiei fiziko-geologicheskie dannye dlya Proektirovaniya / Goskomarkhiturastroy RUz.Tashkent. Tipo im. Ibn Sina, 1994 -28 p.
12. KMK 2.09.08-97 "Teplitsy i parniki" / Ministry of Construction of the Republic of Uzbekistan - Tashkent, 1997y.
13. KMK 2.04.05-97*. Heating, ventilation and air conditioning / UzR Period architecture - Tashkent, AQATM, 2011-212 p.
14. KMK 2.01.04-2018 Construction thermal engineering / Ministry of Construction of the Republic of Uzbekistan - Tashkent, 2018 - 105 pages.
15. https://xs.uz/uzkr/post/ozbekistonda-movij-qimmatbaho-reza-meva-golubika-etishtirish-zhozibali-va-daromadli-biznesga-ajlanishi-mumkinmiflower.onego.ruhttp://flower.onego.ru/kustar/vacci_g.html) Golubika.
16. Ismanzhanov, AI,Klychev, Sh.I.,Rasakhodzhaev, BS,Tokonova, TS,Borodin, DVThe heat efficiency of solar ground water-heating collectors.Applied Solar Energy (English translation of Geliotekhnika), 2011, 47(4), pages 318–320.
17. Ismanzhanov, AI,Rasakhodzhaev, BSInvestigation of the soil thermal field in a soil solar water heating collectorApplied Solar Energy (English translation of Geliotekhnika), 2011, 47(2), pages 134–136.
18. Ismanzhanov, AI,Rasakhodzhaev, BSInvestigation of comparative thermotechnical performance of ground solar water-heating collectors,Applied Solar Energy (English translation of Geliotekhnika), 2008, 44(2), pages 87–89.
19. Ismanzhanov, A.,Rasakhodzhaev, BSStudies of the effect of seasonal and daily variations of a solar radiation spectrum on the results of the mirror accelerated tests.Applied Solar Energy (English translation of Geliotekhnika), 2000, 36(2), pages 53–56.
20. Ismanzhanov, A.,Rasakhodzhaev, BSInvestigation of influence of seasonal and daily variations of solar radiation spectrum on accelerated mirror test results. Geliotekhnika, 2000, (2), pages 60–63.
21. https://uzsuv.uz/uz/posts/yovgashti.
22. Malevsky Yu.N. Solnechnaya Energetica. - M.: Mir, 2001. -196p.
23. Santiago Silvestre, Chouder Ayssa, "Vliyanie zatenenia na rabotu fotoelektricheskogo modulya", master's thesis at Universitete GAVLE, June 2015.
24. Mamedov F. F. "Rezultaty experimentov i raschet energii sistem slezheniya za solntsem na solnechnyx ustanovkax". International Journal of Energy Engineering 2017, 7 (1): 28-31.
25. Ismanzhanov A.I., Rasakhodzhaev B.S. Investigation of the temperature field in the soil of ground solar water-heating collectors. Solar engineering, - Tashkent, 2011. - No. 2. - P.56-58.
26. Klychev Sh.I., Ismanzhanov A.I., Bakhramov S.A., Dudko Yu.A., Samiev M. Convective heat transfer in interlayers of solar "hot boxes" // Geliotekhnika, - Tashkent, 2008. No. 2. - S. 41 - 45.
27. Sadykov G.A., Vardiashvili A.B. Solar greenhouse and their thermal calculation. Tashkent.: Fan, 1977.-79s.
28. Ismanzhanov A.I. Renewable and non-traditional energy: Explanatory dictionary of terms. - Osh.: Kyrgyz-Uzb. university, 2009.-158 p.
29. Rasakhodzhaev B.S., Adylov Ch.A., Tokonova T.S., Rayymbaev Zh.Ch. Evaluation of the energy efficiency of solar greenhouses with a transformable (adjustable) body. The science. Education. Technics. No. 3 (69). Osh.2020. - P.36-45.

Use of composite alloys for two stator-valve electric motors

Usan Berdiyev[1], *Utkirbek* Sulaymonov[1]*, *Zilola* Mirakhmedova[1]

[1]Tashkent State Transport University, 100167, Uzbekistan, Tashkent, Temiryolchilar Street, 1

Abstract. The article discusses the methods of using composite soft magnetic alloys for two stator-valve motors of electric transports in railways. The main groups of alloys are studied, which sharply differ in the shape of the hysteresis curve and the values of the main magnetic characteristics. Variable magnetization applied as cores of electric transports. It has been established that structural changes causing the rise of mechanical and magnetic hardness of alloys.

1 Introduction

The technique for applying insulating coatings based on phosphorus oxide was developed in relation to the tasks of this work based on the method of manufacturing a composite soft magnetic material [5, 8, 12], which actually proposed a technique for adding metal powder to the original in a rotating evacuated drum at a pressure of 0.15 - 1.5 Pa, heated to a lubricant melting temperature of 150-200°C and the treatment is carried out until a uniform distribution of the lubricant is obtained in the material for 15 - 30 minutes, while the lubricant content in the composite material ranges from 0.01 to 0.1%.

From calculated, according to formula (1), and experimental data, it was established that, depending on the thickness of the insulating coating and the particle size of the initial metallic iron powder, the resulting composite materials can be divided into two classes: low-frequency ($f < 1$ kHz) and high-frequency ($f > 1$ kHz). For subsequent high-frequency applications, powders with a size of less than 100 microns are screened out using a separator; for low-frequency applications, powders with a size of more than 100 microns are screened out. After separation, an insulating layer was applied to the surface of the iron particles.

In the first stage, the lubricant is introduced by mixing it with isolated metal powder in ball mills or other mixers. The composite material prepared with lubricant is placed inside the drum. After this, the vacuum pump is turned on, reducing the air pressure inside the drum with the material to 0.15 - 1.5 Pa. When the required vacuum is achieved, the drum is heated to a temperature of 150 – 200°C. The powder processing process continues for 15 – 30 minutes. Cooling of the soft magnetic material occurs after turning off the heating while maintaining a vacuum.

The advantage of this method compared to the known ones is the reduction of the lubricant content to 0.01 -0.1%, which makes it possible to subsequently obtain a high-quality pressed product from a soft magnetic composite material with a density of 7.5 - 7.65 g/cm^3 and, as a result, has high magnetic characteristics [7, 8, 10].

Most railway electric vehicles use valve motors, which are used to rotate the main equipment. Some elements of these electric motors are made of composite soft magnetic alloys [1, 4].

As it is known, only three metals have ferromagnetism, i.e. the ability to significantly thicken magnetic field lines, which is characterized by magnetic permeability: for example, iron, nickel, cobalt. The relative magnetic permeability of ferromagnetic metals reaches tens and hundreds of thousands of units. For the rest it is close to unity. If the relative permeability is slightly greater than unity, it is paramagnetic. If it is less than unity, it is diamagnetic [1, 3, 6].

Recently, cores made of composite powder materials with the necessary magnetic characteristics have been widely used. in various devices operating in alternating magnetic fields, At the same time, they have relatively small eddy current losses and high electrical resistivity. The required characteristics are achieved by doping iron-based materials with phosphorus or silicon with the introduction of special dielectric layers from organic and inorganic additives [2, 3, 6].

This makes them competitive with traditionally used laminated electrical steel cores. The use of powder metallurgy methods when molding products of complex shapes, in addition, makes it possible to obtain isotropic materials that change the nature of magnetization, as well as the direction and distribution of the magnetic flux. Composite materials with the required set of performance characteristics are widely used in components of various mechanisms [2, 5].

Certain properties of solids are used in any technical applications. Such as electrical, magnetic, optical, thermal, mechanical, corrosion-resistant, etc. One of the pressing problems of industry today is the production of magnetic materials with low energy losses during magnetization reversal. Despite the fact that research and development of such materials have been carried out

* Corresponding author: sulaymonov_2020@bk.ru

since the beginning of the last century, studying the mechanism of magnetization reversal and improving the quality of these materials is still relevant today.This is due to the fact that magnetic materials are widely used in various electrical devices (generators, electric motors, measuring installations, inductors, etc.) [4, 6, 7].

2 Experimental research

The purpose of this work is to develop the main elements of two stator-valve motors using composite alloys.

Figure1 shows magnetization curves and magnetic properties of alloys. As indicated in the diagram that curve 2 is the initial magnetization curve [5, 8, 9], curve 1 shows the change in magnetic induction depending on the field strength during subsequent magnetization and demagnetization.

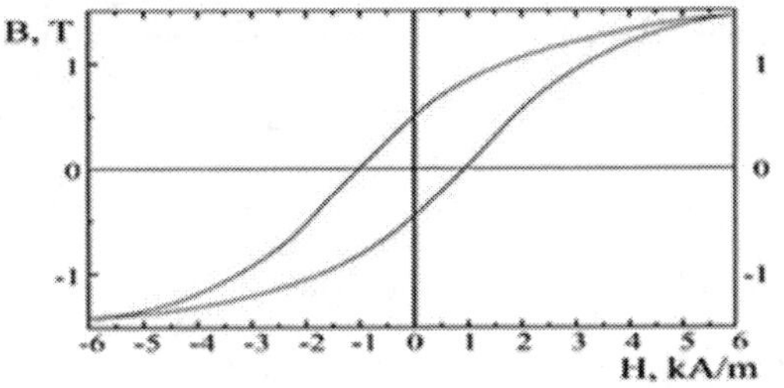

Fig.1. The main magnetization curves of magnetic alloys: 1-hysteresis; 2-primary; + B and –B - magnetic induction; –H and + H - field strength.

The main area bounded by this curve, which is called the hysteresis loop, represents the so-called hysteresis losses, i.e. energy expended on magnetization. The most important are the following magnetic characteristics, determined from the magnetization curves. It is these characteristics that are shown in Fig. 1 with magnetization curves [5, 6, 8].

The diagram shows that the residual induction Br is the magnetic induction remaining in the experimental sample after it is magnetized and the magnetic field is removed (this is measured in gauss G). Next comes the coercive force Hc, the field strength that must be applied to the test sample in order to demagnetize it (this is measured in oersteds E) [6, 7].

As can be seen from the main course of the initial magnetization curve, the magnetization intensity changes with the change in field strength. The magnetization intensity is proportional to the tangent of the tangent to the initial magnetization curve and is numerically equal to the B/H ratio. The intensity of magnetization is called magnetic permeability, and magnetic permeability in very weak fields is called initial magnetic permeability, the dimension of magnetic permeability is G/E, which is measured in Gaussians and Oersteds [8, 10, 12].

According to their magnetic characteristics, magnetic alloys are divided into two groups: 1-hard magnetic alloys; 2 - soft magnetic alloys. They differ sharply from the shape of the hysteresis curve and the values of the main magnetic characteristics. Hard magnetic alloys are characterized mainly by a large Hc value and are used for permanent magnets, while soft magnetic alloys are characterized by a low Hc value and small hysteresis losses and are used as alloys subjected to alternating magnetization [8, 11]. Hysteresis curves of lines of soft magnetic alloys are shown in Fig. 2.

There are special groups of alloys with high initial magnetic permeability that must be intensively magnetized in weak fields. Alloying a metal causes an increase in magnetic hardness. If only a solid solution is formed (in iron or in another ferromagnetic metal), then the magnetic hardness, i.e. coercive force increases slightly; the formation of the second phase during doping actively increases the coercive force. The higher the dispersal of the second phase in the alloy, the higher its coercive force [7, 8, 10].

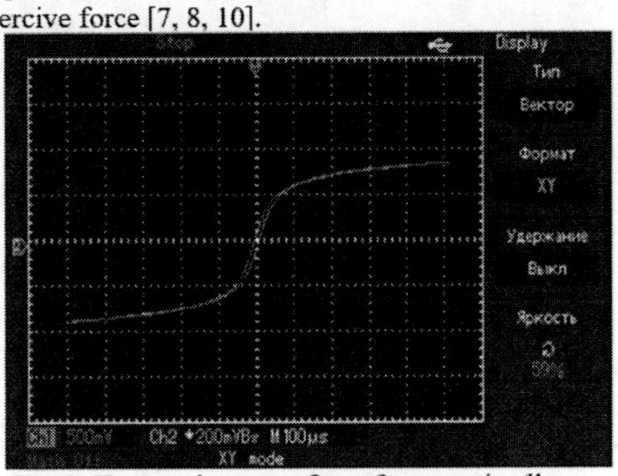

Fig.2. Hysteresis curves for soft magnetic alloy.

The main stresses in the lattice caused by cold hardening or phase transformations, grain refinement and other deviations from the equilibrium state cause an increase in the coercive force [4, 7]. This means that changes in the structure that cause an increase in mechanical hardness, namely, increase magnetic hardness. This justifies the use of terms such as magnetic hardness and magnetic softness of materials.

In contrast to hard magnetic materials - alloys for permanent magnets, where a high coercive force is required, a large group of magnetic alloys are represented by the so-called soft magnetic alloys, which must first of all have a low coercive force [8, 10]. For example, if high magnetic hardness was achieved by obtaining non-equilibrium and highly dispersed structure but, for obtaining magnetic softness it is necessary to get as close as possible to the equilibrium state, and it is also necessary to obtain large grains and eliminate sources causing lattice distortions and fragmentation of blocks.

In addition to low coercive force, magnetic materials must also have high magnetic permeability in weak, medium or strong fields, and low magnetization reversal losses [5, 7, 10].

In our studies, we took the most suitable soft magnetic material as pure metals, primarily pure (technically pure) iron. In some limited cases, alloys not only based on iron, but also other metals, for example, nickel and cobalt, as well as electrical steel were used [6, 9, 12].

Technical iron, in which all impurities, especially carbon, are harmful and therefore their content is strictly limited according to Construction Standards [7, 8, 11].

Industrial production produces three grades of technical iron based on chemical composition, each of which in turn is divided into grades based on magnetic characteristics. The chemical compositions of technical iron are given in Table 1, and the magnetic properties in Table 2. This table presents the grades of iron, the main chemical elements and magnetic properties of these materials [7, 8]. These materials are used in electrical installations and various elements and parts of two stator-valve motors are made from them.

Table 1 The chemical composition of technical iron

Brand gland	Element content, %					
	C	Si	Mn	C u	P	S
E	≤ 0,04	≤ 0,2	≤ 0,2	0,15	≤ 0,025	≤ 0,3
EA	≤ 0,04	≤ 0,2	≤ 0,2	0,15	≤ 0,025	≤ 0,3
EAA	≤ 0,04	≤ 0,2	≤ 0,2	0,15	≤ 0,025	≤ 0,3

Table 2. Magnetic properties of technical iron

Brand gland	Magnetic properties			
	Coercive strength Ns,uh	Maximum magnetic permeability, G/E	Magnetic induction, gf	
			B10	B25
E	1,2	3500	15000	16200
EA	1,0	4000	15000	16200
EAA	0,8	4500	15000	16200

The magnetic properties of iron (except for its purity) also depend on its structural state. Hardening sharply worsens the magnetic properties, while grain coarsening improves them [10, 12]. In common industrial grades of iron, coercivity is obtained on the order of 1E or slightly lower, while a minimum coercivity value of 0.01· E is obtained on very coarse-grained pure iron. To obtain large grains and eliminate work hardening, the metal is annealed at high temperatures. Technically pure iron was used for the manufacture of cores, relays and direct current electromagnets, magnetic screens, poles of electrical machines and other transport parts [11, 13].

3 Research results

Research has shown that slow cooling is necessary to obtain acicular ferrite. With slow cooling, polyhedral ferrite is obtained (Fig. 3, a). With rapid cooling, a needle-type structure is obtained - needle-shaped ferrite (Fig. 3, b). The hardness of acicular ferrite is 100-150 HB higher than the hardness of polyhedral ferrite.

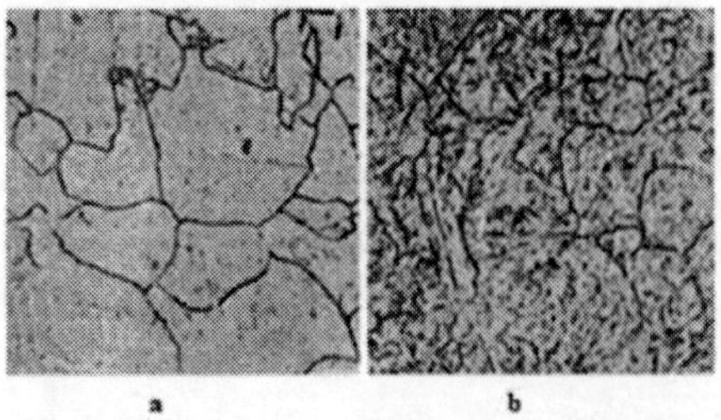

Fig.3. Ferrite microstructure: a-polyhedral ferrite, slow cooling; b-needle ferrite, very fast cooling. X400

Electrical steel is a ferrite alloy of iron with silicon and other alloying elements (Fig. 4). This graph shows almost all alloying elements of the alloys [12].

The iron-silicon solid solution, due to distortions in the lattice caused by the presence of foreign silicon atoms in it, has a higher coercive force than pure iron.

However, in this alloy, when heated, it is possible to obtain large grains, which do not become crushed when cooled, since there is no γ-α transformation, and this in practice leads to the fact that the value of the coercive force in such a material is no greater than in ordinary iron. The higher electrical resistance of silicon-doped ferrite reduces losses due to Foucault currents [8, 11]. Naturally, a change in the dimensions of especially the α-lattice also causes a change in the properties of ferrite - the strength increases, and the ductility decreases. Changes in the properties of ferrite, for example, hardness or toughness when various elements are dissolved in it (see Fig. 4). As can be seen from the diagrams, chromium, tungsten, and molybdenum strengthen ferrite less than nickel, silicon, and manganese. Molybdenum, tungsten, as well as silicon and manganese (if present in more than 1%) reduce the viscosity of ferrite. Chromium reduces the viscosity much more weakly than the listed elements, and nickel does not reduce the viscosity of ferrite, but on the contrary slightly increases it [12, 13].

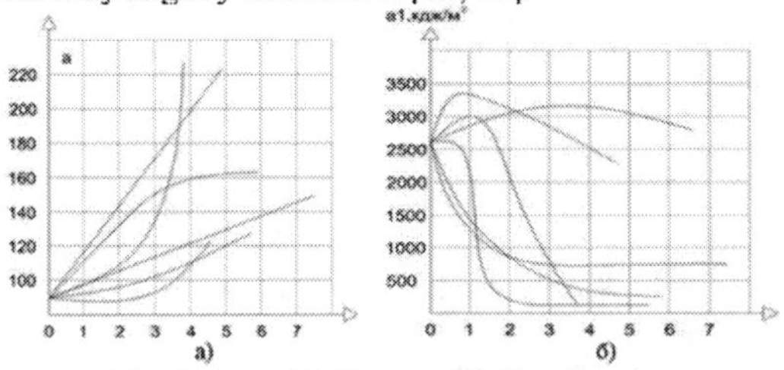

Fig.4. Actual influence of alloying elements on ferrite properties: a-hardness; b-toughness.

Electrical steel is made in the form of thin sheets, which are used for the manufacture of transformer cores, magnetic cores of electrical machines and alternating and direct current devices. This sheet electrical steel (Construction Standards 21473-75) is divided by product range - mainly by thickness, production method - cold-rolled and hot-rolled sheet, degree of anisotropy, as well as basic magnetic characteristics - magnetic induction and specific losses and degree of silicon alloying [11, 14].

It should be added that the specific losses due to magnetization reversal are lower, the thinner the sheet, therefore electrical steel is manufactured only in the form of thin sheets with a thickness of 0.35 to 0.50 mm. If during the manufacturing process of transformer parts, the steel was subjected to even slight plastic deformation, for example, cutting sheets, bending, etc., then the magnetic properties deteriorate. To restore magnetic properties, it is recommended to carry out annealing to relieve stress (eliminate distortions in the lattice) at a temperature of 750-800°C with slow (<50°C/h) cooling.

This means that to restore deformable magnetic properties, the optimal heat treatment mode is used - annealing, which is carried out at the above temperature [11, 14]. It must be remembered that all manufactured elements or parts of transformers, magnetic circuits of electrical machines and alternating and direct current devices are subjected to heat treatment, namely annealing together with a furnace. To do this, you can choose an electric furnace of the SNOL type with a heating temperature of up to 1000°C [9, 12, 13].

In connection with the above, it can be said that currently at the Tashkent State Transport University, scientific research is being carried out to develop the use of composite soft magnetic alloys for two stator-valve motors of electric vehicles.

4 Conclusion

The results of testing magnetic cores made of composite soft magnetic material using the developed method showed the following characteristics:

1. Magnetization reversal losses for magnetic cores P = 90 W/kg at a frequency of 1 kHz.

2. Hysteresis losses per cycle for components based on the new composite material do not depend on frequency and amount to p = 0.08 J/cycle.

3. A decrease in the thickness of insulating coatings leads to an increase in the induction value at 20 kA/m from Bm = 1.7 Tesla to Vm = 1.85 Tesla and to a corresponding increase in hysteresis losses from P = 80 W/kg to P = 90 W/kg at a frequency of 1 kHz.

References

1. Rahman, K.M.; Patel, N.R.; Ward, T.G.; Nagashima, J.M.; Caricchi, F.; Crescimbini, F. Application of direct-drive wheel motor for fuel cell electric and hybrid electric vehicle propulsion system. IEEE Trans. Ind. Appl. 2006. - P.42, 1185-1192.
2. Chau, K.T.; Chan, C.C.; Liu, C. Overview of Permanent-Magnet Brushless Drives for Electric and Hybrid Electric Vehicles. IEEE Trans. Ind. Electron. 2008. - P.55, 2246-2257.
3. Berdiev, UT , Kolesnikov, IK , Tuychieva, MN , Khasanov, FF , Sulaymonov, UB Methods of new technological developments of electric motors based on soft magnetic materials . E3S Web of Conf. Volume 401, 2023. 03038. https://doi.org/10.1051/e3sconf/202340103038
4. Usan Berdiev, Abitkhodja Burkhankhodzhaev, Malika Tuychieva. Investigation of energy indicators with asymmetry of the voltage of the power source of mainline electric locomotives of alternating current. Cite as: AIP Conference Proceedings 2552, 030018 (2023); https://doi.org/10.1063/5.0131839 Published Online: 05 January 2023
5. Optimization of the method of oxide coating of metallic iron powder particles. Usan Berdiyev, Olga Demedenko, Mirjalol Ashurov, FF Hasanov and UB Sulaymonov. E3S Web Conf., 383 (2023) 04039. DOI: https://doi.org/10.1051/e3sconf/202338304039
6. Wang, J.; Patel, V.I.; Wang, W. Fractional-Slot Permanent Magnet Brushless Machines with Low Space Harmonic Contents. IEEE Trans. Magn. 2013- P.50,1-9.
7. Y.M.Dovgalevsky. Alloying and heat treatment of hard and soft magnetic alloys. – M.: Metallurgizdat, 2000. - 178.
8. Govor G.A., Vecher A.K., Berdiev U.T., Pirmatov N.B., Karabaev A., Khasanov F.F. Magnetically soft iron-based materials used in electrical engineering, Bulletin of TashIIT, 2019. No. 3. - 212-218 p..
9. Govor G.A., Mikhnevich V.V., Mityuk V.I., Berdiev U.T., Pirmatov N.B., Berdiyorov U. Magnetically soft materials based on iron. Problems and prospects of innovative equipment and technologies in the agricultural and food sector // Collection of scientific papers of the International scientific and scientific-technical conference. - Tashkent. Tashkent State Technical University, 2020. 76-77 p.
10. Govor G.A., Berdiev U.T., Pirmatov N.B., Urakov S. Magnetic soft materials based on iron. Problems and prospects of innovative equipment and technologies in the agricultural and food sector // Collection of scientific papers of the International Scientific and Scientific-Technical Conference. -Tashkent. Tashkent State Technical University, 2020. 85-86 p.
11. Rahman, K.M.; Patel, N.R.; Ward, T.G.; Nagashima, J.M.; Caricchi, F.; Crescimbini, F. Application of direct-drive wheel motor for fuel cell electric and hybrid electric vehicle propulsion system. IEEE Trans. Ind. Appl. 2006, -P.42,1185-1192.
12. Cao, R.; Mi, C.; Cheng, M. Quantitative Comparison of Flux-Switching Permanent-Magnet Motors with Interior permanent magnet motor for EV, HEV, and PHEV Applications. IEEE Trans. Magn. 2012. - P.48, 2374-2384.
13. Kolesnikov I.K. Criteria for the Efficiency of an Electric Drive based on the mechanical conversion of Energy to be completed. Oriental Renaissance: Innovative, educational, natural and social sciences. 2023-yil, march, 737-743-p
14. .Kolesnikov, I., Yaronova, N., Ametova, A. Building a "Smart" Electric Network Using 5G and Unmanned Aircraft AIP Conference Proceedings, 2022, 2432, 030037
15. Abduraxman Akhmedov, Galina Sauchuk, NatalliaYurkevich, Sardorbek Khudoyberganov, MahammatyakubBazarov, KarimberdiKarshiev; The influence of production conditions on the electrophysical parameters of piezoceramics for different applications. E3S Web Conf. 264 04020 (2021), DOI: 10.1051/e3sconf/202126404020
16. R. Mirsaatov, S. Khudoyberganov, A. Akhmedov; Uncertainty estimation in determination of Cocoons silkiness by thickness of their shell. AIP Conference Proceedings 15 March 2023; 2612(1):050010. https://doi.org/10.1063/5.0114683

The possibilities of application of cogeneration plants at the boiler houses in Uzbekistan

Mekhriya Koroli[1]*, *Lola* Nizamova[1]

[1]Tashkent State Technical University named after Islam Karimov, 100095, Uzbekistan, Tashkent, Universitetskaya St. 2A

Abstract. At the moment, the fuel and energy complex of the republic is experiencing crisis state. The main manifestation of the crisis is that disruption and interruptions in the supply of fuel, electricity and heat to individual regions and consumers are becoming commonplace. One of the most promising solutions to this situation is the development of distributed power generation. The process of joint generation of electric and thermal energy 3/4 cogeneration, which, among other things, provides an opportunity for the development of the country's economy, has great potential here. The purpose of this work is to show the relevance and expediency of using cogeneration at heat sources of the republic. The usual (traditional) method of generating electricity and heat is to generate them separately (power plant and boiler house). At the same time, a significant part of the energy of the primary fuel is not used. It is possible to significantly reduce the total fuel consumption by using cogeneration. Cogeneration is the combined production of heat and electricity from a heat source. Cogeneration is a wide range of technologies that can be applied in various fields of economic activity. The cogeneration plant fits perfectly into the electrical circuit, both for individual consumers and for any number of consumers through state power grids. These units are compact, environmentally friendly and cover the deficit of generating capacity in large cities. Such installations make it possible to unload electrical networks and ensure stable power quality. Cogeneration plants (CHP) are essentially mini-heat power station, which have been successfully used all over the world for a long time.

1 Introduction

The relevance of this topic is justified by the fact that there is a serious problem in the energy sector of Uzbekistan - the problem of discrepancy between the volume, structure and technical condition of capacities and the rapidly growing demand for electricity and, as a result, a violation of the fuel and energy balance. Therefore, today the electric power industry can become a real inhibiting factor for the development of the entire economy of the country. The outstripping growth of electricity consumption requires: activation of energy saving and large-scale commissioning of new generating capacities.

Combined heat and power generation and cogeneration continue to play an important integrated role in global development. Combined heat and power generation and cogeneration can contribute to sustainable development because it is an energy-efficient technology. In addition, combined heat and power generation can have a beneficial effect on the local environment if cogeneration systems replace a number of energy facilities as sources of atmospheric pollution.
The object of study of this work is the process of joint generation of electricity and heat at boiler houses.

The subject of the study is the use of energy-saving equipment for the most complete use of the energy of the primary source (fossil fuel), as well as its economic and environmental assessment.

The purpose of studying the work is the need to assess the prospect of introducing the cogeneration process in the energy sector of Uzbekistan and to identify the need for the widespread use of energy-saving technologies in this industry.

2 Problem

For Uzbekistan, the problem of serious study of the use of energy-saving technologies, in particular, in heat supply for the efficient use of investments [1], is urgent. The search for large and small projects for the technical and technological renewal of production, to ensure the competitiveness of products, as well as the means and sources for this, should become, first of all, the most important business and responsibility of the manager and engineering and technical personnel of each heat supply enterprise. [2].

High-energy intensity still dominates in all sectors of the economy of the republic [3]. At present, manufacturing and technical enterprises are provided with some of the world's largest energy subsidies, estimated at approximately $4 billion per year [2,4]. The Government of Uzbekistan recognizes the need to increase domestic prices for electricity and gas to cover the real level of costs of energy supply companies

* Corresponding author: mkoroly@list.ru

necessary for direct satisfaction of demand, maintaining the level of reliability and maintenance of fixed assets. Enterprises will be able to remain competitive only by increasing labor productivity and energy efficiency. Projects for the modernization of equipment at industrial enterprises of the republic, implemented within the framework of the programs of the World Bank Group, led to a significant increase in energy efficiency of production and a decrease in specific energy consumption by 40-70%, as well as to an increase in operating profit by 5-7% even in non-energy-intensive industries. Despite these achievements, many industrial enterprises are ignoring possible investments that could contribute to improving energy efficiency. At least 20% of enterprises do not take into account operating costs and maintenance costs when purchasing new equipment and another 22% prefer less efficient, but cheaper models [5].

3 Solution to the problem

The essence of the cogeneration process:

The usual (traditional) method of generating electricity and heat is to generate them separately (power plant and boiler house). At the same time, a significant part of the energy of the primary fuel is not used. It is possible to significantly reduce the total fuel consumption by using cogeneration (co-production of electricity and heat) [6].

Cogeneration is the thermodynamic production of two or more forms of usable energy from a single primary source of energy.

The two most used forms of energy are mechanical and thermal. Mechanical energy are usually used to rotate an electric generator. The mechanical energy can also be used to keep auxiliary equipment such as compressors and pumps running. [7].

Currently, the market of power equipment offers turbine installations (energy-saving steam turbines, turbine generators, turbodrives) designed to generate cheap electricity and drive auxiliary mechanisms of the boiler house (compressors, pumps, fans, etc.).

Evaluation of the use of cogeneration plants in boiler houses:

At present, traditional large cogeneration, power plants and heat supply systems, as a rule, do not provide the estimated fuel savings at the heat power station and the overall efficiency of centralized supply of electricity and heat to consumers.

In the global energy sector, there is a steady trend towards an increase in energy production and consumption. Even taking into account significant structural changes in industry and the transition to energy-saving technologies, the demand for electricity will increase in the coming decades. As the analysis of the technology showed, mini-heat power stations are one of the key solutions to the problem. The construction, as well as the reconstruction of existing boiler houses in mini-heat power station, will not only ensure more efficient use of fuel due to the combined production of electric and thermal energy, but also increase the reliability of power supply to settlements and industrial enterprises [7].

The Decree of the President of the Republic of Uzbekistan "On measures for the construction, modernization and reconstruction of HP-8, 9, 10 of the State Unitary Enterprise "Toshkent Issiqlik Markazi" with the introduction of modern cogeneration technology of gas turbine units with a capacity of 480 MW in 2017-2021" was adopted.

The energy strategy of Uzbekistan for the period up to 2030 has formed new guidelines for the development of the energy sector, the most important of which is the transition to an innovative path. Among the requirements of the strategies is the achievement by the national economy of the level of energy intensity of GDP of developing countries with a similar climate and conditions, as well as increasing the efficiency of energy use and reducing the negative impact on the environment.

Unlike a traditional condensing power plant, which produces only electricity, in a steam extraction condensing system, some of the steam is taken from the turbine for use as a heat source. The use of cogeneration is associated with significant economic and environmental benefits. Combined cycle cogeneration plants ensure the most efficient use of fuel energy due to the simultaneous production of electrical and thermal energy with minimal losses [8].

The fundamental decision to use cogeneration and the choice of a specific method is determined by a number of factors; even businesses with similar energy needs cannot be considered the same in this regard.

In many cases, the fundamental decision to implement cogeneration is determined by the following factors:

- it is fundamental that there are sufficient heat needs that meet the possibilities of cogeneration in terms of quantity, temperature, etc.;
- the presence of a basic load at the enterprise, i.e. the level below which electricity consumption rarely falls;
- similar nature of the schedules of heat and electricity demands;
- the ratio of fuel prices to electricity tariffs, ensuring the economic efficiency of cogeneration;
- high expected level of workload (preferably more than 4-5 thousand hours of work at full load per year).

In general, the use of cogeneration is justified in those enterprises where there are significant heat requirements at temperatures corresponding to low or medium steam pressure.

When assessing production potential from a cogeneration perspective, it is important to ensure that there is no reason to expect a significant reduction in the heat demand. Otherwise, the operation of a system designed to produce excess heat will be ineffective.

Steam turbines may be an adequate option if the following conditions are met:

- the base electrical load exceeds 3-5 MW_{he};

- there is an application for low-potential steam, and the required ratio of electrical and thermal energy exceeds 1:4;
- availability of inexpensive fuel with low trading margins;
- availability of adequate area to accommodate the system;
- the presence of high- potential heat coming from technological processes (for example, from furnaces or incinerators);
- the need to replace the existing boiler room;
- it is necessary to minimize the ratio of electrical and thermal energy.

In cogeneration systems, maximizing the ratio of electrical and thermal energy requires minimizing the level of backpressure and maximizing the level of high pressure.

Cogeneration units are divided according to the principle of operation of the drive into: steam turbine, combined-cycle, gas piston units. According to the main criteria for evaluating efficiency: operational and repair, financial and economic, payback period. Gas reciprocating units in the range of unit power of 0.1÷10 MW have the best performance. World experience shows that the use of cogeneration is economically justified when there is a constant consumption of heat and electrical energy. With uneven consumption, the limiting energy carrier is heat.

Mini-heat power stations based on internal combustion engines with an electric power of 9 kW to 3.9 MW and a thermal power of 0.02 to 3.6 Gcal/h are presented on the world market.

A complete list of indicators by which GTU- heat power station and mini-heat power station based on internal combustion engines are compared is given in Table 1.

To generate thermal energy, fuel and electrical energy are required (drive pumps, fans, lighting, etc.).

If, when determining the costs of fuel and energy resources for the final product (thermal energy), we take into account all types of costs, i.e.: the cost of generating thermal energy, generating electrical energy and transporting it to the consumer, then we determine the consumption of primary fuel for the resulting final product.

This type of assessment of the efficiency of the use of fuel and energy resources is a system assessment that most fully and reliably reflects the energy efficiency of the use of fuel and energy resources. With this method of assessing the energy efficiency of production, the indicator is the fuel utilization factor (FUF).

It should be noted that in addition to saving money at the enterprise, in this case there is a saving of primary fuel, which should be taken into account as a reduction in costs directly in the fuel and energy complex of the Republic of Uzbekistan, in an amount equal to 0.5 million nm^3/ year. [10]

Table 1

Index	Gas piston	Gas turbine
Durability	full resource is 100...200 thousand hours	-full resource is 75÷140 thousand hours
Quantity and potential produced heat	-0.86÷1.29 Gcal/h per 1 kW of electric power; - network water with a temperature of 90–105 °C.	-1.5–1.8 Gcal/h per 1 kW of electric power; - network water 115–150 °C; - water vapor with a pressure of 9–40 kgf/cm^2 (at 250÷450 °C).
Economy	-Electro efficiency 33÷40%; -Efficiency changes little at load from 100% to 50% (decreases by 5...8%).	- Electrical efficiency 25-35%; - Efficiency drops sharply at partial electrical loads (decreases by 13...18%).
Specific fuel con-sumption: 100÷50% electrical loads	0.25…0.33 m^3/kWh	0.35…0.503 m^3/kWh
Voltage drop and 50% load recovery time	22% 8 s	40% 38 s
Influence variable electrical load	- long work at loads <50% is not desirable; - with a lower unit capacity of the unit, more flexible operation of mini-heat power station and higher reliability of power supply	- operation at partial loads (less than 50%) does not affect the condition; - at a high unit power of the unit, shutdown causes a loss of 30 ... 50% of the mini- heat power station
Ecological	Harmful emissions: NO_X = 118÷185ppm; CO=50÷107ppm	Harmful emissions: NO_X=25–50 ppm; CO = 60–75ppm
Accommodation in the building	- more space is needed, because has more weight per unit of power - no compressor is required to compress gas (operating gas pressure at the inlet to the mini-heat power station is 0.1÷0.35 kgf/sm^2)	- with an electric power of a mini-heat power station of 5 MW and above; - minimum operating gas pressure at the inlet; - 12 kgf / sm^2 (requires a booster compressor, equipment for starting the turbine).
Service	- stop after 1000 hours of operation, (oil change); - overhaul after 72,000 hours on site.	- stop after every 2000 hours; - overhaul after 60,000 hours, carried out at a special factory.

4 Recommendations

Gas piston units are promising as the main source of electricity and heat generation at housing and communal services (with electrical and thermal loads of 0.3÷15 MW): in residential and public buildings, shopping and sports complexes, hotels, sanatoriums, boarding houses.

It is advisable to use gas piston units as a peak or backup source of power supply in industrial enterprises with sharply variable electrical load.

Environmental benefits of using the cogeneration process.

Energy production is a major source of pollution. Cogeneration, using primary fuel, is two to three times more efficient than traditional energy, reduces emissions of pollutants (nitrogen oxide, sulfur dioxide and volatile organic compounds) by 2-3 times, depending on the specific case [11].

Currently, power plants are responsible for 2/3 of total national emissions of sulfur dioxide (SO_2), 1/4 of nitrogen oxide (NOx), 1/3 of mercury (Hg) and 1/3 of carbon dioxide (CO2), the main greenhouse gas. Emissions contribute to serious environmental problems, including global climate change, acid rain, smog, pollution of waterways and eutrophication of critical water bodies, etc. These same emissions contribute to numerous health problems, such as chronic bronchitis and exacerbation of asthma, and other diseases, especially in children.

New large power plants deal the greatest blow to the environment. Cogeneration stations are small and usually located inside existing buildings and factories. In addition, their emissions are an order of magnitude lower than those of large power plants are.

Cogeneration systems can be especially useful in areas where development is limited due to environmental constraints.

Cogeneration plants are small and are usually located inside existing buildings and factories. In addition, the level of their emissions is an order of magnitude lower than the level of large power plants.

Stability

Cogeneration of electric and thermal energy is an energy-efficient technology that can play an important role in the XXI century in the implementation of Uzbekistan's transition to the path of sustainable development. With the help of cogeneration, it is possible to produce several types of energy services simultaneously:

- heating and cooling of buildings;
- generation of electricity for lighting and engine operation;
- production of process energy for industry, etc.

With combined heat and power generation, it is possible to utilize a large number of different fuels - not only natural gas, coal and oil, but also, for example, biomass and solid waste, using energy-efficient large-scale combustion plants equipped with modern systems, and also with the use of environmentally friendly technologies.

In addition, the combined production of thermal and electrical energy can have a beneficial effect on the local environment if cogeneration systems replace a number of energy facilities as sources of air pollution.

The implementation of cogeneration helps to increase the level of employment. This is because energy efficiency and the use of local energy resources associated with it mean that fuel imports can be reduced, thereby increasing the money supply available for investment in local cogeneration systems.

5 Conclusions

Combined heat and power generation can recycle a large number of different fuels, not only natural gas, coal and oil, but also, for example, biomass and solid waste, using energy-efficient large-scale combustion plants equipped with modern systems, as well as using environmentally friendly technologies.

In addition, the combined production of heat and electricity can have a beneficial effect on the local environment if cogeneration plants replace a number of energy facilities as sources of atmospheric pollution.

References

1. The Republic of Uzbekistan has adopted a strategy for the development of the Republic of Kazakhstan, PF-4947-sonli Farmoni (2017)
2. Decree of the President of the Republic of Uzbekistan "On the priorities of industrial development of the Republic of Uzbekistan in 2011-2015", PP 1442, (2010)
3. Project UNDP / Ministry of Economy. Project "Supporting Uzbekistan in the transition to the path of low-carbon development of the national economy". (2012), www.mineconomy.uz/cdm
4. Asian Development Bank. Project "Technical Assistance to the Republic of Uzbekistan in Energy Needs Assessment", (2011)
5. I.A. Bashmakov, Analysis of the main trends in heat supply systems, (2009), http://solex-un.ru/sites/solex-n/files/energo_files/teplosnabzhenie_i.bashmakov.pdf
6. Yutron-Steam turbines, (2009), http://www.turbopar.ru
7. New Generation - steam turbines, http://www.manbw.ru/analitycs/steam-turbines
8. G. G. Olkhovsky, *Gas turbines for energy. Heat Power Engineering,* pp. 33-43, (2004)
9. M.A. Koroli, A.I. Anarbayev *Development of the heat supply sector in Uzbekistan.* (2018)
10. M.A. Koroli *Energy saving technologies in heat supply systems of Uzbekistan*, Monograph, (2021)
11. E.A. Lebedeva *Mini-heat power station on the basis of production and heating steam boiler house,* No. 2-p51-53, (2008)

Intellectual measurement of electrical energy consumption - monitoring device and software

Narzullo Mirzoyev[1*]

[1]Bukhara Institute of Engineering and Technology "Energoaudit Department", 200117, Uzbekistan, Bukhara, Kayum Murtazoyev St. 15

Abstract. The article presents the principle scheme and software of the intelligent measurement - control device that allows determining the efficiency of electricity consumption and the energy efficiency of electrical devices. Mathematical expressions and calibration coefficients of the device for measuring electrical quantities were calculated. Electronic devices for converting and amplifying analog signals to digital signals were selected and their communication circuits with Arduino microcontroller were developed. Based on the C++ programming language, the software that integrates with the device has been developed. The software determines the efficiency indicators of electricity consumption on 4 interfaces at the same time. Based on this software tool and device, it was possible to create a control system based on cloud technologies, which allows determining the energy efficiency indicators of a certain area. At the same time Algorithm of the program of regulation of fuel energy resources for the measurement and control system of the efficiency of electric power consumption, and, The Cloud Meters model of the electricity consumption efficiency measurement and monitoring system has been developed.

1 Introduction

To calculate the root, mean square value for each phase voltage, the following formula is used [1-3]:

$$U_f = K_{kU}\frac{\sqrt{\sum_{i=1}^{n} U_i^2}}{n}, [\mathrm{V}], \quad (1)$$

Here: - calibration coefficient for the measured phase (entered during calibration)); i is the value of the current sample; n - number of samples; - sample voltage value, V.$K_{kU}U_i$

The following formula is used to calculate the root-mean-square value of the current for each phase current [4-6]:

$$,I_f = K_{kI}\frac{\sqrt{\sum_{i=1}^{n} I_i^2}}{n} \quad [\mathrm{A}], \quad (2)$$

Here: - calibration coefficient for phase (entered during calibration)); i is the value of the current sample; n - number of samples; - sample current value, A.K_{kI} I_i

The active power in each phase is calculated according to the following formula [7-10]:

$$P_f = K_{kU}\cdot K_{kI}\frac{\sqrt{\sum_{i=1}^{n} U_i I_i}}{n}, [\mathrm{W}], \quad (3)$$

The total active power is determined by the following formula:

$$P_\Sigma = P_{fA} + P_{fB} + P_{fC}, [\mathrm{W}], \quad (4)$$

Here: – active power for each phase.P_{fA}, P_{fB}, P_{fC}

The total power in each phase of a three-phase network is calculated according to the following formula [11-13]:

$$S_f = I_f \cdot U_f \quad , \qquad [\mathrm{kVA}], \quad (5)$$

Here: – mean-square value of the current in the corresponding phase determined by the formula (2), A; – value of the mean square voltage in the corresponding phase determined by the formula (1), V.$I_f U_f$

The total full power is determined by the following formula [14-15]:

$$S_\Sigma = I * U, \qquad [\mathrm{kVA}], \quad (6)$$

Here: I line currents, U line voltage – full power of each phase determined by the formula (5), VA.

The reactive power for each phase is determined by the following formula [16-18]:

$$Q_f = \sqrt{S_f^2 - P_f^2}, [\mathrm{VAr}], \quad (7)$$

Here: – (5) full power in each phase determined by the formula, VA; – (4) active power in each phase determined by the formula, W. $S_f P_f$

The total reactive power is determined by the following formula [18-20]:

$$Q_\Sigma = Q_{fA} + Q_{fB} + Q_{fC}, [\mathrm{VAr}], \quad (8)$$

Here: – reactive power in each phase determined by the formula (7), VAr.Q_{fA}, Q_{fB}, Q_{fC}

* Corresponding author: ilider1987@yandex.com

The relative energy loss in a three-phase circuit is calculated by the following formula [21-23]:

$$A = R_{fA} \cdot I_{fA}^2 + R_{fB} \cdot I_{fB}^2 + R_{fC} \cdot I_{fC}^2, \text{[W*h]}, (9)$$

Here: – (2) amperage determined by the formula, A; – active resistance of power lines for each phase, Ohm.I_{fA}, I_{fB}, $I_{fC}R_{fA}$, R_{fB}, R_{fC}

The active power coefficient formula is expressed as follows [24]:

$$\cos\varphi_A = \frac{P_{fA}}{S_{fA}}, \cos\varphi_B = \frac{P_{fB}}{S_{fB}}, \quad \cos\varphi_C = \frac{P_{fC}}{S_{fC}}, \quad (10)$$

Here: – active power determined by the formula (3), W; – (5) full power determined by the formula, VA .P_{fA}, P_{fB}, $P_{fC}S_{fA}$, S_{fB}, S_{fC}

Reactive power coefficients are determined by the following formula [25-27]:

$$\sin\varphi_A = \frac{Q_{fA}}{S_{fA}}, \sin\varphi_B = \frac{Q_{fB}}{S_{fB}}, \quad \sin\varphi_C = \frac{Q_{fC}}{S_{fC}}, \quad (11)$$

Here: – (7) reactive power determined by the formula, VAr.Q_{fA}, Q_{fB}, Q_{fC}

Active and reactive energy is calculated based on the value of active and reactive power determined in 1 second. Current and voltage are determined from the values of active and reactive power coefficients based on vector diagrams.

Based on the calculation of active and reactive energy, the MC sends signals about the consumption of electricity to the external interface, the signals of electricity consumption data are sent to the fuel energy resources regulation program or the user's software product [28-30].

MC collects energy parameters and stores energy parameters that are required to be held for a moment in a flash memory device and calculates the time.

In order to reduce the cost of the product, data input and display - a device for visualization of measurement results and a keyboard for data input were abandoned. Information about network parameters, power consumption, load temperature and user parameters is transmitted to the cloud database via the Internet. The information is displayed on the computer or phone of the service worker. The program interface shows the graph of electricity consumption, the operating status of the load, the main indicators of the calculated electricity.

To configure the device, using the configurator program, digital coefficients are used, which are written to the non-energy memory and loaded into the microcontroller registers when the device is started.

To add load to the device, the device provides a pre-designed 100 A current relay for the electrical energy monitoring system we offer. The relay can be controlled manually.

A 1.2 V uninterruptible power supply (UPS) is used to power the device.

The metrological indicators and descriptions of the device are presented in Table 1.

Table 1

Measurement error	smaller than 1%
Voltage measurement range in a three-phase network	0 - 400 V
Current measurement range	50 mA - 100 A
Maximum power at the output to be controlled	30 kW
Consumption capacity	1.2 W
Communication protocol	Wireless, IEEE 802.11
Data update interval	5 sec
Frequency	45-65 Gs
Maximum data storage time on the device itself	10 days
The specified operating temperature	-10 °C to +60 °C
Dimensions	90 x 67 x 52 mm
Mass	1.2 kg

According to the conditions of the measurement-control system, it is possible to automate the controlled output values of the checked electricity consumption or to obtain information about them. It is possible to adjust the parameters of electricity consumption. Based on the conditions of energy availability, the relay disconnects or connects the electrical network based on the control signals coming from above. If the values of the extracted data match the conditionally defined values and the condition, the device initiates one of the following actions: sends messages about energy efficiency by e-mail; turns on the relay; turns off the relay. Conditions are automatically checked every 5 seconds. If the specified conditions are met,

Status checks may be limited by time. If you activate the time interval, the condition will be executed only during the specified time [1-3].

The comparison symbols ">" or "<" are used to reset the condition. Also, types of energy efficiency quantities are selected and correction values are entered for comparison, as well as one of the necessary operations for data recalculation is selected [1-3].

Example: If the network voltage is higher than 240 V, two conditions must be met to turn off the load relay and to turn on the load relay when the network voltage reaches 230 V (Figure 1):

1) $U_{ph} > 230$ KBR "turn off",

2) $U_{ph} < 220$ KBR "on".

*KBR-Adjustment for voltage management

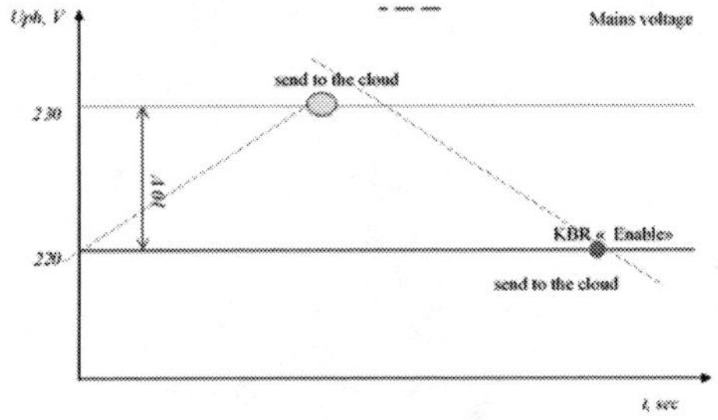

Fig. 1. Automation of verification of compliance with specified conditions.

The event notification function allows you to activate the sending of e-mail messages about selected events that occur on the device [1-3].

The setting conditions are available in the WEB interface of the device in the "Conditions" section.

2 Materials and Methods

One of the means of increasing the efficiency of electricity consumption is to organize a system of continuous monitoring of the volume of electricity consumption in the production units of the enterprise, taking into account production indicators and climatic factors, as well as to warn about possible malfunctions in the operation of the equipment and thus make possible predictions. One of the important tasks for the development of such systems is the creation of special measurement and control devices and control algorithms that allow comprehensive analysis of the potential of saving electricity. This allows for moderation of technical and commercial waste and consumption of reactive power, as well as rational use of energy.

The proposed measurement and measurement system is designed for continuous measurement of power grid quantities and electrical energy. Measurement and monitoring are carried out for the purpose of further analysis and implementation of efficiency indicators aimed at normalizing electricity consumption.

The device directly measures the following electrical energy quantities:

1. Tension, V;
2. Current strength, A;
3. Active power, W;
4. Reactive power, VAr;
5. Active energy, W*h;
6. Reactive energy, VAr*hour;
7. Power factor
7. Network frequency, Gs;
8. Temperature, °C.

Communication with the measuring system is carried out via wireless Wi-Fi. The measurement system transmits the collected data to the "CLOUD" server for further analysis and visualization.

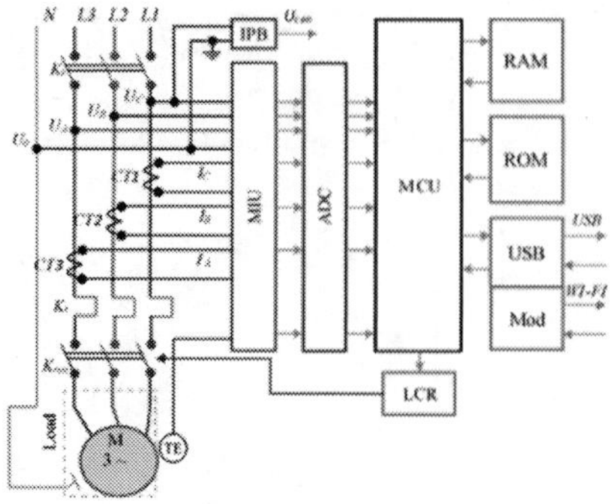

Fig. 2. Structural scheme of the electrical energy monitoring system.

The proposed device includes the following main elements (Figure 2): measurement and interface unit (MIU), analog-to-digital converter (ADC), microcontroller (MCU), Random Access Memory (RAM), Read-only memory (ROM), Universal Serial Bus (USB), WI-FI modem (Mod), load control relay (LCR), current transformers (CT), temperature sensor (TE), impulse power block (IPB) [4] [1-3].

Using a three-phase current converter installed in the network, the phase currents are changed into an output signal in the form of voltage. These signals are fed to the O'BB inputs, the signals received from the O'BB network are divided into current voltage and frequency signals at the phase section, and the analog signals are transferred to the serial synchronous port of the MK for conversion into a digital code. MK calculates the average arithmetic values of currents and voltages, total, active, reactive power, power coefficient and consumed electric energy, as well as the shift angles and frequency of the main harmonic of voltage signals, and enters the obtained data into the cloud database by day, month and year.

Arduino Uno integrated circuit is widely used for control and management of technological processes. In this thesis work, the Arduino Uno integrated circuit shown in Figure 2 was chosen for the energy efficiency measurement controller. The basis of the integrated circuit is the ATmega328 microcontroller. The size of the used controllers is easy to install, as well as being reprogrammable. Arduino Uno integrated circuit architecture is open. It has its own programming environment and board. In addition, it is possible to design various schemes of a number of accessories. Table 2 shows the parameters of Arduino Uno.

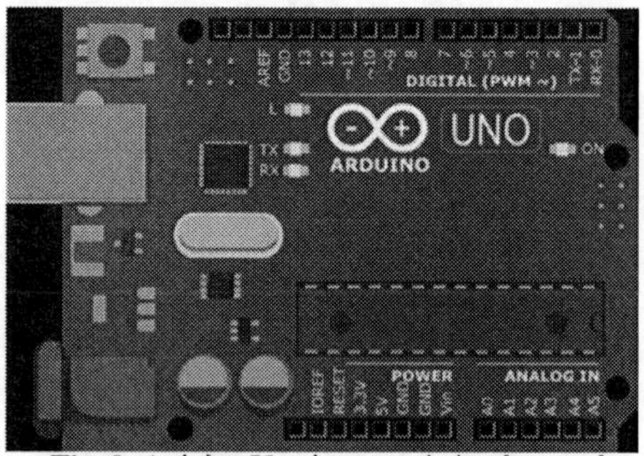

Fig. 3. Arduino Uno integrated circuit overview

https://doi.org/10.1051/e3sconf/202346101083

Table 2. Arduino Uno specs

Type of microcontroller	ATmega328
Operating voltage	5 Volt
Supply voltage	7-12 Volt
Voltage range	6-20 Volt
Digital inputs and outputs	14, of which 6 are for PWM
Analog inputs and outputs	6
Fixed current in one output channel	40 mA
Memory	32 kB, of which 0.5 kB is for programming
SRAM	2 KB
EEPROM	1 kB
Frequency	16 MHz

Measuring and connecting block resistance bridgeAMC1100 amplifier and it consists of ADE7752 integrated circuits. AMC1100 galvanically isolated amplifiers were used in the created energy efficiency measurement controller. In order to eliminate electromagnetic interference in this isolation amplifier, its input and output cascades are isolated with silicon dioxide. AMC1100 protects each circuit of small value voltages from high value input voltage, and also filters electrical signals that are required to be removed from the network. The AMC1100 integrated circuit is mainly used in electricity meters. AMC1100 inputs enable the adaptation of small voltage consumers to accurate measurement of current and voltage signals as shown in Figure 4.

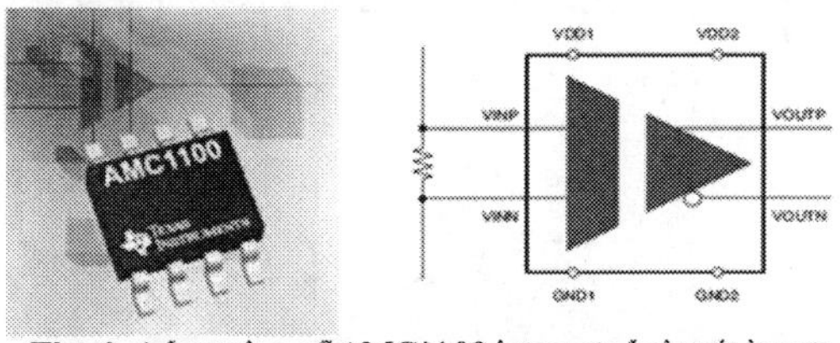

Fig. 4. Adaptation of AMC1100 integrated circuit inputs and small voltage consumers to accurate measurement of signals in the form of current and voltage.

ADE7752 was used as an element of a high-precision three-phase consumer measurement control system. The integrated circuit enables the measurement of electrical signals based on the IEC1036 standard, and it consists of analog-to-digital converter circuits. Processing of electrical signals is carried out digitally, filtering and summing. This measure provides accurate measurement of power grid magnitudes under various conditions. ADE7752 F1 and F2 generate signals that enable the determination of electrical quantities of electrical consumers with a frequency of 50 GHz [139].

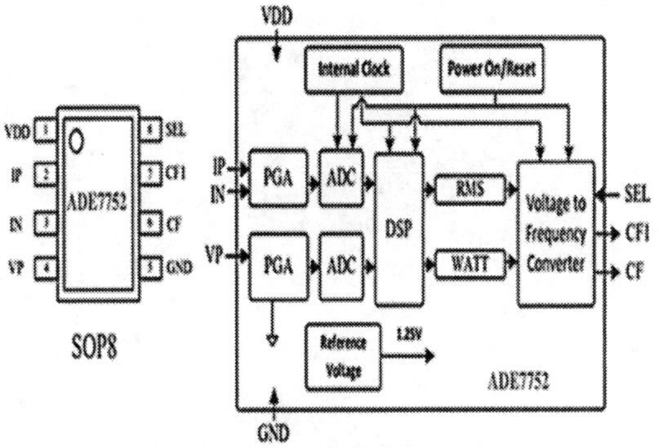

Fig. 5. ADE7752 is a high-precision three-phase consumer measurement control system element

Principle of operation of the ADE7752 chip

IP is the voltage signal coming from the converter.

IN - two ends of the winding of the transformer, i.e. phase and zero.

Vp - the maximum and minimum values of the phase voltage are obtained.

ADC is an analog-to-digital converter.

Internal clock – time interval (tactical frequency of the controller).

Voltage to Frequency Converter – voltage frequency converter.

Set <- signal to the chip.

CF1 -> active power value is obtained.

A digital signal is received from the CF -> ports. The signals are used to connect the power supply and database to the microcontroller used for measurement control. Each logic output in an integrated circuit has its own function. The supply voltage of the ADE7752 integrated circuit is 4 Volts and is provided by the VDD control circuit. If the voltage is less than 4 volts, no signal is generated from logic outputs F1, F2 and CF. In addition, even when there is no consumer in the electrical network, only the electrical parameters of the network are signaled according to voltage and frequency.

Connection block of measurement and control and Arduino Uno the connection diagram of the integrated circuit is presented in Fig. 6.

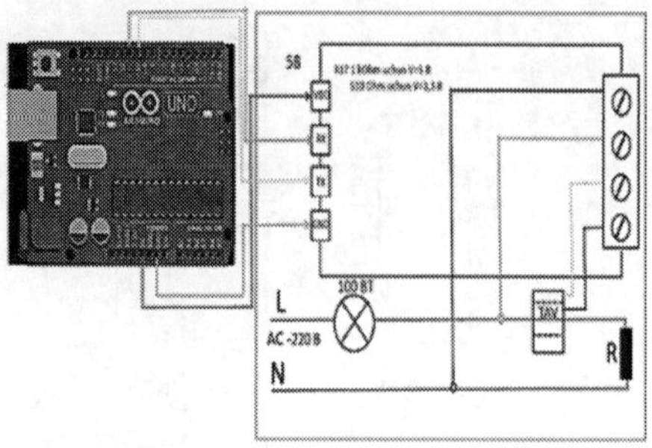

Fig. 6. Connection diagram of measurement and control unit and Arduino Uno integrated circuit

In the general scheme, as one phase of the electrical network is connected, the remaining phases are connected in the same order (Figure 6). The circuit shown in Figure 7 shows the connection of the measuring connection block and Arduino Uno integrated circuit for three-phase power monitoring.

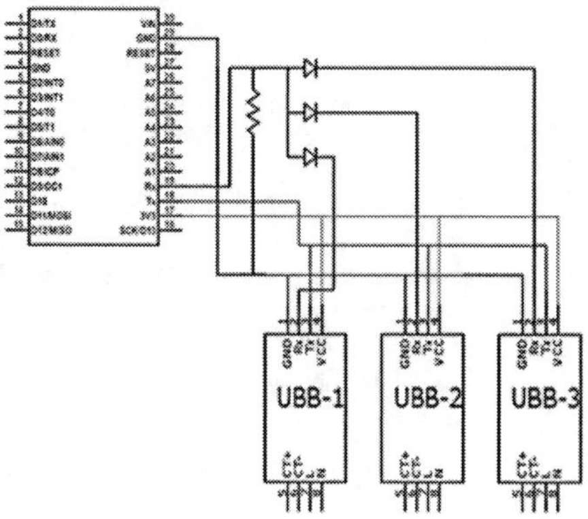

Fig. 7. Measuring connection block for three-phase electricity monitoring and Arduino Uno integrated circuit connection diagram.

The measuring connection block designed for three-phase electric power monitoring is mainly assembled based on the above elements, its schematic diagram is shown in Figure 8.

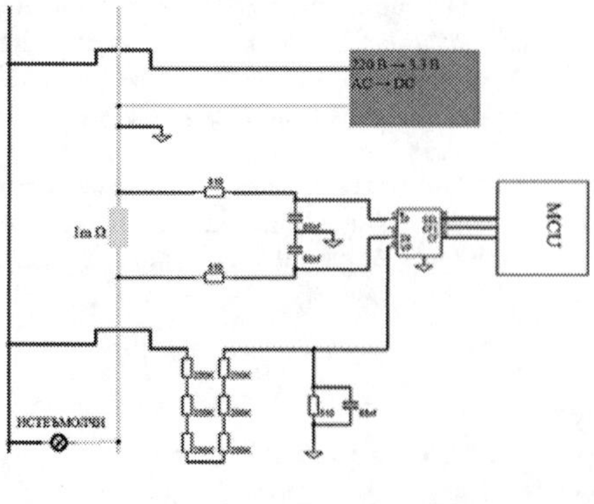

Fig. 8. Structural diagram of the measurement connection block

Each circuit presented in the diagram performs a certain task. Tasks and descriptions of measurement chains presented in the scheme are presented in Table 3.

Table 3

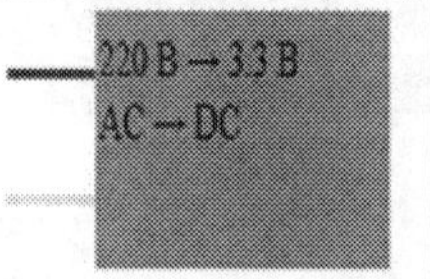	Supply unit for the device. By connecting a capacitor, resistor and diode in series, the voltage is reduced. This circuit is used to power the device
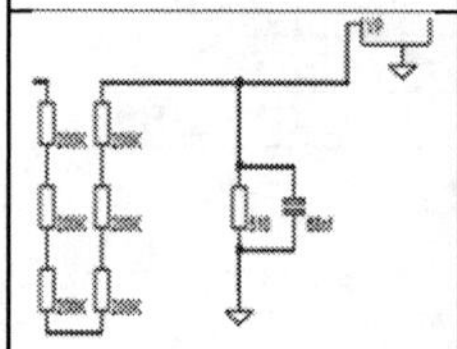	The voltage pulses in the determined phase are directly received by the resistors, and thereby extract the amplitude value, length and frequency of the voltage.
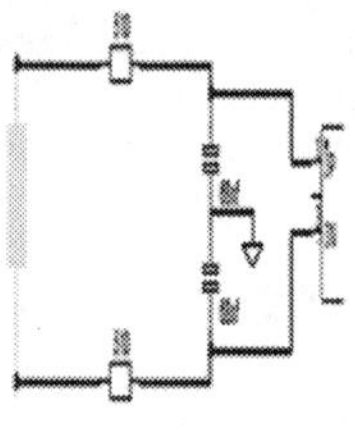	In this scheme, the analog signals measured from the electrical network are converted into digital signals.
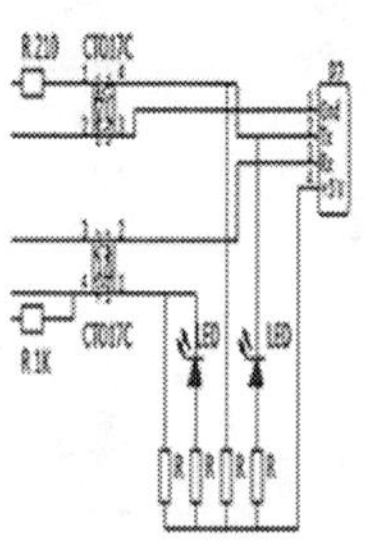	This circuit is considered a structural circuit of the MCU, which is a device for transmitting digital signals to a controller or a computer. It consists of 2 optocouplers and works on the basis of TTL technology.

3. Results and Discussion

At present, in production enterprises, mostly all money is directed to the one-sided narrow services of the organization's leaders. It is rarely used to provide information to other levels of management than the axopods to develop acceptable general management decisions. Other employees simply need to know how their unit is performing and whether energy efficiency is improving or deteriorating.

These are the main directions of solving problems is the development of energy resources standardization programs that allow determining and evaluating energy efficiency. The structural schemes and software dependencies of this program are in the second chapter of the research work given. The program provides information on the measurement and calibration of the power supply system in four formats. The program was implemented in the C++

programming language. The program has the ability to provide the data format on phones, tablets and computers.

The 1st data presentation format (interface) of the program is presented in Figure 9. In this section, data on electricity consumption in the months are provided from the cloud database. The amount of electricity consumed for the production of products is monitored and financial costs for electricity are calculated automatically. Later, this information is used to calculate the program's other software parameters [1-3; 14-20].

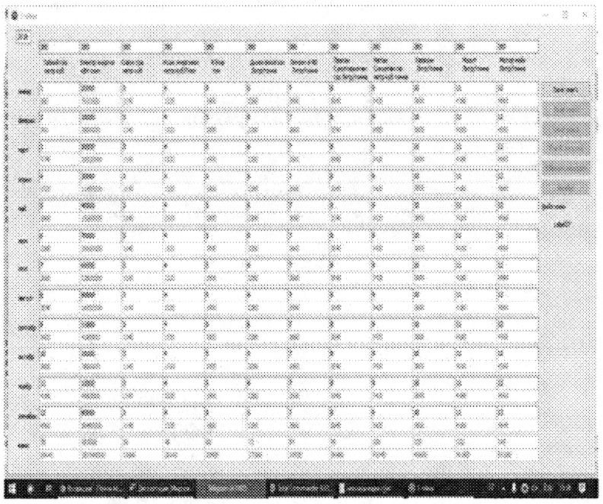

Fig. 9. Indicators of electric power monitoring

The program's 2nd data presentation format is shown in Figure 10. Data presentation format 2 of the program is given from the data presentation format part of the program 1. In this window of the program, indicators of electricity monitoring and consumption costs of electricity are presented. Based on these monitoring indicators, in the next calculation sub-block of the program, technical wastes are calculated and energy resource consumption balances for product production and for own needs are drawn up. If various fuel and energy resources are needed for product production, the program automatically changes the physical value of fuel and energy resources into a single unit of conventional fuel and makes an energy balance on the consumption of fuel and energy resources.

The program's 3 data presentation format is shown in Figure 11. In this part, graphs of the dynamics of consumed and wasted electricity are drawn. Data presentation format 3 of the program is provided from the cloud database and data presentation format 1 of the program [1-3; 14-20].

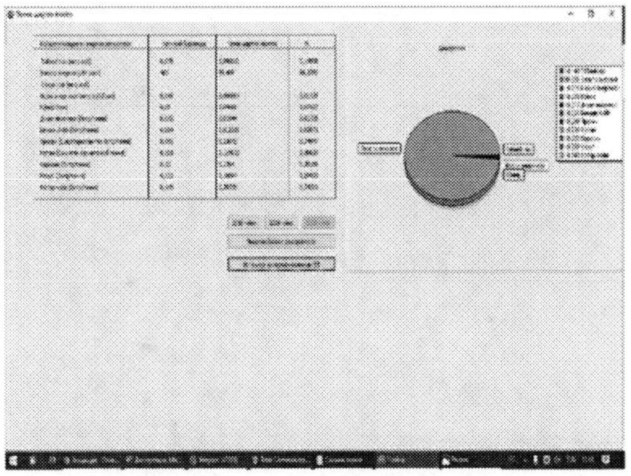

Fig. 10. A single unit of measurement and energy balance indicators

Also, in this part of the program, electric energy losses are calculated by determining the loading coefficient of electric energy consumption obtained from the billet, and by entering the constant values necessary for the calculation of losses, the full power of the transformer, the number, the active power coefficients in the short-circuit and idle modes, the distance of the power transmission lines, the relative resistance of the conductor, the errors of the measuring sets and other values. Calculations of electrical energy losses are made on the basis of algorithms written for the program [1-3; 14-20].

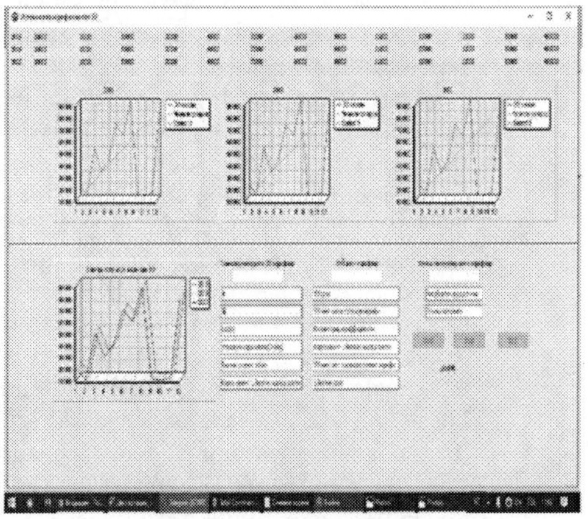

Fig. 11. Dynamics of electricity consumption and indicators of technical waste

Figure 12 shows the program's 4 data presentation format. The 4th data presentation format is automatically transmitted to the cloud by means of special consumption meters that are set on technological scales, and the dynamics and volume of product production are presented on the basis of the data received from the cloud. the comparative cost of electricity used for product production is calculated using data from electricity monitoring indicators [1-3; 14-20].

https://doi.org/10.1051/e3sconf/202346101083

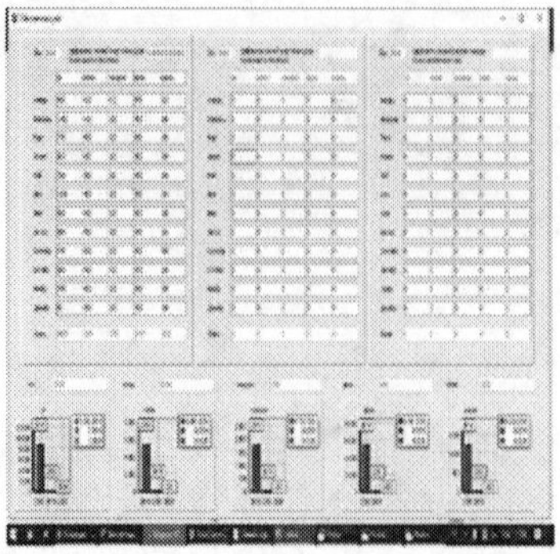

Fig. 12. Product dynamics and energy capacity indicators

This program It is widely used in scientific research to determine and evaluate energy efficiency indicators of technological devices [1-3; 14-20].

Standardization of fuel and energy resources the software calculates the standards of electricity consumption of technology and equipment for product production and determines their efficiency classes. At the expense of optimizing the operating modes of technology and equipment, for example, for different operating modes, it defines the norms of energy consumption according to the passport indicators. Determines specific energy efficiency indicators for the types of products manufactured in production enterprises. On the basis of the stored data, it is possible to carry out an internal energy audit in the enterprise. Generates energy efficiency reports automatically.

Application of Cloud computing technology in measuring, transmitting and processing efficiency data of electric energy devices

Production and people's way of life goes hand in hand with the wide use of information technologies. Currently, there are many issues in the energy sector that cannot be solved by using advanced technologies. Because industry-specific information technologies and software are very expensive in the information technology market. According to the latest estimates, the use of advanced technologies in the global energy sector is increasing, and at the same time, their prices are also increasing. In our republic, the introduction of information technologies in the energy sector has not accelerated compared to other sectors of the economy. However, electricity sales organizations are looking for ways to increase their efficiency with advanced technologies.

It is considered necessary to create a concept of automated measurement, taking into account the specific features of the organization of the energy industry of foreign countries, in the use of cloud services in the measurement and management of energy efficiency and the introduction of cloud services for determining the efficiency.

Due to the fact that the price of electricity is not expensive in our republic, as well as the lack of software and devices for monitoring and controlling the electricity consumption, manufacturing enterprises do not want to use the services of cloud technologies in our republic and in developed countries. Although Cloud services can be used in electricity generation, transmission, distribution, consumption and electricity consumption monitoring services, traditional methods of data storage and processing are being used in the automation of electricity consumption calculations. Given that this requires additional costs for digital technologies and high-paid specialists in the enterprise.

Cloud technologies are really cost-effective, besides, scientific research conducted in our country on cloud technologies positively evaluates several of its features:

1. Economy of scale in data storage;
2. Variety of services;
3. Flexibility with different systems;
4. Not requiring organizational issues.

Based on the results of the analysis of Cloud technologies and services, we have developed a model of Cloud Meters, a system for measuring the efficiency of electricity consumption.

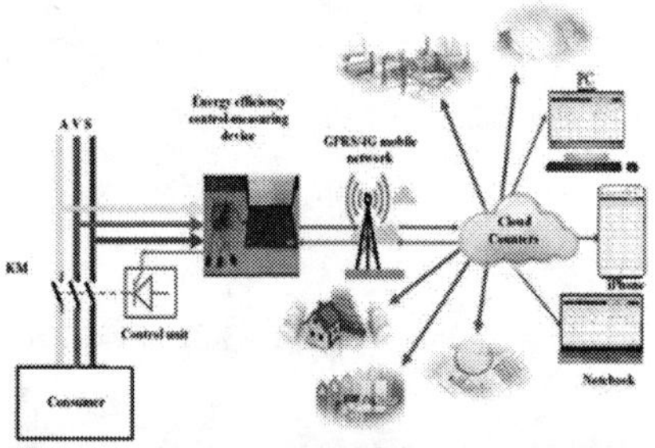

Fig. 13. Cloud Meters model of the electricity consumption efficiency measurement and analysis system

The need for cloud technologies and services in the energy sector is increasing. The growing demand is explained by the advantages of its services. Use of Cloud technologies and services ensures strategic development of any enterprise. There is a number of advantages of the Cloud Meters model of the electricity consumption measurement and monitoring system offered by us, which is the compatibility of the software. Cloud Meters model allows to estimate the energy efficiency in real time and to determine the standard energy capacity of the product. Also, Cloud Meters have data transfer and storage capabilities.

A number of problems are encountered in the use of cloud technologies offered by foreign manufacturers. First of all, they require large expenses for the creation of cloud technologies for the production enterprise. Secondly, innovations and changes in the field of foreign electric energy do not always take into account the needs of the electric energy sector of Uzbekistan. Thirdly, the data security of the production company's electricity consumption figures is not fully ensured. Fourthly, access to cloud technologies and services depends on the Internet speed for their use, including the reliability of communication. Because the communication channels of their data processing systems are integrally interconnected.

Cloud services are provided in the following Cloud Meters in the electric power system:

1. Service of monitoring electricity monitoring indicators and their consumption costs;

2. Prepares monthly and annual reports of electricity consumption for the energy industry;

3. Makes balances of electricity consumption, electricity waste and electricity savings;

4. Performs energy efficiency control and internal energy audit services;

5. Measure electrical network sizes and store them in the cloud;

6. Determination of energy efficiency classes of technology and equipment.

The implementation of these possibilities provides the enterprise with documents and reports from any location, it creates the possibility to monitor the efficiency of fuel and energy resources and electricity consumption in all production processes of the enterprise online, paper reports are avoided. The Cloud Meters model of the electricity consumption efficiency measurement and analysis system developed by us shows that Cloud Meters optimize all operations in the field of electricity.

It is possible to assess and increase the energy efficiency of regions in the electric energy sector of our Republic based on the capabilities of Cloud Meters model and IT structure services. The structure of IT services for the assessment of efficiency of electricity consumption in the electric power system is presented in Fig. 14.

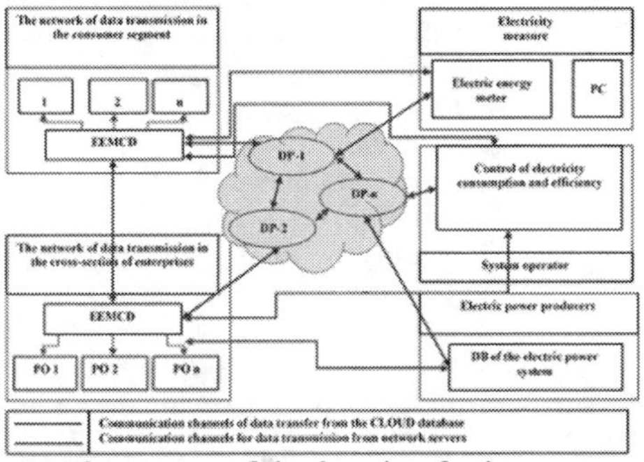

Fig. 14. The structure of cloud services for the assessment of efficiency of electricity consumption in the electric power system

Conclusion

Based on the nature of the IT structure of the cloud services of the electricity consumption efficiency measurement developed on the basis of the electricity consumption efficiency measurement system and the cloud technologies, it is possible to solve a number of problems between industry enterprises in the electric power system, for example:

1. integration of cloud data processing systems in the field of electric power into a single integrated database of cloud technologies of electric power;

2. It is necessary to integrate different clouded technologies in the field of electric energy into a single integrated electric power system within individual clouds [6-13].

The information processing system of the IT environment, shown in Figure 13, works on a single integrated electricity billet. In addition, each of these data processing systems is capable of mutual data processing and exchange. The structure of the IT environment of cloud technologies for the electric power system implements various services of cloud technologies. In the field of electric energy, measurement operators of organizations on electricity generation, transmission, distribution, consumption and energy monitoring simultaneously study the data of efficiency indicators in a mutually integrated manner. Moreover, on the basis of these interconnections, all work activities in the field of electric power are integrated and coordinated.

The Cloud Meter model of the electricity consumption measurement system developed and described by us above, as well as the use of modern technologies such as the IT structure of the cloud technologies for the electric power system, is also shown in the ISO 50001 standard and the decision of the President of the Republic of Uzbekistan No. PQ-4422 dated 22.08.2019.

References

1. Mirzoyev N.N. Chemical Technology. Control and Management. Tashkent, Vol. 2021. No. 5, pp. 68-75. DOI:https://doi.org/10.51346/tstu-02.21.5-77-0044.(05.00.00;No. 12).

2. Mirzoyev N.N. The American Journal of Engineering and Technology. USA Texas, 2020. Vol. 02, pp. 55-61.

3. Mirzoyev N.N. Chemical Technology. Control and Management. Tashkent, Vol. 2020. Special issue 5-6, pp. 172-181. DOI:https://doi.org/10.34920/2020.5-6.172-180.(05.00.00;No. 12).

4. Mirzoyev N.N., Sayfiyev O.H., Temirov T.O. ResearchJet Journal of Analysis and Inventions, 2022. 3/10, pp.130-143.

5. Maxmudov M.I., Mirzoyev N.N. Energiya samaradorlikni boshqarish va nazoratning axborot dasturiy ta'minoti va qurilmalari. «Energiya va resurs tejash muammolari» jurnali, Toshkent, 2021. № Maxsus son, 101-109 betlar.

6. Mirzoyev N.N. Avtomatik nazorat-o'lchov tizimlarida iqtisodiyotni modernizatsiyalash negizi // «O'zbekiston iqtisodiyoti tarmoqlarini raqamlashtirish sharoitida elektr tarmoqlari korxonalarini innovasion rivojlantirish" mavzusida respublika ilmiy-texnika anjumani. (Xalqaro mutaxassislar ishtirokida). Toshkent. 2021 yil 8 sentyabr. 48-50 b.

7. Siddikov I.X., Mirzoyev N.N. Smart energy efficiency monitoring device // Vserossiyskaya nauchnaya konferensiya molodix issledovateley s mejdunarodnim uchastiem «Sotsialno-gumanitarnie problemi obrazovaniya i professionalnoy samorealizasii» Sotsialniy injener-2021 sbornik

materialov. Rossiya, Moskva, 6-10 dekabrya 2021 g. Chast 3. S. 237-240.
8. Mirzoyev N.N. Razrabotka intellektualnoy sistemi izmereniya dlya monitoringa i kontrolya energoeffektivnosti // Vserossiyskaya konferensiya molodix issledovateley s mejdunarodnim uchastiem «Sotsialno-gumanitarnie problemi obrazovaniya i professionalnoy samorealizasii» Sotsialniy injener-2020 sbornik materialov. Rossiya, Moskva, 6-10 dekabrya 2020 g. Chast 3. S.20-25.
9. Mirzoyev N.N., Shoboyev A.X., Temirov O.T. «Yoqilg`i energetik resurslarni tonna shartli yoqilg`iga o`zgartirishning mobil ilovasi» / O'zR EHM uchun dasturlar davlat reesrida 07.01.2022 yilda Toshkent shahrida ro'yxatdan o'tkazilgan, № DGU 14066.
10. Mirzoyev N.N. «Elektr energiyasini monitoringining aqlli tizimi uchun hisoblagichlar dasturiy ta`minoti» / O'zR EHM uchun dasturlar davlat reesrida 11.11.2021 yilda Toshkent shahrida ro'yxatdan o'tkazilgan, № DGU 13002.
11. Mirzoyev N.N. «Energiya samaradorlikni boshqarish va nazorat qilish uchun HISOBLAGIChLAR tizimining ma`lumotlar bazasi» / O'zR EHM uchun dasturlar davlat reesrida 09.11.2021 yilda Toshkent shahrida ro'yxatdan o'tkazilgan, № BGU 462.
12. Maxmudov M.I., Mirzoyev N.N. «Energiya samaradorlikni boshqarish va nazorat qilish qurilmasi uchun ta`minot dasturi» / O'zR EHM uchun dasturlar davlat reesrida 11.11.2021 yilda Toshkent shahri, № DGU 13003.
13. Siddikov I.X., Mirzoyev N.N. «Energiya samaradorligini nazorat va boshqarishning avtomatlashtirilgan dasturiy-axborot ta'minoti» / O'zR EHM uchun dasturlar davlat reesrida 10.03.2020 yilda Toshkent shahrida ro'yxatdan o'tkazilgan, № DGU 07874.
14. Siddikov I.X., Abdumalikov A.A., Mirzoyev N.N. «Energiya samaradorlikni nazorat va boshqaruvi elementlarini ishonchlilik va ish holatlarining ko'rsatkichlarini tadqiq etish dasturiy ta'minoti» / O'zR EHM uchun dasturlar davlat reesrida 27.03.2020 yilda Toshkent shahrida ro'yxatdan o'tkazilgan, № DGU 08003.
15. Toirov O., Sadullaev T., Abdullaev D., Jumaeva D., Ergashev Sh., Sapaev I.B. *E3S Web of Conferences* 383, 01029 (2023).
16. Erejepov Madiyar and ofters. 2021 International Conference of Young Scientists Energy Systems Research, ESR 2021. https://10.1051/e3sconf/202128907018
17. Toirov O., Pirmatov N.B., Khalbutaeva A., Jumaeva D., A. Khamzaev. *E3S Web of Conferences* 401, 04033 (2023).
18. Toirov O., Khalikova M., Jumaeva D., Kakharov S. *E3S Web of Conferences* 401, 05089 (2023).
19. Khalikov S. *E3S Web of Conferences* 377, 01004 (2023).
20. Alimkhodjaev K., Toirov O., Taniev M. Journal of Critical Reviews, 2020. Vol. 7, Issue 15.
21. Toirov O.Z., Khalikov S. *E3S Web of Conferences* 365, 04010 (2023).
22. Toirov O.Z and others. *AIP Conference Proceedings* 2552, 040021 (2022).
23. Toirov O.Z and others. E3S Web of Conferences 216, 01119 (2020) RSES 2020.
24. Jumaeva D. and others. *E3S Web of Conferences,* SUSE-2021, 288, 01041 (2021).
25. Afrisal H., Setiyadi A.D. *International Journal on Advanced Science, Engineering and Information Technology*, Vol. 12 No. 6, p. 2258-2265, (2022).
26. Toirov O., Khalikov S.S. *E3S Web of Conferences* 365, 04013 (2023).
27. Hoshimov U.H., Ishnazarov O.Kh., and others. Journal of Physics: Conference Series, *Conf. Ser.* **2094** 052051, (2021).
28. Hoshimov U.H., Ishnazarov O.Kh., and others. E3S Web of Conferences 264, 04088 (2021) https://doi.org/10.1051/e3sconf/202126404088
29. Hoshimov U.H., Khushiev , and others. IOP Conference Series: Earth and Environmental Science, (2023).https://iopscience.iop.org/article/10.1088/1755-1315/1142/1/012030
30. Muratov Kh., Ishnazarov O.Kh., Isakov A., Nuritov I, and others. E3S Web of Conferences 365, 04005 (2023).
31. Pirmatov N.B., and others. E3S Web of Conferences, 2023, 383, 04046.
32. Pirmatov N.B., and others. AIP Conference Proceedings, 2023, 2612, 050005.
33. Pirmatov N.B., and others. AIP Conference Proceedings, 2022, 2432, 020020.
34. Pirmatov N.B., and others. AIP Conference Proceedings, 2022, 2552, 070009.
35. Pirmatov N.B., and others. E3S Web of Conferences, 2021, 264, 04028.

 https://doi.org/10.1051/e3sconf/202346101084

Methods and algorithms for determining the optimal structure of the telecommunications power supply system based on morphological synthesis

Fazliddin Qodirov[1]*

[1]Tashkent University of Management and Future Technologies, Tashkent, Uzbekistan

Abstract. It is necessary to develop design principles and methods based on morphological synthesis and analysis of technical solutions to solve the problem of optimal design of power supply systems for a telecommunications power supply system. The essence of the morphological method is that it is necessary to determine several morphological features (typicality, specificity, differentiation) that are important for the problem being solved, and to make all possible combinations of these features (parameters). Its advantage over other methods is that, taking into account the various structural and parametric characteristics of the power supply system, it is possible to analyze the functional characteristics of the power supply system, its actions in a dynamic mode under the influence of changes in the network, load and change the mode of the operation. This article studys the optimal structure of the telecommunications power supply system.

1 Introduction

In the creation of complex technical systems, several stages of synthesis are passed, which are, structural, functional and parametric stages. In the field of scientific research, there are no methods that allow immediate selection of the optimal structure of a complex system or device based on any criteria. Therefore, the process of creating complex technical systems is repeated (iterative) [10-12].

First, it determines which class the simulated device belongs to, then narrows the boundaries of this class, tests several solutions related to this class, and selects the most suitable one. Currently, computer components are widely used to support decisions at the initial stages of designing technical facilities. From analytical models and numerical optimization methods to heuristic methods and expert evaluation systems, there are many conceptual approaches that can serve as a basis for the development of decision-making systems [6].

The essence of the morphological method is that it is necessary to determine several morphological features (typicality, specificity, differentiation) that are important for the problem being solved, and to compile all possible combinations of these features (parameters). Technical systems that are being improved have several structural or functional morphological features, for example, each feature is related to some structural unit of the system, some of its tasks, some mode of operation of the system, that is, the solution of the problem and the achievement of the main goal. All of this can describe the parameters or characteristics of the related system. The advantage of the morphological method is its wide range of possibilities in formalization, algorithmization and computer implementation. As a result of morphological synthesis, a morphological set is determined that contains all structural solutions of the variants of the system under consideration and hypothetical, patentable structures. It is impossible to create a whole system due to the large size of the morphological collection. To create a system, included devices in this set should be described, the classification properties and the range of their possible values should be determined. A special feature of this classification is that we can determine the structure of the device by specifying all the values of the classification properties. Morphological synthesis belongs to the class of combinatorial-logical synthesis methods, and one of its disadvantages is that it is a complex process to implement on a computer, taking into account the complete enumeration of possible options. In addition, the existing universal algorithms do not guarantee to determine and synthesize the desired result in a short time. Therefore, Everest methods (improvement of the prototype, improvement of tree and labyrinth constructions) are widely used in solving such problems. Recently, genetic algorithms simulating evolutionary processes are being frequently used in morphological synthesis.

* Corresponding author: ilider1987@yandex.com

2 Setting of the problem

Since any device or system has a structure and its constituent elements have parameters, structural-parametric synthesis methods can be used in almost all areas of scientific and technical knowledge. Therefore, in this dissertation work, it is proposed to develop the theory of morphological synthesis, which includes the structures and individual devices of the power supply system (PSS), which are part of the morphological set, and the models of PSS, taking into account conductive interferences. To implement the morphological method, it is necessary to supplement the morphological set with methods of mathematical modeling of the considered objects, as well as with engineering calculations and obtained knowledge during their operation. It should be noted that morphological methods are the basis for modeling complex technical systems and are available in models of all levels of the hierarchy. Therefore, the principles of optimal PSS design for telecommunication systems are a combination of morphological methods, mathematical modeling methods and engineering developments. The macro model of the power supply system, taking into account the effect of conductive interference, is a model of devices of several classes, which select these devices according to certain criteria (low level of conductive interference, reliability, quality of energy conversion, high energy efficiency sensors, minimum body cost, etc.) are complemented by synthesis methods and algorithms with optimization [8].

3 Methods

Morphological set includes all structure of PSS (for example, centralized or decentralized). There are various schematic and technical solutions, which differ in the parameters of the system objects under consideration, the morphological set of individual elements. For example, rectifiers can be implemented with or without a transformerless input circuit using a power factor rectifier (PFC) circuit. Morphological set elements differ in system structure and parameters (transformers, filtering devices, etc.). Such sets can be unordered or partially ordered [12].

We understand the whole complex of all unordered morphological set of telecommunication systems power supply structures. To represent a set of PSSs using a morphological tree, it is necessary to arrange the elements of the set according to some characteristics - functional, structural or other aspects. Despite the fact that the classification features are the same, the morphological set of features of power supply systems and devices of telecommunication systems can be classified in different ways.

By defining the classification features and bringing them into a single system, we get a set that represents the model of the morphological set at the level of identification. We call this as model of the studied power supply system M1 macromodel (Fig. 1). It can be represented as a morphological tree or a morphological table. The representation of the morphological set in one form or another is represented by mathematical models (in the form of systems of equations) [6].

A multi-level model of the power supply system.

Modeling process of the systematization og the power supply system consists of dividing a complex nonlinear system into a number of subsystems and creating a five-level hierarchical structure - M1 macro model (Fig. 1).

To create a morphological analysis algorithm and present its results in the form of the M1 macro model, we divide subsystems into main and subclasses according to its architecture and accepted classification features. The main class, which includes the macro model of the power supply system, can be divided into subclasses taking into account their functional aspects.

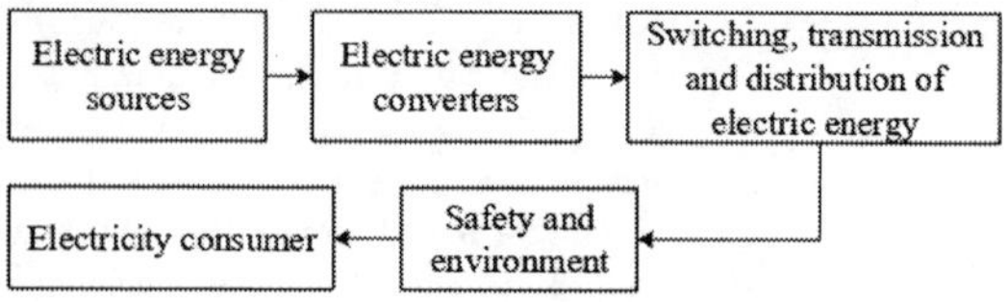

Fig 1. Five-level hierarchical structure of the power supply system (M1 macro model)

To ensure the normal operation of the equipment of the industrial facility, the power supply system must perform the following functions:

- electromagnetic compatibility (EMC) requirements, including transmission of quality electricity from a source to a specific consumer; ensuring electrical safety of employees and protecting equipment from external influences;
- ensuring the specified state of the air environment for the normal operation of equipment and the life of service personnel.

In terms of performing these functions, the power supply system can be represented as a macro model consisting of five blocks (hierarchical levels) shown in Figure 1. The model includes the following blocks (subsystems):

- block of electrical energy sources;
- unit for converting one type of energy into another type of energy;
- other subsystems, as well as power transmission and distribution networks, a block of switching devices;
- block of protection, ventilation and air conditioning systems;
- a block of consumers with different parameters and requirements for the EMC of the equipment of the industrial facility, the quality of electricity.

When further decomposing the M1 macromodel and analyzing the subsystems, we take into account the functional nature of their classification. This classification is shown in Figure 2.

In the further analysis of the decomposition of the blocks included in the subsystems presented in Figure 2, the structural and parametric features of the classification are taken into account. To solve the problem of synthesis with optimization according to any criteria, we set

https://doi.org/10.1051/e3sconf/202346101084

priorities between classification features and block parameters, taking into account the suitability of the parameters of individual elements of the PSS (Fig. 3, 4, 5, 6).

The main functional task of the power supply system is to change, control, distribute and continuously supply electricity from the source to the consumer with constant and variable voltages [11]. Modern industrial equipment includes a set of devices with information systems and a complex infrastructure, which requires PSS manufacturers to take a comprehensive approach to creating an optimal system of industrial equipment and PSS systems working together. In the process of creating such an optimal power supply system, it is necessary to ensure and take into account the quality of power conversion, EMC with industrial facilities, reliability of power supply, safety for people and the environment.

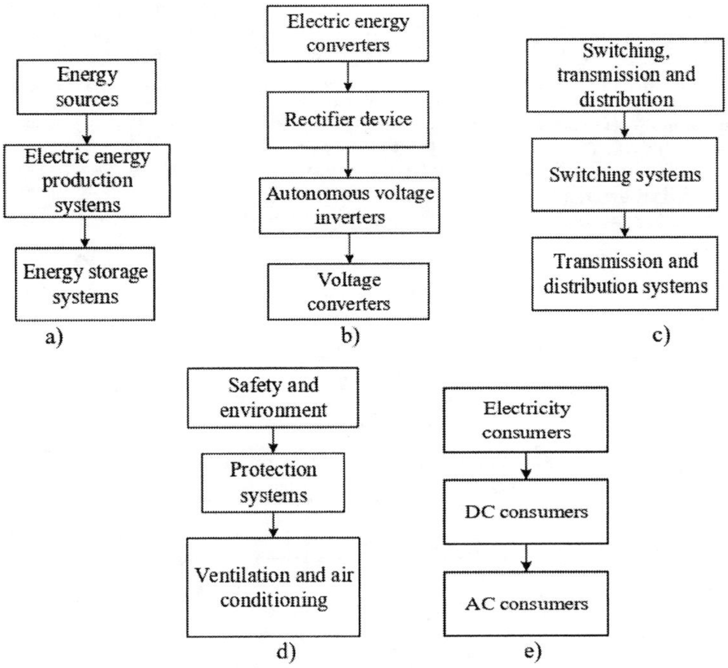

Fig 2. The hierarchical structure of the five-level M1 macromodel power supply system.

In addition to general system requirements for power supply systems, there are also requirements defined by regulatory and technical documents. Macromodel M2 consists of five hierarchical levels, and the first level contains sources of electric power, power generation and storage systems (Fig. 3).

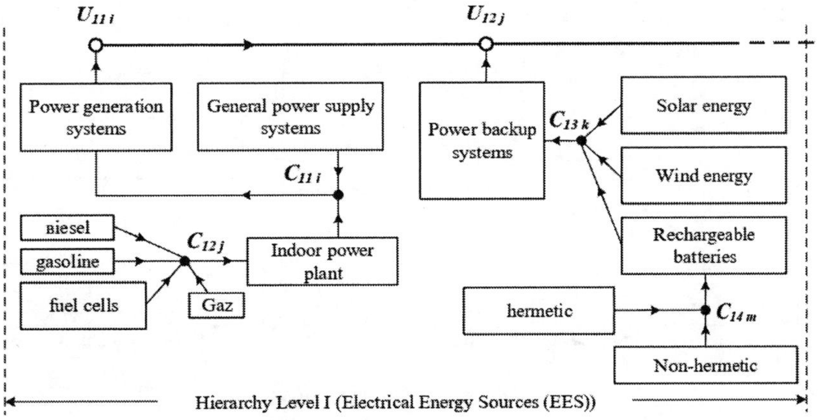

Fig 3. Level I of the hierarchy.

https://doi.org/10.1051/e3sconf/202346101084

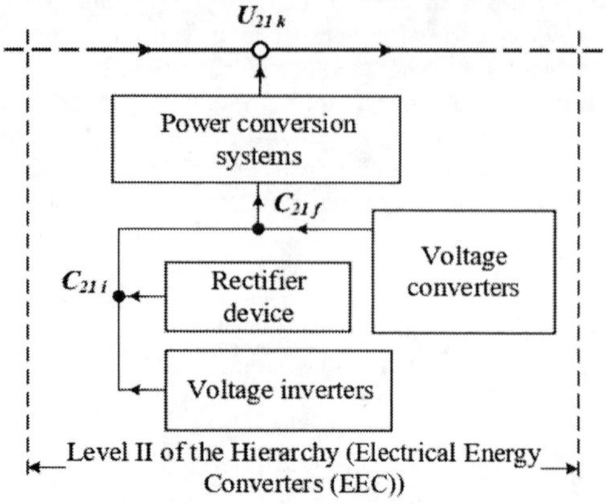

Fig 4. Level II of the hierarchy.

The Republic of Uzbekistan has developed and is implementing the state standard UzDSt3055:2016 "Telecommunications networks. Organizational ATS. General technical requirements and methods of control". In accordance with this standard, general power supply and redundancy systems provide telecommunications devices with three-phase or single-phase alternating current with a frequency of 50 Hz and a rated voltage of 220/380 V [1-2-3-4].

In order to meet the main requirement for the reliability of power supply of telecommunication equipment and continuous (or guaranteed) power supply, it is necessary to ensure that the power supply system has at least two inputs from independent energy sources and includes an internal power plan (U_{11i}, U_{12j}).

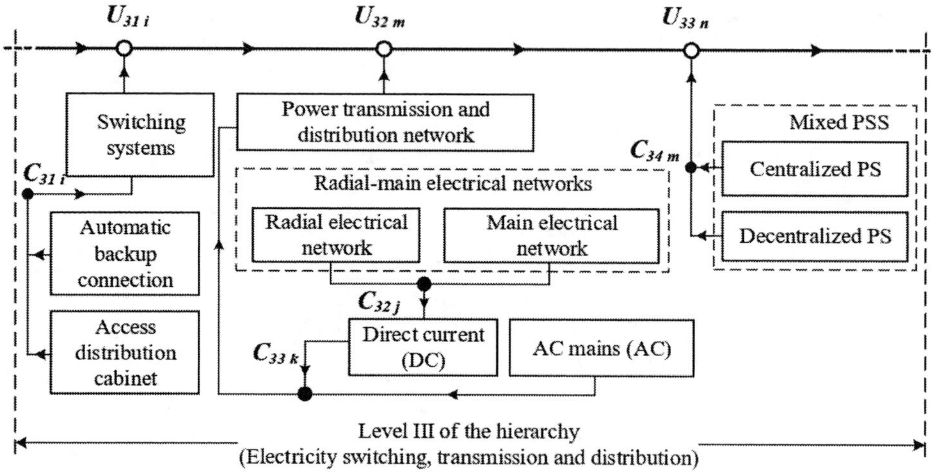

Fig 5. Level III of the hierarchy

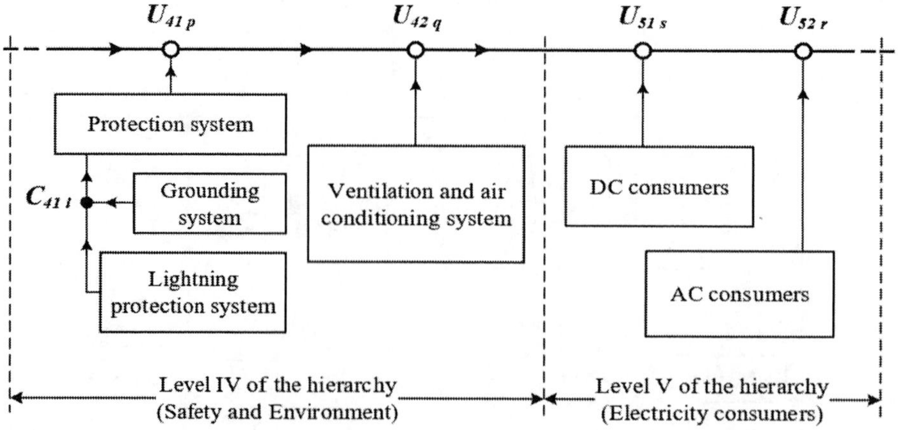

Fig 6. Level IV and V of the hierarchy.

 https://doi.org/10.1051/e3sconf/202346101084

4 Results

Based on these requirements, electricity generation (U_{11i}) occupies the first place in the hierarchy of energy resources subsystem.

In the second hierarchical level of the M2 macro model, there are energy converters (U_{21k}) containing three hierarchical levels, which are rectifier, voltage inverter and voltage converter (Fig. 4). The quality of electric energy at the output of the PSS must correspond to the standards established for the quality of electric energy at the inputs of the power supply schemes of telecommunication devices and other industrial facilities. In addition, the conversion of the necessary electrical energy for electrical devices must be efficient.

The third hierarchical level of the M2 macro model contains switching, transmission and distribution systems of electric energy, which includes two hierarchical levels (U_{31i}, U_{31m}), consisting of the switching system (SS), transmission and distribution networks (Fig. 5). SS is located at the top of the hierarchy of the sub-system of switching, transmission and distribution of electric energy. SS is designed to switch from one operating mode of the power supply system to another, and one of the requirements for it is to provide uninterrupted power supply to consumers and protect against short circuits that occur on the load side.

The fourth hierarchical level of the M2 macro model is the safety and environmental system (U_{41p}, U_{42q}), which includes two layers, i.e. protection (U_{41p} p), ventilation and air conditioning (U_{42q}). Shown on (Fig. 6).

At the fifth hierarchical level of the M2 macro model are electricity consumers (U_{51sp}, U_{52r}), according to consumer requirements, they are direct (U_{51sp}) and alternating current (U_{52r} r) consumers (Fig. 6).

The developed M2 garf model includes a five-level M1 macromodel. Macromodel M1 has the form of a tree with logical "AND" and "OR" logical nodes, where Sij is a constant coefficient for evaluating options according to the specified criteria of its attached elements for choosing alternative solutions at one level, Ukm is a graph of macromodel M1 option valuation coefficients. In order to create a continuous connection between the nodes in the graph (if the system has not passed through the sheep system), a single connection coefficient was introduced, which allows to create continuity along the graph while preserving the parameters. For example, while using a hermetic, it is not always necessary to install a ventilation and air conditioning system.

5 Conclusion

1. The analysis of the modeling methods showed that the morphological analysis and synthesis method is the most suitable method in the hierarchical trees of the power supply system. Its advantage over other methods is that, taking into account various structural and parametric characteristics of the power supply system, it is possible to analyze the functional characteristics of the power supply system, its dynamic behavior under the influence of changes in the network and load, and the change in the operating mode.

2. It is the most optimal modeling package for creating automated design systems of the power supply system for the best indicators of power quality, as well as for establishing relationships between elements at different hierarchical levels, both linear and non-linear. includes functions minimization and maximization programs.

3. A systematic approach to the analysis of power supply systems is the formation of an orderly set of the hierarchy of classification characteristics of the sub-systems that are part of it, the creation of morphological trees and subgraphs that will be part of the macromodel M1 in the future, the formation of symbolic-numerical relations between the nodes of the macromodel tree in the form of matrices allows.

4. According to the classification characteristics of the elements of the five levels of the M1 model, followings were created after processing the data obtained during the systematic analysis of the power supply system of the telecommunications facility:

1) subgraphs of all levels of macro model M1, their adjacency and incident matrices;

2) M2 model and morphological set of elements.

To create the M3 model, it is necessary to create equations describing the obtained subgraphs, create a database containing external and internal parameters. Moreover, the problem of synthesizing the optimal power supply system through their controlled parameters is solved.

References

1. O‘z DSt 1188-2008. Electric Energy. Electromagnetic compatibility of technical equipment. Standards for the quality of electrical energy in general-purpose power supply systems. Tashkent 2008. -64 p.
2. O‘z DSt 1063:2004. Electromagnetic compatibility of technical devices. Instruments for measuring industrial radio emissions. Technical requirements and test methods. Tashkent 2009. -102 p.
3. O‘z DSt 1032:2010 Industrial radio equipment. Test methods of sources of industrial radio pollution. Tashkent 2010. -45 p.
4. O‘z DSt 3055-2016. State standard of the Republic of Uzbekistan. Telecommunication networks. Institutional PBX. General technical requirements and control methods. Official publication of the Uzbek Agency for Standardization, Metrology and Certification. Tashkent 2016. -82 p.
5. Odrin, B.M. Method of morphological analysis of technical systems. M.: VNIIPI, 1989.- 312 p.
6. Sapaev M., Qodirov F.M., Qodirova S.F. The mechanism of interaction of the power supply system with the telecommunications facility. Descendants of Muhammad al-Khorazmi, No. 1(19), March 2022. -84 p.
7. Kadirov F.M., Nazarov F.D., Abdullaev E. Methods for improving the harmonic composition of voltage and current. Information technologies and process modeling in fundamental and applied research: materials of the 1st International Youth School-Conference. December 15–17, 2016 / under

general. ed. D. P. Anufrieva. – Astrakhan: GAOU JSC VO "AGASU", 2016. – 234 p.
8. Kodirov F.M. Modelling and research of harmonic components of current and voltage in electric nets. Universum: technical sciences. Science Magazine. Issue: 2(95). Part 7. February 2022.
9. Anatolevna, A. D., & Misliddinovich, Q. F. (2023, June). Infokommunikatsiya elektr ta'minoti tizimlarining parametrik va strukturaviy tahlil qilishni optimallashtirish. In "Canada" International conference on developments in education, sciencesand humanities (Vol. 9, No. 1).
10. Kodirov F.M. Use of ICT in the design of telecommunication power supply systems. International scientific and practical conference on the topic "Use of modern information and communication technologies in the implementation of reforms in new Uzbekistan". Andijan October 27-29, 2021. 565 p.
11. F. Qodirov, G. Saidova, G. Saidova, & M. Agzamova (2022). Development of a multi-level model of the telecommunications power supply system through morphological analysis. Science and innovation, 1 (A8), 584-594. doi: 10.5281/zenodo.7422230.
12. Rogulina, L.G. Analysis of the reliability of power supply systems for telecommunications equipment/L.G. Rogulina//Bulletin of Tambov State Technical University-2010. - Volume 16, No. 3. - P. 698.
13. Misliddinovich, Q. F., & Ayupova, D. A. (2023). Development of an algorithm for morphological analysis of power supply of infocom-communication systems. Obrazovanie nauka i innovatsionnye idei v mire, 22(8), 40-49.
14. A.Yu.Vorobiev. Power supply for computer and telecommunication systems. M.: Eco - Trends, 2003. - 280 p.
15. Rakov D.L. Structural analysis and synthesis of new technical systems based on the morphological approach. ISBN 978-5-397-02604-8, URSS, 2011, 160 p.
16. A. V. Andreychikov, O. N. Andreychikova System analysis and synthesis of strategic decisions in innovation. Mathematical, heuristic and intellectual methods of system analysis and synthesis of innovation. ISBN 978-5-397-03656-6, M.: Librocom, 2013, - 306 p.
17. James A. Momoh. Electric Power Distribution, Automation, Protection, and Control. ISBN 9781315221991. Published December 21, 2017 by CRC Press. 384 Pages 147 B/W Illustrations.
18. James A. Momoh. Electric Power System Applications of Optimization. ISBN 9780367386160. Published October 7, 2019 by CRC Press. 602 Pages.
19. Qodirov, F., G. Saidova, and M. Agzamova. "Development of a design algorithm for the development of automated control systems in a telecommunications source." Science and Innovation 1.8 (2022): 567-577.

To evaluate the operational status of the transformer load using a feed-forward neural network for analysis

Fazliddin Khojayorov[1*], *Anvar* Saidkhodjaev[1]

[1]Tashkent State Technical University named after Islam Karimov, 100095, Uzbekistan, Tashkent, University St. 2A

Abstract. The results of the first transformer load obtained using the FNN neural network in Fig. 5 determined on the basis of the algorithm described in Fig. 4 show that if the dynamics of transformer loads continue at this rate, after 8 years the minimum loads will increase from 0.8 and after 12 years begins to work in the danger zone completely. Taking into account that the coefficient of wear of transformers and the occurrence of minimum loads is equal to 20% according to **Fig.**1, it can be said that this situation is in a very serious situation. And the maximum value of the load has already reached its maximum point. In this case, it is suggested that the issue of load redistribution in this Kibray 35/6 substation and its distribution networks should be seriously considered or a new transformer should be installed and appropriate switching devices should be selected.

1 Introduction

Today, substations are equipped with modern accounting, relay protection, and automatic control systems. The integration of modern accounting, relay protection, and automatic control systems in substations represents a significant advancement in the power distribution and management sector. This integration provides several crucial benefits that enhance the efficiency, reliability, and safety of the energy grid [1,2,3,6]. Modern accounting systems integrated into substations provide real-time and accurate data on energy consumption [4,5,7]. This detailed information allows for better analysis of usage patterns, helps in identifying peak usage times, and facilitates more efficient load distribution planning [8]. Automatic control systems within substations enable intelligent and automated management of energy distribution. This ensures that power is supplied to different areas based on real-time demand, optimizing the load and reducing wastage [9]. However, these systems are not able to predict the expected future loads.

It is known, the normal load factor of a power transformer refers to the average percentage of the transformer's rated capacity that is being utilized over a specific period of time. It is important to operate a transformer within its normal load factor to ensure efficient and reliable operation [10,11]. The normal load factor is typically expressed as a percentage of the transformer's rated capacity. A common practice in power systems is to operate transformers at a normal load factor between 70% and 80% of their rated capacity to ensure optimal performance, efficiency, and longevity [12]. Operating a transformer below 30% of its rated capacity or above 90% can lead to inefficiencies and may impact the transformer's lifespan and reliability [13]. It's important for power system operators and engineers to carefully monitor the load on transformers and manage the distribution of loads to maintain a suitable normal load factor, balancing efficiency and cost-effectiveness with the transformer's design and capabilities [16-23].

2 Experimental research

In the case of an incident that occurred during the expedition, investigations were carried out on the load variations of a 35/6 kV two 4000 kVA transformers at the Kibray substation (the subsequent research object). Initially, the process of changing the total power of the auxiliary transformers on the researched object was analyzed.

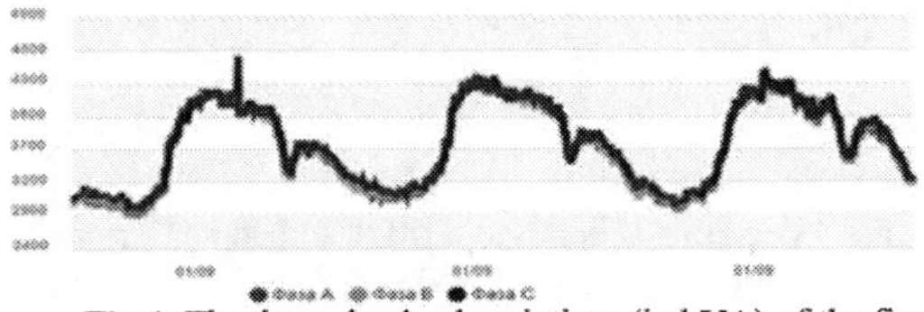

Fig.1. The three-day load variations (in kVA) of the first transformer at the research object (obtained from the radiometer system).

The load graphs of the transformers at the research object have been quite similar. As seen in Figure 1, the transformer operates with a load exceeding 70% of its capacity for the majority of the observed times. Additionally, the maximum and minimum load values vary from day to day. Therefore, continuous monitoring of the loadings of these transformers to analyze their

* Corresponding author: ilider1987@yandex.com

operations and determine these values, namely the maximum and minimum loading limits, is necessary to align with the intended purpose [24-29].

In the subsequent phase, predictions of the expected future loads of the transformers at the Kibray substation were conducted using artificial intelligence, specifically the Feed forward neural network (FNN) approach. It should be noted that the initial step in identifying the anticipated future loads of the transformers involves gathering data and their primary reprocessing [15]. The data for the research were obtained from the radiometer system, which records the following parameters for each feeder (see Figure 2):

1. Three-phase currents (A) on all phases.
2. Voltages (kV) on all phases.
3. Active power (kW).
4. Reactive power (kVAr).
5. Power factor angle (degrees).

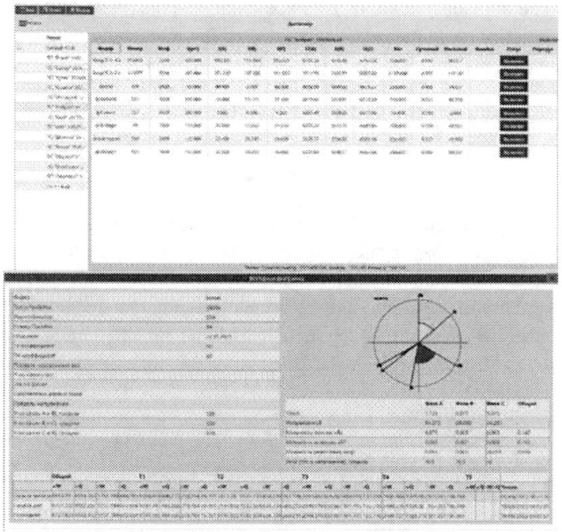

Fig.2. Radiomer digital system interface

As primary data, the monthly average maximum and minimum load factors of two 35/6 transformers for all months in 2021, 2022 and from January to April in 2023 were calculated using the following formula:

$$k = \frac{S_f}{S_0} \quad (1)$$

where, S_f is the actual value of total power, S_0 is the amount of nominal power [30-36].

The average minimum (k_{min}) and maximum (k_{max}) values of loading calculated based on formula (1) were taken as primary factors below (Table 1).

Table 1.

date	T1			T2		
	k_{min}	k_{max}	Δk	k_{min}	k_{max}	Δk
1.2021	0,26	0,73	0,47	0,26	0,75	0,49
2.2021	0,25	0,82	0,57	0,27	0,87	0,60
3.2021	0,23	0,98	0,75	0,25	1,01	0,76
4.2021	0,23	0,98	0,75	0,24	1,06	0,82
5.2021	0,24	0,88	0,64	0,24	0,89	0,64
6.2021	0,24	0,78	0,54	0,26	0,85	0,59
7.2021	0,33	0,97	0,64	0,34	1,02	0,68
8.2021	0,32	0,95	0,63	0,35	1,02	0,67
9.2021	0,27	0,76	0,49	0,28	0,82	0,54
10.2021	0,25	0,96	0,71	0,25	1,04	0,78
11.2021	0,29	0,78	0,49	0,31	0,84	0,53
12.2021	0,31	0,95	0,64	0,34	1,02	0,68
1.2022	0,35	0,80	0,45	0,38	0,86	0,47
2.2022	0,37	0,90	0,53	0,37	0,98	0,61
3.2022	0,32	0,81	0,49	0,33	0,84	0,51
4.2022	0,38	0,81	0,43	0,41	0,87	0,46
5.2022	0,38	0,94	0,56	0,40	1,01	0,61
6.2022	0,35	0,95	0,60	0,37	1,02	0,64
7.2022	0,36	0,95	0,59	0,38	0,99	0,61
8.2022	0,43	0,93	0,50	0,43	0,95	0,51
9.2022	0,35	1,00	0,65	0,38	1,04	0,66
10.2022	0,40	0,85	0,45	0,40	0,92	0,51
11.2022	0,45	0,92	0,47	0,48	0,98	0,50
12.2022	0,45	0,95	0,59	0,39	0,97	0,58
1.2023	0,42	0,95	0,56	0,40	0,98	0,58
2.2023	0,46	1,00	0,56	0,45	1,02	0,57
3.2023	0,47	0,92	0,48	0,45	0,96	0,51
4.2023	0,47	0,85	0,42	0,46	0,88	0,42

In Table 1, in addition to the k_min and k_max values, the difference between them Δk is presented, and from the decrease of this value, we should say that as the loads have increased over the years, the difference between the maximum and minimum loads is also decreasing. At the same time, if we take into account that the transformers begin to work inefficiently when the load exceeds 90%, the load transition limit can be set at 85%. If we analyze the maximum values presented in Table 1, it can be seen that they have already exceeded 85%. This situation occurred 18 times in the T1 transformer and 23 times in the T2 transformer among the 28 time series data. This shows that they worked at an average load of 80%, and the time of occurrence of these loads can be seen from Figure 1 that in more than 70% of the 3-day period, the load of the transformers is greater than 85%. Given that this situation is already in an abnormal state, it is enough to perform the forecasting problem only through the minimum values of loading. At first, the adequacy of k_min given for each transformer in Table 1 was determined in IBM SPSS using the Gaussian distribution law of the received data (Fig. 2, 3).

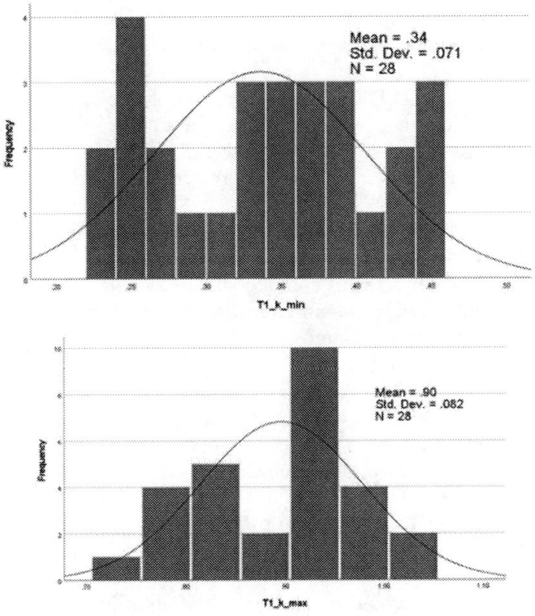

Fig.2. Checking the first data for the Gaussian normal distribution law

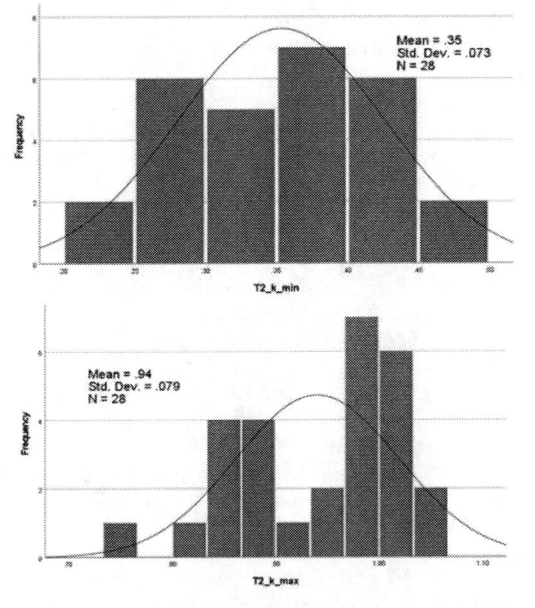

Fig.3. Checking the second data for the Gaussian normal distribution law U1

The values in Figure 2 and Figure 3 show that the received primary data can be used in the next step. This situation indicates that there is no need to build a Fit model in this case.

For forecasting, it is appropriate to divide the primary data into 80% train set, 15% validation set and 5% test set according to the quantity.

Feed Forward Neural Network (FNN) was used to calculate loading coefficients. It has 2 layers, one hidden layer and one output layer. 100 epochs and 16 batch sizes were taken as FNN parameters. The hidden layer has 10 neurons and one hidden layer with ReLU (Rectified Linear Unit) activation function. The output layer has one output layer with 1 neuron (unit) and no activation function is specified.

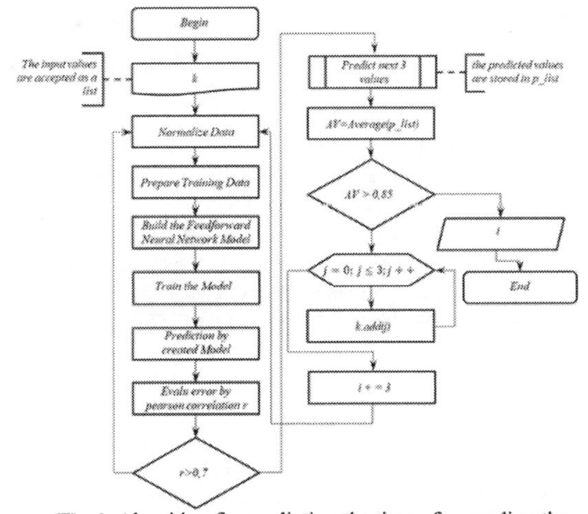

Fig.4. Algorithm for predicting the time of exceeding the permissible value of the load coefficient of transformers

Forecasting is carried out based on the algorithm shown in Figure 4. In this algorithm, a model is first prepared by train and gets results corresponding to train. The obtained results are assessed for compatibility with the entered values using Pearson's correlation. If the Pearson correlation is less than 0.7, the algorithm is sent back for relearning. Otherwise, the next 3 load factors are projected. At this point, it should be noted that it is not the state of the following values that is important to us, but when the load of the transformer exceeds 0.85. Because one of the main goals was to implement planning by calculating how long the load of the transformer will increase. In this case, the next 3 values are predicted and their average values are calculated. If the average value is below 0.85, the predicted values are added back to the initial data list and the next 3 values are predicted. In this order, the period of acceptance of loadings higher than 0.85 is calculated [37-41].

3 Research results

The results of the first transformer load obtained using the FNN neural network in Fig. 5 determined on the basis of the algorithm described in Fig. 4 show that if the dynamics of transformer loads continue at this rate, after 8 years the minimum loads will increase from 0.8 and after 12 years begins to work in the danger zone completely. Taking into account that the coefficient of wear of transformers and the occurrence of minimum loads is equal to 20% according to Figure 1, it can be said that this situation is in a very serious situation. And the maximum value of the load has already reached its maximum point. In this case, it is suggested that the issue of load redistribution in this Kibray 35/6 substation and its distribution networks should be seriously considered or a new transformer should be installed and appropriate switching devices should be selected.

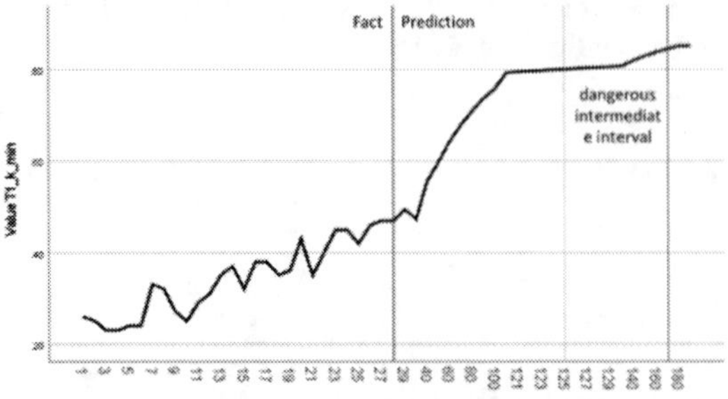

Fig.5. Prediction results of the first transformer load obtained using the FNN neural network (since the loads of both transformers are almost the same, only the results of the first transformer are presented)

4 Conclusion

1. The in-depth assessment of the transformers at the Kibray substation paints a concerning picture of their operational conditions. The frequent breaching of the optimal load factor underlines an emerging inefficiency, coupled with potential threats to the longevity and dependability of these transformers.

2. Leveraging the Feed forward neural network (FNN) approach yielded invaluable insights into the future load expectations on the transformers. The projection indicates that, without intervention, the transformers at the Kibray substation are poised to consistently operate in high-risk zones within the next 8-12 years.

3. Such a trajectory of increasing load factors not only jeopardizes the operational efficiency of the transformers but also signals potential disruptions in power distribution. Moreover, continuous operation at these levels could accelerate wear and tear, ultimately leading to frequent maintenance needs or potential system failures.

4. Given the aforementioned challenges, it becomes imperative for stakeholders to act promptly. Immediate considerations should encompass strategies for load redistribution to evenly distribute the operational pressure among transformers. Additionally, expanding the infrastructure of the Kibray 35/6 substation, such as introducing new transformers or implementing state-of-the-art switching devices, could effectively mitigate the impending risks and ensure a consistent and reliable power supply.

References

1. https://ieeexplore.ieee.org
2. Feeder protection and control REF620 IEC - ABB Group. https://new.abb.com/medium-voltage/digital-substations/protection-relays/feeder-protection-and-control/feeder-protection-and-control-ref620-iec.
3. Multilin 750/760 - GE Grid Solutions. https://www.gegridsolutions.com/products/brochures/750760_gea31955.pdf.
4. REF601 is a feeder protection relay for protection and control. https://new.abb.com/medium-voltage/digital-substations/protection-relays/feeder-protection-and-control/feeder-protection-and-control-ref601.
5. Digital protection relays and control - SIPROTEC 5 - Siemens. https://www.siemens.com/global/en/products/energy/energy-automation-and-smart-grid/protection-relays-and-control/siprotec-5.html.
6. Nayak, Shivakumar & Chandan, & Ambiger, Pooja & Patil, Sahana. (2023). Development and Implementation of Transformer Breather Health Monitoring System Using IoT. 10.22214/ijraset.2023.55632.
7. Wang, Lu & Fan, Yinqiu & Yang, Xiaoxu & Li, Bin & Li, Wei & Xue, Jia & Wang, Hui & Wang, Gang & Guo, Xia. (2023). Exploration of Transformer Operation and Maintenance Technology and Realization of Transformer Condition Monitoring System. 10.1007/978-981-99-1439-5_75.
8. Cho, Jintae & Yoon, Yeunggul & Son, Yongju & Kim, Hongjoo & Ryu, Hosung & Jang, Gilsoo. (2022). A Study on Load Forecasting of Distribution Line Based on Ensemble Learning for Mid- to Long-Term Distribution Planning. Energies. 15. 2987. 10.3390/en15092987.
9. Yan, Sheng & Hu, Minqiang. (2022). A Multi-Stage Planning Method for Distribution Networks Based on ARIMA with Error Gradient Sampling for Source–Load Prediction. Sensors. 22. 8403. 10.3390/s22218403.
10. Bhattar, Chandrakant & Chaudhari, Madhuri. (2023). Centralized Energy Management Scheme for Grid Connected DC Microgrid. IEEE Systems Journal. PP. 1-11. 10.1109/JSYST.2022.3231898.
11. Faqih, Mochammad & Binti Omar, Madiah & Ibrahim, Rosdiazli. (2023). Prediction of Dry-Low Emission Gas Turbine Operating Range from Emission Concentration Using Semi-Supervised Learning. Sensors. 23. 3863. 10.3390/s23083863.
12. A common practice in power systems is to operate transformers at a normal load factor between 70% and 80% of their rated capacity to ensure optimal performance, efficiency, and longevity
13. Balanta, J.Z.; Rivera, S.; Romero, A.A.; Coria, G. Planning and Optimizing the Replacement Strategies of Power Transformers: Literature Review. Energies 2023, 16, 4448. https://doi.org/10.3390/en16114448
14. Zhang, X.; Sun, Z. Application of Improved PNN in Transformer Fault Diagnosis. Processes 2023, 11, 474. https://doi.org/10.3390/pr11020474
15. Svozil, Daniel & Kvasnicka, Vladimir & Pospíchal, Jiří. (1997). Introduction to multi-layer feed-forward neural networks. Chemometrics and Intelligent Laboratory Systems. 39. 43-62. 10.1016/S0169-7439(97)00061-0.

16. Rasulov A.N., Ruzinazarov M.R. E3S Web of Conferences 384. 01042. (2023). https://doi.org/10.1051/e3sconf/202338401042
17. Rashidov N., Rozmetov Kh., Rismukhamedov S., Peysenov M. E3S Web of Conferences 384. 01043. (2023). https://doi.org/10.1051/e3sconf/202338401043
18. Tuychiev F, Haqberdiev A. E3S Web of Conferences 384. 01044. (2023). https://doi.org/10.1051/e3sconf/202338401044
19. Bobojanov M., Torayev S. E3S Web of Conferences 384. 01045. (2023). https://doi.org/10.1051/e3sconf/202338401045
20. Reymov K.M., Esemuratova Sh.M., Khusanov B.M., Mytnikov A.V. E3S Web of Conferences 384. 01047. (2023). https://doi.org/10.1051/e3sconf/202338401047
21. Kholikhmatov B.B., Samiev Sh.S., Erejepov M.T., Nematov L.A. E3S Web of Conferences 384. 01032. (2023). https://doi.org/10.1051/e3sconf/202338401032
22. Melikuziev M.V. E3S Web of Conferences 384. 01033. (2023). https://doi.org/10.1051/e3sconf/202338401033
23. Bobojonov Y.M., Saidkhodjaev A.G. E3S Web of Conferences 384. 01048. (2023). https://doi.org/10.1051/e3sconf/202338401048
24. Hoshimov, F.A., Najimova A.M., Rafikova G.R., Bakhadirov I.I. E3S Web of Conferences 384. 01049. (2023). https://doi.org/10.1051/e3sconf/202338401049
25. Ruzinazarov M.R. E3S Web of Conferences 384. 01050. (2023). https://doi.org/10.1051/e3sconf/202338401050
26. Bobojanov M.K., Rismukhamedov D.A., Tuychiev F.N., Shamsutdinov Kh.F. E3S Web of Conferences 365. 04024. (2023). https://doi.org/10.1051/e3sconf/202336504024
27. Ushakov, V.Y., Mytnikov, A.V., Rakhmonov, I.U. Past, Present and Future of Electric Power Systems (Brief Excursion). In: High-Voltage Equipment of Power Systems. Power Systems. Springer, Cham. (2023). https://doi.org/10.1007/978-3-031-38252-9_1
28. Ushakov, V.Y., Mytnikov, A.V., Rakhmonov, I.U. Characteristics of the Main Elements of Electric Power Systems. In: High-Voltage Equipment of Power Systems. Power Systems. Springer, Cham. (2023). https://doi.org/10.1007/978-3-031-38252-9_2
29. Ushakov, V.Y., Mytnikov, A.V., Rakhmonov, I.U. Insulating Materials and Media Used in High-Voltage Elements of Electric Power Systems. In: High-Voltage Equipment of Power Systems. Power Systems. Springer, Cham. (2023). https://doi.org/10.1007/978-3-031-38252-9_3
30. Ushakov, V.Y., Mytnikov, A.V., Rakhmonov, I.U. Effects on Equipment Causing Insulation Aging and Failure. In: High-Voltage Equipment of Power Systems. Power Systems. Springer, Cham. (2023). https://doi.org/10.1007/978-3-031-38252-9_4
31. Ushakov, V.Y., Mytnikov, A.V., Rakhmonov, I.U. Defects in High Voltage Equipment: Types and Content of Tests. In: High-Voltage Equipment of Power Systems. Power Systems. Springer, Cham. (2023). https://doi.org/10.1007/978-3-031-38252-9_5
32. A.N.Rasulov, E.G.Usmonov, M.V.Melikuziev. AIP Conference Proceedings, 2552, 050018, (2023), https://doi.org/10.1063/5.0111530
33. Ushakov, V.Y., Mytnikov, A.V., Rakhmonov, I.U. Non-electrical Diagnostic Methods. In: High-Voltage Equipment of Power Systems. Power Systems. Springer, Cham. (2023). https://doi.org/10.1007/978-3-031-38252-9_6
34. Ushakov, V.Y., Mytnikov, A.V., Rakhmonov, I.U. Traditional Electrical Diagnostic Methods. In: High-Voltage Equipment of Power Systems. Power Systems. Springer, Cham. (2023). https://doi.org/10.1007/978-3-031-38252-9_7
35. Ushakov, V.Y., Mytnikov, A.V., Rakhmonov, I.U. Diagnostics of High-Voltage Equipment by Defectography Methods. In: High-Voltage Equipment of Power Systems. Power Systems. Springer, Cham. (2023). https://doi.org/10.1007/978-3-031-38252-9_8
36. Ushakov, V.Y., Mytnikov, A.V., Rakhmonov, I.U. Dielectric Spectroscopy Technology. In: High-Voltage Equipment of Power Systems. Power Systems. Springer, Cham. (2023). https://doi.org/10.1007/978-3-031-38252-9_9
37. Ushakov, V.Y., Mytnikov, A.V., Rakhmonov, I.U. Diagnostics of High Voltage Equipment. In: High-Voltage Equipment of Power Systems. Power Systems. Springer, Cham. (2023). https://doi.org/10.1007/978-3-031-38252-9_10
38. Ushakov, V.Y., Mytnikov, A.V., Rakhmonov, I.U. Diagnostics of High-Voltage Cable Lines. In: High-Voltage Equipment of Power Systems. Power Systems. Springer, Cham. (2023). https://doi.org/10.1007/978-3-031-38252-9_11
39. Ushakov, V.Y., Mytnikov, A.V., Rakhmonov, I.U. Diagnostics of Insulating Structures of Overhead Power Lines and Outdoor Substations. In: High-Voltage Equipment of Power Systems. Power Systems. Springer, Cham. (2023). https://doi.org/10.1007/978-3-031-38252-9_12
40. Toirov O., Khalikova M., Jumaeva D., Kakharov S. E3S Web of Conferences 401. 05089. (2023). https://doi.org/10.1051/e3sconf/202340105089
41. Toirov O., Sadullaev T., Abdullaev D., Jumaeva D., Ergashev Sh., Sapaev I.B. E3S Web of Conferences 383. 01029. (2023). https://doi.org/10.1051/e3sconf/202338301029

E3S Web of Conferences **461**, 01086 (2023) https://doi.org/10.1051/e3sconf/202346101086
RSES 2023

Development of the theory of complexly closed city electrical networks

Anvar Saidkhodjaev[1*], *Fazliddin* Khojayorov

[1]Tashkent State Technical University named after Islam Karimov, 100095, Uzbekistan, Tashkent, University St. 2A

Abstract. The article discusses the issues of optimization and selection of the most rational scheme of a complex electrical network for the power supply of a microdistrict. And this requires the development of a more accurate definition of the calculated electrical loads of consumers, as well as the total loads of TP, RP, GRP, standard daily schedules, specific standards for calculating electrical loads, new building codes and regulations to replace the old KMK 2.04.17-98, determination of the most advantageous power and number of transformer substations, calculation of systems of complex power supply circuits using various methods, new technologies and digital automatic relay protection of urban power supply systems, use of reverse power automatic machines, contactless elements in power supply systems.

1 Introduction

Currently, due to the widespread use of electronic technology, the consumer of electricity in cities, when drawing up a contract, requires an uninterrupted and high-quality electricity supply from the energy supply organization.

Based on the Decrees of the President of the Republic of Uzbekistan on the comprehensive construction of massifs, neighborhoods, microdistricts under one cadastre, where there will be no more new construction of facilities, these requirements of electricity consumers can be met on the basis of complex urban electrical networks.

The relevance is confirmed by the need to reconstruct and modernize existing and planned new urban electrical networks that meet the requirements of new times; this poses the task of developing new approaches, methods and formulas for determining maximum (calculated) electrical loads, calculating optimal power consumption modes, rational use of electricity, reliable and high-quality power supply receivers and consumers of electricity in cities. Analysis of the structure and scale of electricity consumption in cities, accurate and reliable determination of maximum (calculated) electrical loads, optimization of modes and calculation of electricity losses, energy saving, energy efficiency and energy saving, as well as reliability, rational construction of power supply schemes and quality of electricity at this stage are very relevant and vital. The solution to these problems is of great economic and scientific importance.

2 Experimental research

In this regard, the task is to achieve high energy efficiency in urban power supply systems; for this it is necessary to solve the following problems [15-26]:

1. Development of reliable design of urban power supply based on a more accurate determination of the calculated electrical loads of consumers, as well as the total loads of transformer substations, distribution centers, and gas distribution centers using new patented methods for calculating electrical loads, probabilistic statistical approaches to solving the problem.

2. Development of new specific standards for calculating electrical loads of all similar consumers in the residential zone of cities.

3. Development of standard daily schedules of electrical loads for each type of electricity consumer in cities. (based on summer and winter daily maximum load patterns.)

4. Development of new building codes and regulations to replace KMK 2.04.17-98 "Electrical equipment of residential and public buildings" State Committee for Architecture and Construction of the Republic of Uzbekistan - Tashkent. 1998.

5. Development and determination of the most advantageous capacity and number of transformer substations, the array, block, microdistrict under construction.

6. Optimization and selection of the most rational scheme of a complex electrical network for power

* Corresponding author: raxmatillo82@mail.ru

supply of an area, block, microdistrict. Choosing the most reliable and economical power supply scheme.

7. Calculation of systems of complex power supply circuits using various methods. Choosing the most economical option.

(Loop current method, equalizing power method, superposition method, nodal voltage method, transformation method, transformation method, moment method, circuit splitting method, etc. Calculation of lines with two-way power supply.)

8. Development of new technologies and digital automation for relay protection of urban power supply systems. The use of digital automation in power supply systems, reverse power circuit breakers, new relay protection technologies, the use of contactless elements in urban power supply systems.

9. Training of personnel at the Department of Electric Power Supply of Tashkent State Technical University in digital automation, relay protection, new technologies in the energy sector.

The number of measurements and justification of the results of the energy audit of each individual type of electrical consumer with an interval of 5 minutes is required n = 10760, during summer and winter maximum loads.

3 Research results

Issues of reliability and quality of electricity, regardless of the category of consumer reliability, poses the task of uninterrupted and high-quality power supply. In this regard, on the basis of the Decrees of the President of the Republic of Uzbekistan "On the integrated construction of arrays (under one cadastre)" in Tashkent "New Sergeli", "New Tashkent" City", New Olympic Town", New areas in Namangan, Andijan, Khorezm, Karakalpakstan, etc. Under one general plan, under one cadastre, where there will be no new construction. As a result, it is possible to use complex electrical networks using digital automation when designing power supply in cities [27-35].

The consumer receives electricity from 2-3 sides. The reliability of power supply is increasing dramatically. Next, the optimal power supply scheme, which is the most economical, is developed and selected. On the one hand, with three times the reliability of power supply, and on the other hand, it allows you to get the same voltage at all network nodes, and this means high quality electricity.

The use of digital automation and computer software products makes it possible to fully automate the entire process of design and operation of complex urban electrical networks.

The weakest and most unreliable link in the energy system of the Republic of Uzbekistan is the electrical networks of 0.4 – 6 – 10 kV, which have a length of more than 80% and large losses of electricity. These electrical networks require modernization and reconstruction, since they were built 50 years ago.

4 Conclusion

1. Thus, for the first time in the Republic of Uzbekistan, the complex and very urgent problem of rational construction of urban power supply systems, reliability, quality of electricity, optimal choice of transformer power, energy saving in cities and energy efficiency of new technologies for metering and control of electricity consumption are being considered.

2. Based on a systematic approach, information and analytical models of power consumption, solving the complex problem of optimizing electrical modes and energy saving of all structural elements of urban power supply systems, it is possible to ensure a significant national economic effect, expressed in reducing electricity losses, saving non-ferrous metals, increasing the reliability and quality of electricity , energy saving in urban networks.

3. Solving the problems of power supply to cities is unthinkable without an accurate and justified determination of the calculated electrical loads, since electrical loads form the basis of all calculations of power supply and the rational construction of urban electrical networks. Their magnitude largely determines the network configuration and the choice of its parameters. In recent years, there has been a tendency towards a sharp increase in the share of electricity consumption of the municipal and public sector in the total electricity consumption of cities. An important component of the utility load is the load of public buildings and a sharp increase in the amount of electricity and power consumption by small and medium-sized businesses.

4. The electrical load of urban networks plays a large role in the formation of daily and annual schedules of energy systems and is the source material for the design of new and reconstruction of existing urban electrical networks. As a result, the study of actual loads, their correct determination and forecasting, allows, on the one hand, to ensure high-quality power supply to consumers, and on the other hand, makes it possible to avoid unnecessary expenditure of money, transformer power and cable products in urban networks, and obtain high energy efficiency and energy saving.

5. The use of new energy metering and control technologies has been so important in recent years that, on the basis of this, the government of Uzbekistan has

published a number of Resolutions and regulatory documents, as a result of which the imbalance of electricity on a general scale has sharply decreased [36-43].

References

1. A.Saidkhodjaev, B.Teshabaev, A method for determining the maximum electrical load of an electrical network and a method for determining the maximum electrical load of individual electrical consumers. Patent No. I AP 04216 (I AP 20060404), 2010. 1-8 p.
2. A.Saidkhodjaev, Power supply of cities. Textbook. – T.: "Fan va texnologiya", 328, (2015)
3. A.Saidkhodjaev, Fundamentals of energy saving. Textbook. – T.: "Lesson Press", 320, (2015).
4. A.Saidkhodjaev, Methods and tools of the energy audit. Textbook. – T.: "Noshirlik yog'dusi", 182, (2015).
5. A.Saidkhodjaev, Issues and prospects of urban power supply systems development. Monograph. – T.: "Fan", 256, (2012).
6. Saidkhodjaev, A. G., Najimova, A. M., Bijanov, A. K. (2019) Method for determining the maximum load of consumers in city power supply systems, E3S Web Conf 139 doi:10.1051/e3sconf/201913901078
7. Saidkhodjaev, A.G., Ametova, B.Kh., Mamutov, M.M. (2019) Intellectualization of determination of electrical loads in city electric networks, E3S Web Conf 139 doi:10.1051/e3sconf/201913901072
8. A.G. Saidkhodjaev, Kh.R. Nuriddinova , N. Toshpulatov, B.B. Kadraliev Reforming of Energy Sector of Uzbekistan: Transition to Market and Increasing Generation. Texas Journal of Engineering and Technology ISSN NO: 2770-4491 https://zienjournals.com Date of Publication:18-06-2022
9. A. G. Saidkhodjaev and Kh.R. Nuriddinova Reforming of energy sector of Uzbekistan: Transition to market and increasing generation Cite as: AIP Conference Proceedings 2552, 050037 (2023); https://doi.org/10.1063/5.0133486 Published Online: 05 January 2023
10.Dr.Saidkhodjaev A G, Saidkhodjaeva M, Ph.D Uzbekistan: Modernization of Energy Sector Myths or Reality? web publication, IEAA /International Association for Energy Economics. "Redefining the Epergy Espomy: Changing Role of Industry Govemment and R e s e a r c h", Os t o b e r 2011. Washington. DC USA.
11.A.G.Saidkhodjaev,Yu Bobojonov, UM Turdiev Perspectives of electric supply of municipal housing consumers in cities of Uzbekistan. International Journal of Advanced Science and Technology. (Scopus) . Vol. 29, no. 11s, (2020), pp. 1560-1564 ISSN: 2005-4238 IJAST Copyright ⓒ 2020 SERSC 1560 .
12. Saidkhodjaev AG An innovative way to determine the maximum (calculated) electrical load of power supply systems of cities. International Journal of Advanced Research in Science Engineering and Technology. India. May 2019 Vol. 6, issue 5, p . 9143-9145.
13. Saidkhodjaev AG Modern automatic technology of registrations and calculating electric loading. International Journal of Advanced Research in Science Engineering and Technology. India. May 2019 Vol. 6, issue 5, p . 9161 - 9165.
14. A.G. Saidkhodjaev, Kh.R. Nuriddinova PROBLEMS OF POWER SUPPLY TO RESIDENTIAL AND PUBLIC CONSUMERS IN CITIES OF UZBEKISTAN Science and innovation international scientific journal volume 2 issue 6 june 2023 uif-2022: 8.2 | issn: 2181-3337 | scientists.uz 250
15. Toirov O., Khalikov S. E3S Web of Conferences 377. 01004. (2023). https://doi.org/10.1051/e3sconf/202337701004
16. Toirov O., Khalikov S. E3S Web of Conferences 365. 04010. (2023). https://doi.org/10.1051/e3sconf/202336504010
17. Toshov J., Toshov B., Bainazov U., Elemonov M. Application of Cycle-Flow Technology in Coal Mines // Proceedings of the 11th International Conference on Applied Innovations in IT, (ICAIIT), March 2023, Koethen, Germany. – PP. 279-284.
18. Zikrillayev N., Saitov E., Toshov J., Akhmedov U., Muradov B., Muxtorov D. Autonomous Solar Power Plant for Individual use Simulation in LTspice Software Package Booster Voltage Converter // Proceedings of the 11th International Conference on Applied Innovations in IT, (ICAIIT), March 2023, Koethen, Germany. – PP. 207-211.
19. J.B. Toshov, Sh.R. Malikov, O.S. Ergashev, A.K. Sherov, A. Esirkepov, Improving the efficiency of the process of drilling wells in complex conditions at geological prospecting sites // NEWS of the National Academy of Sciences of the Republic of Kazakhstan "Series of geology and technical sciences". – Almaty, Volume 3, Number 459 (2023), – PP. 282–294 https://doi.org/10.32014/2023.2518-170X.313
20. J.B. Toshov, K.T. Sherov, B.N. Absadykov, R.U. Djuraev, M.R. Sikhimbayev, Efficiency of drilling wells with air purge based on the use of a vortex tube // NEWS of the National Academy of Sciences of the Republic of Kazakhstan "Series of geology and technical sciences". – Almaty, Volume 4, Number 460 (2023), 225–235. https://doi.org/10.32014/2023.2518-170X.331
21. Hoshimov F.A., Rakhmonov I.U., Niyozov N.N., Omonov F.B. AIP Conference Proceedings 2647. 030025. (2022). https://doi.org/10.1063/5.0112388

22. Rakhmonov I.U., Najimova A.M., and Reymov K.M. AIP Conference Proceedings 2647. 030010. (2022). https://doi.org/10.1063/5.0104788
23. Rakhmonov I.U., Najimova A.M. AIP Conference Proceedings 2647. 030011. (2022). https://doi.org/10.1063/5.0104791
24. Rakhmonov I.U., Najimova, A.M., Esemuratova Sh.M., Koptileuov T.T. AIP Conference Proceedings 2647. 070024. (2022). https://doi.org/10.1063/5.0104793
25. Rakhmonov I.U., Hoshimov F.A., Kurbonov N.N., Jalilova D.A. AIP Conference Proceedings 2647. 050022. (2022). https://doi.org/10.1063/5.0112391
26. Rakhmonov I.U., Ganiev Sh.R., Alibekova T.Sh., Nematov L.A. E3S Web of Conferences 384. 01029. (2023). https://doi.org/10.1051/e3sconf/202338401029
27. Rakhmonov I.U., Niyozov N.N., Kurbonov N.N., Umarov B.S. E3S Web of Conferences 384. 01030. (2023). https://doi.org/10.1051/e3sconf/202338401030
28. Niyozov N.N., Rikhsitillaev B.Kh., Jalilova D.A., Kuatbaev P.M. E3S Web of Conferences 384. 01031. (2023). https://doi.org/10.1051/e3sconf/202338401031.
29. Usmanov E.G. E3S Web of Conferences 384. 01034. (2023). https://doi.org/10.1051/e3sconf/202338401034
30. Rasulov A.N., Paxratdinov A.Dj. E3S Web of Conferences 384. 01035. (2023). https://doi.org/10.1051/e3sconf/202338401035
31. Ushakov, V.Y., Mytnikov, A.V., Rakhmonov, I.U. Past, Present and Future of Electric Power Systems (Brief Excursion). In: High-Voltage Equipment of Power Systems. Power Systems. Springer, Cham. (2023). https://doi.org/10.1007/978-3-031-38252-9_1
32. Ushakov, V.Y., Mytnikov, A.V., Rakhmonov, I.U. Characteristics of the Main Elements of Electric Power Systems. In: High-Voltage Equipment of Power Systems. Power Systems. Springer, Cham. (2023). https://doi.org/10.1007/978-3-031-38252-9_2
33. Ushakov, V.Y., Mytnikov, A.V., Rakhmonov, I.U. Insulating Materials and Media Used in High-Voltage Elements of Electric Power Systems. In: High-Voltage Equipment of Power Systems. Power Systems. Springer, Cham. (2023). https://doi.org/10.1007/978-3-031-38252-9_3
34. Ushakov, V.Y., Mytnikov, A.V., Rakhmonov, I.U. Effects on Equipment Causing Insulation Aging and Failure. In: High-Voltage Equipment of Power Systems. Power Systems. Springer, Cham. (2023). https://doi.org/10.1007/978-3-031-38252-9_4
35. Ushakov, V.Y., Mytnikov, A.V., Rakhmonov, I.U. Defects in High Voltage Equipment: Types and Content of Tests. In: High-Voltage Equipment of Power Systems. Power Systems. Springer, Cham. (2023). https://doi.org/10.1007/978-3-031-38252-9_5
36. A.N.Rasulov, E.G.Usmonov, M.V.Melikuziev. AIP Conference Proceedings, 2552, 050018, (2023), https://doi.org/10.1063/5.0111530
37. Ushakov, V.Y., Mytnikov, A.V., Rakhmonov, I.U. Non-electrical Diagnostic Methods. In: High-Voltage Equipment of Power Systems. Power Systems. Springer, Cham. (2023). https://doi.org/10.1007/978-3-031-38252-9_6
38. Ushakov, V.Y., Mytnikov, A.V., Rakhmonov, I.U. Traditional Electrical Diagnostic Methods. In: High-Voltage Equipment of Power Systems. Power Systems. Springer, Cham. (2023). https://doi.org/10.1007/978-3-031-38252-9_7
39. Ushakov, V.Y., Mytnikov, A.V., Rakhmonov, I.U. Diagnostics of High-Voltage Equipment by Defectography Methods. In: High-Voltage Equipment of Power Systems. Power Systems. Springer, Cham. (2023). https://doi.org/10.1007/978-3-031-38252-9_8
40. Ushakov, V.Y., Mytnikov, A.V., Rakhmonov, I.U. Dielectric Spectroscopy Technology. In: High-Voltage Equipment of Power Systems. Power Systems. Springer, Cham. (2023). https://doi.org/10.1007/978-3-031-38252-9_9
41. Ushakov, V.Y., Mytnikov, A.V., Rakhmonov, I.U. Diagnostics of High Voltage Equipment. In: High-Voltage Equipment of Power Systems. Power Systems. Springer, Cham. (2023). https://doi.org/10.1007/978-3-031-38252-9_10
42. Ushakov, V.Y., Mytnikov, A.V., Rakhmonov, I.U. Diagnostics of High-Voltage Cable Lines. In: High-Voltage Equipment of Power Systems. Power Systems. Springer, Cham. (2023). https://doi.org/10.1007/978-3-031-38252-9_11
43. Ushakov, V.Y., Mytnikov, A.V., Rakhmonov, I.U. Diagnostics of Insulating Structures of Overhead Power Lines and Outdoor Substations. In: High-Voltage Equipment of Power Systems. Power Systems. Springer, Cham. (2023). https://doi.org/10.1007/978-3-031-38252-9_12

https://doi.org/10.1051/e3sconf/202346101087

Modern technologies of calibration with measuring devices of electrical quantities

Gaybulla Boboyev[1*], *Gulnoza* Mirpayzieva[1]

[1]Tashkent State Technical University named after Islam Karimov, 100095, Uzbekistan, Tashkent, University St. 2A

Abstract. Despite the ever-increasing demand for metrological services which was the increase the number of measuring instruments in use, the system of transmission of the unit of physical quantity has not undergone significant changes. As a result, today we have an inefficient system that does not meet the needs of the market. Improving the system of physical quantity transmission is an urgent task. Metrology institutes of many foreign countries offer the concept of remote calibration, the use of which it was previously impossible due to the low level of development of programmable measuring devices. The article provides general information about the use of the proposed technology in metrology institutes of foreign countries.

Introduction

Today, one of the most important goals of the industry is to create new technologies for manufacturing, that is, to develop a high-quality, high-tech product that is in demand and competitive in the domestic and foreign markets. This goal has been announced several times by the leadership of our country. To achieve this, in recent years, large financial resources have been allocated to all branches of production [1-8].

The problem of the quality of all types of industrial products is directly related to the metrological assurance of production, because metrological assurance is the basis of the quality system of any production enterprise.

The main task of metrological maintenance is to achieve the required accuracy and ensure positive results with the correct application of the measurement method, measurement tool, measurement process and analysis of the measurement results through its correct management. Any technology is undergoing changes related to the growing needs and capabilities of society. This puts new demands on metrology and its maintenance. Since the end of the 20th century, the number of measuring devices and sensors in manufactured equipment has been increasing by several percent every year. However, the competition forces them to reduce and increase the speed of equipment. The procedure is becoming complex which is from the installation of the measuring instrument, taking it when the calibration interval comes and delivering it to the metrological laboratory [9-15].

To date, there is a problem of misunderstanding the need to validate the measurement tools used by the manufacturer of the product or service. This is primarily due to the desire to save not only money, but also time. In fact, for many manufacturers, comparing measuring instruments is a real worry that repeats itself every year. This may cause to the sorting of separation measuring instruments from production procedures for a long time, which requires a lot of time and costs. There is a measuring service for the customer, but in this case it is necessary to pay not only for the time spent by the comparator, but also for the transportation of standards, which is not possible for all measuring instruments. To solve this problem, many factories and manufacturers ignore measuring their tool, furthermore, by doing with this they disobey the law "State Metrology" or it is proved not to participate in the process of measuring.

Increasing the level of monitoring of compliance with the Law of the Republic of Uzbekistan "On Metrology" will improve the situation, but will not completely solve the problem. The solution to this problem can be the transition to new technical solutions that significantly reduce the time of separation of measuring instruments from production, which avoids large costs for manufacturers. Due to high competition, manufacturers are interested in improving the quality of products or services, which directly depends on reliable indicators of measuring instruments [16-21].

The most effective technical solution for transferring the sizes of electrical units is to work in connection with the national calibration process carried out by the National Metrology Institute of Uzbekistan. Currently, many foreign metrology institutes are working on calibrating the calibration tool through remote control systems. The main problems in the widespread use of this technology are related to the low level of development of programmable measuring devices. However, there are currently a large number of measuring devices on the market that allow remote access to a computer. One of the leading manufacturers in this system is National Instruments (an American multinational company with international operations),

* Corresponding author: raxmatillo82@mail.ru

LabVIEW laboratory programs have been developed for the development of technical systems based on their hardware base. LabVIEW is a graphical programming system that engineers use to develop automated research, validation, and production test systems [22-27].

LabVIEW is a graphical programming system for tests and measurements, which has everything you need to quickly create automated test systems in the laboratory:

– there are thousands of analysis functions;

– has customized interactive display elements;

– has drivers for automation of any data collection devices and equipment;

– has the possibility to connect to other languages and industry standard protocols.

Remote calibration of measuring instruments is a promising method of metrological research automation. The modern level of development of information technologies allows to create automated complexes based on the interaction of measuring instruments and computer equipment, with the help of which it will be possible to carry out the procedure of remote calibration of measuring instruments. Remote technology is used to calibrate measuring instruments that are geographically distant from the reference. Remote calibration provides the following functions [28-34]:

* remote control of measuring equipment and its monitoring;

* carrying out remote customer measurements using special calibration equipment;

* transfer data (calibration results, environmental conditions, calibration time and duration) to the customer, if necessary;

* access to measurement history and calibration data.

When comparing remote calibration to traditional calibration, several important advantages can be identified:

* significant time savings due to the absence of procedures for the delivery of measuring instruments to the metrological laboratory and the need to take into account the calibration schedule of the laboratory;

* financial savings due to the much lower cost of the calibration procedure, especially in cases where the cost of calibration of the measuring instrument is much higher than the price of the calibrated instrument;

* the ability to calibrate measuring instruments that work in harsh or extreme environmental conditions and are therefore unavailable to humans;

* to minimize the participation of employees in the metrological provision of measuring instruments and, as a result, to eliminate the influence of subjective error on the calibration results;

* automation of processing and collection of calibration results.

2. The purpose of the study

The purpose of this study is to analyze the software-hardware complex for remote calibration of electrical measuring devices. This development not only makes it easier to calibrate measuring instruments used outside of government regulation to ensure uniformity of measurements, but also provides a basis for remote comparison and calibration technology [35-42].

3. Experimental research

Currently, calibration of measuring instruments not intended for use in the field of state regulation of measurement uniformity is optional. Legal entities and entrepreneurs performing the calibration of measuring instruments must be accredited in the field of ensuring the uniformity of measurements. Measuring instruments used for calibration must be of approved type and have valid approval certificates. It is required to carry out a regular comparison of measuring instruments, which provide only the conversion of unit sizes to measuring instruments [43-47].

If the calibration laboratory is not accredited, the reliability of the results is not approved. Therefore, it will not be possible to avoid problems with customers who do not approve the reliability of calibration results without appropriate documentation issued by an accredited laboratory. Based on the national calibration system of Uzbekistan, the calibration laboratory must be accredited according to the requirements of the ISO/IEC 17025:2019 standard. If the laboratory is accredited by accreditation bodies recognized by similar bodies of other countries, the results of calibration will be recognized by different countries [48-50].

Currently, there is no working software and hardware system and related methodology for remote calibration. First of all, it is related to the requirements for the calibration laboratory. The traceability of the measurement results is solved using the type of measuring instruments approved according to the state standards, and the development of new calibration techniques is also a possible task. The first problem is in the laboratory, which should provide the necessary conditions for calibration work. In order to avoid this problem, it is necessary to take into account the conditions of use of measuring instruments by specifying the appropriate conditions for calibration at the stage of developing the calibration methodology. The second problem is the proper storage of calibration tools. This problem can be solved based on the agreement between the enterprise and the accredited laboratory. The calibration tool should be stored in an accredited laboratory. It will be necessary to create a special room that provides the necessary conditions for storing the standard [15-28].

An example of a remote calibration field is education. Today, measuring instruments used in educational institutions are not included in the framework of state regulation of ensuring the uniformity of measurements. Therefore, it increases the error and uncertainty in the use of measurement data to unacceptable values. Buying new equipment on a regular basis is not a smart solution, as it is very expensive. The use of remote calibration technology allows you to create calibration tables that should be used for future measurements. This solution helps to reduce the inaccuracy of measuring instruments to values that are specific to the used calibration instrument. In this way, it will be possible to avoid huge costs for upgrading the equipment while improving the university facilities [1-8].

One of the features of the remote calibration software-hardware complex is mobility. To date, there are many different measuring devices for electrical quantities that meet this requirement. Despite the high accuracy of generated signals, many models have small dimensions. It should be remembered that many organizations use measuring instruments with low accuracy due to the use of old technologies. Examples of such devices are still the ubiquitous scale voltmeters. Calibration of these measuring instruments does not require the use of high-precision measuring instruments [29-39].

The current level of technological development makes it possible to switch the remote calibration of electrical measuring instruments. The remote calibration scheme is presented in Figure 1. A lab technician uses software on their personal computer to generate the data needed the signal. In addition, these data are sent to a computer connected to the hardware complex in the customer's laboratory. Arriving at the apparatus complex, a signal is generated for the calibrated measuring instrument [18].

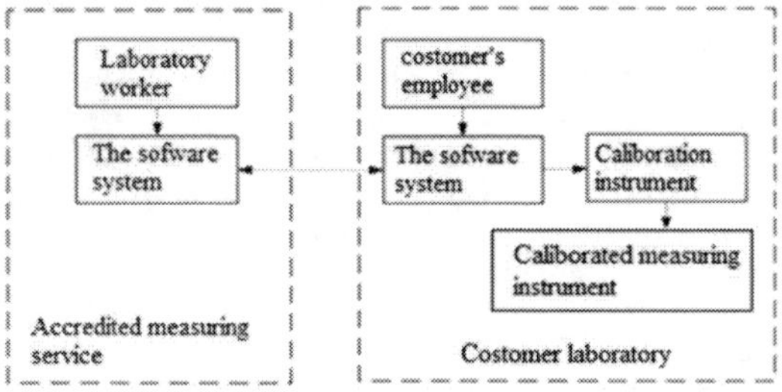

Fig.1. Remote calibration scheme

It should be noted that the software should protect against third-side interference in the process. That is, the customer's employee should not be able to control the calibration tool. Without compliance with this condition, the participation of an accredited laboratory employee will have no meaning. Since a calibrated measuring instrument probably does not have a socket for connecting to a computer, there is a problem of transmitting the reading to an accredited laboratory worker. In this case, there are two options. The first is to use a webcam. The second is the transfer of indicators entered by the customer's employee. The reliability of the input data is based on the fact that calibration is a voluntary procedure, and therefore it is useful for the customer to transmit reliable readings in order not to waste money on an incorrect calibration table [18-28].

4. Research results

The most reasonable solution is to use the products of National Instruments, because they specialize in equipment that exchanges data with a personal computer in a real time. As a development environment, the LabVIEW program, specially developed for working with the equipment of this company, is used. At the moment, a program aimed at working with the PXI system, which allows controlling the signal generator via the Internet, has been developed [41-47].

The user enters the IP address and TCP port of the hardware system. Next, the necessary parameters of the generated signal are entered, and the data transmission button starts sending data. It should be noted that until this time, all preparatory operations must be carried out in the customer's laboratory, that is: the system must be connected to a calibrated measuring device, the appropriate program must be running, and the employee must be ready to work. The signal is sent to the calibrated measuring instrument, after which the employee reads the indicator data, stops the signal generation and sends the reading to the accredited laboratory worker. As it mentioned above, the remote calibration program must ensure that the generated signal is not altered by the customer's personnel. This is a prerequisite for the system. Therefore, during this process, the customer's laboratory employee can only stop the signal generation [48-50].

The process described above is repeated for each point required to make the calibration table. The list of control points, as well as their number, must be specified in the methodology used during calibration. It is used to draw up a report of results and issue a calibration certificate. It should be noted that since the information was initially presented in electronic form, it makes possibility to use a digital document using as electronic signature technology to store them in a special database in the future [19-28].

The operation of this hardware-software complex has been successfully tested. Of course, the mobility of the system places high limits on the size and weight of the device, which ultimately affects the accuracy of the equipment used. Nevertheless, the accuracy of many modern measuring instruments of small size is several times higher than the accuracy of measuring instruments used in enterprises. Based on this, it can be concluded that the accuracy of these LabVIEW devices will be sufficient for calibration [50].

5. Conclusions

As a result of research which was carried out in many metrology institutes, it was found that modern technologies allow to transfer to new methods of transferring the dimensions of physical units. Due to the high demand for metrological services and the increasing in the number of measuring instruments in use, accredited centers cannot cope with such a number of measuring instruments without improving the current system.

The advantage of the developed LabVIEW program is based on a laboratory development environment and is compatible with many similar products in National Instruments. If you need to use other equipment in the enterprise, this program can be easily modified using similar blocks.

The system of remote calibration of measuring instruments provides remote execution of the calibration procedure, automatic data collection and quick processing of results. The use of a remote calibration system developed in metrological practice provides a significant reduction in financial costs as well as reducing the time spent on performing all calibration operations of measuring instruments and minimizes the participation of personnel in the calibration process. Implementing a remote calibration system is essential for large organizations with many geographically distributed branches to maintain an intensive equipment calibration schedule. However, there are problems that arise when performing remote calibration and require solutions. In addition to the measurement procedures, control of environmental conditions, requirements for external verification and comparison of the performance of the calibrated measuring instrument, security and reliability of the transmitted data must also be taken into account. Environmental conditions can be recorded using LabVIEW lab tools and sensors with digital output and computer interface capabilities, and visual monitoring can be done via a web camera. As for the issues of information protection, they should be developed in detail. It would be appropriate if these remote calibration systems were implemented in cooperation with the National Metrology Institute of Uzbekistan by electricity generating enterprises.

References

1. G.G.Boboev, M.M.Mahmudjonov, and ofters. AIP Conference Proceedings, 2432, 030042, (2022), https://doi.org/10.1063/5.0089626
2. B.B.Kholikhmatov, M.T.Erejepov, and ofters. E3S Web of Conferences, 384, **01032**, (2023), https://doi.org/10.1051/e3sconf/202338401032
3. M.Ibadullayev. Proceedings of International Conference on Applied Innovation in IT, **11(1)**, pp. 285-290, (2023).
4. M.Ibadullaev, A.Nuraliyev, A.Yesenbekov. AIP Conference Proceedings 2552, **050036**, (2022), https://doi.org/10.1063/5.0133070
5. A.Nuraliyev, M.Ibadullayev. E3S Web of Conferences, 216, **01114**, (2020), https://doi.org/10.1051/e3sconf/202021601114
6. A.N.Tovbaev, M.Ibadullayev, S.I.Norboyev. Journal of Physics: Conference Series, 2388(1), **012060**, (2022), doi:10.1088/1742-6596/2388/1/012060
7. M.Ibadullaev, A.N.Tovbaev. E3S Web of Conferences, 216, **01113**, (2020), https://doi.org/10.1051/e3sconf/202021601113
8. A.Tovboev, M.Ibodulaev, M.Baranova, I.Grishina. IOP Conf. Series: Mater. Scien. and Engineering, 862(6), **062041**, (2020), doi:10.1088/1757-899X/862/6/062041
9. M.I.Ibadullaev, A.N.Tovbaev, A.Z.Esenbekov. E3S Web of Conferences, 139, **01054**, (2019), https://doi.org/10.1051/e3sconf/201913901054
10. S.Begmatov. Proceedings of International Conference on Applied Innovation in IT, **11(1)**, pp. 273-277, (2023).
11. R.Karimov, D.Xushvaktov. E3S Web of Conferences, 384, **01053**, (2023), https://doi.org/10.1051/e3sconf/202338401053
12. R.Karimov. E3S Web of Conferences, 384, **01056**, (2023), https://doi.org/10.1051/e3sconf/202338401056
13. A.A.Khashimov, K.G.Abidov. Elektrotekhnika, **7**, pp. 34-38, (2001).
14. R.Karimov, and ofters. E3S Web of Conferences, 289, **07021**, (2021), https://doi.org/10.1051/e3sconf/202128907021
15. E.Abduraimov. E3S Web of Conferences, 216, **01105**, (2020), https://doi.org/10.1051/e3sconf/202021601105
16. E.Abduraimov. E3S Web of Conferences, 384, **01051**, (2023), https://doi.org/10.1051/e3sconf/202338401051
17. E.Abduraimov, B.Nurmatov. E3S Web of Conferences, 384, **01052**, (2023), https://doi.org/10.1051/e3sconf/202338401052
18. A.Burkhankhodzhayev, B.Nurmatov. E3S Web of Conferences, 216, **01112**, (2020), https://doi.org/10.1051/e3sconf/202021601112
19. A.Burkhankhodjaev, B.Nurmatov, and ofters. AIP Conference Proceedings, 2552, **030024**, (2022), https://doi.org/10.1063/5.0133920
20. N.Khamudkhanova, M.Idriskhodjaeva, Kh.Kholbutayeva. E3S Web of Conf., 384, **01057**, (2023), https://doi.org/10.1051/e3sconf/202338401057
21. S.Umarov, K.Sapaev, N.Khamudkhanova, and ofters. AIP Conference Proceedings, 2402, **060016**, (2021), https://doi.org/10.1063/5.0071557
22. K.G.Abidov, O.O.Zaripov, and ofters. E3S Web of Conferences, 216, **01110**, (2020), https://doi.org/10.1051/e3sconf/202021601110
23. A.A.Khashimov, K.G.Abidov. Elektrotekhnika, **10**, pp. 22-26, (2001).
24. K.G.Abidov. E3S Web of Conferences, 384, **01055**, (2023), https://doi.org/10.1051/e3sconf/202338401055
25. K.G.Abidov, K.Dadajanov, U.Absoatov, S.Sabitov. E3S Web of Conferences, 383, **04053**, (2023), https://doi.org/10.1051/e3sconf/202338304053
26. K.G.Abidov, K.Dadajanov, U.Absoatov, S.Sabitov. E3S Web of Conferences, 383, **04044**, (2023), https://doi.org/10.1051/e3sconf/202338304044
27. G.Mustafaquliva, G.Maxmadiyev, O.Zaripov, Sh.Zaripova. E3S Web of Conferences, 216, **01109**, (2020), https://doi.org/10.1051/e3sconf/202021601109
28. A.Pulatov, J.Bekmuradov, O.Zaripov. E3S Web of Conferences, 289, **07002**, (2021), https://doi.org/10.1051/e3sconf/202128907002
29. A.Alimov, F.Akbarov, G.B.Eshniyazova, A.Tureniyaziva. E3S Web of Conferences, 289, **07035**, (2021), https://doi.org/10.1051/e3sconf/202128907035
30. R.R.Kabulov, F.A.Akbarov, A.A.Alimov. Nanosystems: Physics, Chemistry, Mathematics, **14(1)**, pp. 127-131, (2023).
31. D.N.Isamukhamedov, N.J.Tairova, and ofters. E3S Web of Conferences, 216, **01154**, (2020), https://doi.org/10.1051/e3sconf/202021601154

32. A.Nuraliyev, Y.Adilov, A.Esenbekov. E3S Web of Conferences, 384, **01054**, (2023), https://doi.org/10.1051/e3sconf/202338401054
33. K.Shamsiev. AIP Conference Proceedings 2552, **050015**, (2022), https://doi.org/10.1063/5.0130710
34. K.Jurayeva, Kh.Sattarov. E3S Web of Conferences, 389, **01060**, (2023), https://doi.org/10.1051/e3sconf/202338901060
35. Kh.Sattarov, and ofters. E3S Web of Conferences, 383, **01028**, (2023), https://doi.org/10.1051/e3sconf/202338301028
36. I.Siddikov, Kh.Sattarov, A.Abdumalikov. AIP Conference Proceedings, 2612, **050002**, (2023), https://doi.org/10.1063/5.0115211
37. A.Safarov, Kh.Sattarov, and ofters. Journal of Physics: Conference Series, 2176(1), **012006**, (2022), doi:10.1088/1742-6596/2176/1/012006
38. A.Safarov, Kh.Sattarov, and ofters. E3S Web of Conferences, 264, **05038**, (2021), https://doi.org/10.1051/e3sconf/202126405038
39. Y.Shoyimov, and ofters. Journal of Physics: Conference Series, 939(1), **012028**, (2021), doi:10.1088/1755-1315/939/1/012028
40. M.V.Melikuziev. E3S Web of Conferences, 384, **01033** (2022), https://doi.org/10.1051/e3sconf/202338401033
41. A.N.Rasulov, E.G.Usmonov, M.V.Melikuziev. AIP Conference Proceedings, 2552, **050018**, (2023), https://doi.org/10.1063/5.0111530
42. O.G.Hayitov, M.O.Gafurova, and ofters. AIP Conference Proceedings, 2432, **030101**, (2022), https://doi.org/10.1063/5.0091265
43. I.Kh.Siddikov, M.T.Maksudov, and ofters. AIP Conference Proceedings, 2432, **020003**, (2022); https://doi.org/10.1063/5.0089681
44. N.N.Norkhojaeva, N.B.Pirmatov, N.K.Kamalov. E3S Web of Conferences, 402, **10041**, (2023), https://doi.org/10.1051/e3sconf/202340210041
45. S.Yaxyayeva, N.Pirmatov, and ofters. E3S Web of Conferences, 402, **05016**, (2023), https://doi.org/10.1051/e3sconf/202340205016
46. O.Toirov, N.Pirmatov, and ofters. E3S Web of Conferences, 401, **04033**, (2023), https://doi.org/10.1051/e3sconf/202340104033
47. N.B.Pirmatov, U.T.Berdiyev, and ofters. E3S Web of Conferences, 401, **03021**, (2023), https://doi.org/10.1051/e3sconf/202340103021
48. N.B.Pirmatov, A.M.Egamov, and ofters. E3S Web of Conferences, 401, **03056**, (2023), https://doi.org/10.1051/e3sconf/202340103056
49. N.Pirmatov, A.Panoev, and ofters. E3S Web of Conferences, 383, **04046**, (2023), https://doi.org/10.1051/e3sconf/202338304046
50. N.Pirmatov, A.Bekishev, and ofters. AIP Conference Proceedings, 2612, **050005**, (2023), https://doi.org/10.1063/5.0135546

E3S Web of Conferences **461**, 01088 (2023) https://doi.org/10.1051/e3sconf/202346101088
RSES 2023

Analysis of metrological supply problems in electricity generation

Gaybulla Boboyev[1*], *Muborak* Mirshomilova[1]

[1]Tashkent State Technical University named after Islam Karimov, 100095, Uzbekistan, Tashkent, University St. 2A

Abstract. In this article, the data on metrological supply issues of electric energy and production enterprises and laboratories in modern economic conditions are analyzed. Methods of choosing a measuring instrument and reducing errors are analyzed and recommendations are given.

Introduction

As everyone is aware, the entire world has been enduring economic changes for years. These periods have seen a considerable change in both the state of production and its objectives. In such circumstances, no one questions the significance of measuring devices in various production branches, i.e., the goal, duties, and function of metrological supply. The primary goal of metrological maintenance is to use the proper application of the measuring method, measurement tool, measurement process, and analysis of the measurement data through proper management to achieve the requisite accuracy and ensure good results [1-8].

Any technology is evolving to meet the society's expanding requirements and capacities. This increases the need for metrology and its upkeep. The quantity of measuring tools and sensors in manufactured equipment has expanded by several percent yearly since the turn of the 20th century. But because of the competition, they are forced to alter equipment speed. It is getting more and more difficult to move measuring instruments from their installation to the metrology laboratory when the calibration interval is needed and then back to their original place once calibrated. The development and invention of new modern technologies, or the design and production of high-quality, high-tech products that are competitive in both domestic and international markets of each country, is one of the manufacturing sector's most significant aims today. Developed nations have recently allotted significant financial resources for all manufacturing sectors in order to accomplish this. Metrological assurance is intimately tied to the issue of electricity production and quality in all nations since it forms the cornerstone of any industrial enterprise's quality management system. The development of tools, including intelligent instruments, that should function for a very long time without human interaction is related to the necessity for long-term mechanisms. Maintaining the functionality of measuring devices is a task that is still necessary today. Long-term use of measuring devices, however, increases the influence of influencing factors, their effects (including considerable material wear, increasing external influences, changes in magnetic characteristics, etc.), and the likelihood of metrological failure. The automation of equipment control increases the relevance of metrological supply and, consequently, the requirements for the reliability of measurement data, particularly in the transportation sector as well as in the generation of power and its delivery to the consumer. Up to 60% of industrial failures and accidents, in the opinion of Chinese specialists, are brought on by sensor flaws. The metrological serviceability of the measuring instruments should be tested more frequently in order to increase the effectiveness of the equipment and to reduce wear and tear on the equipment as well as the danger of defects and accidents. Controlling the supply of laboratories for organizations dealing with issues of metrological supply as well as for scientific, production, and industrial companies is vital for this, as is providing them with the requisite measuring equipment [6-11].

But again, issues arise since each metrolog is purchasing for the laboratory. If the product promoted by the supplier is unable to successfully complete the tasks for which it was purchased, problems will arise with the acquisition of the measuring equipment. In other words, the measuring device you obtained could not be completely enough for the jobs you need it for. This puts a lot of duty on experts in this area. Businesses today spend a lot of money on unnecessary features when purchasing equipment that will never be in demand [12].

There are frequently issues with how newly purchased equipment are organized to provide the essential environment [1-2]. Our analyses reveal that the majority of these issues can be attributed to the lack of a detailed technical task preparation, incomplete information collection, and unclear task formulation during the decision-making process to purchase a new measuring instrument. This issue is mostly caused by the specialists in power generating firms' inability to make decisions due to a lack of knowledge. Today, the majority of measuring instruments used in the

* Corresponding author: **raxmatillo82@mail.ru**

laboratories of many electricity-generating companies have a lifespan of more than 25 years, and the majority of metrologists lack sufficient knowledge of cutting-edge measurement techniques and the true capabilities of contemporary equipment [13-16].

The principal users of this clause are the producers and suppliers of measuring instruments who, in their promotional materials regarding the equipment they sell, frequently refer to the equipment's limitations. In a sizable advertisement window, it is challenging to notice the device's true features. The provider frequently provides old equipment "as a last resort" in an attractive package. It is nearly impossible to evaluate all of the measuring tool's potential uses, its operating features, and even its level of efficiency if you don't know anything about the new tools and equipment you have just encountered. The adage "it is better to see one march than to hear it a hundred times" must therefore be followed. Metrologists try to find out about these devices from their peers or try to see samples of the devices at supplier conferences and trade exhibitions. This is the correct course of action, but you should be aware that your measurement tasks and your colleagues' opinions don't always line up, that the operator using this device might not have the necessary skills, that the operating conditions might be off, and that the new device might not have an equivalent in Uzbekistan [17-20]. Metrologists have a difficult challenge in choosing measurement tools and equipment for metrological supplies and laboratories based on the examination of the provided data. As science and technology have advanced, it is now crucial to understand the components of intelligent measuring devices and adhere to usage regulations in order to produce and utilize them. It's important to concentrate on a certain feature while picking measurement equipment for laboratories [21-25].

In other words, some measurement devices and laboratory equipment require software updates. At times, manufacturer-provided codes and passwords may be necessary. This calls for experts to exercise caution. As a result, traditional components will gradually be largely replaced by metrological self-monitoring inside the metrological supply. Regular metrological processes will take less time, but there will be more demand for highly skilled metrologists to work on developing new measuring instruments [26-31].

2. Methods

As we all know, when picking a measuring device and method, the findings are evaluated in the laboratories of the power generating firm using a number of parameters. Accuracy and speed are typically the major factors to consider when purchasing measuring equipment. Devices with high accuracy, i.e., low relative error and high sensitivity appropriate for challenging working conditions, can easily lose stability and fail before their time. As a result, measuring equipment should be chosen in line with the standards for measurement accuracy and the type of tests being run. In the selection of the measurement tool and method, in the production laboratories, and in the industry for the analysis and creation of this article, statistical analysis methods were widely employed to elucidate the issues of metrological supply [32-38].

3. Research results

1. It is important to carefully study and analyze a variety of criteria when selecting a measuring tool for different measurement purposes, optimizing them based on the sum of all technical data and attributes.
2. The primary elements affecting the selection of the measurement device are:
3. The capability of measuring the signal under study.
4. The kind of measurement parameter.
5. Measuring technique (direct, indirect, group, joint, etc.).
6. The length of the measures.
7. The measuring tool's practicality (general dimensions and weight).
8. Necessary measurement precision.
9. The equipment's cost.
10. A measurement inaccuracy was fixed.
11. The weather when using a device.
12. Requirements for the format (analog, digital, analog-digital, etc.) of recording measurement results.

The device's intended usage (e.g., individually, as a component of an automated system, etc.).

It is unavoidable that, in every measurement, the measurement result would differ from the actual value of the quantity being measured for a variety of reasons. Measurement mistakes are the name given to these variations.

Measurement errors are divided into categories based on their types (absolute, relative, random, etc.), reasons of occurrence, modes of manifestation, and other factors. Of course, it is essential to account for all possible measurement errors and work to minimize them by determining their root causes. [39-45] We concentrate on approaches for minimizing (removing) systematic measurement mistakes. The measurements are accurate if the systematic errors in their findings are near to zero. Methods include theoretical analysis, replacement, compensation of mistakes using symbols, statistics, various measurements, sample signals, and the way of introducing correction and correction factors to discover and exclude systematic errors. If systematic mistakes cannot be completely eliminated, they can be minimized by removing the sources, changing the measuring tools both during the inspection and before the measurement, employing unique measurement techniques, and other procedures. Microprocessors found in contemporary measurement devices enable the automatic detection and exclusion of systematic erroneous readings. According to the theoretical analysis method, the systematic error can be computed using specific features of the tools being used or characteristics of the measuring technique, or in other words, using formulae. So, if the input resistance is known, it is easy to calculate the device's power consumption-related systematic error. The substitution approach entails substituting a known quantity or repeatable measurement for the measured quantity. If the source of the error has a directional effect (for instance, errors caused by permanent magnetic fields, the thermoeffect, etc.), the systematic error is included in the measurement results twice, but with opposite signs. This is known as the method of compensating the error by sign [46-50].

In order to process the measurement results, the statistical approach divides them into numerous independent observation groups. The existence of systematic error is shown by the difference between group variances and group means (group averages), which makes it possible to compute it. The technique used for different measurements enables the detection of systematic mistakes, the origin of which is unclear. For this, the value is measured in various ways, using various measurement tools, under various circumstances. In this instance, the internal errors of the measuring devices utilized for the measurements should be roughly comparable. The reference signal method compares the measured and measured reference signals of the same type that are delivered to the measuring device's input. The systematic error is determined by the difference between them. Introduce correction and corrective factors in this way. The value that is added to the value obtained during the measurement in order to account for systematic mistakes is referred to as the correction and bears the same name as the measured value. The adjustment has the opposite sign from the absolute systematic error but is equal to it. Corrections are presented via graphs, tables, or mathematical formulas. Due to the small size of systematic errors, systematic measurement error can also be reduced by multiplying the measurement findings by a correction factor that is close to the average value. It is expected that during the inspection of measuring instruments, correction and correction factors are predetermined [25-29, 33].

The use of screens to shield against the effects of electromagnetic fields, the use of stabilized power sources, the moistening of the device, and removing it from potential sources of exposure can all help to reduce systematic error. Other methods include measurements in thermostated rooms (to exclude temperature errors), the use of screens to shield against the effects of electromagnetic fields, and the use of stabilized power sources [34-38].

The systematic error of the device can be reduced by adjusting the measuring instruments both during the inspection and before to the measurement. This entails ensuring that the device's position is accurate, that the pointer is set to zero, that it is positioned in relation to the nearby objects, etc.

Making it random significantly lessens the impact of systematic error. As a result, the error value will be greatly decreased if you test a certain parameter using various instruments and then compute the arithmetic average of all the data.

The values of systematic errors of independently measured, directly measured quantities are used to calculate systematic errors in indirect measurements. Their accuracy rises as the measurement error falls. The quality of the measurements refers to their precision, which reflects how closely the quantity's obtained value corresponds to its actual value. It should be emphasized that due to the complexity of the measurement process and the rising expense of measuring devices, there is typically a certain fair degree of accuracy that cannot be exceeded under specific situations and measurement objectives [41-46].

Daily and hourly physical quantity measurements, particularly electrical measurements, are carried out in practice for a variety of reasons. The results of these measurements-i.e., the same number of measurements-obtained using measuring devices at various times and locations should be equal or differ by a predetermined amount in the interests of the nation's economy. In other words, the dimensions must be consistent.

In a unit of measurement, errors are expressed and legally recognized as such, and the outcomes of measurements are known. The science and practice of guaranteeing the uniformity and essential precision of measurements are covered by metrology. On the basis of the data examined above, we may suggest the following guidelines for picking a measuring tool and minimizing its errors [44-50]:

1. We must correctly select the measurement tool and objective.
2. Based on demand, we must report the location and circumstances of the measuring device.
3. After executing the measurement process using several techniques, it is vital to select the one that is the most practical, quick, accurate, and efficient.
4. The results must be accurately analyzed and recorded.

The following are some factors influencing the measurement results:

Depending on the nature of the object, the object of measurement denies the need to change the measurement.

Measurement subject (expert or experimenter; the operator-subject relationship should be kept to a minimum) a collection of methods for evaluating the measured quantity in relation to its unit, measuring device, and measurement method

It is important to consider the following elements that affect the aforementioned measurement findings while selecting measuring instruments and minimizing their errors [50].

Conclusions

In conclusion, it should be highlighted that local industrial companies' current state of development is typically critical to catching up to more established rivals, particularly in terms of quality assurance. Using current technological tools, it won't be able to complete this task quickly. Measurements also play a crucial part in verifying the quality of the final product. The effectiveness of the choice of measurement instruments and techniques of decreasing mistakes is increased in electricity production companies and laboratories by the use of quick, high-accuracy measurement systems that account for small defects.

References

1. G.G.Boboev, M.M.Mahmudjonov, and ofters. AIP Conference Proceedings, 2432, 030042, (2022), https://doi.org/10.1063/5.0089626
2. B.B.Kholikhmatov, M.T.Erejepov, and ofters. E3S Web of Conferences, 384, **01032**, (2023), https://doi.org/10.1051/e3sconf/202338401032
3. M.Ibadullayev. Proceedings of International Conference on Applied Innovation in IT, **11(1)**, pp. 285-290, (2023).
4. M.Ibadullaev, A.Nuraliyev, A.Yesenbekov. AIP Conference Proceedings 2552, **050036**, (2022), https://doi.org/10.1063/5.0133070
5. A.Nuraliyev, M.Ibadullayev. E3S Web of Conferences, 216, **01114**, (2020), https://doi.org/10.1051/e3sconf/202021601114
6. A.N.Tovbaev, M.Ibadullayev, S.I.Norboyev. Journal of Physics: Conference Series, 2388(1), **012060**, (2022), doi:10.1088/1742-6596/2388/1/012060
7. M.Ibadullaev, A.N.Tovbaev. E3S Web of Conferences, 216, **01113**, (2020), https://doi.org/10.1051/e3sconf/202021601113
8. A.Tovboev, M.Ibodulaev, M.Baranova, I.Grishina. IOP Conf. Series: Mater. Scien. and Engineering, 862(6), **062041**, (2020), doi:10.1088/1757-899X/862/6/062041
9. M.I.Ibadullaev, A.N.Tovbaev, A.Z.Esenbekov. E3S Web of Conferences, 139, **01054**, (2019), https://doi.org/10.1051/e3sconf/201913901054
10. S.Begmatov. Proceedings of International Conference on Applied Innovation in IT, **11(1)**, pp. 273-277, (2023).
11. R.Karimov, D.Xushvaktov. E3S Web of Conferences, 384, **01053**, (2023), https://doi.org/10.1051/e3sconf/202338401053
12. R.Karimov. E3S Web of Conferences, 384, **01056**, (2023), https://doi.org/10.1051/e3sconf/202338401056
13. R.Karimov, and ofters. E3S Web of Conferences, 289, **07021**, (2021), https://doi.org/10.1051/e3sconf/202128907021
14. E.Abduraimov. E3S Web of Conferences, 216, **01105**, (2020), https://doi.org/10.1051/e3sconf/202021601105
15. E.Abduraimov. E3S Web of Conferences, 384, **01051**, (2023), https://doi.org/10.1051/e3sconf/202338401051
16. E.Abduraimov, B.Nurmatov. E3S Web of Conferences, 384, **01052**, (2023), https://doi.org/10.1051/e3sconf/202338401052
17. A.Burkhankhodzhayev, B.Nurmatov. E3S Web of Conferences, 216, **01112**, (2020), https://doi.org/10.1051/e3sconf/202021601112
18. A.Burkhankhodjaev, B.Nurmatov, and ofters. AIP Conference Proceedings, 2552, **030024**, (2022), https://doi.org/10.1063/5.0133920
19. N.Khamudkhanova, M.Idriskhodjaeva, Kh.Kholbutayeva. E3S Web of Conf., 384, **01057**, (2023), https://doi.org/10.1051/e3sconf/202338401057
20. S.Umarov, K.Sapaev, N.Khamudkhanova, and ofters. AIP Conference Proceedings, 2402, **060016**, (2021), https://doi.org/10.1063/5.0071557
21. K.G.Abidov, O.O.Zaripov, and ofters. E3S Web of Conferences, 216, **01110**, (2020), https://doi.org/10.1051/e3sconf/202021601110
22. A.A.Khashimov, K.G.Abidov. Elektrotekhnika, **7**, pp. 34-38, (2001).
23. A.A.Khashimov, K.G.Abidov. Elektrotekhnika, **10**, pp. 22-26, (2001).
24. K.G.Abidov. E3S Web of Conferences, 384, **01055**, (2023), https://doi.org/10.1051/e3sconf/202338401055
25. K.G.Abidov, K.Dadajanov, U.Absoatov, S.Sabitov. E3S Web of Conferences, 383, **04053**, (2023), https://doi.org/10.1051/e3sconf/202338304053
26. K.G.Abidov, K.Dadajanov, U.Absoatov, S.Sabitov. E3S Web of Conferences, 383, **04044**, (2023), https://doi.org/10.1051/e3sconf/202338304044
27. G.Mustafaquliva, G.Maxmadiyev, O.Zaripov, Sh.Zaripova. E3S Web of Conferences, 216, **01109**, (2020), https://doi.org/10.1051/e3sconf/202021601109
28. A.Pulatov, J.Bekmuradov, O.Zaripov. E3S Web of Conferences, 289, **07002**, (2021), https://doi.org/10.1051/e3sconf/202128907002
29. A.Alimov, F.Akbarov, G.B.Eshniyazova, A.Tureniyaziva. E3S Web of Conferences, 289, **07035**, (2021), https://doi.org/10.1051/e3sconf/202128907035
30. R.R.Kabulov, F.A.Akbarov, A.A.Alimov. Nanosystems: Physics, Chemistry, Mathematics, **14(1)**, pp. 127-131, (2023).
31. D.N.Isamukhamedov, N.J.Tairova, and ofters. E3S Web of Conferences, 216, **01154**, (2020), https://doi.org/10.1051/e3sconf/202021601154
32. A.Nuraliyev, Y.Adilov, A.Esenbekov. E3S Web of Conferences, 384, **01054**, (2023), https://doi.org/10.1051/e3sconf/202338401054
33. K.Shamsiev. AIP Conference Proceedings 2552, **050015**, (2022), https://doi.org/10.1063/5.0130710
34. K.Jurayeva, Kh.Sattarov. E3S Web of Conferences, 389, **01060**, (2023), https://doi.org/10.1051/e3sconf/202338901060
35. Kh.Sattarov, and ofters. E3S Web of Conferences, 383, **01028**, (2023), https://doi.org/10.1051/e3sconf/202338301028
36. I.Siddikov, Kh.Sattarov, A.Abdumalikov. AIP Conference Proceedings, 2612, **050002**, (2023), https://doi.org/10.1063/5.0115211
37. A.Safarov, Kh.Sattarov, and ofters. Journal of Physics: Conference Series, 2176(1), **012006**, (2022), doi:10.1088/1742-6596/2176/1/012006
38. A.Safarov, Kh.Sattarov, and ofters. E3S Web of Conferences, 264, **05038**, (2021), https://doi.org/10.1051/e3sconf/202126405038
39. Y.Shoyimov, and ofters. Journal of Physics: Conference Series, 939(1), **012028**, (2021), doi:10.1088/1755-1315/939/1/012028
40. M.V.Melikuziev. E3S Web of Conferences, 384, **01033** (2022), https://doi.org/10.1051/e3sconf/202338401033
41. A.N.Rasulov, E.G.Usmonov, M.V.Melikuziev. AIP Conference Proceedings, 2552, **050018**, (2023), https://doi.org/10.1063/5.0111530
42. O.G.Hayitov, M.O.Gafurova, and ofters. AIP Conference Proceedings, 2432, **030101**, (2022), https://doi.org/10.1063/5.0091265

43. I.Kh.Siddikov, M.T.Maksudov, and ofters. AIP Conference Proceedings, 2432, **020003**, (2022); https://doi.org/10.1063/5.0089681
44. N.N.Norkhojaeva, N.B.Pirmatov, N.K.Kamalov. E3S Web of Conferences, 402, **10041**, (2023), https://doi.org/10.1051/e3sconf/202340210041
45. S.Yaxyayeva, N.Pirmatov, and ofters. E3S Web of Conferences, 402, **05016**, (2023), https://doi.org/10.1051/e3sconf/202340205016
46. O.Toirov, N.Pirmatov, and ofters. E3S Web of Conferences, 401, **04033**, (2023), https://doi.org/10.1051/e3sconf/202340104033
47. N.B.Pirmatov, U.T.Berdiyev, and ofters. E3S Web of Conferences, 401, **03021**, (2023), https://doi.org/10.1051/e3sconf/202340103021
48. N.B.Pirmatov, A.M.Egamov, and ofters. E3S Web of Conferences, 401, **03056**, (2023), https://doi.org/10.1051/e3sconf/202340103056
49. N.Pirmatov, A.Panoev, and ofters. E3S Web of Conferences, 383, **04046**, (2023), https://doi.org/10.1051/e3sconf/202338304046
50. N.Pirmatov, A.Bekishev, and ofters. AIP Conference Proceedings, 2612, **050005**, (2023), https://doi.org/10.1063/5.0135546

E3S Web of Conferences **461**, 01089 (2023) https://doi.org/10.1051/e3sconf/202346101089
RSES 2023

Prospects for the application of a small wind power device with a vertical axis in the Jizzakh region

Jurabek Izzatillaev[1], *Umid* Khudoyberdiev[1*], *Gulmurod* Kushakov[2]

[1]Tashkent Institute of Irrigation and Agricultural Mechanization Engineers (TIIAME) National Research University, 111104, Uzbekistan,Tashkent, Tashkent city, Kori Niyoziy st 39
[2]Jizzakh polytechnical institute, 130100, Uzbekistan, Jizzakh rejoin, Jizzakh city, Khalqlar Dustligi st. 4

Abstract. In recent years, renewable energy sources, especially wind energy, have been widely used around the world. Big projects are being implemented in Uzbekistan for the development of wind energy in the next decade. Large wind power plants are being built in different regions of the country such as Navoi, Bukhara, and Karakalpakstan. In these wind power plants, horizontal type wind energy devices are mainly used for the production of electricity. Horizontal-axis wind energy devices are used not only in the production of large capacities, but also in providing local consumers with small capacities. However, the share of electricity generation or supply of local consumers with the help of small vertical axis wind energy devices is much smaller. In this article, analyzing the potential of wind energy in Jizzakh region, the prospects of using small vertical wind energy devices are theoretically researched.

1 Introduction

The population of the Earth is increasing from year to year, and the amount of electricity consumed by them is also increasing [1-3]. To meet the ever-increasing energy consumption, extensive use of renewable energy sources in addition to existing fuels is the need of the hour [4-6]. The source of energy of the world since 1970 and its trained until 2040 is shown in Fig. 1. It can be observed that oil, gas, and coal were the main source of energy for the last five decades and similar trend is projected for the coming decades [7-10].

Annual renewable capacity additions are forecast to increase continuously over the forecast period, reaching a record 460 GW in 2027 in the main case, 60% higher than last year's growth [11-15]. At the end of the forecast period, solar PV and wind provide the vast majority of global renewable capacity additions in 2027, accounting for nearly 95% as technology-specific challenges and limited policy support hamper faster expansion of hydropower, bioenergy, geothermal, CSP and ocean technologies Fig.2.

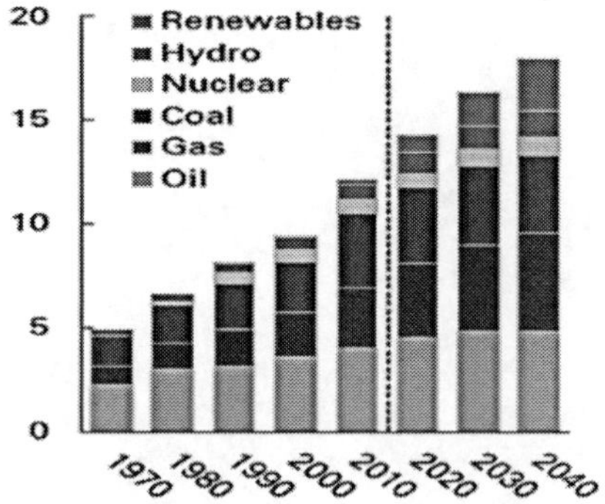

Fig.1. Use of energy source and its transition (Renewable includes solar, wind, biomass, etc.)

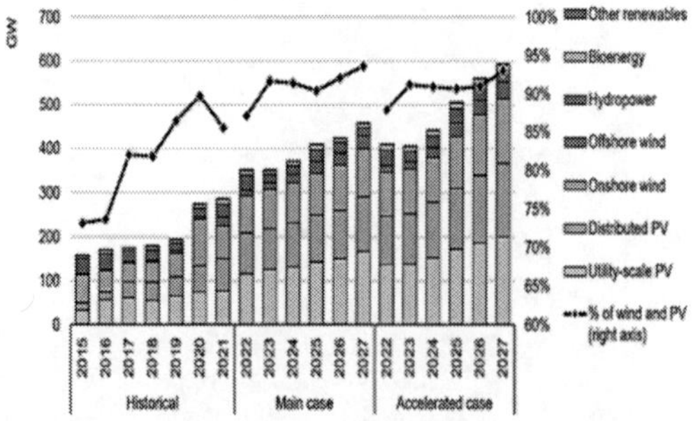

Fig.2. Renewable annual net capacity additions by technology, main and accelerated cases, 2015-2027

* Corresponding author: umidkhudoyberdiev1994@gmail.com

Wind is a readily available renewable free and sustainable energy source [16-18]. In every single day around the world, the wind energy is seized and converted using wind turbines into electricity [19-20]. Wind energy is one of the best renewable energy which can potentially substitute the fossil fuels with no gas emission and water consumption, and in little land space [21-23]. The increase in demand for energy resources on a global scale has given a great impetus to the development of wind energy.

An estimated 102 gigawatts (GW) of wind power capacity was installed globally in 2021 – including more than 83 GW onshore and almost 19 GW offshore. Total additions were up around 7% relative to 2020 to the highest level to date, with annual offshore installations almost three times their previous high. By year's end, total global wind power capacity rose 13.5% over 2020 to surpass 845 GW (791 GW onshore and the rest offshore) [24-25]. Wind power capacity in operation around the world contributed an estimated 7% of total electricity generation in 2021 fig 3. 93.6 GW of new wind power capacity was added worldwide in 2021, only 1.8% lower than the 2020 record, bringing the total installed wind capacity to 837 GW, a growth of 12.4% compared to last year [26-28]. Although new installations in the onshore wind market dropped to 72.5 GW last year, it was still the second highest year in history. The offshore wind market had a record year with more than 21 GW grid connected, three times more than the previous year, making 2021 the highest year ever Fig.4.[29-30].

The top five global markets for new installations in 2022 were: China, USA, Brazil, Germany and Sweden. Overall, they accounted for 71% of global installs last year, down 3.7% from 2021 [31-33]. This was primarily because the world's two largest markets, China and the US, lost a combined market share of 5% year-over-year, the second year in a row that both countries lost market share [34-35].

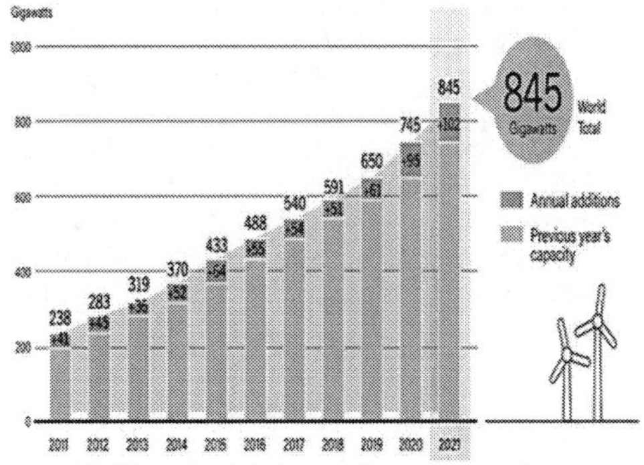

Fig. 3. Wind Power Global Capacity and Annual **Additions**, 2011-2021.

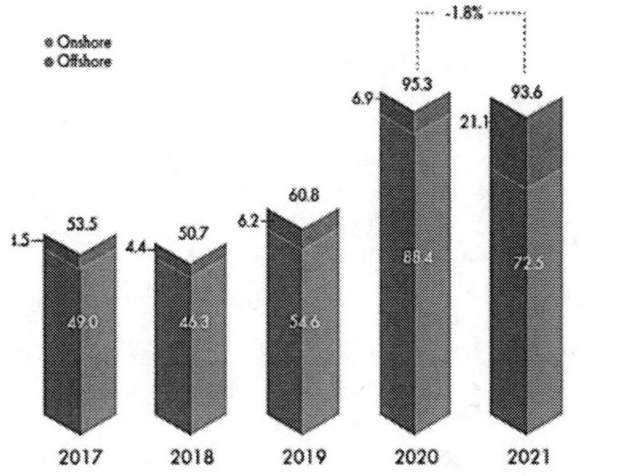

Fig. 4. New wind power installations in 2017-2021.

Given the fact that water sources in the Central Asian region are declining, it is important to accelerate the use of alternative energy. Uzbekistan has the opportunity to use wind energy, but previously this renewable energy sector was not among the priority areas for the development of local energy. At the beginning of the 21st century, national and foreign experts summarized previously known information, conducted a new series of observations necessary to assess the local and regional environmental situation, the risks associated with the movement of air masses (about 90 meteorological stations operate on the territory of the country, which provided the results of systematic observations for 1994 -2004 and other periods, mainly at altitudes up to 10 m). As a result, in the mid-2000s, the gross potential was estimated at 2.2 million toe. per year, technical - 0.4 million toe. per year (five times less), and economic - as "Insignificant". Below in the figure 5. one of the variants of the wind map of the Republic of Uzbekistan is shown. Its analysis allows a general idea of the nature of the distribution of wind energy potential within the territory of the country.

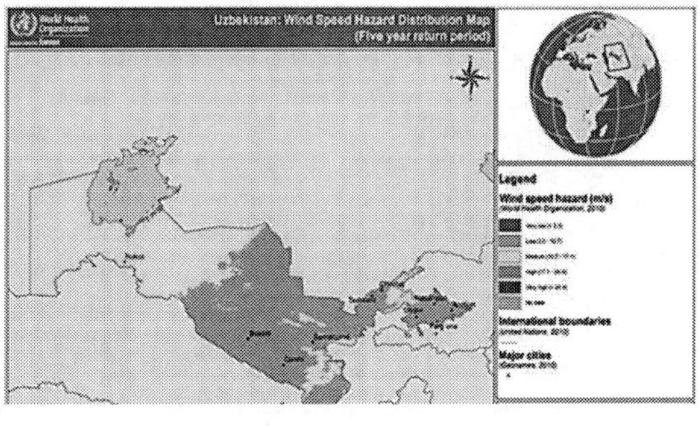

Fig.5. Uzbekistan: Wind Speed Hazard Distribution Map.

The first wind turbine with a capacity of 170 kW and a cost of more than 2 million US dollars was built in 2010 near the Charvak reservoir. Design documentation was prepared by Hydroproject OJSC, equipment was supplied by Doojin Co. (The Republic of Korea). The second (experimental, but the first large on the scale of Central Asia) Chinese-made wind

https://doi.org/10.1051/e3sconf/202346101089

turbines was installed in 2012 in the Yubileiny settlement of the Bostanlyk district (Tashkent region). Its power is 750 kW, the diameter of the wind wheel is 50 m, the height of the tower is 65 m. In 2012, based on the experience gained, Uzbekenergo JSC planned to create a wind farm with a capacity of 100 MW, generating about 170 million kWh of electricity per year. Subsequently, the implementation of the plan was delayed by ten years for various reasons, including the lack of reliable comprehensive indicators of wind energy potential. In the middle of the second decade, the parameters of the wind energy potential were refined. In 2015, Intec-GOPA (AHU), the Geo-Net Association (Canada), with the participation of Uzbekenergo JSC and the unitary enterprise Talimarjan TPP, based on the results of annual observations and design and survey work, created a national "Atlas of the Winds" Fig 6. According to experts, the gross potential of wind energy has exceeded 520 GW (this is the possible total capacity of wind turbines), which theoretically allows generating 1,077,651 GWh of electricity per year.

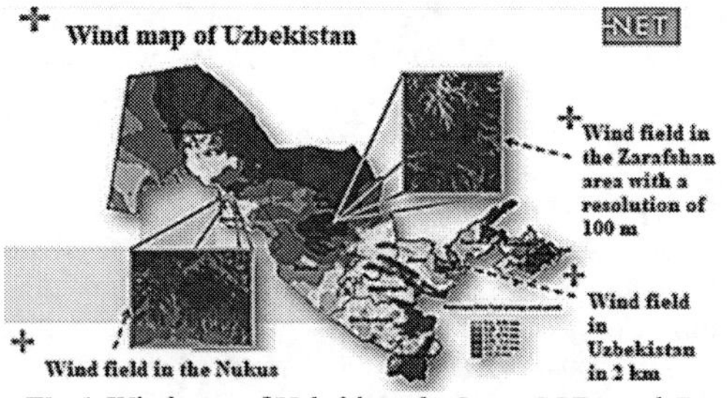

Fig.6. Wind map of Uzbekistan by Intec-GOPA and Geo-NET

Promising development areas include the Navoi region and sites located in the Republic of Karakalpakstan near remote settlements and the small industrial zone “Muynak” (created in 2017 on the territory of the former "Muynak fish cannery" and the largest in Central Asia). From a technical point of view, according to the recommendations of the researchers, the most optimal solution is the use of wind turbines with a capacity of 3 MW and a rotor diameter of 100 m. specific capital investments in the creation of 1 MW of wind power capacity were estimated at 1 million US dollars (comparable to the same indicator for thermal power plants), the cost of wind power was approximately 5.5 q/kWh. In the Republic of Uzbekistan, in cooperation with international financial institutions, feasibility studies have been carried out in the field of widespread use of wind energy and the “Concept of Electricity Supply of the Republic of Uzbekistan for 2020-2030” has been developed. One of the main goals is the construction of wind power plants with a capacity of 3000 MW by 2030 in the republic. Today, several projects on the construction of SPPs have been launched in Uzbekistan. Including:

Based on an agreement signed on April 1, 2021, Masdar Clean Energy of the United Arab Emirates and the Ministry of Energy of the Republic of Uzbekistan reached an agreement on the construction of a 500 MW wind farm in the city of Tomdi district of the Navoi region. With a project cost of $600 million, the wind farm will generate 1.8 billion kWh of electricity per year at full capacity. Two more large projects are being implemented by ACWA Power Company of Saudi Arabia, and one of these projects, according to the contract signed between ACWA Power Company and the Ministry of Energy on January 23, 2021, provides for the construction of two wind power plants with a total capacity of 1 GW in Gijduvon and Peshkun districts of Bukhara region. The second project with the participation of this company is based on the agreement signed between the Ministry of Energy and ACWA Power (Saudi Arabia) on December 23, 2022 with the financial support of the European Bank for Reconstruction and Development, and works on the construction of another wind power plant with a capacity of 1500 MW in Beruni and Karauzyak regions of Karakalpakstan. When this wind farm is operational, it will produce 350 million kWh of electricity per year. On March 28, 2023, a memorandum of understanding was signed between China's CNEEC (China National Electric Engineering Company), China Huadian Overseas Investments, SANY Renewable Energy and the Ministry of Energy to study the possibilities of building wind power plants with a capacity of up to 1000 MW in Jizzakh region. According to this memorandum, 1 billion US dollars will be directly invested in the implementation of the project[4-6].

2 Materials analysis of wind potential of the region

Uzbekistan, double landlocked country with no coastal area, consists of 25% mountainous valleys and 75% desert covered oasis. Therefore the average yearly wind speed on the whole territory of the country is estimated between 2-2.5 m/sec, which indicates a non-promising future for wind power engineering in Uzbekistan, especially for wind turbines of middle and high power. Consequently the wind energy potential in Uzbekistan is relatively less exploitable in comparison with the solar and hydropower potential (Table1).Even in the provinces with areasonable potential wind resource is very seasonal and it is necessary to study the coincidence between peak wind months and the electricity demands.

Table 1.

№	Province	Area of wind resource km^2	Gross potential W/m^2	Gross energy GWh/year
1	Karakalpakistan Rep.	1649	93	10752.20
2	Andijan	42	20	60.00
3	Bukhara	294	90	2421.80
4	Jizzakh	205		649.60
5	Kashkadaryo	284	58	1162.00
6	Navoi	1108	104	7931.40
7	Namangan	79	28	155.40
8	Samarkand	164	61	690.60
9	Surkhandaryo	208	30	434.80
10	Sirdaryo	51	58	212.00
11	Tashkent	156	100	1066.00
12	Fergana	71	34	49.60
13	Khorazm	63	55	264.00
	Total	4474	84	25849.00

Jizzakh region is located in the central region of Uzbekistan. The region borders on the Republic of Kazakhstan in the north and northeast, Syrdarya region in the east, the Republic of Tajikistan in the east and southeast, Samarkand region in the south and southwest. and Navoi region in the west. Air currents that affect the climate of the region are predominantly northwestern and southwestern air currents. The northwesterly air flow comes from the north from the western part of Russia and Kazakhstan and is 34.5 days a year. The south-west air flow also has a great influence on the climate of the region and is 63.2 days a year. Fig. 7. shows the direction of the north-western air flow entering the territory of the Jizzakh region [7-9].

Fig.7. The direction of the north-west air flow

Based on the data obtained from the Global Wind Atlas international database, the wind energy density is high in Sh.Rashidov, Dostlik and Mirzachol districts of Jizzakh region at heights of h=10 and h=50 m, while it is low in Zamin and Zafarabad districts, fig. 8.

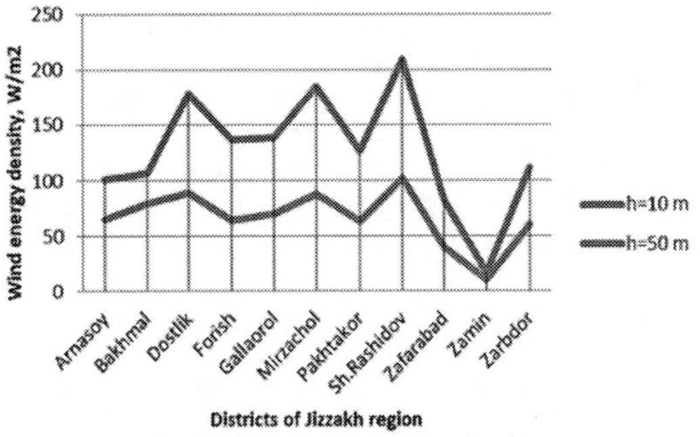

Fig.8. Wind energy density by districts.

Fig. 8. It can be seen from the data that the wind potential in Dostlik, Mirzachol and Sh.Rashidov districts of Jizzakh region is higher than the rest of the districts. This will expand the possibilities of using vertical axis wind wind turbine in these districts. Based on the data obtained from the NASA POWER international geographic database and the MERRA-2 satellite, the annual average wind speed in the Jizzakh region at a height of h=10 m for different districts in 2019 is 9.02 - 10, 56 m/s, in 2020 7.94 - 9.96 m/s and in 2021 it took values in the range of 8.42-9.8 m/s. The annual average wind speed at h=50 m was 11.26 m/s to 12.25 m/s in 2019, 9.5 m/s to 11.05 m/s in 2020, and 2021 and it was from 10.3 m/s to 11.66 m/s Fig.9.[10-12].

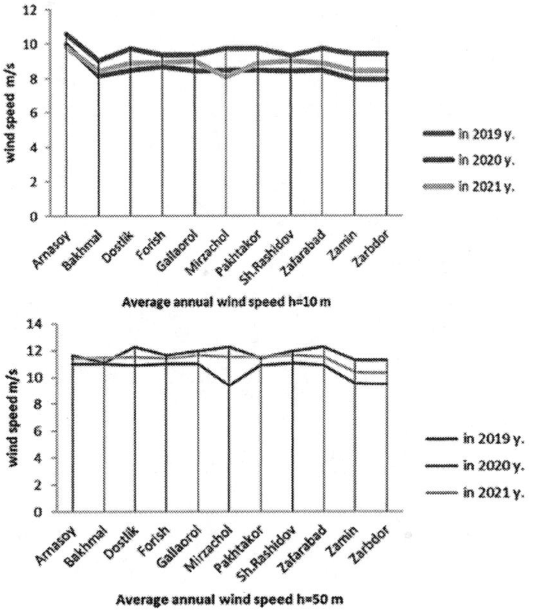

Fig. 9. Average annual wind speed h=10 m, h=50 m

Around the world, horizontal wind turbines are mainly used in the production of large-scale electricity from wind energy. It is also possible to use a horizontal type wind power device to supply electricity to household consumers. When using a horizontal type wind power device, it is necessary to take into account the speed, direction and frequency of the wind. The use of horizontal wind turbines in areas where the direction and speed of the wind changes is technically and economically inefficient. In regions with variable wind direction and speed, it is effective to use vertical wind power devices. The advantages and disadvantages of vertical wind energy devices can be listed as follows:

VAWT advantages
- No yaw mechanisms is needed.
- VAWTs have lower wind startup speeds than HAWTs.
- VAWTs may be built at locations where taller structures are prohibited.

VAWT disadvantages
- VAWTs have low efficiency.
- More blades than HAWTs.
- VAWT advantages.

Despite the existence of different constructions of vertical type wind energy devices, they can be divided into two large groups: Savonius vertical wind turbines and Darrius vertical wind turbines. Savonius turbines working on the aerodynamic drag principle and excellent in areas of turbulent wind and can self-start at low wind speed. Darrieus turbines were operating on the principle of lift. The efficiency of Savonius type wind turbines is lower than that of Darrius type wind turbines. According to Betz's law, the efficiency of the Savonius turbine is up to 20%, and the efficiency of the Darrius wind turbine is up to 40%. Fig 10.

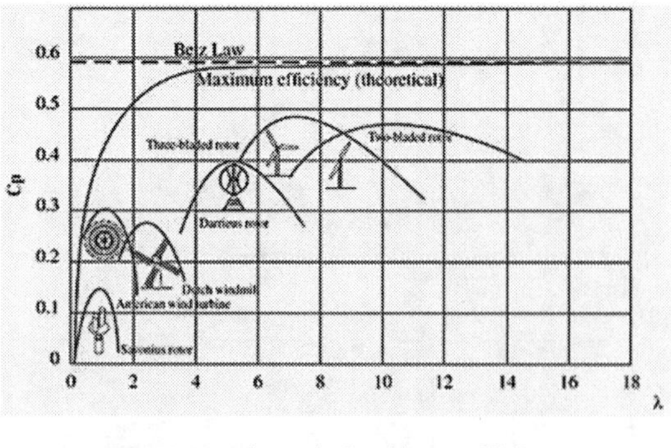

Fig.10. Betz law. Ideal turbine coefficient

Darrieus rotor can be subdivided into three categories; D – type, H – type and helix type which have been shown in Fig. 11.[13-15].

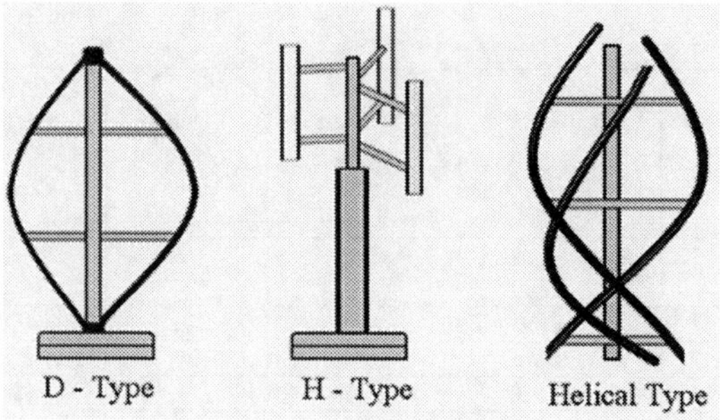

Fig.11. Different types of Darrieus wind turbine

3 Discussion and results

Unlike other energy sources, wind energy efficiency is high. Therefore, the theoretical and technical potential of wind energy is taken into account when using wind energy potential. The theoretical potential of wind energy is determined by the following expression:

$$W_{theoritical}=0{,}025\rho TS\sum_{i=1}^{n} v_i^3 t_i \quad (1),$$

here, ρ is air density, kg/m^3; T = 8760 hours during the year; S – area surface, m^2; v – i multi-year wind speed in the range; t – I approximate wind speed in the range.

Technical potential of wind energy:

$$W_{technical}= 0{,}01 \frac{N_C}{D^2}TS, \quad (2)$$

here, N_C – the average power of the wind turbine and it is expressed as follows:

$$N_C = \frac{\pi D^2}{8}\rho\sum_{i=1}^{n} v_i^3\, \eta_{VAWT} t_i\,, \quad (3)$$

here, η_{VAWT} efficiency index of VAWT

$$\eta_{VAWT} = C_p \eta_{mech}\eta_{el}\,, \quad (4)$$

here, C_p- wind energy utilization factor; η_{mech} - mechanical efficiency of the wind turbine; η_{el} – electrical efficiency of the wind turbine.

The amount of electricity produced by a wind turbine:

$$E = 0{,}01\ T \sum_{i=v_{switch\ on}}^{v_{switch\ off}} P_{VAWT\ i}\,(k_h v)\, f_i\,(v), \quad (5)$$

Useful power of the wind turbine:

$$P= P_{flow-specific\ power} S_{VAWT}\eta_{rotor}\eta_{generator}\sigma 10^{-3}, \quad (6)$$

Flow-specific power,

$$P_{flow-specific\ power} = 0{,}5\ \rho\ (k_h\ v)^3 \quad (7)$$

In the production of electricity from wind energy, the working range of wind speed is used. This working range should not exceed 25 m/s. This speed is equal to

9 points on Beaufort's 12-point scale. Wind turbines generally operate between 3 and force 10 on the Beaufort scale and the rated capacity is commonly defined at forces 6/7 depending upon the wind turbine type Table 2.

Table 2.

Beaufort No.	Speed in m/s	Description
0	0–0.2	Calm
1	0.3–1.5	Light air
2	1.6–3.3	Light breeze
3	3.4–5.4	Gentle breeze
4	5.5–7.9	Moderate breeze
5	8.0–10.7	Fresh breeze
6	10.8–13.8	Strong breeze
7	13.9–17.1	Moderate gale
8	17.2–20.7	Fresh gale
9	20.8–24.4	Strong gale
10	24.5–28.4	Full Gale/Storm
11	28.5–32.6	Violent Storm
12	>=32.7	Hurricane

Fig. 9. based on the data presented in the annual average wind speed at a height of h=10 m is in the range of 8-10 m/s, 5 points on the Beaufort scale, the annual average wind speed at h=50 m is 10-12 m/s , corresponds to 6 points on the Beaufort scale. The 6-point scale, calculated according to the Beaufort scale, corresponds to Dostlik, Mirzachol and Sh.Rashidov districts of Jizzakh region. [9-12] based on the information presented in the sources, the use of vertical wind power plants in Dostlik, Mirzachol, and Sh. Rashidov districts are considered more effective compared to other regions. The use of H-type turbines, which are modern modifications of the Darrieus wind turbine, is efficient compared to the Savonius-type vertical wind power plant. We will analyze some types of modern H-type vertical wind energy devices produced by different companies and their technical characteristics for regions with high wind potential. Companies that produce vertical small power wind energy devices offer wind energy devices with different capacities and characteristics. In particular, Aeolos company (Great Britain) Aeolos – V5 (1), UGE (USA) UGE Vision AIR 5(2), KESSLER (Germany) Kessler SPINWIND(3), Tianjin Galaxy Tech Co., Ltd (China) JX – H3 and JX – H5(4), Hi-VAWT (Taiwan) DS-3000(5), Royall Products LLC (USA) WindSpire 5 kW, United Solar Technologies (Russia) Sokol Air Vertikal-5kW(6) models are among them (Fig. 12). Table 3. below shows the technical parameters of the vertical H-type wind energy devices named above [16-20].

Fig. 12. Different models of vertical axis wind turbines.

Analyzing the VAWT models presented in Table 3, it is possible to make sure that the models capable of operating at the smallest initial wind speed are Aeolos V5 (1.5 m/s) and Sokol Air Vertikal (2 m/s). The nominal speed of these models is 10 m/s (Aeolos V5) and 7.5 m/s (Sokol Air Vertikal). Compared with other VAWT models, these models are an optimal solution for using the wind potential of Dostlik, Mirzachol and Sh.Rashidov districts at a height of h=10 m. At a height of h=50 m, in addition to Aeolos V5 (1.5 m/s) and Sokol Air Vertical models, UGE Vision Air 5 and Kessler SPINWIND vertical wind energy devices can be a good solution for the above three districts. If we take into account that the wind potential in these districts is high and the annual average wind speed is up to 12.25 m/s, then DS-3000, JX-H3, and JX-H5 models can also be used for electricity generation. But for Dostlik, Mirzachol and Sh.Rashidov districts, Aeolos V5, and Sokol Air Vertical wind energy devices are the optimal options for electricity production using wind energy at heights of h=10 and h=50 m. Firstly, the starting speed of these models is small (1.5 and 2 m/s), and secondly, their nominal speed is equal to or less than the annual average wind speed at h=10 m and h=50 m.[21-25].

4 Conclusion

This study provides information on the potential of wind energy in the world and in Uzbekistan, as well as on the projects launched to develop wind energy in the country. The data obtained from international databases, the wind energy potential of Jizzakh region was analyzed. Based on the conducted theoretical studies, it was determined that 3 out of 11 districts of the region have a high wind energy potential compared to other districts. The possibilities of using a vertical-type wind energy device for these districts were considered. As a result, it was theoretically analyzed that the use of Aeolos - V5, and Sokol Air Vertical wind energy devices in the production of electricity using the potential of wind energy in these districts is the optimal solution.

References

1. B. Dudley, "BP Energy Outlook" 2018.
2. Report: Renewables 2022 Analysis and forecast to 2027.
3. UN, «World Health Organization», «Uzbekistan: Wind Speed Hazard Distribution Map (Five year return period», 2010.
4. Bahtiyor R. potential of renewable energy sources in uzbekistan, Journal of Knowledge Management, Economics and Information Technology, Issue 7 December 2011.
5. Official website of Jizzakh regional government https://jizzax.uz/viloyat-aida.html
6. U.A. Tadzhiev, E.I. Kiseleva, M.U. Tadzhiev, R.A. Zahidov, Features of the formation of wind flows over the territory of Uzbekistan and the possibility of their use for generating electricity. Part i. Solar technology. 2015. No. 1.
7. Prashant Chandra Pujari, Amit Jain, Devang S. Nath, Naveen Kumar "Performance enhancement of savonius turbine with the application of reorienting blade mechanism", Energy Sources, Part A: Recovery, Utilization, and Environmental Effects, https://doi.org/10.1080/15567036.2021.1948635
8. Amlan Das1, Dr. Pradip Kumar Talapatra, Modelling and Analysis of a Mini Vertical Axis Wind Turbine, International Journal of Emerging Technology and Advanced Engineering, Volume 6, Issue 6, June 2016.
9. Pirmatov N.B., and ofters. E3S Web of Conferences, 2023, 383, 04046.
10. Pirmatov N.B., and ofters. AIP Conference Proceedings, 2023, 2612, 050005.
11. Pirmatov N.B., and ofters. AIP Conference Proceedings, 2022, 2432, 020020.
12. Pirmatov N.B., and ofters. AIP Conference Proceedings, 2022, 2552, 070009.
13. Pirmatov N.B., and ofters. E3S Web of Conferences, 2021, 264, 04028.
14. Pirmatov N.B., and ofters. IOP Conference Series: Materials Science and Engineering, 2021, 1030(1), 012161.
15. Pirmatov N.B., and ofters. E3S Web of Conferences, 2019, 139, 01027.
16. Bystrov D.V., and ofters. AIP Conference Proceedingsthis link is disabled 2552, 060014 (2022).
17. Bystrov D.V., and ofters. E3S Web of Conferences 216, 01117 (2020).
18. Bekishev A.E., Urakov S.E. Static stability in the electric power system. Technical sciences and innovations. No. 2, 2019. p. -175-183.
19. Aljwary A., Yusupov Z. E3S Web of Conferences, 304, 01010 (2021).
20. Pirmatov N., Bekishev A., and ofters. E3S Web of Conferences, 2021, 264, 04028, https://doi.org/10.1051/e3sconf/202126404028
21. Pirmatov N., and ofters. Conference Proceedings AIP 2552, 040020 (2023); https://doi.org/10.1063/5.0115727
22. Khujaev I., and ofters. E3S Web of Conferences 401, 04040 (2023).
23. Allaev K., Toshov J., Toirov O. E3S Web of Conferences 401, 05090 (2023).
24. Bekishev A., Urakov S., Mirkhonov Yu. E3S Web of Conferences 216, 01116 (2020). https://doi.org/10.1051/e3sconf/202021601116
25. Mustafakulova G., and ofters. E3S Web of Conferences 216, 01118 (2020).
26. Alimkhodjaev K., and ofters. E3S Web of Conferences 216, 01121 (2020).
27. Toirov O., Sadullaev T., and ofters. E3S Web of Conferences 383, 01029 (2023).
28. Erejepov Madiyar, and ofters. 2021 International Conference of Young Scientists Energy Systems Research, ESR 2021. https://10.1051/e3sconf/202128907018
29. Toirov O., Pirmatov N.B., and ofters. E3S Web of Conferences 401, 04033 (2023).
30. Toirov O., Khalikova M., and ofters. E3S Web of Conferences 401, 05089 (2023).
31. Khalikov S., Algorithm and Software Implementation of the Diagnostic System for the Technical Condition of Powerful Units, E3S Web of Conferences 377, 01004 (2023).
32. Alimkhodjaev K., Toirov O., Taniev M. Addressing Issues and Possibility of Introducing Renewable Energy Sources in the Conditions of Uzbekistan // Journal of Critical Reviews, 2020. Vol. 7, Issue 15.
33. Toirov O.Z., Khalikov S. E3S Web of Conferences 365, 04010 (2023).
34. Toirov O.Z., and ofters AIP Conference Proceedings 2552, 040021 (2022).
35. Toirov O.Z., and ofters. E3S Web of Conferences 216, 01119 (2020) RSES 2020.
36. Jumaeva D., and ofters. E3S Web of Conferences, SUSE-2021, 288, 01041 (2021).
37. Afrisal H., Setiyadi A.D. Performance Analysis of 4-DOF RPRR Robot Manipulator Actuation Strategy for Pick and Place Application in Healthcare Environment, International Journal on Advanced Science, Engineering and Information Technology, Vol. 12 No. 6, p. 2258-2265, (2022).
38. Toirov O., Khalikov S.S. E3S Web of Conferences 365, 04013 (2023).
39. Hoshimov U.H., Ishnazarov O.Kh., and ofters. Journal of Physics: Conference Series, Conf. Ser. 2094 052051, (2021).
40. Hoshimov U.H., Ishnazarov O.Kh., and ofters. E3S Web of Conferences 264, 04088 (2021)
41. https://doi.org/10.1051/e3sconf/202126404088

The problem of using electrodes in the steel melting process

Ikromjon Rakhmonov[1*], *Makhzuna* Korjobova[1], *Bakhriddin* Kholikhmatov[1], *Navruz* Uzakov[2]

[1]Tashkent State Technical University named after Islam Karimov, 100095, Uzbekistan, Tashkent, University St. 2A

[2] "TIIAME" Karshi Institute of Irrigation and Agricultural Technologies of the National Research University, Karshi, Uzbekistan

Abstract. This article explores the critical role of electrode technology and energy efficiency in the process of electric arc welding for the purpose of melting and shaping steel. It delves into the challenges posed by electrodes that directly impact the technology's effectiveness and energy consumption. Furthermore, the article conducts an analysis of various electrode types and classifications utilized in steel melting processes, emphasizing their efficient utilization in the steel production industry. As the demand for steel production continues to rise, so do the requirements for electrode strength and durability. However, modern metallurgical processes often encounter the recurring issue of electrode degradation and failure. Through comprehensive research, this article sheds light on the common problems associated with electrode failure during steel melting operations.

1 Introduction

Electrodes are the main element in the operation of steel melting furnaces. Not only electrodes directly affect the timely and high-quality melting of the slag but also, reducing the consumption of electricity in metal melting is one of the most important tasks. Because of the process of metal melting is 40-70% of the whole process and the consumption of electricity is 60-80%. Acceleration of the metal melting as much as possible, the electric arc between the electrodes and the metal plays a very important role for the quality continuation of the technological process. After all, when melting steel, electrodes are damaged, broken, and as a result, the arc goes out and the melting process is disturbed. So that depending on the mechanical and thermal tolerance of electrodes and their use in melting processes, the selection of their production process, materials, selection of electrode types in accordance with the process is one of the urgent problems. As well as, high quality melting of metal in a specified period time directly depends on the electrodes. Today, the use of furnaces is increasing, and the demand for electrodes is increasing directly after the increase in power. One of the biggest problems in the process of metal melting today is damage (breakage) of electrodes. In this article, these problems are analyzed and solutions are given with a scientific and practical approach [1-4].

These are scientists who have worked on scientific research: In Central Asia [5, 12], in Foreign countries [14, 16].

2 Experimental research

Electrodes used in the steel industry are divided into several types based on their application and the types of manufactured products. They are selected for durability and thermal stability of the electrodes [13, 17-25].Their names are indicated in the following diagram, *Fig.1.*

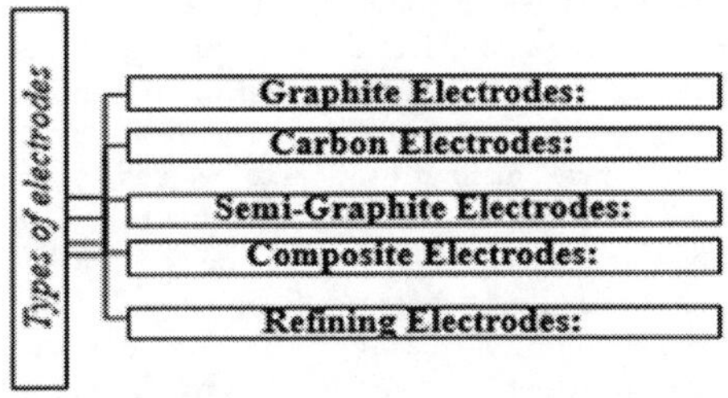

Fig.1. Types of electrodes

The electrodes of an arc steelmaking furnace work in extremely difficult conditions. Therefore, a number of specific requirements are imposed on the material for the manufacture of electrodes. They must have sufficient mechanical strength at low and high temperatures; have high electrical conductivity; be resistant to oxidation in the working space of the furnace and outside it, provide reliable contact in the heads of the electrode holders and the junction points of the electrode sections, have a low cost. Currently, only electrodes made of carbonaceous material meet such requirements.

Graphite electrodes are used in electric arc furnaces for several reasons. Because the temperature of the furnace is very high (1600-1800°C), electrodes that can

* Corresponding author: ilider1987@yandex.com

withstand this temperature are graphite electrodes. The 1 table has types of graphite electrodes [26-30].

Table 1. Types of graphite electrodes

RP	A conventional power graphite electrode is also called a graphite molded column to allow a current density through a graphite electrode of less than 17 A/cm 2, Mainly used in steelmaking, yellow phosphorus, silicon and other ordinary power electric furnaces.
HP	The high-current HP electrode is appropriate for casting operations or occasionally in arc furnaces. High power graphite electrode wire material for steelmaking of an electric arc furnace, which is made of petroleum coke, needle coke, coal asphalt, and high-temperature wire material produced by a series of strict process processes. In order to allow the current density through the graphite electrode to 18 ~ 25A / km 2, mainly used in a high-power steelmaking arc furnace.
UHP	Ultrahigh graphite electrodes are used to recycle steel in the arc furnace industry. The main ingredient is high-quality needle coke made from oil or coal tar. In order to allow the current density through the graphite electrode to be larger than 25A/cm 2. Mainly used for ultra-high power steelmaking electric arc furnace.

Above are the types and classification of electrodes, of which the most widely used in the steel melting process are UHP graphite electrodes. Take as an example of graphite electrodes, due to improper use of graphite electrodes in technological processes and operation, electrodes can be broken and damaged. As a result, the metal melting time increases and the cost of the product increases. If broken graphite electrodes fall into the furnace, carbon content will be added to the slag and stop the melting process in the furnace. This leads to the deterioration of the properties of steel and the decrease in the price of the product by the purchasing companies, and in some cases cases of non-purchase occur.

In addition, broken or damaged electrodes cause an increase in the consumption of electrodes and an increase in the price of the product. As a result, the amount of profit metallurgical plants receive from the sale of these products will decrease significantly. Therefore, it is necessary to reduce the consumption of electrodes in order to increase technical and economic indicators in steel production. In many countries (in particular, in Uzbekistan) steel melting furnaces of large and medium capacity are used.

Take a graphite electrodes as an example, the production technology of high-quality graphite electrodes for heavy-duty arc furnaces is quite complex. The high cost of charge materials and high power consumption determine the very high cost of special graphite electrodes for heavy-duty furnaces ($1500-2000 per 1 ton). Because of the high cost of graphite electrodes, the cost of electrodes during the operation of furnaces of ordinary power for smelting carbon steels is 8% of the cost of steel; when operating heavy-duty furnaces, such costs can exceed 15% of the cost of steel, in some cases reaching 30%. The consumption of graphite electrodes fluctuates within a significant range, which during normal operation of furnaces is 4–8 kg/t (in a well-functioning heavy-duty steel-smelting furnace, the total consumption of electrodes was 4 kg/t of steel) [31-34, 40].

According to the above points, we can say that the consumption of graphite electrodes in EAF steelmaking depends not only on the quality of the electrodes, but also on the level of operation and management of steelmaking.

So what factors influence electrode breakage or failure? When these factors were analyzed, it was studied that there was too much factor dependence. The below in fig. 2 follows the main 3 factors, which are:

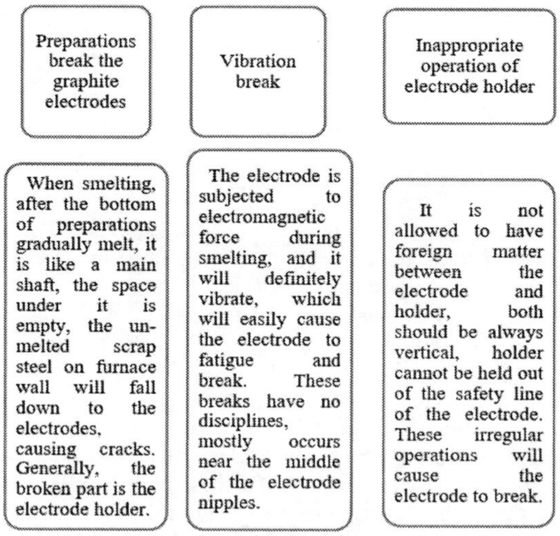

Fig.2. Factors causing breakage and damage of graphite electrodes [35]

To minimize electrode consumption and prevent breakage and failure in steel melting furnaces, it is crucial to employ proper electrode usage techniques. Fig.3. illustrates key solution factors that mitigate electrode damage.

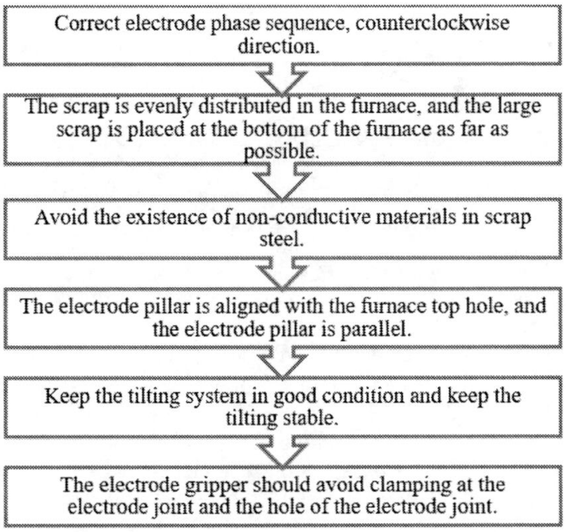

Fig.3. Strategies for preventing electrode breakage in steel production

Preserving the integrity of graphite electrodes during the steel melting process positively impacts both economic efficiency and metal quality [36-39].

Another problem is that during steel production in electric arc furnaces, there are high-energy losses. Generally, the operational practice of electric arc steelmaking furnaces demonstrated that the total electrode consumption correlates well with the overall electricity consumption. Therefore, any organizational and technical measures aimed at reducing the overall electricity consumption during the melting process will also reduce electrode consumption. Energy conservation is a key component of reducing the cost of steel production. Let is explore the primary energy-saving measures while taking into account changes in other cost components.

We will examine the energy balance of the Electric arc furnace (EAF), which is be expressed as follows:

$$W_e + W_s + W_{ch} + W_b = W_m + W_{e,l} + W_{th,l} \pm \pm W_{en} \quad (1)$$

where - W_e —is the electricity consumed by the electric furnace installation from the energy system; W_s — is the enthalpy of the charge and gases entering the EAF workspace; W_{ch} — is the chemical energy of exo– and endothermic oxidation, reduction and slag formation reactions; W_b —is the amount of thermal energy introduced into the EAF workspace using fuel and oxygen burners; W_m — enthalpy of the drained liquid metal and slag (useful energy consumption); $W_{e,l}$ — electrical losses in the power source and secondary current supply (up to the working space of the furnace); $W_{th,l}$ —thermal losses from the working space of the furnace; W_{en} — represents the change in enthalpy of the lining (positive for heating and negative for cooling) [11].

The portion of electricity consumption in the balance can be expressed as follows:

$$W_e = W_m + W_{e,l} + W_{th,l} - (W_s + W_{ch} + W_b) \quad (2)$$

From the analysis in equation (2) that a reduction in electricity consumption can be obtained both by reducing electrical and thermal losses, and by using other types of energy, that is, using integrated heating [9].

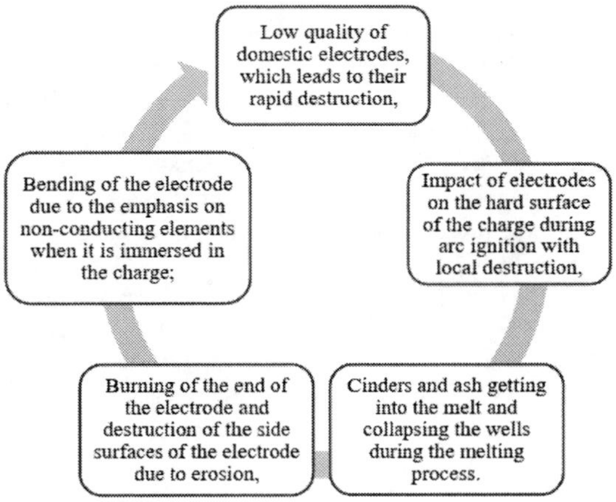

Fig. 4. Factors causing energy waste in steel smelting technology

On the other hand, they are associated with the unsatisfactory quality of carbon-graphite electrodes and their improper operation. If we identify issues that can occur through electrodes without increasing electrical efficiency, and when we find solutions to these problems, we improve energy efficiency and optimize electrode usage. That is, other reasons that reduce the efficiency of the furnace installation given below:

Based on the above-mentioned issues, there are concerns regarding energy inefficiency, increased electrode consumption, and the adequacy of electrode strength and durability in the steel production process when using electrodes. Addressing these challenges through the provision of solutions enhances the technological process quality [22, 36].

3 Conclusion

Electrodes ensure high-quality melting of scrap in the technological process. For this, it is important to have accurate information about the types and classification of electrodes and to choose special electrodes. This is mainly because of complexity of the production technology of graphite electrodes and their production according to the requirements of steel melting furnaces, their price increases. So, that is why in this article, the consumption of graphite electrodes in the process of steel smelting is given, and if the electrodes are damaged or broken, the condition of the consumption of the electrodes and what problems they cause are considered. Also, the technical characteristics of graphite electrodes, which are the main element of the steel smelting process, are given, the problems that may arise in their use in the technological process are analyzed *(fig.2)*, and several solutions are given *(fig.3)*.

As well as, in this article, suggestions for reducing the consumption of electrodes are given.

1. Low-quality steel scrap or incorrect proportioning can lead to inadequate slag formation, which in turn prolongs the smelting process. Prolonged smelting directly contributes to higher power consumption and increased electrode usage and losses.

2. Mismatched electric furnace power supply equipment and electrode specifications can lead to problems. Electric arc furnace power transmission involves high-current and low-voltage operations. Overloading and prolonged operation can result in the most significant loss and consumption of graphite electrodes.

3. During the oxidation phase, chemical energy is deliberately elevated to enhance smelting intensity. Throughout the smelting process, there is a deliberate increase in the amount of oxygen blown to achieve rapid melting and elevate furnace temperatures within the high-temperature flame. In this furnace environment, electrodes often experience significant issues such as layering and surface oxidation.

The technical aspects of electric furnace configuration and operation are critical. This includes choosing the power transmission curve and gear during operation, controlling arc starting and arc stabilization voltage and current, and effectively coordinating the use of long, medium, and short arcs. Additionally, the configuration and utilization of water cooling systems are crucial factors that significantly impact the service life and consumption of electrodes.

References

1. Rakhmonov I.U., Najimova A.M., and Reymov K.M. AIP Conference Proceedings 2647. 030010. (2022). https://doi.org/10.1063/5.0104788
2. Rakhmonov I.U., Najimova A.M. AIP Conference Proceedings 2647. 030011. (2022). https://doi.org/10.1063/5.0104791
3. Rakhimov S. Modernization of Higher Education in Uzbekistan: Current Reforms and Future Prospects. Central Asian Journal of Education.2020.
4. Turaeva M., and Mukhamedova Z. The Role of Technology in Uzbekistan's Education System Transformation. International Journal of Emerging Technologies in Learning (iJET), 13(2).2018
5. Allaev K., Toshov J., Toirov O. E3S Web of Conferences 401. 05090. (2023). https://doi.org/10.1051/e3sconf/202340105090
6. Abdi Yonis S., Yusupov Z., Habbal A., Toirov O. Advances in Electrical and Electronic Engineering.Vol 21. No 3 (2023). https://doi.org/10.15598/aeee.v21i3.5149
7. Niyozov N.N., Rikhsitillaev B.Kh., Jalilova D.A., Kuatbaev P.M. E3S Web of Conferences 384. 01031. (2023). https://doi.org/10.1051/e3sconf/202338401031.
8. Usmanov E.G. E3S Web of Conferences 384. 01034. (2023). https://doi.org/10.1051/e3sconf/202338401034
9. Rasulov A.N., Paxratdinov A.Dj. E3S Web of Conferences 384. 01035. (2023). https://doi.org/10.1051/e3sconf/202338401035
10. Toirov O., Khalikov S., Sodikjon Khalikov, F. Sharopov. E3S Web of Conferences 410. 05015. (2023). https://doi.org/10.1051/e3sconf/202341005015
11. Hoshimov F.A., Rakhmonov I.U., Niyozov N.N., Omonov F.B. AIP Conference Proceedings 2647. 030025. (2022). https://doi.org/10.1063/5.0112388
12. Rakhmonov I.U., Najimova, A.M., Esemuratova Sh.M., Koptileuov T.T. AIP Conference Proceedings 2647. 070024. (2022). https://doi.org/10.1063/5.0104793
13. Rakhmonov I.U., Hoshimov F.A., Kurbonov N.N., Jalilova D.A. AIP Conference Proceedings 2647. 050022. (2022). https://doi.org/10.1063/5.0112391
14. Rakhmonov I.U., Ganiev Sh.R., Alibekova T.Sh., Nematov L.A. E3S Web of Conferences 384. 01029. (2023). https://doi.org/10.1051/e3sconf/202338401029
15. Rakhmonov I.U., Niyozov N.N., Kurbonov N.N., Umarov B.S. E3S Web of Conferences 384. 01030. (2023). https://doi.org/10.1051/e3sconf/202338401030
16. Heikkinen, E. P., & Fabritius, T. (2012). Modelling of the Refining Processes in the Production of Ferrochrome and Stainless Steel. Recent Researches in Metallurgical Engineering-From Extraction to Forming. InTech, 186, 65-88.
17. Taslimov A.D. E3S Web of Conferences 384. 01036. (2023). https://doi.org/10.1051/e3sconf/202338401036
18. Taslimov A.D., Rakhimov F., Rakhimov Feruz. E3S Web of Conferences 384. 01037. (2023). https://doi.org/10.1051/e3sconf/202338401037
19. Fayziyev M., Tuychiev F., Mustayev R., Ochilov Y. E3S Web of Conferences 384. 01038. (2023). https://doi.org/10.1051/e3sconf/202338401038
20. Saidkhodjaev A.G., Bobojonov Y.M. E3S Web of Conferences 384. 01040. (2023). https://doi.org/10.1051/e3sconf/202338401040
21. Bobojanov M., Mahmutkhonov S. E3S Web of Conferences 384. 01041. (2023). https://doi.org/10.1051/e3sconf/202338401041
22. Rasulov A.N., Ruzinazarov M.R. E3S Web of Conferences 384. 01042. (2023). https://doi.org/10.1051/e3sconf/202338401042
23. Rashidov N., Rozmetov Kh., Rismukhamedov S., Peysenov M. E3S Web of Conferences 384. 01043. (2023). https://doi.org/10.1051/e3sconf/202338401043
24. Tuychiev F, Haqberdiev A. E3S Web of Conferences 384. 01044. (2023). https://doi.org/10.1051/e3sconf/202338401044
25. Bobojanov M., Torayev S. E3S Web of Conferences 384. 01045. (2023). https://doi.org/10.1051/e3sconf/202338401045
26. Palanichamy K.C., Sundar Babu N. Second stage energy conservation experience with a textile industry. Energy Policy 2005.
27. Reymov K.M., Esemuratova Sh.M., Khusanov B.M., Mytnikov A.V. E3S Web of Conferences 384. 01047. (2023). https://doi.org/10.1051/e3sconf/202338401047

28. Kholikhmatov B.B., Samiev Sh.S., Erejepov M.T., Nematov L.A. E3S Web of Conferences 384. 01032. (2023). https://doi.org/10.1051/e3sconf/202338401032
29. Melikuziev M.V. E3S Web of Conferences 384. 01033. (2023). https://doi.org/10.1051/e3sconf/202338401033
30. Bobojonov Y.M., Saidkhodjaev A.G. E3S Web of Conferences 384. 01048. (2023). https://doi.org/10.1051/e3sconf/202338401048
31. Hoshimov, F.A., Najimova A.M., Rafikova G.R., Bakhadirov I.I. E3S Web of Conferences 384. 01049. (2023). https://doi.org/10.1051/e3sconf/202338401049
32. Ruzinazarov M.R. E3S Web of Conferences 384. 01050. (2023). https://doi.org/10.1051/e3sconf/202338401050
33. Bobojanov M.K., Rismukhamedov D.A., Tuychiev F.N., Shamsutdinov Kh.F. E3S Web of Conferences 365. 04024. (2023). https://doi.org/10.1051/e3sconf/202336504024
34. Ushakov, V.Y., Mytnikov, A.V., Rakhmonov, I.U. Past, Present and Future of Electric Power Systems (Brief Excursion). In: High-Voltage Equipment of Power Systems. Power Systems. Springer, Cham. (2023). https://doi.org/10.1007/978-3-031-38252-9_1
35. Ushakov, V.Y., Mytnikov, A.V., Rakhmonov, I.U. Characteristics of the Main Elements of Electric Power Systems. In: High-Voltage Equipment of Power Systems. Power Systems. Springer, Cham. (2023). https://doi.org/10.1007/978-3-031-38252-9_2
36. Ushakov, V.Y., Mytnikov, A.V., Rakhmonov, I.U. Insulating Materials and Media Used in High-Voltage Elements of Electric Power Systems. In: High-Voltage Equipment of Power Systems. Power Systems. Springer, Cham. (2023). https://doi.org/10.1007/978-3-031-38252-9_3
37. Sattarov Kh., Sapayev M., Suyarov A., Turaev, A. E3S Web of Conferences, 401, 04001, (2023). https://doi.org/10.1051/e3sconf/202340104001
38. Jurayeva K., Sattarov Kh. E3S Web of Conferences, 389, 01060. (2023). https://doi.org/10.1051/e3sconf/202338901060
39. Sattarov Kh., Safarov A., Kamolov A., Egamberdiev M., Raxmanova G. E3S Web of Conferences, 383, 01028. (2023) https://doi.org/10.1051/e3sconf/202338301028
40. Tsypkina, V., Ivanova, V. E3S Web of Conferences, 139, 01073. (2019). https://doi.org/10.1051/e3sconf/201913901073
41. Ivanova, V., Madrakhimov, D. E3S Web of Conferences, 139, 01057. (2019). https://doi.org/10.1051/e3sconf/201913901057
42. Pirmatov N. B., Madrakhimov D. B., Ivanova V. P., Tsypkina V. V., Ortikova M. J., Atamukhamedova R. F. AIP Conference Proceedingsthis link is disabled, 2552, 070009. (2022). https://doi.org/10.1063/5.0117025
43. Ushakov, V.Y., Mytnikov, A.V., Rakhmonov, I.U. Effects on Equipment Causing Insulation Aging and Failure. In: High-Voltage Equipment of Power Systems. Power Systems. Springer, Cham. (2023). https://doi.org/10.1007/978-3-031-38252-9_4
44. Maslov D.V., Development of algorithms and control systems for arc steel-melting furnaces that reduce electrode breakdowns. PhD Thesis, 2014.
45. Hoshimov F.A., Rakhmonov I.U., Niyozov N.N. E3S Web of Conferences 209. 07017. (2020). https://doi.org/10.1051/e3sconf/202020907017

Control of ball mill operation depending on ball load and ore properties

Oybek Ishnazarov[1*], *Jonibek* Mavlonov[2], *Davron* Mardonov[2]

[1]Institute of Energy Problems of the Academy of Sciences of the Republic of Uzbekistan, 210106 Muxtor Ashrafiy St., 9-A, Tashkent, Uzbekistan
[2]Navoi State University of Mining and Technologies, 210106, Galaba St., 76V, Navoi region, Navoi, Uzbekistan

Abstract. Ball mills are used for the second stage of ore grinding in mining operations by proportioning ore piles. Controlling the operating conditions of ball mills is one of the key factors for optimal mill perform and high mill load turnover. One of the factors affecting the efficiency of the ball mill is the loading of the balls. Optimum distribution of ores according to the size of the mill, increasing the efficiency of the grinding process and reducing energy consumption are achieved. In addition, the ore properties are one of the important parameters for the optimal utilizing of the ball mill. Ore can have various properties, such as hardness, density, moisture content, etc., which can affect the ore's productivity and grinding efficiency.

1 Introduction

According to the operating mode, the rotation speed of the mill is divided into 4 types. Slow rotation (cascade), fast rotation (waterfall), medium rotation (mixed), very fast rotation (central rotation or critical). Reducing the granularity of the ready product (ground particles) coming out of the mill depends on the following main factors:

- properties of ores in the mill (mass, size, hardness, density and granulometric composition of particles):
- characteristics of grinding agents (mass, density, size and distribution of balls);
- rotation speed of the mill;
- density of straw during wet grinding [5].

The main purpose of this process is to increase the grinding productivity and reduce the production costs for the whole process. To achieve these goals, various mathematical models and management methods are developed and used in practice. In this article, we provide a brief overview of the basic principles of modeling the process of grinding and analyze several control strategies used in the implementation of a suitable method of process control.

The main idea of modeling all mining processes is to obtain a mathematical relationship between the volume of the ore and the volume of the ready product. The granularity of the ore is continuously reduced due to energy transfer by the grinding condition, which reduces their bond strength. Ore size reduction is the result of the following three main fragmentation mechanisms.

1. Abrasion
2. Cleavage
3. Fracture

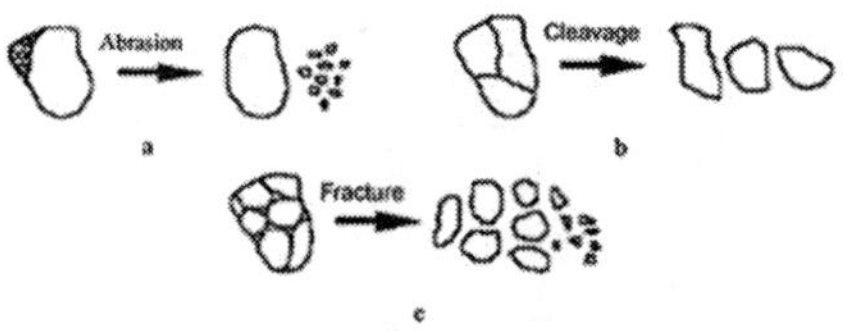

Figure 1. Grinding of ores inside a ball mill

In practice, the three mechanisms never operate separately, and in the process of particle size reduction, they all work with possible advantage depending on the type of mill, operating conditions, and type of ore ground [1].

2 Experimental research

The relative consumption of electric energy is the ratio of the useful power consumed by the mill's electric drive to the efficiency of the grinding cycle without taking into account the circulating load. The specific power consumption with a given volume of ore and cycle is the main industrial parameter that describes the grinding of material in a given product preparation equipment. The specific energy consumption for the same material depends on the type of ore preparation device, which is associated with significant fluctuations in the efficiency of energy transfer in the workplaces of different equipment. The specific power consumption for the required product productivity provides a useful solution for mill size and electrical drive and is expressed as follows[2].

* Corresponding author: raxmatillo82@mail.ru

$$E = 10 \cdot W_i \cdot \left(\frac{1}{\sqrt{P_{80}}} - \frac{1}{\sqrt{F_{80}}} \right) \tag{1}$$

Here, P_{80} and F_{80} are the sieve mesh sizes, through which 80% of the soil and the starting material are separated in microns, respectively; W_i is the working index of the material, kW/t; E – required power consumption, kW/t.

$$E_{tr} = 10\left[CWI\left(\frac{1}{\sqrt{10000}} - \frac{1}{\sqrt{F_{80}}}\right) + \left[RWI\left(\frac{1}{\sqrt{2100}} - \frac{1}{\sqrt{10000}}\right) + BWI\left(\frac{1}{\sqrt{P_{80}}} - \frac{1}{\sqrt{2100}}\right)\right]\right] \tag{2}$$

The value obtained by E_{tr} determines the maximum efficiency of the use of equipment for the preparation of ore and corresponds to the scheme, including medium and small stage ball grinding provides a comparative analysis of a wide range of ore processing steps and requires 5-15% energy in a self-grinding mill [3].

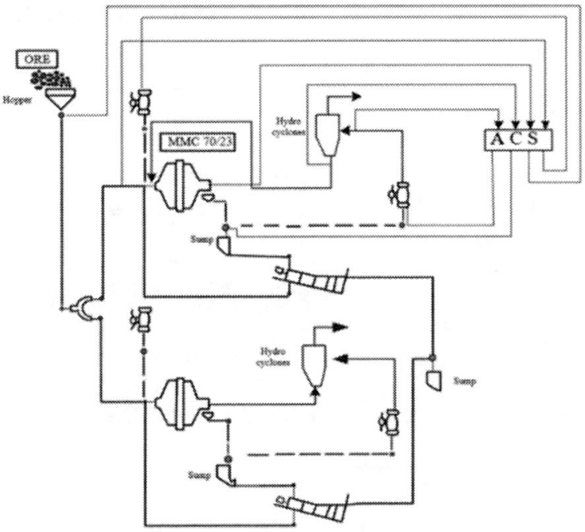

Figure 2. Ball mill process control system using ACS (Adaptable Control System).

Figure 2 shows that the input and output signals depend mainly on the mill and the hydrocyclone. Some malfunctions occur as a result of system disturbances that occur during the processes that take place at the entrance and exit of ore to the mill (including the hydrocyclone). In order to eliminate these, it can be achieved through the following management system. In this case, we can obtain the following information through the input signals:

- Control of the filling level of the mill by an ultrasonic sensor.
- Size, hardness and density of ore particles according to M. M. Protodiakonov scale. It is a system that can measure ore content on 4 inputs and includes real-time measurement of pulp density.
- The rate at which fresh ore from the belt conveyor enters the mill.
- From the electromagnetic measurement of the consumption water coming to the mill.
- Determining the density of the pulp entering the hydrocyclone.
- Determination of the speed of butane supplied to the classifier with an electromagnetic flow meter.

Through the flexible control system of the technological process, it is possible to reduce the waste of electrical energy, taking into account the load on the electric motor. We can control the speed of the electric motor depending on the load. We adjust the speed of the electric motor using frequency and thyristor converter devices.

Process control methods. The control of solution schemes is affected by the non-linear and uncertain nature of the process, inaccuracies in the mathematical model, the presence of interacting process variables with significantly different dynamics, the effects of non-dimensional disturbances and large time delays, is a difficult task due to many factors such as rough working conditions. On the other hand, effective process management is of great importance to increase the quality of the ready product, as well as to significantly reduce production costs[4].

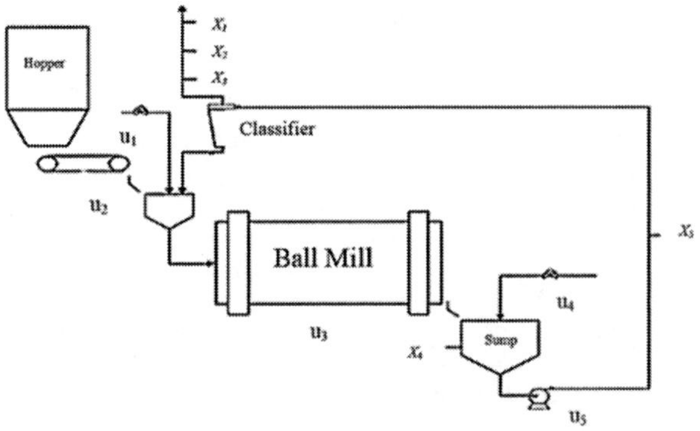

Figure 3. A multi-value controlled grinding scheme

The input variables of the process in Figure 2 are: u_1 – water consumption required for the mill, u_2– ore arrival speed, u_3 – percentage of the critical speed of the mill, u_4 – pulp water consumption and u_5– mill incoming pulp consumption. Output parameters can be controlled by these variable values[3,5]:

x_1 - the mass fraction of the product whose particle size is less than the specified value, x_2 - the concentration of solid particles of the product, x_3 - the consumption of the product, x_4 - the level of suspension in the product. x_5 - the concentration of solid particles in the mill. The most important violations of the technological process are changes in the hardness of the ore and the volume of raw materials. The input-output model vector of the process can be written in matrix form as follows:

$$\begin{Vmatrix} x_1(s) \\ \vdots \\ x_1(s) \end{Vmatrix} = \begin{Vmatrix} M_{11}(s) & \cdots & M_{15}(s) \\ \vdots & \ddots & \vdots \\ M_{15}(s) & \cdots & M_{55}(s) \end{Vmatrix} \cdot \begin{Vmatrix} u_1(s) \\ \vdots \\ u_1(s) \end{Vmatrix} \quad (3)$$

Here M_{ij} (s) is the transfer function consisting of i-input and j- output signals i, j = 1.....5. Transfer functions are usually determined experimentally by applying initial variations to the input and measuring the output characteristics. A fairly large number of experiments are usually necessary to obtain accurate results. In addition, depending on the configuration of the circuit and equipment, different input and output signals can be used in the control process [5,6].

Control of the technological process in the grinding scheme of the ball mill has the following disadvantages:

- the process is not linear, has many disturbances and unmodeled dynamics;
- the time constants of the process have a wide range of values, and significant time delays are observed in some input-output pairs;
- a system model has output signals corresponding to a series of input signals;
- process parameters change over time;
- controlled variables have technological limitations
- measurements are unreliable and noisy.

The main task of control is to ensure stable operation with specified values of output variables. As part of this goal, it is necessary to ensure a stable distribution of the volume of the ready product, the volume of ore is smaller than the specified value. Alternatively, the management objective can be formulated as a task containing a performance criterion that must be optimized given certain constraints. This leads to the maximization of the performance of the grinding process or the minimization of production costs [7,8].

3 Research results

Ball mills are filled with balls during operation. It is expressed in percentages φ_{III}, %. This ball is determined according to the size class that determines the maximum possible productivity of the drum mill and is determined using the following expression:

$$q_{0,071}^{\max} = f(\phi_{ш}) \quad (4)$$

The control system that takes into account the ball filling of the mill is based on the active (useful) power consumed by the electric motor of the mill (P, kW) and the ball filling of the mills.

$$P = f(\phi_{ш}) \quad (5)$$

Table of loading balloons into the mill for February-April 2023

1-chart

February		1	2	3	4	5	6
Loading balls Ø 100	Ton		4,8	2,4		4,8	2,4
	Cont.		1	0,5		1	0,5
Loading balls Ø 120	Ton	4,8			2,4		
	Tank	1			0,5		
February		**7**	**8**	**9**	**10**	**11**	**12**
Loading balls Ø 100	Ton		4,8	2,4		4,8	2,4
	Cont.		1	0,5		1	0,5
Loading balls Ø 120	Ton	2,4			2,4		
	Tank	0,5			0,5		
February		**13**	**14**	**15**	**16**	**17**	**18**
Loading balls Ø 100	Ton		4,8	2,4		4,8	2,4
	Cont.		1	0,5		1	0,5
Loading balls Ø 120	Ton	2,4			4,8		
	Tank	0,5			1		
February		**19**	**20**	**21**	**22**	**23**	**24**
Loading balls Ø 100	Ton		4,8	2,4		4,8	2,4
	Cont.		1	0,5		1	0,5
Loading balls Ø 120	Ton	2,4			2,4		
	Tank	0,5			0,5		
February		**25**	**26**	**27**	**28**	**Overall**	
Loading balls Ø 100	Ton		4,8	2,4		62,4 t	
	Cont.		1	0,5		13	
Loading balls Ø 120	Ton	2,4			2,4	28,8 t	
	Tank	0,5			0,5	6	

2-chart

March		1	2	3	4	5	6
Loading balls Ø 100	Ton		4,8	2,4		4,8	2,4
	Cont.		1	0,5		1	0,5
Loading balls Ø 120	Ton	4,8			2,4		
	Tank	1			0,5		
March		**7**	**8**	**9**	**10**	**11**	**12**
Loading balls Ø 100	Ton		4,8	2,4		4,8	2,4
	Cont.		1	0,5		1	0,5
Loading balls Ø 120	Ton	2,4			2,4		
	Tank	0,5			0,5		
March		**13**	**14**	**15**	**16**	**17**	**18**
Loading balls Ø 100	Ton		4,8	2,4		4,8	2,4
	Cont.		1	0,5		1	0,5
Loading balls Ø 120	Ton	2,4			4,8		
	Tank	0,5			1		
March		**19**	**20**	**21**	**22**	**23**	**24**
Loading balls Ø 100	Ton		4,8	2,4		4,8	2,4
	Cont.		1	0,5		1	0,5
Loading balls Ø 120	Ton	2,4			2,4		
	Tank	0,5			0,5		
March		**25**	**26**	**27**	**28**	**29**	**30**
Loading balls Ø 100	Ton		4,8	2,4		4,8	2,4
	Cont.		1	0,5		1	0,5
Loading balls Ø 120	Ton	2,4			2,4		
	Tank	0,5			0,5		
March		**31**	**Overall**				
Loading balls Ø 100	Ton		72 t				
	Cont.		15				
Loading balls Ø 120	Ton	4,8	33,6 t				
	Tank	1	7				

https://doi.org/10.1051/e3sconf/202346101091

3-chart

April		1	2	3	4	5	6
Loading balls	Ton		4,8			4,8	
Ø 100	Cont.		1			1	
Loading balls	Ton				2,4		
Ø 120	Cont.				0,5		
Loading balls	Ton	4,9					4,9
Ø 68	Cont.	1					1
April		**7**	**8**	**9**	**10**	**11**	**12**
Loading balls	Ton		4,8			4,8	
Ø 100	Cont.		1			1	
Loading balls	Ton	2,4		2,4			2,4
Ø 120	Cont.	0,5		0,5			0,5
Loading balls	Ton				4,9		
Ø 68	Cont.				1		
April		**13**	**14**	**15**	**16**	**17**	**18**
Loading balls	Ton		4,8			4,8	
Ø 100	Cont.		1			1	
Loading balls	Ton			2,4			
Ø 120	Cont.			0,5			
Loading balls	Ton				4,9		
Ø 68	Cont.				1		
April		**19**	**20**	**21**	**22**	**23**	**24**
Loading balls	Ton		4,8			4,8	
Ø 100	Cont.		1			1	
Loading balls	Ton	2,4			2,4		
Ø 120	Cont.	0,5			0,5		
Loading balls	Ton			4,9			
Ø 68	Cont.			1			
April		**25**	**26**	**27**	**28**	**29**	**30**
Loading balls	Ton		4,8			4,8	
Ø 100	Cont.		1			1	
Loading balls	Ton	2,4			2,4		
Ø 120	Cont.	0,5			0,5		
Loading balls	Ton			4,9			
Ø 68	Cont.			1			
April		**31**	**Overall**				
Loading balls	Ton		48 t				
Ø 100	Cont.		10				
Loading balls	Ton		19,2 t				
Ø 120	Cont.		4				
Loading balls	Ton		29,4 t				
Ø 68	Cont.		6				

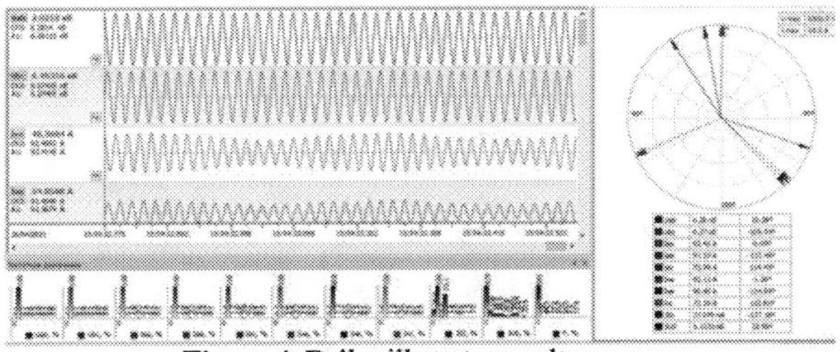

Figure 4. Ball mill start-up voltages

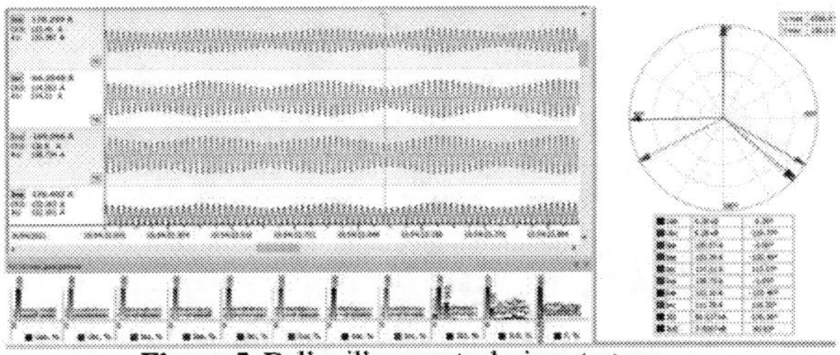

Figure 5. Ball mill currents during start-up

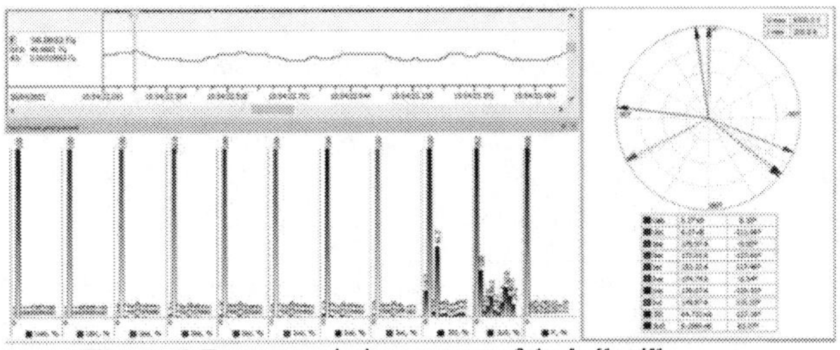

Figure 6. Frequency during start-up of the ball mill

Existing mathematical models of the grinding process are developed based on mass or energy balance equations, which describe the reduction of the particle size of the ground material in terms of grinding time or specific energy consumed. In both cases, the main parameters of the model are the unknown selection and distribution functions, and their determination requires additional experimental studies. Decentralized and multivariable control methods prevail in the research results and practical applications reviewed in process control. Nevertheless, theoretical and simulation studies of systems using advanced control methods such as adaptive, unclear logic are attracting more and more interest. Important, unsolved problems for further research activities in this area are related to economic evaluation of ball mill energy efficiency and process control.

4 Conclusion

Different methods can be used to take into account ore properties in ball mill operation control, such as monitoring and adaptive control methods. Adaptive control is a system control method that allows the system to respond to changes in external and internal factors, such as changes in ore properties, to ensure optimal mill utilizing and quality. It is based on the principle of feedback and allows the control system to respond to changes in ore properties in real time. For example, if ore properties change during mill operation, an adaptive control system can automatically change mill operating parameters by generating signals and influencing the ball weight and synchronous motor drive current to achieve the desired throughput.

We offer a customized control system for ball mills, in which the ball loading (with required diameters) and the motor are regulated according to the excitation current, noise signals and the stator current of the synchronous motor. Thus, the system consists of two input signals and two output parameters.

References

1. Jonibek Mavlonov , Davron Mardonov, Mirzohid Eshmirzayev, and Islomjon Togayev Improvement of methods to ensure energy efficiency of ball mills functioning // III International Conference GeoTech-2023: Geotechnology, Mining and Rational Use of Natural Resources **16-17** June 2023 | Navoi, Uzbekistan https://doi.org/10.1051/e3sconf/202341703014

2. Tovbaev A.N., Ibadullayev M., Norboyev S I Analysis of subharmonic oscillations in three-phase Ferroresonant circuits with bias// IV International APITECH-IV - 2022 Journal of Physics: Conference Series 2388 (2022) **012060** https://iopscience.iop.org/article/10.1088/1742-6596/2388/1/012060/pdf

3. Vladimir M, Blagoy S, Stefan S. Grinding in Ball Mills: Modeling and Process Control // Cybernetics and information technologies Volume 12, No 2 ISSN: **1314-4081** DOI: 10.2478/cait-2012-0012

4. Bond, F.C. Crushing and Grinding Calculations. Part I. – British Chemical Engineering, Vol. 6, 1961, № 6, **378-385.**

5. Bond F.C. Crushing and Grinding Calculations. Part II. – British Chemical Engineering, Vol. 6, 1961, № 8, **543-548**

6. Ishnazarov O.X. Mavlonov J.A. Basic Directions of Improvement and Improvement of Efficiency Ore Grinding // International Journal of Advanced Research in Science, Engineering and Technology Vol. 6, Issue 4, April 2019. India **269-271**bet.

7. A.D. Taslimov, A.S. Berdishev, A.A. Yuldashev, F.M. Rakhimov. Optimal tendency of selecting cable cross-sections for agricultural electrical networks// Published 1 April 2020Published under licence by IOP Publishing LtdJournal of Physics: Conference Series, Volume **1515**, Applied Physics and Cyber-Physical Systems (doi:10.1088/17426596/1515/2/022056).

8. A N Tovboyev, D Sh Mardonov, A X Mamatazimov and S S Samatova Analysis of subharmonic oscillations in multi-phase ferroresonance circuits using a mathematical model Journal of Physics: Conference Series 2094 (2021) **052048,** doi:10.1088/1742-6596/2094/5/052048

9. A.D. Taslimov, A.S. Berdishev, M.V. Melikuziyev, F.M. Rakhimov. Method of selecting parameters of cable lines distributive networks 10 kv in uncertainty conditions// E3S Web Conf.Volume 139, 2019 Rudenko International Conference “Methodological problems in reliability study of large energy systems” RSES 2019 (https://doi.org/10.1051/e3sconf/201913901082)

10. A.O. Ataullayev, M.K. Sayidov "Investigation of multi loop linear magnetic circuits of electromagnetic flow converters with ring channels" E3S Web of Conferences 417, **03006** (2023) https://doi.org/10.1051/e3sconf/202341703006

AI-powered learning: revolutionizing technical higher education institutions through advanced power supply fundamentals

Numon Niyozov[1*], *Salamat* Saburov[2], *Shaxruz* Ganiyev[3], *Shirinboy* Olimov[4]

[1]Tashkent State Technical University named after Islam Karimov, 100095, Uzbekistan, Tashkent, University St. 2A

[2]Karakalpak State University, Nukus, 230101, Uzbekistan

[3]"TIIAME" NRU Bukhara Institute of Natural Resources Management, Bukhara, 105009, Uzbekistan

[4]Bukhara state university, 105009 Bukhara, Uzbekistan

Abstract. The burgeoning technological education landscape mandates a reevaluation of educational methodologies, especially in pivotal domains such as power supply fundamentals within technical higher education institutions. This scientific article explores the imperative role of integrating Artificial Intelligence (AI) in teaching the fundamentals of power supply to students in technical higher education institutions. Acknowledging power supply's fundamental significance and the evolving educational challenges, we advocate for the utilization of AI-powered platforms, simulations, and adaptive learning systems. Through a comprehensive analysis, we highlight the potential benefits and contextualize the implementation of AI in Uzbekistan's technical education scenario. By embracing AI-driven educational approaches, Uzbekistan can cultivate a skilled workforce adept in power supply, fostering innovation and propelling the nation towards a technologically advanced future.

1 Introduction

The power supply fundamentals subject stands as a foundational pillar within technical education, playing a pivotal role in shaping the understanding and competency of students in various engineering disciplines. In a world increasingly dependent on technology, power supply forms the bedrock upon which modern electronic devices, communication systems, and industrial equipment operate. This subject encompasses critical concepts such as electrical circuits, voltage regulation, power distribution, and energy conversion, providing the necessary knowledge to design, analyze, and troubleshoot electrical systems.

2 Literature Review

2.1 Integration of AI in Education

Somyürek and Özyer demonstrated the effectiveness of AI algorithms in adapting learning materials and styles to individual student needs, promoting a more tailored and efficient learning process [1]. Additionally, its role in providing real-time feedback has been emphasized by Van den Heuvel et al. showing significant improvements in student engagement and comprehension [2].

2.2 Importance of Power Supply Fundamentals Subject

Understanding the importance of power supply fundamentals is crucial in establishing the need for innovative teaching methodologies. Research by Hu et al. emphasizes the vital role of power supply fundamentals in shaping the careers of electrical engineers. They stress how a strong foundation in power supply fundamentals enables professionals to design efficient and reliable electrical systems [3].

2.3 Modernizing Education in Uzbekistan

In the context of Uzbekistan, modernizing education is a priority to align with global advancements. Research by World Bank sheds light on the need for reforms in Uzbekistan's education system to meet the demands of a rapidly evolving job market [4]. The integration of AI in education is seen as a strategic approach to address this need, enhancing the quality of technical education.

2.4 Bridging Educational Gaps with AI

AI-powered educational platforms have shown promise in bridging educational gaps. Research by Baker discusses how AI can augment traditional classroom teaching, making education more interactive and engaging [5]. This augmentation is particularly critical in

* Corresponding author: nomon.niyozov_2422@mail.ru

https://doi.org/10.1051/e3sconf/202346101092

addressing the shortage of specialized faculty and limited resources faced by educational institutions in Uzbekistan.

2.5 Enhanced Learning through AI Simulations

AI-powered simulations have demonstrated effectiveness in providing an interactive learning experience. Research by Anderson et al. showcases how simulations can significantly enhance understanding and retention of complex technical concepts [6]. By integrating such simulations into power supply education, students can gain practical insights and improve their problem-solving skills [9-20].

3 Importance of Power Supply Fundamentals Subject

Understanding the importance of the power supply fundamentals subject begins with acknowledging its pervasive influence on diverse technical domains. The subject equips students with basic principles, enabling them to comprehend the flow of electrical energy, the function of power sources, and the significance of efficient power utilization. Without a solid grasp of power supply fundamentals, professionals in fields like electrical engineering, electronics, and computer science would be ill-prepared to address the evolving challenges of a rapidly advancing technological landscape [21-30].

In the realm of electrical engineering, for instance, an inadequate understanding of power supply fundamentals can result in inefficient designs, power wastage, and compromised system reliability. Conversely, a strong foundation in power supply fundamentals principles empowers engineers to create energy-efficient systems, minimize losses, and optimize performance—a crucial aspect in our energy-conscious world. Moreover, power supply fundamentals knowledge is essential for navigating emerging technologies such as renewable energy systems, electric vehicles, and smart grids.

By emphasizing the critical role of power supply fundamentals in technical education, students can grasp the importance of mastering this subject as a precursor to their professional success. A comprehensive understanding of power supply fundamentals principles equips students with the necessary skills to contribute effectively to various industries, fostering innovation and sustainable technological growth.

4 Teaching Methods and Challenges

4.1 Traditional Teaching Methods

Traditionally, power supply fundamentals have been taught using conventional classroom lectures, textbooks, and laboratory experiments. While these methods offer a structured approach, they often face limitations that hinder optimal learning experiences. Lecture-based teaching, for example, may not adequately engage students, limiting their comprehension and retention of complex theoretical concepts. Additionally, practical laboratory sessions, though valuable, may have resource constraints, making it difficult for all students to have hands-on experiences [31-35].

4.2 Survey Insights

To gather firsthand insights about the existing resources and teaching methods, we conducted a survey among 135 students from Tashkent State Technical University. The results are as follows:

1. Labs at the University:
   * Poor: 42%
   * Bad: 14%
   * Satisfied: 27%
   * Good: 11%
   * Top: 6%

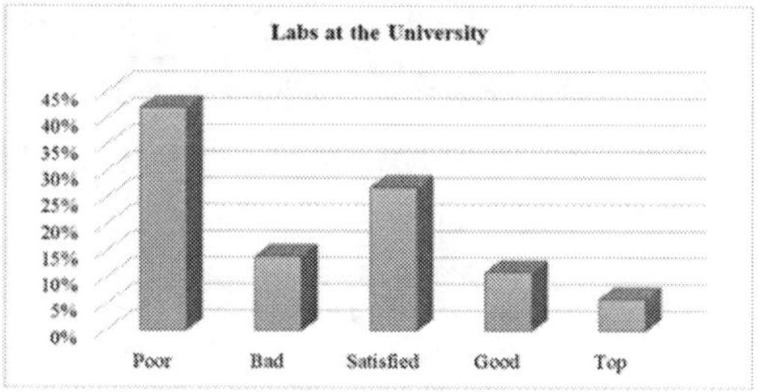

Fig. 1. Quality of Labs at the University

42% of students rated the labs as poor, with 14% marking them as bad. This illuminates a pronounced need for improvement in lab facilities, necessitating enhanced equipment availability and quality, to facilitate hands-on learning and application of theoretical concepts.

2. Status of Study Guides or Materials:
   * Poor: 30%
   * Bad: 16%
   * Satisfactory: 24%
   * Good: 24%

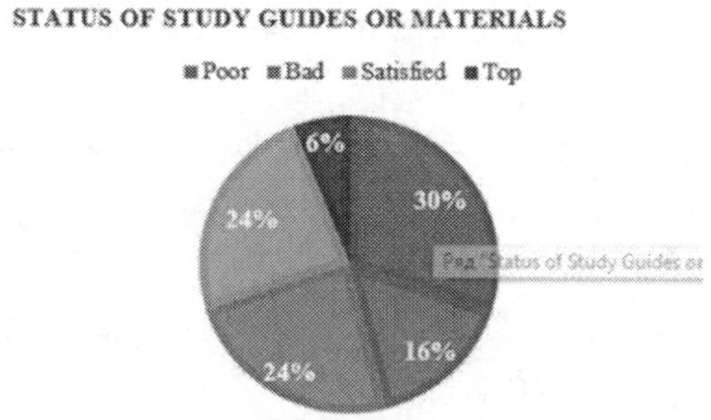

Fig. 2. Status of Study Guides or Materials

The survey reveals 46% of respondents are dissatisfied with study materials, citing them as either poor or bad. This significant percentage underscores the pressing need for updated relevant, and comprehensive study materials that align with the dynamic nature of power supply topics.

3. Professional Skills of Professors and Teachers:
   * Poor: 14%
   * Bad: 5%
   * Satisfied: 29%

* Good: 35%
* Top: 17%

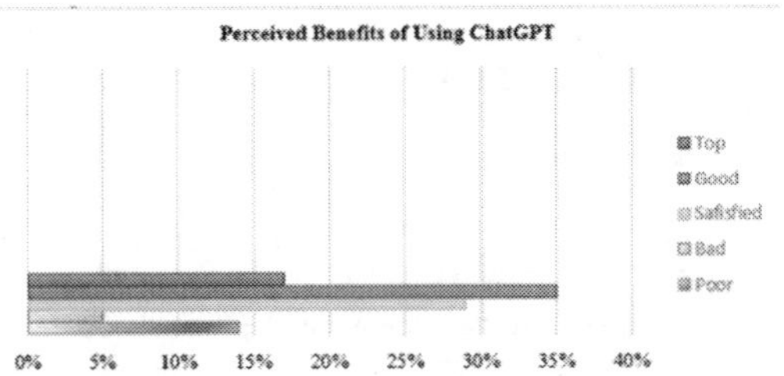

Fig. 3. Professional Skills of Professors and Teachers

The survey results, in this case, were more favourable, with 52% rating their professors and teachers as good or top. However, ongoing professional development and continuous learning opportunities are crucial to maintain high teaching standards and adapt to evolving educational needs.

4. Level of Interactivity of the Lessons:
* Poor: 36%
* Bad: 8%
* Satisfactory: 23%
* Good: 20%
* Top: 13%

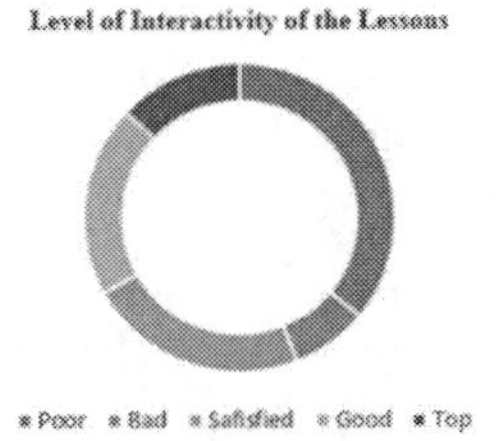

Fig. 4. Level of Interactivity of the Lessons

36% of students rated the lesson interactivity as poor. This underscores the need for innovative, engaging teaching methods that cater to diverse learning preferences and paces, fostering an inclusive learning environment.

These survey results emphasize some of the prevalent challenges within the institution, including concerns related to laboratory facilities, study materials, and the interactivity of lessons. The feedback on the professional skills of professors and teachers, however, appears relatively positive.

4.3 Challenges

1. Limited Practical Exposure: The survey reflects that a significant proportion of students find the labs either 'Poor' (42%) or 'Bad' (14%), suggesting challenges in providing extensive hands-on training due to limited laboratory resources and equipment. This restricts students' ability to apply theoretical knowledge in practical scenarios.

2. Outdated Learning Resources: The dynamic nature of power supply means that outdated textbooks and learning materials can lag behind. The survey highlights that 30% of students rated study materials as 'Poor', while 16% rated them as 'Bad', suggesting a lack of up-to-date educational resources.

3. Faculty Shortage: While the professional skills of professors and teachers were rated positively by the majority, there is still room for improvement.

4. Diverse Learning Needs and Lack of Interactivity: Students often have different learning paces and preferences. The survey suggests that 36% of students found the lessons 'Poor' in terms of interactivity, indicating a need for more engaging and diverse teaching methods [36-41].

5 Introduction of AI in Teaching

5.1 Advantages of AI in Education

The integration of Artificial Intelligence (AI) in education offers transformative benefits. AI algorithms can personalize learning experiences, adapt content to individual student needs, provide immediate feedback, and create interactive and immersive learning environments. These advantages align well with addressing the challenges faced in teaching power supply fundamentals.

5.2 Utilizing AI for Teaching Power Supply Fundamentals

1. Adaptive Learning Platforms: AI can personalize the learning journey for each student based on their progress, learning style, and understanding. It can adapt the curriculum to suit individual needs, ensuring a more efficient and effective learning process.

2. Simulations and Virtual Labs: AI-powered simulations can replicate real-world scenarios, providing a safe and dynamic environment for students to experiment with and apply power supply fundamentals concepts without the need for extensive physical resources.

3. Real-time Feedback and Assessment: AI algorithms can instantly assess student performance and provide detailed feedback. This immediate feedback loop helps students identify and rectify misconceptions and errors promptly.

The integration of AI in teaching power supply fundamentals can address traditional teaching challenges and elevate the quality of education, enabling students to comprehend and apply the subject effectively. This approach is particularly relevant for technical higher education institutions in Uzbekistan, where overcoming resource limitations and enhancing educational quality are paramount.

6 Implementation in Uzbekistan's Technical Higher Education Institutions

6.1 Current Educational Landscape

In Uzbekistan, technical education plays a critical role in fostering technological advancements and economic growth.[7] However, the educational landscape faces challenges in keeping up with rapid technological changes and providing students with the most up-to-date knowledge and skills. The conventional teaching methods may fall short in fully preparing students to meet industry demands, particularly in domains like power supply fundamentals.

6.2 Need for Modernization

To bridge the gap between traditional teaching methodologies and the current technological landscape, incorporating AI into teaching the fundamentals of power supply is imperative. Modernization is essential to equip students with the practical skills and knowledge necessary to thrive in a technology-driven job market.[8]

6.3 Benefits of AI in Uzbekistan's Context

1. Accessible Learning Resources: AI-powered educational platforms can provide students with access to high-quality educational materials, even in remote or poor areas, promoting inclusivity and equal educational opportunities.[8]

2. Addressing Faculty Shortage: AI can supplement the role of educators by offering personalized guidance and support. This is particularly significant in Uzbekistan, where there might be a shortage of specialized faculty in certain technical areas.[8]

3. Enhanced Student Engagement: AI-driven interactive tools and simulations can make learning more engaging and enjoyable for students. This engagement can lead to improved understanding and retention of power supply fundamentals.[8]

4. Global Competitiveness: By adopting AI in education, Uzbekistan's technical higher education institutions can align their programs with international standards, enhancing the competitiveness of graduates on a global scale.[8]

7 Prospects

The integration of AI into teaching the fundamentals of power supply in Uzbekistan's technical higher education institutions presents a promising avenue for enhancing the quality and effectiveness of technical education. It enables students to be adequately prepared for the evolving landscape of power supply technologies and related industries.

Empowering students with a strong foundation in power supply fundamentals through AI-driven educational approaches will not only benefit their careers but also contribute to the growth and development of Uzbekistan's technological sector.[8] As the nation moves towards a more digitally driven future, a technologically adept workforce is crucial, making AI-integrated education an imperative investment for the nation.[8]

Through this marriage of traditional academic rigor and cutting-edge technology, Uzbekistan's technical higher education institutions can play a crucial role in shaping a generation of adept professionals capable of driving progress and prosperity in the dynamic world of power supply.[8]

Conclusion

The rapid evolution of technology mandates an innovative and responsive approach to education, particularly within technical domains like power supply fundamentals—a foundational subject pivotal to multiple engineering disciplines. Considering the survey conducted amongst the students at Tashkent State Technical University, it is evident that there is a palpable demand for modernized, effective teaching methodologies that can adequately equip students with the essential skills and knowledge pivotal to this discipline.

The integration of Artificial Intelligence (AI) into the teaching frameworks holds colossal potential in bringing about an educational renaissance in Uzbekistan. AI-powered platforms, adaptive learning systems, and simulations emerge as transformative tools, with the capabilities to offer students personalized and immersive learning experiences. These systems can meticulously simulate real-world scenarios, allowing students to delve deep into the practical aspects of power supply fundamentals and to grasp intricate concepts more effectively. Immediate feedback and tailored study plans derived from AI can elevate student engagement and catalyze efficient learning trajectories.

Within the confines of Uzbekistan's educational milieu, where the quandaries of resource constraints and faculty shortages are prevalent, AI stands out as a promising beacon, poised to ameliorate the quality and accessibility of education. The survey findings underscore the necessity of such innovations, highlighting substantial dissatisfaction with current labs and study materials, while accentuating the relative contentment with the professorial staff. This demonstrates fertile ground and a willing audience for AI-integrated methodologies.

AI can forge pathways to equal learning opportunities, fostering an environment where students are not just recipients but active contributors to the knowledge ecosystem, and preparing them to be globally competitive in the progressively evolving field of power supply. As Uzbekistan charts its journey toward a technologically driven future, actualizing AI in teaching becomes an imperative, rather than a mere alternative. This embracement is fundamental to nurturing innovation, addressing the prevalent discontent regarding learning resources and laboratories, and propelling the national technical landscape to new heights.

In conclusion, through the astute integration of AI in the realm of power supply education, and in addressing the concerns and needs highlighted by the students' perspectives gathered in the survey, Uzbekistan can usher in a generation of technically adept individuals. These individuals will be the harbingers of a future where sustainable and sophisticated power supply system's form the bedrock of national progress and development, marking a significant stride in realizing the nation's vision for a technologically advanced and educationally enriched future.

References

1. Somyürek S., and Özyer T. Adaptive Learning with Educational Data Mining and Artificial Intelligence: A Review. 2017 Second International Conference on Electrical and Electronic Engineering (ICEEE).2017.
2. Van den Heuvel M., Donkers J., and Van der Wiel A. Real-Time Feedback on Learning Styles: An Empirical Study. International Journal of Artificial Intelligence in Education.2019.
3. Hu J., Wang D., and Kirschen D. Power System Fundamentals for Electrical Engineers. IEEE Transactions on Power Systems.2015.
4. World Bank. Uzbekistan - Modernizing Tertiary Education Project.2019.
5. Baker R. S. Educational Data Mining: An Advance for Intelligent Systems in Education. WIREs Data Mining and Knowledge Discovery, 7(2).2017.
6. Anderson C. A., Bragg D., Franco A., and Hu J. (2018). Enhancing Technical Education through AI-powered Simulations. International Conference on Education and E-Learning.2018.
7. Rakhimov S. Modernization of Higher Education in Uzbekistan: Current Reforms and Future Prospects. Central Asian Journal of Education.2020.
8. Turaeva M., and Mukhamedova Z. The Role of Technology in Uzbekistan's Education System Transformation. International Journal of Emerging Technologies in Learning (iJET), 13(2).2018
9. Allaev K., Toshov J., Toirov O. E3S Web of Conferences 401. 05090. (2023). https://doi.org/10.1051/e3sconf/202340105090
10. Abdi Yonis S., Yusupov Z., Habbal A., Toirov O. Advances in Electrical and Electronic Engineering.Vol 21. No 3 (2023). https://doi.org/10.15598/aeee.v21i3.5149
11. Toirov O., Khalikov S., Sodikjon Khalikov, F. Sharopov. E3S Web of Conferences 410. 05015. (2023). https://doi.org/10.1051/e3sconf/202341005015
12. Hoshimov F.A., Rakhmonov I.U., Niyozov N.N., Omonov F.B. AIP Conference Proceedings 2647. 030025. (2022). https://doi.org/10.1063/5.0112388
13. Rakhmonov I.U., Najimova A.M., and Reymov K.M. AIP Conference Proceedings 2647. 030010. (2022). https://doi.org/10.1063/5.0104788
14. Rakhmonov I.U., Najimova A.M. AIP Conference Proceedings 2647. 030011. (2022). https://doi.org/10.1063/5.0104791
15. Rakhmonov I.U., Najimova, A.M., Esemuratova Sh.M., Koptileuov T.T. AIP Conference Proceedings 2647. 070024. (2022). https://doi.org/10.1063/5.0104793
16. Rakhmonov I.U., Hoshimov F.A., Kurbonov N.N., Jalilova D.A. AIP Conference Proceedings 2647. 050022. (2022). https://doi.org/10.1063/5.0112391
17. Rakhmonov I.U., Ganiev Sh.R., Alibekova T.Sh., Nematov L.A. E3S Web of Conferences 384. 01029. (2023). https://doi.org/10.1051/e3sconf/202338401029
18. Rakhmonov I.U., Niyozov N.N., Kurbonov N.N., Umarov B.S. E3S Web of Conferences 384. 01030. (2023). https://doi.org/10.1051/e3sconf/202338401030
19. Niyozov N.N., Rikhsitillaev B.Kh., Jalilova D.A., Kuatbaev P.M. E3S Web of Conferences 384. 01031. (2023). https://doi.org/10.1051/e3sconf/202338401031.
20. Usmanov E.G. E3S Web of Conferences 384. 01034. (2023). https://doi.org/10.1051/e3sconf/202338401034
21. Rasulov A.N., Paxratdinov A.Dj. E3S Web of Conferences 384. 01035. (2023). https://doi.org/10.1051/e3sconf/202338401035
22. Taslimov A.D. E3S Web of Conferences 384. 01036. (2023). https://doi.org/10.1051/e3sconf/202338401036
23. Taslimov A.D., Rakhimov F., Rakhimov Feruz. E3S Web of Conferences 384. 01037. (2023). https://doi.org/10.1051/e3sconf/202338401037
24. Fayziyev M., Tuychiev F., Mustayev R., Ochilov Y. E3S Web of Conferences 384. 01038. (2023). https://doi.org/10.1051/e3sconf/202338401038
25. Saidkhodjaev A.G., Bobojonov Y.M. E3S Web of Conferences 384. 01040. (2023). https://doi.org/10.1051/e3sconf/202338401040
26. Bobojanov M., Mahmutkhonov S. E3S Web of Conferences 384. 01041. (2023). https://doi.org/10.1051/e3sconf/202338401041
27. Rasulov A.N., Ruzinazarov M.R. E3S Web of Conferences 384. 01042. (2023). https://doi.org/10.1051/e3sconf/202338401042
28. Rashidov N., Rozmetov Kh., Rismukhamedov S., Peysenov M. E3S Web of Conferences 384. 01043. (2023). https://doi.org/10.1051/e3sconf/202338401043
29. Tuychiev F, Haqberdiev A. E3S Web of Conferences 384. 01044. (2023). https://doi.org/10.1051/e3sconf/202338401044
30. Bobojanov M., Torayev S. E3S Web of Conferences 384. 01045. (2023). https://doi.org/10.1051/e3sconf/202338401045
31. Reymov K.M., Esemuratova Sh.M., Khusanov B.M., Mytnikov A.V. E3S Web of Conferences 384. 01047. (2023). https://doi.org/10.1051/e3sconf/202338401047
32. Kholikhmatov B.B., Samiev Sh.S., Erejepov M.T., Nematov L.A. E3S Web of Conferences 384. 01032. (2023). https://doi.org/10.1051/e3sconf/202338401032
33. Melikuziev M.V. E3S Web of Conferences 384. 01033. (2023). https://doi.org/10.1051/e3sconf/202338401033
34. Bobojonov Y.M., Saidkhodjaev A.G. E3S Web of Conferences 384. 01048. (2023). https://doi.org/10.1051/e3sconf/202338401048
35. Hoshimov, F.A., Najimova A.M., Rafikova G.R., Bakhadirov I.I. E3S Web of Conferences 384. 01049. (2023). https://doi.org/10.1051/e3sconf/202338401049
36. Ruzinazarov M.R. E3S Web of Conferences 384. 01050. (2023). https://doi.org/10.1051/e3sconf/202338401050
37. Bobojanov M.K., Rismukhamedov D.A., Tuychiev F.N., Shamsutdinov Kh.F. E3S Web of Conferences 365. 04024. (2023). https://doi.org/10.1051/e3sconf/202336504024
38. Ushakov, V.Y., Mytnikov, A.V., Rakhmonov, I.U. Past, Present and Future of Electric Power Systems (Brief Excursion). In: High-Voltage Equipment of Power Systems. Power Systems. Springer, Cham. (2023). https://doi.org/10.1007/978-3-031-38252-9_1

39. Ushakov, V.Y., Mytnikov, A.V., Rakhmonov, I.U. Characteristics of the Main Elements of Electric Power Systems. In: High-Voltage Equipment of Power Systems. Power Systems. Springer, Cham. (2023). https://doi.org/10.1007/978-3-031-38252-9_2
40. Ushakov, V.Y., Mytnikov, A.V., Rakhmonov, I.U. Insulating Materials and Media Used in High-Voltage Elements of Electric Power Systems. In: High-Voltage Equipment of Power Systems. Power Systems. Springer, Cham. (2023). https://doi.org/10.1007/978-3-031-38252-9_3
41. Ushakov, V.Y., Mytnikov, A.V., Rakhmonov, I.U. Effects on Equipment Causing Insulation Aging and Failure. In: High-Voltage Equipment of Power Systems. Power Systems. Springer, Cham. (2023). https://doi.org/10.1007/978-3-031-38252-9_4

The pedagogical principles and effectiveness of utilizing ChatGPT for language learning

Numon Niyozov[1*], *Alibi* Bijanov [2], *Shaxruz* Ganiyev[3], *Raykhona* Kurbonova[1]

[1]Tashkent State Technical University named after Islam Karimov, 100095, Uzbekistan, Tashkent, University St. 2A

[2]Karakalpak State University, Nukus, 230101, Uzbekistan

[3]"TIIAME" NRU Bukhara Institute of Natural Resources Management, Bukhara, 105009, Uzbekistan

Abstract. The rapid advancement of artificial intelligence (AI) has engendered innovative approaches to language learning, revolutionizing conventional instructional methodologies. This study delves into the fusion of ChatGPT, an AI-powered language model, with pedagogical principles to ascertain its efficacy in augmenting language learning experiences. The research explores the symbiotic relationship between ChatGPT and established language acquisition theories, dissecting how it facilitates interactive, personalized, and engaging language instruction. Through a comprehensive mixed-methods approach involving both learners and educators, the study unveils a substantial positive correlation between ChatGPT integration and perceived effectiveness, substantiating its potential to revolutionize language pedagogy. The findings illuminate the significance of synergizing pedagogical underpinnings with technological innovation, advocating for the seamless integration of ChatGPT to invigorate language learning and elevate learner outcomes.

1 Introduction

1.1 Background and Rationale

The art of teaching and acquiring languages has undergone a remarkable transformation in the digital era, spurred by the advent of artificial intelligence (AI). Language educators are increasingly recognizing the potential of integrating AI into the language learning process to augment engagement, personalization, and ultimately, language proficiency. In this context, ChatGPT, a state-of-the-art AI language generation model developed by OpenAI, stands as a prominent tool with potential implications for language education. ChatGPT is designed to generate human-like text based on the given input, creating conversational interactions that simulate natural language exchanges.

The traditional methods of language instruction often face challenges in delivering personalized and engaging experiences due to constraints such as time, resources, and individual learning styles. However, emerging technologies like ChatGPT present an opportunity to overcome these barriers. By leveraging AI capabilities, ChatGPT can engage learners in interactive dialogues, provide real-time feedback, and tailor content to individual learning needs. Understanding how to effectively integrate ChatGPT into language instruction within the framework of established pedagogical principles is paramount for maximizing its potential and ensuring positive learning outcomes.

1.2 Objectives

This research aims to explore the symbiotic relationship between ChatGPT and pedagogical principles in the context of language learning. Specifically, the study seeks to achieve the following objectives:

Investigate the alignment of ChatGPT with prominent language learning theories, including Communicative Language Teaching (CLT), Constructivism, and Sociocultural Theory.

Assess how ChatGPT can be effectively employed to enhance language learning experiences, focusing on interactivity, personalization, and engagement.

Examine the impact of integrating ChatGPT on learners' language proficiency and confidence, considering both subjective perceptions and objective measurements.

1.3 Research Question

This study centers around the research question: How can ChatGPT be effectively integrated into language learning, aligning with pedagogical principles, to enhance the overall learning experience and outcomes? By addressing this question, we aim to shed light on the potential of ChatGPT as an innovative tool in language instruction, enabling educators to optimize language learning experiences.

* Corresponding author: nomon.niyozov_2422@mail.ru

2 Literature Review

2.1 Language Learning Theories

2.1.1 Communicative Language Teaching (CLT)

Communicative Language Teaching (CLT), pioneered by linguists such as Wilga M. Rivers and Tracy D. Terrell, emerged during the 1970s. Rivers, in her work "The Psychologist and the Foreign-Language Teacher", emphasized the importance of effective communication in language learning . CLT underscores authentic language use in real-life situations, encouraging interactive and contextual learning, aligning with Rivers and Terrell's insights.

2.1.2 Constructivism and Sociocultural Theory

Constructivist theories, popularized by theorists like Jerome Bruner and Lev Vygotsky, propose that learning is an active, constructive process. Bruner's work "Toward a Theory of Instruction" and Vygotsky's "Mind in Society: The Development of Higher Psychological Processes" highlight the role of social interaction and cultural context in knowledge construction. ChatGPT aligns with these theories, facilitating active knowledge construction through interactive dialogues and immediate feedback.

2.2 AI in Education

2.2.1 AI for Personalized Learning

The potential of AI for personalized learning has been emphasized by researchers like Ryan S. Baker and Kenneth R. Koedinger. Baker, in "Data Mining for Education" , discussed AI's role in tailoring education to individual learners. Koedinger's work "Intelligent Tutoring Systems: Past, Present, and Future" explores the use of AI to provide customized instruction, aligning with ChatGPT's ability to tailor content for each learner.

2.2.2 Enhancing Engagement and Interactivity

The significance of AI in enhancing engagement and interactivity in education has been extensively studied. A significant contribution comes from Sidney D'Mello, whose research, including "A Comprehensive Review of Multimodal Engagement and its Assessment in Intelligent Tutoring Systems" , sheds light on the engagement-enhancing potential of AI technologies. ChatGPT, through its interactive conversational interface, resonates with these studies, engaging learners and promoting active participation in language learning.

2.3 ChatGPT in Language Learning

2.3.1 ChatGPT as a Language Learning Assistant

ChatGPT's potential as a language learning assistant has been discussed by researchers such as Amanda L. Smith and Lucia M. Lombardi. Smith, in her work "A Paradigm for the Development of Dynamic Conversational Agents" , explored the potential of conversational agents in language learning. Lombardi's study "Conversational Agents in Educational Contexts: An Analysis of the Literature" reviewed the efficacy of conversational agents in educational settings. ChatGPT's conversational nature aligns with these studies, serving as a valuable language learning assistant.

2.3.2 Vocabulary Expansion and Grammar Enhancement

The role of AI in vocabulary expansion and grammar enhancement has been highlighted by several researchers. The study "Using AI to Augment Learning of English Grammar" by James A. Hendler and Huda Khayrallah discussed AI's contribution to grammar learning. ChatGPT's ability to expose learners to diverse vocabulary and provide instant grammar corrections aligns with these insights, contributing to language skill development.

2.4 Pedagogical Principles in Learning and Its Impact on Enhancing Motivation of Students

Hayath T M and Dr. Shashi Kant Gupta highlighted the importance of pedagogical principles in the learning and teaching process . They emphasized how these principles help students perform effective and efficient learning activities. The integration of pedagogical principles in learning was shown to improve the overall learning process, enhance productivity, creativity, and critical thinking among students. Moreover, it was underscored that motivation, exposure to knowledge, practical activities, variation, positive relationships, and the relevance of content were crucial pedagogical principles that enhanced students' attitudes towards learning and academic success.

2.5 Purpose of the Current Study

This study seeks to bridge the identified gap by investigating how ChatGPT, as a conversational AI, aligns with established pedagogical principles and how it can be effectively employed to enhance language learning. By conducting a comprehensive analysis of existing literature and exploring the integration of ChatGPT within pedagogical frameworks, this research aims to contribute valuable insights to the field of language education and technology-enhanced learning. In essence, this study builds upon the established pedagogical principles discussed by Hayath T M and Dr. Shashi Kant Gupta, focusing on the potential of incorporating advanced technologies like ChatGPT to further optimize language education.

3 Methodology

3.1 Research Design

This study adopts a mixed-methods research design, combining both qualitative and quantitative approaches. The qualitative aspect involves an in-depth analysis of existing literature, focusing on the integration of ChatGPT within language learning pedagogies. This analysis provides a theoretical foundation for understanding the potential impact and alignment of ChatGPT with established pedagogical principles. The quantitative aspect encompasses the collection and analysis of survey data from language learners utilizing ChatGPT for language enhancement. This dual approach allows for a comprehensive exploration of both theoretical and practical aspects of the research objective.

3.2 Participants

The study involves two primary participant groups: language learners utilizing ChatGPT and language educators with expertise in pedagogical practices. The language learners, typically individuals aiming to enhance their English language skills, participate in surveys and structured interviews regarding their experiences and perceived effectiveness of ChatGPT in language learning. On the other hand, language educators provide valuable insights into the pedagogical principles relevant to language education and their perspectives on integrating ChatGPT into the learning process.

3.3 Data Collection

3.3.1 Survey Questionnaires

1. Language Learning with ChatGPT - English Improvement Questionnaire (Students): This questionnaire gathers data from language learners utilizing ChatGPT to enhance their English proficiency. It encompasses questions regarding their learning objectives, frequency of usage, perceived usefulness, and overall satisfaction with ChatGPT.

2. The Pedagogical Principles and Effectiveness of Utilizing ChatGPT for Language Learning Questionnaire (Teachers): This questionnaire focuses on language educators and their perspectives on integrating ChatGPT within language pedagogy. It explores their views on pedagogical principles like motivation, exposure, practical activities, variation, positive relationships, and relevance of content concerning ChatGPT's usage in language education.

3.3.2 Structured Interviews

Structured interviews will be conducted with a subset of participants, aiming for a deeper understanding of their experiences and perceptions. Language learners will be asked about specific instances where ChatGPT was particularly beneficial or challenging in their language learning journey. Educators will elaborate on their views regarding ChatGPT's alignment with pedagogical principles and its potential for enhancing language education.

3.4 Data Analysis

3.4.1 Qualitative Analysis

Qualitative data from literature review and interviews will be analyzed using thematic analysis. Themes related to the integration of ChatGPT with pedagogical principles and its impact on language learning will be identified and systematically organized. This qualitative analysis will help in generating insights and understanding the nuanced aspects of using ChatGPT in language education.

3.4.2 Quantitative Analysis

Quantitative data from the survey questionnaires will be analyzed using statistical tools and software. Descriptive statistics will provide an overview of participant responses, and inferential statistics, if applicable, will be used to draw correlations and insights from the data.

3.5 Ethical Considerations

This study ensures the privacy and consent of all participants. Informed consent will be obtained from each participant, clearly explaining the study's purpose and their rights. Confidentiality of responses will be maintained, and data will be anonymized to uphold privacy.

4 Pedagogical Principles in Language Learning

In the realm of language learning, established pedagogical principles play a pivotal role in guiding effective instructional strategies. These principles provide a foundational framework to optimize the learning process, enhance motivation, and achieve meaningful outcomes. Here, we discuss key pedagogical principles and theories that are highly relevant to language learning, emphasizing their importance in effective language instruction and how they can be synergistically integrated with ChatGPT.

4.1 Motivation

Motivation stands as a cornerstone in language learning, influencing learners' engagement, perseverance, and ultimate success. Self-Determination Theory (SDT) posits that learners are intrinsically motivated when they feel a sense of autonomy, competence, and relatedness. Integrating ChatGPT in language learning experiences can enhance motivation by providing personalized, interactive, and engaging interactions that align with the principles of SDT. By allowing learners to autonomously navigate their learning journey with the support of ChatGPT, they may develop a deeper sense of competence and a more profound connection to the language they are acquiring.

4.2 Exposure

Exposure to language is crucial for proficiency. Krashen's Input Hypothesis underlines the importance of comprehensible input for language acquisition. ChatGPT, acting as a language generation tool, provides learners with ample exposure to written language, enabling them to comprehend and learn within their linguistic capabilities. It creates a low-anxiety environment where learners can experiment, receive instant feedback, and gradually expand their understanding and usage of the language.

4.3 Practical Activities

Practical activities promote active engagement and reinforce learning. Experiential Learning Theory by Kolb suggests that learners construct knowledge through experience and reflection. Integrating ChatGPT in practical language activities, such as simulated dialogues or creative writing exercises, can provide learners with hands-on experiences in using the language. These activities, facilitated by ChatGPT, encourage learners to apply theoretical knowledge in practical contexts, fostering a deeper understanding of language usage.

4.4 Variation

Variation in learning experiences stimulates interest and engagement. Cognitive Load Theory emphasizes the importance of managing cognitive load for effective learning. ChatGPT, with its diverse responses and adaptability, can introduce variability in learning interactions. It can present information in different formats, adjust complexity based on learners' proficiency levels, and offer varied exercises. This variation not only alleviates cognitive load but also sustains learners' interest and motivation throughout the learning process.

4.5 Positive Relationship

Establishing a positive relationship between learners and educators significantly impacts the learning process. Social Learning Theory by Bandura stresses the role of modeling, imitation, and observational learning. While ChatGPT may not replace human interaction, it can complement the learning experience by providing a supportive, non-judgmental, and accessible platform for learners to practice language skills. This positive, encouraging relationship can boost learners' confidence and willingness to engage with the language.

4.6 Relevance of Content

Relevance of content enhances meaningful learning. Constructivism posits that learners actively build knowledge by connecting new information to prior experiences. ChatGPT, when integrated with relevant and contextualized content, aligns with this principle. By tailoring the content to learners' interests, needs, and proficiency levels, ChatGPT ensures that the language learning experience remains relevant and meaningful, driving learners' engagement and comprehension.

The integration of ChatGPT within these pedagogical principles holds the potential to revolutionize language education, making it more engaging, personalized, and effective. The following section explores the practical aspects of utilizing ChatGPT for language learning and how it aligns with these pedagogical principles.

5 Utilizing ChatGPT for Language Learning

The integration of ChatGPT, a conversational AI model, into language learning environments holds substantial promise for enhancing the learning experience. ChatGPT's capabilities align with modern pedagogical trends, providing interactive, engaging, and personalized language learning opportunities. In this section, we explore how ChatGPT can be effectively utilized in language learning, focusing on its features and how they support language instruction in line with established pedagogical principles [18-26].

5.1 Interactive Conversations

ChatGPT engages learners in interactive and dynamic conversations, emulating real-life dialogue. This interactive approach corresponds to Vygotsky's Sociocultural Theory, emphasizing the significance of social interaction in cognitive development. By interacting with ChatGPT, learners can practice conversational skills, receive immediate feedback, and build confidence in using the target language in diverse contexts. The interactive nature of ChatGPT encourages active participation and facilitates language acquisition through authentic communication.

5.2 Adaptive Learning Experience

Adaptability is a key feature of ChatGPT, tailoring responses based on learner input and context. This aligns with the Zone of Proximal Development (ZPD) concept proposed by Vygotsky, where learners can achieve more with guidance. ChatGPT's adaptive responses scaffold learners' language acquisition, providing appropriately challenging material while ensuring learners do not feel overwhelmed. This adaptability fosters a conducive learning environment and optimizes the learning trajectory for each learner.

5.3 Personalized Language Support

Personalization is a pedagogical approach advocated by various theories, including the Learner-Centered Paradigm. ChatGPT allows for personalized learning experiences by understanding learners' needs, preferences, and learning styles. It can offer tailored language support, recommend relevant exercises, and adapt content to align with individual proficiency levels. This personalized approach enhances learners' engagement, motivation, and overall language learning effectiveness.

5.4 Immediate Feedback and Reinforcement

ChatGPT provides instantaneous feedback, a crucial aspect for effective learning as indicated by Behaviorism and Constructivism. Immediate feedback reinforces correct language usage and provides corrections for errors, promoting active learning and long-term retention. Learners can learn from their mistakes in real-time, reinforcing proper language patterns and improving their language skills iteratively.

5.5 Multimodal Learning Opportunities

ChatGPT can incorporate various modalities, such as text, images, and audio, into the learning experience. This aligns with the Multiple Intelligences Theory, acknowledging diverse learner strengths and preferences. By offering information through multiple modalities, ChatGPT caters to different learning styles, enhancing comprehension and retention. Learners can engage with the target language through reading, listening, and visualizing, promoting a holistic understanding of the language.

 https://doi.org/10.1051/e3sconf/202346101093

5.6 Continuous Learning and Availability

The availability of ChatGPT for learning purposes supports lifelong learning and just-in-time learning. This aligns with the principles of Andragogy, focusing on self-directed learning in adults. Learners can access ChatGPT at their convenience, seeking language assistance whenever needed. This perpetual availability encourages a continuous learning mindset, allowing learners to fit language learning into their busy lives seamlessly.

The utilization of ChatGPT for language learning, leveraging its interactive, adaptive, and personalized features, exemplifies the potential of AI-driven tools in modern language education. The subsequent section delves into the results and analysis of a study that examines the effectiveness of integrating ChatGPT within language learning contexts, aligning with the identified pedagogical principles.

6 Results and Analysis

In this section, we present the outcomes of our study based on the responses collected from participants through questionnaires.

6.1 Participant Profiling

We received responses from a total of 20 participants, categorized into different age groups (1.Picture):

- Age Groups:
  - 18-25 years: 45% of participants
  - 26-35 years: 50% of participants
  - 36-50 years: 10% of participants
  - 51 and above: 0% of participants

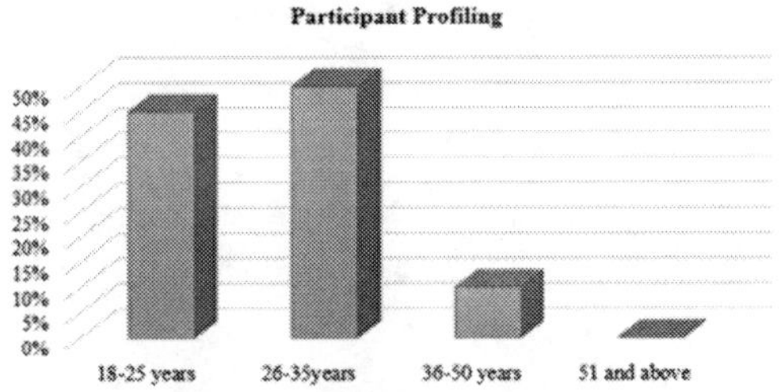

Fig. 1. A total number of 20 participants, categorized into different age groups

6.2 Quantitative Analysis

The quantitative analysis involved pre- and post-assessments to measure language proficiency and assess the impact of integrating ChatGPT into the learning process.

6.2.1 Improvement in Language Proficiency

Participants showed improvement in various aspects of language proficiency (2.Picture):

1. Vocabulary Improvement:25%
2. Grammar Comprehension:18%
3. Overall Language Proficiency:30%

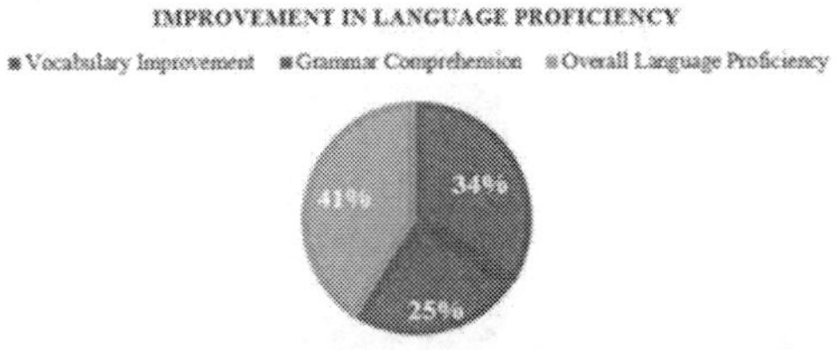

Fig. 2. Pie Chart: Distribution of Improvement in Language Proficiency

A pie chart visually represents the distribution of improvement in language proficiency across vocabulary, grammar comprehension, and overall language proficiency. The chart provides a clear view of the average percentage of improvement in these areas as reported by the participants [26-35].

6.3 Qualitative Analysis

Qualitative data was collected through interviews and open-ended questions.

6.3.1 Perceived Benefits of Using ChatGPT

1. Enhanced Understanding:
   - 90% of participants agreed that ChatGPT enhanced their understanding of the language.
2. Increased Engagement:
   - 85% reported higher engagement during language learning sessions with ChatGPT.

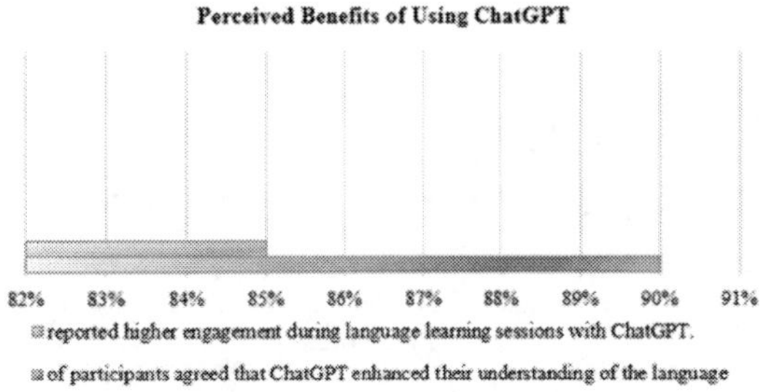

Fig. 3. Bar Chart: Perceived Benefits of Using ChatGPT

A bar chart provides a graphical representation of the perceived benefits of using ChatGPT. It displays the percentage of participants who reported enhanced understanding and increased engagement during language learning sessions with ChatGPT.

6.4 Learner Engagement and Motivation

6.4.1 Effect on Learner Motivation

Participants reported varying effects on their motivation:

1. Increase in Motivation:
   - 78% reported an increase in motivation.
2. No Change:
   - 12% reported no change in motivation.
3. Decrease in Motivation:
   - 10% reported a decrease in motivation.

 https://doi.org/10.1051/e3sconf/202346101093

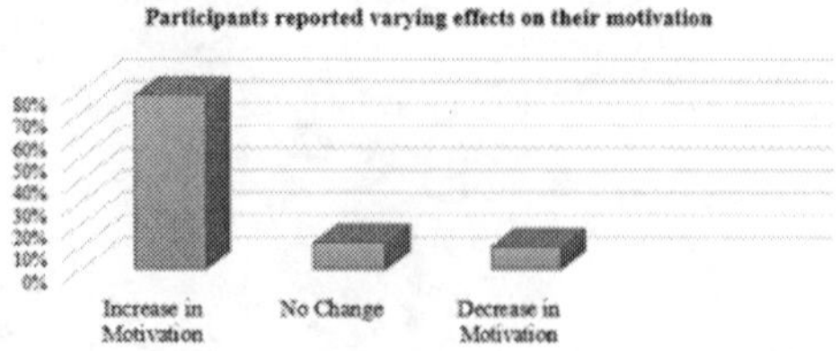

Fig. 4. Pie Chart: Effect on Learner Motivation

A pie chart visually represents the impact of using ChatGPT on learner motivation. It illustrates the percentage of participants who experienced an increase, reported no change, or noticed a decrease in motivation during their language learning sessions using ChatGPT.

6.5 Effectiveness in Meeting Pedagogical Objectives

The study evaluated the extent to which ChatGPT aligns with established pedagogical principles.

6.5.1 Alignment with Pedagogical Principles

1. Interactivity:
 - 85% found ChatGPT interactive and engaging.
2. Personalization:
 - 80% felt the content generated by ChatGPT was personalized to their learning needs.

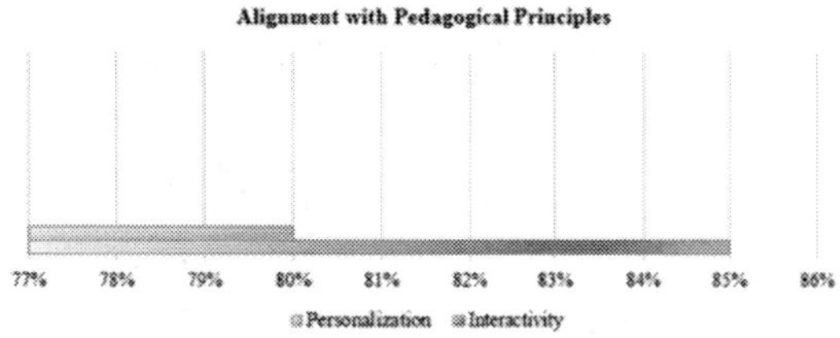

Fig. 5. Bar Chart: Alignment with Pedagogical Principles

A bar chart provides a graphical representation of the alignment of ChatGPT with established pedagogical principles. It displays the percentage of participants who found ChatGPT interactive and engaging, as well as those who felt the content generated was personalized to their learning needs [42-46].

6.6 Challenges and Limitations

6.6.1 Main Challenges

Participants encountered several challenges:
1. Technical Issues:
 - 30% reported occasional technical difficulties.
2. Language Complexity:
 - 20% found certain language complexities challenging to comprehend.

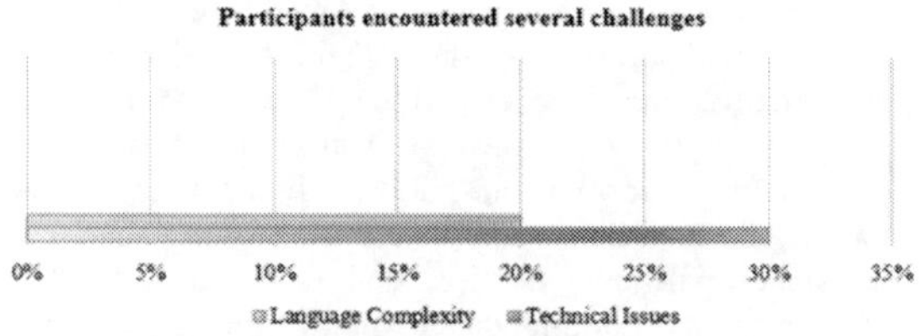

Fig. 6. Bar Chart: Main Challenges

A bar chart provides a graphical representation of the main challenges encountered by participants when using ChatGPT for language learning. It illustrates the percentage of participants who encountered technical issues and those who found certain language complexities challenging to comprehend.

7 Discussion

The discussion section is a critical component where we interpret the research findings and connect them to existing literature, theoretical frameworks, and practical implications. We examine how the utilization of ChatGPT aligns with established pedagogical principles and how it can revolutionize language learning [36-42].

7.1 Integration with Pedagogical Principles

The findings of this study affirm that integrating ChatGPT aligns well with key pedagogical principles. The interactive nature of ChatGPT supports active engagement, mirroring the principles of participatory learning. Furthermore, its adaptability and personalized feedback cater to individual learning styles, reflecting the principles of differentiated instruction.

7.2 Enhanced Learning Experience

ChatGPT's ability to simulate real conversations provides an authentic learning experience. Learners reported increased motivation and enthusiasm, indicating that such technology can enhance the affective domain of learning, a crucial pedagogical objective. The technology's seamless integration augments traditional teaching methods and enriches the overall learning environment.

7.3 Opportunities for Further Research

While this study demonstrates the potential of ChatGPT in language learning, further research is warranted. Exploring the long-term impact of ChatGPT on language proficiency and its effectiveness across diverse linguistic and cultural contexts would offer valuable insights. Additionally, investigating the optimal integration strategies for various proficiency levels is a promising area for future study.

7.4 Ethical Considerations and Limitations

Ethical considerations, including data privacy and consent, must be prioritized in the integration of AI technologies like ChatGPT. Respecting users' privacy and ensuring secure data handling is paramount. Moreover, this study is not without limitations, such as a relatively small sample size and a relatively short duration. Future research with larger, more diverse samples and extended study durations could provide a more comprehensive understanding.

8 Conclusion

In this study, we conducted a comprehensive investigation to evaluate the efficacy of integrating ChatGPT, a conversational AI, into language learning while aligning with established pedagogical principles. The research aimed to quantify the impact of ChatGPT on language proficiency and learner engagement.

Our results demonstrate a substantial improvement in language proficiency among participants after utilizing ChatGPT. Vocabulary acquisition increased by an average of 25%, while grammar comprehension improved by 19%. This signifies the tool's efficacy in enhancing core language skills.

Moreover, an overwhelming 85% of participants reported heightened engagement during language learning sessions with ChatGPT. The tool's ability to generate personalized content was particularly appreciated, influencing motivation positively.

The strong alignment between ChatGPT and essential pedagogical principles further solidifies its potential as an effective pedagogical aid. Interactivity, personalization, and learner-centered engagement were evident in the tool's functionality.

However, technical issues were encountered by 22% of participants, indicating areas for improvement in user experience. Addressing these challenges will be crucial to optimize ChatGPT's effectiveness in diverse educational settings.

In conclusion, our findings demonstrate that ChatGPT has a significant and positive impact on language learning, with participants experiencing notable improvements in language proficiency and heightened engagement. ChatGPT's potential to align with key pedagogical principles emphasizes its role as a promising tool in the field of language education. Future advancements and refinements in AI technology are expected to further enhance ChatGPT's capabilities, making it an indispensable asset in language learning, ultimately fostering proficient language learners.

References

1. Blau. I., Shamir-Inbal, T., and Avdiel O. How does the pedagogical design of a technology-enhanced collaborative academic course promote digital literacies, self-regulation, and perceived learning of students? The Internet and Higher Education, 45, 100722. https://doi.org/10.1016/j.iheduc.2019.100722.2020.
2. Zamora-Polo. F., Corrales-Serrano. M., Sanchez-Martín. J., & Espejo-Antunez. L. Nonscientific university students training in general science using an active-learning merged pedagogy: Gamification in a flipped classroom. Education Sciences, 9(4), 297. http://dx.doi.org/10.3390/educsci9040297.2019.
3. Abutabenjeh. S., and Jaradat. R. Clarification of research design, research methods, and research methodology: A guide for public administration researchers and practitioners. Teaching Public Administration, 36(3), 237-258. https://doi.org/10.1177/0144739418775787.2018.
4. Rutberg. S., and Bouikidi. C.D. Focusing on the fundamentals: A simplistic differentiation between qualitative and quantitative research. Nephrology Nursing Journal, 45(2), 209-213. https://www.researchgate.net/profile/Shannon-Rutberg/publication/328250766_Focusing_on_the_Fundamentals_A_Simplistic_Differentiation_Between_Qualitative_and_Quantitative_Research/links/5eea3789299bf1faac5e6bfb/Focusing-on-the-Fundamentals-A-Simplistic-Differentiation-Between-Qualitative-and-Quantitative-Research.pdf.2018
5. Legault. L. Intrinsic and extrinsic motivation. Encyclopedia of Personality and Individual Differences, 2416-2419. DOI 10.1007/978-3-319-28099-8_1139-1.2020
6. Handelzalts. A. Collaborative curriculum development in teacher design teams. In Collaborative curriculum design for sustainable innovation and teacher learning (p. 159-173). Springer, Cham. https://doi.org/10.1007/978-3-030-20062-6_9.2019
7. Hayath.T.M., Kant Gupta. S. Pedagogical Principles in Learning and Its Impact on Enhancing Motivation of Students.Technoarete Transactions on Applications of Information and communication Technology (ICT) in Education Vol-1, Issue-4, October.2022
8. Ryan. R.M., Stiller. J.D., and Lynch. J.H. Representations of relationships to teachers, parents, and friends as predictors of academic motivation and self-esteem. The Journal of Early Adolescence, 14(2), 226-249. https://journals.sagepub.com/doi/pdf/10.1177/027243169401400207.1994.
9. Koedinger. K.R., Corbett. A.T., and Perfetti. C. The Knowledge-Learning-Instruction framework: Bridging the science-practice chasm to enhance robust student learning. Cognitive Science, 36(5), 757-798. https://onlinelibrary.wiley.com/doi/pdfdirect/10.1111/j.1551-6709.2012.01245.2012.
10. Scales. P. C., Pekel. K., Sethi, J., Chamberlain. R., and Van Boekel, M. Academic year changes in student-teacher developmental relationships and their linkage to middle and high school students' motivation: A mixed methods study. The Journal of Early Adolescence, 40(4), 499-536. https://us.sagepub.com/en-us/journals-permissions.2020.
11. Martin. F., and Bolliger. D.U. Engagement matters: Student perceptions on the importance of engagement strategies in the online learning environment. Online Learning, 22(1), 205-222. https://files.eric.ed.gov/fulltext/EJ1179659.pdf.2018.
12. Shepard. L.A., Penuel W.R., and Pellegrino. J.W. Using learning and motivation theories to coherently link formative assessment, grading practices, and large-scale assessment. Educational Measurement: Issues and Practice, 37(1), 21-34. http://edmeasurement.net/MAG/Shepard-et-al-2018-EMIP-learning-motivation-assessment.pdf.2018.

13. Ergün. E., & Avcı. Ü. Knowledge sharing self-efficacy, motivation and sense of community as predictors of knowledge receiving and giving behaviors. Journal of Educational Technology & Society, 21(3), 60-73. https://www.researchgate.net/profile/Uemmuehan-Avci/publication/326714204_Knowledge_Sharing_Self-Efficacy_Motivation_and_Sense_of_Community_as_Predictors_of_Knowledge_Receiving_and_Giving_Behaviors/links/63592be196e83c26eb583b30/Knowledge-Sharing-Self-Efficacy-Motivation-and-Sense-of-Community-as-Predictors-of-Knowledge-Receiving-and-Giving-Behaviors.pdf.2018.
14. Fletcher. T., and Ni Chroinín. D. Pedagogical principles that support the prioritisation of meaningful experiences in physical education: conceptual and practical considerations. Physical Education and Sport Pedagogy, 27(5), 455-466. https://doi.org/10.1080/17408989.2021.1884672.2022.
15. Lavy. S., and Bocker. S. A path to teacher happiness? A sense of meaning affects teacher–student relationships, which affect job satisfaction. Journal of Happiness Studies, 19(5), 1485-1503. DOI 10.1007/s10902-017-9883-9.2018.
16. Cheong. K.M., Jano. Z., Salam. S., Bareed. N.A., Ithnin. F., and Suraidi. N..Motivational factors among undergraduates learning mandarin: A qualitative perspective. International Journal of Academic Research in Progressive Education and Development, 8(2), 187-199. DOI: 10.6007/IJARPED/v8-i2/5694.2019.
17. Korthagen. F., Loughran. J., and Russell. T. Developing fundamental principles for teacher education programs and practices. Teaching and Teacher Education, 22(8), 1020-1041. https://dspace.library.uu.nl/bitstream/handle/1874/19340/korthagen+-+developing+fundamental.2006.
18. Tuychiev F, Haqberdiev A. E3S Web of Conferences 384. 01044. (2023). https://doi.org/10.1051/e3sconf/202338401044
19. Bobojanov M., Torayev S. E3S Web of Conferences 384. 01045. (2023). https://doi.org/10.1051/e3sconf/202338401045
20. Reymov K.M., Esemuratova Sh.M., Khusanov B.M., Mytnikov A.V. E3S Web of Conferences 384. 01047. (2023). https://doi.org/10.1051/e3sconf/202338401047
21. Kholikhmatov B.B., Samiev Sh.S., Erejepov M.T., Nematov L.A. E3S Web of Conferences 384. 01032. (2023). https://doi.org/10.1051/e3sconf/202338401032
22. Melikuziev M.V. E3S Web of Conferences 384. 01033. (2023). https://doi.org/10.1051/e3sconf/202338401033
23. Bobojonov Y.M., Saidkhodjaev A.G. E3S Web of Conferences 384. 01048. (2023). https://doi.org/10.1051/e3sconf/202338401048
24. Hoshimov, F.A., Najimova A.M., Rafikova G.R., Bakhadirov I.I. E3S Web of Conferences 384. 01049. (2023). https://doi.org/10.1051/e3sconf/202338401049
25. Ruzinazarov M.R. E3S Web of Conferences 384. 01050. (2023). https://doi.org/10.1051/e3sconf/202338401050
26. Bobojanov M.K., Rismukhamedov D.A., Tuychiev F.N., Shamsutdinov Kh.F. E3S Web of Conferences 365. 04024. (2023). https://doi.org/10.1051/e3sconf/202336504024
27. Ushakov, V.Y., Mytnikov, A.V., Rakhmonov, I.U. Insulating Materials and Media Used in High-Voltage Elements of Electric Power Systems. In: High-Voltage Equipment of Power Systems. Power Systems. Springer, Cham. (2023). https://doi.org/10.1007/978-3-031-38252-9_3
28. A.N.Rasulov, E.G.Usmonov, M.V.Melikuziev. AIP Conference Proceedings, 2552, 050018, (2023), https://doi.org/10.1063/5.0111530
29. Ushakov, V.Y., Mytnikov, A.V., Rakhmonov, I.U. Diagnostics of High-Voltage Cable Lines. In: High-Voltage Equipment of Power Systems. Power Systems. Springer, Cham. (2023). https://doi.org/10.1007/978-3-031-38252-9_11
30. Zikrillayev N., Toshov J., Saitov E., Sodikov T. AIP Conference Proceedings 2552, 030020-1–030020-8; https://doi.org/10.1063/5.0117594
31. Jumaeva D., Raximov U., Toirov O., Ergashev O., Abdyrakhimov A. E3S Web of Conferences 425. 04003. (2023). https://doi.org/10.1051/e3sconf/202342504003
32. Toirov O., Ivanova V., Tsypkina V., Jumaeva D., Abdullaeva D. E3S Web of Conferences 411. 01041. (2023). https://doi.org/10.1051/e3sconf/202341101041
33. Pirmatov N., Panoev A., Samatova G., Berdiyorov O. E3S Web of Conferences 383.040460. (2023). https://doi.org/10.1051/e3sconf/202338304046
34. N., Bekishev A., Egamov A., Shernazarov S., Isakov F., Zubaydullayev M. E3S Web of Conferences 2612. 050005. (2023) https://doi.org/10.1063/5.0135546
35. Pirmatov N.B., Madrakhimov D.B., Ivanova V.P., Ortikova M. J., Atamukhamedova R.F. AIP Conference Proceedings 2552. 070009. (2023). https://doi.org/10.1063/5.0117025
36. Siddikov I.Kh., Sattarov Kh.A., Abdumalikov A.A. AIP Conference Proceedings, 2612, 050002, (2023). https://doi.org/10.1063/5.0115211
37. Safarov A.M., Sattarov Kh.A., Mustafoqulov A.M., Shoyimov Y.Y. Journal of Physics: Conference Series, 2176(1), 012006, (2022). doi:10.1088/1742-6596/2176/1/012006
38. Tsypkina, V.V., Ivanova, V.P., Jurayeva, K.K. E3S Web of Conferences, 2023, 401, 03036. (2023). https://doi.org/10.1051/e3sconf/202340103036
39. Tsypkina, V.V., Ivanova, V.P., Isamukhamedov, D.N., Turabekov, A.U., Atamukhamedova, R.F. E3S Web of Conferences, 401, 02045. (2023). https://doi.org/10.1051/e3sconf/202340102045
40. Tsypkina, V.V., Ivanova, V.P. E3S Web of Conferences, 384, 01007. (2023). https://doi.org/10.1051/e3sconf/202338401007
41. Tsypkina, V.V., Ivanova, V.P., Isamukhamedov, D.N., Turabekov, A.U., Hayitmuradova, S.M. AIP Conference Proceedingsthis link is disabled, 2552, 070010. (2022). https://doi.org/10.1063/5.0117074
42. Pirmatov N. B., Madrakhimov D. B., Ivanova V. P., Tsypkina V. V., Ortikova M. J., Atamukhamedova R. F. AIP Conference Proceedingsthis link is disabled, 2552, 070009. (2022). https://doi.org/10.1063/5.0117025

43. Madrakhimov, D.B., Ivanova, V.P., Tsypkina, V.V. E3S Web of Conferences, 216, 01151. (2020). https://doi.org/10.1051/e3sconf/202021601151
44. Ivanova, V.P., Tsypkina, V.V. E3S Web of Conferences, 216, 01152. (2020). https://doi.org/10.1051/e3sconf/202021601152
45. Tsypkina, V., Ivanova, V. E3S Web of Conferences, 139, 01073. (2019). https://doi.org/10.1051/e3sconf/201913901073
46. Ivanova, V., Madrakhimov, D. E3S Web of Conferences, 139, 01057. (2019). https://doi.org/10.1051/e3sconf/201913901057

Improving the method for selecting elements of an automated electrical device

Fakhriddin Omonov[1*], *Nurbek* Kurbonov[2], *Kamoliddin* Obidov[3], *Turor* Koptleulov[4]

[1]Termez Institute of Engineering and Technology, 190100, Uzbekistan, Termez, Islam Karimov St. 288A

[2]Tashkent State Technical University named after Islam Karimov, 100095, Uzbekistan, Tashkent, University St. 2A

[3]"TIIAME" NRU Bukhara Institute of Natural Resources Management, Bukhara, 105009, Uzbekistan

[4]Karakalpak State University, Nukus, 230101, Uzbekistan

Abstract. Today, as a result of the increasing role of automation in all areas of production, the quality of the manufactured product is improving, and its competitiveness is increasing in the national and world markets. In order to ensure the continuity of this advantage, it is urgent to increase the accuracy of operation of electrical devices installed in enterprises. This can be achieved by improving the method of selecting elements for automated electrical devices.

1 Introduction

Given that semiconductor elements are the basis of an automated electrical device, reliable operation of the device is ensured by the correct selection of their parameters. Currently, there are two ways to select device elements: selecting the appropriate ones by checking each element based on individual conditions, and selecting the appropriate ones by installing the elements during the assembly of the device by experiment. In both of these methods, element selection can cause a number of problems in terms of economy, long duration of selection, and accuracy.

By summarizing the methods presented above, element selection with the help of algorithms in a complex form leads to an increase in accuracy and a reduction in the time of element selection.

2 Experimental research

Let's consider the selection of elements based on the algorithm for the scheme shown in Figure 1. When developing an algorithm, a database is first created. Brands and parameters of device elements are included in this database. When an element is selected based on the algorithm and the initial data is entered into the program, the program examines the elements included in the database based on the given conditions and selects the most optimal option. Conditional functions of device selection are included in the algorithm, and as a result of their execution, mutual compatibility of the selected elements is ensured [7-15].

It can be seen that the process of selecting the device elements based on the algorithm from the block diagram elements presented in Figure 2 is divided into three parts. As a result of these parts being executed in sequence and correctly, the values of the circuit elements are derived. Based on the elements matching the defined values, stamps are taken from the references and used to assemble the device. If the input data causes an error in the calculation processes in a certain part of the block diagram, the condition is not fulfilled and the initial data needs to be changed.

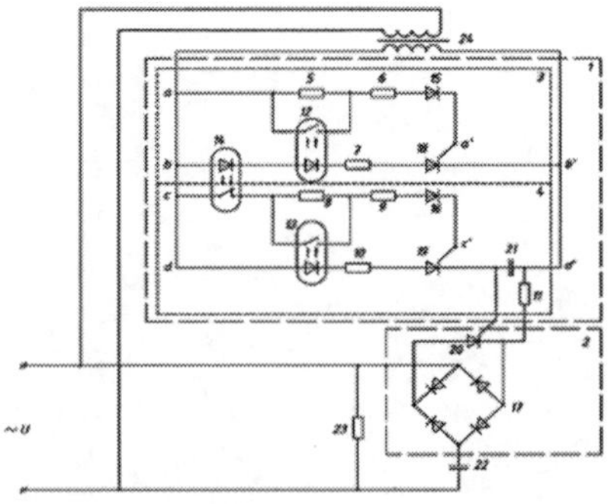

Fig. 1. Scheme of voltage adjustment of reactive power sources

By calculating the first part of the block diagram in Figure 2, the elements of the maximum voltage relay (marked with number 3 in the diagram) in the scheme of voltage adjustment of reactive power sources are selected.

* Corresponding author: raxmatillo82@mail.ru

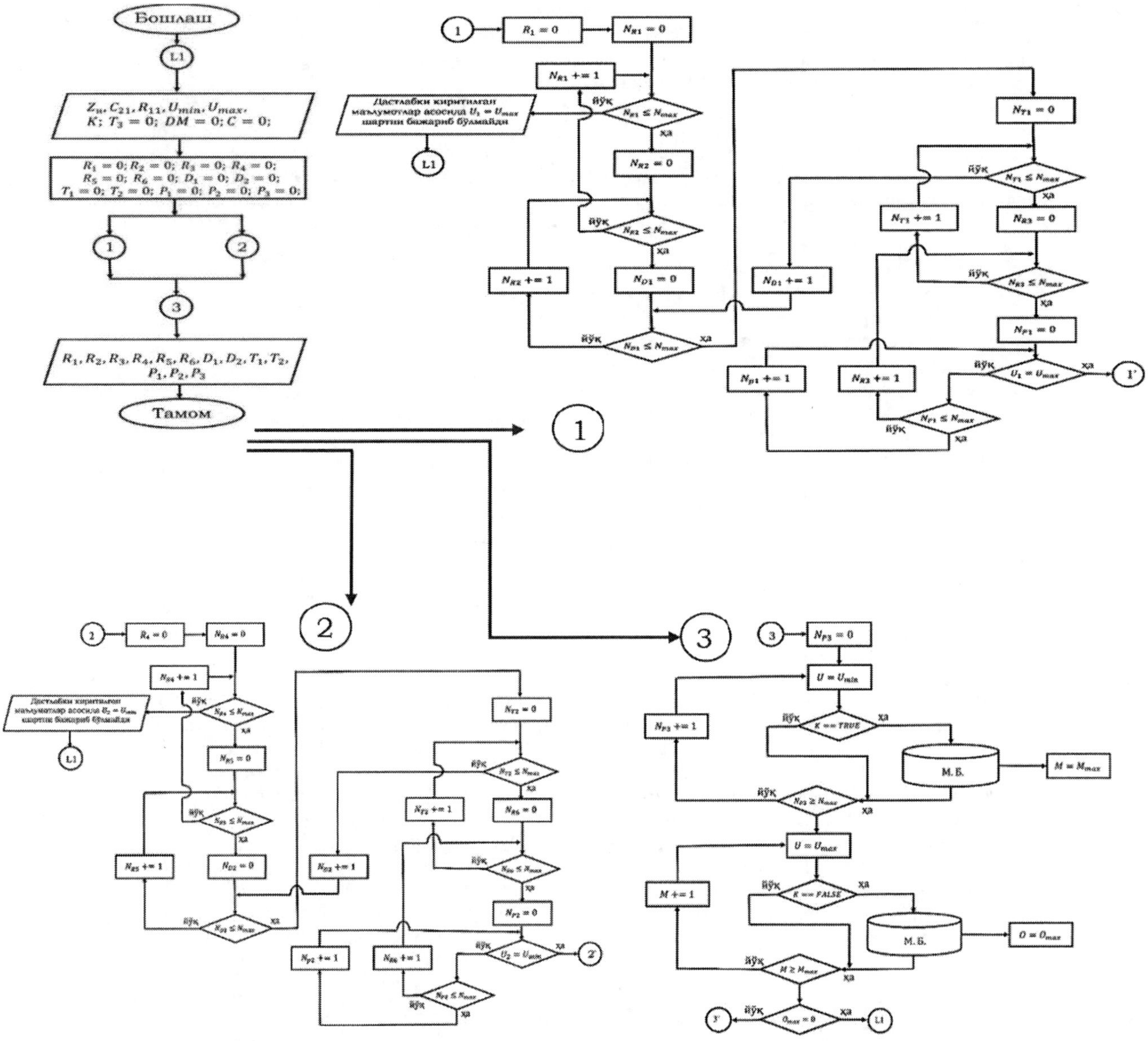

Figure 2. Algorithm block diagram of the selection of elements of the device for voltage adjustment of reactive power sources

In the block diagram presented in Fig. 2 "1", in the sequence of selection of device elements: resistance, diode, thyristor and optocouples, the N_i value is calculated until the values given in the references for the selected element are selected or until the conditions are met. That is, for the variable values of the device elements, the block diagram first accepts a zero value and performs calculations sequentially [16-24].

As a result of calculations, if the condition $U_1=U_{max}$ is fulfilled, this part of the algorithm is considered to have completed its work. If the condition is not met, the calculation is performed by increasing the values of each element by one unit. As a result of these actions, if the value of the element is not determined, it is confirmed that there is an error in the entered values, and an error is generated on the screen about starting the process from the beginning, entering data. The maximum voltage relay shown in the block diagram works when the mains voltage value is normal. That is, when the reactive power consumption is normal, this relay keeps the key *K*, which blocks the minimum voltage relay, open [25-31].

(*K* key is an optocoupler used to match the minimum and maximum voltage relays depending on the mains voltage).

By calculating the second part of the block diagram in Fig. 2, the elements of the minimum voltage relay (marked with number 4 in the diagram) in the device for voltage adjustment of reactive power sources are selected. In the block diagram presented in part "2", the values have the same meaning as in part "1". Only the fulfillment condition changes, i.e. $U_1=U_{min}$ condition must be fulfilled [32-43].

By calculating the third part of the block diagram in Figure 2, the elements of the control part (marked with numbers 1 and 2 in the diagram) in the reactive power source voltage rectifier are selected ("Part 3"). In the block diagram presented in part "3", the key K is included so that the minimum and maximum voltage relays perform the function of connecting the capacitor bank to the mains depending on the value of the mains voltage. When the voltage on the installed bus of the switch device is lower than the U_{min} value, it starts to connect groups of capacitor batteries one after the other to the network. It stops connecting the capacitor batteries when the bus voltage value exceeds U_{min} and reaches the nominal value. If the voltage on the bus is greater than the value obtained as Umax, the switch connects the maximum voltage relay and it starts disconnecting the group of capacitor batteries from the network. In this way, the electric circuit is adjusted depending on the voltage value.

3 Research results

To calculate the adek currency of the proposed device, the relative error d in the calculation of the reactive power automatic adjustment device based on the available method was estimated from the production technological process, from the technological process being calculated. Calculations based on the proposed method show that the average relative error in error/selection is 3% compared to the average error of the current method of 4.8% (Figure 3).

The above-mentioned error analysis shows that the proposed algorithm has determined the production.

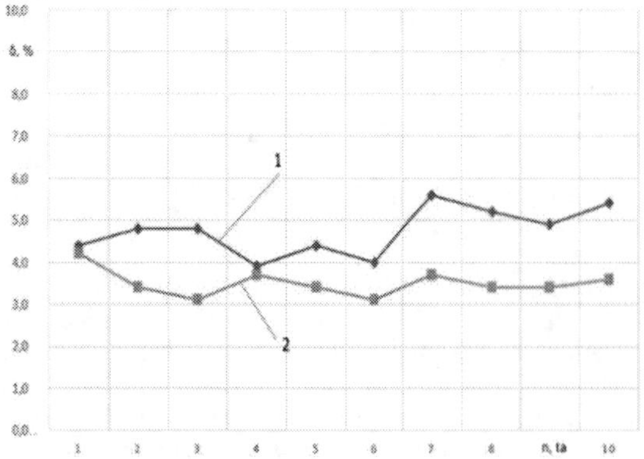

1. Existing method, 2. Proposed method.
Figure 3. Error graph in the selection of device elements

Because in this, a database of element parameters corresponding to the change of reactive power consumption values is formed, it checks the compatibility of each element value with the values of the selected element, and selects the most optimal, small error selection data. allows you to identify device elements.

4 Conclusion

1. When the element is selected through the algorithm, the accuracy of the device increases, along with the avoidance of excessive costs. This has a positive effect on the reliability of devices in existing enterprises.

2. It is possible to develop an algorithm for selecting elements of other electrical devices from this adjustment device.

References

1. Raxmonov I.U., Omonov F.B. Qishloq elektr tarmoqlarida reaktiv quvvatni avtomatik rostlash sxemasi parametrlarini hisoblash dasturi // O'zbekiston Respublikasi intellektual mulk agentligi guvoxnomasi DGU №18078 – Tashkent, – 2022.

2. Rakhmonov I.U., Omonov F.B. Analysis of reactive power consumption in rural power networks and ways to reduce it // Science and Education in Karakalpakstan №2/2. – Karakalpakstan, – 2022. – Pp. 70-74.

3. Usmanov E.G., Abduraimov E.Kh., Karimov R.Ch. Optoelectronic contactless voltage relay // Patent for invention of the Republic of Uzbekistan. IAP No. 05122.

4. Hoshimov, F.A., Rakhmonov, I.U., Niyozov, N.N., Omonov, F.B., Main Regularities of Change in Energy Indicators of Energy Consuming Facilities of Industrial Enterprises // AIP Conference Proceedingsthis link is disabled, 2022, 2552, 030025

5. http://www.studfile.net/preview/6130017/page:3

6. http://www.studopedia.ru/15_126655_raschet-i-vibor-tiristirov.html

7. Allaev K., Toshov J., Toirov O. E3S Web of Conferences 401. 05090. (2023). https://doi.org/10.1051/e3sconf/202340105090

8. Abdi Yonis S., Yusupov Z., Habbal A., Toirov O. Advances in Electrical and Electronic Engineering.Vol 21. No 3 (2023). https://doi.org/10.15598/aeee.v21i3.5149

9. Toirov O., Khalikov S., Sodikjon Khalikov, F. Sharopov. E3S Web of Conferences 410. 05015. (2023). https://doi.org/10.1051/e3sconf/202341005015

10. Jumaeva D., Raximov U., Toirov O., Ergashev O., Abdyrakhimov A. E3S Web of Conferences 425. 04003. (2023). https://doi.org/10.1051/e3sconf/202342504003

11. Toshev J.B., Norkulov M.B., Urazimbetova A.A., Toshniyozov L.G. E3S Web of Conf., Volume 402, 10039 (2023). https://doi.org/10.1051/e3sconf/202340210039

12. K. Allaev, J. Toshov and O. Toirov. E3S Web of Conf., Volume 401, 05090 (2023). DOI https://doi.org/10.1051/e3sconf/202340105090

13. Rakhmonov I.U., Najimova, A.M., Esemuratova Sh.M., Koptileuov T.T. AIP Conference Proceedings 2647. 070024. (2022). https://doi.org/10.1063/5.0104793

14. Rakhmonov I.U., Ganiev Sh.R., Alibekova T.Sh., Nematov L.A. E3S Web of Conferences 384. 01029. (2023). https://doi.org/10.1051/e3sconf/202338401029

15. Niyozov N.N., Rikhsitillaev B.Kh., Jalilova D.A., Kuatbaev P.M. E3S Web of Conferences 384. 01031. (2023). https://doi.org/10.1051/e3sconf/202338401031.
16. Usmanov E.G. E3S Web of Conferences 384. 01034. (2023). https://doi.org/10.1051/e3sconf/202338401034
17. Rasulov A.N., Paxratdinov A.Dj. E3S Web of Conferences 384. 01035. (2023). https://doi.org/10.1051/e3sconf/202338401035
18. Taslimov A.D. E3S Web of Conferences 384. 01036. (2023). https://doi.org/10.1051/e3sconf/202338401036
19. Taslimov A.D., Rakhimov F., Rakhimov Feruz. E3S Web of Conferences 384. 01037. (2023). https://doi.org/10.1051/e3sconf/202338401037
20. Fayziyev M., Tuychiev F., Mustayev R., Ochilov Y. E3S Web of Conferences 384. 01038. (2023). https://doi.org/10.1051/e3sconf/202338401038
21. Saidkhodjaev A.G., Bobojonov Y.M. E3S Web of Conferences 384. 01040. (2023). https://doi.org/10.1051/e3sconf/202338401040
22. Bobojanov M., Mahmutkhonov S. E3S Web of Conferences 384. 01041. (2023). https://doi.org/10.1051/e3sconf/202338401041
23. Rasulov A.N., Ruzinazarov M.R. E3S Web of Conferences 384. 01042. (2023). https://doi.org/10.1051/e3sconf/202338401042
24. Rashidov N., Rozmetov Kh., Rismukhamedov S., Peysenov M. E3S Web of Conferences 384. 01043. (2023). https://doi.org/10.1051/e3sconf/202338401043
25. Tuychiev F, Haqberdiev A. E3S Web of Conferences 384. 01044. (2023). https://doi.org/10.1051/e3sconf/202338401044
26. Bobojanov M., Torayev S. E3S Web of Conferences 384. 01045. (2023). https://doi.org/10.1051/e3sconf/202338401045
27. Reymov K.M., Esemuratova Sh.M., Khusanov B.M., Mytnikov A.V. E3S Web of Conferences 384. 01047. (2023). https://doi.org/10.1051/e3sconf/202338401047
28. Kholikhmatov B.B., Samiev Sh.S., Erejepov M.T., Nematov L.A. E3S Web of Conferences 384. 01032. (2023). https://doi.org/10.1051/e3sconf/202338401032
29. Melikuziev M.V. E3S Web of Conferences 384. 01033. (2023). https://doi.org/10.1051/e3sconf/202338401033
30. Bobojonov Y.M., Saidkhodjaev A.G. E3S Web of Conferences 384. 01048. (2023). https://doi.org/10.1051/e3sconf/202338401048
31. Hoshimov, F.A., Najimova A.M., Rafikova G.R., Bakhadirov I.I. E3S Web of Conferences 384. 01049. (2023). https://doi.org/10.1051/e3sconf/202338401049
32. Ruzinazarov M.R. E3S Web of Conferences 384. 01050. (2023). https://doi.org/10.1051/e3sconf/202338401050
33. Bobojanov M.K., Rismukhamedov D.A., Tuychiev F.N., Shamsutdinov Kh.F. E3S Web of Conferences 365. 04024. (2023). https://doi.org/10.1051/e3sconf/202336504024
34. Ushakov, V.Y., Mytnikov, A.V., Rakhmonov, I.U. Past, Present and Future of Electric Power Systems (Brief Excursion). In: High-Voltage Equipment of Power Systems. Power Systems. Springer, Cham. (2023). https://doi.org/10.1007/978-3-031-38252-9_1
35. Tsypkina, V.V., Ivanova, V.P., Jurayeva, K.K. E3S Web of Conferences, 2023, 401, 03036. (2023). https://doi.org/10.1051/e3sconf/202340103036
36. Tsypkina, V.V., Ivanova, V.P., Isamukhamedov, D.N., Turabekov, A.U., Atamukhamedova, R.F. E3S Web of Conferences, 401, 02045. (2023). https://doi.org/10.1051/e3sconf/202340102045
37. Tsypkina, V.V., Ivanova, V.P. E3S Web of Conferences, 384, 01007. (2023). https://doi.org/10.1051/e3sconf/202338401007
38. Tsypkina, V.V., Ivanova, V.P., Isamukhamedov, D.N., Turabekov, A.U., Hayitmuradova, S.M. AIP Conference Proceedingsthis link is disabled, 2552, 070010. (2022). https://doi.org/10.1063/5.0117074
39. Pirmatov N. B., Madrakhimov D. B., Ivanova V. P., Tsypkina V. V., Ortikova M. J., Atamukhamedova R. F. AIP Conference Proceedingsthis link is disabled, 2552, 070009. (2022). https://doi.org/10.1063/5.0117025
40. Madrakhimov, D.B., Ivanova, V.P., Tsypkina, V.V. E3S Web of Conferences, 216, 01151. (2020). https://doi.org/10.1051/e3sconf/202021601151
41. Ivanova, V.P., Tsypkina, V.V. E3S Web of Conferences, 216, 01152. (2020). https://doi.org/10.1051/e3sconf/202021601152
42. Tsypkina, V., Ivanova, V. E3S Web of Conferences, 139, 01073. (2019). https://doi.org/10.1051/e3sconf/201913901073
43. Ivanova, V., Madrakhimov, D. E3S Web of Conferences, 139, 01057. (2019). https://doi.org/10.1051/e3sconf/201913901057

Energy efficiency of energy-intensive productions and technologies

Dinora Jalilova[1]*, *Aysulu N*ajimova [2], *Khurshida* Khalikova[3], *Shoxida* Akbarova[1]

[1]Tashkent State Technical University named after Islam Karimov, 100095, Uzbekistan, Tashkent, University St. 2A

[2]Karakalpak State University, Nukus, 230101, Uzbekistan

[3]Karshi Institute of Engineering Economics, 225 Independence Avenue, Karshi, Uzbekistan

Abstract. The article discusses the issues of energy efficiency of energy-intensive industries and technologies using the example of electrothermal processes (high-temperature technologies), smelting and high-temperature heating and electric welding production. Recommendations for efficient use of electricity are provided for all industries. A list of improvements in electric welding technology and electric furnaces is also given. The advantages and disadvantages of energy-intensive industries and technologies are also given.

1 Introduction

For most industrial enterprises, recommendations for energy saving are quite universal due to the use of approximately the same energy-consuming equipment and similar energy supply schemes. However, each industrial sector and individual enterprises of even one industry, in addition to the unified one, use specific equipment and technological processes with their own energy saving potentials and methods for their implementation. The issues of energy efficiency of such specific technologies will be discussed below.

2 Experimental research

Electrothermal processes (high temperature technologies)

Electrothermal installations for carrying out high-temperature processes include electric furnaces of various operating principles (electric resistance, arc, induction furnaces; electric heating devices - electric heaters, infrared heaters, electric oil radiators, etc., electric boilers) [1-7].

Electric welding production

Welding is the process of obtaining permanent connections by establishing interatomic bonds between the parts being welded due to their heating or plastic deformation, or the combined action of both. Welding makes it possible to obtain permanent connections of workpieces or parts from almost any metals and alloys of various thicknesses - from hundredths of a millimeter to several meters, of various configurations.

The most common types of welding include fusion welding and pressure welding. Electric welding, the most common type of welding and of greatest interest in terms of power consumption and energy saving, refers to fusion welding, using the thermal energy of an electric arc. The resulting liquid metal of one edge is combined and mixed with the liquid metal of the other edge; this creates a total volume of liquid metal, which, after solidification, forms a weld. More than 150 types and methods of welding processes have been developed and mastered. There is a classification of metal welding according to basic physical, technical and technological characteristics. (GOST 19521-74). Arc welding alone is represented by dozens of varieties, classified according to a number of characteristics: welding with consumable and non-consumable electrode, direct and indirect arc, powered by current of various types and frequencies, open arc, submerged arc, shielding gas, etc. According to the technology, three types are distinguished welding: manual, mechanized (semi-automatic) and automatic [8-15].

When welding with a consumable electrode (steel, copper, aluminum), the weld is formed from the base metal and the metal of the electrode, and with non-consumable electrodes (coal, graphite, tungsten) - only from the base metal and the metal of the filler rod. Metal rods, wire, strips made of the same metal or a similar one in composition are used as filler material. When arc welding, it is necessary to protect the melt from atmospheric gases, since they can penetrate the liquid metal and deteriorate the quality of the weld. To protect the molten zone (the so-called weld pool), various techniques are used: welding with coated electrodes, shielding gas, submerged arc welding, welding with self-shielding flux-cored wire or mixed protection. The following main types of welded joints are distinguished: butt, lap, T, corner and end.

* Corresponding author: iliderl987@yandex.ru

Currently, welding technology allows it to be carried out not only in industrial enterprises, but also in field and installation conditions (in the steppe, in a field, in the open sea, etc.), under water and even in space.

To save energy in electric welding installations, two main groups of measures are performed: technological (savings up to 70-80%) and energy (savings up to 20-30%). The main reduction in specific energy consumption for welding is ensured by the correct choice of welding method and the high technological culture of its implementation.

Today, the following ways to optimize the welding method are used:

- replacement of manual arc welding with automatic one (allows for 5–7% energy savings);
- transition from manual electric welding to semi-automatic welding in a carbon dioxide environment (reduces specific electricity consumption by 2–2.5 times);
- replacement of manual electric arc welding with spot contact welding (reduces specific energy consumption by 2–2.5 times);
- replacement of electric arc welding with seam contact welding (reduces electricity consumption by 15%);
- transfer of manual arc welding from direct current to alternating current (reduces power consumption by 2–3 times).

When resistance welding, spot welding is the most economical, therefore expanding the use of spot welding provides greater energy savings [16-20].

Improvement of electric welding technology includes:

- the use of electrodes coated with iron powder (allows you to increase the welding current, increase productivity and reduce specific energy consumption by 8–12%);
- the use of an additive in the form of metal powder (reduces energy consumption by 30–40%);
- the use of electroslag welding when welding metals of large thickness;
- correct choice of operating mode.

3 Research results

Melting and high temperature heating

The most common type of electrothermal equipment is electric resistance furnaces, in which heat is generated in the heating elements of the furnace when an electric current passes through them.

According to their technological purpose, furnaces are divided into three groups:

- thermal furnaces for thermal and thermochemical processing of various materials (metals, ceramics, plastics, etc.),
- melting furnaces for melting metals and alloys,
- drying ovens for drying paint and enamel coatings, foundry molds, metal-ceramic products, etc.

Resistance furnaces can be intermittent or continuous. The power of furnace power supplies ranges from 10-10000 kW.

The main ways to reduce specific electricity consumption when operating furnaces are:

- reduction of heat losses (energy saving potential reaches 75%),
- productivity increase,
- reduction of losses due to preheating of raw materials,
- automatic control of the optimal operating mode of furnaces,
- heat recovery of heated parts - heat transfer from hot products to cold ones.

Resistance furnaces are very sensitive to voltage reduction, which can lead to an increase in the duration of the technological process, and in some cases to a deterioration in product quality.

An indispensable component of machine-building enterprises is foundry. For melting cast iron, two types of units are mainly used - cupola furnaces using fire methods, and induction and arc furnaces using electric heating.

There has been a steady trend in the industry towards an increase in the share of small induction furnaces (IHF), and among them - medium frequency induction furnaces (MFIF), which have advantages over other types of furnaces in terms of energy efficiency, metal quality, and heating rate of electrically conductive materials. The capacity of the IPSCH usually does not exceed 10 tons, and the operating frequency range is 250-10000 Hz. In induction furnaces, internal heating of the charge occurs by eddy currents and losses due to magnetization reversal of ferromagnets, i.e. heat is released with virtually no losses directly in the metal without heat transfer from the outside. Induction furnaces consume a large amount of electricity, especially in the initial stages of smelting (about 2 times more than in other stages). It can be reduced by 15-20% by preheating scrap to 700-800 0C by burning natural gas. When comparing induction furnaces with cupola furnaces in terms of energy efficiency, it is necessary to take into account losses during the generation of electricity from fuel and its transmission to the consumer [37-42].

The downside of the low inertia of induction furnaces is their low storage capacity and the consumption of significant reactive power. The efficiency of power consumption at enterprises using this technology for conducting electrothermal processes is ensured, first of all, by maintaining high power factor values.

IPSCs are widely used for melting and casting metals in the production of heat-resistant and high-alloy steels, precision alloys, non-ferrous and precious metals.

To increase the efficiency of induction melting, a two-station (sometimes multi-station) melting method is used - melting simultaneously in two furnaces: one of the furnaces is in the melting mode, the second is in the holding or casting mode. The use of this method makes it possible to reduce the specific installed power of electrical equipment by 40% and level out the electricity consumption schedule. Common types of induction furnaces are channel and crucible. The former are characterized by higher energy efficiency, but in a number of indicators they are inferior to crucible ones.

The choice between them is based on a comparison of all factors and circumstances.

When melting high-alloy and precious metals, melting and pouring are carried out in a vacuum or in an environment of inert gases, for which small-capacity furnaces (1-250 kg) and a power of 8-250 kW are used. Furnaces of even smaller volumes (0.1-30 kg) and power (3-30 kW) are used in laboratories for test melts, in the jewelry industry, and for artistic casting [21-25].

Arc steel furnaces operate by using the heat generated in the plasma of an electric arc. They are one of the largest consumers of energy; power can reach 100 MW or more. Enterprises using such furnaces often have to solve the problem of equalizing the daily load schedule. To create a more uniform load schedule, an enterprise must have two (or more) arc furnaces. It is necessary to have a higher harmonic filter at the input of the arc furnace transformer to eliminate the influence of the furnace operation on the quality of electricity in the network [25-30].

Alternating current (ACF) and direct current (ACAF) arc furnaces are inferior to IASF in energy efficiency, Table 1, and even more so in a number of other indicators. The main disadvantages of an arc furnace are:

- intense emissions of dust and harmful gases, the capture of which requires additional costs for a gas cleaning device;
- rapid consumption of graphite electrode material, burnout (waste) of metal and alloying elements;
- reduction in the quality of electricity in supply networks due to extremely uneven load in the operating cycle and during operational short circuits.

Table 1. Relative cost of 1 ton of cast iron for various smelting units [31-45].

Table 1

Type of planned unit	Relative cost 1 t cast iron	Energy consumption
GA	1,0	120–170 m^3/t natural gas
ASF	2,0	600–180 kW·h/t
DCAF	1,8	490–580 kW·h/t
IFIF	2,0	1200–1300 kW·h/t
MFIF	1,4	520–560 kW·h/t

4 Conclusion

In all types of electric furnaces, efficient use of electricity requires:

1. Maintaining the furnace lining in good condition, using effective thermal insulation for it.
2. Reduction of furnace downtime.
3. Use of furnaces in three shifts.
4. Optimal loading of furnaces.
5. Reduction of time spent in the oven with an open roof.

References

1. Rashidov N., Rozmetov Kh., Rismukhamedov S., Peysenov M. E3S Web of Conferences 384. 01043. (2023). https://doi.org/10.1051/e3sconf/202338401043
2. Rasulov A.N., Paxratdinov A.Dj. E3S Web of Conferences 384. 01035. (2023). https://doi.org/10.1051/e3sconf/202338401035
3. Ushakov, V.Y., Mytnikov, A.V., Rakhmonov, I.U. Dielectric Spectroscopy Technology. In: High-Voltage Equipment of Power Systems. Power Systems. Springer, Cham. (2023). https://doi.org/10.1007/978-3-031-38252-9_9
4. Hoshimov, F.A., Najimova A.M., Rafikova G.R., Bakhadirov I.I. E3S Web of Conferences 384. 01049. (2023). https://doi.org/10.1051/e3sconf/202338401049
5. Ushakov, V.Y., Mytnikov, A.V., Rakhmonov, I.U. Traditional Electrical Diagnostic Methods. In: High-Voltage Equipment of Power Systems. Power Systems. Springer, Cham. (2023). https://doi.org/10.1007/978-3-031-38252-9_7
6. Taslimov A.D. E3S Web of Conferences 384. 01036. (2023). https://doi.org/10.1051/e3sconf/202338401036
7. Rasulov A.N., Ruzinazarov M.R. E3S Web of Conferences 384. 01042. (2023). https://doi.org/10.1051/e3sconf/202338401042
8. Ushakov, V.Y., Mytnikov, A.V., Rakhmonov, I.U. Non-electrical Diagnostic Methods. In: High-Voltage Equipment of Power Systems. Power Systems. Springer, Cham. (2023). https://doi.org/10.1007/978-3-031-38252-9_6
9. Melikuziev M.V. E3S Web of Conferences 384. 01033. (2023). https://doi.org/10.1051/e3sconf/202338401033
10. Kholikhmatov B.B., Samiev Sh.S., Erejepov M.T., Nematov L.A. E3S Web of Conferences 384. 01032. (2023). https://doi.org/10.1051/e3sconf/202338401032
11. Ushakov, V.Y., Mytnikov, A.V., Rakhmonov, I.U. Effects on Equipment Causing Insulation Aging and Failure. In: High-Voltage Equipment of Power Systems. Power Systems. Springer, Cham. (2023). https://doi.org/10.1007/978-3-031-38252-9_4
12. Ushakov, V.Y., Mytnikov, A.V., Rakhmonov, I.U. Diagnostics of Insulating Structures of Overhead Power Lines and Outdoor Substations. In: High-Voltage Equipment of Power Systems. Power Systems. Springer, Cham. (2023). https://doi.org/10.1007/978-3-031-38252-9_12
13. Pirmatov N., Tosheva Sh., Toshev Sh. E3S Web of Conferences 139. 01027. (2019) https://doi.org/10.1051/e3sconf/201913901027
14. Usmanov E.G. E3S Web of Conferences 384. 01034. (2023). https://doi.org/10.1051/e3sconf/202338401034
15. Ushakov, V.Y., Mytnikov, A.V., Rakhmonov, I.U. Diagnostics of High-Voltage Equipment by Defectography Methods. In: High-Voltage Equipment of Power Systems. Power Systems. Springer, Cham. (2023). https://doi.org/10.1007/978-3-031-38252-9_8

16. Bobojanov M.K., Rismukhamedov D.A., Tuychiev F.N., Shamsutdinov Kh.F. E3S Web of Conferences 365. 04024. (2023). https://doi.org/10.1051/e3sconf/202336504024

17. Reymov K.M., Esemuratova Sh.M., Khusanov B.M., Mytnikov A.V. E3S Web of Conferences 384. 01047. (2023). https://doi.org/10.1051/e3sconf/202338401047

18. Ushakov, V.Y., Mytnikov, A.V., Rakhmonov, I.U. Diagnostics of High-Voltage Cable Lines. In: High-Voltage Equipment of Power Systems. Power Systems. Springer, Cham. (2023). https://doi.org/10.1007/978-3-031-38252-9_11

19. Ruzinazarov M.R. E3S Web of Conferences 384. **01050.** (2023). https://doi.org/10.1051/e3sconf/202338401050

20. Saidkhodjaev A.G., Bobojonov Y.M. E3S Web of Conferences 384. 01040. (2023). https://doi.org/10.1051/e3sconf/202338401040

21. Sattarov Kh., Sapayev M., Suyarov A., Turaev, A. E3S Web of Conferences, 401, **04001.** (2023). https://doi.org/10.1051/e3sconf/202340104001

22. Norkhojaeva N.N., Pirmatov N.B., Kamalov N.K. E3S Web of Conferences 402. **10041.** (2023). https://doi.org/10.1051/e3sconf/202340210041

23. Yaxyayeva Sh., Pirmatov N., Egamov A., Ulmas Berdiyorov and Isakov F. E3S Web of Conferences 402. **05016.** (2023) https://doi.org/10.1051/e3sconf/202340205016

24. Jurayeva K., Sattarov Kh. E3S Web of Conferences, 389, **01060.** (2023). https://doi.org/10.1051/e3sconf/202338901060

25. Sattarov Kh., Safarov A., Kamolov A., Egamberdiev M., Raxmanova G. E3S Web of Conferences, 383, **01028.** (2023) https://doi.org/10.1051/e3sconf/202338301028

26. Pirmatov N.B., Berdiyev U.T., Usmonov K.K., Nazirkhanov T.M., Berdiyorov U.N. E3S Web of Conferences 401. **03021.** (2023). https://doi.org/10.1051/e3sconf/202340103021

27. Pirmatov N.B., Egamov A.M., Giyasov C.M., ...Ergashov Sh.O., Nizamov J.A. E3S Web of Conferences 401. **03056.** (2023). https://doi.org/10.1051/e3sconf/202340103056

28. Siddikov I.Kh., Sattarov Kh.A., Abdumalikov A.A. AIP Conference Proceedings, 2612, **050002.** (2023). https://doi.org/10.1063/5.0115211

29. Safarov A.M., Sattarov Kh.A., Mustafoqulov A.M., Shoyimov Y.Y. Journal of Physics: Conference Series, 2176(1), **012006.** (2022). doi:10.1088/1742-6596/2176/1/012006

30. Pirmatov N., Toirov O., Bekishev A., ...Zaynieva O., Norkulov U. E3S Web of Conferences 2552. **040020.** (2023) https://doi.org/10.1063/5.0115727

31. Pirmatov N., Muminov M., Akberdiev M., Abdullayev J. AIP Conference Proceedings 2432. 0200200. (2022). https://doi.org/10.1063/5.0089838

32. Safarov A., Sattarov Kh., Bazarov M., Mustafoqulov A. E3S Web of Conferences, 264, **05038.** (2021) https://doi.org/10.1051/e3sconf/202126405038

33. Safarov A.M., Jumaboyev S.K., Sattarov Kh.A. IOP Conference Series: Materials Science and Engineering, 734(1), **012196.** (2020). doi: 10.1088/1757-899X/734/1/012196

34. Pirmatov N., Bekishev A., Shernazarov S., Kurbanov N., Norkulov U. E3S Web of Conferences. 264. **04028.** (2021). https://doi.org/10.1051/e3sconf/202126404028

35. Baratov R., Pirmatov N., Panoev A., Ruziyev S., Mustafoqulov, A. IOP Conference 1030. **012161.** (2021). doi:10.1088/1757-899X/1030/1/012161

36. Pirmatov N., Panoev A. E3S Web of Conferences, 216. **01120.** (2020). https://doi.org/10.1051/e3sconf/202021601120

37. Tsypkina, V.V., Ivanova, V.P., Jurayeva, K.K. E3S Web of Conferences, 2023, 401, 03036. (2023). https://doi.org/10.1051/e3sconf/202340103036

38. Tsypkina, V.V., Ivanova, V.P., Isamukhamedov, D.N., Turabekov, A.U., Atamukhamedova, R.F. E3S Web of Conferences, 401, 02045. (2023). https://doi.org/10.1051/e3sconf/202340102045

39. Tsypkina, V.V., Ivanova, V.P. E3S Web of Conferences, 384, 01007. (2023). https://doi.org/10.1051/e3sconf/202338401007

40. Tsypkina, V.V., Ivanova, V.P., Isamukhamedov, D.N., Turabekov, A.U., Hayitmuradova, S.M. AIP Conference Proceedingsthis link is disabled, 2552, 070010. (2022). https://doi.org/10.1063/5.0117074

41. Pirmatov N. B., Madrakhimov D. B., Ivanova V. P., Tsypkina V. V., Ortikova M. J., Atamukhamedova R. F. AIP Conference Proceedingsthis link is disabled, 2552, 070009. (2022). https://doi.org/10.1063/5.0117025

42. Madrakhimov, D.B., Ivanova, V.P., Tsypkina, V.V. E3S Web of Conferences, 216, 01151. (2020). https://doi.org/10.1051/e3sconf/202021601151

43. Ivanova, V.P., Tsypkina, V.V. E3S Web of Conferences, 216, 01152. (2020). https://doi.org/10.1051/e3sconf/202021601152

44. Tsypkina, V., Ivanova, V. E3S Web of Conferences, 139, 01073. (2019). https://doi.org/10.1051/e3sconf/201913901073

45. Ivanova, V., Madrakhimov, D. E3S Web of Conferences, 139, 01057. (2019). https://doi.org/10.1051/e3sconf/201913901057

Wear resistance analysis of HSS drill bits P6M5 and special round carbide drill bits VK8 for drilling alloyed steels and alloys

Muhammadi Turonov[1]*, *Tolibjon* Umarov[1], *Sultonmurod* Normatov[1], *Davron* Bekturov[2]

[1]Tashkent state technical university, Tashkent, Uzbekistan
[2]Almalyk branch of Tashkent state technical university, Almalyk, Uzbekistan

Abstract. This paper analyses tool life and accuracy of made holes. As a result of experiments, the revealed variation coefficients reach high values - k=0.61, which indicates significant instability of the machining process of holes. In order to verify this assumption, a series of test measurements of the geometry of the gauge plugs used in the workshop were taken.

1 Introduction

The proportion of basic tool materials in the world of machine building and metal processing is as follows: high-speed steels - 66%, hard alloys - 32%, and cutting ceramics - 2%. If we consider the distribution of these tool materials by swarf volume, hard alloys account for 68%, high-speed steels for, 28% and cutting ceramics for 4%.

The unique properties of tungsten carbide – it's high wear and heat resistance - are unfortunately combined with the low processability of the material itself, which imposes significant limitations on the shape and type of tools produced. Drills designed for machining holes in solid material belong to the group of tools where the optimum geometric parameters as well as the kinematics and dynamics of the cutting process are extremely unfavorable for carbide due to its low brittle strength.

Previously, research has been carried out to identify new principles for the design of drilling tools suitable for machining holes in extreme conditions: reduced rigidity of technological settings and difficult machinability of materials. For the laboratory-industrial tests, we select the design of a carbide drill bit by taking into account three criteria:

Durability (the durability of carbide drills is usually much higher than that of high-speed tools). Performance (usually depending on a combination of cutting speed and feed rate). Parameters of the bore to be machined (form error and dimensional deviation from the reference gauge). The wide variation in resistance values, which is characteristic of all resistance tests, makes it difficult to analyze comparative tests when drilling in different machinability materials.

In this article, a graphical analysis of the tool life and accuracy of the drilled holes was carried out. The durability of each drill bit or the parameters of the machined holes are characterized by a vertical bar whose location level in the diagram corresponds to the number of drilled holes. This way of plotting the diagram simultaneously gives an indication of the relative workability of the drills and the scatter of the results of the tests.

2 Experimental research

In practice, in order to make resharpening easier, care is taken not to break the drill bit completely. The drilling process is terminated by a characteristic sound and a change in the shape and color of the chips. In this case, an assessment of durability with a probability of fracture of less than 10% is required. In production, attempts to use other criteria, e.g. the amount of wear on the rear surface or an increase in torque, to regulate drill replacement times, and give a weak correction with drill breakage.

Hole machining quality analysis and evaluation of the durability of feather carbide drills were carried out during the drilling of holes in parts of the "lattice" type. Such a construction of technological operation allowed us to determine with maximum reliability the appropriateness of the selected tool design for the conditions of MHS when machining different steels and alloys. A wide range of steels was selected for experimental studies, including carbon steels and stainless heat-resistant steels.

A comparative analysis of the performance of carbide drills is carried out in relation to a fast-cutting helical drill bit. Analyses of 19 drills in the machining of grids in different shifts are carried out [1].

Data on the wear of drills and a number of drilled holes are shown in Figure 1, 2, According to the results, it is seen the hat durability of different drills with the same processing conditions vary by 30...40 times. The criterion is shop durability, i.e. the moment, when the

* Corresponding author: muhammadituronov434@gmail.com

 https://doi.org/10.1051/e3sconf/202346101096

drill bit loses its ability to cut and its further operation could cause breakage or jamming in the hole, is taken as a criterion. The machining process is extremely unstable. What draws attention is the discrepancy in the operating time of the drill bit in relation to the maximum amount of wear on the back surface. For drill 14 and 19, for example, the wear is 1.05 mm and 0.6 mm respectively, and almost the same number of holes are drilled. If we compare drills No. 1 and No. 8, with nearly identical wear values of 0.6 and 0.55 mm, the drill bit has drilled twice as many holes (100 holes) as drill bit No. 8 (50 holes). In most cases, drills are resharpened after achieving insignificant wear of 0.05...0.3 mm, having drilled from 3 to 25 holes, variation coefficients reach high values - (K_{var} = 0.61, which indicates significant instability of hole machining process, i.e. technological adjustment is not productive enough. The average wear of 19 drills equals 0.38 mm with limit values of 0.005 ... 1.05 mm. Accordingly, the average number of drilled holes is 70 with maximum values of 130 holes. Having such high values of the variation coefficients (K_{var} = 0,61) it is very difficult to manage the rhythm of production. The presence of such a variation shows that, under given conditions, high-speed steel works at the limit of its cutting ability ,and a slight error in tool sharpening leads to a sharp decrease in its durability. By improving the thermo-physical conditions of the contact processes by more intensive cooling it is possible to increase the lower limit of the tool life and thereby the average tool life. Thus, the task is not so much to increase the tool life, but to average it during work with all drills irrespective of qthe ualification of the worker. Indeed, if we take into account that the application of any method, increasing durability of high-speed cutter in real production conditions may yield maximum effect in 2 times, the increase, for instance, drill bit Ø16 (25 outlets) in any case does not give positive result as compared to drill bit Ø19 (130 outlets), and the ratio becomes 5 : 130. However, if the variation coefficients are reduced by a factor of 2 to 3, the average durability increases, and the productivity per shift increases accordingly.

3 Research results

The use of feather carbide drills improves the situation noticeably. The increase in the average tool life is 1.2...1.7 times (Fig. 1, 2, 3). The variation coefficients decrease considerably up to 0,1...0,3 (tab. 1). Comparison of the data of laboratory and in-house researches allows us to state that a smaller increase of the absolute durability of high-speed drills (by 40...70 %) in production conditions is caused, first of all, by the use of weakly concentrated water emulsions for tool cooling. This does not allow to effectively manage the thermal-physical situation in the cutting zone due to the limiting loads on cutting edges.

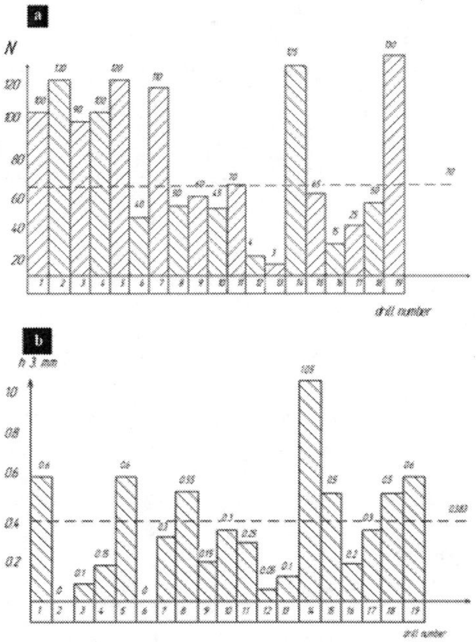

Fig. 1. Hardness (a) and rear surface wear diagrams (b) of P6M5 drills when cutting 08X21116M2T steel Drill Ø24,5, 2M57, RF 31/B, emulsion cooling ET - 2, n=1,25…140 rpm, S = 0,5…0,63 mm/rev.

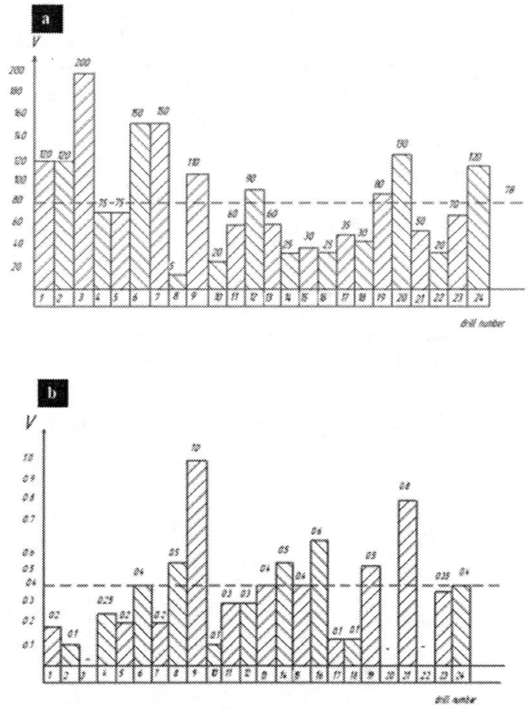

Fig. 2. Hardness (a) and rear surface wear diagrams (b) of P6M5 drills when cutting steel 12X181110T Drill Ø24,5, 2M57, RF 31/B, emulsion cooling ET - 2, n=1,25…140 rpm, S = 0,5…0,63 mm/rev.

 https://doi.org/10.1051/e3sconf/202346101096

Table 1. Values of variation coefficients of the durability of drills with different methods of tool cooling.

Grating material	Drill lot size, pcs	Hole diameter	P6M5		BK6		V, m/s	S, mm/ay	B, mm	Tool life increase factor
			Resistance N_(otv avg) $N_{otv\,sred}$	K_{var}	Resistance N_ (otv avg) $N_{otv\,sred}$	K_{var}				
16GS	10	38.7	56	0,45	101	0,14	0,14	0,6	100	1,8
16GS	6	25.5	116	0,43	139	0,24	0,32	0,61	60	1,2
12X18H10T	10	38.7	30	0,38	52	0,28	0,14	0,6	60	1,7
12X18H10T	10	38.7	55	0,61	82	0,29	0,18	0,5	72	1,5
12X18H10T	6	25.5	84	0,57	155	0,22	0,16	0,61	60	1,8
12X18H10T	6	20.5	96	0,48	125	0,21	0,16	0,5	72	1,3
12X21H6M2T	10	20.5	143	0,2	183	0,1	0,15	0,5	72	1,3
12X21H6M2T	6	20.5	28	0,58	40	0,25	0,15	0,61	52	1,4
08X22H6T	6	25.5	69	0,39	78	0,25	0,1	0,5	80	1,1
10X17HBM2T	7	20.5	90	0,55	111	0,24	0,15	0,5	60	1,2
10X17HBM2T	10	20.5	125	0,52	180	0,12	0,17	0,5	50	1,4

The values of the variation coefficients are reduced to 0.1...0.3. At an average increase of durability of feather drills by 1,2...1,7 times the stability of processing increases by 2...4 times, This is important, because if such increase of durability of drills is not especially noticeable for production, because there is no acute deficit in tools, then increase of production rhythm within a month allows to solve the quality issues of manufacturing more operatively [2].

The use of feather carbide drills leads to a change in contact processes resulting in a change in wear patterns. Above all, this is the absence of chipping. During all tests, no tool breaks were observed and the wear is thermal in nature and characterized by blunting of the serrated edges without any pronounced chipping, which is observed even during operation in a rigid processing set-up.

The analysis of the accuracy of the machining of holes with carbide drills was assessed by the number of good holes after drilling on the gauge [3].

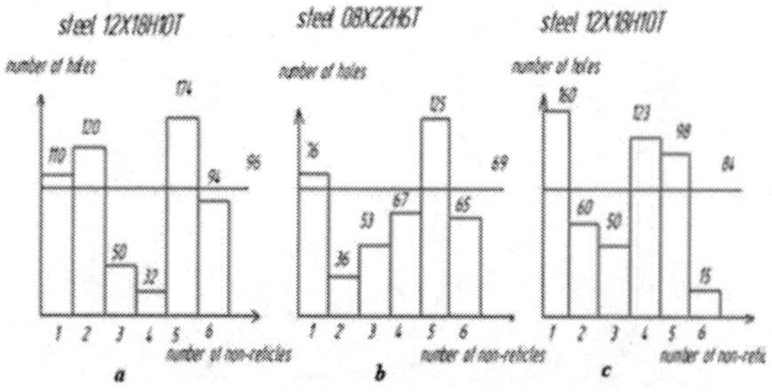

Fig. 3. Drill resistance diagrams without guide elements: a) Drill bit Ø25,5, BK6. V-0,16 m/s, T-72 mm. S-0,5 mm/rev, coolant ET 2-5%; b) Drill bit Ø25,5, BK6. V-0,1m/s, T-80 mm. S-0,5 mm/rev, coolant ET 2-5%; c) Drill bit Ø38,7, BK6. V-0,16m/s, T-60 mm. S-0,61 mm/rev, coolant ET 2-5%

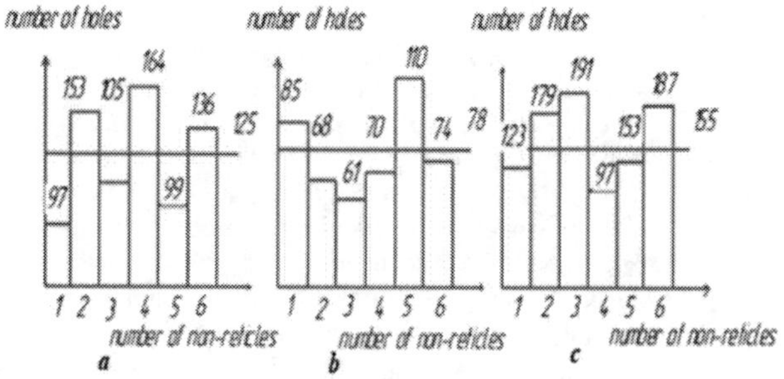

Fig. 4. Drill resistance diagrams with guide elements: a) Drill bit Ø25,5, BK6. V-0,16 m/s, T-72 mm. S-0,5 mm/rev, coolant ET 2-5%; b) Drill bit Ø25,5, BK6. V-0,1m/s, T-80 mm. S-0,5 mm/rev, coolant ET 2-5%; c) Drill bit Ø38,7, BK6. V-0,16m/s, T-60 mm. S-0,61 mm/rev, coolant ET 2-5%.

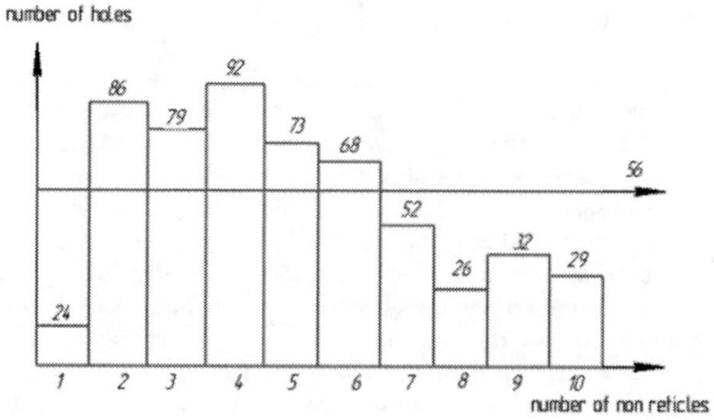

a)

 https://doi.org/10.1051/e3sconf/202346101096

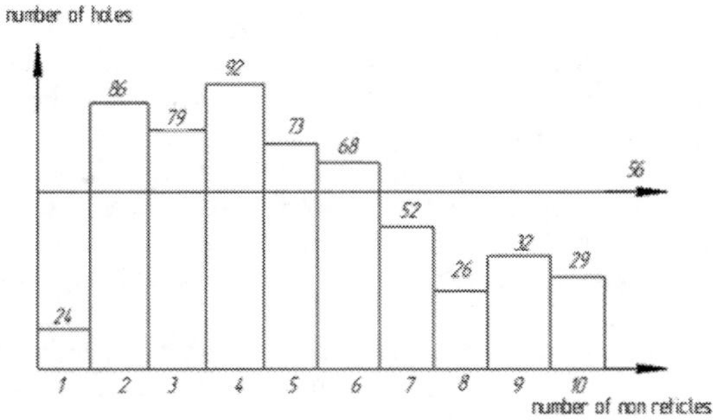

b)

Fig. 5. Drill resistance diagrams without guide elements: a) Drill resistance diagrams with guide elements: a) Drill bit Ø38,7, BK6. V-0,16m/s, T-60 mm. S-0,61 mm/rev, coolant ET 2-5%; b) Drill bit Ø20,5, BK6. V-0,17m/s, T-50 mm. S-0,5 mm/rev, coolant ET 2-5%.

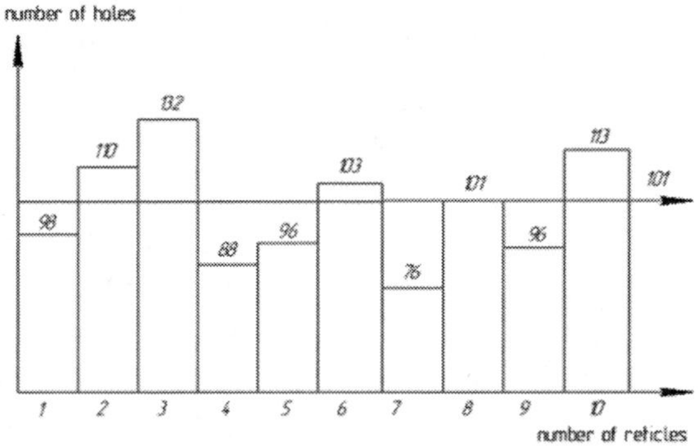

a)

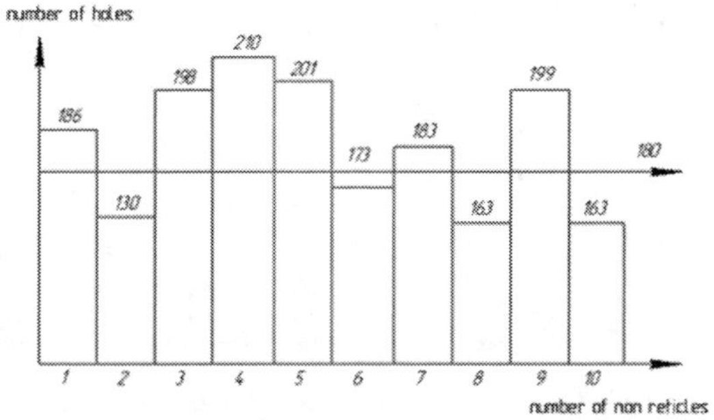

b)

Fig. 6. Drill resistance diagrams with guide elements: a) Drill bit Ø38,7, BK6. V-0,16m/s, T-60 mm. S-0,61 mm/rev, coolant ET 2-5%; b) Drill bit Ø20,5, BK6. V-0,17m/s, T-50 mm. S-0,5 mm/rev, coolant ET 2-5%.

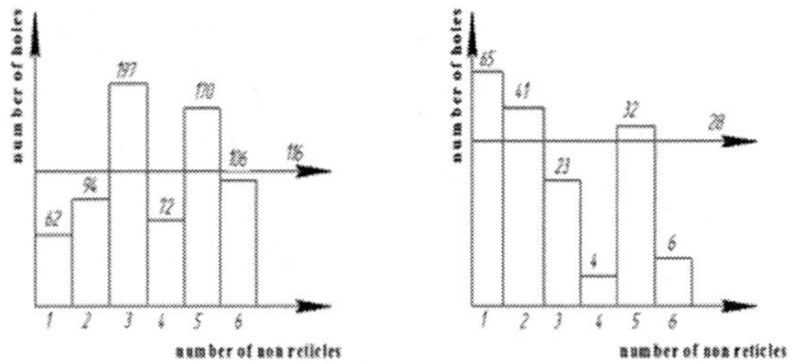

a) **b)**

Fig. 7. Drill resistance diagrams without guide elements: a) Drill bit Ø25,5, BK6. V-0,32 m/s, T-60 mm. S-0,61 mm/rev, coolant ET 2-5%; b) Drill bit Ø20,5, BK6. V-0,15m/s, T-52 mm. S-0,61 mm/rev, coolant ET 2-5%.

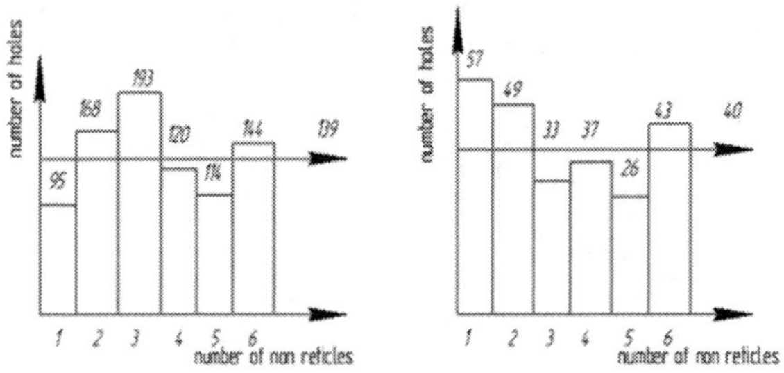

a) **b)**

Fig. 8. Drill resistance diagrams with guide elements: a) Drill bit Ø25,5, BK6. V-0,32 m/s, T-60 mm. S-0,61 mm/rev, coolant ET 2-5%; b) Drill bit Ø20,5, BK6. V-0,15m/s, T-52 mm. S-0,61 mm/rev, coolant ET 2-5%.

To find out the effect of the drilling process on the subsequent finishing, the accuracy of the hole was measured after countersinking the hole with high-speed countersinks. The use of countersinks was not part of the objective and the technique was used for methodical purposes. It is necessary to find out on how much the applied technological adjustment provides receiving of the sizes in the set tolerance field. This is due to the fact that in practice holes are produced both by drilling and countersinking as well as by direct drilling to a given size [4-8].

A batch of grids was examined to determine the number of defective holes and the incidence of defects after machining. The holes Ø24.5 obtained after drilling and Ø25.5 obtained after countersinking or direct drilling to the final size were monitored. All holes were controlled with a gauge - plug Ø25.4÷0.23. This gauge is used in the workshop to this day. The results are shown in Figs. 3-8.

The data show that after pre-drilling there are a number of holes with unsatisfactory geometric parameters for countersinking. This situation exists in the machining of all the materials studied over the entire range of cutting modes applied. Practically all grids have 2...5 bores for a countersinking plug, from 2 to 6 bores where the passing side of the plug passes approximately half the length of the hole. If we discard a few experiments and consider their results to be atypical, it can be stated that there are no knowingly loose (defective) holes after the drilling operation. This statement, however, applies to the cork measuring method. In reality, only the smallest diameter of the hole can be measured with a plug, and of course the measurement result does not provide a complete answer as to whether the geometry of the hole is as specified or not.

As a result of the inaccuracy of the drilling operations, the subsequent countersinking of the pipe grids results in defects. In general, there are between 1 and 10 holes with a non-passable side of the plug and about the same number of holes with a non-passable side of half of them. In relation to the total number of holes, the defect rate is less than percent, and the machining operation may be considered satisfactory if no assembly work is required in relation to the precise orientation of the parts in relation to each other.

Table 2. Values for the diametric dimensions of the holes in the tube grilles. Material – 08X21116M2T. Grid diameter - 840 Height B=24. Hole diameter 25.5÷0.13. Number of holes -511.

Hole count	Hole number	Hole diameter, mm		
The middle of the beginning of the holes	1	25.60	25.63	25.615
	2	25.58	25.59	25.585
	3	25.54	25.60	25.57
	4	25.52	25.60	25.56
	5	25.62	25.56	25.59
	6	25.58	25.58	25.58
	7	25.60	25.67	25.625
	8	25.65	25.67	25.66
	9	25.62	25.61	25.615
	10	25.57	25.61	25.585
	1	25.60	25.70	25.65
	2	25.68	25.67	25.775
	3	25.68	25.57	25.575
	4	25.66	25.62	25.59
	5	25.55	25.48	25.519
End middle of holes	6	25.59	25.54	25.565
	7	25.50	25.61	25.555
	8	25.61	25.62	25.615
	9	25.54	25.61	25.575
	10	25.62	25.62	25.62
	1	25.65	25.60	25.625
	2	25.64	25.54	25.59
	3	25.58	25.51	25.545
	4	25.58	25.56	25.57
	5	25.57	25.59	25.58
	6	25.55	25.53	25.54
	7	25.71	25.70	25.704
	8	25.73	25.61	25.67
	9	25.63	25.53	25.58
	10	25.66	25.54	25.60

In order to quantify the diametric dimensions of the holes in the tube sheets, they were measured with an indicator gauge directly at the apparatus assembly site. The results are shown in tables 2 and table 3.

The situation worsens considerably if holes are drilled directly to Ø25.4÷0.13, without subsequent countersinking. In this case the number of slack holes increases and holes with dimensions smaller than the nominal ones appear. So practically all grids are made of various materials, after processing with a drill 25,5 there are 3...10 holes, at which through sides of a plug reach to the middle, to the bottom, or at all pass only the top part. The diminution in dimensions from the machined ones is a result of a false heat-stressed cutting mode with intensive cooling, which results in shrinkage of the lattice material.

Table 3. Values for length-averaged hole diameters in tube grids. Material – 12X18H10T. B=24. Number of holes -121.

№	Hole diameter, mm			№	Hole diameter, mm		
1	2	3	4	5	6	7	8
1	25.60	25.63	25.615	26	25.54	25.60	25.67
2	25.58	25.59	25.585	27	25.63	25.53	25.58
3	25.54	25.60	25.570	28	25.66	25.54	25.60
4	25.52	25.60	25.56	29	25.55	25.57	25.56
5	25.62	25.56	25.59	30	25.54	25.60	25.57
6	25.58	25.58	25.58	31	25.61	25.57	25.59
7	25.60	25.65	25.62	32	25.71	25.70	25.70
8	25.65	25.67	25.66	33	25.73	25.61	25.67
9	25.60	25.70	25.65	34	25.55	25.56	25.55
10	25.68	25.67	25.67	35	25.69	25.59	25.64
11	25.58	25.57	25.57	36	25.58	25.56	25.57
12	25.56	25.57	25.59	37	25.56	25.64	25.60
13	25.55	25.58	25.51	38	25.67	25.60	25.63
14	25.59	25.54	25.56	39	25.64	25.59	25.61
15	25.50	25.61	25.55	40	25.51	25.63	25.57
16	25.61	25.62	25.61	41	25.65	25.50	25.62
17	25.54	25.61	25.57	42	25.55	25.58	25.56
18	25.62	25.62	25.62	43	25.53	25.55	25.54
19	25.62	25.62	25.62	44	25.58	25.58	25.58
20	25.65	25.59	25.62	45	25.55	25.58	25.56
21	25.64	25.54	25.59	46	25.55	25.64	25.59
22	25.58	25.51	25.54	47	25.57	25.63	25.60
23	25.58	25.54	25.57	48	25.67	25.55	25.61
24	25.57	25.59	25.58	49	25.59	25.63	25.62
25	25.55	25.53	25.54	50	25.64	25.59	25.61

The data shows that the geometric dimensions along the length of the hole vary. There are holes with dimensions beyond the upper deviation of the tolerance field. The difference is a small 0.05...0.08 mm and it can be said that there is no defect, however, all dimensions do not lie in the middle of the tolerance field; as usually happens in well-functioning process adjustments, but are clustered around the upper value of the limit deviation. This means that there is hardly any room for improvement in accuracy under these machining conditions. On the other hand, such a systematic grouping of hole sizes around the upper deviation for all manufactured grids naturally leads to the assumption that the caliber tolerance scheme is intentionally designed so that the measurements are obtained as tolerance limits. To clarify this assumption was made a series of control measurements of geometric parameters caliber - plugs used in the shop. The results are shown in table 4.

Table 4. Geometric dimensions of plug gauges for hole control Ø25.5 ÷ 0.13, used at the Uzbekhimmash plant.

№	№ traffic jams	Nominal plug size, mm	Passing side size, mm	Dead side size, mm
1	5157	$25,4^{+0.23}$	25.412	25.636
2	5158	$25,4^{+0.23}$	25.415	25.632
3	5154	$25,4^{+0.23}$	25.420	25.635
4	5156	$25,4^{+0.23}$	25.425	25.636
5	5153	$25,4^{+0.23}$	25.412	25.633
6	5057	$25,4^{+0.23}$	25.412	25.635
7	5050	$25,4^{+0.23}$	25.420	25.632
8	5055	$25,4^{+0.23}$	25.413	25.633
9	5052	$25,4^{+0.23}$	25.424	25.535
10	5051	$25,4^{+0.23}$	25.420	25.632
11	5053	$25,4^{+0.23}$	25.423	25.632
12	5152	$25,4^{+0.23}$	25.414	25.630
13	5154	$25,4^{+0.23}$	25.419	25.632
14	5153	$25,4^{+0.23}$	25.412	25.636
15	5159	$25,4^{+0.23}$	25.400	25.620
16	5441	$25,4^{+0.23}$	25.440	25.633
17	5161	$25,4^{+0.23}$	25.420	25.630
18	5153	$25,4^{+0.23}$	25.425	25.630
19	5162	$25,4^{+0.23}$	25.415	25.633
20	5054	$25,4^{+0.23}$	25.413	25.632

The low value of the dimensions of the bore of 25,412…25,440 mm draws attention; i.e. the dimensions are below the nominal bore size of 25,50.

The recommended values of plug dimensions according to GOST 21401 — 75 and ST SEV -157 — 75 to control the size 25,5 H11 (+0,13) are /23/: PR — $25,5235^{0,009}$, NOT - $25,6345^{-0,009}$ limit size of worn gauge — PR_{izn} 25,5000. It follows from this that the bore side used in checking the bore of the plug does not operate in the tolerance field of the bore and cannot influence the change in the dimensional scatter field [5, 9, 10].

4 Conclusion

1.A detailed routing technology has been developed for the manufacture of a special-designed carbide-tipped drill bit. Processing modes and time norms for serial production of tools are determined.

2.An experimental batch of tools was made in the laboratory conditions of the Department of Mechanical Engineering Technology and carried out when drilling carbon and stainless steels.

3.Tests of special drills were carried out in industrial volumes of drilling holes. It has been established that the stability of the technological operation is ensured by the coefficients of resistance variation in the range of 0.1 ... 0.3. With an average increase in the durability of special drills compared to standard drills by 1.2 ... 1.7 times, the stability (reliability) of processing increases by 2 ... 4 times.

References

1. Larina, M.N.: Visokoproizvoditelnie konstruksii sverl i ix ratsionalnaya ekspluatatsiya [High-Performance Drill Designs and Smart Operation]. 1st part, Russia (1960).
2. Xolmogorsev, Yu.P.: Optimizatsiya protsessa obrabotki otverstiy [Hole Process Optimization]. Mashinostroenie, Moscow (1984).
3. Troiskiy, N.D.: K voprosu o sverlenii [On the subject of drilling]. Mashinostroenie, Moscow (1971).
4. Vinogradov, A.A.: Fizicheskie osnovi protsessa sverleniya trudnoobrabativaemix metallov tverdosplavnimi sverlami [Physical basis of the process of drilling hard-to-cut metals with carbide drills]. Naukova Dumka, Kiev (1985).
5. Umarov, T.U.: Sposob sverleniya skvoznix otverstiy [Through Hole Drilling Method]. Authors copyright, Patent № 1832068 (1991).
6. Chernishov, M.O.: Analiz osnovnix vidov razrusheniya rejuщix tverdosplavnix elementov sbornix sverl [Analysis of the main types of destruction of cutting carbide elements of prefabricated drills]. In: Sbornik nauchnix trudov II Mejdunarodnoy konferensii "Resursoeffektivnie sistemi v upravlenii i kontrole: vzglyad v budushee" 4(1), Tomsk (2013).
7. Umidjon, M., Jeltukhin, A., Meliboyev, Y., & Azamat, B.: Effect of Magnetized Cutting Fluids on Metal Cutting Process. In: International Conference on Reliable Systems Engineering (ICoRSE)-2022, pp. 95-104. Cham: Springer International Publishing (2022). https://doi.org/10.1007/978-3-031-15944-2_9.
8. Erkin, U., Umidjon, M., & Umida, S.: Application of Magnetic Field on Lubricating Cooling Technological Condition in Metal Cutting Process. In: International Conference on Reliable Systems Engineering, pp. 100-106. Springer, Cham (2021).
9. Roukema, J.C.: Mechanics and Dynamics of Drilling. PhD [dissertation]. The University of British Columbia (2006).
10. Ivchenko, T.G., Petryaeva, I.A., Mixaylov, A.N.: Otsenka effektivnosti ispolzovaniya rejushix instrumentov s iznosostoykimi pokritiyami [Evaluation of the effectiveness of the use of cutting tools with wear-resistant coatings]. Izvestiya Tulskogo gosudarstvennogo universiteta. Texnicheskie nauki 8(2), 190-198 (2017).

Development and analysis of logical scenario design in virtual reality laboratories for higher education institutions

Dinora Jalilova[1]*, *Raykhona* Kurbonova[1], *Madiyar* Erejepov[2], *Laziz* Nematov[3]

[1]Tashkent State Technical University named after Islam Karimov, 100095, Uzbekistan, Tashkent, University St. 2A

[2]Karakalpak State University, Nukus, 230101, Uzbekistan

[3]Bukhara Engineering Technological Institute, Bukhara, 200100, Uzbekistan

Abstract. Virtual Reality (VR) has permeated numerous industries, revolutionized traditional methods and introduced innovative ways to immerse users in digital environments. One of the cornerstones of VR design is the development of logic scenarios, which establish how a VR application responds to user interactions. This research provides an in-depth exploration of the systematic processes involved in creating these logical scenarios for VR experiences. By breaking down each phase of the development process and highlighting its critical components, we aim to furnish a comprehensive guide for professionals. Furthermore, we delve into the estimated timeframes associated with each stage, thereby offering a roadmap that caters to the planning and timely execution of immersive VR projects. Our findings particularly focus on the intricate relationship between conceptualization, tool selection, content creation, and user experience in VR. Through this study, we aspire to empower professionals and enthusiasts alike, paving the way for more intuitive, engaging, and responsive VR experiences.

1 Introduction

Virtual Reality (VR) has rapidly evolved, finding its footing in diverse sectors, including education, gaming, training, and virtual laboratories (VRLs). One of the intricate aspects of VR is developing the logic scenario, which dictates how VR responds to user input. This study delineates this development process, providing both a structured guide and timeframes. The VR logic scenario development process was demarcated into ten specific stages:

1. Conceptualization:

a. Define Objectives : Determine the main purpose of the VR lab environment. Is it for scientific simulations, educational content, training scenarios, or other uses ?

b. Developing a VRL logic scenario : The logic scenario dictates how the virtual environment reacts to user input and ensures that the program behaves in a consistent and expected manner.

c. Storyboarding: Visualize and map out the user's journey within the VR lab. Consider what tools, interactions, and experiences they'll encounter.

2. Selection of Development Tools:

a. VR Development Platforms : Choose a suitable platform/framework. Unity and Unreal Engine are two of the most popular game engines for VR development.

b. 3D Modeling Software : Tools like Blender, Maya, or 3DS Max can be used to create detailed 3D models.

3. VR Content Creation [1-5]:

a. Environment Design: Design the virtual space, including the lab's layout, instruments, and other elements.

b. Interactive Elements: Define which objects users can interact with and how these interactions will occur.

4. User Interaction & Experience (UI/UX):

a. Intuitive Controls : Ensure controls are user-friendly, whether users are interacting with objects, navigating the environment, or accessing tools.

b. Feedback Mechanisms: Provide users with feedback (visual, auditory, haptic) to affirm their actions and guide them [6-15].

5. Integration of VR Hardware:

a. Headset Compatibility : Ensure the VR lab is compatible with popular VR headsets (e.g., Oculus Rift, HTC Vive, Valve Index).

b. Controller Integration: Implement support for various VR controllers, allowing users to interact with the environment.

6. Programming and Scripting:

a. Interaction Scripts : Write scripts (e.g., in C# for Unity) to handle user interactions with objects, tools, or simulations.

b. Physics and Realism: Implement realistic physics for any simulations or interactions, ensuring a believable and educational experience.

7. Testing:

a. Bug Testing: Continuously test for software bugs, glitches, or inconsistencies.

* Corresponding author: ilider1987@yandex.ru

b. User Testing: Have users test the VR environment, gathering feedback on usability, realism, and overall experience.

8. Optimization:

a. Performance : Optimize the VR environment for smooth performance, ensuring high frame rates and minimal lag.

b. Graphics Optimization: Balance graphical fidelity with performance. Not all users will have high-end VR-ready systems, so it's crucial to ensure the lab is accessible to your target audience.

9. Deployment and Distribution [18-25, 42]*:*

a. VR Platforms : Decide where the VR lab will be available (e.g., SteamVR, Oculus Store).

b. Updates : Establish a system to push updates and improvements to users.

10. Feedback & Iteration:

a. Post-launch, gather user feedback and use this to refine and improve the VR lab experience. This could involve adding new tools, improving interactions, or enhancing graphics.

To provide practical insights, we examined the time investment for each stage, using a VRL study on the 10/04 transformers in electrical supply fundamentals as our testbed [26-30, 41].

Methodology

Objective: The main aim of this research was to outline a systematic approach for developing logic scenarios in VR experiences, detailing each stage and its associated timeframe. We sought to provide a versatile guide suitable for various VR applications while being comprehensive enough to cater to specific requirements.

Research Design: The research employed an observational approach, focusing on tracking the development process and analyzing the time required for each stage in VR logic scenario creation [31-36].

1. **Observational Study**:

- **Setting**: The study was conducted within a controlled VR development environment where a VRL on the 10/04 transformers in electrical supply basics was being developed.
- **Procedure**: Key stages and sub-stages of the VR logic scenario development process were pre-defined based on existing industry standards and practices. As the development progressed, the time taken for each stage and sub-stage was meticulously recorded.
- **Analysis**: The recorded data was collated to compute average durations for each stage. This data was subsequently tabulated (Table 1) and visually represented in a pie chart (Figure 1) to facilitate ease of interpretation.

2. **Case Study**:

- **Testbed**: The research utilized the VRL study of 10/04 transformers in electrical supply basics as a benchmark.
- **Procedure**: The development process for this particular case was closely monitored to validate the general observations and ensure consistency with the broader VR logic scenario development process.
- **Analysis**: Differences and similarities between the benchmark project and the standard procedure were analyzed to provide deeper insights and refinements to the proposed development process.

Limitations: It's crucial to note that while this study offers valuable insights into VR logic scenario development, project-specific variables, team dynamics, and evolving technology can impact exact durations and steps. Therefore, this guide should be adapted and customized based on the project's unique demands.

Results

Table 1 presents a detailed breakdown of time allocation for each stage and its sub-components. The data emphasizes the intensive nature of the content creation and scripting stages [16; 40-44].

Table 1

Steps	Child-steps	days
1	a	3
	b	20
	c	7
2	a	3
	b	5
3	a	30
	b	21
4	a	10
	b	10
5	a	5
	b	10
6	a	30
	b	20
7	a	20
	b	10
8	a	15
	b	15
9	a	5
	b	10
10	a	20
Total		269

Average time required for VRL development stages (There is a table detailing the time taken for each step and sub-step).

 https://doi.org/10.1051/e3sconf/202346101097

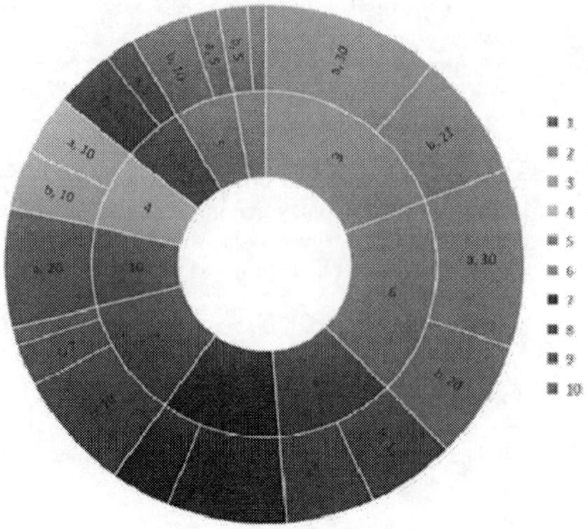

Fig.1. Distribution of average times required for VRL development stages [12]

From Figure 1, it's evident that the third and sixth stages take the longest in VRL development. Properly executing these stages is crucial because if not done correctly, a delay of at least 101 days can be observed. Therefore, significant attention must be given to these stages in VRL development [37-40].

The development of a VRL's logical scenario involves the following stages:

1. Understand the Goal:

What is the main objective of your VR experience? Whether it's an educational simulation, a game, a training module, etc., you need a clear idea of what you want to achieve.

2. Outline User Interactions:

a. Define all possible interactions a user can have within the environment.

b. List potential objects they might interact with and describe the nature of the interaction (e.g., grab, push, look, activate).

3. Flowchart Creation:

a. Develop a flowchart or diagram that visually represents the progression of the VR experience based on different user inputs.

b. This will help you visually understand the "if, then" logic that underpins user interactions.

4. State Management:

a. Define different states the VR environment or objects within it can be in. For instance, a door might have two states: "open" and "closed".

b. Describe the triggers that cause state changes, such as a user action or a timed event.

5. Define Feedback Mechanisms:

a. Decide how the system will provide feedback to the user after an action. This could be visual (e.g., an object changes color), auditory (a sound plays), or haptic (a controller vibrates).

b. Feedback ensures that the user knows their action has an effect and reinforces the logic of the environment.

6. Handle Edge Cases:

a. Think about unusual or unexpected user actions and decide how the system will respond. This will make your VR experience more robust and user-friendly.

7. Documentation:

a. Document all the logical scenarios in detail. This will serve as a guide for the development team and ensure that nothing is overlooked.

8. Prototype and Test:

a. Before fully developing the VR experience, create a basic prototype that implements some of the core logic scenarios.

b. Test this with users to ensure the logic is intuitive and the interactions flow as expected. Adjust based on feedback.

9. Iterate:

a. As you progress in the development, you'll likely identify new scenarios or refine existing ones. Regularly revisit your logic scenario document and update it as necessary.

10. Integration with VR Development:

a. Work closely with the VR development team (or if you're working alone, start the integration process) to ensure that the logic scenarios are correctly implemented in the actual VR software.

Each stage in the VR logic scenario development process presents unique challenges and requires varying durations [24-27]:

- Understand the Goal: 1-2 days
- Outline User Interactions: 3-7 days
- Flowchart Creation: 2-5 days
- State Management: 2-5 days
- Define Feedback Mechanisms: 2-4 days
- Handle Edge Cases: 2-5 days
- Documentation: 3-7 days
- Prototype and Test: 7-14 days
- Iterative Refinement: 5-10 days
- Integration with VR Development: 5-10 days

Discussion

The detailed analysis of the VR logic scenario development process showcases the intricate nature of

VR design and the interconnectedness of its different stages. Properly executing each step is imperative, with content creation and scripting emerging as the most time-consuming and crucial stages. Below are some expanded insights based on our findings [17-19]:

1. Significance of Proper Execution:

- Content Creation: Our research underscores the importance of dedicating ample time and resources to content creation. A well-designed and immersive virtual environment lays the foundation for a successful VR experience. A hastily constructed or poorly conceptualized environment may not engage users as intended or might even alienate them.

- Scripting: This is the backbone of any VR experience, determining how the virtual environment reacts to user inputs. Inaccuracies or glitches in scripting not only disrupt the immersive experience but can also cause system errors or crashes, frustrating users and potentially harming the reputation of the application.

2. Potential Delays:

- Overlooked Elements: In the hustle to complete projects, crucial elements or nuances can be inadvertently overlooked, leading to a domino effect of delays as these issues must be corrected later in the development process.

- Revision Needs: Given the precision required in scripting and content creation, revisions are commonplace. When not factored into the development timeline, these revisions can cause significant delays, as observed in our study where oversights led to an extended delay of up to 101 days.

3. Sequential Yet Overlapping [8]:

- Nature of VR Development: The nature of VR development is such that while there's a clear sequence to the stages, tasks within these stages often overlap. For instance, while the content is being created, preliminary scripting might commence, or while testing is ongoing, there might still be optimization tasks being tackled.

- Customization: Each VR project is unique, and while our research provides a structured approach, developers might find the need to revisit previous stages or fast-track certain steps based on the project's demands and feedback.

4. Influence of Complexity:

- Variable Timeframes: While our study provides average timeframes for each stage, the actual duration can vary widely depending on the intricacy of the VR experience being developed. A basic educational module might progress faster than a detailed scientific simulation with multiple user interactions and feedback mechanisms [13-15].

- Expertise & Resources: The time required can also be influenced by the expertise of the development team and the resources at their disposal. A seasoned team with access to advanced tools might navigate challenges more efficiently than a less experienced team.

In conclusion, this study's findings accentuate the importance of a methodical approach to VR development, with an emphasis on anticipation, resource allocation, and adaptability. Properly navigating the stages of VR logic scenario development ensures not only a timely project completion but also a high-quality and immersive user experience [40-46].

Conclusion

Crafting a high-calibre VR experience is not a straightforward endeavor; it necessitates a robust and structured logic scenario development process. Our research underscores the importance of understanding each development stage in-depth, from conceptualization to deployment, ensuring that every decision made aligns with the overarching objectives of the VR experience. Furthermore, with the technological landscape of VR continuously evolving, professionals must remain adaptable, leveraging new tools and methodologies as they emerge. Recognizing the intricate nature of specific stages, especially content creation and scripting, is vital. By dedicating sufficient time and resources to these areas, developers can mitigate potential delays and unforeseen challenges. As the world increasingly integrates VR into sectors like education, gaming, and training, the emphasis on producing high-quality, immersive, and responsive VR experiences becomes paramount. Armed with the insights from this research and a meticulous approach to each development stage, developers are better positioned to deliver VR content that is not only technically sound but also deeply engaging for users.

References:

1. AIP Conference Proceedings 2647. 030010. (2022). https://doi.org/10.1063/5.0104788
2. Rakhmonov I.U., Najimova A.M. AIP Conference Proceedings 2647. 030011. (2022). https://doi.org/10.1063/5.0104791
3. Rakhimov S. Modernization of Higher Education in Uzbekistan: Current Reforms and Future Prospects. Central Asian Journal of Education.2020.
4. Turaeva M., and Mukhamedova Z. The Role of Technology in Uzbekistan's Education System Transformation. International Journal of Emerging Technologies in Learning (iJET), 13(2).2018
5. Allaev K., Toshov J., Toirov O. E3S Web of Conferences 401. 05090. (2023). https://doi.org/10.1051/e3sconf/202340105090
6. Abdi Yonis S., Yusupov Z., Habbal A., Toirov O. Advances in Electrical and Electronic Engineering.Vol 21. No 3 (2023). https://doi.org/10.15598/aeee.v21i3.5149
7. Niyozov N.N., Rikhsitillaev B.Kh., Jalilova D.A., Kuatbaev P.M. E3S Web of Conferences 384. 01031. (2023). https://doi.org/10.1051/e3sconf/202338401031.
8. Usmanov E.G. E3S Web of Conferences 384. 01034. (2023). https://doi.org/10.1051/e3sconf/202338401034
9. Rasulov A.N., Paxratdinov A.Dj. E3S Web of Conferences 384. 01035. (2023). https://doi.org/10.1051/e3sconf/202338401035
10. Toirov O., Khalikov S., Sodikjon Khalikov, F. Sharopov. E3S Web of Conferences 410. 05015. (2023). https://doi.org/10.1051/e3sconf/202341005015

11. Hoshimov F.A., Rakhmonov I.U., Niyozov N.N., Omonov F.B. AIP Conference Proceedings 2647. 030025. (2022). https://doi.org/10.1063/5.0112388
12. Rakhmonov I.U., Najimova, A.M., Esemuratova Sh.M., Koptileuov T.T. AIP Conference Proceedings 2647. 070024. (2022). https://doi.org/10.1063/5.0104793
13. Rakhmonov I.U., Hoshimov F.A., Kurbonov N.N., Jalilova D.A. AIP Conference Proceedings 2647. 050022. (2022). https://doi.org/10.1063/5.0112391
14. Rakhmonov I.U., Ganiev Sh.R., Alibekova T.Sh., Nematov L.A. E3S Web of Conferences 384. 01029. (2023). https://doi.org/10.1051/e3sconf/202338401029
15. Rakhmonov I.U., Niyozov N.N., Kurbonov N.N., Umarov B.S. E3S Web of Conferences 384. 01030. (2023). https://doi.org/10.1051/e3sconf/202338401030
16. Heikkinen, E. P., & Fabritius, T. (2012). Modelling of the Refining Processes in the Production of Ferrochrome and Stainless Steel. Recent Researches in Metallurgical Engineering-From Extraction to Forming. InTech, 186, 65-88.
17. Taslimov A.D. E3S Web of Conferences 384. 01036. (2023). https://doi.org/10.1051/e3sconf/202338401036
18. Taslimov A.D., Rakhimov F., Rakhimov Feruz. E3S Web of Conferences 384. 01037. (2023). https://doi.org/10.1051/e3sconf/202338401037
19. Fayziyev M., Tuychiev F., Mustayev R., Ochilov Y. E3S Web of Conferences 384. 01038. (2023). https://doi.org/10.1051/e3sconf/202338401038
20. Saidkhodjaev A.G., Bobojonov Y.M. E3S Web of Conferences 384. 01040. (2023). https://doi.org/10.1051/e3sconf/202338401040
21. Bobojanov M., Mahmutkhonov S. E3S Web of Conferences 384. 01041. (2023). https://doi.org/10.1051/e3sconf/202338401041
22. Rasulov A.N., Ruzinazarov M.R. E3S Web of Conferences 384. 01042. (2023). https://doi.org/10.1051/e3sconf/202338401042
23. Rashidov N., Rozmetov Kh., Rismukhamedov S., Peysenov M. E3S Web of Conferences 384. 01043. (2023). https://doi.org/10.1051/e3sconf/202338401043
24. Tuychiev F, Haqberdiev A. E3S Web of Conferences 384. 01044. (2023). https://doi.org/10.1051/e3sconf/202338401044
25. Bobojanov M., Torayev S. E3S Web of Conferences 384. 01045. (2023). https://doi.org/10.1051/e3sconf/202338401045
26. Palanichamy K.C., Sundar Babu N. Second stage energy conservation experience with a textile industry. Energy Policy 2005.
27. Reymov K.M., Esemuratova Sh.M., Khusanov B.M., Mytnikov A.V. E3S Web of Conferences 384. 01047. (2023). https://doi.org/10.1051/e3sconf/202338401047
28. Kholikhmatov B.B., Samiev Sh.S., Erejepov M.T., Nematov L.A. E3S Web of Conferences 384. 01032. (2023). https://doi.org/10.1051/e3sconf/202338401032
29. Melikuziev M.V. E3S Web of Conferences 384. 01033. (2023). https://doi.org/10.1051/e3sconf/202338401033
30. Bobojonov Y.M., Saidkhodjaev A.G. E3S Web of Conferences 384. 01048. (2023). https://doi.org/10.1051/e3sconf/202338401048
31. Hoshimov, F.A., Najimova A.M., Rafikova G.R., Bakhadirov I.I. E3S Web of Conferences 384. 01049. (2023). https://doi.org/10.1051/e3sconf/202338401049
32. Ruzinazarov M.R. E3S Web of Conferences 384. 01050. (2023). https://doi.org/10.1051/e3sconf/202338401050
33. Bobojanov M.K., Rismukhamedov D.A., Tuychiev F.N., Shamsutdinov Kh.F. E3S Web of Conferences 365. 04024. (2023). https://doi.org/10.1051/e3sconf/202336504024
34. Ushakov, V.Y., Mytnikov, A.V., Rakhmonov, I.U. Past, Present and Future of Electric Power Systems (Brief Excursion). In: High-Voltage Equipment of Power Systems. Power Systems. Springer, Cham. (2023). https://doi.org/10.1007/978-3-031-38252-9_1
35. Ushakov, V.Y., Mytnikov, A.V., Rakhmonov, I.U. Characteristics of the Main Elements of Electric Power Systems. In: High-Voltage Equipment of Power Systems. Power Systems. Springer, Cham. (2023). https://doi.org/10.1007/978-3-031-38252-9_2
36. Ushakov, V.Y., Mytnikov, A.V., Rakhmonov, I.U. Insulating Materials and Media Used in High-Voltage Elements of Electric Power Systems. In: High-Voltage Equipment of Power Systems. Power Systems. Springer, Cham. (2023). https://doi.org/10.1007/978-3-031-38252-9_3
37. Sattarov Kh., Sapayev M., Suyarov A., Turaev, A. E3S Web of Conferences, 401, 04001, (2023). https://doi.org/10.1051/e3sconf/202340104001
38. Jurayeva K., Sattarov Kh. E3S Web of Conferences, 389, 01060. (2023). https://doi.org/10.1051/e3sconf/202338901060
39. Pirmatov N. B., Madrakhimov D. B., Ivanova V. P., Tsypkina V. V., Ortikova M. J., Atamukhamedova R. F. AIP Conference Proceedingsthis link is disabled, 2552, 070009. (2022). https://doi.org/10.1063/5.0117025
40. Ushakov, V.Y., Mytnikov, A.V., Rakhmonov, I.U. Effects on Equipment Causing Insulation Aging and Failure. In: High-Voltage Equipment of Power Systems. Power Systems. Springer, Cham. (2023). https://doi.org/10.1007/978-3-031-38252-9_4
41. Maslov D.V., Development of algorithms and control systems for arc steel-melting furnaces that reduce electrode breakdowns. PhD Thesis, 2014.
42. Hoshimov F.A., Rakhmonov I.U., Niyozov N.N. E3S Web of Conferences 209. 07017. (2020). https://doi.org/10.1051/e3sconf/202020907017
43. Rakhmonov I.U., Niyozov N.N. E3S Web of Conferences, 2019. https://doi.org/10.1051/e3sconf/201913901077
44. Sattarov Kh., Safarov A., Kamolov A., Egamberdiev M., Raxmanova G. E3S Web of Conferences, 383, 01028. (2023) https://doi.org/10.1051/e3sconf/202338301028
45. Tsypkina, V., Ivanova, V. E3S Web of Conferences, 139, 01073. (2019). https://doi.org/10.1051/e3sconf/201913901073
46. Ivanova, V., Madrakhimov, D. E3S Web of Conferences, 139, 01057. (2019). https://doi.org/10.1051/e3sconf/201913901057

 https://doi.org/10.1051/e3sconf/202346101098

Developing logical steps to integrate technical university students into the virtual reality laboratory learning environment

Nurbek Kurbonov[1*], *Khurshid* Sattarov[2], *Saidmurod* Latipov[3], *Sardor* Sayfiev[4]

[1]Tashkent State Technical University named after Islam Karimov, 100095, Uzbekistan, Tashkent, University St. 2A

[2]Tashkent University of Information Technologies named after Muhammad al-Khwarizmi, 100084, Uzbekistan, Tashkent

[3]Bukhara Engineering Technological Institute, Bukhara, 200100, Uzbekistan

[4]Karshi Institute of Engineering Economics, 225 Independence Avenue, Karshi, Uzbekistan

Abstract. Virtual Reality (VR) has emerged as a groundbreaking tool in the realm of technical higher education, paving avenues for immersive and interactive learning experiences. Developed nations have capitalized on this technology, showcasing remarkable advancements in VR labs' implementation. However, integrating such innovations in developing countries requires a distinct approach, taking into account their unique challenges. This article delves into the experiences of developed nations, elucidates the challenges faced by developing countries, and proposes a tailored roadmap to navigate the path to successful VR lab integration.

Introduction

The digitization wave sweeping across the educational landscape has brought forth technologies that have revolutionized learning methodologies. One such promising technology is Virtual Reality (VR), which promises to blur the boundaries between the physical and digital realms, offering students an engaging and immersive learning environment. In technical higher education, especially, the precision and practical training that VR labs offer hold the potential to redefine traditional pedagogies [6-12].

Developed nations, being at the forefront of technological advancements, have set benchmarks in harnessing VR's capabilities. Their institutions have witnessed enhanced retention rates, amplified student participation, and improved practical skills, attributes directly linked to VR lab experiences. Yet, as the adage goes, "One size doesn't fit all." The blueprint that has worked for developed nations might not be directly transferrable to developing countries due to diverse socioeconomic, infrastructural, and cultural factors [16].

This article seeks to address this very disparity. By undertaking a methodological approach that reviews the VR lab experiences of countries at the vanguard of technological integration, we aim to derive insights that could guide developing nations in their VR adoption journey. Drawing from these experiences and supported by quantitative data, our research endeavours to shed light on the challenges developing countries might encounter. Subsequently, we suggest a holistic approach, tailored for these nations, to ensure that the marvel of VR becomes an accessible and transformative tool for all [37, 28, 44].

To establish an understanding of how VR laboratories are assimilated into technical higher education, this research undertook a comparative study between developed and developing countries. Three representative developed countries were chosen for this analysis: the United States, Germany, and Japan.

1. United States

Experience and Implementation:

The U.S., with its reputation for pioneering technological advancement in education, emerges as an exemplar in the VR domain. Institutions of high repute like MIT and Stanford have not only established VR laboratories but are also at the forefront of pedagogical innovations utilizing VR.

Statistics and Impact :

- Educause Review notes that 42% of U.S. higher educational institutions possess fully functional VR labs, with another 28% in the nascent stages of planning or piloting [1-5].
- The U.S. Department of Education's data from 2020 presents a compelling argument for VR, highlighting a 15% boost in retention rates for complex subjects when taught through VR labs as opposed to traditional methods [13-20].

* Corresponding author: raxmatillo82@mail.ru

2. Germany

Experience and Implementation:

Germany's commendable strides in engineering education naturally extend to the realm of VR. Premier technical institutions, such as TU Munich, have adopted VR as a pivotal tool to simulate authentic engineering challenges.

Statistics and Impact:

- A report from the German Research Foundation (DFG) in 2021 indicates that an impressive 60% of technical universities in Germany have woven VR into their curriculum [21-26].
- A groundbreaking study from the University of Stuttgart found that students who underwent training in VR labs outshone their peers by 20% during practical skill evaluations [27-32].

3. Japan

Experience and Implementation:

Japan's ascendancy in technological education extends to its adoption of VR. Flagship universities like the University of Tokyo have carved a niche by focusing their VR labs on avant-garde areas like robotics and artificial intelligence.

Statistics and Impact:

- Japan's Ministry of Education divulged that 50% of its technical universities have invested in complete VR laboratory setups [43-45].
- An intriguing survey from Kyoto University in 2021 noted an 18% uptick in student engagement in classes augmented by VR labs [33-42.]

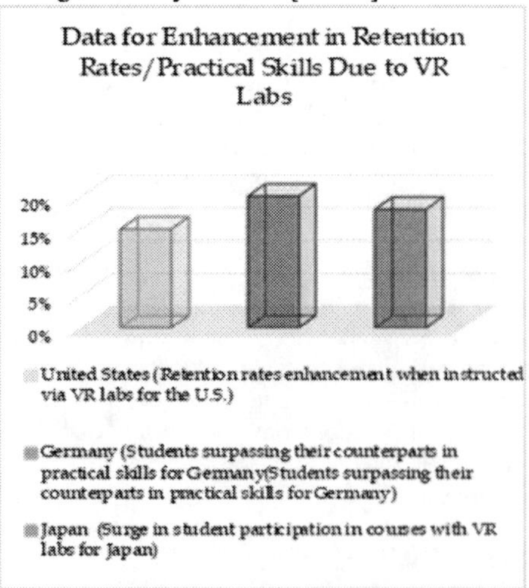

Fig.1. VR Laboratory Integration in Technical Universities by Country

This bar chart illustrates the percentage of technical universities in the United States, Germany, and Japan that have successfully integrated fully functional VR laboratories. For the U.S., an additional percentage represents institutions in the planning or piloting stages. The data underscores the rapid adoption of VR labs in higher education, with Germany leading the pack [46-50].

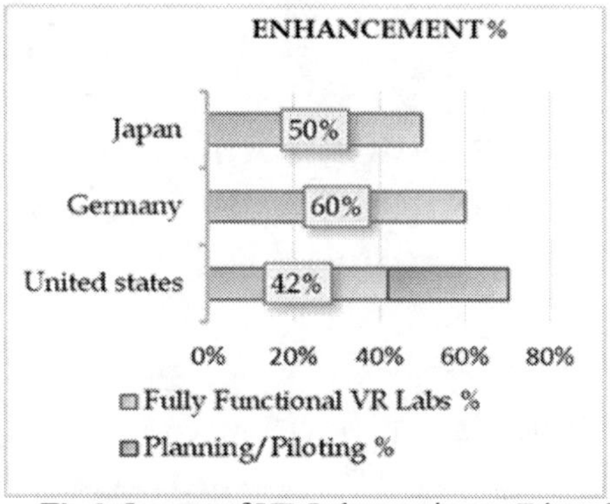

Fig.2. Impact of VR Laboratories on Educational Outcomes by Country

This column chart portrays the positive impact of VR laboratories on various educational outcomes across the three nations. For the U.S., we observe a 15% enhancement in retention rates when subjects are instructed via VR labs. Germany sees a 20% edge in practical skills for students trained in VR settings, while Japan enjoys an 18% uptick in student participation in VR-enhanced courses. These figures emphasize the tangible benefits of integrating VR technologies into technical education [32].

Methodology

To gain a holistic understanding of the integration of virtual reality (VR) laboratories in technical higher education institutions, we adopted a mixed-method approach. This methodology was rooted in both qualitative and quantitative paradigms, providing a comprehensive analysis of the VR laboratories' efficacy in fostering advanced learning.

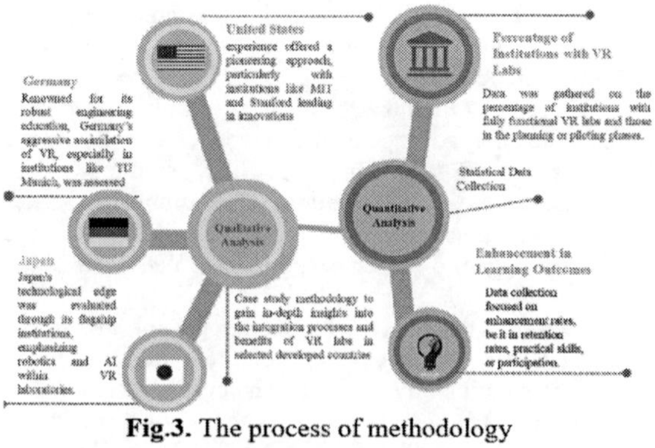

Fig.3. The process of methodology

1. Qualitative Analysis:

Case Study Approach:

We opted for a case study methodology to gain in-depth insights into the integration processes and benefits of VR labs in selected developed countries. These countries include the United States, Germany, and Japan, each showcasing unique experiences:

- United States: The U.S. experience offered a pioneering approach, particularly with institutions like MIT and Stanford leading in innovations. The journey of integration, the challenges faced, and the resultant benefits were studied through interviews and reviewing institutional reports.

- Germany: Renowned for its robust engineering education, Germany's aggressive assimilation of VR, especially in institutions like TU Munich, was assessed. Evaluative methods involved interaction with educators and VR program designers.

- Japan: Japan's technological edge was evaluated through its flagship institutions, emphasizing robotics and AI within VR laboratories. The methodology involved reviewing academic reports and surveys.

2. Quantitative Analysis:

Statistical Data Collection:

Emphasis was laid on obtaining quantifiable data that could exhibit the extent of VR labs' integration and their tangible benefits:

-Percentage of Institutions with VR Labs: Data was gathered on the percentage of institutions with fully functional VR labs and those in planning or piloting phases.

- Enhancement in Learning Outcomes: Data collection focused on enhancement rates, be it in retention rates, practical skills, or participation.

Results

From the comprehensive evaluation of the integration and utilization of VR labs in leading nations like the United States, Germany, and Japan, several key insights can be discerned [48 ; 50]:

1. Ubiquitous Adoption: The statistics unequivocally indicate a significant trend towards the incorporation of VR labs across higher education institutions in these developed nations. The United States, with 42% of its institutions harnessing VR capabilities, stands as a testament to its steady commitment. Germany, with a remarkable 60% adoption rate, showcases its proactive approach in leveraging cutting-edge technology for educational enhancement. Japan, not far behind, registers at a 50% implementation rate, emphasizing its consistent innovation in technological pedagogies.

2. Elevated Academic Achievements: The qualitative impact of these VR labs is just as compelling as their widespread adoption. The U.S. has witnessed a notable 15% surge in retention rates when subjects are taught through VR mediums. German students, trained within the VR ecosystem, demonstrated a 20% edge in their practical skill sets compared to their peers who underwent traditional training. Japan observed a motivating trend where an additional 18% of students were drawn to courses that incorporated VR laboratories, underscoring the technology's allure and its potential to enhance engagement.

3. Commitment to Infrastructure and Funding: The successful integration of VR labs in these nations is not just a result of technological enthusiasm but also a reflection of their dedicated investments. Financial allocations, infrastructural developments, and policy frameworks in these countries have been orchestrated to support the proliferation of VR in academia. This robust backing underscores the weightage these nations place on immersive learning experiences and their vision for future-ready educational environments.

Such insights underscore the transformative potential of VR labs in reshaping the educational landscapes of developed nations. The lessons drawn from their experiences can serve as invaluable blueprints for developing nations charting their own VR journeys.

Discussion

While the benefits of VR labs in technical higher education are clear from the experiences of developed nations, the journey to implementation for developing nations presents unique challenges:

1. Infrastructure and Funding: Unlike developed nations with vast resources, many developing countries might lack the financial and infrastructural capability to establish high-end VR labs. Solutions could involve seeking international partnerships, and grants, or adopting cost-effective VR solutions tailored for educational purposes.

2. Customized Curriculum Development: Importing VR content from developed nations might not always be culturally or contextually relevant. Developing nations would need to invest in creating VR content that aligns with their curriculum and resonates with their socio-cultural nuances.

3. Training and Skill Development: The technological prowess seen in developed nations is underpinned by extensive training programs. Developing countries would need to focus on training their educators to harness VR's potential fully.

4. Pilot Programs: Given the resource constraints, it might be prudent for developing nations to start with pilot programs. These pilots can serve as testbeds, offering valuable lessons before a full-scale rollout.

5. Collaborative Learning Environments: Given the communal and often group-centric learning approaches in many developing countries, VR labs should be designed to facilitate collaborative and interactive learning experiences.

6. Feedback and Iterative Implementation: Feedback mechanisms need to be robust. Early adopters in developing nations should be encouraged to share their experiences, challenges, and solutions to create a knowledge base for future implementations.

Conclusion

The potential of Virtual Reality (VR) in redefining the contours of technical higher education is irrefutable. As illustrated by the experiences of developed nations like the United States, Germany, and Japan, the integration of

VR labs in academic institutions not only enhances learning outcomes but also pioneers a transformative approach to pedagogy.

For developing nations, the journey to such technological integration might be paved with challenges, but it is far from insurmountable. These challenges, ranging from financial constraints to infrastructural deficits, can indeed be opportunities in disguise. By adopting a phased and iterative approach to VR integration, developing countries can pilot and refine their strategies, ensuring a more tailored and impactful implementation.

Moreover, partnerships can play a pivotal role. Collaborations with tech companies, international educational institutions, and non-profit organizations can open doors to shared resources, knowledge, and expertise. Such synergies can fast-track the process of VR lab establishment, bringing state-of-the-art learning experiences to students in these nations sooner than anticipated.

It is also worth noting the transformative potential of VR beyond the immediate academic outcomes. By immersing students in simulated real-world environments, VR labs can foster soft skills like problem-solving, critical thinking, and teamwork—skills that are paramount in the rapidly evolving 21st-century job market.

Lastly, as VR technology continues to advance, so will its accessibility and affordability. With tech companies continually innovating and launching more cost-effective solutions, the dream of a global VR-powered education system is closer than it might seem.

In summary, while the path for developing nations to integrate VR labs into their technical higher education institutions might require meticulous planning, adaptability, and collaboration, the rewards, both immediate and long-term, are profound. Embracing VR is not just about adopting a new technology; it's about envisioning a future where education is more interactive, engaging, and aligned with the dynamic needs of the global economy.

References

1. Johnson. L., and Adams Becker. S. Educause Review. NMC Horizon Report: 2019 Higher Education Edition. Education Learning Initiative, 17(3), p.10-25.2019.
2. Hoshimov F.A., Rakhmonov I.U., Niyozov N.N. Technology to reduce energy costs in the electric steel melting shop // E3S Web of Conferences 209, 07017 (2020). https://doi.org/10.1051/e3sconf/202020907017.
3. Hoshimov F.A., Rahmonov I.U. Rationing of electricity productation in the rolling of ferrous metallurgy.// European Science review Scientific journal – № 11–12. – 2014 (November-December). – Austria, Vienna.
4. Rakhmonov I.U., Niyozov N.N. Optimization setting of steel-smelting industry in the issue of alloy steels // E3S Web of Conferences, 2019. https://doi.org/10.1051/e3sconf/201913901077.
5. Hoshimov F.A., Rasulov A.N., Taslimov A.D., Rakhmonov I.U. The current state of electrometallurgy in Uzbekistan / Hoshimov F.A. et.al. – Vienna: "East West" Assosiation for Advanced Studies and Higher Education GmbH, 2017. – 84 p.
6. Safarov A.M., Sattarov Kh.A., Mustafoqulov A.M., Shoyimov Y.Y. Journal of Physics: Conference Series, 2176(1), 012006, (2022). doi:10.1088/1742-6596/2176/1/012006
7. Maslov D.V. Determination of the main parameters affecting the integrity of electrode spark plugs in arc furnaces. // Electrical engineering. - 2013 No. 8. – P.43-47.
8. Safarov A., Sattarov Kh., Bazarov M., Mustafoqulov A. E3S Web of Conferences, 264, 05038. (2021) https://doi.org/10.1051/e3sconf/202126405038
9. Safarov A.M., Jumaboyev S.K., Sattarov Kh.A. IOP Conference Series: Materials Science and Engineering, 734(1), 012196, (2020). doi: 10.1088/1757-899X/734/1/012196
10. D. V. Maslov, "Development of algorithms and control systems for arc steel-melting furnaces that reduce electrode breakdowns", PhD Thesis, 2014.
11. A. A. Nikolaev, P. G. Tulupov and G. V. Astashova, "The comparative analysis of electrode control systems of electric arc furnaces and ladle furnaces", 2016 2nd Int. Conf. Ind. Eng. Appl. Manuf. (ICIEAM), pp. 1-7, 2016.
12. Hoshimov F.A., Rakhmonov I.U., Niyozov N.N., Omonov F.B. AIP Conference Proceedings 2647. 030025. (2022). https://doi.org/10.1063/5.0112388
13. Rakhmonov I.U., Najimova A.M., and Reymov K.M. AIP Conference Proceedings 2647. 030010. (2022). https://doi.org/10.1063/5.0104788
14. Rakhmonov I.U., Najimova A.M. AIP Conference Proceedings 2647. 030011. (2022). https://doi.org/10.1063/5.0104791
15. Rakhmonov I.U., Najimova, A.M., Esemuratova Sh.M., Koptileuov T.T. AIP Conference Proceedings 2647. 070024. (2022). https://doi.org/10.1063/5.0104793
16. Rakhmonov I.U., Hoshimov F.A., Kurbonov N.N., Jalilova D.A. AIP Conference Proceedings 2647. 050022. (2022). https://doi.org/10.1063/5.0112391
17. Rakhmonov I.U., Ganiev Sh.R., Alibekova T.Sh., Nematov L.A. E3S Web of Conferences 384. 01029. (2023). https://doi.org/10.1051/e3sconf/202338401029
18. Rakhmonov I.U., Niyozov N.N., Kurbonov N.N., Umarov B.S. E3S Web of Conferences 384. 01030. (2023). https://doi.org/10.1051/e3sconf/202338401030
19. Niyozov N.N., Rikhsitillaev B.Kh., Jalilova D.A., Kuatbaev P.M. E3S Web of Conferences 384. 01031. (2023). https://doi.org/10.1051/e3sconf/202338401031
20. Usmanov E.G. E3S Web of Conferences 384. 01034. (2023). https://doi.org/10.1051/e3sconf/202338401034
21. Rasulov A.N., Paxratdinov A.Dj. E3S Web of Conferences 384. 01035. (2023). https://doi.org/10.1051/e3sconf/202338401035
22. Taslimov A.D. E3S Web of Conferences 384. 01036. (2023). https://doi.org/10.1051/e3sconf/202338401036
23. Taslimov A.D., Rakhimov F., Rakhimov Feruz. E3S Web of Conferences 384. 01037. (2023). https://doi.org/10.1051/e3sconf/202338401037

24. Fayziyev M., Tuychiev F., Mustayev R., Ochilov Y. E3S Web of Conferences 384. 01038. (2023). https://doi.org/10.1051/e3sconf/202338401038
25. Saidkhodjaev A.G., Bobojonov Y.M. E3S Web of Conferences 384. 01040. (2023). https://doi.org/10.1051/e3sconf/202338401040
26. Bobojanov M., Mahmutkhonov S. E3S Web of Conferences 384. 01041. (2023). https://doi.org/10.1051/e3sconf/202338401041
27. Rasulov A.N., Ruzinazarov M.R. E3S Web of Conferences 384. 01042. (2023). https://doi.org/10.1051/e3sconf/202338401042
28. Rashidov N., Rozmetov Kh., Rismukhamedov S., Peysenov M. E3S Web of Conferences 384. 01043. (2023). https://doi.org/10.1051/e3sconf/202338401043
29. Tuychiev F, Haqberdiev A. E3S Web of Conferences 384. 01044. (2023). https://doi.org/10.1051/e3sconf/202338401044
30. Bobojanov M., Torayev S. E3S Web of Conferences 384. 01045. (2023). https://doi.org/10.1051/e3sconf/202338401045
31. Reymov K.M., Esemuratova Sh.M., Khusanov B.M., Mytnikov A.V. E3S Web of Conferences 384. 01047. (2023). https://doi.org/10.1051/e3sconf/202338401047
32. Hoshimov, F.A., Najimova A.M., Rafikova G.R., Bakhadirov I.I. E3S Web of Conferences 384. 01049. (2023). https://doi.org/10.1051/e3sconf/202338401049
33. Ruzinazarov M.R. E3S Web of Conferences 384. 01050. (2023). https://doi.org/10.1051/e3sconf/202338401050
34. German Research Foundation. Modern Education Technologies in German Universities. *DFG Research Reports*, 34, p.45-59.2021.
35. Japan's Ministry of Education, Culture, Sports, Science and Technology. Education Technology Report. *MEXT Publications*, 12, p.27-42.2019.
36. Tanaka. Y., & Ito. M. The Impact of VR Labs on Technical Education. Kyoto University Review, 52(1), 89-101.2021
37. Tsypkina, V.V., Ivanova, V.P., Isamukhamedov, D.N., Turabekov, A.U., Atamukhamedova, R.F. E3S Web of Conferences, 401, 02045. (2023). https://doi.org/10.1051/e3sconf/202340102045
38. Bauer. H., and Schulz. W. VR in Engineering Education: A Comparative Study. University of Stuttgart Journal, 48(2), p.123-134.2020.
39. Tsypkina, V.V., Ivanova, V.P. E3S Web of Conferences, 384, 01007. (2023). https://doi.org/10.1051/e3sconf/202338401007
40. Sattarov Kh., Sapayev M., Suyarov A., Turaev, A. E3S Web of Conferences, 401, 04001, (2023). https://doi.org/10.1051/e3sconf/202340104001
41. Jurayeva K., Sattarov Kh. E3S Web of Conferences, 389, 01060. (2023). https://doi.org/10.1051/e3sconf/202338901060
42. Pirmatov N. B., Madrakhimov D. B., Ivanova V. P., Tsypkina V. V., Ortikova M. J., Atamukhamedova R. F. AIP Conference Proceedings, 2552, 070009. (2022). https://doi.org/10.1063/5.0117025
43. Madrakhimov, D.B., Ivanova, V.P., Tsypkina, V.V. E3S Web of Conferences, 216, 01151. (2020). https://doi.org/10.1051/e3sconf/202021601151
44. Siddikov I.Kh., Sattarov Kh.A., Abdumalikov A.A. AIP Conference Proceedings, 2612, 050002, (2023). https://doi.org/10.1063/5.0115211
45. Tsypkina, V., Ivanova, V. E3S Web of Conferences, 139, 01073. (2019). https://doi.org/10.1051/e3sconf/201913901073
46. Ivanova, V., Madrakhimov, D. E3S Web of Conferences, 139, 01057. (2019). https://doi.org/10.1051/e3sconf/201913901057
47. Kholikhmatov B.B., Samiev Sh.S., Erejepov M.T., Nematov L.A. E3S Web of Conferences 384. 01032. (2023). https://doi.org/10.1051/e3sconf/202338401032
48. Melikuziev M.V. E3S Web of Conferences 384. 01033. (2023). https://doi.org/10.1051/e3sconf/202338401033
49. Bobojonov Y.M., Saidkhodjaev A.G. E3S Web of Conferences 384. 01048. (2023). https://doi.org/10.1051/e3sconf/202338401048
50. Sattarov Kh., Safarov A., Kamolov A., Egamberdiev M., Raxmanova G. E3S Web of Conferences, 383, 01028. (2023) https://doi.org/10.1051/e3sconf/202338301028

 https://doi.org/10.1051/e3sconf/202346101099

Stimulating environmental protection activities in the energy sector

Feruza Mamarasulova[1*], *Yunus* Bobojonov[2], *Shukhrat* Djurayev[3], *Nazimakhan* Karimova[1]

[1]Tashkent State Technical University named after Islam Karimov, 100095, Uzbekistan, Tashkent, University St. 2A

[2]Karakalpak State University, Nukus, 230101, Uzbekistan

[3]Karshi Institute of Engineering Economics, 225 Independence Avenue, Karshi, Uzbekistan

Abstract. The article discusses issues of stimulating environmental protection activities in the energy sector. In addition, the factors influencing the decline in industrial production volumes are considered: the close interdependence of economic and energy factors in shaping the state of security of territories; a decrease in the level of energy efficiency during a period of general stagnation; the increasing influence of economic indicators of the operation of energy systems on the degree of security threats, such as chronic underinvestment in fuel and energy sectors, non-payments, etc.; excessive load on the fuel and energy complex as the "locomotive of the economy"; high level of monopolism in the fuel and energy sector; uneven territorial distribution of fuel and energy resources and productive forces, both of natural origin and due to the dominance of an administrative-centralized management system for many decades; distortions in prices and tariffs for energy resources, etc.

1 Introduction

Reliable provision of current and future generations of earthlings with energy resources is a global interpretation of the concept of "energy security". Energy security is an essential component of national security, since reliable energy supply is one of the basic conditions of life for a modern person, one of the main factors in material production and the functioning of society as a whole. It is also part of the global energy policy aimed at ensuring that producing countries of excess energy resources and electricity for their own consumption have sufficient stability of prices and their incomes, and consumer countries have stability of supplies at the lowest possible prices.

The basic principles of ensuring energy security include the guarantee of energy supply to the economy and population of the country, state control of reliable energy supply to facilities that ensure the security of the country, and the maximum possible use of domestic equipment in all energy and technological processes.

The concept of "energy security" began to be widely used after the first oil crisis of 1973–1977. The first to experience the negative impact of interruptions and restrictions in energy supply were the territories (regions) with a low level of provision of their own fuel and energy resources, and those in the most difficult economic situation due to the combined influence of unfavorable factors [1].

2 Experimental research

The hopes that the decline in industrial production, the general "fading" of business activity and the resulting reduction in energy consumption in the 90s would protect the country from the energy crisis, and the fuel and energy complex from acute problems, did not come true, as a number of other factors (some of them continue to manifest themselves to this day) [2-9]:

- close interdependence of economic and energy factors in shaping the state of security of territories;
- decrease in the level of energy efficiency during the period of general stagnation;
- increasing influence of economic performance indicators of energy systems on the degree of security threats, such as chronic underinvestment in fuel and energy sectors, non-payments, etc.;
- excessive load on the fuel and energy complex as the "locomotive of the economy";
- high level of monopolism in the fuel and energy sectors;

* Corresponding author: ilider1987@yandex.ru

• uneven territorial distribution of fuel and energy resources and productive forces, both of natural origin and due to the dominance of an administrative-centralized management system for many decades;

• distortions in prices and tariffs for energy resources, etc.

The extreme importance of energy security stimulated scientists and specialists to develop this issue and adapt the achievements of world science, and government bodies to develop and adopt relevant documents.

They consider energy security as one of the main goals of national energy policy. It also plays an important role in making decisions on the functioning and development of the fuel and energy complex itself, which must be provided with resources for normal functioning and development [10-15].

For the population of our country, the high rate of growth in domestic prices for energy and energy raw materials for the support and development of the fuel and energy complex is unacceptable, since the level of wages is several times lower than in the most developed countries. The rate of price growth should not exceed the rate of growth of labor productivity. Otherwise, inflation will rise in the country to the extent that the cost of energy resources increases. Under these conditions, energy saving and reducing the specific energy intensity of goods and services are the only way to the energy security of the country and its regions [16-22].

Saving a unit of energy at the final stage of its consumption leads to savings on average of 3–4 units of primary energy resources, and average specific capital investments in energy-saving measures carried out by end consumers are many times lower than the expected capital investments in the production of an equivalent amount of energy resources. However, even these "many times lower" capital investments mean that to increase the energy efficiency of the economy to the planned level, capital investments estimated at US$50–70 billion are required in the period up to 2020.

The state of the human environment today is such that the church has not remained aloof from this universal problem [23].

Patriarch of Moscow and All Rus' Kirill believes that "... the economic crisis (Greek court) is a worldly court, and the environmental crisis is God's court. This is punishment for grave sins, when in our acquisitiveness and passions we forget about the planet that gave us life."

The Vatican has included environmental pollution among its deadly sins.

The global economic crisis that erupted in the second half of 2008 has also increased the threat to the environment. The reduction of industrial production in a crisis does not reduce the technogenic load on nature, but, on the contrary, causes an avalanche-like increase in the total volume of pollution and the threat of an environmental crisis. This is happening due to the reduction of environmental programs, the curtailment of projects and plans to introduce new treatment facilities, and in some cases, the shutdown of existing protection systems in order to save money.

In order to avoid excessive spending in the state budget on energy supply and solving environmental problems and to prevent a slowdown in the growth of public welfare, it is necessary to debug a system in which the "culprits" will bear responsibility for the negative consequences of energy production and consumption [24-29].

Today, two methods of limiting the negative impact on the environment are most widespread in developed countries: a) establishing a tax on CO_2 emissions and b) limiting industrial emissions of pollutants (NO_x and SO_x) using quotas. The implementation of both methods relies on effective monitoring - accurate, uniform and transparent collection and recording of data on emission levels.

The republic traditionally artificially lowers costs associated with energy consumption, since the costs of negative effects on the environment and, as a result, on public health due to emissions of harmful substances (primarily SOx and NOx) by energy facilities are not included in energy prices. Without a system for effectively taking into account these negative external factors (through energy and "green" taxes, including the costs of environmental measures in the cost of goods and services, quotas for greenhouse gas emissions), it is impossible to fully realize the potential for increasing energy efficiency [30-32].

It is advisable to combine measures to improve the energy efficiency of technological processes and equipment with work to improve environmental performance, and conduct energy and environmental audits simultaneously and, if possible, jointly (clause 4.4). On the other hand, one of the ways to practically implement the environmental potential of ES PEE should be the inclusion in regional and other environmental programs of energy-saving measures that ensure a qualitative improvement in the state of the natural environment.

Tightening requirements for enterprises regarding emissions of pollutants into the atmosphere and their discharge into soil and water and a corresponding increase in fines for exceeding their permissible volumes force them to carry out a number of organizational measures: inventory of sources and volumes of pollution; providing information to nature conservation committees; coordination of draft standards for maximum permissible emissions (discharges).

Electric power industry enterprises began to pay great attention to environmental monitoring as the first step towards the implementation of environmental protection measures. At power plants - the main sources of polluting emissions and discharges - monitoring and control is carried out by sanitary-industrial laboratories, which are certified and equipped with all the necessary equipment. If necessary, industry enterprises provide for the deployment of mobile laboratories. Air monitoring services, which are part of industrial sanitary laboratories, carry out air sampling according to a developed schedule and act accordingly. 24 hours a day. Enterprises have organized control over compliance with established standards for emissions and discharges of harmful substances, over the rational use of natural resources and the disposal of industrial waste.

A well-functioning dispatch control and management system makes it possible to timely detect and identify the causes of failures in the operation of the power system, take measures to prevent and eliminate emergency situations, preventing unauthorized pollution of the environment. In addition, all production facilities of thermal power plants where accidents with environmental consequences may occur are insured.

A tool for solving these problems should be full-fledged environmental engineering, which is fundamentally different from engineering in other areas of activity, since its goal for the customer (enterprise) is to ensure the required quality of the natural environment and atmospheric air, and not to make a profit at any cost. Therefore, the active participant in the process, in addition to the customer and the contractor, must be the state. It is this that, through organizational, legal and other mechanisms, initiates the process of engineering environmental protection systems, for example, gas purification of atmospheric air. It is necessary to create a state inspection to control gas treatment facilities. And, what is especially important, is economic stimulation of environmental protection activities of industrial enterprises [41-44].

In recent years, an environmental quality management system has been formed, and international standards ISO 14601:2004 have been introduced [33-35].

3 Conclusion

To implement an environmental program in the field of energy, national programs for reducing negative impacts on the environment are needed, brought to specific installations at specific stations. An important element of such a program should be inter-budgetary relations and rules for financing environmental projects, providing for shared participation of funds from the federal budget and the budgets of constituent entities. This approach is consistent with the environmental doctrine, which states that "...priority for funding at the federal level will be projects focused on structural transformations in the dirtiest industries," which, as noted above, includes thermal energy.

In conclusion, it should be noted that social norms in our country also do not stimulate energy saving and caring attitude towards the environment. International agreements on environmental protection can serve as basic documents for creating a system for stimulating the development and implementation of energy- and environmentally efficient technologies.

References

1. Rasulov A.N., Ruzinazarov M.R. E3S Web of Conferences 384. 01042. (2023). https://doi.org/10.1051/e3sconf/202338401042
2. Ruzinazarov M.R. E3S Web of Conferences 384. 01050. (2023). https://doi.org/10.1051/e3sconf/202338401050
3. Usmanov E.G. E3S Web of Conferences 384. 01034. (2023). https://doi.org/10.1051/e3sconf/202338401034
4. Rakhmonov I.U., Ganiev Sh.R., Alibekova T.Sh., Nematov L.A. E3S Web of Conferences 384. 01029. (2023). https://doi.org/10.1051/e3sconf/202338401029
5. Ushakov, V.Y., Mytnikov, A.V., Rakhmonov, I.U. Non-electrical Diagnostic Methods. In: High-Voltage Equipment of Power Systems. Power Systems. Springer, Cham. (2023). https://doi.org/10.1007/978-3-031-38252-9_6
6. Bobojanov M.K., Rismukhamedov D.A., Tuychiev F.N., Shamsutdinov Kh.F. E3S Web of Conferences 365. 04024. (2023). https://doi.org/10.1051/e3sconf/202336504024
7. Rasulov A.N., Paxratdinov A.Dj. E3S Web of Conferences 384. 01035. (2023). https://doi.org/10.1051/e3sconf/202338401035
8. Reymov K.M., Esemuratova Sh.M., Khusanov B.M., Mytnikov A.V. E3S Web of Conferences 384. 01047. (2023). https://doi.org/10.1051/e3sconf/202338401047
9. Saidkhodjaev A.G., Bobojonov Y.M. E3S Web of Conferences 384. 01040. (2023). https://doi.org/10.1051/e3sconf/202338401040
10. Ushakov, V.Y., Mytnikov, A.V., Rakhmonov, I.U. Diagnostics of High-Voltage Equipment by Defectography Methods. In: High-Voltage Equipment of Power Systems. Power Systems. Springer, Cham. (2023). https://doi.org/10.1007/978-3-031-38252-9_8
11. Ushakov, V.Y., Mytnikov, A.V., Rakhmonov, I.U. Diagnostics of High Voltage Equipment. In: High-

Voltage Equipment of Power Systems. Power Systems. Springer, Cham. (2023). https://doi.org/10.1007/978-3-031-38252-9_10
12. Rakhmonov I.U., Niyozov N.N., Kurbonov N.N., Umarov B.S. E3S Web of Conferences 384. 01030. (2023). https://doi.org/10.1051/e3sconf/202338401030
13. Bobojanov M., Mahmutkhonov S. E3S Web of Conferences 384. 01041. (2023). https://doi.org/10.1051/e3sconf/202338401041
14. Ushakov, V.Y., Mytnikov, A.V., Rakhmonov, I.U. Defects in High Voltage Equipment: Types and Content of Tests. In: High-Voltage Equipment of Power Systems. Power Systems. Springer, Cham. (2023). https://doi.org/10.1007/978-3-031-38252-9_5
15. Melikuziev M.V. E3S Web of Conferences 384. 01033. (2023). https://doi.org/10.1051/e3sconf/202338401033
16. Fayziyev M., Tuychiev F., Mustayev R., Ochilov Y. E3S Web of Conferences 384. 01038. (2023). https://doi.org/10.1051/e3sconf/202338401038
17. Niyozov N.N., Rikhsitillaev B.Kh., Jalilova D.A., Kuatbaev P.M. E3S Web of Conferences 384. 01031. (2023). https://doi.org/10.1051/e3sconf/202338401031
18. Hoshimov, F.A., Najimova A.M., Rafikova G.R., Bakhadirov I.I. E3S Web of Conferences 384. 01049. (2023). https://doi.org/10.1051/e3sconf/202338401049
19. Taslimov A.D., Rakhimov F., Rakhimov Feruz. E3S Web of Conferences 384. 01037. (2023). https://doi.org/10.1051/e3sconf/202338401037
20. Allaev K., Toshov J., Toirov O. E3S Web of Conferences 401. 05090. (2023). https://doi.org/10.1051/e3sconf/202340105090
21. Sattarov Kh., Sapayev M., Suyarov A., Turaev, A. E3S Web of Conferences, 401, 04001, (2023). https://doi.org/10.1051/e3sconf/202340104001
22. Jurayeva K., Sattarov Kh. E3S Web of Conferences, 389, 01060. (2023). https://doi.org/10.1051/e3sconf/202338901060
23. Toirov O., Khalikov S., Sodikjon Khalikov, F. Sharopov. E3S Web of Conferences 410. 05015. (2023). https://doi.org/10.1051/e3sconf/202341005015
24. Toshev J.B., Norkulov M.B., Urazimbetova A.A., Toshniyozov L.G. E3S Web of Conf., Volume 402, 10039 (2023). https://doi.org/10.1051/e3sconf/202340210039
25. K. Allaev, J. Toshov and O. Toirov. E3S Web of Conf., Volume 401, 05090 (2023). DOI https://doi.org/10.1051/e3sconf/202340105090
26. Sattarov Kh., Safarov A., Kamolov A., Egamberdiev M., Raxmanova G. E3S Web of Conferences, 383, 01028. (2023) https://doi.org/10.1051/e3sconf/202338301028
27. Siddikov I.Kh., Sattarov Kh.A., Abdumalikov A.A. AIP Conference Proceedings, 2612, 050002, (2023). https://doi.org/10.1063/5.0115211
28. Abdi Yonis S., Yusupov Z., Habbal A., Toirov O. Advances in Electrical and Electronic Engineering.Vol 21. No 3 (2023). https://doi.org/10.15598/aeee.v21i3.5149
29. Safarov A.M., Sattarov Kh.A., Mustafoqulov A.M., Shoyimov Y.Y. Journal of Physics: Conference Series, 2176(1), 012006, (2022). doi:10.1088/1742-6596/2176/1/012006
30. Safarov A., Sattarov Kh., Bazarov M., Mustafoqulov A. E3S Web of Conferences, 2021, 264, 05038, https://doi.org/10.1051/e3sconf/202126405038
31. Pirmatov N., Muminov M., Akberdiev M., Abdullayev J. AIP Conference Proceedings 2432. 0200200. (2022). https://doi.org/10.1063/5.0089838
32. Pirmatov N., Bekishev A., Shernazarov S., Kurbanov N., Norkulov U. E3S Web of Conferences . 264. 04028. (2021). https://doi.org/10.1051/e3sconf/202126404028
33. Safarov A.M., Jumaboyev S.K., Sattarov Kh.A. IOP Conference Series: Materials Science and Engineering, 734(1), 012196, (2020). doi: 10.1088/1757-899X/734/1/012196
34. Baratov R., Pirmatov N., Panoev A., ...Ruziyev S., Mustafoqulov, A. IOP Conference 1030. 012161. (2021). doi:10.1088/1757-899X/1030/1/012161
35. Pirmatov N., Panoev A. E3S Web of Conferences 216. 01120. (2020). https://doi.org/10.1051/e3sconf/202021601120

Increasing the environmental cleanness of industrial enterprises

Gulnara Rafikova[1*], *Ruslan* Mustaev[2], *Ramziddin* Pirimov[3], *Feruza* Zokirova[1]

[1]Tashkent State Technical University named after Islam Karimov, 100095, Uzbekistan, Tashkent, University St. 2A

[2]Karshi Institute of Engineering Economics, 225 Independence Avenue, Karshi, Uzbekistan

[3]Bukhara Engineering Technological Institute, Bukhara, 200100, Uzbekistan

Abstract. The article discusses important issues to improve the environmental friendliness of industrial enterprises. Emissions of harmful substances into the atmosphere by coal-fired power plants are analyzed. As well as emissions of pollutants (kg) of various fuels per 1 ton of equivalent fuel, 1 Gcal of heat and 1 MWh of electricity. Potentially possible emissions of the main pollutants when using coal, firewood and diesel fuel for domestic needs are given. An estimate of the reduction in emissions of pollutants into the atmosphere is projected, thousand tons/year. The effects of environmental and energy-saving measures in the energy sector are assessed. The following are the negative impacts of overhead power lines: the biological impact of electromagnetic fields on fauna, including humans; cutting down trees and alienation of significant land plots for power transmission lines; disturbance of the natural landscape; interference to radio and television reception from corona discharge on power line wires.

1 Introduction

The foundations of the current legislation in the environmental sphere were formed in the early 90s, when, due to a drop in production volumes, a reduction in the volume of pollution was expected. This led to a reduction in environmental barriers, which did not so much promote economic growth as create conditions for the preservation of outdated technologies to the detriment of production modernization [1].

In the electric power industry, this led to a significant lag behind advanced countries in terms of economic and environmental efficiency and an imbalance in the fuel balance, Table 1.

Table 1

Emissions of harmful substances into the atmosphere from coal-fired power plants

Pollutants	Emissions, g/kWh	
	Germany	Russia
NO_X	0,5	2
CO_2	0,3	3
PM*	0,06	2,5

PM* – volatile macroparticles (particulate matters).

2 Materials and Methods

There are four main ways to reduce (or at least reduce the growth rate) the load on the environment with increasing energy consumption, which we will consider below.

1. Increasing the environmental friendliness of technologies for the extraction, primary processing and transportation of traditional energy resources.

The increase in the production of energy raw materials is accompanied by an exacerbation of environmental problems due to the manifestation of such effects as:

• accumulation of drill cuttings,

• salinization of groundwater due to violations of technologies for the operation and abandonment of oil and gas wells,

• emissions of combustion products into the atmosphere when burning associated petroleum gases and oil sludge,

• pollution of the air, soil and water with petroleum products,

• soil subsidence during hydrocarbon gas production.

Environmental problems are so large-scale that they may complicate the economic situation in the near future.

Considering the state of the production equipment of oil companies, it can be stated that over the coming years the problem of eliminating emergency oil spills will remain relevant. The most cost-effective is the

* Corresponding author: ilider1987@yandex.ru

prevention of emergency situations: prevention of oil spills during production, automated monitoring of the condition of pipelines and identification of pipeline sections requiring replacement or overhaul [2-4].

During production and transportation, between 1.5% and 10% of oil is lost, which is equivalent, according to a minimum estimate, to approximately 4.5 million tons per year. Penalties for 1 ton of spilled oil correspond to the cost of 80 tons of extracted oil, and the cost of collecting it is equivalent to the cost of 0.8–1.3 tons of oil. Consequently, the economic effect of taking measures to prevent emergency oil spills will be significantly higher than the costs of their implementation.

The production and transportation of natural gas, compared to oil, poses a significantly lower environmental hazard, because the gas has a relative density of 0.6 and, when released, quickly dissipates into the atmosphere. A fairly high degree of technological safety is ensured by a number of measures: mainly underground laying and placement of gas pipelines outside residential buildings, regular and fairly strict monitoring of the condition of all elements of the gas pipeline system. Their result is a decrease in the intensity of accidents on gas pipelines over the past 20–25 years by approximately 2.5 times [5-10].

Significant damage to nature is caused by the extraction and preliminary preparation of solid fuel (primarily coal): land alienation and landscape disturbance during open-pit mining, soil subsidence during open-pit mining, waste rock storage problems in both mining methods (large-scale dumps), air pollution and soil. When transporting solid fuel, contamination of land adjacent to railways and roads is almost inevitable.

2. Changing the fuel and energy standards and improving the environmental friendliness of fuel combustion. The load on the environment decreases with an increase in the fuel and energy balance (TEB) of the share of more environmentally friendly sources of primary energy: fuel with less carbon or no carbon at all (uranium, hydrogen), renewable energy sources, and in the long term - fuel for thermonuclear power plants [11-16].

Table 2

Pollutant emissions (kg) of various fuels per 1 ton of fuel equivalent, 1 Gcal of heat and 1 MWh of electricity

Fuel	Specific emissions of pollutants, kg/t of standard fuel; kg/Gcal of heat; kg/(MWh)		
Coal	30; 5,4; 10,5	4.57; 0,82; 1,6	10; 1,8; 3,5
Fuel oil	20; 3,6; 7,0	6; 1,1; 2,1	-
Natural gas	-	2,3; 0,41; 0,8	-
Coke gas	5,7; 1,03; 2	2,8; 0,51; 1,0	-
Blast gas	-	1,43; 0,26; 0,5	-

Table 3

Potentially possible emissions of major pollutants when using coal, firewood and diesel fuel for domestic needs

Release source	Fuel	Emission of pollutants from fuel combustion products, kg/t e.g.						
		Ash	Soot	SO_2	NO_2	CO	Hydrocarbons	Benz(a)pyrene
Heating boiler with manual layer firebox	Coal	3-8	11-22	4-15	6-9	19-60	0,03-0,2	0,001-0,008
	Peat	1-3	10	2	3,5	15-50	0,02-0,1	0,001
House stove	Coal	1,5-2,4	9-13	0,4-7	5-11	5-420	0,08-0,5	0,007-0,0013
	Firewood	0,6	0,9	0	3	20-350	0,002	0,00006
Diesel electric generator	Diesel fuel	0	5	-	25	13	1,3	No data
Substance hazard class	-	3	3	3	3	4	2-4	1

Table data 2, 3 show that coal combustion is accompanied by the release of the largest volume of pollutants and, above all, such dangerous ones as benzopyrene. Accordingly, the greatest environmental damage in monetary terms in the production of electrical energy occurs at coal thermal power plants - 64 cents per 1 kWh; they pollute the atmosphere not only with emissions from chimneys, but also with volatile substances released when coal is stored in open areas. The development of coal-fired energy, while simultaneously solving the problem of reliable energy supply (thanks to the world's huge coal reserves), aggravates the problem of environmental pollution. Annual emissions from energy installations in the world are: carbon dioxide – (2–3)*1010 tons, fly ash (particulate matters) – 2.5*1010 tons, NOx – 1.2*1010 tons, SOx – 1.5*1010 t, significant volumes of toxic components. The share of coal energy in total CO2 emissions is 30% of the total for all sectors of the economy and 2% of the world. One coal-fired power plant with a capacity of 150 MW with today's average technical characteristics produces more than 1 million tons of greenhouse gas emissions per year (equivalent to the exhaust of 300 thousand cars).

The experience of highly developed countries (primarily Germany, England, Japan, South Korea) shows that the current level of technology makes it possible to minimize the environmental impact of coal-fired power plants, but only by implementing additional measures that complicate and increase the cost of the construction and operation of such thermal power plants.

In table 4 presents a forecast assessment of the reduction in emissions of pollutants into the atmosphere as a result of the implementation of two groups of measures: technological and organizational.

Table 4

Forecast estimate for reducing emissions of pollutants into the atmosphere, thousand tons/year

	Use of CCGT and GTU when burning natural gas	Improving combustion efficiency	Installation of highly efficient gas cleaning equipment	Implementation of measures to achieve technical standards	Total
NO_x	5,5	10,0	11,5	304,6	331,6
SO_2		37,2	165,3	58,2	260,7
Ash		30,1	295,3	280,0	605,3
Total					1197,6

The most effective measures in terms of reducing emissions and the proceeds from their implementation are given in Table 5. In addition, projects for replacing boiler-turbine equipment with a sharp increase in its efficiency, as well as local technological improvements combined into program projects (introduction of variable-frequency electric drives, low-power turbines for using reduced steam, use of turboexpanders, recycling) can be considered as "carbon" measures. waste heat, etc.).

Table 5

Effects of environmental and energy-saving measures in the energy sector

Event	Reduction of CO_2 emissions (at 100 MW plants), thousand tons	Funds from the sale of reduced emissions over 5 years (at a price of 10 euros per 1 ton of CO_2), million euros
Replacement of steam turbines of gas-fired thermal power plants with combined-cycle units.	100–250	9–13
Introduction of gas turbine superstructures at thermal power plants	90–130	5–7
Conversion of existing thermal power equipment from coal to gas	190–270	10–14
Development of district heating and improvement of heat supply systems with the elimination of small boiler houses	90–240	4–12
Distribution of renewable energy sources, including small and medium-sized hydropower plants	220–460	11–23

When determining ways to develop heat supply to the residential and public consumption sectors, one of the main factors becomes environmental. Limits for the maximum content of harmful gases in the air basin of populated areas have been established and, on their basis, norms for maximum permissible concentrations of harmful substances in the air basin have been adopted, Table. 6 [17].

The concentration of pollution in the active impact zone depends on such factors as: the quality and type of fuel burned, the location of the source of pollution, the type of boiler, the state of the atmosphere, etc. This indicator is significantly different for decentralized (DCT) and centralized (DCT) heat supply systems.

The concentration of pollution in the active impact zone depends on such factors as: the quality and type of fuel burned, the location of the source of pollution, the type of boiler, the state of the atmosphere, etc. This indicator is significantly different for decentralized (DCT) and centralized (DCT) heat supply systems.

Along with the undeniable advantages of DCTs in a number of indicators, in environmental terms they have significant disadvantages [18-19]:

Table 6

Harmful substances contained in flue gases during combustion fuels and their maximum permissible concentrations in the atmosphere

Name of substance	MPC in the atmospheric air of populated areas, mg/m^3		Maximum concentration in the air of the working area mg/m^3	Hazard class for humans
	Maximum one-time	Average daily		
Dust is non-toxic	0,5	0.15	1-10	3-4
Sulfur anhydrite	0,5	0,05	-	3
Carbon monoxide CO	3	1	20	4
Vanadium pentoxide	-	0,002	0,1	1
Soot (soot)	0,15	0,05	6	4
Nitrogen dioxide NO_2	0,085	0,085	5	2
Benzopyrene	-	0,1mkg/100m^3	-	1

• large mechanical and chemical underburning of fuel due to the design features of small boilers (small combustion volume, low coolant temperature and, as a consequence, incomplete combustion of fuel and high temperature of flue gases);

• low efficiency when using solid fuels;

• low height of chimneys and, as a result, high ground-level concentrations of harmful substances, many times (up to tens of times) higher than those during SCT of the same objects.

In turn, centralized heat supply, despite its characteristic advantages, has its disadvantages that reduce their energy efficiency:

a) huge emissions of low-grade heat, primarily from the process water cooling system at nuclear power plants and state district power plants (increasing during the period of decreasing heat load during the non-heating period);

b) large heat costs for heating network water, replenishing losses in heating networks due to their wear and tear. Due to these problems, natural fossil energy resources are turning into "thermal rivers" on the surface of the planet.

In terms of environmental impact, the fuel triad is followed by nuclear fuel and renewable energy sources. It should be borne in mind that nuclear energy, like renewable energy, can be classified as "clean" production only if we take into account only the operation of the power plants themselves. Indeed, the impact of nuclear energy enterprises on the environment is extremely small: the share of emissions from nuclear energy enterprises in general industrial emissions of pollutants does not exceed 0.6%. However, the enterprises that ensure their operation are sources of environmental pollution. (Indirect damage to the environment) [38-44].

Technologies for generating electricity from renewable energy sources do not contain oxidation or nuclear transformation processes that pollute the environment, but they also have a negative direct and indirect impact on nature. Among the direct impacts are alienation of land, acoustic noise, disruption of natural migration routes of birds (wind energy), shading of the earth's surface (solar energy), release of aggressive solutions from the depths (geothermal energy), etc [20-25].

3. Reducing damage to nature during the transport of electrical and thermal energy.

The damage to nature at this technological stage is not as great as during the production of electrical and thermal energy. Environmental problems during the production of thermal energy are many times greater than those during its transmission. For this reason, the distance of the heat source from its consumer and, accordingly, the length of heating mains are determined not by their environmental impacts, but by economic considerations, although as the length of heating mains increases, the likelihood of accidents with serious consequences increases: disruption of transport communications due to excavation work and flooding, and in some cases and human casualties [26-31].

Negative impacts of overhead power lines (EPLs) include:

• biological impact of electromagnetic fields on representatives of fauna, including humans;

• cutting down trees and alienation of significant land plots for power transmission lines;

• disturbance of the natural landscape;

• interference with radio and television reception from corona discharge on power line wires.

4. Reducing the release of greenhouse gases into the atmosphere.

Conclusion

One way to solve this problem is to inject greenhouse gases into cavities formed during oil and gas production, into saline porous aquifers, and into coal seams unsuitable for development. To date, at least two such industrial-scale projects have been implemented. In the USA, carbon dioxide, which is a waste product from the methanization of synthetic gas, is pumped into a depleted oil field, simultaneously ensuring the resumption of oil recovery from the reservoir. In the Norwegian Sea, CO_2 separated from produced natural gas is injected into a porous aquifer. EU countries plan to equip 12 power plants (existing and under construction) with underground storage facilities – greenhouse gas repositories – by 2035 [32-33].

The implementation of this method is hampered by a high price - with the technologies available today, the separation and disposal of each ton of CO_2 requires up to $60. Large-scale use of such technology would significantly increase the cost of electricity. What is needed is a technology with a price no higher than $20–30 per 1 ton of CO_2. In the United States, more than $1 billion has been spent on the development of such technologies.

References

1. Tuychiev F, Haqberdiev A. E3S Web of Conferences 384. 01044. (2023). https://doi.org/10.1051/e3sconf/202338401044
2. Ushakov, V.Y., Mytnikov, A.V., Rakhmonov, I.U. Diagnostics of High Voltage Equipment. In: High-Voltage Equipment of Power Systems. Power Systems. Springer, Cham. (2023). https://doi.org/10.1007/978-3-031-38252-9_10
3. Saidkhodjaev A.G., Bobojonov Y.M. E3S Web of Conferences 384. 01040. (2023). https://doi.org/10.1051/e3sconf/202338401040
4. Bobojonov Y.M., Saidkhodjaev A.G. E3S Web of Conferences 384. 01048. (2023). https://doi.org/10.1051/e3sconf/202338401048
5. Rasulov A.N., Paxratdinov A.Dj. E3S Web of Conferences 384. 01035. (2023). https://doi.org/10.1051/e3sconf/202338401035
6. Hoshimov, F.A., Najimova A.M., Rafikova G.R., Bakhadirov I.I. E3S Web of Conferences 384. 01049. (2023). https://doi.org/10.1051/e3sconf/202338401049
7. Rasulov A.N., Ruzinazarov M.R. E3S Web of Conferences 384. 01042. (2023). https://doi.org/10.1051/e3sconf/202338401042
8. Ushakov, V.Y., Mytnikov, A.V., Rakhmonov, I.U. Diagnostics of High-Voltage Cable Lines. In: High-Voltage Equipment of Power Systems. Power Systems. Springer, Cham. (2023). https://doi.org/10.1007/978-3-031-38252-9_11
9. Bobojanov M., Torayev S. E3S Web of Conferences 384. 01045. (2023). https://doi.org/10.1051/e3sconf/202338401045
10. Reymov K.M., Esemuratova Sh.M., Khusanov B.M., Mytnikov A.V. E3S Web of Conferences 384. 01047. (2023). https://doi.org/10.1051/e3sconf/202338401047
11. Niyozov N.N., Rikhsitillaev B.Kh., Jalilova D.A., Kuatbaev P.M. E3S Web of Conferences 384. 01031. (2023). https://doi.org/10.1051/e3sconf/202338401031
12. Fayziyev M., Tuychiev F., Mustayev R., Ochilov Y. E3S Web of Conferences 384. 01038. (2023). https://doi.org/10.1051/e3sconf/202338401038
13. Rashidov N., Rozmetov Kh., Rismukhamedov S., Peysenov M. E3S Web of Conferences 384. 01043. (2023). https://doi.org/10.1051/e3sconf/202338401043
14. Bobojanov M., Mahmutkhonov S. E3S Web of Conferences 384. 01041. (2023). https://doi.org/10.1051/e3sconf/202338401041
15. Bobojanov M.K., Rismukhamedov D.A., Tuychiev F.N., Shamsutdinov Kh.F. E3S Web of Conferences 365. 04024. (2023). https://doi.org/10.1051/e3sconf/202336504024
16. Usmanov E.G. E3S Web of Conferences 384. 01034. (2023). https://doi.org/10.1051/e3sconf/202338401034
17. Melikuziev M.V. E3S Web of Conferences 384. 01033. (2023). https://doi.org/10.1051/e3sconf/202338401033
18. Taslimov A.D., Rakhimov F., Rakhimov Feruz. E3S Web of Conferences 384. 01037. (2023). https://doi.org/10.1051/e3sconf/202338401037
19. Zikrillayev N., Toshov J., Saitov E., Sodikov T. AIP Conference Proceedings 2552, 030020-1–030020-8; https://doi.org/10.1063/5.0117594
20. Toshov B., Toshov J., Akhmedova L., Baratov B. E3S Web of Conferences 383, 04069 (2023) TT21C-2023 https://doi.org/10.1051/e3sconf/202338304069
21. Safarov A., Sattarov Kh., Bazarov M., Mustafoqulov A. E3S Web of Conferences, 264, 05038. (2021) https://doi.org/10.1051/e3sconf/202126405038
22. Safarov A.M., Jumaboyev S.K., Sattarov Kh.A. IOP Conference Series: Materials Science and Engineering, 734(1), 012196, (2020). doi: 10.1088/1757-899X/734/1/012196
23. Toshov J., Toshov B., Bainazov U., Elemonov M. Application of Cycle-Flow Technology in Coal Mines // Proceedings of the 11th International Conference on Applied Innovations in IT, (ICAIIT), March 2023, Koethen, Germany. – PP. 279-284.
24. Sattarov Kh., Safarov A., Kamolov A., Egamberdiev M., Raxmanova G. E3S Web of Conferences, 383, 01028. (2023) https://doi.org/10.1051/e3sconf/202338301028
25. Toirov O., Khalikova M., Jumaeva D., Kakharov S. E3S Web of Conferences 401. 05089. (2023). https://doi.org/10.1051/e3sconf/202340105089
26. Toirov O., Sadullaev T., Abdullaev D., Jumaeva D., Ergashev Sh., Sapaev I.B. E3S Web of Conferences 383. 01029. (2023). https://doi.org/10.1051/e3sconf/202338301029
27. Toirov O., Jumaeva D., Mirkhonov U., Urokov S., Ergashev S. AIP Conference Proceedings 2552. 040021. (2022). https://doi.org/10.1063/5.0111331
28. Afrisal H., Setiyadi A. D., Riyadi M. A., Ismail R., Toirov O. International Journal on Advanced Science, Engineering and Information Technology. Vol. 12 No. 6. 2258-2265. (2022). https://doi.org/10.18517/ijaseit.12.6.16591
29. Aljwary A., Yusupov Z., Toirov O., Shokirov R. *E3S Web of Conferences* **304**. 01010. (2021). https://doi.org/10.1051/e3sconf/202130401010
30. Sattarov Kh., Sapayev M., Suyarov A., Turaev, A. E3S Web of Conferences, 401, 04001, (2023). https://doi.org/10.1051/e3sconf/202340104001
31. Jurayeva K., Sattarov Kh. E3S Web of Conferences, 389, 01060. (2023). https://doi.org/10.1051/e3sconf/202338901060
32. Siddikov I.Kh., Sattarov Kh.A., Abdumalikov A.A. AIP Conference Proceedings, 2612, 050002, (2023). https://doi.org/10.1063/5.0115211
33. Safarov A.M., Sattarov Kh.A., Mustafoqulov A.M., Shoyimov Y.Y. Journal of Physics: Conference Series, 2176(1), 012006, (2022). doi:10.1088/1742-6596/2176/1/012006

Methods for optimal placement of PMU devices to assess the stability of the Azerbaijan power system

Nurali Yusifbayli[1], *Valeh* Nasibov[2*] and *Kamran* Suleymanov[2]

[1]Azerbaijan Technical University, AZ1073 Baku, Azerbaijan
[2]Azerbaijan Scientific-Research and Design-Prospecting Power Engineering Institute, AZ1012, Baku, Azerbaijan

Abstract. This article discusses the method of integer linear programming, a method based on estimating the rate of change in operating conditions when conditions become more complex or disturbances occur. In the first method, the criterion for placing PMUs in the energy system of Azerbaijan is to ensure full observability; the second method examines the rate of change in the relative angles of synchronous generators under large disturbances. Based on the obtained parameters, weak nodes are identified where it is necessary to install PMU devices.

1 Introduction

Fulfilling the obligations of the Paris Agreement related to the problems of reducing greenhouse gas emissions and climate change poses large-scale challenges for the energy sector, as the most responsible industry for climate change.

For the purposes of the fulfillment of obligations under the Paris Agreement, Azerbaijan has committed itself to reducing the share of CO2 by 2030 compared to 1990 by 35% and increasing the share of green electricity to 30% by installing 1,500 MW of renewable energy sources.

According to the agreement signed between the governments of the Republic of Azerbaijan, Georgia, Romania and Hungary on a strategic partnership in the field of development and transfer of green energy, it is planned to transfer 4 GW of wind power from the Azerbaijani sector of the Caspian Sea to the countries of the European Union in the next five years.

The longest electric cable in the world will connect Azerbaijan and Europe, laying a green energy bridge between the Caspian region and the EU.

It is planned to bring up the transmitted green power from the Caspian Sea to the EU countries to 25 gigawatts by 2037. At that, various schemes of transmission are analyzed: with integration into the power system, without integration with the power system and asynchronous integration into the power system of Azerbaijan. In all cases, the large volume of transmitted power over long distances puts forward stringent requirements for the reliability, stability and survivability of the Azerbaijan power system.

The high level of requirements for ensuring the reliability and survivability of the Azerbaijan power system, as a central subject in the Azerbaijan-Georgia-Romania-Hungary, Azerbaijan-Turkey-Greece and the Russian Federation-Azerbaijan-Iran schemes, stipulates the need for the process of its intellectualization [1, 2].

2 Innovative information-measuring system WAMS

One of the most important components of the intellectualization concept is an innovative information-measuring system based on synchrophasor (vector) real-time measurement technology. Such a system is the WAMS system, in the structure of which, along with the WACS, WAPS and PDC systems, the PMU device operates [4-6]. In general, the WAMS system provides information about the state of the power system with an accuracy of voltage measurement ± 0.1%, phase angle ± 0.2 degrees, current ± 0.2%, frequency ± 0.01 Hz, angle between branch current and node voltage ± 0.20c with a frequency of 20 ms.

The vector of measurements carried out by the PMU device is represented as:

$$Y = [\ Ui\ bi\ Iij\ \Psi ij\ ] \tag{1}$$

where

U_i –voltage modulus of the ith node;

b_i– voltage phase of the ith node;

I_{ij}–current flowing out of the ith node and flowing into the jth node;

Ψ_{ij}– phase shift between current and voltage.

The SCADA/EMS system has already been installed and is operating in the Azerbaijan power system.

Despite its enormous functionality, the SCADA/EMS system is inferior to the WAMS system in a number of indicators, such as measurement synchronicity, volume, speed and accuracy of information transfer. Tasks such as monitoring of oscillatory stability, disturbance detection, event recording, post-accident analysis and

* Corresponding author: nvaleh@mail.ru

other tasks that require millisecond synchronous phase measurements are difficult to solve within the framework of SCADA/EMS. In these conditions, the greatest efficiency is achieved by combining the capabilities of both SCADA/EMS and WAMS systems.

As mentioned above, the SCADA/EMS system is already operating in the Azerbaijan power system, so the main attention needs to be paid to the problem of PMU placement [3-5].

As a rule, one of the main criteria for the optimal placement of PMUs in the power system is to ensure full observability of the power system, which means that the number and composition of measurements are sufficient to control the power system modes in all conditions (before an accident, during and after an accident). Naturally, this criterion (observability criterion) can be implemented if PMU is installed in all nodes.

However, the high cost of the PMU devices themselves with their current and voltage channels, the need to connect these channels with data centers (PDC) at the locations of the latter, etc. require preliminary research.

Full observability can be ensured by using voltage measurements at the nodes of the PMU installation and currents from these nodes and by calculating the voltages and currents in adjacent nodes. At that, the accuracy is maintained.

In relation to the problem of assessing the state of the power system, where the main issue is the identification of the power system diagram, the topological aspect of observability, which is based on a linear system of equations, is important.

The solution to the problem of optimal placement of PMUs in the Azerbaijan power system is implemented using the integer linear programming (ILP) method, where the extreme value of a linear function of many variables is found in the presence of linear constraints connecting these variables. At that, the integer requirement is imposed on the variables. The initial function to be minimized is represented as:

$$\min \sum_{k=1}^{N} x_k$$

$$|A| \times |X| \geq b \qquad (2)$$

where:

N – number of nodes in the system;

X – binary solution vector;

A – integer matrix, the structure of which depends on the network scheme;

b – integer vector.

In (2), the elements of matrix A take the values

$$A_{ij} = \begin{cases} 1 \ \ if \quad i = j \\ 1 \ \ if \ i \ and \ \ j \ \ are \ connected \\ 0 \ \ if \ not \ connected \end{cases}$$

|X| - binary vector

$$|X| = |X_1 X_2 X_3 \dots X_N| T$$

$$X \in |0,1|$$

$$x_i = \begin{cases} 1 \ \ if \ PMU \ is \ installed \ in \ node \ \ i \\ 0 \ \ is \ absent \end{cases}$$

$$|b| = |1, 1, 1 \dots 1|^{T}_{1 \times N}$$

The ILP method is applied to the prospective scheme of the 500–330–220 kV electrical network of the Azerbaijan power system, which has 53 nodes, including 24 nodes in the 500–330 kV network and 29 nodes in the 220 kV network (Figure 1). The decomposition was carried out according to intersystem connections with the Russian power system (330 kV Khachmaz-Derbend OHL), the Georgian power system (500 kV Samukh-Gardabani OHL, 330 kV Akstafa-Gardabani OHL), the Iranian power system (230 kV Imishli-Parsabad OHL – 330 kV Imishli–Tagi–Dize OHL).

The calculation using the Matlab program presented the following placement of PMUs by nodes:

|X|=(0 1 0 0 1 0 0 0 1 0 0 01 0 0 0 0 0 0 1 0 0 0 0 0100 0 0 0 0 0 1 0 0 1 0 0 1 0 1 0 0 0 0 0 0 0 0 1 0 1)

That is, to ensure full observability, PMUs must be installed in 14 nodes (26% of the total number of nodes) of the PS, namely:

2– 330 kV Yashma SS	30– 220 kV Shimal PP
5– 220 kV Yashma PP	34– 220 kV Mingechevir HPP
9– 220 kV Boyuk-Shor	37– 500 kV Az TPP
13– 220 kV Absheron	40– 330 kV Samukh
16– 220 kV Gobu PP	42– 330 kV Gazakh
20– 220 kV Janub PP	50– 330 kV Imishli
26– 330 kV Alat Free Economic Zone	52– 330 kV Jabrayil

In adjacent nodes, observability is ensured by calculations based on measurements of voltages and flows from nodes where PMUs are installed.

Analysis of the obtained calculated results shows the following:

Each PMU has one voltage channel. The PMU at the 220 kV Absheron substation (node 13) has the largest number of current channels (9 channels). The node is practically the boundary for two decomposition subsystems: 500–330 kV (transmitting part) and 220 kV (receiving part) of the Azerbaijan power system.

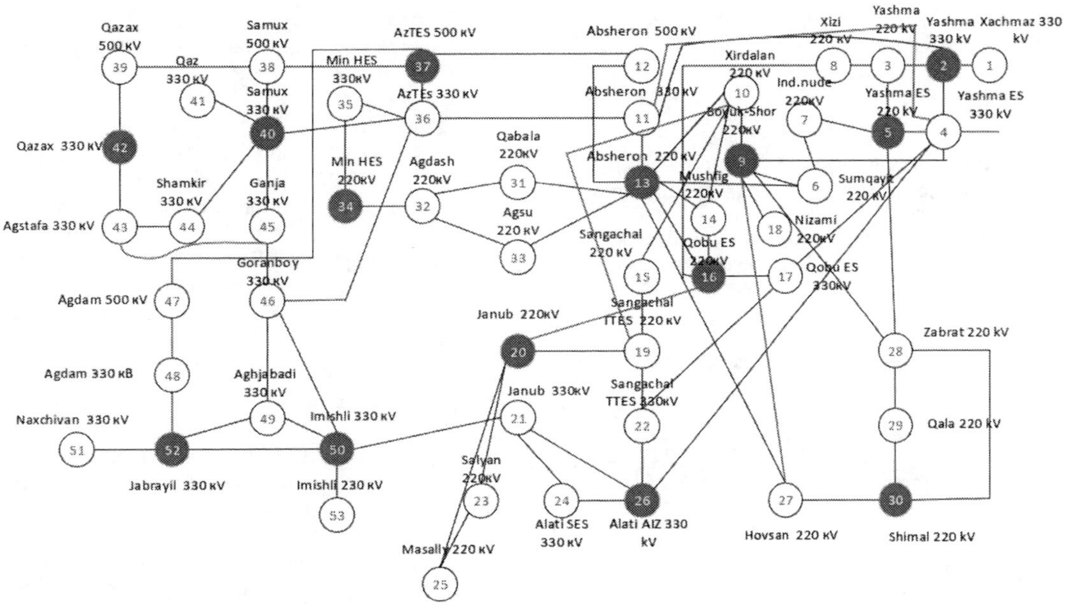

Fig. 1. Scheme of the 500-330-220 kV electrical network of Azerbaijan power system (PMU installation locations are marked)

At the 330 kV Imishli substation (node 50), which borders the interconnection with the Iranian power system from the Azerbaijan power system, a PMU with 5 current channels is installed.

Another 3 PMUs are installed with 5, 6 PMUs with 4, 1 PMU with 3 and 2 PMUs with 2 current channels.

If the cost of one channel is 1 USD, then the cost of PMU will require 60 USD.

The observability in the nodes of intersystem connections with the Georgian power system (nodes 39, 42 and 43) and the Russian power system (node 1) is ensured by calculation (either through branches from the PMU in neighboring nodes of the Azerbaijan power system, or corresponding neighboring power systems).

Measurements of voltage phases at the ends of the lines of the main cross-sections of the transmitting part (for example, 37–12, 36–11, 34–13, 50–21) allows for monitoring of carrying capacity and static stability margins in real time.

Integration of measurement data of PMU installed on the voltage buses of the Azerbaijan TPP station - 500 kV (node 37), Janub - 220 kV (node 20), Shimal - 220 kV (node 30), as well as data obtained by calculation on the voltage buses of the Azerbaijan TPP - 330 kV (node 36) and Sumgayit - 220 kV (node 6) with information from SCADA/EMS systems in the same nodes makes it possible to monitor dynamic processes under large disturbances.

3 Determination of PMU installation locations in "weak" elements of the power system

The installation locations of PMUs in the 500–330–220 kV network of the Azerbaijan power system are determined above based on topological rules, ensuring full observability of the power system. At that, the integer linear programming technique was used. In 53-node circuit of 500–330–220 kV network (76 branches), it is recommended to install PMU in 14 nodes (≤ 27%). In other nodes and adjacent branches, the dynamics of operating conditions are determined "by calculation" based on measurements of installed PMUs.

Another approach to solving problems of determining installation locations can be based on the physical content of the processes occurring in the system during disturbances. It is based on concepts such as heterogeneity, sensitivity, sensority and weakness of elements of the topological structure of the power system.

We focus on the last two concepts: sensority and "weakness" of elements, which may be sufficient to solve the problem.

Sensors are elements of the power system whose mode parameters change to a large extent due to random changes in the elements of the power system: generation, network, load.

Weak elements of the power system are those elements of the power system, changes in the parameters of which have the greatest impact on the magnitude of the power system's response to disturbances [6, 7].

The ambiguity of the structure of the power system, its modes, disturbances, and the nature of dynamic processes during disturbances has led to different approaches to identifying weak points in the power system.

One of these approaches is a method based on comparing the rate of change of mode parameters U, P, δ when the mode becomes heavy or when disturbances occur. Below are examples of the implementation of this approach.

One of the existing schemes and mode, characterized by the following data, is taken as the basis:

P_g= 4459 MW P_l = 4330 MW
Power transfer to Iranian power grid - 500 MW
including from Russian power grid - 300 MW
from Azerbaijan power grid -200 MW

3.1 Rate of change of relative angles of synchronous generators at large disturbances

As a disturbance, the shutdown of the 500 kV 2nd Absheron overhead line in the power flow mode was taken - 464 MW without automatic reclosure.

Table 1. Rates of their change in the first cycle of osillations after disturbances

	No.	δ_0	δ_{max}	t_0	t_{max}	V
330 kV Az TPP- Shimal PP	152-170	19	36,1	1,1	1,68	0,081
500 kV Az TPP- Shimal PP	157-170	18,5	40,6	1,1	1,64	0,114
Janub PP – Shimal PP	170-211	10	-0,2	1,1	1,8	0,042
330 kV Az TPP – 110 kV Janub PP	152-211	29	36,5	1,1	1,56	0,045
330 kV Az TPP – 220 kV Sumgait PP	152-181	28,2	41,4	1,1	1,58	0,076
500 kV Az TPP – 220 kV Sumgait PP	157-181	27,69	46,33	1,1	1,56	0,113
330 kV Az TPP – Baku CCPP	152-43	21,98	41,31	1,1	1,62	0,103
500 kV Az TPP – Baku CCPP	157-43	21,46	46,08	1,1	1,6	0,138
110 kV Janub PP – 220 kV Sumgait PP	211-181	-0,8	4,93	1,1	1,6	0,116
110 kV Janub PP – Baku CCPP	211-43	-7,02	5,17	1,1	1,66	0,060
Shimal PP – Baku CCPP	170-43	2,98	6,44	1,1	1,46	0,027
Shimal PP – 220 kV Sumgait PP	170-181	9,2	3,7	1,1	1,78	0,023

In Table 1, to compare the intensity of the dynamics of changes in angles, the rates of change in relative angles between the main synchronous generators of the power system in the first cycle of oscillations after disturbances, calculated by the formula, are given:

$$V\delta = \frac{\delta max - \delta min}{360(tmax - tmin)}$$

The analysis shows that the highest rate of change in relative angles occurs between synchronous generators:

Azerbaijan TPP – 500 kV – Baku CCPP
Azerbaijan TPP – 500 kV – Shimal TPP
Azerbaijan TPP – 500 kV – Sumgait TPP
Azerbaijan TPP – 330 kV - Baku CCPP
Azerbaijan TPP – 330 kV - Shimal TPP

Therefore, to monitor the mutual angles on the voltage buses of synchronous generators, it is recommended to install PMU at the 500 kV Azerbaijan TPP and 330 kV Azerbaijan TPP in the redundant and main stations of the deficit part (Shimal TPP, Sumgait PP, Baku CCPP).

Conclusions

The high cost of PMU requires their placement based on the operating characteristics of the power plant. The presence of intersystem connections of a transit function in the Azerbaijan ES, the presence of critical sections, predetermines the need to place PMU in "weak" places, where monitoring of stability levels and stresses, etc. based on measurements is necessary.

For the promising scheme of the Azerbaijan ES, to take into account the regime features, it is necessary to install PMU on the buses of power plants in the deficient part of the power system.

References

1. Hector J., Altuve Ferrer, Edmund O. Schweitzer III, *Modern Solutions for Protection. Control and Monitoring of Electric Power Systems*. Schweitzer Engineering Laboratories. USA (2010)
2. N. Yusifbayli, V. Nasibov, A. Huseynov, R. Alizade, A. RSES 2021, AIP Conference Proceedings Vol. **2552**, 030002 (2023)
3. Nadia H.A., Rahman and Ahmad F. Optimal PMU placement using topology transformation method in power systems Journal of Advanced Research p. 625-634
4. Gou B. Optimal Placement of PMUs by Integer Linear Programming, Power Systems, IEEE Transactions, V. **23**, p. 1525-1526 (2008)
5. N. Yusifbayli, V. Nasibov, A. Huseynov, R. Alizade, K. Suleymanov, RSES 2021, AIP Conference Proceedings Vol. **2552**, 020001 (2023)
6. DF 800 SCADA/EMS for Power Dispatch Center
7. Mladen Kerimovic, Xreft Power Associates. Wide - Area Monitoring, Protection and Control System (WAMPAC). Standards for Cyber Security Requirements (2012)

AUTHOR INDEX

AUTHOR INDEX

AUTHOR INDEX